KU-572-819

NINTH EDITION

Chemistry for Today

General, Organic, and Biochemistry

Spencer L. Seager

University of South Dakota
Weber State University

Michael R. Slabaugh

University of South Dakota
Weber State University

Maren S. Hansen

West High School, Salt Lake City, UT

CENGAGE
Learning·

Australia • Brazil • Mexico • Singapore • United Kingdom • United States

CENGAGE
Learning®

Chemistry for Today: General, Organic, and Biochemistry, **Ninth Edition**
Spencer L. Seager, Michael R. Slabaugh

Product Director: Dawn Giovanniello

Product Manager: Courtney Heilman

Content Developer: Peter McGahey

Product Assistant: Anthony Bostler

Media Developer: Elizabeth Woods

Marketing Manager: Ana Albinson

Content Project Manager: Teresa L Trego

Art Director: Sarah B. Cole

Manufacturing Planner: Judy Inouye

Production Service: MPS Limited

Photo Researcher: Lumina Datamatics

Text Researcher: Lumina Datamatics

Copy Editor: MPS Limited

Text Designer: Hespenheide Design

Cover Designer: Delgado and Company

Cover Image: Paul Souders/Getty Images

Compositor: MPS Limited

For product information and technology assistance, contact us at
Cengage Learning Customer & Sales Support, 1-800-354-9706

For permission to use material from this text or product,
submit all requests online at **www.cengage.com/permissions**
Further permissions questions can be e-mailed to
permissionrequest@cengage.com

Library of Congress Control Number: 2016952183

Student Edition:
ISBN: 978-1-305-96006-0

Loose-leaf Edition:
ISBN: 978-1-305-96870-7

Cengage Learning
20 Channel Center Street
Boston, MA 02210
USA

Cengage Learning is a leading provider of customized learning solutions with employees residing in nearly 40 different countries and sales in more than 125 countries around the world. Find your local representative at **www.cengage.com**

Cengage Learning products are represented in Canada by Nelson Education, Ltd.

To learn more about Cengage Learning Solutions, visit **www.cengage.com**

Purchase any of our products at your local college store or at our preferred online store **www.cengagebrain.com**

Printed in the United States of America
Print Number: 02 Print Year: 2017

About the Authors

Spencer L. Seager

Spencer L. Seager retired from Weber State University in 2013 after serving for 52 years as a chemistry department faculty member. He served as department chairman from 1969 until 1993. He taught general and physical chemistry at the university. He was also active in projects designed to help improve chemistry and other science education in local elementary schools. He received his B.S. in chemistry and Ph.D. in physical chemistry from the University of Utah. He currently serves as an adjunct professor at Weber State and the University of South Dakota where he teaches online courses in general chemistry, elementary organic chemistry, and elementary biochemistry.

Michael R. Slabaugh

Michael R. Slabaugh is an adjunct professor at the University of South Dakota and at Weber State University, where he teaches the yearlong sequence of general chemistry, organic chemistry, and biochemistry. He received his B.S. degree in chemistry from Purdue University and his Ph.D. degree in organic chemistry from Iowa State University. His interest in plant alkaloids led to a year of postdoctoral study in biochemistry at Texas A&M University. His current professional interests are chemistry education and community involvement in science activities, particularly the State Science and Engineering Fair in Utah. He also enjoys the company of family, hiking in the mountains, and fishing the local streams.

Maren S. Hansen

Maren S. Hansen is a science teacher at West High School, where she teaches honors biology. She has also taught AP biology and biology in the International Baccalaureate Program. She received her B.A. and master of education degrees from Weber State University. Her professional interests have focused upon helping students participate in Science Olympiad and Science Fair. Other interests include adventure travel, mountain hiking, gardening, and the company of friends and family. She hopes to share her love of science with her two children.

Brief Contents

Contents

Chapter 7

Solutions and Colloids 210

Chapter 8

Reaction Rates and Equilibrium 250

Chapter 9

Acids, Bases, and Salts 276

Chapter 13

Alcohols, Phenols, and Ethers 424

Chapter 14

Aldehydes and Ketones 458

Chapter 15

Carboxylic Acids and Esters 488

Chapter 22

Nutrition and Energy for Life 702

Chapter 23

Carbohydrate Metabolism 732

Chapter 24

Lipid and Amino Acid Metabolism 760

Chapter 25

Body Fluids 788

Preface

The Image of Chemistry

We, as authors, are pleased that the acceptance of the previous eight editions of this text-book by students and their teachers has made it possible to publish this ninth edition. In the earlier editions, we expressed our concern about the negative image of chemistry held by many of our students, and their genuine fear of working with chemicals in the laboratory. Unfortunately, this negative image not only persists, but seems to be intensifying. Reports in the media related to chemicals or to chemistry continue to be primarily negative, and in many cases seem to be designed to increase the fear and concern of the general public. With this edition, we continue to hope that those who use this book will gain a more positive understanding and appreciation of the important contributions that chemistry makes in their lives.

Theme and Organization

This edition continues the theme of the positive and useful contributions made by chemistry in our world.

This text is designed to be used in either a two-semester or three-quarter course of study that provides an introduction to general chemistry, organic chemistry, and biochemistry. Most students who take such courses are majoring in nursing, other health professions, or the life sciences, and consider biochemistry to be the most relevant part of the course of study. However, an understanding of biochemistry depends upon a sound background in organic chemistry, which in turn depends upon a good foundation in general chemistry. We have attempted to present the general and organic chemistry in sufficient depth and breadth to make the biochemistry understandable.

The decisions about what to include and what to omit from the text were based on our combined 75-plus years of teaching, input from numerous reviewers and adopters, and our philosophy that a textbook functions as a personal tutor to each student. In the role of a personal tutor, a text must be more than just a collection of facts, data, and exercises. It should also help students relate to the material they are studying, carefully guide them through more difficult material, provide them with interesting and relevant examples of chemistry in their lives, and become a reference and a resource that they can use in other courses or their professions.

New to This Edition

In this ninth edition of the text, we have some exciting new features, including Ask a Pharmacist boxes written by Marvin Orrock and Chemistry Tips for Living Well. We have also retained features that received a positive reception from our own students, the students of other adopters, other teachers, and reviewers. The retained features are Case Studies, which begin each chapter, including 8 new to this edition; 45 Chemistry Around Us boxes, including 19 new to this edition; 23 Study Skills boxes; 4 How Reactions Occur boxes; and 10 Ask an Expert boxes. The 12 Ask a Pharmacist boxes reflect coverage of both prescription and nonprescription health-related products. The 25 Chemistry Tips for Living Well contain current chemistry-related health issues and suggestions. In addition, approximately 10% of the end-of-chapter exercises have been changed.

Also new to this edition are many new photographs and updated art to further enhance student comprehension of key concepts, processes, and preparation.

Revision Summary of Ninth Edition:

Chapter 1:
- New Case Study
- New Case Study Follow-up
- Several revised figures
- New photography
- New Ask an Expert: *Does Food Density Matter When You're Trying to Lose Weight?*
- New Chemistry Around Us: *Are Chemicals Getting a Bad Rap?*
- New Chemistry Tips for Living Well: *Choose Wisely for Health Information*
- 10% new Exercises

Chapter 2:
- Several revised figures
- New photography
- Updated element table
- New Chemistry Around Us: *Chemical Elements in the Human Body*
- New Ask a Pharmacist: *Uprooting Herbal Myths*
- New Chemistry Tips for Living Well: *Take Care of Your Bones*
- 10% new Exercises

Chapter 3:
- Several revised figures
- New photography
- New Chemistry Tips for Living Well: *Watch the Salt*
- New Chemistry Around Us: *A Solar Future*
- New Chemistry Around Us: *Transition and Inner-Transition Elements in Your Smart Phone*
- 10% new Exercises

Chapter 4:
- Several revised figures
- New photography
- New Ask a Pharmacist: *Are All Iron Preparations Created Equal?*
- New Chemistry Tips for Living Well: *Consider the Mediterranean Diet*
- New Chemistry Around Us: *Ozone: Good up High, Bad Nearby*
- 10% new Exercises

Chapter 5:
- New Case Study
- New Case Study Follow-up
- Several revised figures
- New photography
- New Chemistry Tips for Living Well: *Add Color to Your Diet*
- New Chemistry Around Us: *Teeth Whitening*
- New Chemistry Around Us: *Electric Cars*
- 10% new Exercises

Chapter 6:
- Several revised figures
- New photography
- New Ask a Pharmacist: *Zinc for Colds?*
- New Chemistry Tips for Living Well: *Get an Accurate Blood Pressure Reading*
- New Chemistry Around Us: *Air Travel*
- 10% new Exercises

Chapter 7:

- New Case Study
- New Case Study Follow-up
- Several revised figures
- New photography
- New Chemistry Around Us: *Health Drinks*
- New Chemistry Around Us: *CO₂ Emissions: A Blanket around the Earth*
- New Chemistry Tips for Living Well: *Stay Hydrated*
- 10% new Exercises

Chapter 8:

- Several revised figures
- New photography
- New Ask a Pharmacist: *Energy for Sale*
- New Chemistry Around Us: *Why "Cold" Does Not Exist*
- New Chemistry Tips for Living Well: *Use Your Phone to Help You Stay Healthy*
- 10% new Exercises

Chapter 9:

- Several revised figures
- New photography
- New Chemistry Tips for Living Well: *Beware of Heartburn*
- New Chemistry Around Us: *Sinkholes*
- 10% new Exercises

Chapter 10:

- New Case Study
- New Case Study Follow-up
- Several revised figures
- New photography
- New Ask a Pharmacist: *Medications to Avoid on Test Day*
- New Chemistry Tips for Living Well: *Check the Radon Level in Your Home*
- 10% new Exercises

Chapter 11:

- Several revised figures
- New photography
- New Chemistry Around Us: *Fracking Oil Wells*
- New Chemistry Around Us: *Reducing Your Carbon Footprint*
- New Chemistry Tips for Living Well: *Take Care of Dry Skin*
- 10% new Exercises

Chapter 12:

- Several revised figures
- New photography
- New Ask a Pharmacist: *Controlled Substances*
- New Chemistry Tips for Living Well: *Think before Getting Brown*
- New Chemistry Around Us: *Three-Dimensional Printers*
- New Chemistry Around Us: *Polycarbonate—The Lucky Polymer*
- New Chemistry Around Us: *Graphene*
- 10% new Exercises

Chapter 13:

- Several revised figures
- New photography

- New Ask a Pharmacist: Marijuana: *A Gateway Drug*
- New Chemistry Tips for Living Well: *Take Advantage of Hand Sanitizers*
- 10% new Exercises

Chapter 14:

- Several revised figures
- New photography
- New Chemistry Tips for Living Well: *Get the Right Dose of Exercise*
- 10% new Exercises

Chapter 15:

- New Case Study
- New Case Study Follow-up
- Several revised figures
- New photography
- New Chemistry Tips for Living Well: *Consider Low-Dose Aspirin*
- 10% new Exercises

Chapter 16:

- Several revised figures
- New photography
- New Ask a Pharmacist: *A Wake-Up Call for Treating Insomnia*
- New Chemistry Tips for Living Well: *Try a Little Chocolate*
- 10% new Exercises

Chapter 17:

- Several revised figures
- New photography
- New Chemistry Tips for Living Well: *Put Fiber into Snacks and Meals*
- 10% new Exercises

Chapter 18:

- Several revised figures
- New photography
- New Chemistry Tips for Living Well: *Consider Olive Oil*
- New Chemistry Around Us: *Biofuels Move into the Kitchen*
- 10% new Exercises

Chapter 19:

- New Case Study
- New Case Study Follow-up
- Several revised figures
- New photography
- New Ask a Pharmacist: *Who Really Needs Gluten-Free Food?*
- New Chemistry Around Us: *A Milk Primer*
- New Chemistry Tips for Living Well: *Go for the Good Grains*
- 10% new Exercises

Chapter 20:

- Several revised figures
- New photography
- New Ask a Pharmacist: *Treatment Options for the Common Cold*
- New Chemistry Around Us: *No Milk Please*
- New Chemistry Tips for Living Well: *Cut Back on Processed Meat*
- 10% new Exercises

Chapter 21:

- New Case Study
- New Case Study Follow-up
- Several revised figures
- New photography
- New Chemistry Tips for Living Well: *Reduce Your Chances for Developing Cancer*
- 10% new Exercises

Chapter 22:

- Several revised figures
- New photography
- New Ask a Pharmacist: Sports Supplements: *Where Is My Edge?*
- New Chemistry Tips for Living Well: *Select a Heart-Healthful Diet*
- 10% new Exercises

Chapter 23:

- Several revised figures
- New photography
- New Chemistry Tips for Living Well: *Choose Complex Carbohydrates*
- 10% new Exercises

Chapter 24:

- New Case Study
- New Case Study Follow-up
- New photography
- New Chemistry Tips for Living Well: *Pick the Right Fats*
- 10% new Exercises

Chapter 25:

- New photography
- New Ask a Pharmacist: *Performance-Enhancing Drugs*
- New Chemistry Tips for Living Well: *Select the Right Pre-Exercise Foods*
- 10% new Exercises

Features

Each chapter has features especially designed to help students study effectively, as well as organize, understand, and enjoy the material in the course.

Case Studies. These scenarios introduce you the students to diverse situations a health care professional might encounter. The purpose of the case studies is to stimulate inquiry; for that reason, we've placed them at the beginning of each chapter of the book. Vocabulary and scenarios may be unfamiliar to you who are studying these course materials, but our intent is to raise questions and pique your curiosity. Medicine has long been described as an art. The questions raised by these case studies rarely have a single correct answer. With the knowledge that you gain from this text and your future training,

 Case Study

Purpose: The case study scenarios introduce diverse situations that a health care professional might encounter. Their purpose is to stimulate inquiry; for that reason, we've placed them at the beginning of each chapter. Vocabulary and scenarios may be unfamiliar, but our intention is to stimulate questions and to pique curiosity. Medicine has long been described as an art as well as a science. The questions raised by these case studies rarely have a single correct answer. With the knowledge that you gain from this text, and your future training, acceptable answers to the questions raised in our scenarios will become apparent.

Disclaimer: Some of the case studies are based on real-life situations. In such cases, names have been changed to protect the individual's anonymity.

acceptable answers to the questions raised in our scenarios will become apparent. A Case Study Follow-up to each Case Study can be found at the end of each chapter before the Concept Summary.

Chapter Outlines and Learning Objectives. At the beginning of each chapter, a list of learning objectives provides students with a convenient overview of what they should gain by studying the chapter. In order to help students navigate through each chapter and focus on key concepts, these objectives are repeated at the beginning of the section in which the applicable information is discussed. The objectives are referred to again in the concept summary at the end of each chapter along with one or two suggested end-of-chapter exercises. By working the suggested exercises, students get a quick indication of how well they have met the stated learning objectives. Thus, students begin each chapter with a set of objectives and end with an indication of how well they satisfied the objectives.

Key Terms. Identified within the text by the use of bold type, key terms are defined in the margin near the place where they are introduced. Students reviewing a chapter can quickly identify the important concepts on each page with this marginal glossary. A full glossary of key terms and concepts appears at the end of the text.

Ask a Pharmacist. These boxed features written by Marvin Orrock, Pharm.D., contain useful information about health-related products that are readily available to consumers with or without a prescription. The information in each box provides a connection between the chemical behavior of the product and its effect on the body.

ASK A PHARMACIST 12.1

Controlled Substances

So what are *controlled substance,* anyway, and why do we have them? Before we answer those questions, let's briefly review the major legislation that pertains to products used as medicines. Prior to the 1900s there were no governmental regulations on foods or drugs. As a result, some products were contaminated and some not labeled accurately. Consequently, the U.S. Congress passed the Pure Food and Drug Act of 1906. It proved to be helpful, but opiates and cocaine were not regulated. A significant percentage of the population became addicted, and many deaths were attributed to the use of products that were "pure" and "labeled" correctly but still contained addicting materials. In 1914 the Harrison Act was passed. It regulated heroin and cocaine sales. During the

United States, or a currently accepted medical use with severe restrictions. Abuse of the substance might lead to severe psychological or physical dependence (e.g., Percocet, Demerol, Ritalin).

Schedule III: The substance has a potential for abuse less than the compounds in Schedules I and II. The substance has a currently accepted medical use for treatment in the United States. Abuse of the substance might lead to moderate or low physical dependence or high psychological dependence (e.g., Tylenol with codeine used for pain, anabolic steroids).

Schedule IV: The substance has a low potential for abuse relative to the compounds in Schedule III. The substance has a currently accepted medical use for treatment in the United States. Abuse of the substance might lead to limited physical dependence or psychological dependence relative to the sub-

Chemistry Around Us. These boxed features present everyday applications of chemistry that emphasize in a real way the important role of chemistry in our lives. Thirty percent of these are new to this edition and emphasize health-related applications of chemistry.

Chemistry Tips for Living Well. These boxed features contain current chemistry-related health issues such as "Add Color to Your Diet," and suggestions for maintaining good health such as "Consider the Mediterranean Diet," "Cut Back on Processed Meat," and "Try a Little Chocolate."

 CHEMISTRY TIPS FOR LIVING WELL **14.1**

Get the Right Dose of Exercise

Experts agree that exercise is one of the best preventative "medicines" available. It increases energy, stamina, and one's sense of well-being. In the long term it also reduces the risk of premature death from cardiovascular disease. Put simply, it makes you feel better and live longer. We expect medicines to make us feel better when we are ill. But exercise acts as a powerful medicine to prevent illness. How do you know what the proper dose is? Do you need to exercise on a daily basis or will a weekly dose provide the desired health benefits? Just how little can you get away with and stay healthy?

Researchers arrive at the proper dose by examining

times the recommended amount), health benefits are comparable to those achieved by people who merely meet the minimum requirements. In other words, many extra hours of exercise do not equate to huge gains in longevity. On the other hand, many times the recommended exercise level is not considered to be harmful. It is difficult to overdose on moderate exercise.

Intensity, as well as frequency, should be considered when calculating the ideal exercise dose. People who spend part of their daily exercise time in vigorous activity, rather than moderate activity alone (e.g., running instead of walking) reap additional health benefits. People who spent up to

Ask an Expert. These boxed features, written by Melina B. Jampolis, M.D., engage students by presenting questions and answers about nutrition and health, as related to chemistry, that are relevant and important in today's world.

Examples. To reinforce students in their problem-solving skill development, complete step-by-step solutions for numerous examples are included.

Learning Checks. Short self-check exercises follow examples and discussions of key or difficult concepts. A complete set of solutions is included in Appendix C. These allow students to measure immediately their understanding and progress.

Study Skills. Most chapters contain a *Study Skills* feature in which a challenging topic, skill, or concept of the chapter is addressed. Study suggestions, analogies, and approaches are provided to help students master these ideas.

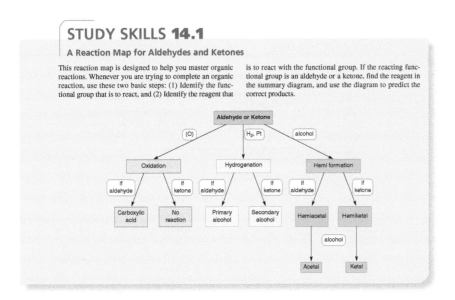

STUDY SKILLS 14.1

A Reaction Map for Aldehydes and Ketones

This reaction map is designed to help you master organic reactions. Whenever you are trying to complete an organic reaction, use these two basic steps: (1) Identify the functional group that is to react, and (2) Identify the reagent that is to react with the functional group. If the reacting functional group is an aldehyde or a ketone, find the reagent in the summary diagram, and use the diagram to predict the correct products.

How Reactions Occur. The mechanisms of representative organic reactions are presented in four boxed inserts to help students dispel the mystery of how these reactions take place.

Concept Summary. Located at the end of each chapter, this feature provides a concise review of the concepts and includes suggested exercises to check achievement of the learning objectives related to the concepts.

Concept Summary

Symbols and Formulas Symbols based on names have been assigned to every element. Most consist of a single capital letter followed by a lowercase letter. A few consist of a single capital letter. Compounds are represented by formulas made up of elemental symbols. The number of atoms of each element in a molecule is shown by subscripts.
Objective 1 (Section 2.1), Exercise 2.4

Inside the Atom Atoms are made up of numerous smaller particles, of which the most important to chemical studies are the proton, neutron, and electron. Positively charged protons and neutral neutrons have a relative mass of 1 u each and

are located in the nuclei of atoms. Negatively charged electrons with a mass of 1/1836 u are located outside the nuclei of atoms.
Objective 2 (Section 2.2), Exercises 2.10 and 2.12

Isotopes Most elements in their natural state are made up of more than one kind of atom. These different kinds of atoms of a specific element are called isotopes and differ from one another only in the number of neutrons in their nuclei. A symbol incorporating atomic number, mass number, and elemental symbol is used to represent a specific isotope.
Objective 3 (Section 2.3), Exercises 2.16 and 2.22

Key Terms and Concepts. These are listed at the end of the chapter for easy review, with a reference to the chapter section in which they are presented.

Key Equations. This feature provides a useful summary of general equations and reactions from the chapter. This feature is particularly helpful to students in the organic chemistry chapters.

Exercises. Nearly 1,700 end-of-chapter exercises are arranged by section. Approximately half of the exercises are answered in the back of the text. Complete solutions to these answered exercises are included in the Student Study Guide. Solutions and answers to the remaining exercises are provided in the Instructor's Manual. We have included a significant number of clinical and other familiar applications of chemistry in the exercises.

Chemistry for Thought. Included at the end of each chapter are special questions designed to encourage students to expand their reasoning skills. Some of these exercises are based on photographs found in the chapter, while others emphasize clinical or other useful applications of chemistry.

Allied Health Exam Connection. These examples of chemistry questions from typical entrance exams used to screen applicants to allied health professional programs help students focus their attention on the type of chemical concepts considered important in such programs.

Allied Health Exam Connection

The following questions are from these sources:

- *Nursing School Entrance Exam* © 2005, Learning Express, LLC.
- *McGraw-Hill's Nursing School Entrance Exams* by Thomas A. Evangelist, Tamara B. Orr, and Judy Unrein © 2009, The McGraw-Hill Companies, Inc.
- *NSEE Nursing School Entrance Exams,* 3rd edition © 2009, Kaplan Publishing.
- *Cliffs Test Prep: Nursing School Entrance Exams* by Fred N. Grayson © 2004, Wiley Publishing, Inc.
- *Peterson's Master the Nursing School and Allied Health Entrance Exams,* 18th edition by Marion F. Gooding © 2008, Peterson's, a Nelnet Company.

9.137 An acid is a substance that dissociates in water into one or more _____ ions and one or more _____.
 a. hydrogen . . . anions
 b. hydrogen . . . cations
 c. hydroxide . . . anions
 d. hydroxide . . . cations

9.138 A base is a substance that dissociates in water into one or more _____ ions and one or more _____.
 a. hydrogen . . . anions
 b. hydrogen . . . cations
 c. hydroxide . . . anions

9.143 Dissolving H_2SO_4 in water creates an acid solution by increasing the:
 a. sulfate ions.
 b. water ions.
 c. hydrogen ions.
 d. oxygen ions.

9.144 When a solution has a pH of 7, it is:
 a. a strong base.
 b. a strong acid.
 c. a weak base.
 d. neutral.

Possible Course Outlines

This text may be used effectively in either a two-semester or three-quarter course of study:

First semester: Chapters 1–13 (general chemistry and three chapters of organic chemistry)

Second semester: Chapters 14–25 (organic chemistry and biochemistry)

First semester: Chapters 1–10 (general chemistry)

Second semester: Chapters 11–21 (organic chemistry and some biochemistry)

First quarter: Chapters 1–10 (general chemistry)

Second quarter: Chapters 11–18 (organic chemistry)

Third quarter: Chapters 19–25 (biochemistry)

Supporting Materials

Please visit **http://www.cengage.com/chemistry/seager/gob9e** for information about student and instructor resources for this text.

Acknowledgments

We express our sincere appreciation to the following reviewers, who helped us revise the many editions:

Hugh Akers
Lamar University–Beaumont

Johanne I. Artman
Del Mar College

Gabriele Backes
Portland Community College

Bruce Banks
University of North Carolina–Greensboro

David Boykin
Georgia State University

Deb Breiter
Rockford College

Lorraine C. Brewer
University of Arkansas

Martin Brock
Eastern Kentucky University

Jonathan T. Brockman
College of DuPage

Kathleen Brunke
Christopher Newport University

Christine Brzezowski
University of Utah

Sybil K. Burgess
University of North Carolina–Wilmington

Sharmaine S. Cady
East Stroudsburg University

Linda J. Chandler
Salt Lake Community College

Tom Chang
Utah State University

Ngee Sing Chong
Middle Tennessee State University

Judith Ciottone
Fitchburg State University

Caroline Clower
Clayton State University

Sharon Cruse
Northern Louisiana University

Thomas D. Crute
Augusta College

Jack L. Dalton
Boise State University

Lorraine Deck
University of New Mexico

Kathleen A. Donnelly
Russell Sage College

Jan Fausset
Front Range Community College

Patricia Fish
The College of St. Catherine

Harold Fisher
University of Rhode Island

John W. Francis
Columbus State Community College

Wes Fritz
College of DuPage

Jean Gade
Northern Kentucky University

Galen George
Santa Rosa Junior College

Anita Gnezda
Ball State University

Meldath Govindan
Fitchburg State University

Jane D. Grant
Florida Community College

James K. Hardy
University of Akron

Leland Harris
University of Arizona

Robert H. Harris
University of Nebraska–Lincoln

David C. Hawkinson
University of South Dakota

Jack Hefley
Blinn College

Claudia Hein
Diablo Valley College

John Henderson
Jackson Community College

Mary Herrmann
University of Cincinnati

Arthur R. Hubscher
Brigham Young University–Idaho

Kenneth Hughes
University of Wisconsin–Oshkosh

Jeffrey A. Hurlbut
Metropolitan State College of Denver

Jim Johnson
Sinclair Community College

Richard. F. Jones
Sinclair Community College

Frederick Jury
Collin County Community College

Lidija Kampa
Kean College of New Jersey

Laura Kibler-Herzog
Georgia State University

Margaret G. Kimble
*Indiana University–Purdue University
 Fort Wayne*

James F. Kirby
Quinnipiac University

Peter J. Krieger
Palm Beach Community College

Terrie L. Lampe
De Kalb College–Central Campus

Carol Larocque
Cambrian College

Richard Lavallee
Santa Monica College

Donald Linn
*Indiana University—Purdue University
 Fort Wayne*

Leslie J. Lovett
Fairmont State College

James Luba
University of Arkansas at Little Rock

Regan Luken
University of South Dakota

Gregory Marks
Carroll University

Armin Mayr
El Paso Community College

James McConaghy
Wayne College

Evan McHugh
Pikes Peak Community College

Trudy McKee
Thomas Jefferson University

Melvin Merken
Worcester State College

W. Robert Midden
Bowling Green State University

Pamela S. Mork
Concordia College

Phillip E. Morris, Jr.
University of Alabama–Birmingham

Robert N. Nelson
Georgia Southern University

Marie Nguyen
Highline Community College

Elva Mae Nicholson
Eastern Michigan University

H. Clyde Odom
Charleston Southern University

Howard K. Ono
California State University–Fresno

Jeff Owens
Highline Community College

Dwight Patterson
Middle Tennessee State University

James A. Petrich
San Antonio College

Thomas G. Richmond
University of Utah

James Schreck
University of Northern Colorado

William Scovell
Bowling Green State University

Jean M. Shankweiler
El Camino Community College

Francis X. Smith
King's College

J. Donald Smith
University of Massachusetts–Dartmouth

Malcolm P. Stevens
University of Hartford

Eric R. Taylor
University of Southwestern Louisiana

Krista Thomas
Johnson County Community College

Linda Thomas-Glover
Guilford Technical Community College

James A. Thomson
University of Waterloo

Mary Lee Trawick
Baylor University

Katherin Vafeades
University of Texas–San Antonio

John Vincent
University of Alabama

Scott White
Southern Arkansas University

Cary Willard
Grossmont College

Don Williams
Hope College

Les Wynston
California State University–Long Beach

Jean Yockey
University of South Dakota

We also give special thanks to Dawn Giovanniello, Product Director, and Peter McGahey, Senior Content Developer for Cengage Learning, who guided and encouraged us in the preparation of this ninth edition. We would also like to thank Teresa Trego, Senior Content Project Manager; Elizabeth Woods, Content Developer and Ana Albinson, Associate Marketing Manager. All were essential to the team and contributed greatly to the success of the project. We are very grateful for the superb work of Prashant Kumar Das of MPS Limited for his outstanding coordination of production, and Erika Mugavin, IP Project Manager, for coordinating the excellent photos. We are especially pleased with the new feature Ask a Pharmacist and wish to thank Marvin Orrock for his excellent work. We appreciate the significant help of four associates: Monica Linford, who did an excellent job writing 8 new case studies, Mary Ann Francis, who helped with submitting the manuscript, Kimberly Francis, who helped write the Chemistry Around Us features, and David Shinn of the U.S. Merchant Marine Academy for assistance with accuracy checking.

Finally, we extend our love and heartfelt thanks to our families for their patience, support, encouragement, and understanding during a project that occupied much of our time and energy.

Spencer L. Seager

Michael R. Slabaugh
Maren S. Hansen

1

Matter, Measurements, and Calculations

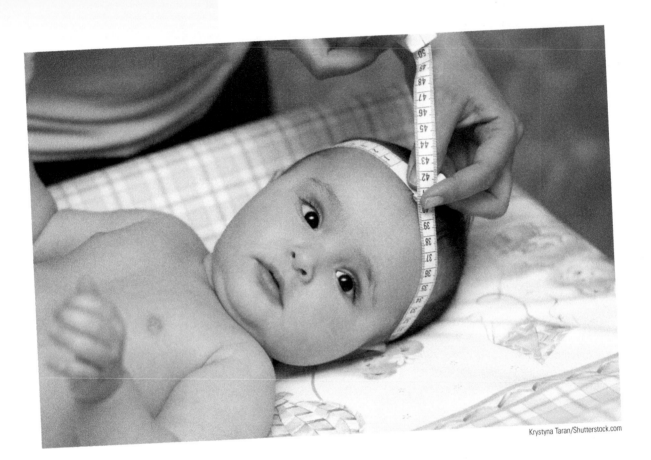

Krystyna Taran/Shutterstock.com

▶ Case Study

Purpose: The case study scenarios introduce diverse situations that a health care professional might encounter. Their purpose is to stimulate inquiry; for that reason, we've placed them at the beginning of each chapter. Vocabulary and scenarios may be unfamiliar, but our intention is to stimulate questions and to pique curiosity. Medicine has long been described as an art as well as a science. The questions raised by these case studies rarely have a single correct answer. With the knowledge that you gain from this text, and your future training, acceptable answers to the questions raised in our scenarios will become apparent.

 Disclaimer: Some of the case studies are based on real-life situations. In such cases, names have been changed to protect the individual's anonymity.

Any resemblance to a particular person is purely coincidental. Models are used in all photos illustrating the cases. No photos of actual people experiencing particular medical scenarios are ever used in this text.

Case Study: Katie enjoyed well-child appointments at the military clinic. Because of the remote location, several doctors operated the clinic in turn. Katie liked the positive feedback at Norah's two-week checkup, where doctors praised Katie for her attentive mothering and congratulated her on Norah's impressive weight gain on what one doctor called "high-octane" milk. Today, at Norah's nine-month check, the nurse recorded important measurements of weight, length, temperature, and head circumference. Doctor Watson pondered these for a disconcertingly long time. He asked questions, including "Does she crawl?" and "Can she say ten words?" Dr. Watson admitted his concern about microcephaly and directed that Norah should be returned every two weeks for head measurements. Katie felt sure of her daughter's intelligence, but perhaps she was just a proud parent. Two months later, a different pediatrician examined Norah and reassured Katie that hats come in different sizes for a reason. Now, thirty years later, Norah's name is followed by Ph.D.

▶ **What other factors should the doctor consider when microcephaly is suspected? How important is it for medical professionals to consider the emotional impact of their diagnoses on family members (e.g., the mother's anxiety)?**

Follow-up to this Case Study appears at the end of the chapter before the *Concept Summary*.

Learning Objectives

When you have completed your study of this chapter, you should be able to:

1 Explain what matter is. **(Section 1.1)**

2 Explain the difference between the terms *physical* and *chemical* as applied to the properties of matter and changes in matter. **(Section 1.2)**

3 Describe matter in terms of the accepted scientific model. **(Section 1.3)**

4 On the basis of observation or information given to you, classify matter into the correct category of each of the following pairs: heterogeneous or homogeneous, solution or pure substance, and element or compound. **(Section 1.4)**

5 Recognize the use of measurement units in everyday activities. **(Section 1.5)**

6 Recognize units of the metric system, and convert measurements done using the metric system into related units. **(Section 1.6)**

7 Express numbers using scientific notation, and do calculations with numbers expressed in scientific notation. **(Section 1.7)**

8 Express the results of measurements and calculations using the correct number of significant figures. **(Section 1.8)**

9 Use the factor-unit method to solve numerical problems. **(Section 1.9)**

10 Do calculations involving percentages. **(Section 1.10)**

11 Do calculations involving densities. **(Section 1.11)**

Chemistry is often described as the scientific study of matter. In a way, almost any study is a study of matter, because matter is the substance of everything. Chemists, however, are especially interested in matter; they study it and attempt to understand it from nearly every possible point of view.

The chemical nature of all matter makes an understanding of chemistry useful and necessary for individuals who are studying in a wide variety of areas, including the

health sciences, the natural sciences, home economics, education, environmental science, and law enforcement.

Matter comes in many shapes, sizes, and colors that are interesting to look at and describe. Early chemists did little more than describe what they observed, and their chemistry was a descriptive science that was severely limited in scope. It became a much more useful science when chemists began to make quantitative measurements, do calculations, and incorporate the results into their descriptions. Some fundamental ideas about matter are presented in this chapter, along with some ideas about quantitative measurement, the scientific measurement system, and calculations.

1.1 | What Is Matter?

Learning Objective

1. Explain what matter is.

Definitions are useful in all areas of knowledge; they provide a common vocabulary for both presentations to students and discussions between professionals. You will be expected to learn a number of definitions as you study chemistry, and the first one is a definition of *matter*. Earlier, we said that matter is the substance of everything. That isn't very scientific, even though we think we know what it means. If you stop reading for a moment and look around, you will see a number of objects that might include people, potted plants, walls, furniture, books, windows, and a TV set or radio. The objects you see have at least two things in common: Each one has mass, and each one occupies space. These two common characteristics provide the basis for the scientific definition of matter. **Matter** is anything that has mass and occupies space. You probably understand what is meant by an object occupying space, especially if you have tried to occupy the same space as some other object. The resulting physical bruises leave a lasting mental impression.

You might not understand the meaning of the term *mass* quite as well, but it can also be illustrated "painfully." Imagine walking into a very dimly lit room and being able to just barely see two large objects of equal size on the floor. You know that one is a bowling ball and the other is an inflated plastic ball, but you can't visually identify which is which. However, a hard kick delivered to either object easily allows you to identify each one. The bowling ball resists being moved much more strongly than does the inflated ball. Resistance to movement depends on the amount of matter in an object, and **mass** is an actual measurement of the amount of matter present.

The term *weight* is probably more familiar to you than *mass*, but the two are related. All objects are attracted to each other by gravity, and the greater their mass, the stronger the attraction between them. The **weight** of an object on Earth is a measurement of the gravitational force pulling the object toward Earth. An object with twice the mass of a second object is attracted with twice the force, and therefore has twice the weight of the second object. The mass of an object is constant no matter where it is located (even if it is in a weightless condition in outer space). However, the weight of an object depends on the strength of the gravitational attraction to which it is subjected. For example, a rock that weighs 16 pounds on Earth would weigh about 2.7 pounds on the moon because the gravitational attraction is only about one-sixth that of Earth. However, the rock contains the same amount of matter and thus has the same mass whether it is located on Earth or on the moon.

Despite the difference in meaning between mass and weight, the determination of mass is commonly called "weighing." We will follow that practice in this book, but we will use the correct term *mass* when referring to an amount of matter.

matter Anything that has mass and occupies space.

mass A measurement of the amount of matter in an object.

weight A measurement of the gravitational force acting on an object.

A Central Science

Chemistry is often referred to as the "central science" because it serves as a necessary foundation for many other scientific disciplines. Regardless of which scientific field you are interested in, every single substance you will discuss or work with is made up of chemicals. Also, many processes important to those fields will be based on an understanding of chemistry.

As you read this text, you will encounter chapter opening photos dealing with applications of chemistry in the health-care professions. Within the chapters, other Chemistry Around Us boxes focus on specific substances that play essential roles in meeting the needs of society.

```
Health sciences  ←→  Nutrition
         ↘       ↗
        Chemistry
         ↙       ↘
Microbiology  ←   →  Physiology
         ↓
        Botany
```

Chemistry is the foundation for many other scientific disciplines.

We also consider chemistry a central science because of its crucial role in responding to the needs of society. We use chemistry to discover new processes, develop new sources of energy, produce new products and materials, provide more food, and ensure better health.

Chemicals are present in everything we can touch, smell, or see. Chemistry is all around us.

1.2 | Properties and Changes

Learning Objective

2. Explain the difference between the terms *physical* and *chemical* as applied to the properties of matter and changes in matter.

When you looked at your surroundings earlier, you didn't have much trouble identifying the various things you saw. For example, unless the decorator of your room had unusual tastes, you could easily tell the difference between a TV set and a potted plant by observing such characteristics as shape, color, and size. Our ability to identify objects or materials and discriminate between them depends on such characteristics. Scientists prefer to use the term *property* instead of *characteristic*, and they classify properties into two categories, physical and chemical.

Physical properties are those that can be observed or measured without changing or trying to change the composition of the matter in question—no original substances are destroyed, and no new substances appear. For example, you can observe the color or measure the size of a sheet of paper without attempting to change the paper into anything else. Color and size are physical properties of the paper. **Chemical properties** are the properties matter demonstrates when attempts are made to change it into other kinds of matter. For example, a sheet of paper can be burned; in the process, the paper is changed into new substances. On the other hand, attempts to burn a piece of glass under similar conditions fail. The ability of paper to burn is a chemical property, as is the inability of glass to burn.

physical properties Properties of matter that can be observed or measured without trying to change the composition of the matter being studied.

chemical properties Properties that matter demonstrates when attempts are made to change it into new substances.

Are Chemicals Getting a Bad Rap?

The following question was overheard in a grocery store when a customer approached a sales clerk and asked, "Are there any chemicals in this yogurt?" If the clerk had been a chemistry student, a correct answer would have been "Of course, the yogurt itself is made of chemicals."

The reason for this correct answer is that all matter, including yogurt, is made up of atoms of the elements. This means that any sample of any kind of matter contains atoms, and therefore contains chemicals. What the customer really should have asked is a more specific question, such as "Does this yogurt contain any chemical preservative?" or, if the customer had a condition such as lactose intolerance, "Does this yogurt contain any lactose?"

However, unfortunately, in today's world the word "chemical" is often used in a negative way, as illustrated by this conversation. Hopefully, students using this textbook will be taking a course that will eliminate the negative feeling toward chemistry and chemicals.

All matter, including yogurt, is comprised of chemicals.

physical changes Changes matter undergoes without changing composition.

chemical changes Changes matter undergoes that involve changes in composition.

You can easily change the size of a sheet of paper by cutting off a piece. The paper sheet is not converted into any new substance by this change, but it is simply made smaller. **Physical changes** can be carried out without changing the composition of a substance. However, there is no way you can burn a sheet of paper without changing it into new substances. Thus, the change that occurs when paper burns is called a **chemical change. Figure 1.1** shows an example of a chemical change, the burning of magnesium metal. The bright light produced by this chemical change led to the use of magnesium in the flash powder used in early photography. Magnesium is still used in fireworks to produce a brilliant white light.

Example **1.1** Classifying Changes as Physical or Chemical

Classify each of the following changes as physical or chemical: (a) a match is burned; (b) iron is melted; (c) limestone is crushed; (d) limestone is heated, producing lime and carbon dioxide; (e) an antacid seltzer tablet is dissolved in water; and (f) a rubber band is stretched.

Solution

Changes b, c, and f are physical changes because no composition changes occurred and no new substances were formed.

The others are chemical changes because new substances were formed. A match is burned—combustion gases are given off, and matchstick wood is converted to ashes. Limestone is heated—lime and carbon dioxide are the new substances. A seltzer tablet is dissolved in water—the fizzing that results is evidence that at least one new material (a gas) is produced.

✔ **LEARNING CHECK 1.1** Classify each of the following changes as physical or chemical, and, in the cases of chemical change, describe one observation or test that indicates new substances have been formed: (a) milk sours, (b) a wet handkerchief dries, (c) fruit ripens, (d) a stick of dynamite explodes, (e) air is compressed into a steel container, and (f) water boils.

1 A strip of magnesium metal.

2 After being ignited with a flame, the magnesium burns with a blinding white light.

3 The white ash of magnesium oxide from the burning of several magnesium strips.

Figure 1.1 A chemical change occurs when magnesium metal burns.

Among the most common physical changes are changes in state, such as the melting of solids to form liquids, the sublimation of solids to form gases, or the evaporation of liquids to form gases. These changes take place when heat is added to or removed from matter, as represented in **Figure 1.2**. We will discuss changes in state in more detail in Chapter 6.

Figure 1.2 Examples of physical change.

A Solid iodine becomes gaseous iodine when heated.

B Liquid benzene becomes solid benzene when cooled.

1.3 | A Model of Matter

Learning Objective

3. Describe matter in terms of the accepted scientific model.

Model building is a common activity of scientists, but the results in many cases would not look appropriate on a fireplace mantle. **Scientific models** are explanations for observed behavior. Some, such as the well-known representation of the solar system, can easily be depicted in a physical way. Others are so abstract that they can be represented only by mathematical equations.

scientific models Explanations for observed behavior in nature.

Our present understanding of the nature of matter is a model that has been developed and refined over many years. Based on careful observations and measurements of the properties of matter, the model is still being modified as more is learned. In this book, we will concern ourselves with only some very basic concepts of this model, but even these basic ideas will provide a powerful tool for understanding the behavior of matter.

The study of the behavior of gases—such as air, oxygen, and carbon dioxide—by some of the earliest scientists led to a number of important ideas about matter. The volume of a gas kept at a constant temperature was found to change with pressure. An increase in pressure caused the gas volume to decrease, whereas a decrease in pressure permitted the gas volume to increase. It was also discovered that the volume of a gas maintained at constant pressure increased as the gas temperature was increased. Gases were also found to have mass and to mix rapidly with one another when brought together.

A simple model for matter was developed that explained these gaseous properties, as well as many properties of solids and liquids. Some details of the model are discussed in Chapter 6, but one conclusion is important to us now. All matter is made up of particles that are too small to see (see **Figure 1.3**). The early framers of this model called the small particles *molecules*. It is now known that molecules are the constituent particles of many, but not all, substances. In this chapter, we will limit our discussion to substances made up of molecules. Substances that are not made of molecules are discussed in Sections 4.3 and 4.11.

The results of some simple experiments will help us formally define the term *molecule*. Suppose you have a container filled with oxygen gas and you perform a number of experiments with it. You find that a glowing splinter of wood bursts into flames when placed in the gas. A piece of moist iron rusts much faster in the oxygen than it does in air. A mouse or other animal can safely breathe the gas.

Now suppose you divide another sample of oxygen the same size as the first into two smaller samples. The results of similar experiments done with these samples would be the same as before. Continued subdivision of an oxygen sample into smaller and smaller samples does not change the ability of the oxygen in the samples to behave just like the oxygen in the original sample. We conclude that the physical division of a sample of oxygen gas into smaller and smaller samples does not change the oxygen into anything else—it is still oxygen. Is there a limit to such divisions? What is the smallest sample of oxygen that will behave like the larger sample? We hope you have concluded that the smallest sample must be a single molecule. Although its very small size would make a one-molecule sample difficult to handle, it would nevertheless behave just as a larger sample would—it could be stored in a container, it would make wood burn rapidly, it would rust iron, and it could be breathed safely by a mouse.

We are now ready to formally define the term *molecule*. A **molecule** is the smallest particle of a pure substance that has the properties of that substance and is capable of a stable independent existence. Alternatively, a molecule is defined as the limit of physical subdivision for a pure substance.

In less formal terms, these definitions indicate that a sample of pure substance—such as oxygen, carbon monoxide, or carbon dioxide—can be physically separated into smaller and smaller samples only until there is a single molecule. Any further separation cannot be done physically, but if it were done (chemically), the resulting sample would no longer have the same properties as the larger samples.

The idea that it might be possible to chemically separate a molecule into smaller particles grew out of continued study and experimentation by early scientists. In modern terminology, the smaller particles that make up molecules are called atoms. John Dalton (1766–1844) is generally credited with developing the first atomic theory containing ideas that are still used today. The main points of his theory, which he proposed in 1808, can be summarized in the following five statements:

1. All matter is made up of tiny particles called atoms.
2. Substances called elements are made up of atoms that are all identical.
3. Substances called compounds are combinations of atoms of two or more elements.
4. Every molecule of a specific compound always contains the same number of atoms of each kind of element found in the compound.
5. In chemical reactions, atoms are rearranged, separated, or combined, but are never created nor destroyed.

Figure 1.3 A hang glider soars far above the ground. How does this feat confirm that air is matter?

Alexandra Lande/Shutterstock.com

molecule The smallest particle of a pure substance that has the properties of that substance and is capable of a stable independent existence. Alternatively, a molecule is the limit of physical subdivision for a pure substance.

Early scientists used graphic symbols such as circles and squares to represent the few different atoms that were known at the time. Instead of different shapes, we will use representations such as those in **Figure 1.4** for oxygen, carbon monoxide, and carbon dioxide molecules.

Figure 1.4 Symbolic representations of molecules.

The three pure substances just mentioned illustrate three types of molecules found in matter. Oxygen molecules consist of two oxygen atoms, and are called **diatomic molecules** to indicate that fact. Molecules such as oxygen that contain only one kind of atom are also called **homoatomic molecules** to indicate that the atoms are all of the same kind. Carbon monoxide molecules also contain two atoms and therefore are diatomic molecules. However, in this case the atoms are not identical, a fact indicated by the term **heteroatomic molecule**. Carbon dioxide molecules consist of three atoms that are not all identical, so carbon dioxide molecules are described by the terms **triatomic** and heteroatomic. The words *diatomic* and *triatomic* are commonly used to indicate two- or three-atom molecules, but the word **polyatomic** is usually used to describe molecules that contain more than three atoms.

diatomic molecules Molecules that contain two atoms.

homoatomic molecules Molecules that contain only one kind of atom.

heteroatomic molecules Molecules that contain two or more kinds of atoms.

triatomic molecules Molecules that contain three atoms.

polyatomic molecules Molecules that contain more than three atoms.

Example 1.2 Classifying Molecules

Use the terms *diatomic, triatomic, polyatomic, homoatomic,* or *heteroatomic* to classify the following molecules correctly:

Solution
A. Polyatomic and heteroatomic (more than three atoms, and the atoms are not all identical)
B. Polyatomic and homoatomic (more than three atoms, and the atoms are identical)
C. Diatomic and homoatomic (two atoms, and the atoms are identical)
D. Triatomic and heteroatomic (three atoms, and the atoms are not identical)
E. Diatomic and heteroatomic (two atoms, and the atoms are not identical)

The subdivision of molecules into smaller particles is a chemical change. How far can such subdivisions of molecules go? You are probably a step ahead of us and have guessed that the answer is atoms. In fact, this provides us with a definition of atoms. An **atom** is the limit of chemical subdivision. In less formal terms, atoms are the smallest particles of matter that can be produced as a result of chemical changes. However, all chemical changes do not necessarily break molecules into atoms. In some cases, chemical changes might just divide a large molecule into two or more smaller molecules. Also, as we will see later, some chemical changes form larger molecules from smaller ones. The important point is that only chemical changes will produce a division of molecules, and the smallest particles of matter that can possibly be produced by such a division are called atoms.

atom The limit of chemical subdivision for matter.

1.4 | Classifying Matter

Learning Objective

4. On the basis of observation or information given to you, classify matter into the correct category of each of the following pairs: heterogeneous or homogeneous, solution or pure substance, and element or compound.

Unknown substances are often analyzed to determine their compositions. An analyst, upon receiving a sample to analyze, will always ask an important question: Is the sample a pure substance or a mixture? Any sample of matter must be one or the other. Pure water and sugar are both pure substances, but you can create a mixture by stirring together some sugar and pure water.

What is the difference between a pure substance and a mixture? Two differences are that a **pure substance** has a constant composition and a fixed set of physical and chemical properties. Pure water, for example, always contains the same proportions of hydrogen and oxygen and freezes at a specific temperature. A **mixture** of sugar and water, however, can vary in composition, and the properties will be different for the different compositions. For example, a glass of sugar water could contain a few crystals of sugar or several spoonfuls. Properties such as the sweetness and freezing point would vary depending on the amount of sugar present in the mixture.

Another difference is that a pure substance cannot be physically separated into simpler substances, whereas a mixture can theoretically be separated into its components. For example, if we heat a sugar-and-water mixture, the water evaporates, and the sugar remains. We say mixtures can theoretically be separated, but some separations are very difficult to achieve. **Figure 1.5** summarizes these ideas.

pure substance Matter that has a constant composition and fixed properties.

mixture A physical blend of matter that can theoretically be physically separated into two or more components.

Figure 1.5 Mixtures and pure substances.

Pure substances, and mixtures such as sugar water, are examples of **homogeneous matter**—matter that has a uniform appearance and the same properties throughout. Homogeneous mixtures such as sugar water are called **solutions** (see **Figure 1.6**). Mixtures in which the properties and appearance are not uniform throughout the sample are examples of **heterogeneous matter**. The mixture of rock salt and sand that is spread on snowy roads during the winter is an example.

Commonly, the word *solution* is used to describe homogeneous liquid mixtures such as sugar water, but solutions of gases and solids also exist. The air around us is a gaseous solution, containing primarily nitrogen and oxygen. The alloys of some metals are solid solutions. For example, small amounts of copper are often added to the gold used in making jewelry. The resulting solid solution is harder than gold and has greater resistance to wear.

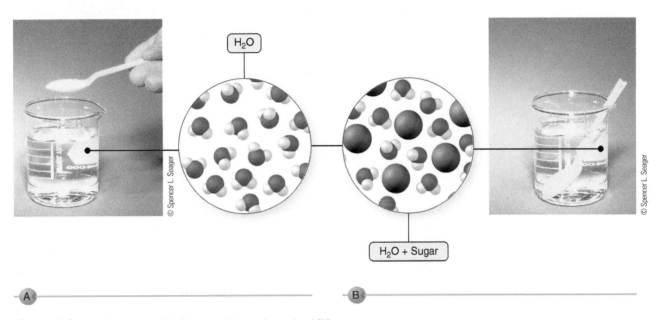

Figure 1.6 Sugar and water **(A)** form a solution when mixed **(B)**.

Most matter is found in nature in the form of heterogeneous mixtures. The properties of such mixtures depend on the location from which samples are taken. In some cases, the heterogeneity is obvious. In a slice of tomato, for example, the parts representing the skin, juice, seeds, and pulp can be easily seen and identified because they look different. Thus, at least one property (e.g., color or texture) is different for the different parts. However, a sample of clean sand from a seashore must be inspected very closely before slight differences in appearance can be seen for different grains. At this point, you might be thinking that even the solutions described earlier as homogeneous mixtures would appear to be heterogeneous if they were looked at closely enough. We could differentiate between sugar and water molecules if sugar solutions were observed under sufficient magnification. We will generally limit ourselves to differences normally visible when we classify matter as heterogeneous on the basis of appearance.

Earlier, we looked at three examples of pure substances—oxygen, carbon monoxide, and carbon dioxide—and found that the molecules of these substances are of different types. Oxygen molecules are diatomic and homoatomic, carbon monoxide molecules are diatomic and heteroatomic, and carbon dioxide molecules are triatomic and heteroatomic. Many pure substances have been found to consist of either homoatomic or heteroatomic molecules—a characteristic that permits them to be classified into one of two categories. Pure substances made up of homoatomic molecules are called **elements**, and those made up of heteroatomic molecules are called **compounds**. Thus, oxygen is an element, whereas carbon monoxide and carbon dioxide are compounds.

It is useful to note a fact here that is discussed in more detail later in Section 4.11. The smallest particles of some elements and compounds are individual atoms rather than molecules. However, in elements of this type, the individual atoms are all of the same kind, whereas in compounds, two or more kinds of atoms are involved. Thus, the classification

of a pure substance as an element or a compound is based on the fact that only one kind of atom is found in elements and two or more kinds are found in compounds. In both cases, the atoms may be present individually or in the form of homoatomic molecules (elements) or heteroatomic molecules (compounds). Some common household materials are pure substances (elements or compounds), such as aluminum foil, baking soda, and table salt.

✔ **LEARNING CHECK 1.3** Classify the molecules represented below as those of an element or a compound:

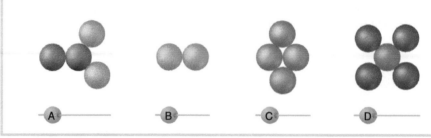

The characteristics of the molecules of elements and compounds lead us to some conclusions about their chemical behavior. Elements cannot be chemically subdivided into simpler pure substances, but compounds can. Because elements contain only one kind of atom and the atom is the limit of chemical subdivision, there is no chemical way to break an element into any simpler pure substance—the simplest pure substance is an element. On the other hand, because the molecules of compounds contain more than one kind of atom, breaking such molecules into simpler pure substances is possible. For example, a molecule of table sugar can be chemically changed into two simpler molecules (which are also sugars) or into atoms or molecules of the elements carbon, hydrogen, and oxygen. Thus, compounds can be chemically subdivided into simpler compounds or elements. **Figure 1.7** summarizes these ideas, and **Figure 1.8** illustrates a classification scheme for matter based on the ideas we have discussed.

Figure 1.7 Elements and compounds.

Example 1.3 Classifying Substances

When sulfur, an element, is heated in air, it combines with oxygen to form sulfur dioxide. Classify sulfur dioxide as an element or a compound.

Solution
Because sulfur and oxygen are both elements and they combine to form sulfur dioxide, the molecules of sulfur dioxide must contain atoms of both sulfur and oxygen. Thus, sulfur dioxide is a compound because its molecules are heteroatomic.

✔ **LEARNING CHECK 1.4** Suppose an element and a compound combine to form only one product. Classify the product as an element or a compound.

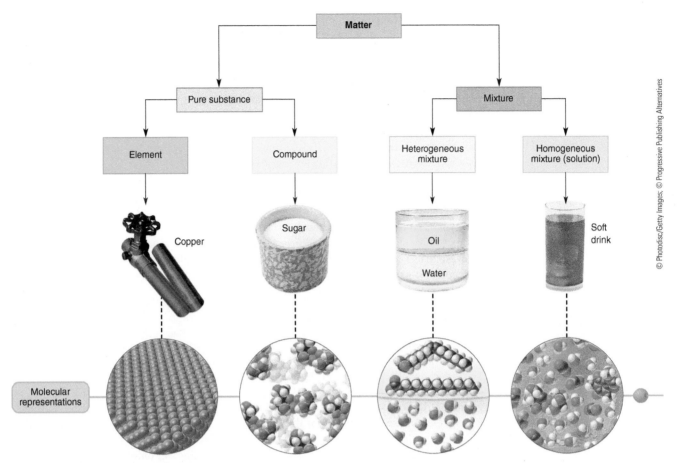

Figure 1.8 A classification scheme for matter.

1.5 | Measurement Units

Learning Objective

5. Recognize the use of measurement units in everyday activities.

Matter can be classified and some physical or chemical properties can be observed without making any measurements. However, the use of quantitative measurements and calculations greatly expands our ability to understand the chemical nature of the world around us. A measurement consists of two parts, a number and an identifying unit. A number expressed without a unit is generally useless, especially in scientific work. We constantly make and express measurements in our daily lives. We measure the gallons of gasoline put into our cars, the time it takes to drive a certain distance, and the temperature on a hot or cold day. In some of our daily measurements, the units might be implied or understood. For example, if someone said the temperature outside was 39, you would probably assume this was 39 degrees Fahrenheit if you lived in the United States, but in most other parts of the world, it would be 39 degrees Celsius. Such confusion is avoided by expressing both the number and the unit of a measurement.

All measurements are based on units agreed on by those making and using the measurements. When a measurement is made in terms of an agreed-on unit, the result is expressed as some multiple of that unit. For example, when you purchase 10 pounds of potatoes, you are buying a quantity of potatoes equal to 10 times the standard quantity called 1 pound. Similarly, 3 feet of string is a length of string 3 times as long as the standard length that has been agreed on and called 1 foot.

The earliest units used for measurements were based on the dimensions of the human body. For example, the foot was the length of some important person's foot, and the

biblical cubit was the length along the forearm from the elbow to the tip of the middle finger. One problem with such units is obvious; the size of the units changed when the person on whom they were based changed because of death, change in political power, and so on.

As science became more quantitative, scientists found that the lack of standard units became more and more of a problem. A standard system of units was developed in France about the time of the French Revolution and was soon adopted by scientists throughout the world. This system, called the *metric system*, has since been adopted and is used by almost all nations of the world. The United States adopted the system but has not yet put it into widespread use.

In an attempt to further standardize scientific measurements, an international agreement in 1960 established certain basic metric units, and units derived from them, as preferred units to be used in scientific measurements. Measurement units in this system are known as SI units after the French Système International d'Unités. SI units have not yet been totally put into widespread use. Many scientists continue to express certain quantities, such as volume, in non-SI units. The metric system in this book is generally based on accepted SI units but also includes a few of the commonly used non-SI units.

1.6 | The Metric System

Learning Objective

6. Recognize units of the metric system, and convert measurements done using the metric system into related units.

The metric system has a number of advantages compared with other measurement systems. One of the most useful of these advantages is that the metric system is a decimal system in which larger and smaller units of a quantity are related by factors of 10. See **Table 1.1** for a comparison between the metric and English units of length—a meter is slightly longer than a yard. Notice in Table 1.1 that the units of length in the metric system are related by multiplying a specific number of times by 10—remember, $100 = 10 \times 10$ and $1000 = 10 \times 10 \times 10$. The relationships between the units of the English system show no such pattern.

TABLE 1.1 Metric and English Units of Length

	Base Unit	Larger Unit	Smaller Unit
Metric	1 meter	1 kilometer = 1000 meters	10 decimeters = 1 meter 100 centimeters = 1 meter 1000 millimeters = 1 meter
English	1 yard	1 mile = 1760 yards	3 feet = 1 yard 36 inches = 1 yard

basic unit of measurement A specific unit from which other units for the same quantity are obtained by multiplication or division.

derived unit of measurement A unit obtained by multiplication or division of one or more basic units.

The relationships between units of the metric system that are larger or smaller than a **basic** (defined) **unit** are indicated by prefixes attached to the name of the basic unit. Thus, 1 kilometer (km) is a unit of length that is 1000 times longer than the basic unit of 1 meter (m), and a millimeter (mm) is only $\frac{1}{1000}$ the length of 1 m. Some commonly used prefixes are given in **Table 1.2**.

Area and volume are examples of **derived units** of measurement; they are obtained or derived from the basic unit of length:

$$\text{area} = (\text{length})(\text{length}) = (\text{length})^2$$

$$\text{volume} = (\text{length})(\text{length})(\text{length}) = (\text{length})^3$$

The unit used to express an area depends on the unit of length used.

TABLE 1.2 Common Prefixes of the Metric System

Prefix[a]	Abbreviation	Relationship to Basic Unit	Exponential Relationship to Basic Unit[b]
mega-	M	$1{,}000{,}000 \times$ basic unit	$10^6 \times$ basic unit
kilo-	k	$1000 \times$ basic unit	$10^3 \times$ basic unit
deci-	d	$1/10 \times$ basic unit	$10^{-1} \times$ basic unit
centi-	c	$1/100 \times$ basic unit	$10^{-2} \times$ basic unit
milli-	m	$1/1000 \times$ basic unit	$10^{-3} \times$ basic unit
micro-	μ	$1/1{,}000{,}000 \times$ basic unit	$10^{-6} \times$ basic unit
nano-	n	$1/1{,}000{,}000{,}000 \times$ basic unit	$10^{-9} \times$ basic unit
pico-	p	$1/1{,}000{,}000{,}000{,}000 \times$ basic unit	$10^{-12} \times$ basic unit

[a]The prefixes in boldface (heavy) type are the most common ones. [b]The use of exponents to express large and small numbers is discussed in Section 1.7.

Example 1.4 Calculating Areas

Calculate the area of a rectangle that has sides of 1.5 and 2.0 m. Express the answer in units of square meters and square centimeters.

Solution

$$area = (length)(length)$$

In terms of meters, area $= (1.5 \text{ m})(2.0 \text{ m}) = 3.0 \text{ m}^2$. Note that m^2 represents meter squared, or square meters. In terms of centimeters, area $= (150 \text{ cm})(200 \text{ cm}) = 30{,}000 \text{ cm}^2$. The lengths expressed in centimeters were obtained by remembering that 1 m = 100 cm.

✔ **LEARNING CHECK 1.5** The area of a circle is given by the formula $A = \pi r^2$, where r is the radius and $\pi = 3.14$. Calculate the area of a circle that has a radius of 3.5 cm.

The unit used to express volume also depends on the unit of length used in the calculation. Thus, a volume could have such units as cubic meters (m^3), cubic decimeters (dm^3), or cubic centimeters (cm^3). The abbreviation cc is also used to represent cubic centimeters, especially in medical work. The liter (L), a non-SI unit of volume, has been used as a basic unit of volume by chemists for many years (see **Figure 1.9**). For all practical purposes, 1 L and 1 dm^3 are equal volumes. This also means that 1 milliliter (mL) is equal to 1 cm^3. Most laboratory glassware is calibrated in liters or milliliters.

Example 1.5 Calculating Liquid Volumes

A circular petri dish with vertical sides has a radius of 7.50 cm. You want to fill the dish with a liquid medium to a depth of 2.50 cm. What volume of medium in milliliters and liters will be required?

Solution

The volume of medium required will equal the area of the circular dish (in square centimeters, cm^2) multiplied by the liquid depth (in centimeters, cm). Note that the unit of this product will be cubic centimeters (cm^3). According to Learning Check 1.5 above, the area of a circle is equal to πr^2, where $\pi = 3.14$. Thus, the liquid volume will be

$$V = (3.14)(7.50 \text{ cm})^2(2.50 \text{ cm}) = 442 \text{ cm}^3$$

© Spencer L. Seager

Figure 1.9 A liter is slightly larger than a quart.

Because 1 cm³ = 1 mL, the volume equals 442 mL. Also, because 1 L = 1000 mL, the volume can be converted to liters:

$$(442 \text{ mL})\left(\frac{1 \text{ L}}{1000 \text{ mL}}\right) = 0.442 \text{ L}$$

Notice that the milliliter units canceled in the calculation. This conversion to liters is an example of the factor-unit method of problem solving, which is discussed in Section 1.9.

✔ **LEARNING CHECK 1.6** A rectangular aquarium has sides with lengths of 30.0 cm and 20.0 cm, and a height of 15.0 cm. Calculate the volume of the aquarium, and express the answer in milliliters and liters.

The basic unit of mass in the metric system is 1 kilogram (kg), which is equal to about 2.2 pounds in the English system. A kilogram is too large to be conveniently used in some applications, so it is subdivided into smaller units. Two of these smaller units that are often used in chemistry are the gram (g) and milligram (mg) (see **Figure 1.10**). The prefixes *kilo-* (k) and *milli-* (m) indicate the following relationships between these units:

$$1 \text{ kg} = 1000 \text{ g}$$

$$1 \text{ g} = 1000 \text{ mg}$$

$$1 \text{ kg} = 1{,}000{,}000 \text{ mg}$$

0.4-g paperclip 3.0-g razor blade

3.1-g penny 4.7-g nickel

© West

Figure 1.10 Metric masses of some common items as found in a 0.4-g paperclip, 3.0-g razor blade, 3.1-g penny, and 4.7-g nickel.

Example 1.6 Expressing Measurements in Metric Units

All measurements in international track and field events are made using the metric system. Javelins thrown by female competitors must have a mass of no less than 600 g. Express this mass in kilograms and milligrams.

Solution
Because 1 kg = 1000 g, 600 g can be converted to kilograms as follows:

$$600 \text{ g} \times \frac{1 \text{ kg}}{1000 \text{ g}} = 0.600 \text{ kg}$$

Also, because 1 g = 1000 mg,

$$600 \text{ g} \times \frac{1000 \text{ mg}}{1 \text{ g}} = 600{,}000 \text{ mg}$$

Once again, the units of the original quantity (600 g) were canceled, and the desired units were generated by this application of the factor-unit method (see **Section 1.9**).

✔ **LEARNING CHECK 1.7** The javelin thrown by male competitors in track and field meets must have a minimum mass of 0.800 kg. A javelin is weighed and found to have a mass of 0.819 kg. Express the mass of the weighed javelin in grams.

Temperature is difficult to define but easy for most of us to measure—we just read a thermometer. However, thermometers can have temperature scales that represent different units. For example, a temperature of 293 would probably be considered quite high until it was pointed out that it is just room temperature as measured using the Kelvin temperature scale. Temperatures on this scale are given in kelvins, K. (Notice that the abbreviation is K, not °K.)

The Celsius scale (formerly known as the centigrade scale) is the temperature scale used in most scientific work. On this scale, water freezes at 0°C and boils at 100°C under normal atmospheric pressure. A Celsius degree (division) is the same size as a kelvin of the Kelvin scale, but the two scales have different zero points. **Figure 1.11** compares the two scientific temperature scales and the familiar Fahrenheit scale. There are 100 Celsius degrees (divisions) between the freezing point (0°C) and the boiling point (100°C) of water. On the Fahrenheit scale, these same two temperatures are 180 degrees (divisions) apart (the freezing point is 32°F and the boiling point is 212°F). Readings on these two scales are related by the following equations:

$$°C = \frac{5}{9}(°F - 32°) \tag{1.1}$$

$$°F = \frac{9}{5}(°C) + 32° \tag{1.2}$$

As mentioned, the difference between the Kelvin and Celsius scales is simply the zero point; consequently, readings on the two scales are related as follows:

$$°C = K - 273 \tag{1.3}$$

$$K = °C + 273 \tag{1.4}$$

Notice that Equation 1.2 can be obtained by solving Equation 1.1 for Fahrenheit degrees, and Equation 1.4 can be obtained by solving Equation 1.3 for kelvins. Thus, you need to remember only Equations 1.1 and 1.3, rather than all four.

Figure 1.11 Fahrenheit, Celsius, and Kelvin temperature scales. The lowest temperature possible is absolute zero, 0 K.

Example 1.7 Converting Fahrenheit Temperatures to Celsius

A temperature reading of 77°F is measured with a Fahrenheit thermometer. What reading would this temperature give if a Celsius thermometer were used?

Solution

The change is from a Fahrenheit reading to a Celsius reading, so Equation 1.1 is used:

$$°C = \frac{5}{9}(°F - 32°) = \frac{5}{9}(77° - 32°) = \frac{5}{9}(45°) = 25°$$

Thus, the reading on a Celsius thermometer would be 25°C.

✔ **LEARNING CHECK 1.8** What Kelvin thermometer reading would correspond to the 77°F reading described in Example 1.7?

The last units discussed at this point are derived units of energy. Other units will be introduced later in the book as they are needed. The metric system unit of energy is a joule (J), pronounced "jewl." A joule is quite small, as shown by the fact that a 50-watt light-bulb uses 50 J of energy every second. A typical household in the United States uses several billion joules of electrical energy in a month.

The calorie (cal), a slightly larger unit of energy, is sometimes used by chemists. One calorie is the amount of heat energy required to increase the temperature of 1 g of water by 1°C. The calorie and joule are related as follows:

$$1 \text{ cal} = 4.184 \text{ J}$$

The nutritional calorie of the weight watcher is actually 1000 scientific calories, or 1 kcal. It is represented by writing *calorie* with a capital C (Calorie, abbreviated Cal). **Table 1.3** contains a list of the commonly used metric units, their relationship to basic units, and their relationship to English units.

TABLE 1.3 Commonly Used Metric Units

Quantity	Metric Unit	Relationship to Metric Basic Unit	Relationship to English Unit
Length	meter (m)	Basic unit	1 m = 1.094 yd
	centimeter (cm)	100 cm = 1 m	1 cm = 0.394 in.
	millimeter (mm)	1000 mm = 1 m	1 mm = 0.0394 in.
	kilometer (km)	1 km = 1000 m	1 km = 0.621 mi
Volume	cubic decimeter (dm^3)	Basic unit	1 dm^3 = 1.057 qt
	cubic centimeter (cm^3 or cc)	1000 cm^3 = 1 dm^3	1 cm^3 = 0.0338 fl oz
	liter (L)	1 L = 1 dm^3	1 L = 1.057 qt
	milliliter (mL)[a]	1000 mL = 1 dm^3	1 mL = 0.0338 fl oz
Mass	gram (g)	1000 g = 1 kg	1 g = 0.035 oz
	milligram (mg)	1,000,000 mg = 1 kg	1 mg = 0.015 grain
	kilogram (kg)	Basic unit	1 kg = 2.20 lb
Temperature	degree Celsius (°C)	1°C = 1 K	1°C = 1.80°F
	kelvin (K)	Basic unit	1 K = 1.80°F
Energy	calorie (cal)	1 cal = 4.184 J	1 cal = 0.00397 BTU[b]
	kilocalorie (kcal)	1 kcal = 4184 J	1 kcal = 3.97 BTU
	joule (J)	Basic unit	1 J = 0.000949 BTU
Time	second (s)	Basic unit	Same unit used

[a]1 mL = 1 cm^3. [b]A BTU (British thermal unit) is the amount of heat required to increase the temperature of 1 pound of water 1°F.

Effects of Temperature on Body Function

The human body has the ability to remain at a relatively constant temperature even when the surrounding temperature increases or decreases. Because of this characteristic, we humans are classified as warm-blooded. In reality, our body temperature varies over a significant range depending on the time of day and the temperature of our surroundings. "Normal" body temperature is considered to be 37.0°C when measured orally. However, this "normal" value can fluctuate between a low of 36.1°C for an individual just waking up in the morning to a value as high as 37.2°C just before bedtime in the late evening.

In addition to this regular variation, our body temperature fluctuates in response to extremes in the temperature of our surroundings. In extremely hot environments, the capacity of our perspiration-based cooling mechanism can be overtaxed, and our body temperature will increase. Body temperatures more than 3.5°C above normal begin to interfere with bodily functions. Body temperatures above 41.1°C can cause convulsions and can result in permanent brain damage, especially in children.

Hypothermia occurs when the body's internal heat generation is not sufficient to balance the heat lost to very cold surroundings. As a result, the body temperature decreases, and at 28.5°C the afflicted person appears pale and might have an irregular heartbeat. Unconsciousness usually results if the body temperature gets lower than 26.7°C. At these low temperatures respiration also slows and becomes shallow, resulting in a decrease in the delivery of oxygen to body tissues.

Thus, the body temperature of we warm-blooded humans is really not constant. Even though we have built-in heating and cooling systems, their capacity to maintain a constant, normal body temperature is limited.

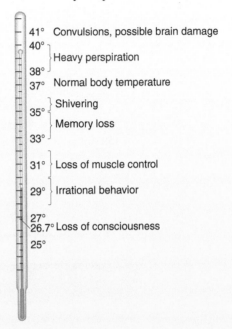

The effects of body temperature on body function using the Celsius scale.

1.7 | Large and Small Numbers

Learning Objective

7. Express numbers using scientific notation, and do calculations with numbers expressed in scientific notation.

Numbers are used in all measurements and calculations. Many numbers are readily understood and represented because of common experience with them. A price of 10 dollars, a height of 7 feet, a weight of 165 pounds, and a time of 40 seconds are examples of such numbers. But how do we handle numbers like the diameter of a hydrogen atom (about one hundred-millionth of a centimeter) or the distance light travels in 1 year—a light-year (about 6 trillion miles)? These numbers are so small and large, respectively, that they defy understanding in terms of relationships to familiar distances. Even if we can't totally relate to them, it is important in scientific work to be able to conveniently represent and work with such numbers.

Scientific notation provides a method for conveniently representing any number including those that are very large or very small. In **scientific notation**, numbers are represented as the product of a nonexponential term and an exponential term in the general form $M \times 10^n$. The nonexponential term M is a number between 1 and 10 (but not equal to 10) written with a decimal to the right of the first nonzero digit in the number. This position of the decimal is called the **standard position**. The exponential term is a 10 raised to a whole number exponent n that may be positive or negative. The value of n is the number of places the decimal must be moved from the standard position in M to be at the original position in the number when the number is written normally without using scientific notation. If n is positive, the original decimal position is to the right of the standard position. If n is negative, the original decimal position is to the left of the standard position.

scientific notation A way of representing numbers consisting of a product between a nonexponential number and 10 raised to a whole-number exponent that may be positive or negative.

standard position for a decimal In scientific notation, the position to the right of the first nonzero digit in the nonexponential number.

Example 1.8 Converting Scientific Notation to Non-scientific

The following numbers are written using scientific notation. Write them without using scientific notation.

a. 3.72×10^5 b. 8.513×10^{-7}

Solution

a. The exponent 5 indicates that the original position of the decimal is located 5 places to the right of the standard position. Zeros are added to accommodate this change:

$$3.72 \times 10^5 = 372,000. = 372,000$$

standard original
position position

b. The exponent -7 indicates that the original position of the decimal is 7 places to the left of the standard position. Again, zeros are added as needed:

$$8.513 \times 10^{-7} = 0.0000008513$$

original standard
position position

Example 1.9 Converting Non-scientific Notation to Scientific

Write the following numbers using scientific notation:

a. 8725.6 b. 0.000729

Solution

a. The standard decimal position is between the 8 and 7: 8.7256. However, the original position of the decimal is 3 places to the right of the standard position. Therefore, the exponent must be +3:

$$8725.6 = 8.7256 \times 10^3$$

b. The standard decimal position is between the 7 and 2: 7.29. However, the original position is 4 places to the left of standard. Therefore, the exponent must be -4:

$$0.000729 = 7.29 \times 10^{-4}$$

The zeros to the left of the 7 are dropped because they are not significant figures (see **Section 1.8**); they only locate the decimal in the nonscientific notation and are not needed in the scientific notation.

✔ **LEARNING CHECK 1.9** Some of the following numbers are written using scientific notation, and some are not. In each case, rewrite the number using the notation in which it is not written.

a. 5.88×10^2 c. 3.915×10^{-4} e. 36.77
b. 0.000439 d. 9870 f. 0.102

Example 1.10 Recognizing Scientific Notation

Determine which of the following numbers are written correctly using scientific notation. For those that are not, rewrite them correctly.

a. 001.5×10^{-3} b. 28.0×10^2 c. 0.35×10^4

Solution

a. Incorrect; the zeros to the left are not needed. The correct answer is 1.5×10^{-3}.
b. Incorrect; the decimal is not in the standard position. Move the decimal one position to the left and increase the exponent by 1 to give the correct answer of 2.80×10^3.
c. Incorrect; the decimal is not in the standard position. Move the decimal one position to the right and decrease the exponent by 1 to give the correct answer of 3.5×10^3.

✔ **LEARNING CHECK 1.10** Determine which of the following numbers are written correctly using scientific notation. For those that are not, rewrite them correctly.

a. 62.5×10^4 b. 0.0098 c. 0.0041×10^{-3} d. 7.85×10^2

The multiplication and division of numbers written in scientific notation can be done quite simply by using some characteristics of exponentials. Consider the following multiplication:

$$(a \times 10^y)(b \times 10^z)$$

The multiplication is done in two steps. First, the nonexponential terms a and b are multiplied in the usual way. The exponential terms 10^y and 10^z are multiplied by adding the exponents y and z and using the resulting sum as a new exponent of 10. Thus, we can write

$$(a \times 10^y)(b \times 10^z) = (a \times b)(10^{y+z})$$

Division is done similarly. The nonexponential terms are divided in the usual way, and the exponential terms are divided by subtracting the exponent of the bottom term from that of the top term. The final answer is then written as a product of the resulting nonexponential and exponential terms:

$$\frac{a \times 10^y}{b \times 10^z} = \left(\frac{a}{b}\right)(10^{y-z})$$

Multiplication and division calculations involving scientific notation are easily done using a calculator. **Table 1.4** gives the steps, the typical calculator procedures (buttons to press), and typical calculator readout or display for the division of 7.2×10^{-3} by 1.2×10^4.

TABLE 1.4 Using a Calculator for Scientific Notation Calculations		
Step	**Procedure**	**Calculator Display**
1. Enter 7.2	Press buttons 7, ., 2	7.2
2. Enter 10^{-3}	Press button that activates exponential mode (EE, Exp, etc.)	7.2^{00}
	Press 3	7.2^{03}
	Press change-sign button (\pm, etc.)	7.2^{-03}
3. Divide	Press divide button (\div)	7.2^{-03}
4. Enter 1.2	Press buttons 1, ., 2	1.2
5. Enter 10^4	Press button that activates exponential mode (EE, Exp, etc.)	1.2^{00}
	Press 4	1.2^{04}
6. Obtain answer	Press equals button ($=$)	$6.^{-07}$

Example 1.11 More Working with Scientific Notation

Do the following operations:

a. $(3.5 \times 10^4)(2.0 \times 10^2)$ c. $(4.6 \times 10^{-7})(5.0 \times 10^3)$

b. $\dfrac{3.8 \times 10^5}{1.9 \times 10^2}$ d. $\dfrac{1.2 \times 10^3}{3.0 \times 10^{-2}}$

Solution

a. $(3.5 \times 10^4)(2.0 \times 10^2) = (3.5 \times 2.0)(10^4 \times 10^2)$
$$= (7.0)(10^{4+2}) = 7.0 \times 10^6$$

b. $\dfrac{3.8 \times 10^5}{1.9 \times 10^2} = \dfrac{3.8}{1.9} \times \dfrac{10^5}{10^2} = (2.0)(10^{5-2}) = 2.0 \times 10^3$

c. $(4.6 \times 10^{-7})(5.0 \times 10^3) = (4.6 \times 5.0)(10^{-7} \times 10^3)$
$$= (23)(10^{-7+3}) = 23 \times 10^{-4}$$

To get the decimal into the standard position, move it one place to the left. This changes the exponent from -4 to -3, so the final result is 2.3×10^{-3}. (This number in decimal form, 0.0023, can be written correctly as either 23×10^{-4} or 2.3×10^{-3}, but scientific notation requires that the decimal point be to the right of the first nonzero number.)

d. $\dfrac{1.2 \times 10^3}{3.0 \times 10^{-2}} = \dfrac{1.2}{3.0} \times \dfrac{10^3}{10^{-2}} = (0.40)(10^{3-(-2)}) = 0.40 \times 10^5$

Adjust the decimal to the standard position and get 4.0×10^4. If these examples were done using a calculator, the displayed answers would normally be given in correct scientific notation.

✔ **LEARNING CHECK 1.11** Perform the following operations, and express the result in correct scientific notation:

a. $(2.4 \times 10^3)(1.5 \times 10^4)$ c. $\dfrac{6.3 \times 10^5}{2.1 \times 10^3}$

b. $(3.5 \times 10^2)(2.0 \times 10^{-3})$ d. $\dfrac{4.4 \times 10^{-2}}{8.8 \times 10^{-3}}$

The diameter of a hydrogen atom, mentioned earlier as one hundred-millionth of a centimeter, is written in scientific notation as 1.0×10^{-8} cm. Similarly, 1 light-year of 6 trillion miles is 6.0×10^{12} mi.

1.8 | Significant Figures

Learning Objective

8. Express the results of measurements and calculations using the correct number of significant figures.

Every measurement contains an uncertainty that is characteristic of the device used to make the measurement. These uncertainties are represented by the numbers used to record the measurement. Consider the small square in **Figure 1.12**:

Figure 1.12 Measurements using two different rulers.

Ruler in A is divided into centimeters.

Ruler in B is divided into tenths of centimeters.

In (A), the length of one side of the square is measured with a ruler divided into centimeters. It is easy to see that the length is greater than 1 cm, but not quite 2 cm. The length is recorded by writing the number that is known with certainty to be correct (the 1) and writing an estimate for the uncertain number. The result is 1.9 cm, where the .9 is the estimate. In (B), the ruler is divided into tenths of centimeters. It is easy to see that the length is at least 1.8 cm, but not quite 1.9 cm. Once again the certain numbers (1.8) are written, and an estimate is made for the uncertain part. The result is 1.86 cm. When measurements are recorded this way, the numbers representing the certain measurement plus one number representing the estimate are called **significant figures**. Thus, the first measurement of 1.9 cm contains two significant figures, and the second measurement of 1.86 cm contains three significant figures.

The maximum number of significant figures possible in a measurement is determined by the design of the measuring device and cannot be changed by expressing the measurement in different units. The 1.8-cm length determined earlier can also be represented in terms of meters and millimeters as follows:

significant figures The numbers in a measurement that represent the certainty of the measurement, plus one number representing an estimate.

$$1.8 \text{ cm} = 0.018 \text{ m} = 18 \text{ mm}$$

In this form, it appears that the length expressed as 0.018 m contains four significant figures, but this is impossible; a measurement made with a device doesn't become more certain simply by changing the unit used to express the number. Thus, the zeros are not significant figures; their only function is to locate the correct position for the decimal. Zeros located to the left of nonzero numbers, such as the two zeros in 0.018 cm, are never considered to be significant. Thus, 12.5 mg, 0.0125 g, and 0.0000125 kg all represent the same measured mass, and all contain three significant figures.

Zeros located between nonzero numbers are considered significant. Thus, 2055 μL, 2.055 mL, and 0.002055 L all represent the same volume measurement, and all contain four significant figures. Trailing zeros at the end of a number are significant if there is a decimal point present in the number. Otherwise, trailing zeros are not significant.

The rule that covers trailing zeros is generally followed by scientists, but sometimes there is ambiguity. For example, suppose you read in a newspaper that the population of a city is 1,250,000 people. Should the four trailing zeros be considered significant? If they are, it means that the population is known with certainty to the nearest 10 people and that the measurement has an uncertainty of only plus or minus 1 person. A more reasonable conclusion is that the census is correct to the nearest 1000 people. This could be represented using scientific notation, 1.250×10^6. In this case, only four significant figures are used. In scientific notation, the correct number of significant figures is used in the nonexponential term, and the location of the decimal is determined by the exponent. In this book, large numbers will always be represented by scientific notation. However, you are likely to encounter nonsignificant trailing zeros in other reading materials. The rules for determining the significance of zeros are summarized as follows:

1. Zeros not preceded by nonzero numbers are not significant figures. These zeros are sometimes called *leading zeros.*
2. Zeros located between nonzero numbers are significant figures. These zeros are sometimes called *buried* or *confined zeros.*
3. Zeros located at the end of a number are significant figures if there is a decimal point present in the number. These zeros are sometimes called *trailing zeros.* If no decimal point is present, trailing zeros are not significant.

Example 1.12 Significant Figures and Scientific Notation

Determine the number of significant figures in each of the following measurements, and use scientific notation to express each measurement using the correct number of significant figures:

a. 24.6°C b. 0.036 g c. 15.0 mL d. 0.0020 m

Solution

a. All the numbers are significant: three significant figures, $2.46 \times 10^1 \, °C$.
b. The leading zeros are not significant: two significant figures, $3.6 \times 10^{-2} \, g$.
c. The trailing zero is significant: three significant figures, $1.50 \times 10^1 \, mL$.
d. The leading zeros are not significant, but the trailing zero is: two significant figures, $2.0 \times 10^{-3} \, m$.

✔ **LEARNING CHECK 1.12** Determine the number of significant figures in each of the following measurements:

a. 250 mg
b. 18.05 mL
c. 0.0108 kg
d. 37°C
e. 0.001 mm
f. 101.0 K

✔ **LEARNING CHECK 1.13** Use scientific notation to express each of the following measurements using the correct number of significant figures:

a. 101 m
b. 1200 g
c. 0.00230 kg
d. 1296°C
e. 21.65 mL
f. 0.015 km

Most measurements that are made do not stand as final answers. Instead, they are usually used to make calculations involving multiplication, division, addition, or subtraction. The answer obtained from such a calculation cannot have more certainty than the least certain measurement used in the calculation. It should be written to reflect an uncertainty equal to that of the most uncertain measurement. This is accomplished by the following rules:

1. The answer obtained by multiplication or division must contain the same number of significant figures as the quantity with the fewest significant figures used in the calculation.
2. The answer obtained by addition or subtraction must contain the same number of places to the right of the decimal as the quantity in the calculation with the fewest number of places to the right of the decimal.

To follow these rules, it is often necessary to reduce the number of significant figures by rounding answers. The following are rules for rounding:

1. If the first of the nonsignificant figures to be dropped from an answer is 5 or greater, all the nonsignificant figures are dropped, and the last significant figure is increased by 1.
2. If the first of the nonsignificant figures to be dropped from an answer is less than 5, all nonsignificant figures are dropped, and the last significant figure is left unchanged.

Remember, if you use a calculator, it will often express answers with too few or too many figures (see **Figure 1.13**). It will be up to you to determine the proper number of significant figures to use and to round the calculator answer correctly.

Figure 1.13 Calculators usually display the sum of 4.362 and 2.638 as 7 (too few figures), and the product of 0.67 and 10.14 as 6.7938 (too many figures).

© Jeffrey M. Seager

Example 1.13 Significant Figures in Multiplication and Division

Do the following calculations, and round the answers to the correct number of significant figures:

a. $(4.95)(12.10)$ b. $\dfrac{3.0954}{0.0085}$ c. $\dfrac{(9.15)(0.9100)}{3.117}$ d. $\dfrac{320}{4.00}$

Solution

All calculations are done with a calculator, and the calculator answer is written first. Appropriate rounding is done to get the final answer.

a. Calculator answer: 59.895

The number 4.95 has three significant figures, and 12.10 has four. Thus, the answer must have three significant figures:

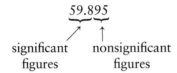

The first of the nonsignificant figures to be dropped is 9, so after both are dropped, the last significant figure is increased by 1. The final answer containing three significant figures is 59.9.

b. Calculator answer: 364.16471

The number 3.0954 has five significant figures, and 0.0085 has two. Thus, the answer must have only two:

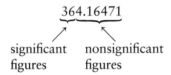

The first of the nonsignificant figures to be dropped is 4, so the last significant figure remains unchanged after the nonsignificant figures are dropped. The correct answer then is 360, which contains two significant figures. The answer also can be written with the proper number of significant figures by using scientific notation, 3.6×10^2.

c. Calculator answer: 2.6713186

The number 9.15 has three significant figures, 0.9100 has four, and 3.117 has four. Thus, the answer must have only three:

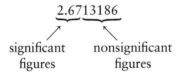

Appropriate rounding gives 2.67 as the correct answer.

d. Calculator answer: 80

The number 320 has two significant figures and 4.00 contains three significant figures, so the answer should have two. The correct answer should be expressed as 80., which contains two significant figures.

✔ **LEARNING CHECK 1.14** Do the following calculations, and round the answers to the correct number of significant figures:

a. $(0.0019)(21.39)$ b. $\dfrac{8.321}{4.1}$ c. $\dfrac{(0.0911)(3.22)}{1.379}$

Example 1.14 Significant Figures in Addition and Subtraction

Do the following additions and subtractions, and write the answers with the correct number of significant figures:

a. $1.9 + 18.65$ b. $15.00 - 8.0$ c. $1500 + 10.9 + 0.005$ d. $5.1196 - 5.02$

Solution

In each case, the numbers are arranged vertically with the decimals in a vertical line. The answer is then rounded so it contains the same number of places to the right of the decimal as the smallest number of places in the quantities added or subtracted.

a.
$$\begin{array}{r} 1.9 \\ \underline{18.65} \\ 20.55 \end{array}$$

The answer must be expressed with one place to the right of the decimal to match the one place in 1.9.

Correct answer: 20.6 (Why was the final 5 increased to 6?)

b.
$$\begin{array}{r} 15.00 \\ \underline{-8.0} \\ 7.0 \end{array}$$

The answer must be expressed using one place to the right of the decimal to match the 8.0. A typical calculator answer would probably be 7, requiring that a zero be added to the right of the decimal to provide the correct number of significant figures.

c.
$$\begin{array}{r} 1500. \\ 10.9 \\ \underline{0.005} \\ 1510.905 \end{array}$$

The answer must be expressed with no places to the right of the decimal to match the 1500.

Correct answer: 1511 (Why was the final 0 of the answer increased to 1?)

d.
$$\begin{array}{r} 5.1196 \\ \underline{-5.02} \\ 0.0996 \end{array}$$

The answer must be expressed with two places to the right of the decimal to match the 5.02.

Correct answer: 0.10 (What rounding rule was followed?)

Notice that the answer to (a) has more significant figures than the least significant number used in the calculation. The answer to (d), on the other hand, has fewer significant figures than either number used in the calculation. This happened because the rule for dealing with addition and subtraction focuses on the number of figures located to the right of the decimal and is not concerned with the figures to the left of the decimal. Thus, the number of significant figures in the answer is sometimes increased as a result of addition and decreased as a result of subtraction. You should be aware of this and not be confused when it happens.

✔ **LEARNING CHECK 1.15** Do the following additions and subtractions, and round the answers to the correct number of significant figures:

a. $8.01 + 3.2$ c. $4.33 - 3.12$
b. $3000. + 20.3 + 0.009$ d. $6.023 - 2.42$

exact numbers Numbers that have no uncertainty; numbers from defined relationships, counting numbers, and numbers that are part of simple fractions.

Some numbers used in calculations are **exact numbers** that have no uncertainty associated with them and are considered to contain an unlimited number of significant trailing zeros. Such numbers are not used when the appropriate number of significant figures is determined for calculated answers. In other words, exact numbers do not limit the number of significant figures in calculated answers. One kind of exact number is a number used as part of a defined relationship between quantities. For example, 1 m contains exactly 100 cm:

$$1 \text{ m} = 100 \text{ cm}$$

Thus, the numbers 1 and 100 are exact. A second kind of exact number is a counting number obtained by counting individual objects. A dozen eggs contains exactly 12 eggs,

not 11.8 and not 12.3. The 12 is an exact counting number. A third kind of exact number is one that is part of a simple fraction such as $\frac{1}{4}$, $\frac{2}{3}$, or the $\frac{5}{9}$ used in Equation 1.1 to convert Fahrenheit temperature readings into Celsius readings.

1.9 | Using Units in Calculations

Learning Objective

9. Use the factor-unit method to solve numerical problems.

Some beginning chemistry students are concerned about not being able to solve numerical chemistry problems. They may say, "I can work the problems, I just can't set them up." What they are really saying is "I can do the arithmetic once the numbers are properly arranged, but I can't do the arranging."

This section presents a method for arranging numbers that will work for most of the numerical problems you will encounter in this course. This method has a number of names, including the factor-unit method, the factor-label method, and dimensional analysis. We will call it the factor-unit method. It is a systematic approach to solving numerical problems and consists of the following steps:

Step 1. Write down the known or given quantity. Include both the numerical value and units of the quantity.

Step 2. Leave some working space and set the known quantity equal to the units of the unknown quantity.

Step 3. Multiply the known quantity by one or more factors, such that the units of the factor cancel the units of the known quantity and generate the units of the unknown quantity. These **factors** are fractions derived from numerical relationships between quantities. These relationships can be definitions or experimentally measured quantities. For example, the defined relationship 1 m = 100 cm provides the following two factors:

factors used in the factor-unit method Fractions obtained from numerical relationships between quantities.

$$\frac{1 \text{ m}}{100 \text{ cm}} \quad \text{and} \quad \frac{100 \text{ cm}}{1 \text{ m}}$$

Step 4. After you get the desired units, do the necessary arithmetic to produce the final answer.

Example 1.15 Factor-Unit Calculations

Use the factor-unit method and numerical relationships from Table 1.3 (see page 18) to calculate the number of yards in 10.0 m.

Solution
The known quantity is 100 m, and the unit of the unknown quantity is yards (yd).

Step 1. 10.0 m

Step 2. 10.0 m \qquad = yd

Step 3. $10.0 \text{ m} \times \dfrac{1.094 \text{ yd}}{1 \text{ m}} = \text{yd}$

The factor $\dfrac{1.094 \text{ yd}}{1 \text{ m}}$ came from the numerical relationship 1 m = 1.094 yd found in Table 1.3.

Step 4. $\dfrac{(10.0)(1.094)\text{yd}}{1} = 10.94 \text{ yd}$

This answer should be rounded to 10.9 yd, an answer that contains three significant figures, just as 10.0 m does. The 1 m in the factor is an exact number used as part of a defined relationship, so it doesn't influence the number of significant figures in the answer.

Example 1.16 More Factor-Unit Calculations

A laboratory technician uses a micropipet to measure a 5.0-μL (5.0-microliter) sample of blood serum for analysis. Express the sample volume in liters (L).

Solution

The known quantity is 5.0 μL, and the unit of the unknown quantity is liters.

Step 1. 5.0 μL

Step 2. 5.0 μL $\qquad\qquad\qquad\qquad$ = L

Step 3. 5.0 $\mu\cancel{L} \times \dfrac{1 \times 10^{-6}\,L}{1\,\mu\cancel{L}}$ = L

The factor $\dfrac{1 \times 10^{-6}\,L}{1\,\mu L}$ came from the numerical relationship 1 μL = 1 $\times 10^{-6}$ L

described in Table 1.2.

Step 4. $\dfrac{(5.0)(1 \times 10^{-6}\,L)}{1} = 5.0 \times 10^{-6}\,L$

The answer is expressed using the same number of significant figures as the 5.0 μL because 1 and 1 $\times 10^{-6}$ are exact numbers by definition.

✔ **LEARNING CHECK 1.16** Creatinine is a substance found in the blood. An analysis of a blood serum sample detected 1.1 mg of creatinine. Express this amount in grams by using the factor-unit method. Remember, the prefix milli- means $\frac{1}{1000}$, so 1 g = 1000 mg.

Example 1.17 More Factor-Unit Calculations

One of the fastest-moving nerve impulses in the body travels at a speed of 405 feet per second (ft/s). What is the speed in miles per hour?

Solution

The known quantity is 405 ft/s, and the unit of the unknown quantity is miles per hour (mi/h).

Step 1. $\dfrac{405\,ft}{s}$

Step 2. $\dfrac{405\,ft}{s} \qquad\qquad\qquad\qquad\qquad\qquad = \dfrac{mi}{h}$

Step 3. $\left(\dfrac{405\,\cancel{ft}}{\cancel{s}}\right)\left(\dfrac{1\,mi}{5280\,\cancel{ft}}\right)\left(\dfrac{60\,\cancel{s}}{1\,\cancel{min}}\right)\left(\dfrac{60\,\cancel{min}}{1\,h}\right) = \dfrac{mi}{h}$

The factors came from the following well-known numerical relationships: 1 mi = 5280 ft, 1 min = 60 s, 1 h = 60 min. All numbers in these factors are exact numbers based on definitions.

Step 4. $\dfrac{(405)(1)(60)(60)\,mi}{(5280)(1)(1)\,h} = 276.1\,\dfrac{mi}{h}$

Rounding to three significant figures, the same as in 405 ft/s, gives 276 mi/h.

✔ **LEARNING CHECK 1.17** A world-class sprinter can run 100 m in 10.0 s. This corresponds to a speed of 10.0 m/s. Convert this speed to miles per hour. Use information from Table 1.3 (see page 18).

1.10 | Calculating Percentages

Learning Objective

10. Do calculations involving percentages.

The word *percent* literally means per one hundred. It is the number of specific items in a group of 100 such items. Since items are seldom found in groups of exactly 100, we usually have to calculate the number of specific items that would be in the group if it did contain exactly 100 items. This number is the percentage, and the calculation follows:

$$\text{percent} = \frac{\text{number of specific items}}{\text{total items in the group}} \times 100 \qquad (1.5)$$

$$\% = \frac{\text{part}}{\text{total}} \times 100 \qquad (1.6)$$

In Equation 1.6, the word *part* is used to represent the number of specific items included in the total.

Example 1.18 Percentage Calculations

A college has 4517 female and 3227 male students enrolled. What percentage of the student body is female?

Solution

The total student body consists of 7744 people, of which 4517 are female.

$$\% \text{ female} = \frac{\text{number of females}}{\text{total number of students}} \times 100$$

$$\% \text{ female} = \frac{4517}{7744} \times 100 = 58.33$$

Rearrangement of Equation 1.6 gives another useful percent relationship. Because $\% = \text{part}/\text{total} \times 100$,

$$\text{part} = \frac{(\%)(\text{total})}{100} \qquad (1.7)$$

According to Equation 1.7, the number of specific items corresponding to a percentage can be calculated by multiplying the percentage and total, then dividing the product by 100.

Example 1.19 More Percentage Calculations

The human body is approximately 65% water by mass. What is the mass of water in a 170-pound (lb) person?

Solution

In this problem, what we would classify as the part is the mass of water in a person who weighs a total of 170 lb. Substitution into Equation 1.7 gives

$$\text{mass of water} = \frac{(\%)(\text{total})}{100} = \frac{(65)(170 \text{ lb})}{100} = 110.5 \text{ lb}$$

We should round our answer to only two significant figures to match the two in the 65%. The answer is 110 lbs, where the trailing zero is not significant.

a. A student is saving money to buy a computer that will cost a total of $1200. The student has saved $988. What percentage of the purchase price has been saved?

b. In a chemistry class of 83 students, 90.4% voted not to have the final exam. How many students wanted to take the exam?

STUDY SKILLS **1.1**

Help with Calculations

Many students feel uneasy about working chemistry problems that involve the use of mathematics. The uneasiness is often increased if the problem to be solved is a story problem. One tip that will help you solve such problems in this textbook is to remember that almost all of these problems are one of two types: those for which a specific formula applies and those where the factor-unit method is used. When you do homework or take quizzes or examinations and encounter a math-type problem, your first task should be to decide which type of problem it is, formula or factor-unit.

In this chapter, formula-type problems were those that dealt with percentage calculations (see **Examples 1.18** and **1.19**) and the conversion from one temperature scale to another (see **Example 1.7**). If you decide a problem is a formula type, Step 1 in solving it is to write down the formula that applies. This important step makes it easier to do the next step because a formula is like a road map that tells you how to proceed from one point to another. Step 2 is to identify the given quantity. In Step 3 put its number and units into the formula. This will leave only the answer missing in most formula-type problems. You can then obtain the answer by doing the appropriate calculations in Step 4.

In this textbook, the factor-unit method discussed in Section 1.9 is used for most problems that require mathematical calculations. This method simplifies problem solving and should be mastered so it can be used where it applies. The beauty of this method is that it mimics your natural, everyday way of solving problems. This real-life method usually involves identifying where you are, where you want to go, and how to get there. The factor-unit method follows the same pattern: Step 1, identify the given number and its units; Step 2, write down the unit of the desired answer; Step 3, put in factors that will convert the units of the given quantity into the units of the desired answer; Step 4, do calculations to get answer.

Thus, we see that working story problems is not as difficult a task as it might first appear. A key is to see through all the words and find what is given (number and units). Then look for what is wanted by focusing on key words or phrases like "how much," "what is," and "calculate." Finally, use one of the two methods, formula or factor-unit, to solve the problem.

The steps to follow in solving both types of problems are summarized in the following flow charts.

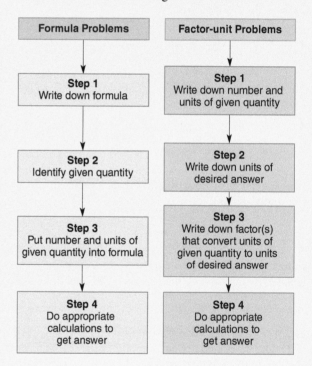

1.11 | **Density**

Learning Objective

11. Do calculations involving densities.

A discussion of the density of matter will conclude the topics of this chapter. Density is a physical property of matter, so it can be measured without changing the composition of

the sample of matter under investigation. **Density** is the number obtained by dividing the mass of a sample of matter by the volume of the same sample.

density The number given when the mass of a sample of a substance is divided by the volume of the same sample.

$$\text{density} = \frac{\text{mass}}{\text{volume}} \qquad (1.8)$$

$$d = \frac{m}{V} \qquad (1.9)$$

We see from these equations that once a numerical value has been obtained for the density, two factors are available that relate mass and volume, and these may be used to solve problems.

Example 1.20 Using Density in Calculations

The density of iron metal has been determined to be 7.2 g/cm³.

a. Use the density value to calculate the mass of an iron sample that has a volume of 35.0 cm³.
b. Use the density value to calculate the volume occupied by 138 g of iron.

Solution

The value of the density tells us that one cubic centimeter (cm³) of iron has a mass of 7.2 g. This may be written as 7.2 g = 1.0 cm³, using two significant figures for the volume. This relationship gives two factors that can be used to solve our problems:

$$\frac{7.2 \text{ g}}{1.0 \text{ cm}^3} \qquad \text{and} \qquad \frac{1.0 \text{ cm}^3}{7.2 \text{ g}}$$

a. The sample volume is 35.0 cm³, and we wish to use a factor to convert this to grams. The first factor given above will work.

$$35.0 \text{ cm}^3 \times \frac{7.2 \text{ g}}{1.0 \text{ cm}^3} = 252 \text{ g (calculator answer)}$$

$$= 2.5 \times 10^2 \text{ g (properly rounded answer)}$$

b. The sample mass is 138 g, and we wish to convert this to cubic centimeters (cm³). The second factor given above will work.

$$138 \text{ g} \times \frac{1.0 \text{ cm}^3}{7.2 \text{ g}} = 19.17 \text{ cm}^3 \text{ (calculator answer)}$$

$$= 19 \text{ cm}^3 \text{ (properly rounded answer)}$$

✔ **LEARNING CHECK 1.19** Aluminum metal has a density of 2.7 g/cm³.

a. Calculate the mass of an aluminum sample with a volume of 60.0 cm³.
b. Calculate the volume of an aluminum sample that has a mass of 98.5 g.

For some substances, density is rather easily determined experimentally by direct measurement. The mass of a sample is obtained by weighing the sample. The sample volume can be calculated if the sample is a regular solid such as a cube. If the sample is a liquid or an irregular solid, the volume can be measured by using volumetric apparatus such as those shown in **Figure 1.14**. Densities of solids are often given in units of g/cm³, and those of liquids in units of g/mL because of the different ways the volumes are determined. However, according to Table 1.3 (see page 18), 1 cm³ = 1 mL, so the numerical value is the same regardless of which of the two volume units is used.

Figure 1.14 Glassware for measuring volumes of liquids. Clockwise from top center: buret, graduated cylinder, syringe, pipet, and volumetric flask.

CHEMISTRY TIPS FOR LIVING WELL **1.1**

Choose Wisely for Health Information

The World Wide Web contains more than three million sites that are related to health issues or health problems or that sell health products. Many of these sites are very good resources for individuals concerned about such topics. Unfortunately, however, there are also sites run by scam artists who are simply interested in trying to make money at the expense of gullible or uneducated web surfers. The information provided by these sites is not only useless but might also be dangerous. There have been reports of sites run by individuals claiming to have a "miracle cure" for serious diseases. The sites encourage individuals to stop taking their prescription medication and instead buy the new miracle product. Another characteristic of a site to be avoided is one that claims to have a physician who will diagnose or treat you without requiring you to have a proper examination and consultation. As a general rule, you should use common sense. Another good idea is to find websites that are already linked with organizations you are familiar with or recognize as being legitimate. The following sites are very helpful and contain accurate and complete information.

Healthfinder: *www.healthfinder.gov*
USDA's Food & Nutrition Information Center: *www.fnic*
.nal.usda.gov
National Cancer Institute: *www.cancer.gov*
Centers for Disease Control and Prevention: *www.cdc*
.gov
U.S. Department of Health & Human Services, Agency for Healthcare Research and Quality: *www.ahrq.gov*
/research/data/infoqual.html
U.S. National Library of Medicine: *www.nlm.nih.gov*
/medlineplus/

New England Journal of Medicine: *www.nejm.org*
American Medical Association: *www.ama-assn.org/*
American Cancer Society: *www.cancer.org*
Harvard Health Publications: *www.health.harvard.edu*
Mayo Clinic: *www.mayoclinic.org*
American Academy of Family Physicians:
www.familydoctor.org
Medical Library Association: *www.mlanet.org*
NOAH: New York Online Access to Health:
www.noah-health.org

An enormous amount of health information is available through home computers and the Internet.

Example **1.21** More Use of Density in Calculations

a. A hypodermic syringe was used to deliver 5.0 cc (cm^3) of alcohol into an empty container that had a mass of 25.12 g when empty. The container with the alcohol sample weighed 29.08 g. Calculate the density of the alcohol.

b. A cube of copper metal measures 2.00 cm on each edge and weighs 71.36 g. What is the density of the copper sample?

c. According to its owner, a chain necklace is made of pure gold. In order to check this, the chain was weighed and found to have a mass of 19.21 g. The chain was then put into a graduated cylinder that contained 20.8 mL of water. After the chain was put into the cylinder, the water level rose to 21.9 mL. The density of pure gold was looked up in a handbook and found to be 19.2 g/cm^3. Is the chain made of pure gold?

Solution

a. According to Table 1.3, 1 cc (or cm^3) = 1 mL, so the volume is 5.0 mL. The sample mass is equal to the difference between the mass of the container with the sample inside and the mass of the empty container:

$$m = 29.08 \text{ g} - 25.12 \text{ g} = 3.96 \text{ g}$$

The density of the sample is equal to the sample mass divided by the sample volume:

$$d = \frac{m}{V} = \frac{3.96 \text{ g}}{5.0 \text{ mL}} = 0.79 \text{ g/mL (rounded value)}$$

b. The volume of a cube is equal to the product of the three sides:

$$V = (2.00 \text{ cm})(2.00 \text{ cm})(2.00 \text{ cm}) = 8.00 \text{ cm}^3$$

The density is equal to the mass divided by the volume:

$$d = \frac{m}{V} = \frac{71.36 \text{ g}}{8.00 \text{ cm}^3} = 8.92 \text{ g/cm}^3 \text{ (rounded value)}$$

c. The volume of the chain is equal to the difference in the water level in the cylinder with and without the chain present:

$$V = 21.9 \text{ mL} - 20.8 \text{ mL} = 1.1 \text{ mL} = 1.1 \text{ cm}^3$$

The density is equal to the mass divided by the volume:

$$d = \frac{m}{V} = \frac{19.21 \text{ g}}{1.1 \text{ cm}^3} = 17 \text{ g/cm}^3 \text{ (rounded value)}$$

The experimentally determined density is less than the density of pure gold, so the chain is not made of pure gold.

✔ LEARNING CHECK 1.20

a. A pipet was used to put a 10.00-mL sample of a liquid into an empty container. The empty container weighed 51.22 g, and the container with the liquid sample weighed 64.93 g. Calculate the density of the liquid in g/mL.

b. A box of small irregular pieces of metal was found in a storage room. It was known that the metal was either nickel or chromium. A 35.66-g sample of the metal was weighed and put into a graduated cylinder that contained 21.2 mL of water (**Figure 1.15**). The water level after the metal was added was 25.2 mL (**Figure 1.15, right**). Was the metal nickel (density = 8.9 g/cm^3) or chromium (density = 7.2 g/cm^3)?

Figure 1.15 Measuring the volume of irregular metal pieces.

© Jeffrey M. Seager

ASK AN EXPERT 1.1

Q: **Does food density matter when you're trying to lose weight?**

Melina B. Jampolis, M.D., is a Physician Nutrition Specialist focusing on nutrition for weight loss and disease prevention and treatment.

A: When it comes to losing weight, it seems like there is a new fad diet out every month. Low fat, low carb, high protein, vegan, vegetarian low carb—the choices are endless. What really works for weight loss? The answer varies among individuals and likely depends on a combination of genes, food preferences, and lifestyle. But one of the most important concepts for achieving permanent weight loss is the energy density of the food you consume. Energy density refers to the number of calories per gram in a given type of food. Eating foods that have a lower energy density allows you to eat larger, more satisfying, portions while consuming fewer calories. Research has shown that this is important both mentally and physically when it comes to weight loss. Beginning at about the age of 3 or 4, we begin to eat in response to visual cues in addition to physiological cues such as feeling full. Studies show that when presented with larger portions, most people eat significantly more yet do not report feeling more full. By lowering the energy density of the foods you consume, your portions will be visually more substantial and will also physically take up more space in your stomach.

Energy density depends on three factors: water, fiber, and fat. Water has the strongest relationship to energy density. Foods high in water content like fruits, vegetables, soups, hot cereal, and low-fat or fat-free dairy have a lower energy density because water contributes to the weight and size of a serving but not the calories. Dried fruit and dry snacks like pretzels and crackers are higher in energy density, even if they are fat free, because they don't contain water.

High-fiber foods, like whole grains, beans, and, again, fruits and vegetables, have a lower energy density because fiber contributes to the weight of the food but is not well absorbed and digested by the body, and so it contributes far fewer calories. Fat, on the other hand, is very dense calorically. It contains 9 calories per gram compared to approximately 4 calories per gram for both protein and carbohydrates. So foods that are high in fat, like oils, nuts, cheese, salad dressing, high-fat meat, cream-based sauces, and desserts have a higher energy density. Oil is particularly energy dense because it is not only high in fat, it also lacks fiber and contains no water.

Although there are no standard categories for energy density, one of the leading researchers in the field, Dr. Barbara Rolls, suggests the following:

Very low energy density = 0.0–0.6 kcal/g
Low energy density = 0.6–1.5 kcal/g
Moderate energy density = 1.5–4 kcal/g
High energy density = >4 kcal/g

By eating more foods that have lower energy density, restricting portions of foods that have high energy density, and modifying recipes to decrease the energy density (substituting apple sauce for oil in baking, combining chopped mushrooms with ground meat before grilling, making soups instead of casseroles), you can lose or maintain your weight with far less effort and more success than following the "diet of the month."

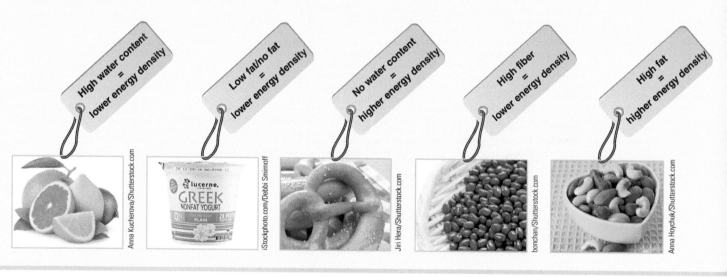

High water content = lower energy density · Low fat/no fat = lower energy density · No water content = higher energy density · High fiber = lower energy density · High fat = higher energy density

Anna Kucherova/Shutterstock.com · iStockphoto.com/Debbi Smirnoff · Jiri Hera/Shutterstock.com · bonchan/Shutterstock.com · Anna Hoychuk/Shutterstock.com

▶ Case Study Follow-up

"Microcephaly" refers to the condition of children whose head circumference falls within three standard deviations below the norm. It is caused by a variety of either genetic or environmental factors and can develop both before and after birth. Since head circumference indicates brain volume, microcephaly correlates with abnormal brain development. Causes of microcephaly include severe malnutrition, uncontrolled phenylketonuria (PKU) in the mother, infection of the fetus during pregnancy, exposure to drugs or alcohol, decreased oxygen during pregnancy or delivery, genetic abnormalities, or craniosynostosis (premature fusing of the cranial sutures). In addition to these causes, the Centers for Disease Control and Prevention (CDC) is investigating a possible new cause—the

mosquito-borne Zika virus. This highly suspected connection may explain an unusual increase in microcephaly cases in Brazil and other tropical areas. Complications of microcephaly include developmental delays in speech or movement, poor coordination, dwarfism, facial distortions, hyperactivity, and mental retardation. Head measurements can give some indication of future abilities, but head circumference alone should not be used to establish a prognosis for intellectual development. Microcephaly may be one factor in a number of important physical, neurological, developmental, and metabolic symptoms. No cure is available for microcephaly; treatment focuses on preventing and minimizing deformities and helping the child reach his or her full potential in the community.

Concept Summary

What Is Matter? Matter, the substance of everything, is defined as anything that has mass and occupies space. Mass is a measurement of the amount of matter present in an object. Weight is a measure of the gravitational force pulling on an object.
Objective 1 (Section 1.1), Exercise 1.2

Properties and Changes Chemical properties cannot be determined without attempting to change one kind of matter into another. Physical properties can be determined without attempting such composition changes. Any change in matter that is accompanied by a composition change is a chemical change. Physical changes take place without the occurrence of any composition changes.
Objective 2 (Section 1.2), Exercises 1.8 a & b and 1.10 b & c

A Model of Matter Scientific models are explanations for observed behavior. The results of many observations led scientists to a model for matter in which all matter is composed of tiny particles. In many substances, these particles are called molecules, and they represent the smallest piece of such substances that is capable of a stable existence. Molecules, in turn, are made up of atoms, which represent the limit of chemical subdivision for matter. The terms *diatomic, triatomic, polyatomic, homoatomic,* and *heteroatomic* are commonly used to describe the atomic composition of molecules.
Objective 3 (Section 1.3), Exercise 1.12

Classifying Matter It often simplifies things to classify the items being studied. Some useful categories into which matter can be classified are heterogeneous, homogeneous, solution, pure substance, element, and compound. All matter is either heterogeneous or homogeneous. Heterogeneous matter

is a mixture in which the properties and appearance are not uniform. Homogeneous matter is either a mixture of two or more pure substances (and the mixture is called a solution), or it is a pure substance. If it is a pure substance, it is either an element (containing atoms of only one kind) or a compound (containing two or more kinds of atoms).
Objective 4 (Section 1.4), Exercises 1.18, 1.22, and 1.24

Measurement Units All measurements are based on standard units that have been agreed on and adopted. The earliest measurements were based on human body dimensions, but the changeable nature of such basic units made the adoption of a worldwide standard desirable.
Objective 5 (Section 1.5), Exercise 1.28

The Metric System The metric system of measurement is used by most scientists worldwide and all major nations except the United States. It is a decimal system in which larger and smaller units of a quantity are related by factors of 10. Prefixes are used to designate relationships between the basic unit and larger or smaller units of a quantity.
Objective 6 (Section 1.6), Exercises 1.30 and 1.40

Large and Small Numbers Because of difficulties in working with very large or very small numbers in calculations, a system of scientific notation has been devised to represent such numbers. In scientific notation, numbers are represented as products of a nonexponential number and 10 raised to some power. The nonexponential number is always written with the decimal in the standard position (to the right of the first nonzero digit in the number). Numbers written in scientific notation can be manipulated in calculations by following a few rules.
Objective 7 (Section 1.7), Exercises 1.48 and 1.60

Significant Figures In measured quantities, the significant figures are the numbers representing the part of the measurement that is certain, plus one number representing an estimate. The maximum number of significant figures possible in a measurement is determined by the design of the measuring device. The results of calculations made using numbers from measurements can be expressed with the proper number of significant figures by following simple rules.

Objective 8 (Section 1.8), Exercises 1.64 and 1.66

Using Units in Calculations The factor-unit method for doing calculations is based on a specific set of steps. One crucial step involves the use of factors that are obtained from fixed numerical relationships between quantities. The units of the factor must always cancel the units of the known quantity and generate the units of the unknown or desired quantity.

Objective 9 (Section 1.9), Exercise 1.82

Calculating Percentages The word *percent* means per one hundred, and a percentage is literally the number of specific items contained in a group of 100 items. Few items always occur in groups of exactly 100, so a calculation can be done that determines how many specific items would be in a group if the group actually did contain exactly 100 items.

Objective 10 (Section 1.10), Exercise 1.92

Density The density of a substance is the number obtained by dividing the mass of a sample by the volume of the same sample. Measured values of density provide two factors that can be used with the factor-unit method to calculate the mass of a substance if the volume is known, or the volume if the mass is known.

Objective 11 (Section 1.11), Exercise 1.98

Key Terms and Concepts

Atom (1.3)
Basic unit of measurement (1.6)
Chemical changes (1.2)
Chemical properties (1.2)
Compound (1.4)
Density (1.11)
Derived unit of measurement (1.6)
Diatomic molecules (1.3)
Element (1.4)
Exact numbers (1.8)

Factors used in the factor-unit method (1.9)
Heteroatomic molecules (1.3)
Heterogeneous matter (1.4)
Homoatomic molecules (1.3)
Homogeneous matter (1.4)
Mass (1.1)
Matter (1.1)
Mixture (1.4)
Molecule (1.3)
Physical changes (1.2)

Physical properties (1.2)
Polyatomic molecules (1.3)
Pure substance (1.4)
Scientific models (1.3)
Scientific notation (1.7)
Significant figures (1.8)
Solutions (1.4)
Standard position for a decimal (1.7)
Triatomic molecules (1.3)
Weight (1.1)

Key Equations

1. Conversion of temperature readings from one scale to another (Section 1.6)	$^\circ C = \dfrac{5}{9}(^\circ F - 32^\circ)$	Equation 1.1
	$^\circ F = \dfrac{9}{5}(^\circ C) + 32^\circ$	Equation 1.2
	$^\circ C = K - 273$	Equation 1.3
	$K = {}^\circ C + 273$	Equation 1.4
2. Calculation of percentage (Section 1.10)	$\text{percent} = \dfrac{\text{number of specific items}}{\text{total items in the group}} \times 100$	Equation 1.5
	$\% = \dfrac{\text{part}}{\text{total}} \times 100$	Equation 1.6
3. Calculation of number of items representing a specific percentage of a total (Section 1.10)	$\text{part} = \dfrac{(\%)(\text{total})}{100}$	Equation 1.7
4. Calculation of density from mass and volume data (Section 1.11)	$d = \dfrac{m}{V}$	Equation 1.9

Exercises

What Is Matter? (Section 1.1)

1.1 A heavy steel ball is suspended by a thin wire. The ball is hit from the side with a hammer but hardly moves. Describe what you think would happen if this identical experiment were carried out on the moon.

1.2 Explain how the following are related to each other: matter, mass, and weight.

1.3 Tell how you would try to prove to a doubter that air is matter.

1.4 Which of the following do you think is likely to change the most when done on Earth and then on the moon? Carefully explain your reasoning.

 a. The distance you can throw a bowling ball.

 b. The distance you can roll a bowling ball on a flat, smooth surface.

1.5 The attractive force of gravity for objects near Earth's surface increases as you move toward Earth's center. Suppose you are transported from a deep mine to the top of a tall mountain.

 a. How would your mass be changed by the move?

 b. How would your weight be changed by the move?

1.6 Earth's rotation causes it to bulge at the equator. How would the weights of people of equal mass differ when one was determined at the equator and one at the North Pole? (See Exercise 1.5.)

Properties and Changes (Section 1.2)

1.7 Classify each of the following as a physical or chemical change, and give at least one observation, fact, or reason to support your answer.

 a. A plum ripens.

 b. Water boils.

 c. A glass window breaks.

 d. Food is digested.

1.8 Classify each of the following as a physical or chemical change, and give at least one observation, fact, or reason to support your answer.

 a. A stick is broken into two pieces.

 b. A candle burns.

 c. Rock salt is crushed by a hammer.

 d. Tree leaves change color in autumn.

1.9 Classify each of the following properties as physical or chemical. Explain your reasoning in each case.

 a. Iron melts at 1535°C.

 b. Alcohol is very flammable.

 c. The metal used in artificial hip-joint implants is not corroded by body fluids.

 d. A 1-in. cube of aluminum weighs less than a 1-in. cube of lead.

 e. An antacid tablet neutralizes stomach acid.

1.10 Classify each of the following properties as physical or chemical. Explain your reasoning in each case.

 a. Mercury metal is a liquid at room temperature.

 b. Sodium metal reacts vigorously with water.

 c. Water freezes at 0°C.

 d. Gold does not rust.

 e. Chlorophyll molecules are green in color.

A Model of Matter (Section 1.3)

1.11 A sample of liquid alcohol is frozen to a solid, then allowed to melt back to a liquid. Have the alcohol molecules been changed by the process? Explain your answer.

1.12 Succinic acid, a white solid that melts at 182°C, is heated gently, and a gas is given off. After the gas evolution stops, a white solid remains that melts at a temperature different from 182°C.

 a. Have the succinic acid molecules been changed by the process? Explain your answer.

 b. Is the white solid that remains after heating still succinic acid? Explain your answer.

 c. In terms of the number of atoms contained, how do you think the size of succinic acid molecules compares with the size of the molecules of the white solid produced by this process? Explain your answer.

 d. Classify molecules of succinic acid by using the term *homoatomic* or *heteroatomic*. Explain your reasoning.

1.13 A sample of solid elemental phosphorus that is deep red in color is burned. While the phosphorus is burning, a white smoke is produced that is actually a finely divided solid that is collected.

 a. Have the molecules of phosphorus been changed by the process of burning? Explain your answer.

 b. Is the collected white solid a different substance from the phosphorus? Explain your answer.

 c. In terms of the number of atoms contained, how do you think the size of the molecules of the white solid compares with the size of the molecules of phosphorus? Explain your answer.

 d. Classify molecules of the collected white solid using the term *homoatomic* or *heteroatomic*. Explain your reasoning.

1.14 Oxygen gas and solid carbon are both made up of homoatomic molecules. The two react to form a single substance, carbon dioxide. Use the term *homoatomic* or *heteroatomic* to classify molecules of carbon dioxide. Explain your reasoning.

1.15 Under appropriate conditions, hydrogen peroxide can be changed to water and oxygen gas. Use the term *homoatomic* or *heteroatomic* to classify molecules of hydrogen peroxide. Explain your reasoning.

1.16 Water can be decomposed to hydrogen gas and oxygen gas by passing electricity through it. Use the term *homoatomic* or *heteroatomic* to classify molecules of water. Explain your reasoning.

1.17 Methane gas, a component of natural gas, is burned in pure oxygen. The only products of the process are water and carbon dioxide. Use the term *homoatomic* or *heteroatomic* to classify molecules of methane. Explain your reasoning.

Classifying Matter (Section 1.4)

1.18 Classify each pure substance represented below by a capital letter as an element or a compound. Indicate when such a classification cannot be made, and explain why.

 a. Substance A is composed of heteroatomic molecules.

 b. Substance D is composed of homoatomic molecules.

 c. Substance E is changed into substances G and J when it is heated.

1.19 Classify each pure substance represented below by a capital letter as an element or a compound. Indicate when such a classification cannot be made, and explain why.

 a. Two elements when mixed combine to form only substance L.

 b. An element and a compound when mixed form substances M and Q.

 c. Substance X is not changed by heating it.

1.20 Consider the following experiments, and answer the questions pertaining to classification:

 a. A pure substance R is heated, cooled, put under pressure, and exposed to light but does not change into anything else. What can be said about classifying substance R as an element or a compound? Explain your reasoning.

 b. Upon heating, solid pure substance T gives off a gas and leaves another solid behind. What can be said about classifying substance T as an element or compound? Explain your reasoning.

 c. What can be said about classifying the solid left in part b as an element or compound? Explain your reasoning.

1.21 Early scientists incorrectly classified calcium oxide (lime) as an element for a number of years. Discuss one or more reasons why you think they might have done this.

1.22 Classify each of the following as homogeneous or heterogeneous:

 a. a pure gold chain

 b. liquid eyedrops

 c. chunky peanut butter

 d. a slice of watermelon

 e. cooking oil

 f. Italian salad dressing

 g. window glass

1.23 Classify each of the following as homogeneous or heterogeneous:

 a. muddy flood water

 b. gelatin dessert

 c. normal urine

 d. smog-filled air

 e. an apple

 f. mouthwash

 g. petroleum jelly

1.24 Classify as pure substance or solution each of the materials of Exercise 1.22 that you classified as homogeneous.

1.25 Classify as pure substance or solution each of the materials of Exercise 1.23 that you classified as homogeneous.

Measurement Units (Section 1.5)

1.26 Briefly discuss why a system of measurement units is an important part of our modern society.

1.27 In the distant past, 1 in. was defined as the length resulting from laying a specific number of grain kernels such as corn in a row. Discuss the disadvantages of such a system.

1.28 An old British unit used to express weight is a stone. It is equal to 14 lb. What sort of weighings might be expressed in stones? Suggest some standard that might have been used to establish the unit.

The Metric System (Section 1.6)

1.29 Which of the following quantities are expressed in metric units?

 a. The amount of aspirin in a tablet: 5 grains

 b. The distance between two cities: 55 km

 c. The internal displacement of an auto engine: 5 L

 d. The time for a race: 4 min, 5.2 s

 e. The area of a field: 3.6 acres

 f. The temperature on a hot day: 104°F

1.30 Which of the following quantities are expressed in metric units?

 a. Normal body temperature: 37°C

 b. The amount of soft drink in a bottle: 2 L

 c. The height of a ceiling in a room: 8.0 ft

 d. The amount of aspirin in a tablet: 81 mg

 e. The volume of a cooking pot: 4 qt

 f. The time for a short race to be won: 10.2 s

1.31 Referring to Table 1.3, suggest an appropriate metric system unit for each nonmetric unit in Exercise 1.29.

1.32 Referring to Table 1.3, suggest an appropriate metric system unit for each nonmetric unit in Exercise 1.30.

1.33 Referring only to Table 1.2, answer the following questions:

 a. A computer has 12 megabytes of memory storage. How many bytes of storage is this?

 b. A 10-km race is 6.2 mi long. How many meters long is it?

 c. A chemical balance can detect a mass as small as 0.1 mg. What is this detection limit in grams?

 d. A micrometer is a device used to measure small lengths. If it lives up to its name, what is the smallest metric length that could be measured using a micrometer?

1.34 Referring only to Table 1.2, answer the following questions:

 a. Devices are available that allow liquid volumes as small as one microliter (μL) to be measured. How many microliters would be contained in 1.00 liter?

 b. Electrical power is often measured in kilowatts. How many watts would equal 75 kilowatts?

 c. Ultrasound is sound of such high frequency that it cannot be heard. The frequency is measured in hertz

(vibrations per second). How many hertz correspond to 15 megahertz?

 d. A chlorine atom has a diameter of 200 picometers. How many meters is this diameter?

1.35 One inch is approximately equal to 2.54 cm. Express this length in millimeters and meters.

1.36 Cookbooks are going metric. In such books, 1 cup is equal to 240 mL. Express 1 cup in terms of liters and cubic centimeters.

1.37 Two cities in Germany are located 25 km apart. What is the distance in miles?

1.38 The shotput used by female track and field athletes has a mass of 4.0 kg. What would be the weight of such a shotput in pounds?

1.39 Referring to Table 1.3, answer the following questions:

 a. Which is longer, a centimeter or an inch?

 b. How many milliliters are in a quart?

 c. How many grams are in an ounce?

1.40 Referring to Table 1.3, answer the following questions:

 a. Approximately how many inches longer is a meter stick than a yardstick?

 b. A temperature increases by 65°C. How many kelvins would this increase be?

 c. You have a 5-lb bag of sugar. Approximately how many kilograms of sugar do you have?

1.41 Do the following, using appropriate values from Table 1.3:

 a. Calculate the area in square meters of a circular skating rink that has a 12.5-m radius. For a circle, the area (A) is related to the radius (r) by $A = \pi r^2$, where $\pi = 3.14$.

 b. Calculate the floor area and volume of a rectangular room that is 5.0 m long, 2.8 m wide, and 2.1 m high. Express your answers in square meters and cubic meters (meters cubed).

 c. A model sailboat has a triangular sail that is 25 cm high (h) and has a base (b) of 15 cm. Calculate the area (A) of the sail in square centimeters. $A = \dfrac{(b)(h)}{2}$ for a triangle.

1.42 Using appropriate values from Table 1.3, answer the following questions:

 a. One kilogram of water has a volume of 1.0 dm³. What is the mass of 1.0 cm³ of water?

 b. One quart is 32 fl oz. How many fluid ounces are contained in a 2.0-L bottle of soft drink?

 c. Approximately how many milligrams of aspirin are contained in a 5-grain tablet?

1.43 The weather report says the temperature is 23°F. What is this temperature on the Celsius scale? On the Kelvin scale?

1.44 Recall from Chemistry Around Us 1.3 (see page 19) that a normal body temperature might be as low as 36.1°C in the morning and as high as 37.2°C at bedtime. What are these temperatures on the Fahrenheit scale?

1.45 One pound of body fat releases approximately 4500 kcal of energy when it is metabolized. How many joules of energy is this? How many BTUs?

Large and Small Numbers (Section 1.7)

1.46 Which of the following numbers are written using scientific notation correctly? For those that are not, explain what is wrong.

 a. 02.7×10^{-3}

 b. 4.1×10^2

 c. 71.9×10^{-6}

 d. 10^3

 e. $.0405 \times 10^{-2}$

 f. 0.119

1.47 Which of the following numbers are written using scientific notation correctly? For those that are not, explain what is wrong.

 a. 4.2×10^3

 b. 6.8^4

 c. 202×10^{-3}

 d. 0.026×10^{-2}

 e. 10^{-2}

 f. 74.5×10^5

1.48 Write each of the following numbers using scientific notation:

 a. 14 thousand

 b. 365

 c. 0.00204

 d. 461.8

 e. 0.00100

 d. 9.11 hundred

1.49 Write each of the following numbers using scientific notation:

 a. three hundred

 b. 4003

 c. 0.682

 d. 91.86

 e. six thousand

 f. 400

1.50 The speed of light is about 186 thousand mi/s, or 1100 million km/h. Write both numbers using scientific notation.

1.51 A sheet of paper is 0.0106 cm, or 0.0042 in., thick. Write both numbers using scientific notation.

1.52 A single copper atom has a mass of 1.05×10^{-22} g. Write this number in a decimal form without using scientific notation.

1.53 In 2.0 g of hydrogen gas, there are approximately 6.02×10^{23} hydrogen molecules. Write this number without using scientific notation.

1.54 Do the following multiplications, and express each answer using scientific notation:

 a. $(8.2 \times 10^{-3})(1.1 \times 10^{-2})$

 b. $(2.7 \times 10^2)(5.1 \times 10^4)$

 c. $(3.3 \times 10^{-4})(2.3 \times 10^2)$

 d. $(9.2 \times 10^{-4})(2.1 \times 10^4)$

 e. $(4.3 \times 10^6)(6.1 \times 10^5)$

1.55 Do the following multiplications, and express each answer using scientific notation:

a. $(6.3 \times 10^5)(4.2 \times 10^{-8})$

b. $(2.8 \times 10^{-3})(1.4 \times 10^{-4})$

c. $(8.6 \times 10^2)(6.4 \times 10^{-3})$

d. $(9.1 \times 10^4)(1.4 \times 10^3)$

e. $(3.7 \times 10^5)(6.1 \times 10^{-3})$

1.56 Express each of the following numbers using scientific notation, then carry out the multiplication. Express each answer using scientific notation.

a. $(144)(0.0876)$

b. $(751)(106)$

c. $(0.0422)(0.00119)$

d. $(128,000)(0.0000316)$

1.57 Express each of the following numbers using scientific notation, then carry out the multiplication. Express each answer using scientific notation.

a. $(538)(0.154)$

b. $(600)(524)$

c. $(22.8)(341)$

d. $(23.6)(0.047)$

1.58 Do the following divisions, and express each answer using scientific notation:

a. $\dfrac{3.1 \times 10^{-3}}{1.2 \times 10^2}$

b. $\dfrac{7.9 \times 10^4}{3.6 \times 10^2}$

c. $\dfrac{4.7 \times 10^{-1}}{7.4 \times 10^2}$

d. $\dfrac{0.00229}{3.16}$

e. $\dfrac{119}{3.8 \times 10^3}$

1.59 Do the following divisions, and express each answer using scientific notation:

a. $\dfrac{223}{1.67}$

b. $\dfrac{6.7 \times 10^3}{4.2 \times 10^4}$

c. $\dfrac{8.7 \times 10^{-4}}{2.3 \times 10^{-2}}$

d. $\dfrac{6.8 \times 10^3}{2.7 \times 10^{-4}}$

e. $\dfrac{1.8 \times 10^{-2}}{6.5 \times 10^4}$

1.60 Do the following calculations, and express each answer using scientific notation:

a. $\dfrac{(5.3)(0.22)}{(6.1)(1.1)}$

b. $\dfrac{(3.8 \times 10^{-4})(1.7 \times 10^{-2})}{6.3 \times 10^3}$

c. $\dfrac{4.8 \times 10^6}{(7.4 \times 10^3)(2.5 \times 10^{-4})}$

d. $\dfrac{5.6}{(0.022)(109)}$

e. $\dfrac{(4.6 \times 10^{-3})(2.3 \times 10^2)}{(7.4 \times 10^{-4})(9.4 \times 10^{-5})}$

1.61 Do the following calculations, and express each answer using scientific notation:

a. $\dfrac{(7.4 \times 10^{-3})(1.3 \times 10^4)}{5.5 \times 10^{-2}}$

b. $\dfrac{6.4 \times 10^5}{(8.8 \times 10^3)(1.9 \times 10^{-4})}$

c. $\dfrac{(6.4 \times 10^{-2})(1.1 \times 10^{-8})}{(2.7 \times 10^{-4})(3.4 \times 10^{-4})}$

d. $\dfrac{(963)(1.03)}{(0.555)(412)}$

e. $\dfrac{1.15}{(0.12)(0.73)}$

Significant Figures (Section 1.8)

1.62 Indicate to what decimal position readings should be estimated and recorded (nearest 0.1, 0.01, etc.) for measurements made with the following devices:

a. A ruler with a smallest scale marking of 0.1 cm

b. A measuring telescope with a smallest scale marking of 0.1 mm

c. A protractor with a smallest scale marking of 1°

d. A tire pressure gauge with a smallest scale marking of 1 lb/in.2

1.63 Indicate to what decimal position readings should be estimated and recorded (nearest 0.1, 0.01, etc.) for measurements made with the following devices:

a. A buret with a smallest scale marking of 0.1 mL

b. A graduated cylinder with a smallest scale marking of 1 mL

c. A thermometer with a smallest scale marking of 0.1°C

d. A barometer with a smallest scale marking of 1 torr

1.64 Write the following measured quantities as you would record them, using the correct number of significant figures based on the device used to make the measurement.

a. Exactly 6 mL of water measured with a graduated cylinder that has a smallest scale marking of 0.1 mL

b. A temperature that appears to be exactly 37 degrees using a thermometer with a smallest scale marking of 1°C

c. A time of exactly nine seconds measured with a stopwatch that has a smallest scale marking of 0.1 second

d. Fifteen and one-half degrees measured with a protractor that has 1-degree scale markings

1.65 Write the following measured quantities as you would record them, using the correct number of significant figures based on the device used to make the measurements.

Even-numbered exercises answered in Appendix B.
Blue-numbered exercises **are more challenging.**

a. A length of two and one-half centimeters measured with a measuring telescope with a smallest scale marking of 0.1 mm

b. An initial reading of exactly 0 for a buret with a smallest scale marking of 0.1 mL

c. A length of four and one-half centimeters measured with a ruler that has a smallest scale marking of 0.1 cm

d. An atmospheric pressure of exactly 690 torr measured with a barometer that has a smallest scale marking of 1 torr

1.66 In each of the following, identify the measured numbers and exact numbers. Do the indicated calculation, and write your answer using the correct number of significant figures.

 a. A bag of potatoes is found to weigh 5.06 lb. The bag contains 16 potatoes. Calculate the weight of an average potato.

 b. The foul-shooting percentages for the five starting players of a women's basketball team are 71.2%, 66.9%, 74.1%, 80.9%, and 63.6%. What is the average shooting percentage of the five players?

1.67 In each of the following, identify the measured numbers and exact numbers. Do the indicated calculations, and write your answer using the correct number of significant figures.

 a. An individual has a job of counting the number of people who enter a store between 1 p.m. and 2 p.m. each day for 5 days. The counts were 19, 24, 17, 31, and 40. What was the average number of people entering the store per day for the 5-day period?

 b. The starting five members of a women's basketball team have the following heights: 6′9″, 5′8″, 5′6″, 5′1″, and 4′11″. What is the average height of the starting five?

1.68 Determine the number of significant figures in each of the following:

 a. 0.0400

 b. 309

 c. 4.006

 d. 4.4×10^{-3}

 e. 1.002

 f. 255.02

1.69 Determine the number of significant figures in each of the following:

 a. 0.040

 b. 11.91

 c. 2.48×10^2

 d. 149.1

 e. 10.003

 f. 148.67

1.70 Do the following calculations and use the correct number of significant figures in your answers. Assume all numbers are the results of measurements.

 a. $(3.71)(1.4)$

 b. $(0.0851)(1.2262)$

 c. $\dfrac{(0.1432)(2.81)}{(0.7762)}$

 d. $(3.3 \times 10^4)(3.09 \times 10^{-3})$

 e. $\dfrac{(760)(2.00)}{6.02 \times 10^{20}}$

1.71 Do the following calculations and use the correct number of significant figures in your answers. Assume all numbers are the results of measurements.

 a. $(1.21)(3.2)$

 b. $(6.02 \times 10^{23})(0.220)$

 c. $\dfrac{(0.023)(1.1 \times 10^{-3})}{100}$

 d. $\dfrac{(365)(7.000)}{60}$

 e. $\dfrac{(810)(3.1)}{8.632 \times 10^{-1}}$

1.72 Do the following calculations and use the correct number of significant figures in your answers. Assume all numbers are the results of measurements.

 a. $0.208 + 4.9 + 1.11$

 b. $228 + 0.999 + 1.02$

 c. $8.543 - 7.954$

 d. $(3.2 \times 10^{-2}) + (5.5 \times 10^{-1})$

 (HINT: Write in decimal form first, then add.)

 e. $336.86 - 309.11$

 f. $21.66 - 0.02387$

1.73 Do the following calculations and use the correct number of significant figures in your answers. Assume all numbers are the results of measurements.

 a. $2.1 + 5.07 + 0.119$

 b. $0.051 + 8.11 + 0.02$

 c. $4.337 - 3.211$

 d. $(2.93 \times 10^{-1}) + (6.2 \times 10^{-2})$

 (HINT: Write in decimal form first, then add.)

 e. $471.19 - 365.09$

 f. $17.76 - 0.0479$

1.74 Do the following calculations and use the correct number of significant figures in your answers. Assume all numbers are the results of measurements. In calculations involving both addition/subtraction and multiplication/division, it is usually better to do additions/subtractions first.

 a. $\dfrac{(0.0267 + 0.0019)(4.626)}{28.7794}$

 b. $\dfrac{212.6 - 21.88}{86.37}$

 c. $\dfrac{27.99 - 18.07}{4.63 - 0.88}$

 d. $\dfrac{18.87}{2.46} - \dfrac{18.07}{0.88}$

 (HINT: Do divisions first, then subtract.)

 e. $\dfrac{(8.46 - 2.09)(0.51 + 0.22)}{(3.74 + 0.07)(0.16 + 0.2)}$

 f. $\dfrac{12.06 - 11.84}{0.271}$

1.75 Do the following calculations and use the correct number of significant figures in your answers. Assume all numbers are the results of measurements. In calculations involving both addition/subtraction and multiplication/division, it is usually better to do additions/subtractions first.

a. $\dfrac{132.15 - 32.16}{87.55}$

b. $\dfrac{(0.0844 + 0.1021)(7.174)}{19.1101}$

c. $\dfrac{(2.78 - 0.68)(0.42 + 0.4)}{(1.058 + 0.06)(0.22 + 0.2)}$

d. $\dfrac{27.65 - 21.71}{4.97 - 0.36}$

e. $\dfrac{12.47}{6.97} - \dfrac{203.4}{201.8}$
(HINT: Do divisions first, then subtract.)

f. $\dfrac{19.37 - 18.49}{0.822}$

1.76 The following measurements were obtained for the length and width of a series of rectangles. Each measurement was made using a ruler with a smallest scale marking of 0.1 cm.

Black rectangle: $l = 12.00$ cm, $w = 10.40$ cm
Red rectangle: $l = 20.20$ cm, $w = 2.42$ cm
Green rectangle: $l = 3.18$ cm, $w = 2.55$ cm
Orange rectangle: $l = 13.22$ cm, $w = 0.68$ cm

a. Calculate the area (length × width) and perimeter (sum of all four sides) for each rectangle and express your results in square centimeters and centimeters, respectively, and give the correct number of significant figures in the result.

b. Change all measured values to meters and then calculate the area and perimeter of each rectangle. Express your answers in square meters and meters, respectively, and give the correct number of significant figures.

c. Does changing the units used change the number of significant figures in the answers?

Using Units in Calculations (Section 1.9)

1.77 Determine a single factor derived from Table 1.3 that could be used as a multiplier to make each of the following conversions:

a. 3.4 lb to kilograms
b. 3.0 yd to meters
c. 1.5 oz to grams
d. 40. cm to inches

1.78 Determine a single factor from Table 1.3 that could be used as a multiplier to make each of the following conversions:

a. 20. mg to grains
b. 350 mL to fl oz
c. 4 qt to liters
d. 5 yd to meters

1.79 Obtain a factor from Table 1.3 and calculate the number of liters in 1.00 gal (4 qt) by using the factor-unit method of calculation.

1.80 A marathon race is about 26 miles. Obtain a factor from Table 1.3 and use the factor-unit method to calculate the distance of a marathon in kilometers.

1.81 A metric cookbook calls for a baking temperature of 200°C. Your oven settings are in degrees Fahrenheit. What Fahrenheit setting should you use?

1.82 A metric cookbook calls for 250 mL of milk. Your measuring cup is in English units. About how many cups of milk should you use? (*Note*: You will need two factors, one from Table 1.3 and one from the fact that 1 cup = 8 fl oz.)

1.83 An Olympic competitor threw the javelin 62.75 m. What is this distance in feet?

1.84 You have a 40-lb baggage limit for a transatlantic flight. When your baggage is put on the scale, you think you are within the limits because it reads 18.0. But then you realize that weight is in kilograms. Do a calculation to determine whether your baggage is overweight.

1.85 You need 3.00 lb of meat that sells for $3.41/lb (i.e., 1 lb = $3.41). Use this price to determine a factor to calculate the cost of the meat you need using the factor-unit method.

1.86 During a glucose tolerance test, the serum glucose concentration of a patient was found to be 131 mg/dL. Convert the concentration to grams per liter.

Calculating Percentages (Section 1.10)

1.87 Retirement age is 65 years in many companies. What percentage of the way from birth to retirement is a 55-year-old person?

1.88 A salesperson made a sale of $467.80 and received a commission of $25.73. What percent commission was paid?

1.89 After drying, 140 lbs of grapes yields 32 lbs of raisins. What percentage of the grapes' mass was lost during the drying process?

1.90 The recommended daily intake of thiamin is 1.4 mg for a male adult. Suppose such a person takes in only 1.0 mg/day. What percentage of the recommended intake is he receiving?

1.91 The recommended daily caloric intake for a 20-year-old woman is 2000. How many Calories should her breakfast contain if she wants it to be 45% of her recommended daily total?

1.92 Immunoglobulin antibodies occur in five forms. A sample of serum is analyzed with the following results. Calculate the percentage of total immunoglobulin represented by each type.

Type:	IgG	IgA	IgM	IgD	IgE
Amount (mg):	987.1	213.3	99.7	14.4	0.1

Density (Section 1.11)

1.93 Calculate the density of the following materials for which the mass and volume of samples have been measured. Express the density of liquids in g/mL, the density of solids in g/cm^3, and the density of gases in g/L.

a. 250 mL of liquid mercury metal (Hg) has a mass of 3400 g.
b. 500. mL of concentrated liquid sulfuric acid (H_2SO_4) has a mass of 925 g.
c. 5.00 L of oxygen gas has a mass of 7.15 g.
d. A 200-cm^3 block of magnesium metal (Mg) has a mass of 350 g.

1.94 Calculate the density of the following materials for which the mass and volume of samples have been measured. Express the density of liquids in g/mL, the density of solids in g/cm^3, and the density of gases in g/L.

 a. A 50.0-mL sample of liquid acetone has a mass of 39.6 g.

 b. A 1.00-cup (236-mL) sample of homogenized milk has a mass of 243 g.

 c. 20.0 L of dry carbon dioxide gas (CO_2) has a mass of 39.54 g.

 d. A 25.0-cm^3 block of nickel metal (Ni) has a mass of 222.5 g.

1.95 Calculate the volume and density of a rectangular block of metal with edges of 5.50 cm, 12.0 cm, and 4.00 cm. The block weighs 929.5 g.

1.96 Calculate the volume and density of a cube of lead metal (Pb) that has a mass of 718.3 g and has edges that measure 3.98 cm.

1.97 The volume of an irregularly shaped solid can be determined by immersing the solid in a liquid and measuring the volume of liquid displaced. Find the volume and density of the following:

 a. An irregular piece of the mineral quartz is found to weigh 12.4 g. It is then placed into a graduated cylinder that contains some water. The quartz does not float. The water in the cylinder was at a level of 25.2 mL before the quartz was added and at 29.9 mL afterward.

 b. The volume of a sample of lead shot is determined using a graduated cylinder, as in part (a). The cylinder readings are 16.3 mL before the shot is added and 21.7 mL after. The sample of shot weighs 61.0 g.

 c. A sample of coarse rock salt is found to have a mass of 11.7 g. The volume of the sample is determined by the graduated-cylinder method described in (part a), but kerosene is substituted for water because the salt will not dissolve in kerosene. The cylinder readings are 20.7 mL before adding the salt and 26.1 mL after.

1.98 The density of ether is 0.736 g/mL. What is the volume in mL of 280. g of ether?

1.99 Calculate the mass in grams of 100.0 mL of chloroform (d = 1.49 g/mL).

Additional Exercises

1.100 Do the following metric system conversions by changing only the power of 10. For example, convert 2.5 L to mL: 2.5 L = 2.5 × 10^3 mL

 a. Convert 4.5 km to mm

 b. Convert 6.0 × 10^6 mg to g

 c. Convert 9.86 × 10^{15} m to km

 d. Convert 1.91 × 10^{-4} kg to mg

 e. Convert 5.0 ng to mg

1.101 A single water molecule has a mass of 2.99 × 10^{-23} g. Each molecule contains two hydrogen atoms that together make up 11.2% of the mass of the water molecule. What is the mass in grams of a single hydrogen atom?

1.102 It has been found that fat makes up 14% of a 170-lb person's body weight. One pound of body fat provides 4500 kcal of energy when it is metabolized. The person requires 2000 kcal of energy per day to survive. Assume all the energy needed for survival comes from the metabolism of body fat, and calculate the number of days the person could survive without eating before depleting the entire body fat reserve.

1.103 Cooking oil has a density of 0.812 g/mL. What is the mass in grams of 1.00 quart of cooking oil? Use Table 1.3 for any necessary factors.

1.104 A 175-lb patient is to undergo surgery and will be given an anesthetic intravenously. The safe dosage of anesthetic is 12 mg/kg of body weight. Determine the maximum dose of anesthetic in mg that should be used.

1.105 At 4.0°C, pure water has a density of 1.00 g/mL. At 60.0°C, the density is 0.98 g/mL. Calculate the volume in mL of 1.00 g of water at each temperature, and then calculate the percentage increase in volume that occurs as water is heated from 4.0°C to 60.0°C.

Chemistry for Thought

1.106 The following pairs of substances represent heterogeneous mixtures. For each pair, describe the steps you would follow to separate the components and collect them.

 a. wood sawdust and sand

 b. sugar and sand

 c. iron filings and sand

 d. sand soaked with oil

1.107 Explain why a bathroom mirror becomes foggy when someone takes a hot shower. Classify any changes that occur as physical or chemical.

1.108 A 20-year-old student was weighed and found to have a mass of 44.5 kg. She converted this to pounds and got an answer of 20.2 lb. Describe the mistake she probably made in doing the calculation.

1.109 Liquid mercury metal freezes to a solid at a temperature of −38.9°C. Suppose you want to measure a temperature that is at least as low as −45°C. Can you use a mercury thermometer? If not, propose a way to make the measurement.

1.110 Answer the question contained in Figure 1.3 on page 8. How does hang gliding confirm that air is an example of matter?

1.111 Show how the factor-unit method can be used to prepare an oatmeal breakfast for 27 guests at a family reunion. The directions on the oatmeal box say that 1 cup of dry oatmeal makes 3 servings.

1.112 A chemist is brought a small solid figurine. The owner wants to know if it is made of silver but doesn't want it damaged during the analysis. The chemist decides to determine the density, knowing that silver has a density of 10.5 g/mL. The figurine is put into a graduated cylinder that contains 32.6 mL of water. The reading while the figurine is in the water is 60.1 mL. The mass of the figurine is 240.8 g. Is the figurine made of silver? Explain your reasoning.

1.113 Refer to Chemistry Around Us 1.2 and explain what is meant by the following statement: All matter contains chemicals.

1.114 Refer to Figure 1.6, then use the model of matter described in Section 1.3 to propose an explanation for the following observation. When two teaspoons of sugar are dissolved in a small glass of water, the volume of the resulting solution is not significantly larger than the original volume of the water.

Allied Health Exam Connection

The following questions are from these sources:

- *Nursing School Entrance Exam* © 2005, Learning Express, LLC.

- *McGraw-Hill's Nursing School Entrance Exams* by Thomas A. Evangelist, Tamara B. Orr, and Judy Unrein © 2009, The McGraw-Hill Companies, Inc.

- *NSEE Nursing School Entrance Exams,* 3rd edition © 2009, Kaplan Publishing.

- *Cliffs Test Prep: Nursing School Entrance Exams* by Fred N. Grayson © 2004, Wiley Publishing, Inc.

- *Peterson's Master the Nursing School and Allied Health Entrance Exams,* 18th edition by Marion F. Gooding © 2008, Peterson's, a Nelnet Company.

1.115 Which of the following properties is considered a physical property?

 a. flammability

 b. boiling point

 c. reactivity

 d. osmolarity

1.116 Which of the following depicts a chemical process?

 a. Helium is combined with neon

 b. Iron forms rust

 c. Water causes soil erosion

 d. Ice melts

1.117 Which of the following is a mixture?

 a. sodium chloride

 b. rice and beans

 c. magnesium sulfate

 d. water

1.118 If it is 92°F, approximately what temperature is it on the Celsius scale?

 a. 18°C

 b. 33°C

 c. 58°C

 d. 104°C

1.119 If the temperature is 25°C, what is the temperature in °F?

 a. 25°F

 b. 298°F

 c. 0°F

 d. 77°F

1.120 The correct formula for converting Fahrenheit to Celsius is given by: $°C = \frac{5}{9}(°F - 32°)$. Convert 72°F into temperature in Celsius.

 a. 72°C

 b. 40°C

 c. 25°C

 d. 22°C

1.121 In degrees Kelvin, the freezing pointing of water is:

 a. −273°

 b. 0°

 c. 100°

 d. 273°

1.122 The number of degrees on the Fahrenheit thermometer between the freezing point and the boiling point of water is:

 a. 100 degrees

 b. 180 degrees

 c. 212 degrees

 d. 273 degrees

1.123 A calorie is a form of:

 a. light

 b. heat

 c. darkness

 d. sound

1.124 How many millimeters are there in one centimeter?

 a. 10,000

 b. 1000

 c. 100

 d. 10

1.125 Convert 4.50×10^2 nm into pm.

 a. 4.50×10^2 pm

 b. 4.50×10^{-2} pm

 c. 4.50×10^{11} pm

 d. 4.50×10^5 pm

1.126 One millimeter contains how many μm?

 a. 10

 b. 100

 c. 1000

 d. 10,000

1.127 Convert 4.50×10^2 nm into m.

 a. 4.50×10^2 m

 b. 4.50×10^{11} m

 c. 4.50×10^{-7} m

 d. 4.50×10^8 m

1.128 The quantity 6185 meters can be rewritten as:

 a. 6.185×10^3 meters.

 b. 6185 kilometers.

 c. 6185×10^3 meters.

 d. 185×10^3 meters.

1.129 The number 1,000,000 is what power of 10?

 a. 10^{-6}

 b. 10^6

 c. 1^6

 d. 0.000001

1.130 What exponent or power of ten would you use to express how many meters are in a kilometer?

 a. 10^5

 b. 10^3

 c. 10^4

 d. 10^2

1.131 Express 0.0562 in exponential notation.

 a. 0.056×10^{-3}

 b. 56×10^{-3}

 c. 562×10^{-4}

 d. 5.62×10^{-2}

1.132 Write the correct answer (correct number of significant figures) for the following calculation: $(27 + 93) \times 5.1558$.

 a. 618.697

 b. 618.7

 c. 619

 d. 618.6970

1.133 The oxidation of 1 gram of CHO (carbohydrate) produces 4 calories. How much CHO must be oxidized in the body to produce 36 calories?

 a. 4 grams

 b. 7 grams

 c. 9 grams

 d. 12 grams

1.134 The percentage of oxygen by weight in $Al_2(SO_4)_3$ (atomic weights: Al = 27, S = 32, O = 16) is approximately:

 a. 19.

 b. 21.

 c. 56.

 d. 92.

1.135 A 10-percent solution of glucose will contain:

 a. 1 gram of glucose per 1,000 milliliters of solution.

 b. 1 gram of glucose per 100 milliliters of solution.

 c. 1 gram of glucose per 10 microliters of solution.

 d. 10 grams of glucose per 100 milliliters of solution.

1.136 The density of gold (Au) is 19.3g/cm^3 and that of iron (Fe) is 7.9g/cm^3. A comparison of the volumes (V) of 50 gram samples of each metal would show that:

 a. $V_{Au} = V_{Fe}$

 b. $V_{Au} < V_{Fe}$

 c. $V_{Au} > V_{Fe}$

 d. There is no predictable relationship between volumes.

Atoms and Molecules

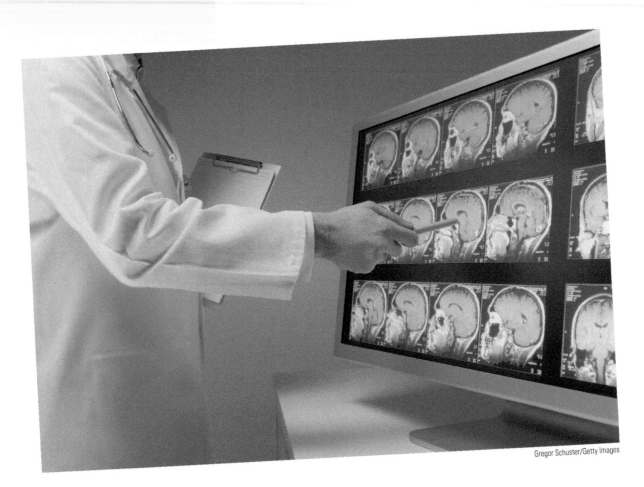

Gregor Schuster/Getty Images

▶ Case Study

Raju, a 15-year-old high-school student, was small for his age. Perhaps he took after his mother, who was quite petite. In the middle of his school year, Raju began having severe headaches, feeling them mostly behind his eyes. Over-the-counter remedies didn't seem to help much. Raju's mother had suffered from migraine headaches for some time. His doctor was inclined to diagnose migraine disorder, but Raju's symptoms didn't seem to completely support that diagnosis. Motivated by intuition, the doctor ordered an MRI scan. An MRI scan produces clear images of soft tissues such as the brain. Raju's scan revealed a golf ball–sized tumor impinging on the pituitary gland, located close to the hypothalamus, at the base of the brain. The tumor was likely the cause of Raju's growth issues as well as the headaches, because the pituitary gland regulates

hormones related to growth. Fortunately, Raju's tumor was operable. He is now leading a normal life, and annual MRI tests have shown no recurrence of the tumor.

▶ **How does an MRI scan work? What element in the body does an MRI scan detect? How important is it for health professionals to consider unlikely explanations for seemingly straightforward symptoms?**

Follow-up to this Case Study appears at the end of the chapter before the *Concept Summary*.

Learning Objectives

When you have completed your study of this chapter, you should be able to:

1 Use symbols for chemical elements to write formulas for chemical compounds. **(Section 2.1)**

2 Identify the characteristics of protons, neutrons, and electrons. **(Section 2.2)**

3 Use the concepts of atomic number and mass number to determine the number of subatomic particles in isotopes and to write correct symbols for isotopes. **(Section 2.3)**

4 Use atomic weights of the elements to calculate molecular weights of compounds. **(Section 2.4)**

5 Use isotope percent abundances and masses to calculate atomic weights of elements. **(Section 2.5)**

6 Use the mole concept to obtain relationships between number of moles, number of grams, and number of atoms for elements, and use those relationships to obtain factors for use in factor-unit calculations. **(Section 2.6)**

7 Use the mole concept and molecular formulas to obtain relationships between number of moles, number of grams, and number of atoms or molecules for compounds, and use those relationships to obtain factors for use in factor-unit calculations. **(Section 2.7)**

We introduced some fundamental ideas about matter, atoms, molecules, measurements, and calculations in Chapter 1. In this chapter, these ideas are applied, the mole is defined, and the quantitative nature of chemistry becomes more apparent. A system of symbols is introduced that simplifies the way atoms and molecules are represented.

2.1 | Symbols and Formulas

Learning Objective

1. Use symbols for chemical elements to write formulas for chemical compounds.

In Chapter 1, we defined elements as homogeneous pure substances made up of identical atoms. At least 118 different elements are known to exist. This leads to the conclusion that a minimum of 118 different kinds of atoms exist. Eighty-eight of the elements are naturally occurring and therefore are found in Earth's crust, oceans, or atmosphere. The others are synthetic elements produced in the laboratory.

Each element can be characterized and identified by its unique set of physical and chemical properties, but it would be very cumbersome to list all these properties each time a specific element is discussed. For this reason, each element has been assigned a unique name and symbol. Many elements have been named by their discoverer, and as a result the names are varied. Some are based on elemental properties, others come from the names of famous scientists or places, while others are derived from the names of astronomical bodies or mythological characters.

An **elemental symbol** is based on the element's name and consists of a single capital letter or a capital letter followed by a lowercase letter. The symbols for 11 elements are

elemental symbol A symbol assigned to an element based on the name of the element, consisting of one capital letter or a capital letter followed by a lowercase letter.

based on the element's name in Latin or German. An elemental symbol is sometimes used to represent an element in a general way or to represent a single atom of an element. **Table 2.1** lists the elements whose names and symbols have been agreed on.

TABLE 2.1 The Chemical Elements and Their Symbols

Ac	actinium	Es	einsteinium	Mt	meitnerium	Sb	antimony (*stibium*)[a]
Ag	silver (*argentum*)[a]	Eu	europium	N	nitrogen	Sc	scandium
Al	aluminum	F	fluorine	Na	sodium (*natrium*)[a]	Se	selenium
Am	americium	Fe	iron (*ferrum*)[a]	Nb	niobium	Sg	seaborgium
Ar	argon	Fl	flerovium	Nd	neodymium	Si	silicon
As	arsenic	Fm	fermium	Ne	neon	Sm	samarium
At	astatine	Fr	francium	Nh	nihonium	Sn	tin (*stannum*)[a]
Au	gold (*aurum*)[a]	Ga	gallium	Ni	nickel	Sr	strontium
B	boron	Gd	gadolinium	No	nobelium	Ta	tantalum
Ba	barium	Ge	germanium	Np	neptunium	Tb	terbium
Be	beryllium	H	hydrogen	O	oxygen	Tc	technetium
Bh	bohrium	He	helium	Og	oganesson	Te	tellurium
Bi	bismuth	Hf	hafnium	Os	osmium	Th	thorium
Bk	berkelium	Hg	mercury (*hydrargyrum*)[a]	P	phosphorus	Ti	titanium
Br	bromine	Ho	holmium	Pa	protactinium	Tl	thallium
C	carbon	Hs	hassium	Pb	lead (*plumbum*)[a]	Tm	thulium
Ca	calcium	I	iodine	Pd	palladium	Ts	tennessine
Cd	cadmium	In	indium	Pm	promethium	U	uranium
Ce	cerium	Ir	iridium	Po	polonium	V	vanadium
Cf	californium	K	potassium (*kalium*)[a]	Pr	praseodymium	W	tungsten (*wolfram*)[a]
Cl	chlorine	Kr	krypton	Pt	platinum	Xe	xenon
Cm	curium	La	lanthanum	Pu	plutonium	Y	yttrium
Cn	copernicium	Li	lithium	Ra	radium	Yb	ytterbium
Co	cobalt	Lr	lawrencium	Rb	rubidium	Zn	zinc
Cr	chromium	Lu	lutetium	Re	rhenium	Zr	zirconium
Cs	cesium	Lv	livermorium	Rf	rutherfordium		
Cu	copper (*cuprum*)[a]	Mc	moscovium	Rg	roentgenium		
Db	dubnium	Md	mendelevium	Rh	rhodium		
Ds	darmstadtium	Mg	magnesium	Rn	radon		
Dy	dysprosium	Mn	manganese	Ru	ruthenium		
Er	erbium	Mo	molybdenum	S	sulfur		

[a]Elements with symbols not derived from their English names.

compound formula A representation of the molecule of a compound, consisting of the symbols of the atoms found in the molecule. Atoms present in numbers greater than one have the number indicated by a subscript.

Compounds are pure substances made up of two or more different kinds of atoms. The atoms found in compounds are the same ones found in elements. Thus, the symbols used to represent elements can be combined and used to represent compounds. A molecular compound is depicted by a **compound formula**, in which each atom is represented by an appropriate elemental symbol. When more than one atom of an element is present in a molecule, a subscript is used to indicate the number.

Notice the similarity between this practice and the molecular representations in Figure 1.4. The carbon dioxide molecules in Figure 1.4 are represented by the formula CO_2, where C represents an atom of carbon and O an atom of oxygen. The subscript 2 on

Chemical Elements in the Human Body

As indicated in the diagram, the human body, like all matter, is made up of chemical elements. The elements are in the form of compound molecules or ions. The most abundant element in terms of mass is oxygen, because the body contains a lot of water-based fluids and water is 89% oxygen and 11% hydrogen by mass.

Almost 99% of the total mass of the human body consists of six elements: oxygen, carbon, hydrogen, nitrogen,

calcium, and phosphorus. About 1% is made up of another four elements: potassium, sulfur, sodium, and chlorine. All of these ten elements are necessary for life. In addition, more than a dozen other elements that are present in small amounts are thought to be necessary for life or to play a role in good health.

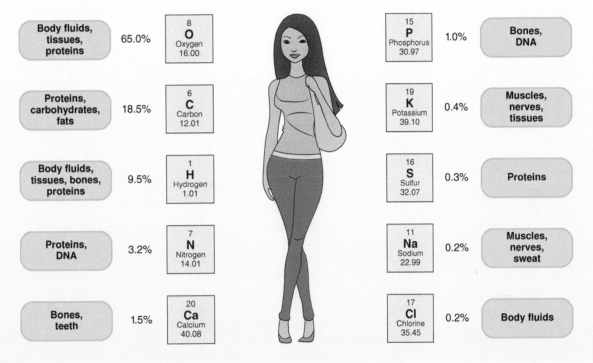

Body fluids, tissues, proteins	65.0%	8 O Oxygen 16.00		15 P Phosphorus 30.97	1.0%	Bones, DNA
Proteins, carbohydrates, fats	18.5%	6 C Carbon 12.01		19 K Potassium 39.10	0.4%	Muscles, nerves, tissues
Body fluids, tissues, bones, proteins	9.5%	1 H Hydrogen 1.01		16 S Sulfur 32.07	0.3%	Proteins
Proteins, DNA	3.2%	7 N Nitrogen 14.01		11 Na Sodium 22.99	0.2%	Muscles, nerves, sweat
Bones, teeth	1.5%	20 Ca Calcium 40.08		17 Cl Chlorine 35.45	0.2%	Body fluids

the oxygen indicates that the molecule contains two atoms of oxygen. Notice that the single carbon atom in the molecule is not indicated by a subscript 1. Subscript 1 is never used in molecular formulas; it is understood. Formulas for molecular compounds are sometimes used to represent a compound in a general way, or to represent a single molecule of a compound. See **Table 2.2** and **Figure 2.1** for other examples of compound formulas.

Cengage Learning Steven Hyatt

Figure 2.1 Iron pyrite is a mineral that contains iron and sulfur atoms in a 1:2 ratio, respectively. The golden crystals, called "fool's gold" by experienced miners, have caused (temporary) excitement for many novice prospectors. What is the formula for iron pyrite?

TABLE 2.2 **Examples of Compound Formulas**		
Compound Name	**Molecular Representation**	**Molecular Formula**
Methane		CH_4
Water		H_2O
Carbon monoxide		CO
Hydrogen peroxide		H_2O_2

Example 2.1 Writing Compound Formulas

Write formulas for the following compounds:

a. Nitrogen dioxide: one nitrogen (N) atom and two oxygen (O) atoms
b. Sulfuric acid: two hydrogen (H) atoms, one sulfur (S) atom, and four oxygen (O) atoms

Solution

a. The single N atom will not have a subscript because ones are understood and never written. The two O atoms will be represented by writing a subscript 2. The molecular formula is NO_2.
b. Using similar reasoning, the H atom will have a subscript 2, the S atom will have no subscript, and the O atom will have a subscript 4. Therefore, the molecular formula is H_2SO_4.

✔ **LEARNING CHECK 2.1** Write molecular formulas for the following compounds:

a. Phosphoric acid: three hydrogen (H) atoms, one phosphorus (P) atom, and four oxygen (O) atoms
b. Sulfur trioxide: one sulfur (S) atom and three oxygen (O) atoms
c. Glucose: six carbon (C) atoms, twelve hydrogen (H) atoms, and six oxygen (O) atoms

2.2 | Inside the Atom

Learning Objective

2. Identify the characteristics of protons, neutrons, and electrons.

An atom has been defined as the limit of chemical subdivision for matter. On the basis of the characteristics of atoms that have been discussed, you probably have a general (and correct) idea that atoms can be considered to be the units from which matter is made. However, the question of how atoms interact to form matter has not yet been addressed. This interesting topic is discussed in Chapters 3 and 4, but a bit more must be learned about atoms first.

Extensive experimental evidence collected since the middle of the 19th century indicates that atoms are made up of many smaller particles. More than 100 of these subatomic particles have been discovered, and the search for more continues. As yet there is no single theory that can explain all observations involving subatomic particles, but three fundamental particles are included in all current theories: the proton, the neutron, and the electron. Most chemical behavior of matter can be explained in terms of a few of the well-known characteristics of these particles. These important characteristics are mass, electrical charge, and location in atoms. They are summarized in **Table 2.3**. The atomic mass unit (u) shown in the table is discussed in Section 2.4.

TABLE 2.3 Characteristics of the Fundamental Subatomic Particles

Particle	Common Symbols	Characteristics			
		Charge (±)	Mass (g)	Mass (u)	Location
Electron	e^-	1−	9.07×10^{-28}	1/1836	Outside nucleus
Proton	p, p^+, H^+	1+	1.67×10^{-24}	1	Inside nucleus
Neutron	n	0	1.67×10^{-24}	1	Inside nucleus

For many years, scientists believed in the existence of atoms based primarily on the successful use of the atomic theory to explain the observed properties of matter. Direct visual evidence of atoms, however, eluded scientists until 1970, when an electron microscope, which uses electron beams instead of light, obtained images that showed the locations of large, heavy atoms of thorium and uranium.

In 1981, the scanning tunneling microscope (STM) was introduced. A different type of microscope, the STM does not have lenses and a light source like the microscopes with which you are familiar. Instead, an STM uses a very thin, needlelike probe that moves horizontally over the surface of a sample. A high-voltage electrical current passes between the sharp tip of the probe and the sample under investigation. The magnitude of the current passing between the sample and the probe depends on the distance between the tip of the probe and the surface of the sample. As the probe passes over the atoms making up the sample surface, the STM keeps the magnitude of the current constant by moving the probe up and down to maintain a very tiny but constant separation distance between the atoms and the probe tip. It moves up when it goes over a protruding atom and moves down when it is above a space between atoms. A computer detects and processes these tiny up-and-down motions and converts them into a greatly enlarged, three-dimensional image. Because the motion resulted from the atomic bumps on the surface, the image shows the location of the bump-producing atoms on the surface. The inventors of the STM, Gerd Binnig, Heinrich Rohrer, and Ernst Ruska, shared the 1986 Nobel Prize in physics for their accomplishment.

A view from a scanning tunneling microscope (STM) shows the locations of atoms protruding from a short strand of a DNA double helix.

In Chapter 1, atoms were described as being very tiny particles. The masses of single atoms of the elements are now known to fall within the range of 1.67×10^{-24} g for the least massive to 5.00×10^{-22} g for the most massive. Thus, it is not surprising to find that the particles that make up atoms have very small masses as well. Even though atomic masses are very small, the mass information given in Table 2.3 indicates that most of the mass of an atom comes from the protons and neutrons it contains. The protons and neutrons are tightly bound together to form the central portion of an atom called the **nucleus**. Because protons each have a +1 electrical charge, and neutrons have no charge, the nucleus of an atom has a positive electrical charge that is equal to the number of protons it contains.

Electrons are negatively charged particles located outside the nucleus of an atom. Protons and electrons carry equal but opposite electrical charges, so a neutral atom that has no electrical charge must have the same number of protons in its nucleus as it has electrons outside the nucleus. The electrons of an atom are thought to move very rapidly throughout a relatively large volume of space surrounding the small but very heavy nucleus (see **Figure 2.2**).

Even though subatomic particles exist, the atom itself is the particle of primary interest in chemistry, because subatomic particles do not lead an independent existence for any appreciable length of time. The only way they gain long-term stability is by combining with other particles to form an atom.

nucleus The central core of atoms that contains protons, neutrons, and most of the mass of atoms.

Figure 2.2 Electrons move rapidly around a massive nucleus. This figure is not drawn to scale. For a nucleus of the size shown, the closest electrons would be at least 80 m away.

2.3 | Isotopes

Learning Objective

3. Use the concepts of atomic number and mass number to determine the number of subatomic particles in isotopes and to write correct symbols for isotopes.

atomic number of an atom
A number equal to the number of protons in the nucleus of an atom. Symbolically it is represented by Z.

Atoms of elements have no electrical charge and so must contain identical numbers of positive protons and negative electrons. However, because neutrons have no electrical charge, their numbers in an atom do not have to be the same as the numbers of protons or electrons. The number of protons in the nucleus of an atom is given by the **atomic number** for the atom. Atomic numbers are represented by the symbol Z. All atoms of a specific element must have the same atomic number. The atomic numbers for each element are the numbers above the elemental symbols of the periodic table inside the front cover of this book. Remember, this is also the number of electrons in the atoms of each element.

The possibility of having different numbers of neutrons combined with one given number of protons to form nuclei leads to some interesting results. For example, three different kinds of hydrogen atoms are known to exist. Each kind of atom contains one proton and one electron, so all have an atomic number of 1. However, the nuclei of the different kinds of hydrogen atoms contain different numbers of neutrons. The most common kind has no neutrons in the nucleus, the next most common has one, and the least common kind has two. The sum of the number of protons and the number of neutrons in a nucleus is called the **mass number** and is represented by the symbol A. Thus, the three kinds of hydrogen atoms all have atomic numbers of 1 and mass numbers of 1, 2, and 3, respectively. Atoms that have the same atomic number but different mass numbers are called **isotopes**. Most elements are made up of mixtures of two or more isotopes. When it is important to distinguish between them, the following notation is used for each isotope: $_Z^A E$, where E is the symbol for the element.

mass number of an atom
A number equal to the sum of the number of protons and neutrons in the nucleus of an atom. Symbolically it is represented by A.

isotopes Atoms that have the same atomic number but different mass numbers. That is, they are atoms of the same element that contain different numbers of neutrons in their nuclei.

The three isotopes of hydrogen are represented as follows using this notation: $_1^1 H$, $_1^2 H$, and $_1^3 H$. When these symbols are not convenient to use, as in written or spoken references to the isotopes, the elemental name followed by the mass number is used. Thus, the three hydrogen isotopes are hydrogen-1, hydrogen-2, and hydrogen-3. These three isotopes have specific names: protium ($A = 1$), deuterium ($A = 2$), and tritium ($A = 3$).

Example 2.2 Using the Periodic Table

Use the periodic table inside the front cover to answer the following questions about isotopes:

a. What are the mass number, atomic number, and isotope symbol ($_Z^A E$) for an atom that contains 7 protons and 8 neutrons?
b. How many neutrons are contained in an atom of nickel-60?
c. How many protons and how many neutrons are contained in an atom with a mass number of 26 and the symbol Mg?

Solution

a. The mass number, A, equals the sum of the number of protons and the number of neutrons: $A = 7 + 8 = 15$. The atomic number, Z, equals the number of protons: $Z = 7$. According to the periodic table, the element with an atomic number of 7 is nitrogen, with the symbol N. The isotope symbol is $_7^{15}N$.
b. According to the periodic table, nickel has the symbol Ni, and an atomic number, Z, of 28. The mass number, 60, is equal to the sum of the number of protons and the number of neutrons. The number of protons is equal to the atomic number, 28. Therefore, the number of neutrons is $60 - 28 = 32$. The atom contains 32 neutrons.

c. According to the periodic table, the element with the symbol Mg is magnesium, which has an atomic number of 12. Therefore, the atom contains 12 protons. Since A, the number of protons plus neutrons is equal to 26, the number of neutrons is $26 - 12$, or 14. The atom contains 14 neutrons.

✔ **LEARNING CHECK 2.2** Use the periodic table inside the front cover to answer the following questions about isotopes:

a. What are the atomic number, mass number, and isotope symbol for an atom that contains 4 protons and 5 neutrons?
b. How many neutrons are contained in an atom of chlorine-37?
c. How many protons and how many neutrons are contained in an atom with a mass number of 28 and the symbol Si?

2.4 | Relative Masses of Atoms and Molecules

Learning Objective

4. Use atomic weights of the elements to calculate molecular weights of compounds.

Because of their extremely small size, it is very inconvenient to use the actual masses of atoms when the atoms are being characterized or when quantitative calculations are done. In fact, the earliest chemists had no way of determining the actual masses of atoms. For this reason, a system was devised that utilized relative or comparative masses for the atoms. These relative masses are the numbers that are given beneath the symbol and name for each element in the periodic table inside the front cover.

Relative masses provide a simple way of comparing the masses of atoms. For example, the mass of neon atoms, Ne, from the periodic table is given as 20.18. Similarly, the mass of calcium atoms, Ca, is given as 40.08. These numbers simply indicate that calcium atoms have a mass that is about twice the mass of neon atoms. The exact relationship between the two masses is calculated as follows, using the correct number of significant figures:

$$\frac{\text{Ca atom mass}}{\text{Ne atom mass}} = \frac{40.08}{20.18} = 1.986$$

In a similar way, we arrive at the conclusion that helium atoms are about four times as massive as hydrogen atoms:

$$\frac{\text{He atom mass}}{\text{H atom mass}} = \frac{4.003}{1.008} = 3.971$$

In each case, we have been able to determine the relationship between the masses of the atoms without using the actual masses.

Modern instruments called mass spectrometers allow the actual masses of individual atoms to be measured. These measured masses show the same relationships to each other as do the relative masses:

$$\frac{\text{Ca atom mass}}{\text{Ne atom mass}} = \frac{6.655 \times 10^{-23} \text{ g}}{3.351 \times 10^{-23} \text{ g}} = 1.986$$

$$\frac{\text{He atom mass}}{\text{H atom mass}} = \frac{6.647 \times 10^{-24} \text{ g}}{1.674 \times 10^{-24} \text{ g}} = 3.971$$

ASK A PHARMACIST 2.1

Uprooting Herbal Myths

Truth is seldom final or absolute; it depends on the kind and amount of evidence that substantiates it. Just how substantial, and how valid, does the evidence need to be before something is believable? Too often, the selection of herbs used as medications is based on common myths. See if any of these sound familiar. Are they fact or fallacy?

MYTH: Herbs are natural, so they are "safe and effective."

TRUTH: Many herbs are "safe," but never consider all natural substances as safe. There are many natural compounds that are very dangerous—think of hemlock, cyanide, and botulinum. Herbs are not required by law to be proven safe and effective before they are marketed. Adverse reactions could be from the direct effect, side effect, overdose, or chronic use. The "effectiveness" of a compound could be influenced by how it was grown, harvested, and prepared for use. Usually, herbs are less effective than traditional dosage forms. For example, it takes 10 cups of white willow bark tea (a source of aspirin) to be equivalent to one 5-grain aspirin tablet.

MYTH: Herbs must be safe and effective because they have been used for hundreds of years.

TRUTH: Some toxicities might take years to develop and might need large studies or a case report registry to identify infrequent adverse side effects. There might be no proven benefit from cultural or traditional use. At the present time, no adverse event reporting is required for herbal therapies.

MYTH: You don't need to tell your doctor that you're taking herbs.

TRUTH: Don't romanticize a "natural benefit" and ignore potential risks. If an herb has enough potency to cause a desired effect, it might have enough potency to cause harm or confound a treatment plan. Taking ginkgo for cognitive improvement might interfere with blood coagulation. Ma huang taken to suppress appetite might aggravate mood/anxiety disorders. There are many well-described drug-herb interactions that need to be avoided.

MYTH: If it is on the label, it must be true.

TRUTH: Herbal therapies are not governed by the FDA, and consequently do not have to conform to the labeling requirements for prescription (Rx) or over-the-counter (OTC) medications. Herbal supplements fall under the Dietary Supplement Health and Education Act (DSHEA) of 1994, which allows structure and function claims on the label.

MYTH: I don't need special training to take herbs.

TRUTH: Since there is less reliable information available about herbs, and there are few safeguards in a relatively unregulated industry, it is even more important to have special training to allow for an objective, informed decision to be made.

Reliable information can be obtained from the National Center for Complementary and Alternative Medicine funded by the National Institutes of Health.

Be sure to tell your doctor if you are taking herbal products.

Thus, we see that the relative masses used in the periodic table give the same results as the actual masses when the masses of atoms are compared with one another.

Actual atomic masses are given in mass units such as grams, but the relative values used in the periodic table are given in units referred to as **atomic mass units**. The symbol u will be used to represent atomic mass units throughout the text. The actual mass represented by a single atomic mass unit is $\frac{1}{12}$ the mass of a single carbon-12 atom or 1.661×10^{-24} g.

The relative masses of the elements as given in the periodic table are referred to as atomic masses or atomic weights. We will use the term **atomic weight** in this book. In those cases where the naturally occurring element exists in the form of a mixture of isotopes, the recorded atomic weight is the average value for the naturally occurring isotope mixture. This idea is discussed in Section 2.5.

atomic mass unit (u) A unit used to express the relative masses of atoms. One u is equal to $\frac{1}{12}$ the mass of an atom of carbon-12.

atomic weight The mass of an average atom of an element expressed in atomic mass units.

Take Care of Your Bones

Osteoporosis, the abnormal thinning of bones that often accompanies aging, may lead to bone fractures, disability, and even death. While women are most susceptible, this serious condition also affects men, but usually at a more advanced age than women. A number of significant risk factors have been identified, including the following:

1. A poor diet, especially one low in calcium
2. Advanced age
3. The onset of menopause (or having had ovaries removed)
4. A sedentary lifestyle
5. Smoking
6. A family history of osteoporosis or hip fracture
7. Heavy drinking
8. The long-term use of certain steroid drugs
9. Vitamin D deficiency

Ninety-nine percent of the calcium found in the body is located in the skeleton and teeth, so it is not surprising that the behavior of this metallic element in the body plays a central role in a number of these risk factors for osteoporosis. For example, it is known that in later life the body's ability to absorb calcium from food in the small intestine decreases (factor 2), and such absorption requires the presence of vitamin D (factor 9).

It is well recognized that the best insurance against developing osteoporosis in later life is to build and strengthen as much bone as possible during the first 25 to 35 years of life. Two essential components of this process are eating a healthful diet containing adequate amounts of calcium and vitamin D, and following a healthful lifestyle that includes regular weight-bearing

exercise such as walking, jogging, weight lifting, stair climbing, or physical labor. Dietary calcium supplements provide an additional way to enhance calcium intake, especially for individuals who are at risk to develop osteoporosis.

A bone density test provides an effective way to diagnose the presence or extent of osteoporosis in an individual. Such tests measure the absorption of X-rays by bones, and are not invasive or uncomfortable. Bone density tests were not routinely recommended in the past, but some health and wellness organizations now suggest that women who are at high risk should have such a test by the age of 50, and all women over age 65 should be routinely tested.

Sally Ho/Getty Images

An active lifestyle is an essential component of building and strengthening bones.

A convenient way to visualize the concept of relative masses, and to compare the relative masses of atoms, involves the use of the simple child's toy called a seesaw or teeter-totter, shown below.

When the masses on both sides of the central pivot are equal, the seesaw will be in balance. When the masses on each side are different, the seesaw will be out of balance, and the side with the greatest mass will go down. This characteristic lends itself to comparing the masses of objects to each other without knowing their actual values in mass units such as grams.

Example 2.3 Atomic Weight Calculations

Round the atomic weights of the periodic table to the nearest whole number, and answer the following questions:

a. How many calcium atoms, Ca, would have to be put on one side of a seesaw to balance one atom of bromine, Br, that was placed on the other side?
b. How many helium atoms, He, would balance one oxygen atom on a seesaw?
c. How many hydrogen atoms, H, would balance one carbon atom, C, on a seesaw?

Solution

a. Calcium atoms have a periodic table atomic weight of 40.08 u, which rounds to 40 u. Bromine atoms have a periodic table atomic weight of 79.90 u, which rounds to 80 u. Thus, two calcium atoms would be required to balance one bromine atom as indicated below:

b. Helium atoms have a periodic table atomic weight of 4.003 u, which rounds to 4 u. Oxygen atoms have a periodic table atomic weight of 16.00 u, which rounds to 16 u. Thus, four helium atoms will balance one oxygen atom:

c. Hydrogen atoms have a periodic table atomic weight of 1.008 u, which rounds to 1 u. Carbon atoms have a periodic table atomic weight of 12.01 u, which rounds to 12 u. Twelve hydrogen atoms will balance one carbon atom:

✔ **LEARNING CHECK 2.3** Round atomic weights of the periodic table to the nearest whole number, and answer the following questions using the seesaw balance idea:

a. How many nitrogen atoms, N, will balance one atom of iron, Fe?
b. How many carbon atoms, C, will balance six helium atoms, He?
c. How many calcium atoms, Ca, will balance two argon atoms, Ar?

Molecules are made up of atoms, so the relative mass of a molecule can be calculated by adding together the atomic weights of the atoms that make up the molecule. Relative molecular masses calculated in this way are called **molecular weights** and are also given in atomic mass units.

molecular weight The relative mass of a molecule expressed in atomic mass units and calculated by adding together the atomic weights of the atoms in the molecule.

Example 2.4 Atomic Weights and Molecular Weights

Use atomic weights from the periodic table inside the front cover to determine the molecular weight of urea, CH_4N_2O, the chemical form in which much nitrogenous body waste is excreted in the urine.

Solution

According to the formula given, a urea molecule contains one carbon atom, C, four hydrogen atoms, H, two nitrogen atoms, N, and one oxygen atom, O. The molecular weight is calculated as follows:

$$MW = 1(\text{at. wt. C}) + 4(\text{at. wt. H}) + 2(\text{at. wt. N}) + 1(\text{at. wt. O})$$

$$MW = 1(12.01 \text{ u}) + 4(1.008 \text{ u}) + 2(14.01 \text{ u}) + 1(16.00 \text{ u}) = 60.062 \text{ u}$$

Rounded to four significant figures, the correct answer is 60.06 u.

✔ **LEARNING CHECK 2.4**

a. Use atomic weights from the periodic table to determine the molecular weight of sulfuric acid. Each molecule contains two hydrogen (H) atoms, one sulfur (S) atom, and four oxygen (O) atoms.

b. Determine the molecular weight of isopropyl alcohol (C_3H_8O), the active ingredient in most rubbing alcohol sold commercially.

2.5 | Isotopes and Atomic Weights

Learning Objective

5. Use isotope percent abundances and masses to calculate atomic weights of elements.

The atomic weights discussed in Section 2.4 were defined as the relative masses of atoms of the elements. It would have been more correct to define them as the relative masses of average atoms of the elements. Why include the idea of an average atom? Remember, the mass number of an isotope is the sum of the number of protons and neutrons in the nucleus of atoms of the isotope. Also, protons and neutrons both have masses of 1 u (see **Table 2.3**). The masses of electrons are quite small, so the atomic weights of isotopes are very nearly equal to their mass numbers. For example, all naturally occurring phosphorus is made up of atoms containing 15 protons and 16 neutrons. The mass number of this isotope is 31, and its symbol is $^{31}_{15}P$. The relative mass (atomic weight) of atoms of this isotope is 30.97 u, the same as the atomic weight of the element given in the periodic table. The atomic weight of the isotope and the listed atomic weight of the element are the same because all atoms contained in the element are identical and have the same relative mass.

Naturally occurring chlorine is different. It is a mixture of two isotopes, $^{35}_{17}Cl$ and $^{37}_{17}Cl$ Chlorine-35 has a mass number of 35 and a relative mass of 34.97 u. Chlorine-37 has a mass number of 37 and a relative mass of 36.97 u. Note that the mass numbers and relative masses (atomic weights) of the atoms of an isotope are essentially identical. However, naturally occurring chlorine is a mixture containing both isotopes. Thus, the determined relative mass of chlorine atoms will be the average relative mass of the atoms found in the mixture (see **Figure 2.3**).

The average mass of each particle in a group of particles is simply the total mass of the group divided by the number of particles in the group. This calculation requires that the total number of particles be known. However, the percentage of each isotope in a mixture of atoms is easier to determine than the actual number of atoms of each isotope present. The percentages can be used to calculate average masses. Remember that percent means per 100, so we use an imaginary sample of an element containing 100 atoms in the calculation. On this basis, the number of atoms of each isotope in the 100-atom sample will be the percentage of that isotope in the sample. The mass contributed to the sample by each isotope will be the product of the number of atoms of the isotope (the percentage of the isotope) and the mass of the isotope. The total mass of the sample will be the sum of the masses contributed by each isotope. This total mass divided by 100, the number of atoms in our imaginary sample, gives the mass of an average atom, which is the atomic weight of the element.

Figure 2.3 The pieces of fruit in a bowl are somewhat like atoms of the isotopes of an element. Each piece of fruit may have the same color, taste, and texture, but it is unlikely that any two have *exactly* the same mass. The 12 oranges in the bowl weigh a total of 2.36×10^3 g. What is the average mass of each orange in the bowl?

Example 2.5 Isotope and Atomic Weight Relationships

Calculate the atomic weight of chlorine, given that the naturally occurring element consists of 75.53% chlorine-35 (mass = 34.97 u) and 24.47% chlorine-37 (mass = 36.97 u).

Solution

$$\text{atomic weight} = \frac{(\%\ \text{chlorine-35})(\text{mass chlorine-35}) + (\%\ \text{chlorine-37})(\text{mass chlorine-37})}{100}$$

$$= \frac{(75.53)(34.97\ \text{u}) + (24.47)(36.97\ \text{u})}{100}$$

$$= \frac{2641.28\ \text{u} + 904.66\ \text{u}}{100} = \frac{3545.94\ \text{u}}{100} = 35.4594\ \text{u}$$

$$= 35.46\ \text{u (rounded value)}$$

This result is slightly different from the periodic table atomic weight value of 35.45 because of slight errors introduced in rounding the isotope masses to four significant figures.

✔ LEARNING CHECK 2.5

a. Naturally occurring fluorine consists of a single isotope, fluorine-19, with a mass of 19.00 u (using four significant figures). Determine the atomic weight of fluorine, and compare your answer with the value given in the periodic table.

b. Naturally occurring magnesium has three isotopes: magnesium-24, magnesium-25, and magnesium-26. Their relative masses and percent abundances are, respectively, 23.99 u (78.70%), 24.99 u (10.13%), and 25.98 u (11.17%). Determine the atomic weight of magnesium, and compare it with the value given in the periodic table.

2.6 | Avogadro's Number: The Mole

Learning Objective

6. Use the mole concept to obtain relationships between number of moles, number of grams, and number of atoms for elements, and use those relationships to obtain factors for use in factor-unit calculations.

Figure 2.4 Samples of magnesium and carbon with masses of 24.31 g and 12.01 g, respectively.

The atomic weights of the elements given in the periodic table have far more uses than simply comparing the masses of the atoms of various elements. However, these uses are not apparent until some additional ideas are developed. According to the periodic table, the atomic weight of magnesium, Mg, is 24.31 u, and the atomic weight of carbon, C, is 12.01 u. As we have seen, this means that an average magnesium atom has about twice the mass of an average carbon atom. Modern instruments allow us to determine that the actual mass of an average magnesium atom is 4.037×10^{-23} g, and the actual mass of an average carbon atom is 1.994×10^{-23} g.

Even though the actual masses of magnesium and carbon atoms are extremely small, it should still be possible to collect together enough atoms of either element to give a sample of any desired mass. Suppose we wish to collect together enough atoms of each element to give a sample with a mass in grams equal to the atomic weight of each element. That is, we want to collect enough magnesium atoms to give a sample with a total mass of 24.31 g, and enough carbon atoms to give a sample with a total mass of 12.01 g (see **Figure 2.4**). How many atoms of each element will be required?

We know the mass of one atom of each element, and this fact will provide us with a factor that will allow us to convert the desired sample mass into the number of atoms required. The given mass of one atom of magnesium can be written as

$$1 \text{ Mg atom} = 4.037 \times 10^{-23} \text{ g Mg}$$

This known relationship provides two factors that can be used to solve factor-unit problems:

$$\frac{1 \text{ Mg atom}}{4.037 \times 10^{-23} \text{ g Mg}} \quad \text{and} \quad \frac{4.037 \times 10^{-23} \text{ g Mg}}{1 \text{ Mg atom}}$$

Our task is to convert the desired sample mass of 24.31 g Mg into the number of Mg atoms in the sample. The first factor will cancel the units "g Mg" and will generate the units "Mg atoms."

$$(24.31 \text{ g Mg}) \times \frac{1 \text{ Mg atom}}{4.037 \times 10^{-23} \text{ g Mg}} = 6.022 \times 10^{23} \text{ Mg atoms}$$

In a similar way, the number of C atoms needed to produce a sample with a mass of 12.01 g is calculated:

$$(12.01 \text{ g C}) \times \frac{1 \text{ C atom}}{1.994 \times 10^{-23} \text{ g C}} = 6.023 \times 10^{23} \text{ C atoms}$$

The result that the same number of atoms is required for each sample might seem surprising, but it is a consequence of the sample sizes we wanted to produce. If we collected just one atom of each element, the ratio of the mass of Mg to the mass of C would be equal to the atomic weight of Mg divided by the atomic weight of C:

$$\frac{\text{Mg mass}}{\text{C mass}} = \frac{4.037 \times 10^{-23} \text{ g}}{1.994 \times 10^{-23} \text{ g}} = 2.024 = \frac{\text{At. wt. Mg}}{\text{At. wt. C}} = \frac{24.31 \text{ u}}{12.01 \text{ u}}$$

This result reflects the fact that magnesium atoms have about twice the mass of carbon atoms. If we collected samples of 100 atoms of each element, the ratio of the mass of the Mg sample to the mass of the C sample would still be 2.024 because each sample would have a mass 100 times greater than the mass of a single atom:

$$\frac{\text{Mg mass}}{\text{C mass}} = \frac{(100)(4.037 \times 10^{-23} \text{ g})}{(100)(1.994 \times 10^{-23} \text{ g})} = 2.024$$

It follows that if we collected samples containing 6.022×10^{23} atoms of each element, the ratio of the mass of the Mg sample to the mass of the C sample would still be 2.024 because each sample would have a mass that would be 6.022×10^{23} times greater than the mass of a single atom:

$$\frac{\text{Mg mass}}{\text{C mass}} = \frac{(6.023 \times 10^{23})(4.037 \times 10^{-23} \text{ g})}{(6.023 \times 10^{23})(1.994 \times 10^{-23} \text{ g})} = 2.024$$

These results lead to the following conclusion: Any samples of magnesium and carbon that have mass ratios equal to 2.024 will contain the same number of atoms.

Example 2.6 Mass Ratios and Atomic Weights

Show that samples of magnesium and carbon with masses of 9.663 g and 4.774 g, respectively, have a mass ratio of 2.024 and contain the same number of atoms.

Solution
The mass ratio is obtained by dividing the mass of the magnesium sample by the mass of the carbon sample:

$$\frac{\text{Mg mass}}{\text{C mass}} = \frac{9.663 \text{ g}}{4.774 \text{ g}} = 2.024$$

The number of atoms in each is calculated using the factors obtained earlier from the mass of one atom of each element:

$$(9.663 \text{ g Mg}) \times \frac{1 \text{ Mg atom}}{4.037 \times 10^{-23} \text{ g Mg}} = 2.394 \times 10^{23} \text{ Mg atoms}$$

$$(4.774 \text{ g C}) \times \frac{1 \text{ C atom}}{1.994 \times 10^{-23} \text{ g C}} = 2.394 \times 10^{23} \text{ C atoms}$$

✔ **LEARNING CHECK 2.6** Show that samples of magnesium and carbon with masses of 13.66 g and 6.748 g, respectively, have a mass ratio of 2.024 and contain the same number of atoms.

The preceding results for magnesium and carbon may be generalized for all the elements of the periodic table as follows. Any samples of two elements that have a mass ratio equal to the ratio of their atomic weights will contain identical numbers of atoms. In addition, we have seen that if the number of grams of sample is equal numerically to the atomic weight of an element, the number of atoms in the sample is equal to 6.022×10^{23}. **Figure 2.5** illustrates these ideas for particles that are familiar to most of us.

Figure 2.5 An average jelly bean has a mass that is 1.60 times the mass of an average dry bean. Each jar contains the same number of beans. The total mass of jelly beans is 472 g. What is the total mass of the dry beans?

© Spencer L. Seager

With modern equipment, it is possible to determine the number of atoms in any size sample of an element. However, before this was possible, the practice of focusing on samples with a mass in grams equal to the atomic weights of the elements became well established. It continues today. We have learned that the number of atoms in such samples is 6.022×10^{23}.

We saw in Section 2.4 that molecular weights, the relative masses of molecules, are calculated by adding the atomic weights of the atoms contained in the molecules. The resulting molecular weights are expressed in atomic mass units, just as are the atomic weights of the atoms. The same ideas we used to discuss the actual and relative masses of atoms can be applied to molecules. We will not go through the details but simply state the primary conclusion: A sample of a compound with a mass in grams equal to the molecular weight of the compound contains 6.022×10^{23} molecules of the compound.

The number 6.022×10^{23} is called *Avogadro's number* in honor of Amedeo Avogadro (1776–1856), an Italian scientist who made important contributions to the concept of atomic weights. As we have seen, this number represents the number of atoms or molecules in a specific sample of an element or compound. Because of its importance in calculations, the number of particles represented by Avogadro's number is given a specific name; it is called a **mole**, abbreviated mol.

mole The number of particles (atoms or molecules) contained in a sample of an element or compound with a mass in grams equal to the atomic or molecular weight, respectively. Numerically, 1 mol is equal to 6.022×10^{23} particles.

It is sometimes helpful to remember that the word *mole* represents a specific number, just as the word *dozen* represents 12, regardless of the objects being counted. Thus, 6.022×10^{23} people would be called 1 mol of people, just as 12 people would be called 1 dozen people. The immensity of Avogadro's number is illustrated by the results of a few calculations based on 1 mol of people. One mol of people would be enough to populate about 1×10^{14} Earths at today's level. That is 100 trillion Earths. Or, put another way, the present population of Earth is $1 \times 10^{-12}\%$ (0.000000000001%) of 1 mol.

In the development of these ideas to this point, we have used four significant figures for atomic weights, molecular weights, and Avogadro's number to minimize the introduction of rounding errors. However, in calculations throughout the remainder of the book, three significant figures will generally be sufficient and will be used.

In a strict sense, 1 mol is a specific number of particles. However, in chemistry it is customary to follow the useful practice of also letting 1 mol stand for the mass of a sample of element or compound that contains Avogadro's number of particles. Thus, the application of the mole concept to sulfur (at. wt. = 32.1 u) gives the following relationships:

$$1 \text{ mol S atoms} = 6.02 \times 10^{23} \text{ S atoms} = 32.1 \text{ g S}$$

When written individually, these three relationships can be used to generate six factors for use in factor-unit calculations involving sulfur:

$$1 \text{ mol S atoms} = 6.02 \times 10^{23} \text{ S atoms}$$

$$6.02 \times 10^{23} \text{ S atoms} = 32.1 \text{ g S}$$

$$1 \text{ mol S atoms} = 32.1 \text{ g S}$$

Example 2.7 Factor-Unit Calculations for Sulfur

Determine the following using the factor-unit method of calculation and factors obtained from the preceding three relationships given for sulfur (S):

a. The mass in grams of 1.35 mol of S
b. The number of moles of S atoms in 98.6 g of S
c. The number of S atoms in 98.6 g of S
d. The mass in grams of one atom of S

Solution

a. The known quantity is 1.35 mol of S, and the unit of the unknown quantity is grams of S. The factor comes from the relationship 1 mol S atoms = 32.1 g S.

$$(1.35 \; \text{mol S atoms}) \left(\frac{32.1 \; \text{g S}}{1 \; \text{mol S atoms}} \right) = 43.3 \; \text{g S}$$

b. The known quantity is 98.6 g of S, and the unit of the unknown quantity is moles of S atoms. The factor comes from the same relationship used in (a).

$$(98.6 \; \text{g S}) \left(\frac{1 \; \text{mol S atoms}}{32.1 \; \text{g S}} \right) = 3.07 \; \text{mol S atoms}$$

c. The known quantity is, again, 98.6 g of S, and the unit of the unknown is the number of S atoms. The factor comes from the relationship 6.02×10^{23} S atoms = 32.1 g S.

$$(98.6 \; \text{g S}) \left(\frac{6.02 \times 10^{23} \; \text{S atoms}}{32.1 \; \text{g S}} \right) = 1.85 \times 10^{24} \; \text{S atoms}$$

d. The known quantity is one S atom, and the unit of the unknown is grams of S. The factor comes from the same relationship used in (c), 6.02×10^{23} S atoms = 32.1 g S. Note that the factor is the inverse of the one used in (c) even though both came from the same relationship. Thus, we see that each relationship provides two factors.

$$(1 \; \text{S atom}) \left(\frac{32.1 \; \text{g S}}{6.02 \times 10^{23} \; \text{S atoms}} \right) = 5.33 \times 10^{-23} \; \text{g S}$$

✔ **LEARNING CHECK 2.7** Calculate the mass of a single oxygen atom in grams. How does the ratio of this mass divided by the mass of a carbon atom given earlier (1.994×10^{-23} g) compare with the ratio of the atomic weights of oxygen and carbon given in the periodic table?

The mole concept can also be applied to particles that are molecules instead of atoms. The compound carbon dioxide consists of molecules that contain one carbon atom, C, and two oxygen atoms, O. The formula for the molecule is CO_2. The molecular weight of the molecule is calculated as shown earlier by adding together the atomic weight of one carbon atom and the atomic weight of two oxygen atoms:

$$MW = 1 \; (\text{at. wt. C}) + 2 \; (\text{at. wt. O}) = 1 \; (12.0 \; \text{u}) + 2 \; (16.0 \; \text{u}) = 44.0 \; \text{u}$$

Thus, the relative mass of a CO_2 molecule is 44.0 u, and application of the mole concept to CO_2 molecules gives the following relationships:

$$1 \; \text{mol CO}_2 \; \text{molecules} = 6.02 \times 10^{23} \; \text{CO}_2 \; \text{molecules} = 44.0 \; \text{g CO}_2$$

When written individually, these three relationships can be used to generate six factors for use in factor-unit calculations involving CO_2.

$$1 \text{ mol } CO_2 \text{ molecules} = 6.02 \times 10^{23} \text{ } CO_2 \text{ molecules}$$

$$6.02 \times 10^{23} \text{ } CO_2 \text{ molecules} = 44.0 \text{ g } CO_2$$

$$1 \text{ mol } CO_2 \text{ molecules} = 44.0 \text{ g } CO_2$$

Example 2.8 Factor-Unit Calculations for Carbon Dioxide

Determine the following using the factor-unit method of calculation and factors obtained from the preceding three relationships given for carbon dioxide, CO_2:

a. The mass in grams of 1.62 mol of CO_2
b. The number of moles of CO_2 molecules in 63.9 g of CO_2
c. The number of CO_2 molecules in 63.9 g of CO_2
d. The mass in grams of one molecule of CO_2

Solution

a. The known quantity is 1.62 mol of CO_2, and the unit of the unknown quantity is g CO_2. The factor comes from the relationship 1 mol CO_2 molecules = 44.0 g CO_2.

$$(1.62 \text{ mol } CO_2 \text{ molecules})\left(\frac{44.0 \text{ g } CO_2}{1 \text{ mol } CO_2 \text{ molecules}}\right) = 71.3 \text{ g } CO_2$$

b. The known quantity is 63.9 g of CO_2, and the unit of the unknown quantity is mol of CO_2 molecules. The factor comes from the same relationship used in (a).

$$(63.9 \text{ g } CO_2)\left(\frac{1 \text{ mol } CO_2 \text{ molecules}}{44.0 \text{ g } CO_2}\right) = 1.45 \text{ mol } CO_2 \text{ molecules}$$

c. The known quantity is, again, 63.9 g of CO_2, and the unit of the unknown is the number of CO_2 molecules. The factor comes from the relationship 6.02×10^{23} CO_2 molecules = 44.0 g CO_2.

$$(63.9 \text{ g } CO_2)\left(\frac{6.02 \times 10^{23} \text{ } CO_2 \text{ molecules}}{44.0 \text{ g } CO_2}\right) = 8.74 \times 10^{23} \text{ } CO_2 \text{ molecules}$$

d. The known quantity is one CO_2 molecule, and the unit of the unknown is g CO_2. The factor comes from the same relationship used in (c), but the factor is the inverse of the one used in (c).

$$(1 \text{ } CO_2 \text{ molecule})\left(\frac{44.0 \text{ g } CO_2}{6.02 \times 10^{23} \text{ } CO_2 \text{ molecules}}\right) = 7.31 \times 10^{-23} \text{ g } CO_2$$

✔ **LEARNING CHECK 2.8** Calculate the molecular weight of carbon monoxide, CO, in u, then calculate the mass of a single CO molecule in grams. How does the ratio of this mass divided by the mass of a single CO_2 molecule (from Example 2.8 d) compare to the ratio of the molecular weight of CO in u divided by the molecular weight of CO_2 in u (from Example 2.8)?

2.7 | The Mole and Chemical Formulas

Learning Objective

7. Use the mole concept and molecular formulas to obtain relationships between number of moles, number of grams, and number of atoms or molecules for compounds, and use those relationships to obtain factors for use in factor-unit calculations.

According to Section 2.1 and as demonstrated in Example 2.8, the formula for a compound is made up of the symbols for each element present. Subscripts following the elemental symbols indicate the number of each type of atom in the molecule represented. Thus, chemical formulas represent the numerical relationships that exist among the atoms in a compound. Application of the mole concept to the atoms making up the formulas provides additional useful information.

Consider water as an example. The formula H_2O represents a 2:1 ratio of hydrogen atoms to oxygen atoms in a water molecule. Since this ratio is fixed, the following statements can be written:

1. 2 H_2O molecules contain 4 H atoms and 2 O atoms.
2. 10 H_2O molecules contain 20 H atoms and 10 O atoms.
3. 100 H_2O molecules contain 200 H atoms and 100 O atoms.
4. 6.02×10^{23} H_2O molecules contain 12.04×10^{23} H atoms and 6.02×10^{23} O atoms.

Statement 4 is significant because 6.02×10^{23} particles is 1 mol. Thus, Statement 4 can be changed to Statement 5:

5. 1 mol of H_2O molecules contains 2 mol of H atoms and 1 mol of O atoms.

Figure 2.6 contains another example of this concept.

© Jeffrey M. Seager

Figure 2.6 Liquid carbon disulfide (CS_2) is composed of carbon (left) and sulfur (right), elements that are solids. How many moles of sulfur atoms would be contained in 1.5 mol of CS_2 molecules?

Example 2.9 Mole Calculations Using Rabbits

How many moles of ears, tails, and legs are contained in 1 mol of normal rabbits?

Solution

This example is nonchemical, but it might help you grasp the relationships that exist between the individual parts of a formula and the formula as a whole. The parts of a rabbit are related to a rabbit just as the parts of a formula are related to the entire formula.

Ears. Each rabbit has two ears and a 2:1 ratio exists between the number of ears and the number of rabbits. Therefore, 1 mol of rabbits contains 2 mol of ears.

Tails. The 1:1 ratio of tails to rabbits leads to the result that 1 mol of rabbits contains 1 mol of tails.

Legs. The 4:1 ratio of legs to rabbits leads to the result that 1 mol of rabbits contains 4 mol of legs.

Example 2.10 Mole Calculations Using Molecular Formulas

How many moles of each type of atom are contained in 1 mol of chloroform ($CHCl_3$)?

Solution

Just as each rabbit has two ears, each chloroform molecule contains one C atom, one H atom, and three Cl atoms. Therefore, 1 mol of $CHCl_3$ contains 1 mol of C atoms, 1 mol of H atoms, and 3 mol of Cl atoms.

✔ **LEARNING CHECK 2.9** How many moles of each type of atom would be contained in 0.50 mol of glucose ($C_6H_{12}O_6$)?

STUDY SKILLS 2.1

Help with Mole Calculations

Problems involving the use of the mole often strike fear into the hearts of beginning chemistry students. The good news is that problems involving the use of moles, atoms, molecules, and grams are made easier by using the factor-unit method. The method focuses your attention on the goal of eliminating the unit of the known, or given, quantity and of generating the unit of the answer, or unknown, quantity. Remember, Step 1 is to write down the number and unit of the given quantity. In Step 2, write down the unit of the answer. In Step 3, multiply the known quantity by a factor whose units will cancel that of the known quantity and will generate the unit of the answer. In Step 4, obtain the answer by doing the required arithmetic using the numbers that were introduced in Steps 1–3.

The ability to write the necessary factors for use in Step 3 is essential if you are to become proficient in solving mole problems using the factor-unit method. The factors come from numerical relationships between quantities that are obtained from definitions, experimental measurements, or combinations of the two. The definition of the mole, coupled with experimentally determined atomic and molecular weights, gives the following numerical relationships:

Atom Y: 1 mol Y atoms = 6.02×10^{23} Y atoms = y g Y where y is the atomic weight of element Y.

Molecule Z: 1 mol Z molecules = 6.02×10^{23} Z molecules = z g Z

where z is the molecular weight of compound Z. Each of these sets of three related quantities will give six factors

that can be used in factor-unit problems. Each factor is simply a ratio between any two of the three related quantities such as

$$\frac{1 \text{ mol } Y \text{ atoms}}{6.02 \times 10^{23} \text{ } Y \text{ atoms}}$$

Because each ratio can be inverted, six different factors result from each set of three quantities. See if you can write the five other factors for atom Y. The steps given in the flow charts below will help you.

The usefulness of this approach can be increased by remembering and using the mass relationships of the mole concept. Thus, Statement 5 written earlier for water can be changed to Statement 6:

6. 18.0 g of water contains 2.0 g of H and 16.0 g of O.

Or, in a more concise form:

$$18.0 \text{ g H}_2\text{O} = 2.0 \text{ g H} + 16.0 \text{ g O}$$

Mass relationships such as these allow percent compositions to be calculated easily.

Example 2.11 Mass Percentage Calculations

Ammonia (NH_3) and ammonium nitrate (NH_4NO_3) are commonly used agricultural fertilizers. Which one of the two contains the higher mass percentage of nitrogen (N)?

Solution
In each case, the mass percentage of N is given by

$$\% \text{ N} = \frac{\text{part}}{\text{total}} \times 100 = \frac{\text{mass of N}}{\text{mass of compound}} \times 100$$

We will use 1 mol of each compound as a sample because the mass in grams of 1 mol of compound and the mass in grams of N in the 1 mol of compound are readily determined. One mol of NH_3 weighs 17.0 g and contains 1 mol of N atoms, which weighs 14.0 g.

$$\% \text{ N} = \frac{14.0 \text{ g}}{17.0 \text{ g}} \times 100 = 82.4\%$$

Similarly, 1 mol of NH_4NO_3 weighs 80.0 g and contains 2 mol of N atoms, which weigh 28.0 g.

$$\% \text{ N} = \frac{28.0 \text{ g}}{80.0 \text{ g}} \times 100 = 35.0\%$$

✔ **LEARNING CHECK 2.10** Determine the mass percentage of carbon in carbon dioxide (CO_2) and carbon monoxide (CO).

▶ Case Study Follow-up

During an MRI (magnetic resonance imaging) scan, large magnets cause the hydrogen atoms in the body to line up in relation to the magnet. About half point in one direction, and the other half point in the opposite direction, mostly canceling each other out. A radio-frequency pulse specific to hydrogen is then applied that causes any unmatched atoms to spin in a particular direction at a particular frequency. When the pulse is turned off, the spinning atoms emit energy that is detected by the MRI device. A powerful computer uses mathematical formulae to convert this energy information into a meaningful image.

Anyone who drives a car knows that a series of complex behaviors can become automatic: A driver may arrive at home with hardly any recollection of the journey there. Health professionals who diagnose common complaints can fall into a routine and automatically diagnose patients with simple conditions when in fact their case is complex or serious. When this occurs, it can result in increased cost to and prolonged suffering for the patient and can delay a timely resolution of the health concern. It can even become life threatening. Busy health professionals have to be diligent in considering a range of possible explanations for the conditions with which each individual presents. Accurate diagnosis depends on focused attention concerning all of a patient's symptoms, family history, behavior, and environment.

Concept Summary

Symbols and Formulas Symbols based on names have been assigned to every element. Most consist of a single capital letter followed by a lowercase letter. A few consist of a single capital letter. Compounds are represented by formulas made up of elemental symbols. The number of atoms of each element in a molecule is shown by subscripts.

Objective 1 (Section 2.1), Exercise 2.4

Inside the Atom Atoms are made up of numerous smaller particles, of which the most important to chemical studies are the proton, neutron, and electron. Positively charged protons and neutral neutrons have a relative mass of 1 u each and

are located in the nuclei of atoms. Negatively charged electrons with a mass of 1/1836 u are located outside the nuclei of atoms.

Objective 2 (Section 2.2), Exercises 2.10 and 2.12

Isotopes Most elements in their natural state are made up of more than one kind of atom. These different kinds of atoms of a specific element are called isotopes and differ from one another only in the number of neutrons in their nuclei. A symbol incorporating atomic number, mass number, and elemental symbol is used to represent a specific isotope.

Objective 3 (Section 2.3), Exercises 2.16 and 2.22

Relative Masses of Atoms and Molecules Relative masses called atomic weights have been assigned to each element and are tabulated in the periodic table. The units used are atomic mass units, abbreviated u. Relative masses for molecules, called molecular weights, are determined by adding the atomic weights of the atoms making up the molecules.

Objective 4 (Section 2.4), Exercise 2.32

Isotopes and Atomic Weights The atomic weights measured for elements are average weights that depend on the percentages and masses of the isotopes in the naturally occurring element. If the isotope percent abundances and isotope masses are known for an element, its atomic weight can be calculated.

Objective 5 (Section 2.5), Exercise 2.38

Avogadro's Number: The Mole Avogadro's number of the atoms of an element has a mass in grams equal to the atomic weight of the element. Avogadro's number of molecules has a mass in grams equal to the molecular weight. Avogadro's number of particles is called a mole, abbreviated mol.

Objective 6 (Section 2.6), Exercises 2.44 a & b and 2.46 a & b

The Mole and Chemical Formulas The mole concept when applied to molecular formulas gives numerous relationships that yield useful factors for factor-unit calculations.

Objective 7 (Section 2.7), Exercises 2.50 b and 2.52 b

Key Terms and Concepts

Atomic mass unit (u) (2.4)
Atomic number of an atom (2.3)
Atomic weight (2.4)
Compound formula (2.1)

Elemental symbol (2.1)
Isotopes (2.3)
Mass number of an atom (2.3)
Mole (2.6)

Molecular weight (2.4)
Nucleus (2.2)

Exercises

Even-numbered exercises are answered in Appendix B.

Blue-numbered exercises are more challenging.

You will find it useful to refer to Table 2.1 and the periodic table inside the front cover as you work these exercises.

Symbols and Formulas (Section 2.1)

2.1 Draw a "formula" for each of the following molecules using circular symbols of your choice to represent atoms:

 a. A diatomic molecule of an element

 b. A diatomic molecule of a compound

 c. A triatomic molecule of an element

 d. A molecule of a compound containing one atom of one element and four atoms of another element

2.2 Draw a "formula" for each of the following molecules using circular symbols of your choice to represent atoms:

 a. A triatomic molecule of a compound

 b. A molecule of a compound containing two atoms of one element and two atoms of a second element

 c. A molecule of a compound containing two atoms of one element, one atom of a second element, and four atoms of a third element

 d. A molecule containing two atoms of one element, six atoms of a second element, and one atom of a third element

2.3 Write formulas for the following molecules using elemental symbols from Table 2.1 and subscripts. Compare these formulas with those of Exercise 2.1.

 a. A diatomic molecule of chlorine gas

 b. A diatomic molecule of hydrogen fluoride (one hydrogen atom and one fluorine atom)

 c. A triatomic molecule of ozone (a molecular form of the element oxygen)

 d. A molecule of carbon tetrachloride (one atom of carbon and four atoms of chlorine)

2.4 Write formulas for the following molecules using elemental symbols from Table 2.1 and subscripts. Compare these formulas with those of Exercise 2.2.

 a. A molecule of water (two hydrogen atoms and one oxygen atom)

 b. A molecule of hydrogen peroxide (two hydrogen atoms and two oxygen atoms)

 c. A molecule of sulfuric acid (two hydrogen atoms, one sulfur atom, and four oxygen atoms)

 d. A molecule of ethyl alcohol (two carbon atoms, six hydrogen atoms, and one oxygen atom)

2.5 Determine the number of each type of atom in molecules represented by the following formulas:

 a. ammonia (NH_3)

 b. acetic acid ($C_2H_4O_2$)

 c. boric acid (H_3BO_3)

 d. ethane (C_2H_6)

2.6 Determine the number of each type of atom in molecules represented by the following formulas:

 a. sulfur dioxide (SO_2)

 b. butane (C_4H_{10})

 c. chlorous acid ($HClO_2$)

 d. boron trifluoride (BF_3)

2.7 Tell what is wrong with each of the following molecular formulas and write a correct formula:

 a. H_3PO_3 (phosphorous acid)

 b. $SICl_4$ (silicon tetrachloride)

 c. SOO (sulfur dioxide)

 d. 2HO (hydrogen peroxide—two hydrogen atoms and two oxygen atoms)

2.8 Tell what is wrong with each of the following formulas and write a correct formula:

 a. HSH (hydrogen sulfide)

 b. $HCLO_2$ (chlorous acid)

 c. $2HN_2$ (hydrazine—two hydrogen atoms and four nitrogen atoms)

 d. C_2H_6 (ethane)

Inside the Atom (Section 2.2)

2.9 Determine the charge and mass (in u) of nuclei made up of the following particles:

 a. 6 protons and 6 neutrons

 b. 8 protons and 9 neutrons

 c. 20 protons and 25 neutrons

 d. 52 protons and 78 neutrons

2.10 Determine the charge and mass (in u) of nuclei made up of the following particles:

 a. 4 protons and 5 neutrons

 b. 9 protons and 10 neutrons

 c. 20 protons and 23 neutrons

 d. 47 protons and 60 neutrons

2.11 Determine the number of electrons that would have to be associated with each nucleus described in Exercise 2.9 to produce a neutral atom.

2.12 Determine the number of electrons that would have to be associated with each nucleus described in Exercise 2.10 to produce a neutral atom.

Isotopes (Section 2.3)

2.13 Determine the number of electrons and protons contained in an atom of the following elements:

 a. sulfur

 b. As

 c. element number 24

2.14 Determine the number of electrons and protons contained in an atom of the following elements:

 a. silicon

 b. Sn

 c. element number 74

2.15 Determine the number of protons, number of neutrons, and number of electrons in atoms of the following isotopes:

 a. 3_2He

 b. 9_4Be

 c. $^{235}_{92}U$

2.16 Determine the number of protons, number of neutrons, and number of electrons in atoms of the following isotopes:

 a. $^{34}_{16}S$

 b. $^{91}_{40}Zr$

 c. $^{131}_{54}Xe$

2.17 Write symbols like those given in Exercises 2.15 and 2.16 for the following isotopes:

 a. cadmium-110

 b. cobalt-60

 c. uranium-235

2.18 Write symbols like those given in Exercises 2.15 and 2.16 for the following isotopes:

 a. silicon-28

 b. argon-40

 c. strontium-88

2.19 Determine the mass number and atomic number for atoms containing the nuclei described in Exercise 2.9. Write symbols for each atom like those given in Exercises 2.15 and 2.16.

2.20 Determine the mass number and atomic number for neutral atoms containing the nuclei described in Exercise 2.10. Write symbols for each atom like those given in Exercises 2.15 and 2.16.

2.21 Write isotope symbols for neutral atoms with the following characteristics:

 a. Contains 20 electrons and 20 neutrons

 b. Contains 1 electron and 2 neutrons

 c. A magnesium atom that contains 14 neutrons

2.22 Write isotope symbols for neutral atoms with the following characteristics:

 a. Contains 17 electrons and 20 neutrons

 b. A copper atom with a mass number of 65

 c. A zinc atom that contains 36 neutrons

Relative Masses of Atoms and Molecules (Section 2.4)

2.23 Write the symbols and names for two elements whose average atoms have masses that are within 0.3 u of each other. Don't look beyond element number 83.

2.24 Round atomic weights to the nearest whole number, and determine how many helium atoms would balance one carbon atom on a seesaw.

2.25 Round atomic weights to the nearest whole number, and determine how many lithium atoms would balance two nitrogen atoms on a seesaw.

2.26 What are the symbol and name for an element whose average atoms have a mass that is 77.1% of the mass of an average chromium atom?

2.27 In the first 36 elements, 6 elements have atoms whose average mass is within 0.2 u of being twice the atomic number of the element. Write the symbols and names for these 6 elements.

2.28 What are the symbol and name of the element whose average atoms have a mass very nearly half the mass of an average silicon atom?

2.29 Determine the molecular weights of the following in u:
 a. fluorine (F_2)
 b. carbon disulfide (CS_2)
 c. sulfurous acid (H_2SO_3)
 d. ethyl alcohol (C_2H_6O)
 e. ethane (C_2H_6)

2.30 Determine the molecular weights of the following in u:
 a. nitrogen dioxide (NO_2)
 b. ammonia (NH_3)
 c. glucose ($C_6H_{12}O_6$)
 d. ozone (O_3)
 e. ethylene glycol ($C_2H_6O_2$)

2.31 The molecular weight was determined for a gas that is known to be an oxide of nitrogen. The value obtained experimentally was 43.98 u. Which of the following is most likely to be the formula of the gas? NO, N_2O, NO_2.

2.32 A flammable gas is known to contain only carbon and hydrogen. Its molecular weight is determined and found to be 28.05 u. Which of the following is the likely identity of the gas? Acetylene (C_2H_2), ethylene (C_2H_4), ethane (C_2H_6).

2.33 Glycine, an amino acid found in proteins, has a molecular weight of 75.07 u and is represented by the formula $C_2H_xNO_2$. What number does x stand for in the formula?

2.34 Serine, an amino acid found in proteins, has a molecular weight of 105.10 u and is represented by the formula $C_yH_7NO_3$. What number does y stand for in the formula?

Isotopes and Atomic Weights (Section 2.5)

2.35 Naturally occurring sodium has a single isotope. Determine the following for the naturally occurring atoms of sodium:
 a. The number of neutrons in the nucleus
 b. The mass (in u) of the nucleus (to three significant figures)

2.36 Naturally occurring aluminum has a single isotope. Determine the following for the naturally occurring atoms of aluminum:
 a. The number of neutrons in the nucleus
 b. The mass (in u) of the nucleus (to three significant figures)

2.37 Calculate the atomic weight of lithium on the basis of the following percent composition and atomic weights of the naturally occurring isotopes. Compare the calculated value with the atomic weight listed for lithium in the periodic table.

$$\text{lithium-6} = 7.42\% \ (6.0151 \text{ u})$$
$$\text{lithium-7} = 92.58\% \ (7.0160 \text{ u})$$

2.38 Calculate the atomic weight of boron on the basis of the following percent composition and atomic weights of the naturally occurring isotopes. Compare the calculated value with the atomic weight listed for boron in the periodic table.

$$\text{boron-10} = 19.78\% \ (10.0129 \text{ u})$$
$$\text{boron-11} = 80.22\% \ (11.0093 \text{ u})$$

2.39 Calculate the atomic weight of silicon on the basis of the following percent composition and atomic weights of the naturally occurring isotopes. Compare the calculated value with the atomic weight listed for silicon in the periodic table.

$$\text{silicon-28} = 92.21\% \ (27.9769 \text{ u})$$
$$\text{silicon-29} = 4.70\% \ (28.9765 \text{ u})$$
$$\text{silicon-30} = 3.09\% \ (29.9738 \text{ u})$$

2.40 Calculate the atomic weight of copper on the basis of the following percent composition and atomic weights of the naturally occurring isotopes. Compare the calculated value with the atomic weight listed for copper in the periodic table.

$$\text{copper-63} = 69.09\% \ (62.9298 \text{ u})$$
$$\text{copper-65} = 30.91\% \ (64.9278 \text{ u})$$

Avogadro's Number: The Mole (Section 2.6)

2.41 Refer to the periodic table and determine how many grams of sulfur contain the same number of atoms as 3.10 grams of phosphorus.

2.42 Refer to the periodic table and determine how many grams of fluorine contain the same number of atoms as 1.60 g of oxygen.

2.43 Write three relationships (equalities) based on the mole concept for each of the following elements:
 a. beryllium
 b. lead
 c. sodium

2.44 Write three relationships (equalities) based on the mole concept for each of the following elements:
 a. phosphorus
 b. aluminum
 c. krypton

2.45 Use a factor derived from the relationships written in Exercise 2.43 and the factor-unit method to determine the following:
 a. The number of moles of beryllium atoms in a 10.0 g sample of beryllium
 b. The number of lead atoms in a 2.00 mol sample of lead
 c. The number of sodium atoms in a 50.0 g sample of sodium

2.46 Use a factor derived from the relationships written in Exercise 2.44 and the factor-unit method to determine the following:

 a. The mass in grams of one phosphorus atom

 b. The number of grams of aluminum in 1.65 mol of aluminum

 c. The total mass in grams of one-fourth Avogadro's number of krypton atoms

The Mole and Chemical Formulas (Section 2.7)

2.47 Refer to the periodic table and calculate the molecular weights for the compounds PH_3 and SO_2. Then, determine how many grams of PH_3 contain the same number of molecules as 6.41 g of SO_2.

2.48 Refer to the periodic table and calculate the molecular weights for the compounds BF_3 and H_2S. Then, determine how many grams of BF_3 contain the same number of molecules as 0.34 g of H_2S.

2.49 For each formula given below, write statements equivalent to Statements 1–6 (see Section 2.7):

 a. carbon dioxide (CO_2)

 b. ethane (C_2H_6)

 c. glucose ($C_6H_{12}O_6$)

2.50 For each formula given below, write statements equivalent to Statements 1–6 (see Section 2.7):

 a. ethyl ether ($C_4H_{10}O$)

 b. fluoroacetic acid ($C_2H_3O_2F$)

 c. aniline (C_6H_7N)

2.51 Answer the following questions based on information contained in the statements you wrote for Exercise 2.49.

 a. How many moles of oxygen atoms are contained in 1 mol of CO_2 molecules?

 b. How many grams of carbon are contained in 1.00 mol of C_2H_6?

 c. What is the mass percentage of oxygen in $C_6H_{12}O_6$?

2.52 Answer the following questions based on information contained in the statements you wrote for Exercise 2.50.

 a. How many moles of hydrogen atoms are contained in 0.50 mol of ethyl ether?

 b. How many carbon atoms are contained in 0.25 mol of $C_2H_3O_2F$?

 c. How many grams of hydrogen are contained in 2.00 mol of C_6H_7N?

2.53 How many moles of N_2O_5 contain the same number of nitrogen atoms as 3.0 mol of NO_2?

2.54 How many grams of C_2H_6O contain the same number of oxygen atoms as 0.75 mol of H_2O?

2.55 Determine the mass percentage of carbon in CO and CO_2.

2.56 Determine the mass percentage of hydrogen in CH_4 and C_2H_6.

2.57 Any of the statements based on a mole of substance (Statements 4–6) can be used to obtain factors for problem solving by the factor-unit method. Write statements equivalent to 4, 5, and 6 for nitrophenol ($C_6H_5NO_3$). Use a single factor obtained from the statements to solve each of the following. A different factor will be needed in each case.

 a. How many grams of nitrogen are contained in 70.0 g of $C_6H_5NO_3$?

 b. How many moles of oxygen atoms are contained in 1.50 mol of $C_6H_5NO_3$?

 c. How many atoms of carbon are contained in 9.00×10^{22} molecules of $C_6H_5NO_3$?

2.58 Any of the statements based on a mole of substance (Statements 4–6) can be used to obtain factors for problem solving by the factor-unit method. Write statements equivalent to 4, 5, and 6 for fructose ($C_6H_{12}O_6$). Use a single factor obtained from the statements to solve each of the following. A different factor will be needed in each case.

 a. How many grams of oxygen are contained in 43.5 g of $C_6H_{12}O_6$?

 b. How many moles of hydrogen atoms are contained in 1.50 mol of $C_6H_{12}O_6$?

 c. How many atoms of carbon are contained in 7.50×10^{22} molecules of $C_6H_{12}O_6$?

2.59 Urea (CH_4N_2O) and ammonium sulfate ($N_2H_8SO_4$) are both used as agricultural fertilizers. Which one contains the higher mass percentage of nitrogen?

2.60 Two iron ores that have been used as sources of iron are magnetite (Fe_3O_4) and hematite (Fe_2O_3). Which one contains the higher mass percentage of iron?

2.61 Both calcite ($CaCO_3$) and dolomite ($CaMgC_2O_6$) are used as dietary calcium supplements. Calculate the mass percentage of calcium in each mineral.

Additional Exercises

2.62 The two major isotopes of uranium metal found in nature are U-235 and U-238. Which isotope has the greater density? Explain your reasoning.

2.63 About one billion (1.0×10^9) peas can fit into a railroad car. What percentage of one mole of peas is this number?

2.64 The mass of a single carbon-12 atom is 1.99×10^{-23} g. What is the mass in grams of a single carbon-14 atom?

2.65 One mole of water molecules, H_2O, has a mass of 18.0 g. What would be the mass in grams of one mole of heavy water molecules, D_2O, where D represents the isotope?

2.66 According to the caption of Figure 2.2, atoms are composed primarily of empty space (99.999% empty). What would happen to the density of matter (increase or decrease) if the electrons were actually located at the distance from the nucleus shown in the atom drawing of Figure 2.2? Explain your reasoning.

Chemistry for Thought

2.67 **a.** Explain how atoms of different elements differ from one another.

 b. Explain how atoms of different isotopes of the same element differ from one another.

2.68 The atomic weight of aluminum is 26.98 u, and the atomic weight of nickel is 58.69 u. All aluminum atoms have a mass of 26.98 u, but not a single atom of nickel has a mass of 58.69 u. Explain.

2.69 Answer the question in the caption of Figure 2.3 (see page 57). Would you expect any orange in the bowl to have the exact mass you calculated as an average? Explain.

2.70 Answer the question in the caption of Figure 2.5 (see page 60). Use your answer and the fact that an average jelly bean has a mass of 1.18 g to calculate the number of beans in each jar.

2.71 Answer the question in the caption of Figure 2.6 (see page 63). How many CS_2 molecules would be required to contain 0.25 mol of sulfur atoms?

2.72 In Section 2.4 it was pointed out that an atomic mass unit, u, is equal to $\frac{1}{12}$ the mass of an atom of carbon-12. Suppose one atomic mass unit was redefined as being equal to $\frac{1}{24}$ the mass of a carbon-12 atom. How would this change influence the value of the atomic weight of magnesium?

2.73 How would the change of question 2.72 influence the ratio of the atomic weight of magnesium divided by the atomic weight of hydrogen?

2.74 How would the change of question 2.72 influence the value of Avogadro's number?

Allied Health Exam Connection

The following questions are from these sources:

- *Nursing School Entrance Exam* © 2005, Learning Express, LLC.

- *McGraw-Hill's Nursing School Entrance Exams* by Thomas A. Evangelist, Tamara B. Orr, and Judy Unrein © 2009, The McGraw-Hill Companies, Inc.

- *NSEE Nursing School Entrance Exams,* 3rd edition © 2009, Kaplan Publishing.

- *Cliffs Test Prep: Nursing School Entrance Exams* by Fred N. Grayson © 2004, Wiley Publishing, Inc.

- *Peterson's Master the Nursing School and Allied Health Entrance Exams,* 18th edition by Marion F. Gooding © 2008, Peterson's, a Nelnet Company.

2.75 The symbol K on the periodic table stands for

 a. potassium.

 b. calcium.

 c. carbon.

 d. phosphorus.

2.76 Which one of the following substances is a chemical compound?

 a. blood

 b. water

 c. oxygen

 d. air

2.77 Which of the following is true about compounds?

 a. Compounds are pure substances that are composed of two or more elements in a fixed proportion.

 b. Compounds can be broken down chemically to produce their constituent elements or other compounds.

 c. Both A and B are correct.

 d. Neither A nor B is correct.

2.78 $^{34}_{17}Cl$ has

 a. 17 protons, 17 electrons, and 17 neutrons.

 b. 17 protons, 19 electrons, and 17 neutrons.

 c. 17 protons, 18 electrons, and 17 neutrons.

 d. 34 protons, 34 electrons, and 17 neutrons.

2.79 If two atoms are isotopes, they will

 a. have the same number of protons and neutrons.

 b. have the same number of neutrons, but different numbers of protons.

 c. have the same number of protons, but different numbers of neutrons.

 d. have the same number of neutrons and electrons.

2.80 Copper (Cu) has an atomic number of 29 and a mass number of 64. One copper atom, therefore, has how many protons?

 a. 27

 b. 29

 c. 31

 d. 35

2.81 Atoms are electrically neutral. This means that an atom will contain

 a. more protons than neutrons.

 b. more electrons than protons.

 c. an equal number of protons and electrons.

 d. None of the above.

2.82 The negative charged particle found within the atom is the

 a. proton.

 b. electron.

 c. nucleus.

 d. neutron.

2.83 Two atoms, L and M, are isotopes. Which of the following properties would they NOT have in common?

 a. atomic number

 b. atomic weight

 c. chemical reactivity

 d. the number of protons in the nucleus

2.84 The major portion of an atom's mass consists of:

 a. neutrons and protons.

 b. electrons and protons.

 c. electrons and neutrons.

 d. neutrons and positrons.

2.85 The mass of an atom is almost entirely contributed by its

a. nucleus.

b. protons.

c. electrons and protons.

d. neutrons.

2.86 Which of the following is the chemical symbol for the species that has 16 protons, 17 neutrons, and 18 electrons?

a. $^{33}_{16}S$

b. $^{33}_{17}Cl$

c. $^{35}_{17}Cl$

d. $^{33}_{16}S^{2-}$

2.87 An atom with an atomic number of 58 and an atomic mass of 118 has

a. 58 neutrons.

b. 176 neutrons.

c. 60 neutrons.

d. 116 neutrons.

2.88 What is the mass number of an atom with 60 protons, 60 electrons, and 75 neutrons?

a. 120

b. 135

c. 75

d. 195

2.89 Which of the following represents Avogadro's number?

a. 1.66×10^{-24}

b. 1.0×10^{24}

c. 6.022×10^{23}

d. 3.011×10^{23}

2.90 Which of the following has the greatest number of atoms?

a. 1.0 mol N

b. 1.0 g N

c. 1.0 mol NO_2

d. 0.5 mol NH_3

2.91 The formula of carbon dioxide is CO_2. Its molecular weight is 44 u. A sample of 11 grams of CO_2 contains

a. 1.0 mole of carbon dioxide.

b. 1.5 grams of carbon.

c. 3.0 grams of carbon.

d. 6.0 grams of oxygen.

2.92 What is the molar mass of calcium oxide, CaO?

a. 56

b. 28

c. 640

d. 320

2.93 How many grams are contained in 0.200 mol of calcium phosphate, $Ca_3(PO_4)_2$?

a. 6.20

b. 62.0

c. 124

d. 31.0

2.94 How many moles are contained in a 54.0 g sample of Al?

a. 1.0

b. 2.0

c. 0.5

d. 4.0

3 Electronic Structure and the Periodic Law

▶ Case Study

Maggie was 18 months old. She was very fair, but she was a redhead, so her pale skin didn't seem unusual. Maggie's mother, Suzanne, was a "chocoholic," especially for very dark chocolate. Suzanne was given some 80%-cocoa dark chocolate from Germany. Maggie incessantly asked for a bite of the chocolate. Suzanne gave her a tiny taste, thinking that she surely wouldn't like it. It wasn't even sweet. Maggie loved it and launched into a full-fledged tantrum to be given more. Suzanne was flabbergasted by her daughter's behavior. Maggie's older sister had just taken a nutrition class and suggested that Maggie might be anemic. Suzanne considered her daughter's suggestion and took Maggie to a pediatrician. Sure enough, she was anemic. Her hemoglobin level was 10.5 g/dL. The doctor prescribed 3 mg of ferrous sulfate per kilogram of body weight and recommended another hematocrit in three months.

▶ How likely is it that Maggie's chocolate craving was related to the anemia? Was it a coincidence? Could this be an example of pica (odd cravings due to anemia)?

Follow-up to this Case Study appears at the end of the chapter before the *Concept Summary*.

Thinkstock/Stockbyte/Getty Images

When you have completed your study of this chapter, you should be able to:

1 Locate elements in the periodic table on the basis of group and period designations. **(Section 3.1)**

2 Determine the number of electrons in designated atomic orbitals, subshells, or shells. **(Section 3.2)**

3 Determine the number of valence shell electrons and the electronic structure for atoms, and relate this information to the location of elements in the periodic table. **(Section 3.3)**

4 Determine the following for elements: the electronic configuration of atoms, the number of unpaired electrons

in atoms, and the identity of atoms based on provided electronic configurations. **(Section 3.4)**

5 Determine the shell and subshell locations of the distinguishing electrons in elements, and based on their location in the periodic table, classify elements into the categories given in Figures 3.10 and 3.12. **(Section 3.5)**

6 Recognize property trends of elements within the periodic table, and use the trends to predict selected properties of the elements. **(Section 3.6)**

In Chapter 1, we defined atoms as particles that represent the limit of chemical subdivision. According to this idea, atoms of a specific element cannot be divided into smaller particles or converted into atoms of another element by any physical or chemical change. Then in Chapter 2, we introduced the idea that atoms are, in fact, made up of particles that are smaller than the atoms themselves. Two of these particles, protons and neutrons, form the nuclei of atoms, whereas electrons are located outside the nuclei.

The idea that atoms are made up of subatomic particles implies that it should be possible to obtain even smaller particles from atoms. Scientists have found that it is possible. During chemical changes, electrons are transferred from one atom to another or are shared between atoms. Some details of these processes are given in Chapter 4, but it is known that they depend on the arrangements of the electrons around the nuclei of atoms. These electronic arrangements are one of the major topics of this chapter.

3.1 | The Periodic Law and Table

Learning Objective

1. Locate elements in the periodic table on the basis of group and period designations.

By the early 19th century, detailed studies of the elements known at that time had produced an abundance of chemical information. Scientists looked for order in these facts, with the hope of providing a systematic approach to the study of chemistry. Two scientists independently, and almost simultaneously, made the same important contribution to this end. Julius Lothar Meyer, a German, and Dimitri Mendeleev, a Russian, each produced classification schemes for the elements in 1869. Both schemes were based on the **periodic law**, which in its present form is stated as follows: When all the elements are arranged in order of increasing atomic numbers, elements with similar chemical properties will occur at regular (periodic) intervals.

A convenient way to compactly represent such behavior is to use tables. The arrangement of the elements in a table based on the periodic law is called a *periodic table*. In a modern periodic table, such as the one inside the front cover of this book, elements with similar chemical properties are found in vertical columns called **groups** or **families**.

In this book, the groups are designated in two ways. In the U.S. system, a Roman numeral and letter, such as IIA, is used. In the IUPAC system, a simple number, such as 2, is used.

periodic law A statement about the behavior of the elements when they are arranged in a specific order. In its present form, it is stated as follows: Elements with similar chemical properties occur at regular (periodic) intervals when the elements are arranged in order of increasing atomic numbers.

group or family of the periodic table A vertical column of elements that have similar chemical properties.

Both designations appear on the periodic table inside the front cover of the book, with the IUPAC number in parentheses. We will generally refer to groups in the text by using both the U.S. system and the IUPAC system. Thus, a reference to the second group would be given as group IIA(2), where we have put the IUPAC number in parentheses. The horizontal rows in the table are called **periods** and are numbered from top to bottom. Thus, each element belongs to both a period and a group of the periodic table.

Example 3.1 Groups and Periods of the Periodic Table

Identify the group and period to which each of the following belongs:

a. P b. Cr c. element number 30 d. element number 53

Solution
a. Phosphorus (P) is in group VA(15) and period 3.
b. Chromium (Cr) is in group VIB(6) and period 4.
c. The element with atomic number 30 is zinc (Zn), which is found in group IIB(12) and period 4.
d. Element number 53 is iodine (I), found in group VIIA(17) and period 5.

✔ **LEARNING CHECK 3.1** Write the symbol for the element found in the following places of the periodic table:

a. Group IVA(14) and period 4 b. Group VIIB(7) and period 6

It should be noted that the periodic table as given inside the front cover appears to violate the practice of arranging the elements according to increasing atomic number. Element 72 follows element 57, and 104 follows 89, whereas elements 58–71 and 90–103 are arranged in two rows at the bottom of the table. Technically, these two rows should be included in the body of the table as shown in **Figure 3.1** (top). To save horizontal space, they are placed in the position shown in Figure 3.1 (bottom). This exception presents no problem, as long as it is understood.

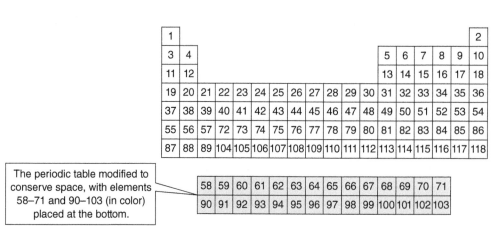

The periodic table modified to conserve space, with elements 58–71 and 90–103 (in color) placed at the bottom.

Figure 3.1 Forms of the periodic table.

Example 3.2 More Groups and Periods of the Periodic Table

a. How many elements are found in period 6 of the periodic table?
b. How many elements are found in group VA(15) of the periodic table?

Solution

a. Period 6 includes elements 55–86, even though elements 58–71 are shown below the main table. Therefore, period 6 contains 32 elements.
b. A count shows group VA(15) contains 6 elements: N, P, As, Sb, Bi, and number 115.

✔ **LEARNING CHECK 3.2** How many elements are found in the following?

a. Period 1 of the periodic table
b. Group IIB(12) of the periodic table

3.2 | Electronic Arrangements in Atoms

Learning Objective

2. Determine the number of electrons in designated atomic orbitals, subshells, or shells.

In 1913 Niels Bohr, a young Danish physicist, made an important contribution to our understanding of atomic structure. He was working under the direction of Ernest Rutherford, a British scientist, who had proposed a solar system model for atoms in which negative electrons moved in circular orbits around the positive nucleus, much like the planets move around the sun. Bohr built on this model by proposing that the single electron of a hydrogen atom could occupy orbits only at specific distances from the nucleus, and thus the electron could have only specific energies (see **Figure 3.2**). Bohr further proposed that the electron changed orbits only by absorbing or releasing energy. The addition of energy to a hydrogen atom elevated the electron to a higher-energy orbit located farther from the nucleus. The energy released when an electron dropped from a higher- to a lower-energy orbit appeared as emitted light.

Research into the nature of atoms continued after Bohr's proposal, and in 1926 a revised model of atomic structure was proposed by Erwin Schrödinger, an Austrian physicist, who received the Nobel Prize in physics in 1933 in recognition of this achievement. According to Schrödinger's quantum mechanical model, the precise paths of electrons cannot be determined accurately, as Bohr's model required. It was found that the location and energy of electrons around a nucleus can be specified using three terms: shell, subshell, and orbital. This is somewhat like locating an individual in a city by specifying a street, building, and apartment.

The location of electrons in a **shell** is indicated by assigning a number, n, to the shell and to all the electrons within the shell. The n value of the lowest-energy shell is 1, that of the next higher energy is 2, the next is 3, and so on. Higher n values for a shell correspond to higher energies and greater distances from the nucleus for the electrons of the shell. Electrons in the third shell all have an n value of 3, all have an energy higher than the energies of electrons in shells 1 and 2, and also are located farther from the nucleus than those of shells 1 and 2.

Each shell is made up of **subshells** that are designated by a letter from the group s, p, d, and f. Because all subshells are designated by one of these letters regardless of the shell in which the subshell is found, a combination of both shell number and subshell letter is used to identify subshells clearly. Thus, a p subshell in shell number 2 is referred to as a $2p$ subshell. The number of subshells found in a shell is the same as the value of n for the shell. Thus, shell number 2 ($n = 2$) contains two subshells. The subshells are the $2p$ mentioned earlier and a $2s$. Electrons located in specific subshells are often referred to in terms of the same number and letter as the subshell. For example, we might refer to an atom as having three $2p$ electrons. All electrons within a specific subshell have the same energy.

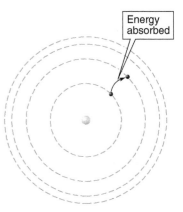

Figure 3.2 A diagram of the Bohr hydrogen atom (not drawn to scale; the orbits are actually much larger than the nucleus). The electron is elevated to a higher-energy orbit when energy is absorbed.

shell A location and energy of electrons around a nucleus that is designated by a value for n, where $n = 1, 2, 3$, etc.

subshell A component of a shell that is designated by a letter from the group s, p, d, and f.

Watch the Salt

New York City's Department of Health has required that all chain restaurants add a salt-shaker-like warning symbol next to foods that contain more than the recommended daily limit of 2300 milligrams of sodium per serving (the amount in about one teaspoon of table salt).

An average adult in the United States consumes about 3400 milligrams of sodium each day in the form of table salt. Only about one in ten Americans meets the one teaspoon guideline suggested by New York's Department of Health. The overconsumption of sodium increases the risk of developing high blood pressure, a leading cause of cardiovascular disease that, in turn, accounts for two-thirds of all strokes and half of the heart disease in the United States. Some research has even linked the excessive intake of salt or salty foods to an increase in the incidence of stomach cancer. Research has also found that the amount of calcium your body loses via urination increases with the amount of salt you consume. If calcium is in short supply in the blood, it can be leached out of the bones. So a diet high in sodium could have an additional unwanted effect, the bone-thinning disease of osteoporosis.

Salt is often present in processed and prepared foods such as pizza, bread, cold cuts, and sandwiches. A fast-food breakfast sandwich can contain as much as 50% of the recommended daily limit of salt in one food item alone. Make it a practice to read labels of items you buy at the grocery store. When you eat at home, be aware of the amount of salt you use.

Increase the amount of potassium in your diet: bananas, yogurt, baked potatoes, salmon, and avocado are among the tasty foods rich in potassium. Dietary potassium helps relax blood vessels, decreases blood pressure, and helps eliminate excess sodium from your body.

© Cengage Learning

Salt is often present in processed foods such as pizza, bread, and cold cuts.

atomic orbital A volume of space around atomic nuclei in which electrons of the same energy move. Groups of orbitals with the same *n* value form subshells.

The description of the location and energy of electrons moving around a nucleus is completed in the quantum mechanical model by specifying an orbital. Each subshell consists of one or more **atomic orbitals**, which are specific volumes of space around nuclei in which electrons move. These atomic orbitals must not be confused with the fixed electron orbits of the original Bohr theory; they are *not* the same. These volumes of space around nuclei have different shapes, depending on the energy of the electrons they contain (see **Figure 3.3**). All *s* subshells consist of a single orbital that is also designated by the letter *s* and further identified by the *n* value of the shell to which the subshell belongs. Thus, the 2*s* subshell mentioned earlier consists of a single 2*s* orbital. All *p* subshells consist of three *p* orbitals that also carry the *n* value of the shell. Thus, the 2*p* subshell of shell number 2 consists of three 2*p* orbitals. Since all the electrons in a subshell have the same energy, an electron in any one of the three 2*p* orbitals has the same energy, regardless of which orbital of the three it occupies. All *d* subshells contain five orbitals, and all *f* subshells contain seven orbitals. According to the quantum mechanical model, each orbital within a subshell can contain a maximum of two electrons.

Figure 3.3 Shapes of typical *s*, *p*, and *d* orbitals.

s

p

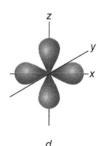

d

The shapes of orbitals such as those given in Figure 3.3 must not be interpreted incorrectly. The fact that s orbitals are spherical in shape does not mean that the electrons move around on the spherical surface. According to the quantum mechanical model, electrons in s orbitals move around inside the spherical volume of the orbital in paths that cannot be determined. All that can be determined about their behavior within the orbital is the probability of finding them in a specific location, which is what Figure 3.3 depicts. Thus, if it were determined that an electron had a 2% probability of being at a specific location, it would simply mean that the electron could be found at that location within an orbital 2 times out of every 100 times we looked for it there. Similarly, electrons in p or d orbitals do not move on the surface of the dumbbell- or cloverleaf-shaped orbitals; they move *within* the three-dimensional dumbbell- or cloverleaf-shaped volumes.

The energy of electrons located in an orbital within a subshell and shell is determined by two factors. As mentioned earlier, the higher the value of n, the higher the energy. In addition, if n is the same, but the subshell is different, the energy of the contained electrons increases in the order s, p, d, f. Thus, a $3p$ electron (an electron in a $3p$ subshell) has a higher energy than a $3s$ electron (an electron in a $3s$ subshell). **Figure 3.4** is a diagrammatic representation of the fourth shell of an atom completely filled with electrons.

Figure 3.4 The fourth shell of an atom, completely filled with electrons.

Example 3.3 Shells and Subshells of Atoms

Determine the following for the third shell of an atom:

a. The number of subshells
b. The designation for each subshell
c. The number of orbitals in each subshell
d. The maximum number of electrons that can be contained in each subshell
e. The maximum number of electrons that can be contained in the shell

Solution

a. The number of subshells is the same as the number used to designate the shell. Therefore, the third shell contains three subshells.
b. Subshells increase in energy according to the order s, p, d, f. The subshells in the shell are therefore designated $3s, 3p,$ and $3d$.
c. The number of orbitals in the subshells is 1, 3, and 5 because s subshells always contain a single orbital, p subshells always contain three orbitals, and d subshells always contain five orbitals.
d. Each atomic orbital can contain a maximum of two electrons, independent of the type of orbital under discussion. Therefore, the $3s$ subshell (one orbital) can hold a maximum of two electrons, the $3p$ subshell (three orbitals) a maximum of 6 electrons, and the $3d$ subshell (five orbitals) a maximum of 10 electrons.
e. The maximum number of electrons that can be contained in the shell is simply the sum of the maximum number for each subshell, $2 + 6 + 10 = 18$.

✔ **LEARNING CHECK 3.3** In each of the following, what is the maximum number of electrons that can be found?

a. a $4p$ orbital b. a $5d$ subshell c. shell number 1

It might seem to you at this point that the modifications to the Bohr theory have created a number of hard-to-remember relationships between shells, subshells, orbitals, and electrons. However, some patterns help make the relationships easy to remember. **Table 3.1** summarizes them for the first four shells of an atom.

TABLE 3.1 Relationships between Shells, Subshells, Orbitals, and Electrons

Shell Number (n)	Number of Subshells in Shell	Subshell Designation	Number of Orbitals in Subshell	Orbital Designation	Maximum Number of Electrons in Subshell	Maximum Number of Electrons in Shell
1	1	1s	1	1s	2	2
2	2	2s	1	2s	2	
		2p	3	2p	6	8
3	3	3s	1	3s	2	
		3p	3	3p	6	
		3d	5	3d	10	18
4	4	4s	1	4s	2	
		4p	3	4p	6	
		4d	5	4d	10	
		4f	7	4f	14	32

3.3 | The Shell Model and Chemical Properties

Learning Objective

3. Determine the number of valence shell electrons and the electronic structure for atoms, and relate this information to the location of elements in the periodic table.

The arrangement of electrons into orbitals, subshells, and shells provides an explanation for the similarities in chemical properties of various elements. **Table 3.2** gives the number of electrons in each shell for the first 20 elements of the periodic table.

TABLE 3.2 Electron Occupancy of Shells

Element	Belongs to Group	Symbol	Atomic Number	Electrons in Shell Number 1	2	3	4
Hydrogen	IA(1)	H	1	1			
Helium	Noble gas(18)	He	2	2			
Lithium	IA(1)	Li	3	2	1		
Beryllium	IIA(2)	Be	4	2	2		
Boron	IIIA(13)	B	5	2	3		
Carbon	IVA(14)	C	6	2	4		
Nitrogen	VA(15)	N	7	2	5		
Oxygen	VIA(16)	O	8	2	6		
Fluorine	VIIA(17)	F	9	2	7		
Neon	Noble gas(18)	Ne	10	2	8		
Sodium	IA(1)	Na	11	2	8	1	
Magnesium	IIA(2)	Mg	12	2	8	2	
Aluminum	IIIA(13)	Al	13	2	8	3	
Silicon	IVA(14)	Si	14	2	8	4	
Phosphorus	VA(15)	P	15	2	8	5	
Sulfur	VIA(16)	S	16	2	8	6	
Chlorine	VIIA(17)	Cl	17	2	8	7	
Argon	Noble gas(18)	Ar	18	2	8	8	
Potassium	IA(1)	K	19	2	8	8	1
Calcium	IIA(2)	Ca	20	2	8	8	2

In Table 3.2, notice that the third shell stops filling when 8 electrons are present, even though the shell can hold a maximum of 18 electrons. The reasons for this are discussed in Section 3.4. Also, note that all the elements in a specific group of the periodic table have the same number of electrons in the outermost occupied shell. This outermost occupied shell (the one of highest energy) is called the **valence shell**. Similarities in elemental chemical properties result from identical numbers of electrons in the valence shells of the atoms (see **Figure 3.5**).

valence shell The outermost (highest-energy) shell of an element that contains electrons.

Figure 3.5 Left to right: Magnesium, calcium, and strontium, members of group IIA(2) of the periodic table, have similar chemical properties and appearances.

Example 3.4 Number of Electrons in Valence Shells for Atoms

Referring to Table 3.2, indicate the number of electrons in the valence shell of elements in groups IA(1), IIA(2), IIIA(13), and IVA(14).

Solution

According to Table 3.2, the elements in group IA(1) are hydrogen, lithium, sodium, and potassium. Each element has one electron in the valence shell. Hydrogen belongs in group IA(1) on the basis of its electronic structure, but its properties differ significantly from other group members.

The group IIA(2) elements are beryllium, magnesium, and calcium. Each has two electrons in the valence shell. The group IIIA(13) elements are boron and aluminum. Both have three electrons in the valence shell. The group IVA(14) elements are carbon and silicon; each has four valence-shell electrons.

✔ **LEARNING CHECK 3.4** Referring to Table 3.2, indicate the number of electrons in the valence shell of elements in groups VA(15), VIA(16), and VIIA(17), and the noble gases(18).

Example 3.4 and Learning Check 3.4 emphasize the fact that elements belonging to the same periodic table group have the same number of electrons in the valence shell (helium, the first element in the noble gases, is an exception). Notice also that the number of electrons in the valence shell is identical to the Roman numeral that designates the group number. Elements of group IIIA(13), for example, have three electrons in the valence shell. It is also apparent that the n value for the valence shell increases by 1 with each heavier member of a group, that is, $n = 2$ for Li, $n = 3$ for Na, and $n = 4$ for K.

Example 3.5 Valence Shell Relationships for Atoms

Using the periodic table and Example 3.4 and Learning Check 3.4, determine the n value for the valence shell and the number of electrons in the valence shell for the following elements:

a. Ba b. Br c. element number 50
d. The third element of group VIA(16)

Solution

a. Ba is the fifth element of group IIA(2), and because $n = 4$ for Ca (the third element of the group), $n = 6$ for the fifth element. The number of valence-shell electrons is two, the same as the group number.
b. In a similar way, Br, the third element of group VIIA(17), has $n = 4$ and seven valence-shell electrons.
c. Element number 50 is tin (Sn), which is the fourth element of group IVA(14). Therefore, $n = 5$ and the number of valence-shell electrons is four.
d. The third element of group VIA(16) is selenium (Se), for which $n = 4$ and the number of valence-shell electrons is six.

✔ **LEARNING CHECK 3.5** How many electrons will be found in the following?

a. The valence shell of Sr
b. The valence shell of the third element in group IVA(14)
c. The valence shell of the fifteenth element in period 4

3.4 | Electronic Configurations

Learning Objective

4. Determine the following for elements: the electronic configuration of atoms, the number of unpaired electrons in atoms, and the identity of atoms based on provided electronic configurations.

According to Section 3.3, similarities in chemical properties between elements are related to the number of electrons that occupy the valence shells of their atoms. We now look at the electronic arrangements of atoms in more detail. These detailed arrangements, called **electronic configurations**, indicate the number of electrons that occupy each subshell and orbital of an atom.

Imagine that electrons are added one by one to the orbitals that belong to subshells and shells associated with a nucleus. The first electron will go to as low an energy state as possible, which is represented by the $1s$ orbital of the $1s$ subshell of the first shell. The second electron will join the first and completely fill the $1s$ orbital, the $1s$ subshell, and the first shell (remember, an orbital is filled when it contains two electrons). The third electron will have to occupy the lowest-energy subshell ($2s$) of the second shell. The fourth electron will also occupy (and fill) the $2s$ subshell. The fifth electron must seek out the next–highest-energy subshell, which is the $2p$. The $2p$ subshell contains three $2p$ orbitals, so the sixth electron can either join the fifth in one of the $2p$ orbitals or go into an empty $2p$ orbital. It will go into an empty orbital in compliance with **Hund's rule**, which states: Electrons will not join other electrons in an orbital if an empty orbital of the same energy is available.

It has been found that electrons behave as if they spin on an axis, and only electrons spinning in opposite directions (indicated by ↑ and ↓) can occupy the same orbital. This principle, known as the **Pauli exclusion principle**, explains why orbitals can contain a maximum of two electrons. Hund's rule and the Pauli exclusion principle can be combined: Electrons will pair with other electrons in an orbital only if there is no empty orbital of the same energy available and if there is one electron with opposite spin already in the orbital.

The seventh added electron will occupy the last empty $2p$ orbital, and the eighth, ninth, and tenth electrons will pair up with electrons already in the $2p$ orbitals. The tenth electron fills the $2p$ subshell, thus completing the second shell. This filling order is illustrated in **Figure 3.6**.

electronic configurations The detailed arrangement of electrons indicated by a specific notation, $1s^2 2s^2 2p^4$, etc.

Hund's rule A statement of the behavior of electrons when they occupy orbitals: Electrons will not join other electrons in an orbital if an empty orbital of the same energy is available for occupancy.

Pauli exclusion principle A statement of the behavior of electrons when they occupy orbitals: Only electrons spinning in opposite directions can simultaneously occupy the same orbital.

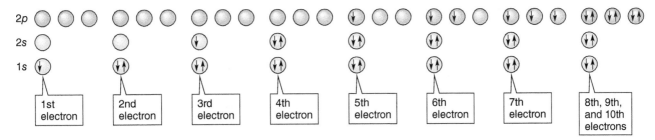

Figure 3.6 The filling order for the first 10 electrons.

The eleventh electron, with no empty orbital available that has the same energy as a 2p and no partially filled orbitals, will occupy the empty lowest-energy subshell of the third shell. The filling order for electrons beyond the tenth follows the pattern given in **Figure 3.7**, where shells are indicated by large rectangles, subshells by colored rectangles, and orbitals by circles. The filling order is obtained by following the arrows.

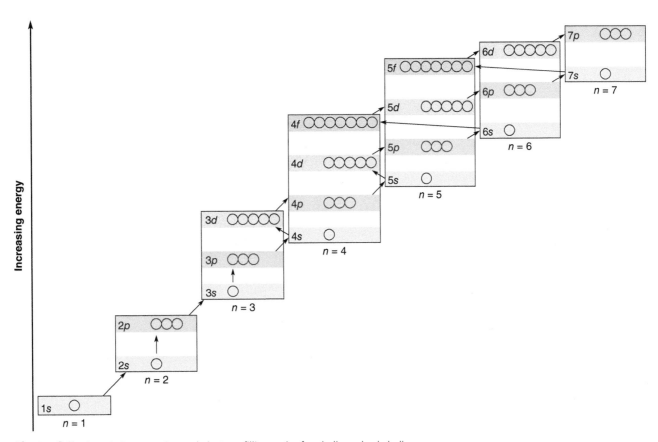

Figure 3.7 The relative energies and electron-filling order for shells and subshells.

As shown in Figure 3.7, some low-energy subshells of a specific shell have energies lower than the upper subshell of a preceding shell. For example, the 4s subshell has a lower energy than, and fills before, the 3d subshell. Figure 3.7 indicates that the third shell will not accept more than 8 electrons until the 4s subshell is complete. Thus, electrons 21 through 30 go into the 3d subshell and complete the filling of the third shell.

It is often convenient to represent the electronic configuration of an atom in a concise way. This is done by writing the subshells in the correct filling order and then indicating the number of electrons in each subshell by a superscript.

Example 3.6 Electronic Configurations and Unpaired Electrons

Write the electronic configurations for the following, and indicate the number of unpaired electrons in each case:

a. An atom that contains 7 electrons
b. An atom that contains 17 electrons
c. An atom of element number 22
d. An atom of arsenic (As)

Solution

The correct filling order of subshells from Figure 3.7 is

$$1s, 2s, 2p, 3s, 3p, 4s, 3d, 4p, 5s, 4d, 5p, 6s, 4f, 5d, 6p, 7s, 5f, 6d, 7p$$

a. Even though s subshells can hold 2 electrons, p subshells 6, d subshells 10, and f subshells 14, only enough subshells to hold 7 electrons will be used. Therefore, the configuration is written as shown below, starting on the left, with circles representing orbitals and arrows representing electrons. It is apparent that no subshells beyond $2p$ are needed because that subshell contains only 3 electrons. Note that both the $1s$ and $2s$ subshells are full; the $2p$ subshell is half-full, with one electron in each of the three orbitals (Hund's rule). These three electrons are unpaired.

$$1s^2 \quad 2s^2 \quad 2p^3$$

b. Similarly, the configuration for 17 electrons is shown below. Here the $1s$, $2s$, $2p$, and $3s$ subshells are full, as are the first and second shells. The $3p$ subshell is not full and can accept 1 more electron. One electron in the $3p$ subshell is unpaired.

$$1s^2 \quad 2s^2 \quad 2p^6 \quad 3s^2 \quad 3p^5$$

c. An atom of element number 22 contains 22 protons in the nucleus and must therefore contain 22 electrons. The electronic configuration is

$$1s^2 \quad 2s^2 \quad 2p^6 \quad 3s^2 \quad 3p^6 \quad 4s^2 \quad 3d^2$$

Note here that only two of the 3d orbitals of the 3d subshell are occupied, and each of these orbitals contains a single electron. Thus, there are two unpaired electrons.

d. Arsenic (As) is element number 33 and therefore contains 33 electrons. The electronic configuration is

$$1s^2 2s^2 2p^6 3s^2 3p^6 4s^2 3d^{10} 4p^3$$

This time, the circles and arrows have not been used to indicate orbitals and electrons. You should satisfy yourself that the following facts are clear: The first, second, and third shells are full. The fourth shell is partially full, with the $4s$ subshell being full and the $4p$ subshell being half-full. The $4p$ subshell contains three unpaired electrons.

✔ **LEARNING CHECK 3.6** Write the electronic configurations for the following, and indicate the number of unpaired electrons in each case:

a. Element number 9
b. Mg
c. The element found in group VIA(16) and period 3
d. An atom that contains 23 protons

CHEMISTRY AROUND US 3.1

A Solar Future

Energy to power human activities comes from numerous sources, but it often takes the form of heat released by burning fuel. In some cases the fuel is burned directly and the released energy is used to move an object like an automobile or airplane. In other cases the heat energy of the burning fuel is converted into other forms, such as the electricity provided by a power plant.

All energy obtained from burning fuels is really solar energy that was stored in the fuels. Advances in technology are making it more and more possible to convert solar energy directly into electrical energy using a device called a solar cell. Solar cells are made by covering an electrically conductive surface such as a metal with a semiconductive material such as silicon. When light strikes the semiconductor, electrons are released that flow through the metal and create an electrical current that can be used to power machines or other devices.

Another approach to harnessing solar energy to produce electricity uses mirrors rather than solar cells. The Ivanpah system located in the Mojave Desert uses more than 300,000 mirrors to reflect sunlight onto giant boilers located on towers. The sunlight heats water in the boilers, producing high-pressure, superheated steam. The steam then flows through pipes to a turbine where electricity is generated for approximately 140,000 homes in California.

Central tower of a solar-thermal plant, Ivanpah, California.

Ethan Miller/Getty Images News/Getty Images

Although Figure 3.7 gives the details of subshell-filling order, a more concise diagram is available and easy to remember. It is shown in **Figure 3.8**, where the subshells are first arranged as shown on the left and then diagonal arrows are drawn as shown on the right. To get the correct subshell-filling order, follow the arrows from top to bottom, going from the head of one arrow to the tail of the next.

1s			
2s	2p		
3s	3p	3d	
4s	4p	4d	4f
5s	5p	5d	5f
6s	6p	6d	
7s			

Figure 3.8 An aid to remembering subshell-filling order.

The electronic configurations described to this point provide details of the shells, subshells, and orbitals involved but are somewhat cumbersome. In some applications, these details are not needed, and simplified representations are used that emphasize the electrons in the valence shell.

We see from Table 3.2 that the noble gases neon and argon both have electronic configurations that end with a completely filled p subshell. This is true for all noble gases except helium, which ends with a filled 1s subshell. In Chapter 4, we will see that these **noble gas configurations** are important in understanding the bonding that occurs between atoms.

noble gas configuration An electronic configuration in which the last eight electrons occupy and fill the s and p subshells of the highest-occupied shell.

Noble gas configurations can be used to write abbreviated electronic configurations. Instead of writing the configurations in their entirety, the symbols for the noble gases are used in brackets to represent the electrons found in their configurations. Electrons that are present in addition to those of the noble gases are written following the symbol. For example, the electronic configuration for sodium can be represented as $1s^2 2s^2 2p^6 3s^1$ or $[Ne]3s^1$.

Example 3.7 Abbreviated Electronic Configurations

Write abbreviated electronic configurations for the following:

a. An atom that contains 7 electrons
b. An atom that contains 17 electrons
c. An atom of element number 22
d. An atom of arsenic (As)

Solution

All these configurations were written in conventional form in Example 3.6.

a. From Example 3.6, the configuration is $1s^2 2s^2 2p^3$. We see that the two electrons in the $1s$ subshell represent the noble gas configuration of helium (He), so we can write the configuration as $[He]2s^2 2p^3$.
b. From Example 3.6, the configuration is $1s^2 2s^2 2p^6 3s^2 3p^5$. We see that the $1s^2 2s^2 2p^6$ portion is the configuration of neon (Ne), so we can write the configuration as $[Ne]3s^2 3p^5$.
c. From Example 3.6, the configuration is $1s^2 2s^2 2p^6 3s^2 3p^6 4s^2 3d^2$. We see that the first 18 electrons represented by $1s^2 2s^2 2p^6 3s^2 3p^6$ correspond to the electrons of argon (Ar). Thus, we can write the configuration as $[Ar]4s^2 3d^2$.
d. From Example 3.6, the configuration of arsenic is $1s^2 2s^2 2p^6 3s^2 3p^6 4s^2 3d^{10} 4p^3$. Once again, the first 18 electrons can be represented by the symbol for argon. The configuration is $[Ar]4s^2 3d^{10} 4p^3$.

✔ **LEARNING CHECK 3.7** Write abbreviated electronic configurations for the following. These are the same elements used in Learning Check 3.6.

a. An atom of element number 9
b. An atom of magnesium (Mg)
c. An atom of the element found in group VIA(16) and period 3
d. An atom that contains 23 protons

3.5 | Another Look at the Periodic Table

Learning Objective

5. Determine the shell and subshell locations of the distinguishing electrons in elements, and based on their location in the periodic table, classify elements into the categories given in Figures 3.10 and 3.12.

Now that you know more about electronic configurations, you can better understand the periodic law and table. First, consider the electronic configurations of the elements belonging to the same group of the periodic table. Group IA(1), for example, contains

Li, Na, K, Rb, Cs, and Fr, with electronic configurations for the first four elements as shown below:

Element Symbol	Conventional Form	Abbreviated Form
Li	$1s^2 2s^1$	$[He]2s^1$
Na	$1s^2 2s^2 2p^6 3s^1$	$[Ne]3s^1$
K	$1s^2 2s^2 2p^6 3s^2 3p^6 4s^1$	$[Ar]4s^1$
Rb	$1s^2 2s^2 2p^6 3s^2 3p^6 4s^2 3d^{10} 4p^6 5s^1$	$[Kr]5s^1$

Notice that each of these elements has a single electron in the valence shell. Further, each of these valence-shell electrons is located in an *s* subshell. Elements belonging to other groups also have valence-shell electronic configurations that are similar for all members of the group, and all members have similar chemical properties. It has been determined that the similar chemical properties of elements in the same group result from similar valence-shell electronic configurations.

The periodic table becomes more useful when we interpret it in terms of the electronic configurations of the elements in various areas. One relationship is shown in **Figure 3.9**, where the periodic table is divided into four areas on the basis of the type of subshell occupied by the highest-energy electron in the atom. This last electron added to an atom is called the **distinguishing electron**. Note that the *s* area is 2 columns (elements) wide, the *p* area is 6 columns wide, the *d* area is 10 columns wide, and the *f* area is 14 columns wide—exactly the number of electrons required to fill the *s*, *p*, *d*, and *f* subshells, respectively.

distinguishing electron The last and highest-energy electron found in an element.

Figure 3.9 The periodic table divided by distinguishing electrons of the elements.

Electronic configurations are also used to classify elements of the periodic table, as shown in **Figure 3.10**.

The *noble gases* make up the group of elements found on the extreme right of the periodic table. They are all gases at room temperature and are unreactive with most other substances (hence, the group name). With the exception of helium, the first member of the group, noble gases are characterized by filled *s* and *p* subshells of the highest occupied shell.

Figure 3.10 Elemental classification on the basis of electronic configurations.

| | Representative elements | | | | | | Transition elements | | | | | | | | Representative elements | | | | | Noble gases |
|---|
| | (1) IA | (2) IIA | (3) IIIB | (4) IVB | (5) VB | (6) VIB | (7) VIIB | (8) VIIIB | (9) VIIIB | (10) | (11) IB | (12) IIB | (13) IIIA | (14) IVA | (15) VA | (16) VIA | (17) VIIA | (18) VIIIA |
| 1 | 1 H | | | | | | | | | | | | | | | | | 2 He |
| 2 | 3 Li | 4 Be | | | | | | | | | | | 5 B | 6 C | 7 N | 8 O | 9 F | 10 Ne |
| 3 | 11 Na | 12 Mg | | | | | | | | | | | 13 Al | 14 Si | 15 P | 16 S | 17 Cl | 18 Ar |
| 4 | 19 K | 20 Ca | 21 Sc | 22 Ti | 23 V | 24 Cr | 25 Mn | 26 Fe | 27 Co | 28 Ni | 29 Cu | 30 Zn | 31 Ga | 32 Ge | 33 As | 34 Se | 35 Br | 36 Kr |
| 5 | 37 Rb | 38 Sr | 39 Y | 40 Zr | 41 Nb | 42 Mo | 43 Tc | 44 Ru | 45 Rh | 46 Pd | 47 Ag | 48 Cd | 49 In | 50 Sn | 51 Sb | 52 Te | 53 I | 54 Xe |
| 6 | 55 Cs | 56 Ba | 57 La | 72 Hf | 73 Ta | 74 W | 75 Re | 76 Os | 77 Ir | 78 Pt | 79 Au | 80 Hg | 81 Tl | 82 Pb | 83 Bi | 84 Po | 85 At | 86 Rn |
| 7 | 87 Fr | 88 Ra | 89 Ac | 104 Rf | 105 Db | 106 Sg | 107 Bh | 108 Hs | 109 Mt | 110 Ds | 111 Rg | 112 Cn | 113 Nh | 114 Fl | 115 Mc | 116 Lv | 117 Ts | 118 Og |

Period (left axis)

Inner-transition elements:

58 Ce	59 Pr	60 Nd	61 Pm	62 Sm	63 Eu	64 Gd	65 Tb	66 Dy	67 Ho	68 Er	69 Tm	70 Yb	71 Lu
90 Th	91 Pa	92 U	93 Np	94 Pu	95 Am	96 Cm	97 Bk	98 Cf	99 Es	100 Fm	101 Md	102 No	103 Lr

representative element An element in which the distinguishing electron is found in an *s* or a *p* subshell.

transition element An element in which the distinguishing electron is found in a *d* subshell.

inner-transition element An element in which the distinguishing electron is found in an *f* subshell.

Representative elements are those found in the *s* and *p* areas of the periodic table, not including the noble gases. The distinguishing electrons of representative elements partially or completely fill an *s* subshell—groups IA(1) and IIA(2)—or partially fill a *p* subshell—groups IIIA(13), IVA(14), VA(15), VIA(16), and VIIA(17). Most of the common elements are representative elements.

The *d* area of the periodic table contains the **transition elements** (Figure 3.10), in which the distinguishing electron is found in a *d* subshell. Some transition elements are used for everyday applications (**Figure 3.11**). **Inner-transition elements** are those in the *f* area of the periodic table, and the distinguishing electron is found in an *f* subshell.

Figure 3.11 Transition elements, such as gold, silver, copper, nickel, platinum, and zinc, are often used in coins and medals. List some chemical and physical properties that would be desirable in metals used for such purposes.

iStockphoto.com/Cveltri

STUDY SKILLS 3.1

The Convention Hotels Analogy

A new concept is often made easier to understand by relating it to something familiar. The concept of electronic configurations is very likely new to you, but you are probably familiar with hotels. The way electrons fill up orbitals, subshells, and shells around a nucleus can be compared to the way rooms, floors, and hotels located near a convention center will fill with convention delegates. To make our analogy work, imagine that the hotels are located on a street that runs uphill from the convention center, as shown. Further imagine that none of the hotels has elevators, so the only way to get to upper floors is to climb the stairs.

In this analogy, the convention center is equivalent to the nucleus of an atom, and each hotel represents a shell, each floor represents a subshell, and each room represents an orbital. If you were a delegate who wanted to describe to a friend where you were staying, you would indicate the hotel (shell), floor (subshell), and room (orbital) assigned to you. Electronic configurations such as $1s^2 2s^2 2p^1$ give similar information for each electron: The numbers preceding each letter indicate the shells, the letters indicate the subshells, and the superscripts coupled with Hund's rule indicate which orbitals are occupied.

Three more assumptions allow us to extend the analogy: (1) No more than two delegates can be assigned to a room, (2) delegates prefer not to have roommates if an empty room on the same floor is available, and (3) all delegates want to use as little energy as possible when they walk from the convention center to their rooms (remember, there are no elevators).

With these assumptions in mind, it is obvious that the first delegate to check in will choose to stay in the single room of the s floor of Hotel One (a small but very exclusive hotel). The second delegate will also choose this same floor and room of Hotel One and will fill the hotel to capacity. The third delegate will have to go to Hotel Two and will choose the one room on the s floor. The fourth delegate will also choose the one room remaining on the s floor, and fill that floor. The fifth delegate will choose a room on the p floor of Hotel Two because the s floor is full. The sixth delegate will also choose a room on the p floor of Hotel Two but will choose one that is not occupied. The seventh delegate will also choose an empty room on the p floor of Hotel Two. Delegates eight, nine, and ten can either pair up with the delegates already in the rooms of the p floor of Hotel Two or go uphill to Hotel Three. They choose to save energy by walking up the stairs and staying in rooms with roommates on the p floor of Hotel Two. Additional arriving delegates will occupy the floors of Hotels Three and Four in the order dictated by these same energy and pairing considerations. Thus, we see that the delegates in their rooms are analogous to the electrons in their orbitals.

Example 3.8 Distinguishing Electrons and Element Classifications

Use the periodic table and Figures 3.9 and 3.10 to determine the following for Ca, Fe, S, and Kr:

a. The type of distinguishing electron
b. The classification based on Figure 3.10

Solution

a. On the basis of the location of each element in Figure 3.9, the distinguishing electrons are of the following types:

$$\text{Ca: } s \quad \text{Fe: } d \quad \text{S: } p \quad \text{Kr: } p$$

b. The classifications based on Figure 3.10 are: Ca, representative element; Fe, transition element; S, representative element; Kr, noble gas.

✔ **LEARNING CHECK 3.8** Determine the following for element numbers 38, 47, 50, and 86:

a. The type of distinguishing electron
b. The classification according to Figure 3.10

The elements can also be classified into the categories of metals, nonmetals, and metalloids. This approach, used in **Figure 3.12**, shows that most elements are classified as metals. It is also apparent from Figure 3.10 that most nonmetals and all metalloids are representative elements.

	(1) IA	(2) IIA	(3) IIIB	(4) IVB	(5) VB	(6) VIB	(7) VIIB	(8) VIIIB	(9) VIIIB	(10) VIIIB	(11) IB	(12) IIB	(13) IIIA	(14) IVA	(15) VA	(16) VIA	(17) VIIA	(18) VIIIA
1	1 H																	2 He
2	3 Li	4 Be											5 B	6 C	7 N	8 O	9 F	10 Ne
3	11 Na	12 Mg											13 Al	14 Si	15 P	16 S	17 Cl	18 Ar
4	19 K	20 Ca	21 Sc	22 Ti	23 V	24 Cr	25 Mn	26 Fe	27 Co	28 Ni	29 Cu	30 Zn	31 Ga	32 Ge	33 As	34 Se	35 Br	36 Kr
5	37 Rb	38 Sr	39 Y	40 Zr	41 Nb	42 Mo	43 Tc	44 Ru	45 Rh	46 Pd	47 Ag	48 Cd	49 In	50 Sn	51 Sb	52 Te	53 I	54 Xe
6	55 Cs	56 Ba	57 La	72 Hf	73 Ta	74 W	75 Re	76 Os	77 Ir	78 Pt	79 Au	80 Hg	81 Tl	82 Pb	83 Bi	84 Po	85 At	86 Rn
7	87 Fr	88 Ra	89 Ac	104 Rf	105 Db	106 Sg	107 Bh	108 Hs	109 Mt	110 Ds	111 Rg	112 Cn	113 Nh	114 Fl	115 Mc	116 Lv	117 Ts	118 Og

Metals Metalloids Nonmetals

58 Ce	59 Pr	60 Nd	61 Pm	62 Sm	63 Eu	64 Gd	65 Tb	66 Dy	67 Ho	68 Er	69 Tm	70 Yb	71 Lu
90 Th	91 Pa	92 U	93 Np	94 Pu	95 Am	96 Cm	97 Bk	98 Cf	99 Es	100 Fm	101 Md	102 No	103 Lr

Figure 3.12 Locations of metals, nonmetals, and metalloids in the periodic table of the elements.

metals Elements found in the left two-thirds of the periodic table. Most have the following properties: high thermal and electrical conductivities, high malleability and ductility, and a metallic luster.

Most **metals** have the following properties. (Are these properties physical or chemical?)

High thermal conductivity—they transmit heat readily.
High electrical conductivity—they transmit electricity readily.
Ductility—they can be drawn into wires.
Malleability—they can be hammered into thin sheets.
Metallic luster—they have a characteristic "metallic" appearance.

nonmetals Elements found in the right one-third of the periodic table. They often occur as brittle, powdery solids or gases and have properties generally opposite those of metals.

Nonmetals, the elements found in the right one-third of the periodic table, generally have chemical and physical properties opposite those of metals. Under normal conditions, they often occur as brittle, powdery solids or as gases.

metalloids Elements that form a narrow diagonal band in the periodic table between metals and nonmetals. They have properties somewhat between those of metals and nonmetals.

Metalloids, such as boron (B) and silicon (Si), are the elements that form a diagonal separation zone between metals and nonmetals in the periodic table. Metalloids have properties somewhat between those of metals and nonmetals, and they often exhibit some of the characteristic properties of each type.

✔ **LEARNING CHECK 3.9** Classify each of the following elements as metal, nonmetal, or metalloid:

a. Xe b. As c. Hg d. Ba e. Th

Transition and Inner-Transition Elements in Your Smart Phone

The smart phones carried and used by a significant percentage of people today could not function without tiny amounts of transition and inner-transition elements (rare earths) used in the construction of their electrical circuitry. Refer to the periodic table and classify each of the nine elements shown in the diagram below as transition or inner-transition.

Color Screen
Yttrium
Lanthanum
Praseodymium
Europium
Gadolinium
Terbium
Dysprosium

Glass Polishing
Lanthanum
Cerium
Praseodymium

Phone Circuitry
Lanthanum
Praseodymium
Neodymium
Gadolinium
Dysprosium

Speakers
Praseodymium
Neodymium
Gadolinium
Terbium
Dysprosium

Vibration Unit
Neodymium
Terbium
Dysprosium

Alexey Boldin/Shutterstock.com

3.6 | Property Trends within the Periodic Table

Learning Objective

6. Recognize property trends of elements within the periodic table, and use the trends to predict selected properties of the elements.

In Figure 3.12, the elements are classified into the categories of metal, metalloid, or nonmetal according to their positions in the table and properties such as thermal conductivity and metallic luster. It is generally true that within a period of the periodic table, the elements become less metallic as we move from left to right. Within a group, the metallic character increases from top to bottom.

Consider group VA(15) as an example. We see that the elements of the group consist of the two nonmetals nitrogen (N) and phosphorus (P), the two metalloids arsenic (As) and antimony (Sb), and the metal bismuth (Bi). We see the trend toward more metallic character from top to bottom that was mentioned above. According to what has been discussed concerning the periodic law and periodic table, these elements should have some similarities in chemical properties because they belong to the same group of the periodic table. Studies of their chemical properties show that they are similar but not identical. Instead, they follow trends according to the location of the elements within the group.

Certain physical properties also follow such trends, and some of these are easily observed, as shown in **Figure 3.13**. At room temperature and ordinary atmospheric pressure, nitrogen is a colorless gas, phosphorus is a white nonmetallic solid, arsenic is a brittle gray solid with a slight metallic luster, antimony is a brittle silver-white solid with a metallic luster, and bismuth is a lustrous silver-white metal.

Figure 3.13 The elements of group VA(15) of the periodic table. Phosphorus must be stored under water because it will ignite when exposed to the oxygen in air. It has a slight color because of reactions with air.

We see that these physical properties are certainly not identical, but we do see that they change in a somewhat regular way (they follow a trend) as we move from element to element down the group. Nitrogen is a gas, but the others are solids. Nitrogen is colorless, phosphorus is white, then the other three become more and more metallic-looking. Also, as mentioned previously, we note a change from nonmetal to metalloid to metal as we come down the group.

The trends in these properties occur in a regular way that would allow us to predict some of them for one element if they were known for the other elements in the group. For example, if the properties of bismuth were unknown, we would have predicted that it would have a silvery white color and a metallic luster based on the appearance of arsenic and antimony. The three elements from group VIIA(17) shown in **Figure 3.14** also demonstrate some obvious predictable trends in physical properties.

Figure 3.14 Chlorine, bromine, and iodine (left to right) all belong to group VIIA(17) of the periodic table. Their atoms all have the same number of electrons in the valence shell and therefore similar chemical properties. However, they do not have similar appearances. At room temperature and under normal atmospheric pressure, chlorine is a pale yellow gas, bromine is a dark red liquid that readily changes into a gas, and iodine is a gray-black solid that changes into a purple gas when heated slightly. Astatine is the next member of the group after iodine. Try to think like Mendeleev and predict its color and whether it will be a liquid, solid, or gas under normal conditions.

Trends in properties occur in elements that form periods across the periodic table as well as among those that form vertical groups. We will discuss two of these properties and their trends for representative elements. Our focus will be on the general trends, recognizing that some elements show deviations from these general behaviors. We will also propose explanations for the trends based on the electronic structure of atoms discussed in the chapter.

The first property we will consider is the size of the atoms of the representative elements. The size of an atom is considered to be the radius of a sphere extending from the center of the nucleus of the atom to the location of the outermost electrons around the nucleus. The behavior of this property across a period and down a group is shown in **Figure 3.15**. We see that the size increases from the top to the bottom of each group, and decreases from left to right across a period.

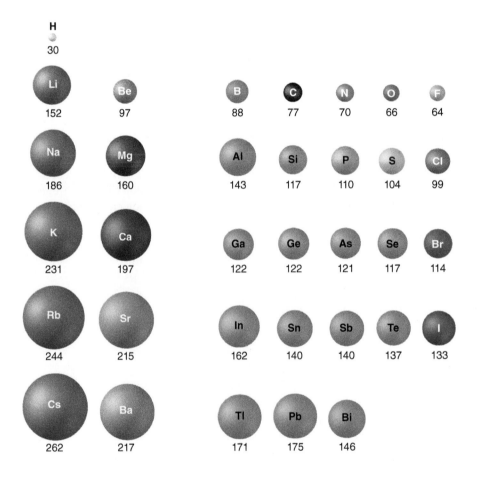

Figure 3.15 Scale drawings of the atoms of some representative elements enlarged about 60 million times. The numbers are the atomic radii in picometers (10^{-12} m).

Remember that when we go from element to element down a group, a new electronic shell is being filled. Consider group IA(1). In lithium atoms (Li), the second shell ($n = 2$) is beginning to fill, in sodium atoms (Na), the third shell ($n = 3$) is beginning to fill, and so on for each element in the group until the sixth shell ($n = 6$) begins to fill in cesium atoms (Cs). As n increases, the distance of the electrons from the nucleus in the shell designated by n increases, and the atomic radius as defined above also increases.

The decrease in atomic radius across a period can also be understood in terms of the outermost electrons around the nucleus. Consider period 3, for example. The abbreviated electronic structure for sodium atoms (Na) is [Ne]$3s^1$, and the outermost electron is in the third shell ($n = 3$). In magnesium atoms (Mg), the electronic structure is [Ne]$3s^2$, and we see that the outermost electron is still in the third shell. In aluminum atoms (Al), the electronic structure is [Ne]$3s^23p^1$, and we see that, once again, the outermost electron is in the third shell. In fact, the outermost electrons in all the atoms across period 3 are in the third shell.

Because all these electrons are in the same third shell, they should all be the same distance from the nucleus, and the atoms should all be the same size. However, each time another electron is added to the third shell of these elements, another positively charged proton is added to the nucleus. Thus, in sodium atoms there are 11 positive nuclear charges attracting the electrons, but in aluminum atoms there are 13. The effect of the increasing nuclear charge attracting electrons in the same shell is to pull all the electrons of the shell closer to the nucleus and cause the atomic radii to decrease as the nuclear charge increases.

The chemical reactivity of elements is dependent on the behavior of the electrons of the atoms of the elements, especially the valence electrons. One property that is related to the behavior of the electrons of atoms is the ionization energy.

The *ionization energy* of an element is the energy required to remove an electron from an atom of the element in the gaseous state. The removal of one electron from an atom leaves the atom with a net 1+ charge because the nucleus of the resulting atom contains one more proton than the number of remaining electrons. The resulting charged atom is called an *ion*. We will discuss ions and the ionization process in more detail in Sections 4.2 and 4.3. A reaction for the removal of one electron from an atom of sodium is

$$Na(g) \rightarrow Na^+(g) + e^-$$

first ionization energy The energy required to remove the first electron from a neutral atom.

Because this process represents the removal of the first electron from a neutral sodium atom, the energy necessary to accomplish the process is called the **first ionization energy**. If a second electron were removed, the energy required would be called the second ionization energy, and so forth. We will focus on only the first ionization energy for representative elements. **Figure 3.16** contains values for the first ionization energy of a number of representative elements.

Figure 3.16 First ionization energies for selected representative elements. The values are given in kJ/mol. (Source: Data from Bard, A. J.; Parsons, R.; Jordan, J. *Standard Potentials in Aqueous Solution.* New York: Dekker, 1985, pp. 24–27.)

IA							VIIIA
H 1311	IIA	IIIA	IVA	VA	VIA	VIIA	**He** 2370
Li 521	**Be** 899	**B** 799	**C** 1087	**N** 1404	**O** 1314	**F** 1682	**Ne** 2080
Na 496	**Mg** 737	**Al** 576	**Si** 786	**P** 1052	**S** 1000	**Cl** 1245	**Ar** 1521
K 419	**Ca** 590	**Ga** 576	**Ge** 784	**As** 1013	**Se** 939	**Br** 1135	**Kr** 1351
Rb 402	**Sr** 549	**In** 559	**Sn** 704	**Sb** 834	**Te** 865	**I** 1007	**Xe** 1170
Cs 375	**Ba** 503	**Tl** 590	**Pb** 716	**Bi** 849	**Po** 791	**At** 926	**Rn** 1037

The general trend is seen to be a decrease from the top to the bottom of a group, and an increase from left to right across a period. The higher the value of the ionization energy, the more difficult it is to remove an electron from the atoms of an element. Thus, we see that in general, the electrons of metals are more easily removed than are the electrons of nonmetals. Also, the farther down a group a metal is located, the easier it is to remove an electron.

The metals of group IA(1) all react with ethyl alcohol, C_2H_5OH, to produce hydrogen gas, H_2, as follows, where M is a general representation of the metals of the group:

$$2M + 2C_2H_5OH \rightarrow 2C_2H_5OM + H_2$$

In this reaction, electrons are removed from the metal and transferred to the hydrogen. The reaction is shown in progress for the first three members of the group in **Figure 3.17**. The rate (speed) of the reaction is indicated by the amount of hydrogen gas being released.

© Jeffrey M. Seager

Figure 3.17 The reaction of group IA(1) elements with ethyl alcohol. Left to right: Lithium (Li), sodium (Na), potassium (K). Each metal sample was wrapped in a wire screen to keep it from floating.

✔ **LEARNING CHECK 3.10** Refer to Figure 3.17, and do the following:

a. Arrange the three metals of the group vertically in order of the rate of the reaction with ethyl alcohol. Put the slowest reaction at the top and the fastest at the bottom.
b. Compare the trend in reaction rate coming down the group with the trend in ionization energy coming down the group.
c. Compare the trend in reaction rate coming down the group with the trend in ease of removing an electron from an atom of the metal coming down the group.

The trends in ionization energy and ease of removing an electron can be explained using arguments similar to those for explaining the sizes of atoms. As we come down a group, the valence electrons are located farther and farther away from the nucleus because they are located in higher-energy shells. The farther the electrons are away from the nucleus, the weaker is the attraction of the positively charged nucleus for the negatively charged electrons, and the easier it is to pull the electrons away.

Similarly, as we go from left to right in a period, the valence electrons are going into the same shell and therefore should be about the same distance from the nucleus. But as we saw before, the nuclear charge increases as well as the number of electrons. As a result, there is a greater nuclear charge attracting the electrons the farther we move to the right. Thus, a valence electron of an atom farther to the right is more difficult to remove than a valence electron of an atom farther to the left.

The general trends we have discussed are summarized in **Figure 3.18**. It is easiest to remember the trends by always going in the same direction in the table and noting how the properties change in that direction. We have chosen to summarize by always going from top to bottom for groups, and left to right for periods.

Figure 3.18 General trends for atomic size and ionization energy of representative elements.

► Case Study Follow-up

The element iron (Fe), occupying the position of period 4, group VIIIB(8) on the periodic table and having the atomic number 26, is present in every cell in the human body. Iron is part of many enzymes and is required for numerous cell functions.

A conservative definition of pica is "the ingestion of non-food substances frequently associated with a nutritional deficiency such as iron-deficiency anemia." Some health-care practitioners include unusual food items in the definition. Pica is common among pregnant women and varies by culture. In the United States, pica is most often observed in African American women living in rural areas who also have a family history of the condition. Craving ice has been associated with anemia

and is frequently classified as pica. In the same way, craving what is essentially plain cocoa by a toddler could be considered pica. Since it was an isolated instance in this case study, a meaningful correlation could not be made. The craving and the anemia might be a coincidence. However, the behavior served as a signal to the mother to assess her child. This case study highlights the importance of family members or closely associated caregivers having knowledge of a child's typical behavior and the caregiver's ability to detect unusual symptoms. It also calls attention to the role of education in public health. The fact that the older sister had some education in nutrition was an important factor in the diagnosis of the anemia.

Concept Summary

The Periodic Law and Table The chemical properties of the elements tend to repeat in a regular (periodic) way when the elements are arranged in order of increasing atomic numbers. This periodic law is the basis for the arrangement of the elements called the periodic table. In this table, each element belongs to a vertical grouping, called a group or family, and a horizontal grouping, called a period. All elements in a group or family have similar chemical properties.

Objective 1 (Section 3.1), Exercise 3.4

Electronic Arrangements in Atoms Niels Bohr proposed a theory for the electronic structure of hydrogen based on the idea that the electrons of atoms move around atomic nuclei in fixed circular orbits. Electrons change orbits only when they absorb or release energy. The Bohr model was modified as a result of continued research. It was found that precise Bohr orbits for electrons could not be determined. Instead, the energy and location of electrons could be specified in terms of shells, subshells, and orbitals, which are indicated by a notation system of numbers and letters.

Objective 2 (Section 3.2), Exercise 3.12

The Shell Model and Chemical Properties The modified Bohr model, or shell model, of electronic structure provides an explanation for the periodic law. The rules governing electron occupancy in shells, subshells, and orbitals result in a repeating pattern of valence-shell electron arrangements. Elements with similar chemical properties turn out to be elements with identical numbers and types of electrons in their valence shells.

Objective 3 (Section 3.3), Exercises 3.18 and 3.22

Electronic Configurations The arrangements of electrons in orbitals, subshells, and shells are called electronic configurations. Rules and patterns have been found that allow these configurations to be represented in a concise way. Written electronic configurations allow details of individual orbital, subshell, and shell electron occupancies to be seen readily. Also, the number of unpaired electrons in elements is easily determined. Electronic configurations can be represented in an abbreviated form by using noble gas symbols to represent some of the inner electrons.

Objective 4 (Section 3.4), Exercises 3.24 and 3.28

Another Look at the Periodic Table Correlations between electronic configurations for the elements and the periodic table arrangement of elements make it possible to determine a number of details of electronic structure for an element simply on the basis of the location of the element in the periodic table. Special attention is paid to the last or distinguishing electron in an element. Elements are classified according to the type of subshell (*s, p, d, f*) occupied by this electron. The elements are also classified on the basis of other properties as metals, nonmetals, or metalloids.

Objective 5 (Section 3.5), Exercises 3.34 and 3.36

Property Trends within the Periodic Table Chemical and physical properties of elements follow trends within the periodic table. These trends are described in terms of changes in properties of elements from the top to the bottom of groups, and from the left to the right of periods. The sizes of atoms and first ionization energies are two properties that show distinct trends.

Objective 6 (Section 3.6), Exercises 3.40 and 3.42

Key Terms and Concepts

Atomic orbital (3.2)
Distinguishing electron (3.5)
Electronic configurations (3.4)
First ionization energy (3.6)
Group or family of the periodic table (3.1)
Hund's rule (3.4)
Inner-transition element (3.5)

Metals (3.5)
Metalloids (3.5)
Noble gas configuration (3.4)
Nonmetals (3.5)
Pauli exclusion principle (3.4)
Periodic law (3.1)
Period of the periodic table (3.1)

Representative element (3.5)
Shell (3.2)
Subshell (3.2)
Transition element (3.5)
Valence shell (3.3)

Exercises

Even-numbered exercises are answered in Appendix B.

Blue-numbered exercises are more challenging.

The Periodic Law and Table (Section 3.1)

3.1 Identify the group and period to which each of the following elements belongs:

 a. Si

 b. element number 21

 c. zinc

 d. element number 35

3.2 Identify the group and period to which each of the following elements belongs:

 a. element number 27

 b. Pb

 c. arsenic

 d. Ba

3.3 Write the symbol and name for the elements located in the periodic table as follows:

 a. Belongs to group VIA(16) and period 3

 b. The first element (reading down) in group VIB(6)

 c. The fourth element (reading left to right) in period 3

 d. Belongs to group IB(11) and period 5

3.4 Write the symbol and name for the elements located in the periodic table as follows:

 a. The noble gas belonging to period 4

 b. The fourth element (reading down) in group IVA(14)

 c. Belongs to group VIB(6) and period 5

 d. The sixth element (reading left to right) in period 6

3.5 **a.** How many elements are located in group VIIB(7) of the periodic table?

 b. How many elements are found in period 5 of the periodic table?

 c. How many total elements are in groups IVA(14) and IVB(4) of the periodic table?

3.6 **a.** How many elements are located in group VIIB(7) of the periodic table?

 b. How many total elements are found in periods 1 and 2 of the periodic table?

 c. How many elements are found in period 5 of the periodic table?

3.7 The following statements either define or are closely related to the terms *periodic law*, *period*, and *group*. Match the terms to the appropriate statements.

 a. This is a vertical arrangement of elements in the periodic table.

 b. The chemical properties of the elements repeat in a regular way as the atomic numbers increase.

 c. The chemical properties of elements 11, 19, and 37 demonstrate this principle.

 d. Elements 4 and 12 belong to this arrangement.

3.8 The following statements either define or are closely related to the terms *periodic law*, *period*, and *group*. Match the terms to the appropriate statements.

 a. This is a horizontal arrangement of elements in the periodic table.

 b. Element 11 begins this arrangement in the periodic table.

 c The element nitrogen is the first member of this arrangement.

 d. Elements 9, 17, 35, and 53 belong to this arrangement.

Electronic Arrangements in Atoms (Section 3.2)

3.9 According to the Bohr theory, which of the following would have the higher energy?

 a. An electron in an orbit close to the nucleus

 b. An electron in an orbit located farther from the nucleus

3.10 What particles in the nucleus cause the nucleus to have a positive charge?

3.11 What is the maximum number of electrons that can be contained in each of the following?

 a. A 3*d* orbital

 b. A 3*d* subshell

 c. The third shell

3.12 What is the maximum number of electrons that can be contained in each of the following?

 a. A 2*p* orbital

 b. A 2*p* subshell

 c. The second shell

3.13 How many orbitals are found in the fourth shell? Write designations for the orbitals.

3.14 How many orbitals are found in the second shell? Write designations for the orbitals.

3.15 How many orbitals are found in a $3d$ subshell? What is the maximum number of electrons that can be located in this subshell?

3.16 How many orbitals are found in a $4f$ subshell? What is the maximum number of electrons that can be located in this subshell?

3.17 Identify the subshells found in the fourth shell; indicate the maximum number of electrons that can occupy each subshell and the total number of electrons that can occupy the shell.

The Shell Model and Chemical Properties (Section 3.3)

3.18 Look at the periodic table and tell how many electrons are in the valence shell of the following elements:

 a. element number 54

 b. the first element (reading down) in group VA(15)

 c. Sn

 d. The fourth element (reading left to right) in period 3

3.19 Look at the periodic table and tell how many electrons are in the valence shell of the following elements:

 a. element number 35

 b. Zn

 c. strontium

 d. the second element in group VA(15)

3.20 What period 6 element has chemical properties most like sodium? How many valence-shell electrons does this element have? How many valence-shell electrons does sodium have?

3.21 What period 5 element has chemical properties most like silicon? How many valence-shell electrons does this element have? How many valence-shell electrons does silicon have?

3.22 If you discovered an ore deposit containing copper, what other two elements might you also expect to find in the ore? Explain your reasoning completely.

3.23 Radioactive isotopes of strontium were produced by the explosion of nuclear weapons. They were considered serious health hazards because they were incorporated into the bones of animals that ingested them. Explain why strontium would be likely to be deposited in bones.

Electronic Configurations (Section 3.4)

3.24 Write an electronic configuration for each of the following elements, using the form $1s^2 2s^2 2p^6$, and so on. Indicate how many electrons are unpaired in each case.

 a. element number 37

 b. Si

 c. titanium

 d. Ar

3.25 Write an electronic configuration for each of the following elements, using the form $1s^2 2s^2 2p^6$, and so on. Indicate how many of the electrons are unpaired in each case.

 a. K

 b. chromium

 c. element number 33

 d. Ti

3.26 Write electronic configurations and answer the following:

 a. How many total s electrons are found in magnesium?

 b. How many unpaired electrons are in nitrogen?

 c. How many subshells are completely filled in Al?

3.27 Write electronic configurations and answer the following:

 a. How many total electrons in Ge have a number designation (before the letters) of 4?

 b. How many unpaired p electrons are found in sulfur? What is the number designation of these unpaired electrons?

 c. How many $3d$ electrons are found in tin?

3.28 Write the symbol and name for each of the elements described. More than one element will fit some descriptions.

 a. Contains only two $2p$ electrons

 b. Contains an unpaired $3s$ electron

 c. Contains two unpaired $3p$ electrons

 d. Contains three $4d$ electrons

 e. Contains three unpaired $3d$ electrons

3.29 Write the symbol and name for each of the elements described. More than one element will fit some descriptions.

 a. Contains one unpaired $5p$ electron

 b. Contains a half-filled $5s$ subshell

 c. Contains a half-filled $6p$ subshell

 d. The last electron completes the $4d$ subshell

 e. The last electron half fills the $4f$ subshell

3.30 Write abbreviated electronic configurations for the following:

 a. arsenic

 b. an element that contains 25 electrons

 c. silicon

 d. element number 53

3.31 Write abbreviated electronic configurations for the following:

 a. an element that contains 24 electrons

 b. element number 21

 c. iodine

 d. copper

3.32 Refer to the periodic table and write abbreviated electronic configurations for all elements in which the noble gas symbol used will be [Ne].

3.33 Refer to the periodic table and determine how many elements have the symbol [Kr] in their abbreviated electronic configurations.

Another Look at the Periodic Table (Section 3.5)

3.34 Classify each of the following elements into the s, p, d, or f area of the periodic table on the basis of the distinguishing electron:

 a. lead

 b. element 27

 c. Tb

 d. Rb

3.35 Classify each of the following elements into the *s*, *p*, *d*, or *f* area of the periodic table on the basis of the distinguishing electron:

 a. Kr

 b. tin

 c. Pu

 d. element 40

3.36 Classify the following elements as representative, transition, inner-transition, or noble gases:

 a. iron

 b. element 15

 c. U

 d. xenon

 e. tin

3.37 Classify the following elements as representative, transition, inner-transition, or noble gases:

 a. W

 b. Cm

 c. element 10

 d. helium

 e. barium

3.38 Classify the following as metals, nonmetals, or metalloids:

 a. argon

 b. element 3

 c. Ge

 d. boron

 e. Pm

3.39 Classify the following as metals, nonmetals, or metalloids:

 a. rubidium

 b. arsenic

 c. element 50

 d. S

 e. Br

Property Trends within the Periodic Table (Section 3.6)

3.40 Use trends within the periodic table to predict which member of each of the following pairs is more metallic:

 a. K or Ti

 b. As or Bi

 c. Mg or Sr

 d. Sn or Ge

3.41 Use trends within the periodic table to predict which member of each of the following pairs is more metallic:

 a. C or Sn

 b. Sb or In

 c. Ca or As

 d. Al or Mg

3.42 Use trends within the periodic table and indicate which member of each of the following pairs has the larger atomic radius:

 a. Ga or Se

 b. N or Sb

 c. O or C

 d. Te or S

3.43 Use trends within the periodic table and indicate which member of each of the following pairs has the larger atomic radius:

 a. Mg or Sr

 b. Rb or Ca

 c. S or Te

 d. I or Sn

3.44 Use trends within the periodic table and indicate which member of each of the following pairs gives up one electron more easily:

 a. Li or K

 b. C or Sn

 c. Mg or S

 d. Li or N

3.45 Use trends within the periodic table and indicate which member of each of the following pairs gives up one electron more easily:

 a. Mg or Al

 b. Ca or Be

 c. S or Al

 d. Te or O

Additional Exercises

3.46 How would you expect the chemical properties of isotopes of the same element to compare to each other? Explain your answer.

3.47 Bromine (Br) and mercury (Hg) are the only elements that are liquids at room temperature. All other elements in the periodic group that contains mercury are solids. Explain why mercury and bromine are not in the same group.

3.48 What would be the mass in mg of 3.0×10^{20} atoms that all have the same electronic configuration of $1s^2\, 2s^2\, 2p^4$?

3.49 Refer to Figure 3.15 and predict what would happen to the density of the metallic elements (purple color) as you go from left to right across a period of the periodic table. Explain your reasoning.

3.50 A 10.02-g sample of an element contains 0.250 mol of the element. Classify the element into the correct category of representative, transition, inner-transition, or noble gas. Will the element conduct electricity?

Chemistry for Thought

3.51 Samples of three metals that belong to the same group of the periodic table are shown in Figure 3.5. When magnesium reacts with bromine, a compound with the formula $MgBr_2$ results. What would be the formulas of the compounds formed by reactions of bromine with each of the other metals shown? Explain your reasoning.

3.52 Answer the problem posed in Figure 3.14, then predict the same things for fluorine, the first member of the group. Explain the reasoning that led you to your answers.

3.53 Answer the problem posed in Figure 3.11. What property that makes gold suitable for coins and medals also makes it useful in electrical connectors for critical electronic parts such as computers in spacecraft?

3.54 Calcium metal reacts with cold water as follows:

$$Ca + 2H_2O \rightarrow Ca(OH)_2 + H_2$$

Magnesium metal does not react with cold water. What behavior toward cold water would you predict for strontium and barium? Write equations to represent any predicted reactions.

3.55 Refer to the hotels analogy in Study Skills 3.1 and determine the number of floors and the number of rooms on the top floor of Hotel Five.

3.56 A special sand is used by a company as a raw material. The company produces zirconium metal that is used to contain the fuel in nuclear reactors. What other metals are likely to be produced from the same raw material? Explain your answer.

Allied Health Exam Connection

The following questions are from these sources:

- *Nursing School Entrance Exam* © 2005, Learning Express, LLC.
- *McGraw-Hill's Nursing School Entrance Exams* by Thomas A. Evangelist, Tamara B. Orr, and Judy Unrein © 2009, The McGraw-Hill Companies, Inc.
- *NSEE Nursing School Entrance Exams*, 3rd edition © 2009, Kaplan Publishing.

- *Cliffs Test Prep: Nursing School Entrance Exams* by Fred N. Grayson © 2004, Wiley Publishing, Inc.
- *Peterson's Master the Nursing School and Allied Health Entrance Exams*, 18th edition by Marion F. Gooding © 2008, Peterson's, a Nelnet Company.

3.57 The arrangement of the modern periodic table is based on atomic:

 a. mass

 b. number

 c. radius

 d. electronegativity

3.58 The horizontal rows of the periodic table are called:

 a. families

 b. groups

 c. representative elements

 d. periods

3.59 Which of the following is an example of a transition element?

 a. aluminum

 b. astatine

 c. nickel

 d. rubidium

3.60 Which two elements have chemical properties that are similar?

 a. H and He

 b. Fe and W

 c. Li and Be

 d. Mg and Ca

3.61 Which statement below is false?

 a. Hydrogen is a nonmetal.

 b. Aluminum is a semimetal.

 c. Calcium is a metal.

 d. Argon is a gas at room temperature.

3.62 Where on the periodic table are the nonmetals located?

 a. upper right

 b. upper left

 c. lower right

 d. lower left

3.63 What does the number 36 represent on the periodic table entry for krypton?

 a. atomic number

 b. relative atomic mass

 c. group number

 d. electron configuration

3.64 Which of the following is an alkali metal (group IA)?

 a. calcium

 b. sodium

 c. aluminum

 d. alkanium

3.65 Which of the following is an alkaline earth metal?

 a. Na

 b. Mg

 c. Sc

 d. Ti

3.66 What is the maximum number of electrons that each p orbital can hold?

 a. 8

 b. 2

 c. 6

 d. 4

3.67 From the periodic table, which of K and Br is larger?

 a. K is larger.

 b. Br is larger.

 c. They are the same size.

 d. We cannot know which one is larger.

Even-numbered exercises answered in Appendix B.

3.68 The element with the smallest atomic radius of the following is:

 a. Sr.

 b. Mg.

 c. Ba.

 d. Ra.

3.69 Ionization energy is:

 a. The energy required to completely remove an electron from an atom or ion.

 b. The energy created by an ion.

 c. The same as kinetic energy.

 d. The attraction between a proton and neutron.

3.70 Which of the following has the largest first ionization energy?

 a. Cs

 b. Rb

 c. Ba

 d. Sr

3.71 Which elements conduct electricity?

 a. metals

 b. nonmetals

 c. metalloids

 d. ions

3.72 What term describes the electrons in the outermost principal energy level of an atom?

 a. vector

 b. core

 c. kernel

 d. valence

3.73 If the electron configuration of an element is written $1s^2\,2s^2\,2p_x^2\,2p_y^2\,2p_z^2\,3s^1$, the element's atomic:

 a. number is 11.

 b. number is 12.

 c. weight is 11.

 d. weight is 12.

3.74 Identify the two atoms with the same number of electrons in their outermost energy level.

 a. Na/K

 b. K/Ca

 c. Na/Mg

 d. Ca/Na

3.75 The number of unpaired electrons in the outer subshell of a phosphorus atom (atomic number 15) is:

 a. 2

 b. 0

 c. 3

 d. 1

3.76 How many valence electrons are needed to complete the outer valence shell of sulfur?

 a. 1

 b. 2

 c. 3

 d. 4

3.77 An atom that has five $3p$ electrons in its ground state is:

 a. Si

 b. P

 c. Cl

 d. O

4 Forces between Particles

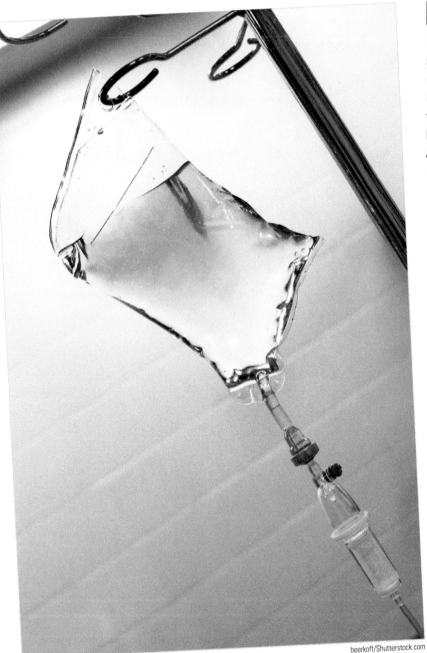

beerkoff/Shutterstock.com

▶ Case Study

Isabelle was a perfectionist and a straight-A student. She kept her room clean, followed the rules, and never gave her parents any trouble. Isabelle was a pretty girl; but having gone through puberty early, by sixth grade she was larger than her peers. Isabelle was not emotionally prepared to accept the weight gain that accompanies puberty. By the beginning of seventh grade she began counting calories. By eighth grade she started to binge and purge. She was 5'6" tall and felt that it was unacceptable to be over 100 lbs. Her parents weren't concerned. They saw Isabelle sometimes consuming large amounts of food—but they just thought she had a high metabolic rate. Isabelle moved on to college, and her parents saw her infrequently. They were shocked to receive a call one day from a hospital informing them that Isabelle had collapsed during a choir rehearsal. Her heart rate was elevated, her blood pressure was low, and her eyes appeared sunken. Isabelle was dehydrated and had an electrolyte imbalance. She was given an IV and referred to the Behavioral Health Unit to receive treatment for her eating disorder.

▶ How do anorexia and bulimia differ? Can a patient have both at the same time? How does family involvement influence patient outcomes?

Follow-up to this Case Study appears at the end of the chapter before the *Concept Summary*.

When you have completed your study of this chapter, you should be able to:

1 Draw correct Lewis structures for atoms of representative elements. **(Section 4.1)**

2 Use electronic configurations to determine the number of electrons gained or lost by atoms as they achieve noble gas electronic configurations. **(Section 4.2)**

3 Use the octet rule to correctly predict the ions formed during the formation of ionic compounds, and write correct formulas for binary ionic compounds containing a representative metal and a representative nonmetal. **(Section 4.3)**

4 Correctly name binary ionic compounds. **(Section 4.4)**

5 Determine formula weights for ionic compounds. **(Section 4.5)**

6 Draw correct Lewis structures for covalent molecules. **(Section 4.6)**

7 Draw correct Lewis structures for polyatomic ions. **(Section 4.7)**

8 Use VSEPR theory to predict the shapes of molecules and polyatomic ions. **(Section 4.8)**

9 Use electronegativities to classify covalent bonds of molecules, and determine whether covalent molecules are polar or nonpolar. **(Section 4.9)**

10 Write correct formulas for ionic compounds containing representative metals and polyatomic ions, and correctly name binary covalent compounds and compounds containing polyatomic ions. **(Section 4.10)**

11 Relate melting and boiling points of pure substances to the strength and type of interparticle forces present in the substances. **(Section 4.11)**

I n the discussion to this point, we have emphasized that matter is composed of tiny particles. However, we have not yet discussed the forces that hold these particles together to form the matter familiar to us. It is these forces that produce many of the properties associated with various types of matter—properties such as the electrical conductivity of copper wire, the melting point of butter, and the boiling point of water.

4.1 | Noble Gas Configurations

Learning Objective

1. Draw correct Lewis structures for atoms of representative elements.

The octet rule, proposed in 1916 by G. N. Lewis (an American) and Walther Kossel (a German), was the first widely accepted theory to describe bonding between atoms. The basis for the rule was the observed chemical stability of the noble gases. Because the noble gases did not react readily with other substances and because chemical reactivity depends on electronic structure, chemists concluded that the electronic structure of the noble gases represented a stable (low-energy) configuration. This noble gas configuration (see **Section 3.4**) is characterized by two electrons in the valence shell of helium and eight valence-shell electrons for the other members of the group (Ne, Ar, Kr, Xe, and Rn).

Example 4.1 Working with Electronic Configurations

Use the techniques described in Section 3.4 to write the electronic configurations for the noble gases. Identify the valence-shell electrons, and verify the noble gas configurations just mentioned.

Solution

The subshell-filling order in Figure 3.7 gives the electronic configurations shown below. Valence-shell electrons are in darker type, and the noble gas configurations are verified.

		Number of Valence-Shell Electrons
He:	$1s^2$	2
Ne:	$1s^2 2s^2 2p^6$	8
Ar:	$1s^2 2s^2 2p^6 3s^2 3p^6$	8
Kr:	$1s^2 2s^2 2p^6 3s^2 3p^6 4s^2 3d^{10} 4p^6$	8
Xe:	$1s^2 2s^2 2p^6 3s^2 3p^6 4s^2 3d^{10} 4p^6 5s^2 4d^{10} 5p^6$	8
Rn:	$1s^2 2s^2 2p^6 3s^2 3p^6 4s^2 3d^{10} 4p^6 5s^2 4d^{10} 5p^6 6s^2 4f^{14} 5d^{10} 6p^6$	8

✔ **LEARNING CHECK 4.1** Write the electronic configurations for F and K. What would have to be done to change these configurations to noble gas configurations (how many electrons would have to be added or removed)?

Lewis structure A representation of an atom or ion in which the elemental symbol represents the atomic nucleus and all but the valence-shell electrons. The valence-shell electrons are represented by dots arranged around the elemental symbol.

A simplified way to represent the valence-shell electrons of atoms was invented by G. N. Lewis. In these representations, called electron-dot formulas or **Lewis structures**, the symbol for an element represents the nucleus and all electrons around the nucleus except those in the valence shell. Valence-shell electrons are shown as dots around the symbol. Thus, rubidium is represented as Rb·.

When a Lewis structure is to be written for an element, it is necessary to determine the number of valence-shell electrons in the element. This can be done by writing the electronic configuration using the methods described in Chapter 3 and identifying the valence-shell electrons as those with the highest n value. A simpler alternative for representative elements is to refer to the periodic table and note that the number of valence-shell electrons in the atoms of an element is the same as the number of the group in the periodic table to which the element belongs. The group number used must be the one that precedes the letter A, not the one in parentheses as given in the periodic table inside the front cover of this book. Using this method, we see that rubidium belongs to group IA(1) and therefore has one valence-shell electron.

Example 4.2 Drawing Lewis Structures

Draw Lewis structures for atoms of the following:

a. element number 4
b. cesium (Cs)
c. aluminum (Al)
d. selenium (Se)

The accepted procedure is to write the element's symbol and put a dot for each valence electron in one of four equally spaced locations around the symbol. Imagine a square around the symbol. Each side of the square represents one of the four locations. An element with four valence electrons would have one dot in each of the four locations. A fifth electron would be represented by one additional dot in one of the locations. Each location can have a maximum of two dots.

Solution

a. Element number 4 is beryllium; it belongs to group IIA(2) and thus has two valence-shell electrons. The Lewis structure is Be·.
b. Cesium is in group IA(1), has one valence-shell electron, and has the Lewis structure Cs·.
c. Aluminum is in group IIIA(13). It has three valence-shell electrons and is represented by the following Lewis structure: Al·.
d. Selenium is in group VIA(16) and thus has six valence-shell electrons. The Lewis structure is : Se·.

✔ **LEARNING CHECK 4.2** Draw Lewis structures for atoms of the following:

a. element number 9
b. magnesium (Mg)
c. sulfur (S)
d. krypton (Kr)

It is important to clearly understand the relationship between Lewis structures and electronic configurations for atoms. The following example and learning check will help you review this relationship.

Example 4.3 Using Abbreviated Electronic Configurations

Represent the following using abbreviated electronic configurations and Lewis structures:

a. F b. K c. Mg d. Si

Solution

a. Fluorine (F) contains nine electrons, with seven of them classified as valence electrons (those in the $2s$ and $2p$ subshells). The configuration and Lewis structure are

$$[He]2s^22p^5 \qquad \text{and} \qquad :\overset{\cdot\cdot}{\underset{\cdot\cdot}{F}}\cdot$$

b. Potassium (K) contains nineteen electrons, with one of them classified as a valence electron (the one in the $4s$ subshell). The configuration and Lewis structure are

$$[Ar]4s^1 \qquad \text{and} \qquad K\cdot$$

c. Magnesium (Mg) contains twelve electrons, with two of them classified as valence electrons (the two in the $3s$ subshell). The configuration and Lewis structure are

$$[Ne]3s^2 \qquad \text{and} \qquad \overset{\cdot}{Mg}\cdot$$

d. Silicon (Si) contains fourteen electrons, with four of them classified as valence electrons (those in the $3s$ and $3p$ subshells). The configuration and Lewis structure are

$$[Ne]3s^23p^2 \qquad \text{and} \qquad \cdot\overset{\cdot\cdot}{\underset{\cdot}{Si}}\cdot$$

✔ **LEARNING CHECK 4.3** Using the two methods illustrated in Example 4.3, write electronic configurations and Lewis structures for the following atoms:

a. Li b. Br c. Sr d. S

4.2 | Ionic Bonding

Learning Objective

2. Use electronic configurations to determine the number of electrons gained or lost by atoms as they achieve noble gas electronic configurations.

According to the **octet rule** of Lewis and Kossel, atoms tend to interact through electronic rearrangements that produce a noble gas electronic configuration for each atom involved in the interaction. Except for the lowest-energy shell, this means that each atom ends up with eight electrons in the valence shell. There are exceptions to this octet rule, but it is still used because of the amount of information it provides. It is especially effective in describing reactions between the representative elements of the periodic table (see **Figure 3.10**).

During some chemical interactions, the octet rule is satisfied when electrons are transferred from one atom to another. As a result of the transfers, neutral atoms acquire net positive or negative electrical charges and become attracted to one another. These charged atoms are called

octet rule A rule for predicting electron behavior in reacting atoms. It says that atoms will gain or lose sufficient electrons to achieve an outer electron arrangement identical to that of a noble gas. This arrangement usually consists of eight electrons in the valence shell.

simple ions, and the attractive force between oppositely charged atoms constitutes an **ionic bond**. A second type of interaction that also satisfies the octet rule is discussed in Section 4.6.

Example 4.4 Ions and Noble Gas Configurations

Show how the following atoms can achieve a noble gas configuration and become ions by gaining or losing electrons:

a. Na b. Cl

Solution

a. The electronic structure of sodium (Na) is represented below using an abbreviated configuration and a Lewis structure:

$$[Ne]3s^1 \qquad \text{and} \qquad Na\cdot$$

The first representation makes it obvious that the Ne configuration with 8 electrons in the valence shell would result if the Na atom lost the single electron located in the $3s$ subshell. The loss is represented by the following equation:

$$Na \rightarrow Na^+ + 1e^-$$

Notice that the removal of a single negative electron from a neutral Na atom leaves the atom with 11 positive protons in the nucleus and 10 negative electrons. This gives the atom a net positive charge. The atom has become a positive ion.

b. The electronic structure for chlorine (Cl) is shown below using an abbreviated configuration and a Lewis structure:

$$[Ne]3s^2 3p^5 \qquad \text{and} \qquad :\overset{\cdot\cdot}{\underset{\cdot\cdot}{Cl}}\cdot$$

The Cl atom can achieve an Ne configuration by losing the 7 valence electrons. However, it is energetically much more favorable to achieve the configuration of argon (Ar) by adding 1 electron to the valence shell. This electron would complete an octet and change a Cl atom into a negative chloride ion as represented by the following equation, where Cl^- represents a Cl atom with 17 protons and 18 electrons, giving the atom a net negative charge:

$$Cl + 1e^- \rightarrow Cl^-$$

✔ **LEARNING CHECK 4.4** In Learning Check 4.1 (see page 102), you described what had to be done to change the electronic configurations of F and K to noble gas configurations. Write equations to illustrate these changes.

As a general rule, metals lose electrons and nonmetals gain electrons during ionic bond formation. The number of electrons lost or gained by a single atom rarely exceeds three, and this number can be predicted accurately for representative elements by using the periodic table. The number of electrons easily lost by a representative metal atom is the same as the group number (represented by Roman numerals preceding the A in the periodic table). The number of electrons that tend to be gained by a representative nonmetal atom is equal to eight minus the group number. However, the nonmetal hydrogen is an exception. In most reactions, it loses one electron, consistent with its placement in group IA(1). In other, less common, reactions it gains one electron just as group VIIA(17) elements do.

Example 4.5 Losing and Gaining Electrons

Use the periodic table to predict the number of electrons lost or gained by atoms of the following elements during ionic bond formation. Write an equation to represent the process in each case.

a. Li
b. Any group IIA(2) element, represented by the symbol M

c. element number 15
d. carbon

Solution

a. Lithium (Li), a metal, is in group IA(1); therefore it will lose one electron per atom:

$$Li \rightarrow Li^+ + 1e^-$$

b. Any group IIA(2) metal will lose two electrons; therefore,

$$M \rightarrow M^{2+} + 2e^-$$

c. Element number 15 is phosphorus (P), a nonmetal of group VA(15). It will gain $8 - 5 = 3$ electrons. The equation is

$$P + 3e^- \rightarrow P^{3-}$$

d. Carbon (C) is in group IVA(14) and is a nonmetal. It should, therefore, gain $8 - 4 = 4$ electrons. However, no more than three electrons generally become involved in ionic bond formation, so we conclude that carbon will not react readily to form ionic bonds.

✔ **LEARNING CHECK 4.5** Predict the number of electrons that would be gained or lost during ionic bond formation for each of the following elements. Write an equation to represent each predicted change.

a. element number 34 b. Rb c. element number 18 d. In

While your attention is focused on noble gas configurations, it is appropriate to emphasize the following point: The attainment of a noble gas electronic configuration by an atom does not mean that the atom is converted into a noble gas. Instead, as seen above, the atoms are converted into simple ions (charged atoms). This point is emphasized by the comparison given in **Table 4.1**. Notice that the electronic configuration of Mg^{2+} is the same as that of Ne, but the number of protons (the atomic number) of Mg^{2+} is still the same as that of Mg. Atoms and simple ions that have identical electronic configurations are said to be **isoelectronic**.

isoelectronic A term that literally means "same electronic," used to describe atoms or ions that have identical electronic configurations.

TABLE 4.1 A Metal Atom, a Metal Ion, and a Noble Gas Atom Compared

Particle	Symbol	Number of Protons in Nucleus	Number of Electrons around Nucleus	Net Charge on Particle
Magnesium atom	Mg	12	12	0
Magnesium ion	Mg^{2+}	12	10	+2
Neon atom	Ne	10	10	0

4.3 | Ionic Compounds

Learning Objective

3. Use the octet rule to correctly predict the ions formed during the formation of ionic compounds, and write correct formulas for binary ionic compounds containing a representative metal and a representative nonmetal.

So far the discussion has focused on the electron-transfer process as it occurs for isolated atoms. In reality, the electrons lost by a metal are the same ones gained by the nonmetal with which it is reacting. Substances formed from such reactions are called *ionic*

compounds. No atom can lose electrons unless another atom is available to accept them, as shown in the formulas used to represent ionic compounds. These formulas represent the combining ratio of the positive and negative ions found in the compounds. This ratio is determined by the charges on the ions, which are determined by the number of electrons transferred.

Example 4.6 Electron Transfer Processes

Represent the electron-transfer process that takes place when the following pairs of elements react ionically. Determine the formula for each resulting ionic compound.

a. Na and Cl b. Mg and F

Solution

a. Sodium (Na) is in group IA(1), and we can write:

$$Na \rightarrow Na^+ + 1e^-$$

Chlorine (Cl), in group VIIA(17), reacts as follows:

$$Cl + 1e^- \rightarrow Cl^-$$

The resulting ions, Na^+ and Cl^-, will combine in a 1:1 ratio because the total positive and total negative charges in the final formula must add up to zero. The formula then is NaCl. Note that the metal is written first in the formula, a generally followed practice. Also, the formula is written using the smallest number of each ion possible—in this case, one of each.

The actual electron-transfer process and the achievement of octets can more easily be visualized as follows:

$$[Ne]3s^1 + [Ne]3s^23p^5 \rightarrow [Ne] + [Ne]3s^23p^6$$
$$\quad Na \qquad\quad Cl \qquad\quad Na^+ \qquad Cl^-$$

In this representation, it must be remembered that the $[Ne]3s^23p^6$ electronic configuration of Cl^- is the same configuration as argon, the noble gas that follows chlorine in the periodic table.

b. Magnesium (Mg) of group IIA(2) will lose two electrons per atom, whereas fluorine (F) of group VIIA(17) will gain one electron per atom. Therefore, we can write

$$Mg \rightarrow Mg^{2+} + 2e^-$$
$$F + 1e^- \rightarrow F^-$$

It is apparent that two fluorine atoms will be required to accept the electrons from one magnesium atom. From another point of view, two F^- ions will be needed to balance the charge of a single Mg^{2+} ion. Both observations lead to the formula MgF_2 for the compound. Note the use of subscripts to indicate the number of ions involved in the formula. The subscript 1, on Mg, is understood and never written.

✔ **LEARNING CHECK 4.6** Write equations to represent the formation of ions for each of the following pairs of elements. Write a formula for the ionic compound that would form in each case.

a. Mg and O b. K and S c. Ca and Br

binary compound A compound made up of two different elements.

Ionic compounds of the types used in Example 4.6 and Learning Check 4.6 are called **binary compounds** because each contains only two different kinds of atoms, Na and Cl in one case (see **Figure 4.1**) and Mg and F in the other.

Consider the Mediterranean Diet

If you are looking for a heart-healthy eating plan, the Mediterranean diet might be right for you. The Mediterranean diet incorporates the basics of healthy eating—plus a splash of flavorful olive oil and perhaps even an occasional glass of red wine. These components characterize the traditional cooking style of countries that border the Mediterranean Sea.

Most healthful diets include fruits, vegetables, fish, and whole grains, and limit unhealthy fats. While these parts of a healthful diet remain tried and true, subtle variations or differences in the proportions of certain foods might make a difference in your risk for heart disease.

Research has shown that the traditional Mediterranean diet reduces the risk of heart disease. In fact, an analysis of more than 1.5 million healthy adults demonstrated that following a Mediterranean diet was associated with a reduced risk of death from heart disease and cancer, as well as a reduced incidence of Parkinson's and Alzheimer's diseases.

The *Dietary Guidelines for Americans* recommends the Mediterranean diet as an eating plan that can help promote health and prevent disease. In addition, it is one your entire family can follow for good health.

The Mediterranean diet emphasizes the following dietary practices:

1. Eating primarily plant-based foods, such as fruits, vegetables, whole grains, legumes, and nuts.

2. Replacing butter with healthy fats such as olive oil.
3. Using herbs and spices instead of salt to flavor foods.
4. Limiting red meat to no more than a few times a month.
5. Eating fish and poultry at least twice a week.
6. Drinking red wine in moderation (optional).

The Mediterranean diet also recognizes the importance of being physically active and enjoying meals with family and friends.

YuliiaHolovchenko/Shutterstock.com

The Mediterranean diet incorporates a variety of healthy foods.

Figure 4.1 The reaction of sodium metal and chlorine gas. The vigorous reaction releases energy that heats the flask contents to a high temperature. Would you expect a similar reaction to occur between potassium metal and fluorine gas?

Na

Cl$_2$

NaCl

Sodium metal and chlorine gas.

The combination reaction of sodium and chlorine. This reaction releases significant amounts of energy.

4.4 | Naming Binary Ionic Compounds

Learning Objective

4. Correctly name binary ionic compounds.

Names for binary ionic compounds are easily assigned when the names of the two elements involved are known. The name of the metallic element is given first, followed by the stem of the nonmetallic elemental name to which the suffix -*ide* has been added.

$$\text{name} = \text{metal} + \text{nonmetal stem} + \text{-ide}$$

The stem of the name of a nonmetal is the name of the nonmetal with the ending dropped. **Table 4.2** gives the stem of the names of some common nonmetallic elements.

TABLE 4.2 Stem Names and Ion Formulas of Common Nonmetallic Elements

Element	Stem	Formula of Ion
Bromine	brom-	Br^-
Chlorine	chlor-	Cl^-
Fluorine	fluor-	F^-
Iodine	iod-	I^-
Nitrogen	nitr-	N^{3-}
Oxygen	ox-	O^{2-}
Phosphorus	phosph-	P^{3-}
Sulfur	sulf-	S^{2-}

Example 4.7 Naming Binary Ionic Compounds

Name the following binary ionic compounds:

a. KCl b. SrO c. Ca_3N_2

Solution

a. The metal is potassium (K), and the nonmetal is chlorine (Cl). Thus, the compound name is potassium <u>chlor</u>ide (the stem of the nonmetallic name is underlined).
b. Similarly, strontium (Sr) and oxygen (O) give the compound name strontium <u>ox</u>ide.
c. The elements are calcium (Ca) and nitrogen (N), hence the name calcium <u>nitr</u>ide.

✔ **LEARNING CHECK 4.7** Assign names to the binary compounds whose formulas you wrote in Learning Check 4.6.

Names for the constituent ions of a binary compound are obtained in the same way as the compound name. Thus, K^+ is a potassium ion, whereas Cl^- is a chloride ion.

Some metal atoms, especially those of transition and inner-transition elements, form more than one type of charged ion. Copper, for example, forms both Cu^+ and Cu^{2+}, and iron forms Fe^{2+} and Fe^{3+}. The names of ionic compounds containing such elements must indicate which ion is present in the compound. A nomenclature system that does this well indicates the ionic charge of the metal ion by a Roman numeral in parentheses following the name of the metal. Thus, CuCl is copper(I) chloride and $CuCl_2$ is copper(II) chloride. These names are expressed verbally as "copper one chloride" and "copper two chloride."

An older system is still in use but works only for naming compounds of metals that can form only two different charged ions. In this method, the endings -*ous* and -*ic* are

attached to the stem of the metal name. For metals with elemental symbols derived from non-English names, the stem of the non-English name is used. The *-ous* ending is always used with the ion of lower charge and the *-ic* ending with the ion of higher charge. Thus, CuCl is cuprous chloride, and $CuCl_2$ is cupric chloride. In the case of iron, $FeCl_2$ is ferrous chloride, and $FeCl_3$ is ferric chloride (see **Figure 4.2**). Notice that the Cu^{2+} ion was called cupric, whereas the Fe^{2+} was called ferrous. The designations do not tell the actual ionic charge, but only which of the two is higher (Cu^{2+} and Fe^{3+}) or lower (Cu^+ and Fe^{2+}) for the metal in question.

© Jeffrey M. Seager

Figure 4.2 Chloride compounds of copper and iron. Top: Cuprous and cupric chloride. Bottom: Ferrous and ferric chloride.

Example 4.8 Formulas for Ionic Compounds

Write formulas for ionic compounds that would form between the following simple ions. Note that the metal forms two different simple ions, and name each compound two ways.

a. Cr^{2+} and S^{2-} b. Cr^{3+} and S^{2-}

Solution
In each case, the metal ion is from the metal chromium (Cr).

a. Because the ionic charges are equal in magnitude but opposite in sign, the ions will combine in a 1:1 ratio. The formula is CrS. Chromium forms simple ions with 2+ and 3+ charges. This one is 2+, the lower of the two; thus, the names are chromium(II) sulfide and chromous sulfide.

b. In this case, the charges on the combining ions are 3+ and 2−. The smallest combining ratio that balances the charges is two Cr^{3+} (a total of 6+ charge) and three S^{2-} (a total of 6− charge). The formula is Cr_2S_3. The names are chromium(III) sulfide and chromic sulfide.

✔ **LEARNING CHECK 4.8** Write formulas for the ionic compounds that would form between the following simple ions. The metal is cobalt, and it forms only the two simple ions shown. Name each compound, using two methods.

a. Co^{2+} and Br^- b. Co^{3+} and Br^-

4.5 | The Smallest Unit of Ionic Compounds

Learning Objective

5. Determine formula weights for ionic compounds.

In Section 1.3 (see page 8), a molecule was defined as the smallest unit of a pure substance capable of a stable, independent existence. As you will see later, some compound formulas are used to represent single molecules. True molecular formulas represent the precise numbers of atoms of each element that are found in a molecule. However, as we have seen, formulas for ionic compounds represent only the simplest combining ratio of the ions in the compounds.

The stable form of an ionic compound is not a molecule, but a crystal in which many ions of opposite charge occupy **lattice sites** in a rigid three-dimensional arrangement called a **crystal lattice.** For example, the lattice of sodium chloride (ordinary table salt) represented in **Figure 4.3** is the stable form of pure sodium chloride.

Figure 4.3 Crystal lattice for sodium chloride (table salt).

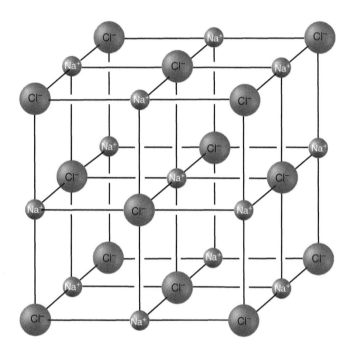

Even though the formulas of ionic compounds do not represent true molecular formulas, they are often used as if they did. This is especially true when equations representing chemical reactions are written, or when the mole concept is applied to chemical formulas (see **Section 2.7**, page 63). When the atomic weights of the atoms making up a true molecular formula are added together, the result is called the molecular weight of the compound (see **Section 2.4**, page 56). A similar quantity obtained by adding up the atomic weights of the atoms shown in the formula of an ionic compound is called a **formula weight.** The mole concept is applied to formula weights in a manner similar to the way it is applied to molecular weights.

Example 4.9 Comparing Molecular and Ionic Compounds

Carbon dioxide, CO_2, is a molecular compound, whereas magnesium chloride, $MgCl_2$, is an ionic compound.

a. Determine the molecular weight for CO_2 and the formula weight for $MgCl_2$ in atomic mass units.
b. Determine the mass in grams of 1.00 mol of each compound.
c. Determine the number of CO_2 molecules in 1.00 mol and the number of Mg^{2+} and Cl^- ions in 1.00 mol of $MgCl_2$.

Solution

a. For CO_2, the molecular weight is the sum of the atomic weights of the atoms in the formula:

$$MW = (1)(\text{at. wt. C}) + (2)(\text{at. wt. O}) = (1)(12.0 \text{ u}) + (2)(16.0 \text{ u})$$
$$MW = 44.0 \text{ u}$$

For $MgCl_2$, the formula weight is also equal to the sum of the atomic weights of the atoms in the formula:

$$FW = (1)(\text{at. wt. Mg}) + (2)(\text{at. wt. Cl}) = (1)(24.3 \text{ u}) + (2)(35.5 \text{ u})$$
$$FW = 95.3 \text{ u}$$

b. According to Section 2.6, 1.00 mol of a molecular compound has a mass in grams equal to the molecular weight of the compound. Thus, 1.00 mol CO_2 = 44.0 g CO_2.

For ionic compounds, 1.00 mol of compound has a mass in grams equal to the formula weight of the compound. Thus, 1.00 mol $MgCl_2$ = 95.3 g $MgCl_2$.

c. In Section 2.6 we learned that 1.00 mol of a molecular compound contains Avogadro's number of molecules. Thus, 1.00 mol CO_2 = 6.02×10^{23} molecules of CO_2.

In the case of ionic compounds, 1.00 mol of compound contains Avogadro's number of formula units. That is, 1.00 mol of magnesium chloride contains Avogadro's number of $MgCl_2$ units, where each unit represents one Mg^{2+} ion and two Cl^- ions. Thus,

$$1.00 \text{ mol } MgCl_2 = 6.02 \times 10^{23} \text{ } MgCl_2 \text{ units}$$

or

$$1.00 \text{ mol } MgCl_2 = 6.02 \times 10^{23} \text{ } Mg^{2+} \text{ ions} + 1.20 \times 10^{24} \text{ } Cl- \text{ ions}$$

Note that the number of Cl^- ions is simply twice Avogadro's number.

✔ **LEARNING CHECK 4.9** Determine the same quantities that were determined in Example 4.9, but use the molecular compound hydrogen sulfide, H_2S, and the ionic compound calcium oxide, CaO.

4.6 | Covalent Bonding

Learning Objective

6. Draw correct Lewis structures for covalent molecules.

In Example 4.5, we found that carbon, a nonmetal, would have to gain four electrons in order to form ionic bonds. We concluded that carbon does not generally form ionic bonds. Yet carbon is known to form many compounds with other elements. Also, electron transfers would not be expected to take place between two atoms of the same element because such an exchange would not change the electronic configuration of the atoms involved. Yet molecules of elements containing two atoms, such as chlorine (Cl_2), oxygen (O_2), and nitrogen (N_2), are known to exist and, in fact, represent the stable form in which these elements occur in nature. What is the nature of the bonding in these molecules?

Again, G. N. Lewis provided an answer by suggesting that the valence-shell electrons of the atoms in such molecules are shared in a way that satisfies the octet rule for each of the atoms. This process is symbolized below for fluorine, (F_2), using Lewis structures:

$$:\ddot{\text{F}}\cdot + \cdot\ddot{\text{F}}: \longrightarrow :\ddot{\text{F}}:\ddot{\text{F}}:$$

shared pair (counted in octet of each atom)

Sometimes the shared electron pairs are shown as straight lines, and sometimes the nonshared pairs are not shown, as illustrated in Example 4.10.

Example 4.10 Lewis Structures and Reactions

Represent the following reactions using Lewis structures:

a. $Cl + Cl \rightarrow Cl_2$ b. $H + H \rightarrow H_2$

Solution

a. Each chlorine (Cl) atom has seven valence-shell electrons. Because one electron from each atom is shared, the octet rule is satisfied for each atom

$$:\ddot{\underset{\cdot\cdot}{Cl}}\cdot + \cdot\ddot{\underset{\cdot\cdot}{Cl}}: \longrightarrow :\ddot{\underset{\cdot\cdot}{Cl}}:\ddot{\underset{\cdot\cdot}{Cl}}:$$

shared pair

The structure of Cl_2 can also be represented as

$$:\ddot{\underset{\cdot\cdot}{Cl}} - \ddot{\underset{\cdot\cdot}{Cl}}: \quad or \quad Cl - Cl$$

b. Each hydrogen (H) atom has one valence-shell electron. By sharing the two electrons, each H atom achieves the helium (He) structure:

$$H \cdot + \cdot H \rightarrow H:H \quad or \quad H - H$$

✔ **LEARNING CHECK 4.10** Represent the following reactions using Lewis structures:

a. $N + N \rightarrow N_2$ (more than two electrons will be shared) b. $Br + Br \rightarrow Br_2$

To understand the origin of the attractive force between atoms that results from electron sharing, consider the H_2 molecule and the concept of atomic orbital overlap. Suppose two H atoms are moving toward each other. Each atom has a single electron in a spherical $1s$ orbital. While the atoms are separated, the orbitals are independent of each other; but as the atoms get closer together, the orbitals overlap and blend to create an orbital common to both atoms called a *molecular orbital*. The two shared electrons then move throughout the overlap region but have a high probability of being found somewhere between the two nuclei. As a result, both of the positive nuclei are attracted toward the negative pair of electrons and hence toward each other. This process, represented in **Figure 4.4**, produces a net attractive force between the nuclei. This attractive force is the **covalent bond** that holds the atoms together. Because forces cannot be seen, we represent only the presence of the shared electron pair between the atoms (by a pair of dots or a line) and remember that an attractive force is also there.

covalent bond The attractive force that results between two atoms that are both attracted to a shared pair of electrons.

Figure 4.4 Orbital overlap during covalent bond formation.

Covalent bonding between like atoms has been described, but it also occurs between unlike atoms. Examples of both types are given in **Table 4.3**. Lewis structures, like those

An adequate supply of clean water is essential to our health and well-being. We can live without food for many days, but life would end in only a few days without water. Our circulatory system is an aqueous stream that distributes an amazing variety of substances throughout the body. Our cells are filled with water solutions in which the chemical reactions of life take place. Without water and its unique properties, life would not be possible on Earth.

Three-fourths of Earth's surface is covered with water, which gives the planet a blue color when it is viewed from space. The oceans represent the world's largest liquid solution. Water participates in most of the chemical reactions that occur in nature.

Most water used for human consumption comes from reservoirs, lakes, rivers, and wells. Much of this water is used and reused by numerous cities as it travels downstream. Such reused water may become seriously polluted with waste from previous users and with pathogenic microorganisms. As a safety precaution, most of the water we use undergoes chemical and physical treatment to purify it. This treatment process includes settling, whereby specific materials are added to bring down suspended solids. This is followed by filtration through sand and gravel to remove still more suspended matter. The filtered water may then be aerated by spraying it into the air. This part of the treatment process removes some odors and improves the taste of the water. Charcoal filtration, a treatment process that is gaining in use, also removes odors along with colored materials. In a final step, chlorine or another disinfectant is added to kill any remaining bacteria.

The treatment of water before use is an important safety precaution.

in Table 4.3, are most easily drawn by using a systematic approach such as the one represented by the following steps:

1. Use the molecular formula to determine how many atoms of each kind are in the molecule.
2. Use the given connecting pattern of the atoms to draw an initial structure for the molecule, with the atoms arranged properly.
3. Determine the total number of valence-shell electrons contained in the atoms of the molecule.
4. Put one pair of electrons between each bonded pair of atoms in the initial structure drawn in Step 2. Subtract the number of electrons used in this step from the total number determined in Step 3. Use the remaining electrons to complete the octets of all atoms in the structure, beginning with the atoms that are present in greatest number in the molecule. Remember, hydrogen atoms require only one pair to achieve the electronic configuration of helium.
5. If all octets cannot be satisfied with the available electrons, move nonbonding pairs (those that are not between bonded atoms) to positions between bonded atoms to complete octets. This will create **double** or **triple bonds** between some atoms.

double and triple bonds The bonds resulting from the sharing of two and three pairs of electrons, respectively.

TABLE 4.3 Examples of Covalent Bonding

Molecule	Atomic Lewis Structure	Sharing Pattern	Molecular Lewis Structure
Nitrogen gas (N_2)	$\cdot\ddot{N}\cdot$ ($1s^2 2s^2 2p^3$)	:N ⇄ N⋅:	:N::N:
Carbon dioxide (CO_2) (each O is bonded to the C)	$\cdot\dot{C}\cdot$ ($1s^2 2s^2 2p^2$) $:\dot{O}\cdot$ ($1s^2 2s^2 2p^4$)	:O C O:	:O::C::O:
Formaldehyde (H_2CO) (each H and the O are bonded to the C)	$\cdot\dot{C}\cdot$ ($1s^2 2s^2 2p^2$) $:\ddot{O}\cdot$ ($1s^2 2s^2 2p^4$) H⋅ ($1s^1$)	H⋅ C O: H⋅	H. C::O: H
Methane (CH_4) (each H is bonded to the C)	$\cdot\dot{C}\cdot$ ($1s^2 2s^2 2p^2$) H⋅ ($1s^1$)	H H⋅ C ⋅H H	H H:C:H H
Ammonia (NH_3) (each H is bonded to the N)	$\cdot\ddot{N}\cdot$ ($1s^2 2s^2 2p^3$) H⋅ ($1s^1$)	H⋅ N ⋅H H	H:N:H H
Water (H_2O) (each H is bonded to the O)	$:\ddot{O}\cdot$ ($1s^2 2s^2 2p^4$) H⋅ ($1s^1$)	:O⋅ ⋅H H	:O:H H
Oxygen difluoride (OF_2) (each F is bonded to the O)	$:\ddot{O}\cdot$ ($1s^2 2s^2 2p^4$) $:\dot{F}:$ ($1s^2 2s^2 2p^5$)	:O⋅ ⋅F: :F:	:O:F: :F:

Example 4.11 Drawing Lewis Structures

Draw Lewis structures for the following molecules:

a. NH_3 b. SO_3 c. C_2H_2

Solution

a. **Step 1.** The formula indicates that the molecule contains one nitrogen (N) atom and three hydrogen (H) atoms.

Step 2. The connecting pattern will have to be given to you in some form. Table 4.3 shows that each H atom is connected to the N atom; thus, we draw

$$H\ N\ H$$
$$H$$

Step 3. Nitrogen is in group VA(15) of the periodic table and so has five valence electrons; hydrogen is in group IA(1) and has one valence electron. The total is eight (five from one N atom and three from the three H atoms).

Step 4. We put one pair of electrons between each H atom and the N atom in the initial structure drawn in Step 2:

$$H:N:H$$
$$\ddot{H}$$

This required six of the eight available electrons. The remaining pair is used to complete the octet of nitrogen. Remember, hydrogen achieves the noble gas configuration of helium with only two electrons:

$$H\!:\!\overset{\cdot\cdot}{N}\!:\!H \qquad \text{or} \qquad H-\overset{\cdot\cdot}{\underset{|}{N}}-H$$
$$\overset{\cdot\cdot}{H} \qquad\qquad\qquad\qquad H$$

Step 5. All octets are satisfied by Step 4, so nothing more needs to be done.

b. **Step 1.** The formula indicates that one sulfur (S) and three oxygen (O) atoms are present in a molecule.

Step 2. Each O atom is bonded only to the S atom; thus, we draw

$$O\ S\ O$$
$$O$$

Step 3. Sulfur and oxygen are both in group VIA(16), and so each atom has 6 valence electrons. The total is 24 (6 from the one S atom and 18 from the three O atoms).

Step 4. We put one pair of electrons between each O atom and the S atom:

$$O\!:\!S\!:\!O$$
$$\overset{\cdot\cdot}{O}$$

This required 6 of the 24 available electrons. The remaining 18 are used to complete the octets, beginning with the O atoms:

$$:\!\overset{\cdot\cdot}{O}\!:\!S\!:\!\overset{\cdot\cdot}{O}\!:$$
$$:\!\overset{\cdot\cdot}{O}\!:$$

Step 5. We see that the octet of sulfur is not completed, even though all available electrons have been used. If one nonbonding pair of any one of the three O atoms is moved to a bonding position between the oxygen and sulfur, it will help satisfy the octet of both atoms. The final Lewis structure is given below. Note that it contains a double bond between one of the oxygens and the sulfur:

$$:\!\overset{\cdot\cdot}{O}\!:\!\overset{\cdot\cdot}{S}\!:\!:\!\overset{\cdot\cdot}{O}\!: \qquad \text{or} \qquad :\!\overset{\cdot\cdot}{O}\!-\!\overset{|}{S}\!=\!\overset{\cdot\cdot}{O}\!:$$
$$:\!\overset{\cdot\cdot}{O}\!: \qquad\qquad\qquad\qquad :\!\overset{\cdot\cdot}{O}\!:$$

If a nonbonding pair of electrons from either of the other two oxygens is used, the following structures result. These are just as acceptable as the preceding structures:

$$:\!\overset{\cdot\cdot}{O}\!=\!\overset{|}{S}\!-\!\overset{\cdot\cdot}{O}\!: \qquad\qquad :\!\overset{\cdot\cdot}{O}\!-\!\overset{||}{S}\!-\!\overset{\cdot\cdot}{O}\!:$$
$$:\!\overset{\cdot\cdot}{O}\!: \qquad\qquad\qquad\qquad O\!:$$

c. **Step 1.** The formula indicates two carbon (C) atoms and two hydrogen (H) atoms per C_2H_2 molecule.

Step 2. The C atoms are bonded to each other, and one H atom is bonded to each C atom.

$$H\ C\ C\ H$$

Step 3. Carbon is in group IVA(14) and has four valence electrons; hydrogen is in group IA(1) and has one valence electron. The total electrons available is ten (eight from the two C atoms and two from the two H atoms).

Step 4. We put one pair of electrons between the two C atoms and between each C atom and H atom.

$$H\!:\!C\!:\!C\!:\!H$$

This required six of the available ten electrons. The remaining four electrons are used to complete the C octets (remember, hydrogen needs only two electrons):

$$H\!:\!C\!:\!\overset{\cdot\cdot}{C}\!:\!H$$

Step 5. All electrons are used in Step 4, but the octet of one C atom is still incomplete. It needs two more pairs. If both nonbonding pairs on the second C atom were shared with the first C atom, both carbons would have complete octets. The result is

$$H:C:::C:H \qquad \text{or} \qquad H-C\equiv C-H$$

This molecule contains a triple bond (three shared pairs) between the C atoms.

✔ **LEARNING CHECK 4.11** Draw Lewis structures for the following molecules:

a. CH_4 (each H atom is bonded to the C atom)
b. H_2CO (the O atom and two H atoms are each bonded to the C atom)
c. HNO_3 (the O atoms are each bonded to the N atom, and the H atom is bonded to one of the O atoms)

4.7 | Polyatomic Ions

Learning Objective

7. Draw correct Lewis structures for polyatomic ions.

polyatomic ions Covalently bonded groups of atoms that carry a net electrical charge.

An interesting combination of ionic and covalent bonding is found in compounds that contain **polyatomic ions**. These ions are covalently bonded groups of atoms that carry a net electrical charge. With the exception of the ammonium ion, NH_4^+, the common polyatomic ions are negatively charged.

Lewis structures can be drawn for polyatomic ions with just a slight modification in the steps given earlier for covalently bonded molecules. In Step 3, the total number of electrons available is obtained by first determining the total number of valence-shell electrons contained in the atoms of the ion. To this number are added electrons representing the number required to give the negative charge to the ion.

For example, in the SO_4^{2-} ion, the total number of valence-shell electrons is thirty, with six coming from the sulfur atom and six from each of the four oxygen atoms. To this number we add two, because two additional electrons are required to give the group of neutral atoms the charge of 2− found on the ion. This gives thirty-two total available electrons.

The only exception to this procedure for the common polyatomic ions is for the positively charged ammonium ion, NH_4^+. In this case, the number of electrons available is the total number of valence electrons minus one, because one electron would have to be removed from the neutral group of atoms to produce the + charge on the ion.

Example 4.12 Lewis Structures for Polyatomic Ions

Draw Lewis structures for the following polyatomic ions. The connecting patterns of the atoms to each other are indicated.

a. SO_4^{2-} (each O atom is bonded to the S atom)
b. NO_3^- (each O atom is bonded to the N atom)
c. $H_2PO_4^-$ (each O atom is bonded to the P atom, and each H atom is bonded to an O atom)

Solution

a. The only difference between drawing Lewis structures for molecules and polyatomic ions comes in Step 3, where the total number of valence-shell electrons is determined.

Step 1. The formula indicates the ion contains one sulfur (S) atom and four oxygen (O) atoms.

Step 2. The given bonding relationships lead to the following initial structure:

$$\begin{matrix} & O & \\ O & S & O \\ & O & \end{matrix}$$

Step 3. Both the S and O atoms are in group VIA(16), so each atom contributes six valence electrons. The total number of valence-shell electrons from the atoms is therefore $5 \times 6 = 30$. However, the ion has a $2-$ charge. This charge comes from the presence of two electrons in the ion in addition to the electrons from the constituent atoms. These two additional electrons must be included in the total valence-shell electrons. Thus, the total number of electrons is $30 + 2 = 32$.

Step 4. The following is obtained when the 32 electrons are distributed among the atoms of the initial structure of Step 2:

$$
\begin{array}{c}
:\ddot{O}: \\
:\ddot{O}:S:\ddot{O}: \\
:\ddot{O}:
\end{array}
$$

Step 5. All octets are satisfied by Step 4. However, it is conventional to enclose Lewis structures of polyatomic ions in brackets and to indicate the net charge on the ion. Thus, the Lewis structures are

$$
\left[\begin{array}{c}
:\ddot{O}: \\
:\ddot{O}:\ddot{S}:\ddot{O}: \\
:\ddot{O}:
\end{array}\right]^{2-}
\quad \text{or} \quad
\left[\begin{array}{c}
O \\
| \\
O-S-O \\
| \\
O
\end{array}\right]^{2-}
$$

where the second structure does not show nonbonding (unshared) electron pairs, and the bonding pairs are represented by lines.

b. **Step 1.** Three oxygen (O) atoms and one nitrogen (N) atom are found in the ion.

Step 2.

$$
\begin{array}{c}
O \\
O \; N \; O
\end{array}
$$

Step 3. Electrons from oxygen $= 3 \times 6 = 18$.
Electrons from nitrogen $= 1 \times 5 = 5$.
One electron comes from the ionic charge $= 1$.
Total electrons $= 18 + 5 + 1 = 24$.

Step 4.

$$
\begin{array}{c}
:\ddot{O}: \\
:\ddot{O}:N:\ddot{O}:
\end{array}
$$

Step 5. The octet of nitrogen is not satisfied by Step 4, so an unshared pair of electrons on an O atom must be shared with the N atom. This will form a double bond in the ion. The resulting Lewis structures are

$$
\left[\begin{array}{c}
:\ddot{O}: \\
:\ddot{O}:N::\ddot{O}:
\end{array}\right]^{-}
\quad \text{or} \quad
\left[\begin{array}{c}
O \\
| \\
O-N=O
\end{array}\right]^{-}
$$

c. **Step 1.** Two hydrogen (H) atoms, one phosphorus (P) atom, and four oxygen (O) atoms are found in the ion.

Step 2.

$$
\begin{array}{c}
O \\
O \; P \; O \; H \\
O \\
H
\end{array}
$$

Step 3. Electrons from hydrogen $= 2 \times 1 = 2$.
Electrons from oxygen $= 4 \times 6 = 24$.
Electrons from phosphorus $= 1 \times 5 = 5$.
Electron from ionic charge $= 1$.
Total electrons $= 2 + 24 + 5 + 1 = 32$.

Step 4.

$$\begin{matrix} & :\overset{..}{O}: & \\ :\overset{..}{O}:\overset{..}{P}:\overset{..}{O}:H \\ & :\overset{..}{O}: & \\ & H & \end{matrix}$$

Step 5. All octets are satisfied by Step 4 (remember, H atoms are satisfied by a single pair of electrons). The resulting Lewis structures are

$$\left[\begin{matrix} & :\overset{..}{O}: & \\ :\overset{..}{O}:\overset{..}{P}:\overset{..}{O}:H \\ & :\overset{..}{O}: & \\ & H & \end{matrix} \right]^{-} \quad \text{or} \quad \left[\begin{matrix} & O & \\ & | & \\ O - P - O - H \\ & | & \\ & O & \\ & | & \\ & H & \end{matrix} \right]^{-}$$

✔ **LEARNING CHECK 4.12** Draw Lewis structures for the following polyatomic ions:

a. PO_4^{3-} (each O atom is bonded to the P atom)
b. SO_3^{2-} (each O atom is bonded to the S atom)
c. NH_4^{+} (each H atom is bonded to the N atom)

4.8 | Shapes of Molecules and Polyatomic Ions

Learning Objective

8. Use VSEPR theory to predict the shapes of molecules and polyatomic ions.

Most molecules and polyatomic ions do not have flat, two-dimensional shapes like those implied by the molecular Lewis structures of Table 4.3. In fact, the atoms of most molecules and polyatomic ions form distinct three-dimensional shapes. Being able to predict the shape is important because the shape contributes to the properties of the molecule or ion. This can be done quite readily for molecules composed of representative elements.

To predict molecular or ionic shapes, first draw Lewis structures for the molecules or ions using the methods discussed in Sections 4.6 and 4.7. Once you have drawn the Lewis structure for a molecule or ion, you can predict its shape by applying a simple theory to the structure. The theory is called the *valence-shell electron-pair repulsion theory*, or **VSEPR theory** (sometimes pronounced "vesper" theory). According to VSEPR theory, electron pairs in the valence shell of an atom are repelled by other electron pairs and get as far away from one another as possible. Any atom in a molecule or ion that is bonded to two or more other atoms is called a *central atom*. When the VSEPR theory is applied to the valence-shell electrons of central atoms, the shape of the molecule or ion containing the atoms can be predicted. Two rules are followed:

VSEPR theory A theory based on the mutual repulsion of electron pairs. It is used to predict molecular shapes.

1. All valence-shell electron pairs around the central atom are counted equally, regardless of whether they are bonding or nonbonding pairs.
2. Double or triple bonds between atoms are treated like a single pair of electrons when predicting shapes.

The electron pairs around a central atom will become oriented in space to get as far away from one another as possible. Thus, two pairs will be oriented with one pair on each opposite side of the central atom. Three pairs will form a triangle around the central atom, and four pairs will be located at the corners of a regular tetrahedron with the central atom in the center (see **Figure 4.5** and **Figure 4.6**). The VSEPR theory can be used to predict shapes of molecules and ions with five or more pairs on the central atom, but we will not go beyond four pairs in this book.

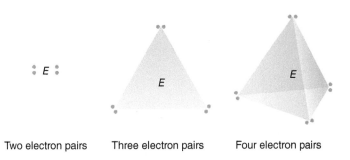

Two electron pairs Three electron pairs Four electron pairs

Figure 4.5 Arrangements of electron pairs around a central atom (E).

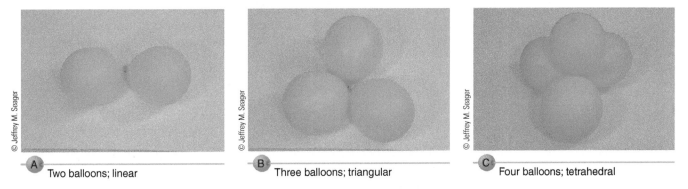

A — Two balloons; linear B — Three balloons; triangular C — Four balloons; tetrahedral

Figure 4.6 When balloons of the same size and shape are tied together, they will assume positions like those taken by pairs of valence electrons around a central atom. Where would the central atom be located in these balloon models?

Example 4.13 Applications of VSEPR Theory

Draw Lewis structures for the following molecules, apply the VSEPR theory, and predict the shape of each molecule:

a. CO_2 (each O atom is bonded to the C atom)
b. C_2Cl_4 (the C atoms are bonded together, and two Cl atoms are bonded to each C atom)
c. NH_3 (each H atom is bonded to the N atom)
d. H_2O (each H atom is bonded to the O atom)

Solution

a. The Lewis structure is $\ddot{O}=C=\ddot{O}$. The central atom is carbon (C) and it has two pairs of electrons in the valence shell surrounding it (remember, each double bond is treated like a single pair of electrons in the VSEPR theory). The two pairs will be located on opposite sides of the C atom, so the molecule has the shape drawn above with the O, C, and O atoms in a line; it is a linear molecule.

b. The Lewis structure is

Both C atoms are central atoms in this molecule, and each one has three electron pairs in the valence shell (again, the double bond is treated as only one pair). The three pairs around each C atom will be arranged in the shape of a triangle:

Thus, the molecule has the shape of two flat triangles in the same plane connected together at one point:

c. The Lewis structure is

$$H - \overset{\displaystyle ..}{N} - H$$
$$|$$
$$H$$

The central atom is nitrogen (N), and it has four electron pairs in the valence shell. The four pairs will be located at the corners of a tetrahedron with the N atom in the middle:

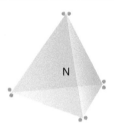

The shape of the molecule is determined only by the positions of the atoms, not by the position of the unshared pair of electrons. Thus, the NH_3 molecule has the shape of a pyramid with a triangular base. The N atom is at the peak of the pyramid, and an H atom is at each corner of the base:

d. The Lewis structure is $H : \overset{\displaystyle ..}{\underset{\displaystyle ..}{O}} : H$. The central atom is oxygen (O), and it is seen to have four pairs of electrons in its valence shell. The four pairs will be located at the corners of a tetrahedron with the O atom in the middle:

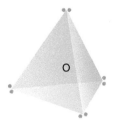

Once again, the shape of the molecule is determined by the locations of the atoms. Thus, H_2O is seen as having a bent or angular shape:

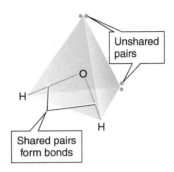

✔ **LEARNING CHECK 4.13** Predict the shapes of the following molecules by applying the VSEPR theory:

a. BF_3 (the Lewis structure is

$$:\ddot{F}: \\ B:\ddot{F}: \\ :\ddot{F}:$$

for this molecule, which does not obey the octet rule).
b. CCl_4 (each Cl atom is bonded to the C atom)
c. SO_2 (each O atom is bonded to the S atom)
d. CS_2 (each S atom is bonded to the C atom)

Example 4.14 More Applications of VSEPR Theory

Use the VSEPR theory to predict the shapes of the following polyatomic ions. The Lewis structures were drawn in Example 4.12.

a. SO_4^{2-}
b. NO_3^-

Solution

a. The Lewis structure from Example 4.12 is

$$\left[\begin{array}{c} :\ddot{O}: \\ :\ddot{O}:S:\ddot{O}: \\ :\ddot{O}: \end{array} \right]^{2-}$$

We see that sulfur is the central atom, and it has four electron pairs around it. The four pairs will be located at the corners of a tetrahedron with the S in the middle.

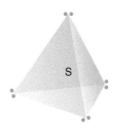

The shape of the ion is determined by the location of the oxygen atoms. Thus, the ion has a tetrahedral shape:

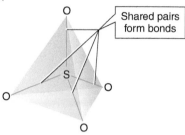

b. The Lewis structure from Example 4.12 is

$$\left[\begin{array}{c} :\ddot{O}: \\ :\ddot{O}:N::\ddot{O}: \end{array} \right]^{-}$$

We see that nitrogen is the central atom, and it has three electron pairs around it. Remember, the double bond formed by the two shared electron pairs between the nitrogen and one of the oxygens counts as only one pair in the VSEPR theory. The three pairs will form a triangle around the N.

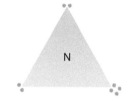

An oxygen is located at each corner of the triangle, so the ion has a flat triangular shape:

✔ **LEARNING CHECK 4.14** Use the VSEPR theory to predict the shapes of the following polyatomic ions. The Lewis structures were drawn in Learning Check 4.12.

a. $PO_4{}^{3-}$ b. $SO_3{}^{2-}$ c. $NH_4{}^{+}$

4.9 | The Polarity of Covalent Molecules

Learning Objective

9. Use electronegativities to classify covalent bonds of molecules, and determine whether covalent molecules are polar or nonpolar.

Molecules of chlorine (Cl—Cl) and hydrogen chloride (H—Cl) have a number of similarities. For example, they contain two atoms each and are therefore diatomic molecules, and the atoms are held together by a single covalent bond. Some differences also exist, the most obvious being the homoatomic nature of the chlorine molecule (an element) and the heteroatomic nature of the hydrogen chloride molecule (a compound). This difference influences the distribution of the shared electrons in the two molecules.

ASK A PHARMACIST 4.1

Are All Iron Preparations Created Equal?

The short answer to the above question is no. Neither are the people who take them—at least in terms of their ability to absorb the iron. Of all the elements in the body, iron is one of the most important. It is so important that the body has developed several mechanisms for transporting and storing it. One of the most important functions of iron is its role as part of the hemoglobin molecule in transporting oxygen to the body cells and exchanging other gases.

About two-thirds of the iron in the body is in the form of the ferrous ion Fe^{2+} that is located in the hemoglobin molecule. Oxygen reversibly bonds to the ferrous ion of hemoglobin and is transported through the body in this form. If iron is in the ferric state Fe^{3+}, it loses its ability to carry oxygen. Another important function of iron in the body is its role in many enzyme-catalyzed reactions.

There are numerous causes of iron deficiency anemia that require patients to take an iron supplement. Among these causes are blood loss from injury, ulcers, absorption problems, pregnancy, rapid growth demands, menstruation, etc. Iron deficiencies have been linked to a wide variety of health problems, ranging from metabolism issues to learning and behavior disorders in children. If it is determined by a health professional that a patient needs to take supplemental iron orally, what form should be used?

There are several different forms available "over the counter," and they are all labeled "iron." In most over-the-counter products, the iron is in the ferrous or 2+ ionic form. The other part of the over-the-counter products will typically be sulfate, gluconate, or fumarate. Which product you choose will determine how much iron you are getting per dosage for your money. Ferrous sulfate contains only about 20% iron, ferrous gluconate only about 12% iron, and ferrous fumarate about 33% iron. You should also check with your health professional to make certain you are not overdosing.

Also, it is good to know that iron is best absorbed from an acidic environment. So, it is not a good idea to take any antacid products along with your iron supplements. Again, your health professional should be consulted if you need direction.

Iron preparations vary in the amount of iron present.

An electron pair shared by two identical atoms is attracted equally to each of the atoms, and the probability of finding the electrons close to an atom is equal for the two atoms. Thus, on average, the electrons spend exactly the same amount of time associated with each atom. Covalent bonds of this type are called **nonpolar covalent bonds**.

Different atoms generally have different tendencies to attract the shared electrons of a covalent bond. A measurement of this tendency is called **electronegativity**. The electronegativity of the elements is a property that follows trends in the periodic table. It increases from left to right across a period and decreases from top to bottom of a group. These trends are shown in **Table 4.4**, which contains electronegativity values for the common representative elements. As a result of electronegativity differences, the bonding electrons shared by two different atoms are shared unequally. The electrons spend more of their time near the atom with the higher electronegativity. The resulting shift in average location of bonding electrons is called **bond polarization** and produces a **polar covalent bond**.

As a result of the unequal electron sharing, the more electronegative atom acquires a partial negative charge ($\delta-$), and the less electronegative atom has a partial positive charge ($\delta+$). The molecule as a whole has no net charge, just an uneven charge distribution.

nonpolar covalent bond A covalent bond in which the bonding pair of electrons is shared equally by the bonded atoms.

electronegativity The tendency of an atom to attract shared electrons of a covalent bond.

bond polarization A result of shared electrons being attracted to the more electronegative atom of a bonded pair of atoms.

polar covalent bond A covalent bond that shows bond polarization; that is, the bonding electrons are shared unequally.

TABLE 4.4 Electronegativities for the Common Representative Elements

Increasing electronegativity →

				H 2.1		
Li 1.0	Be 1.5	B 2.0	C 2.5	N 3.0	O 3.5	F 4.0
Na 0.9	Mg 1.2	Al 1.5	Si 1.8	P 2.1	S 2.5	Cl 3.0
K 0.8	Ca 1.0	Ga 1.6	Ge 1.8	As 2.0	Se 2.4	Br 2.8
Rb 0.8	Sr 1.0	In 1.7	Sn 1.8	Sb 1.9	Te 2.1	I 2.5
Cs 0.7	Ba 0.9					

Decreasing electronegativity ↓

Example 4.15 The Periodic Table and Properties

Use only the periodic table to determine the following for the diatomic covalent molecules listed below: (1) the more electronegative element, (2) the direction of bond polarization, and (3) the charge distribution resulting from the polarization.

a. I—Cl b. Br—Br c. C≡O

Solution

a. Both chlorine (Cl) and iodine (I) belong to group VIIA(17); chlorine is higher in the group and therefore is the more electronegative. The bond polarization (shift in average bonding-electron location) will be toward chlorine, as indicated by the arrow shown below the bond. The result is a partial negative charge on Cl and a partial positive charge on I:

$$\overset{\delta+}{I} \!-\! \overset{\delta-}{Cl}$$
$$\rightarrow$$

b. Since both bromine (Br) atoms have the same electronegativity, no bond polarization or unequal charge distribution results. The molecule is nonpolar covalent and is represented as Br—Br.
c. Oxygen (O) and carbon (C) both belong to the second period. Oxygen, being located farther to the right in the period, is the more electronegative. Thus, the molecule is represented as

$$\overset{\delta+}{C} \!\equiv\! \overset{\delta-}{O}$$
$$\rightarrow$$

✔ LEARNING CHECK 4.15 Use only the periodic table and show the direction of any bond polarization and resulting charge distribution in the following molecules:

a. N≡N b. I—Br c. H—Br

The extent of bond polarization depends on electronegativity differences between the bonded atoms and forms one basis for a classification of bonds. When the electronegativity difference (ΔEN) between the bonded atoms is 0.0, the bond is classified as nonpolar covalent (the electrons are shared equally). When ΔEN is 2.1 or greater, the bond is classified as ionic (the electrons are transferred). When ΔEN is between 0.0 and 2.1, the bond is classified as polar covalent (the electrons are unequally shared, and the bond is polarized). Note, however, that the change from purely covalent to completely ionic compounds is gradual and continuous as ΔEN values change. Thus, a bond between hydrogen (H)

and boron (B) ($\Delta EN = 0.1$) is classified as polar, but the electron sharing is only slightly unequal and the bond is only slightly polar (see **Figure 4.7**). As the values in **Table 4.4** indicate, compounds formed by a reaction of a metal with a nonmetal are generally ionic, while those between nonmetals are either nonpolar or polar covalent.

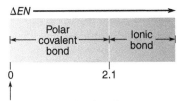

Nonpolar covalent bond

Figure 4.7 The darker the shading, the larger ΔEN is for the bonded atoms. Larger ΔEN values correspond to greater inequality in the sharing of the bonding electrons.

Example 4.16 Classifying Covalent Bonds

Using Table 4.4, classify the bonds in the following compounds as nonpolar covalent, ionic, or polar covalent:

a. ClF b. MgO c. PI$_3$

Solution

a. The electronegativities of chlorine (Cl) and fluorine (F) are 3.0 and 4.0, respectively. Obtain the ΔEN value by subtracting the smaller from the larger, regardless of the order of the elements in the formula. Thus,

$$\Delta EN = 4.0 - 3.0 = 1.0$$

and the bond is polar covalent.

b. Similarly,

$$\Delta EN = 3.5 - 1.2 = 2.3$$

and the bond is ionic.

c. The ΔEN is determined for each phosphorus–iodine bond:

$$\Delta EN = 2.5 - 2.1 = 0.4$$

Thus, each bond is classified as polar covalent.

✔ **LEARNING CHECK 4.16** Using Table 4.4, classify the bonds in the following compounds as nonpolar covalent, ionic, or polar covalent:

a. KF b. NO c. AlN d. K$_2$O

You can now predict the polarity of bonds within molecules, but it is also important to be able to predict the polar nature of the molecules themselves. The terms *polar* and *nonpolar*, when used to describe molecules, indicate their electrical charge symmetry (see **Figure 4.8**). In **polar molecules** (often called dipoles), the charge distribution resulting

polar molecule A molecule that contains polarized bonds and in which the resulting charges are distributed nonsymmetrically throughout the molecule.

Figure 4.8 Polar and nonpolar molecules are affected differently by electrical charges. The ΔEN between C and Cl is 0.5. Propose a reason why CCl_4 molecules are nonpolar.

A. A stream of polar water molecules is attracted by a charged balloon.

B. A stream of nonpolar carbon tetrachloride (CCl_4) molecules is not affected by a charged balloon.

from bond polarizations is nonsymmetric. In **nonpolar molecules**, any charges caused by bond polarization are symmetrically distributed through the molecule. **Table 4.5** illustrates polar and nonpolar molecules. The molecular shapes were determined by using the VSEPR theory.

TABLE 4.5 The Polarity of Molecules

Molecular Formula	Geometric Structure	Bond Polarization and Molecular Charge Distribution	Geometry of Charge Distribution	Classification of Charge Distribution	Classification of Molecule
H_2	H — H	H — H	No charges	No charges	Nonpolar
HCl	H — Cl	$\overset{\delta+}{H} \Rightarrow \overset{\delta-}{Cl}$	+ −	Nonsymmetric	Polar
CO_2	O = C = O	$\overset{\delta-}{O} \rightleftharpoons \overset{\delta 2+}{C} \Rightarrow \overset{\delta-}{O}$	− 2+ −	Symmetric	Nonpolar
BF_3	F, B, F, F (trigonal)	B with δ3+, F atoms δ−	− / 3+ / − −	Symmetric	Nonpolar
N_2O	N ≡ N — O	$N \equiv \overset{\delta+}{N} \Rightarrow \overset{\delta-}{O}$	+ −	Nonsymmetric	Polar
H_2O	H — O, H (bent)	$\overset{\delta+}{H} \Rightarrow \overset{\delta 2-}{O}$, $\overset{\delta+}{H}$	+ 2− +	Nonsymmetric	Polar

4.10 | More about Naming Compounds

Learning Objective

10. Write correct formulas for ionic compounds containing representative metals and polyatomic ions, and correctly name binary covalent compounds and compounds containing polyatomic ions.

In Section 4.4, we discussed the rules for naming binary ionic compounds. In this section, we expand the rules to include the naming of binary covalent compounds and ionic compounds that contain polyatomic ions.

It is apparent from Table 4.4 that covalent bonds (including polar covalent) form most often between representative elements classified as nonmetals. It is not possible to predict formulas for these compounds in a simple way, as was done for ionic compounds in Section 4.4. However, simple rules do exist for naming binary covalent compounds on the basis of their formulas. The rules are similar to those used to name binary ionic compounds: (1) Give the name of the less electronegative element first (the element given first in the formula), (2) give the stem of the name of the more electronegative element next and add the suffix *-ide*, and (3) indicate the number of each type of atom in the molecule by means of the Greek prefixes listed in **Table 4.6**.

TABLE 4.6 Greek Numerical Prefixes

Number	Prefix
1	*mono-*
2	*di-*
3	*tri-*
4	*tetra-*
5	*penta-*
6	*hexa-*
7	*hepta-*
8	*octa-*
9	*nona-*
10	*deca-*

Example 4.17 Naming Binary Covalent Compounds

Name the following binary covalent compounds:

a. CO_2 b. CO c. NO_2 d. N_2O_5 e. CS_2

Solution

a. The elements are carbon (C) and oxygen (O), and carbon is the less electronegative. Because the stem of oxygen is *ox-*, the two portions of the name will be carbon and oxide. Only one C atom is found in the molecule, but the prefix *mono-* is not used

when it appears at the beginning of a name. The two O atoms in the molecule are indicated by the prefix *di-*. The name therefore is carbon dioxide.

b. Similarly, we arrive at the name carbon monoxide for CO.

c. NO_2 is assigned the name nitrogen dioxide.

d. N_2O_5 is assigned the name dinitrogen pentoxide. *Note:* The a is dropped from the *penta-* prefix for oxygen. This is done to avoid the pronunciation problem created by having two vowels next to each other in a name (dinitrogen pentaoxide).

e. CS_2 is named carbon disulfide.

✔ **LEARNING CHECK 4.17** Name the following binary covalent compounds:

a. SO_3 b. BF_3 c. S_2O_7 d. CCl_4

The formulas and names of some common polyatomic ions are given in **Table 4.7**. From this information, the formulas and names for compounds containing polyatomic ions can be written. The rules are essentially the same as those used earlier for binary ionic compounds. In the formulas, the metal (or ammonium ion) is written first, the positive and negative charges must add up to zero, and parentheses are used around the polyatomic ions if more than one is used. In names, the positive metal (or ammonium) ion is given first, followed

STUDY SKILLS 4.1

Help with Polar and Nonpolar Molecules

When concepts are closely related, it is very easy to become confused. In this chapter, the terms *polar* and *nonpolar* are used to describe both bonds and molecules. What is the correct way to apply the terms?

It may be helpful to remember the numbers 3 and 2: three types of bonds, but only two types of molecules. The bonds are nonpolar covalent ($\Delta EN = 0$), polar covalent ($\Delta EN = 0.1-2.0$), and ionic ($\Delta EN = 2.1$ or greater). Remember that compounds bonded by ionic bonds do not form molecules, so there are only two types of molecules: nonpolar and polar.

The polarity of a molecule depends on two factors: (1) the polarity of the bonds between the atoms of the molecule, and (2) the geometric arrangement of the atoms in space. First, determine whether any of the bonds between atoms are polar. If only nonpolar bonds are present, the molecule will be nonpolar regardless of the geometric arrangement of the atoms. However, if one or more polar bonds are found, the arrangement of the atoms must be considered as a second step.

For diatomic (two-atom) molecules, a polar bond between the atoms results in a polar molecule:

$A-A$ $\Delta EN = 0$ because atoms are identical; nonpolar
$A-B$ If $\Delta EN = 0$; nonpolar
$A-B$ If $\Delta EN = 0.1-2.0$; polar

Thus, we see that for diatomic molecules, the type of bond between atoms and the resulting type of molecule is the same.

For a molecule that contains polar bonds and three or more atoms, the molecular geometry must be known and taken into account before the polar nature of the molecule can be determined. Remember, the unequal sharing of electrons between two atoms bonded by a polar bond will cause one of the bonded atoms to have a partial positive charge and one a partial negative charge. If these atoms with their charges are symmetrically arranged in space to form the molecule, the molecule will be nonpolar. Thus, each of the following planar (flat) molecules will be nonpolar even if the bonds between the atoms are polar:

$$B-A-B$$

On the other hand, each of the following molecules will be polar if they contain polar bonds, because the resulting charges on the atoms are not distributed symmetrically in space:

TABLE 4.7 Some Common Polyatomic Ions

Very Common		Common	
NH_4^+	ammonium	CrO_4^{2-}	chromate
$C_2H_3O_2^-$	acetate	$Cr_2O_7^{2-}$	dichromate
CO_3^{2-}	carbonate	NO_2^-	nitrite
ClO_3^-	chlorate	MnO_4^-	permanganate
CN^-	cyanide	SO_3^{2-}	sulfite
HCO_3^-	hydrogen carbonate (bicarbonate)	ClO^-	hypochlorite
OH^-	hydroxide	HPO_4^{2-}	hydrogen phosphate
NO_3^-	nitrate	$H_2PO_4^-$	dihydrogen phosphate
PO_4^{3-}	phosphate	HSO_4^-	hydrogen sulfate (bisulfate)
SO_4^{2-}	sulfate	HSO_3^-	hydrogen sulfite (bisulfite)

by the name of the negative polyatomic ion (see **Figure 4.9**). No numerical prefixes are used except where they are a part of the polyatomic ion name. Names and formulas of acids (compounds in which hydrogen is bound to polyatomic ions) will be given in Chapter 9.

Figure 4.9 Examples of compounds that contain polyatomic ions. The samples are shown as solids and as solids dissolved in water. Referring to Table 4.7, write formulas for these compounds (clockwise from top): potassium carbonate, potassium chromate, potassium phosphate, and potassium permanganate.

© Jeffrey M. Seager

Example 4.18 Writing Compound Formulas and Names

Write formulas and names for compounds composed of ions of the following metals and the polyatomic ions indicated:

a. Na and NO_3^- c. K and HPO_4^{2-}
b. Ca and ClO_3^- d. NH_4^+ and NO_3^-

Solution

a. Sodium (Na) is a group IA(1) metal and forms Na^+ ions. Electrical neutrality requires a combining ratio of one Na^+ for one NO_3^-. The formula is $NaNO_3$. The name is given by the metal name plus the polyatomic ion name: sodium nitrate.

b. Calcium (Ca), a group IIA(2) metal, forms Ca^{2+} ions. Electrical neutrality therefore requires a combining ratio of one Ca^{2+} for two ClO_3^- ions. The formula is $Ca(ClO_3)_2$. (*Note:* The use of parentheses around the polyatomic ion prevents the confusion resulting from writing $CaClO_{32}$, which implies that there are 32 oxygen atoms in the formula. Parentheses are always used when multiples of a specific polyatomic ion are used in a formula and are indicated by a subscript.) The name is calcium chlorate.

c. Potassium (K), a group IA(1) metal, forms K^+ ions. The required combining ratio of 2:1 gives the formula K_2HPO_4. The name is potassium hydrogen phosphate.

d. The NH_4^+ is not a metallic ion, but it behaves like one in numerous compounds. The 1:1 combining ratio gives the formula NH_4NO_3. (*Note:* The polyatomic ions are written separately, and the nitrogen atoms are not grouped to give a formula such as $N_2H_4O_3$.) The name is ammonium nitrate.

✔ **LEARNING CHECK 4.18** Write formulas and names for compounds containing ions of the following metals and the polyatomic ions indicated:

a. Ca and HPO_4^{2-}

b. Mg and PO_4^{3-}

c. K and MnO_4^-

d. NH_4^+ and $Cr_2O_7^{2-}$

4.11 | Other Interparticle Forces

Learning Objective

11. Relate melting and boiling points of pure substances to the strength and type of interparticle forces present in the substances.

Ionic and covalent bonding can account for certain properties of many substances. However, some experimental observations can be explained only by proposing the existence of other types of forces between particles.

Earlier in this chapter, a crystal lattice was described in connection with ionic bonding (see **Figure 4.3**). Most pure substances (elements or compounds) in the solid state also exist in the form of a crystal lattice. However, in some of these solids, neutral atoms or molecules occupy the lattice sites instead of ions. When solids are melted, the forces holding the lattice particles in place are overcome, and the particles move about more freely in what is called the liquid state. The addition of more heat overcomes the attractive interparticle forces to a still greater extent, and the liquid is converted into a gas or vapor; the liquid boils. Particles in the vapor state move about very freely and are influenced only slightly by interparticle attractions. Thus, the temperatures at which melting and boiling take place give an indication of the strength of the interparticle forces that are being overcome. (These states of matter—solid, liquid, and gas—are discussed in more detail in Chapter 6.)

Suppose an experiment is carried out with several pure substances. Some are familiar to you, but others you probably have never seen in the solid state. The substances are listed in **Table 4.8**, together with some pertinent information.

TABLE 4.8 Some Characteristics of Selected Pure Substances

Substance	Formula or Symbol	Classification	Particles Occupying Lattice Sites
Sodium chloride	NaCl	Compound	Na^+ and Cl^- ions
Water	H_2O	Compound	H_2O molecules
Carbon monoxide	CO	Compound	CO molecules
Quartz (pure sand)	SiO_2	Compound	Si and O atoms
Copper metal	Cu	Element	Cu atoms
Oxygen	O_2	Element	O_2 molecules

The experiment is started at the very low temperature of $-220°C$ to have all substances in the solid state. The temperature is then increased slowly and uniformly to

2600°C, at which point all the substances will be in the gaseous state. See **Table 4.9**, which gives only those temperatures corresponding to a specific change in one of the substances.

TABLE 4.9 The Behavior of Selected Pure Substances in Response to Heating

| Temperature (°C) | Behavior or State of Substance | | | | | |
	Oxygen (O_2)	Carbon Monoxide (CO)	Water (H_2O)	Salt (NaCl)	Copper (Cu)	Quartz (SiO_2)
−220	Solid	Solid	Solid	Solid	Solid	Solid
−218	**Melts**	Solid	Solid	Solid	Solid	Solid
−199	Liquid	**Melts**	Solid	Solid	Solid	Solid
−192	Liquid	**Boils**	Solid	Solid	Solid	Solid
−183	**Boils**	Gas	Solid	Solid	Solid	Solid
0	Gas	Gas	**Melts**	Solid	Solid	Solid
100	Gas	Gas	**Boils**	Solid	Solid	Solid
801	Gas	Gas	Gas	**Melts**	Solid	Solid
1083	Gas	Gas	Gas	Liquid	**Melts**	Solid
1413	Gas	Gas	Gas	**Boils**	Liquid	Solid
1610	Gas	Gas	Gas	Gas	Liquid	**Melts**
2230	Gas	Gas	Gas	Gas	Liquid	**Boils**
2595	Gas	Gas	Gas	Gas	**Boils**	Gas
2600	Gas	Gas	Gas	Gas	Gas	Gas

The melting points of the substances used in this experiment show that the weakest interparticle forces are found in solid oxygen. The forces then increase in the order CO, H_2O, NaCl, Cu, and SiO_2. With two exceptions, the boiling points follow the same order. How do these melting and boiling points relate to the lattice particles of these substances?

The lattice particles in solid silicon dioxide are individual atoms of silicon and oxygen. They are held together in the lattice by covalent bonds. Solids of this type are called **network solids**, and when such solids are melted or vaporized, strong covalent bonds must be broken.

The individual atoms that are the lattice particles of copper metal are held together in the lattice by what is called a **metallic bond**. As described in Section 4.2, the atoms of metals lose valence-shell electrons readily. Imagine a large number of metal atoms occupying lattice sites. Now imagine that the valence-shell electrons of each of the atoms move readily throughout the lattice. The attraction of the positive kernels (the metal atom nuclei plus low-level electrons) to the mobile electrons, and hence to one another, constitutes a metallic bond. The mobile electrons of the metallic bond are responsible for a number of the observed properties of metals, including high thermal conductivity, high electrical conductivity, and the characteristic metallic luster.

The bonds broken when sodium chloride (NaCl) melts or boils are ionic bonds resulting from attractions between positive (Na^+) and negative ($Cl−$) ions, the lattice particles in Figure 4.3. As shown by Table 4.9, these bonds are generally quite strong.

The lattice particles of solid carbon monoxide (CO) are the nonsymmetric CO molecules. These molecules are polar because of the nonsymmetric distribution of charges

network solids Solids in which the lattice sites are occupied by atoms that are covalently bonded to each other.

metallic bond An attractive force responsible for holding solid metals together. It originates from the attraction between positively charged atomic kernels that occupy the lattice sites and mobile electrons that move freely through the lattice.

Ozone: Good up High, Bad Nearby

Ozone, O_3, is a gas that occurs both in the Earth's upper atmosphere and at ground level. Ozone is an allotrope of oxygen. Allotropes are different molecular forms of the same element. The elements carbon and sulfur also occur naturally in allotropic forms. Carbon, for example, occurs as diamond and as charcoal. Ozone can be good or bad for your health and the environment, depending on its location in the atmosphere.

Ozone occurs in two layers of the atmosphere. The layer closest to the Earth's surface is the troposphere. Here, ground-level or "bad" ozone is an air pollutant that is harmful to breathe and damages crops, trees, and other vegetation. It is a main ingredient of urban smog. The troposphere generally extends to a level about 6 miles up from the Earth's surface. The second layer of the atmosphere, called the stratosphere, extends from the top of the troposphere to a height of about 30 miles. Ozone in the stratosphere is "good" ozone. It protects life on Earth by absorbing harmful ultraviolet (UV) radiation that is a part of sunlight.

Ozone is produced naturally in the stratosphere, but this "good" ozone is gradually being destroyed by human-produced substances referred to as ozone-depleting substances (ODS). Examples of ODS include chlorofluorocarbons (CFCs), hydrochlorofluorocarbons (HCFCs), halons, methyl bromide, carbon tetrachloride, and methyl chloroform. These ODS materials were formerly used and sometimes are still used as air conditioner coolants, as foaming agents to make products like styrofoam, as solvents, as pesticides, and as aerosol propellants. Once released into the air, these ODS materials degrade very slowly. Some remain unchanged as they move up through the troposphere until they reach the stratosphere. In the stratosphere they are broken down by the intense ultraviolet rays in the sunlight, and release chlorine and bromine atoms. The released chlorine and bromine atoms react with and destroy the good ozone of the stratosphere. Scientists estimate that one chlorine atom can destroy 100,000 good ozone molecules.

Even though the use of ODS compounds has been reduced or eliminated, their past use continues to influence the protective ozone layer around the earth. The good news is that research indicates that the depletion rate of the "good" ozone layer is being reduced worldwide. The changes in the protective ozone layer can be observed using satellite measurements, particularly over the polar regions of the Earth.

The slowing down of the depletion of "good" ozone in the stratosphere is good health news. Overexposure to UV radiation is believed to increase the chances of developing melanoma, the most fatal skin cancer. Other serious health conditions associated with overexposure to UV radiation are cataracts and impaired immune systems. Since 1990, the risk of developing melanoma has more than doubled.

Ozone is the main ingredient in urban smog.

between the carbon and oxygen atoms. They are held in the solid lattice by **dipolar forces** resulting from the attraction of the positive end of one polar molecule to the negative end of another polar molecule. These forces are usually weak, and melting and vaporization take place at very low temperatures. Such substances are usually thought of as gases because that is their normal state at room temperature.

Water molecules, the lattice particles of ice, are also held in place by dipolar forces. However, these forces are stronger than those of the dipoles for CO. In water molecules, the hydrogens carry a partial positive charge, and the oxygen has a partial negative charge (see **Table 4.5**). Thus, the hydrogens of one molecule are attracted to the oxygens of other molecules. This attraction, called **hydrogen bonding**, is stronger than most other dipolar attractions because of the small size of the hydrogen atom and the high electronegativity of oxygen (see **Figure 4.10**). Hydrogen bonding occurs in gases, liquids, and solids composed of polar molecules in which hydrogen atoms are covalently bonded to highly electronegative elements (generally O, N, or F).

dipolar force The attractive force that exists between the positive end of one polar molecule and the negative end of another.

hydrogen bonding The result of attractive dipolar forces between molecules in which hydrogen atoms are covalently bonded to very electronegative elements (O, N, or F).

Polar water molecule

B Hydrogen bonding in liquid water

C Hydrogen bonding in solid water

Figure 4.10 Hydrogen bonding in water. Hydrogen bonds are dotted, and covalent bonds are solid.

Figure 4.11 Icebergs move as a result of wind and ocean currents. They can be very dangerous to ships, so their positions are reported and their probable courses estimated by an International Ice Patrol. The patrol was established in 1914 following the sinking of the *Titanic*.

Water is the most well-known substance in which hydrogen bonding dramatically influences the properties. These properties make water useful in many processes, including those characteristic of living organisms. Because of this widespread use, water is not often thought of as being a peculiar substance. However, it is often the peculiarities that make it so useful.

As we saw in Example 4.13d, water molecules are angular. Because of this shape and the large difference in electronegativity between hydrogen and oxygen, water molecules are polar, with partial positive charges on the hydrogens and a partial negative charge on the oxygen, as shown in Figure 4.10A. The attractions between these oppositely charged parts of the molecules result in hydrogen bonding. This is represented for the liquid and solid states of water in Figures 4.10B and C.

Let's look at some of the peculiar properties of water. As shown in **Table 4.9**, water has a normal boiling point of 100°C. This is much higher than would be predicted from the measured boiling points of compounds containing hydrogen and the other members of group VIA(16) of the periodic table. The boiling point of H_2Te is -2.2°C, of H_2Se is -41.3°C, and of H_2S is -60.3°C. Thus, it appears that the boiling point decreases with decreasing compound molecular weight. On the basis of this trend, water should boil at approximately -64°C. The 164-degree difference between predicted and measured boiling points is caused by strong hydrogen bonds between water molecules in the liquid. The other three compounds do not have strong hydrogen bonds between molecules because the electronegativites of the elements to which hydrogen is covalently bonded are too low. This deviation from prediction makes water a liquid at normal temperatures and allows it to be used in many of the ways familiar to us.

It is also common knowledge that solid water (ice) floats on liquid water (see **Figure 4.11**). Nothing peculiar here—or is there? In fact, there is. Water, like most other liquids, increases in density as it is cooled. This means that a specific mass of liquid decreases in volume as its temperature is lowered. Unlike most liquids, water behaves this way only until it reaches a temperature of about 4°C, as shown by the data in **Table 4.10**. Then, its density decreases as it is cooled further. When its temperature reaches 0°C, water freezes and, in the process, undergoes a dramatic decrease in density. Put another way, a quantity of liquid water at 0°C expands significantly when it freezes to solid at 0°C. Thus, a specific volume of ice has a lower mass than the same volume of liquid water, so the ice floats on the water. Hydrogen bonds between water molecules are, once again, responsible. The strong hydrogen bonds orient water molecules into a very open three-dimensional crystal lattice when it freezes (see **Figure 4.10C**). This open lattice of the solid occupies more space than is occupied by the molecules in the liquid state.

Even though containers or engines might be ruptured or cracked if water is allowed to freeze in them, the overall effect of this characteristic of water is beneficial. If ice did not float, it would form on natural waters in the winter and sink to the bottom. Gradually, ponds, lakes, and other bodies of water would fill with ice. In the spring, the ice would melt from the top down because the heavier solid ice would stay under the water. As a result, most natural waters in areas that experience freezing winters would contain significant amounts of ice year-round. It would be a strange (and possibly hostile) environment for us and for wildlife.

The amount of heat required to melt a quantity of solid and the amount of heat required to vaporize a quantity of liquid also depend on the attractive forces between molecules. As expected, these are high for water, compared with the hydrogen compounds of the three other elements in group VIA(16). The beneficial results of these high values for water will be discussed in Chapter 6.

The forces between O_2 molecules, the lattice particles of solid oxygen, do not fit into any of the classifications we have discussed to this point. The O_2 molecule is not polar and contains no ionic or metallic bonds, and solid oxygen melts at much too low a temperature to fit into the category of a network solid. The forces between O_2 molecules are called **dispersion forces** and result from momentary nonsymmetric electron distributions in the molecules.

The nonbonding electrons in an O_2 molecule can be visualized as being uniformly distributed. However, there is a small statistical chance that, during their normal movement, more of the electrons will momentarily be on one side of a molecule than on the other. This condition causes the molecule to become dipolar for an instant. The resulting negative side of the dipole will tend to repel electrons of adjoining molecules and induce them also to become dipolar (they become induced dipoles). The original (statistical) dipole and all induced dipoles are then attracted to one another. This happens many times per second throughout the solid or liquid oxygen. The net effect is a weak dispersion force of attraction between the molecules. The net force is weak because it represents the average result of many weak, short-lived attractions per second between molecules.

Dispersion forces exist in all matter, but because they are very weak, their contribution is negligible when other, stronger forces are also present. Thus, solid water is held together by dispersion forces and hydrogen bonds, but the properties are almost entirely the result of the hydrogen bonds. The ease with which a dipole can be induced increases with the size of the particle (atoms or molecules). The larger the particle, the stronger the resulting dipolar attraction (dispersion force), and the harder to separate the particles by melting or boiling. Thus, we expect to find the melting and boiling points of the elements increase as we move down a group of the periodic table.

TABLE 4.10 Water Density as a Function of Temperature	
Temperature (°C)	Water Density (g/mL)
100	0.9586
80	0.9719
60	0.9833
40	0.9923
20	0.9982
10	0.9997
5	0.9999
4 (actually 3.98)	1.0000
2	0.9999
0 (liquid H_2O)	0.9998
0 (solid H_2O)	0.9170

dispersion forces Very weak attractive forces acting between the particles of all matter. They result from momentary nonsymmetric electron distributions in molecules or atoms.

Example 4.19 The Behavior of Dispersion Forces

Illustrate the behavior of dispersion forces by using the information in **Table 4.11**.

Solution

The molecules increase in size in the order F_2, Cl_2, Br_2, I_2. The strength of dispersion forces increases in the same order, as shown in Table 4.11 by the increases in melting and boiling points.

✔ **LEARNING CHECK 4.19** Using only the periodic table, predict which member of each of the following pairs of elements would have the higher melting and boiling points:

a. O and Se b. Sb and P c. He and Ne

TABLE 4.11 The Normal Melting and Boiling Points of Group VIIA(17) Elements		
Substance	Melting Point (°C)	Boiling Point (°C)
F_2	−223	−188
Cl_2	−103	−34.6
Br_2	−7.2	58.8
I_2	113.9	184.3

The melting and boiling points in Example 4.19 indicate the variation in magnitude of dispersion forces. Similar variations are found for the other forces described in this section; sometimes their strengths overlap. For example, some metallic bonds are weaker than ionic bonds, whereas others are stronger. This is illustrated in **Figure 4.12**, which summarizes the relative strengths of interparticle forces.

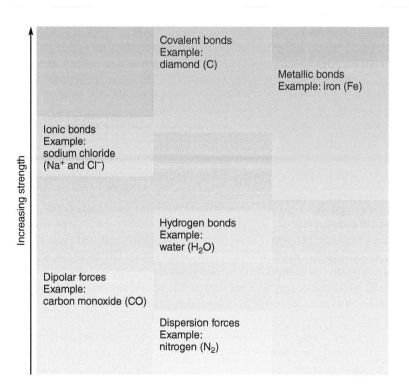

Figure 4.12 The relative strengths of interparticle forces.

▶ Case Study Follow-up

Anorexia is characterized by failure to maintain a body weight of at least 85% of normal, whereas bulimia involves the intake of large quantities of food and then eliminating it by one or more of the following methods: forced vomiting, excessive use of laxatives, fasting, and exercise. Bulimic behaviors can result in electrolyte imbalances (see **Section 25.5**). Many people with bulimia maintain a normal or slightly elevated weight. A patient who may "binge and purge" but also fasts and exercises to the point of failing to maintain 85% of recommended body weight would be considered anorexic, whereas a person who mainly binges and purges but maintains a normal weight would be considered bulimic.

A 5′ 6″ tall woman who weighs below 102 lbs, with no other explanation for her low weight, would be considered anorexic. Both eating disorders occur more prevalently in females who come from highly controlling families where nurturance is lacking. When there is a family-environment component in the etiology of the eating disorder, it is helpful for the entire family to receive counseling in order to improve the functioning of family relationships and to create an environment where healthy thinking and behavior can thrive. If the person with the eating disorder receives treatment and subsequently returns to an unchanged family environment, chances of maintaining recovery are reduced.

Concept Summary

Noble Gas Configurations The lack of reactivity for noble gases led to the proposal that the electronic configurations of the noble gases represented stable configurations. These configurations, usually consisting of eight electrons in the valence shell, can be represented in several useful ways.
Objective 1 (Section 4.1), Exercise 4.2

Ionic Bonding Ionic compounds are formed when reacting atoms gain or lose electrons to achieve a noble gas configuration of eight electrons in the valence shell. This octet rule predicts that atoms will be changed into charged particles

called simple ions. Ions of opposite charge are attracted to each other; the attractive force is called an ionic bond.
Objective 2 (Section 4.2), Exercise 4.12

Ionic Compounds Oppositely charged ions group together to form compounds in ratios determined by the positive and negative charges of the ions. The formulas representing these ratios contain the lowest number of each ion possible in a proportion such that the total positive charges and total negative charges used are equal.
Objective 3 (Section 4.3), Exercises 4.20 and 4.22

Naming Binary Ionic Compounds Binary ionic compounds contain a metal and a nonmetal. They are named by naming the metal, then adding the suffix *-ide* to the stem of the nonmetal.

Objective 4 (Section 4.4), Exercise 4.30

The Smallest Unit of Ionic Compounds Ionic compounds do not exist in the form of molecules but as three-dimensional arrangements of oppositely charged ions. The sum of the atomic weights of the elements in the formula of ionic compounds is called the formula weight and is used in calculations somewhat like the molecular weight of molecular compounds.

Objective 5 (Section 4.5), Exercise 4.38

Covalent Bonding Elements with little or no tendency to gain or lose electrons often react and achieve noble gas electronic configurations by sharing electrons. Lewis structures are useful in representing electron sharing. Shared pairs of electrons exert an attractive force on both atoms that share them. The atoms are held together by this attraction to form a covalent bond.

Objective 6 (Section 4.6), Exercise 4.48

Polyatomic Ions Polyatomic ions are groups of two or more covalently bonded atoms that carry a net electrical charge. They are conveniently represented using Lewis structures.

Objective 7 (Section 4.7), Exercise 4.50

Shapes of Molecules and Polyatomic Ions The shapes of many molecules and polyatomic ions can be predicted by using the valence-shell electron-pair repulsion theory (VSEPR). According to the VSEPR theory, electron pairs in the valence shell of the central atom of a molecule or ion repel one another and become arranged so as to maximize their separation distances. The resulting arrangement determines the molecular or ionic shape when one or all of the electron pairs involved form bonds between the central atom and other atoms.

Objective 8 (Section 4.8), Exercises 4.52 and 4.54

The Polarity of Covalent Molecules Shared electrons may be shared equally, or they may be attracted more strongly to one of the atoms they bond together. The tendency of a covalently bonded atom to attract shared electrons is called the electronegativity of the atom. Unequally shared bonding-electron pairs form polar covalent bonds. The extent of bond polarization can be estimated from the electronegativity differences between the bonded atoms. The higher the electronegativity difference, the more polar (or ionic) is the bond. Polar covalent bonds cause partial positive and negative charges to form within molecules. When these charges are symmetrically distributed in the molecule, it is said to be nonpolar. An unsymmetric distribution gives rise to a polar molecule.

Objective 9 (Section 4.9), Exercises 4.58 and 4.64

More about Naming Compounds Binary covalent compounds are named using the name of the less electronegative element first, followed by the stem plus *-ide* of the more electronegative element. Greek prefixes are used to represent the number of each type of atom in molecules of the compounds. Ionic compounds that contain a metal ion (or ammonium ion) plus a polyatomic ion are named by first naming the metal (or ammonium ion) followed by the name of the polyatomic ion.

Objective 10 (Section 4.10), Exercises 4.66, 4.70, and 4.72

Other Interparticle Forces Forces other than ionic and covalent bonds are also known to hold the particles of some pure substances together in the solid and liquid states. These forces include metallic bonds, dipolar attractions, hydrogen bonds, and dispersion forces. The strength of the predominant force acting in a substance is indicated by the melting and boiling points of the substance.

Objective 11 (Section 4.11), Exercises 4.78 and 4.80

Key Terms and Concepts

Binary compound (4.3)
Bond polarization (4.9)
Covalent bond (4.6)
Crystal lattice (4.5)
Dipolar force (4.11)
Dispersion force (4.11)
Double bond (4.6)
Electronegativity (4.9)
Formula weight (4.5)

Hydrogen bonding (4.11)
Ionic bond (4.2)
Isoelectronic (4.2)
Lattice site (4.5)
Lewis structure (4.1)
Metallic bond (4.11)
Network solids (4.11)
Nonpolar covalent bond (4.9)
Nonpolar molecule (4.9)

Octet rule (4.2)
Polar covalent bond (4.9)
Polar molecule (4.9)
Polyatomic ion (4.7)
Simple ion (4.2)
Triple bond (4.6)
VSEPR theory (4.8)

Exercises

Even-numbered exercises are answered in Appendix B.

Blue-numbered exercises are more challenging.

Noble Gas Configurations (Section 4.1)

4.1 Refer to the group numbers of the periodic table and draw Lewis structures for atoms of the following:

 a. potassium

 b. barium

 c. aluminum

 d. bromine

4.2 Refer to the group numbers of the periodic table and draw Lewis structures for atoms of the following:

 a. iodine

 b. strontium

 c. tin

 d. sulfur

4.3 Write abbreviated electronic configurations for the following:

 a. fluorine

 b. element 37

 c. selenium

 d. silicon

4.4 Write abbreviated electronic configurations for the following:

 a. tin

 b. Cs

 c. element number 49

 d. strontium

4.5 Draw Lewis structures for the elements given in Exercise 4.3.

4.6 Draw Lewis structures for the elements given in Exercise 4.4.

4.7 Use the symbol E to represent an element in a general way and draw Lewis structures for atoms of the following:

 a. Any group IIIA(13) element

 b. Any group VIA(16) element

4.8 Use the symbol E to represent an element in a general way and draw Lewis structures for atoms of the following:

 a. Any group IIA(2) element

 b. Any group VA(15) element

Ionic Bonding (Section 4.2)

4.9 Indicate both the minimum number of electrons that would have to be added and the minimum number that would have to be removed to change the electronic configuration of each element listed in Exercise 4.3 to a noble gas configuration.

4.10 Indicate both the minimum number of electrons that would have to be added and the minimum number that would have to be removed to change the electronic configuration of each element listed in Exercise 4.4 to a noble gas configuration.

4.11 Use the periodic table and predict the number of electrons that will be lost or gained by the following elements as they change into simple ions. Write an equation using elemental symbols, ionic symbols, and electrons to represent each change.

 a. Mg

 b. silicon

 c. element 53

 d. sulfur

4.12 Use the periodic table and predict the number of electrons that will be lost or gained by the following elements as they change into simple ions. Write an equation using elemental symbols, ionic symbols, and electrons to represent each change.

 a. Cs

 b. oxygen

 c. element number 7

 d. iodine

4.13 Write a symbol for each of the following ions:

 a. A bromine atom that has gained one electron

 b. A sodium atom that has lost one electron

 c. A sulfur atom that has gained two electrons

4.14 Write a symbol for each of the following ions:

 a. A selenium atom that has gained two electrons

 b. A rubidium atom that has lost one electron

 c. An aluminum atom that has lost three electrons

4.15 Identify the element in period 2 that would form each of the following ions. E is used as a general symbol for an element.

 a. E^-

 b. E^{2+}

 c. E^{3-}

 d. E^+

4.16 Identify the element in period 3 that would form each of the following ions. E is used as a general symbol for an element.

 a. E^{2-}

 b. E^{3+}

 c. E^+

 d. E^-

4.17 Identify the noble gas that is isoelectronic with each of the following ions:

 a. Mg^{2+}

 b. Te^{2-}

 c. N^{3-}

 d. Be^{2+}

4.18 Identify the noble gas that is isoelectronic with each of the following ions:

 a. Li^+

 b. I^-

 c. S^{2-}

 d. Sr^{2+}

Ionic Compounds (Section 4.3)

4.19 Write equations to represent positive and negative ion formation for the following pairs of elements. Then write a formula for the ionic compound that results when the ions combine.

 a. Ca and S

 b. Mg and N

 c. elements number 19 and 17

4.20 Write equations to represent positive and negative ion formation for the following pairs of elements. Then write a formula for the ionic compound that results when the ions combine.

 a. Ca and Cl

 b. lithium and bromine

 c. elements number 12 and 16

4.21 Write the formula for the ionic compound formed from Ca^{2+} and each of the following ions:

 a. Br^-

 b. O^{2-}

 c. S^{2-}

 d. N^{3-}

4.22 Write the formula for the ionic compound formed from Ba^{2+} and each of the following ions:

 a. Se^{2-}

 b. P^{3-}

 c. I^-

 d. As^{3-}

4.23 Classify each of the following as a binary compound or not a binary compound:

 a. HF

 b. OF_2

 c. H_2SO_4

 d. H_2S

 e. $MgBr_2$

4.24 Classify each of the following as a binary compound or not a binary compound:

 a. PbO_2

 b. $CuCl_2$

 c. KNO_3

 d. Be_3N_2

 e. $CaCO_3$

Naming Binary Ionic Compounds (Section 4.4)

4.25 Name the following metal ions:

 a. Ca^{2+}

 b. K^+

 c. Al^{3+}

 d. Rb^+

4.26 Name the following metal ions:

 a. Li^+

 b. Mg^{2+}

 c. Ba^{2+}

 d. Cs^+

4.27 Name the following nonmetal ions:

 a. Cl^-

 b. N^{3-}

 c. S^{2-}

 d. Se^{2-}

4.28 Name the following nonmetal ions:

 a. Br^-

 b. O^{2-}

 c. P^{3-}

 d. F^-

4.29 Name the following binary ionic compounds:

 a. Na_2O

 b. $CaCl_2$

 c. Al_2S_3

 d. LiF

 e. Ba_3N_2

4.30 Name the following binary ionic compounds:

 a. SrS

 b. CaF_2

 c. $BaCl_2$

 d. Li_2O

 e. MgO

4.31 Name the following binary ionic compounds, using a Roman numeral to indicate the charge on the metal ion:

 a. $CrCl_2$ and $CrCl_3$

 b. CoS and Co_2S_3

 c. FeO and Fe_2O_3

 d. $PbCl_2$ and $PbCl_4$

4.32 Name the following binary ionic compounds, using a Roman numeral to indicate the charge on the metal ion:

 a. PbO and PbO_2

 b. CuCl and $CuCl_2$

 c. Au_2S and Au_2S_3

 d. CoO and Co_2O_3

4.33 Name the binary compounds of Exercise 4.31 by adding the endings *-ous* and *-ic* to indicate the lower and higher ionic charges of the metal ion in each pair of compounds. The non-English root for lead (Pb) is *plumb-*.

4.34 Name the binary compounds of Exercise 4.32 by adding the endings *-ous* and *-ic* to indicate the lower and higher ionic charges of the metal ion in each pair of compounds. The non-English root for gold (Au) is *aur-*, and that of lead (Pb) is *plumb-*.

4.35 Write formulas for the following binary ionic compounds:

 a. manganese(II) chloride

 b. iron(III) sulfide

 c. chromium(II) oxide

 d. iron(II) bromide

 e. tin(II) chloride

4.36 Write formulas for the following binary ionic compounds:

 a. mercury(I) oxide

 b. lead(II) oxide

 c. platinum(IV) iodine

 d. copper(I) nitride

 e. cobalt(II) sulfide

The Smallest Unit of Ionic Compounds (Section 4.5)

4.37 Determine the formula weight in atomic mass units for each of the following binary ionic compounds:

 a. Na_2O

 b. FeO

 c. PbS_2

 d. $AlCl_3$

4.38 Determine the formula weight in atomic mass units for each of the following binary ionic compounds:

 a. NaBr

 b. CaF_2

 c. Cu_2S

 d. Li_3N

4.39 Identify the ions that would occupy lattice sites in a solid sample of each compound given in Exercise 4.37.

4.40 Identify the ions that would occupy lattice sites in a solid sample of each compound given in Exercise 4.38.

4.41 Calculate the mass in grams of positive ions and negative ions contained in 1 mol of each compound given in Exercise 4.37.

4.42 Calculate the mass in grams of positive ions and negative ions contained in 1 mol of each compound given in Exercise 4.38.

4.43 Calculate the number of positive ions and negative ions contained in 1.00 mol of each compound given in Exercise 4.37.

4.44 Calculate the number of positive ions and negative ions contained in 1.00 mol of each compound given in Exercise 4.38.

Covalent Bonding (Section 4.6)

4.45 Represent the following reaction using Lewis structures:

$$I + I \rightarrow I_2$$

4.46 Represent the following reaction using Lewis structures:

$$8S \rightarrow S_8 \text{ (the atoms form a ring)}$$

4.47 Represent the following molecules by Lewis structures:

 a. HF

 b. IBr

 c. PH_3 (each H atom is bonded to the P atom)

 d. $HClO_2$ (the O atoms are each bonded to the Cl, and the H is bonded to one of the O atoms)

4.48 Represent the following molecules by Lewis structures:

 a. CH_4 (each H atom is bonded to the C atom)

 b. CO_2 (each O atom is bonded to the C atom)

 c. H_2Se (each H atom is bonded to the Se atom)

 d. NH_3 (each H atom is bonded to the N atom)

Polyatomic Ions (Section 4.7)

4.49 Draw Lewis structures for the following polyatomic ions:

 a. ClO_3^- (each O atom is bonded to the Cl atom)

 b. CN^-

 c. CO_3^{2-} (each O atom is bonded to the C atom)

4.50 Draw Lewis structures for the following polyatomic ions:

 a. NH_4^+ (each H atom is bonded to the N atom)

 b. PO_4^{3-} (each O atom is bonded to the P atom)

 c. SO_3^{2-} (each O atom is bonded to the S atom)

Shapes of Molecules and Polyatomic Ions (Section 4.8)

4.51 Draw Lewis structures for the following molecules:

 a. O_3 (the O atoms are bonded together, like beads on a string)

 b. CS_2 (each S atom is bonded to the C atom)

 c. SeO_2 (each O atom is bonded to the Se atom)

 d. H_2SO_3 (each O atom is bonded to the S atom, and one H atom is bonded to each of two O atoms)

4.52 Predict the shape of each of the following molecules by first drawing a Lewis structure, then applying the VSEPR theory:

 a. H_2S (each H atom is bonded to the S atom)

 b. PCl_3 (each Cl atom is bonded to the P atom)

 c. OF_2 (each F atom is bonded to the O atom)

 d. SnF_4 (each F atom is bonded to the Sn atom)

4.53 Predict the shape of each of the following molecules by first drawing a Lewis structure, then applying the VSEPR theory:

 a. O_3 (see Exercise 4.51 for Lewis structure)

 b. SeO_2 (see Exercise 4.51 for Lewis structure)

 c. PH_3 (each H atom is bonded to the P atom)

 d. SO_3 (each O atom is bonded to the S atom)

4.54 Predict the shape of each of the following polyatomic ions by first drawing a Lewis structure, then applying the VSEPR theory:

 a. NO_2^- (each O is bonded to N)

 b. ClO_3^- (each O is bonded to Cl)

 c. CO_3^{2-} (each O is bonded to C)

 d. H_3O^+ (each H is bonded to O)

 Note the positive charge; compare with NH_4^+.

4.55 Predict the shape of each of the following polyatomic ions by first drawing a Lewis structure, then applying the VSEPR theory:

 a. NH_2^- (each H is bonded to N)

 b. PO_3^{3-} (each O is bonded to P)

 c. $BeCl_4^{2-}$ (each Cl is bonded to Be)

 d. ClO_4^- (each O is bonded to Cl)

The Polarity of Covalent Molecules (Section 4.9)

4.56 Use the periodic table and Table 4.4 to determine which of the following bonds will be polarized. Show the resulting

charge distribution in those molecules that contain polarized bonds.

a. H—I

b.
S=O
/
O

c. O—O
\\
O

4.57 Use the periodic table and Table 4.4 to determine which of the following bonds will be polarized. Show the resulting charge distribution in those molecules that contain polarized bonds.

a. Cl—F

b. H—Se
\
H

c.
H H
\ /
B — B
/ \
H H

4.58 Use Table 4.4 and classify the bonds in the following compounds as nonpolar covalent, polar covalent, or ionic:

a. LiBr

b. HCl

c. PH_3 (each H is bonded to P)

d. SO_2 (each O is bonded to S)

e. CsF

4.59 Use Table 4.4 and classify the bonds in the following compounds as nonpolar covalent, polar covalent, or ionic:

a. MgI_2 (each I is bonded to Mg)

b. NCl_3 (each Cl is bonded to N)

c. H_2S (each H is bonded to S)

d. RbF

e. SrO

4.60 On the basis of the charge distributions you drew for the molecules of Exercise 4.56, classify each of the molecules as polar or nonpolar.

4.61 On the basis of the charge distributions you drew for the molecules of Exercise 4.57, classify each of the molecules as polar or nonpolar.

4.62 Use Table 4.4 and predict the type of bond you would expect to find in compounds formed from the following elements:

a. magnesium and chlorine

b. carbon and hydrogen

c. phosphorus and hydrogen

4.63 Use Table 4.4 and predict the type of bond you would expect to find in compounds formed from the following elements:

a. C and Br

b. aluminum and chlorine

c. nitrogen and oxygen

4.64 Show the charge distribution in the following molecules, and predict which are polar molecules:

a. C≡O

b. H—Se
|
H

c.
I
|
Al
/ \
I I

4.65 Show the charge distribution in the following molecules, and predict which are polar molecules:

a. S=C=S

b. H—C≡N

c. F—O
\
F

More about Naming Compounds (Section 4.10)

4.66 Name the following binary covalent compounds:

a. PCl_3

b. N_2O_5

c. CCl_4

d. BF_3

e. CS_2

4.67 Name the following binary covalent compounds:

a. SiO_2

b. SiF_4

c. P_2O_5

d. $AlBr_3$

e. CBr_4

4.68 Write formulas for the following binary covalent compounds:

a. chlorine dioxide

b. dinitrogen monoxide

c. sulfur dioxide

d. carbon tetrachloride

4.69 Write formulas for the following binary covalent compounds:

a. disulfur monoxide

b. sulfur hexafluoride

c. silicon tetrachloride

d. carbon diselenide

4.70 Write the formulas and names for compounds composed of ions of the following metals and the indicated polyatomic ions:

a. calcium and hypochlorite ion

b. cesium and nitrite ion

c. Mg and SO_3^{2-}

d. K and $Cr_2O_7^{2-}$

4.71 Write the formulas and names for compounds composed of ions of the following metals and the indicated polyatomic ions:

a. calcium and carbonate ion

b. sodium and sulfate ion

c. K^+ and PO_4^{3-}

d. Mg^{2+} and NO_3^-

4.72 Write formulas for the following compounds:

 a. barium hydroxide

 b. magnesium sulfite

 c. calcium carbonate

 d. ammonium sulfate

 e. lithium hydrogen carbonate

4.73 Write formulas for the following compounds:

 a. potassium hydroxide

 b. lithium carbonate

 c. ammonium chloride

 d. sodium phosphate

 e. calcium nitrate

4.74 Write a formula for the following compounds, using M with appropriate charges to represent the metal ion:

 a. Any group IA(1) element and SO_3^{2-}

 b. Any group IA(1) element and $C_2H_3O_2^-$

 c. Any metal that forms M^{2+} ions and $Cr_2O_7^{2-}$

 d. Any metal that forms M^{3+} ions and PO_4^{3-}

 e. Any metal that forms M^{3+} ions and NO_3^-

4.75 Write a formula for the following compounds, using M with appropriate charges to represent the metal ion:

 a. Any group IIA(2) element and PO_4^{3-}

 b. Any group IIA(2) element and NO_3^-

 c. Any metal that forms M^+ ions and SO_4^{2-}

 d. Any metal that forms M^{3+} ions and OH^-

 e. Any metal that forms M^{2+} ions and HPO_4^{2-}

Other Interparticle Forces (Section 4.11)

4.76 The covalent compounds ethyl alcohol and dimethyl ether both have the formula C_2H_6O. However, the alcohol melts at $-117.3°C$ and boils at $78.5°C$, whereas the ether melts at $-138.5°C$ and boils at $-23.7°C$. How could differences in forces between molecules be used to explain these observations?

4.77 The following structural formulas represent molecules of ethyl alcohol and dimethyl ether. Assign the correct name to each formula and explain how your choice is consistent with your answer to Exercise 4.76.

$$
\begin{array}{ccc}
& \text{H} & & \text{H} \\
& | & & | \\
\text{H} - \text{C} - & \text{O} & - \text{C} - \text{H} \\
& | & & | \\
& \text{H} & & \text{H}
\end{array}
\qquad
\begin{array}{ccc}
\text{H} & \text{H} & \\
| & | & \\
\text{H} - \text{C} - & \text{C} - \text{O} - \text{H} \\
| & | & \\
\text{H} & \text{H} &
\end{array}
$$

4.78 Describe the predominant forces that exist between molecules of the noble gases. Arrange the noble gases in a predicted order of increasing boiling point (lowest first) and explain the reason for the order.

4.79 Use the concept of interparticle forces to propose an explanation for the fact that CO_2 is a soft, low-melting solid (dry ice), whereas SiO_2 is a hard solid (sand). Focus on the nature of the particles that occupy lattice sites in the solid.

4.80 Table sugar, sucrose, melts at about $185°C$. Which interparticle forces do you think are unlikely to be the predominant ones in the lattice of solid sucrose?

4.81 The formula for sucrose is $C_{12}H_{22}O_{11}$, where many of the hydrogens and oxygens are combined to form OH groups that are bonded to carbon atoms. What type of predominant interparticle bonding would you now propose for solid sucrose (see Exercise 4.80)?

Additional Exercises

4.82 Three atoms with an electronic configuration of $1s^1$ are covalently bonded to one atom with an electronic configuration of $1s^2\,2s^2\,2p^3$ to form a molecule. What is the formula of this molecule? Make a drawing to show how two of these molecules would hydrogen bond with each other.

4.83 Suppose an element from group IIA(2) and period 3 of the periodic table forms an ionic compound with the element that has an electronic configuration of $1s^2\,2s^2\,2p^5$. What would be the formula of the compound, and what would be the name of the compound?

4.84 What would be the mass in grams of 0.200 moles of the ionic compound formed when magnesium metal reacts with oxygen?

4.85 The ampere unit is used to describe the flow of electricity in an electrical circuit. One ampere is an amount of electricity corresponding to the flow of 6.2×10^{18} electrons past a point in a circuit in 1 second. In a hydrogen fuel cell, hydrogen atoms are dissociated into H^+ ions and electrons ($H \rightarrow H^+ + 1e^-$). How many grams of hydrogen must be dissociated each second in a fuel cell to produce 1.0 ampere of electricity?

4.86 If one atom of oxygen reacted with two atoms of nitrogen to form a molecule, what would be the formula of the molecule? Use Table 4.4 and Figure 4.7 to determine if the bond between the atoms is ionic or covalent. Also, name the compound that is formed.

Chemistry for Thought

4.87 Refer to Figure 4.1, and answer the question in the caption. What other metals and nonmetals would you predict might react in a similar way?

4.88 The colors of some compounds, such as those shown in Figure 4.2, result from the presence of water in the compounds. Propose an experiment you could perform to see if this was true for the compounds shown in Figure 4.2.

4.89 Refer to Figure 4.6, and answer the question in the caption. Propose the shapes that would be assumed by a group of five balloons and by a group of six balloons.

4.90 Refer to Figure 4.9. Two of the compounds are highly colored (other than white). All the compounds consist of potassium and a polyatomic ion. If you have not yet done so, write formulas for the compounds and see if you can find a characteristic of the polyatomic ions of the colored compounds that is not found in the white compounds. Then refer to Table 4.7 and predict which of the other polyatomic anions would form colored compounds of potassium.

4.91 Recall how a metal atom changes to form a positively charged metal ion. How do you think the sizes of a metal atom and a positive ion of the same metal will compare?

4.92 Recall how a nonmetal atom such as chlorine changes to form a negatively charged ion. How do you think the size of a nonmetal atom and a negatively charged ion of the same nonmetal will compare?

4.93 Neon atoms do *not* combine to form Ne_2 molecules. Explain.

4.94 Refer to Figure 4.8, and answer the question in the caption. The balloon carries a negative charge. What would happen if a positively charged object was used in place of the balloon?

4.95 It has been found that NO is a vital biological molecule. Explain how NO forms when a fuel such as natural gas, CH_4, is burned in air at a high temperature.

Allied Health Exam Connection

The following questions are from these sources:

• *Nursing School Entrance Exam* © 2005, Learning Express, LLC.

• *McGraw-Hill's Nursing School Entrance Exams* by Thomas A. Evangelist, Tamara B. Orr, and Judy Unrein © 2009, The McGraw-Hill Companies, Inc.

• *NSEE Nursing School Entrance Exams,* 3rd edition © 2009, Kaplan Publishing.

• *Cliffs Test Prep: Nursing School Entrance Exams* by Fred N. Grayson © 2004, Wiley Publishing, Inc.

• *Peterson's Master the Nursing School and Allied Health Entrance Exams,* 18th edition by Marion F. Gooding © 2008, Peterson's, a Nelnet Company.

4.96 Noble gases:

 a. have low boiling points.

 b. are all gases at room temperature.

 c. are also called inert gases.

 d. all of the above.

4.97 The elements in group zero of the periodic table are considered inert gases because each has how many electrons in its outermost energy level?

 a. 8

 b. 7

 c. 4

 d. 2

4.98 Name the type of bond that is formed when *electrons are shared* between two atoms.

 a. shared bond

 b. ionic bond

 c. covalent bond

 d. multiple bond

4.99 Which of the following is the correct name for Li_2SO_3?

 a. lithium sulfite

 b. lithium sulfide

 c. lithium sulfate

 d. lithium disulfate

4.100 An atom becomes an ion that possesses a negative charge. The atom must have:

 a. gained protons.

 b. lost protons.

 c. lost electrons.

 d. gained electrons.

4.101 When calcium reacts with chlorine to form calcium chloride, it:

 a. shares two electrons.

 b. gains two electrons.

 c. loses two electrons.

 d. gains one electron.

4.102 Which of the following is the probable charge for an ion formed from Ca?

 a. +1

 b. +2

 c. −1

 d. −2

4.103 A covalent bond is believed to be caused by a:

 a. transfer of electrons.

 b. sharing of electrons.

 c. release of energy.

 d. none of the above.

4.104 Which molecule below has a nonpolar bond in which the electrons are being shared equally?

 a. H_2O

 b. NH_3

 c. Cl_2

 d. CH_4

4.105 What is the formula for bismuth(III) hydroxide?

 a. Bi_3OH

 b. $Bi(OH)_3$

 c. $Bi(OH)_2$

 d. BiOH

4.106 Which of the following species will combine with a chloride ion to produce ammonium chloride?

 a. NH_3

 b. K^+

 c. NH_4^+

 d. Al^{+++}

4.107 What type of bond is created when bromine and magnesium are reacted to form $MgBr_2$?

 a. polar covalent

 b. metallic

 c. ionic

 d. nonpolar covalent

4.108 The parts of an atom directly involved in ionic bonding are the:

 a. protons in the nucleus.

 b. neutrons in the nucleus.

 c. electrons in the outer energy shell.

 d. electrons in the innermost energy shell.

4.109 In forming an ionic bond with an atom of chlorine, a sodium atom will:

 a. receive one electron from the chlorine atom.

 b. receive two electrons from the chlorine atom.

 c. give up one electron to the chlorine atom.

 d. give up two electrons to the chlorine atom.

4.110 In bonding, what would happen between the electrons of K and Br?

 a. They would be transferred.

 b. They would be shared.

 c. Neither of the above.

 d. They would be both transferred and shared.

4.111 Which compound contains a bond with **no** ionic character?

 a. CO

 b. CaO

 c. K_2O

 d. Na_2O

4.112 Which of the following molecules is nonpolar?

 a. NH_3

 b. H_2O

 c. PCl_3

 d. N_2

4.113 Which molecule is nonpolar and contains a nonpolar covalent bond?

 a. F_2

 b. HI

 c. KCl

 d. NH_3

4.114 Which of the following is a nonpolar covalent bond?

 a. the bond between two carbons

 b. the bond between sodium and chloride

 c. the bond between two water molecules

 d. the bond between nitrogen and hydrogen

4.115 The type of bond that forms between two molecules of water is a:

 a. polar covalent bond.

 b. hydrogen bond.

 c. nonpolar covalent bond.

 d. peptide bond.

4.116 Which of the following is an example of hydrogen bonding?

 a. The bond between O and H in a single molecule of water

 b. The bond between the O of one water molecule and the H of a second water molecule

 c. The bond between the O of one water molecule and the O of a second water molecule

 d. The bond between the H of one water molecule and the H of a second water molecule

5

Chemical Reactions

Excellent backgrounds/Shutterstock.com

▶ Case Study

In spite of extensive laboratory-safety training, Ron suffered serious injuries in a work-related accident at an explosives company. Ron wanted to repeat a new experiment just once more before he left for the day. He used a towel to quickly wipe a beaker to run the test. This caused a static electricity spark to form. The spark caused titanium dust in the air to ignite. The resulting explosion caused third-degree burns on both of Ron's hands. Because the ambulance arrived from a chemical laboratory, the hospital staff assumed Ron's burns resulted from exposure to chemicals. The staff then followed the protocol for chemical burns, including a painful 15-minute shower. Ron tried to explain that his injury was caused by fire and not by chemical exposure, but the hospital staff did not grasp the difference. Ron was grateful that safety gear had

protected his face from injury. However, his treatment required skin grafts, physical therapy, and compression therapy for more than a year. Once a gifted piano player, Ron lost his beloved pastime in one moment of haste, and common, everyday tasks like typing, dressing, and driving inflicted pain.

▶ **What is the difference between a chemical burn, a heat burn, and a radiation burn? What does the word _debridement_ mean?**

Follow-up to this Case Study appears at the end of the chapter before the _Concept Summary_.

Learning Objectives

When you have completed your study of this chapter, you should be able to:

1 Identify the reactants and products in written reaction equations, and balance the equations by inspection. **(Section 5.1)**

2 Assign oxidation numbers to elements in chemical formulas, and identify the oxidizing and reducing agents in redox reactions. **(Section 5.3)**

3 Classify reactions into the categories of redox or nonredox, then into the categories of decomposition, combination, single replacement, or double replacement. **(Sections 5.4–5.6)**

4 Write molecular equations in total ionic and net ionic forms. **(Section 5.7)**

5 Classify reactions as exothermic or endothermic. **(Section 5.8)**

6 Use the mole concept to do calculations based on chemical reaction equations. **(Section 5.9)**

7 Use the mole concept to do calculations based on the limiting-reactant principle. **(Section 5.10)**

8 Use the mole concept to do percentage-yield calculations. **(Section 5.11)**

I n previous chapters, we introduced the terms _molecule_, _element_, _compound_, and _chemical change_. In Chapter 1, you learned that chemical changes result in the transformation of one or more substances into one or more new substances. The processes involved in such changes are called _chemical reactions_. In this chapter, you will learn to write and read chemical equations that represent chemical reactions, to classify reactions, and to do calculations based on the application of the mole concept to chemical equations.

5.1 | Chemical Equations

Learning Objective

1. Identify the reactants and products in written reaction equations, and balance the equations by inspection.

A simple chemical reaction between elemental hydrogen and oxygen has been used to power the engines of a number of spacecraft, including the space shuttle. The products are water and much heat. For the moment, we will focus only on the substances involved; we will deal with the heat later. The reaction can be represented by a word equation:

$$\text{hydrogen} + \text{oxygen} \rightarrow \text{water} \tag{5.1}$$

The word equation gives useful information, including the reactants and products of the reaction. The **reactants** are the substances that undergo the chemical change; by convention,

reactants of a reaction The substances that undergo chemical change during the reaction. They are written on the left side of the equation representing the reaction.

products of a reaction The substances produced as a result of the reaction taking place. They are written on the right side of the equation representing the reaction.

law of conservation of matter Atoms are neither created nor destroyed in chemical reactions.

balanced equation An equation in which the number of atoms of each element in the reactants is the same as the number of atoms of that same element in the products.

these are written on the left side of the equation. The **products**, or substances produced by the chemical change, are written on the right side of the equation. The reactants and products are separated by an arrow that points to the products. A plus sign (+) is used to separate individual reactants and products.

Word equations convey useful information, but the chemical equations used by chemists convey much more:

$$2H_2(g) + O_2(g) \rightarrow 2H_2O(\ell) \tag{5.2}$$

In this equation, the reactants and products are represented by molecular formulas that tell much more than the names used in the word equation. For example, it is apparent that both hydrogen and oxygen molecules are diatomic. The equation is also consistent with a fundamental law of nature called the **law of conservation of matter**. According to this law, atoms are neither created nor destroyed in chemical reactions but are rearranged to form new substances. Thus, atoms are conserved in chemical reactions, but molecules are not. The numbers written as coefficients to the left of the molecular formulas make the equation consistent with this law by making the total number of each kind of atom equal in the reactants and products. Note that coefficients of 1 are never written but are understood. Equations written this way are said to be **balanced**. In addition, the symbol in parentheses to the right of each formula indicates the state or form in which the substance exists. Thus, in Equation 5.2, the reactants hydrogen and oxygen are both in the form of gases (g), and the product water is in the form of a liquid (ℓ). Other common symbols you will encounter are (s) to designate a solid and (aq) to designate a substance dissolved in water. The symbol (aq) comes from the first two letters of *aqua,* the Latin word for water.

Example 5.1 Interpreting Equations

Determine the number of atoms of each type on each side of Equation 5.2.

Solution
The coefficient 2 written to the left of H_2 means that two hydrogen molecules with the formula H_2 are reacted. Since each molecule contains two hydrogen atoms, a total of four hydrogen atoms is represented. The coefficient to the left of O_2 is 1, even though it is not written in the equation. The oxygen molecule contains two atoms, so a total of two oxygen atoms is represented. The coefficient 2 written to the left of H_2O means that two molecules of H_2O are produced. Each molecule contains two hydrogen atoms and one oxygen atom. Therefore, the total number of hydrogen atoms represented is four, and the number of oxygen atoms is two. These are the same as the number of hydrogen and oxygen atoms represented in the reactants.

✔ **LEARNING CHECK 5.1** Determine the number of atoms of each type on each side of the following balanced equation:

$$N_2(g) + 3H_2(g) \rightarrow 2NH_3(g)$$

When the identity and formulas of the reactants and products of a reaction are known, the reaction can be balanced by applying the law of conservation of matter.

Example 5.2 Writing Balanced Equations

Nitrogen dioxide (NO_2) is an air pollutant that is produced in part when nitric oxide (NO) reacts with oxygen gas (O_2). Write a balanced equation for the production of NO_2 by this reaction.

Solution
The reactants written to the left of the arrow will be NO and O_2. The product is NO_2:

$$NO(g) + O_2(g) \rightarrow NO_2(g) \tag{5.3}$$

A quick inspection shows that the reactants contain three oxygen atoms, whereas the product has just two. The number of nitrogen atoms is the same in both the reactants and the products. A practice sometimes resorted to by beginning chemistry students is to change the formula of oxygen gas to O by changing the subscript and to write

$$NO(g) + O(g) \rightarrow NO_2(g) \qquad (5.4)$$

This is not allowed. The natural molecular formulas of any compounds or elements cannot be adjusted; they are fixed by the principles of chemical bond formation described in Chapter 4. All that can be done to balance a chemical equation is to change the coefficients of the reactants and products. In this case, inspection reveals that the following is a balanced form of the equation:

$$2NO(g) + O_2(g) \rightarrow 2NO_2(g) \qquad (5.5)$$

Notice that the following forms of the equation are also balanced:

$$4NO(g) + 2O_2(g) \rightarrow 4NO_2(g) \qquad (5.6)$$

and

$$NO(g) + \frac{1}{2} O_2(g) \rightarrow NO_2(g) \qquad (5.7)$$

In Equation 5.6, the coefficients are double those used in Equation 5.5; in Equation 5.7, one coefficient is a fraction. The lowest possible whole-number coefficients are used in balanced equations. Thus, Equation 5.5 is the correct form.

✔ **LEARNING CHECK 5.2** Write and balance an equation that represents the reaction of sulfur dioxide (SO_2) with oxygen gas (O_2) to give sulfur trioxide (SO_3).

5.2 Types of Reactions

A large number of chemical reactions are known to occur; only a relatively small number of them will be studied in this book. This study is made easier by classifying the reactions according to certain characteristics. Such a classification scheme could be developed in a number of ways, but we have chosen to first classify reactions as being either oxidation–reduction (redox) reactions or nonredox reactions. As you will see, redox reactions are very important in numerous areas of study, including metabolism. Once reactions are classified as redox or nonredox, many can be further classified into one of several other categories, as shown in **Figure 5.1**. Notice that according to Figure 5.1, single-replacement, or substitution, reactions are redox reactions, whereas double-replacement, or metathesis, reactions are nonredox. Combination and decomposition reactions can be either redox or nonredox.

Figure 5.1 A classification of chemical reactions.

5.3 | Redox Reactions

Learning Objective

2. Assign oxidation numbers to elements in chemical formulas, and identify the oxidizing and reducing agents in redox reactions.

Almost all elements react with oxygen to form oxides. The process is so common that the word *oxidation* was coined to describe it. Some examples are the rusting of iron,

$$4Fe(s) + 3O_2(g) \rightarrow 2Fe_2O_3(s) \tag{5.8}$$

and the burning of hydrogen,

$$2H_2(g) + O_2(g) \rightarrow 2H_2O(\ell) \tag{5.9}$$

The reverse process, *reduction*, originally referred to the technique of removing oxygen from metal oxide ores to produce the free metal. Some examples are

$$CuO(s) + H_2(g) \rightarrow Cu(s) + H_2O(\ell) \tag{5.10}$$

and

$$2Fe_2O_3(s) + 3C(s) \rightarrow 4Fe(s) + 3CO_2(g) \tag{5.11}$$

Today, the words **oxidation** and **reduction** are used in a rather broad sense. **Table 5.1** contains most of the common meanings. To understand oxidation and reduction in terms of electron transfer and oxidation number (O.N.) change, you must become familiar with the concept of oxidation numbers.

Oxidation numbers, sometimes called **oxidation states**, are positive or negative numbers assigned to the elements in chemical formulas according to a specific set of rules. The following rules will be used; be sure to note that Rule 1 applies only to uncombined elements—that is, elements in their free state. Rules 2 through 7 apply to elements combined to form compounds or ions.

Rule 1. The oxidation number (O.N.) of any uncombined element is 0.
Examples: Al(0), $O_2(0)$, $Br_2(0)$, and Na(0)

Rule 2. The O.N. of a simple ion is equal to the charge on the ion.
Examples: $Na^+(+1)$, $Mg^{2+}(+2)$, $S^{2-}(-2)$, and $Br^-(-1)$

Rule 3. The O.N.s of combined group IA(1) and IIA(2) elements are +1 and +2, respectively.
Examples: $Na_2CO_3(Na = +1)$, $Sr(NO_3)_2$ (Sr = +2), and $CaCl_2$ (Ca = +2)

Rule 4. The O.N. of combined hydrogen is +1.
Example: HCl (H = +1) and H_3PO_4 (H = +1)

Rule 5. The O.N. of combined oxygen is −2 except in peroxides, where it is −1.
Examples: CaO (O = −2), H_2SO_4 (O = −2), H_2O (O = −2), and H_2O_2 (O = −1)

Rule 6. The algebraic sum of the O.N.s of all atoms in a complete compound formula equals zero.

Examples: K_2CO_3: 2(O.N. of K) + (O.N. of C) + 3(O.N. of O) = 0

$$
\begin{array}{cccc}
2(+1) & +4 & +3(-2) & = 0 \\
+2 & +4 & +(-6) & = 0
\end{array}
$$

HNO_2: (O.N. of H) + (O.N. of N) + 2(O.N. of O) = 0

$$
\begin{array}{cccc}
+1 & +3 & +2(-2) & = 0 \\
+1 & +3 & +(-4) & = 0
\end{array}
$$

oxidation Originally, a process involving a reaction with oxygen. Today it means a number of things, including a process in which electrons are given up, hydrogen is lost, or an oxidation number increases.

reduction Originally, a process in which oxygen was lost. Today it means a number of things, including a process in which electrons are gained, hydrogen is accepted, or an oxidation number decreases.

oxidation numbers or oxidation states Positive or negative numbers assigned to the elements in chemical formulas according to a specific set of rules.

TABLE 5.1 Common Uses of the Terms *Oxidation* and *Reduction*

Term	Meaning
Oxidation	To combine with oxygen
	To lose hydrogen
	To lose electrons
	To increase in oxidation number
Reduction	To lose oxygen
	To combine with hydrogen
	To gain electrons
	To decrease in oxidation number

Rule 7. The algebraic sum of the O.N.s of all atoms in a polyatomic ion is equal to the charge on the ion.

Examples: MnO_4^-: (O.N. of Mn) + 4(O.N. of O) = −1

$$
\begin{array}{ccc}
+7 & +4(-2) & = -1 \\
+7 & +(-8) & = -1
\end{array}
$$

HPO_4^{2-}: (O.N. of H) + (O.N. of P) + 4(O.N. of O) = −2

$$
\begin{array}{cccc}
+1 & +5 & +4(-2) & = -2 \\
+1 & +5 & +(-8) & = -2
\end{array}
$$

Example 5.3 Assigning Oxidation Numbers

a. Assign O.N.s to the blue element in each of the following:

$$CO_2, \quad NO_2^-, \quad H_2O, \quad N_2, \quad K^+$$

b. Assign O.N.s to each element in the following:

$$CO_2, \quad Mg(NO_3)_2, \quad NO_3^-, \quad CH_2O$$

Solution

a. CO_2, O = −2 (Rule 5); NO_2^-, O = −2 (Rule 5); H_2O, H = +1 (Rule 4); N_2, N = 0 (Rule 1); K^+, K = +1 (Rule 2).

b. CO_2: The O.N. of O is −2 (Rule 5), and the O.N. of C can be calculated by using Rule 6 as follows:

$$
\begin{aligned}
&(\text{O.N. of C}) + 2(\text{O.N. of O}) = 0 \\
&(\text{O.N. of C}) + 2(-2) = 0 \\
&(\text{O.N. of C}) + (-4) = 0
\end{aligned}
$$

Therefore,

$$\text{O.N. of C} = +4$$

$Mg(NO_3)_2$: The O.N. of Mg is +2 (Rule 3), the O.N. of O is −2 (Rule 5), and the O.N. of N can be calculated by using Rule 6 as follows:

$$
\begin{aligned}
&(\text{O.N. of Mg}) + 2(\text{O.N. of N}) + 6(\text{O.N. of O}) = 0 \\
&(+2) + 2(\text{O.N. of N}) + 6(-2) = 0 \\
&2(\text{O.N. of N}) + 2 + (-12) = 0 \\
&2(\text{O.N. of N}) - 10 = 0
\end{aligned}
$$

Therefore,

$$\text{O.N. of N} = +5$$

NO_3^-: The O.N. of O is −2 (Rule 5), and the O.N. of N can be calculated by using Rule 7 as follows:

$$
\begin{aligned}
&(\text{O.N. of N}) + 3(\text{O.N. of O}) = -1 \\
&(\text{O.N. of N}) + 3(-2) = -1 \\
&(\text{O.N. of N}) + (-6) = -1
\end{aligned}
$$

Therefore,

$$\text{O.N. of N} = +5$$

Note that N has the same O.N. in NO_3^- and in $Mg(NO_3)_2$. This is expected because the polyatomic NO_3^- ion is present in $Mg(NO_3)_2$.

CH$_2$O: The O.N. of H is +1 (Rule 4), the O.N. of O is −2 (Rule 5), and the O.N. of C can be calculated by using Rule 6 as follows:

$$(\text{O.N. of C}) + 2(\text{O.N. of H}) + (\text{O.N. of O}) = 0$$
$$(\text{O.N. of C}) + 2(+1) + (-2) = 0$$
$$(\text{O.N. of C}) + 2 - 2 = 0$$

Therefore,

$$\text{O.N. of C} = 0$$

This example shows that an O.N. of 0 may be found for some combined elements as well as those in an uncombined state (Rule 1).

✔ **LEARNING CHECK 5.3** Assign O.N.s to each element in the following:

a. SO$_3$ b. Ca(ClO$_3$)$_2$ c. ClO$_4^-$

Now that you can assign O.N.s to the elements involved in reactions, you are ready to look at redox processes in terms of O.N.s and electron transfers. The reaction when sulfur is burned in oxygen is represented by the equation

$$S(s) + O_2(g) \rightarrow SO_2(g) \tag{5.12}$$

You can see that this reaction represents an oxidation because sulfur has combined with oxygen (remember Table 5.1, page 158). When oxidation numbers are assigned to sulfur, the reactant S has an oxidation number of 0, while the S in the SO$_2$ product has an oxidation number of +4. Thus, the oxidation that has taken place resulted in an increase in the oxidation number of sulfur. Oxidation always corresponds to an increase in oxidation number.

The same conclusion is arrived at by thinking of oxidation numbers as charges. Thus, a sulfur atom with 0 charge is oxidized as it acquires a +4 charge. This change in charge results when the sulfur atom releases four electrons that have a −1 charge each. This leads to another definition of oxidation: Oxidation takes place when electrons are lost.

The oxygen of Equation 5.12 undergoes an oxidation number change from 0 in O$_2$ to −2 in SO$_2$. This decrease in oxidation number corresponds to a reduction of the oxygen. In terms of electrons, this same change is accomplished when each uncharged oxygen atom in the O$_2$ molecule accepts two electrons. Thus, the four electrons lost by sulfur as it was oxidized is the same total number accepted by the oxygen as it was reduced.

An oxidation process must always be accompanied by a reduction process, and all the electrons released during oxidation must be accepted during reduction. Thus, oxidation and reduction processes always take place simultaneously, hence the term redox. In redox reactions, the substance that is oxidized (and releases electrons) is called the **reducing agent** because it is responsible for reducing another substance. Similarly, the substance that is reduced (accepts electrons) is called the **oxidizing agent** because it is responsible for oxidizing another material (see **Figure 5.2**). These characteristics are summarized in **Table 5.2.**

© Cengage Learning/Charles D. Winters

Figure 5.2 Combustion, the first reaction known to be carried out by humans, is a rapid redox reaction. Identify the oxidizing and reducing agents of the reaction.

reducing agent The substance that contains an element that is oxidized during a chemical reaction.

oxidizing agent The substance that contains an element that is reduced during a chemical reaction.

TABLE 5.2 Properties of Oxidizing and Reducing Agents	
Oxidizing Agent	**Reducing Agent**
Gains electrons	Loses electrons
Oxidation number decreases	Oxidation number increases
Becomes reduced	Becomes oxidized

A word of caution is appropriate here. Although electron transfers are a useful concept for understanding redox reactions involving covalent substances, it must be remembered that such transfers actually take place only during the formation of ionic compounds (see **Section 4.3**). In covalent substances, the electrons are actually shared (see **Section 4.6**). The oxidation number assignment rules that were given, as well as the electron-transfer idea, are based on the arbitrary practice of assigning shared electrons to the more electronegative element sharing them. However, it must be remembered that none of the atoms in covalent molecules actually acquire a net charge.

Example 5.4 Identifying Oxidizing and Reducing Agents

Determine oxidation numbers for each atom represented in the following equations and identify the oxidizing and reducing agents:

a. $4Al(s) + 3O_2(g) \rightarrow 2Al_2O_3(s)$
b. $CO(g) + 3H_2(g) \rightarrow H_2O(g) + CH_4(g)$
c. $S_2O_8^{2-}(aq) + 2I^-(aq) \rightarrow I_2(aq) + 2SO_4^{2-}(aq)$

Solution

The O.N. under each elemental symbol was calculated by using the methods demonstrated in Example 5.3.

a. $4Al + 3O_2 \rightarrow 2Al_2\ O_3$
 $00+3\ -2$

 The O.N. of Al has changed from 0 to +3. Therefore, Al has been oxidized and is the reducing agent. The O.N. of O has decreased from 0 to −2. Oxygen has been reduced and is the oxidizing agent.

b. $C\ \ O + 3H_2 \rightarrow H_2\ O + C\ H_4$
 $+2\ -20+1\ -2-4\ +1$

 The O.N. of H_2 increased from 0 to +1. H_2 has been oxidized and is the reducing agent. The O.N. of C has decreased from +2 to −4. Carbon has been reduced and could be called the oxidizing agent. However, when one element in a molecule or ion is the oxidizing or reducing agent, the convention is to refer to the entire molecule or ion by the appropriate term. Thus, carbon monoxide (CO) is the oxidizing agent.

c. $S_2\ O_8^{2-} + 2I^- \rightarrow I_2 + 2S\ \ O_4^{2-}$
 $+7\ -2-10+6\ -2$

 The O.N. of I has increased from −1 to 0, and the I^- ion is the reducing agent. The O.N. of S in $S_2O_8^{2-}$ has decreased from +7 to +6. Thus, sulfur has been reduced, and $S_2O_8^{2-}$ is the oxidizing agent.

✔ **LEARNING CHECK 5.4** Assign oxidation numbers to each atom represented in the following equations and identify the oxidizing and reducing agents:

a. $Zn(s) + 2H^+(aq) \rightarrow Zn^{2+}(aq) + H_2(g)$
b. $2KI(aq) + Cl_2(aq) \rightarrow 2KCl(aq) + I_2(aq)$
c. $IO_3^-(aq) + 3HSO_3^-(aq) \rightarrow I^-(aq) + 3HSO_4^-(aq)$

5.4 | Decomposition Reactions

Learning Objective

3. Classify reactions into the categories of redox or nonredox, then into the categories of decomposition, combination, single replacement, or double replacement.

In **decomposition reactions**, a single substance is broken down to form two or more simpler substances, as shown in **Figure 5.3**. In this and other "box" representations of reactions, the number of molecules in the boxes will not match the coefficients of the reaction equation. However, they will be in the correct proportions. In Figure 5.3, for example,

$2H_2O_2 \longrightarrow 2H_2O + O_2$

Figure 5.3 A decomposition reaction.

decomposition reaction A chemical reaction in which a single substance reacts to form two or more simpler substances.

the box on the left contains the same number of H_2O_2 molecules as the number of H_2O molecules on the right. Also, the number of H_2O molecules on the right is twice the number of O_2 molecules. The general form of the equation for a decomposition reaction is

$$A \rightarrow B + C \tag{5.13}$$

Some decomposition reactions are also redox reactions, whereas others are not. Examples of decomposition reactions are given by Equations 5.14 and 5.15:

$$2HgO(s) \rightarrow 2Hg(\ell) + O_2(g) \tag{5.14}$$

$$CaCO_3(s) \rightarrow CaO(s) + CO_2(g) \tag{5.15}$$

Equation 5.14 represents the redox reaction that takes place when mercury(II) oxide (HgO) is heated. Mercury metal (Hg) and oxygen gas (O_2) are the products. This reaction was used by Joseph Priestley in 1774 when he discovered oxygen. Equation 5.15 represents a nonredox reaction used commercially to produce lime (CaO) by heating limestone ($CaCO_3$) to a high temperature. The decomposition of H_2O_2 represented by Figure 5.3 is shown in **Figure 5.4**.

Figure 5.4 The decomposition of hydrogen peroxide. A solution of hydrogen peroxide (H_2O_2) in water (left) does not decompose rapidly until an enzyme catalyst from a piece of freshly cut potato is added (right). How can you tell that one of the products of the reaction is a gas?

© Jeffrey M. Seager © Jeffrey M. Seager

5.5 | Combination Reactions

Learning Objective

3. Classify reactions into the categories of redox or nonredox, then into the categories of decomposition, combination, single replacement, or double replacement.

combination reaction A chemical reaction in which two or more substances react to form a single substance.

Combination reactions are sometimes called *addition* or *synthesis reactions*. Their characteristic is that two or more substances react to form a single substance (see **Figure 5.5**). The reactants can be any combination of elements or compounds, but the product is always a compound. The general form of the equation is

$$A + B \rightarrow C \tag{5.16}$$

H₂ + F₂ ───────────→ 2HF

Figure 5.5 A combination reaction.

At high temperatures, a number of metals will burn and give off very bright light. This burning is a redox combination reaction, represented for magnesium metal by Equation 5.17 and shown in **Figure 5.6**.

$$2Mg(s) + O_2(g) \rightarrow 2MgO(s) \qquad (5.17)$$

A nonredox combination reaction that takes place in the atmosphere contributes to the acid rain problem. An air pollutant called sulfur trioxide (SO_3) reacts with water vapor and forms sulfuric acid. The reaction is represented by Equation 5.18:

$$SO_3(g) + H_2O(\ell) \rightarrow H_2SO_4(aq) \qquad (5.18)$$

5.6 | **Replacement Reactions**

Learning Objective

3. Classify reactions into the categories of redox or nonredox, then into the categories of decomposition, combination, single replacement, or double replacement.

Single-replacement reactions, also called *substitution reactions*, are always redox reactions and take place when one element reacts with a compound and displaces another element from the compound. A single-replacement reaction is represented in **Figure 5.7** and demonstrated in **Figure 5.8**. The general equation for the reaction is shown in Equation 5.19.

$$A + BX \rightarrow B + AX \qquad (5.19)$$

Figure 5.6 Magnesium metal burns in air to form magnesium oxide.

single-replacement reaction A chemical reaction in which an element reacts with a compound and displaces another element from the compound.

$$H_2 \qquad + \qquad CuO \longrightarrow Cu + H_2O$$

Figure 5.7 A single-replacement reaction.

Figure 5.8 When a piece of copper wire (Cu) is placed in a solution of silver nitrate ($AgNO_3$) in water, crystals of silver metal (Ag) form on the wire, and the liquid solution that was originally colorless turns blue as copper nitrate [$Cu(NO_3)_2$] forms in solution. Write a balanced equation for this single-replacement reaction.

This type of reaction is useful in a number of processes used to obtain metals from their oxide ores. Iron, for example, can be obtained by reacting iron(III) oxide ore (Fe_2O_3) with carbon. The carbon displaces the iron from the oxide, and carbon dioxide is formed. The equation for the reaction is

$$3C(s) + 2Fe_2O_3(s) \rightarrow 4Fe(s) + 3CO_2(g) \tag{5.20}$$

double-replacement reaction A chemical reaction in which two compounds react and exchange partners to form two new compounds.

Double-replacement reactions, also called *metathesis reactions*, are never redox reactions. These reactions, represented in **Figure 5.9**, often take place between substances dissolved in water. The general Equation 5.21 shows the partner-swapping characteristic of these reactions:

$$AX + BY \rightarrow BX + AY \tag{5.21}$$

NaOH + HF ⟶ NaF + H₂O

Figure 5.9 A double-replacement, or metathesis, reaction.

The reaction that takes place when a base is used to neutralize an acid is a good example of a double-replacement reaction:

$$HCl(aq) + NaOH(aq) \rightarrow NaCl(aq) + H_2O(\ell) \tag{5.22}$$

Example 5.5 Classifying Reactions

Classify each of the reactions represented by the following equations as redox or nonredox. Further classify them as decomposition, combination, single-replacement, or double-replacement reactions.

a. $SO_2(g) + H_2O(\ell) \rightarrow H_2SO_3(aq)$
b. $2K(s) + 2H_2O(\ell) \rightarrow 2KOH(aq) + H_2(g)$
c. $N_2(g) + 3H_2(g) \rightarrow 2NH_3(g)$
d. $BaCl_2(aq) + Na_2CO_3(aq) \rightarrow BaCO_3(s) + 2NaCl(aq)$

Solution
The O.N. under each elemental symbol was calculated by using the methods demonstrated in Example 5.3.

a. S O₂ + H₂ O → H₂ S O₃
 +4 −2 +1 −2 +1 +4 −2
 No O.N. changes take place; therefore, the reaction is nonredox. Because two substances combine to form a third, the reaction is a combination reaction.
b. 2K + 2H₂ O → 2K O H + H₂
 0 +1 −2 +1 −2 +1 0
 The O.N. of K increases from 0 to +1 and that of H decreases from +1 to 0. The reaction is a redox reaction. Because K displaces H, it is a single-replacement reaction.
c. N₂ + 3H₂ → 2N H₃
 0 0 −3 +1
 The O.N.s of both N and H change; therefore, the reaction is a redox reaction. Two substances combine to form a third, so it is a combination reaction.
d. Ba Cl₂ + Na₂ C O₃ → Ba C O₃ + 2Na Cl
 +2 −1 +1 +4 −2 +2 +4 −2 +1 −1
 No changes in O.N. occur. The reaction is nonredox and is an example of a double-replacement, or partner-swapping, reaction.

✔ LEARNING CHECK 5.5 Classify each of the reactions represented by the following equations as redox or nonredox. Further classify them as decomposition, combination, single-replacement, or double-replacement reactions.

a. $2HI(g) \rightarrow H_2(g) + I_2(g)$
b. $2H_2O_2(aq) \rightarrow 2H_2O(\ell) + O_2(g)$ (note: H_2O_2 is a peroxide)
c. $NaCl(aq) + AgNO_3(aq) \rightarrow AgCl(s) + NaNO_3(aq)$
d. $4P(s) + 5O_2(g) \rightarrow 2P_2O_5(s)$
e. $2NaI(aq) + Cl_2(aq) \rightarrow 2NaCl(aq) + I_2(aq)$

5.7 | Ionic Equations

Learning Objective

4. Write molecular equations in total ionic and net ionic forms.

Many of the reactions of interest in the course you are taking occur between compounds or elements dissolved in water. Ionic compounds and some polar covalent compounds break apart (dissociate) into ions when they are dissolved in water. Thus, a water solution of sodium hydroxide (NaOH), an ionic compound, does not contain molecules of NaOH but, rather, contains equal numbers of sodium ions (Na^+) and hydroxide ions (OH^-). Covalently bonded hydrogen chloride, HCl, dissolves readily in water to form H^+ and Cl^- ions. Equations for reactions between substances that form ions in solution can be written in several ways. For example, Equation 5.22 contains three substances that form ions, HCl, NaOH, and NaCl. Equation 5.22 is written in the form of a **molecular equation** in which each compound is represented by its formula. This same reaction, when represented by a **total ionic equation**, becomes

$$H^+(aq) + Cl^-(aq) + Na^+(aq) + OH^-(aq) \rightarrow$$
$$Na^+(aq) + Cl^-(aq) + H_2O(\ell) \quad (5.23)$$

In this equation each ionic compound is shown dissociated into ions, the form it takes when it is dissolved in water. Some of the ions appear as both reactants and products. These so-called **spectator ions** do not actually undergo any changes in the reaction. Because of this, they are dropped from the equation when it is written as a **net ionic equation**:

$$H^+(aq) + OH^-(aq) \rightarrow H_2O(\ell) \quad (5.24)$$

The net ionic equation makes the partner-swapping characteristics of this double-replacement reaction less obvious, but it does emphasize the actual chemical changes that take place. **Figure 5.10** shows an experiment in which an ionic reaction occurs.

molecular equation An equation written with each compound represented by its formula.

total ionic equation An equation written with all soluble ionic substances represented by the ions they form in solution.

spectator ions The ions in a total ionic reaction that are not changed as the reaction proceeds. They appear in identical forms on the left and right sides of the equation.

net ionic equation An equation that contains only un-ionized or insoluble materials and ions that undergo changes as the reaction proceeds. All spectator ions are eliminated.

Figure 5.10 The liquid in the large container is a solution of solid sodium chloride (NaCl) in water. The liquid being added is a solution of solid silver nitrate ($AgNO_3$) in water. When the two liquids are mixed, an insoluble white solid forms. The solid is silver chloride (AgCl). Write the molecular, total ionic, and net ionic equations for the reaction.

© Mike Slabaugh

Add Color to Your Diet

Scientific evidence accumulated during the 1990s suggested that diets rich in fruits and vegetables had a protective effect against a number of different types of cancer. Studies showed that simply increasing the levels of vitamins and minerals in the diet did not provide the increased protection. This led to research into the nature of other substances found in fruits and vegetables that are important for good health. As a result of this research, a number of chemical compounds found in plants and called *phytonutrients* have been shown to be involved in the maintenance of healthy tissues and organs. The mechanism for their beneficial action in the body is not understood for all phytonutrients, but a significant number are known to work as antioxidants that stop harmful oxidation reactions from occurring.

The colors of fruits and vegetables help identify those containing beneficial compounds. The table below contains a list of some of the more well-known phytonutrients together with sources, colors, and beneficial actions. The amount of evidence supporting the existence of benefits from phytonutrients is not the same for all those listed in the table. In some cases, the experimental evidence is extensive (e.g., the cancer-blocking behavior of isothiocyanates), while in other cases the listed benefits are based on a limited amount of research and more studies are being done (e.g., the contribution to eye health by anthocyanins).

Fruit/Vegetable Color	Fruit/Vegetable Examples	Phytonutrients	Possible Benefits
Red	Tomatoes, watermelon, pink grapefruit	Lycopene (a carotenoid)	Protect against prostate, cervical, and pancreatic cancer and heart and lung disease
Red/purple	Red and blue grapes, blueberries, strawberries, beets, eggplant, red cabbage, red peppers, plums, red apples	Anthocyanins (flavonoids)	Antioxidants; block formation of blood clots and help maintain good eye health
Orange	Carrots, mangoes, sweet potatoes, cantaloupe, winter squash	Alpha- and beta-carotenes	Cancer fighters; protect skin against free radicals, promote repair of damaged DNA
Orange/yellow	Oranges, peaches, papaya, nectarines	Beta-cryptoxanthin	May help prevent heart disease
Yellow/green	Spinach, collards, corn, green peas, avocado, honeydew	Lutein and zeaxanthin (both are carotenoids)	Reduce risk of cataracts and age-related macular degeneration
Green	Broccoli, brussels sprouts, cabbage, kale, bok choy	Sulforaphane, isothiocyanates, and indoles	Cancer blocking
White/green	Onions, leeks, garlic, celery, asparagus, pears, green grapes, cauliflower, mushrooms	Allicin (in onions) and flavonoids	Antitumor agent; antioxidants

Example 5.6 Writing Reaction Equations

Write equations for the following double-replacement reaction in total ionic and net ionic forms. Note that barium sulfate ($BaSO_4$) does not dissolve in water and should not be written in dissociated form. All other compounds are ionic and soluble in water.

$$Na_2SO_4(aq) + BaCl_2(aq) \rightarrow BaSO_4(s) + 2NaCl(aq)$$

Solution

Total ionic:

$$2Na^+(aq) + SO_4^{2-}(aq) + Ba^{2+}(aq) + 2Cl^-(aq) \rightarrow$$
$$BaSO_4(s) + 2Na^+(aq) + 2Cl^-(aq)$$

Net ionic:

$$SO_4^{2-}(aq) + Ba^{2+}(aq) \rightarrow BaSO_4(s)$$

The Na^+ and Cl^- ions appear in equal numbers as reactants and products. Thus, they are spectator ions and are not shown in the net ionic equation.

✔ **LEARNING CHECK 5.6** Write equations for the following reactions in total ionic and net ionic forms. Consider all ionic compounds to be soluble except $CaCO_3$ and $BaSO_4$, and remember that covalent molecules do not form ions when they dissolve.

a. $2NaI(aq) + Cl_2(aq) \rightarrow 2NaCl(aq) + I_2(aq)$
b. $CaCl_2(aq) + Na_2CO_3(aq) \rightarrow 2NaCl(aq) + CaCO_3(s)$
c. $Ba(OH)_2(aq) + H_2SO_4(aq) \rightarrow 2H_2O(\ell) + BaSO_4(s)$

5.8 | Energy and Reactions

Learning Objective

5. Classify reactions as exothermic or endothermic.

Besides changes in composition, energy changes accompany all chemical reactions. The equation for the reaction between hydrogen and oxygen given at the beginning of this chapter (Equation 5.2) can also be written

$$2H_2(g) + O_2(g) \rightarrow 2H_2O(g) + energy \qquad (5.25)$$

Most of the energy of this reaction appears as heat, but the energy released or absorbed during chemical changes can take many forms including sound, electricity, light, high-energy chemical bonds, and motion. Some of these forms are discussed in detail in later chapters, but here the focus will be on energy in general, and it will be expressed in calories or joules, as if it all took the form of heat.

On the basis of heat, Equation 5.25 can be written

$$2H_2(g) + O_2(g) \rightarrow 2H_2O(g) + 115.6 \text{ kcal } (483.7 \text{ kJ}) \qquad (5.26)$$

This equation assumes that the hydrogen and oxygen are gases when they react and that the water produced is also a gas (vapor). According to Equation 5.26, whenever 2 mol of water vapor is formed from 2 mol of hydrogen gas and 1 mol of oxygen gas, 115.6 kcal, or 483.7 kJ, of energy is also released.

The reaction represented by Equation 5.26 is an example of an **exothermic** (heat out) **reaction**, in which heat is released as the reaction takes place. In **endothermic** (heat in) **reactions**, heat is absorbed. A reaction of this type is utilized in emergency cold packs, which consist of a small plastic pouch of liquid sealed inside a larger pouch that also

exothermic reaction A reaction that liberates heat.

endothermic reaction A reaction that absorbs heat.

contains a solid substance. When the large pouch is squeezed firmly, the small pouch of liquid breaks, the liquid and solid react, heat is absorbed, and the liquid contents of the larger pouch become cold. This reaction is described in more detail in Chapter 7.

5.9 | The Mole and Chemical Equations

Learning Objective

6. Use the mole concept to do calculations based on chemical reaction equations.

stoichiometry The study of mass relationships in chemical reactions.

The mole concept introduced in Section 2.6 (see page 58) and applied to chemical formulas in Section 2.7 (see page 63) can also be used to calculate mass relationships in chemical reactions. The study of such mass relationships is called **stoichiometry**, a word derived from the Greek *stoicheion* (element) and *metron* (measure).

Stoichiometry calculations require that balanced equations be used for the reactions being studied. Consider the following equation for the redox reaction that takes place when methane (CH_4), a major constituent of natural gas, is burned:

$$CH_4(g) + 2O_2(g) \rightarrow CO_2(g) + 2H_2O(\ell) \tag{5.27}$$

Remember, coefficients in balanced equations refer to the formula that follows the coefficient. With this in mind, note that the following statement is consistent with this balanced equation:

1. 1 CH_4 molecule + 2 O_2 molecules \rightarrow 1 CO_2 molecule + 2 H_2O molecules

This statement indicates that one molecule of CH_4 gas reacts with two molecules of O_2 gas and produces one molecule of CO_2 gas plus two molecules of H_2O liquid. Put another way, according to this reaction equation, the number of O_2 molecules that react with CH_4 will be equal to twice the number of CH_4 molecules that react. We remember that one mole of any kind of molecule is equal to 6.02×10^{23} molecules. Thus, one mole of CH_4 is equal to 6.02×10^{23} CH_4 molecules. As shown earlier, the number of O_2 molecules that will react with CH_4 will be twice the number of CH_4 molecules that react. Thus, the number of O_2 molecules that will react with 6.02×10^{23} CH_4 molecules is $2(6.02 \times 10^{23})$. Remembering that 6.02×10^{23} molecules is 1 mole of molecules leads to the conclusion that 1 mole of CH_4 molecules will react with 2 moles of O_2 molecules. We note that the numbers 1 and 2 are identical to the coefficients of the CH_4 and O_2 in the balanced equation. This leads to the conclusion that the coefficients in reaction equations can be interpreted as the number of moles of reactants and products involved in the reaction, which leads to the following statement for the reaction of CH_4 with O_2:

2. 1 mol $CH_4(g)$ + 2 mol $O_2(g) \rightarrow$ 1 mol $CO_2(g)$ + 2 mol $H_2O(\ell)$

In general, any balanced reaction equation can be interpreted this way, where the coefficients of the equation become the number of moles of reactants and products involved in the reaction.

The fact used earlier that 1 mole of any substance is equal to 6.02×10^{23} particles of the substance leads to statement 3 given below for the reaction of CH_4 with O_2:

3. 6.02×10^{23} CH_4 molecules + $2(6.02 \times 10^{23})$ O_2 molecules \rightarrow
 6.02×10^{23} CO_2 molecules + $2(6.02 \times 10^{23})$ H_2O molecules

Application of the fact that one mole of a compound has a mass in grams equal to the molecular weight of the compound leads to statement 4 given below for the reaction:

4. 16.0 g CH_4 + 2(32.0) g $O_2 \rightarrow$ 44.0 g CO_2 + 2(18.0) g H_2O

Statements 2, 3, and 4 are all based on the mole definition, and are very useful in solving numerical problems involving balanced reaction equations and the factor-unit method described earlier in Section 1.9 (see page 29). Any two quantities from statements 2, 3, and 4 can be used to form factors that can be used to solve problems. For example, the following

CHEMISTRY AROUND US 5.1

Teeth Whitening

Tooth enamel is the hardest substance in the human body. At a microscopic level, tooth enamel is made up of tiny rods of hard calcium phosphate. The enamel occurs in two layers of a tooth. The top white layer covers a second yellow layer, so a normal tooth appears to be white in color. Teeth become more yellow in color when substances are applied to them that cause the inner yellow layer to darken. The antibiotic tetracycline and large amounts of fluoride stain teeth by this mechanism. Teeth also become stained by the frequent use of tobacco or the use of highly colored beverages such as coffee, tea, or colored soft drinks. Such stains become permanent when colored organic material fills the space between the calcium phosphate rods that make up the enamel.

Perhaps it is ego-driven, but the process called "teeth whitening" has become quite popular. Four options for this process are now available: whitening done by professionals, the use of dentist-prescribed medications, over-the-counter medications, and so-called "natural methods." With the exception of the natural method, all the methods involve the use of hydrogen peroxide (H_2O_2), which is an oxidizing agent that breaks down the colored organic material that fills the spaces between the calcium phosphate rods of tooth enamel. The natural method uses malic acid found in fruit to whiten the teeth. This method includes the requirement to eat an apple or some strawberries each day.

The American Dental Association says that teeth-whitening methods are safe and effective when 3.6% hydrogen peroxide concentrations or equivalent are used. In order to protect enamel, many treatment mixtures also include fluoride. Some methods result in temporary tooth sensitivity or gum irritation, but they do not cause any permanent damage. Eat fruits to maintain whiteness, especially after drinking coffee, tea, or dark-colored soda.

Spotmatik Ltd/Shutterstock.com

Teeth whitening has become a popular procedure.

factors are just four of the many that can be obtained from statements 2, 3, and 4 by combining various quantities from the statements:

$$\frac{2 \text{ mol O}_2}{1 \text{ mol CH}_4} \qquad \frac{16.0 \text{ g CH}_4}{44.0 \text{ g CO}_2} \qquad \frac{2(18.0) \text{ g H}_2\text{O}}{6.02 \times 10^{23} \text{ CH}_4 \text{ molecules}} \qquad \frac{1 \text{ mol CH}_4}{44.0 \text{ g CO}_2}$$

Suppose you were asked this question: How many moles of CO_2 could be formed by reacting together 2 mol of CH_4 and 4 mol of O_2? Statement 2 can be used to solve this problem quickly. We see that the amounts reacted correspond to twice the amounts represented by Statement 2. Thus, 2 mol of CO_2 would be formed, which is also twice the amount represented by Statement 2. However, many stoichiometric problems cannot be solved quite as readily, so it is helpful to learn a general approach that works well for many problems of this type. This approach is based on the factor-unit method described earlier in Section 1.9. The needed factors are obtained from Statements 2, 3, and 4.

Example 5.7 Interpreting Equations

The following questions are based on Equation 5.27.

a. How many moles of oxygen gas (O_2) will be required to react with 1.72 mol of methane (CH_4)?

b. How many grams of H_2O will be produced when 1.09 mol of CH_4 reacts with an excess of O_2?

c. How many grams of O_2 must react with excess CH_4 to produce 8.42 g of carbon dioxide (CO_2)?

Solution

a. The steps to follow in the factor-unit method are as follows (see Section 1.9):

 1. Write down the known or given quantity. It will include both a number and a unit.
 2. Leave working space, and set the known quantity equal to the unit of the unknown quantity.
 3. Multiply the known quantity by a factor such that the unit of the known quantity is canceled and the unit of the unknown quantity is generated.
 4. After the units on each side of the equation match, do the necessary arithmetic to get the final answer.

In this problem, the known quantity is 1.72 mol of CH_4, and the unknown quantity has the unit of mol O_2.

Step 1. 1.72 mol CH_4

Step 2. 1.72 mol CH_4 $\qquad\qquad$ = mol O_2

Step 3. 1.72 $\cancel{\text{mol }CH_4} \times \dfrac{2 \text{ mol } O_2}{1 \cancel{\text{ mol }CH_4}}$ = mol O_2

The factor

$$\frac{2 \text{ mol } O_2}{1 \text{ mol } CH_4}$$

was used because it properly canceled the unit of the known and generated the unit of the unknown.

Step 4. $(1.72)\left(\dfrac{2 \text{ mol } O_2}{1}\right) = 3.44 \text{ mol } O_2$

The answer contains the proper number of significant figures because the 2 and 1 of the factor are exact numbers (see Section 1.8).

b.

Step 1. 1.09 mol CH_4

Step 2. 1.09 mol CH_4 $\qquad\qquad\qquad$ = g H_2O

Step 3. 1.09 $\cancel{\text{mol }CH_4} \times \dfrac{2(18.0) \text{ g } H_2O}{1 \cancel{\text{ mol }CH_4}}$ \qquad = g H_2O

Step 4. 1.09 $\times \dfrac{2(18.0) \text{ g } H_2O}{1} = 39.24 \text{ g } H_2O = 39.2 \text{ g } H_2O$

The answer was rounded to three significant figures to match the three in 1.09 and 18.0 g of H_2O. The 1 and 2 are exact counting numbers.

c.

Step 1. 8.42 g CO_2

Step 2. 8.42 g CO_2 $\qquad\qquad\qquad\qquad$ = g O_2

Step 3. 8.42 $\cancel{\text{g }CO_2} \times \dfrac{2(32.0) \text{ g } O_2}{44.0 \cancel{\text{ g }CO_2}}$ \qquad = g O_2

Step 4. 8.42 $\times \dfrac{2(32.0) \text{ g } O_2}{44.0} = 12.2472 \text{ g } O_2 = 12.2 \text{ g } O_2$

The answer was rounded to three significant figures to match the three in 8.42, 32.0 g of O_2, and the 44.0. The 2 is an exact counting number.

✔ **LEARNING CHECK 5.7** The following reaction equation is balanced:

$$N_2(g) + 3H_2(g) \rightarrow 2NH_3(g)$$

a. Write statements equivalent to statements 2, 3, and 4 based on this equation.
b. Calculate the number of mol of NH_3 that would be produced if 2.11 mol of H_2 was reacted with excess N_2.
c. Calculate the number of g of N_2 required to react with 9.47 g H_2.

5.10 | The Limiting Reactant

Learning Objective

7. Use the mole concept to do calculations based on the limiting-reactant principle.

Nearly everyone is familiar with two characteristics of automobiles. They stop running when they run out of gasoline, and they stop running when the air intake to the engine becomes clogged. Most people also know that gasoline and air are mixed, and the mixture is burned inside the automobile engine. However, the combustion reaction occurs only as long as both reactants, gasoline and air, are present. The reaction (and engine) stops when gasoline is absent, despite the presence of ample amounts of air. On the other hand, the reaction stops when air is absent, even though plenty of gasoline is available.

The behavior of the auto engine illustrates the **limiting-reactant principle** that a chemical reaction will take place only as long as all necessary reactants are present. The reaction will stop when one of the reactants, the **limiting reactant**, is used up. According to this principle, the amount of product formed depends on the amount of limiting reactant present because the reaction will stop (and no more product will form) when the limiting reactant is used up. This is represented in **Figure 5.11** for the reaction $F_2(g) + H_2(g) \rightarrow 2HF(g)$.

limiting-reactant principle The maximum amount of product possible from a reaction is determined by the amount of reactant present in the least amount, based on its reaction coefficient and molecular weight.

limiting reactant The reactant present in a reaction in the least amount, based on its reaction coefficients and molecular weight. It is the reactant that determines the maximum amount of product that can be formed.

$$H_2 \quad + \quad F_2 \longrightarrow \text{Mixture of reactants} \longrightarrow HF + \text{unreacted } H_2$$

Figure 5.11 The limiting reactant is used up in a reaction.

Example 5.8 Identifying Limiting Reactants

A mixture containing 20.0 g of CH_4 and 100. g of O_2 is ignited and burns according to Equation 5.27, which is repeated here: $CH_4(g) + 2O_2(g) \rightarrow CO_2(g) + 2H_2O(\ell)$. What substances will be found in the mixture after the reaction stops?

Solution

The task is to identify the limiting reactant. It will be completely used up, and the other reactant will be found mixed with the products (CO_2 and H_2O) after the reaction stops. A simple way to identify the limiting reactant is to find out which reactant will give the least amount of products. To do this, two problems must be solved, one in which the known quantity is 20.0 g of CH_4 and one in which the known quantity is 100. g of O_2. In each case, the theoretical amount of CO_2 produced is calculated, although the same information would be obtained by solving for the amount of water produced.

$$20.0 \text{ g } CH_4 \times \frac{44.0 \text{ g } CO_2}{16.0 \text{ g } CH_4} = 55.0 \text{ g } CO_2$$

$$100. \text{ g } O_2 \times \frac{44.0 \text{ g } CO_2}{2(32.0) \text{ g } O_2} = 68.75 \text{ g } CO_2 = 68.8 \text{ g } CO_2$$

Since the 20.0 g of CH_4 produces the smaller amount of CO_2, CH_4 is the limiting reactant. Therefore, the final mixture will contain the products CO_2 and H_2O and the leftover O_2.

CHEMISTRY AROUND US 5.2

Electric Cars

Electricity is a form of energy that was considered as an energy source for some of the earliest automobiles. However, the internal combustion engine won out. The typical battery used to power early electric cars was the heavy lead-acid battery still used today for starting the engines of gasoline-fueled cars.

Today, with air pollution and global warming being topics for serious consideration, the idea of electrically powered cars has once again become a serious topic for consideration. However, lead-acid batteries are simply too heavy to use as power sources in the electric cars of today. The best current candidate for powering electric cars is called a lithium-ion battery pack.

Lithium-ion batteries consist of three main components: a positive electrode, a negative electrode, and an electrolyte solution. The positive electrode (cathode) is made from a type of layered lithium oxide, such as lithium cobalt oxide ($LiCoO_2$). The negative electrode, or anode, is made of a pure form of carbon called graphite. The electrolyte is a non-aqueous solution of a lithium salt. A permeable membrane, called a separator, plays a critical role between the positive and negative electrodes. The resulting separation of charge is what allows the battery to generate usable electricity. Otherwise the battery would short-circuit and explode.

The major drawbacks for electric cars are their range, which limits their use to relatively short distances, and their high prices. Also, since most electricity in the United States is generated by burning coal, the use of electric cars continues to contribute to air pollution by carbon dioxide. However, it is thought that their range will improve over time as better-performing lithium-ion batteries are developed.

Electric cars use rechargeable lithium-ion batteries for power.

Example 5.9 Working with Limiting Reactants

The reaction shown in Figure 5.8 was carried out by a student who wanted to produce some silver metal. The equation for the reaction is

$$Cu(s) + 2AgNO_3(aq) \rightarrow 2Ag(s) + Cu(NO_3)_2(aq)$$

The student dissolved 17.0 g of $AgNO_3$ in distilled water, then added 8.50 g of copper metal, Cu, to the resulting solution. What mass of silver metal, Ag, was produced?

Solution

The limiting reactant must be identified because it will determine the amount of silver produced. To find the limiting reactant, we calculate the theoretical amount of silver produced by each reactant. The reactant that will produce the smallest mass of silver is the limiting reactant, and the mass of silver calculated for it is the amount produced by the reaction.

$$17.0 \text{ g AgNO}_3 \times \frac{2(108) \text{ g Ag}}{2(170) \text{ g AgNO}_3} = 10.8 \text{ g Ag}$$

$$8.50 \text{ g Cu} \times \frac{2(108) \text{ g Ag}}{63.6 \text{ g Cu}} = 28.8679 \text{ g Ag} = 28.9 \text{ g Ag}$$

We see that the $AgNO_3$ is the limiting reactant, and it will produce 10.8 g of Ag.

✔ **LEARNING CHECK 5.8** Refer to the reaction used for Learning Check 5.7. Assume that 2.00 mol of H_2 and 15.5 g of N_2 are reacted. What is the maximum mass of ammonia (NH_3) that can be produced? Which reactant is the limiting reactant?

5.11 | Reaction Yields

Learning Objective

8. Use the mole concept to do percentage-yield calculations.

In Sections 5.9 and 5.10, calculations were done to determine how much product could be obtained from the reaction of specified amounts of reactants. In many instances, the amount determined by such calculations would be greater than the amount produced by reacting the specified reactants in a laboratory. Does this mean matter is destroyed when reactions are actually done in the laboratory? No, this would violate the law of conservation of matter, a fundamental law of nature. Less product might be obtained because some of the reactants form compounds other than the desired product. Such reactions are called **side reactions**. Carbon dioxide gas, CO_2, is produced when carbon-containing fuels burn in an ample supply of oxygen. However, a side reaction produces toxic carbon monoxide gas, CO, when the oxygen supply is limited.

side reactions Reactions that do not give the desired product of a reaction.

Another cause of less product than expected might be poor laboratory technique resulting in such things as losses when materials are transferred from one container to another.

STUDY SKILLS 5.1

Help with Oxidation Numbers

Assigning oxidation numbers is a task that seems to be straightforward, but mistakes caused by carelessness often show up when it is done on exams. You can minimize such mistakes by using a systematic approach to the task. First, have the seven rules for assigning oxidation numbers (see page 148) well in mind. Then, use an organized method such as the boxlike approach (shown below to assign an oxidation number to the Mn in Mn_2O_3). In this approach, we begin with an oxidation number that is given by one of the rules (O is –2 according to Rule 5):

Each Mn must be +3 because two Mn give +6 total.

$$Mn_2 \quad O_3$$
$$+3 \qquad$$

Rule 5: O.N. of O = −2.

−2

Rule 6: Sum of total O.N. of Mn and O = 0.

$$+6 \leftarrow -6$$

Total O.N. for three O's = −6

Follow the arrows in a clockwise direction to solve for the O.N. of each atom of Mn (+3 is correct). Remember that oxidation numbers are reported for individual atoms. For example, the total oxidation number due to Mn atoms is +6, but because there are two atoms of Mn in the formula, the oxidation number for each atom is just +3.

Whatever the cause, the mass of product obtained in an experiment is called the *actual yield*, and the mass calculated according to the methods of Sections 5.9 and 5.10 is called the *theoretical yield*. The **percentage yield** is the actual yield divided by the theoretical yield multiplied by 100 (see **Section 1.10**):

percentage yield The percentage of the theoretical amount of a product actually produced by a reaction.

$$\% \text{ yield} = \frac{\text{actual yield}}{\text{theoretical yield}} \times 100 \qquad (5.28)$$

Example 5.10 Calculating Percentage Yields

A chemist wants to produce urea (N_2CH_4O) by reacting ammonia (NH_3) and carbon dioxide (CO_2). The balanced equation for the reaction is

$$2NH_3(g) + CO_2(g) \rightarrow N_2CH_4O(s) + H_2O(\ell)$$

The chemist reacts 5.11 g of NH_3 with excess CO_2 and isolates 3.12 g of solid N_2CH_4O. Calculate the percentage yield of the experiment.

Solution

First, the theoretical yield must be calculated. Because this involves only masses of reactants and products, statements 2 and 4 based on the balanced equation will be used:

$$2NH_3(g) + CO_2(g) \rightarrow N_2CH_4O(s) + H_2O(\ell)$$

2. \qquad 2 mol $NH_3(g)$ + 1 mol $CO_2(g) \rightarrow$ 1 mol $N_2CH_4O(s)$ + 1 mol $H_2O(\ell)$

4. \qquad 2(17.0) g NH_3 + 44.0 g $CO_2 \rightarrow$ 60.0 g N_2CH_4O + 18.0 g H_2O

The known quantity is 5.11 g of NH_3, and the unit of the unknown quantity is g N_2CH_4O. The needed factor will come from statement 4.

Step 1. \quad 5.11 g NH_3

Step 2. \quad 5.11 g NH_3 $\hspace{8cm}$ = g N_2CH_4O

Step 3. \quad $5.11 \text{ g } \cancel{NH_3} \times \dfrac{60.0 \text{ g } N_2CH_4O}{2(17.0) \text{ g } \cancel{NH_3}}$ $\hspace{4cm}$ = g N_2CH_4O

Step 4. \quad $5.11 \times \dfrac{60.0 \text{ g } N_2CH_4O}{2(17.0)} = 9.0176$ g N_2CH_4O = 9.02 g N_2CH_4O

The actual yield was 3.12 g, so the percentage yield is

$$\% \text{ yield} = \frac{\text{actual yield}}{\text{theoretical yield}} \times 100 = \frac{3.12 \text{ g}}{9.02 \text{ g}} \times 100 = 34.5898\% = 34.6\%$$

✔ **LEARNING CHECK 5.9**

a. A chemist isolates 17.43 g of product from a reaction that has a calculated theoretical yield of 21.34 g. What is the percentage yield?

b. Lime (CaO) is produced by heating calcium carbonate. The equation for the reaction is $CaCO_3(s) \rightarrow CaO(s) + CO_2(g)$. Suppose 510. g of $CaCO_3$ is heated, and 235 g of CaO is isolated after the reaction mixture cools. What is the percentage yield for the reaction?

▶ **Case Study Follow-up**

Burns are categorized as first-, second-, or third-degree by the extent of tissue damage. First-degree burns affect the first layer of the skin, called the epidermis. Nearly everyone experiences a first-degree burn sometime in his or her life in the form of a sunburn. Second-degree burns extend into the second layer of skin, called the dermis. These burns appear wet and can be exceedingly painful. Third-degree burns involve the underlying fat layer below the skin. These burns appear to be dry and are either white-colored or charred black.

In general, burns range in severity between being mildly irritating to being life threatening. Severe burns are defined as those covering at least 25% of the body. Severe burns or burns involving the hands, feet, face, neck, or groin require emergency care, often at a special burn center. Treatment for severe burns might include skin grafts (removing areas of healthy skin and applying them to the burned area), administration of IV fluids, removal of damaged skin (debridement), massage, physical therapy, and compression therapy to minimize scarring.

The term "burn" describes tissue damage, and is not always associated with heat. Chemicals, radiation, and even freezing temperatures can cause damage the same as burning caused by heat. Burns occur quickly and might have long-lasting or permanent effects including disfigurement and death.

Concept Summary

Chemical Equations Chemical reactions are conveniently represented by equations in which reacting substances, called reactants, and the substances produced, called products, are written in terms of formulas. Coefficients are placed before reactant and product formulas to balance the equation. A balanced equation satisfies the law of conservation of matter.

Objective 1 (Section 5.1), Exercises 5.2 and 5.6

Types of Reactions To facilitate their study, reactions are classified into the general category redox or nonredox, and further classified as decomposition, combination, single-replacement, or double-replacement reactions.

Redox Reactions Reactions in which reactants undergo oxidation or reduction are called redox reactions. Oxidation and reduction are indicated conveniently by the oxidation number changes that occur. Oxidation numbers are assigned according to a specific set of rules. A substance is oxidized when the oxidation number of a constituent element increases, and it is reduced when the oxidation number of a constituent element decreases.

Objective 2 (Section 5.3), Exercises 5.10 and 5.15

Decomposition Reactions Decomposition reactions are characterized by one substance reacting to give two or more products. Decomposition reactions can be redox or nonredox.

Objective 3 (Section 5.4), Exercise 5.20

Combination Reactions Combination reactions, also called addition or synthesis reactions, are characterized by two or more reactants that form a single compound as a product. Combination reactions can be redox or nonredox.

Objective 3 (Section 5.5), Exercise 5.20

Replacement Reactions Single-replacement reactions, sometimes called substitution reactions, are always

redox reactions. One element reacts with a compound and displaces another element from the compound. Double-replacement reactions, also called metathesis reactions, are always nonredox reactions. They can be recognized by their partner-swapping characteristics.

Objective 3 (Section 5.6), Exercise 5.20

Ionic Equations Many water-soluble compounds separate (dissociate) into ions when dissolved in water. Reactions of such materials can be represented by molecular equations in which no ions are shown, total ionic equations in which all ions are shown, or net ionic equations in which only ions actually undergoing a change are shown.

Objective 4 (Section 5.7), Exercise 5.30a, b, c

Energy and Reactions Energy changes accompany all chemical reactions. The energy can appear in a variety of forms, but heat is a common one. Exothermic reactions liberate heat, and endothermic reactions absorb it.

Objective 5 (Section 5.8), Exercise 5.34

The Mole and Chemical Equations The mole concept, when applied to chemical equations, yields relationships that can be used to obtain factors for doing factor-unit calculations.

Objective 6 (Section 5.9), Exercise 5.42

The Limiting Reactant The limiting reactant is the reactant present in a reaction in an amount that determines the maximum amount of product that can be made. Factor-unit calculations can be used to determine which reactant is limiting.

Objective 7 (Section 5.10), Exercise 5.52

Reaction Yields The mass of product isolated after a reaction is often less than the mass theoretically possible. The ratio of the actual isolated mass (actual yield) to the calculated theoretical yield multiplied by 100 is the percentage yield.

Objective 8 (Section 5.11), Exercise 5.56

Key Terms and Concepts

Balanced equation (5.1)
Combination reaction (5.5)
Decompostion reaction (5.4)
Double-replacement reaction (5.6)
Endothermic reaction (5.8)
Exothermic reaction (5.8)
Law of conservation of matter (5.1)
Limiting reactant (5.10)

Limiting-reactant principle (5.10)
Molecular equation (5.7)
Net ionic equation (5.7)
Oxidation (5.3)
Oxidation number (oxidation state) (5.3)
Oxidizing agent (5.3)
Percentage yield (5.11)
Products of a reaction (5.1)

Reactants of a reaction (5.1)
Reducing agent (5.3)
Reduction (5.3)
Side reactions (5.11)
Single-replacement reaction (5.6)
Spectator ions (5.7)
Stoichiometry (5.9)
Total ionic equation (5.7)

Key Equations

1.	Decomposition reaction—one substance changes to two or more new substances (Section 5.4):	$A \rightarrow B + C$	Equation 5.13
2.	Combination reaction—two or more substances react to produce one new substance (Section 5.5):	$A + B \rightarrow C$	Equation 5.16
3.	Single-replacement reaction—one element reacts with a compound to produce a new compound and new element (Section 5.6):	$A + BX \rightarrow B + AX$	Equation 5.19
4.	Double-replacement reaction—partner-swapping reaction (Section 5.6):	$AX + BY \rightarrow BX + AY$	Equation 5.21
5.	Percentage yield (Section 5.11):	$\% \text{ yield} = \dfrac{\text{actual yield}}{\text{theoretical yield}} \times 100$	Equation 5.28

Exercises

Even-numbered exercises are answered in Appendix B.

Blue-numbered exercises are more challenging.

Chemical Equations (Section 5.1)

5.1 Identify the reactants and products in each of the following reaction equations:

 a. $CH_4(g) + 2O_2(g) \rightarrow CO_2(g) + 2H_2O(g)$

 b. $2Al(s) + 3Cl_2(g) \rightarrow 2AlCl_3(s)$

 c. methane + water → carbon monoxide + hydrogen

 d. copper(II) oxide + hydrogen → copper + water

5.2 Identify the reactants and products in each of the following reaction equations:

 a. $H_2(g) + Cl_2(g) \rightarrow 2HCl(g)$

 b. $2KClO_3(s) \rightarrow 2KCl(s) + 3O_2(g)$

 c. magnesium oxide + carbon → magnesium + carbon monoxide

 d. ethane + oxygen → carbon dioxide + water

5.3 Identify which of the following are consistent with the law of conservation of matter. For those that are not, explain why they are not.

 a. $Fe(s) + O_2(g) \rightarrow Fe_2O_3(s)$

 b. $2Na_3PO_4(aq) + 3MgCl_2(aq) \rightarrow Mg_3(PO_4)_2(s) + 6NaCl(aq)$

 c. 3.20 g oxygen + 3.21 g sulfur → 6.41 g sulfur dioxide

 d. $CH_4(g) + 2O_2(g) \rightarrow CO_2(g) + 2H_2O(g)$

5.4 Identify which of the following are consistent with the law of conservation of matter. For those that are not, explain why they are not.

 a. $ZnS(s) + O_2(g) \rightarrow ZnO(s) + SO_2(g)$

 b. $Cl_2(aq) + 2I^-(aq) \rightarrow I_2(aq) + 2Cl^-(aq)$

 c. 1.50 g oxygen + 1.50 g carbon → 2.80 g carbon monoxide

 d. $2C_2H_6(g) + 7O_2(g) \rightarrow 4CO_2(g) + 6H_2O(g)$

5.5 Determine the number of atoms of each element on each side of the following equations and decide which equations are balanced:

 a. $3Pb(NO_3)_2(aq) + 2AlBr_3(aq) \rightarrow 3PbBr_2(s) + 2Al(NO_3)_3(aq)$

 b. $K(s) + H_2O(\ell) \rightarrow KOH(aq) + H_2(g)$

 c. $CaCO_3(s) \rightarrow CaO(s) + CO_2(g)$

 d. $Ba(ClO_3)_2(aq) + H_2SO_4(aq) \rightarrow 2HClO_3(aq) + BaSO_4(s)$

5.6 Determine the number of atoms of each element on each side of the following equations and decide which equations are balanced:

 a. $Ag(s) + Cu(NO_3)_2(aq) \rightarrow Cu(s) + AgNO_3(aq)$

 b. $2N_2O(g) + 3O_2(g) \rightarrow 4NO_2(g)$

 c. $Mg(s) + O_2(g) \rightarrow 2MgO(s)$

 d. $H_2SO_4(aq) + Ca(OH)_2(aq) \rightarrow CaSO_4(s) + 2H_2O(\ell)$

5.7 Balance the following equations:

 a. $Li_3N(s) \rightarrow Li(s) + N_2(g)$

 b. $Cu_2O(s) + O_2(g) \rightarrow CuO(s)$

 c. $Zn(s) + HNO_3(aq) \rightarrow Zn(NO_3)_2(aq) + H_2(g)$

 d. $K_2SO_4(aq) + BaCl_2(aq) \rightarrow BaSO_4(s) + KCl(aq)$

 e. dinitrogen monoxide → nitrogen + oxygen

 f. dinitrogen pentoxide → nitrogen dioxide + oxygen

 g. $P_4O_{10}(s) + H_2O(\ell) \rightarrow H_4P_2O_7(aq)$

 h. $CaCO_3(s) + HCl(aq) \rightarrow CaCl_2(aq) + H_2O(\ell) + CO_2(g)$

5.8 Balance the following equations:

 a. $KClO_3(s) \rightarrow KCl(s) + O_2(g)$

 b. $C_2H_6(g) + O_2(g) \rightarrow CO_2(g) + H_2O(\ell)$

c. nitrogen + oxygen → dinitrogen pentoxide

d. $MgCl_2(s) + H_2O(g) \rightarrow MgO(s) + HCl(g)$

e. $CaH_2(s) + H_2O(\ell) \rightarrow Ca(OH)_2(s) + H_2(g)$

f. $Al(s) + Fe_2O_3(s) \rightarrow Al_2O_3(s) + Fe(s)$

g. aluminum + bromine → aluminum bromide

h. $Hg_2(NO_3)_2(aq) + NaCl(aq) \rightarrow Hg_2Cl_2(s) + NaNO_3(aq)$

Redox Reactions (Section 5.3)

5.9 Assign oxidation numbers to the blue element in each of the following formulas:

a. $CaCO_3$

b. H_2Se

c. Ca_2^+

d. Cl_2

e. Na_2CrO_4

f. PO_4^{3-}

5.10 Assign oxidation numbers to the blue element in each of the following formulas:

a. $HClO$

b. $NaNO_2$

c. N_2

d. $Ca(ClO)_2$

e. Al^{3+}

f. N_2O_5

5.11 Find the element with the highest oxidation number in each of the following formulas:

a. N_2O_5

b. $KHCO_3$

c. $NaOCl$

d. $NaNO_3$

e. $HClO_4$

f. $Ca(NO_3)_2$

5.12 Find the element with the highest oxidation number in each of the following formulas:

a. $Na_2Cr_2O_7$

b. $K_2S_2O_3$

c. HNO_3

d. P_2O_5

e. $Mg(ClO_4)_2$

f. $HClO_2$

5.13 For each of the following equations, indicate whether the blue element has been oxidized, reduced, or neither oxidized nor reduced.

a. $2Mg(s) + O_2(g) \rightarrow 2MgO(s)$

b. $CuO(s) + H_2(g) \rightarrow Cu(s) + H_2O(g)$

c. $Ag^+(aq) + Cl^-(aq) \rightarrow AgCl(s)$

d. $BaCl_2(aq) + H_2SO_4(aq) \rightarrow BaSO_4(s) + 2HCl(aq)$

e. $Zn(s) + 2H^+(aq) \rightarrow Zn^{2+}(aq) + H_2(g)$

5.14 For each of the following equations, indicate whether the blue element has been oxidized, reduced, or neither oxidized nor reduced.

a. $4Al(s) + 3O_2(g) \rightarrow 2Al_2O_3(s)$

b. $SO_2(g) + H_2O(\ell) \rightarrow H_2SO_3(aq)$

c. $2KClO_3(s) \rightarrow 2KCl(s) + 3O_2(g)$

d. $2CO(g) + O_2(g) \rightarrow 2CO_2(g)$

e. $2Na(s) + 2H_2O(\ell) \rightarrow 2NaOH(aq) + H_2(g)$

5.15 Assign oxidation numbers to each element in the following equations and identify the oxidizing and reducing agents:

a. $H_2(g) + Cl_2(g) \rightarrow 2HCl(g)$

b. $H_2O(g) + CH_4(g) \rightarrow CO(g) + 3H_2(g)$

c. $CuO(s) + H_2(g) \rightarrow Cu(s) + H_2O(g)$

d. $B_2O_3(s) + 3Mg(s) \rightarrow 2B(s) + 3MgO(s)$

e. $Fe_2O_3(s) + CO(g) \rightarrow 2FeO(s) + CO_2(g)$

f. $Cr_2O_7^{2-}(aq) + 2H^+(aq) + 3Mn^{2+}(aq) \rightarrow$
$$2Cr^{3+}(aq) + 3MnO_2(s) + H_2O(\ell)$$

5.16 Assign oxidation numbers to each element in the following equations and identify the oxidizing and reducing agents:

a. $2Cu(s) + O_2(g) \rightarrow 2CuO(s)$

b. $Cl_2(aq) + 2KI(aq) \rightarrow 2KCl(aq) + I_2(aq)$

c. $3MnO_2(s) + 4Al(s) \rightarrow 2Al_2O_3(s) + 3Mn(s)$

d. $2H^+(aq) + 3SO_3^{2-}(aq) + 2NO_3^-(aq) \rightarrow$
$$2NO(g) + H_2O(\ell) + 3SO_4^{2-}(aq)$$

e. $Mg(s) + 2HCl(aq) \rightarrow MgCl_2(aq) + H_2(g)$

f. $4NO_2(g) + O_2(g) \rightarrow 2N_2O_5(g)$

5.17 The tarnish of silver objects is a coating of silver sulfide (Ag_2S), which can be removed by putting the silver in contact with aluminum metal in a dilute solution of baking soda or salt. The equation for the cleaning reaction is

$$3Ag_2S(s) + 2Al(s) \rightarrow 6Ag(s) + Al_2S_3(s)$$

The sulfur in these compounds has a -2 oxidation number. What are the oxidizing and reducing agents in the cleaning reaction?

5.18 Aluminum metal reacts rapidly with highly basic solutions to liberate hydrogen gas and a large amount of heat. This reaction is utilized in a popular solid drain cleaner that is composed primarily of lye (sodium hydroxide) and aluminum granules. When wet, the mixture reacts as follows:

$$6NaOH(aq) + 2Al(s) \rightarrow 3H_2(g) + 2Na_3AlO_3(aq) + heat$$

The liberated H_2 provides agitation that, together with the heat, breaks the drain stoppage loose. What are the oxidizing and reducing agents in the reaction?

5.19 Identify the oxidizing and reducing agents in the Haber process for producing ammonia from elemental nitrogen and hydrogen. The equation for the reaction is

$$N_2(g) + 3H_2(g) \rightarrow 2NH_3(g)$$

Decomposition, Combination, and Replacement Reactions (Sections 5.4–5.6)

5.20 Classify each of the reactions represented by the following equations, first as a redox or nonredox reaction. Then further classify each redox reaction as a decomposition, single-replacement, or combination reaction, and each nonredox reaction as a decomposition, double-replacement, or combination reaction.

 a. $K_2CO_3(s) \rightarrow K_2O(s) + CO_2(g)$

 b. $Ca(s) + 2H_2O(\ell) \rightarrow Ca(OH)_2(s) + H_2(g)$

 c. $BaCl_2(aq) + H_2SO_4(aq) \rightarrow BaSO_4(s) + 2HCl(aq)$

 d. $SO_2(g) + H_2O(\ell) \rightarrow H_2SO_3(aq)$

 e. $2NO(g) + O_2(g) \rightarrow 2NO_2(g)$

 f. $2Zn(s) + O_2(g) \rightarrow 2ZnO(s)$

5.21 Classify each of the reactions represented by the following equations, first as a redox or nonredox reaction. Then further classify each redox reaction as a decomposition, single-replacement, or combination reaction, and each nonredox reaction as a decomposition, double-replacement, or combination reaction:

 a. $N_2O_5(g) + H_2O(\ell) \rightarrow 2HNO_3(aq)$

 b. $Cr_2O_3(s) + 2Al(s) \rightarrow 2Cr(s) + Al_2O_3(s)$

 c. $CaO(s) + SiO_2(s) \rightarrow CaSiO_3(s)$

 d. $H_2CO_3(aq) \rightarrow CO_2(g) + H_2O(\ell)$

 e. $PbCO_3(s) \rightarrow PbO(s) + CO_2(g)$

 f. $Zn(s) + Cl_2(g) \rightarrow ZnCl_2(s)$

5.22 Baking soda ($NaHCO_3$) can serve as an emergency fire extinguisher for grease fires in the kitchen. When heated, it liberates CO_2, which smothers the fire. The equation for the reaction is

$$2NaHCO_3(s) \xrightarrow{\text{Heat}} Na_2CO_3(s) + H_2O(g) + CO_2(g)$$

Classify the reaction into the categories used in Exercises 5.20 and 5.21.

5.23 Baking soda may serve as a source of CO_2 in bread dough. It causes the dough to rise. The CO_2 is released when $NaHCO_3$ reacts with an acidic substance:

$$NaHCO_3(aq) + H^+(aq) \rightarrow Na^+(aq) + H_2O(\ell) + CO_2(g)$$

Classify the reaction as redox or nonredox.

5.24 Many homes are heated by the energy released when natural gas (represented by CH_4) reacts with oxygen. The equation for the reaction is

$$CH_4(g) + 2O_2(g) \rightarrow CO_2(g) + 2H_2O(g)$$

Classify the reaction as redox or nonredox.

5.25 Hydrogen peroxide will react and liberate oxygen gas. In commercial solutions, the reaction is prevented to a large degree by the addition of an inhibitor. The equation for the oxygen-liberating reaction is

$$2H_2O_2(aq) \rightarrow 2H_2O(\ell) + O_2(g)$$

Classify the reaction as redox or nonredox.

5.26 Chlorine, used to treat drinking water, undergoes the reaction in water represented by the following equation:

$$Cl_2(aq) + H_2O(\ell) \rightarrow HOCl(aq) + HCl(aq)$$

Classify the reaction as redox or nonredox.

5.27 Triple superphosphate, an ingredient of some fertilizers, is prepared by reacting rock phosphate (calcium phosphate) and phosphoric acid. The equation for the reaction is

$$Ca_3(PO_4)_2(s) + 4H_3PO_4(aq) \rightarrow 3Ca(H_2PO_4)_2(s)$$

Classify the reaction into the categories used in Exercises 5.20 and 5.21.

Ionic Equations (Section 5.7)

5.28 Consider all of the following ionic compounds to be water soluble, and write the formulas of the ions that would be formed if the compounds were dissolved in water. Table 4.7 will be helpful.

 a. $LiNO_3$

 b. Na_2HPO_4

 c. $Ca(ClO_3)_2$

 d. KOH

 e. $MgBr_2$

 f. $(NH_4)_2SO_4$

5.29 Consider all of the following ionic compounds to be water soluble, and write the formulas of the ions that would be formed if the compounds were dissolved in water. Table 4.7 will be helpful.

 a. Na_2SO_4

 b. $CaCl_2$

 c. $(NH_4)_3PO_4$

 d. $NaOH$

 e. KNO_3

 f. $NaMnO_4$

5.30 Reactions represented by the following equations take place in water solutions. Write each molecular equation in total ionic form, then identify spectator ions and write the equations in net ionic form. Solids that do not dissolve are designated by (s), gases that do not dissolve are designated by (g), and substances that dissolve but do not dissociate appear in blue.

 a. $SO_2(aq) + H_2O(\ell) \rightarrow H_2SO_3(aq)$

 b. $CuSO_4(aq) + Zn(s) \rightarrow Cu(s) + ZnSO_4(aq)$

 c. $2KBr(aq) + 2H_2SO_4(aq) \rightarrow$
 $$Br_2(aq) + SO_2(aq) + K_2SO_4(aq) + 2H_2O(\ell)$$

 d. $AgNO_3(aq) + NaOH(aq) \rightarrow AgOH(s) + NaNO_3(aq)$

 e. $BaCO_3(s) + 2HNO_3(aq) \rightarrow$
 $$Ba(NO_3)_2(aq) + CO_2(g) + H_2O(\ell)$$

 f. $N_2O_5(aq) + H_2O(\ell) \rightarrow 2HNO_3(aq)$

5.31 Reactions represented by the following equations take place in water solutions. Write each molecular equation in total ionic form, then identify spectator ions and write the equations in net ionic form. Solids that do not dissolve are designated by (s), gases that do not dissolve are designated by (g), and substances that dissolve but do not dissociate appear in blue.

 a. $H_2O(\ell) + Na_2SO_3(aq) + SO_2(aq) \rightarrow 2NaHSO_3(aq)$

 b. $3Cu(s) + 8HNO_3(aq) \rightarrow$
 $$3Cu(NO_3)_2(aq) + 2NO(g) + 4H_2O(\ell)$$

 c. $2HCl(aq) + CaO(s) \rightarrow CaCl_2(aq) + H_2O(\ell)$

 d. $CaCO_3(s) + 2HCl(aq) \rightarrow CaCl_2(aq) + CO_2(aq) + H_2O(\ell)$

 e. $MnO_2(s) + 4HCl(aq) \rightarrow MnCl_2(aq) + Cl_2(aq) + 2H_2O(\ell)$

 f. $2AgNO_3(aq) + Cu(s) \rightarrow Cu(NO_3)_2(aq) + 2Ag(s)$

5.32 The following molecular equations all represent neutralization reactions of acids and bases. These reactions will be discussed further in Chapter 9. Write each equation in total ionic form, identify the spectator ions, then write the net ionic equation. Water is the only substance that does not dissociate. What do you notice about all the net ionic equations?

a. $HNO_3(aq) + KOH(aq) \rightarrow KNO_3(aq) + H_2O(\ell)$

b. $H_3PO_4(aq) + 3NH_4OH(aq) \rightarrow (NH_4)_3PO_4(aq) + 3H_2O(\ell)$

c. $HI(aq) + NaOH(aq) \rightarrow NaI(aq) + H_2O(\ell)$

5.33 The following molecular equations all represent neutralization reactions of acids and bases. These reactions will be discussed further in Chapter 9. Write each equation in total ionic form, identify the spectator ions, then write the net ionic equation. Water is the only substance that does not dissociate. What do you notice about all the net ionic equations?

a. $HCl(aq) + LiOH(aq) \rightarrow LiCl(aq) + H_2O(\ell)$

b. $H_3PO_4(aq) + 3KOH(aq) \rightarrow K_3PO_4(aq) + 3H_2O(\ell)$

c. $H_2SO_4(aq) + 2NH_4OH(aq) \rightarrow (NH_4)_2SO_4(aq) + 2H_2O(\ell)$

Energy and Reactions (Section 5.8)

5.34 In addition to emergency cold packs, emergency hot packs are available that heat up when water is mixed with a solid. Is the process that takes place in such packs exothermic or endothermic? Explain.

5.35 In refrigeration systems, the area to be cooled has pipes running through it. Inside the pipes, a liquid evaporates and becomes a gas. Is the evaporation process exothermic or endothermic? Explain.

5.36 An individual wants to keep some food cold in a portable picnic cooler. A piece of ice is put into the cooler, but it is wrapped in a thick insulating blanket to slow its melting. Comment on the effectiveness of the cooler in terms of the direction of heat movement inside the cooler.

5.37 The human body cools itself by the evaporation of perspiration. Is the evaporation process endothermic or exothermic? Explain.

The Mole and Chemical Equations (Section 5.9)

5.38 For the reactions represented by the following equations, write statements equivalent to Statements 2, 3, and 4 given in Section 5.9.

a. $S(s) + O_2(g) \rightarrow SO_2(g)$

b. $Sr(s) + 2H_2O(\ell) \rightarrow Sr(OH)_2(s) + H_2(g)$

c. $2H_2S(g) + 3O_2(g) \rightarrow 2H_2O(g) + 2SO_2(g)$

d. $4NH_3(g) + 5O_2(g) \rightarrow 4NO(g) + 6H_2O(g)$

e. $CaO(s) + 3C(s) \rightarrow CaC_2(s) + CO(g)$

5.39 For the reactions represented by the following equations, write statements equivalent to Statements 2, 3, and 4 given in Section 5.9.

a. $2SO_2(g) + O_2(g) \rightarrow 2SO_3(g)$

b. $4HCl(g) + O_2(g) \rightarrow 2Cl_2(g) + 2H_2O(g)$

c. $Fe_2O_3(s) + 3C(s) \rightarrow 2Fe(s) + 3CO(g)$

d. $2H_2O_2(aq) \rightarrow 2H_2O(\ell) + O_2(g)$

e. $2C_3H_6(g) + 9O_2(g) \rightarrow 6CO_2(g) + 6H_2O(g)$

5.40 For the following equation, write statements equivalent to Statements 2, 3, and 4 given in Section 5.9. Then write at least six factors (including numbers and units) that could be used to solve problems by the factor-unit method.

$$2SO_2(g) + O_2(g) \rightarrow 2SO_3(g)$$

5.41 Calculate the number of grams of SO_2 that must react to produce 350. g of SO_3. Use the statements written in Exercise 5.40 and express your answer using the correct number of significant figures.

5.42 Calculate the mass of limestone ($CaCO_3$) that must be decomposed to produce 500. g of lime (CaO). The equation for the reaction is

$$CaCO_3(s) \rightarrow CaO(s) + CO_2(g)$$

5.43 Calculate the number of moles of CO_2 generated by the reaction of Exercise 5.42 when 500. g of CaO is produced.

5.44 Calculate the number of grams of bromine (Br_2) needed to react exactly with 50.1 g of aluminum (Al). The equation for the reaction is

$$2Al(s) + 3Br_2(\ell) \rightarrow 2AlBr_3(s)$$

5.45 Calculate the number of moles of $AlBr_3$ produced by the process of Exercise 5.44.

5.46 How many grams of $AlBr_3$ are produced by the process in Exercise 5.44?

5.47 In Exercise 5.17 you were given the following equation for the reaction used to clean tarnish from silver:

$$3Ag_2S(s) + 2Al(s) \rightarrow 6Ag(s) + Al_2S_3(s)$$

a. How many grams of aluminum would need to react to remove 0.250 g of Ag_2S tarnish?

b. How many moles of Al_2S_3 would be produced by the reaction described in (a)?

5.48 Pure titanium metal is produced by reacting titanium(IV) chloride with magnesium metal. The equation for the reaction is

$$TiCl_4(s) + 2Mg(s) \rightarrow Ti(s) + 2MgCl_2(s)$$

How many grams of Mg would be needed to produce 1.00 kg of pure titanium?

5.49 An important metabolic process of the body is the oxidation of glucose to water and carbon dioxide. The equation for the reaction is

$$C_6H_{12}O_6(aq) + 6O_2(aq) \rightarrow 6CO_2(aq) + 6H_2O(\ell)$$

a. What mass of water in grams is produced when the body oxidizes 1.00 mol of glucose?

b. How many grams of oxygen are needed to oxidize 1.00 mol of glucose?

5.50 Caproic acid is oxidized in the body as follows:

$$C_6H_{12}O_2(aq) + 8O_2(aq) \rightarrow 6CO_2(aq) + 6H_2O(\ell)$$

How many grams of oxygen are needed to oxidize 1.00 mol of caproic acid?

The Limiting Reactant (Section 5.10)

5.51 A sample of 4.00 g of methane (CH_4) is mixed with 15.0 g of chlorine (Cl_2).

a. Determine which is the limiting reactant according to the following equation:

$$CH_4(g) + 4Cl_2(g) \rightarrow CCl_4(\ell) + 4HCl(g)$$

b. What is the maximum mass of CCl_4 that can be formed?

5.52 Nitrogen and oxygen react as follows:

$$N_2(g) + 2O_2(g) \rightarrow 2NO_2(g)$$

Suppose 1.25 mol of N_2 and 50.0 g of O_2 are mixed together.

a. Which one is the limiting reactant?

b. What is the maximum mass in grams of NO_2 that can be produced from the mixture?

5.53 Suppose you want to use acetylene (C_2H_2) as a fuel. You have a cylinder that contains 5.00×10^2 g of C_2H_2 and a cylinder that contains 2.00×10^3 g of oxygen (O_2). Do you have enough oxygen to burn all the acetylene? The equation for the reaction is

$$2C_2H_2(g) + 5O_2(g) \rightarrow 4CO_2(g) + 2H_2O(g)$$

5.54 Ammonia, carbon dioxide, and water vapor react to form ammonium bicarbonate as follows:

$$NH_3(aq) + CO_2(aq) + H_2O(\ell) \rightarrow NH_4HCO_3(aq)$$

Suppose 50.0 g of NH_3, 80.0 g of CO_2, and 2.00 mol of H_2O are reacted. What is the maximum number of grams of NH_4HCO_3 that can be produced?

5.55 Chromium metal (Cr) can be prepared by reacting the oxide with aluminum. The equation for the reaction is

$$Cr_2O_3(s) + 2Al(s) \rightarrow 2Cr(s) + Al_2O_3(s)$$

What three substances will be found in the final mixture if 150 g of Cr_2O_3 and 150 g of Al are reacted?

Reaction Yields (Section 5.11)

5.56 The actual yield of a reaction was 11.74 g of product, while the calculated theoretical yield was 16.09 g. What was the percentage yield?

5.57 A product weighing 14.37 g was isolated from a reaction. The amount of product possible according to a calculation was 17.55 g. What was the percentage yield?

5.58 For a combination reaction, it was calculated that 7.59 g of A would exactly react with 4.88 g of B. These amounts were reacted, and 9.04 g of product was isolated. What was the percentage yield of the reaction?

5.59 A sample of calcium metal with a mass of 2.00 g was reacted with excess oxygen. The following equation represents the reaction that took place:

$$2Ca(s) + O_2(g) \rightarrow 2CaO(s)$$

The isolated product (CaO) weighed 2.26 g. What was the percentage yield of the reaction?

5.60 Upon heating, mercury(II) oxide undergoes a decomposition reaction:

$$2HgO(s) \rightarrow 2Hg(\ell) + O_2(g)$$

A sample of HgO weighing 7.22 g was heated. The collected mercury weighed 5.95 g. What was the percentage yield of the reaction?

Additional Exercises

5.61 Would you expect argon, Ar, to be involved in a redox reaction? Explain your reasoning.

5.62 Rewrite the following word equation using chemical formulas. Balance the equation and write the balanced equation in net ionic form:

barium chloride (aq) + sodium sulfate (aq) →
sodium chloride (aq) + barium sulfate (s)

5.63 The element with an electron configuration of $1s^22s^22p^63s^1$ undergoes a combination reaction with the element that has 15 protons in its nucleus. Assume the product of the reaction consists of simple ions, and write a balanced chemical equation for the reaction.

5.64 Assuming a 100% reaction yield, it was calculated that 6.983 g of naturally occurring elemental iron would be needed to react with another element to form 18.149 g of product. If iron composed only of the iron-60 isotope was used to form the product, how many grams of iron-60 would be required?

5.65 The decomposition of a sample of a compound produced 1.20×10^{24} atoms of nitrogen and 80.0 grams of oxygen atoms. What was the formula of the sample that was decomposed? What is the correct name of the decomposed sample?

Chemistry for Thought

5.66 In experiments where students prepare compounds by precipitation from water solutions, they often report yields of dry product greater than 100%. Propose an explanation for this.

5.67 Refer to Figure 5.2 and follow the instructions. Then explain how the concept of limiting reactant is used to extinguish a fire.

5.68 Refer to Figure 5.4 and answer the question. Suggest another material that might provide the enzyme catalyst.

5.69 What do the observations of Figure 5.10 indicate to you about the abilities of the following solids to dissolve in water: NaCl, $AgNO_3$, and AgCl? What would you expect to observe if you put a few grams of solid AgCl into a test tube containing 3 mL of water and shook the mixture? What would you expect to observe if you repeated this experiment but used a few grams of $NaNO_3$ instead of AgCl?

5.70 The concentration of alcohol (CH_3CH_2OH) in the breath of an individual who has been drinking is measured by an instrument called a Breathalyzer. The breath sample is passed through a solution that contains the orange-colored $Cr_2O_7^{2-}$ ion. The equation for the net ionic redox reaction that occurs is:

$$2Cr_2O_7^{2-}(aq) + 3CH_3CH_2OH(aq) + 16H^+(aq) \rightarrow$$
$$4Cr^{3+}(aq) + 3CH_3COOH(aq) + 11H_2O(\ell)$$

The Cr^{3+} ion has a pale violet color in solution. Explain how color changes in the solution could be used to indicate the amount of alcohol in the breath sample.

5.71 Certain vegetables and fruits, such as potatoes and apples, darken quickly when sliced. Submerging the slices in water slows this process. Explain.

5.72 In an ordinary flashlight battery, an oxidation reaction and a reduction reaction take place at different locations to produce an electrical current that consists of electrons. In one of the reactions, the zinc container of the battery slowly dissolves as it is converted into zinc ions. Is this the oxidation or the reduction reaction? Is this reaction the source of electrons, or are electrons used to carry out the reaction? Explain your answers with a reaction equation.

Allied Health Exam Connection

The following questions are from these sources:

- *Nursing School Entrance Exam* © 2005, Learning Express, LLC.

- *McGraw-Hill's Nursing School Entrance Exams* by Thomas A. Evangelist, Tamara B. Orr, and Judy Unrein © 2009, The McGraw-Hill Companies, Inc.

- *NSEE Nursing School Entrance Exams,* 3rd edition © 2009, Kaplan Publishing.

- *Cliffs Test Prep: Nursing School Entrance Exams* by Fred N. Grayson © 2004, Wiley Publishing, Inc.

- *Peterson's Master the Nursing School and Allied Health Entrance Exams,* 18th edition by Marion F. Gooding © 2008, Peterson's, a Nelnet Company.

5.73 Balance the following redox reaction:

$Mg(s) + H_2O(g) \rightarrow Mg(OH)_2(s) + H_2(g)$

- **a.** $Mg(s) + H_2O(g) \rightarrow Mg(OH)_2(s) + H_2(g)$
- **b.** $Mg(s) + 4H_2O(g) \rightarrow Mg(OH)_2(s) + H_2(g)$
- **c.** $Mg(s) + 2H_2O(g) \rightarrow Mg(OH)_2(s) + H_2(g)$
- **d.** $Mg(s) + H_2O(g) \rightarrow Mg(OH)_2(s) + \frac{1}{2}H_2(g)$

5.74 Which one of the following equations is balanced?

- **a.** $H_2O \rightarrow H_2 + O_2$
- **b.** $Al + H_2SO_4 \rightarrow Al_2(SO_4)_3 + H_2$
- **c.** $S + O_2 \rightarrow SO_3$
- **d.** $2HgO \rightarrow 2Hg + O_2$

5.75 Which of the following equations is balanced?

- **a.** $Mg + N_2 \rightarrow Mg_3N_2$
- **b.** $Fe + O_2 \rightarrow Fe_2O_3$
- **c.** $C_{12}H_{22}O_{11} \rightarrow 12C + 11H_2O$
- **d.** $Ca + H_2O \rightarrow Ca(OH)_2 + H_2$

5.76 The coefficient of O_2 after the following equation is balanced is: $_CH_4 + _O_2 \rightarrow _CO_2 + _H_2O$

- **a.** 1
- **b.** 2
- **c.** 3
- **d.** 4

5.77 Ammonia can be produced from nitrogen and hydrogen, according to the unbalanced equation

$$N_2(g) + H_2(g) = NH_3(g)$$

After balancing the equation, the coefficient before ammonia should be:

- **a.** 1
- **b.** 2
- **c.** 3
- **d.** 4

5.78 What is the oxidation number of sodium in the following reaction? $Pb(NO_3)_2(aq) + 2NaI(aq) \rightarrow PbI_2(s) + 2NaNO_3(aq)$

- **a.** $+1$
- **b.** $+2$
- **c.** -1
- **d.** -2

5.79 What is the oxidation number for nitrogen in HNO_3?

- **a.** -2
- **b.** $+5$
- **c.** -1
- **d.** -5

5.80 The oxidation number of sulfur in the ion SO_4^{-2} is:

- **a.** -2
- **b.** $+2$
- **c.** $+6$
- **d.** $+10$

5.81 Which of the following is the oxidation number of sulfur in the compound sodium thiosulfate, $Na_2S_2O_3$?

- **a.** $+1$
- **b.** -1
- **c.** $+2$
- **d.** -2

5.82 Which best describes the following redox reaction:

$10Br^-(aq) + 2MnO_4^-(aq) + 16H^+(aq) \rightarrow$
$$5Br_2(\ell) + 2Mn^{2+}(aq) + 8H_2O(\ell)$$

- **a.** Br and Mn are both reduced
- **b.** Br is oxidized and Mn is reduced
- **c.** Br is oxidized and O is reduced
- **d.** Br is reduced and Mn is oxidized

5.83 Which reactant is oxidized and which is reduced in the following reaction?

$C_2H_4(g) + 3O_2(g) \rightarrow 2CO_2(g) + 2H_2O(g)$

- **a.** oxidized: $C_2H_4(g)$, reduced: $3O_2(g)$
- **b.** oxidized: $C_2H_4(g)$, reduced: $2H_2O(g)$
- **c.** oxidized: $C_2H_4(g)$, reduced: $2CO_2(g)$
- **d.** oxidized: $2CO_2(g)$, reduced: $C_2H_4(g)$

5.84 Which of the following species is being oxidized in this redox reaction?

$Zn(s) + Cu^{2+}(aq) \rightarrow Zn^{2+}(aq) + Cu(s)$

- **a.** $Zn(s)$
- **b.** $Cu^{2+}(aq)$
- **c.** $Zn^{2+}(aq)$
- **d.** $Cu(s)$

5.85 Identify the oxidizing agent and the reducing agent in the following reaction:

$8H^+(aq) + 6Cl^-(aq) + Sn(s) + 4NO_3^-(aq) \rightarrow$
$SnCl_6^{2-}(aq) + 4NO_2(g) + 4H_2O(\ell)$

a. oxidizing agent: $8H^+(aq)$, reducing agent: $Sn(s)$

b. oxidizing agent: $4NO_3^-(aq)$, reducing agent: $Sn(s)$

c. oxidizing agent: $4NO_3^-(aq)$, reducing agent: $4NO_2(g)$

d. oxidizing agent: $4NO_3^-(aq)$, reducing agent: $8H^+(aq)$

5.86 The chemical reaction: $2Zn + 2HCl \rightarrow 2ZnCl + H_2$ is an example of:

a. double displacement.

b. synthesis.

c. analysis.

d. single displacement.

5.87 Identify the following as an oxidation, a reduction, a decomposition, or a dismutation reaction:
$Cl_2 + 2e^- \rightarrow 2Cl^-$

a. a reduction

b. an oxidation

c. a decomposition

d. a dismutation

5.88 The reaction $2NaI + Cl_2 \rightarrow 2NaCl + I_2$ demonstrates a:

a. decomposition reaction.

b. synthesis reaction.

c. single-replacement reaction.

d. double-replacement reaction.

5.89 Classify the reaction as to the reaction type:
$Mg(OH)_2 + 2HCl \rightarrow MgCl_2 + 2H_2O$

a. synthesis

b. decomposition

c. single replacement

d. double replacement

5.90 Identify the statement which is NOT characteristic of exergonic reactions:

a. They are downhill reactions.

b. They have a negative energy change ($-H$).

c. They are uphill reactions.

d. The products have less energy than the reactants.

5.91 What is the net ionic equation of the following transformations?
$Ca(NO_3)_2(aq) + 2KCl(aq) \rightarrow CaCl_2(s) + 2KNO_3(aq)$

a. $2NO_3^-(aq) + 2K^+(aq) \rightarrow 2KNO_3(aq)$

b. $Ca^{2+}(aq) + 2Cl^-(aq) \rightarrow CaCl_2(s)$

c. $Ca^{2+}(aq) + 2NO_3^-(aq) + 2K^+(aq) + 2Cl^-(aq) \rightarrow$
$CaCl_2(s) + 2K^+(aq) + 2NO_3^-(aq)$

d. $Ca^{2+}(aq) + 2NO_3^-(aq) + 2K^+(aq) + 2Cl^-(aq) \rightarrow$
$Ca^{2+}(aq) + 2Cl^-(aq) + 2K^+(aq) + 2NO_3^-(aq)$

5.92 In the Haber process, ammonia is produced according to the following equation:

$$N_2(g) + 3H_2(g) = 2NH_3(g)$$

How many moles of hydrogen gas are needed to react with one of nitrogen?

a. 1

b. 3

c. 6

d. 22.4

5.93 What mass of product would you expect given that you started with 17.0 g of NH_3 and 36.5 g of HCl?

$$NH_3 + HCl \rightarrow NH_4Cl$$

a. 17.0 g

b. 36.5 g

c. 53.5 g

d. 19.5 g

5.94 The number of grams of hydrogen formed by the action of 6 grams of magnesium (atomic weight = 24) on an appropriate quantity of acid is:

a. 0.5

b. 8

c. 22.4

d. 72

5.95 In the reaction $CaCl_2 + Na_2CO_3 \rightarrow CaCO_3 + 2NaCl$, if 0.5 mole of NaCl is to be formed, then:

a. 1 mole of Na_2CO_3 is needed.

b. 0.5 mole of $CaCO_3$ is also formed.

c. 0.5 mole of Na_2CO_3 is needed.

d. 0.25 mole of $CaCl_2$ is needed.

5.96 In the reaction $4Al + 3O_2 \rightarrow 2Al_2O_3$, how many grams of O_2 are needed to completely react with 1.5 moles of Al?

a. 24 g

b. 36 g

c. 48 g

d. 60 g

6

The States of Matter

▶ Case Study

The combined sound of a seal barking and a child crying was coming from 3-year-old Adrian's room at 2:00 a.m. His father, Kyle, turned on the light to investigate. Adrian was blue around the lips, and when he inhaled his stomach looked as if a vacuum was pulling it in. When Kyle took his son to the car, he seemed to improve in the cold winter air. At the ER, Adrian was diagnosed with a classic case of croup. The doctor called for a respiratory therapist to administer a treatment of an aerosol containing racemic epinephrine and a shot of corticosteroids. After two hours in the ER, the respiratory therapist recommended oxygen and humidity treatment administered through a tent. Adrian was admitted to Pediatrics for treatment. Kyle asked if he could stay with his son, and, to his relief, the hospital encouraged parental accompaniment.

▶ **How important is it to allow parents to stay with hospitalized youngsters? What is an acceptable O$_2$ saturation level for a 3-year-old? What are the signs that a cough is a medical emergency?**

Follow-up to this Case Study appears at the end of the chapter before the *Concept Summary*.

Learning Objectives

When you have completed your study of this chapter, you should be able to:

1 Do calculations based on the property of density. **(Section 6.1)**

2 Demonstrate an understanding of the kinetic molecular theory of matter. **(Section 6.2)**

3 Use the kinetic molecular theory to explain and compare the properties of matter in different states. **(Sections 6.3–6.5)**

4 Do calculations to convert pressure and temperature values into various units. **(Section 6.6)**

5 Do calculations based on Boyle's law, Charles's law, and the combined gas law. **(Section 6.7)**

6 Do calculations based on the ideal gas law. **(Section 6.8)**

7 Do calculations based on Dalton's law. **(Section 6.9)**

8 Do calculations based on Graham's law. **(Section 6.10)**

9 Classify changes of state as exothermic or endothermic. **(Section 6.11)**

10 Demonstrate an understanding of the concepts of vapor pressure and evaporation. **(Section 6.12)**

11 Demonstrate an understanding of the process of boiling and the concept of boiling point. **(Section 6.13)**

12 Demonstrate an understanding of the processes of sublimation and melting. **(Section 6.14)**

13 Do calculations based on energy changes that accompany heating, cooling, or changing the state of a substance. **(Section 6.15)**

I f you live in an area that has cold winters, you have probably seen water in the three different forms used to categorize the states in which matter occurs. On a cold day, you can usually find solid water (ice) floating in a pool of cold liquid water, and at the same time you can see a small cloud of tiny water droplets that forms when gaseous water condenses as you exhale into the cold air.

Most matter is not as easily observed in all three states as the water in the preceding example. In fact, most matter is classified as a solid, a liquid, or a gas on the basis of the form in which it is commonly observed. However, according to Section 4.11, the state of a substance depends on temperature. You will see in this chapter that the state also depends on pressure. Therefore, when a substance is classified as a solid, a liquid, or a gas, we are usually simply stating its form under normal atmospheric pressure and at a temperature near 25°C. **Figure 6.1** gives the states of the elements under those conditions.

In this chapter, we will study the characteristics of the common states of matter, along with a theory that relates these states to molecular behavior. We will also investigate the energy relationships that accompany changes of state.

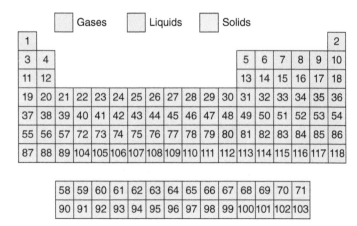

Figure 6.1 The common states of the elements at normal atmospheric pressure and 25°C.

6.1 | Observed Properties of Matter

Learning Objective

1. Do calculations based on the property of density.

Solids, liquids, and gases can easily be recognized and distinguished by differences in physical properties. Four properties that can be used are density, shape, compressibility, and thermal expansion.

Density was defined in Section 1.11 as the mass of a sample of matter divided by the volume of the same sample. A failure to remember this is the basis for the wrong answers given to the children's riddle, "Which is heavier, a pound of lead or a pound of feathers?" Of course, they weigh the same, but many people say lead is heavier because they think in terms of samples having the same volume. Of course, if you have lead and feather samples of the same volume, the lead sample will be much heavier because its density is much greater.

shape Shape depends on the physical state of matter.

The **shape** of matter is sometimes independent of a container (for solids), or it may be related to the shape of the container (for liquids and gases). In the case of liquids, the shape depends on the extent to which the container is filled. Gases always fill the container completely (see **Figure 6.2**).

1 Solids have a shape and volume that does not depend on the container.

2 Liquids take the shape of the part of the container they fill. Each sample above has the same volume.

3 Gases completely fill and take the shape of their container. When the valve separating the two parts of the container is opened, the gas fills the entire container volume (bottom photo).

Figure 6.2 Characteristics of the states of matter.

compressibility The change in volume of a sample resulting from a pressure change acting on the sample.

Compressibility, the change in volume resulting from a pressure change, is quite high for gases. Compressibility allows a lot of gas to be squeezed into a small volume if the gas is put under sufficient pressure—think of automobile tires, or a tank of compressed helium used to fill toy balloons.

thermal expansion The change in volume of a sample resulting from a change in temperature of the sample.

Thermal expansion, the change in volume resulting from temperature changes, is a property used in liquid-in-glass thermometers. As the temperature of the liquid increases, the liquid expands and fills more of the fine capillary tube on which the temperature scale is engraved. You see the liquid move up the tube and know the temperature is increasing.

These four properties are compared for the three states of matter in **Table 6.1**.

TABLE 6.1 Physical Properties of Solids, Liquids, and Gases

Property	State		
	Solid	Liquid	Gas
Density	High	High—usually lower than that of corresponding solid	Low
Shape	Definite	Indefinite—takes shape of container to the extent it is filled	Indefinite—takes shape of container it fills
Compressibility	Small	Small—usually greater than that of corresponding solid	Large
Thermal expansion	Very small	Small	Moderate

Example 6.1 Calculating Densities

Samples of plumber's solder, rubbing alcohol, and air are collected. The volume and mass of each sample are determined at 20°C as follows: solder—volume = 28.6 mL, mass = 268.8 g; alcohol—volume = 100.0 mL, mass = 78.5 g; air—volume = 500.0 mL, mass = 0.602 g. Calculate the density of each substance in grams per milliliter.

Solution

The density (d) is obtained by dividing the sample mass (in grams) by the sample volume (in milliliters):

solder: $\quad d = \dfrac{268.8}{28.6 \text{ mL}} = 9.40 \text{ g/mL}$

alcohol: $\quad d = \dfrac{78.5 \text{ g}}{100.0 \text{ mL}} = 0.785 \text{ g/mL}$

air: $\quad d = \dfrac{0.602 \text{ g}}{500.0 \text{ mL}} = 0.00120 \text{ g/mL}$

The calculated densities follow the pattern given in Table 6.1.

✔ **LEARNING CHECK 6.1** Samples of copper, glycerin, and helium are collected and weighed at 20°C. The results are as follows: copper—volume = 12.8 mL, mass = 114.2 g; glycerin—volume = 50.0 mL, mass = 63.0 g; helium—volume = 1500 mL, mass = 0.286 g.

a. Calculate the density of each sample at 20°C.
b. Describe qualitatively (increase, decrease, little change, no change, etc.) what happens to the following properties for each sample when the pressure on the sample is doubled: sample volume, sample mass, sample density (see Table 6.1).

The density calculations done in Example 6.1 used Equation 1.9, which is repeated here:

$$d = \frac{m}{V}$$

Remember, in this equation d is density, m is mass, and V is volume. This useful equation can be rearranged so that any one quantity in it can be calculated if the other two are known.

Example 6.2 Calculating Volumes Using Densities

The sample of rubbing alcohol used in Example 6.1 is heated to 50°C. The density of the sample is found to decrease to a value of 0.762 g/mL. What is the volume of the sample at 50°C?

Solution

It is assumed that no alcohol is lost by evaporation, so the sample mass has the same value of 78.5 g it had in Example 6.1. Equation 1.9 is rearranged and solved for volume:

$$V = \frac{m}{d} \qquad (6.1)$$

The new volume is

$$V = \frac{m}{d} = \frac{78.5 \text{ g}}{0.762 \text{ g/mL}} = 103 \text{ mL}$$

Thus, the heating causes the liquid to expand from a volume of 100 mL to 103 mL.

✔ **LEARNING CHECK 6.2** Calculate the mass in grams of a 1200-mL air sample at a temperature where the air has a density of 1.18×10^{-3} g/mL. *Note:* Equation 6.1 can be rearranged to $m = d \times V$.

6.2 | The Kinetic Molecular Theory of Matter

Learning Objective

2. Demonstrate an understanding of the kinetic molecular theory of matter.

The long title of this section is the name scientists have given to a model or theory used to explain the behavior of matter in its various states. Some theories, including this one, are made up of a group of generalizations or postulates. This useful practice makes it possible to study and understand each postulate individually, instead of the entire theory.

The postulates of the kinetic molecular theory are

1. Matter is composed of tiny particles called molecules.
2. The particles are in constant motion and therefore possess kinetic energy.
3. The particles possess potential energy as a result of attracting or repelling each other.
4. The average particle speed increases as the temperature increases.
5. The particles transfer energy from one to another during collisions in which no net energy is lost from the system.

These postulates contain two new terms, kinetic energy and potential energy. **Kinetic energy** is the energy a particle has as a result of its motion. Mathematically, kinetic energy is calculated as

$$KE = \frac{1}{2}mv^2 \qquad (6.2)$$

where m is the particle mass and v is its velocity. Thus, if two particles of different mass are moving at the same velocity, the heavier particle will possess more kinetic energy than the other particle. Similarly, the faster moving of two particles with equal masses will have more kinetic energy.

kinetic energy The energy a particle has as a result of its motion. Mathematically, it is $KE = \frac{1}{2}mv^2$.

Example 6.3 Calculating Kinetic Energies

Calculate the kinetic energy of two particles with masses of 2.00 g and 3.00 g if they are both moving with a velocity of 15.0 cm/s.

Solution

The kinetic energy of the 2.00-g particle is

$$\text{KE} = \frac{1}{2}(2.00 \text{ g})\left(15.0 \frac{\text{cm}}{\text{s}}\right)^2 = 225 \frac{\text{g cm}^2}{\text{s}^2}$$

The kinetic energy of the 3.00-g particle is

$$\text{KE} = \frac{1}{2}(3.00 \text{ g})\left(15.0 \frac{\text{cm}}{\text{s}}\right)^2 = 337.5 \frac{\text{g cm}^2}{\text{s}^2}$$

Rounding gives 338 g cm^2/s^2; thus, the more massive particle has more kinetic energy.

✔ **LEARNING CHECK 6.3** Calculate the kinetic energy of two 3.00-g particles if one has a velocity of 10.0 cm/s and the other has a velocity of 20.0 cm/s.

Potential energy results from attractions or repulsions of particles. A number of these interactions are familiar, such as the gravitational attraction of Earth and the behavior of the poles of two magnets brought near each other. In each of these examples, the size of the force and the potential energy depend on the separation distance. The same behavior is found for the potential energy of atomic-sized particles. The potential energy of attraction increases as separation distance increases, whereas the potential energy of repulsion decreases with increasing separation.

The kinetic molecular theory provides reasonable explanations for many of the observed properties of matter. An important factor in these explanations is the relative influence of cohesive forces and disruptive forces. **Cohesive forces** are the attractive forces associated with potential energy, and **disruptive forces** result from particle motion (kinetic energy). Disruptive forces tend to scatter particles and make them independent of each other; cohesive forces have the opposite effect. Thus, the state of a substance depends on the relative strengths of the cohesive forces that hold the particles together and the disruptive forces tending to separate them. Cohesive forces are essentially temperature-independent because they involve interparticle attractions of the type described in Chapter 4. Disruptive forces increase with temperature because they arise from particle motion, which increases with temperature (Postulate 4). This explains why temperature plays such an important role in determining the state in which matter is found.

potential energy The energy a particle has as a result of attractive or repulsive forces acting on it.

cohesive forces The attractive force between particles; they are associated with potential energy.

disruptive forces The force resulting from particle motion; they are associated with kinetic energy.

6.3 | The Solid State

Learning Objective

3. Use the kinetic molecular theory to explain and compare the properties of matter in different states.

In the solid state, the cohesive forces are stronger than the disruptive forces (see **Figure 6.3A**). Each particle of a crystalline solid occupies a fixed position in the crystal lattice (see **Section 4.11**). Disruptive kinetic energy causes the particles to vibrate about their fixed positions, but the strong cohesive forces prevent the lattice from breaking down. The properties of solids in Table 6.1 are explained by the kinetic theory as follows:

High density. The particles of solids are located as closely together as possible. Therefore, large numbers of particles are contained in a small volume, resulting in a high density.
Definite shape. The strong cohesive forces hold the particles of solids in essentially fixed positions, resulting in a definite shape.

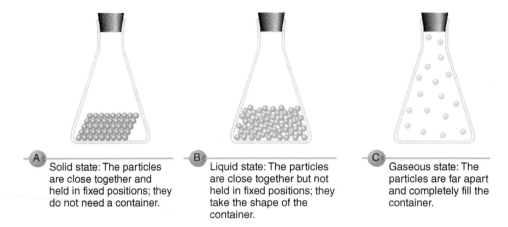

Figure 6.3 A kinetic molecular view of solids, liquids, and gases.

A. Solid state: The particles are close together and held in fixed positions; they do not need a container.

B. Liquid state: The particles are close together but not held in fixed positions; they take the shape of the container.

C. Gaseous state: The particles are far apart and completely fill the container.

Small compressibility. Because there is very little space between particles of solids, increased pressure cannot push them closer together, and it will have little effect on the volume.

Very small thermal expansion. Increased temperature increases the vibrational motion of the particles and the disruptive forces acting on them. Each particle vibrates with an increased amplitude and "occupies" a slightly larger volume. However, there is only a slight expansion of the solid because the strong cohesive forces prevent this effect from becoming very large.

6.4 | The Liquid State

Learning Objective

3. Use the kinetic molecular theory to explain and compare the properties of matter in different states.

Particles in the liquid state are randomly packed and relatively close to each other (see **Figure 6.3B**). They are in constant, random motion, sliding freely over one another, but without sufficient kinetic energy to separate completely from each other. The liquid state is a situation in which cohesive forces dominate slightly. The characteristic properties of liquids are also explained by the kinetic theory:

High density. The particles of liquids are not widely separated; they essentially touch each other. There will, therefore, be a large number of particles per unit volume and a high density.

Indefinite shape. Although not completely independent of each other, the particles in a liquid are free to move over and around each other in a random manner, limited only by the container walls and the extent to which the container is filled.

Small compressibility. Because the particles in a liquid essentially touch each other, there is very little space between them. Therefore, increased pressure cannot squeeze the particles much more closely together.

Small thermal expansion. Most of the particle movement in a liquid is vibrational because the particles can move only a short distance before colliding with a neighbor. Therefore, the increased particle velocity that accompanies a temperature increase results only in increased vibration. The net effect is that the particles push away from each other a little more, thereby causing a slight volume increase in the liquid.

6.5 | The Gaseous State

Learning Objective

3. Use the kinetic molecular theory to explain and compare the properties of matter in different states.

Disruptive forces completely overcome cohesive forces between particles in the gaseous or vapor state. As a result, the particles of a gas move essentially independently of one another in a totally random way (see **Figure 6.3C**). Under ordinary pressure, the particles

are relatively far apart except when they collide with each other. Between collisions with each other or with the container walls, gas particles travel in straight lines. The particle velocities and resultant collision frequencies are quite high for gases, as shown in **Table 6.2**.

TABLE 6.2 **Some Numerical Data Related to the Gaseous State**			
Gas	Average Speed at 0°C	Average Distance Traveled between Collisions at 1 atm of Pressure	Number of Collisions of 1 Molecule in 1 s at 1 atm of Pressure
Hydrogen (H_2)	169,000 cm/s (3700 mi/h)	1.12×10^{-5} cm	1.60×10^{10}
Nitrogen (N_2)	45,400 cm/s (1015 mi/h)	0.60×10^{-5} cm	0.80×10^{10}
Carbon dioxide (CO_2)	36,300 cm/s (811 mi/h)	0.40×10^{-5} cm	0.95×10^{10}

The kinetic theory explanation of gaseous-state properties follows the same pattern seen earlier for solids and liquids:

Low density. The particles of a gas are widely separated. There are relatively few of them in a given volume, which means there is little mass per unit volume.

Indefinite shape. The forces of attraction between particles have been overcome by kinetic energy, and the particles are free to travel in all directions. The particles, therefore, completely fill the container and assume its inner shape.

Large compressibility. The gas particles are widely separated, so that a gas sample is mostly empty space. When pressure is applied, the particles are easily pushed closer together, decreasing the amount of empty space and the gas volume, as shown by **Figure 6.4**.

Moderate thermal expansion. Gas particles move in straight lines except when they collide with each other or with container walls. An increase in temperature causes the particles to collide with more energy. Thus, they push each other away more strongly, and at constant pressure this causes the gas itself to occupy a significantly larger volume.

It must be understood that the size of the particles is not changed during expansion or compression of gases, liquids, or solids. The particles are merely moving farther apart or closer together, and the space between them is changed.

Gas at low pressure | Gas at higher pressure

Figure 6.4 The compression of a gas.

6.6 | **The Gas Laws**

Learning Objective

4. Do calculations to convert pressure and temperature values into various units.

The states of matter have not yet been discussed in quantitative terms. We have pointed out that solids, liquids, and gases expand when heated, but the amounts by which they expand have not been calculated. Such calculations for liquids and solids are beyond the intended scope of this text, but gases obey relatively simple quantitative relationships that were discovered during the 17th, 18th, and 19th centuries. These relationships, called **gas laws**, describe in mathematical terms the behavior of gases as they are mixed, subjected to pressure or temperature changes, or allowed to diffuse.

To use the gas laws, you must clearly understand the units used to express pressure, and you must utilize the Kelvin temperature scale introduced in Section 1.6. **Pressure** is defined as force per unit area. However, most of the units commonly used to express pressure reflect a relationship to barometric measurements of atmospheric pressure. The mercury barometer was invented by Italian physicist Evangelista Torricelli (1608–1647). Its essential components are shown in **Figure 6.5**. A glass tube, sealed at one end, is filled with mercury, stoppered, and inverted so the stoppered end is under

gas laws Mathematical relationship that describe the behavior of gases as they are mixed, subjected to pressure or temperature changes, or allowed to diffuse.

pressure A force per unit area of surface on which the force acts. In measurements and calculations involving gases, it is often expressed in units related to measurements of atmospheric pressure.

the surface of a pool of mercury. When the stopper is removed, the mercury in the tube falls until its weight is just balanced by the weight of air pressing on the mercury pool. The pressure of the atmosphere is then expressed in terms of the height of the supported mercury column.

Figure 6.5 Setting up a simple mercury barometer. Standard atmospheric pressure is 760 mm of mercury.

Mercury

Mercury cup

760 mm

Despite attempts to standardize measurement units, you will probably encounter a number of different units of pressure in your future studies and employment. Some of the more common are standard atmosphere (atm), torr, millimeters of mercury (mmHg), inches of mercury (in. Hg), pounds per square inch (psi), bar, and kilopascals (kPa). One **standard atmosphere** is the pressure needed to support a 760-mm column of mercury in a barometer tube, and 1 **torr** (named in honor of Torricelli) is the pressure needed to support a 1-mm column of mercury in a barometer tube. The relationships of the various units to the standard atmosphere are given in **Table 6.3**. Note that the values of 1 atm and 760 torr (or 760 mmHg) are exact numbers based on definitions and do not limit the number of significant figures in calculated numbers.

standard atmosphere The pressure needed to support a 760-mm column of mercury in a barometer tube.

torr The pressure needed to support a 1-mm column of mercury in a barometer tube.

TABLE 6.3 Units of Pressure

Unit	Relationship to One Standard Atmosphere	Typical Application
Atmosphere	—	Gas laws
Torr	760 torr = 1 atm	Gas laws
Millimeters of mercury	760 mmHg = 1 atm	Gas laws
Pounds per square inch	14.7 psi = 1 atm	Compressed gases
Bar	1.01 bar = 29.9 in. Hg = 1 atm	Meteorology
Kilopascal	101 kPa = 1 atm	Gas laws

Example 6.4 Units of Gas Pressure

The gauge on a cylinder of compressed oxygen gas reads 1500 psi. Express this pressure in terms of (a) atm, (b) torr, and (c) mmHg.

Solution

We will use the factor-unit method of calculation from Section 1.9, and the necessary factors from Table 6.3.

a. It is seen from Table 6.3 that 14.7 psi = 1 atm. Therefore, the known quantity, 1500 psi, is multiplied by the factor atm/psi in order to generate units of atm. The result is

$$1500 \text{ psi} \left(\frac{1 \text{ atm}}{14.7 \text{ psi}} \right) = 1.0 \times 10^2 \text{ atm}$$

b. Table 6.3 does not give a direct relationship between psi and torr. However, both psi and torr are related to atm. Therefore, the atm is used somewhat like a "bridge" between psi and torr:

$$(1500 \text{ psi}) \left(\frac{1 \text{ atm}}{14.7 \text{ psi}} \right) \left(\frac{760 \text{ torr}}{1 \text{ atm}} \right) = 77,551 \text{ torr} = 7.8 \times 10^4 \text{ torr}$$

Note how the "bridge" term canceled out.

c. Because the torr and mmHg are identical, the problem is worked as it was in part b:

$$(1500 \text{ psi}) \left(\frac{1 \text{ atm}}{14.7 \text{ psi}} \right) \left(\frac{760 \text{ mmHg}}{1 \text{ atm}} \right) = 7.8 \times 10^4 \text{ mmHg}$$

✔ **LEARNING CHECK 6.4** A barometer has a pressure reading of 670 torr. Convert this reading into (a) atm and (b) psi.

According to the kinetic molecular theory, a gas expands when it is heated at constant pressure, because the gaseous particles move faster at the higher temperature. It makes no difference to the particles what temperature scale is used to describe the heating. However, gas law calculations are based on the Kelvin temperature scale (**Section 1.6**) rather than the Celsius or Fahrenheit scales. The only apparent difference between the Kelvin and Celsius scales is the location of the zero reading (see **Figure 1.11**). It should be noted that a 0 reading on the Kelvin scale has a great deal of significance. It is the lowest possible temperature and is called **absolute zero**. It represents the temperature at which particles have no kinetic energy because all motion stops. The Kelvin and Celsius temperature scales have the same size degree, but the 0 of the Kelvin scale is 273 degrees below the freezing point of water, which is the 0 of the Celsius scale. Thus, a Celsius reading can be converted to a Kelvin reading simply by adding 273.

absolute zero The temperature at which all motion stops; a value of 0 on the Kelvin scale.

Example 6.5 Converting Temperature Values

Refer to Figure 1.11 for the necessary conversion factors, and make the following temperature conversions (remember, Kelvin temperatures are expressed in kelvins, K):

a. 37°C (body temperature) to K b. −50.°C to K c. 400. K to °C

Solution

a. Celsius is converted to Kelvin by adding 273, K = °C + 273. Therefore,

$$K = 37°C + 273 = 310. \text{ K (body temperature)}$$

b. Again, 273 is added to the Celsius reading. However, in this case the Celsius reading is negative, so the addition must be done with the signs in mind. Therefore,

$$K = −50.°C + 273 = 223 \text{ K}$$

Get an Accurate Blood Pressure Reading

The measurement of blood pressure seems to be a very straightforward process. Stick your arm in a cuff for a few seconds, and read the two values that appear. However, blood pressure is a moving target with a value that can vary from hour to hour or even minute to minute depending on a number of variables. Getting an accurate reading requires some knowledge of what these variables are.

Stress, body posture (sitting or standing), eating a meal, exciting or soothing music or video, and even the time of day can influence blood pressure. Because it changes so much, you are more likely to obtain a "normal" reading if you check blood pressure at home rather than have it taken in a doctor's office. The American Heart Association recommends this practice for individuals with high blood pressure or the risk of developing it.

To get an accurate blood pressure reading, you should become educated about factors that influence it. The following are six tips for getting the most accurate readings:

1. **Take your blood pressure readings at the same time each day.** Levels are usually lowest in the morning and increase through the day.
2. **Go to the bathroom.** A full bladder can raise systolic pressure by 10–15 mmHg. Diuretics (a common blood pressure medication) lower blood pressure by causing the kidneys to remove more sodium and water from the body: this helps to relax the blood vessel walls.
3. **Rest for 5 minutes before taking blood pressure readings.** Exertion can raise your blood pressure.

4. **Avoid exercise, tobacco, and caffeine for a least an hour before taking a reading.** All of these can make blood pressure spike temporarily.
5. **Use proper technique.** Feet should be flat on the floor, legs uncrossed, and cuff at heart level. These factors can each influence the systolic pressure by 2–10 mmHg.
6. **Put the cuff on bare skin.** Putting the cuff over clothes can increase the systolic pressure by 10–15 mmHg.

Know your numbers. Ask your doctor what range is ideal for you, and what reading should prompt an immediate call to the medical office. Contact your doctor if you have any unusual or persistent increases in your pressure.

A sphygmomanometer reports blood pressure as a ratio such as 100/80 (the higher or systolic blood pressure is shown here).

c. Because K = °C + 273, an algebraic rearrangement shows that °C = K − 273. Therefore,

$$°C = 400. K − 273 = 127°C$$

✔ **LEARNING CHECK 6.5** Convert the following Celsius temperatures into kelvins, and the Kelvin (K) temperatures into degrees Celsius:

a. 27°C b. 0.°C c. 0.00 K d. 100. K

6.7 | Pressure, Temperature, and Volume Relationships

Learning Objective

5. Do calculations based on Boyle's law, Charles's law, and the combined gas law.

Experimental investigations into the behavior of gases as they were subjected to changes in temperature and pressure led to several gas laws that could be expressed by simple mathematical equations. In 1662 Robert Boyle, an Irish chemist, reported his discovery of

a relationship between the pressure and volume of a gas sample kept at constant temperature. This relationship, known as **Boyle's law**, is

$$P = \frac{k}{V} \tag{6.3}$$

or

$$PV = k \tag{6.4}$$

where k is an experimentally determined constant, and (remember) the measurements are made without changing the temperature of the gas. Mathematically, the pressure and volume are said to be related by an inverse relationship because the change in one is in a direction opposite (inverse) to the change in the other. That is, as the pressure on a gas sample is increased, the volume of the gas sample decreases.

Another gas law was discovered in 1787 by Jacques Charles, a French scientist. He studied the volume behavior of gas samples kept at constant pressure as they were heated. He found that at constant pressure, the volume of a gas sample was directly proportional to its temperature expressed in kelvins. In other words, if the temperature was doubled, the sample volume doubled as long as the pressure was kept constant (see **Figure 6.6**). This behavior, known as **Charles's law**, is represented mathematically as

$$\frac{V}{T} = k' \tag{6.5}$$

Figure 6.6 A balloon collapses when the gas it contains is cooled in liquid nitrogen (temperature = −196°C, or 77 K). Assume the inflated yellow balloon had a volume of 4.0 L at room temperature (25°C) and the prevailing pressure. Assume the pressure remained constant and calculate the volume of the gas in the balloon at the temperature of liquid nitrogen.

© Cengage Learning/Charles D. Winters

© Cengage Learning/Charles D. Winters

or

$$V = k'T \tag{6.6}$$

where k' is another experimentally determined constant.

Boyle's law and Charles's law can be combined to give a single law called the combined gas law, which provides a relationship between the pressure, volume, and temperature of gases. The **combined gas law** is written as

$$\frac{PV}{T} = k'' \tag{6.7}$$

where k'' is another experimentally determined constant.

Equation 6.7 can be put into a useful form by the following line of reasoning. Suppose a gas sample is initially at a pressure and temperature of P_i and T_i and has a volume of V_i. Now suppose that the pressure and temperature are changed to some new (final) values represented by P_f and T_f and that the volume changes to a new value of V_f. According to Equation 6.7,

$$\frac{P_i V_i}{T_i} = k''$$

and

$$\frac{P_f V_f}{T_f} = k''$$

Since both quotients on the left side are equal to the same constant, they can be set equal to each other, and we get

$$\frac{P_i V_i}{T_i} = \frac{P_f V_f}{T_f} \tag{6.8}$$

Example 6.6 Gas Law Calculations

a. A sample of helium gas has a volume of 5.00 L at 25°C and a pressure of 0.951 atm. What volume will the sample have if the temperature and pressure are changed to 50.°C and 1.41 atm?
b. A sample of gas has a volume of 3.75 L at a temperature of 25°C and a pressure of 1.15 atm. What will the volume be at a temperature of 35°C and a pressure of 620. torr?

Solution

a. Equation 6.8 will be used to solve this problem. An important step is to identify the quantities that are related and will thus have the same subscripts in Equation 6.8. In this problem, the 5.00-L volume, 25°C temperature, and 0.951 atm pressure are all related and will be used as the initial conditions. The final conditions are the 50.°C temperature, the 1.41 atm pressure, and the final volume V_f, which is to be calculated. Substitution of these quantities into Equation 6.8 gives

$$\frac{(0.951\ \text{atm})(5.00\ \text{L})}{(298\ \text{K})} = \frac{(1.41\ \text{atm})V_f}{(323\ \text{K})}$$

Note that the Celsius temperatures have been changed to kelvins. The desired quantity, V_f, can be isolated by multiplying both sides of the equation by 323 K and dividing both sides by 1.41 atm:

$$\frac{(323\ \text{K})(0.951\ \cancel{\text{atm}})(5.00\ \text{L})}{(298\ \text{K})(1.41\ \cancel{\text{atm}})} = \frac{(\cancel{323\ \text{K}})(\cancel{1.41\ \text{atm}})V_f}{(\cancel{323\ \text{K}})(\cancel{1.41\ \text{atm}})}$$

or

$$V_f = \frac{(323\ \text{K})(0.951\ \cancel{\text{atm}})(5.00\ \text{L})}{(298\ \text{K})(1.41\ \cancel{\text{atm}})} = 3.66\ \text{L}$$

b. This is the same as the problem in part a, except that the initial and final pressures are given in different units. As we saw in part a, the pressure units must cancel in the final calculation, so the initial and final units must be the same. The units of pressure used make no difference as long as they are the same, so we will convert the initial pressure into torr. The necessary conversion factor was obtained from Table 6.3:

$$P_i = (1.15\ \cancel{\text{atm}})\left(\frac{760\ \text{torr}}{1\ \cancel{\text{atm}}}\right) = 874\ \text{torr}$$

The quantities are substituted into Equation 6.8 to give

$$\frac{(874\ \text{torr})(3.75\ \text{L})}{(298\ \text{K})} = \frac{(620.\ \text{torr})(V_f)}{(308\ \text{K})}$$

or

$$V_f = \frac{(308\ \cancel{K})(874\ \cancel{torr})(3.75\ \text{L})}{(298\ \cancel{K})(620.\ \cancel{torr})} = 5.46\ \text{L}$$

Example 6.7 More Gas Law Calculations

a. A steel cylinder has a bursting point of 1.00×10^4 psi. It is filled with gas at a pressure of 2.50×10^3 psi when the temperature is 20.°C. During transport, the cylinder is allowed to sit in the hot sun, and its temperature reaches 100.°C. Will the cylinder burst?
b. A sample of gas has a volume of 3.00 ft^3 at a temperature of 30.°C. If the pressure on the sample remains constant, at what Celsius temperature will the volume be half the volume at 30.°C?

Solution

a. The volume of the gas is constant (unless the cylinder bursts), so $V_i = V_f$. With this in mind, Equation 6.8 is used with the following values: $P_i = 2.50 \times 10^3$ psi, $V_i = V_f$, $T_i = 20.°C + 273$ K, $P_f =$ unknown pressure, and $T_f = 100.°C + 273$ K. Substitution into Equation 6.8 gives

$$\frac{(2.50 \times 10^3\ \text{psi})(V_i)}{293\ \text{K}} = \frac{(P_f)(V_f)}{373\ \text{K}}$$

or

$$P_f = \frac{(2.50 \times 10^3\ \text{psi})(\cancel{V_f})(373\ \cancel{K})}{(\cancel{V_f})(293\ \cancel{K})} = 3.18 \times 10^3\ \text{psi}$$

The cylinder will not burst.

b. In this case, $P_i = P_f$ because the pressure remains constant. Also, $V_i = 3.00$ ft^3; $V_f = \frac{1}{2}V_i$, or 1.50 ft^3; $T_i = 30.°C + 273 = 303$ K; and T_f is to be determined. Substitution into Equation 6.8 gives

$$\frac{(P_i)(3.00\ \text{ft}^3)}{(303\ \text{K})} = \frac{(P_f)(1.50\ \text{ft}^3)}{T_f}$$

The quantity to be calculated, T_f, is in the denominator of the right side. We put it into the numerator by inverting both sides of the equation:

$$\frac{(303\ \text{K})}{(P_i)(3.00\ \text{ft}^3)} = \frac{T_f}{(P_f)(1.50\ \text{ft}^3)}$$

The desired quantity, T_f, is now isolated by multiplying both sides of the equation by $(P_f)(1.50\ \text{ft}^3)$:

$$\frac{(P_f)(1.50\ \cancel{\text{ft}^3})(303\ \text{K})}{(P_i)(3.00\ \cancel{\text{ft}^3})} = \frac{(\cancel{P_f})(\cancel{1.50\ \text{ft}^3})(T_f)}{(\cancel{P_f})(\cancel{1.50\ \text{ft}^3})}$$

or

$$T_f = \frac{(\cancel{P_f})(1.50\ \cancel{\text{ft}^3})(303\ \text{K})}{(\cancel{P_i})(3.00\ \cancel{\text{ft}^3})} = 152\ \text{K}$$

Zinc for Colds?

The common cold is the number one medical reason for missed work and school absenteeism. While there is no known cure, some evidence from studies done during the past 20 years indicates that certain zinc-containing chemical compounds, when taken internally, have the ability to shorten the duration and reduce the severity of cold symptoms.

In a 100-patient study done at the Cleveland Clinic, patients were given lozenges within 24 hours of first showing symptoms of the common cold. Half the patients received lozenges that contained the compound zinc gluconate. The other half received placebo lozenges that contained no zinc compounds. During their waking hours, all patients dissolved a lozenge in their mouth every 2 hours. The results of this study showed that cold symptoms lasted an average of 4.4 days in patients who received the zinc gluconate, compared to an average duration of 7.6 days in patients who received the non-zinc placebos.

The sales of over-the-counter zinc-containing medications, especially lozenges, skyrocketed following the publication of these results in 1996. However, this enthusiasm was diminished somewhat by research results published in 1998. In these studies it was concluded that zinc-containing lozenges were not effective against cold symptoms in children and teenagers.

Researchers continue to study this topic today. As a result of these studies, there is fair evidence that the use of zinc gluconate lozenges does marginally reduce the duration of the common cold when compared to a placebo. However, the design of these studies has been called into question, so the conclusions are still considered to be somewhat tentative.

It must also be noted that the reported positive results of treating a cold with zinc-containing lozenges applies only to the use of lozenges. The use of zinc-containing nasal or throat sprays has been found to have serious negative side effects, and such use should be avoided.

Zinc-containing lozenges may marginally reduce the duration of the common cold.

Remember, P_f and P_i canceled because they were equal (the pressure was kept constant). The answer is given in kelvins, but we want it in degrees Celsius:

$$K = {}^\circ C + 273$$

or

$$^\circ C = K - 273 = 152 - 273 = -121^\circ C$$

✔ LEARNING CHECK 6.6

a. A sample of argon gas is confined in a 10.0-L container at a pressure of 1.90 atm and a temperature of 30.°C. What volume would the sample have at 1.00 atm and $-10.2^\circ C$?

b. A sample of gas has a volume of 5.00×10^2 mL at a temperature and pressure of 300. K and 800. torr. It is desired to compress the sample to a volume of 250. mL at a pressure of 900. torr. What temperature in both kelvins and Celsius degrees will be required?

6.8 | The Ideal Gas Law

Learning Objective

6. Do calculations based on the ideal gas law.

The combined gas law (Equation 6.8) works for samples only in which the mass of gas remains constant during changes in temperature, pressure, and volume. However, it is often useful to work with situations in which the amount of gas varies. The foundation for this kind of work was proposed in 1811 by Amadeo Avogadro, an Italian scientist. According to his proposal, which is now known as **Avogadro's law**, equal volumes of different gases measured at the same temperature and pressure contain equal numbers of gas molecules. According to this idea, two identical compressed gas cylinders of helium and oxygen at the same pressure and temperature would contain identical numbers of molecules of the respective gases. However, the mass of gas in the cylinders would *not* be the same because the molecules of the two gases have different molecular weights.

The actual temperature and pressure used do not influence the validity of Avogadro's law, but it is convenient to specify a standard set of values. Chemists have chosen 0°C (273 K) and 1.00 atm to represent what are called **standard conditions** for gas measurements. These conditions are often abbreviated STP (standard temperature and pressure).

As we have seen, the mole (defined in Section 2.6, page 60) is a convenient quantity of matter to work with. What volume does 1.00 mol of a gas occupy according to Avogadro's law? Experiments show that 1.00 mol of any gas molecules has a volume of 22.4 L at STP (see **Figure 6.7**).

A combination of Boyle's, Charles's, and Avogadro's laws leads to another gas law that includes the quantity of gas in a sample as well as the temperature, pressure, and volume of the sample. This law, known as the **ideal gas law**, is written as

$$PV = nRT \qquad (6.9)$$

In this equation, *P*, *V*, and *T* are defined as they were in the gas laws given earlier. The symbol *n* stands for the number of moles of gas in the sample being used, and *R* is a constant known as the **universal gas constant**. The measured value for the volume of 1.00 mol of gas at STP allows *R* to be evaluated by substituting the values into Equation 6.9 after rearrangement to isolate *R*:

$$R = \frac{PV}{nT} = \frac{(1.00 \text{ atm})(22.4 \text{ L})}{(1.00 \text{ mol})(273 \text{ K})} = 0.0821 \frac{\text{L atm}}{\text{mol K}}$$

This value of *R* is the same for all gases under any conditions of temperature, pressure, and volume.

Example 6.8 Ideal Gas Law Calculations

Use the ideal gas law to calculate the volume of 0.413 mol of hydrogen gas at a temperature of 20.°C and a pressure of 1.20×10^3 torr.

Solution

To use the ideal gas law (Equation 6.9), it is necessary that all units match those of *R*. Thus, the temperature will have to be expressed in kelvins and the pressure in atmospheres:

$$K = °C + 273 = 20.°C + 273 = 293 \text{ K}$$

$$P = (1.20 \times 10^3 \text{ torr})\left(\frac{1 \text{ atm}}{760 \text{ torr}}\right) = 1.58 \text{ atm}$$

Avogadro's law Equal volumes of gases measured at the same temperature and pressure contain equal numbers of molecules.

standard conditions (STP) A set of specific temperature and pressure values used for gas measurements.

Figure 6.7 The ball has a volume of 22.4 L, the volume of 1 mole of any gas at STP. The bottle has a volume of 1.00 quart (0.946 L). How many moles of gas would the bottle contain at STP?

ideal gas law A gas law that relates the pressure, volume, temperature, and number of moles in a gas sample. Mathematically, it is *PV = nRT*.

universal gas constant The constant that relates pressure, volume, temperature, and number of moles of gas in the ideal gas law.

© Spencer L. Seager

Because volume is the desired quantity, we isolate it by dividing both sides of the ideal gas law by P:

$$\frac{PV}{P} = \frac{nRT}{P} \qquad \text{or} \qquad V = \frac{nRT}{P}$$

Substitution of quantities gives

$$V = \frac{(0.413 \text{ mol})\left(0.0821\dfrac{\text{L atm}}{\text{mol K}}\right)(293 \text{ K})}{(1.58 \text{ atm})} = 6.29 \text{ L}$$

✔ **LEARNING CHECK 6.7** A 2.15-mol sample of sulfur dioxide gas (SO_2) occupies a volume of 12.6 L at 30.°C. What is the pressure of the gas in atm?

Because R is a constant for all gases, it follows that if any three of the quantities P, V, T, or n are known for a gas sample, the fourth quantity can be calculated by using Equation 6.9. An interesting application of this concept makes it possible to determine the molecular weights of gaseous substances. If the mass of a sample is known, the number of moles in the sample is the mass in grams divided by the molecular weight. This fact is represented by Equation 6.10, where n is the number of moles in a sample that has a mass in grams of m and a molecular weight of MW:

$$n = \frac{m}{\text{MW}} \tag{6.10}$$

When $\dfrac{m}{\text{MW}}$ is substituted for n in Equation 6.9, the result is

$$PV = \frac{mRT}{\text{MW}} \tag{6.11}$$

Example 6.9 More Ideal Gas Law Calculations

A gas sample has a volume of 2.74 L, a mass of 16.12 g and is stored at a pressure of 4.80 atm at a temperature of 25°C. The gas might be CH_4, C_2H_6, or C_3H_{10}. Which gas is it?

Solution
The molecular weight of the gas is determined using Equation 6.11 and compared with the molecular weights of the three possibilities calculated from their formulas. Rearrangement of Equation 6.11 to isolate molecular weight gives

$$\text{MW} = \frac{mRT}{PV}$$

Substitution of the known quantities after conversion to proper units gives

$$\text{MW} = \frac{mRT}{PV} = \frac{(16.12 \text{ g})\left(0.0821\dfrac{\text{L atm}}{\text{mol K}}\right)(298 \text{ K})}{(4.80 \text{ atm})(2.74 \text{ L})} = 30.0 \text{ g/mol}$$

Thus, the molecular weight is 30.0 u. This matches the molecular weight of C_2H_6 calculated from atomic weights and the molecular formula.

✔ **LEARNING CHECK 6.8** A sample of unknown gas has a mass of 3.35 g and occupies 2.00 L at 1.21 atm and 27°C. What is the molecular weight of the gas? Is the gas H_2S, HBr, or NH_3?

The gas laws discussed in this chapter apply only to gases that are ideal, but interestingly, no ideal gases actually exist. If they did exist, ideal gases would behave exactly as predicted by the gas laws at all temperatures and pressures. Real gases deviate from the behavior predicted by the gas laws, but under normally encountered temperatures and pressures, the deviations are small for many real gases. This fact allows the gas laws to be used for real gases. Interparticle attractions tend to make gases behave less ideally. Thus, the gas laws work best for gases in which such forces are weak, that is, those made up of single atoms (the noble gases) or nonpolar molecules (O_2, N_2, etc.). Highly polar molecules such as water vapor, hydrogen chloride, and ammonia deviate significantly from ideal behavior.

6.9 | Dalton's Law

Learning Objective

7. Do calculations based on Dalton's law.

John Dalton (1766–1844), an English schoolteacher, made a number of important contributions to chemistry. Some of his experiments led to the law of partial pressures, also called **Dalton's law**. According to this law, the total pressure exerted by a mixture of different gases kept at a constant volume and temperature is equal to the sum of the partial pressures of the gases in the mixture. The **partial pressure** of each gas in such mixtures is the pressure each gas would exert if it were confined alone under the same temperature and volume conditions as the mixture.

Imagine you have four identical gas containers, as shown in **Figure 6.8**. Place samples of three different gases (represented by △, ○, and □) into three of the containers, one to a

> **Dalton's law of partial pressures** The total pressure exerted by a mixture of gases is equal to the sum of the partial pressures of the gases in the mixture.
>
> **partial pressure** The pressure an individual gas of a mixture would exert if it were in the container alone at the same temperature as the mixture.

Figure 6.8 Dalton's law of partial pressures.

container, and measure the pressure exerted by each sample. Then place all three samples into the fourth container and measure the total pressure (P_t) exerted. The result is a statement of Dalton's law:

$$P_t = P_\triangle + P_\bigcirc + P_\square \qquad (6.12)$$

where P_\triangle, P_\bigcirc, and P_\square are the partial pressures of gases △, ○, and □, respectively.

1 A mylar balloon (left) and rubber balloon freshly filled with helium.

2 The same balloons the next day.

Figure 6.9 Graham's law in action. Does helium diffuse more rapidly through rubber or plastic?

effusion A process in which a gas escapes from a container through a small hole.

diffusion A process that causes gases to spontaneously intermingle when they are brought together.

Graham's law A mathematical expression that relates rates of effusion or diffusion of two gases to the masses of the molecules of the two gases.

Example 6.10 Partial Pressure Calculations

A sample of air is collected when the atmospheric pressure is 742 torr. The partial pressures of nitrogen and oxygen in the sample are found to be 581 torr and 141 torr, respectively. Assume water vapor to be the only other gas present in the air sample, and calculate its partial pressure.

Solution

Dalton's law says $P_t = P_{O_2} + P_{N_2} + P_{H_2O}$. The total pressure of the sample is the atmospheric pressure of 742 torr. Therefore,

$$742 \text{ torr} = 141 \text{ torr} + 581 \text{ torr} + P_{H_2O}$$

and

$$P_{H_2O} = 742 - 141 - 581 = 20. \text{ torr}$$

✔ **LEARNING CHECK 6.9** A mixture is made of helium, nitrogen, and oxygen. Their partial pressures are, respectively, 310 torr, 0.200 atm, and 7.35 psi. What is the total pressure of the mixture in torr?

6.10 | Graham's Law

Learning Objective

8. Do calculations based on Graham's law.

Effusion and diffusion are processes by which gases move and mix. **Effusion** is the escape of a gas through a small hole in its container. **Diffusion** is the process in which two or more gases spontaneously intermingle when brought together. Even though the processes appear to be different, they are related, and both follow a law proposed in 1828 by Thomas Graham to describe effusion (see **Figure 6.9**). For two gases, represented by A and B, **Graham's law** is

$$\frac{\text{effusion rate } A}{\text{effusion rate } B} = \sqrt{\frac{\text{molecular mass of } B}{\text{molecular mass of } A}} \tag{6.13}$$

Example 6.11 Graham's Law Calculations

Oxygen molecules weigh 16 times as much as hydrogen molecules. Which molecule will diffuse faster and how much faster?

Solution

Applying Equation 6.13 but recognizing that diffusion follows the equation as well as effusion, we get

$$\frac{\text{rate } H_2}{\text{rate } O_2} = \sqrt{\frac{\text{mass } O_2}{\text{mass } H_2}} = \sqrt{\frac{32}{2.0}} = \sqrt{16} = 4.0$$

Therefore, we conclude that hydrogen will diffuse four times faster than oxygen.

✔ **LEARNING CHECK 6.10** Which will diffuse faster, He molecules or Ne molecules? How much faster?

6.11 | Changes in State

Learning Objective

9. Classify changes in state as exothermic or endothermic.

Matter can be changed from one state into another by processes such as heating, cooling, or changing the pressure. Heating and cooling are the processes most often used, and a change in state that requires an input of heat is called endothermic, whereas one in which

heat is given up (or removed) is called exothermic (see **Section 5.8**). (These terms come from the Greek *endo* = in and *exo* = out). Endothermic changes are those in which particles are moved farther apart as cohesive forces are being overcome, such as the change of a solid to a liquid or gas. Exothermic changes are those in the opposite direction. See **Figure 6.10**.

Figure 6.10 Endothermic and exothermic changes of state.

✔ **LEARNING CHECK 6.11** Classify the following processes as endothermic or exothermic:

a. Alcohol evaporates.
b. Water freezes.
c. Iron melts in a furnace.

6.12 | **Evaporation and Vapor Pressure**

Learning Objective

10. Demonstrate an understanding of the concepts of vapor pressure and evaporation.

The evaporation of liquids is a familiar process. Water in a container will soon disappear (evaporate) if the container is left uncovered. **Evaporation**, or **vaporization**, is an endothermic process that takes place as a result of molecules leaving the surface of a liquid. The rate of evaporation depends on the temperature of the liquid and the surface area from which the molecules can escape. Temperature is an important factor because it is related directly to the speed and kinetic energy of the molecules and hence to their ability to break away from the attractive forces present at the liquid surface.

Evaporating molecules carry significant amounts of kinetic energy away from the liquid, and as a result, the temperature of the remaining liquid will drop unless heat flows in from the surroundings. This principle is involved in all evaporative cooling processes, including evaporative coolers for homes, the cooling of the human body by perspiration, and the cooling of a panting dog by the evaporation of saliva from the mucous membranes of its mouth.

Compare evaporation in a closed container (see **Figure 6.11**) with that in an open container. Evaporation occurs in both containers, as indicated by a drop in liquid level. But, unlike that in an open container, the liquid level in the closed container eventually stops dropping and becomes constant.

evaporation or vaporization
An endothermic process in which a liquid is changed to a gas.

Figure 6.11 Liquid evaporation in a closed container. The drop in liquid level is greatly exaggerated for emphasis.

| Initially | After some time | At equilibrium |

What would explain this behavior? In a closed container the molecules of liquid that go into the vapor (gaseous) state are unable to move completely away from the liquid surface as they do in an open container. Instead, the vapor molecules are confined to a space immediately above the liquid, where they have many random collisions with the container walls, other vapor molecules, and the liquid surface. Occasionally, their collisions with the liquid result in condensation, and they are recaptured by the liquid. **Condensation** is an exothermic process in which a gas or vapor is converted to either a liquid or a solid. Thus, two processes—evaporation (escape) and condensation (recapture)—actually take place in the closed container. Initially, the rate of evaporation exceeds that of condensation, and the liquid level drops. However, the rates of the two processes eventually become equal, and the liquid level stops dropping because the number of molecules that escape in a given time is the same as the number recaptured.

A system in which two opposite processes take place at equal rates is said to be in equilibrium (see **Chapter 8**). Under the equilibrium conditions just described, the number of molecules in the vapor state remains constant. This constant number of molecules will exert a constant pressure on the liquid surface and the container walls. This pressure exerted by a vapor in equilibrium with a liquid is called the **vapor pressure** of the liquid. The magnitude of a vapor pressure depends on the nature of the liquid (molecular polarity, mass, etc.) and the temperature of the liquid. These dependencies are illustrated in **Tables 6.4** and **6.5**.

condensation An exothermic process in which a gas or vapor is changed to a liquid or solid.

vapor pressure The pressure exerted by vapor that is in equilibrium with its liquid.

TABLE 6.4 The Vapor Pressure of Various Liquids at 20°C

Liquid	Molecular Weight (u)	Polarity	Vapor Pressure (torr)
pentane (C_5H_{12})	72	Nonpolar	414.5
hexane (C_6H_{14})	86	Nonpolar	113.9
heptane (C_7H_{16})	100	Nonpolar	37.2
ethanol (C_2H_5—OH)	46	Polar (hydrogen bonds)	43.9
1-propanol (C_3H_7—OH)	60	Polar (hydrogen bonds)	17.3
1-butanol (C_4H_9—OH)	74	Polar (hydrogen bonds)	7.1

TABLE 6.5 Vapor Pressure of Water at Various Temperatures

Temperature (°C)	Vapor Pressure (torr)
0	4.6
20	17.5
40	55.3
60	149.2
80	355.5
100	760.0

The effect of molecular mass on vapor pressure is seen in both the pentane–heptane and ethanol–1-butanol series of compounds in **Table 6.4**. Within each series, the molecular polarities are the same for each member, but vapor pressure decreases as molecular weight increases. This shows the effect of stronger dispersion forces between heavier molecules. A comparison of pentane and 1-butanol shows the effects of polar attractions between molecules. Molecules of these liquids have similar molecular weights, but the hydrogen-bonded 1-butanol has a much lower vapor pressure.

The effect of increasing the kinetic energy of molecules by heating is clearly evident by the vapor pressure behavior of water shown in **Table 6.5**.

✔ **LEARNING CHECK 6.12** Decide which member of each of the following pairs of compounds would have the higher vapor pressure. Explain your choice in each case.

a. Methyl alcohol (CH_3OH) and propyl alcohol (C_3H_7OH)
b. Liquid helium (He) and liquid nitrogen (N_2)
c. Liquid hydrogen fluoride (HF) and liquid neon (Ne)

The Earth is sometimes called the "water planet" because more than 70% of its surface is covered by an ocean of water in liquid or solid form. Above this ocean, the earth is covered by another ocean. This second ocean, called the atmosphere, is made up of gases and consists of about 78.1% nitrogen gas, 20.9% oxygen gas, and trace amounts of other gases such as water vapor and carbon dioxide. Just as the pressure at the bottom of a liquid water ocean is greater than at the top, the pressure of the gaseous atmospheric ocean is greatest at the bottom, near the surface of the Earth, and decreases as the altitude above the Earth's surface increases.

At the 29,029 foot top of Mount Everest, the atmospheric pressure is only about one-third as high as the atmospheric pressure at sea level. The cruising altitude for commercial airliners averages from 30,000 feet to 40,000 feet. This altitude is above the clouds, so cruising commercial aircraft encounter less inclement weather and turbulence than aircraft flying at lower altitudes. Because the density of the atmosphere is less at high altitudes, the correspondingly fewer air molecules offer less resistance to the moving aircraft. Thus, flying at high altitude increases the fuel efficiency of the aircraft.

When an aircraft cruises at 30,000 feet, the outside air pressure is only 4.4 pounds per square inch (psi), so the passenger cabin must be pressurized. Normal sea-level atmospheric pressure of 14.7 psi exceeds the strength of a typical commercial aircraft's fuselage, so the accepted value for cabin pressure in most commercial aircraft is 10.9 psi, a pressure equal to the atmospheric pressure at an altitude of 8,000 feet. Most passengers can tolerate breathing air at

this reduced pressure and do not even notice the decreased oxygen levels.

Passengers often feel their ears "pop" when taking off or landing in an aircraft. The middle ear, next to the eardrum, contains a pocket of air that is normally under the same pressure as the surrounding air. If the surrounding air pressure decreases rapidly, the higher pressure in the middle ear painfully pushes against the eardrum, giving the sensation of the ear popping.

During landings, the surrounding air pressure increases, but passengers' ears are now filled with air that is under lower pressure than the surrounding air. Some surrounding air rushes into the middle ear to equalize the pressure and the ears "pop." The popping can be avoided during takeoffs and landings by pinching your nostrils closed and gently exhaling through your mouth.

Aircraft flying at high altitudes have pressurized passenger cabins.

6.13 | **Boiling and the Boiling Point**
Learning Objective

11. Demonstrate an understanding of the process of boiling and the concept of boiling point.

As a liquid is heated, its vapor pressure increases, as shown for water in **Table 6.5**. If the liquid's temperature is increased enough, its vapor pressure will reach a value equal to that of the prevailing atmospheric pressure. Up to that temperature, all vaporization appears to take place at the liquid surface. However, when the vapor pressure becomes equal to atmospheric pressure, vaporization begins to occur beneath the surface as well. When bubbles of vapor form and rise rapidly to the surface, where the vapor escapes, the liquid is boiling. The **boiling point** of a liquid is the temperature at which the vapor pressure of a liquid is equal to the atmospheric pressure above the liquid. The **normal** or **standard boiling point** of a liquid is the temperature at which the vapor pressure is equal to 1 standard atmosphere (760 torr). Normal values were used in all examples of boiling points given earlier in this book.

Liquids boil at temperatures higher than their normal boiling points when external pressures are greater than 760 torr. When external pressures are less than 760 torr, liquids boil at temperatures below their normal boiling points. Thus, the boiling point

boiling point The temperature at which the vapor pressure of a liquid is equal to the prevailing atmospheric pressure.

normal or standard boiling point The temperature at which the vapor pressure of a liquid is equal to 1 standard atmosphere (760 torr).

of water fluctuates with changes in atmospheric pressure. Such fluctuations seldom exceed 2°C at a specific location, but there can be striking variations at different elevations, as shown by **Table 6.6**. (This is why cooking directions are sometimes different for different altitudes.)

TABLE 6.6 Variations in the Boiling Point of Water with Elevation

Location	Elevation (feet above sea level)	Boiling Point of Water (°C)
San Francisco, CA	Sea level	100.0
Salt Lake City, UT	4,390	95.6
Denver, CO	5,280	95.0
La Paz, Bolivia	12,795	91.4
Mount Everest	29,029	76.5

TABLE 6.7 The Boiling Point of Water in a Pressure Cooker

Pressure above Atmospheric		Boiling Point of Water (°C)
psi	torr	
5	259	108
10	517	116
15	776	121

The increase in boiling point caused by an increase in pressure is the principle used in the ordinary pressure cooker. Increasing the pressure inside the cooker causes the boiling point of the water in it to rise, as shown in **Table 6.7**. It then becomes possible to increase the temperature of the food-plus-water in the cooker above 100°C. Such an increase of just 10°C will make the food cook approximately twice as fast (see **Figure 6.12**).

Figure 6.12 A pressure cooker shortens the time required to cook food. A carrot cooks completely in boiling water (100°C) in 8 minutes. About how long would it take to cook in a pressure cooker that raised the boiling point of water to 110°C?

sublimation The endothermic process in which a solid is changed directly to a gas without first becoming a liquid.

6.14 | Sublimation and Melting

Learning Objective

12. Demonstrate an understanding of the processes of sublimation and melting.

Solids, like liquids, have vapor pressures. Although the motion of particles is much more restricted in solids, particles at the surface can escape into the vapor state if they acquire sufficient energy. However, the strong cohesive forces characteristic of the solid state usually cause the vapor pressures of solids to be quite low.

As expected, vapor pressures of solids increase with temperature. When the vapor pressure of a solid is high enough to allow escaping molecules to go directly into the vapor state without passing through the liquid state, the process is called **sublimation** (see **Figure 6.13**). Sublimation is characteristic of materials such as solid carbon dioxide (dry ice) and naphthalene (moth crystals). Frozen water also sublimes under appropriate conditions. Wet laundry hung out in freezing weather eventually dries as the frozen water sublimes. Freeze drying, a technique based on this process, is used to remove water from materials that would be damaged by heating (e.g., freeze-dried foods).

Even though solids have vapor pressures, most pure substances in the solid state melt before appreciable sublimation takes place. Melting involves the breakdown of a rigid,

Figure 6.13 Solid CO_2 (left) and solid H_2O (right) both sublime. Contrast the endothermic processes involved when each is used in the most common way to keep things cold. Are the same processes involved?

orderly, solid structure into a mobile, disorderly liquid state. This collapse of the solid structure occurs at a characteristic temperature called the **melting point**. At the melting point, the kinetic energy of solid particles is large enough to partially overcome the strong cohesive forces holding the particles together, and the solid and liquid states have the same vapor pressure.

In some instances, solids cannot be changed into liquids, or liquids into gases, by heating. The atoms making up the molecules of some solids acquire enough kinetic energy on heating to cause bonds within the molecules to break before the solid (or liquid) can change into another state. This breaking of bonds within molecules changes the composition of the original substance. When this **decomposition** occurs, the original substance is said to have decomposed. This is why cotton and paper, when heated, char rather than melt.

melting point The temperature at which a solid changes to a liquid; the solid and liquid have the same vapor pressure.

decomposition A change in chemical composition that can result from heating.

6.15 | Energy and the States of Matter

Learning Objective

13. Do calculations based on energy changes that accompany heating, cooling, or changing the state of a substance.

A pure substance in the gaseous state contains more energy than in the liquid state, which in turn contains more energy than in the solid state. Before we look at this, note the following relationships. Kinetic energy, the energy of particle motion, is related to heat. In fact, temperature is a measurement of the average kinetic energy of the particles in a system. Potential energy, in contrast, is related to particle separation distances rather than motion. Thus, we conclude that an increase in temperature on adding heat corresponds to an increase in kinetic energy of the particles, whereas no increase in temperature on adding heat corresponds to an increase in the potential energy of the particles.

Now let's look at a system composed of 1 g of ice at an initial temperature of −20°C. Heat is added at a constant rate until the ice is converted into 1 g of steam at 120°C. The atmospheric pressure is assumed to be 760 torr throughout the experiment. The changes in the system take place in several steps, as shown in **Figure 6.14**.

Figure 6.14 The temperature behavior of a system during changes in state.

The solid is first heated from −20°C to the melting point of 0°C (line *AB*). The temperature increase indicates that most of the added heat causes an increase in the kinetic energy of the molecules. Along line *BC*, the temperature remains constant at 0°C while the solid melts. The constant temperature during melting reveals that the added heat has increased the potential energy of the molecules without increasing their kinetic energy; the molecules are moved farther apart, but their motion is not increased. The addition of more heat warms the liquid water from 0°C to the normal boiling point of 100°C (line *CD*). At 100°C, another change of state occurs as heat is added, and the liquid is converted into vapor (steam) at 100°C (line *DE*). The constant temperature of this process again indicates an increase in potential energy. Line *EF* represents heating the steam from 100°C to 120°C by adding more heat.

The addition of heat to this system resulted in two obviously different results: The temperature of water in a specific state was increased, or the water was changed from one state to another at constant temperature. The amount of heat energy required to change the temperature of a specified amount of a substance by 1°C is called the **specific heat** of the substance. In scientific work, this is often given in units of calories or joules (J) per gram degree. Thus, the specific heat of a substance is the number of calories or joules required to raise the temperature of 1 g of the substance by 1°C. (*Note:* 1 cal = 4.184 J.) The specific heat is related to the amount of heat required to increase the temperature of a sample of substance by Equation 6.14.

specific heat The amount of heat energy required to raise the temperature of exactly 1 g of a substance by exactly 1°C.

$$\text{Heat} = (\text{sample mass}) (\text{specific heat}) (\text{temp. change}) \tag{6.14}$$

Specific heats for a number of substances in various states are listed in **Table 6.8**. A substance with a high specific heat is capable of absorbing more heat with a small temperature change than substances with lower specific heats. Thus, substances to be used as heat transporters (e.g., in cooling and heating systems) should have high specific heats—note the value for liquid water.

Example 6.12 Specific Heat Calculations

You notice that your car is running hot on a summer day. Someone tells you to drain out the ethylene glycol antifreeze and replace it with water. Will this allow the engine to run cooler?

Solution

The engine will run cooler because a given mass of water has a greater ability to carry heat from the engine to the radiator than does an equal mass of ethylene glycol. The coolant in your car is a mixture of ethylene glycol and water, but to simplify the following comparison, let us imagine it is pure ethylene glycol. We calculate the amount of

heat absorbed by 1000 g each of pure ethylene glycol and pure water as the temperature changes from 20°C to 80°C.

$$\text{Heat absorbed} = (\text{mass})(\text{specific heat})(\text{temp. change})$$

ethylene glycol: $\text{Heat absorbed} = (1000 \text{ g})\left(\dfrac{0.57 \text{ cal}}{\text{g} \, °\cancel{C}}\right)(60° \, \cancel{C})$

$$= 34{,}200 \text{ cal} = 34.2 \text{ kcal}$$

water: $\text{Heat absorbed} = (1000 \text{ g})\left(\dfrac{1.00 \text{ cal}}{\text{g} \, °\cancel{C}}\right)(60° \, \cancel{C})$

$$= 60{,}000 \text{ cal} = 60.0 \text{ kcal}$$

Thus, the same amount of water will absorb (and transport) nearly twice as much heat for the same temperature increase.

✔ **LEARNING CHECK 6.13** Some nuclear reactors are cooled by gases. Calculate the number of calories that 1.00 kg of helium gas will absorb when it is heated from 25°C to 700°C. See **Table 6.8** for the specific heat of He.

TABLE 6.8 Specific Heats for Selected Substances

Substance and State	Specific Heat	
	cal/g Degree	J/g Degree
Aluminum (solid)	0.24	1.0
Copper (solid)	0.093	0.39
Ethylene glycol (liquid)	0.57	2.4
Helium (gas)	1.25	5.23
Hydrogen (gas)	3.39	14.2
Lead (solid)	0.031	0.13
Mercury (liquid)	0.033	0.14
Nitrogen (gas)	0.25	1.1
Oxygen (gas)	0.22	0.92
Sodium (solid)	0.29	1.2
Sodium (liquid)	0.32	1.3
Water (solid)	0.51	2.1
Water (liquid)	1.00	4.18
Water (gas)	0.48	2.0

The amount of heat required to change the state of 1 g of a substance at constant temperature is called the **heat of fusion** (for melting a solid) and the **heat of vaporization** (for boiling a liquid). The units used for these quantities are just calories or joules per gram because no temperature changes are involved. These heats represent the amount of energy required to change 1 g of a substance to the liquid or vapor state at the characteristic melting or boiling point. The heats of fusion and vaporization for water are 80 and 540 cal/g, respectively. This explains why a burn caused by steam at 100°C is more severe than one caused by water at 100°C. Vaporization has added 540 cal to each gram of steam, and each gram will release these 540 cal when it condenses on the skin. Liquid water at 100°C would not have this extra heat with which to burn the skin (see **Figure 6.15**).

heat of fusion The amount of heat energy required to melt exactly 1 g of a solid substance at constant temperature.

heat of vaporization The amount of heat energy required to vaporize exactly 1 g of a liquid substance at constant temperature.

Figure 6.15 Water can be boiled in a paper cup. Explain why heat from the burner does not increase the temperature of the water-containing cup to the ignition temperature.

1 A paper cup filled with water does not reach the ignition temperature when heated by a burner, but the water boils.

2 An empty paper cup burns when heated by a burner because the heat from the burner increases the temperature of the cup to the ignition point.

STUDY SKILLS 6.1

Which Gas Law to Use

For many students, the biggest challenge of this chapter is solving problems using the gas laws. Once you recognize that you are faced with a gas law problem, you must decide which of the six gas laws will work to solve the problem. One aid to selecting the appropriate law is to look for key words, phrases, or ideas that are often associated with specific laws. Some of these are given here:

Gas Law	Equation	Key
Boyle's law	$PV = k$	T is constant
Charles's law	$\dfrac{V}{T} = k'$	P is constant
Combined gas law	$\dfrac{P_iV_i}{T_i} = \dfrac{P_fV_f}{T_f}$	$P, V,$ and T all change
Dalton's law	$P_t = P_\triangle + P_\bigcirc + P_\square$	Two or more different gases
Graham's law	$\dfrac{\text{effusion rate } A}{\text{effusion rate } B}$ $= \sqrt{\dfrac{\text{molecular mass of } B}{\text{molecular mass of } A}}$	Effusion and diffusion
Ideal gas law	$PV = nRT$	Moles of gas

For example, if you see the key word *diffusion*, it is likely that you are dealing with a Graham's law problem.

The first three gas laws (Boyle's, Charles's, and the combined) are very similar in that they all use the symbols *P, V,* and *T*. If you have narrowed down a gas law problem to one of these three, a simple approach is to use the combined gas law. It works in all three cases. If *T* is constant,

$$\frac{P_iV_i}{T_i} = \frac{P_fV_f}{T_f}$$

simplifies to $P_iV_i = P_fV_f$, a form of Boyle's law. If *P* is constant, it simplifies to

$$\frac{V_i}{T_i} = \frac{V_f}{T_f}$$

a form of Charles's law. If *V* is constant, the combined law becomes

$$\frac{P_i}{T_i} = \frac{P_f}{T_f}$$

an equation that is not named but is useful for certain problems.

A final point to remember is that temperatures must be expressed in kelvins to get correct answers using any of the gas laws that involve temperature.

Example 6.13 More Specific Heat Calculations

Calculate the heat released when 5.0 kg of steam at 120.°C condenses to water at 100.°C in a radiator of a steam heating system.

Solution

Consider the process as taking place in two steps. The steam must cool from 120.°C to 100.°C and then condense to liquid water at 100.°C. When the steam cools to 100.°C,

$$\text{Heat released} = (\text{mass})(\text{specific heat})(\text{temp. change})$$

$$= (5.0 \times 10^3 \, \cancel{g})\left(\frac{0.48 \, \text{cal}}{\cancel{g} \, °\cancel{C}}\right)(20.° \, \cancel{C})$$

$$= 4.8 \times 10^4 \, \text{cal} = 48 \, \text{kcal}$$

When the steam condenses to liquid water at 100.°C,

$$\text{Heat released} = (\text{mass})(\text{heat of vaporization})$$

$$= (5.0 \times 10^3 \, \cancel{g})\left(\frac{540 \, \text{cal}}{\cancel{g}}\right)$$

$$= 2.7 \times 10^6 \, \text{cal} = 2.7 \times 10^3 \, \text{kcal}$$

Note that the heat of vaporization was used even though the water was changing from the vapor to the liquid state. The only difference between vaporization and condensation is the direction of heat flow; the amount of heat involved remains the same for a specific quantity of material. (Accordingly, 1 g of liquid water will release 80 cal when it freezes.)

These results make it clear that most of the transported heat (98%) was carried in the form of potential energy, which was released when the steam condensed.

✔ **LEARNING CHECK 6.14** Suppose the radiator of your car overheats and begins to boil. Calculate the number of calories absorbed by 5.00 kg of water (about 1 gallon) as it boils and changes to steam.

▶ Case Study Follow-up

Coughing is considered an emergency if it is accompanied by cyanosis (appearance of bluish skin), apnea (cessation of breathing), intercostal retractions (inward movement of muscles between the ribs), seizures or convulsions, high fever, or dehydration. During an emergency visit for a serious cough, a pulse oximeter (a device that clips onto the finger) may be used to measure actual oxygen saturation levels of the blood. Levels below 90% are considered indicative of hypoxemia (low blood-oxygen).

Parental presence in hospitals has been standard for many years. It alleviates the child's anxiety and reduces psychological suffering of the parents as well. Since 2000, parental presence has been supported even during resuscitation of pediatric patients. However, controversy exists over parental presence in some situations. If a parent is anxious, his or her anxiety may serve to increase the child's fears, and concerns about a distressed parent needing treatment and adding to the workload of pediatric nurses have also been raised. In the absence of complicating factors, a calm parent's presence can serve to alleviate the patient's fear and decrease the workload of pediatric nurses.

Concept Summary

Observed Properties of Matter Matter in the solid, liquid, or gaseous state shows differences in physical properties such as density, shape, compressibility, and thermal expansion.
Objective 1 (Section 6.1), Exercise 6.2

The Kinetic Molecular Theory of Matter Much of the behavior of matter in different states can be explained by the kinetic molecular theory, according to which all matter is composed of tiny molecules that are in constant motion and are attracted or repelled by each other.
Objective 2 (Sections 6.2–6.4), Exercise 6.8

The Solid State In the solid state, cohesive forces between particles of matter are stronger than disruptive forces. As a result, the particles of solids are held in rigid three-dimensional lattices in which the particles' kinetic energy takes the form of vibrations about each lattice site.
Objective 3 (Section 6.5), Exercises 6.12 and 6.16

The Liquid State In the liquid state, cohesive forces between particles slightly dominate disruptive forces. As a result, particles of liquids are randomly arranged but relatively close to each other and are in constant random motion, sliding freely over each other but without enough kinetic energy to become separated.
Objective 3 (Section 6.5), Exercises 6.12 and 6.16

The Gaseous State In the gaseous state, disruptive forces dominate and particles move randomly, essentially independent of each other. Under ordinary pressure, the particles are separated from each other by relatively large distances except when they collide.
Objective 3 (Section 6.5), Exercises 6.12 and 6.16

The Gas Laws Mathematical relationships, called gas laws, describe the observed behavior of gases when they are mixed, subjected to pressure or temperature changes, or allowed to diffuse. When these relationships are used, it is necessary to express the volume and pressure in consistent units and the temperature in kelvins.
Objective 4 (Section 6.6), Exercises 6.20 and 6.22

Pressure, Temperature, and Volume Relationships Gas laws discovered by Robert Boyle and Jacques Charles led to the development of the combined gas law. This law allows calculations to be made that relate temperature, pressure, and volume changes for gases.
Objective 5 (Section 6.7), Exercises 6.24, 6.32, and 6.34

The Ideal Gas Law The application of a gas law discovered by Amadeo Avogadro to the combined gas law led to the ideal gas law, which allows calculations to be made that account for the number of moles of gas in a sample.
Objective 6 (Section 6.8), Exercise 6.46

Dalton's Law John Dalton discovered experimentally that the total pressure exerted by a mixture of gases is equal to the sum of the partial pressures of the individual gases in the mixture.
Objective 7 (Section 6.9), Exercise 6.58

Graham's Law Thomas Graham proposed a relationship that mathematically relates the rate of effusion or diffusion of a gas to the mass of the gas molecules:

$$\frac{\text{effusion rate } A}{\text{effusion rate } B} = \sqrt{\frac{\text{molecular mass of } B}{\text{molecular mass of } A}}$$

Objective 8 (Section 6.10), Exercise 6.60

Changes in State Most matter can be changed from one state to another by heating, cooling, or changing pressure. State changes that give up heat are called exothermic, and those that absorb heat are called endothermic.
Objective 9 (Section 6.11), Exercise 6.64

Evaporation and Vapor Pressure The evaporation of a liquid is an endothermic process and as a result is involved in many cooling processes. In a closed container, evaporation takes place only until the rate of escape of molecules from the liquid is equal to the rate at which they return to the liquid. The pressure exerted by the vapor, which is in equilibrium with the liquid, is called the vapor pressure of the liquid. Liquid vapor pressures increase as the liquid temperature increases.
Objective 10 (Section 6.12), Exercise 6.68

Boiling and the Boiling Point At the boiling point of a liquid, its vapor pressure equals the prevailing atmospheric pressure, and bubbles of vapor form within the liquid and rise to the surface as the liquid boils. The boiling point of a liquid changes as the prevailing atmospheric pressure changes.
Objective 11 (Section 6.13), Exercise 6.70

Sublimation and Melting Solids, like liquids, have vapor pressures that increase with temperature. Some solids have high enough vapor pressures to allow them to change to vapor without first becoming a liquid, a process called sublimation. Most solids change to the liquid state before they change to the vapor state. The temperature at which solids change to liquids is called the melting point.
Objective 12 (Section 6.14), Exercise 6.74

Energy and the States of Matter Energy is absorbed or released when matter is changed in temperature or changed from one state to another. The amount of heat energy required to produce temperature changes is called the specific heat of the matter involved. For phase changes, the amount of heat required is called the heat of fusion, or of vaporization.
Objective 13 (Section 6.15), Exercises 6.76 and 6.78

Key Terms and Concepts

Absolute zero (6.6)	Disruptive forces (6.2)	Potential energy (6.2)
Avogadro's law (6.8)	Effusion (6.10)	Pressure (6.6)
Boiling point (6.13)	Evaporation or vaporization (6.12)	Shape (6.1)
Boyle's law (6.7)	Gas laws (6.6)	Specific heat (6.15)
Charles's law (6.7)	Graham's law (6.10)	Standard atmosphere (6.6)
Cohesive forces (6.2)	Heat of fusion (6.15)	Standard conditions (STP) (6.8)
Combined gas law (6.7)	Heat of vaporization (6.15)	Sublimation (6.14)
Compressibility (6.1)	Ideal gas law (6.8)	Thermal expansion (6.1)
Condensation (6.12)	Kinetic energy (6.2)	Torr (6.6)
Dalton's law of partial pressures (6.9)	Melting point (6.14)	Universal gas constant (6.8)
Decomposition (6.14)	Normal or standard boiling point (6.13)	Vapor pressure (6.12)
Diffusion (6.10)	Partial pressure (6.9)	

Key Equations

1. Calculation of volume from mass and density (Section 6.1):	$V = \dfrac{m}{d}$	Equation 6.1
2. Calculation of kinetic energy of particles in motion (Section 6.2):	$KE = \dfrac{1}{2}\, mv^2$	Equation 6.2
3. Boyle's law (Section 6.7):	$P = \dfrac{k}{V}$ $PV = k$	Equation 6.3 Equation 6.4
4. Charles's law (Section 6.7):	$\dfrac{V}{T} = k'$ $V = k'T$	Equation 6.5 Equation 6.6
5. Combined gas law (Section 6.7):	$\dfrac{PV}{T} = k''$ $\dfrac{P_i V_i}{T_i} = \dfrac{P_f V_f}{T_f}$	Equation 6.7 Equation 6.8
6. Ideal gas law (Section 6.8):	$PV = nRT$	Equation 6.9
7. Molecular weight determination (Section 6.8):	$PV = \dfrac{mRT}{MW}$	Equation 6.11
8. Dalton's law (Section 6.9):	$P_t = P_\triangle + P_\bigcirc + P_\square$	Equation 6.12
9. Graham's law of effusion (Section 6.10):	$\dfrac{\text{effusion rate } A}{\text{effusion rate } B} = \sqrt{\dfrac{\text{molecular mass of } B}{\text{molecular mass of } A}}$	Equation 6.13
10. Heat calculation (Section 6.15):	Heat = (sample mass)(specific heat)(temp. change)	Equation 6.14

Exercises

Even-numbered exercises are answered in Appendix B.

Blue-numbered exercises are more challenging.

Observed Properties of Matter (Section 6.1)

6.1 Calculate the volume of 125 g of the following liquids:

 a. Acetone ($d = 0.792$ g/mL)

 b. Olive oil ($d = 0.918$ g/mL)

 c. Chloroform ($d = 1.49$ g/mL)

6.2 Calculate the volume of 125 g of the following liquids:

 a. Sea water ($d = 1.03$ g/mL)

 b. Methyl alcohol ($d = 0.792$ g/mL)

 c. Concentrated sulfuric acid ($d = 1.84$ g/mL)

6.3 Copper metal has a density of $8.92 \, g/cm^3$ at $20.0°C$ and $8.83 \, g/cm^3$ at $100.0°C$. Calculate the change in volume that occurs when a $10.0\text{-}cm^3$ piece of copper is heated from $20.0°C$ to $100.0°C$.

6.4 Liquid water has a density of $1.00 \, g/mL$ at $10.0°C$ and $0.996 \, g/mL$ at $30.0°C$. Calculate the change in volume that occurs when $500. \, mL$ of water is heated from $10.0°C$ to $30.0°C$.

6.5 Gallium metal melts at $29.8°C$. At the melting point, the density of the solid is $5.90 \, g/mL$, and that of the liquid is $6.10 \, g/mL$.

 a. Does solid gallium expand or contract when it is melted? Explain.

 b. What is the change in volume when $5.00 \, mL$ (cm^3) of solid gallium is melted?

6.6 A $1.50\text{-}L$ rubber balloon is filled with carbon dioxide gas at a temperature of $0.00°C$ and a pressure of $1.00 \, atm$. The density of the carbon dioxide gas under these conditions is $1.98 \, g/L$.

 a. Will the density of the carbon dioxide gas increase or decrease when the balloon is heated?

 b. At $50.0°C$, the balloon has a volume of $1.78 \, L$. Calculate the carbon dioxide density at this temperature.

The Kinetic Molecular Theory of Matter (Section 6.2)

6.7 Describe the changes in form of energy (kinetic changes to potential, etc.) that occur for the energy of a rock dropped to the ground from a cliff. What form or forms do you suppose the energy takes when the rock hits the ground?

6.8 Suppose a toy ball is thrown into the air such that it goes straight up, then falls and is caught by the person who threw it. Describe the changes in the form of energy that occur for the ball from the time it is thrown until it is caught.

6.9 Suppose a 180-lb (81.8-kg) halfback running at a speed of 8.0 m/s collides head-on with a 260-lb (118.2-kg) tackle running at 3.0 m/s. Which one will be pushed back? That is, which one has less kinetic energy? If you're not familiar with football, check with someone who is for definition of terms.

6.10 At $25.0°C$, helium molecules (He) have an average velocity of 1.26×10^5 cm/s, and methane molecules (CH_4) have an average velocity of 6.30×10^4 cm/s. Calculate the kinetic energy of each type of molecule at $25.0°C$ and determine which is greater. Express molecular masses in u for this calculation.

6.11 Which have the greater kinetic energy, hydrogen molecules traveling with a velocity of $2v$, or helium molecules traveling with a velocity of v? Express molecular masses in u.

The Solid, Liquid, and Gaseous States (Sections 6.3–6.5)

6.12 Explain each of the following observations using the kinetic molecular theory of matter:

 a. A liquid takes the shape, but not necessarily the volume, of its container.

 b. Solids and liquids are practically incompressible.

 c. A gas always exerts uniform pressure on all walls of its container.

6.13 Explain each of the following observations using the kinetic molecular theory of matter:

 a. Gases have low densities.

 b. The densities of a substance in the solid and liquid states are nearly identical.

 c. Solids, liquids, and gases all expand when heated.

6.14 Discuss differences in kinetic and potential energy of the constituent particles for a substance in the solid, liquid, and gaseous states.

6.15 The following statements are best associated with the solid, liquid, or gaseous states of matter. Match the statements to the appropriate state of matter.

 a. This state is characterized by the lowest density of the three.

 b. This state is characterized by an indefinite shape and a high density.

 c. In this state, disruptive forces prevail over cohesive forces.

 d. In this state, cohesive forces are most dominant.

6.16 The following statements are best associated with the solid, liquid, or gaseous states of matter. Match the statements to the appropriate state of matter.

 a. Temperature changes influence the volume of this state substantially.

 b. In this state, constituent particles are less free to move about than in other states.

 c. Pressure changes influence the volume of this state more than that of the other two states.

 d. This state is characterized by an indefinite shape and a low density.

The Gas Laws (Section 6.6)

6.17 What is a gas law?

6.18 A weather reporter on TV reports the barometer pressure as 28.6 inches of mercury. Calculate this pressure in the following units:

 a. atm

 b. torr

 c. psi

 d. bars

6.19 The pressure of a gas sample is recorded as 585 torr. Calculate this pressure in the following units:

 a. atm

 b. in. Hg

 c. psi

 d. bars

6.20 An engineer reads the pressure gauge of a boiler as 210 psi. Calculate this pressure in the following units:

 a. atm

 b. bars

 c. mmHg

 d. in. Hg

6.21 A chemist reads a pressure from a manometer attached to an experiment as 16.8 cmHg. Calculate this pressure in the following units:

 a. atm

 b. mmHg

c. torr

d. psi

6.22 Convert each of the following temperatures from the unit given to the unit indicated:

 a. The melting point of potassium metal, 63.7°C, to kelvins.

 b. The freezing point of liquid hydrogen, 14.1 K, to degrees Celsius.

 c. The boiling point of liquid helium, −268.9°C, to kelvins.

6.23 Convert each of the following temperatures from the unit given to the unit indicated:

 a. The melting point of gold, 1337.4 K, to degrees Celsius.

 b. The melting point of tungsten, 3410°C, to kelvins.

 c. The melting point of tin, 505 K, to degrees Celsius.

Pressure, Temperature, and Volume Relationships (Section 6.7)

6.24 Use the combined gas law (Equation 6.8) to calculate the unknown quantity for each gas sample described in the following table.

	Sample		
	A	**B**	**C**
P_i	1.50 atm	2.35 atm	9.86 atm
V_i	2.00 L	1.97 L	11.7 L
T_i	300. K	293 K	500. K
P_f	?	1.09 atm	5.14 atm
V_f	3.00 L	?	9.90 L
T_f	450. K	310 K	?

6.25 A 200.-mL sample of oxygen gas is collected at 26.0°C and a pressure of 690. torr. What volume will the gas occupy at STP (0°C and 760 torr)?

6.26 A 200.-mL sample of nitrogen gas is collected at 45.0°C and a pressure of 610. torr. What volume will the gas occupy at STP (0°C and 760 torr)?

6.27 A 3.00-L sample of helium at 0.00°C and 1.00 atm is compressed into a 0.50-L cylinder. What pressure will the gas exert in the cylinder at 50.°C?

6.28 A 2.50-L sample of neon at 0.00°C and 1.00 atm is compressed into a 0.75-L cylinder. What pressure will the gas exert in the cylinder at 30.°C?

6.29 What volume (in liters) of air measured at 1.00 atm would have to be put into a bicycle tire with a 1.00-L volume if the pressure in the bike tire is to be 65.0 psi? Assume the temperature of the gas remains constant.

6.30 What volume (in liters) of air measured at 1.00 atm would have to be put into a car tire with a volume of 14.5 L if the pressure in the car tire is to be 32.0 psi? Assume the temperature of the gas remains constant.

6.31 A sample of gas has a volume of 500. mL at a pressure of 640. torr. What volume in mL will the gas occupy at the same temperature, but at standard atmospheric pressure, 760 torr?

6.32 A sample of gas has a volume of 800. mL at a pressure of 650. torr. What volume in mL will the gas occupy at the same temperature, but at standard atmospheric pressure, 760. torr?

6.33 A 3.0-L sample of gas at 1.0 atm and 0.0°C is heated to 85°C. Calculate the gas volume at the higher temperature if the pressure remains at 1.0 atm.

6.34 A 3.8-L sample of gas at 1.0 atm and 20.°C is heated to 85°C. Calculate the gas volume in liters at the higher temperature if the pressure remains constant at 1.0 atm.

6.35 A sample of gas has a volume of 375 mL at 27°C. The gas is heated at a constant pressure until the volume is 500. mL. What is the new temperature of the gas in degrees Celsius?

6.36 What volume of gas in liters at 120.°C must be cooled to 40.°C if the gas volume at constant pressure and 40.°C is to be 2.0 L?

6.37 A 4.50-L gas sample is collected at a temperature and pressure of 27.0°C and 1.20 atm. The gas is to be transferred to a 3.00-L container at a pressure of 1.00 atm. What must the Celsius temperature of the gas in the 3.00-L container be?

6.38 A 2.50×10^3-L sample of oxygen gas is produced at 1.00 atm pressure. It is to be compressed and stored in a 25.0-L steel cylinder. Assume it is produced and stored at the same temperature and calculate the pressure in atm of the stored oxygen in the cylinder.

6.39 A steel tank with a volume of 6.25 L is full of gas at a pressure of 3.25 atm. What volume in liters would the gas occupy at a pressure of 0.250 atm if its temperature did not change?

6.40 A helium balloon was partially filled with 8.00×10^3 ft^3 of helium when the atmospheric pressure was 0.98 atm and the temperature was 23°C. The balloon rose to an altitude where the atmospheric pressure was 400. torr and the temperature was 5.3°C. What volume did the helium occupy at this altitude?

6.41 You have a 1.50-L balloon full of air at 30.°C. To what Celsius temperature would you have to heat the balloon to double its volume if the pressure remained unchanged?

6.42 A gas occupies 250. mL at a pressure of 2.10 atm. What volume would it occupy at the same temperature and a pressure of 60.0 kPa?

6.43 What minimum pressure would a 250.-mL aerosol can have to withstand if it were to contain 2.00 L of gas measured at 700. torr? Assume constant temperature.

6.44 A 2.00-L sample of nitrogen gas at 760. torr and 0.0°C weighs 2.50 g. The pressure on the gas is increased to 4.00 atm at 0.0°C. Calculate the gas density at the new pressure in g/L.

The Ideal Gas Law (Section 6.8)

6.45 Use the ideal gas law and calculate the following:

 a. The pressure exerted by 2.00 mol of oxygen confined to a volume of 400. mL at 20.0°C.

 b. The volume of hydrogen gas in a steel cylinder if 0.525 mol of the gas exerts a pressure of 3.00 atm at a temperature of 15.0°C.

 c. The temperature (in degrees Celsius) of a nitrogen gas sample that has a volume of 2.25 L and a pressure of 300. torr and contains 0.100 mol.

6.46 Use the ideal gas law and calculate the following:

 a. The number of moles of argon in a gas sample that occupies a volume of 400. mL at a temperature of 90.0°C and has a pressure of 735 torr.

 b. The pressure in atm exerted by 0.750 mol of hydrogen gas confined to a volume of 2.60 L at 50.0°C.

 c. The volume in liters of a tank of nitrogen if 1.50 mol of the gas exerts a pressure of 4.32 atm at 20.0°C.

6.47 Suppose 0.156 mol of SO_2 gas is compressed into a 0.750-L steel cylinder at a temperature of 27°C. What pressure in atmospheres is exerted by the gas?

6.48 Suppose 10.0 g of SO_2 gas was compressed into a 1.25-L steel cylinder at a temperature of 38.0°C. What pressure in atm would the gas exert on the walls of the cylinder?

6.49 Calculate the volume occupied by 8.75 g of oxygen gas (O_2) at a pressure of 0.890 atm and a temperature of 35.0°C.

6.50 The pressure gauge of a steel cylinder of methane gas (CH_4) reads 400. psi. The cylinder has a volume of 1.50 L and is at a temperature of 30.0°C. How many grams of methane does the cylinder contain?

6.51 Suppose 12.0 g of dry ice (solid CO_2) was placed in an empty 400.-mL steel cylinder. What pressure would develop if all the solid sublimed at a temperature of 35.0°C?

6.52 An experimental chamber has a volume of 60. L. How many moles of oxygen gas will be required to fill the chamber at STP?

6.53 How many molecules of nitrogen gas (N_2) are present in a sample that fills a 10.0-L tank at STP?

6.54 A sample of gaseous methyl ether has a mass of 8.12 g and occupies a volume of 3.96 L at STP. What is the molecular weight of methyl ether?

6.55 A sample of a gaseous nitrogen oxide is found to weigh 0.525 g. The sample has a volume of 300. mL at a pressure of 708 torr and a temperature of 25.7°C. Is the gas NO or NO_2?

6.56 A sample of gas weighs 0.176 g and has a volume of 114.0 mL at a pressure and temperature of 640. torr and 20.°C. Determine the molecular weight of the gas, and identify it as CO, CO_2, or O_2.

6.57 A 2.00-g sample of gas has a volume of 1.12 L at STP. Calculate its molecular weight and identify it as He, Ne, or Ar.

Dalton's Law (Section 6.9)

6.58 A steel cylinder contains a mixture of nitrogen, oxygen, and carbon dioxide gases. The total pressure in the tank is 2100. torr. The pressure exerted by the nitrogen and oxygen is 810 and 920 torr, respectively. What is the partial pressure in torr of the carbon dioxide in the mixture?

6.59 A 250-mL sample of oxygen gas is collected by water displacement. As a result, the oxygen is saturated with water vapor. The partial pressure of water vapor at the prevailing temperature is 22 torr. Calculate the partial pressure of the oxygen if the total pressure of the sample is 720. torr.

Graham's Law (Section 6.10)

6.60 Hydrogen gas (H_2) is found to diffuse approximately four times as fast as oxygen gas (O_2). Using this information, determine how the masses of hydrogen molecules and oxygen molecules compare. How do they compare based on information in the periodic table?

6.61 The mass of a bromine molecule is 160. u, and the mass of an argon molecule is 40. u. Compare the rates at which these gases will diffuse.

6.62 Two identical rubber balloons were filled with gas—one with helium and the other with nitrogen. After a time, it was noted that one of the balloons appeared to be going "flat." Which one do you think it was? Explain.

6.63 Assume the balloon in Exercise 6.62 that went flat first showed signs of "flatness" 10 hours after it was filled. How long would it take for the other balloon to begin to show signs of going flat?

Changes in State (Section 6.11)

6.64 Classify each of the following processes as endothermic or exothermic:

a. condensation

b. liquefaction

c. boiling

6.65 Classify each of the following processes as endothermic or exothermic:

a. freezing

b. sublimation

c. vaporization

6.66 Discuss what is meant by a change in state.

Evaporation and Vapor Pressure (Section 6.12)

6.67 The following are all nonpolar liquid hydrocarbon compounds derived from petroleum: butane (C_4H_{10}), pentane (C_5H_{12}), hexane (C_6H_{14}), and heptane (C_7H_{16}). Arrange these compounds in order of increasing vapor pressure (lowest first, highest last) and explain how you arrived at your answer.

6.68 Methylene chloride (CH_2Cl_2) was used at one time as a local anesthetic by dentists. It was sprayed onto the area to be anesthetized. Propose an explanation for how it worked.

6.69 Suppose a drop of methyl ether (C_2H_6O) was put on the back of one of your hands and a drop of ethyl ether ($C_4H_{10}O$) was put on your other hand. Propose a way you could tell which compound was which without smelling them.

Boiling and the Boiling Point (Section 6.13)

6.70 Each of two glass containers contains a clear, colorless, odorless liquid that has been heated until it is boiling. One liquid is water (H_2O) and the other is ethylene glycol ($C_2H_6O_2$). Explain how you could make one measurement of each boiling liquid, using the same device, and tell which liquid was which.

6.71 Suppose a liquid in an open container was heated to a temperature just 1 or 2 degrees below its boiling point, then insulated so it stayed at that temperature. Describe how the liquid would behave (what you would see happen) if the hot sample was suspended beneath a helium balloon and taken rapidly to higher altitudes.

6.72 Suppose you were on top of Mount Everest and wanted to cook a potato as quickly as possible. You left your microwave oven at home, so you could either boil the potato in water or throw it into a campfire. Explain which method you would use and why.

Sublimation and Melting (Section 6.14)

6.73 List three common substances that will sublime.

6.74 Solid iodine readily sublimes without melting when moderately heated. The hot vapor will condense back to the solid state when it cools. Describe a method that could be used to obtain pure solid iodine from a mixture of solid iodine and sand. Explain your reasoning.

6.75 A mixture was made of pure water and ice. The mixture was allowed to come to a constant temperature of 0.0°C, the melting point of solid water. The vapor pressure of the water was measured and found to be 4.58 torr. What is the vapor pressure of the ice in torr? Explain your answer.

Energy and the States of Matter (Section 6.15)

6.76 Using the specific heat data of Table 6.8, calculate the amount of heat (in calories) needed to increase the temperature of the following:

 a. 50. g of aluminum from 25°C to 55°C

 b. 2.50×10^3 of ethylene glycol from 80.°C to 85°C

 c. 500. g of steam from 110.°C to 120.°C

6.77 Using the specific heat data of Table 6.8, calculate the amount of heat (in calories) needed to increase the temperature of the following:

 a. 210 g of copper from 40.°C to 95.°C

 b. 150 g of mercury from 120.°C to 300.°C

 c. 2.50×10^3 g of helium gas from 250.°C to 900.°C

6.78 For solar energy to be effective, collected heat must be stored for use during periods of decreased sunshine. One proposal suggests that heat can be stored by melting solids that, upon solidification, would release the heat. Calculate the heat that could be stored by melting 1000. kg of each of the following solids. (Note: The water in each formula is included in the molecular weight.)

 a. Calcium chloride ($CaCl_2 \cdot 6H_2O$): melting point = 30.2°C, heat of fusion = 40.7 cal/g

 b. Lithium nitrate ($LiNO_3 \cdot 3H_2O$): melting point = 29.9°C, heat of fusion = 70.7 cal/g

 c. Sodium sulfate ($Na_2SO_4 \cdot 10H_2O$): melting point = 32.4°C, heat of fusion = 57.1 cal/g

6.79 Why wouldn't a solid such as K_2SO_4 (melting point = 1069°C, heat of fusion = 50.3 cal/g) be suitable for use in a solar heat storage system? (See Exercise 6.78.)

6.80 Liquid Freon (CCl_2F_2) is used as a refrigerant. It is circulated inside the cooling coils of older refrigerators or freezers. As it vaporizes, it absorbs heat. How much heat can be removed by 2.00 kg of Freon as it vaporizes inside the coils of a refrigerator? The heat of vaporization of Freon is 38.6 cal/g.

6.81 Calculate the total amount of heat needed to change 0.500 kg of ice at −10.°C into 0.500 kg of steam at 120.°C. Do this by calculating the heat required for each of the following steps and adding to get the total:

 Step 1. Ice (−10.°C) → ice (0.°C)

 Step 2. Ice (0.°C) → water (0.°C)

 Step 3. Water (0.°C) → water (100.°C)

 Step 4. Water (100.°C) → steam (100.°C)

 Step 5. Steam (100.°C) → steam (120.°C)

Additional Exercises

6.82 What is the density of argon gas in g/mL at STP?

6.83 Explosives react very rapidly and produce large quantities of heat and gaseous products. When nitroglycerine explodes, several gases are produced:

$$4C_3H_5O_9N_3(\ell) \rightarrow 12CO_2(g) + O_2(g) + 6N_2(g) + 10H_2O(g)$$

Suppose 11 g of nitroglycerine was sealed inside a 1.0-L soda bottle and detonated. Assume the bottle would not break, and the temperature immediately after detonation was 750 K. Calculate the pressure of the gases inside the bottle in atmospheres.

6.84 How many liters of oxygen gas, O_2, will it take to completely react with 2.31 L of hydrogen gas, H_2, to produce water? Assume both gases are at the same pressure and temperature. What type of reaction is this?

6.85 Review Table 6.4; then propose an explanation for the fact that hydrogen bonding decreases the vapor pressure of liquids made up of molecules with similar molecular weights.

6.86 Some people say they have a hard time breathing on top of a tall mountain because there is less oxygen in the air they breathe. Is this a true statement? Explain your answer.

Chemistry for Thought

6.87 As solids are heated, they melt. Explain this in terms of the effect of temperature on cohesive and disruptive forces.

6.88 Explain how a hot-air balloon works in terms of gas densities.

6.89 Which of the following gases would you expect to behave most ideally: He, Ar, or HCl? Explain.

6.90 Refer to Figure 6.6 and do the calculation. When a gas is heated, it expands. Explain how the gas in a hot-air balloon remains at constant pressure as the gas is heated. Hot-air balloons do *not* stretch.

6.91 Refer to Figure 6.7 and answer the question. Would a quart bottle full of gas actually contain the number of moles you calculated? Explain.

6.92 Refer to Figure 6.9 and answer the question. Propose an explanation for your answer in terms of the distances between molecules in rubber and plastic.

6.93 Refer to Figure 6.12 and answer the question. Water is sometimes made safe to drink by boiling. Explain why this might not work if you attempted to do it in an open pan on the summit of Mount Everest.

6.94 Refer to Figure 6.15 and answer the question. Suppose a sample of water was heated to the boiling point in a glass beaker, using a single burner. What would happen to the temperature of the boiling water if a second burner was added to help with the heating? Explain.

6.95 Suppose you put four 250-mL bottles of water into an ice chest filled with crushed ice. If the bottles were initially at a temperature of 23°C, calculate the number of grams of ice that would have to melt in order to cool the water in the bottles to 5°C. Assume the density of water is 1.00 g/mL.

Allied Health Exam Connection

The following questions are from these sources:

• *Nursing School Entrance Exam* © 2005, Learning Express, LLC.

• *McGraw-Hill's Nursing School Entrance Exams* by Thomas A. Evangelist, Tamara B. Orr, and Judy Unrein © 2009, The McGraw-Hill Companies, Inc.

• *NSEE Nursing School Entrance Exams,* 3rd edition © 2009, Kaplan Publishing.

• *Cliffs Test Prep: Nursing School Entrance Exams* by Fred N. Grayson © 2004, Wiley Publishing, Inc.

• *Peterson's Master the Nursing School and Allied Health Entrance Exams,* 18th edition by Marion F. Gooding © 2008, Peterson's, a Nelnet Company.

6.96 Definite shape and definite volume best describes a sample of:

 a. $I_2(s)$

 b. $Br_2(\ell)$

 c. $Cl(g)$

 d. $F_2(g)$

6.97 Which of the following is NOT characteristic of gases?

 a. They have a definite volume and shape

 b. They are low in density

 c. They are highly compressible

 d. They mix rapidly

6.98 Which of the following states has the highest average translational kinetic energy?

 a. solid

 b. liquid

 c. gas

 d. none of the above

6.99 In which of the following states of matter are molecules most likely to move freely?

 a. solid

 b. liquid

 c. gas

 d. all have similar freedom of movement

6.100 Which of the following indicates the relative randomness of molecules in the three states of matter?

 a. solid > liquid < gas

 b. liquid < solid < gas

 c. liquid > gas > solid

 d. gas > liquid > solid

6.101 In the kinetic molecular theory of gases, which of the following statements concerning average speeds is true?

 a. Most of the molecules are moving at the average speed.

 b. Any given molecule moves at the average speed most of the time.

 c. When the temperature increases, more of the molecules will move at the new average speed.

 d. When the temperature increases, fewer molecules will move at the new average speed.

6.102 All of the following statements underlie the kinetic molecular theory of gases EXCEPT:

 a. Gas molecules have no intermolecular forces.

 b. Gas particles are in random motion.

 c. Gas particles have no volume.

 d. The average kinetic energy is proportional to the temperature ($°C$) of the gas.

6.103 What are the differentiating factors between potential and kinetic energy?

 a. Properties—physical or chemical

 b. State—solid or liquid

 c. Temperature—high or low

 d. Activity—in motion or in storage

6.104 The transformation of a solid directly into a gas is called:

 a. vaporization.

 b. ionization.

 c. sublimation.

 d. polarization.

6.105 Evaporation can best be described as:

 a. the process in which molecules may have enough energy to leave the liquid phase and escape into the gaseous phase.

 b. a heating process.

 c. fusion.

 d. the process in which molecules in the solid phase absorb enough energy to begin the liquid phase.

6.106 When a vapor condenses into a liquid:

 a. it absorbs heat.

 b. it generates heat.

 c. its temperature rises.

 d. its temperature drops.

6.107 When solid iodine becomes gaseous iodine with no apparent liquid phase in between, the process is called:

 a. evaporation.

 b. condensation.

 c. sublimation.

 d. precipitation.

6.108 What does the term *pressure* mean when applied to a gas?

 a. weight

 b. how heavy the gas is

 c. mass divided by volume

 d. force exerted per unit area

6.109 When one liquid evaporates much faster than another liquid, the first liquid is said to be more:

 a. volatile.

 b. transient.

 c. viscous.

 d. evaporative.

6.110 Which of the following laws is related to this expression: $P_T = P_1 + P_2 + P_3$?

 a. Boyle's law

 b. Charles's law

 c. Gay-Lussac's law

 d. Dalton's law

6.111 Which law predicts that if the temperature (in Kelvin) doubles, the pressure will also double?

 a. Boyle's law

 b. Charles's law

c. Gay-Lussac's law

d. Dalton's law

6.112 The inhaling and exhaling of air by the human lungs is mainly an application of:

 a. Boyle's law—the inverse relationship between the pressure and the volume of a gas.

 b. The volume of a gas at standard temperature and pressure (STP).

 c. Charles's law—the direct relationship between the temperature and the volume of a gas.

 d. The number of O_2 and CO_2 particles per mole.

6.113 How much heat is required to raise the temperature of 100. grams of water from 25°C (near room temperature) to 100.°C (its boiling point)? The specific heat of water is approximately 4.2 J per g-K.

 a. 3.2×10^4 J

 b. 32 J

 c. 4.2×10^4 J

 d. 76 J

6.114 How many calories are required to change the temperature of 2.0×10^3 grams of H_2O from 20.°C to 38°C?

 a. 36 calories

 b. 24 calories

 c. 18 calories

 d. 12 calories

6.115 If temperature and pressure are held constant for a sample of gas, and the number of moles is doubled, in what manner will the volume change?

 a. It will double.

 b. It will quadruple.

 c. It will be halved.

 d. There will be no change.

6.116 A gas has a volume of 0.25 liter at a pressure of 1.0 atmosphere. If the volume increases to 0.50 liter and the temperature remains constant, the new pressure will be:

 a. 1.0 atmosphere

 b. 0.50 atmosphere

 c. 0.25 atmosphere

 d. 2.0 atmospheres

6.117 Given 3.00 moles of krypton gas, Kr(g), how many liters will this sample occupy at STP?

 a. 11.2

 b. 22.4

 c. 44.8

 d. 67.2

6.118 A sample of helium at 25°C occupies a volume of 725 mL at 730. mmHg. What volume will it occupy at 25°C and 760. mmHg?

 a. 755 mL

 b. 760. mL

 c. 696 mL

 d. 730. mL

6.119 One liter of a certain gas, under standard conditions, weighs 1.16 grams. A possible formula for the gas is:

 a. C_2H_2

 b. CO

 c. NH_3

 d. O_2

6.120 A mixture consisting of 8.0 g of oxygen and 14 g of nitrogen is prepared in a container such that the total pressure is 750 mmHg. The partial pressure of oxygen in the mixture is:

 a. 125 mmHg.

 b. 500 mmHg.

 c. 135 mmHg.

 d. 250 mmHg.

6.121 A sample of nitrogen at 20°C in a volume of 875 mL has a pressure of 730. mmHg. What will be its pressure at 20°C if the volume is changed to 955 mL?

 a. 750. mmHg

 b. 658 mmHg

 c. 797 mmHg

 d. 669 mmHg

6.122 The weight in grams of 22.4 liters of nitrogen (atomic weight = 14) is:

 a. 3.

 b. 7.

 c. 14.

 d. 28.

6.123 Water at sea level boils at what temperature?

 a. 100°F

 b. 180°F

 c. 212°C

 d. 373 K

6.124 Which gas law states that the volume of a gas is inversely proportional to the pressure?

 a. Ideal gas law

 b. Boyle's law

 c. Combined gas law

 d. Charles's law

6.125 When a liquid is at its boiling point, the vapor pressure of the liquid:

 a. is less than the external pressure on the liquid.

 b. is equal to the external pressure on the liquid.

 c. is greater than the external pressure on the liquid.

 d. can be either less or greater than the external pressure on the liquid.

7

Solutions and Colloids

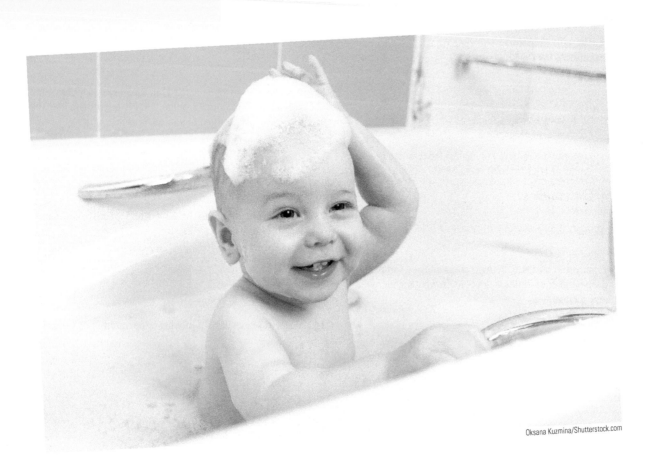

Oksana Kuzmina/Shutterstock.com

▶ Case Study

Joel noticed red patches on his 18-month-old daughter's arms and legs where her elbows and knees bent. He suspected she had acquired an infection by playing on the floor or in the playground sand. Joel felt upset that he had not been more diligent about Clara's hygiene. In an effort to make certain that she was "extra clean," he added liquid bubble bath to her bath water and scrubbed her with a sponge. After 6 days of this treatment, Clara's symptoms got worse. Dry, scaly patches appeared on her skin. The patches itched, and Clara responded by scratching them incessantly. Her grandmother claimed that Clara had eczema. Grandmother also suggested that Joel stop bathing Clara with soap and water and give her an "oatmeal bath." This sounded like an old wives' tale, but Joel consulted a dermatologist who applauded his mother's

old-fashioned wisdom. The doctor explained that Clara had what is medically called "atopic dermatitis" and that colloidal oatmeal baths are helpful in alleviating the symptoms.

▶ **What risk factors are associated with atopic dermatitis? How common is the condition? Can it be cured? How does colloidal oatmeal help?**

Follow-up to this Case Study appears at the end of the chapter before the *Concept Summary*.

Learning Objectives

When you have completed your study of this chapter, you should be able to:

1 Classify mixtures as solutions or nonsolutions based on their appearance. **(Section 7.1)**

2 Demonstrate your understanding of terms related to the solubility of solutes in solution. **(Section 7.2)**

3 Predict in a general way the solubilities of solutes in solvents on the basis of molecular polarity. **(Section 7.3)**

4 Calculate solution concentrations in units of molarity, weight/weight percent, weight/volume percent, and volume/volume percent. **(Section 7.4)**

5 Describe how to prepare solutions of specific concentration using pure solutes and solvent, or solutions of greater concentration than the one desired. **(Section 7.5)**

6 Do stoichiometric calculations based on solution concentrations. **(Section 7.6)**

7 Do calculations based on the colligative solution properties of boiling point, freezing point, and osmotic pressure. **(Section 7.7)**

8 Describe the characteristics of colloids. **(Section 7.8)**

9 Describe the process of dialysis, and compare it to the process of osmosis. **(Section 7.9)**

Earlier, homogeneous matter was classified into two categories—pure substances and mixtures (see **Figure 1.5**, page 5). Since then, the discussion has been limited to pure substances. We now look at homogeneous mixtures called solutions and their distant relatives, colloidal suspensions.

Solutions and colloidal suspensions are very important in our world. They bring nutrients to the cells of our bodies and carry away waste products. The ocean is a solution of water, sodium chloride, and many other substances (even gold). Many chemical reactions take place in solution—including most of those discussed in this book.

7.1 | Physical States of Solutions

Learning Objective

1. Classify mixtures as solutions or nonsolutions based on their appearance.

Solutions are homogeneous mixtures of two or more substances in which the components are present as atoms, molecules, or ions. These uniformly distributed particles are too small to reflect light, and as a result solutions are transparent (clear); light passes through them. In addition, some solutions are colored. The component particles are in constant motion (remember the kinetic theory, Section 6.2) and do not settle under the influence of gravity.

In most solutions, a larger amount of one substance is present compared to the other components. This most abundant substance in a solution is called the **solvent**, and any other components are called **solutes**. Most people normally think of solutions as liquids, but solutions in solid and gaseous forms are known as well. The state of a solution is often the same as the state of the solvent. This is illustrated by **Table 7.1**, which lists examples

solution A homogeneous mixture of two or more substances in which the components are present as atoms, molecules, or ions.

solvent The substance present in a solution in the largest amount.

solute One or more substances present in a solution in amounts less than that of the solvent.

dissolving A term used to describe the process of solution formation when one or more solutes are dispersed in a solvent to form a homogeneous mixture.

of solutions in various states. The original states of the solvents and solutes are given in parentheses. Solution formation takes place when one or more solutes **dissolve** in a solvent.

TABLE 7.1 Solutions in Various States

Solution	Solution State	Solvent	Solute
Salt water	Liquid	Water (liquid)	Sodium chloride (solid)
Alcoholic beverage	Liquid	Water (liquid)	Alcohol (liquid)
Carbonated water	Liquid	Water (liquid)	Carbon dioxide (gas)
Gold alloy (jewelry)	Solid	Gold (solid)	Copper (solid)
Gold amalgam	Solid	Gold (solid)	Mercury (liquid)
Hydrogen in palladium	Solid	Palladium (solid)	Hydrogen (gas)
Air	Gaseous	Nitrogen (gas)	Oxygen (gas)
Humid oxygen	Gaseous	Oxygen (gas)	Water (liquid)
Camphor in nitrogen	Gaseous	Nitrogen (gas)	Camphor (solid)

Example 7.1 Identifying Solution, Solvent, and Solutes

Identify the solvent and solute(s) in each of the following solutions:

a. A sample of natural gas contains 97% methane (CH_4), 1.5% ethane (C_2H_6), 1% carbon dioxide (CO_2), and 0.5% nitrogen (N_2).
b. The label on a bottle of Scotch whiskey says, among other things, 86 proof. This means the contents contain 43% ethyl alcohol (C_2H_5OH). Assume water to be the only other component.
c. The physiological saline solution used in hospitals contains 0.9 g NaCl for each 100 g water.

Solution

a. Methane is present in the largest amount; it is the solvent. All other components are solutes.
b. The whiskey is 43% alcohol and 57% water. Thus, water is the solvent and alcohol is the solute. The whiskey actually contains small amounts of many other components, which impart flavor and color. These components are also solutes.
c. Water is the component present in the larger amount and is therefore the solvent.

✔ **LEARNING CHECK 7.1** Identify the solvent and solute(s) in the following solutions:

a. White gold is a solid solution containing 58.3% gold, 17.0% copper, 7.7% zinc, and 17.0% nickel.
b. Clean, dry air contains about 78.1% nitrogen, 21.0% oxygen, 0.9% argon, and very small amounts of other gases.

7.2 | Solubility

Learning Objective

2. Demonstrate your understanding of terms related to the solubility of solutes in solution.

A few experiments with water, sugar, cooking oil, and rubbing alcohol (isopropyl alcohol) illustrate some important concepts associated with solution formation. Imagine you have three drinking glasses, each containing 100 mL (100 g) of pure water.

You begin by adding small amounts of the other three substances to the glasses. The mixtures are stirred well, with results shown in **Figure 7.1A**. The sugar and alcohol form homogeneous mixtures (solutions) with the water, while the oil forms a two-layer (heterogeneous) mixture. **Soluble substances** such as the sugar and alcohol dissolve completely in the solvent and form solutions. **Insoluble substances** do not dissolve in the solvent. The term **immiscible** is used to describe a liquid solute that does not dissolve in a liquid solvent.

soluble substance A substance that dissolves to a significant extent in a solvent.

insoluble substance A substance that does not dissolve to a significant extent in a solvent.

immiscible A term used to describe liquids that are insoluble in each other.

A Small amounts of substance added

B Larger amounts of substance added

Figure 7.1 The homogeneity of solutions and mixtures of sugar (limited solubility), cooking oil (immiscible), and alcohol (completely soluble).

The experiments are continued by adding more sugar, alcohol, and oil to the water samples. This ultimately leads to the situation shown in **Figure 7.1B**. Regardless of the amount of alcohol added, a solution forms. In fact, other experiments could be done that show that water and isopropyl alcohol are completely soluble in each other and will mix in any proportion. Sugar behaves differently. About 204 g can be dissolved in the 100 mL of water (at 20°C), but any additional sugar simply sinks to the bottom of the glass and remains undissolved. Thus, the sugar has a solubility in water of 204 g/100 g H_2O. The term **solubility** refers to the maximum amount of solute that can be dissolved in a specific amount of solvent at a specific temperature. With the oil–water mixture, any additional oil simply floats on the surface of the water along with the oil added initially.

The solubilities of a number of solutes in water are given in **Table 7.2**. The use of specific units, such as grams of solute per 100 g of water, makes it possible to compare solubilities precisely. However, such precision is often unnecessary, and when this is the case, the approximate terms defined in **Table 7.3** will be used.

solubility The maximum amount of solute that can be dissolved in a specific amount of solvent under specific conditions of temperature and pressure.

TABLE 7.2 Examples of Solute Solubilities in Water (0°C)

Solute Name	Formula	Solubility (g solute/100 g H_2O)
Ammonium chloride	NH_4Cl	29.7
Ammonium nitrate	NH_4NO_3	118.3
Ammonium orthophosphate	$NH_4H_2PO_4$	22.7
Ammonium sulfate	$(NH_4)_2SO_4$	70.6
Calcium carbonate	$CaCO_3$	0.0012
Calcium chloride	$CaCl_2$	53.3
Calcium sulfate	$CaSO_4$	0.23
Potassium carbonate	K_2CO_3	101
Potassium chloride	KCl	29.2
Sodium bicarbonate	$NaHCO_3$	6.9
Sodium bromide	$NaBr$	111
Sodium carbonate	Na_2CO_3	7.1
Sodium chloride	$NaCl$	35.7
Sodium iodide	NaI	144.6
Ascorbic acid (vitamin C)	$C_6H_8O_6$	33
Ethyl alcohol	C_2H_5OH	∞^a
Ethylene glycol (antifreeze)	$C_2H_4(OH)_2$	∞
Glycerin	$C_3H_5(OH)_3$	∞
Sucrose (table sugar)	$C_{12}H_{22}O_{11}$	179.2

aSoluble in all proportions.

TABLE 7.3 Approximate Solubility Terms

Solute Solubility (g solute/ 100 g H_2O)	Solubility Term
Less than 0.1	Insoluble
0.1–1	Slightly soluble
1–10	Soluble
Greater than 10	Very soluble

saturated solution A solution that contains the maximum amount possible of dissolved solute in a stable situation under the prevailing conditions of temperature and pressure.

supersaturated solution An unstable solution that contains an amount of solute greater than the solute solubility under the prevailing conditions of temperature and pressure.

A solution in which the maximum amount of solute has been dissolved in a quantity of solvent is called a **saturated solution**. The final sugar solution described in the earlier experiments (204 g in 100 g H_2O) was a saturated solution. Solutions in which the amount of solute dissolved is greater than the solute solubility are called **supersaturated solutions**. Supersaturated solutions are usually prepared by forming a nearly saturated solution at a high temperature and then cooling the solution to a lower temperature at which the solubility is lower. Such solutions are not stable. The addition of a small amount of solid solute (or even a dust particle) will usually cause the excess solute to crystallize out of solution until the solution becomes saturated (see **Figure 7.2**). The temperature dependence of solute solubility is illustrated in **Figure 7.3**.

1. A supersaturated solution just after adding a small seed crystal.

2. Rapid crystallization begins on the seed crystal.

3. The excess solute has all crystallized, leaving behind a saturated solution.

Figure 7.2 Crystallization converts a supersaturated solution to a saturated solution.

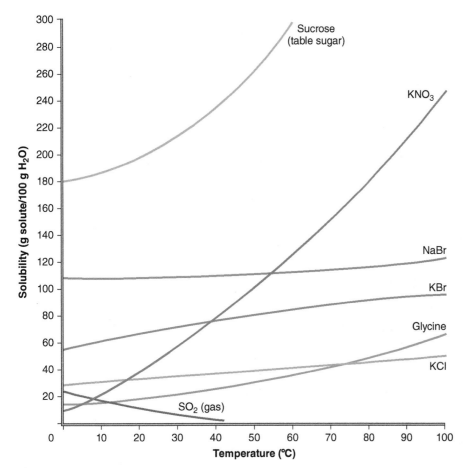

Figure 7.3 The effect of temperature on solute solubility.

Whereas the solubility of most liquids and solids in water increases with temperature, the solubility of most gases in water decreases as the temperature increases (see the SO_2 curve in **Figure 7.3**). This is easily demonstrated for gaseous CO_2 by opening both a cold and a warm carbonated beverage. The solubility of gaseous solutes is also influenced significantly by pressure; the effect on liquid or solid solutes is minimal. It has been found that the solubility of many gases is directly proportional to the pressure of the gas above the solution at constant temperature. Thus, if the gas pressure is doubled, the solubility doubles.

The pressure dependence of gas solubility provides the "sparkle" for carbonated beverages. The cold beverage is saturated with CO_2 and capped under pressure. When the bottle is opened, the pressure is relieved, and the gas, now less soluble, comes out of solution as fine bubbles. A similar effect sometimes takes place in the bloodstream of deep-sea divers. While submerged, they inhale air under pressure that causes nitrogen to be more soluble in the blood than it is under normal atmospheric pressure. If the diver is brought to the lower pressure on the surface too quickly, the excess dissolved nitrogen comes out of solution and forms bubbles in the blood and joints. The result, called decompression sickness or the bends, is painful and dangerous. The chances of getting the bends are decreased by breathing a mixture of oxygen and helium rather than air (oxygen and nitrogen), because helium is less soluble in the body fluids than nitrogen.

Example 7.2 Temperature and Solute Solubilities

A 260-g sample of table sugar is added to 100 g of water at 70°C. The sugar dissolves completely. The resulting solution is allowed to cool slowly. At 30°C, crystals of sugar begin to form and increase in size as the solution cools to 20°C. Refer to **Figure 7.3** and describe the nature of the solution at 70°, 60°, 50°, 40°, 30°, and 20°C.

Solution

The solubility of sugar is greater than 260 g/100 g H_2O at all temperatures above 50°C. Therefore, the solution is unsaturated at 70°C and 60°C. At 50°C, the solubility is equal to the amount dissolved, so the solution is saturated. At 40°C, the solution contains more dissolved sugar than it should on the basis of solubility, and the solution is supersaturated. At 30°C, the excess sugar crystallizes from solution, and the resulting solution becomes saturated. From that point, excess sugar continues to crystallize from the solution, and the solution remains saturated to 20°C.

✔ **LEARNING CHECK 7.2** Refer to Figure 7.3. Saturated solutions of potassium nitrate (KNO_3) and sodium bromide (NaBr) are made at 80°C. Which solution contains more solute per 100 g of H_2O? The solutions are cooled to 50°C, and excess solute crystallizes out of each solution. Which solution at 50°C contains more solute per 100 g of H_2O? Are the solutions saturated at 50°C?

7.3 | The Solution Process

Learning Objective

3. Predict in a general way the solubilities of solutes in solvents on the basis of molecular polarity.

The how and why of solution formation are the topics of this section. How are solute particles removed from the bulk solute and uniformly distributed throughout the solvent? Why are some solutes very soluble whereas others are not?

Consider the formation of a saltwater solution. Earlier we pointed out (see **Section 4.2**, page 103) that solid ionic compounds are collections of ions held together by attractions between the opposite ionic charges. When an ionic compound dissolves, the orderly ionic arrangement is destroyed as the interionic attractions are overcome. Thus, the attractive forces between water molecules and ions must be stronger than the interionic attractions within the crystal.

The solution-forming process for an ionic solute is represented in **Figure 7.4**. When the solid ionic crystal is placed in water, the polar water molecules become oriented so that the negative oxygen portion points toward positive sodium ions, and the positive

Figure 7.4 The dissolving of an ionic substance in water.

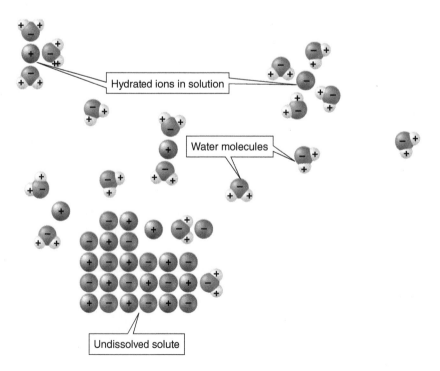

Hydrated ions in solution

Water molecules

Undissolved solute

hydrogen portion points toward negative chloride ions. As the polar water molecules begin to surround ions on the crystal surface, they tend to create a shielding effect that reduces the attraction between the ion and the remainder of the crystal. As a result, the ion breaks away from the crystal surface and is surrounded by water molecules. Ions surrounded by water molecules in solution are called **hydrated ions**. As each ion leaves the surface, others are exposed to the water, and the crystal is picked apart ion by ion. Once in solution, the hydrated ions are uniformly distributed by stirring or by random collisions with other molecules or ions.

The random motion of solute ions in solution causes them to collide with one another, with solvent molecules, and occasionally with the surface of any undissolved solute. Ions undergoing such collisions occasionally stick to the solid surface and thus leave the solution. When the number of ions in solution is low, the chances for collision with the undissolved solute are low. However, as the number of ions in solution increases, so do the chances for such collisions, and more ions leave the solution and become attached once again to the solid. Eventually, the number of ions in solution reaches a level at which ions return to the undissolved solute at the same rate as other ions leave. At this point the solution is saturated, an equilibrium condition. Even though the processes of leaving and returning continue, no net changes in the number of ions in solution or the amount of undissolved solute can be detected as time passes, and an experimenter would observe that no more solid solute dissolves.

Supersaturated solutions form when there are no sites with which the excess solute particles can collide. The addition of such sites in the form of a seed crystal of solute causes the excess solute to crystallize from solution very quickly.

Polar but non-ionic solutes such as sugar dissolve in water in much the same way as ionic solutes. The only difference is the attraction of polar water molecules to both poles of the solute molecules. The process is represented in **Figure 7.5**.

This solution-forming process also explains the low solubility of some solutes. A solute will not dissolve in a solvent if (1) the forces between solute particles are too strong to be overcome by interactions with solvent particles or (2) the solvent particles are more strongly attracted to each other than to solute particles.

The cooking oil of the earlier experiment did not dissolve in water because the polar water molecules were attracted to each other more strongly than they were to the nonpolar oil molecules. The cooking oil will dissolve in a nonpolar solvent such as gasoline

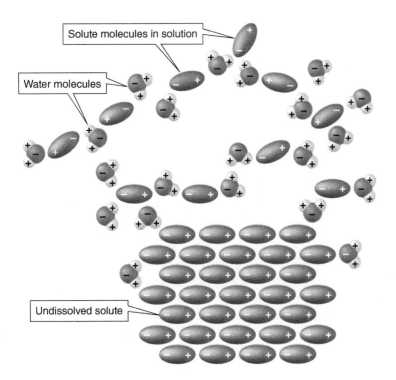

Figure 7.5 The dissolving of a polar solute in water.

Solute molecules in solution

Water molecules

Undissolved solute

or carbon tetrachloride (CCl_4). In these solvents, the weak forces between oil molecules and solvent molecules are no stronger than the weak forces between nonpolar solvent molecules. A good rule of thumb is "like dissolves like." Thus, polar solvents will dissolve polar or ionic solutes, and nonpolar solvents will dissolve nonpolar or non-ionic solutes.

These generalizations apply best to non-ionic compounds. Some ionic compounds (such as $CaCO_3$ and $CaSO_4$) have very low solubilities in water (see **Table 7.2**). Both the attractive forces between ions and the attraction of polar solvent molecules for ions are electrical and depend on such characteristics as ionic charge and size. Changes in ionic compounds that increase the attractive forces between ions also increase the forces between ions and polar solvent molecules, but not always by the same amount. Thus, simple rules for quantitatively predicting the water solubility of ionic compounds are not available. (We can't easily predict how many grams will dissolve in a specific quantity of water.) However, ionic compounds are generally insoluble in nonpolar solvents; they usually follow the solubility guidelines given in **Table 7.4** when water is the solvent.

TABLE 7.4 General Solubilities of Ionic Compounds in Water

Compounds	Solubility	Exceptions
Group IA (Na^+, K^-, etc.) and NH_4^+	Soluble	
Nitrates (NO_3^-)	Soluble	
Acetates ($C_2H_3O_2^-$)	Soluble	
Chlorides (Cl^-)	Soluble	Chlorides of Ag^+, Pb^{2+}, Hg^+ (Hg_2^{2+})
Sulfates (SO_4^{2-})	Soluble	Sulfates of Ba^{2+}, Sr^{2+}, Pb^{2+}, Hg^+ (Hg_2^{2+})
Carbonates (CO_3^{2-})	Insoluble[a]	Carbonates of group IA and NH_4^+
Phosphates (PO_4^{3-})	Insoluble[a]	Phosphates of group IA and NH_4^+

[a]Many hydrogen carbonates (HCO_3^-) and phosphates (HPO_4^{2-}, $H_2PO_4^-$) are soluble.

Example 7.3 Predicting Solute Solubilities

Predict the solubility of the following solutes in the solvent indicated:

a. ammonia gas (NH_3) in water
b. oxygen gas (O_2) in water
c. $Ca(NO_3)_2$ in water
d. $Mg_3(PO_4)_2$ in water
e. paraffin wax (nonpolar) in CCl_4

Solution

a. Soluble: NH_3 is polar—like dissolves like. The actual solubility at 20°C is 51.8 g/100 g H_2O.
b. Insoluble: O_2 is nonpolar. The actual solubility at 20°C is 4.3×10^{-3} g/100 g H_2O.
c. Soluble: nitrates are soluble in water (Table 7.4).
d. Insoluble: phosphates are insoluble except those of group IA(1) and NH_4^+ (Table 7.4).
e. Soluble: CCl_4 is nonpolar—like dissolves like.

✔ **LEARNING CHECK 7.3**

a. Refer to Table 7.4 and decide which of the following salts you would use as a solute if you wanted to prepare a solution that contained as many barium ions (Ba^{2+}) as possible: $BaSO_4$, $Ba(NO_3)_2$, or $BaCO_3$. Explain your answer.
b. A common sight on the evening TV news in the late 1960s and 1970s was sea birds soaked with crude oil being cleaned by concerned people. Which of the following do you think was used to clean the oil from the birds? Explain your answer. Light mineral oil, purified water, seawater, or gasoline.

Be careful not to confuse solute solubility with the rate at which a solute dissolves. Under some conditions, even a very soluble solute will dissolve slowly—for example, a lump of rock candy (sugar) dissolves much more slowly than an equal weight of granulated sugar. The dissolving rate can be increased in a number of ways (see **Figure 7.6**):

1. *Crushing or grinding the solute*—small particles provide more surface area for solvent attack and dissolve more rapidly than larger particles.
2. *Heating the solvent*—solvent molecules move faster and have more frequent collisions with solute at higher temperatures.
3. *Stirring or agitating the solution*—stirring removes locally saturated solution from the vicinity of the solute and allows unsaturated solvent to take its place.

Heat is usually absorbed or released when a solute dissolves in a solvent. When heat is absorbed, the process is endothermic, and the solution becomes cooler. The fact that heat absorption leads to cooling might sound strange, and it needs to be explained. The heat is absorbed by the interacting solvent and solute molecules, and it is removed from the solvent molecules that are not involved in the actual attack on the solute. Since most of the solvent is in the latter category, the entire solution becomes cooled.

When heat is released, the process is exothermic, and the solution temperature increases, this time because the heat released by the interacting molecules is absorbed by the uninvolved solvent. This behavior is the more common one. These processes can be represented by equations:

Figure 7.6 Heat and agitation increase the rate at which solutes dissolve. Will sugar dissolve faster in hot tea or iced tea?

$$\text{Endothermic:} \qquad \text{Solute + solvent + heat} \rightarrow \text{solution} \qquad (7.1)$$

$$\text{Exothermic:} \qquad \text{Solute + solvent} \rightarrow \text{solution + heat} \qquad (7.2)$$

An endothermic solution process is the basis for commercially available instant cold packs. Water is sealed in a thin plastic bag and placed inside a larger, stronger bag together with a quantity of solid solute (NH_4Cl or NH_4NO_3). When the inner bag is broken by squeezing, the solid dissolves in the water, heat is absorbed, and the mixture becomes quite cold (see **Figure 7.7**).

Figure 7.7 Exothermic (top) and endothermic (bottom) solution processes.

1 Solid NaOH not yet added to water.

2 Solution becomes hot when NaOH dissolves in water.

1 Solid NH_4NO_3 not yet added to water.

2 Solution becomes cool when NH_4NO_3 dissolves in water.

7.4 | Solution Concentrations

Learning Objective

4. Calculate solution concentrations in units of molarity, weight/weight percent, weight/volume percent, and volume/volume percent.

In Chapter 5, quantitative calculations were done using equations for reactions such as the following:

$$CH_4(g) + 2O_2(g) \rightarrow CO_2(g) + 2H_2O(g) \tag{7.3}$$

The reaction represented by this equation, and most of the reactions we have discussed to this point in the book, involve pure substances as reactants and products. However, many of the reactions done in laboratories, and most of those that go on in our bodies, take place between substances dissolved in a solvent to form solutions. In our bodies, the solvent is almost always water. A double-replacement reaction of this type done in laboratories is represented by the following equation:

$$2AgNO_3(aq) + Na_2CO_3(aq) \rightarrow Ag_2CO_3(s) + 2NaNO_3(aq) \tag{7.4}$$

We remember from Chapter 5 that the coefficients in equations such as Equation 7.3 allow the relative number of moles of pure reactants and products involved in the reaction to be determined. These relationships coupled with the mole definition in terms of masses then yield factors that can be used to solve stoichiometric problems involving the reactants and products. Similar calculations can be done for reactions that take place between the solutes of solutions if the amount of solute contained in a specific quantity of the reacting solutions is known. Such relationships are known as solution **concentrations**. Solution concentrations may be expressed in a variety of units, but only two, molarity and percentage, will be discussed at this time.

The **molarity (M)** of a solution expresses the number of moles of solute contained in exactly 1 L of the solution:

concentration The relationship between the amount of solute and the specific amount of solution in which it is contained.

molarity (M) A solution concentration expressed in terms of the number of moles of solute contained in a liter of solution.

$$M = \frac{\text{moles of solute}}{\text{liters of solution}} \tag{7.5}$$

It is important to note that even though a concentration in molarity expresses the number of moles contained in 1 L of solution, the molarity of solutions that have total volumes different from 1 L can be calculated using Equation 7.5. We simply determine the number of moles of solute contained in a specified volume of solution, then express that volume in liters before substituting the values into Equation 7.5.

Example 7.4 Calculating Solution Concentrations

Express the concentration of each of the following solutions in terms of molarity:

a. 2.00 L of solution contains 1.50 mol of solute.
b. 150. mL of solution contains 0.210 mol of solute.
c. 315 mL of solution contains 10.3 g of isopropyl alcohol, C_3H_7OH.

Solution

a. Because the data are given in terms of moles of solute and liters of solution, the data may be substituted directly into Equation 7.5:

$$M = \frac{1.50 \text{ mol solute}}{2.00 \text{ L solution}} = 0.750 \frac{\text{mol solute}}{\text{L solution}}$$

The solution is 0.750 molar, or 0.750 M.

b. In this problem, the number of moles of solute is given, but the volume of solution is given in milliliters rather than liters. The volume must first be converted into liters, then the data may be substituted into Equation 7.5:

$$(150. \text{ mL solution})\left(\frac{1 \text{ L}}{1000 \text{ mL}}\right) = 0.150 \text{ L solution}$$

$$M = \frac{0.210 \text{ mol solute}}{0.150 \text{ L solution}} = 1.40 \frac{\text{mol solute}}{\text{L solution}}$$

The solution is 1.40 molar, or 1.40 M.

c. Before Equation 7.5 can be used, the number of moles of solute (isopropyl alcohol) must be determined. This is done as follows, where the factor

$$\frac{1 \text{ mol alcohol}}{60.1 \text{ g alcohol}}$$

comes from the calculated molecular weight of 60.1 u for isopropyl alcohol:

$$(10.3 \text{ g alcohol})\left(\frac{1 \text{ mol alcohol}}{60.1 \text{ g alcohol}}\right) = 0.171 \text{ mol alcohol}$$

Next, this number of moles of solute and the solution volume expressed in liters are substituted into Equation 7.5:

$$M = \frac{0.171 \text{ mol alcohol}}{0.315 \text{ L solution}} = 0.543 \frac{\text{mol alcohol}}{\text{L solution}}$$

The solution is 0.543 molar, or 0.543 M.

✔ **LEARNING CHECK 7.4** Express the concentrations of each of the following solutions in terms of molarity:

a. 2.50 L of solution contains 1.25 mol of solute.
b. 225 mL of solution contains 0.486 mol of solute.
c. 100 mL of solution contains 2.60 g of NaCl solute.

Sometimes a detailed knowledge of the actual stoichiometry of a process involving solutions is not needed, but some information about the solution concentrations would be useful. When this is true, solution concentrations are often expressed as percentages. In general, a concentration in **percent** gives the amount of solute contained in 100 parts of solution. You saw in Section 1.10 (see page 4), that it is convenient to use a formula for percentage calculations because we seldom work with exactly 100 units of anything. The general formula used is:

$$\% = \frac{\text{part}}{\text{total}} \times 100$$

Three different percent concentrations for solutions are used. A **weight/weight percent** (abbreviated w/w) is the mass of solute contained in 100 mass units of solution. Thus, a 12.0% (w/w) sugar solution contains 12.0 grams of sugar in each 100 g of solution. In terms of this concentration, the general formula for percent calculations becomes

$$\%(\text{w/w}) = \frac{\text{solute mass}}{\text{solution mass}} \times 100 \qquad (7.6)$$

Any mass units may be used, but the mass of solute and solution must be expressed in the same units.

A more commonly used percent concentration is **weight/volume percent** (abbreviated w/v), which is the grams of solute contained in 100 mL of solution. In these units, a 12.0% (w/v)

percent A solution concentration that expresses the amount of solute in 100 parts of solution.

weight/weight percent A concentration that expresses the mass of solute contained in 100 mass units of solution.

weight/volume percent A concentration that expresses the grams of solute contained in 100 mL of solution.

Stay Hydrated

Getting enough to drink? Staying hydrated is always important, but it becomes even more so as temperatures increase. Hydration is critical to your heart health and general body function. Proper hydration makes your heart's job of delivering blood to your muscles and brain easier. This helps your muscles work more efficiently, and provides a general feeling of being healthy. Dehydration can cause unpleasant symptoms such as headache and fatigue, or even life-threatening conditions of *heat exhaustion* and *heat stroke*.

The amount of water a person needs to stay adequately hydrated depends on factors such as climatic conditions, exercise intensity, clothing choices, and general health. Also, because perspiration rates vary between individuals, someone who perspires more must drink more fluids than someone who perspires less. Certain medical conditions, such as diabetes, cystic fibrosis, obesity, or heart disease also influence the amount one must drink to maintain proper hydration. Some medications and caffeinated beverages can act as diuretics, causing the body to lose fluid. People over the age of 50 should pay special attention to hydration.

Although you can choose from a variety of beverages, plain water is the best and least expensive drink to use for proper hydration. Some fruits and vegetables contain a high percentage of water, and eating them can contribute to proper body hydration. Sports drinks containing electrolytes are also often high in sugar and calories, and should be reserved for use in activities involving vigorous exercise or activities done in very hot weather. Drinking water before you exercise or go out in hot weather is also important to alleviate heart strain. If you do become dehydrated, avoid sugary drinks, which can lead to you becoming nauseated.

One way to know how much fluid you need is to weigh yourself before and after exercising. This practice is especially important for athletes training during hot weather. Every pound lost during exercise equates to a pint of fluid that should be replenished. Some athletes have been known to lose as much as 5 pounds during a training session. Sweat is an indicator of proper hydration. If you are not sweating during a workout, you might be in danger of developing heat exhaustion. Thirst isn't the best indicator that you need a drink. Thirst signals that dehydration is *already occurring*. One indicator of proper hydration levels is the color of your urine. Pale, clear urine indicates proper hydration, whereas dark urine is a signal to drink more fluids. Simply sitting outside on a hot day, even without exercising, can cause dehydration. Be aware of your hydration status to avoid health problems.

Dangerous results of dehydration fall into two categories called "heat exhaustion" and the more severe "heatstroke." What is the difference?

Heat exhaustion (also called *heat prostration* or *heat collapse*) is caused by the loss of too much fluid by perspiration. This results in a low body fluid volume and low sodium levels in body fluids. Symptoms include intense thirst, weakness, paleness, dizziness, nausea, fainting, and confusion. The condition is corrected by the intake of fluids containing electrolytes (such as sports drinks) and resting in a cool location.

Heatstroke (also called *sunstroke*) is the most serious heat-related condition. It is characterized by hot dry skin, labored breathing, rapid pulse, nausea, and blurred vision. These symptoms sometimes progress to irrational behavior and coma. During an episode of heatstroke, body temperature regulation fails and the body temperature might exceed 105°F. Prolonged exposure to environmental heat or strenuous physical activity cause heatstroke. Heatstroke is a medical emergency that requires immediate professional care. If you suspect heatstroke is taking place, use first aid measures such as removing the victim's clothing and soaking the victim in ice water until help arrives. If the victim is conscious, give him or her fluids.

As described above, failure to properly hydrate causes complications that follow a path of increasingly serious symptoms leading from dehydration to heatstroke. Physiologically, the following take place: first, blood volume declines, followed by a decrease in the amount of water in and around body cells. Sweat, flushing, and dry mouth signal dehydration. As the body temperature increases, muscle weakness and cramps set in (due to increased electrolyte concentrations). If this *heat exhaustion* is not addressed quickly, the body temperature increases sharply; heart rate increases and life-threatening *heatstroke* might result. Awareness of your hydration status is vital to your health and safety.

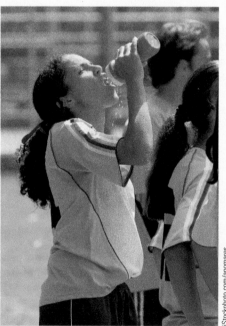

Regular fluid intake is a must for participants in vigorous physical activities.

sugar solution would contain 12.0 g of sugar in each 100 mL of solution. This percent concentration is normally used when the solute is a solid and the solvent and resulting solutions are liquids. The general formula for the calculation of percent concentrations in these units is

$$\%(w/v) = \frac{\text{grams of solute}}{\text{milliliters of solution}} \times 100 \qquad (7.7)$$

In weight/volume percent calculations, the solute amount is always given in grams, and the solution volume is always given in milliliters.

A percent concentration that is useful when the solute and solvent are either both liquids or both gases is **volume/volume percent** (abbreviated v/v). Concentrations given in these units express the number of volumes of solute found in 100 volumes of solution. For these units, the general percentage equation becomes

volume/volume percent
A concentration that expresses the volume of solute contained in 100 volumes of solution.

$$\%(v/v) = \frac{\text{solute volume}}{\text{solution volume}} \times 100 \qquad (7.8)$$

Any volume units may be used, but they must be the same for both the solute and the solution.

Example 7.5 More Solution Calculations

a. A solution contains 100. g of water and 1.20 g of solute. What is the %(w/w) concentration?
b. A solution is made by mixing 90.0 mL of alcohol with enough water to give 250. mL of solution. What is the %(v/v) concentration of alcohol in the solution?
c. A 150.-mL sample of saltwater is evaporated to dryness. A residue of salt weighing 27.9 g is left behind. Calculate the %(w/v) of the original saltwater.

Solution
a. Equation 7.6 is used:

$$\%(w/w) = \frac{\text{solute mass}}{\text{solution mass}} \times 100 = \frac{1.20 \text{ g}}{101.2 \text{ g}} \times 100 = 1.19\% \ (w/w)$$

Note that the solution mass is the sum of the solvent (water) and solute masses.
b. Equation 7.8 is used:

$$\%(v/v) = \frac{\text{solute volume}}{\text{solution volume}} \times 100 = \frac{90. \text{ mL}}{250. \text{ mL}} \times 100 = 36.0\% \ (v/v)$$

c. Equation 7.7 is used after checking to make certain the amount of solute is given in grams and the solution volume is in mL:

$$\%(w/v) = \frac{\text{grams of solute}}{\text{milliliters of solution}} \times 100$$

$$= \frac{27.9 \text{ g}}{150. \text{ mL}} \times 100$$

$$= 18.6\% \ (w/v)$$

✔ LEARNING CHECK 7.5

a. A solution is made by dissolving 0.900 g of salt in 100.0 mL of water. Assume that each milliliter of water weighs 1.00 g and that the final solution volume is 100.0 mL. Calculate the %(w/w) and %(w/v) for the solution using the assumptions as necessary.
b. An alcoholic beverage is labeled 90 proof, which means the alcohol concentration is 45% (v/v). How many milliliters of pure alcohol would be present in 1 oz (30. mL) of the beverage?

7.5 | Solution Preparation

Learning Objective

5. Describe how to prepare solutions of specific concentration using pure solutes and solvent, or solutions of greater concentration than the one desired.

Solutions are usually prepared by mixing together proper amounts of solute and solvent or by diluting a concentrated solution with solvent to produce a solution of lower concentration. In the first method, the solute is measured out and placed in a container, and the correct amount of solvent is added. When the concentration is based on solution volume [%(v/v), %(w/v), and M], a volumetric flask or other container is used that, when filled to a specific mark, holds an accurately known volume (see **Figure 7.8**). When the concentration is based on solvent mass [%(w/w)], the correct mass of solvent is added. This mass is usually converted to a volume by using the density, so that a volume of solvent can be measured rather than an amount weighed on a balance.

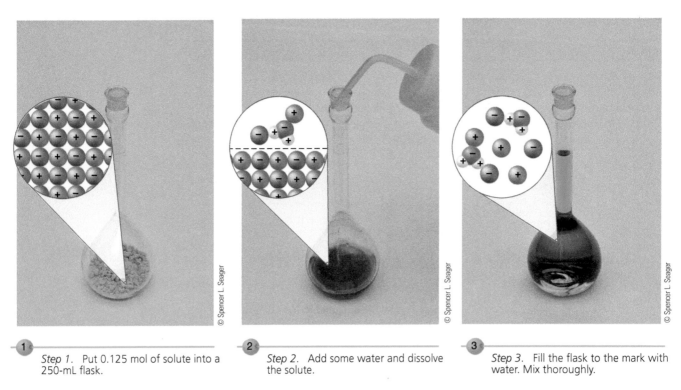

Step 1. Put 0.125 mol of solute into a 250-mL flask.

Step 2. Add some water and dissolve the solute.

Step 3. Fill the flask to the mark with water. Mix thoroughly.

© Spencer L. Seager

Figure 7.8 Preparation of a 0.500 M solution. Use the data given in the figure and show by a calculation that the resulting solution is 0.500 M.

Example 7.6 Preparing Solutions

Describe how you would prepare the following solutions from pure solute and water:

a. 1.00 L of 1.50 M $CoCl_2$ solution
b. 250. mL of 0.900% (w/v) NaCl solution
c. 500. mL of 8.00% (v/v) methyl alcohol solution

Solution

a. According to Equation 7.5, molarity is the moles of solute per (divided by) the liters of solution containing the solute:

$$M = \frac{\text{moles of solute}}{\text{liters of solution}}$$

This equation can be rearranged to calculate the moles of solute required:

$$(M)(\text{liters of solution}) = \text{moles of solute}$$

By substituting the molarity and volume of the desired solution, the required number of moles of solute is determined:

$$\left(1.50 \frac{\text{mol}}{L}\right)(1.00 \, L) = 1.50 \text{ mol}$$

The formula weight of $CoCl_2$ is 129.9 u. The mass of solute is calculated:

$$(1.50 \text{ mol } CoCl_2)\left(\frac{129.9 \text{ g } CoCl_2}{1 \text{ mol } CoCl_2}\right) = 195 \text{ g } CoCl_2$$

The solution is prepared by weighing out 195 g $CoCl_2$, putting it into a 1-L volumetric flask, and adding enough water to fill the flask to the mark (see **Figure 7.8**). It is a good practice to let the solute dissolve completely in part of the solvent before filling the flask to the mark.

b. Similarly, rearrange Equation 7.7, and use the result to calculate the amount of solute needed:

$$\%(w/v) = \frac{\text{grams of solute}}{\text{milliliters of solution}} \times 100$$

Therefore,

$$\frac{(\%)(\text{milliliters of solution})}{100} = \text{grams of solute}$$

Substitute percentage and solution volume:

$$\frac{(0.900\%)(250. \text{ mL})}{100} = 2.25 \text{ g NaCl}$$

Thus, the solution is prepared by putting 2.25 g NaCl into a 250-mL volumetric flask and adding water to the mark.

c. Rearrange Equation 7.8, and use the result to calculate the volume of solute needed:

$$\%(v/v) = \frac{\text{solute volume}}{\text{solution volume}} \times 100$$

Therefore,

$$\frac{(\%)(\text{solution volume})}{100} = \text{solute volume}$$

Substitute percentage and solution volume:

$$\frac{(8.00\%)(500. \text{ mL})}{100} = 40.0 \text{ mL}$$

Thus, 40.0 mL methyl alcohol is put into a 500-mL volumetric flask, and water is added up to the mark.

✔ **LEARNING CHECK 7.6** Describe how you would prepare the following, using pure solute and water:

a. 0.500 L of 1.00 M $MgCl_2$ solution
b. 0.100 L of 12.0% (w/v) $MgCl_2$ solution
c. 1.00 L of 20.0% (v/v) ethylene glycol solution

Solutions are often prepared by diluting a more concentrated solution with solvent (usually water) to produce a solution of lower concentration. Suppose, for example, that you want to prepare 250 mL of 0.100 M NaCl solution using a 2.00 M NaCl solution as the source of NaCl. First, calculate the number of moles of NaCl that would be contained in 250 mL of 0.100 M solution. Remember Equation 7.5,

$$M = \frac{\text{moles of solute}}{\text{liters of solution}}$$

which can be rearranged to

$$(M)(\text{liters of solution}) = \text{moles of solute}$$

So, the number of moles of NaCl contained in the desired solution is

$$\left(0.100\ \frac{\text{mol}}{\text{L}}\right)(0.250\ \text{L}) = 0.0250\ \text{mol}$$

These moles of NaCl must be obtained from the 2.00 M NaCl solution. The volume of this solution that contains the desired number of moles can be obtained by rearranging Equation 7.5 a different way:

$$\text{liters of solution} = \frac{\text{moles of solute}}{M}$$

or

$$\text{liters of solution} = \frac{0.0250\ \text{mol}}{2.00\ \text{mol/L}} = 0.0125\ \text{L} = 12.5\ \text{mL}$$

Thus, the solution is prepared by putting 12.5 mL of 2.00 M NaCl solution into a 250-mL volumetric flask and adding water up to the mark. **Figure 7.9** will help you understand this process.

Step 1. 20.0 mL of 0.200 M $K_2Cr_2O_7$ solution is withdrawn from a beaker using a pipette.

Step 2. The 20.0 mL of the 0.200 M $K_2Cr_2O_7$ solution is put into a 100-mL flask.

Step 3. Water is added (while the solution is swirled) to fill the flask to the mark.

Figure 7.9 Preparation of a 0.0400 M $K_2Cr_2O_7$ solution by dilution of a 0.200 M $K_2Cr_2O_7$ solution. Use the data and show by a calculation that the new solution is 0.0400 M.

The same result can be obtained by using a simplified calculation. Note that the product of concentration in molarity (M) and solution volume in liters will give the number of moles of solute in a sample of solution. In a dilution such as the one done above, the number of moles of solute taken from the concentrated solution and diluted with water is the same as the number of moles of solute in the resulting more dilute solution (see **Figure 7.9**). Thus, the following can be written:

$$M_c V_c = \frac{\text{solute moles in}}{\text{concentrated solution}} = \frac{\text{solute moles in}}{\text{dilute solution}} = M_d V_d$$

or

$$(C_c)(V_c) = (C_d)(V_d) \tag{7.9}$$

where the subscripts c and d refer to the more concentrated and dilute solutions, respectively, and C is used to represent any appropriate concentration. An advantage of using this equation is that any volume units may be used, as long as the same one is used for both V_c and V_d. Also, Equation 7.9 is true for any solution concentration based on volume. This means that M, %(v/v), or %(w/v) concentrations can be used just as long as C_c and C_d are expressed in the same units.

Example 7.7 Preparing Solutions by Dilution

Use Equation 7.9 and describe how to prepare 250. mL of 0.100 M NaCl solution using a 2.00 M NaCl solution as the source of NaCl.

Solution
Equation 7.9 can be used to determine the volume of 2.00 M NaCl needed. The relationships between the terms in Equation 7.9 and the quantities of the problem are $C_c = 2.00$ M, V_c = volume of 2.00 M NaCl needed, $C_d = 0.100$ M, and $V_d = 250.$ mL. Substitution gives

$$(2.00 \text{ M})(V_c) = (0.100 \text{ M})(250 \text{ mL})$$

or

$$V_c = \frac{(0.100 \text{ M})(250. \text{ mL})}{(2.00 \text{ M})} = 12.5 \text{ mL}$$

Thus, we quickly find the volume of 2.00 M NaCl needs to be 12.5 mL, the same result found earlier. This volume of solution is put into a 250-mL flask, and water is added to the mark as before.

✔ **LEARNING CHECK 7.7** Use Equation 7.9 and describe how to prepare 500 mL of a 0.250 M NaOH solution from a 6.00 M NaOH solution.

7.6 | Solution Stoichiometry

Learning Objective
6. Do stoichiometric calculations based on solution concentrations.

The stoichiometry calculations of solution reactions can be done using the factor-unit method. The sources of the needed factors will be the mole interpretation of the reactions introduced as statement 2 in Section 5.9 (see page 158), and the molarities of the solutions involved in the reactions. Each known solution molarity will provide two factors. For example, the following two factors can be obtained based on a 0.400 M HCl solution:

$$\frac{0.400 \text{ mol HCl}}{1 \text{ L HCl soln.}} \quad \text{and} \quad \frac{1 \text{ L HCl soln.}}{0.400 \text{ mol HCl}}$$

Example 7.8 Solution Reaction Calculations

Solutions of NaOH and HCl react according to the following equation:

$$NaOH(aq) + HCl(aq) \rightarrow NaCl(aq) + H_2O(\ell)$$

a. What volume of a 0.250 M NaOH solution contains 0.110 moles of NaOH?
b. What volume of a 0.200 M NaOH solution is needed to exactly react with 0.150 moles of HCl?
c. What volume of a 0.185 M NaOH solution is needed to exactly react with 25.0 mL of 0.225 M HCl solution?

Solution

a. The known quantity is 0.110 mol of NaOH, and the unit of the unknown quantity is a volume that we will express as L NaOH soln.

 Step 1. 0.110 mol NaOH

 Step 2. 0.110 mol NaOH $\qquad\qquad\qquad\qquad$ = L NaOH soln.

 Step 3. $0.110 \; \text{mol NaOH} \times \dfrac{1 \text{ L NaOH soln.}}{0.250 \; \text{mol NaOH}}$ = L NaOH soln.

 The factor $\dfrac{1 \text{ L NaOH soln.}}{0.250 \text{ mol NaOH}}$ came from the known molarity of the NaOH solution as described above.

 Step 4. $0.110 \times \dfrac{1 \text{ L NaOH soln.}}{0.250}$ = 0.44 L NaOH soln. = 0.440 L NaOH soln.

 The calculator answer of 0.44 was rounded by adding a trailing zero so the answer had three significant figures to match the three significant figures in 0.110 and 0.250. The number 1 is an exact counting number.

b. The known quantity is 0.150 mol HCl, and the unit of the unknown quantity is the volume of NaOH solution, which we will express as L NaOH soln.

 Step 1. 0.150 mol HCl

 Step 2. 0.150 mol HCl $\qquad\qquad\qquad\qquad\qquad$ = L NaOH soln.

 Step 3. $0.150 \; \text{mol HCl} \times \dfrac{1 \text{ mol NaOH}}{1 \text{ mol HCl}} \times \dfrac{1 \text{ L NaOH soln.}}{0.200 \; \text{mol NaOH}}$ = L NaOH soln.

 Note that two factors were needed to cancel the unit of the known quantity and generate the unit of the unknown quantity. The first factor came from statement 2 for the reaction:

 $$NaOH(aq) + HCl(aq) \qquad \rightarrow NaCl(aq) + H_2O(\ell)$$

 Statement 2: 1 mol NaOH + 1 mol HCl \rightarrow 1 mol NaCl + 1 mole H_2O

 The second factor came from the 0.200 M concentration of the NaOH solution. Two factors were possible: $\dfrac{0.200 \text{ mol NaOH}}{1 \text{ L NaOH soln.}}$ and $\dfrac{1 \text{ L NaOH soln.}}{0.200 \text{ mol NaOH}}$. The second factor was used because it cancelled the *mol NaOH* unit and generated the *L NaOH soln.* unit.

 Step 4. $0.150 \times \dfrac{1}{1} \times \dfrac{1 \text{ L NaOH soln.}}{0.200}$ = 0.75 L NaOH soln. = 0.750 L NaOH soln.

Once again, a zero was added to the calculator answer of 0.75 to give an answer with three significant figures to match the three in 0.150 and 0.200. Both of the 1 numbers are exact counting numbers.

c. The known quantity is 25.0 mL of a 0.225 M HCl soln., and the unit of the unknown is the volume of a 0.185 M NaOH soln., which we will express as mL NaOH soln. to match the mL unit of the known quantity.

Step 1. 25.0 mL HCl soln.

Step 2. 25.0 mL HCl soln. $=$ mL NaOH soln.

Step 3. $25.0 \text{ mL HCl soln.} \times \dfrac{0.225 \text{ mol HCl}}{1000 \text{ mL HCl soln.}}$

$\times \dfrac{1 \text{ mol NaOH}}{1 \text{ mol HCl}} \times \dfrac{1000 \text{ mL NaOH soln.}}{0.185 \text{ mol NaOH}} = \text{mL NaOH soln.}$

Note that three factors were needed. The first came from the molarity of the HCl solution with the solution volume expressed as 1000 mL rather than 1 L. The second factor came from statement 2 for the reaction as was done in part b. The third factor came from the molarity of the NaOH solution with the solution volume expressed as 1000 mL rather than 1 L.

Step 4. $25.0 \times \dfrac{0.225}{1000} \times \dfrac{1}{1} \times \dfrac{1000 \text{ mL NaOH soln.}}{0.185} = 30.4 \text{ mL NaOH soln.}$

7.7 | Solution Properties

Learning Objective

7. Do calculations based on the colligative solution properties of boiling point, freezing point, and osmotic pressure.

The experiment represented in **Figure 7.10** reveals an interesting property of some solutions. As shown by the figure, the circuit is completed and the light goes on only when the solution conducts electricity. Experiments of this type demonstrate that conductive

1 A solution of a strong electrolyte conducts electricity well.

2 A solution of a weak electrolyte conducts electricity poorly.

3 A solution of a nonelectrolyte does not conduct electricity.

Figure 7.10 Electrical conductivity of solutions.

solutions are formed when soluble ionic materials (see **Section 4.3**), such as sodium chloride, or highly polar covalent materials (see **Section 4.9**), such as hydrogen chloride, dissolve in water. Solutes that form conductive water solutions are called **electrolytes**, whereas solutes that form nonconductive solutions are called **nonelectrolytes**. Conductive solutions result when the dissolved solute dissociates, or breaks apart, to form ions, as shown in the following equations representing the dissociation into ions of polar covalent hydrogen chloride and ionic calcium nitrate:

$$HCl(aq) \rightarrow H^+(aq) + Cl^-(aq) \tag{7.10}$$

$$Ca(NO_3)_2(aq) \rightarrow Ca^{2+}(aq) + 2NO_3^-(aq) \tag{7.11}$$

Some electrolytes, such as HCl and $Ca(NO_3)_2$, dissociate essentially completely in solution and are called *strong electrolytes*. They form strongly conducting solutions. Other electrolytes, such as acetic acid (the acid found in vinegar), dissociate only slightly and form weakly conductive solutions. These solutes are classified as *weak electrolytes*. The solutes in nonconductive solutions dissolve in the solvent but remain in the form of uncharged molecules.

Besides electrical conductivity (or the lack of it), all solutions have properties that depend only on the concentration of solute particles present and not on the actual identity of the solute. Thus, these properties, called **colligative properties**, would be identical for water solutions containing 1 mol of sugar or 1 mol of alcohol per liter.

Three closely related colligative properties are vapor pressure, boiling point, and freezing point. Experiments demonstrate that the vapor pressure of water (solvent) above a solution is lower than the vapor pressure of pure water (see **Figure 7.11**). This behavior causes the boiling point of solutions to be higher than the boiling point of the pure solvent used in the solutions, and the freezing point to be lower (see **Table 7.5**).

Michael R. Slabaugh

1 — Initially, equal volumes of pure water and copper sulfate solution (blue) are placed under a glass dome that prevents water vapor from escaping to the outside.

Michael R. Slabaugh

2 — After some time has passed, the volume of pure water has decreased, while that of the copper sulfate solution has increased.

Figure 7.11 The vapor pressure of water above a solution is lower than the vapor pressure of pure water. Remember that pressure is a force per unit area. Think of this force as "pushing" water molecules in the vapor state, and explain the process shown in the photos.

TABLE 7.5 A Comparison of Colligative Properties of Pure Solvent and Solutions

Property	Pure Solvent	Solution
Vapor pressure	Normal	Lower than pure solvent
Boiling point	Normal	Higher than pure solvent
Freezing point	Normal	Lower than pure solvent

The boiling and freezing point differences between pure solvent and solutions can be calculated by using an equation of the general form

$$\Delta T = nKm \qquad (7.12)$$

In this equation, m is the solution concentration expressed as a molality, a unit we have not discussed. For very dilute solutions, the molality m and molarity M are essentially equal, and M can be used in Equation 7.12 instead of m. This approximation will be used for calculating colligative properties of solutions in this book. The symbol ΔT is the boiling point or freezing point difference between pure solvent and solution. The specific equations used to calculate ΔT for boiling and freezing points are

$$\Delta T_b = nK_b\text{M} \qquad (7.13)$$

$$\Delta T_f = nK_f\text{M} \qquad (7.14)$$

The subscripts b and f refer to boiling or freezing, and K_b and K_f are constants characteristic of the solvent used in the solution (remember, to this point we have focused on water as the solvent). Values for K_b and K_f are given in **Table 7.6** for a number of solvents. In Equations 7.13 and 7.14, M is the molarity of solute in solution, and n is the number of moles of solute particles put into solution when 1 mol of solute dissolves.

TABLE 7.6 Boiling and Freezing Point Constants for Various Solvents

Solvent	Normal Boiling Point (°C)	K_b (°C/M)	Normal Freezing Point (°C)	K_f (°C/M)
Benzene	80.1	2.53	5.5	4.90
Camphor			174.0	40.0
Carbon tetrachloride	76.8	5.03		
Chloroform	61.2	3.63		
Cyclohexane	81.0	2.79	6.5	20.0
Ethyl alcohol	78.5	1.22		
Water	100.0	0.52	0.0	1.86

Example 7.9 Solution Boiling and Freezing Points

Calculate the boiling and freezing points of the following solutions:

a. 171 g of sugar ($C_{12}H_{22}O_{11}$) is dissolved in enough water to give 1.00 L of solution.
b. 13.4 g of NH_4Cl is dissolved in water to form 500. mL of solution.

Solution

In each case, Equations 7.13 and 7.14 are used to calculate the difference between the normal boiling and freezing point of water and the solution. To use these equations,

K_b and K_f are obtained from Table 7.6; the solution molarity, M, is calculated; and n is determined.

a. To find the boiling point, calculate solution molarity.

$$(171 \text{ g } C_{12}H_{22}O_{11})\left(\frac{1 \text{ mol } C_{12}H_{22}O_{11}}{342.0 \text{ g } C_{12}H_{22}O_{11}}\right) = 0.500 \text{ mol } C_{12}H_{22}O_{11}$$

$$M = \frac{\text{moles of solute}}{\text{liters of solution}} = \frac{0.500 \text{ mol}}{1.00 \text{ L}} = 0.500 \frac{\text{mol}}{\text{L}}$$

Determine n: Because sugar does not dissociate on dissolving, $n = 1$. Therefore,

$$\Delta T_b = nK_bM = (1)(0.52°C/M)(0.500 \text{ M}) = 0.26°C$$

Because boiling points are higher in solutions, we add ΔT_b to the normal boiling point of water:

$$\text{solution boiling point} = 100.°C + 0.26°C = 100.26°C = 100.3°C$$

To find the freezing point, calculate ΔT_f:

$$\Delta T_f = nK_fM = (1)(1.86°C/M)(0.500 \text{ M}) = 0.93°C$$

Because freezing points are lower in solutions, we subtract ΔT_f from the normal freezing point of water:

$$\text{solution freezing point} = 0.0°C - 0.93°C = -0.93°C = -0.9°C$$

b. Similarly,

$$(13.4 \text{ g } NH_4Cl)\left(\frac{1 \text{ mol } NH_4Cl}{53.5 \text{ g } NH_4Cl}\right) = 0.250 \text{ mol } NH_4Cl$$

$$M = \frac{\text{moles of solute}}{\text{liters of solution}} = \frac{0.250 \text{ mol}}{0.500 \text{ L}} = 0.500 \frac{\text{mol}}{\text{L}}$$

Because NH_4Cl dissociates in water, 1 mol of solute gives 2 mol of particles (ions):

$$NH_4Cl(aq) \rightarrow NH_4^+(aq) + Cl^-(aq)$$

Thus, we conclude that $n = 2$. Therefore,

$$\Delta T_b = nK_bM = (2)(0.52°C/M)(0.500 \text{ M}) = 0.52°C$$
$$\text{solution boiling point} = 100.0°C + 0.52°C = 100.5°C$$
$$\Delta T_f = nK_fM = (2)(1.86°C/M)(0.500 \text{ M}) = 1.86°C$$
$$\text{solution freezing point} = 0.0°C - 1.86°C = -1.9°C$$

Example 7.9 demonstrates the influence of solute dissociation on colligative properties. Even though the solutions in parts a and b have the same molarity, the NH_4Cl dissociates and produces twice as many solute particles in solution. Hence, it causes twice as much change in the colligative properties.

Osmotic pressure, another important colligative property, can be illustrated by some hypothetical experiments. Consider **Figure 7.12**, where a sugar solution is separated from pure water by a barrier (part A). The barrier is removed, but the mixture is not stirred (part B). After a day or so, the barrier is replaced, with the results shown in (part C). These results

are not surprising; the sugar has diffused throughout the mixture uniformly. It would have been very surprising if the process had not taken place.

Figure 7.12 Diffusion eliminates concentration gradients.

Now consider the experiment shown in **Figure 7.13**, in which two solutions similar to those used before are separated by a semipermeable membrane, which has pores large enough to allow small molecules such as water to pass through but small enough to prevent passage by larger molecules or hydrated ions. In Figure 7.13, the membrane allows water molecules to pass but not sugar molecules.

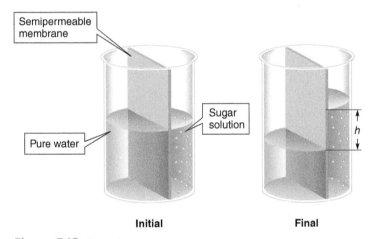

Initial **Final**

Figure 7.13 Osmosis.

osmotic pressure The hydrostatic pressure required to prevent the net flow of solvent through a semipermeable membrane into a solution.

osmosis The process in which solvent molecules flow through a semipermeable membrane into a solution.

In this experiment, a concentration difference between the liquids has been created, but diffusion is prevented from taking place as it did before. Water molecules move through the membrane in both directions, but sugar molecules cannot move into the pure water to equalize the concentration. As a result, the net flow of water through the membrane is into the sugar solution. This has the effect of increasing the volume of the sugar solution and decreasing its concentration to a value closer to that of the pure water. The movement of water creates a difference in liquid levels (*h*), which causes hydrostatic pressure against the membrane. This pressure increases until it becomes high enough to balance the tendency for net water to flow through the membrane into the sugar solution. From that time on, the flow of water in both directions through the membrane is equal, and the volume of liquid on each side of the membrane no longer changes. The hydrostatic pressure required to prevent the net flow of water through a semipermeable membrane into a solution is called the **osmotic pressure** of the solution. The process in which solvent molecules move through semipermeable membranes is called **osmosis**. Although our experiment doesn't illustrate it, solvent molecules will also flow osmotically through semipermeable membranes separating solutions of different concentrations. The net flow of solvent is always from the more dilute solution into the more concentrated (see **Figure 7.14**).

Figure 7.14 Osmosis through carrot membranes. The tissue of a dried shriveled carrot (left) acts as a semipermeable osmotic membrane when placed in water. After some time, the osmotic flow of water into the carrot rehydrates the carrot (right).

STUDY SKILLS 7.1

Getting Started with Molarity Calculations

Knowing where to start or what to do first is a critical part of working any math-type problem. This might be a factor if molarity calculations are difficult for you. First, of course, you must identify a problem as a molarity problem. This can be done by looking for a key word or phrase (molar, molarity, or moles/liter) or the abbreviation M. Second, remember that you have a formula for molarity

$$M = \frac{\text{moles of solute}}{\text{liters of solution}}$$

and the problem might be treated as a formula-type problem. Look for the given numbers and their units, and see if they match the units of the formula. If they do, put them into the formula and do the calculations. For example, if you are asked to calculate the molarity of a solution that contains 0.0856 mol NaCl dissolved in enough water to give 0.100 L of solution, you could put the numbers and their units directly into the formula:

$$M = \frac{0.0856 \text{ mol NaCl}}{0.100 \text{ L solution}} = 0.856 \text{ mol NaCl/L solution}$$

However, if you must calculate the molarity of a solution that contains 5.00 g of NaCl in enough water to give 100. mL of solution, the units of the numbers do not match those of the formula. The factor-unit method can be used to convert each quantity into the units needed by the formula

$$5.00 \text{ g NaCl} \times \frac{1 \text{ mol NaCl}}{58.4 \text{ g NaCl}} = 0.0856 \text{ mol NaCl}$$

$$100 \text{ mL solution} \times \frac{1 \text{ L solution}}{1000 \text{ mL solution}} = 0.100 \text{ L solution}$$

These numbers and units can then be put into the formula

$$M = \frac{0.0856 \text{ mol NaCl}}{0.100 \text{ L solution}} = 0.856 \text{ mol NaCl/L solution}$$

Alternatively, the problem can be solved by remembering the units of the answer (molarity would have the units mol NaCl/L solution), noting the numbers and units of the given quantities (5.00 g NaCl/100 mL solution), and using the factor-unit method to convert the units of the given quantity to those of the answer:

Step 1. $\dfrac{5.00 \text{ g NaCl}}{100.\text{ mL solution}}$

Step 2. $\dfrac{5.00 \text{ g NaCl}}{100.\text{ mL solution}} = \dfrac{\text{mol NaCl}}{\text{L solution}}$

Step 3. $\dfrac{5.00 \text{ g NaCl}}{100.\text{ mL solution}} \times \dfrac{1 \text{ mol NaCl}}{58.4 \text{ g NaCl}}$

$\times \dfrac{1000 \text{ mL solution}}{1 \text{ L solution}} = \dfrac{\text{mol NaCl}}{\text{L solution}}$

Step 4. $\dfrac{(5.00)(1 \text{ mol NaCl})(1000)}{(100.)(58.4)(1 \text{ L solution})} = 0.856 \dfrac{\text{mol NaCl}}{\text{L solution}}$

The final answer, which has the correct units for a molarity, has been rounded to three significant figures.

The osmotic pressure that will develop across a semipermeable membrane separating pure solvent from a solution of molarity M is given by Equation 7.15, which is similar to the ideal gas law given earlier (see **Section 6.8**):

$$\pi = nMRT \tag{7.15}$$

In this equation, π is the osmotic pressure in units that are the same as the pressure units used in the ideal gas constant R. M is the solution molarity, T is the temperature in kelvins, and n, as before, is the number of moles of solute particles obtained when 1 mol of solute dissolves. Scientists in biological and medical fields often call the product of n and M the **osmolarity** of the solution.

osmolarity The product of n and M in the equation $\pi = nMRT$.

Example 7.10 Osmotic Pressure Calculations

Calculate the osmolarity and the osmotic pressure that would develop across a semipermeable membrane if the solutions of Example 7.9 were separated from pure water by the membrane. Assume a solution temperature of 27°C and use an R value of 62.4 L torr/K mol.

Solution

a. The molarity of the sugar solution was found to be 0.500 mol/L. Because sugar does not dissociate, n is equal to 1, and the osmolarity nM is also equal to 0.500 mol/L.

$$\pi = n\text{M}RT = (1)\left(0.500\ \frac{\cancel{\text{mol}}}{\cancel{\text{L}}}\right)\left(\frac{62.4\ \cancel{\text{L}}\ \text{torr}}{\cancel{\text{K}}\ \cancel{\text{mol}}}\right)(300\ \cancel{\text{K}})$$

$$= 9.36 \times 10^3\ \text{torr}$$

Thus, this solution would develop a pressure sufficient to support a column of mercury 9.36×10^3 mm high. This is equal to about 12.3 standard atmospheres of pressure (see **Table 6.3** for pressure units).

b. The molarity of this solution was also found to be 0.500. But, because this solute dissociates into two ions, $n = 2$. This makes the osmolarity equal to (2)(0.500 mol/L) = 1.00 mol/L:

$$\pi = n\text{M}RT = (2)\left(0.500\ \frac{\cancel{\text{mol}}}{\cancel{\text{L}}}\right)\left(\frac{62.4\ \cancel{\text{L}}\ \text{torr}}{\cancel{\text{K}}\ \cancel{\text{mol}}}\right)(300.\ \cancel{\text{K}})$$

$$= 1.87 \times 10^4\ \text{torr, or about 24.6 standard atmospheres}$$

✔ **LEARNING CHECK 7.8** Calculate the boiling point, freezing point, and osmotic pressure of the following solutions. Assume that the osmotic pressure is measured at 27°C.

a. A 0.100 M solution of $CaCl_2$ in water. The $CaCl_2$ is a strong electrolyte.
b. A 0.100 M solution of ethylene glycol in water. Ethylene glycol does not dissociate in solution.

7.8 | Colloids

Learning Objective

8. Describe the characteristics of colloids.

Like solutions, **colloids** (or colloidal dispersions) are homogeneous mixtures of two or more components in which there is more of one component than of the others. In solutions, the terms *solvent* and *solute* are used for the components, but in colloids, the terms **dispersing medium** (for solvent) and **dispersed phase** (for solute) are used.

Solute particles (in solutions) and dispersed phase particles (in colloids) cannot be seen and do not settle under the influence of gravity. However, solute particles do not scatter or reflect light, whereas dispersed phase particles do. This and other variations in properties result from the principal difference between solutions and colloids, the size of the particles making up the solute or dispersed phase. The dissolved solute in a solution is present in the form of tiny particles (small molecules or ions) that are less than about 10^{-7} cm (0.1 μm) in diameter. The dispersed phase of colloids is made up of much larger particles (very large molecules or small pieces of matter) with diameters of 10^{-7} to 10^{-5} cm (0.1–10 μm). As a result of light scattering, colloids often appear to be cloudy. When a beam of light passes through them, they demonstrate the **Tyndall effect** in which the path of the light becomes visible (see **Figure 7.15**).

The word *colloidal* means "gluelike," and some colloids, including some glues, fit this description quite well. However, many, including smoke, shaving cream, and cheese, do not. Colloids are usually differentiated according to the states of the dispersing medium and dispersed phase. Some colloid types are listed in **Table 7.7**, together with examples and specific names.

colloid A homogeneous mixture of two or more substances in which the dispersed substances are present as larger particles than are found in solutions.

dispersing medium The substance present in a colloidal dispersion in the largest amount.

dispersed phase The substance present in a colloidal dispersion in amounts less than the amount of the dispersing medium.

Tyndall effect A property of colloids in which the path of a beam of light through the colloid is visible because the light is scattered.

Figure 7.15 The Tyndall effect.

© Joel Gordon, courtesy of West Publishing Company/CHEMISTRY by Radel & Navidi, 2d ed

TABLE 7.7 Types of Colloids

Colloid Type			
Dispersing Medium	**Dispersed Phase**	**Name**	**Examples**
Gas	Liquid	Aerosol	Fog, aerosol sprays, some air pollutants
Gas	Solid		Smoke, some air pollutants
Liquid	Gas	Foam	Whipped cream, shaving cream
Liquid	Liquid	Emulsion	Milk, mayonnaise
Liquid	Solid	Sol	Paint, ink, gelatin dessert
Solid	Gas	Solid foam	Marshmallow, pumice stone, foam rubber
Solid	Liquid		Butter, cheese
Solid	Solid		Pearls, opals, colored glass, some metal alloys

Sols that become viscous and semisolid are called *gels*. In these colloids, the solid dispersed phase has a very high affinity for the dispersing medium. The gel "sets" by forming a three-dimensional network of solid and dispersing medium. Other examples of gels are fruit jellies and "canned heat" (jellied alcohol).

Much of the interest in colloids is related to their formation or destruction. Ions that are present in the dispersing medium are attracted to colloid particles and stick on their surfaces. The charge (+ or −) of the ions depends on the nature of the colloid, but all colloid particles within a particular system will attract ions of only one charge or the other. In this way, the colloid particles all acquire the same charge and repel each other. This repulsion helps prevent the particles from coalescing into aggregates large enough to settle out.

In the Cottrel precipitator, colloidal solids are removed from gaseous smokestack wastes before they are released into the atmosphere. The precipitator contains a number of highly charged plates or electrodes. As smoke passes over the charged surfaces, the colloid particles lose their charges. The particles then coalesce into larger particles that settle out and are collected for disposal.

Health Drinks

What should you drink to maintain optimal hydration and health? Many drink products boast of providing health benefits and energy-boosting power. With all the new "high-tech" beverages available, it can be difficult to make a choice. Fluids lost through metabolism, breathing, sweating, and waste removal are all best replenished by pure water—H_2O. Water quenches your thirst and rehydrates your body. As long as you have access to clean drinking water, plain water beats all other beverages in health benefits. But, how much do you need?

Since everyone's body is different, there is no single recommendation for water consumption that exactly fits everyone. The broad recommended amount is 125 ounces per day for men, and 91 ounces for women. This recommendation includes water consumed by eating food as well as through drinking the liquid. The old-fashioned recommendation of 8 cups a day is about right for most people. Athletes or people living in hot climates will need to drink more than less active people or those living in cooler climates.

Since water is calorie free and sugar free, you can't make a healthier choice from the point of view of your body weight and dental health. Water is usually the most affordable beverage choice as well. However, since many cultures promote sugary drinks, choosing plain water can sometimes be difficult. Rather than skip water for a sweet alternative, enhance the flavor of your water with healthy additions such as crushed berries, citrus slices, or fresh mint. These types of materials can be used in combinations to create satisfying flavor-infused water. You can keep a pitcher of flavored water in your refrigerator to use throughout the day.

Many people believe that fruit juice or sparkling juice is a healthy alternative to soda. This is a mistake, because juices have as many calories and as much sugar as soda, and are just as detrimental to your dental health and overall body weight. Unlike soda, juice does contribute some vitamins and other nutrients, especially vitamin C. Therefore, adding just a splash of juice to plain water or sparkling water can make the water more palatable and encourage water consumption.

The consumption of high-sugar drinks can cause health problems such as weight gain, increased risk of type 2 diabetes, cardiovascular disease, and gout. The average can of fruit punch or soda contains about 150 calories. Consuming one can of soda per day over the course of a year contributes a weight gain of 5 pounds for the average person. Conversely, eliminating 150 calories per day (or one can of soda) could result in a 5-pound weight reduction (not to mention the reduction in your chances of developing diabetes).

Sports drinks are closely related to soda. They contain a similar number of calories and are designed to deliver carbohydrates and electrolytes to athletes during high-intensity physical workouts. If you are not an athlete burning calories and losing fluid rapidly, sports drinks may not be for you.

Another popular choice is energy drinks. These drinks contain large amounts of sugar and enough caffeine to influence your blood pressure. They also contain additives that produce unknown health effects. If you need a pick-me-up, plain coffee or tea is a better choice. Diet drinks, fruit juice, and milk should also be limited or consumed in moderation. Alcoholic drinks should be limited to no more than one drink per day for women and no more than two drinks per day for men ("one drink" equals 5 ounces of wine, one beer, or 1 ounce of hard liquor). Clean, refreshing, plain water is always the healthiest choice.

mimagephotography/Shutterstock.com

Plain water is always a healthy choice.

Some colloids are stabilized (prevented from coalescing) by substances known as **emulsifying agents** or **stabilizing agents**. Mayonnaise-like salad dressing is a colloid of oil in water, with compounds from egg yolk acting as the emulsifying agent. These compounds form a coating around the oil droplets, which keeps them separated and suspended in the water.

emulsifying agent (stabilizing agent) A substance that when added to colloids prevents them from coalescing and settling.

The cleaning action of soaps and detergents comes from their activity as emulsifying agents. Both soaps and detergents contain long molecules with structures like the following:

$$\underset{\text{Soap}}{H-\overset{\overset{\displaystyle H}{|}}{\underset{\underset{\displaystyle H}{|}}{C}}-\overset{\overset{\displaystyle H}{|}}{\underset{\underset{\displaystyle H}{|}}{C}}-\overset{\overset{\displaystyle H}{|}}{\underset{\underset{\displaystyle H}{|}}{C}}-\overset{\overset{\displaystyle H}{|}}{\underset{\underset{\displaystyle H}{|}}{C}}-\overset{\overset{\displaystyle H}{|}}{\underset{\underset{\displaystyle H}{|}}{C}}-\overset{\overset{\displaystyle H}{|}}{\underset{\underset{\displaystyle H}{|}}{C}}-\overset{\overset{\displaystyle H}{|}}{\underset{\underset{\displaystyle H}{|}}{C}}-\overset{\overset{\displaystyle H}{|}}{\underset{\underset{\displaystyle H}{|}}{C}}-\overset{\overset{\displaystyle H}{|}}{\underset{\underset{\displaystyle H}{|}}{C}}-\overset{\overset{\displaystyle H}{|}}{\underset{\underset{\displaystyle H}{|}}{C}}-\overset{\overset{\displaystyle O}{\|}}{C}-O^-Na^+}$$

$$\underset{\text{Detergent}}{H-C-C-C-C-C-C-C-C-C-C-C-O-\overset{\overset{\displaystyle O}{\|}}{\underset{\underset{\displaystyle O}{\|}}{S}}-O^-Na^+}$$

When placed in water, soaps and detergents dissociate to form ions, as shown in Equations 7.16 and 7.17, where the long carbon chains are represented by a wavy line.

$$\sim\!\!\sim\!\!\sim\overset{\overset{\displaystyle O}{\|}}{C}-O^-Na^+ \longrightarrow \sim\!\!\sim\!\!\sim\overset{\overset{\displaystyle O}{\|}}{C}-O^- + Na^+ \qquad (7.16)$$

$$\sim\!\!\sim\!\!\sim O-\overset{\overset{\displaystyle O}{\|}}{\underset{\underset{\displaystyle O}{\|}}{S}}-O^-Na^+ \longrightarrow \sim\!\!\sim\!\!\sim O-\overset{\overset{\displaystyle O}{\|}}{\underset{\underset{\displaystyle O}{\|}}{S}}-O^- + Na^+ \qquad (7.17)$$

Nonpolar oils and greases are not soluble in water, but they are attracted to the uncharged ends of the soap or detergent ions. As a result of this attraction, the soap or detergent forms a charged layer around the oil droplets, which keeps them separated and suspended (see **Figure 7.16**). Certain compounds (lecithins) of egg yolk act in much the same way to stabilize the mayonnaise-like dressing discussed earlier.

Figure 7.16 The cleaning action of soaps and detergents.

Soap ions dissolved in water Soil particles suspended in water

7.9 | Dialysis

Learning Objective

9. Describe the process of dialysis, and compare it to the process of osmosis.

dialyzing membrane
A semipermeable membrane with pores large enough to allow solvent molecules, other small molecules, and hydrated ions to pass through.

Earlier we discussed semipermeable membranes that selectively allow solvent to pass but retain dissolved solutes during osmosis. **Dialyzing membranes** are semipermeable membranes with larger pores than osmotic membranes. They hold back colloid

CO₂ Emissions: A Blanket around the Earth

In the same way that a greenhouse heats up when radiant heat energy from the sun is trapped as sunlight passes through glass or plastic windows of a greenhouse, the surface of the Earth warms when radiant solar energy passes through the atmosphere and is trapped. Scientists call this heating of the Earth the "greenhouse effect." Most climate scientists agree that one main cause of current observed climate change is *human activity* that enhances this "greenhouse effect." Since the Industrial Revolution, human activity has influenced the climate.

Certain gases, particularly carbon dioxide (CO_2), water vapor (H_2O), methane (CH_4), and nitrous oxide (N_2O) are called "greenhouse gases" because they act like the windows of a greenhouse and let visible light in but prevent heat from radiating back out. As a result of this effect and the presence of these gases in the Earth's atmosphere, the average temperature at the surface of the Earth is a life-supporting 59°F or 15°C. In naturally occurring amounts these gases make life possible on the Earth, but too much of a good thing can create problems.

Carbon dioxide is a by-product of animal respiration, as all animals breathe in oxygen and exhale carbon dioxide. Carbon dioxide is also produced by some geological events such as volcanic eruptions. Other human activities such as deforestation, burning fossil fuels, and agriculture have historically contributed and continue to contribute CO_2 into the atmosphere. It is estimated that since the beginning of the Industrial Revolution human activity has increased the concentration of atmospheric CO_2 by one-third. This is the most significant human contribution to climate change.

Like carbon dioxide, methane has natural sources as well as ones generated by human activity. Methane is produced by ruminant digestion (animals that digest cellulose such as cattle). In addition to natural ruminant digestion, other agricultural practices associated with cattle ranching and manure management as well as decomposition occurring in landfills produces methane. Rice cultivation, which causes decomposition of organic material in flooded fields, also contributes to atmospheric methane levels. Although less methane exists in the atmosphere than CO_2, methane is a more powerful greenhouse gas. Put another way, if a CH_4 molecule and a CO_2 molecule are both heated by 1 degree, the CH_4 molecule would store more heat than the CO_2 molecule.

Nitrous oxide (N_2O) is present in the atmosphere as a result of a natural nitrogen cycle. Animals excrete nitrogen compounds in their waste products. Soil bacteria convert some of these compounds into N_2O. Nitrous oxide is also produced by soil cultivation practices, especially the use of commercial and organic fertilizers. Other sources of atmospheric N_2O include fossil fuel combustion, nitric acid production, and biomass burning. Nitrous oxide is a much more powerful greenhouse gas than carbon dioxide. One pound of N_2O is 300 times more powerful in terms of atmospheric warming than one pound of CO_2.

The consequences of changing the natural atmospheric concentration of greenhouse gases are difficult to predict and controversial, but one thing is certain. During the past century, based on actual measurements, the average surface temperature of the Earth has increased by up to 0.8°C. Although that might not sound like much, a tiny increase in temperature can have enormous consequences. Glaciers, sea ice, and ice sheets around the world are observed to be melting. As the ice melts, sea levels rise. Also, as sea water warms, it expands and sea levels rise even more. The average measured sea level has increased by about 8 inches since about 1870. Again, this might sound insignificant, but weather changes prove otherwise.

Severe storms are occurring more frequently than in the past. Not only has the number of tropical storms increased in recent years, but hurricanes now occur in parts of the world that never previously experienced them. These changes threaten coastal cities with natural disasters. Other significant changes are that extreme cold events in the world are shorter and milder on average, while heat waves are more frequent. This has profound effects on agriculture. Farmers now struggle to grow food in areas that were previously productive. Lack of rain increases the occurrence of wildfires that spread quickly on parched land covered with dried-out vegetation. Some animal species are expected to become extinct because their habitats change more quickly than they can adapt. Only humans can change the amount of greenhouse gases produced by human activity, and minimize these negative environmental effects.

An increase in the rate of breakup of glaciers (calving) as they approach the oceans is one indication of global warming.

particles and large molecules but allow solvent, hydrated ions, and small molecules to pass through. The passage of these ions and small molecules through such membranes is called **dialysis**. Dialysis can be used to separate small particles from colloids, as shown in **Figure 7.17**.

A mixture containing water, ions, small molecules, and colloid particles is placed inside a bag made from a dialyzing membrane. Water flowing around the bag carries away ions and small molecules that pass through the membrane. Water molecules move through the membrane in both directions, but the colloid particles remain inside the bag.

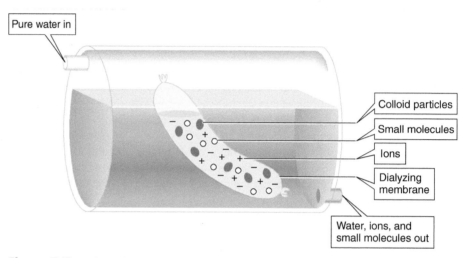

Pure water in

Colloid particles

Small molecules

Ions

Dialyzing membrane

Water, ions, and small molecules out

Figure 7.17 Dialysis. This is one method of dialysis used to purify proteins.

A similar technique is used to clean the blood of people suffering from kidney malfunction. The blood is pumped through tubing made of a dialyzing membrane. The tubing passes through a bath in which impurities collect after passing out of the blood. Blood proteins and other important large molecules remain in the blood.

▶ Case Study Follow-up

Atopic dermatitis, also known as eczema, is an incurable skin condition characterized by alternating remissions and flares. The exact cause is unknown. It is known that atopic dermatitis is related to allergies, but it is not an allergic reaction. Numerous factors that are known to trigger or worsen eczema flares include dry skin resulting from long hot baths or showers, the use of harsh soaps or detergents, exposure to irritants like tobacco smoke, air pollution, dust, and pollen, and exposure to allergens such as eggs, wheat, peanuts, and soy. Dangerous staph infections can develop with flares from excess scratching. Symptoms of atopic dermatitis can be managed somewhat by avoiding the use of harsh washday detergents, applying moisturizers regularly, and avoiding allergens and irritants.

Colloidal oatmeal products such as bath powders or moisturizers are valuable treatments for eczema. The Food and Drug Administration (FDA) regulates such colloidal oatmeal products as over-the-counter medications. These products are manufactured by finely grinding oats, then boiling the resulting powder in water to extract the colloidal materials. The extracted colloidal materials include starches, beta-glucan, phenols, and saponins. The starches and beta-glucan moisturize and protect the skin. The phenols provide anti-inflammatory, antioxidant, and ultraviolet light protection, and the saponins provide cleansing properties. All of these compounds could be helpful for treating atopic dermatitis.

Concept Summary

Physical States of Solutions Solutions, homogeneous mixtures of a solvent and one or more solutes, can be found in any of the three states of matter: solid, liquid, or gas. The physical state of the solution is often the same as the physical state of the solvent.

Objective 1 (Section 7.1), Exercise 7.4

Solubility The amount of solute that will dissolve in a quantity of solvent to form a saturated solution is the solubility of the solute. Solubility depends on similarities in polarities of the solvent and solute; it generally increases with temperature for solid and liquid solutes, but decreases with temperature for gaseous solutes.

Objective 2 (Section 7.2), Exercises 7.6 and 7.12

The Solution Process The solution process of solid solute in a liquid solvent can be thought of in terms of the solvent molecules attracting the solute particles away from the solute crystal lattice. The solute is picked apart and solute particles hydrated as a result of attractions between water molecules and solute particles. Heat is generally absorbed or released when a solute dissolves in a solvent. When heat is released, the solution process is called exothermic, and the solution temperature increases. When heat is absorbed, the solution process is endothermic, and the solution cools as solute dissolves.

Objective 3 (Section 7.3), Exercise 7.16

Solution Concentrations Relationships between the amount of solute and the amount of solution containing the solute are called concentrations. They may be expressed as molarity, weight/weight percent, weight/volume percent, or volume/volume percent.

Objective 4 (Section 7.4), Exercises 7.22b, 7.30c, 7.34a, and 7.38c

Solution Preparation Solutions of specific concentration can be prepared by mixing appropriate amounts of solute and solvent or by diluting a more concentrated solution with solvent.

Objective 5 (Section 7.5), Exercises 7.46 and 7.48b

Solution Stoichiometry When solution concentrations are expressed as molarities, the mole concept can be applied to reactions taking place between substances that are solutes in the solutions.

Objective 6 (Section 7.6), Exercise 7.56

Solution Properties A number of solution properties differ from the properties of pure solvent. Ionic or highly polar solutes that dissociate in solution result in solutions that conduct electricity. Colligative solution properties depend on the concentration of solute particles in the solution and include vapor pressure, boiling point, freezing point, and osmotic pressure.

Objective 7 (Section 7.7), Exercises 7.64a & c and 7.74

Colloids Homogeneous mixtures called colloids, or colloidal dispersions, differ from solutions in terms of the size of the dispersed phase particles. In colloids, the particles are large enough to scatter light and thus show the Tyndall effect. Colloids, like solutions, occur in all three physical states, depending primarily on the physical state of the dispersing medium. In true colloids, the suspension is permanent because dispersed particles acquire similar charges by adsorbing ions from the dispersing medium and repel each other or because emulsifying agents keep the dispersed particles from coalescing.

Objective 8 (Section 7.8), Exercise 7.82

Dialysis Dialyzing membranes are semipermeable but with pores large enough to allow solvent molecules, hydrated ions, and small molecules to pass through in a process called dialysis. The process is important in the removal of impurities from the blood and is applied artificially for people suffering from kidney malfunction.

Objective 9 (Section 7.9), Exercise 7.84

Key Terms and Concepts

Colligative property (7.7)
Colloid (7.8)
Concentration (7.4)
Dialysis (7.9)
Dialyzing membrane (7.9)
Dispersed phase (7.8)
Dispersing medium (7.8)
Dissolving (7.1)
Electrolyte (7.7)
Emulsifying agent (stabilizing agent) (7.8)

Hydrated ions (7.3)
Immiscible (7.2)
Insoluble substance (7.2)
Molarity (M) (7.4)
Nonelectrolyte (7.7)
Osmolarity (7.7)
Osmosis (7.7)
Osmotic pressue (7.7)
Percent (7.4)
Saturated solution (7.2)

Solubility (7.2)
Soluble substance (7.2)
Solute (7.1)
Solution (7.1)
Solvent (7.1)
Supersaturated solution (7.2)
Tyndall effect (7.8)
Volume/volume percent (7.4)
Weight/volume percent (7.4)
Weight/weight percent (7.4)

Key Equations

1.	Solution concentration in terms of molarity (Section 7.4):	$M = \dfrac{\text{moles of solute}}{\text{liters of solution}}$	Equation 7.5
2.	Solution concentration in terms of weight/weight % (Section 7.4):	$\%(w/w) = \dfrac{\text{solute mass}}{\text{solution mass}} \times 100$	Equation 7.6
3.	Solution concentration in terms of weight/volume % (Section 7.4):	$\%(w/v) = \dfrac{\text{grams of solute}}{\text{milliliters of solution}} \times 100$	Equation 7.7
4.	Solution concentration in terms of volume/volume % (Section 7.4):	$\%(v/v) = \dfrac{\text{solute volume}}{\text{solution volume}} \times 100$	Equation 7.8
5.	Dilution of concentrated solution to make less-concentrated solution (Section 7.5):	$(C_c)(V_c) = (C_d)(V_d)$	Equation 7.9
6.	Boiling point elevation of a solution (Section 7.7):	$\Delta T_b = nK_bM$	Equation 7.13
7.	Freezing point depression of a solution (Section 7.7):	$\Delta T_f = nK_fM$	Equation 7.14
8.	Osmotic pressure of a solution (Section 7.7):	$\pi = nMRT$	Equation 7.15

Exercises

Even-numbered exercises are answered in Appendix B.

Blue-numbered exercises are more challenging.

Physical States of Solutions (Section 7.1)

7.1 Many solutions are found in the home. Some are listed below, with the composition as printed on the label. When no percentage is indicated, components are usually given in order of decreasing amount. When water is present, it is often not mentioned on the label or it is included in the inert ingredients. Identify the solvent and solutes of the following solutions:

 a. Antiseptic mouthwash: alcohol 25%, thymol, eucalyptol, methyl salicylate, menthol, benzoic acid, boric acid

 b. Paregoric: alcohol 45%, opium 0.4%

 c. Baby oil: mineral oil, lanolin (there happens to be no water in this solution—why?)

 d. Distilled vinegar: acetic acid 5%

7.2 Many solutions are found in the home. Some are listed below, with the composition as printed on the label. When no percentage is indicated, components are usually given in order of decreasing amount. When water is present, it is often not mentioned on the label or it is included in the inert ingredients. Identify the solvent and solutes of the following solutions:

 a. Liquid laundry bleach: sodium hypochlorite 5.25%, inert ingredients 94.75%

 b. Rubbing alcohol: isopropyl alcohol 70%

 c. Hydrogen peroxide: 3% hydrogen peroxide

 d. Aftershave: SD alcohol, water, glycerin, fragrance, menthol, benzophenone-1, coloring

7.3 Classify the following as being a solution or not a solution. Explain your reasons when you classify one as *not* a solution. For the ones classified as solutions, identify the solvent and solute(s).

 a. Maple syrup

 b. Milk

 c. Eyedrops

 d. Tomato juice

 e. Tap water

7.4 Classify the following as being a solution or not a solution. Explain your reasons when you classify one as *not* a solution. For the ones classified as solutions, identify the solvent and solute(s).

 a. Foggy air

 b. Tears

 c. Freshly squeezed orange juice

 d. Strained tea

 e. Creamy hand lotion

Solubility (Section 7.2)

7.5 Use the term *soluble, insoluble,* or *immiscible* to describe the behavior of the following pairs of substances when they are shaken together:

 a. 25 mL of water and 1 g of salt—the resulting mixture is clear and colorless.

 b. 25 mL of water and 1 g of solid silver chloride—the resulting mixture is cloudy and solid settles out.

 c. 25 mL of water and 5 mL of mineral oil—the resulting mixture is cloudy and gradually separates into two layers.

7.6 Use the term *soluble, insoluble,* or *immiscible* to describe the behavior of the following pairs of substances when they are shaken together:

 a. 25 mL of cooking oil and 25 mL of vinegar—the resulting mixture is cloudy and gradually separates into two layers.

 b. 25 mL of water and 10 mL of rubbing alcohol—the resulting mixture is clear and colorless.

 c. 25 mL of chloroform and 1 g of roofing tar—the resulting mixture is clear but dark brown in color.

7.7 Define the term *miscible.* It is *not* defined in the text.

7.8 Classify the following solutions as unsaturated, saturated, or supersaturated:

 a. A solution to which a small piece of solute is added, and it dissolves.

 b. A solution to which a small piece of solute is added, and much more solute comes out of solution.

 c. The final solution resulting from the process in part b.

7.9 Suppose you put 35.8 g of ammonium sulfate into a flask and add 100 g of water at 0°C. After stirring to dissolve as much solute as possible, will you have a saturated or unsaturated solution? Explain your answer. See Table 7.2.

7.10 Suppose you have a saturated solution that is at room temperature. Discuss how it could be changed into a supersaturated solution without using any additional solute.

7.11 Classify each of the following solutes into the approximate solubility categories of Table 7.3. The numbers in parentheses are the grams of solute that will dissolve in 100 g of water at the temperature indicated.

 a. boric acid, H_3BO_3 (6.35 g at 30°C)

 b. calcium hydroxide, $Ca(OH)_2$ (5.35 g at 30°C)

 c. antimony(III) sulfide, Sb_2S_3 (1.75×10^{-4} g at 18°C)

 d. copper(II) chloride, $CuCl_2$ (70.6 g at 0°C)

 e. iron(II) bromide, $FeBr_2$ (109 g at 10°C)

7.12 Classify each of the following solutes into the approximate solubility categories of Table 7.3. The numbers in parentheses are the grams of solute that will dissolve in 100 g of water at the temperature indicated.

 a. barium nitrate, $Ba(NO_3)_2$ (8.7 g at 20°C)

 b. aluminum oxide, Al_2O_3 (9.8×10^{-5} g at 29°C)

 c. calcium sulfate, $CaSO_4$ (0.21 g at 30°C)

 d. manganese chloride, $MnCl_2$ (72.3 g at 25°C)

 e. lead bromide, $PbBr_2$ (0.46 g at 0°C)

The Solution Process (Section 7.3)

7.13 What is the difference between a nonhydrated ion and a hydrated ion? Draw a sketch using the Cl^- ion to help illustrate your answer.

7.14 Suppose you had a sample of white crystalline solid that was a mixture of barium chloride ($BaCl_2$) and barium sulfate ($BaSO_4$). Describe how you could treat the sample to isolate one of the solids in a pure state. Which solid would it be?

7.15 Ground-up limestone ($CaCO_3$) is used as a gentle abrasive in some powdered cleansers. Why is this a better choice than ground-up soda ash (Na_2CO_3)?

7.16 Indicate which of the following substances (with geometries as given) would be soluble in water (a polar solvent) and in benzene (a nonpolar solvent):

 a.

 (tetrahedral)

 b. Ne

 c.

 (triangular-based pyramid)

 d.

F
|
B (flat triangle)
F F

7.17 Indicate which of the following substances (with geometries as given) would be soluble in water (a polar solvent) and in benzene (a nonpolar solvent):

 a. H—S
 \H

 b. HCl

 c. O—O
 H/ \H

 d. N≡N

7.18 Freons are compounds formerly used in a variety of ways. Explain why Freon-114 was useful as a degreasing agent. The molecular structure is

7.19 Suppose you put a piece of a solid into a beaker that contains water and stir the mixture briefly. You find that the solid does not immediately dissolve completely. Describe three things you might do to try to get the solid to dissolve.

Solution Concentrations (Section 7.4)

7.20 Calculate the molarity of the following solutions:

 a. 1.25 L of solution contains 0.455 mol of solute.

 b. 250 mL of solution contains 0.215 mol of solute.

 c. 0.175 mol of solute is put into a container and enough distilled water is added to give 100 mL of solution.

7.21 Calculate the molarity of the following solutions:

a. 2.50 L of solution that contains 0.860 mol of solute.

b. 400. mL of solution that contains 0.304 mol of solute.

c. 0.120 mol of solute is put into a container and enough distilled water is added to give 250. mL of solution.

7.22 Calculate the molarity of the following solutions:

a. A sample of solid KBr weighing 11.9 g is put in enough distilled water to give 200. mL of solution.

b. A 14.2-g sample of solid Na_2SO_4 is dissolved in enough water to give 500. mL of solution.

c. A 10.0-mL sample of solution is evaporated to dryness and leaves 0.29 g of solid residue that is identified as Li_2SO_4.

7.23 Calculate the molarity of the following solutions:

a. A sample of solid Na_2SO_4 weighing 0.140 g is dissolved in enough water to make 10.0 mL of solution.

b. A 4.50-g sample of glucose ($C_6H_{12}O_6$) is dissolved in enough water to give 150. mL of solution.

c. A 43.5-g sample of K_2SO_4 is dissolved in a quantity of water, and the solution is stirred well. A 25.0-mL sample of the resulting solution is evaporated to dryness and leaves behind 2.18 g of solid K_2SO_4.

7.24 Calculate:

a. How many moles of solute is contained in 1.75 L of 0.215 M solution?

b. How many moles of solute is contained in 250. mL of 0.300 M solution?

c. How many mL of 0.350 M solution contains 0.200 mol of solute?

7.25 Calculate:

a. How many moles of solute is contained in 1.25 L of a 0.350 M solution?

b. How many moles of solute is contained in 200. mL of a 0.750 M solution?

c. What volume of a 0.415 M solution contains 0.500 mol of solute?

7.26 Calculate:

a. How many grams of solid would be left behind if 20.0 mL of a 0.550 M KCl solution was evaporated to dryness?

b. What volume of a 0.315 M HNO_3 solution is needed to provide 0.0410 mol HNO_3?

c. What volume of 1.21 M NH_4NO_3 contains 50.0 g of solute?

7.27 Calculate:

a. How many grams of solid $AgNO_3$ will be needed to prepare 200. mL of a 0.200 M solution?

b. How many grams of vitamin C ($C_6H_8O_6$) would be contained in 25.0 mL of a 1.00 M solution?

c. How many moles of HCl is contained in 250 mL of a 6.0 M solution?

7.28 Calculate the concentration in %(w/w) of the following solutions. Assume water has a density of 1.00 g/mL.

a. 5.3 g of sugar and 100. mL of water

b. 5.3 g of any solute and 100. mL of water

c. 5.3 g of any solute and 100. g of any solvent

7.29 Calculate the concentration in %(w/w) of the following solutions. Assume water has a density of 1.00 g/mL.

a. 6.5 g of table salt and 100. mL of water

b. 6.5 g of any solute and 100. mL of water

c. 6.5 g of any solute and 100. g of any solvent

7.30 Calculate the concentration in %(w/w) of the following solutions. Assume water has a density of 1.00 g/mL.

a. 20.0 g of salt is dissolved in 250. mL of water.

b. 0.100 mol of solid glucose ($C_6H_{12}O_6$) is dissolved in 100. mL of water.

c. 120 g of solid is dissolved in 100. mL of water.

d. 10.0 mL of ethyl alcohol (density = 0.789 g/mL) is mixed with 10.0 mL of water.

7.31 Calculate the concentration in %(w/w) of the following solutions. Assume water has a density of 1.00 g/mL.

a. 5.20 g of $CaCl_2$ is dissolved in 125 mL of water.

b. 0.200 mol of solid KBr is dissolved in 200. mL of water.

c. 50.0 g of solid is dissolved in 250. mL of water.

d. 10.0 mL of ethyl alcohol (density = 0.789 g/mL) is mixed with 10.0 mL of ethylene glycol (density = 1.11 g/mL).

7.32 Calculate the concentration in %(w/w) of the following solutions:

a. 20.0 g of solute is dissolved in enough water to give 150. mL of solution. The density of the resulting solution is 1.20 g/mL.

b. A 10.0-mL sample with a density of 1.10 g/mL leaves 1.18 g of solid residue when evaporated.

c. A 25.0-g sample of solution on evaporation leaves a 1.87-g residue of $MgCl_2$.

7.33 Calculate the concentration in %(w/w) of the following solutions:

a. 424 g of solute is dissolved in enough water to give 1.00 L of solution. The density of the resulting solution is 1.18 g/mL.

b. A 50.0-mL solution sample with a density of 0.898 g/mL leaves 12.6 g of solid residue when evaporated.

c. A 25.0-g sample of solution on evaporation leaves a 2.32-g residue of NH_4Cl.

7.34 Calculate the concentration in %(v/v) of the following solutions:

a. 200. mL of solution contains 15 mL of alcohol.

b. 200. mL of solution contains 15 mL of any soluble liquid solute.

c. 8.0 fluid ounces of oil is added to 2.0 gallons (256 fluid ounces) of gasoline.

d. A solution of alcohol and water is separated by distillation. A 200.-mL sample gives 85.9 mL of alcohol.

7.35 Calculate the concentration in %(v/v) of the following solutions:

a. 250. mL of solution contains 15.0 mL of acetone.

b. 250. mL of solution contains 15.0 mL of any soluble liquid solute.

c. 1.0 quart of acetic acid is put into a 5-gallon container, and enough water is added to fill the container.

d. A solution of acetone and water is separated by distillation. A 300.-mL sample gives 109 mL of acetone.

7.36 Consider the blood volume of an adult to be 5.0 L. A blood alcohol level of 0.50% (v/v) can cause a coma. What volume of pure ethyl alcohol, if consumed in one long drink and assumed to be absorbed completely into the blood, would result in this critical blood alcohol level?

7.37 The blood serum acetone level for a person is determined to be 1.8 mg of acetone per 100 mL of serum. Express this concentration as %(v/v) if liquid acetone has a density of 0.79 g/mL.

7.38 Calculate the concentration in %(w/v) of the following solutions:

 a. 150. mL of solution contains 7.50 g of dissolved solid Na_2SO_4.

 b. 150. mL of solution contains 7.50 g of any dissolved solid solute.

 c. 350. mL of solution contains 30.7 g of dissolved solid solute.

7.39 Calculate the concentration in %(w/v) of the following solutions:

 a. 26.5 g of solute is dissolved in 200. mL of water to give a solution with a density of 1.10 g/mL.

 b. A 30.0-mL solution sample on evaporation leaves a solid residue of 0.38 g.

 c. On analysis for total protein, a blood serum sample of 15.0 mL is found to contain 1.15 g of total protein.

7.40 A saturated solution of KBr in water is formed at 20.0°C. Consult Figure 7.3 and calculate the concentration of the solution in %(w/w).

7.41 Assume the density of the solution prepared in Exercise 7.40 is 1.18 g/mL and express the concentration in %(w/v).

Solution Preparation (Section 7.5)

7.42 Explain how you would prepare the following solutions using pure solute and water. Assume water has a density of 1.00 g/mL.

 a. 200 mL of 0.150 M Na_2SO_4 solution

 b. 250 mL of 0.250 M $Zn(NO_3)_2$ solution

 c. 150 g of 2.25% (w/w) NaCl solution

 d. 125 mL of 0.75% (w/v) KCl solution

7.43 Explain how you would prepare the following solutions using pure solute and water. Assume water has a density of 1.00 g/mL.

 a. 250 mL of a 2.00 M NaOH solution

 b. 500 mL of a 40.0% (v/v) alcohol solution (C_2H_5OH)

 c. 100 mL of a 15.0% (w/v) glycerol solution. Glycerol is a liquid with a density of 1.26 g/mL. Describe two ways to measure out the amount of glycerol needed.

 d. Approximately 50 mL of a normal saline solution, 0.89% (w/w) NaCl

7.44 A solution is prepared by mixing 45.0 g of water and 15.0 g of ethyl alcohol. The resulting solution has a density of 0.952 g/mL. Express the solution concentration in %(w/w) ethyl alcohol and %(w/v) ethyl alcohol.

7.45 Calculate the following:

 a. The number of moles of NaI in 50.0 mL of a 0.400 M solution

 b. The number of grams of KBr in 120. mL of a 0.720 M solution

 c. The number of grams of NaCl in 20.0 mL of a 1.20% (w/v) NaCl solution

 d. The number of milliliters of alcohol in 250. mL of a 20.0% (v/v) solution

7.46 Calculate the following:

 a. The number of grams of Li_2CO_3 in 250. mL of 1.75 M Li_2CO_3 solution

 b. The number of moles of NH_3 in 200. mL of 3.50 M NH_3 solution

 c. The number of mL of alcohol in 250. mL of 12.5% (v/v) solution

 d. The number of grams of $CaCl_2$ in 50.0 mL of 4.20% (w/v) $CaCl_2$ solution

7.47 Explain how you would prepare the following dilute solutions from the more concentrated ones:

 a. 200 mL of 0.500 M HCl from a 6.00 M HCl solution

 b. 50 mL of 2.00 M H_2SO_4 from a 6.00 M H_2SO_4 solution

 c. 100 mL of normal saline solution, 0.89% (w/v) NaCl, from a 5.0% (w/v) NaCl solution

 d. 250 mL of 5.00% (v/v) acetone from 20.5% (v/v) acetone

7.48 Explain how you would prepare the following dilute solutions from the more concentrated ones:

 a. 5.00 L of 6.00 M H_2SO_4 from a 18.0 M H_2SO_4 solution

 b. 250 mL of 0.500 M $CaCl_2$ from a 3.00 M $CaCl_2$ solution

 c. 200 mL of 1.50% (w/v) KBr from a 10.0% (w/v) KBr solution

 d. 500 mL of 10.0% (v/v) alcohol from 50.0% (v/v) alcohol

7.49 What is the molarity of the solution prepared by diluting 25.0 mL of 0.412 M $Mg(NO_3)_2$ solution to each of the following final volumes?

 a. 40.0 mL

 b. 0.100 L

 c. 1.10 L

 d. 350. mL

7.50 What is the molarity of the solution prepared by diluting 50.0 mL of 0.195 M KBr to each of the following volumes?

 a. 1.50 L

 b. 200. mL

 c. 500. mL

 d. 700. mL

Solution Stoichiometry (Section 7.6)

7.51 How many milliliters of 6.00 M HCl solution would be needed to react exactly with 20.0 g of pure solid NaOH?

$$HCl(aq) + NaOH(s) \rightarrow NaCl(aq) + H_2O(\ell)$$

7.52 How many grams of solid Na_2CO_3 will react with 250. mL of a 1.25 M HCl solution?

$$Na_2CO_3(s) + 2HCl(aq) \rightarrow 2NaCl(aq) + CO_2(g) + H_2O(\ell)$$

7.53 How many milliliters of 0.250 M HCl solution will exactly react with 10.5 g of solid $NaHCO_3$?

$$NaHCO_3(s) + HCl(aq) \rightarrow NaCl(aq) + CO_2(g) + H_2O(\ell)$$

7.54 How many milliliters of 0.250 M $AgNO_3$ solution will exactly react with 25.0 mL of a 0.200 M NaCl solution?

$$NaCl(aq) + AgNO_3(aq) \rightarrow NaNO_3(aq) + AgCl(s)$$

7.55 How many milliliters of 0.120 M Na_2S solution will exactly react with 35.0 mL of a 0.150 M $AgNO_3$ solution?

$$2AgNO_3(aq) + Na_2S(aq) \rightarrow Ag_2S(s) + 2NaNO_3(aq)$$

7.56 How many milliliters of 0.225 M NH_3 solution will exactly react with 30.0 mL of a 0.190 M H_2SO_4 solution?

$$2NH_3(aq) + H_2SO_4(aq) \rightarrow (NH_4)_2SO_4(aq)$$

7.57 How many milliliters of 0.124 M NaOH solution will exactly react with 35.0 mL of a 0.210 M H_3PO_4 solution?

$$3NaOH(aq) + H_3PO_4(aq) \rightarrow Na_3PO_4(aq) + 3H_2O(\ell)$$

7.58 How many milliliters of 0.124 M NaOH solution will exactly react with 25.0 mL of a 0.210 M HCl solution?

$$NaOH(aq) + HCl(aq) \rightarrow NaCl(aq) + H_2O(\ell)$$

7.59 How many milliliters of 0.115 M NaOH solution will exactly react with 25.0 mL of a 0.210 M H_2SO_4 solution?

$$2NaOH(aq) + H_2SO_4(aq) \rightarrow Na_2SO_4(aq) + 2H_2O(\ell)$$

7.60 Stomach acid is essentially 0.10 M HCl. An active ingredient found in a number of popular antacids is calcium carbonate, $CaCO_3$. Calculate the number of grams of $CaCO_3$ needed to exactly react with 250. mL of stomach acid.

$$CaCO_3(s) + 2HCl(aq) \rightarrow CO_2(g) + CaCl_2(aq) + H_2O(\ell)$$

7.61 An ingredient found in some antacids is magnesium hydroxide, $Mg(OH)_2$. Calculate the number of grams of $Mg(OH)_2$ needed to exactly react with 200. mL of stomach acid (see Exercise 7.60).

$$Mg(OH)_2(s) + 2HCl(aq) \rightarrow MgCl_2(aq) + 2H_2O(\ell)$$

Solution Properties (Section 7.7)

7.62 Before it is frozen, ice cream is essentially a solution of sugar, flavorings, etc., dissolved in water. Use the idea of colligative solution properties and explain why a mixture of ice (and water) and salt is used to freeze homemade ice cream. Why won't just ice work?

7.63 If you look at the labels of automotive products used to prevent radiator freezing (antifreeze) and radiator boiling, you will find the same ingredient listed, ethylene glycol. Use the idea of colligative properties to explain how the same material can prevent an automobile cooling system from freezing and boiling.

7.64 Calculate the boiling and freezing points of water solutions that are 1.50 M in the following solutes:

a. KCl, a strong electrolyte

b. glycerol, a nonelectrolyte

c. $(NH_4)_2SO_4$, a strong electrolyte

d. $Al(NO_3)_3$, a strong electrolyte

7.65 Calculate the boiling and freezing points of water solutions that are 1.15 M in the following solutes:

a. KBr, a strong electrolyte

b. ethylene glycol, a nonelectrolyte

c. $(NH_4)_2CO_3$, a strong electrolyte

d. $Al_2(SO_4)_3$, a strong electrolyte

7.66 Calculate the boiling and freezing points of the following solutions. Water is the solvent, unless otherwise indicated.

a. A 0.50 M solution of urea, a nonelectrolyte

b. A 0.250 M solution of $CaCl_2$, a strong electrolyte

c. A solution containing 100. g of ethylene glycol ($C_2H_6O_2$), a nonelectrolyte, per 250. mL

7.67 Calculate the boiling and freezing points of the following solutions. Water is the solvent, unless otherwise indicated.

a. A solution containing 50.0 g of H_2SO_4, a strong electrolyte (both Hs dissociate), per 250. mL

b. A solution containing 200. g of table sugar ($C_{12}H_{22}O_{11}$), a nonelectrolyte, per 250. mL

c. A solution containing 75.0 g of octanoic acid ($C_8H_{16}O_2$), a nonelectrolyte, in enough benzene to give 250. mL of solution

7.68 Calculate the osmolarity for the following solutions:

a. A 0.15 M solution of glycerol, a nonelectrolyte

b. A 0.15 M solution of $(NH_4)_2SO_4$, a strong electrolyte

c. A solution containing 25.3 g of LiCl (a strong electrolyte) per liter

7.69 Calculate the osmolarity for the following solutions:

a. A 0.25 M solution of KCl, a strong electrolyte

b. A solution containing 15.0 g of urea (CH_4N_2O), a nonelectrolyte, per 500. mL

c. A solution containing 50.0 mL of ethylene glycol ($C_2H_6O_2$), a nonelectrolyte with a density of 1.11 g/mL, per 250. mL

Note: In Exercises 7.70–7.79, assume the temperature is 25.0°C, and express your answer in torr, mmHg, and atm.

7.70 Calculate the osmotic pressure of any solution with an osmolarity of 0.250.

7.71 Calculate the osmotic pressure of a 0.125 M solution of Na_2SO_4, a strong electrolyte.

7.72 Calculate the osmotic pressure of a 0.200 M solution of Na_2SO_4, a strong electrolyte.

7.73 Calculate the osmotic pressure of a 0.250 M solution of methanol, a nonelectrolyte.

7.74 Calculate the osmotic pressure of a solution that contains 95.0 g of the nonelectrolyte urea, CH_4N_2O, per 500. mL of solution.

7.75 Calculate the osmotic pressure of a solution that contains 1.20 mol of $CaCl_2$, a strong electrolyte, in 1.50 L.

7.76 Calculate the osmotic pressure of a solution that has a freezing point of 20.35°C.

7.77 Calculate the osmotic pressure of a solution that is 0.122 M in solute and has a boiling point 0.19°C above that of pure water.

7.78 Calculate the osmotic pressure of a solution that contains 5.30 g of NaCl and 8.20 g of KCl per 750. mL.

7.79 Calculate the osmotic pressure of a solution that contains 220.0 g of ethylene glycol ($C_2H_6O_2$), a nonelectrolyte, per liter.

7.80 Suppose an osmotic membrane separates a 5.00% sugar solution from a 10.0% sugar solution. In which direction will water flow? Which solution will become diluted as osmosis takes place?

Colloids (Section 7.8)

7.81 Explain how the following behave in a colloidal suspension: dispersing medium, dispersed phase, and colloid emulsifying agent.

7.82 Explain why detergents or soaps are needed if water is to be used as a solvent for cleaning clothes and dishes.

Dialysis (Section 7.9)

7.83 Suppose you have a bag made of a membrane like that in Figure 7.17. Inside the bag is a solution containing water and dissolved small molecules. Describe the behavior of the system when the bag functions as an osmotic membrane and when it functions as a dialyzing membrane.

7.84 Each of the following mixtures was placed in a dialysis bag like the one shown in Figure 7.17. The bag was immersed in pure water. Which substances will pass through the bag into the surrounding water?

 a. NaCl solution and starch (colloid)

 b. Urea (small organic molecule), and starch (colloid)

 c. Albumin (colloid), KCl solution, and glucose solution (small organic molecule)

Additional Exercises

7.85 When 5.0 mL of water and 5.0 mL of rubbing alcohol are mixed together, the volume of the resulting solution is 9.7 mL. Explain in terms of molecules why the final solution volume is not 10.0 mL.

7.86 A student made a 7.00% (w/v) solution by dissolving 7.00 grams of sodium chloride (salt) in an appropriate volume of water. Assume the density of the water was 1.00 g/mL, and that there was no increase in volume when the salt was added to the water. Calculate the density of the resulting solution in g/mL.

7.87 Ethylene glycol is mixed with water and used as an antifreeze in automobile cooling systems. If a 50% (v/v) solution is made using pure ethylene glycol and pure water, it will have a freezing point of about −65°C. What will happen to the freezing point of the mixture if ethylene glycol is added until the concentration is 70% (v/v) ethylene glycol? Explain your answer.

7.88 How many mL of 1.50 M HCl solution contains enough HCl to react completely with 0.500 g of zinc metal? The reaction is $Zn(s) + 2HCl(aq) \rightarrow ZnCl_2(aq) + H_2(g)$.

7.89 At 20°C ethyl alcohol has a vapor pressure of 43.9 torr, and the vapor pressure of water at the same temperature is 17.5 torr. Suppose a 50% (v/v) solution of water and ethyl alcohol was brought to a boil and the first vapors given off were collected and condensed to a liquid. How would the percentage of alcohol in the condensed liquid compare to the percentage of alcohol in the original 50% solution (greater than, less than, equal to)? Explain your reasoning.

Chemistry for Thought

7.90 When a patient has blood cleansed by hemodialysis, the blood is circulated through dialysis tubing submerged in a bath that contains the following solutes in water: 0.6% NaCl, 0.04% KCl, 0.2% NaHCO₃, and 0.72% glucose (all percentages are w/v). Suggest one or more reasons why the dialysis tubing is not submerged in pure water.

7.91 Can the terms *saturated* and *supersaturated* be used to describe solutions made of liquids that are soluble in all proportions? Explain.

7.92 Refer to Figure 7.3 and propose a reason why fish sometimes die when the temperature of the water in which they live increases.

7.93 Small souvenir salt-covered objects are made by forming the object out of wire mesh and suspending the mesh object in a container of water from a salt lake such as the Dead Sea or the Great Salt Lake. As the water evaporates, the wire mesh becomes coated with salt crystals. Describe this process using the key terms introduced in Section 7.2.

7.94 Refer to Figure 7.6 and answer the question. How would the solubility of sugar compare in equal amounts of hot and iced tea?

7.95 Refer to Figure 7.11 and explain the process as requested. Draw simple diagrams showing the initial appearance and appearance after some time for a similar experiment in which the two liquids are 0.20 M copper sulfate solution and 2.0 M copper sulfate solution. Explain your reasoning.

7.96 Strips of fresh meat can be preserved by drying. In one process, the strips are coated with table salt and exposed to the air. Use a process discussed in this chapter and one discussed in Chapter 6 to explain how the drying takes place.

Allied Health Exam Connection

The following questions are from these sources:

• *Nursing School Entrance Exam* © 2005, Learning Express, LLC.

• *McGraw-Hill's Nursing School Entrance Exams* by Thomas A. Evangelist, Tamara B. Orr, and Judy Unrein © 2009, The McGraw-Hill Companies, Inc.

• *NSEE Nursing School Entrance Exams*, 3rd edition © 2009, Kaplan Publishing.

• *Cliffs Test Prep: Nursing School Entrance Exams* by Fred N. Grayson © 2004, Wiley Publishing, Inc.

• *Peterson's Master the Nursing School and Allied Health Entrance Exams,* 18th edition by Marion F. Gooding © 2008, Peterson's, a Nelnet Company.

7.97 If a salt is added to water, which of the following is likely to occur?

 a. The boiling point will increase and the freezing point will decrease.

 b. The boiling point will increase and the freezing point will increase.

 c. The boiling point will decrease and the freezing point will decrease.

 d. The boiling point will decrease and the freezing point will increase.

7.98 A cell is in a solution in which the concentration of solutes is higher inside the cell than outside the cell. The cell will likely:

 a. swell up and possibly burst.

 b. shrivel and shrink.

 c. maintain its size.

 d. grow a cell wall for support.

7.99 What will happen if a semipermeable membrane is placed between two different concentrations of a NaCl solution?

 a. The solute will move toward the higher concentration.

 b. The solute will move toward the lower concentration.

 c. The solvent will move toward the higher concentration.

 d. The solvent will move toward the lower concentration.

7.100 Given a sample of $C_6H_{12}O_6$(aq), which of the following is true?

 a. The glucose is the solvent and water is the solute.

 b. The glucose is the solvent and water is the solvent.

 c. The glucose is the solute and water is the solute.

 d. The glucose is the solute and water is the solvent.

7.101 Which one of the following compounds is a nonelectrolyte when dissolved in water?

 a. KOH

 b. NH_3

 c. NaBr

 d. $CaCl_2$

7.102 An example of a strong electrolyte is:

 a. sugar.

 b. calcium chloride.

 c. glycerin.

 d. boric acid.

7.103 A solution that contains all the solute it can normally dissolve at a given temperature must be:

 a. concentrated.

 b. supersaturated.

 c. saturated.

 d. unsaturated.

7.104 Oil and water are immiscible (do not mix) because:

 a. oil is polar and water is polar.

 b. oil is nonpolar and water is polar.

 c. water is nonpolar and oil is polar.

 d. water is nonpolar and oil is nonpolar.

7.105 Cells that contain more dissolved salts and sugars than the surrounding solution are called:

 a. isotonic.

 b. hypertonic.

 c. hypotonic.

 d. osmosis.

7.106 How many grams of NaOH would be needed to make 250 ml of a 0.200 M solution? (molecular weight of NaOH = 40.0)

 a. 8.00 g

 b. 4.00 g

 c. 2.00 g

 d. 2.50 g

7.107 Ice can be melted most effectively by _____ if 1 mole is used.

 a. sucrose

 b. calcium chloride

 c. sodium chloride

 d. methanol

7.108 Jack has 100 mL of a 12-molar solution of sulfuric acid. How much of it should he put into a graduated cylinder to make 20 mL of a 1.2-molar solution?

 a. 1 mL

 b. 2 mL

 c. 10 mL

 d. 12 mL

7.109 How many grams of sugar are needed to make 500 mL of a 5% (weight/volume) solution of sugar?

 a. 20

 b. 25

 c. 50

 d. 10

7.110 As water is evaporated from a solution, the concentration of the solute in the solution will:

 a. increase.

 b. decrease.

 c. remain the same.

7.111 In a dilute solution of sodium chloride in water, the sodium chloride is the:

 a. solvent.

 b. solute.

 c. precipitate.

 d. reactant.

7.112 A salt solution has a molarity of 1.5 M. How many moles of this salt are present in 2.0 L of this solution?

 a. 1.5 moles

 b. 2.0 moles

 c. 3.0 moles

 d. 0.75 moles

7.113 If 58.5 g of NaCl (1 mole of NaCl) are dissolved in enough water to make 0.500 L of solution, what is the molarity of this solution?

 a. 2.0 M

 b. 11.7 M

 c. 1.0 M

 d. The answer cannot be determined from the information above.

7.114 To prepare 100 mL of a 0.20 M NaCl solution from a stock solution of 1.00 M NaCl, you should mix:

 a. 20 mL of stock solution with 80 mL of water

 b. 40 mL of stock solution with 60 mL of water

 c. 20 mL of stock solution with 100 mL of water

 d. 25 mL of stock solution with 75 mL of water

7.115 If a 2.0 M solution is diluted to 0.5 M, and the final volume is 100 mL, what was the original volume?

 a. 400 mL

 b. 200 mL

 c. 50 mL

 d. 25 mL

7.116 The number of moles of NaCl in 250 mL of a 0.300 M solution of NaCl is:

 a. 0.0750

 b. 0.150

 c. 0.250

 d. 1.15

7.117 One liter of solution is made by dissolving 29.2 g of NaCl in water. What is the molarity of the solution?

 a. 0.5 M

 b. 2.0 M

 c. 1.3 M

 d. 0.82 M

7.118 If a red blood cell is placed in seawater, it will be in what kind of solution?

 a. isotonic

 b. hypotonic

 c. hypertonic

 d. facilitated diffusion

7.119 When placed in distilled water, a human red blood cell:

 a. shrivels up.

 b. neither shrinks nor swells.

 c. takes up more salts to balance all concentrations.

 d. swells to a larger size.

7.120 The movement of substances from a lesser concentration to a higher concentration is called:

 a. osmosis.

 b. diffusion.

 c. active transport.

 d. pinocytosis.

7.121 Primary intermolecular interactions between a K cation and H_2O molecules are:

 a. hydrogen bonds.

 b. dipole-dipole interactions.

 c. ion-dipole interactions.

 d. London forces.

7.122 All of the following are colligative properties of solutions EXCEPT:

 a. vapor pressure.

 b. osmotic pressure.

 c. density.

 d. boiling point elevation.

8 Reaction Rates and Equilibrium

Evgeniy Ayupov/Shutterstock.com

▶ Case Study

Joy wanted to get the kids out of the house on this beautiful end-of-summer day. Stopping at the gas station on the way to the park, she bought corn dogs. Her toddler, Morgan, was eating his corn dog in the stroller while Joy pushed her daughter on the swing. Suddenly, Morgan screamed. Several yellow jackets were flying around him. Apparently, the insects were fond of corn dogs as well! Morgan had been stung on the lip. Joy scraped off the stinger and comforted her son, but almost immediately his face began to swell. They went home and Joy gave Morgan some liquid antihistamine. Half of his face was swollen and his eye was nearly shut and so Joy took Morgan to the clinic. The physician's assistant (PA) examined Morgan and said that Joy had done the right thing, but that her son seemed to be sensitive to yellow jacket stings. The PA added that it would be wise to have a fast-acting treatment on hand, because stings can be dangerous to sensitive individuals. He wrote a prescription for an EpiPen (epinephrine injection) and some cortisone cream.

▶ When is it appropriate to use an epinephrine injection? How can someone distinguish between a mild allergic reaction and a dangerous one?

Follow-up to this Case Study appears at the end of the chapter before the *Concept Summary*.

Learning Objectives

When you have completed your study of this chapter, you should be able to:

1 Use the concepts of energy and entropy to predict the spontaneity of processes and reactions. **(Section 8.1)**

2 Calculate reaction rates from experimental data. **(Section 8.2)**

3 Use the concept of molecular collisions to explain reaction characteristics. **(Section 8.3)**

4 Represent and interpret the energy relationships for reactions by using energy diagrams. **(Section 8.4)**

5 Explain how factors such as reactant concentrations, temperature, and catalysts influence reaction rates. **(Section 8.5)**

6 Relate experimental observations to the establishment of equilibrium. **(Section 8.6)**

7 Write equilibrium expressions based on reaction equations, and do calculations based on equilibrium expressions. **(Section 8.7)**

8 Use Le Châtelier's principle to predict the influence of changes in concentration and reaction temperature on the position of equilibrium for reactions. **(Section 8.8)**

According to calculations, carbon in the form of a diamond will spontaneously change into graphite. Owners of diamonds seem unconcerned by this fact; they know from experience that their investments in gems are secure. In reality, the change does take place, but so slowly that it is not detectable over many human lifetimes. The low rate of the reaction makes the difference.

The element barium is very poisonous when it is taken into the body as barium ions (Ba^{2+}). When barium sulfate ($BaSO_4$) dissolves, Ba^{2+} and SO_4^{2-} ions are formed. However, suspensions of solid $BaSO_4$ are routinely swallowed by patients undergoing diagnostic X-ray photography of the stomach and intestinal tract. The patients are not affected because very little $BaSO_4$ dissolves in the body. This lack of solubility is an equilibrium property of $BaSO_4$.

8.1 | Spontaneous and Nonspontaneous Processes

Learning Objective

1. Use the concepts of energy and entropy to predict the spontaneity of processes and reactions.

Years ago, tourists in Salt Lake City visited "gravity hill," where an optical illusion made a stream of water appear to flow uphill. The fascination with the scene came from the apparent violation of natural laws—everyone knew that water does not flow uphill. Processes that take place naturally with no apparent cause or stimulus are called **spontaneous processes**. Nonspontaneous processes take place only as the result of some cause or stimulus. For example, imagine you are in the positions depicted in **Figure 8.1** and want to move a boulder.

spontaneous process A process that takes place naturally with no apparent cause or stimulus.

Figure 8.1 The problem of moving a boulder.

A — Spontaneous process

B — Spontaneous process once started

C — Nonspontaneous process

In part A, the boulder will roll down the hill as soon as you release it; the process begins and takes place spontaneously. In part B, you must push the boulder a little to get it over the hump, but once started, the boulder spontaneously rolls downhill. The situation in part C is very different. You must continually push on the boulder, or it will not move up the hill; at no time can you stop pushing and expect the boulder to continue moving up. The process is nonspontaneous; it takes place only because of the continuous application of an external stimulus.

As the boulder rolls downhill, it gives up energy; it moves from a state of high potential energy to a lower one. As you push the boulder uphill, it gains energy (from you). Processes that give up energy are called **exergonic** (energy out), whereas those that gain energy are called **endergonic** (energy in). Very often, energy changes in chemical processes involve heat. Those changes that do are referred to as either exothermic (heat out) or endothermic (heat in), two terms you have encountered before (see **Sections 5.8** and **6.11**).

Many spontaneous processes give up energy. A piece of wood, once ignited, spontaneously burns and liberates energy as heat and light. At normal room temperature, steam spontaneously condenses to water and releases heat. However, some spontaneous processes take place and either give up no energy or actually gain energy. These spontaneous processes are always accompanied by another change called an entropy increase. **Entropy** describes the disorder or mixed-up character (randomness) of a system (see **Figure 8.2**). Thus, an entropy increase accompanies all spontaneous processes in which energy remains constant or is gained. A drop of ink placed in a glass of water will eventually become uniformly distributed throughout the water, even though the water is not stirred. No energy change accompanies the diffusion of the ink, but the distribution of ink throughout the water is a more disorderly (higher-entropy) state than when the ink is all together in a single drop. Ice melts spontaneously at 20°C and in the process absorbs heat. The process takes place because the random distribution of water molecules in the liquid is a higher-entropy state than the orderly molecular arrangement found in crystalline ice.

exergonic process A process that gives up energy as it takes place.

endergonic process A process that gains or accepts energy as it takes place.

entropy A measurement or indication of the disorder or randomness of a system. The more disorderly a system is, the higher its entropy.

Figure 8.2 The mixture on the near-right has higher entropy (more randomness, or mixed up character) than the mixture on the far-right.

Energy and entropy changes influence the spontaneity of processes in several ways:

1. A process will always be spontaneous if the energy decreases and the entropy increases. An example of such a process is the burning of a piece of wood in which heat is given up and the gaseous products of combustion are distributed throughout the surroundings, providing the entropy increase.
2. When a spontaneous process is accompanied by an energy increase, a large entropy increase must also occur. Thus, when ice spontaneously melts at 20°C, the increase in entropy is large enough to compensate for the increase in energy that also takes place.
3. A spontaneous process accompanied by an entropy decrease must also be accompanied by a compensating energy decrease. Thus, when water spontaneously freezes at 0°C, the energy loss compensates for the entropy decrease that occurs as the water molecules assume the well-ordered arrangement in ice.

It is useful to think of energy and entropy changes in terms of the directions favoring spontaneity and the directions not favoring spontaneity. It is apparent from the preceding discussion that spontaneity is favored by an energy decrease and/or an entropy increase. Processes in which one of these factors changes in a nonspontaneous direction will be spontaneous only if the other factor changes in a spontaneous direction by a large enough amount to compensate for the nonspontaneous change. The influences listed under 2 and 3 above are examples of this.

A substance that does not undergo spontaneous changes is said to be a **stable substance**. However, stability depends on the surrounding conditions, and a change in those conditions might cause a nonspontaneous process to become spontaneous. Ice is a stable solid at −10°C and 760 torr pressure but spontaneously melts to a liquid at 5°C and 760 torr. Wood is stable at room temperature but spontaneously burns when its temperature equals or exceeds the ignition temperature. In the following discussions, the surrounding conditions are assumed to be those normally encountered. Otherwise, the differences are specified.

stable substance A substance that does not undergo spontaneous changes under the surrounding conditions.

8.2 | Reaction Rates

Learning Objective

2. Calculate reaction rates from experimental data.

The speed of a reaction is called the **reaction rate**. It is determined experimentally as the change in concentration of a reactant or product divided by the time required for the change to occur. This average rate is represented for changes in product concentration by Equation 8.1:

reaction rate The speed of a reaction.

$$\text{Rate} = \frac{\Delta C}{\Delta t} = \frac{C_t - C_0}{\Delta t} \tag{8.1}$$

where the delta symbol (Δ) stands for change. C_t and C_0 are the concentrations of a product at the end and beginning, respectively, of the measured time change, Δt. The time can be measured in any convenient unit.

Example 8.1 Calculating Reaction Rates

The gases NO_2 and CO react as follows:

$$NO_2(g) + CO(g) \rightarrow NO(g) + CO_2(g)$$

Calculate the average rate of the reaction if pure NO_2 and CO are mixed and after 50.0 seconds the concentration of CO_2 is found to be 0.0160 mol/L.

Solution

Because the reaction was started by mixing pure NO_2 and CO, the initial concentration C_0 of CO_2 was 0.000 mol/L. After 50.0 seconds, the concentration C_t of CO_2 was 0.0160 mol/L.

$$\text{Rate} = \frac{C_t - C_0}{\Delta t} = \frac{0.0160 \text{ mol/L} - 0.000 \text{ mol/L}}{50.0 \text{ s}} = 3.20 \times 10^{-4} \frac{\text{mol/L}}{\text{s}}$$

The answer is read 3.20×10^{-4} mole per liter per second. It means that during the 50.0-second time interval, an average of 3.20×10^{-4} mol CO_2 was formed in each liter of reacting mixture each second.

✔ **LEARNING CHECK 8.1** The Ce^{4+} and Fe^{2+} ions react in solution as follows:

$$Ce^{4+}(aq) + Fe^{2+}(aq) \rightarrow Ce^{3+}(aq) + Fe^{3+}(aq)$$

Solutions of Ce^{4+} and Fe^{2+} are mixed and allowed to react. After 75.0 seconds, the concentration of Ce^{3+} is found to be 1.50×10^{-5} mol/L. Calculate the average rate of the reaction.

8.3 | Molecular Collisions

Learning Objective

3. Use the concept of molecular collisions to explain reaction characteristics.

Chemical equations such as those in Example 8.1 and Learning Check 8.1 are useful in identifying reactants and products and in representing the stoichiometry of a reaction. However, they indicate nothing about how a reaction takes place—how reactants get together (or break apart) to form products. The explanation of how a reaction occurs is called a **reaction mechanism**. Reaction mechanisms are not discussed much in this book, but most are based on the following assumptions:

reaction mechanism A detailed explanation of how a reaction actually takes place.

1. Reactant particles must collide with one another in order for a reaction to occur.
2. Particles must collide with at least a certain minimum total amount of energy if the collision is to result in a reaction.
3. In some cases, colliding reactants must be oriented in a specific way if a reaction is to occur.

The validity of the first assumption is fairly obvious. Two molecules cannot react with each other if they are far apart. In order to break bonds, exchange atoms, and form new bonds, they must come in contact. There are, however, some exceptions, such as decompositions in which molecules break into fragments and processes in which molecules react by an internal rearrangement of their atoms.

In general, reactions take place more rapidly in the gaseous or liquid state than in the solid state. This observation verifies the first assumption because molecules of gases and liquids move about freely and can undergo many more collisions than can the rigidly held molecules of solids. Reactions involving solids usually take place only on the solid surface and therefore involve only a small fraction of the total molecules present in the solid. As the reaction proceeds and the products dissolve, diffuse, or fall from the surface, fresh solid is exposed. In this way, the reaction proceeds into the solid. The rusting of iron is an example of such a process.

If collisions were the only factor, however, most gaseous and liquid state reactions would take place almost instantaneously if every collision resulted in a reaction. Such high reaction rates are not observed, a fact that brings us to assumptions 2 and 3.

One of the ways to speed up a chemical reaction is to add energy in the form of heat. The added heat increases both the average speed (kinetic energy) and the internal energy of the molecules. **Internal energy** is the energy associated with molecular vibrations. An increase in internal energy increases the amplitude of molecular vibrations and, if large

internal energy The energy associated with vibrations within molecules.

ASK A PHARMACIST 8.1

Energy for Sale

All individuals, regardless of their age, use up some body energy during a typical day, even when they are sleeping. Many different products are commercially available that claim to replenish the used energy. These products are available in edible forms or as drinks. This feature will focus on energy drinks.

It is difficult to lump all energy drinks together because different products contain different proprietary blends of ingredients, but they all contain some stimulant(s), most often caffeine. Naturally caffeinated drinks (tea or coffee) are usually not considered to be "energy drinks." Typical popular soft drinks that contain 20–50 mg of caffeine per 12 oz serving are also not considered to be energy drinks in this discussion.

Other common ingredients of energy drinks include amino acids (taurine and carnitine), vitamins (in nominal amounts that have little/no benefit), carbohydrates, and herbal extracts (ginseng, guarana, yerba maté, and ginkgo biloba in variable amounts that are unregulated in spite of having numerous interactions with drugs).

Even though the amount of caffeine in a typical energy drink is not more per ounce than brewed coffee, the marketed container size of energy drinks is much larger than a cup of coffee, and most consumers drink the entire can. This means that the total amount of caffeine consumed will have both good and bad pharmacological effects.

The good effects include increased mental alertness at work or school, since caffeine stimulates the central and autonomic nervous systems. This combats fatigue while driving, and might help prevent auto accidents.

The possible bad effects are as follows:

- Causes nervousness, irritability, anxiety, and phobias.
- Causes insomnia. Might be more difficult to fall asleep and stay asleep, and result in less deep sleep.
- Causes rapid heartbeat. Can induce dysrhythmias, heart attacks, and death.
- Causes increased blood pressure, gastrointestinal distress, and muscle tremors.
- Might worsen underlying medical conditions that are influenced by stimulants.

The use of energy drinks should be reviewed by a health care provider.

Contrary to product promotions, there is little credible scientific evidence to support claims that energy drinks increase muscle strength and endurance. Energy drinks are often consumed before exercise or athletic events. The additional fluid intake is healthful, but be aware that caffeine is a diuretic that might increase the risk of dehydration.

Recommendations:

1. If you consistently lack energy, consider healthier alternatives such as scheduling an additional 10–15 minutes of sleep each night.

2. Do not mix an energy drink with an alcoholic beverage. The feeling of intoxication will be blunted, but the consequences are not.

3. Quantify the amount of caffeine and other stimulants consumed per day. Caffeine has its benefits, but too much can cause unwanted effects. More than 250–400 mg per day for an adult appears to have the greatest likelihood of causing adverse effects. But some people are more sensitive to stimulants and might feel better on lower doses or none at all.

4. Avoid caffeine use by children, and limit its use to less than 100 mg per day by adolescents.

5. Eat a healthful diet and follow a regular exercise/recreation program.

6. Evaluate the amount and quality of sleep you get. Chronic sleep deprivation, even in small amounts, will decrease your daytime alertness and might increase your motivation to use an energy drink.

7. Energy drinks that help you get through the day might interfere with your sleep architecture and actually decrease the quality and quantity of your sleep. That leads to a vicious cycle.

8. Assess and reduce stress levels in your life.

9. Check with your health providers and determine if you are taking any medications that can influence your energy levels. Also, explain your symptoms to them, since fatigue can be a symptom of a medical condition such as anemia or hypothyroidism.

10. If you intend to decrease your energy drink consumption for any reason, do not stop abruptly. An abrupt stop might cause withdrawal symptoms such as headache and fatigue. Taper off slowly. Paying attention to the amount of stimulating substances you consume can allow you to assess the risk/benefit ratio, monitor the consequences, and modify your behavior if necessary.

There are both pros and cons to consuming energy drinks.

Bond

Low internal energy

Bond being stretched

Higher internal energy

Bond breaks

Internal energy high enough to break bonds

Figure 8.3 The internal energy of molecules.

enough, breaks bonds (see **Figure 8.3**). Internal energy is also increased by the conversion of some kinetic energy into internal energy during collisions. When heat is added, both the kinetic and the internal energy of molecules is increased. This increases the frequency and speed of collisions and the chances that a collision will cause a sufficient increase in internal energy to break bonds and bring about a reaction (see **Figure 8.4**).

Figure 8.4 The reaction that makes a Cyalume light stick glow takes place more rapidly in hot water (left) than in ice water (right). The light sticks outside the container are at room temperature. Which pair of sticks would glow for a longer time? Explain your reasoning.

In some reaction mixtures, the average total energy of the molecules is too low at the prevailing temperature for a reaction to proceed at a detectable rate; the reaction mixture is stable. However, some reactions can be started by providing **activation energy**. Once the reaction is started, enough energy is released to activate other molecules and keep the reaction going at a good rate. The energy needed to get the boulder over the small hump in **Figure 8.1B** is a kind of activation energy. In many chemical reactions, activation energy causes bonds in reactant molecules to break. When the broken bonds react to form the new bonds of the products, energy is released that can cause bonds in more reactant molecules to break and the reaction to continue. The striking of a kitchen match is a good example. Activation energy is provided by rubbing the match head against a rough surface. Once started, however, the match continues to burn spontaneously.

A number of gases are routinely used in hospitals. Some of these form very flammable or even potentially explosive mixtures. Cyclopropane, a formerly used anesthetic, will burn vigorously in the presence of the oxygen in air. However, a mixture of the two will not react unless activation energy is provided in the form of an open flame or a spark. This is the reason that extreme precautions are taken to avoid open flames and sparks in a hospital operating room. Even sparks from static electricity can set off such gaseous mixtures, so special materials are used in clothing, floor coverings, and so on to prevent static electricity from building up.

Oxygen gas does not burn, but it reacts with anything that is combustible. Oxygen is often used in hospitals in concentrations much higher than the 20% found in air. Thus, precautions must be taken to prevent fire or sparks in any areas where oxygen gas is used in high concentrations.

Orientation effects are related to which side or end of a particle actually hits another particle during a collision. Orientation effects are unimportant in many reactions. For example, the orientation of silver ions (Ag^+) and chloride ions (Cl^-) toward each other during a collision has no effect on the rate of forming AgCl:

$$Ag^+(aq) + Cl^-(aq) \rightarrow AgCl(s) \qquad (8.2)$$

The reason is that Ag^+ and Cl^- are both essentially spherical charged particles. However, collision orientation may be important in reactions that involve nonspherical molecules (assumption 3). Consider the following hypothetical reaction:

$$A—B + C—D \rightarrow A—C + B—D \qquad (8.3)$$

It is clear that the collision orientation of $A—B$ and $C—D$ shown in **Figure 8.5A** is more favorable to reaction than the orientations shown in **Figure 8.5B** and **Figure 8.5C**.

8.4 | Energy Diagrams

Learning Objective

4. Represent and interpret the energy relationships for reactions by using energy diagrams.

Energy relationships for reactions can be represented by energy diagrams like the one in **Figure 8.6**. Notice the similarity to the earlier example of rolling boulders.

Figure 8.6 A typical energy diagram for chemical reactions.

The energy diagrams for most reactions look generally alike, but there are some differences. Typical diagrams for exothermic (exergonic) and endothermic (endergonic) reactions are given in **Figure 8.7**. It is clear that the products of exothermic reactions have lower energy than the reactants and that the products of endothermic reactions have higher energy than the reactants. When exothermic reactions occur, the energy difference between reactants and products is released. This energy often appears as heat, so heat is given up to the surroundings. When endothermic reactions occur, the energy (heat) added to the products is absorbed from the surroundings. Also, different reactions generally have different activation energies, a fact easily represented by energy diagrams.

Favorable orientation: *A* is near *C*, and *B* is near *D* during collision.

Unfavorable orientation: *A* is not near *C*, and *B* is not near *D* during collision.

Unfavorable orientation: *B* is near *D*, but *A* and *C* are far removed during collision.

Figure 8.5 Molecular orientations during collisions.

Figure 8.7 Energy diagrams for exothermic and endothermic reactions.

White phosphorus, a nonmetallic element, spontaneously bursts into flame if left in a rather warm room (34°C):

$$4P(s) + 5O_2(g) \rightarrow P_4O_{10}(s) \tag{8.4}$$

Sulfur, another nonmetallic element, will also burn, but it does not ignite until heated to about 232°C:

$$S(s) + O_2(g) \rightarrow SO_2(g) \tag{8.5}$$

The different activation energies for these two exothermic reactions are represented in **Figure 8.8**.

Figure 8.8 Differences in activation energies.

It should now be clear why we stressed the notion that some substances that are "stable" at normal living conditions undergo spontaneous changes at other conditions. It is simply that a higher temperature would provide the necessary activation energy. In a room hotter than 232°C, for example, both white phosphorus and sulfur would be "unstable" in the presence of oxygen.

8.5 | Factors That Influence Reaction Rates

Learning Objective

5. Explain how factors such as reactant concentrations, temperature, and catalysts influence reaction rates.

Reaction rates are influenced by a number of different factors. Four factors that affect the rates of all reactions are:

1. The nature of the reactants.
2. The concentration of the reactants.
3. The temperature of the reactants.
4. The presence of catalysts.

The formation of insoluble AgCl by mixing together solutions containing Ag^+ and Cl^- ions is represented by Equation 8.2, given earlier. The white solid AgCl forms the instant the two solutions are mixed. This behavior is typical of reactions involving ionic reactants. The high reaction rate results from the attraction of the charged reactants to each other. In contrast, reactions that require covalent bonds to be broken or formed often proceed slowly. The production of water gas, a mixture of hydrogen (H_2) and carbon monoxide (CO), is a reaction involving covalent bonds:

$$H_2O(g) + C(s) \xrightarrow{\text{Heat}} CO(g) + H_2(g) \tag{8.6}$$

In addition, other structural characteristics of reactants, such as bond polarity or molecular size, may also be important factors in reaction rates.

The influence of reactant concentration on reaction rates can be illustrated by using the concept of molecular collisions. Suppose a reaction takes place between the hypothetical molecules A and B, which are mixed in a 1:1 ratio. The proposed reaction is

$$A + B \rightarrow \text{products} \tag{8.7}$$

Collisions with the capability to cause a reaction to occur are called **effective collisions**. Only collisions between A and B molecules can be classified as effective, since collisions between two A molecules or between two B molecules cannot possibly yield products. Imagine the reaction is begun with two A molecules and two B molecules as shown in **Figure 8.9**. The example is simplified by looking only at the collisions of a single A molecule. Initially, two of every three collisions of the A molecule will be effective (part A). On doubling the number (concentration) of B molecules, the number of effective collisions also doubles, and four of every five collisions is effective (part B).

When larger numbers of molecules are used, the results approach more closely those actually observed experimentally for simple chemical reactions. Imagine a mixture containing 1000 A molecules and 1000 B molecules. On the average, 1000 out of every 1999 collisions of a single A molecule will involve a B molecule. (Each A molecule has almost a 50–50 chance of bumping into a B molecule.) This gives essentially a 1:1 ratio of effective to noneffective collisions. When the number of B molecules is doubled to 2000, the ratio becomes 2:1, because 2000 of every 2999 collisions will be effective. Therefore, the reaction rate should double when the concentration of one reactant is doubled. This result has been verified in numerous chemical reactions. Thus, higher concentrations produce a larger number of effective collisions in a given period of time, and this increases the reaction rate.

Gas-phase reactions are easily visualized this way, but reacting liquids and solids must be looked at differently. A large piece of solid contains a large number of molecules, but, as described before, only those on the surface can react with the molecules of another substance. For this reason, the total amount of solid in a sample is not as important as the surface area of solid in contact with other reactants. The effective concentration of a solid therefore depends on its surface area and state of division. A 100-pound sack of flour is difficult to burn when it is in a single pile. The same 100 pounds dispersed in the air as a fine dust burns very rapidly, and a dust explosion results (see **Figure 8.10**). The effective concentrations of reacting liquids must be thought of in a similar way unless the reactants are completely miscible.

effective collision A collision that causes a reaction to occur between the colliding molecules.

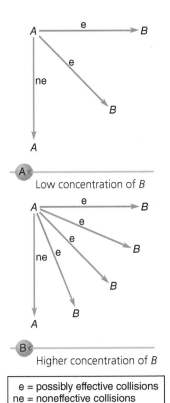

e = possibly effective collisions
ne = noneffective collisions

Figure 8.9 The effect of concentration on reaction rates.

1 A compact pile of lycopodium powder burns sluggishly.

2 A similar sample of lycopodium powder is blown through a blowpipe. The blown powder burns instantly with a bright flash.

Figure 8.10 Surface area influences reaction rates of solids. What is the other reactant in this reaction? How finely divided is it?

The effect of reactant temperature on reaction rates can also be explained by using the molecular collision concept. As was noted earlier, an increase in the temperature of a

system increases the average kinetic energy and internal energy of the reacting molecules. The increased molecular speed (kinetic energy) causes more collisions to take place in a given time. Also, because the kinetic and internal energies of the colliding molecules are greater, a larger fraction of the collisions will be effective because they will provide the necessary activation energy.

As a rough rule of thumb, it has been found that the rate of a chemical reaction doubles for every 10°C increase in temperature. The chemical reactions of cooking take place faster in a pressure cooker because of a higher cooking temperature (see **Section 6.13**). Also, cooling or freezing is used to slow the chemical reactions involved in the spoiling of food, souring of milk, and ripening of fruit.

A **catalyst** is a substance that changes reaction rates without being used up in the reactions. Usually the term *catalyst* is used to describe substances that speed up reactions. A substance that slows reactions is know as an **inhibitor**. Catalysts are used in a number of different forms. Those dispersed uniformly throughout a reaction mixture in the form of individual ions or molecules are called **homogeneous catalysts**. **Heterogeneous** or **surface catalysts** are used in the form of solids; usually they have large surface areas on which reactions take place readily.

Catalysts enhance a reaction rate by providing an alternate reaction pathway that requires less activation energy than the normal pathway. This effect is represented by **Figure 8.11**. According to some theories, a catalyst provides the lower-energy pathway by entering into a reaction and forming an intermediate structure, which then breaks up to produce the final products and regenerate the catalyst:

Uncatalyzed: $A + B \rightarrow$ products (8.8)

Catalyzed: $A + B + \text{catalyst} \rightarrow A\!-\!B \rightarrow \text{products} + \text{catalyst}$ (8.9)

$$\underset{\text{Intermediate structure}}{\overset{| \qquad |}{\text{catalyst}}}$$

According to another proposed mechanism, solid catalysts provide a surface to which reactant molecules become attached with a particular orientation. Reactants attached to these surfaces are sufficiently close to one another and oriented favorably enough to allow the reaction to take place. The products of the reaction then leave the surface and make the attachment sites available for catalyzing other reactant molecules.

catalyst A substance that changes (usually increases) reaction rates without being used up in the reaction.

inhibitor A substance that decreases reaction rates.

homogeneous catalyst A catalytic substance that is distributed uniformly throughout the reaction mixture.

heterogeneous (surface) catalyst A catalytic substance normally used in the form of a solid with a large surface area on which reactions take place.

Figure 8.11 The effect of catalysts on activation energy.

8.6 | Chemical Equilibrium

Learning Objective

6. Relate experimental observations to the establishment of equilibrium.

So far, the focus has been only on the reactants of a reaction. However, all reactions can (in principle) go in both directions, and the products can be looked on as reactants. Suppose equal amounts of gaseous H_2 and I_2 are placed in a closed container and allowed to react and form HI. Initially, no HI is present, so the only possible reaction is

$$H_2(g) + I_2(g) \rightarrow 2HI(g) \tag{8.10}$$

However, after a short time, some HI molecules are produced. They can collide with one another in a way that causes the reverse reaction to occur:

$$2HI(g) \rightarrow H_2(g) + I_2(g) \tag{8.11}$$

The low concentration of HI makes this reaction slow at first, but as the concentration increases, so does the reaction rate. The rate of reaction decreases as the concentrations of H_2 and I_2 decrease. Eventually, the concentrations of H_2, I_2, and HI in the reaction mixture reach levels at which the rates of the forward (Equation 8.10) and reverse (Equation 8.11) reactions are equal. From that time on, the concentrations of H_2, I_2, and HI in the mixture remain constant, since both reactions take place at the same rate and each

substance is being produced as fast as it is used up. When the forward and reverse reaction rates are equal, the reaction is in a **state of equilibrium**, and the concentrations are called **equilibrium concentrations**. The behavior of reaction rates and reactant concentrations for both the forward and reverse reactions is shown graphically in **Figure 8.12**.

Instead of writing separate equations for both the forward and reverse reactions, the usual practice is to represent reversibility by double arrows. Thus, the equation for the reaction between H_2 and I_2 is written

$$H_2(g) + I_2(g) \rightleftharpoons 2HI(g) \qquad (8.12)$$

state of equilibrium A condition in a reaction system when the rates of the forward and reverse reactions are equal.

equilibrium concentrations The unchanging concentrations of reactants and products in a reaction system that is in a state of equilibrium.

8.7 | The Position of Equilibrium

Learning Objective

7. Write equilibrium expressions based on reaction equations, and do calculations based on equilibrium expressions.

position of equilibrium An indication of the relative amounts of reactants and products present at equilibrium.

The **position of equilibrium** for a reaction indicates the relative amounts of reactants and products present at equilibrium. When the position is described as being far to the right, it means that at equilibrium the concentration of products is much higher than the concentration of reactants. A position far to the left means the concentration of reactants is much higher than that of products.

The position of equilibrium can be represented numerically by using the concept of an equilibrium constant. Any reaction that establishes an equilibrium can be represented by a general equation:

$$aA + bB + \cdots \rightleftarrows wW + xX + \cdots \tag{8.13}$$

In this equation, the capital letters stand for substances such as the H_2, I_2, and HI of Equation 8.12, and the lowercase letters are the coefficients in the balanced equation such as the 1, 1, and 2 of Equation 8.12. The dots in Equation 8.13 indicate that any number of reactants and products can be involved in the reaction, but we will limit ourselves to just the four shown.

As we implied earlier, a reaction at equilibrium can be recognized because the concentrations of reactants and products remain constant. That is, they do not change as time passes. For a reaction at equilibrium, the following equation is valid:

$$K = \frac{[W]^w[X]^x}{[A]^a[B]^b} \tag{8.14}$$

equilibrium expression An equation relating the equilibrium constant and reactant and product equilibrium concentrations.

equilibrium constant A numerical relationship between reactant and product concentrations in a reaction at equilibrium.

In this **equilibrium expression**, the brackets [] represent molar concentrations of the reactants (A and B) and the products (W and X). The K is a constant called the **equilibrium constant**, and the powers on each bracket are the coefficients from the balanced equation for the reaction. According to this equation, the product of equilibrium concentrations of products (raised to appropriate powers) divided by the product of the equilibrium concentration of reactants (also raised to appropriate powers) gives a number that does not change with time (the equilibrium constant). The reason K does not change with time is that at equilibrium, the concentrations used to calculate K do not change with time. We could get a different constant by dividing the reactant concentrations by the product concentrations, but it is the accepted practice to calculate K values as we have shown in Equation 8.14.

Example 8.2 Writing Equilibrium Expressions

Write equilibrium expressions for the reactions represented by the following equations:

a. $H_2(g) + I_2(g) \rightleftarrows 2HI(g)$
b. $2SO_2(g) + O_2(g) \rightleftarrows 2SO_3(g)$

Solution

a. The concentration of the product HI goes on top and is raised to the power 2. The concentrations of the reactants H_2 and I_2 go on the bottom, and each is raised to the power 1, which is not written but understood:

$$K = \frac{[HI]^2}{[H_2][I_2]}$$

b. The product is SO_3, and the power is 2. The reactants are SO_2 (power = 2) and O_2 (power = 1).

$$K = \frac{[SO_3]^2}{[SO_2]^2[O_2]}$$

✔ **LEARNING CHECK 8.2** Write equilibrium expressions for the reactions represented by the following equations:

a. $N_2O_4(g) \rightleftarrows 2NO_2(g)$
b. $3NO(g) \rightleftarrows N_2O(g) + NO_2(g)$

For a reaction with an equilibrium position far to the right, the product concentrations [W] and [X] will be much higher than the reactant concentrations [A] and [B]. According to Equation 8.14, such a situation should result in a large K value. For reactions with equilibrium positions far to the left, a similar line of reasoning leads to the conclusion that K values should be small. Thus, the value of the equilibrium constant gives an indication of the position of equilibrium for a reaction.

Example 8.3 Calculating Equilibrium Constants

For each of the following equations, the equilibrium molarity concentration is given below each reactant and product. Calculate K for each reaction and comment on the position of equilibrium.

a. $\quad 2NO_2(g) \quad \rightleftarrows \quad N_2O_4(g)$
$\quad 8.23 \times 10^{-3}$ M $\quad 1.46 \times 10^{-2}$ M at 25°C

b. $CH_4(g) \quad + \quad H_2O(g) \rightleftarrows \quad CO(g) \quad + \quad 3H_2(g)$
$\quad 0.200$ M $\quad 0.150$ M $\quad 1.37 \times 10^{-2}$ M $\quad 4.11 \times 10^{-2}$ M at 900 K

Solution

a. $K = \dfrac{[N_2O_4]}{[NO_2]^2} = \dfrac{(1.46 \times 10^{-2} \text{ mol/L})}{(8.23 \times 10^{-3} \text{ mol/L})^2} = \dfrac{2.16 \times 10^2}{\text{mol/L}}$

We see that K is fairly large, which tells us the equilibrium position is toward the right, or product, side of the reaction. The peculiar sounding "per mole per liter" unit is the result of the way concentration terms are arranged in the equilibrium expression. The unit will not be the same for the K of all reactions. These units are not ordinarily shown, a practice we will follow for the remainder of this chapter.

b. $K = \dfrac{[CO][H_2]^3}{[CH_4][H_2O]}$

$= \dfrac{(1.37 \times 10^{-2} \text{ mol/L})(4.11 \times 10^{-2} \text{ mol/L})^3}{(0.200 \text{ mol/L})(0.150 \text{ mol/L})}$

$= 3.17 \times 10^{-5}$

The small value of K indicates the equilibrium position is toward the left.

✔ **LEARNING CHECK 8.3** The molar concentrations are given below the reactants and products in the equations for the following reactions. Calculate K and comment on the position of each equilibrium.

a. $Br_2(g)$ + $I_2(g)$ ⇌ $2IBr(g)$
1.50×10^{-1} M 5.00×10^{-2} M 1.96×10^{-2} M at 25°C

b. $N_2(g)$ + $3H_2(g)$ ⇌ $2NH_3(g)$
9.23×10^{-3} M 2.77×10^{-2} M 9.00 M at 25°C

Example 8.3 illustrates that equilibrium constants for different reactions may be small or large. In fact, the two reactions used are by no means even close to the extremes encountered for K values. Some are so small (such as 1.1×10^{-36}) that for all practical purposes no products are present at equilibrium. Others are so large (such as 1.2×10^{40}) that the reaction can be considered to go completely to products. For reactions with K values between 10^{-3} and 10^3, the position of equilibrium is not extremely favorable to either side, and significant concentrations of both reactants and products can be detected in the equilibrium mixtures.

You might also have noticed in Example 8.3 that the equilibrium concentrations were given for specific temperatures. The reason for this is that K values are constant for a reaction as long as the temperature remains constant. However, K values will change as the temperature is changed. This is a demonstration of the effect of temperature on the position of equilibrium described in the next section.

8.8 | Factors That Influence Equilibrium Position

Learning Objective

8. Use Le Châtelier's principle to predict the influence of changes in concentration and reaction temperature on the position of equilibrium for reactions.

Le Châtelier's principle The position of an equilibrium shifts in response to changes made in factors of the equilibrium.

A number of factors can change the position of an established equilibrium. The influence of such factors can be predicted by using a concept known as Le Châtelier's principle, in honor of its originator. According to **Le Châtelier's principle**, when a change is made in any factor of an established equilibrium, the position of equilibrium will shift in a direction that will minimize or oppose the change. The factors we will be most concerned with are concentrations of reactants and products, and reaction temperature.

The effect of concentration changes can be illustrated by the reaction of H_2 with I_2 (Equation 8.12):

$$H_2(g) + I_2(g) \rightleftharpoons 2HI(g) \qquad (8.12)$$

Suppose an equilibrium mixture of H_2, I_2, and HI is formed. According to the molecular collision concept, favorable collisions are occurring between H_2 and I_2 molecules to form HI molecules, and favorable collisions are also taking place between HI molecules that cause them to form H_2 and I_2 molecules. Now, suppose some additional I_2 is added to the equilibrium mixture. The chances for favorable collisions between H_2 and I_2 molecules are increased; the rate of formation of HI is increased, and more HI is formed than disappears.

Why "Cold" Does Not Exist

Everyone enjoys a cold drink on a hot day. Just drop a few ice cubes in your drink, and the temperature of your drink decreases. How did this happen? The temperature went down when ice cubes were added, so there are two possible explanations as to what happened. Either the ice cubes added something called "cold" to the drink, or they removed something the opposite of cold, which we will call "heat," from the drink.

According to the kinetic molecular theory of matter discussed in Chapter 6 (see page 178), energy (heat) always naturally flows from a warmer area to a cooler area. So, we conclude that heat energy moved in the natural direction from the warmer drink to the cooler ice, and since heat was removed from the liquid drink, its temperature went down. In the process, the ice melted as it absorbed the heat that was transferred from the liquid. The temperature changes and the ice melting are explained without resorting to the idea that a substance called "cold" was involved.

Blinka/Shutterstock.com

Melting ice cools a glass of water.

The rate of reaction of HI to give H_2 and I_2 increases as the concentration of HI increases, and eventually a new equilibrium position and a new set of equilibrium concentrations will be established. The new equilibrium mixture will contain more HI than the original mixture, but the amounts of the reactants will also be different such that the equilibrium constant remains unchanged. The original equilibrium of the reaction has been shifted toward the right.

Le Châtelier's principle can also be used to predict the influence of temperature on an equilibrium by treating heat as a product or reactant. Consider the following equation for a hypothetical exothermic reaction:

$$A + B \rightleftharpoons \text{products} + \text{heat} \qquad (8.15)$$

If the temperature is increased (by adding more heat, which appears on the right side of the equation), the equilibrium shifts to the left in an attempt to use up the added heat. In the new equilibrium position, the concentrations of A and B will be higher, whereas the chemical product concentrations will be lower than those in the original equilibrium. In this case the value of the equilibrium constant is changed by the change in temperature (see **Figure 8.13**).

Example 8.4 Using Le Châtelier's Principle

Using Le Châtelier's principle, answer the following questions:

a. Ammonia is made from hydrogen and atmospheric nitrogen:

$$N_2(g) + 3H_2(g) \rightleftharpoons 2NH_3(g) + \text{heat}$$

What effect will cooling have on an equilibrium mixture?

b. What effect will removing H_2 have on the equilibrium mixture described in part a?

c. Consider the following equation for a reaction that takes place in solution:

$$\underset{\substack{\text{light}\\\text{brown}}}{Fe^{3+}(aq)} + \underset{\text{colorless}}{6SCN^-(aq)} \rightleftharpoons \underset{\text{dark red}}{Fe(SCN)_6^{3-}(aq)}$$

What effect will the addition of colorless NaSCN solid have on the color of the solution?

Michael C. Slabaugh

Figure 8.13 The effect of temperature on the position of equilibrium for the reaction $N_2O_4 \rightleftharpoons 2NO_2$. *Left:* A sealed tube containing an equilibrium mixture of red-brown NO_2 and colorless N_2O_4 is cooled in an ice bath. *Right:* The same sealed tube is heated in a hot water bath. On which side of the equation will heat appear? Is the reaction exothermic or endothermic?

Solution

a. Heat is removed by cooling. Heat is replenished when the equilibrium shifts to the right. Thus, more NH_3 will be present at equilibrium in the cooler mixture.
b. The equilibrium will shift to the left in an attempt to replenish the H_2. At the new equilibrium position, less NH_3 will be present.
c. When solid NaSCN is added, it will dissolve and form ions:

$$NaSCN(s) \rightarrow Na^+(aq) + SCN^-(aq)$$

The SCN^- concentration in the solution will therefore be increased. In an attempt to use up the added SCN^-, the equilibrium will shift to the right. This shift generates more of the dark red $Fe(SCN)_6{}^{3-}$. Therefore, the new position of equilibrium will be characterized by a darker red color.

✔ **LEARNING CHECK 8.4** Use Le Châtelier's principle and answer the following:

a. A saturated solution of ammonium nitrate is formed as follows. The solution process is endothermic:

$$heat + NH_4NO_3(s) \rightleftarrows NH_4{}^+(aq) + NO_3{}^-(aq)$$

Which way will the equilibrium shift when heat is added? What does this shift mean in terms of the solubility of NH_4NO_3 at the higher temperature?
b. Sulfur trioxide gas is formed when oxygen and sulfur dioxide gas react:

$$2SO_2(g) + O_2(g) \rightleftarrows 2SO_3(g)$$

Which way would the equilibrium shift if O_2 were removed from the system? How would the new equilibrium concentration of SO_3 compare with the earlier equilibrium concentration?

Catalysts cannot change the position of an equilibrium. This fact becomes clear when you remember that a catalyst functions by lowering the activation energy for a reaction. A lowering of the energy barrier for the forward reaction also lowers the barrier for the reverse reaction (see **Figure 8.14**). Hence, a catalyst speeds up both the forward and reverse reactions and cannot change the position of equilibrium. However, the lowered activation energy allows equilibrium to be established more quickly than if the catalyst were absent.

Figure 8.14 The influence of catalysts on forward and reverse activation energies.

STUDY SKILLS 8.1

Le Châtelier's Principle in Everyday Life

Le Châtelier's principle is extremely important in laboratory work and in the chemical industry when the goal is to obtain the maximum amount of product from a reaction. In such cases, increasing reactant concentrations, or adjusting the reaction temperature or pressure to shift the equilibrium position to the product side of the reaction, is a common practice.

To help you understand Le Châtelier's principle, it is often useful to observe familiar situations and events and interpret them in terms of shifting equilibria. For example, when you stand in an upright position, you are in equilibrium with the force of gravity. If a friend leans against you, your equilibrium is upset, and you will lose your balance unless you respond by leaning toward your friend such that the forces on you are once again balanced.

Children playing on a seesaw provide a similar example of forces in equilibrium. If one child slides toward the center of the seesaw, the force acting on that side is reduced, the equilibrium is upset, and the other child will be firmly on the ground. However, if the second child also shifts toward the center, the forces on each side can be made equal again, and play can resume.

The preceding examples involved equilibrium between forces, but we are not limited to only those types of examples. Imagine you are enjoying a warm shower when someone turns on a nearby hot water faucet. Suddenly, your shower turns cold. You respond to this stress by adjusting the shower controls (after shrieking loudly) to let in more hot water in an attempt to restore the pleasant temperature you were enjoying. (However, be aware of what will happen when the other faucet is turned off.)

Be mindful of your surroundings and activities, and you will see many more examples that will remind you of responses to stress that are attempts to restore equilibrium to a previous situation.

► Case Study Follow-up

Allergic reactions are common and range from mild to severe. *Anaphylaxis* is the proper name for a severe allergic reaction. Anaphylaxis is a whole-body reaction to an allergen, and it can be life threatening. Anaphylaxis can occur in response to any allergen, including insect bites and stings, certain foods, and some medicines. Death can occur in as little as 15 minutes—anaphylaxis is a true medical emergency. It is appropriate to call 911 immediately if a person is showing the following symptoms after exposure to an allergen: wheezing, difficulty breathing, palpitations, slurred speech, or difficulty swallowing. Other allergy symptoms that warrant medical attention but might not be an immediate emergency are vomiting, diarrhea, hives, nasal congestion, and cough.

Epinephrine is the treatment of choice for severe allergic reactions. It acts rapidly and has a short duration. Any time a severe allergic reaction is suspected, an epinephrine injection is appropriate for individuals with a prescription for it or if administered by qualified medical personnel. Whenever someone uses an epinephrine injection, he or she should seek additional medical care immediately. Anyone who experiences a moderate to severe allergic reaction should consult his or her health care provider about carrying an epinephrine injector.

Concept Summary

Spontaneous and Nonspontaneous Processes
Spontaneous processes take place naturally with no apparent cause or stimulus. Process spontaneity depends on the energy and entropy changes that accompany the process. Energy decreases and entropy increases favor spontaneity. However, a nonspontaneous change in one of these factors can be compensated for by a large spontaneous change in the other to cause processes to be spontaneous.
Objective 1 (Section 8.1), Exercise 8.6

Reaction Rates The speed of a reaction is called a reaction rate, which can be determined by measuring how fast reactants are used up or products are formed.
Objective 2 (Section 8.2), Exercise 8.14

Molecular Collisions Explanations of how reactions take place are called reaction mechanisms. Most mechanisms are based on three assumptions: (1) Molecules must collide with one another, (2) the collision must involve a

certain minimum of energy, and (3) some colliding molecules must be oriented in a specific way during collision in order to react.

Objective 3 (Section 8.3), Exercise 8.20

Energy Diagrams Energy relationships for reactions can be represented by energy diagrams, in which energy is plotted versus the reaction progress. The concepts of exothermic and endothermic reactions and activation energy are clearly represented by such diagrams.

Objective 4 (Section 8.4), Exercise 8.26

Factors That Influence Reaction Rates Four factors affect the rates of all reactions: (1) the nature of the reactants, (2) reactant concentrations, (3) reactant temperature, and (4) the presence of catalysts.

Objective 5 (Section 8.5), Exercise 8.30

Chemical Equilibrium Reactions are in equilibrium when the rate of the forward reaction is equal to the rate of the reverse reaction. Equilibrium is emphasized in equations

for reactions by writing double arrows pointing in both directions between reactants and products.

Objective 6 (Section 8.6), Exercise 8.38

The Position of Equilibrium The relative amounts of reactants and products present in a system at equilibrium define the position of equilibrium. The equilibrium position is toward the right when a large amount of product is present and toward the left when a large amount of reactant is present. The position is indicated by the value of the equilibrium constant.

Objective 7 (Section 8.7), Exercises 8.40 and 8.46

Factors That Influence Equilibrium Position Factors known to influence the position of equilibrium include changes in amount of reactants and/or products and changes in temperature. The influence of such factors can be predicted by using Le Châtelier's principle. Catalysts cannot change the position of equilibrium.

Objective 8 (Section 8.8), Exercise 8.52

Key Terms and Concepts

Activation energy (8.3)
Catalyst (8.5)
Effective collision (8.5)
Endergonic process (8.1)
Entropy (8.1)
Equilibrium concentrations (8.6)
Equilibrium constant (8.7)

Equilibrium expression (8.7)
Exergonic process (8.1)
Heterogeneous (surface) catalyst (8.5)
Homogeneous catalyst (8.5)
Inhibitor (8.5)
Internal energy (8.3)
Le Châtelier's principle (8.8)

Position of equilibrium (8.7)
Reaction mechanism (8.3)
Reaction rate (8.2)
Spontaneous process (8.1)
Stable substance (8.1)
State of equilibrium (8.6)

Key Equations

1.	Calculation of reaction rate (Section 8.2):	$\text{Rate} = \dfrac{\Delta C}{\Delta t} = \dfrac{C_t - C_0}{\Delta t}$	Equation 8.1
2.	Equilibrium expression for general reaction (Section 8.7):	$aA + bB + \cdots \rightleftarrows wW + xX + \cdots$	Equation 8.13
		$K = \dfrac{[W]^w[X]^x}{[A]^a[B]^b}$	Equation 8.14

Exercises

Even-numbered exercises are answered in Appendix B.

Blue-numbered exercises are more challenging.

Spontaneous and Nonspontaneous Processes (Section 8.1)

8.1 Classify the following processes as spontaneous or nonspontaneous. Explain your answers in terms of whether

energy must be continually supplied to keep the process going.

a. Water is decomposed into hydrogen and oxygen gas by passing electricity through the liquid.

b. An explosive detonates after being struck by a falling rock.

c. A coating of magnesium oxide forms on a clean piece of magnesium exposed to air.

d. A lightbulb emits light when an electric current is passed through it.

e. A cube of sugar dissolves in a cup of hot coffee.

8.2 Classify the following processes as spontaneous or nonspontaneous. Explain your answers in terms of whether energy must be continually supplied to keep the process going.

a. The space shuttle leaves its pad and goes into orbit.

b. The fuel in a booster rocket of the space shuttle burns.

c. Water boils at 100°C and 1 atm pressure.

d. Water temperature increases to 100°C at 1 atm pressure.

e. Your bedroom becomes orderly.

8.3 Classify the following processes as exergonic or endergonic. Explain your answers.

a. Any combustion process

b. Perspiration evaporating from the skin

c. Melted lead solidifying

d. An explosive detonating

e. An automobile being pushed up a slight hill (from point of view of the automobile)

8.4 Classify the following processes as exergonic or endergonic. Explain your answers.

a. An automobile being pushed up a slight hill (from point of view of the one pushing)

b. Ice melting (from point of view of the ice)

c. Ice melting (from point of view of surroundings of the ice)

d. Steam condensing to liquid water (from point of view of the steam)

e. Steam condensing to liquid water (from point of view of surroundings of the steam)

8.5 Describe the energy and entropy changes that occur in the following processes, and indicate whether the processes are spontaneous under the conditions stated:

a. Lumber becomes a house.

b. A seed grows into a tree.

c. On a hot day, water evaporates from a lake.

8.6 Describe the energy and entropy changes that occur in the following processes, and indicate whether the processes are spontaneous under the conditions stated:

a. On a cold day, water freezes.

b. A container of water at 40°C cools to room temperature.

c. The odor from an open bottle of perfume spreads throughout a room.

8.7 Pick the example with the highest entropy from each of the following sets. Explain your answers.

a. Solid ice, liquid water, or steam

b. Leaves on a tree, fallen leaves blown about on the ground, fallen leaves raked and placed in a basket

c. A stack of sheets of paper, a wastebasket containing sheets of paper, a wastebasket containing torn and crumpled sheets of paper

d. A 0.10 M sugar solution, a 1.0 M sugar solution, a 10.0 M sugar solution

e. A banquet table set for dinner, a banquet table during dinner, a banquet table immediately after dinner

8.8 Pick the example with the highest entropy from each of the following sets. Explain your answers.

a. Two opposing football teams just before the ball is snapped, two opposing football teams 1 second after the ball is snapped, two opposing football teams when the whistle is blown, ending the play

b. A 10% copper/gold alloy, a 2% copper/gold alloy, pure gold

c. A purse on which the strap just broke, a purse just hitting the ground, a purse on the ground with contents scattered

d. Coins in a piggy bank, coins in piles containing the same type of coins, coins in stacks of the same type of coins

e. A dozen loose pearls in a box, a dozen pearls randomly strung on a string, a dozen pearls strung on a string in order of decreasing size

8.9 You probably know that on exposure to air silver tarnishes and iron rusts; but gold, stainless steel, and chromium do not change. Explain these facts, using the concept of stability.

Reaction Rates (Section 8.2)

8.10 Classify the following processes according to their rates as very slow, slow, or fast:

a. The souring of milk stored in a refrigerator

b. The cooking of an egg in boiling water

c. The rusting of a shovel left in the garden over the winter

d. The growing of corn during a warm summer

e. The burning of a lighted match.

8.11 Classify the following processes according to their rates as very slow, slow, or fast:

a. The melting of butter put into a hot pan

b. The ripening of a piece of fruit stored at room temperature

c. The cooking of a raw potato in a hot oven

d. The melting of an ice cube in a glass of cool water

e. The combustion of gasoline in the engine of a car

8.12 Describe the observations or measurements that could be made to allow you to follow the rate of the following processes:

a. The melting of a block of ice

b. The setting (hardening) of concrete

c. The burning of a candle

8.13 Describe the observations or measurements that could be made to allow you to follow the rate of the following processes:

a. The diffusion of ink from a drop placed in a pan of quiet, undisturbed water

b. The loss of water from a pan of boiling water

c. The growth of a corn plant

8.14 Consider the following hypothetical reaction:

$$A + B \rightarrow C$$

Calculate the average rate of the reaction on the basis of the following information:

a. Pure A and B are mixed, and after 12.0 minutes the measured concentration of C is 0.396 mol/L.

b. Pure A, B, and C are mixed together at equal concentrations of 0.300 M. After 8.00 minutes, the concentration of C is found to be 0.455 M.

8.15 Consider the following hypothetical reaction:

$$A + B \rightarrow C$$

Calculate the average rate of the reaction on the basis of the following information:

a. Pure A, B, and C are mixed together at concentrations of $A + B = 0.400$ M, $C = 0.150$ M. After 6.00 minutes, the concentration of C is 0.418 M.

b. Pure A and B are mixed together at the same concentration of 0.361 M. After 7.00 minutes, the concentration of A is found to be 0.048 M.

8.16 A reaction generates chlorine gas (Cl_2) as a product. The reactants are mixed and sealed in a 250.-mL container. After 30.0 minutes, 2.97×10^{-2} mol of Cl_2 has been generated. Calculate the average rate of the reaction.

8.17 A reaction generates hydrogen gas (H_2) as a product. The reactants are mixed and sealed in a 250.-mL vessel. After 20.0 minutes, 3.91×10^{-2} mol H_2 has been generated. Calculate the average rate of the reaction.

8.18 Ammonium and nitrite ions react in solution to form nitrogen gas:

$$NH_4^+ \text{ (aq)} + NO_2^- \text{ (aq)} \rightarrow N_2(g) + 2H_2O(\ell)$$

A reaction is run, and the liberated N_2 gas is collected in a previously evacuated 500.-mL container. After the reaction has gone on for 750. seconds, the pressure of N_2 in the 500.-mL container is 2.77×10^{-2} atm, and the temperature of the N_2 is 25.0°C. Use the ideal gas law (Equation 6.9) to calculate the number of moles of N_2 liberated. Then calculate the average rate of the reaction.

8.19 Suppose a small lake is contaminated with an insecticide that decomposes with time. An analysis done in June shows the decomposition product concentration to be 7.8×10^{-4} mol/L. An analysis done 35 days later shows the concentration of decomposition product to be 9.9×10^{-4} mol/L. Assume the lake volume remains constant and calculate the average rate of decomposition of the insecticide.

Molecular Collisions (Section 8.3)

8.20 In each of the following, which reaction mechanism assumption is apparently being violated? Explain your answers.

a. A reaction takes place more rapidly when the concentration of reactants is decreased.

b. A reaction takes place more rapidly when the reaction mixture is cooled.

c. The reaction rate of $A + B \rightarrow A\text{—}B$ increases as the concentration of A is increased but does not change as the concentration of B is increased.

8.21 Which reaction mechanism assumptions are unimportant in describing simple ionic reactions between cations and anions? Why?

8.22 Describe two ways by which an increase in temperature increases a reaction rate.

Energy Diagrams (Section 8.4)

8.23 Sketch energy diagrams to represent each of the following. Label the diagrams completely and tell how they are similar to each other and how they are different.

a. Exothermic (exergonic) reaction with activation energy

b. Exothermic (exergonic) reaction without activation energy

8.24 Sketch energy diagrams to represent each of the following. Label the diagrams completely and tell how they are similar to each other and how they are different.

a. Endothermic (endergonic) reaction with activation energy

b. Endothermic (endergonic) reaction without activation energy

8.25 Use energy diagrams to compare catalyzed and uncatalyzed reactions.

8.26 One reaction occurs at room temperature and liberates 500 kJ/mol of reactant. Another reaction does not take place until the reaction mixture is heated to 150°C. However, it also liberates 500 kJ/mol of reactant. Draw an energy diagram for each reaction and indicate the similarities and differences between the two.

Factors That Influence Reaction Rates (Section 8.5)

8.27 The following reactions are proposed. Make a rough estimate of the rate of each one—rapid, slow, won't react. Explain each answer.

a. $H_2O(\ell) + H^+(aq) \rightarrow H_3O^+(aq)$

b. $H_3O^+(aq) + H^+(aq) \rightarrow H_4O^{2+}(aq)$

c. $3H_2(g) + N_2(g) \rightarrow 2NH_3(g)$

d. $Ba^{2+}(aq) + SO_4^{2-}(aq) \rightarrow BaSO_4(s)$

8.28 The following reactions are proposed. Make a rough estimate of the rate of each one—rapid, slow, won't react. Explain each answer.

a. $CaO(s) + 2HCl(g) \rightarrow CaCl_2(s) + H_2O(\ell)$

b. $2KI(s) + Pb(NO_3)_2(s) \rightarrow PbI_2(s) + 2KNO_3(s)$

c. $Cl^-(aq) + I^-(aq) \rightarrow ICl^{2-}(aq)$

d. $I_2(aq) + I^-(aq) \rightarrow I_3^-(aq)$

8.29 Which reaction mechanism assumption is most important in explaining the following? Why?

a. The influence of concentration on reaction rates

b. The influence of catalysts on reaction rates

8.30 Suppose you are running a reaction and you want to speed it up. Describe three things you might try to do this.

8.31 A reaction is started by mixing reactants. As time passes, the rate decreases. Explain this behavior that is characteristic of most reactions.

8.32 A reaction is run at 10°C and takes 3.7 hours to go to completion. How long would it take to complete the reaction at 30°C?

8.33 What factor is more important than simply the amount of solid reactant present in determining the rate of a reaction? Explain.

8.34 Describe two ways catalysts might speed up a reaction.

Chemical Equilibrium (Section 8.6)

8.35 Describe the establishment of equilibrium in a system represented by a shopper walking up the "down" escalator.

8.36 Describe the observation or measurement result that would indicate when each of the following had reached equilibrium:

a. H_2 $+$ I_2 \rightleftarrows $2HI$
colorless gas violet gas colorless gas

b. solid sugar + water \rightleftarrows sugar solution

c. N_2 $+$ $2O_2$ \rightleftarrows $2NO_2$
colorless gas colorless gas red-brown gas

8.37 Describe the observation or measurement result that would indicate when each of the following had reached equilibrium:

a. $2CO$ $+$ O_2 \rightleftarrows $2CO_2$
colorless gas colorless gas colorless gas
(apply Dalton's law)

b. $LiOH$ $+$ CO_2 \rightleftarrows $LiHCO_3$
colorless solid colorless gas colorless solid

c. paycheck → checking account → checks to pay bills

8.38 Colorless hydrogen gas (H_2) and red-brown colored bromine gas (Br_2) react to form colorless HBr gas. Describe how the concentrations of H_2, Br_2, and HBr would change (increase or decrease) as equilibrium was established in a sealed container that initially contained only H_2 and Br_2. What observation would indicate that equilibrium had been established?

8.39 Colorless N_2O_4 gas decomposes to form red-brown colored NO_2 gas. Describe how the concentrations of N_2O_4 and NO_2 would change (increase or decrease) as equilibrium was established in a sealed container that initially contained only N_2O_4. What observation would indicate that equilibrium had been established?

The Position of Equilibrium (Section 8.7)

8.40 Write an equilibrium expression for each of the following gaseous reactions:

a. $2CO + O_2 \rightleftarrows 2CO_2$
b. $N_2O_4 \rightleftarrows 2NO_2$
c. $2C_2H_6 + 7O_2 \rightleftarrows 4CO_2 + 6H_2O$
d. $2NOCl \rightleftarrows 2NO + Cl_2$
e. $2Cl_2O_5 \rightleftarrows O_2 + 4ClO_2$

8.41 Write an equilibrium expression for each of the following gaseous reactions:

a. $3O_2 \rightleftarrows 2O_3$
b. $COCl_2 \rightleftarrows CO + Cl_2$
c. $CS_2 + 4H_2 \rightleftarrows CH_4 + 2H_2S$
d. $2SO_2 + O_2 \rightleftarrows 2SO_3$
e. $CO + H_2O \rightleftarrows CO_2 + H_2$

8.42 The following equilibria are established in water solutions. Write an equilibrium expression for each reaction.

a. $Fe^{3+} + 6CN^- \rightleftarrows Fe(CN)_6^{3-}$
b. $Ag^+ + 2NH_3 \rightleftarrows Ag(NH_3)_2^+$
c. $Au^{3+} + 4Cl^- \rightleftarrows AuCl_4^-$

8.43 The following equilibria are established in water solutions. Write an equilibrium expression for each reaction.

a. $Ni^{2+} + 6NH_3 \rightleftarrows Ni(NH_3)_6^{2+}$
b. $Sn^{2+} + 2Fe^{3+} \rightleftarrows Sn^{4+} + 2Fe^{2+}$
c. $F_2 + 2Cl^- \rightleftarrows 2F^- + Cl_2$

8.44 Write a reaction equation that corresponds to each of the following equilibrium expressions:

a. $K = \dfrac{[CO_2][H_2O]^2}{[CH_4][O_2]^2}$

b. $K = \dfrac{[CH_4][H_2O]}{[H_2]^3[CO]}$

c. $K = \dfrac{[O_2]^3}{[O_3]^2}$

d. $K = \dfrac{[NH_3]^4[O_2]^7}{[NO_2]^4[H_2O]^6}$

8.45 Write a reaction equation that corresponds to each of the following equilibrium expressions:

a. $K = \dfrac{[PH_3][F_2]^3}{[HF]^3[PF_3]}$

b. $K = \dfrac{[O_2]^7[NH_3]^4}{[NO_2]^4[H_2O]^6}$

c. $K = \dfrac{[O_2][ClO_2]^4}{[Cl_2O_5]^2}$

d. $K = \dfrac{[N_2][H_2O]^2}{[NO]^2[H_2]^2}$

8.46 A sample of gaseous BrCl is allowed to decompose in a closed container at 25°C:

$$2BrCl(g) \rightleftarrows Br_2(g) + Cl_2(g)$$

When the reaction reaches equilibrium, the following concentrations are measured: [BrCl] = 0.38 M, [Cl$_2$] = 0.26 M, [Br$_2$] = 0.26 M. Evaluate the equilibrium constant for the reaction at 25°C.

8.47 At 600°C, gaseous CO and Cl$_2$ are mixed together in a closed container. At the instant they are mixed, their concentrations are CO = 0.79 mol/L and Cl$_2$ = 0.69 mol/L. After equilibrium is established, their concentrations are CO = 0.25 mol/L and Cl$_2$ = 0.15 mol/L. Evaluate the equilibrium constant for the reaction

$$CO(g) + Cl_2(g) \rightleftarrows COCl_2(g)$$

8.48 A mixture of the gases NOCl, Cl$_2$, and NO is allowed to reach equilibrium at 25°C. The measured equilibrium concentrations are [NO] = 0.92 mol/L, [NOCl] = 1.31 mol/L, and [Cl$_2$] = 0.20 mol/L. What is the value of the equilibrium constant at 25°C for the reaction

$$2NOCl(g) \rightleftarrows 2NO(g) + Cl_2(g)$$

8.49 Consider the following equilibrium constants. Describe how you would expect the equilibrium concentrations of reactants and products to compare with each other (larger than, smaller than, etc.) for each case.

 a. $K = 2.1 \times 10^{-6}$

 b. $K = 0.15$

 c. $K = 1.2 \times 10^{8}$

 d. $K = 0.00036$

8.50 Consider the following equilibrium constants. Describe how you would expect the equilibrium concentrations of reactants and products to compare with each other (larger than, smaller than, etc.) for each case.

 a. $K = 5.9$

 b. $K = 3.3 \times 10^{6}$

 c. $K = 2.7 \times 10^{-4}$

 d. $K = 0.0000558$

Factors That Influence Equilibrium Position (Section 8.8)

8.51 Use Le Châtelier's principle to predict the direction of equilibrium shift in the following equilibria when the indicated stress is applied:

 a. $2A + B + \text{heat} \rightleftarrows C$; some A is removed.

 b. $2A + B + \text{heat} \rightleftarrows C + D$; the system is cooled.

 c. $N_2O_4 \rightleftarrows 2NO_2$; some NO_2 is removed.

8.52 Use Le Châtelier's principle to predict the direction of equilibrium shift in the following equilibria when the indicated stress is applied:

 a. $Ag^{+}(aq) + Cl^{-}(aq) \rightleftarrows AgCl(s)$; some Ag^{+} is removed.

 b. $2HI(g) + \text{heat} \rightleftarrows H_2(g) + I_2(g)$; the system is heated.

 c. $6Cu(s) + N_2(g) + \text{heat} \rightleftarrows 2Cu_3N(s)$; the system is cooled and some N_2 is removed.

8.53 Use Le Châtelier's principle to predict the direction of equilibrium shift and the changes that will be observed (color, amount of precipitate, etc.) in the following equilibria when the indicated stress is applied:

 a. $\text{heat} + Co^{2+}(aq) + 4Cl^{-}(aq) \rightleftarrows CoCl_4{}^{2-}(aq)$;
 pink colorless blue
 The equilibrium mixture is heated.

 b. $\text{heat} + Co^{2+}(aq) + 4Cl^{-}(aq) \rightleftarrows CoCl_4{}^{2-}(aq)$;
 pink colorless blue
 Co^{2+} is added to the equilibrium mixture.

 c. $Fe^{3+}(aq) + 6SCN^{-}(aq) \rightleftarrows Fe(SCN)_6{}^{3-}(aq)$;
 brown colorless red
 SCN^{-} is added to the equilibrium mixture.

 d. $Pb^{2+}(aq) + 2Cl^{-}(aq) \rightleftarrows PbCl_2(s) + \text{heat}$;
 colorless colorless white solid
 Pb^{2+} is added to the equilibrium mixture.

 e. $C_2H_4 + I_2 \rightleftarrows C_2H_4I_2 + \text{heat}$;
 colorless gas violet gas colorless gas
 A catalyst is added to the equilibrium mixture.

8.54 Use Le Châtelier's principle to predict the direction of equilibrium shift and the changes that will be observed (color,

amount of precipitate, etc.) in the following equilibria when the indicated stress is applied:

 a. $Cu^{2+}(aq) + 4NH_3(aq) \rightleftarrows Cu(NH_3)_4{}^{2+}(aq)$;
 blue colorless dark purple
 Some NH_3 is added to the equilibrium mixture.

 b. $Pb^{2+}(aq) + 2Cl^{-}(aq) \rightleftarrows PbCl_2(s) + \text{heat}$;
 colorless colorless white solid
 The equilibrium mixture is cooled.

 c. $C_2H_4 + I_2 \rightleftarrows C_2H_4I_2 + \text{heat}$;
 colorless gas violet gas colorless gas
 Some $C_2H_4I_2$ is removed from the equilibrium mixture.

 d. $C_2H_4 + I_2 \rightleftarrows C_2H_4I_2 + \text{heat}$;
 colorless gas violet gas colorless gas
 The equilibrium mixture is cooled.

 e. $\text{heat} + 4NO_2 + 6H_2O \rightleftarrows 7O_2 + 4NH_3$;
 brown colorless colorless colorless
 gas gas gas gas
 A catalyst is added, and NH_3 is added to the equilibrium mixture.

8.55 Tell what will happen to each equilibrium concentration in the following when the indicated stress is applied and a new equilibrium position is established:

 a. $H^{+}(aq) + HCO_3{}^{-}(aq) \rightleftarrows H_2O(\ell) + CO_2(g)$;
 $HCO_3{}^{-}$ is added.

 b. $CO_2(g) + H_2O(\ell) \rightleftarrows H_2CO_3(aq) + \text{heat}$;
 CO_2 is removed.

 c. $CO_2(g) + H_2O(\ell) \rightleftarrows H_2CO_3(aq) + \text{heat}$;
 The system is cooled.

8.56 Tell what will happen to each equilibrium concentration in the following when the indicated stress is applied and a new equilibrium position is established:

 a. $LiOH(s) + CO_2(g) \rightleftarrows LiHCO_3(s) + \text{heat}$; CO_2 is removed.

 b. $2NaHCO_3(s) + \text{heat} \rightleftarrows Na_2O(s) + 2CO_2(g) + H_2O(g)$;
 The system is cooled.

 c. $CaCO_3(s) + \text{heat} \rightleftarrows CaO(s) + CO_2(g)$; The system is cooled.

8.57 The gaseous reaction $2HBr(g) \rightleftarrows H_2(g) + Br_2(g)$ is endothermic. Tell which direction the equilibrium will shift for each of the following:

 a. Some H_2 is added.

 b. The temperature is increased.

 c. Some Br_2 is removed.

 d. A catalyst is added.

 e. Some HBr is removed.

 f. The temperature is decreased, and some HBr is removed.

8.58 The gaseous reaction $N_2(g) + 3H_2(g) \rightleftarrows 2NH_3(g)$ is exothermic. Tell which direction the equilibrium will shift for each of the following:

 a. Some N_2 is added.

 b. The temperature is increased.

 c. Some NH_3 is removed.

 d. Some H_2 is removed.

e. A catalyst is added.

f. The temperature is increased, and some H_2 is removed.

Additional Exercises

8.59 Assume the following reactions take place under identical conditions of concentration and temperature. Which reaction would you expect to take place faster? Explain your answer.

a. $H_2(g) + F_2(g) \rightarrow 2HF(g)$

b. $H_2(g) + I_2(g) \rightarrow 2HI(g)$

8.60 Gases *A* and *B* react as follows: $A(g) + B(g) \rightarrow C(s)$. Suppose gases *A* and *B* are mixed and used to fill a balloon. How could you increase the concentration of *A* and *B*, and speed up the reaction?

8.61 Bacteria found in lakes use dissolved oxygen gas to metabolize organic contaminants and use them as a food source, thus removing the contaminants from the lakes. When lake temperatures increase significantly, the rate of this chemical decontamination decreases. Remember your study of gases and explain this apparent violation of the factors influencing reaction rates studied in this chapter.

8.62 In Section 8.1 of this chapter, three criteria that determine the spontaneity of a process are given. Review these criteria, and determine which one is involved in each of the following spontaneous phase changes:

a. Evaporation of a liquid

b. Condensation of a gas to a liquid

c. Sublimation of a solid to a gas

d. Liquefaction of a gas to a liquid

e. Crystallization of a liquid to a solid

Chemistry for Thought

8.63 Refer to Figure 8.4 and answer the question. How would the total energy released as light for each light stick compare when they have both stopped glowing?

8.64 A mixture of the gases NOCl, NO, and Cl_2 is allowed to come to equilibrium at 400°C in a 1.50-L reaction container. Analysis shows the following number of moles of each substance to be present at equilibrium: NOCl = 1.80 mol, NO = 0.70 mol, and Cl_2 = 0.35 mol. Calculate the value of the equilibrium constant for the reaction at 400°C. The equation for the reaction is

$$2NOCl(g) \rightleftarrows 2NO(g) + Cl_2(g)$$

8.65 The equilibrium constant for the reaction $PCl_5 \rightleftarrows PCl_3 + Cl_2$ is 0.0245 at 250°C. What molar concentration of PCl_5 would be present at equilibrium if the concentrations of PCl_3 and Cl_2 were both 0.250 M?

8.66 At 448°C, the equilibrium constant for the reaction $H_2 + I_2 \rightleftarrows 2HI$ is 50.5. What concentration of I_2 would be found in an equilibrium mixture in which the concentrations of HI and H_2 were 0.500 M and 0.050 M, respectively?

8.67 Refer to Figure 8.10 and answer the questions. Would you expect a crushed antacid tablet (like Alka-Seltzer) to dissolve faster or slower than a whole tablet? Explain.

8.68 Refer to Figure 8.13 and answer the questions. Would the presence of a catalyst in the tube influence the equilibrium concentrations of the two gases? Explain.

8.69 In the blood, both oxygen (O_2) and poisonous carbon monoxide (CO) can react with hemoglobin (Hb). The respective compounds are HbO_2 and HbCO. When both gases are present in the blood, the following equilibrium is established:

$$HbCO + O_2 \rightleftarrows HbO_2 + CO$$

Use Le Châtelier's principle to explain why pure oxygen is often administered to victims of CO poisoning.

8.70 Use the concept of reaction rates to explain why no smoking is allowed in hospital areas where patients are being administered oxygen gas.

8.71 Suppose you have two identical unopened bottles of carbonated beverage. The contents of both bottles appear to be perfectly clear. You loosen the cap of one of the bottles and hear a hiss as gas escapes, and at the same time gas bubbles appear in the liquid. The liquid in the unopened bottle still appears to be perfectly clear. Explain these observations using the concept of equilibrium and Le Châtelier's principle. Remember, a carbonated beverage contains carbon dioxide gas dissolved in a liquid under pressure.

8.72 Someone once suggested that it is impossible to unscramble a scrambled egg. Describe an unscrambled and a scrambled egg in terms of the concept of entropy.

Allied Health Exam Connection

The following questions are from these sources:

- *Nursing School Entrance Exam* © 2005, Learning Express, LLC.

- *McGraw-Hill's Nursing School Entrance Exams* by Thomas A. Evangelist, Tamara B. Orr, and Judy Unrein © 2009, The McGraw-Hill Companies, Inc.

- *NSEE Nursing School Entrance Exams,* 3rd edition © 2009, Kaplan Publishing.

- *Cliffs Test Prep: Nursing School Entrance Exams* by Fred N. Grayson © 2004, Wiley Publishing, Inc.

- *Peterson's Master the Nursing School and Allied Health Entrance Exams,* 18th edition by Marion F. Gooding © 2008, Peterson's, a Nelnet Company.

8.73 A reaction takes place between an acid and 0.5 grams of solid magnesium ribbon. Another reaction takes place between an acid and 0.5 grams of powdered magnesium. Which statement is true?

a. The powdered magnesium reacts faster because the activation energy has been lowered.

b. The magnesium strip reacts faster because it has a higher concentration of magnesium.

c. The powdered magnesium reacts faster because it has a greater surface area.

d. The magnesium strip reacts faster because it will create a higher temperature once the reaction starts.

8.74 If the reaction: $A + B \rightarrow C + D$ is designated as first order, the rate depends on:

a. the concentration of only one reactant.

b. the concentration of each reactant.

c. no specific concentration.

d. the temperature only.

8.75 Reaction kinetics deals with:

a. equilibrium position.

b. reaction rates.

c. molecular reactant size.

d. none of the above.

8.76 A book is held 6 feet above the floor and then dropped. Which statement is true?

a. The potential energy of the book is converted to kinetic energy.

b. The potential energy of the book is destroyed.

c. Kinetic energy is created.

d. The total energy of the system will not be conserved.

8.77 When a crane at a building site lifts a beam to its top height, what type of energy is created?

a. kinetic energy

b. potential energy

c. chemical energy

d. electrical energy

8.78 Stored energy is referred to as:

a. activation energy.

b. kinetic energy.

c. potential energy.

d. electrical energy.

8.79 In exergonic reactions, the energy is:

a. used.

b. stored.

c. released.

d. lost.

8.80 Which is an example of an exothermic change?

a. sublimation

b. condensation

c. melting

d. evaporation

8.81 Which of the following reactions releases heat energy?

a. double replacement

b. decomposition

c. endothermic

d. exothermic

8.82 Which of the following is the best example of potential energy changing to kinetic energy?

a. Pushing a rock off a cliff

b. Sitting in a rocking chair

c. Observing a bird fly

d. Standing on a table

8.83 Which is NOT an example of an endothermic change?

a. melting

b. sublimation

c. freezing

d. evaporation

8.84 Which of the following processes is endothermic?

a. ice melting

b. a piece of paper burning

c. a bomb exploding

d. an organism's metabolism producing a certain amount of heat

8.85 Which sentence best describes the following reaction?

$$2H_2(g) + O_2(g) \rightarrow 2H_2O(\ell) + \text{heat}$$

a. It is an endothermic reaction.

b. It is an exothermic double-replacement reaction.

c. It is a synthesis reaction that is also exothermic.

d. It is a decomposition reaction that is also endothermic.

8.86 By which of the following mechanisms does a catalyst operate?

a. It decreases the activation energy barrier for a reaction.

b. It serves as a reactant and is consumed.

c. It increases the temperature of a reaction.

d. It increases the concentration of reactants.

8.87 Which of the following is NOT true of reversible chemical reactions?

a. A chemical reaction is never complete.

b. The products of the reaction also react to reform the original reactants.

c. When the reaction is finished, both reactants and products are present in equal amounts.

d. The reaction can result in an equilibrium.

8.88 Given the reaction:

$$2CO(g) + O_2(g) \rightleftarrows 2CO_2(g)$$

When there is an increase in pressure to the system one would expect:

a. an increase in the amount of carbon dioxide.

b. an increase in the amount of carbon monoxide and oxygen.

c. a decrease in the amount of carbon dioxide.

d. no change in the system.

8.89 The following reaction is exothermic: $AgNO_3 + NaCl \rightleftarrows AgCl + NaNO_3$. How will the equilibrium be changed if the temperature is increased?

a. Equilibrium will shift to the right.

b. Equilibrium will shift to the left.

c. The reaction will not proceed.

d. Equilibrium will not change.

8.90 Consider the reaction $N_2(g) + 3H_2(g) \rightarrow 2NH_3(g) + heat$. Indicate the incorrect statement.

a. An increase in temperature will shift the equilibrium to the right.

b. An increase in pressure applied will shift the equilibrium to the right.

c. The addition of ammonia will shift the equilibrium to the left.

d. The addition of H_2 will shift the equilibrium to the right.

8.91 In a chemical reaction, [A] and [B] combine to form [C] and [D], as expressed by the reaction [A] + [B] = [C] + [D]. Select the statement that best describes the equilibrium condition.

a. The reaction is shifted to the right.

b. The concentrations of reactants and products are constant.

c. The reaction is shifted to the left.

d. The concentration of products is greater than the concentration of reactants.

8.92 What is the effect of the addition of a catalyst to a reaction in equilibrium?

a. The reaction favors the formation of the products.

b. The reaction favors the formation of the reactants.

c. There is no change in composition of the reaction.

d. The rate of the reaction slows.

8.93 For the reaction: $H_2(g) + Br_2(g) \rightarrow 2HBr(g)$, the reaction can be driven to the left by:

a. increasing the pressure.

b. increasing the hydrogen.

c. increasing hydrogen bromide.

d. decreasing hydrogen bromide.

8.94 What is the equilibrium constant K_c for the following equation?

$$2SO_2(g) + O_2(g) \rightleftarrows 2SO_3(g)$$

a. $K_c = \dfrac{[SO_3]}{[SO_2][O_2]}$

b. $K_c = \dfrac{[SO_2][O_2]}{[SO_3]}$

c. $K_c = \dfrac{[SO_2]^2[O_2]}{[SO_3]^2}$

d. $K_c = \dfrac{[SO_3]^2}{[SO_2]^2[O_2]}$

9

Acids, Bases, and Salts

iStockphoto.com/Fanelie Rosier

▶ Case Study

Dallas grew up on a farm. He learned to work hard and to just tough it out when he didn't feel well. After his 50th birthday, he started having frequent bouts of heartburn. No matter how hard he tried, the heartburn pain could not be ignored. He treated it by drinking baking soda in water, which helped a little, but not much. Then he tried liquid antacids and tablets, but his daughter was concerned because he seemed to be eating them like candy, and she convinced him to see a doctor. The doctor diagnosed Dallas with gastroesophageal reflux disease (GERD). The doctor explained to Dallas that the valve between his stomach and esophagus did not close tightly and that stomach acid was entering his esophagus and causing his pain. The doctor prescribed a proton-pump inhibitor to reduce the amount of acid produced and recommended that Dallas avoid certain foods. She also ordered an endoscopy to check for possible damage to the esophagus.

▶ **Can too many over-the-counter antacids be harmful? How can heartburn symptoms be distinguished from heart attack? What foods aggravate heartburn?**

Follow-up to this Case Study appears at the end of the chapter before the *Concept Summary*.

When you have completed your study of this chapter, you should be able to:

1 Write reaction equations that illustrate Arrhenius acid–base behavior. **(Section 9.1)**

2 Write reaction equations that illustrate Brønsted acid–base behavior, and identify Brønsted acids and bases from written reaction equations. **(Section 9.2)**

3 Name common acids. **(Section 9.3)**

4 Do calculations using the concept of the self-ionization of water. **(Section 9.4)**

5 Do calculations using the pH concept. **(Section 9.5)**

6 Write reaction equations that illustrate the characteristic reactions of acids. **(Section 9.6)**

7 Write reaction equations that represent neutralization reactions between acids and bases. **(Section 9.7)**

8 Write reaction equations that illustrate various ways to prepare salts, and do calculations using the concept of an equivalent of salt. **(Section 9.8)**

9 Demonstrate an understanding of the words *weak* and *strong* as applied to acids and bases. **(Section 9.9)**

10 Demonstrate an understanding of the titration technique used to analyze acids and bases. **(Section 9.10)**

11 Do calculations related to the analysis of acids and bases by titration. **(Section 9.11)**

12 Explain the concept of salt hydrolysis, and write equations to illustrate the concept. **(Section 9.12)**

13 Explain how buffers work, and write equations to illustrate their action. **(Section 9.13)**

Acids, bases, and salts are among the most common and important solutes found in solutions. Until late in the 19th century, these substances were characterized by such properties as taste and color changes induced in certain dyes. Acids taste sour; bases, bitter; and salts, salty. Litmus, a dye, is red in the presence of acids and blue in the presence of bases. These and other observations led to the correct conclusions that acids and bases are chemical opposites, and that salts are produced when acids and bases react with each other. Today, acids and bases are defined in more precise ways that are useful when studying their characteristics.

9.1 | The Arrhenius Theory

Learning Objective

1. Write reaction equations that illustrate Arrhenius acid–base behavior.

In 1887, Swedish chemist Svante Arrhenius proposed a theory dealing with electrolytic dissociation. He defined an **acid** as a substance that dissociates when dissolved in water and produces hydrogen ions (H^+). Similarly, a **base** is a substance that dissociates and releases hydroxide ions (OH^-). Hydrogen chloride (HCl) and sodium hydroxide (NaOH) are examples of an Arrhenius acid and base, respectively. They dissociate in water as follows:

Arrhenius acid Any substance that provides H^+ ions when dissolved in water.

Arrhenius base Any substance that provides OH^- ions when dissolved in water.

$$HCl(aq) \rightarrow H^+(aq) + Cl^-(aq) \tag{9.1}$$

$$NaOH(aq) \rightarrow Na^+(aq) + OH^-(aq) \tag{9.2}$$

Note that the hydrogen ion is a bare proton, the nucleus of a hydrogen atom.

9.2 | The Brønsted Theory

Learning Objective

2. Write reaction equations that illustrate Brønsted acid–base behavior, and identify Brønsted acids and bases from written reaction equations.

Arrhenius did not know that free hydrogen ions cannot exist in water. They covalently bond with water molecules to form hydronium ions (H_3O^+). The water molecules provide both electrons used to form the covalent bond:

$$H^+ + \overset{..}{:}\!\!\underset{\underset{H}{|}}{\overset{..}{O}}\!\!-H \longrightarrow \left[H\!-\!\underset{\underset{H}{|}}{\overset{..}{O}}\!-\!H \right]^+ \tag{9.3}$$

new bond

Or, more simply,

$$H^+(aq) + H_2O(\ell) \rightarrow H_3O^+(aq) \tag{9.4}$$

In 1923, Johannes Brønsted in Denmark and Thomas Lowry in England proposed an acid–base theory that took into account this behavior of hydrogen ions. They defined an **acid** as any hydrogen-containing substance that donates a proton (hydrogen ion) to another substance and a **base** as any substance that accepts a proton.

In conformity with this theory, the acidic behavior of covalently bonded HCl molecules in water is written

$$HCl(aq) + H_2O(\ell) \rightleftharpoons H_3O^+(aq) + Cl^-(aq) \tag{9.5}$$

The HCl behaves as a Brønsted acid by donating a proton to a water molecule. The water molecule, by accepting the proton, behaves as a base.

The double arrows of unequal length in Equation 9.5 indicate that the reaction is reversible, with the equilibrium lying far to the right. In actual water solutions, essentially 100% of the dissolved HCl is in the ionic form at equilibrium. Remember, both the forward and the reverse reactions are taking place at equilibrium (see **Section 8.6**). When the reverse reaction occurs, hydronium ions donate protons to chloride ions to form HCl and H_2O molecules. Thus, H_3O^+ behaves as a Brønsted acid, and Cl^- behaves as a Brønsted base. We see from this discussion that when a substance like HCl behaves as a Brønsted acid by donating a proton, the species that remains (Cl^-) is a Brønsted base. The Cl^- is called the **conjugate base** of HCl. Every Brønsted acid and the base formed when it donates a proton is called a **conjugate acid–base pair**. Thus, in Equation 9.5, we see that HCl and Cl^- form a conjugate acid–base pair, as do H_3O^+ and H_2O for the reverse reaction. Notice that the acid and base in a conjugate acid–base pair differ only by a proton, H^+.

Brønsted acid Any hydrogen-containing substance that is capable of donating a proton (H^+) to another substance.

Brønsted base Any substance that is capable of accepting a proton (H^+) from another substance.

conjugate base The species remaining when a Brønsted acid donates a proton.

conjugate acid–base pair A Brønsted acid and its conjugate base.`

Example 9.1 Identifying Brønsted Acids and Bases

Identify all Brønsted acids, bases, and acid–base conjugate pairs in the reactions represented by the following equations:

a. $HNO_3(aq) + H_2O(\ell) \rightleftharpoons H_3O^+(aq) + NO_3^-(aq)$
b. $HClO_4(aq) + H_2O(\ell) \rightleftharpoons H_3O^+(aq) + ClO_4^-(aq)$
c. $NH_3(aq) + H_2O(\ell) \rightleftharpoons OH^-(aq) + NH_4^+(aq)$

Solution

a. Nitric acid, HNO_3, behaves as a Brønsted acid by donating a proton to H_2O, a base, in the forward reaction. In the reverse reaction, H_3O^+, the conjugate acid of H_2O, donates a proton to NO_3^-, the conjugate base of HNO_3. In summary, the Brønsted

acids are HNO_3 and H_3O^+, the Brønsted bases are H_2O and NO_3^-, and the conjugate acid–base pairs are HNO_3/NO_3^- and H_3O^+/H_2O:

$$HNO_3(aq) + H_2O(\ell) \rightleftharpoons H_3O^+(aq) + NO_3^-(aq)$$

acid base acid base

| conjugate pair |
| conjugate pair |

b. Similarly, perchloric acid ($HClO_4$) is a Brønsted acid, and H_2O is a Brønsted base (forward reaction). Also, H_3O^+ is a Brønsted acid, and the perchlorate ion (ClO_4^-) is a Brønsted base (reverse reaction). The conjugate acid–base pairs are $HClO_4/ClO_4^-$ and H_3O^+/H_2O:

$$HClO_4(aq) + H_2O(\ell) \rightleftharpoons H_3O^+(aq) + ClO_4^-(aq)$$

acid base acid base

| conjugate pair |
| conjugate pair |

c. In this reaction, water donates a proton instead of accepting one. Therefore, H_2O is a Brønsted acid, and ammonia (NH_3) is a Brønsted base (forward reaction). The ammonium ion (NH_4^+) is an acid, and the hydroxide ion (OH^-) is a base (reverse reaction). Note that an Arrhenius base was a substance that released the OH^-, whereas according to Brønsted, the OH^- is a base. The conjugate acid–base pairs are H_2O/OH^- and NH_4^+/NH_3:

$$NH_3(aq) + H_2O(\ell) \rightleftharpoons OH^-(aq) + NH_4^+(aq)$$

base acid base acid

| conjugate pair |
| conjugate pair |

✔ **LEARNING CHECK 9.1** Identify all Brønsted acids, bases, and acid–base conjugate pairs in the reactions represented by the following equations:

a. $HC_2H_3O_2(aq) + H_2O(\ell) \rightleftharpoons H_3O^+(aq) + C_2H_3O_2^-(aq)$
b. $NO_2^-(aq) + H_2O(\ell) \rightleftharpoons HNO_2(aq) + OH^-(aq)$
c. $HS^-(aq) + H_2O(\ell) \rightleftharpoons H_3O^+(aq) + S^{2-}(aq)$

9.3 | Naming Acids

Learning Objective

3. Name common acids.

In Section 4.4 (see page 108), the rules for naming binary ionic compounds were given. The rules for naming binary covalent compounds and ionic compounds that contain polyatomic ions were discussed in Section 4.10 (see page 126). We now conclude our discussion of inorganic nomenclature by giving the rules used to name hydrogen-containing compounds that behave as acids. Examples of the two types of compounds that behave as acids were given in Section 9.2. Acids of the first type, represented by HCl, are compounds in which hydrogen is covalently bonded to a nonmetal. In acids of the second type, hydrogen is covalently bonded to a polyatomic ion. An example of the second type is HNO_3.

You probably recognized that HCl is a binary covalent compound that can be named by the rules given earlier in Section 4.10. According to those rules, HCl should be named hydrogen chloride. In fact, that is the correct name for the compound HCl that has not been dissolved in water and is represented in reaction equations by the notation HCl(g). Such compounds that have not been dissolved in water are said to be anhydrous (without water). However, when the gas is dissolved in water and represented in equations by the

notation HCl(aq), it behaves as an acid and is given another name. The following rules are used to name acidic water solutions of such compounds:

1. The word *hydrogen* in the anhydrous compound name is dropped.
2. The prefix *hydro-* is attached to the stem of the name of the nonmetal that is combined with hydrogen.
3. The suffix *-ide* on the stem of the name of the nonmetal that is combined with hydrogen is replaced with the suffix *-ic*.
4. The word *acid* is added to the end of the name as a separate word.

Example 9.2 Naming Acids and Bases

Determine the name that would be given to each of the following binary covalent compounds in the anhydrous form and in the form of water solutions:

a. HCl (stomach acid)
b. H₂S (a gas produced when some sulfur-containing foods such as eggs decay)

Solution

a. The name of the anhydrous compound was given above as hydrogen chloride. The name of the water solution is obtained by dropping *hydrogen* from the anhydrous compound name and adding the prefix *hydro-* to the stem *chlor*. The suffix *-ide* on the stem *chlor* is replaced by the suffix *-ic* to give the name *hydrochloric*. The word *acid* is added, giving the final name *hydrochloric acid* for the water solution.
b. According to the rules of Section 4.10, the anhydrous compound name is hydrogen sulfide. The first two steps in obtaining the name of the water solution are to drop *hydrogen* and to add the prefix *hydro-* to the stem *sulf*. However, in acids involving sulfur as the nonmetal combined with hydrogen, the stem *sulf* is replaced by the entire name *sulfur* for pronunciation reasons. The next steps involve dropping the suffix *-ide*, adding the suffix *-ic*, and adding the word *acid*. The resulting name of the water solution is *hydrosulfuric acid*.

✔ **LEARNING CHECK 9.2** Determine the name that would be given to each of the following binary covalent compounds in the anhydrous form and in the form of water solutions:

a. HI
b. HBr

Acids of the second type, in which hydrogen is covalently bonded to a polyatomic ion, have the same name in the anhydrous form and in the form of water solutions. The names for these acids are based on the name of the polyatomic ion to which the hydrogen is bonded. The rules are as follows:

1. All hydrogens that are written as the first part of the formula of the acid are removed. The hydrogens are removed in the form of H⁺ ions.
2. The polyatomic ion that remains after the H⁺ ions are removed is named by referring to sources such as **Table 4.7** (see page 128).
3. When the remaining polyatomic ion has a name ending in the suffix *-ate*, the suffix is replaced by the suffix *-ic*, and the word *acid* is added.
4. When the remaining polyatomic ion has a name ending in the suffix *-ite*, the suffix is replaced by the suffix *-ous*, and the word *acid* is added.
5. If the polyatomic ion contains sulfur or phosphorus, the stems *-sulf* or *-phosph* that remain in Steps 3 or 4, when the suffixes *-ate* or *-ite* are replaced, are expanded for pronunciation reasons to *-sulfur* and *-phosphor* before the suffixes *-ic* or *-ous* are added.

Example 9.3 Naming Phosphorus-Containing Acids

Compounds derived from the acid H_3PO_4 serve numerous important functions in the body, including the control of acidity in urine and body cells and the storage of energy in the form of ATP. Name this important acid.

Solution

The removal of the three H^+ ions leaves behind the PO_4^{3-} polyatomic ion. This ion is named the phosphate ion in Table 4.7. According to Rules 1–4, the suffix *-ate* is replaced by the suffix *-ic* to give the name *phosphic acid*. However, Rule 5 must be used for this phosphorus-containing acid. The stem is expanded to *phosphor* to give the final name, *phosphoric acid*.

✔ **LEARNING CHECK 9.3** The acid H_2CO_3 is involved in many processes in the body, including the removal of CO_2 gas produced by cellular metabolism and the control of the acidity of various body fluids. Name this important acid.

9.4 | The Self-Ionization of Water

Learning Objective

4. Do calculations using the concept of the self-ionization of water.

In Examples 9.1a and b, water behaved as a Brønsted base. In Example 9.1c, it was a Brønsted acid. But what happens when only pure water is present? The answer is that water behaves as both an acid and a base and undergoes a self- or auto-ionization. The equation representing this self-ionization is:

$$H_2O(\ell) + H_2O(\ell) \rightleftarrows H_3O^+(aq) + OH^-(aq) \tag{9.6}$$

or

The transfer of a proton from one water molecule (the acid) to another (the base) causes one H_3O^+ and one OH^- to form. Therefore, in pure water the concentrations of H_3O^+ and OH^- must be equal. At 25°C these concentrations are 10^{-7} mol/L (M). Thus, the equilibrium position is far to the left, as indicated by the arrows in Equation 9.6. Unless noted otherwise, all concentrations and related terms in this chapter are given at 25°C.

The term **neutral** is used to describe any water solution in which the concentrations of H_3O^+ and OH^- are equal. Thus, pure water is neutral because each liter of pure water contains 10^{-7} mol H_3O^+ and 10^{-7} mol OH^-, at equilibrium. Although all water solutions are not necessarily neutral, it is true that in any solution that contains water, the product of the molar concentrations of H_3O^+ and OH^- is a constant. This becomes apparent by writing the equilibrium expression for Reaction 9.6:

neutral A term used to describe any water solution in which the concentrations of H_3O^+ and OH^- are equal. Also, a water solution with pH = 7.

$$K = \frac{[H_3O^+][OH^-]}{[H_2O][H_2O]} = \frac{[H_3O^+][OH^-]}{[H_2O]^2} \tag{9.7}$$

This expression contains the square of the concentration of water in the denominator. However, only a tiny amount of water actually reacts to form H_3O^+ and OH^- (10^{-7} mol/L), so the concentration of water remains essentially constant. Rearrangement of Equation 9.7 gives

$$K[H_2O]^2 = [H_3O^+][OH^-] \tag{9.8}$$

This equation may be written as

$$K[H_2O]^2 = K_w = [H_3O^+][OH^-] \tag{9.9}$$

where K_w is a new constant called the **ion product of water**. It is a constant because it is equal to the product of two constants, K and $[H_2O]^2$. At 25°C, K_w can be evaluated from the measured values of $[H_3O^+]$ and $[OH^-]$ in pure water.

$$K_w = [H_3O^+][OH^-] = (1.0 \times 10^{-7}\,\text{mol/L})(1.0 \times 10^{-7}\,\text{mol/L})$$
$$= 1.0 \times 10^{-14}\,(\text{mol/L})^2 \tag{9.10}$$

Equation 9.10 is valid not only for pure water but for any solution in which water is the solvent. Notice that we are including units for equilibrium constants in this discussion. This makes the calculation of concentrations easier to follow and understand.

A solution is classified as **acidic** when the concentration of H_3O^+ is greater than the concentration of OH^-. In a **basic** or **alkaline solution**, the concentration of OH^- is greater than that of H_3O^+. However, the product of the molar concentrations of H_3O^+ and OH^- will be $1.0 \times 10^{-14}\,(\text{mol/L})^2$ in either case. Many acidic and basic materials are found in the home (see **Figure 9.1**).

Figure 9.1 Acidic and basic materials are common in the home. Look at the photos, and note a single category to which most of the acidic materials belong, and one to which most of the basic materials belong.

1 Acidic materials found in the home.

2 Basic or alkaline materials found in the home.

Example 9.4 Solution Classifications and Calculations

Classify each of the following solutions as acidic, basic, or neutral. Calculate the molar concentration of the ion whose concentration is not given.

a. $[H_3O^+] = 1.0 \times 10^{-4}\,\text{mol/L}$
b. $[OH^-] = 1.0 \times 10^{-9}\,\text{mol/L}$
c. $[OH^-] = 1.0 \times 10^{-6}\,\text{mol/L}$

Solution

a. Because $[H_3O^+] = 1.0 \times 10^{-4}\,\text{mol/L}$, a rearrangement of Equation 9.10 gives

$$[OH^-] = \frac{1.0 \times 10^{-14}\,(\text{mol/L})^2}{[H_3O^+]} = \frac{1.0 \times 10^{-14}\,(\text{mol/L})^2}{1.0 \times 10^{-4}\,\text{mol/L}}$$

$$= 1.0 \times 10^{-10}\,\text{mol/L}$$

Thus,

$$[H_3O^+] = 1.0 \times 10^{-4}\,\text{mol/L} \quad \text{and}$$

$$[OH^-] = 1.0 \times 10^{-10}\,\text{mol/L}$$

The solution is acidic because the $[H_3O^+]$ is greater than the $[OH^-]$; 1.0×10^{-4} is greater than 1.0×10^{-10}.

b. Similarly, $[OH^-] = 1.0 \times 10^{-9}$ mol/L; therefore,

$$[H_3O^+] = \frac{1.0 \times 10^{-14}\,(\text{mol/L})^2}{[OH^-]} = \frac{1.0 \times 10^{-14}\,(\text{mol/L})^2}{1.0 \times 10^{-9}\,\text{mol/L}}$$

$$= 1.0 \times 10^{-5}\,\text{mol/L}$$

This solution is also acidic because 1.0×10^{-5} is greater than 1.0×10^{-9}.

c. $[OH^-] = 1.0 \times 10^{-6}$ mol/L; therefore,

$$[H_3O^+] = \frac{1.0 \times 10^{-14}\,(\text{mol/L})^2}{[OH^-]} = \frac{1.0 \times 10^{-14}\,(\text{mol/L})^2}{1.0 \times 10^{-6}\,\text{mol/L}}$$

$$= 1.0 \times 10^{-8}\,\text{mol/L}$$

This solution is basic because the OH^- concentration (1.0×10^{-6} mol/L) is greater than the H_3O^+ concentration (1.0×10^{-8} mol/L).

✔ **LEARNING CHECK 9.4** Classify each of the following solutions as acidic, basic, or neutral. Calculate the molarity of the ion whose concentration is not given.

a. $[OH^-] = 1.0 \times 10^{-5}$ mol/L
b. $[H_3O^+] = 1.0 \times 10^{-9}$ mol/L
c. $[H_3O^+] = 1.0 \times 10^{-2}$ mol/L

9.5 | The pH Concept

Learning Objective

5. Do calculations using the pH concept.

In Section 9.4, you learned that the concentration of H_3O^+ in pure water is 1.0×10^{-7} M. Chemists, technologists, and other laboratory personnel routinely work with solutions in which the H_3O^+ concentration may be anywhere from 10 to 10^{-14} M. Because of the inconvenience of working with numbers that extend over such a wide range, chemists long ago adopted a shortcut notation known as the pH. Mathematically, the **pH** is defined in terms of the negative logarithm (log) of $[H_3O^+]$ by the following two equations, where we now introduce the common practice of substituting H^+ for H_3O^+ to simplify equations:

pH The negative logarithm of the molar concentration of H^+ (H_3O^+) in a solution.

$$pH = -\log[H^+] \tag{9.11}$$

$$[H^+] = 1 \times 10^{-pH} \tag{9.12}$$

Thus, pH is simply the negative of the exponent used to express the hydrogen-ion concentration in moles per liter.

Example 9.5 Calculating pH Values

Express the following concentrations in terms of pH:

a. $[H^+] = 1 \times 10^{-5}$ mol/L
b. $[OH^-] = 1 \times 10^{-9}$ mol/L
c. $[H^+] = 1 \times 10^{-7}$ mol/L
d. $[H^+] = 1 \times 10^{-11}$ mol/L

Solution

a. pH is the negative of the exponent used to express $[H^+]$. Therefore, $pH = -(-5) = 5.0$

b. Here $[OH^-]$ is given, and $[H^+]$ must be calculated. We remember Equation 9.10 and get

$$[H^+] = \frac{1.0 \times 10^{-14}\,(mol/L)^2}{[OH^-]} = \frac{1.0 \times 10^{-14}\,(mol/L)^2}{1 \times 10^{-9}\,mol/L} = 1 \times 10^{-5}\,mol/L$$

Therefore, $pH = -(-5) = 5.0$

c. This $pH = -(-7) = 7.0$. $[H^+] = 1 \times 10^{-7}$ mol/L and corresponds to pure water and neutral solutions. Thus, a pH of 7 represents neutrality.

d. $pH = -(-11) = 11.0$

✔ **LEARNING CHECK 9.5** Express the following concentrations in terms of pH.

a. $[H^+] = 1 \times 10^{-14}$ mol/L
b. $[OH^-] = 1.0$ mol/L
c. $[OH^-] = 1 \times 10^{-8}$ mol/L

The pH value of 5.0 obtained in Example 9.5a corresponds to a solution in which the H^+ concentration is greater than the OH^- concentration. Thus, the solution is acidic. Any solution with a pH less than 7 is classified as acidic. Any solution with a pH greater than 7 is classified as basic or alkaline. The pH values of some familiar solutions are given in **Table 9.1**.

TABLE 9.1 Relationships between [H⁺], [OH⁻], and pH

	$[H^+]$	$[OH^-]$	pH	Examples (solids are dissolved in water)
	10^0	10^{-14}	0	Battery acid
	10^{-1}	10^{-13}	1	Gastric juice
	10^{-2}	10^{-12}	2	Lemon juice
	10^{-3}	10^{-11}	3	Vinegar, carbonated drink Aspirin
	10^{-4}	10^{-10}	4	Orange juice Apple juice
	10^{-5}	10^{-9}	5	Black coffee
	10^{-6}	10^{-8}	6	Normal urine (average value) Milk, liquid dishwashing detergent
Neutral	10^{-7}	10^{-7}	7	Saliva, pure water Blood
	10^{-8}	10^{-6}	8	Soap (not synthetic detergent) Baking soda Phosphate-containing detergent
	10^{-9}	10^{-5}	9	Milk of magnesia Powdered household cleanser
	10^{-10}	10^{-4}	10	Phosphate-free detergent
	10^{-11}	10^{-3}	11	Household ammonia
	10^{-12}	10^{-2}	12	Liquid household cleaner
	10^{-13}	10^{-1}	13	Oven cleaner
	10^{-14}	10^0	14	Liquid drain cleaner

Example 9.6 pH Calculations

Determine the H^+ and OH^- molar concentrations that correspond to the following pH values:

a. pH = 9.0
b. pH = 3.0
c. pH = 11.0

Solution

In each case, the relationships $[H^+] = 1 \times 10^{-pH}$ and $K_w = [H^+][OH^-] = 1.0 \times 10^{-14}$ $(mol/L)^2$ can be used. Note that we have substituted $[H^+]$ for $[H_3O^+]$ in Equation 9.10.

a. Because pH = 9.0, $[H^+] = 1 \times 10^{-pH} = 1 \times 10^{-9}$ mol/L

$$[OH^-] = \frac{K_w}{[H^+]} = \frac{1.0 \times 10^{-14}\,(mol/L)^2}{1 \times 10^{-9}\,mol/L} = 1 \times 10^{-5}\,mol/L$$

b. Because pH = 3.0, $[H^+] = 1 \times 10^{-pH} = 1 \times 10^{-3}$ mol/L

$$[OH^-] = \frac{K_w}{[H^+]} = \frac{1.0 \times 10^{-14}\,(mol/L)^2}{1 \times 10^{-3}\,mol/L} = 1 \times 10^{-11}\,mol/L$$

c. Because pH = 11.0, $[H^+] = 1 \times 10^{-pH} = 1 \times 10^{-11}$ mol/L

$$[OH^-] = \frac{K_w}{[H^+]} = \frac{1.0 \times 10^{-14}\,(mol/L)^2}{1 \times 10^{-11}\,mol/L} = 1 \times 10^{-3}\,mol/L$$

✔ **LEARNING CHECK 9.6** Determine the $[H^+]$ and $[OH^-]$ values that correspond to the following pH values:

a. pH = 10.0
b. pH = 4.0
c. pH = 5.0

It is apparent from Table 9.1 that not all solutions have pH values that are neat whole numbers. For example, the pH of vinegar is about 3.3. How do we deal with such numbers? The H^+ concentration of vinegar could be written $1 \times 10^{-3.3}$, but it is not convenient to work with exponents that are not whole numbers. When exact values are not needed, the pH, $[H^+]$, and so on can be expressed as a range. Thus, the H^+ concentration of vinegar is between 1×10^{-3} and 1×10^{-4} mol/L. Similarly, a solution with $[H^+] = 2 \times 10^{-5}$ mol/L has a pH between 4 and 5.

When more exact values are needed, we must work with logarithms. This is most easily done by using a hand calculator. A hydrogen-ion concentration is converted to pH by taking the logarithm and changing its sign. **Table 9.2** gives the steps of a typical

TABLE 9.2 Calculating pH from Molarity with a Calculator

Step	Calculator Procedure	Calculator Display
1. Enter 3.6	Press buttons 3, ., 6	3.6
2. Enter 10^{-4}	Press button that activates exponential mode (EE, Exp, etc.)	3.6^{00}
	Press 4	3.6^{04}
	Press change-sign button (\pm, etc.)	3.6^{-04}
3. Take logarithm	Press log button (log, etc.)	-3.4437
4. Change sign	Press change-sign button (\pm, etc.)	3.4437

calculator procedure (what button is pushed, etc.) and a typical calculator readout or display for the conversion of $[H^+] = 3.6 \times 10^{-4}$ mol/L into pH. The pH from Table 9.2 would be recorded as 3.44.

✔ **LEARNING CHECK 9.7** Convert the following $[H^+]$ values into pH:

a. $[H^+] = 4.2 \times 10^{-5}$ mol/L
b. $[H^+] = 8.1 \times 10^{-9}$ mol/L

Calculators can also be used to convert pH values into corresponding molar concentrations. The steps are given in **Table 9.3** for the conversion of a pH value of 5.92 into $[H^+]$. As shown in Table 9.3, a pH of 5.92 corresponds to a $[H^+]$ of 1.2×10^{-6} mol/L. Note that the number of figures to the right of the decimal in a pH value should be the same as the number of significant figures in the $[H^+]$ value. Thus, the two figures to the right of the decimal in pH = 5.92 is reflected in the two significant figures in $[H^+] = 1.2 \times 10^{-6}$ mol/L.

TABLE 9.3 Calculating Molarity from pH with a Calculator

Step	Calculator Procedure	Calculator Display
1. Enter 5.92	Press 5, ., 9, 2	5.92
2. Change sign	Press change-sign button (\pm, etc.)	-5.92
3. Take antilog	Press antilog or 10^x button, or (more commonly) press inv. or 2nd function button and then log button	.0000012 or 1.2^{-06}

✔ **LEARNING CHECK 9.8** Convert the following pH values into molar concentrations of H^+:

a. pH = 2.75
b. pH = 8.33

9.6 | Properties of Acids

Learning Objective

6. Write reaction equations that illustrate the characteristic reactions of acids.

Acids and bases are used so often in most laboratories that stock solutions are kept readily available at each work space. The common solutions, their concentrations, and label designations are given in **Table 9.4**.

Example 9.7 Solution Preparation by Dilution

Describe how you would make 250 mL of 1 M HNO_3 by using a dilute HNO_3 stock solution.

Solution
According to Table 9.4, the dilute HNO_3 stock solution is 6 M. The volume of 6 M HNO_3 needed to make 250 mL of 1 M HNO_3 is obtained by using Equation 7.9. It must be remembered that this equation is useful only for dilution problems such as this one. The equation is not useful for calculations involving the volume of one solution that will react with a specific volume of a second solution (see **Sections 9.10** and **9.11**).

$$C_c V_c = C_d V_d$$
$$(6 \text{ M}) V_c = (1 \text{ M})(250 \text{ mL})$$

$$V_c = \frac{(1 \text{ M})(250 \text{ mL})}{6 \text{ M}} = 41.7 \text{ mL}$$

Because the molarity is given using only one significant figure, the solution can be made without too much attention to volumetric flasks and the like. Thus, 42 mL of 6 M HNO_3 is measured with a graduated cylinder, and this is added to 208 mL of distilled water that has also been measured with a graduated cylinder.

✔ **LEARNING CHECK 9.9** Describe how you would prepare 500 mL of 3.0 M aqueous ammonia using a concentrated NH_3 stock solution (see **Table 9.4**).

TABLE 9.4 Common Laboratory Acids and Bases

Name	Formula	Label Concentration	Molarity
Acids			
Acetic acid	$HC_2H_3O_2$	Glacial	18
Acetic acid	$HC_2H_3O_2$	Dilute	6
Hydrochloric acid	HCl	Concentrated	12
Hydrochloric acid	HCl	Dilute	6
Nitric acid	HNO_3	Concentrated	16
Nitric acid	HNO_3	Dilute	6
Sulfuric acid	H_2SO_4	Concentrated	18
Sulfuric acid	H_2SO_4	Dilute	3
Bases			
Aqueous ammonia[a]	NH_3	Concentrated	15
Aqueous ammonia	NH_3	None usually given	6
Sodium hydroxide	NaOH	None usually given	6

[a]Often erroneously called ammonium hydroxide and given the formula NH_4OH.

Different acids have different properties that make some more practical than others for specific uses. However, all acids have certain properties in common. Two of these were mentioned earlier—all acids taste sour and produce H_3O^+ ions when dissolved in water. In addition, all acids undergo characteristic double-replacement reactions with solid oxides, hydroxides, carbonates, and bicarbonates (see **Figure 9.2**).

Figure 9.2 The reaction of hydrochloric acid with three natural forms of $CaCo_3$. What gas is produced? Left to right: limestone, marble, and eggshell.

$$2\text{HCl(aq)} + \text{CuO(s)} \rightarrow \text{CuCl}_2\text{(aq)} + \text{H}_2\text{O}(\ell) \tag{9.13}$$
$$\text{copper}$$
$$\text{oxide}$$

$$2\text{HCl(aq)} + \text{Ca(OH)}_2\text{(s)} \rightarrow \text{CaCl}_2\text{(aq)} + 2\text{H}_2\text{O}(\ell) \tag{9.14}$$
$$\text{calcium}$$
$$\text{hydroxide}$$

$$2\text{HCl(aq)} + \text{CaCO}_3\text{(s)} \rightarrow \text{CaCl}_2\text{(aq)} + \text{CO}_2\text{(g)} + \text{H}_2\text{O}(\ell) \tag{9.15}$$
$$\text{calcium}$$
$$\text{carbonate}$$

$$2\text{HCl(aq)} + \text{Sr(HCO}_3)_2\text{(s)} \rightarrow \text{SrCl}_2\text{(aq)} + 2\text{CO}_2\text{(g)} + 2\text{H}_2\text{O}(\ell) \tag{9.16}$$
$$\text{strontium}$$
$$\text{bicarbonate}$$

Notice that the preceding reactions are written using molecular equations (see **Section 5.7**).
Reactions involving ionic substances can also be written as total ionic equations or net ionic equations. Equation 9.13 is written in total ionic form in Equation 9.17.

$$\underbrace{2\text{H}^+\text{(aq)} + 2\text{Cl}^-\text{(aq)}}_{2\text{HCl}} + \text{CuO(s)} \longrightarrow \underbrace{\text{Cu}^{2+}\text{(aq)} + 2\text{Cl}^-\text{(aq)}}_{\text{CuCl}_2} + \text{H}_2\text{O}(\ell) \tag{9.17}$$

We see that the chloride ions are spectator ions, so a net ionic equation can be written as shown in Equation 9.18.

$$2\text{H}^+\text{(aq)} + \text{CuO(s)} \rightarrow \text{Cu}^{2+}\text{(aq)} + \text{H}_2\text{O}(\ell) \tag{9.18}$$

The general nature of the reaction is emphasized in the net ionic form because the H^+ could come from any acid.

Remember that correctly written molecular equations must have their atoms balanced (see **Section 5.1**). Total ionic and net ionic equations must also have their atoms balanced, but in addition the total charges on each side of the equation must balance. In Equation 9.18, for example, the two H^+ ions provide two positive charges on the left, which are balanced by the two positive charges of Cu^{2+} on the right.

Example 9.8 Writing Ionic Equilibrium Equations

Write Equations 9.14 and 9.15 in total ionic and net ionic forms.

Solution

In Equation 9.14, HCl and CaCl_2 are soluble and ionizable.

Total ionic: $2\text{H}^+\text{(aq)} + 2\text{Cl}^-\text{(aq)} + \text{Ca(OH)}_2\text{(s)} \rightarrow \text{Ca}^{2+}\text{(aq)} + 2\text{Cl}^-\text{(aq)} + 2\text{H}_2\text{O}(\ell)$

The Cl^- is a spectator ion.

Net ionic: $2\text{H}^+\text{(aq)} + \text{Ca(OH)}_2\text{(s)} \rightarrow \text{Ca}^{2+}\text{(aq)} + 2\text{H}_2\text{O}(\ell)$

In Equation 9.15, HCl and CaCl_2 are soluble and ionizable.

Total ionic: $2\text{H}^+\text{(aq)} + 2\text{Cl}^-\text{(aq)} + \text{CaCO}_3\text{(s)} \rightarrow \text{Ca}^{2+}\text{(aq)} + 2\text{Cl}^-\text{(aq)} + \text{CO}_2\text{(g)} + \text{H}_2\text{O}(\ell)$

Again, Cl^- is a spectator ion.

Net ionic: $2\text{H}^+\text{(aq)} + \text{CaCO}_3\text{(s)} \rightarrow \text{Ca}^{2+}\text{(aq)} + \text{CO}_2\text{(g)} + \text{H}_2\text{O}(\ell)$

✔ **LEARNING CHECK 9.10** Write Equation 9.16 in total and net ionic forms. Consider HCl and SrCl_2 as soluble and ionizable.

Another property of acids is their ability to react with (and dissolve) certain metals to yield hydrogen gas. This is a redox reaction (see **Section 5.3**), as evidenced by the change in the oxidation number of hydrogen as the reaction takes place. In compounds such as acids, hydrogen has an oxidation number of $+1$. In hydrogen gas, the oxidation number is 0 (Section 5.3). Thus, hydrogen is reduced during the reaction, and, as shown by Equations 9.19 and 9.20, the metal is oxidized. The ability to reduce hydrogen ions to hydrogen gas is not the same for all metals. Some are such strong reducing agents that they can react with hydrogen ions of very low concentrations such as that found in water. Others are so weak as reducing agents that they cannot react with H^+ at the high concentration found in concentrated acids. These tendencies are represented by the **activity series** of metals shown in **Table 9.5**. The higher a metal is in the series, the more active it is as a reducing agent. Some typical reactions are given in Equations 9.19 and 9.20 (see **Figure 9.3**).

activity series A tabular representation of the tendencies of metals to react with H^+.

TABLE 9.5 The Activity Series of the Metals

Metal	Symbol	Comments
Potassium	K	React violently with cold water
Sodium	Na	
Calcium	Ca	Reacts slowly with cold water
Magnesium	Mg	React very slowly with steam, but quite rapidly in higher H^+ concentrations
Aluminum	Al	
Zinc	Zn	
Chromium	Cr	
Iron	Fe	React in moderately high H^+ concentrations
Nickel	Ni	
Tin	Sn	
Lead	Pb	
Copper	Cu	Do not react with H^+
Mercury	Hg	
Silver	Ag	
Platinum	Pt	
Gold	Au	

© Jeffrey M. Seager

Figure 9.3 Metals vary in their ability to reduce hydrogen ions (H^+) to hydrogen gas (H_2). This difference is apparent when copper, iron, zinc, and magnesium (left to right) are put into hydrochloric acid solution (HCl) of the same molarity. Are these results consistent with Table 9.5? Explain.

Molecular equation: $Zn(s) + 2HCl(aq) \rightarrow ZnCl_2(aq) + H_2(g)$ (9.19)

Net ionic equation: $Zn(s) + 2H^+(aq) \rightarrow Zn^{2+}(aq) + H_2(g)$

Molecular equation: $2K(s) + 2H_2O(\ell) \rightarrow 2KOH(aq) + H_2(g)$ (9.20)

Net ionic equation: $2K(s) + 2H_2O(\ell) \rightarrow 2K^+(aq) + 2OH^-(aq) + H_2(g)$

✔ **LEARNING CHECK 9.11** Write molecular, total ionic, and net ionic equations to represent the following reactions:

a. Calcium (Ca) with cold water (NOTE: $Ca(OH)_2$ is not soluble in water).
b. Mg with H_2SO_4 (NOTE: $MgSO_4$ is soluble and ionizable in water).

9.7 | Properties of Bases

Learning Objective

7. Write reaction equations that represent neutralization reactions between acids and bases.

neutralization reaction A reaction in which an acid and base react completely, leaving a solution that contains only a salt and water.

Solutions containing bases feel soapy or slippery and change the color of litmus from red to blue. Equation 9.14 illustrates their most characteristic chemical property—they react readily with acids. In most of the earliest acid–base reactions studied, the complete reaction of an acid with a base produced a neutral solution. For this reason, such reactions were often called **neutralization reactions** (see below). It is now known that many "neutralization" reactions do not produce neutral solutions (see **Section 9.12**). However, the name for the reactions is still used. More than 20 billion pounds of sodium hydroxide (NaOH) is produced and used in the United States each year. This useful crystalline solid is quite soluble in water and dissociates to form basic solutions.

$$NaOH(s) + H_2O(\ell) \rightarrow Na^+(aq) + OH^-(aq)$$ (9.21)

The neutralization reaction between NaOH and HCl is

Molecular equation: $HCl(aq) + NaOH(aq) \rightarrow NaCl(aq) + H_2O(\ell)$ (9.22)

Total ionic equation: $H^+(aq) + Cl^-(aq) + Na^+(aq) + OH^-(aq) \rightarrow$ (9.23)
$$Na^+(aq) + Cl^-(aq) + H_2O(\ell)$$

Net ionic equation: $H^+(aq) + OH^-(aq) \rightarrow H_2O(\ell)$ (9.24)

The molecular equation (9.22) illustrates the following statement, which is a common definition of neutralization: During a neutralization reaction, an acid and a base combine to form a salt and water. (More is said about salts in the next section.) The net ionic form of the equation (9.24) emphasizes the general nature of neutralization reactions: H^+ ions (from any source) react with OH^- (from any source) to form water.

Bases also react with fats and oils and convert them into smaller, soluble molecules. For this reason, most household cleaning products contain basic substances. For example, lye (NaOH) is the active ingredient in numerous drain cleaners, and many liquid household cleaners contain ammonia.

✔ **LEARNING CHECK 9.12** Write molecular, total ionic, and net ionic equations to represent neutralization reactions between the following acids and bases:

a. HNO_3 and NaOH
b. H_2SO_4 and KOH

Also, because 1 eq = 1000 meq,

$$0.10 \, \cancel{eq} \times \frac{1000 \, meq}{1 \, \cancel{eq}} = 1.0 \times 10^2 \, meq$$

Notice that we could have focused on the negative charges in either part above and still arrived at the same answers. In part a, 0.050 mol KCl produces 0.050 mol Cl^-, or 0.050 mol of negative charge. Similarly, in part b, 0.050 mol $CaCl_2$ produces 2×0.050, or 0.10 mol, Cl^- ions, or 0.10 mol of negative charge. We have arbitrarily chosen to use the positive charges, but either will work. Just remember that you do not count both the negative and positive charges for a salt.

✔ **LEARNING CHECK 9.14** Determine the number of equivalents and milli-equivalents in each of the following:

a. 0.10 mol NaCl
b. 0.10 mol $Mg(NO_3)_2$

Example 9.11 More Equivalent Calculations

A sample of blood serum contains 0.139 equivalents of Na^+ ion per liter of serum. Assume the Na^+ comes from dissolved NaCl, and calculate the number of equivalents, number of moles, and number of grams of NaCl in 250. mL of the serum.

Solution

The Na^+ ion has a single charge, so we may write the following relationships:

$$1.00 \, mol \, NaCl = 1.00 \, mol \, Na^+ = 1.00 \, eq \, Na^+ = 1.00 \, eq \, NaCl$$

These relationships provide numerous factors that can be used in the factor-unit method to solve problems. Two that will be used to solve these problems are:

$$\frac{1.00 \, eq \, NaCl}{1.00 \, eq \, Na^+} \quad and \quad \frac{1.00 \, mol \, NaCl}{1.00 \, eq \, NaCl}$$

The known quantity is 250. mL of serum. This is converted into equivalents of Na^+ by using the given concentration of $\dfrac{0.139 \, eq \, Na^+}{L \, serum}$ as the factor.

$$(0.250 \, \cancel{L \, serum}) \times \frac{0.139 \, eq \, Na^+}{1.00 \, \cancel{L \, serum}} = 0.0347 \, eq \, Na^+$$

This number of equivalents of Na^+ will now be converted into the quantities asked for by using the two factors given above, and a factor obtained from the formula weight for NaCl of 58.44 u.

$$(0.0347 \, \cancel{eq \, Na^+}) \times \frac{1.00 \, eq \, NaCl}{1.00 \, \cancel{eq \, Na^+}} = 0.0347 \, eq \, NaCl$$

$$(0.0347 \, \cancel{eq \, NaCl}) \times \frac{1.00 \, mol \, NaCl}{1.00 \, \cancel{eq \, NaCl}} = 0.0347 \, mol \, NaCl$$

$$(0.0347 \, \cancel{mol \, NaCl}) \times \frac{58.44 \, g \, NaCl}{1.00 \, \cancel{mol \, NaCl}} = 2.03 \, g \, NaCl$$

✔ **LEARNING CHECK 9.15** A sample of blood serum contains 0.103 equivalents of Cl^- ions per liter. Assume the Cl^- comes from dissolved NaCl and calculate the number of equivalents, number of moles, and number of grams of NaCl in 250. mL of serum.

Sinkholes are natural depressions that occur in land surfaces when subterranean caverns collapse. In urban areas, they can swallow roads, cars, homes, larger buildings, ponds, and even swimming pools. Insurance companies in Florida receive an average of seventeen sinkhole-related claims daily. In 2013, a sinkhole in Missouri claimed 15 cars. Deaths from sinkholes are rare, but like other unpredictable catastrophes their spectacular impact can be simultaneously frightening and even awe-inspiring.

Sinkholes result when surface water that lacks a normal surface drainage route instead follows a subsurface route made up of porous rock, especially limestone. Upon exposure to acidic runoff water, limestone slowly dissolves and becomes more porous. When enough of this porous rock dissolves, an underground channel for water drainage is created. When these channels become large enough, the ground above them can no longer bridge them, and they collapse as a sinkhole.

Sinkholes vary in size from a few feet to hundreds of acres, and from less than one to more than a hundred feet deep. One sinkhole in Texas, called the "Devil's sinkhole" is 400 feet deep and opens into a cavern inhabited by thousands of bats.

Certain areas of the United States are more likely to have sinkholes because they contain carbonate rocks, such as limestone and dolomite, beneath the surface. Susceptible states include Alabama, Florida, Kentucky, Missouri, Pennsylvania, Tennessee, and Texas.

The chemistry involved in sinkhole generation is acid–base chemistry. When rain passes through the atmosphere, the rain water and CO_2 in the atmosphere undergo an acid–base reaction:

$$H_2O(\ell) + CO_2(g) \rightarrow H^+(aq) + HCO_3^-(aq)$$

The H^+ ions produced by this reaction react with subsurface limestone and dissolve it:

$$2H^+(aq) + CaCO_3(s) \rightarrow CO_2(g) + H_2O(\ell) + Ca^{2+}(aq)$$

Industrial pollution increases the amount of CO_2 in the atmosphere and the amount of acid rain, which increases the amount of underground limestone that is dissolved, and the likelihood of sinkholes being generated.

Although sinkholes can and do occur naturally, human activity, climate change, and its associated severe weather might increase the occurrence of sinkhole collapses in the future. Warning signs to look for are sudden appearance of standing water, signs or fence posts that appear to slump, crumbling house foundations, windows or doors that close incorrectly, or patches of dying vegetation. Inspectors can confirm the presence of suspected sinkholes using ground-penetrating radar.

Sinkholes, like volcanic eruptions and earthquakes, remind us of the Earth's dynamic nature, and that no such thing as "solid ground" exists on this planet. The Earth is constantly moving and changing. Human activities such as mining, fracking, drilling, groundwater use, and building structures above- and belowground should be done carefully, with consideration of the potent ill impacts on the Earth.

Sinkholes remind us of the Earth's dynamic nature.

9.9 | The Strengths of Acids and Bases

Learning Objective

9. Demonstrate an understanding of the words *weak* and *strong* as applied to acids and bases.

strong acids and strong bases Acids and bases that dissociate (ionize) completely when dissolved to form a solution.

weak (or moderately weak) acids and bases Acids and bases that dissociate (ionize) less than completely when dissolved to form a solution.

When salts dissolve in water, they generally dissociate completely, but this is not true for all acids and bases. The acids and bases that do dissociate almost completely are classified as **strong acids** and **strong bases** (they are also strong electrolytes). Those that dissociate to a much smaller extent are called **weak** or **moderately weak**, depending on the degree of dissociation (they are also weak or moderately weak electrolytes). Examples of strong and weak acids are given in **Table 9.7**.

A 0.10 M solution of hydrochloric acid could be prepared by dissolving 0.10 mol (3.7 g) of HCl gas in enough water to give 1.0 L of solution. According to Table 9.7, 100% of the dissolved gas would dissociate into H^+ and Cl^-. Thus, the concentration of H^+ in a 0.10 M HCl solution is 0.10 mol/L, and the pH is 1.00.

Learning Objective

8. Write reaction equations that illustrate various ways to prepare salts, and do calculations using the concept of an equivalent of salt.

At room temperature, a **salt** is a solid crystalline ionic compound that contains the **cation** (positive ion) of a base and the **anion** (negative ion) of an acid. Thus, ordinary table salt (NaCl) contains Na^+, the cation of NaOH, and Cl^-, the anion of HCl (look again at Equation 9.22). Similarly, $CuSO_4$ is a salt containing the cation of $Cu(OH)_2$ and the anion of H_2SO_4. You must be careful to think of the term *salt* in a general way, and not as representing only table salt (NaCl).

Some acids and bases are not stable enough to be isolated even though their salts are. For example, carbonic acid (H_2CO_3) cannot be isolated in the pure state. When it forms in water, it promptly decomposes:

$$H_2CO_3(aq) \rightleftharpoons H_2O(\ell) + CO_2(g) \qquad (9.25)$$

Despite this characteristic, salts of carbonic acid, such as Na_2CO_3 and $NaHCO_3$, are quite stable.

It is not necessary to identify the parent acid and base in order to write correct salt formulas or names. Just remember that the cation of a salt can be any positive ion except H^+, and it will usually be a simple metal ion or NH_4^+. The salt anion can be any negative ion except OH^-. Most of the polyatomic anions you will use were given earlier in **Table 4.7**. The rules for naming salts were given in **Sections 4.4 and 4.10**.

As previously observed, salts dissolved in solution are dissociated into ions. The salt can be recovered by evaporating away the water solvent. When this is done carefully, some salts retain specific numbers of water molecules as part of the solid crystalline structure. Such salts are called **hydrates**, and the retained water is called the **water of hydration**. Most hydrates lose all or part of the water of hydration when they are heated to moderate or high temperatures. A number of useful hydrates are given in **Table 9.6**.

salt A solid crystalline ionic compound at room temperature that contains the cation of a base and the anion of an acid.

cation A positively charged ion.

anion A negatively charged ion.

hydrate A salt that contains specific numbers of water molecules as part of the solid crystalline structure.

water of hydration Water retained as part of the solid crystalline structure of some salts.

TABLE 9.6 **Some Useful and Common Hydrates**

Formula	Chemical Name	Common Name	Uses
$CaSO_4 \cdot H_2O$	Calcium sulfate monohydrate	Plaster of Paris	Plaster, casts, molds
$CaSO_4 \cdot 2H_2O$	Calcium sulfate dihydrate	Gypsum	Casts, molds, wallboard
$MgSO_4 \cdot 7H_2O$	Magnesium sulfate heptahydrate	Epsom salts	Cathartic
$Na_2B_4O_7 \cdot 10H_2O$	Sodium tetraborate decahydrate	Borax	Laundry
$Na_2CO_3 \cdot 10H_2O$	Sodium carbonate decahydrate	Washing soda	Water softener
$Na_3PO_4 \cdot 12H_2O$	Sodium phosphate dodecahydrate	Trisodium phosphate (TSP)	Water softener
$Na_2SO_4 \cdot 10H_2O$	Sodium sulfate decahydrate	Glauber's salt	Cathartic
$Na_2S_2O_3 \cdot 5H_2O$	Sodium thiosulfate pentahydrate	Hypo	Film photography

Many salts occur in nature, and some are used as industrial raw materials. Examples are sodium chloride, NaCl (a source of Cl_2 and NaOH); calcium carbonate or limestone, $CaCO_3$ (a source of cement and building stone); and calcium phosphate or rock phosphate, $Ca_3(PO_4)_2$ (a source of fertilizer). In the laboratory, salts can be prepared by reacting a solution of an appropriate acid with a metal, a metal oxide, a metal hydroxide, a metal carbonate, or a metal bicarbonate. These reactions, given earlier as examples of acid properties (Equations 9.13, 9.14, 9.15, 9.16), are given below in a general form:

$$\text{acid} + \text{metal} \rightarrow \text{salt} + H_2 \qquad (9.26)$$
$$\text{acid} + \text{metal oxide} \rightarrow \text{salt} + H_2O \qquad (9.27)$$
$$\text{acid} + \text{metal hydroxide} \rightarrow \text{salt} + H_2O \qquad (9.28)$$
$$\text{acid} + \text{metal carbonate} \rightarrow \text{salt} + H_2O + CO_2 \qquad (9.29)$$
$$\text{acid} + \text{metal bicarbonate} \rightarrow \text{salt} + H_2O + CO_2 \qquad (9.30)$$

Example 9.9 Compound Preparation Equations

Write equations to represent the preparation of $Mg(NO_3)_2$, using Reactions 9.26 through 9.30. Use molecular equations to emphasize the salt formation.

Solution

In each case, a dilute solution of nitric acid (HNO_3) is reacted with magnesium metal or the appropriate magnesium compound:

$$2HNO_3(aq) + Mg(s) \rightarrow Mg(NO_3)_2(aq) + H_2(g)$$

$$2HNO_3(aq) + MgO(s) \rightarrow Mg(NO_3)_2(aq) + H_2O(\ell)$$

$$2HNO_3(aq) + Mg(OH)_2(s) \rightarrow Mg(NO_3)_2(aq) + 2H_2O(\ell)$$

$$2HNO_3(aq) + MgCO_3(s) \rightarrow Mg(NO_3)_2(aq) + H_2O(\ell) + CO_2(g)$$

$$2HNO_3(aq) + Mg(HCO_3)_2(s) \rightarrow Mg(NO_3)_2(aq) + 2H_2O(\ell) + 2CO_2(g)$$

✓ **LEARNING CHECK 9.13** Write balanced equations to represent the preparation of $AlCl_3$, using Reactions 9.26 through 9.30. Use molecular equations to emphasize the salt formation.

We have expressed the amounts of materials involved in chemical reactions primarily in terms of mass (grams) or number of moles. However, in certain applications, it is important to express the amount of salt in a solution in terms of the amount of electrical charge represented by the ions of the salt. This is especially true in medical applications, where the levels and balance of electrolytes in various body fluids are extremely important.

A unit that expresses the amount of ionic electrical charge for salts is the equivalent. One **equivalent of salt** is the amount that will produce 1 mol of positive (or negative) charges when dissolved and dissociated. To determine the amount of salt that represents 1 equivalent (eq), you must know what ions are produced when the salt dissociates. For example, potassium chloride (KCl), an electrolyte often administered to patients following surgery, dissociates as follows: $KCl(aq) \rightarrow K^+(aq) + Cl^-(aq)$. Thus, 1 mol, or 74.6 g, of solid KCl provides 1 mol of positive charges (1 mol of K^+ ions) when it dissolves and dissociates. Thus, 1 eq of KCl is equal to 1 mol, or 74.6 g, of KCl.

equivalent of salt The amount that will produce 1 mol of positive (or negative) electrical charges when dissolved and dissociated.

Example 9.10 Equivalents of Salts

Determine the number of equivalents and milliequivalents (meq) of salt contained in the following:

a. 0.050 mol KCl

b. 0.050 mol $CaCl_2$

Solution

a. As shown above, 1 mol KCl = 1 eq KCl. Therefore, 0.050 mol KCl = 0.050 eq KCl. Also, because 1 eq = 1000 meq,

$$0.050 \text{ eq} \times \frac{1000 \text{ meq}}{1 \text{ eq}} = 50 \text{ meq}$$

b. The dissociation reaction is

$$CaCl_2(aq) \rightarrow Ca^{2+}(aq) + 2Cl^-(aq)$$

Thus, we see that 1 mol $CaCl_2$ produces 1 mol Ca^{2+} or 2 mol of positive charges. Thus, 1 mol $CaCl_2$ = 2 eq $CaCl_2$. Therefore,

$$0.050 \text{ mol } CaCl_2 \times \frac{2 \text{ eq } CaCl_2}{1 \text{ mol } CaCl_2} = 0.10 \text{ eq } CaCl_2$$

TABLE 9.7 Some Common Strong and Weak Acids

Name	Formula	% Dissociation[a]	K_a	Classification
Hydrochloric acid	HCl	100	Very large	Strong
Hydrobromic acid	HBr	100	Very large	Strong
Nitric acid	HNO_3	100	Very large	Strong
Sulfuric acid	H_2SO_4	100	Very large	Strong
Phosphoric acid	H_3PO_4	28	7.5×10^{-3}	Moderately weak
Sulfurous acid[b]	H_2SO_3	34	1.5×10^{-2}	Moderately weak
Acetic acid	$HC_2H_3O_2$	1.3	1.8×10^{-5}	Weak
Boric acid	H_3BO_3	0.01	7.3×10^{-10}	Weak
Carbonic acid[b]	H_2CO_3	0.2	4.3×10^{-7}	Weak
Nitrous acid[b]	HNO_2	6.7	4.6×10^{-4}	Weak

[a]Based on dissociation of one proton in 0.1 M solution at 25°C.

[b]Unstable acid.

The strength of acids and bases is shown quantitatively by the value of the equilibrium constant for the dissociation reaction in water solutions. Equation 9.31 represents the dissociation of a general acid, HB, in water, where B^- represents the conjugate base of the acid:

$$HB(aq) + H_2O(\ell) \rightleftharpoons H_3O^+(aq) + B^-(aq) \qquad (9.31)$$

The equilibrium expression for this reaction is

$$K = \frac{[H_3O^+][B^-]}{[HB][H_2O]} \qquad (9.32)$$

In Equation 9.32, the brackets, again, represent molar concentrations of the materials in the solution. Only a tiny amount of the water in the solution actually enters into the reaction, so the concentration of water is considered to be constant (see **Section 9.4**). We can then write Equation 9.33, where K_a is a new constant called the **acid dissociation constant**:

$$K[H_2O] = K_a = \frac{[H_3O^+][B^-]}{[HB]} \qquad (9.33)$$

acid dissociation constant The equilibrium constant for the dissociation of an acid.

In Equation 9.33, $[H_3O^+]$ and $[B^-]$ are, respectively, the equilibrium concentrations of the hydronium ion and the anion conjugate base that is characteristic of the acid. The [HB] represents the concentration of that part of the dissolved acid that remains undissociated in the equilibrium mixture. In solutions of strong acids, $[H_3O^+]$ and $[B^-]$ values are quite large, while [HB] has a value near 0, so K_a is quite large. In weak acids, [HB] has larger values, while $[H_3O^+]$ and $[B^-]$ are smaller, so K_a is smaller. Thus, the larger a K_a value, the stronger the acid it represents. This is illustrated by the K_a values given in Table 9.7. If we simplify by substituting H^+ for H_3O^+, introduced in Section 9.5, we obtain

$$K_a = \frac{[H^+][B^-]}{[HB]} \qquad (9.34)$$

It is important to remember that the terms *weak* and *strong* apply to the extent of dissociation and not to the concentration of an acid or base. For example, gastric juice (0.05% HCl) is a dilute (not weak) solution of a strong acid.

Example 9.12 Dissociation Reactions of Acids

Write dissociation reactions and expressions for K_a for each of the following weak acids:

a. Hydrocyanic acid (HCN)
b. Phosphoric acid (H_3PO_4) (1st H only)
c. Dihydrogen phosphate ion ($H_2PO_4^-$) (1st H only)

Solution

In each case, H^+ has been substituted for H_3O^+.

a. $HCN \leftrightarrows H^+ + CN^-$; $K_a = \dfrac{[H^+][CN^-]}{[HCN]}$

b. $H_3PO_4 \leftrightarrows H^+ + H_2PO_4^-$; $K_a = \dfrac{[H^+][H_2PO_4^-]}{[H_3PO_4]}$

c. $H_2PO_4^- \leftrightarrows H^+ + HPO_4^{2-}$; $K_a = \dfrac{[H^+][HPO_4^{2-}]}{[H_2PO_4^-]}$

✔ **LEARNING CHECK 9.16** Write dissociation reactions and K_a expressions for the following weak acids:

a. Hydrogen phosphate ion (HPO_4^{2-})
b. Nitrous acid (HNO_2)
c. Hydrofluoric acid (HF)

monoprotic acid An acid that gives up only one proton (H^+) per molecule when dissolved.

diprotic acid An acid that gives up two protons (H^+) per molecule when dissolved.

triprotic acid An acid that gives up three protons (H^+) per molecule when dissolved.

Acid behavior is linked to the loss of protons. Thus, acids must contain hydrogen atoms that can be removed to form H^+. **Monoprotic acids** can lose only one proton per molecule, whereas **diprotic** and **triprotic** acids can lose two and three, respectively. For example, HCl is monoprotic, H_2SO_4 is diprotic, and H_3PO_4 is triprotic. Di- and triprotic acids dissociate in steps, as shown for H_2SO_4 in Equations 9.35 and 9.36:

$$H_2SO_4(aq) \rightleftharpoons H^+(aq) + HSO_4^-(aq) \tag{9.35}$$

$$HSO_4^-(aq) \leftarrows H^+(aq) + SO_4^{2-}(aq) \tag{9.36}$$

The second proton is not as easily removed as the first because it must be pulled away from a negatively charged particle, HSO_4^-. Accordingly, HSO_4^- is a weaker acid than H_2SO_4.

The number of ionizable hydrogens cannot always be determined from the molecular formula for an acid. For example, acetic acid ($HC_2H_3O_2$) is monoprotic even though the molecule contains four hydrogen atoms. The dissociation of acetic acid is represented by Equation 9.37, where structural formulas are used to emphasize the different H atoms in the molecule:

$$\begin{array}{c} \text{H} \quad \text{O} \\ | \quad \parallel \\ \text{H}-\text{C}-\text{C}-\text{O}-\text{H} \\ | \\ \text{H} \end{array} \rightleftharpoons \text{H}^+ + \begin{array}{c} \text{H} \quad \text{O} \\ | \quad \parallel \\ \text{H}-\text{C}-\text{C}-\text{O}^- \\ | \\ \text{H} \end{array} \tag{9.37}$$

Only the hydrogen bound to the oxygen is ionizable. Those hydrogens bound to C are too tightly held to be removed. **Table 9.8** contains other examples, with the ionizable hydrogens shown in color.

We have focused our attention on the strength of acids, using the extent of dissociation as a basis. However, all acid dissociations are reversible to some degree, and in the reverse

TABLE 9.8 Examples of Monoprotic, Diprotic, and Triprotic Acids

Name	Formula	Structural Formula	Classification
Butyric acid	$HC_4H_7O_2$		Monoprotic
Carbonic acid	H_2CO_3		Diprotic
Formic acid	$HCHO_2$		Monoprotic
Nitric acid	HNO_3		Monoprotic
Phosphoric acid	H_3PO_4		Triprotic
Phosphorous acid	H_3PO_3		Diprotic

reactions, anions produced by the forward reaction behave as Brønsted bases. What can be said about the strength of these bases? Because Brønsted acid–base behavior is really just competition for protons, we can answer this question by looking again at the simplified form of the equation for dissociation of a general acid:

$$HB(aq) \rightleftharpoons H^+(aq) + B^-(aq) \tag{9.38}$$

We see from this equation that the strength of HB as an acid depends on how tightly the conjugate base B^- holds onto the proton. If HB is a strong acid, the conjugate base holds onto the proton only weakly. If HB is a weak acid, the conjugate base holds on more strongly, depending on the strength of HB. Thus, we have answered our earlier question. If HB is a strong acid, the H^+ is held only weakly by the B^-. We can conclude then that B^- is not strongly attracted to protons—it is a weak base. Conversely, if HB is a weak acid, the H^+ is held tightly by the B^-, and we conclude that B^- is more strongly attracted to protons—it is a stronger base than the B^- from a strong acid.

In general, the conjugate base anions produced by the dissociation of strong Brønsted acids are weak Brønsted bases. The conjugate base anions of weak acids are stronger bases, with their strengths dependent on the strength of the parent acid.

STUDY SKILLS 9.1

Writing Reactions of Acids

As you study this chapter, you will acquire a knowledge of the characteristic reactions of acids and the ability to write balanced equations for the reactions. In the Key Equations section at the end of the chapter, five characteristic reactions are summarized in item 3. One useful way to remember the reactions is to learn them as general word equations, such as

"An acid plus a metal gives a salt plus hydrogen gas"

rather than as specific equations such as $H_2SO_4(aq) + Zn(s) \rightarrow ZnSO_4(aq) + H_2(g)$. If you learn the general word equations and you recognize the starting materials for a reaction (such as an acid and a metal), you will know what the products will be (a salt and H_2 gas).

A second approach is to remember that all five of the general reactions of acids are either single-replacement

$$\text{acid} + \text{metal} \rightarrow \text{salt} + H_2$$

or double-replacement (the remaining four acid reactions in the Key Equations section). The formulas of the products of double-replacement reactions can be predicted by simply breaking each reactant into its positive and negative parts and recombining the parts in the other possible way (the positive part of one reactant with the negative part of the other reactant). For example, let's determine the products and the balanced equation for a reaction between H_2SO_4 and KOH. If you remember that H_2SO_4 is an acid and KOH is a base, the general word equation says the products will be a salt and water:

$$H_2SO_4(aq) + KOH(aq) \rightarrow \text{salt}(aq) + H_2O(\ell)$$

H_2SO_4 breaks apart to give H^+ and SO_4^{2-}; KOH breaks apart to give K^+ and OH^-. If we combine the positive part of the acid (H^+) with the negative part of the base (OH^-), we get the water (H_2O). A similar combination of the positive part

of the base (K^+) with the negative part of the acid (SO_4^{2-}) gives the salt. We must remember that the total charges of the combined parts must add up to 0. Thus, two K^+ will combine with one SO_4^{2-} to give the salt K_2SO_4. The balanced equation is

$$H_2SO_4(aq) + 2KOH(aq) \rightarrow K_2SO_4(aq) + 2H_2O(\ell)$$

As another example, consider a reaction between HBr and KOH. KBr breaks into H^+ and Br^-. KOH breaks into K^+ and OH^-. Now, switch the parts and recombine:

The products are KBr and H_2O, and the balanced equation is

$$HBr(aq) + KOH(aq) \rightarrow KBr(aq) + H_2O(\ell)$$

For a final example, let's try a reaction between HBr and $NaHCO_3$. The word equation predicts that the products should be a salt, water, and carbon dioxide gas. HBr breaks into H^+ and Br^-; $NaHCO_3$ breaks into Na^+ and HCO_3^-. Now, switch the parts and recombine:

The products are NaBr and H_2CO_3. However, H_2CO_3 is not stable; it decomposes to give H_2O and CO_2, the products predicted earlier. The balanced equation is

$$HBr(aq) + NaHCO_3(aq) \rightarrow NaBr(aq) + H_2O(\ell) + CO_2(g)$$

Ammonia (NH_3) is the weak base most often encountered in addition to the anions of strong acids. The dissociation reaction of gaseous NH_3 in water, given earlier in Example 9.1, is

$$NH_3(aq) + H_2O(\ell) \rightleftharpoons NH_4^+(aq) + OH^-(aq) \tag{9.39}$$

The most common strong bases are the hydroxides of group IA(1) metals (NaOH, KOH, etc.) and the hydroxides of group IIA(2) metals ($Mg(OH)_2$, $Ca(OH)_2$, etc.).

Example 9.13 Strengths of Acids

Classify each of the following pairs by identifying the stronger of the pair according to the indicated behavior. Information from Tables 9.7 and 9.8 may be used.

a. H_3PO_4 and $H_2PO_4^-$ (acid)
b. $H_2PO_4^-$ and HPO_4^{2-} (base)
c. HNO_3 and HNO_2 (acid)

Beware of Heartburn

The medical name for acid reflux disease is *gastroesophageal reflux disease,* which is often abbreviated and referred to as GERD. The disease is often mistaken for occasional heartburn and treated with over-the-counter remedies. GERD is the result of a malfunctioning muscle (the lower esophageal sphincter [LES] muscle) located at the bottom of the esophagus, just above the stomach. When operating normally, this muscle relaxes and opens to allow food to pass from the esophagus down into the stomach, then contracts to close the opening and prevent the acidic contents of the stomach from backing up into the esophagus.

When the muscle relaxes at inappropriate times, the acidic stomach contents get into the esophagus and cause the burning chest pain called heartburn. However, when this occurs repeatedly and frequently, the acidic stomach contents can also erode the lining of the esophagus. GERD is a complex condition with many degrees of severity, ranging from only frequent heartburn symptoms to erosive esophagitis, in which the esophagus can suffer different degrees of damage. In extreme cases of erosive esophagitis, ulcers develop in the esophagus and lead to esophageal bleeding that, if persistent and undetected, can lead to iron deficiency and anemia as well as extreme pain and weight loss. In some cases, severe GERD can lead to other serious medical conditions that require hospitalization and even surgery to correct.

Doctors often recommend lifestyle and dietary changes for most GERD patients, including the avoidance of foods and beverages that weaken the LES muscle. These foods include chocolate, peppermint, fatty foods, coffee, and alcoholic beverages. The use of foods and beverages that can irritate a damaged esophageal lining, such as acidic fruits and juices, pepper, and tomato products is also discouraged. GERD symptoms in overweight individuals often diminish when some weight is lost. Smokers who quit also generally gain some relief. Prescription medications that reduce the amount of acid in the stomach are also available. Two types of medication are available. Both types, called H2 blockers and proton (acid) pump inhibitors, decrease the amount of acid secreted into the stomach, but by different mechanisms.

The main symptom that may indicate the presence of GERD is frequent, persistent heartburn that occurs two or more times a week. Other symptoms include difficulty in swallowing and frequent belching and regurgitation. Less common symptoms that occur in some people resemble respiratory conditions and include a persistent sore throat, wheezing, chronic coughing, and hoarseness. Individuals who suspect they might be suffering from some degree of GERD should consult a physician to determine the extent of the disease and proper treatment.

Custom Medical Stock Photo/Alamy Stock Photo

GERD has many degrees of severity.

Solution

a. H_3PO_4 is stronger as an acid. In di- and triprotic acids, each proton in the removal sequence is harder to remove, so the corresponding acid is weaker.

b. The stronger acid produces the weaker anion base. Thus, $H_2PO_4^-$, the anion of the stronger acid H_3PO_4, would be a weaker base than HPO_4^{2-}, the anion of the weaker acid $H_2PO_4^-$. So HPO_4^{2-} is a stronger base than $H_2PO_4^-$.

c. HNO_3 is the stronger acid according to Table 9.7. Generally, when related acids (same atoms, etc.) are compared for strength, the one containing more oxygen atoms will be the stronger.

✔ **LEARNING CHECK 9.17** Classify each of the following according to strength for the indicated behavior. If more than two are compared, list them with the strongest at the top and the weakest at the bottom. Use Tables 9.7 and 9.8 as needed.

a. $HClO$, $HClO_3$, $HClO_2$ (acid)
b. NO_2^- and NO_3^- (base)
c. $HC_2H_3O_2$ and $C_2H_3O_2^-$ (acid)

9.10 | Analyzing Acids and Bases

Learning Objective

10. Demonstrate an understanding of the titration technique used to analyze acids and bases.

The analysis of solutions for the total amount of acid or base they contain is a regular activity in many laboratories. The total amount of acid in a solution is indicated by its capacity to neutralize a base. The pH is related to the acidity or concentration of H^+ in solution, while the capacity to neutralize a base depends on the total amount of H^+ available. For example, a 0.10 M acetic acid solution has an H^+ concentration of about 1.3×10^{-3} M (pH = 2.89). However, 1 L of the solution can neutralize 0.10 mol of OH^-, not just 1.3×10^{-3} mol. The reason is that the dissociation equilibrium of acetic acid is

$$HC_2H_3O_2(aq) \rightleftarrows H^+(aq) + C_2H_3O_2^-(aq) \tag{9.40}$$

As OH^- is added, H^+ reacts to form water (see **Equation 9.24**). The removal of H^+ causes the equilibrium of Reaction 9.40 to shift right in accordance with Le Châtelier's principle. The continued addition of OH^- and removal of H^+ will eventually cause all of the acetic acid molecules to dissociate and react.

A common procedure often used to analyze acids and bases is called **titration** (see **Figure 9.4**). Suppose the total acidity of an unknown acid solution needs to be

> **titration** An analytical procedure in which one solution (often a base) of known concentration is slowly added to a measured volume of an unknown solution (often an acid). The volume of the added solution is measured with a buret.

Figure 9.4 Titration.

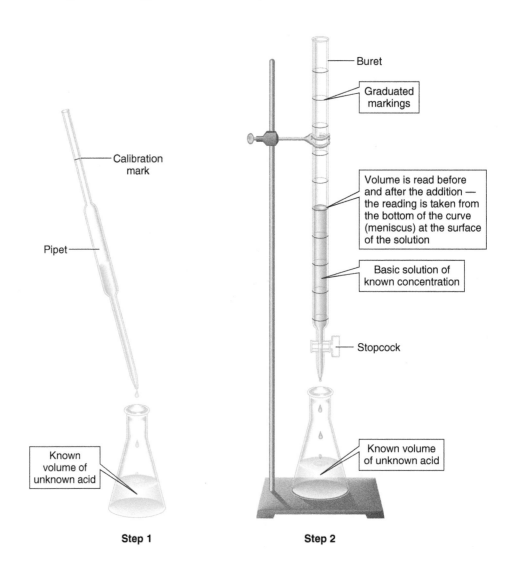

determined. A known volume of the acidic solution is first measured out by drawing it up to the calibration mark of a pipet. This solution is placed in a container (Step 1). A basic solution of known concentration (a standard solution) is added to the acid solution in the container until the **equivalence point of a titration** is reached. This is the point where the unknown acid is completely reacted with base. The volume of base needed to reach the equivalence point is obtained from the buret readings.

equivalence point of a titration The point at which the unknown solution has exactly reacted with the known solution. Neither is in excess.

To successfully complete a titration, the point at which the reaction is completed must somehow be detected. One way to do this is to add an indicator to the solution being titrated. An indicator is an organic compound that changes to different colors depending on the pH of its surroundings. An indicator is selected that will change color at a pH as close as possible to the pH the solution will have at the equivalence point. If the acid and base are both strong, the pH at that point will be 7. However, for reasons discussed later, the pH is not always 7 at the equivalence point. The point at which the indicator changes color and the titration is stopped is called the **endpoint of a titration. Figure 9.5** gives the colors shown by a number of indicators at various pH values.

endpoint of a titration The point at which the titration is stopped on the basis of an indicator color change or pH meter reading.

Figure 9.5 Indicators change color with changes in pH (the numbers across the bottom). Methyl red (top row) goes from red at low pH to yellow at higher pH. Bromthymol blue (middle row) goes from yellow to blue, and phenolphthalein (bottom row) goes from colorless to red. Which of the three indicators would be the best to use to differentiate between the solutions with pH values of 4 and 6?

Indicator papers (litmus paper, pH paper, etc.) are often used to make routine pH measurements. These papers are impregnated with one or more indicators that change to a variety of colors depending on the pH. However, indicators may not be practical under certain conditions. For example, there might not be any indicator that changes color close enough to the equivalence point, or the solution being titrated might be so highly colored that an indicator color change cannot be detected. Under such circumstances, a pH meter can be used. The electrodes of the meter are placed in the solution being titrated, and the pH is read directly (see **Figure 9.6**). The titration is continued until the meter reading matches the pH of the equivalence point.

© Spencer L. Seager

1 ● At the beginning, the pH meter gives the pH of the acid solution being titrated.

© Spencer L. Seager

2 ● Partway through the titration, the pH meter reading is of a solution of unreacted acid and the salt produced by the reaction.

© Spencer L. Seager

3 ● At the end of the titration, the pH meter gives the pH of the salt solution formed by the complete reaction of acid with base.

Figure 9.6 An acid–base titration using a pH meter to detect the equivalence point.

9.11 | Titration Calculations

Learning Objective

11. Do calculations related to the analysis of acids and bases by titration.

Equations 9.41 and 9.42 represent the reactions that occur when solutions of nitric and sulfuric acid are titrated with sodium hydroxide:

$$HNO_3(aq) + NaOH(aq) \rightarrow H_2O(\ell) + NaNO_3(aq) \tag{9.41}$$

$$H_2SO_4(aq) + 2NaOH(aq) \rightarrow 2H_2O(\ell) + Na_2SO_4(aq) \tag{9.42}$$

One mole of HNO_3 requires 1 mol of NaOH for a complete reaction, but 1 mol of H_2SO_4 requires 2 mol of NaOH. Because these are solution reactions, stoichiometric calculations can be done using the methods described in **Section 7.6**, page 227.

Example 9.14 Titration Calculations

Calculate the molarity of the HNO_3 and H_2SO_4 solutions involved in the following titrations:

a. A 25.0-mL sample of an HNO_3 solution requires the addition of 16.3 mL of 0.200 M NaOH to reach the equivalence point.
b. A 25.0-mL sample of an H_2SO_4 solution requires the addition of 32.6 mL of 0.200 M NaOH to reach the equivalence point.

Solution

In each case we know the volume of acid solution reacted. If we also knew the number of moles of acid in the volume of reacted solution, we could calculate the solution molarity by using Equation 7.5:

$$M = \frac{\text{moles of solute}}{\text{liters of solution}}$$

Therefore, our task will be to calculate the number of moles of acid reacted in each case. The methods described in Section 7.6 will be used. We will begin with the volume of NaOH solution used in each titration, and convert that volume to moles of acid using the factor-unit method.

a. Nitric acid

 Step 1. 0.0163 L NaOH solution
 Step 2. 0.0163 L NaOH solution = mol HNO_3

Step 3. $0.0163 \text{ L NaOH solution} \times \dfrac{0.200 \text{ mol NaOH}}{1 \text{ L NaOH solution}} \times \dfrac{1 \text{ mol HNO}_3}{1 \text{ mol NaOH}}$

$$= \text{mol HNO}_3$$

The factor $\dfrac{0.200 \text{ mol NaOH}}{1 \text{ L NaOH solution}}$ came from the molarity of the NaOH solution, and the factor $\dfrac{1 \text{ mol HNO}_3}{1 \text{ mol NaOH}}$ came from the 1 to 1 reaction ratio of HNO_3 with NaOH.

Step 4. $0.0163 \times \dfrac{0.200}{1} \times \dfrac{1 \text{ mol HNO}_3}{1} = 0.00326 \text{ mol HNO}_3$

This result tells us that the 25.0 mL titrated sample of HNO_3 contained 0.00326 moles of HNO_3. After converting the HNO_3 volume to liters, the molarity can be calculated:

$$M = \frac{\text{moles of HNO}_3}{\text{liters of HNO}_3 \text{ solution}} = \frac{0.00326 \text{ mol HNO}_3}{0.0250 \text{ L HNO}_3 \text{ solution}} = \frac{0.130 \text{ mol HNO}_3}{\text{L HNO}_3 \text{ solution}}$$

$$= 0.130 \text{ M HNO}_3$$

b. Sulfuric acid

This calculation follows the same pattern as the HNO_3 calculation, except for the fact that the reaction ratio of NaOH and H_2SO_4 is 2 mol of NaOH reacts with 1 mol of H_2SO_4. This gives a factor of $\dfrac{1 \text{ mol H}_2SO_4}{2 \text{ mol NaOH}}$ that will be used in Step 3.

Step 1. 0.0326 L NaOH solution

Step 2. $0.0326 \text{ L NaOH solution}$ $= \text{mol H}_2SO_4$

Step 3. $0.0326 \text{ L NaOH solution} \times \dfrac{0.200 \text{ mol NaOH}}{1 \text{ L NaOH solution}} \times \dfrac{1 \text{ mol H}_2SO_4}{2 \text{ mol NaOH}}$

$$= \text{mol H}_2SO_4$$

Step 4. $0.0326 \times \dfrac{0.200}{1} \times \dfrac{1 \text{ mol H}_2SO_4}{2} = 0.00326 \text{ mol H}_2SO_4$

The molarity of the solution is calculated as before:

$$M = \frac{\text{moles of H}_2SO_4}{\text{liters of H}_2SO_4 \text{ solution}} = \frac{0.00326 \text{ mol H}_2SO_4}{0.0250 \text{ L H}_2SO_4 \text{ solution}} = \frac{0.130 \text{ mol H}_2SO_4}{\text{L H}_2SO_4 \text{ solution}}$$

$$= 0.130 \text{ M H}_2SO_4$$

Thus, we see that the two acid solutions have the same molar concentrations. Why did the second require twice the volume of NaOH solution for the titration?

✔ **LEARNING CHECK 9.18** Calculate the molarity of a solution of phosphoric acid (H_3PO_4) if a 25.0-mL sample of acid solution requires 14.1 mL of a 0.250 M NaOH solution to titrate to the equivalence point. The equation for the reaction is

$$H_3PO_4(aq) + 3NaOH(aq) \rightarrow 3H_2O(aq) + Na_3PO_4(aq)$$

9.12 | Hydrolysis Reactions of Salts

Learning Objective

12. Explain the concept of salt hydrolysis, and write equations to illustrate the concept.

hydrolysis reaction Any reaction with water. For salts it is a reaction of the acidic cation and/or basic anion of the salt with water.

In general, a **hydrolysis reaction** is a reaction with water. Many types of hydrolysis reactions are known, but at this point we will discuss only one, the hydrolysis of salts. In Section 9.10, we said that the pH at the equivalence point is not 7 for all acid–base titrations. However, we pointed out earlier that the only products of acid–base reactions are water and a salt. Therefore, it seems reasonable to conclude that some salts and water must interact (a hydrolysis reaction) and cause the solution pH to differ from that of pure water.

Suppose three solutions are prepared by dissolving equal molar amounts of sodium acetate, ammonium chloride, and sodium chloride in identical volumes of pure water. Measurement of pH shows that the sodium acetate solution is alkaline (pH higher than 7), the ammonium chloride solution is acidic (pH lower than 7), and the sodium chloride solution is neutral (pH = 7). The alkaline sodium acetate solution must contain more OH^- than H^+ ions. However, the only source of both ions in a solution of salt and water is the dissociation of water, so both should be present in equal amounts. The equation for the water dissociation is

$$H_2O(\ell) \rightleftharpoons H^+(aq) + OH^-(aq) \tag{9.43}$$

The OH^- excess would result if something removed H^+ from this equilibrium, causing it to shift right and produce more OH^-. This is exactly what happens. The acetate anion is the conjugate base of a weak acid (acetic acid), and it is a strong enough Brønsted base to react with the H^+ ions of the above equilibrium:

$$C_2H_3O_2^-(aq) + H^+(aq) \rightleftharpoons HC_2H_3O_2(aq) \tag{9.44}$$

The overall result of this reaction can be seen by adding Equations 9.43 and 9.44:

$$H_2O(\ell) + C_2H_3O_2^-(aq) + H^+(aq) \rightleftharpoons H^+(aq) + OH^-(aq) + HC_2H_3O_2(aq) \tag{9.45}$$

Net reaction: $H_2O(\ell) + C_2H_3O_2^-(aq) \rightleftharpoons OH^-(aq) + HC_2H_3O_2^-(aq) \tag{9.46}$

Now it can be seen that the excess OH^- comes from a reaction between the acetate ion and water. But what about the Na^+ ion that is also in the solution? Does it react with water? The Na^+ cation is the conjugate acid of the strong base NaOH. As a result, Na^+ is an extremely weak Brønsted acid and makes no contribution to the pH of the solution.

In general, salts influence the pH of water solutions as follows: (1) Salts containing an anion of a weak acid and a cation of a strong base will form alkaline water solutions. (2) Salts containing an anion of a strong acid and a cation of a weak base will form acidic solutions. (3) Salts containing an anion and a cation from equal-strength acids and bases (both weak or both strong) will form neutral solutions (see **Figure 9.7**).

The acidic ammonium chloride solution is an example of the second category of salts, where the NH_4^+ cation is the conjugate acid of the weak base NH_3, and the Cl^- anion is the conjugate base of the strong acid HCl. The hydrolysis reaction is given in Equation 9.47, where H_3O^+ is used to emphasize the reaction with water:

$$NH_4^+(aq) + H_2O(\ell) \rightleftharpoons H_3O^+(aq) + NH_3(aq) \tag{9.47}$$

The neutral sodium chloride solution is an example of the third category. Both NaOH and HCl are strong, and an NaCl solution is neutral.

✔ **LEARNING CHECK 9.19** Predict the relative pH value (higher than 7, lower than 7, etc.) for water solutions of the following salts. Table 9.7 may be useful.

a. Sodium nitrate ($NaNO_3$)
b. Sodium nitrite ($NaNO_2$)
c. Potassium borate (K_3BO_3)

Water
(H₂O)

Sodium Acetate
(CH₃COONa)

© Jeffrey M. Seager

Figure 9.7 Samples of pure water (left) and sodium acetate dissolved in water (right) behave differently when phenolphthalein indicator is added. The acetate ion hydrolyzes in water to form a basic solution that turns phenolphthalein to a pink color. Why didn't the Na^+ ion from the sodium acetate hydrolyze?

9.13 | Buffers

Learning Objective

13. Explain how buffers work, and write equations to illustrate their action.

A **buffer** is a solution that has the ability to resist changes in pH when acids or bases are added (see **Figure 9.8**). Most buffers consist of a pair of compounds, one with the ability to react with H^+ and the other with the ability to react with OH^-. An example is a mixture of acetic acid and its salt, sodium acetate. The acetate ion from sodium acetate is the conjugate base of acetic acid and reacts with any added acid (H^+ ions):

$$C_2H_3O_2^-(aq) + H^+(aq) \rightleftharpoons HC_2H_3O_2(aq) \qquad (9.48)$$

Any added base (OH^- ions) reacts with nonionized acetic acid molecules:

$$HC_2H_3O_2(aq) + OH^-(aq) \rightleftharpoons C_2H_3O_2^-(aq) + H_2O(\ell) \qquad (9.49)$$

buffer A solution with the ability to resist changing pH when acids (H^+) or bases (OH^-) are added.

① The solution on the left is not buffered; the one on the right is; universal indicator has been added to each solution.

② Sodium hydroxide has been added to each solution.

③ Hydrochloric acid has been added to two fresh samples that originally looked like the first pair of samples.

Figure 9.8 Buffered solutions resist changing pH when bases and acids are added. What observation of this experiment indicates that the pH of the buffered solutions did not change significantly when base or acid was added?

In this way, the buffer solution resists changes in pH. The amount of H$^+$ or OH$^-$ that a buffer system can absorb without allowing significant pH changes to occur is called its **buffer capacity**.

Buffered solutions are found extensively in the body, where they serve to protect us from the disastrous effects of large deviations in pH. An important buffer system in blood is composed of carbonic acid (H$_2$CO$_3$) and bicarbonate salts such as NaHCO$_3$. The unstable carbonic acid results when dissolved CO$_2$ reacts with water in the blood:

$$H_2O(\ell) + CO_2(aq) \rightleftharpoons H_2CO_3(aq) \tag{9.50}$$

Added H$^+$ is neutralized by reacting with HCO$_3^-$ from the dissolved bicarbonate salts:

$$HCO_3^-(aq) + H^+(aq) \rightleftharpoons H_2CO_3(aq) \tag{9.51}$$

The carbonic acid protects against added OH$^-$:

$$H_2CO_3(aq) + OH^-(aq) \rightleftharpoons HCO_3^-(aq) + H_2O(\ell) \tag{9.52}$$

If large amounts of H$^+$ or OH$^-$ are added to a buffer, the buffer capacity can be exceeded, the buffer system is overwhelmed, and the pH changes. For example, if large amounts of H$^+$ were added to the bicarbonate ion–carbonic acid buffer, Reaction 9.51 would take place until the HCO$_3^-$ was depleted. The pH would then drop as additional H$^+$ ions were added. In blood, the concentration of HCO$_3^-$ is ten times the concentration of H$_2$CO$_3$. Thus, this buffer has a greater capacity against added acid than against bases. This is consistent with the normal functions of the body that cause larger amounts of acidic than basic substances to enter the blood.

The pH of buffers can be calculated by using a form of Equation 9.34 that was given earlier and is repeated here:

$$K_a = \frac{[H^+][B^-]}{[HB]}$$

This equation can be rearranged to give

$$[H^+] = K_a \frac{[HB]}{[B^-]} \tag{9.53}$$

Application of the logarithm concept to Equation 9.53 gives

$$pH = pK_a + \log \frac{[B^-]}{[HB]} \tag{9.54}$$

In this equation, pH $= -\log[H^+]$ (Equation 9.11), and **pK$_a$** $= -\log K_a$. Equation 9.54 is known as the **Henderson–Hasselbalch equation**; it is often used by biologists, biochemists, and others who frequently work with buffers. We see from Equation 9.54 that when the concentrations of a weak acid and its conjugate base (anion) are equal in a solution, the pH of the solution is equal to pK$_a$. When it is desired to produce a buffer with a pH different from the exact pK$_a$, Equation 9.54 indicates it can be done. An acid with a pK$_a$ near the desired pH is selected, and the ratio of the concentrations of conjugate base (anion of the acid) and acid is adjusted to give the desired pH. When a weak acid and its salt (the source of the acid's conjugate base anion) are mixed in solution, you can generally assume that the amount of acid that dissociates is small and may be neglected. This means that the buffer concentrations of the acid and anion (the salt of acid) are "equal" to the made-up concentrations. **Table 9.9** lists some weak acids, together with values of K$_a$ and pK$_a$.

Example 9.15 Buffer Solution Calculations

a. Calculate the pH of a buffer solution that contains 0.10 mol acetic acid (CH$_3$COOH) and 0.10 mol sodium acetate (CH$_3$COONa) per liter.

b. What is the pH of a buffer in which the concentration of NaH$_2$PO$_4$ is 0.10 M and that of Na$_2$HPO$_4$ is 0.50 M?

TABLE 9.9 K_a and pK_a Values for Selected Weak Acids

Name	Formula	K_a	pK_a
Acetic acid[a]	CH_3COOH	1.8×10^{-5}	4.74
Ammonium ion	NH_4^+	5.6×10^{-10}	9.25
Boric acid	H_3BO_3	7.3×10^{-10}	9.14
Dihydrogen borate ion	$H_2BO_3^-$	1.8×10^{-13}	12.75
Hydrogen borate ion	HBO_3^{2-}	1.6×10^{-14}	13.80
Carbonic acid	H_2CO_3	4.3×10^{-7}	6.37
Bicarbonate ion	HCO_3^-	5.6×10^{-11}	10.25
Citric acid[a]	$C_3H_4(OH)(COOH)_3$	8.4×10^{-4}	3.08
Dihydrogen citrate ion[a]	$C_3H_4(OH)(COOH)_2COO^-$	1.8×10^{-5}	4.74
Hydrogen citrate ion[a]	$C_3H_4(OH)(COOH)(COO)_2^{2-}$	4.0×10^{-6}	5.40
Formic acid[a]	$HCOOH$	1.8×10^{-4}	3.74
Lactic acid[a]	$C_2H_4(OH)COOH$	1.4×10^{-4}	3.85
Nitrous acid	HNO_2	4.6×10^{-4}	3.33
Phosphoric acid	H_3PO_4	7.5×10^{-3}	2.12
Dihydrogen phosphate ion	$H_2PO_4^-$	6.2×10^{-8}	7.21
Hydrogen phosphate ion	HPO_4^{2-}	2.2×10^{-13}	12.66
Sulfurous acid	H_2SO_3	1.5×10^{-2}	1.82
Bisulfite ion	HSO_3^-	1.0×10^{-7}	7.00

[a]The hydrogen that ionizes in organic acids and ions is a part of a carboxylic acid group, represented by COOH.

c. A buffer system consisting of the ions $H_2PO_4^-$ and HPO_4^{2-} helps control the pH of urine. What value of the ratio $[HPO_4^{2-}]/[H_2PO_4^-]$ would be required to maintain a pH of 6.00, the average value for normal urine?

Solution

a. The acetic acid concentration is 0.10 M and that of its conjugate base, the acetate ion from the sodium acetate salt, is also 0.10 M. The pK_a for acetic acid is 4.74:

$$pH = pK_a + \log\frac{[CH_3COO^-]}{[CH_3COOH]} = 4.74 + \log\left(\frac{0.10 \text{ mol/L}}{0.10 \text{ mol/L}}\right)$$

$$= 4.74 + \log(1) = 4.74 + 0 = 4.74$$

We see that, as mentioned earlier, when the acid and conjugate base concentrations are equal, pH = pK_a.

b. In this case, both compounds are salts that produce the $H_2PO_4^-$ and HPO_4^{2-} ions. In such situations, the ion with more hydrogens will be the acid, and the one with fewer will be the anion or conjugate base. Therefore, the acid is $H_2PO_4^-$ at a concentration of 0.10 M, and the conjugate base or anion is HPO_4^{2-} at a concentration of 0.50 M. The pK_a for $H_2PO_4^-$ is 7.21:

$$pH = pK_a + \log\frac{[HPO_4^{2-}]}{[H_2PO_4^-]} = 7.21 + \log\left(\frac{0.50 \text{ mol/L}}{0.10 \text{ mol/L}}\right)$$

$$= 7.21 + \log(5.00) = 7.21 + 0.70 = 7.91$$

Q: Does diet play a role in peptic ulcer disease?

Melina B. Jampolis, M.D., is a Physician Nutrition Specialist focusing on nutrition for weight loss and disease prevention and treatment.

A: Peptic ulcer disease (PUD), which includes ulcer formation in the stomach or the duodenum (the first portion of the small intestine), affects a significant portion of the U.S. population and is the leading reason for gastrointestinal surgery. Since PUD affects your digestive tract, it makes sense that what you eat on a daily basis should impact the development of ulcer disease. However, direct associations between what many people believe to be ulcer-causing components of the diet and the development of PUD are surprisingly limited and inconsistent. Large observational studies have not established a direct link between the intake of alcohol, caffeine, or spicy foods and ulcer formation. This is true despite the fact that alcohol and caffeine are known to stimulate gastric acid secretion, which in turn makes ulcer symptoms worse and interferes with ulcer healing. Research over the past few decades has found that 80–90% of peptic ulcers are actually caused by a bacterium called *Helicobacter pylori* (*H. pylori*). So while diet might play a role, particularly in terms of ulcer healing, *H. pylori* infection is the primary cause of PUD.

Milk, which many people still mistakenly believe protects against or even helps to treat PUD, might actually increase its severity by causing an increase in acid secretion and masking ulcer symptoms. Masked symptoms allow the disease to progress without treatment. On the other hand, fermented milk products, including yogurt and cheese, have been associated with a reduction in PUD, possibly due to their antimicrobial properties. Research shows that probiotic strains, including *Lactobacillus* and *Bifidobacterium bifidum*, inhibit the growth of the *H. pylori*. These probiotic strains have also been shown to improve the effectiveness of treatment for *H. pylori* infection.

In the past, fiber consumption was discouraged for patients with PUD. Current research suggests that soluble fiber might actually play a protective role. In a large prospective study, those with the highest intakes of soluble fiber, particularly fiber from legumes, including beans, tofu, peanut butter, and nuts, had a 50% lower incidence of duodenal ulcer compared to those consuming the least amount of soluble fiber. Vegetable fiber and vitamin A intake also appear to exert a protective effect. Conversely, higher intakes of refined grains (which are lower in fiber) and refined sugar have been associated with a higher risk of duodenal ulcer disease.

Polyphenols, plant-based nutrients that play a role in decreasing the risk of numerous diseases, also appear to play an important role in ulcer prevention. Animal studies have shown that several different polyphenols help prevent PUD by increasing the formation of gut-protecting mucus; decreasing acid secretion; inhibiting the growth of *H. pylori*; and acting as powerful antioxidants, which help protect the lining of the gut. Promising research in rats has shown a potential beneficial effect of numerous polyphenol-rich plant compounds, including ginger, fennel, and olive leaf extract.

Smoking has been strongly associated with the incidence of PUD through numerous mechanisms. On the other hand, physical activity has been shown to decrease the risk of duodenal ulcer. So, the adoption of a healthy lifestyle, including a diet containing a variety of fruits and vegetables, plenty of soluble fiber, and fermented dairy products with limited added sugar; not smoking; and exercising regularly can play a significant role in the prevention of ulcer disease.

Peptic ulcer disease

Fermented dairy products are good for PUD

Ginger is good for PUD

Fennel is good for PUD

Milk may increase the severity of PUD

Patients with PUD should avoid refined sugars

A B C D E F

Veronika Burmeister/Visuals Unlimited/Corbis · tacar/Shutterstock.com · Reika/Shutterstock.com · amst/Shutterstock.com · Jiri Hera/Shutterstock.com · Olga Danylenko /Shutterstock.com

c. As in part (b), the acid is $H_2PO_4^-$ and the conjugate base is HPO_4^{2-}. The desired pH is 6.00. Substitution into Equation 9.54 gives

$$pH = pK_a + \log \frac{[HPO_4^{2-}]}{[H_2PO_4^-]}$$

$$6.00 = 7.21 + \log \frac{[HPO_4^{2-}]}{[H_2PO_4^-]}$$

$$6.00 - 7.21 = \log \frac{[HPO_4^{2-}]}{[H_2PO_4^-]}$$

$$-1.21 = \log \frac{[HPO_4^{2-}]}{[H_2PO_4^-]}$$

At this point, we know the value of the log of the desired ratio. To get the ratio, we must evaluate the antilog of -1.21. Use Step 3 given in Table 9.3 to get 0.062 as the value of the ratio. This result says the ratio $[HPO_4^{2-}]/[H_2PO_4^-]$ must be 0.062. In other words, the HPO_4^{2-} concentration must be only about 0.06 times the value of the $H_2PO_4^-$ concentration. Any concentrations giving this ratio would be satisfactory. For example, if $[H_2PO_4^-] = 0.50$ M, then $[HPO_4^{2-}]$ would have to equal 0.062×0.5 M, or 0.031 M.

✔ **LEARNING CHECK 9.20**

a. What is the pH of a buffer solution in which formic acid (HCOOH) and sodium formate (HCOONa) are both at a concentration of 0.22 M?
b. What is the pH of a buffer solution that is 0.25 M in sulfurous acid (H_2SO_3) and 0.10 M in sodium bisulfite ($NaHSO_3$)?
c. A buffer system in the blood consists of the bicarbonate ion, HCO_3^-, and carbonic acid, H_2CO_3. What value of the ratio $[HCO_3^-]/[H_2CO_3]$ would be required to maintain blood at the average normal pH of 7.40?

▶ Case Study Follow-up

Over-the-counter medications should be taken as directed. Some antacids contain large amounts of sodium, which might be problematic for someone on a sodium-restricted diet, and calcium-containing antacids might exacerbate existing kidney problems. Antacids should only be used for occasional relief. If a person is taking antacids on a daily basis for extended periods of time, a different medication, such as a proton-pump inhibitor, should be considered. Frequent long-lasting heartburn may indicate gastroesophageal reflux disease (GERD).

Symptoms of heartburn include a burning sensation in the chest area occurring after a meal or while lying down, or a sour taste in the mouth from stomach acid. Heart attack symptoms might be mistaken for heartburn. However, if chest pain occurs during exercise; is accompanied by shortness of breath, nausea, or dizziness; and is radiating to the arm, neck, shoulder, or jaw, heart attack is a possibility—and emergency help should be sought immediately.

Heartburn may be initiated or made worse by certain foods. Someone suffering from heartburn should avoid fatty foods, tomato sauce, chocolate, caffeine, and spicy foods. Other factors that aggravate heartburn are excess body weight, tight clothing, and lying down too soon after a meal. Maintaining a healthy weight and eating frequent small meals instead of infrequent large meals, especially before bedtime, will help control heartburn.

Concept Summary

The Arrhenius Theory Svante Arrhenius defined acids as substances that dissociate in water to provide hydrogen ions (H^+), and bases as substances that dissociate in water to provide hydroxide ions (OH^-).
Objective 1 (Section 9.1), Exercise 9.2

The Brønsted Theory Johannes Brønsted and Thomas Lowry proposed a theory in which acids are defined as any hydrogen-containing substances capable of donating protons to other substances. Bases are substances that accept and form covalent bonds with protons. When a substance behaves as a Brønsted acid by donating a proton, the substance becomes a conjugate base.
Objective 2 (Section 9.2), Exercises 9.6 and 9.10

Naming Acids Two types of acids are named differently. Water solutions of binary covalent compounds containing hydrogen and a nonmetal are named following the pattern hydro(stem)ic acid, where (stem) is the stem of the name of the nonmetal bonded to hydrogen. Acids in which hydrogen is bonded to polyatomic ions have names based on the name of the polyatomic ion to which hydrogen is bonded.
Objective 3 (Section 9.3), Exercise 9.22

The Self-Ionization of Water Water, a hydrogen-containing material, is able to behave as both a Brønsted acid and a Brønsted base. In pure water, a small number of water molecules (10^{-7} mol/L) donate protons to other water molecules.
Objective 4 (Section 9.4), Exercises 9.28a & b and 9.30a & b

The pH Concept The pH is the negative logarithm of the molar H^+ concentration of a solution. Solutions with pH values lower than 7 are acidic, those with pH values higher than 7 are basic or alkaline, and those with a pH value of 7 are neutral.
Objective 5 (Section 9.5), Exercises 9.36 and 9.40

Properties of Acids All acids have certain characteristic properties that include (1) a sour taste; (2) a reaction with water to produce H_3O^+; (3) reactions with solid metallic oxides, hydroxides, carbonates, and bicarbonates; and (4) a reaction with certain metals to give hydrogen gas.
Objective 6 (Section 9.6), Exercise 9.50

Properties of Bases Basic solutions feel soapy or slippery and change the color of litmus from red to blue. Their most characteristic chemical property is a neutralization reaction with acids to produce water and a salt.
Objective 7 (Section 9.7), Exercise 9.60

Salts At room temperature, salts are solid crystalline substances that contain the cation of a base and the anion of an acid. Hydrated salts contain specific numbers of water

molecules as a part of their crystalline structures. Salts can be prepared by reacting an appropriate acid with one of a number of other materials.
Objective 8 (Section 9.8), Exercises 9.68 and 9.74

The Strengths of Acids and Bases Acids and bases that dissociate completely when dissolved in solution are called strong, and those that do not are called weak or moderately weak, depending on the degree of dissociation they undergo. Acid strength is indicated by the value of K_a, the acid dissociation constant. In general, polyprotic acids become weaker as an acid during each successive dissociation reaction. The Brønsted base produced by the dissociation of an acid has a strength opposite that of the acid. The acid strengths of cations produced by the dissociation of bases follow a similar pattern.
Objective 9 (Section 9.9), Exercise 9.86

Analyzing Acids and Bases The neutralization reaction of acids and bases is used in a process called titration to analyze acids and bases. During a typical titration, a base solution of known concentration is added slowly to an acid solution of unknown concentration. The titration is stopped at the endpoint when a color change occurs in an indicator. The volumes of acid and base required are used to calculate the acid concentration.
Objective 10 (Section 9.10), Exercise 9.92

Titration Calculations Titrations are used to determine the total amount of acid or base in solutions. Data collected from titrations are treated like other stoichiometric data in calculations.
Objective 11 (Section 9.11), Exercises 9.98 and 9.100a

Hydrolysis Reactions of Salts The cation and anion of a salt may have the same or different Brønsted acid and base strengths. When they are the same, solutions of the salt have a pH of 7. When they are different, salt solutions have H^+ and OH^- concentrations that are unequal and pH values different than 7. The reactions of salt ions with water that cause these results are called hydrolysis reactions.
Objective 12 (Section 9.12), Exercise 9.108

Buffers Solutions with the ability to maintain essentially constant pH values when acid (H^+) or base (OH^-) are added are called buffers. All buffers have a limit to the amount of acid or base they can absorb without changing pH. This limit is called the buffer capacity.
Objective 13 (Section 9.13), Exercise 9.116

Key Terms and Concepts

Acid dissociation constant (9.9)
Acidic solution (9.4)
Activity series (9.6)
Anion (9.8)
Arrhenius acid (9.1)
Arrhenius base (9.1)
Basic or alkaline solution (9.4)
Brønsted acid (9.2)
Brønsted base (9.2)
Buffer (9.13)
Buffer capacity (9.13)
Cation (9.8)

Conjugate acid–base pair (9.2)
Conjugate base (9.2)
Diprotic acid (9.9)
Endpoint of a titration (9.10)
Equivalence point of a titration (9.10)
Equivalent of salt (9.8)
Henderson–Hasselbalch equation (9.13)
Hydrate (9.8)
Hydrolysis reaction (9.12)
Ion product of water (9.4)
Monoprotic acid (9.9)
Neutral (9.4)

Neutralization reaction (9.7)
pH (9.5)
pK_a (9.13)
Salt (9.8)
Strong acids and strong bases (9.9)
Titration (9.10)
Triprotic acid (9.9)
Water of hydration (9.8)
Weak (or moderately weak) acids and bases (9.9)

Key Equations

1.	Relationship between $[H^+]$ and $[OH^-]$ in water solutions (Section 9.4):	$K_w = [H_3O^+][OH^-] = 1.0 \times 10^{-14}\ (mol/L)^2$	Equation 9.10
2.	Relationships between pH and $[H^+]$ (Section 9.5):	$pH = -\log[H^+]$ $[H^+] = 1 \times 10^{-pH}$	Equation 9.11 Equation 9.12
3.	General reactions of acids (Section 9.8):	acid + metal → salt + H_2 acid + metal oxide → salt + H_2O acid + metal hydroxide → salt + H_2O acid + metal carbonate → salt + H_2O + CO_2 acid + metal bicarbonate → salt + H_2O + CO_2	Equation 9.26 Equation 9.27 Equation 9.28 Equation 9.29 Equation 9.30
4.	Dissociation constant for acids (Section 9.9):	$K_a = \dfrac{[H^+][B^-]}{[HB]}$	Equation 9.34
5.	Calculation of buffer pH, Henderson–Hasselbalch equation (Section 9.13):	$pH = pK_a + \log\dfrac{[B^-]}{[HB]}$	Equation 9.54

Exercises

Even-numbered exercises are answered in Appendix B.

Blue-numbered exercises are more challenging.

The Arrhenius Theory (Section 9.1)

9.1 Write the dissociation equations for the following that emphasize their behavior as Arrhenius acids:

 a. HI

 b. HBrO

 c. HCN

 d. $HClO_2$

9.2 Write the dissociation equations for the following that emphasize their behavior as Arrhenius acids:

 a. $HBrO_2$

 b. HS^-

 c. HBr

 d. $HC_2H_3O_2$ (only the 1st listed H dissociates)

9.3 Each of the following produces a basic solution when dissolved in water. Identify those that behave as Arrhenius bases and write dissociation equations to illustrate that behavior.

 a. CsOH

 b. CH_3NH_2

 c. NH_3

 d. $Ca(OH)_2$

9.4 Each of the following produces a basic solution when dissolved in water. Identify those that behave as Arrhenius bases and write dissociation equations to illustrate that behavior.

 a. $NaNH_2$

 b. RbOH

 c. $C_3H_7NH_2$

 d. $Ba(OH)_2$

The Brønsted Theory (Section 9.2)

9.5 Identify each Brønsted acid and base in the following equations. Note that the reactions are assumed to be reversible.

 a. $HBr(aq) + H_2O(\ell) \rightleftharpoons H_3O^+(aq) + Br^-(aq)$

 b. $H_2O(\ell) + N_3^-(aq) \rightleftharpoons HN_3(aq) + OH^-(aq)$

 c. $H_2S(aq) + H_2O(\ell) \rightleftharpoons H_3O^+(aq) + HS^-(aq)$

 d. $SO_3^{2-}(aq) + H_2O(\ell) \rightleftharpoons HSO_3^-(aq) + OH^-(aq)$

 e. $HCN(aq) + H_2O(\ell) \rightleftharpoons H_3O^+(aq) + CN^-(aq)$

9.6 Identify each Brønsted acid and base in the following equations. Note that the reactions are assumed to be reversible.

 a. $HC_2O_4^-(aq) + H_2O(\ell) \rightleftharpoons H_3O^+(aq) + C_2O_4^{2-}(aq)$

 b. $HNO_2(aq) + H_2O(\ell) \rightleftharpoons H_3O^+(aq) + NO_2^-(aq)$

 c. $PO_4^{3-}(aq) + H_2O(\ell) \rightleftharpoons HPO_4^{2-}(aq) + OH^-(aq)$

 d. $H_2SO_3(aq) + H_2O(\ell) \rightleftharpoons HSO_3^-(aq) + H_3O^+(\ell)$

 e. $F^-(aq) + H_2O(\ell) \rightleftharpoons HF(aq) + OH^-(aq)$

9.7 Identify each conjugate acid–base pair in the equations of Exercise 9.5.

9.8 Identify each conjugate acid–base pair in the equations of Exercise 9.6.

9.9 Write equations to represent the Brønsted acid behavior for each of the following acids in water solution. Remember to represent the reactions as being reversible.

 a. HI

 b. HBrO

 c. HCN

 d. HSe^-

9.10 Write equations to represent the Brønsted acid behavior for each of the following acids in water solution. Remember to represent the reactions as being reversible.

 a. HF

 b. $HClO_3$

 c. HClO

 d. HS^-

9.11 Write a formula for the conjugate base formed when each of the following behaves as a Brønsted acid:

 a. HSO_3^-

 b. HPO_4^{2-}

 c. $HClO_3$

 d. $CH_3NH_3^+$

 e. $H_2C_2O_4$

9.12 Write a formula for the conjugate base formed when each of the following behaves as a Brønsted acid:

 a. HSO_4^-

 b. $CH_3NH_3^+$

 c. $HClO_4$

 d. NH_4^+

 e. HCl

9.13 Write a formula for the conjugate acid formed when each of the following behaves as a Brønsted base:

 a. NH_2^-

 b. CO_3^{2-}

 c. OH^-

 d. $(CH_3)_2NH$

 e. NO_2^-

9.14 Write a formula for the conjugate acid formed when each of the following behaves as a Brønsted base:

 a. HCO_3^-

 b. S^{2-}

 c. HS^-

 d. $HC_2O_4^-$

 e. $HN_2O_2^-$

9.15 The following reactions illustrate Brønsted acid–base behavior. Complete each equation.

 a. $HI(aq) + ? \rightarrow H_3O^+(aq) + I^-(aq)$

 b. $NH_3(\ell) + ? \rightarrow NH_4^+ + NH_2^-$

 c. $H_2C_2O_4(aq) + H_2O(\ell) \rightarrow ? + HC_2O_4^-(aq)$

 d. $H_2N_2O_2(aq) + H_2O(\ell) \rightarrow H_3O^+(aq) + ?$

 e. $? + H_2O(\ell) \rightarrow H_3O^+(aq) + CO_3^{2-}(aq)$

9.16 The following reactions illustrate Brønsted acid–base behavior. Complete each equation.

 a. $H_2AsO_4^-(aq) + NH_3(aq) \rightarrow NH_4^+(aq) + ?$

 b. $? + H_2O(\ell) \rightarrow C_6H_5NH_3^+(aq) + OH^-(aq)$

 c. $S^{2-}(aq) + ? \rightarrow HS^-(aq) + OH^-(aq)$

 d. $? + HBr(aq) \rightarrow (CH_3)_2NH_2^+(aq) + Br^-(aq)$

 e. $CH_3NH_2(aq) + HCl(aq) \rightarrow ? + Cl^-(aq)$

9.17 Write equations to illustrate the acid–base reaction when each of the following pairs of Brønsted acids and bases are combined:

	Acid	*Base*
a.	$HOCl$	H_2O
b.	$HClO_4$	NH_3
c.	H_2O	NH_2^-
d.	H_2O	OCl^-
e.	$HC_2O_4^-$	H_2O

9.18 Write equations to illustrate the acid–base reaction when each of the following pairs of Brønsted acids and bases are combined:

	Acid	*Base*
a.	H_3O^+	NH_2^-
b.	$H_2PO_4^-$	NH_3
c.	$HS_2O_3^-$	OCl^-
d.	H_2O	ClO_4^-
e.	H_2O	NH_3

Naming Acids (Section 9.3)

9.19 A water solution of HF gas is used to etch glass. Name the water solution as an acid.

9.20 Hydrogen cyanide, HCN, behaves in a water solution very much like the binary covalent compounds of hydrogen, but it liberates the cyanide ion, CN^-. Name the acidic water solution by following the rules for binary covalent compounds of hydrogen.

9.21 Name the following acids. Refer to Table 4.7 as needed.

 a. $H_2Se(aq)$

 b. $HClO_3$

 c. H_2SO_4

 d. HNO_3

9.22 Name the following acids. Refer to Table 4.7 as needed.

 a. $H_2Te(aq)$

 b. $HClO$

 c. H_2SO_3

 d. HNO_2

9.23 The acid $H_3C_6H_5O_7$ forms the citrate ion, $C_6H_5O_7^{3-}$, when all three hydrogens are removed. This acid is involved in an important energy-storing process in the body. Name the acid.

9.24 The acid $H_2C_4H_4O_4$ forms the succinate ion, $C_4H_4O_4^{2-}$, when both hydrogens are removed. This acid is involved in the same energy-storing process as the acid of Exercise 9.23. Name $H_2C_4H_4O_4$ as an acid.

9.25 Refer to Table 4.7, and write the formula for chromic acid.

9.26 Refer to Table 4.7, and write the formula for permanganic acid.

The Self-Ionization of Water (Section 9.4)

9.27 Calculate the molar concentration of OH^- in water solutions with the following H_3O^+ molar concentrations:

 a. 1.0×10^{-6}

 b. 4.8×10^{-4}

 c. 6.2×10^{-10}

 d. 1.4

 e. 0.038

9.28 Calculate the molar concentration of OH^- in water solutions with the following H_3O^+ molar concentrations:

 a. 0.044

 b. 1.3×10^{-4}

 c. 0.0087

 d. 7.9×10^{-10}

 e. 3.3×10^{-2}

9.29 Calculate the molar concentration of H_3O^+ in water solutions with the following OH^- molar concentrations:

 a. 1.0×10^{-5}

 b. 3.4×10^{-4}

 c. 8.3×10^{-9}

 d. 0.072

 e. 3.2

9.30 Calculate the molar concentration of H_3O^+ in water solutions with the following OH^- molar concentrations:

 a. 6.9×10^{-5}

 b. 0.074

 c. 4.9

 d. 1.7×10^{-3}

 e. 9.2×10^{-9}

9.31 Classify the solutions represented in Exercises 9.27 and 9.29 as acidic, basic, or neutral.

9.32 Classify the solutions represented in Exercises 9.28 and 9.30 as acidic, basic, or neutral.

The pH Concept (Section 9.5)

9.33 Classify solutions with the following characteristics as acidic, basic, or neutral:

 a. pH = 10

 b. pH = 4

 c. pH = 7.3

 d. pH = 6

9.34 Classify solutions with the following characteristics as acidic, basic, or neutral:

 a. pH = 2.8

 b. pH = 8

 c. pH = 6.9

 d. pH = 12

9.35 Determine the pH of water solutions with the following characteristics. Classify each solution as acidic, basic, or neutral.

 a. $[H^+] = 1.0 \times 10^{-4}$

 b. $[OH^-] = 5.0 \times 10^{-2}$

 c. $[H^+] = [OH^-]$

 d. $[H^+] = 8.0 \times 10^{-9}$

 e. $[OH^-] = 4.0 \times 10^{-10}$

9.36 Determine the pH of water solutions with the following characteristics. Classify each solution as acidic, basic, or neutral.

 a. $[H^+] = 4.1 \times 10^{-9}$

 b. $[OH^-] = 9.4 \times 10^{-4}$

 c. $[OH^-] = 10.[H^+]$

 d. $[H^+] = 2.3 \times 10^{-2}$

 e. $[OH^-] = 5.1 \times 10^{-10}$

9.37 Determine the pH of water solutions with the following characteristics. Classify each solution as acidic, basic, or neutral.

 a. $[H^+] = 6.2 \times 10^{-5}$

 b. $[H^+] = 8.4 \times 10^{-8}$

 c. $[H^+] = 2.2 \times 10^{-10}$

 d. $[OH^-] = 4.9 \times 10^{-2}$

 e. $[OH^-] = 6.2 \times 10^{-7}$

9.38 Determine the pH of water solutions with the following characteristics. Classify each solution as acidic, basic, or neutral.

 a. $[H^+] = 2.2 \times 10^{-3}$

 b. $[H^+] = 3.9 \times 10^{-12}$

 c. $[H^+] = 7.5 \times 10^{-6}$

 d. $[OH^-] = 2.5 \times 10^{-4}$

 e. $[OH^-] = 8.6 \times 10^{-10}$

9.39 Determine the $[H^+]$ value for solutions with the following characteristics:

 a. pH = 3.64

 b. pH = 12.82

 c. pH = 7.14

9.40 Determine the [H$^+$] value for solutions with the following characteristics:

 a. pH = 9.27

 b. pH = 2.55

 c. pH = 5.42

9.41 Convert the following pH values into both [H$^+$] and [OH$^-$] values:

 a. pH = 9.00

 b. pH = 6.27

 c. pH = 3.10

9.42 Convert the following pH values into both [H$^+$] and [OH$^-$] values:

 a. pH = 3.95

 b. pH = 4.00

 c. pH = 11.86

9.43 The pH values listed in Table 9.1 are generally the average values for the listed materials. Most natural materials, such as body fluids and fruit juices, have pH values that cover a range for different samples. Some measured pH values for specific body fluid samples are given below. Convert each one to [H$^+$], and classify the fluid as acidic, basic, or neutral.

 a. Blood, pH = 7.41

 b. Gastric juice, pH = 1.60

 c. Urine, pH = 5.93

 d. Saliva, pH = 6.85

 e. Pancreatic juice, pH = 7.85

9.44 The pH values listed in Table 9.1 are generally the average values for the listed materials. Most natural materials, such as body fluids and fruit juices, have pH values that cover a range for different samples. Some measured pH values for specific body fluid samples are given below. Convert each one to [H$^+$], and classify the fluid as acidic, basic, or neutral.

 a. Bile, pH = 8.05

 b. Vaginal fluid, pH = 3.93

 c. Semen, pH = 7.38

 d. Cerebrospinal fluid, pH = 7.40

 e. Perspiration, pH = 6.23

9.45 The pH values of specific samples of food items are listed below. Convert each value to [H$^+$], and classify the sample as acidic, basic, or neutral.

 a. Milk, pH = 6.39

 b. Coffee, pH = 5.10

 c. Orange juice, pH = 4.07

 d. Vinegar, pH = 2.65

9.46 The pH values of specific samples of food items are listed below. Convert each value to [H$^+$], and classify the sample as acidic, basic, or neutral.

 a. Soft drink, pH = 2.91

 b. Tomato juice, pH = 4.11

 c. Lemon juice, pH = 2.32

 d. Grapefruit juice, pH = 3.07

Properties of Acids (Section 9.6)

9.47 Using the information in Table 9.4, describe how you would prepare each of the following solutions.

 a. About 750 mL of 0.5 M H_2SO_4 from dilute sulfuric acid

 b. About 200 mL of 0.1 M NaOH from stock sodium hydroxide solution

 c. About 1.0 L of 1 M acetic acid from glacial acetic acid

9.48 Using the information in Table 9.4, describe how you would prepare each of the following solutions.

 a. About 2 L of 3.0 M HNO_3 from dilute nitric acid solution

 b. About 500 mL of 1.5 M aqueous ammonia from concentrated aqueous ammonia solution

 c. About 5 L of 0.2 M HCl from concentrated hydrochloric acid solution

9.49 Write balanced molecular equations to illustrate the following characteristic reactions of acids, using nitric acid (HNO_3).

 a. Reaction with water to form hydronium ions

 b. Reaction with the solid oxide CaO

 c. Reaction with the solid hydroxide $Mg(OH)_2$

 d. Reaction with the solid carbonate $CuCO_3$

 e. Reaction with the solid bicarbonate $KHCO_3$

 f. Reaction with Mg metal

9.50 Write balanced molecular equations to illustrate the following characteristic reactions of acids, using sulfuric acid (H_2SO_4).

 a. Reaction with water to form hydronium ions

 b. Reaction with the solid oxide CaO

 c. Reaction with the solid hydroxide $Mg(OH)_2$

 d. Reaction with the solid carbonate $CuCO_3$

 e. Reaction with the solid bicarbonate $KHCO_3$

 f. Reaction with Mg metal

9.51 Write each molecular equation of Exercise 9.49 in total ionic and net ionic form. Use Table 7.4 to decide which products will be soluble.

9.52 Write each molecular equation of Exercise 9.50 in total ionic and net ionic form. Use Table 7.4 to decide which products will be soluble.

9.53 Write balanced molecular equations to illustrate five different reactions that could be used to prepare $BaCl_2$ from hydrochloric acid (HCl) and other appropriate substances.

9.54 Write balanced molecular equations to illustrate five different reactions that could be used to prepare $SrCl_2$ from hydrochloric acid (HCl) and other appropriate substances.

9.55 Write balanced molecular, total ionic, and net ionic equations to illustrate each of the following reactions. All the metals form 2+ ions.

 a. Zinc with H_2SO_4

 b. Magnesium with HCl

 c. Calcium with $HC_2H_3O_2$

9.56 Write balanced molecular, total ionic, and net ionic equations to illustrate each of the following reactions. All the metals form 2+ ions.

 a. Tin with H_2SO_3

 b. Magnesium with H_3PO_4

 c. Calcium with HBr

Properties of Bases (Section 9.7)

9.57 Write balanced molecular, total ionic, and net ionic equations to represent neutralization reactions between RbOH and the following acids. Use all H's possible for each acid.

 a. HCl

 b. HNO_3

 c. H_2SO_4

9.58 Write balanced molecular, total ionic, and net ionic equations to represent neutralization reactions between RbOH and the following acids. Use all H's possible for each acid.

 a. H_3PO_4

 b. $H_2C_2O_4$ (oxalic acid)

 c. $HC_2H_3O_2$

9.59 Some polyprotic acids can form more than one salt, depending on the number of H's that react with base. Write balanced molecular, total ionic, and net ionic equations to represent the following neutralization reactions between KOH and

 a. H_2SO_4 (only react one H).

 b. H_2SO_4 (react both H's).

 c. H_3PO_4 (only react one H).

9.60 Some polyprotic acids can form more than one salt, depending on the number of H's that react with base. Write balanced molecular, total ionic, and net ionic equations to represent the following neutralization reactions between KOH and

 a. H_3PO_4 (react two H's).

 b. H_3PO_4 (react three H's).

 c. $H_2C_2O_4$ (react one H).

Salts (Section 9.8)

9.61 Identify with ionic formulas the cations and anions of the following salts:

 a. LiCl

 b. $Cu(NO_3)_2$

 c. $SrSO_4$

 d. K_3PO_4

 e. K_2HPO_4

 f. $CaCO_3$

9.62 Identify with ionic formulas the cations and anions of the following salts:

 a. NH_4NO_3

 b. $CaCl_2$

 c. $Mg(HCO_3)_2$

 d. $KC_2H_3O_2$

 e. $LiHSO_3$

9.63 Identify with formulas the acid and base from which the anion and cation of each salt in Exercise 9.61 was derived. Pay special attention to salts derived from polyprotic acids and be sure to list the acid formula with all H's.

9.64 Identify with formulas the acid and base from which the anion and cation of each salt in Exercise 9.62 was derived. Pay special attention to salts derived from polyprotic acids and be sure to list the acid formula with all H's.

9.65 Calculate the mass of water that would be released if the water of hydration were completely driven off 1.0 mol of (a) plaster of Paris and (b) gypsum (see Table 9.6). How would the products of these reactions compare?

9.66 Calculate the mass of water that would be released if the water of hydration were completely driven off 1.0 mol of (a) Epsom salts and (b) borax (see Table 9.6). How would the products of these reactions compare?

9.67 Write formulas for the acid and indicated solid that could be used to prepare each of the following salts:

 a. $CuCl_2$ (solid is an oxide)

 b. $MgSO_4$ (solid is a carbonate)

 c. LiBr (solid is a hydroxide)

9.68 Write formulas for the acid and indicated solid that could be used to prepare each of the following salts:

 a. KNO_3 (solid is a bicarbonate)

 b. $ZnCl_2$ (solid is a metal)

 c. LiBr (solid is an oxide)

9.69 Write balanced molecular equations to illustrate each salt preparation described in Exercise 9.67.

9.70 Write balanced molecular equations to illustrate each salt preparation described in Exercise 9.68.

9.71 Determine the number of moles of each of the following salts that would equal 1 eq of salt:

 a. KNO_3

 b. Li_2CO_3

 c. $SrCl_2$

9.72 Determine the number of moles of each of the following salts that would equal 1 eq of salt:

 a. $MgCO_3$

 b. $Zn(HCO_3)_2$

 c. $FeCl_3$

9.73 Determine the number of equivalents and milliequivalents in each of the following:

 a. 0.10 mol KI

 b. 0.25 mol $MgCl_2$

 c. 4.73×10^{-2} mol $AgNO_3$

9.74 Determine the number of equivalents and milliequivalents in each of the following:

 a. 0.22 mol $ZnCl_2$

 b. 0.45 mol CsCl

 c. 3.12×10^{-2} mol $Fe(NO_3)_2$

9.75 Determine the number of equivalents and milliequivalents in 5.00 g of each of the following salts. Include any waters of hydration given in the salt formula when you calculate salt formula weights.

a. NaCl

b. $NaNO_3$

c. Na_3PO_4

d. $MgSO_4 \cdot 7H_2O$

9.76 Determine the number of equivalents and milliequivalents in 5.00 g of each of the following salts. Include any waters of hydration given in the salt formula when you calculate salt formula weights.

a. $Na_2CO_3 \cdot 10H_2O$

b. $CuSO_4 \cdot 5H_2O$

c. Li_2CO_3

d. NaH_2PO_4

9.77 A sample of intracellular fluid contains 45.1 meq/L of Mg^{2+} ion. Assume the Mg^{2+} comes from dissolved $MgCl_2$, and calculate the number of moles and number of grams of $MgCl_2$ that would be found in 250. mL of the intracellular fluid.

9.78 A sample of intracellular fluid contains 133 meq/L of K^+ ion. Assume the K^+ comes from dissolved K_2SO_4, and calculate the number of moles and number of grams of K_2SO_4 that would be found in 150. mL of the intracellular fluid.

The Strengths of Acids and Bases (Section 9.9)

9.79 Illustrate the difference between weak, moderately weak, and strong acids by writing dissociation reactions for the hypothetical acid HB, using arrows of various lengths.

9.80 The K_a values have been determined for four acids and are listed below. Arrange the acids in order of increasing acid strength (weakest first, strongest last).

$$\text{acid } A \ (K_a = 5.6 \times 10^{-5})$$
$$\text{acid } B \ (K_a = 1.8 \times 10^{-5})$$
$$\text{acid } C \ (K_a = 1.3 \times 10^{-4})$$
$$\text{acid } D \ (K_a = 1.1 \times 10^{-3})$$

9.81 Arrange the four acids classified as weak in Table 9.7 in order of increasing strength (weakest first, strongest last).

9.82 K_a values for four weak acids are given below:

$$\text{acid } A \ (K_a = 2.6 \times 10^{-4})$$
$$\text{acid } B \ (K_a = 3.7 \times 10^{-5})$$
$$\text{acid } C \ (K_a = 5.8 \times 10^{-4})$$
$$\text{acid } D \ (K_a = 1.5 \times 10^{-3})$$

a. Arrange the four acids in order of increasing acid strength (weakest first, strongest last).

b. Arrange the conjugate bases of the acids (identify as base A, etc.) in order of increasing base strength (weakest base first, strongest last).

9.83 Write dissociation reactions and K_a expressions for the following weak acids:

a. Hypobromous acid, HBrO

b. Sulfurous acid, H_2SO_3 (1st H only)

c. Hydrogen sulfite ion, HSO_3^-

d. Hydroselenic acid, H_2Se (1st H only)

e. Arsenic acid, H_3AsO_4 (1st H only)

9.84 Write dissociation reactions and K_a expressions for the following weak acids:

a. Hydrogen selenide ion, HSe^-

b. Dihydrogen borate ion, $H_2BO_3^-$ (1st H only)

c. Hydrogen borate ion, HBO_3^{2-}

d. Hydrogen arsenate ion, $HAsO_4^{2-}$

e. Hypochlorous acid, HClO

9.85 Equal molar solutions are made of three monoprotic acids, HA, HB, and HC. The pHs of the solutions are, respectively, 4.82, 3.16, and 5.47. Rank the acids in order of increasing acid strength and explain your reasoning.

9.86 If someone asked you for a weak acid solution, which of the following would you provide according to definitions in this chapter?

a. 0.05 M HCl

b. 20% acetic acid

If the individual really wanted the other solution, what term should have been used instead of *weak*?

9.87 Arsenic acid (H_3AsO_4) is a moderately weak triprotic acid. Write equations showing its stepwise dissociation. Which of the three anions formed in these reactions will be the strongest Brønsted base? Which will be the weakest Brønsted base? Explain your answers.

Analyzing Acids and Bases (Section 9.10)

9.88 Explain the purpose of doing a titration.

9.89 Describe the difference between the information obtained by measuring the pH of an acid solution and by titrating the solution with base.

9.90 Suppose a student is going to titrate an acidic solution with a base and just picks an indicator at random. Under what circumstances will (a) the endpoint and equivalence point be the same? (b) The endpoint and equivalence point be different?

9.91 Determine the number of moles of NaOH that could be neutralized by each of the following:

a. 1.00 L of 0.250 M HCl

b. 500. mL of 0.300 M HNO_3

9.92 Determine the number of moles of NaOH that could be neutralized by each of the following:

a. 250. mL of 0.400 M HBr

b. 750. mL of 0.300 M $HClO_4$

Titration Calculations (Section 9.11)

9.93 Write a balanced molecular equation to represent the neutralization reaction between NaOH and each of the following acids. React all of the acid H's.

a. Molybdic acid, H_2MoO_4

b. Permanganic acid, $HMnO_4$

c. Phosphoric acid, H_3PO_4

9.94 Write a balanced molecular equation to represent the neutralization reaction between NaOH and each of the following acids. React all of the acid H's.

　a. Trichloroacetic acid, $HC_2O_2Cl_3$

　b. Dithionic acid, $H_2S_2O_6$

　c. Hypophosphorous acid, $H_4P_2O_6$

9.95 Write a balanced molecular equation to represent the neutralization reaction between HCl and each of the following bases:

　a. $Cd(OH)_2$

　b. $Cr(OH)_3$

　c. $Fe(OH)_2$

9.96 Write a balanced molecular equation to represent the neutralization reaction between HCl and each of the following bases:

　a. $Zn(OH)_2$

　b. $Tl(OH)_3$

　c. CsOH

9.97 A 25.00-mL sample of gastric juice is titrated with a 0.0210 M NaOH solution. The titration to the equivalence point requires 26.4 mL of NaOH solution. If the equation for the reaction is

$$HCl(aq) + NaOH(aq) \rightarrow NaCl(aq) + H_2O(\ell)$$

what is the molarity of HCl in the gastric juice?

9.98 A 25.00-mL sample of $H_2C_2O_4$ solution required 43.88 mL of 0.1891 M NaOH solution to titrate it to the equivalence point. Calculate the molarity of the $H_2C_2O_4$ solution.

9.99 A 20.00-mL sample of each of the following acid solutions is to be titrated to the equivalence point using a 0.120 M NaOH solution. Determine the number of milliliters of NaOH solution that will be needed for each acid sample.

　a. 0.200 M HCl

　b. 0.200 M H_2SO_4

　c. 0.250 M HCl

　d. 10.00 g of H_3PO_4 in 150. mL of solution

　e. 0.150 mol H_2MoO_4 in 400. mL of solution

　f. 0.215 mol H_2MoO_4 in 600. mL of solution

9.100 A 20.00-mL sample of each of the following acid solutions is to be titrated to the equivalence point using a 0.120 M NaOH solution. Determine the number of milliliters of NaOH solution that will be needed for each acid sample.

　a. 0.200 M $HClO_4$

　b. 0.125 M H_2SO_4

　c. 0.150 M $H_4P_2O_6$

　d. 0.120 mol H_3PO_4 in 500. mL of solution

　e. 6.25 g of H_2SO_4 in 250. mL of solution

　f. 0.500 mol $HClO_3$ in 1.00 L of solution

9.101 The following acid solutions were titrated to the equivalence point with the base listed. Use the titration data to calculate the molarity of each acid solution.

　a. 25.00 mL of HI solution required 27.15 mL of a 0.250 M NaOH solution.

b. 20.00 mL of H_2SO_4 solution required 11.12 mL of a 0.109 M KOH solution.

c. 25.00 mL of gastric juice (HCl) required 18.40 mL of a 0.0250 M NaOH solution.

9.102 The following acid solutions were titrated to the equivalence point with the base listed. Use the titration data to calculate the molarity of each acid solution.

　a. 5.00 mL of dilute H_2SO_4 required 29.88 mL of a 1.17 M NaOH solution.

　b. 10.00 mL of vinegar (acetic acid) required 35.62 mL of a 0.250 M KOH solution.

　c. 10.00 mL of muriatic acid (HCl) used to clean brick and cement required 20.63 mL of a 6.00 M NaOH solution.

9.103 A 20.00-mL sample of diprotic oxalic acid ($H_2C_2O_4$) solution is titrated with a 0.250 M NaOH solution. A total of 26.54 mL of NaOH is required. Calculate:

　a. The number of moles of oxalic acid in the 20.00-mL sample.

　b. The molarity of the oxalic acid solution.

　c. The number of grams of oxalic acid in the 20.00-mL sample.

9.104 A sample of monoprotic benzoic acid weighing 0.5823 g is dissolved in about 25 mL of water. The solution is titrated to the equivalence point using 0.1021 M NaOH. The volume of base required is 46.75 mL. Calculate the molecular weight of the solid acid.

Hydrolysis Reactions of Salts (Section 9.12)

9.105 A solution of solid NH_4Cl dissolved in pure water is acidic (the pH is less than 7). Explain.

9.106 A solution of solid Na_3PO_4 dissolved in pure water is basic (the pH is greater than 7). Explain.

9.107 Predict the relative pH (greater than 7, less than 7, etc.) for water solutions of the following salts. Table 9.9 may be useful. For each solution in which the pH is greater or less than 7, explain why and write a net ionic equation to justify your answer.

　a. Potassium sulfite, K_2SO_3

　b. Lithium nitrite, $LiNO_2$

　c. Sodium carbonate, Na_2CO_3

　d. Methylammonium chloride, CH_3NH_3Cl (CH_3NH_2 is a weak base)

9.108 Predict the relative pH (greater than 7, less than 7, etc.) for water solutions of the following salts. Table 9.9 may be useful. For each solution in which the pH is greater or less than 7, explain why and write a net ionic equation to justify your answer.

　a. Sodium hypochlorite, NaOCl (HOCl is a weak acid)

　b. Sodium formate, $NaCHO_2$

　c. Potassium nitrate, KNO_3

　d. Sodium phosphate, Na_3PO_4

9.109 A chemist has 20.00-mL samples of 0.100 M acid A and 0.100 M acid B in separate flasks. Both acids are monoprotic. Unfortunately, the flasks were not labeled, so the chemist doesn't know which sample is in which flask. But fortunately, it

is known that acid A is strong and acid B is weak. Before thinking about the problem, the chemist adds 20.00 mL of 0.100 M NaOH solution to each flask. Explain how the chemist could use a pH meter (or pH paper) to determine which flask originally contained which acid.

9.110 Explain why the hydrolysis of salts makes it necessary to have available in a laboratory more than one acid–base indicator for use in titrations.

9.111 How would the pH values of equal molar solutions of the following salts compare (highest, lowest, etc.)? NaH_2PO_4, Na_2HPO_4, and Na_3PO_4.

Buffers (Section 9.13)

9.112 Write equations similar to Equations 9.48 and 9.49 of the text to illustrate how a mixture of sodium hydrogen phosphate (Na_2HPO_4) and sodium dihydrogen phosphate (NaH_2PO_4) could function as a buffer when dissolved in water. Remember that phosphoric acid (H_3PO_4) ionizes in three steps.

9.113 Could a mixture of ammonia (NH_3), a weak base, and ammonium chloride (NH_4Cl) behave as a buffer when dissolved in water? Use reaction equations to justify your answer.

9.114 Some illnesses lead to a condition of excess acid (acidosis) in the body fluids. An accepted treatment is to inject solutions containing bicarbonate ions (HCO_3^-) directly into the bloodstream. Write an equation to show how this treatment would help combat the acidosis.

9.115 Calculate the pH of a buffer made by dissolving 1 mol formic acid (HCOOH) and 1 mol sodium formate (HCOONa) in 1 L of solution (see Table 9.9).

9.116 a. Calculate the pH of a buffer that is 0.1 M in lactic acid, $C_2H_4(OH)COOH$, and 0.1 M in sodium lactate, $C_2H_4(OH)COONa$.

 b. What is the pH of a buffer that is 1 M in lactic acid and 1 M in sodium lactate?

 c. What is the difference between the buffers described in parts a and b?

9.117 Which of the following acids and its conjugate base would you use to make a buffer with a pH of 3.00? Explain your reasons: formic acid, lactic acid, nitrous acid.

9.118 Calculate the pH of buffers that contain the acid and conjugate base concentrations listed below.

 a. $[CH_3COOH] = 0.40$ M, $[CH_3COO^-] = 0.25$ M

 b. $[H_2PO_4^-] = 0.10$ M, $[HPO_4^{2-}] = 0.40$ M

 c. $[HSO_3^-] = 1.50$ M, $[SO_3^{2-}] = 0.20$ M

9.119 Calculate the pH of buffers that contain the acid and conjugate base concentrations listed below.

 a. $[HPO_4^{2-}] = 0.33$ M, $[PO_4^{3-}] = 0.52$ M

 b. $[HNO_2] = 0.029$ M, $[NO_2^-] = 0.065$ M

 c. $[HCO_3^-] = 0.50$ M, $[CO_3^{2-}] = 0.15$ M

9.120 What ratio of concentrations of NaH_2PO_4 and Na_2HPO_4 in solution would give a buffer with pH = 7.65?

9.121 A citric acid–citrate buffer has a pH of 3.20. You want to increase the pH to a value of 3.35. Would you add citric acid or sodium citrate to the solution? Explain.

Additional Exercises

9.122 Consider the following reaction equation:

$$Cl^-(aq) + H_3O^+(aq) \rightarrow HCl(aq) + H_2O(\ell)$$

Draw Lewis structures (electron-dot formulas) for all the substances in the reaction equation, identify each reactant as a Brønsted acid or base, and then propose a definition for an acid and a base based on the ability of the substance to accept or donate a pair of electrons to form a covalent bond.

9.123 Consider the following dissociation reaction of a weak acid, HA:

$$HA(aq) + H_2O(\ell) \rightleftharpoons H_3O^+(aq) + A^-(aq)$$

It was determined that 2.63% of the acid in a 0.150 M solution of HA in water was dissociated. Calculate the pH of the 0.150 M solution.

9.124 The value of K_w for water varies with temperature. At 25°C, $K_w = 1.0 \times 10^{-14}$. At 50°C, $K_w = 5.5 \times 10^{-14}$. Calculate the pH of water at 50°C.

9.125 When sodium metal, Na, reacts with water, hydrogen gas, H_2, and sodium hydroxide, NaOH, in solution are produced. Write a balanced equation for the reaction, and explain how a basic solution is produced.

Chemistry for Thought

9.126 In an early industrial method, H_2SO_4 was manufactured in lead-lined chambers. Propose an explanation for this.

9.127 A saturated solution of solid $Ca(OH)_2$ in water has a $[OH^-]$ of only 2.50×10^{-2} M, and yet $Ca(OH)_2$ is a strong base. Explain this apparent contradiction.

9.128 Imagine that a solution of weak acid is being titrated with a strong base. Describe the substances present in the solution being titrated when it has been titrated halfway to the equivalence point. How is the pH of this half-titrated solution related to the pK_a for the weak acid?

9.129 Calculate K_a for the following weak acids based on the equilibrium concentrations and pH values given:

 a. Benzoic acid, represented by HBz:

 pH = 2.61, $[Bz^-] = 2.48 \times 10^{-3}$, $[HBz] = 0.0975$

 b. Abietic acid, represented by HAb:

 pH = 4.16, $[Ab^-] = 6.93 \times 10^{-5}$, $[HAb] = 0.200$

 c. Cacodylic acid, represented by HCc:

 pH = 3.55, $[Cc^-] = 0.000284$, $[HCc] = 0.150$

9.130 Bottles of ketchup are routinely left on the counters of cafés, yet the ketchup does not spoil. Why not?

9.131 Refer to Figure 9.2 and answer the question. What implications does this reaction have for the long-term durability of marble structures when exposed to acid rain?

9.132 Refer to Figure 9.3 and answer the question. Do you think the results would be the same if sulfuric acid (H_2SO_4) were substituted for the HCl? Explain.

9.133 Refer to Figure 9.5. Which of the three indicators would be the best to use to differentiate between the solutions with pH values of 4 and 6?

9.134 Refer to Figure 9.6 and explain why the pH reading of the meter might not be 7.00 at the equivalence point in a titration. Would the pH at the equivalence point be greater or less than 7 if hydrochloric acid were titrated with aqueous ammonia?

9.135 Refer to Figure 9.7 and answer the question. List at least two other ions that would behave like the Na^+ ion.

9.136 Refer to Figure 9.8 and answer the question. What color is the universal indicator in acidic solutions? In basic solutions? Explain how you arrived at your conclusions.

Allied Health Exam Connection

The following questions are from these sources:

• *Nursing School Entrance Exam* © 2005, Learning Express, LLC.

• *McGraw-Hill's Nursing School Entrance Exams* by Thomas A. Evangelist, Tamara B. Orr, and Judy Unrein © 2009, The McGraw-Hill Companies, Inc.

• *NSEE Nursing School Entrance Exams,* 3rd edition © 2009, Kaplan Publishing.

• *Cliffs Test Prep: Nursing School Entrance Exams* by Fred N. Grayson © 2004, Wiley Publishing, Inc.

• *Peterson's Master the Nursing School and Allied Health Entrance Exams,* 18th edition by Marion F. Gooding © 2008, Peterson's, a Nelnet Company.

9.137 An acid is a substance that dissociates in water into one or more _____ ions and one or more _____.

 a. hydrogen . . . anions

 b. hydrogen . . . cations

 c. hydroxide . . . anions

 d. hydroxide . . . cations

9.138 A base is a substance that dissociates in water into one or more _____ ions and one or more _____.

 a. hydrogen . . . anions

 b. hydrogen . . . cations

 c. hydroxide . . . anions

 d. hydroxide . . . cations

9.139 Which is true of alkaline solutions?

 a. More H^+ ion than OH^- ion

 b. Same amount of H^+ ion + OH^- ion

 c. More OH^- ion than H^+ ion

 d. None of the above

9.140 Which of the following is NOT an acid/conjugate base pair?

 a. HCN/CN^-

 b. H_2CO_3/OH^-

 c. H_2SO_4/HSO_4^-

 d. $H_3PO_4/H_2PO_4^-$

9.141 What is the formula of the hydronium ion?

 a. H^+

 b. NH_4^+

 c. H_3O^+

 d. H_2O^+

9.142 Which of the following substances has a pH closest to 7?

 a. ammonia

 b. blood

 c. lemon juice

 d. vinegar

9.143 Dissolving H_2SO_4 in water creates an acid solution by increasing the:

 a. sulfate ions.

 b. water ions.

 c. hydrogen ions.

 d. oxygen ions.

9.144 When a solution has a pH of 7, it is:

 a. a strong base.

 b. a strong acid.

 c. a weak base.

 d. neutral.

9.145 A common detergent has a pH of 11.0, so the detergent is:

 a. neutral.

 b. acidic.

 c. alkaline.

 d. none of the above.

9.146 In a 0.001 M solution of HCl, the pH is:

 a. 2.

 b. −3.

 c. 1.

 d. 3.

9.147 The pH of a blood sample is 7.40 at room temperature. The pOH is therefore:

 a. 6.60.

 b. 7.40.

 c. 6×10^{-6}.

 d. 4×10^{-7}.

9.148 As the concentration of hydrogen ions in a solution decreases:

 a. the pH numerically decreases.

 b. the pH numerically increases.

 c. the product of the concentrations $[H^+] \times [OH^-]$ comes closer to 1×10^{-14}.

 d. the solution becomes more acidic.

9.149 Blood with a pH of 7.4 indicates that the blood sample is:

 a. strongly acidic.

 b. strongly basic.

 c. weakly acidic.

 d. weakly basic.

9.150 Atmospheric moisture (H_2O) combines with oxides of carbon, nitrogen, and sulfur (CO_2, NO_3, and SO_2) to produce:

 a. alkaline precipitation.

 b. acid rain.

 c. alcohol.

 d. none of the above.

9.151 Which of the following substances is most likely to taste sour?

 a. $NaOH$

 b. $NaCl$

 c. NH_3

 d. $HC_2H_3O_2$

9.152 Which of the following compounds would be classified as a salt?

 a. Na_2CO_3

 b. $Ca(OH)_2$

 c. H_2CO_3

 d. CH_3OH

9.153 A substance that functions to prevent rapid, drastic changes in the pH of a body fluid by changing strong acids and bases into weak acids and bases is called a(n):

 a. salt.

 b. buffer.

 c. enzyme.

 d. coenzyme.

9.154 Which one of the following equations represents neutralization?

 a. $2Na + Cl_2 \rightarrow 2NaCl$

 b. $CO_2 + H_2O \rightarrow H_2CO_3$

 c. $HNO_3 + KOH \rightarrow KNO_3 + H_2O$

 d. $CaO + H_2O \rightarrow Ca(OH)_2$

9.155 Which reaction below demonstrates a neutralization reaction?

 a. $^{232}_{90}Th \rightarrow\ ^{228}_{88}Ra +\ ^4_2He$

 b. $NaCl + H_2O \rightarrow HCl + NaOH$

 c. $2HNO_3 + Mg(OH)_2 \rightarrow 2H_2O + Mg(NO_3)_2$

 d. $CH_4 + 2O_2 \rightarrow CO_2 + 2H_2O$

9.156 In a titration of 40.0 mL of 0.20 M NaOH with 0.4 M HCl, what will be the final volume of the solution when the sodium hydroxide is completely neutralized?

 a. 42 mL

 b. 20 mL

 c. 60 mL

 d. 80 mL

9.157 When titrating 50 mL of 0.2 M HCl, what quantity of 0.5 M NaOH is needed to bring the solution to the equivalence point?

 a. 80 mL

 b. 40 mL

 c. 20 mL

 d. 10 mL

10 Radioactivity and Nuclear Processes

Suttha Burawonk/Shutterstock.com

▶ Case Study

Al bragged that he never got sick. He rarely even caught a cold. He attributed this to his ironclad immune system. As a combat veteran, he had developed real physical strength and tenacity. He felt strong.

When Al noticed a persistent, nagging pain in his middle back, he suspected something serious might be causing it. Because these symptoms were unusual for Al, his physician, Dr. Meyers, ordered a CT scan. In a CT scan procedure, powerful X-rays are passed through the body. Bones and organs of the body scatter the X-rays. A powerful computer is used to analyze the scattering pattern, and the pattern is used to construct an image of the bones and internal organs. This image can then be used to detect abnormalities, including the presence of cancers.

Al felt nervous when Dr. Meyers asked him to come in to discuss the CT scan results. Dr. Meyers informed Al that he had bladder cancer. The tumor partially blocked the right ureter, which disrupted kidney drainage and caused pain.

Suddenly, Al's life changed dramatically. The recommended treatment included a total cystectomy (removal of the bladder and prostrate) followed by chemotherapy. Al wondered how he could have developed cancer. Dr. Meyers informed him that cigarette smoking caused most bladder cancers.

▶ How does smoking cause bladder cancer? Is contrast solution dangerous? How does removal of the bladder impact the patient's life?

Follow-up to this Case Study appears at the end of the chapter before the *Concept Summary*.

Learning Objectives

When you have completed your study of this chapter, you should be able to:

1 Describe and characterize the common forms of radiation emitted during radioactive decay and other nuclear processes. **(Section 10.1)**

2 Write balanced equations for nuclear reactions. **(Section 10.2)**

3 Solve problems using the half-life concept. **(Section 10.3)**

4 Describe the effects of radiation on health. **(Section 10.4)**

5 Describe and compare the units used to measure quantities of radiation. **(Section 10.5)**

6 Describe, with examples, medical uses of radioisotopes. **(Section 10.6)**

7 Describe, with examples, nonmedical uses of radioisotopes. **(Section 10.7)**

8 Show that you understand the concept of induced nuclear reactions. **(Section 10.8)**

9 Describe the differences between nuclear fission and nuclear fusion reactions. **(Section 10.9)**

I n Chapters 3 and 4, we learned that the arrangements of electrons around the nuclei of atoms determine the chemical behavior of elements and compounds. We saw that chemical reactions take place between atoms and molecules as electrons rearrange to achieve noble gas configurations for the atoms involved. These noble gas configurations represent a stable arrangement for the electrons. In this chapter, we will see that the achievement of stability is also a driving force that causes certain atomic nuclei to undergo changes that release small subnuclear particles and energy.

10.1 | Radioactive Nuclei

Learning Objective

1. Describe and characterize the common forms of radiation emitted during radioactive decay and other nuclear processes.

In 1896 Henri Becquerel, a French physicist, discovered that uranium compounds emitted rays that could expose and fog photographic plates wrapped in lightproof paper. Research conducted since that time has shown that the penetrating rays originate from changes that occur in the nuclei of some atoms. We saw earlier (see **Section 2.3**, page 52) that isotopes of the same element differ only in the number of neutrons present in their nuclei. Thus, each of the three isotopes of hydrogen, represented by the symbols $_1^1H$, $_1^2H$, and $_1^3H$, contains one proton in the nucleus and one electron outside the nucleus. However, besides the one nuclear proton, $_1^2H$ and $_1^3H$ contain one and two nuclear neutrons, respectively. It has been found that some combinations of neutrons and protons in the nucleus are stable and do not change spontaneously, but some combinations are not stable. It is these unstable nuclei that emit radiation as they become more stable. For example, both $_1^1H$ and $_1^2H$ are stable, but $_1^3H$ emits radiation. Nuclei that emit radiation are said to be **radioactive nuclei**.

Henri Becquerel discovered natural radioactivity by chance. Through research he found that radiation was emitted by any compound of uranium. He further found that the intensity of the radiation was unaffected by factors that normally influence the rates of chemical reactions: the temperature, pressure, and type of uranium compound used.

radioactive nuclei Nuclei that undergo spontaneous changes and emit energy in the form of radiation.

Later studies by other investigators showed that the radiation emitted by uranium, and by other radioactive elements discovered later, could be separated into three types by an electrical or magnetic field (see **Figure 10.1**). The three types had different electrical charges: One was positive, one was negative, and one carried no charge. The types of radiation were given names that are still used today: alpha rays (positive), beta rays (negative), and gamma rays (uncharged).

Figure 10.1 The three types of emitted radiation.

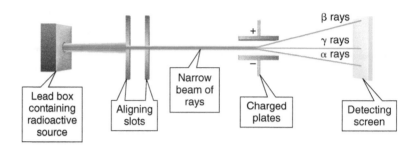

Today it is known that other types of radiation, such as neutrons and positrons, are also emitted by radioactive nuclei. However, alpha, beta, and gamma are the most common. **Table 10.1** summarizes the characteristics of these forms. In this book, we usually use the symbol given first for each form; the symbols in parentheses are alternatives you might find elsewhere. Except for gamma radiation, which is very high-energy radiation somewhat like X-rays, the radiation emitted by radioactive nuclei consists of streams of particles. The emission of radiation by unstable nuclei is often called **radioactive decay**.

radioactive decay A process in which an unstable nucleus changes energy states and in the process emits radiation.

TABLE 10.1 Characteristics of Nuclear Radiation

Type of Radiation	Symbols	Mass Number	Charge	Composition
Alpha	$^4_2\alpha$ (α, 4_2He, He$^{2+}$)	4	+2	Helium nuclei, 2 protons + 2 neutrons
Beta	$^0_{-1}\beta$ (β, β^-, $^0_{-1}e$, e^-)	0	−1	Electrons produced in nucleus and ejected
Gamma	γ ($^0_0\gamma$)	0	0	Electromagnetic radiation
Neutron	1_0n (n)	1	0	Neutrons
Positron	$^0_1\beta$ (β^+, 0_1e, e^+)	0	+1	Positive electrons

We note from Table 10.1 that the particles that make up alpha rays are identical to the nuclei of helium atoms; that is, they are clusters containing two protons and two neutrons. Because they are positively charged, alpha particles are attracted toward the negatively charged plate in Figure 10.1. **Alpha particles** move at a speed of nearly one-tenth the speed of light. They are the most massive particles emitted by radioactive materials, and they have the highest charge. Because of their mass and charge, they collide often with molecules of any matter through which they travel. As a result of these collisions, their energy is quickly dissipated, and they do not travel far. They cannot even penetrate through a few sheets of writing paper or the outer cells of the skin. However, exposure to intense alpha radiation can result in severe burns of the skin.

Beta particles are actually electrons, but they are produced within the nucleus and then emitted. They are not part of the group of electrons moving around the nucleus that is responsible for chemical characteristics of the atoms. Beta particles have a smaller charge than alpha particles and are also 7000 times less massive than alpha particles. They undergo far fewer collisions with molecules of matter through which they pass and therefore

alpha particle The particle that makes up alpha rays. It is identical to the helium nucleus and is composed of two protons and two neutrons.

beta particle The particle that makes up beta rays. It is identical to an electron but is produced in the nucleus when a neutron is changed into a proton and an electron.

are much more penetrating than alpha particles. Because of their negative charge and tiny mass, beta particles are attracted toward the positive plate of Figure 10.1, and are deflected more than alpha particles.

Gamma rays are not streams of particles but rays of electromagnetic radiation similar to X-rays that have higher energies. Like X-rays, gamma radiation is very penetrating and dangerous to living organisms. Adequate shielding from gamma rays is provided only by thick layers of heavy materials like metallic lead or concrete. Because they are not charged, gamma rays are not attracted to either charged plate in Figure 10.1.

gamma ray A high-energy ray that is like an X-ray, but with a higher energy.

10.2 | Equations for Nuclear Reactions

Learning Objective

2. Write balanced equations for nuclear reactions.

Isotopes of elements that emit nuclear radiation are called **radioisotopes**. In nuclear reactions, a specific isotope of an element may behave differently from another isotope of the same element. Thus, all particles involved in nuclear reactions are designated by a symbol, a mass number (the sum of protons and neutrons in the particle), and an atomic number (or charge for electrons or positrons). The symbolism used was introduced in Section 2.3. It is $^A_Z X$, where X is the symbol for the particle, A is the mass number, and Z is the atomic number or charge. When X represents the nucleus of an isotope of an element, the chemical symbol for the element is generally used.

radioisotope An isotope of an element that emits nuclear radiation.

Example 10.1 Writing Symbols for Particles

Write appropriate symbols for the following particles using the $^A_Z X$ symbolism:

a. A lead-214 nucleus
b. An alpha particle
c. A deuteron, a particle containing one proton and one neutron

Solution

a. Lead-214 is an isotope of lead, so the symbol X is replaced by Pb. The mass number is 214 according to rules learned earlier for designating isotopes, so $A = 214$. The atomic number of lead obtained from the periodic table is 82, so $Z = 82$, and the symbol is $^{214}_{82}$Pb.
b. An alpha particle is represented by the symbol α. Since it contains two protons and two neutrons, $A = 4$ and $Z = 2$. The symbol is $^4_2\alpha$, as shown in Table 10.1.
c. A deuteron is not familiar to you, so we will replace its symbol X with D. It contains one proton and one neutron, so $A = 2$ and $Z = 1$. The symbol is $^2_1 D$. This particle is the nucleus of a deuterium or heavy hydrogen atom and so could also be represented as 2_1H.

✔ **LEARNING CHECK 10.1** Write appropriate symbols for the following particles using the $^A_Z X$ symbolism:

a. An iodine-131 nucleus
b. A beta particle
c. A hypothetical particle composed of four protons
d. A hypothetical particle composed of four electrons

A nuclear equation is balanced when the sum of the atomic numbers on the left side equals the sum of the atomic numbers on the right side, and the sum of the mass numbers on the left equals the sum of the mass numbers on the right. Thus, we focus on mass numbers and atomic numbers in nuclear reactions. As shown in the following example, a nucleus of an element that undergoes a nuclear reaction by emitting a particle is changed to a nucleus of a different element.

Example 10.2 Equations for Nuclear Reactions

Uranium-238 nuclei undergo radioactive decay by emitting an alpha particle and changing to the nucleus of another element. Write a balanced nuclear equation for the process.

Solution

Because the uranium-238 nucleus gives up an alpha particle, $_2^4\alpha$, the mass number of the uranium-238 must decrease by 4 and the atomic number must decrease by 2. As a result, the new nucleus, called a **daughter nucleus**, has a mass number of 234 and an atomic number of 90 (remember, the atomic number of uranium can be obtained from a periodic table). The balanced nuclear equation is

$$_{92}^{238}U \rightarrow \; _2^4\alpha + \; _{90}^{234}Th$$

Note that this equation is balanced because

$$238 = 4 + 234 \qquad \text{and} \qquad 92 = 2 + 90$$

The symbol for the daughter (thorium) was obtained by locating element number 90 in the periodic table.

Processes such as this where one alpha particle is ejected can be represented by the general equation

$$_Z^A X \rightarrow \; _2^4\alpha + \; _{Z-2}^{A-4}Y$$

where Y is the symbol of the daughter nucleus.

✔ **LEARNING CHECK 10.2** Thorium, element number 90, exists in a number of isotopic forms. One form, thorium-234, decays by emitting a beta particle. Write a balanced nuclear equation for the process and write a general equation for decays in which one beta particle is ejected.

Because gamma rays have no mass or atomic numbers, they do not enter into the balancing process of nuclear reactions. However, they do represent energy and should be included in balanced equations when they are known to be emitted.

As described earlier, beta particles are electrons ejected from the nucleus of a radioactive atom, and yet, we know from our discussions of atomic theory in Chapter 2 that atomic nuclei contain only protons and neutrons. Where then do the beta particles originate? According to a simplified theory of nuclear behavior, a beta particle (or electron) is produced in the nucleus when a neutron changes into a proton. The balanced equation for the nuclear reaction is

$$_0^1 n \rightarrow \; _1^1 p + \; _{-1}^0\beta \tag{10.1}$$

Note that the criteria for balance in this equation are satisfied; that is, $1 = 1 + 0$ and $0 = 1 + (-1)$. Thus, a nuclear neutron is converted to a proton during beta emission, and the daughter has an atomic number higher by 1 than the decaying nucleus. You found this result if you did Learning Check 10.2 correctly. Similar changes in nuclear particles occur during two other types of nuclear reactions. Some nuclei decay by positron emission. A **positron** is a positively charged electron with the $_1^0\beta$ (see **Table 10.1**). When a nucleus emits a positron, a nuclear proton is changed to a neutron. Positron-emitting isotopes are used in diagnostic PET scans of the brain (see **Figure 10.2**).

✔ **LEARNING CHECK 10.3** Write a general balanced equation for the nuclear process that occurs during positron emission. How do the atomic numbers of daughter nuclei compare with those of the decaying nucleus?

Certain nuclear changes occur when an electron from outside the nucleus is drawn into the nucleus, where it reacts with a proton and converts it to a neutron. This process, called **electron capture**, is not as common as other decay processes.

daughter nuclei The new nuclei produced when unstable nuclei undergo radioactive decay.

positron A positively charged electron.

electron capture A mode of decay for some unstable nuclei in which an electron from outside the nucleus is drawn into the nucleus, where it combines with a proton to form a neutron.

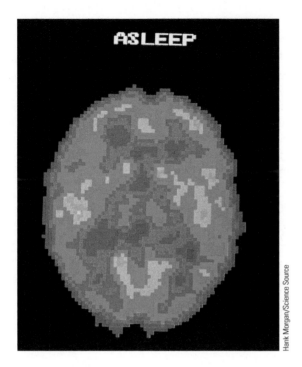

ASLEEP

Hank Morgan/Science Source

Figure 10.2 Positron emission tomography (PET) scan shows normal brain activity during sleep. A radioactive isotope that emits positrons is made into a chemical compound that is absorbed by active areas of the brain. The emitted positrons collide with nearby electrons and produce gamma rays that pass through the skull to detectors surrounding the patient's head. A computer uses the detector data to construct the image.

✔ **LEARNING CHECK 10.4** Write a general balanced equation for the nuclear process that occurs during electron capture. How do the atomic numbers of daughter nuclei compare with those of the decaying nucleus?

Example 10.3 Writing Balanced Nuclear Equations

Write a balanced nuclear equation for the decay of each of the following. The mode of decay is indicated in parentheses.

a. $^{14}_{6}C$ (beta emission)

b. $^{122}_{53}I$ (positron emission)

c. $^{55}_{26}Fe$ (electron capture)

d. $^{212}_{84}Po$ (alpha emission)

Solution

In each case, the atomic number of the daughter is obtained by balancing atomic numbers in the equation. The symbol for the daughter is obtained by matching atomic numbers to symbols in the periodic table.

a. $^{14}_{6}C \rightarrow \, ^{0}_{-1}\beta + \, ^{14}_{7}N$

b. $^{122}_{53}I \rightarrow \, ^{0}_{1}\beta + \, ^{122}_{52}Te$

c. $^{55}_{26}Fe + \, ^{0}_{-1}e \rightarrow \, ^{55}_{25}Mn$

d. $^{212}_{84}Po \rightarrow \, ^{4}_{2}\alpha + \, ^{208}_{82}Pb$

✔ **LEARNING CHECK 10.5** Write a balanced nuclear equation for the decay of each of the following. The mode of decay is indicated in parentheses.

a. $^{50}_{25}Mn$ (positron emission)

b. $^{54}_{25}Mn$ (electron capture)

c. $^{56}_{25}Mn$ (beta emission)

d. $^{224}_{88}Ra$ (alpha emission)

10.3 | Isotope Half-Life

Learning Objective

3. Solve problems using the half-life concept.

half-life The time required for one-half the unstable nuclei in a sample to undergo radioactive decay.

Some radioactive isotopes are more stable than others. The more stable isotopes undergo radioactive decay more slowly than the less stable isotopes. The **half-life** of an isotope is used to indicate stability, and it is equal to the time required for one-half (50%) of the atoms of a sample of the isotope to decay. **Table 10.2** contains examples showing the wide range of half-lives that have been determined.

TABLE 10.2 Examples of Half-Lives

Isotope	Half-Life	Source
$^{238}_{92}U$	4.5×10^9 years	Naturally occurring
$^{40}_{19}K$	1.3×10^9 years	Naturally occurring
$^{226}_{88}Ra$	1600 years	Naturally occurring
$^{14}_{6}C$	5600 years	Naturally occurring
$^{239}_{94}Pu$	24,000 years	Synthetically produced
$^{90}_{38}Sr$	28 years	Synthetically produced
$^{131}_{53}I$	8 days	Synthetically produced
$^{24}_{11}Na$	15 hours	Synthetically produced
$^{15}_{8}O$	2 minutes	Synthetically produced
$^{5}_{3}Li$	10^{-21} seconds	Synthetically produced

Example 10.4 Calculating Half-Lives

Rubidium-79 decays by positron emission and forms krypton-79, which is a gas. A weighed 100.00-mg sample of solid rubidium-79 was allowed to decay for 42 minutes, then weighed again. Its mass was 25.00 mg. What is the half-life of rubidium-79?

Solution

We assume all the gaseous krypton-79 that was formed escaped into the surrounding air and so was not weighed. In the first half-life, one-half the original 100.00-mg sample of rubidium would have been lost, so the sample would have a mass of 50.00 mg. During the second half-life, this 50.00-mg sample would again lose half its mass and would be reduced to 25.00 mg. Thus, we can conclude that two half-lives passed during the 42 minutes that elapsed between the weighings. Because two half-lives equal 42 minutes, one half-life equals 21 minutes.

✔ **LEARNING CHECK 10.6** Potassium-38 decays by positron emission to argon-38, a gas. Potassium-38 has a half-life of 7.7 minutes. How long will it take a 200.00-mg sample of potassium-38 to be reduced to 25.00 mg?

Because radioisotopes are continuously decaying, any found in nature must belong to one of three categories. They may have very long half-lives and decay very slowly ($^{238}_{92}U$), they may be daughters produced by the decay of long-lived isotopes ($^{226}_{88}Ra$), or they may result from natural processes such as cosmic-ray bombardment of stable nuclei ($^{14}_{6}C$).

The fraction of original radioactive atoms remaining in a sample after a specific time has passed can be determined from the half-life. After one half-life has passed, $\frac{1}{2}$ the original number have decayed, so $\frac{1}{2}$ remain. During the next half-life, $\frac{1}{2}$ the remaining $\frac{1}{2}$ decay, so $\frac{1}{4}$ of the original atoms remain. After three half-lives, $\frac{1}{2} \times \frac{1}{2} \times \frac{1}{2} = \frac{1}{8}$ of the original atoms remain undecayed (see **Figure 10.3**).

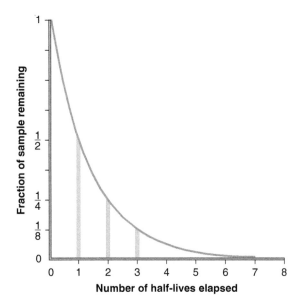

Figure 10.3 A radioactive decay curve (bars give values at 1, 2, and 3 half-lives).

The amount of radiation given off by a sample of radioactive material is proportional to the number of radioactive atoms present in the sample. The amount of radiation given off is called the *activity* of the sample and is measured by devices such as the Geiger–Müller counter described in Section 10.5. Data such as the sample fractions shown in **Figure 10.3** are obtained quite easily. For example, suppose a sample gave an initial reading of 120 units of activity, and it took 35 minutes for the reading to drop to 60 units. The reading of 60 units means that the sample contained just half as many radioactive atoms as it contained when the reading was 120 units. Thus, the 35 minutes represents one half-life, and the fraction remaining from the time of the initial measurement is $\frac{1}{2}$.

Example 10.5 More Half-Life Calculations

The activity (amount of radiation emitted) of a radioactive substance is measured at 9 a.m. At 5 p.m. the activity is found to be only $\frac{1}{16}$ the original (9 a.m.) value. What is the half-life of the radioactive material?

Solution

Because only $\frac{1}{16}$ the original radioactive material is present, four half-lives have passed: $\frac{1}{2} \times \frac{1}{2} \times \frac{1}{2} \times \frac{1}{2} = \frac{1}{16}$. The time between measurements is 8 hours (9 a.m. to 5 p.m.). Therefore, 8 hours is four half-lives, and the half-life is 2 hours.

✔ **LEARNING CHECK 10.7** Iodine-123 is a radioisotope used to diagnose the function of the thyroid gland. It has a half-life of 13.3 hours. What fraction of a diagnostic dose of iodine-123 would be present in a patient 79.8 hours (a little over 3 days) after it was administered?

10.4 | The Health Effects of Radiation

Learning Objective

4. Describe the effects of radiation on health.

You are probably aware of the fact that radiation is hazardous to living organisms. But why? What does radiation do that is so dangerous? The greatest danger to living organisms results from the ability of high-energy or ionizing radiation to knock electrons out of compounds and generate highly reactive particles called **radicals** or **free radicals** in tissue through which the radiation passes. These electron-deficient

radical or free radical An electron-deficient particle that is very reactive.

particles are very reactive. They cause reactions to occur among more stable materials in the cells of living organisms. When these reactions involve genetic materials such as genes and chromosomes, the changes might lead to genetic mutations, cancer, or other serious consequences.

Long-term exposure to low-level radiation is more likely to cause the problems just described than are short exposures to intense radiation. Short-term intense radiation tends to destroy tissue rapidly in the area exposed and can cause symptoms of so-called **acute radiation syndrome**. Some of these symptoms are given in **Table 10.3**. It is this rapid destruction of tissue that makes relatively intense radiation a useful tool in the treatment of some cancers.

TABLE 10.3 The Effects on Humans of Short-Term, Whole-Body Radiation Exposure

Dose (rems)[a]	Effects
0–25	No detectable clinical effects.
25–100	Slight short-term reduction in number of some blood cells, disabling sickness not common.
100–200	Nausea and fatigue, vomiting if dose is greater than 125 rems, longer-term reduction in number of some blood cells.
200–300	Nausea and vomiting first day of exposure, up to a 2-week latent period followed by appetite loss, general malaise, sore throat, pallor, diarrhea, and moderate emaciation. Recovery in about 3 months unless complicated by infection or injury.
300–600	Nausea, vomiting, and diarrhea in first few hours. Up to a 1-week latent period followed by loss of appetite, fever, and general malaise in the second week, followed by hemorrhage, inflammation of mouth and throat, diarrhea, and emaciation. Some deaths in 2–6 weeks. Eventual death for 50% if exposure is above 450 rems; others recover in about 6 months.
600 or more	Nausea, vomiting, and diarrhea in first few hours. Rapid emaciation and death as early as second week. Eventual death of nearly 100%.

[a]The rem, a biological unit of radiation, is defined in Section 10.5.

Source: U.S. Atomic Energy Commission.

The health hazards presented by radiation make it imperative that exposures be minimized. This is especially true for individuals with occupations that continually present opportunities for exposure. One approach is to absorb radiation by shielding. As shown in **Table 10.4**, alpha and beta rays are the easiest to stop. Gamma radiation and X-rays require the use of very dense materials such as lead to provide protection. If the type of radiation is known, a careful choice of shielding materials can provide effective protection.

TABLE 10.4 Penetrating Abilities of Alpha, Beta, and Gamma Rays

Type of Radiation	Depth of Penetration Into		
	Dry Air	Tissue	Lead
Alpha	4 cm	0.05 mm	0
Beta[a]	6–300 cm	0.06–4 mm	0.005–0.3 mm
Gamma[b]	400 m	50 cm	300 mm

[a]Depth depends on energy of rays.

[b]Depth listed is that necessary to reduce initial intensity by 10%.

Another protection against radiation is distance. Because radiation spreads in all directions from a source, the amount falling on a given area decreases the farther that area is from the source. Imagine a lightbulb at the center of a balloon that is being filled slowly with air. The intensity of light on each square centimeter of the balloon will decrease as the surface of the balloon moves away from the light. It has been shown that the intensity is inversely proportional to the square of the distance of the surface from the radiation source. This **inverse square law of radiation** is true only for radiation traveling in a vacuum, but it gives pretty good results when the radiation travels through air. The law is represented by Equation 10.2, where I_x is the intensity at distance d_x and I_y is the intensity at distance d_y:

$$\frac{I_x}{I_y} = \frac{d_y^2}{d_x^2} \qquad (10.2)$$

inverse square law of radiation A mathematical way of saying that the intensity of radiation is inversely proportional to the square of the distance from the source of the radiation.

Example 10.6 Distance and Radiation Intensity

The intensity of radiation is 18.5 units at a distance of 50.0 cm from a source. What is the intensity at a distance of 100. cm?

Solution

Let $I_x = 18.5$ units and $d_x = 50.0$ cm. Then I_y is wanted at $d_y = 100.$ cm. Substitution into Equation 10.2 gives

$$\frac{18.5}{I_y} = \frac{(100. \text{ cm})^2}{(50.0 \text{ cm})^2}$$

or

$$I_y = \frac{(50.0 \text{ cm})^2(18.5)}{(100. \text{ cm})^2} = 4.63 \text{ units}$$

Thus, doubling the distance cuts the intensity to one-fourth the initial intensity.

✔ **LEARNING CHECK 10.8** The intensity of a radioactive source is measured at a distance of 25 feet and equals 10.0 units. What is the intensity 5.0 feet from the source?

10.5 | Measurement Units for Radiation

Learning Objective

5. Describe and compare the units used to measure quantities of radiation.

Two methods, physical and biological, are used to describe quantities of radiation. **Physical units** indicate the activity of a source of radiation, typically in terms of the number of nuclei that decay per unit of time. **Biological units** are related to the damage caused by radiation and account for the fact that a given quantity (or number of particles) of one type of radiation does not have the same damaging effect on tissue as the same quantity of another type of radiation.

The curie and the becquerel are physical units. The **curie** is most widely used at the present time and is equal to 3.7×10^{10} nuclear disintegrations per second. One curie (Ci) represents a large unit, so fractions such as the millicurie (mCi), the microcurie (μCi), and the picocurie (pCi) are often used. The **becquerel** (Bq) is a relatively new unit that has achieved little use among U.S. scientists, even though it is smaller and more manageable than the curie. One becquerel is equal to one disintegration per second.

A number of biological units are used, including the roentgen (R), rad (D), gray (Gy), and rem. The **roentgen** is a biological unit used with X-rays or gamma rays. One roentgen is the quantity of radiation that will generate 2.1×10^9 ion pairs in 1 cm³ of dry air at normal temperature and pressure. The ionization effect of X-rays and gamma rays is greater when the radiation passes through tissue; one roentgen generates about 1.8×10^{12} ion pairs per gram of tissue.

physical unit of radiation A radiation measurement unit indicating the activity of the source of the radiation; for example, the number of nuclear decays per minute.

biological unit of radiation A radiation measurement unit indicating the damage caused by radiation in living tissue.

curie A physical unit of radiation measurement corresponding to 3.7×10^{10} nuclear disintegrations per second.

becquerel A physical unit of radiation measurement corresponding to one nuclear disintegration per second.

roentgen A biological unit of radiation measurement used with X-rays and gamma rays; the quantity of radiation that generates 2.1×10^9 ion pairs per 1 cm³ of dry air or 1.8×10^{12} ion pairs per 1 g of tissue.

CHEMISTRY AROUND US 10.1

Radiation Exposure in Modern Life

Throughout our existence, we humans have been continually exposed to radiation from natural sources such as alpha, beta, or gamma radiation from naturally occurring radioisotopes, sunlight, and cosmic rays from outer space. The discovery of X-rays in 1895 by Wilhelm Roentgen contributed to this exposure when X-rays were introduced as a diagnostic tool for patients. Today, X-rays are used in numerous medical and dental diagnostic tests and procedures.

In the early 1970s, X-rays were integrated with computers to create a diagnostic tool called *computed tomography*, or CT scans. In this procedure, the intensity of X-rays after they pass through the body is interpreted using a computer, and the computer results are used to create a cross-sectional image of the part of the body through which the X-rays passed.

In 1946, the property called *nuclear magnetic resonance* was discovered. Today, this property is widely used medically in the diagnostic procedure called *magnetic resonance imaging*, or MRI. During an MRI procedure, the patient is placed in a very strong magnetic field and radiation in the form of powerful radio waves is passed through the patient's body. The energy of the radio waves is absorbed by hydrogen atoms in the body and then re-emitted. The re-emitted radio waves are interpreted by a powerful computer program that then produces cross-sectional images of the body.

Airline travelers are increasingly being required to pass through full-body scanners before being allowed to board airliners. In this process, the passengers are exposed to low-energy X-rays. Some individuals have received publicity for refusing to go through the scanners in spite of assurances that the radiation dose from such a scan presents, at most, a very tiny risk. An alternative physical search is usually available for such passengers. According to a recent study, an airline passenger receives a radiation dose from flying for one hour at an altitude of 35,000 feet that is at least 100 times more than the dose received by being scanned before the flight. In another comparison, a report from the British Institute of Radiology found that the radiation dose from a single airport scan was 100,000 times lower than the average total annual dose of radiation a person receives from natural background radiation and medical sources.

However, some radiation experts have taken a very cautious approach to the question of scanner safety. One expert urged researchers to carry out investigations that focused on the way airport scans affected specific groups such as children or individuals with gene mutations. The same expert has proposed that the neck and head areas should not be scanned to avoid increasing the risk of a certain type of skin cancer.

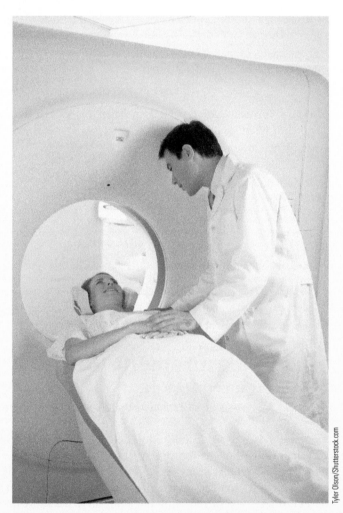

Magnetic resonance imaging (MRI) scans enable medical personnel to view soft tissues in detail.

Tyler Olson/Shutterstock.com

rad A biological unit of radiation measurement corresponding to the transfer of 1×10^{-2} J or 2.4×10^{-3} cal of energy to 1 kg of tissue.

gray A biological unit of radiation measurement corresponding to the transfer of 1 J of energy to 1 kg of tissue.

The rad and the gray are biological units used to describe the effects of radiation in terms of the amount of energy transferred from the radiation to the tissue through which it passes. One **rad** transfers 1×10^{-2} J or 2.4×10^{-3} cal of energy to 1 kg of tissue. One **gray**, a larger unit, transfers 1 J per 1 kg of tissue. From a health point of view, the roentgen and rad are essentially equal. One roentgen of X-rays or gamma rays equals 0.96 rad.

The rem is a biological unit devised to account for health differences in various types of radiation. A 1-rad dose of gamma radiation does not produce the same health effects as a 1-rad dose of alpha radiation. An additive unit was needed so that one unit of alpha had

the same health effects as one unit of gamma. The devised unit is the rem, which stands for roentgen equivalent in man. One **rem** of any type of radiation has the same health effect as 1 roentgen of gamma rays or X-rays. The advantage of such a unit is that individuals exposed to low levels of various types of radiation have to keep track only of the rems absorbed from each type and add them up to get the total health effect. If other units were used, the number of units from each type of radiation would have to be determined and the health effects calculated for each type. The units used for measuring radiation are summarized in **Table 10.5**, together with relationships among the units.

rem A biological unit of radiation measurement corresponding to the health effect produced by 1 roentgen of gamma or X-rays, regardless of the type of radiation involved.

TABLE 10.5 Units for Measuring Radiation

Unit	Type of Unit	Relationships between Units
Curie (Ci)	Physical	$1 \text{ Ci} = 3.7 \times 10^{10} \text{ Bq}$
Becquerel (Bq)	Physical	$1 \text{ Bq} = 2.7 \times 10^{-11} \text{ Ci}$
Roentgen (R)	Biological	$1 \text{ R} = 0.96 \text{ D (tissue)}$
Rad (D)	Biological	$1 \text{ D} = 1 \times 10^{-2} \text{ Gy}$
Gray (Gy)	Biological	$1 \text{ Gy} = 1 \times 10^{2} \text{ D}$
Rem	Biological	1 rem = 1 R X-ray or gamma ray in health effect

Radiation can be detected in a number of ways. Film badges are commonly used by people working in environments where they might be exposed to radiation. The badges contain photographic film that becomes exposed when subjected to radiation. The extent of exposure of the film increases with the amount of radiation. The film is developed after a specific amount of time, and the degree of exposure indicates the radiation dose that has been absorbed during that time (see **Figure 10.4**).

Radiation causes certain substances, called phosphors, to give off visible light. The light appears as brief flashes when radiation strikes a surface coated with a thin layer of phosphors. The number of flashes (called *scintillations*) per unit time is proportional to the amount of radiation striking the surface. A device that employs this approach is called a **scintillation counter**.

A third type of detector, called a **Geiger–Müller tube**, is based on the fact that electricity is conducted through a gas that contains charged particles. A Geiger–Müller tube is represented in **Figure 10.5**. The tube essentially consists of two electrodes charged to high voltages of opposite sign. The tube is filled with a gas at very low pressure. When radiation passes through the tube, it creates a path of charged particles (ions) by knocking electrons off gas molecules. They conduct a brief pulse of current between the electrodes. This brief current is amplified and triggers a counting circuit to record one "count." The number of counts per unit time is proportional to the amount of radiation passing through the Geiger–Müller tube.

Cliff Moore/Science Source

Figure 10.4 Film badges provide a convenient way to monitor the total amount of radiation received during a specific time period.

scintillation counter A radiation-detection device operating on the principle that phosphors give off light when struck by radiation.

Geiger–Müller tube A radiation-detection device operating on the principle that ions form when radiation passes through a tube filled with low-pressure gas.

Figure 10.5 A cross section of a detection tube used in Geiger–Müller counters.

10.6 | Medical Uses of Radioisotopes

Learning Objective

6. Describe, with examples, medical uses of radioisotopes.

Chemically, radioisotopes undergo the same reactions as nonradioactive isotopes of the same element. For example, radioactive iodine-123 is absorbed by the thyroid gland and used to produce the hormone thyroxine just as if it were ordinary nonradioactive iodine. This characteristic, together with the fact that the location of radioisotopes in the body can be readily detected, makes them useful for diagnostic and therapeutic medical applications. In diagnostic applications, radioisotopes are used as **tracers** whose progress through the body or localization in specific organs can be followed.

To minimize the risks to patients from exposure to radiation, radioisotopes used as diagnostic tracers should have as many of the following characteristics as possible:

1. Tracers should have short half-lives so they will decay while the diagnosis is being done but will give off as little radiation as possible after the diagnostic procedure is completed.
2. The daughter produced by the decaying radioisotope should be nontoxic and give off little or no radiation of its own. Ideally, it will be stable.
3. The radioisotope should have a long enough half-life to allow it to be prepared and administered conveniently.
4. The radiation given off by the radioisotope should be penetrating gamma rays, if possible, to ensure that they can be detected readily by detectors located outside the body.
5. The radioisotope should have chemical properties that make it possible for the tissue being studied to either concentrate it in diseased areas and form a **hot spot** or essentially reject it from diseased areas to form a **cold spot**.

Iodine-123 is used diagnostically to determine whether a thyroid gland is properly functioning. As mentioned earlier, the thyroid gland concentrates iodine and uses it in the production of thyroxine, an iodine-containing hormone. Iodine is an excellent tracer to use because the thyroid is the only user of iodine in the body. Iodine-123 is a good tracer because it is a gamma-emitter with a half-life of 13.3 hours. Iodine-131 is also radioactive and has a short half-life (8 days), but it emits gamma and beta radiation. The beta radiation cannot penetrate through the tissue to be detected diagnostically and therefore is simply an added radiation risk to the patient. For this reason, iodine-123 is preferred. An overactive thyroid absorbs more iodine and forms a hot spot, whereas an underactive or nonactive gland absorbs less iodine and may show a cold spot on a thyroid scan (see **Figure 10.6**).

Radioisotopes administered internally for therapeutic use should ideally have the following characteristics:

1. The radioisotope should emit less penetrating alpha or beta radiation to restrict the extent of damage to the desired tissue.
2. The half-life should be long enough to allow sufficient time for the desired therapy to be accomplished.
3. The decay products should be nontoxic and give off little or no radiation.
4. The target tissue should concentrate the radioisotope to restrict the radiation damage to the target tissue.

Iodine-131 is used therapeutically to treat thyroid cancer and hyperthyroidism. The primary advantage of this radioisotope is the ability of thyroid tissue to absorb it and localize its effects.

Until the mid-1950s, when improvements in the generation of penetrating X-rays made them the therapy of choice, cobalt-60 was widely used to treat cancer. This radioisotope emits beta and gamma radiation and has a half-life of 5.3 years. When used therapeutically

tracer A radioisotope used medically because its progress through the body or localization in specific organs can be followed.

hot spot Tissue in which a radioactive tracer concentrates.

cold spot Tissue from which a radioactive tracer is excluded or rejected.

Figure 10.6 A thyroid scan produced after the administration of a radioactive iodine isotope. Is either a hot spot or a cold spot present?

today, a sample of cobalt-60 is placed in a heavy lead container with a window aimed at the cancerous site. The beam of gamma radiation that exits through the window is focused on a small area of the body where the tumor is located. **Table 10.6** contains other examples of radioisotopes used diagnostically and therapeutically.

TABLE 10.6 Examples of Medically Useful Radioisotopes

Isotope	Emission	Half-Life	Applications
$^{3}_{1}H$	Beta	12.3 years	To measure water content of the body
$^{32}_{15}P$	Beta	14.3 days	Detection of tumors, treatment of a form of leukemia
$^{51}_{24}Cr$	Gamma	27.8 days	Diagnosis: size and shape of spleen, gastrointestinal disorders
$^{59}_{26}Fe$	Beta	45.1 days	Diagnosis: anemia, bone marrow function
$^{60}_{27}Co$	Beta, gamma	5.3 years	Therapy: cancer treatment
$^{67}_{31}Ga$	Gamma	78.1 hours	Diagnosis: various tumors
$^{75}_{34}Se$	Beta	120.4 days	Diagnosis: pancreatic scan
$^{81}_{36}Kr$	Gamma	2.1×10^5 years	Diagnosis: lung ventilation scan
$^{85}_{38}Sr$	Gamma	64 days	Diagnosis: bone scan
$^{99}_{43}Tc$	Gamma	6 hours	Diagnosis: brain, liver, kidney, bone, and heart muscle scans
$^{123}_{53}I$	Gamma	13.3 hours	Diagnosis: thyroid cancer
$^{131}_{53}I$	Beta, gamma	8.1 days	Diagnosis and treatment: thyroid cancer
$^{197}_{80}Hg$	Gamma	65 hours	Diagnosis: kidney scan

10.7 | Nonmedical Uses of Radioisotopes

Learning Objective

7. Describe, with examples, nonmedical uses of radioisotopes.

The study of the photosynthetic process in plants has been made much easier because of the use of radioactive carbon-14 as a tracer. The overall photosynthetic process is represented by Equation 10.3:

$$6CO_2(g) + 6H_2O(\ell) \rightarrow C_6H_{12}O_6(aq) + 6O_2(g) \qquad (10.3)$$

However, the conversion of water (H_2O) and carbon dioxide (CO_2) into carbohydrates ($C_6H_{12}O_6$) involves numerous chemical reactions and carbon-containing intermediate compounds. By using CO_2 that contains radioactive carbon-14, scientists have been able to follow the chemical path of CO_2 through the various intermediate compounds.

Interesting uses for radioactive tracers are found in industry. In the petroleum industry, for example, radioisotopes are used in pipelines to indicate the boundary between different products moving through the lines (see **Figure 10.7**). In another application, tracer added to fluids moving in pipes makes it easy to detect leaks. Also, the effectiveness of lubricants has been studied by the use of radioactive metal isotopes as components of metal parts; as the metal part wears, the isotope shows up in the lubricant, and the amount of wear can be determined.

Check the Radon Level in Your Home

Substances that are chemically unreactive are generally not thought to be dangerous to one's health. However, radon, a colorless, odorless, chemically inert noble gas, is considered to be a serious health risk because it is radioactive. It is believed to be the leading cause of lung cancer among nonsmokers. It has been estimated by U.S. officials that up to 8% of the annual lung cancer deaths in the United States can be attributed to indoor exposure to radon-222.

Radon-222 is formed during a multistep process in which naturally occurring uranium-238 decays to stable lead-206. During one step of the process, radium-226, with a half-life of 1600 years, emits alpha and gamma radiation to produce radon-222:

$$^{226}_{88}\text{Ra} \rightarrow {}^{4}_{2}\alpha + {}^{222}_{86}\text{Rn} + \gamma$$

Radon-222, with a half-life of 4 days, is produced in rocks and soil that contain even minute amounts of uranium-238. Because it is a gas, radon migrates readily from the soil into the surrounding air. It seeps into houses and other buildings through openings around pipes, and through cracks in basement floors and walls.

The average level of radon in homes is lower than 1 picocurie/L of air. A picocurie (10^{-12} curie) is equivalent to the radioactive decay of 2 radon nuclei per minute. The Environmental Protection Agency recommends an upper limit of 4 picocuries/L for indoor air, and it has been estimated that 99% of the homes in the United States have levels below this value. However, new studies reported by the World Health Organization have led to recommendations that new upper limits of 2.7 picocuries/L be established. This level is 33% lower than the current upper limit EPA recommendations.

Levels of radiation much higher than 4 picocuries/L have been found in some homes. The highest recorded level of radioactivity from radon in a home was 2700 picocuries/L of air. This was found in a home in Pennsylvania that was built on soil rich in uranium-238. The level of radon in home air can be minimized by sealing the entry points and providing good ventilation.

The hazards associated with breathing radon-222 come not only from the alpha radiation emitted as the radon decays but also from the daughter products of that decay. None of these decay products are gases, but all are radioactive. Because they are not gases, they are not easily exhaled and can become lodged in the lungs, where their radiation can cause serious damage.

It is difficult to decide just how much concern people should have about radon in their homes. Some experts in the field agree with the EPA recommendations, but others question the data on which those recommendations are based. It is quite easy and relatively inexpensive to test for radon in a home, and homes built in locations where the gas has been identified as a problem should be tested.

Radon detectors are available for home use.

Figure 10.7 Radioactive tracers used in industry.

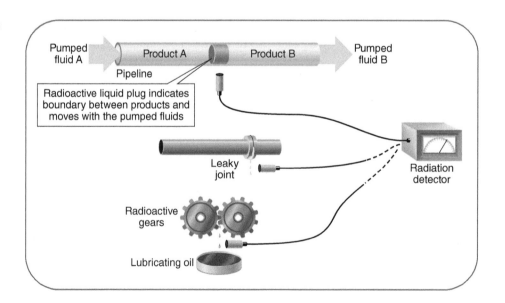

Radioisotopes are also useful in forms other than tracers. The thickness of manufactured materials, such as metal foils or sheets, can be determined continuously by the use of a radioisotope. The amount of radiation that penetrates the material to the detector is related to the thickness of the material.

An interesting application of radioisotopes involves the use of naturally occurring radioactive materials to determine the ages of artifacts and rocks. Perhaps the most widely known example of **radioactive dating** is the use of carbon-14, which is produced in the atmosphere when cosmic-ray neutrons strike nitrogen nuclei. The equation for the reaction is

$$^{14}_{7}\text{N} + {}^{1}_{0}n \rightarrow {}^{14}_{6}\text{C} + {}^{1}_{1}p \qquad (10.4)$$

radioactive dating A process for determining the age of artifacts and rocks, based on the amount and half-life of radioisotopes contained in the object.

The resulting $^{14}_{6}\text{C}$ is converted to carbon dioxide $^{14}_{6}\text{CO}_2$ through a chemical reaction with oxygen. In this form, the radioactive carbon is absorbed by plants and converted into cellulose or other carbohydrates during photosynthesis. As long as the plant lives, it takes in $^{14}_{6}\text{CO}_2$. As a result, an equilibrium is established in which the plant contains the same fraction of carbon-14 as the surrounding air. Thus, as long as the plant lives, a constant fraction of the total carbon present is carbon-14. When the plant dies, $^{14}_{6}\text{CO}_2$ intake stops, and the fractional amount of carbon-14 begins to decrease. From a measurement of this amount in, say, a wooden object, and a knowledge of the half-life of carbon-14 (5600 years), it is possible to calculate the age of the object—how long since the tree was cut down to make the object. An object containing only about one-eighth as much carbon-14 as a fresh wood sample from a living tree would be about three half-lives, or 16,800 years old.

This method is limited to objects less than about 50,000 years old because of difficulties in measuring the small amount of carbon-14 present. For this reason, other dating methods have been developed. One method involves potassium-40 and argon-40. The half-life of potassium-40, which absorbs one of its own inner electrons to produce argon-40, is 1.3×10^9 years. By determining the amount of argon-40 in a potassium-containing mineral, it is possible to estimate the age of the mineral. The equation for the reaction is

$$^{40}_{19}\text{K} + {}^{0}_{-1}e \rightarrow {}^{40}_{18}\text{Ar} \qquad (10.5)$$

10.8 | Induced Nuclear Reactions

Learning Objective

8. Show that you understand the concept of induced nuclear reactions.

Before 1934, the study of radioactivity was limited to reactions of the relatively few radioisotopes found in nature. In that year, Irene and Frederic Joliot-Curie, French physical chemists, found that radioactivity could be induced in nonradioactive nuclei by bombarding them with small, subatomic particles. They produced an artificial radioactive isotope, nitrogen-13, by bombarding boron-10 with alpha particles from a natural radioisotope:

$$^{10}_{5}\text{B} + {}^{4}_{2}\alpha \rightarrow {}^{13}_{7}\text{N} + {}^{1}_{0}n \qquad (10.6)$$

Notice that a neutron, $^{1}_{0}n$, is produced in addition to $^{13}_{7}\text{N}$. Fifteen years earlier, British physicist Ernest Rutherford had used a similar reaction to produce nonradioactive oxygen-17 from nitrogen-14. A small particle, a proton, was produced in addition to oxygen-17:

$$^{14}_{7}\text{N} + {}^{4}_{2}\alpha \rightarrow {}^{17}_{8}\text{O} + {}^{1}_{1}p \qquad (10.7)$$

A great deal of research has been done since these first experiments were performed, and today it is known that a variety of nuclear reactions can be induced by bombarding either stable or naturally radioactive nuclei with high-energy particles. These artificially induced reactions may lead to one of four results. The first possibility is that the bombarded nucleus may be changed into a different nucleus that is stable.

A second possibility is that a new nucleus may be formed that is unstable and undergoes radioactive decay. The other two possibilities are that nuclear fission or nuclear fusion will take place. These are discussed in more detail in **Section 10.9**. The result of bombardment depends on the kind of nuclei bombarded and the nature and energy of the bombarding particles.

Charged or uncharged particles may be used to bombard nuclei, and each type presents unique problems to overcome. It is believed that a bombarding particle will be drawn into the nucleus of an atom by strong nuclear forces if it can get to within 10^{-12} cm of the nucleus. Neutrons are captured readily if their speeds are reduced enough to allow the nuclear forces to work. Neutrons are slowed by passing them through materials known as **moderators**. Graphite is an example of a neutron moderator. Charged particles present a different problem. If they are negatively charged ($_{-1}^{0}\beta$), they are repelled by the electron cloud around the nucleus; for a reaction to take place they must have sufficient energy to overcome this repulsion. Similarly, positive particles must be energetic enough to overcome repulsion by the positively charged nucleus.

Neutrons for bombardment reactions are obtained from other nuclear reactions and, together with proper moderators, are used as they come from the sources. Charged particles are obtained from radioactive materials or from ionization reactions. For example, alpha particles are produced by an ionization reaction in which the two electrons are removed from helium atoms. These charged particles are generally accelerated to high speeds (and energies) before bombarding the intended target nuclei. Two types of particle accelerators, cyclic and linear, are used. The **cyclotron**, one type of cyclic accelerator, is represented in **Figure 10.8**.

moderators Materials capable of slowing down neutrons that pass through them.

cyclotron A cyclic particle accelerator that works by changing electrical polarities as charged particles cross a gap. The particles are kept moving in a spiral path by a strong magnetic field.

Figure 10.8 The cyclotron. Early versions of the cyclotron were about the size of a dinner plate.

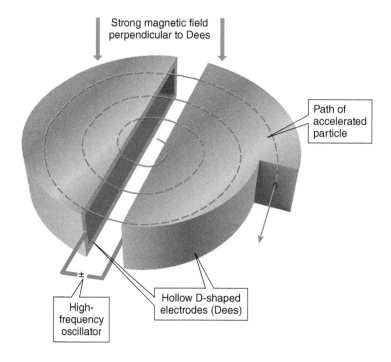

The charged particles to be accelerated enter the evacuated chamber at the center of the cyclotron and, because of the magnetic field, move in a circular path toward the gap between the Dees. Just as the particles reach the gap, the electrical charge on the Dees is adjusted so the particles are repelled by the Dee they are leaving and attracted to the other one. The particles then coast inside the Dee until they again reach a gap, at which point the charges are again adjusted to cause acceleration. This process continues, and the particles are accelerated each time they pass through the gap. As the speed and energy of the particles increase, so does the radius of their circular path, until they finally leave the Dees and strike the target.

In a **linear accelerator**, the particles are accelerated through a series of axially aligned, charged tubes located in an evacuated chamber (see **Figure 10.9**). Each time a particle passes through a gap and goes from one tube to another, it is accelerated in the same way as particles passing between the Dees of a cyclotron. To make the particle accelerate, the charges on the tubes are reversed at the proper times. The tubes get successively longer to allow the particles the same residence or coasting time as they move toward the end at ever-increasing speeds. After acceleration, the particles exit and strike the target.

linear accelerator A particle accelerator that works by changing electrical polarities as charged particles cross gaps between segments of a long tube.

Figure 10.9 A linear particle accelerator.

One of the most interesting results of bombardment is the creation of completely new elements. Four of these elements, produced between 1937 and 1941, filled gaps in the periodic table for which no naturally occurring element had been found. These four are technetium (Tc, number 43), promethium (Pm, number 61), astatine (At, number 85), and francium (Fr, number 87). The equations for the reactions for their production are:

$$^{96}_{42}\text{Mo} + ^{2}_{1}\text{H} \rightarrow ^{1}_{0}n + ^{97}_{43}\text{Tc} \text{ (half-life: } 2.6 \times 10^6 \text{ years)} \tag{10.8}$$

$$^{142}_{60}\text{Nd} + ^{1}_{0}n \rightarrow ^{0}_{-1}\beta + ^{143}_{61}\text{Pm} \text{ (half-life: 265 days)} \tag{10.9}$$

$$^{209}_{83}\text{Bi} + ^{4}_{2}\alpha \rightarrow 3^{1}_{0}n + ^{210}_{85}\text{At} \text{ (half-life: 8.3 hours)} \tag{10.10}$$

$$^{230}_{90}\text{Th} + ^{1}_{1}p \rightarrow 2^{4}_{2}\alpha + ^{223}_{87}\text{Fr} \text{ (half-life: 22 minutes)} \tag{10.11}$$

Synthetic elements heavier than uranium, the heaviest naturally occurring element, have also been produced. All these elements, called **transuranium elements**, are radioactive. The first, neptunium (Np, number 93), was produced in 1940; since then at least 25 more (numbers 94 through 118) have been synthesized.

transuranium elements Synthetic elements with atomic numbers greater than that of uranium.

Significant quantities of only a few of these elements have been produced. Plutonium is synthesized in quantities large enough to make it available for use in atomic weapons and reactors. In 1968 the entire world's supply of californium (10^{-4} g) was gathered and made into a target for bombardment by heavy, accelerated particles. These experiments led to the discovery of several elements heavier than californium. Usually, the amounts of the new elements produced are extremely small—sometimes only hundreds of atoms. When the half-life is short, this small amount quickly disappears. Thus, these elements have to be identified not by chemical properties but by the use of instruments that analyze the characteristic radiation emitted by each new element.

Example 10.7 Writing Equations for Nuclear Reactions

An isotope of gallium (Ga) useful in medical diagnostic work is formed when zinc-66 is bombarded with accelerated protons. Write the equation for this nuclear reaction and determine what isotope of gallium is produced.

Solution
The equation for the reaction is written

$$^{66}_{30}\text{Zn} + ^{1}_{1}p \rightarrow ?$$

where the product is to be determined. By balancing the mass numbers and atomic numbers, we see that the product has a mass number of 67 and an atomic number of 31. Reference to the periodic table shows element number 31 to be gallium, Ga. Therefore, the isotope produced is gallium-67, and the equation for the reaction is:

$$^{66}_{30}\text{Zn} + ^{1}_{1}p \rightarrow ^{67}_{31}\text{Ga}$$

✔ **LEARNING CHECK 10.9** When manganese-55 is bombarded by protons, a neutron is one product. Write the equation for the process and identify the other product.

10.9 | Nuclear Energy

Learning Objective

9. Describe the differences between nuclear fission and nuclear fusion.

In 1903, Ernest Rutherford and Frederick Soddy made observations that led them to conclude that the nuclei of all atoms, not just radioactive ones, must contain large quantities of energy. However, the form it took and the means for releasing it remained a mystery. In 1905, Albert Einstein, who was then working as a patent clerk, provided a partial answer. According to his theory of relativity, matter and energy were equivalent and one could be converted into the other. This fact was expressed in his now-famous equation,

$$E = mc^2 \tag{10.12}$$

According to his equation, a small amount of mass, m, would, upon conversion, yield a huge amount of energy, E. The amount of energy liberated is huge because of the size of the multiplier c^2, the velocity of light squared. The velocity of light is a very large number (3.0×10^8 m/s, or 1.9×10^5 mi/s). The verification of Einstein's theory was not made until 1939, when the work of Otto Hahn, Lise Meitner, Fritz Strassman, and Enrico Fermi led to the discovery of **nuclear fission**. Attempts to produce new heavier transuranium elements by bombarding uranium-235 with neutrons led instead to the production of elements much less massive than uranium-235. The uranium-235 nuclei had undergone fission and split into smaller fragments. In the process, a small amount of matter was converted into energy, just as Einstein had predicted.

Today, we know that uranium-235 is the only naturally occurring isotope that will undergo fission, and a number of different products can result. Equations 10.13–10.16 represent some of the numerous fission reactions that can occur.

$$^{235}_{92}\text{U} + ^{1}_{0}n \nearrow \quad ^{135}_{53}\text{I} + ^{97}_{39}\text{Y} + 4^{1}_{0}n \tag{10.13}$$

$$^{139}_{56}\text{Ba} + ^{94}_{36}\text{Kr} + 3^{1}_{0}n \tag{10.14}$$

$$^{102}_{42}\text{Mo} + ^{132}_{50}\text{Sn} + 2^{1}_{0}n \tag{10.15}$$

$$^{139}_{54}\text{Xe} + ^{95}_{38}\text{Sr} + 2^{1}_{0}n \tag{10.16}$$

For every uranium-235 nucleus undergoing fission, more than one neutron is generated, which opens up the possibility for a chain reaction to take place. If only one of the neutrons produced each time reacts with another uranium-235 nucleus, the process will become a **chain reaction** that continues at a constant rate. If more than one of the neutrons generated per fission reaction produces another reaction, the process becomes an **expanding** or **branching chain reaction** that leads to an explosion. A constant-rate chain reaction is called a **critical reaction**. A branching chain reaction that will lead to an explosion is called **supercritical** (see **Figure 10.10**).

nuclear fission A process in which large nuclei split into smaller, approximately equal-sized nuclei when bombarded by neutrons.

chain reaction A nuclear reaction in which the products of one reaction cause a repeat of the reaction to take place. In the case of uranium fission, neutrons from fission reactions cause other fission reactions to occur.

expanding or branching chain reaction A reaction in which the products of one reaction cause more than one more reaction to occur.

critical reaction A constant-rate chain reaction.

supercritical reaction A branching chain reaction.

Figure 10.10 Nuclear chain reactions.

CRITICAL

SUPERCRITICAL

History's first critical chain reaction was carried out in an atomic pile, which was literally a stack or pile of uranium-containing graphite blocks and blocks of pure graphite that served as a neutron moderator (see **Figure 10.11**). The pile was constructed in a squash court beneath the stands of an athletic fieldhouse at the University of Chicago, and the experiments were carried out by a team of scientists led by Italian physicist Enrico Fermi. In December 1942, the team observed that nuclear reactions in the pile had become self-sustaining, or critical. The control rods, composed of strong neutron absorbers, were pushed into the pile to stop the reaction.

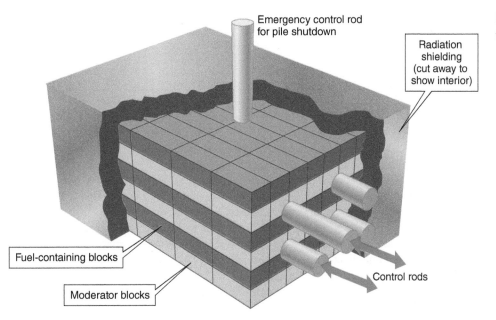

Emergency control rod
for pile shutdown

Radiation
shielding
(cut away to
show interior)

Fuel-containing blocks

Moderator blocks

Control rods

Figure 10.11 An atomic pile, the world's first nuclear reactor.

Once it had been demonstrated that uranium-235 would undergo self-sustaining fission, the construction of an atomic bomb was fairly simple in principle, but it depended on the concept of critical mass. Neutrons released inside a small piece of fissionable material might escape to the outside before they collide with and split another atom. If the majority of neutrons are lost this way, a chain reaction will not take place. However, as the size of the piece of fissionable material is increased, the chance for neutrons to escape without hitting other atoms decreases. The amount of fissionable material needed to cause a

critical mass The minimum amount of fissionable material needed to sustain a critical chain reaction at a constant rate.

supercritical mass The minimum amount of fissionable material that must be present to cause a branching chain reaction to occur.

critical reaction to occur is the **critical mass**. The amount needed to cause a branching chain reaction and explosion to occur is called a **supercritical mass**.

During World War II, one major difficulty encountered in building a nuclear bomb was the size of the device needed to bring together subcritical masses to form a supercritical mass. Any bomb used in combat would have to fit into a B-29, the largest bomber available at the time, and could not be heavier than the load capacity of the airplane. The configurations of the first two devices actually used are represented in **Figure 10.12**. The spherical configuration resulted in a short, wide bomb, while the tubular configuration gave rise to a long, narrow bomb. In each configuration, explosive charges were used to drive subcritical masses together to form a supercritical mass. These nuclear bombs were the first applications of the concepts of subcritical and supercritical mass.

Figure 10.12 Configurations used in fission-type nuclear bombs.

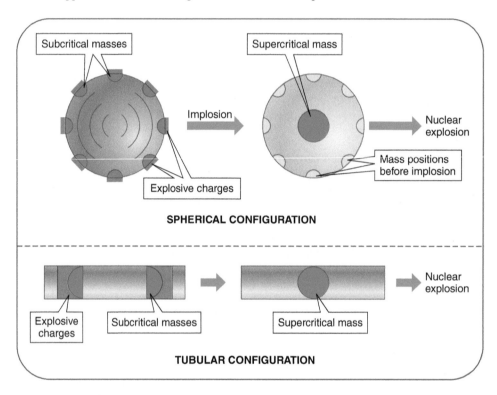

On August 6, 1945, the United States became the first nation to use a nuclear weapon in combat. This bomb used a tubular configuration. Approximately 70,000 inhabitants of Hiroshima, Japan, were killed in the explosion; an equal number were seriously injured. A few days later a second bomb, with a spherical configuration, was dropped on the city of Nagasaki. The casualty figures were similar, and World War II came to an end. The Nagasaki bomb was made of plutonium, an artificial fissionable material produced in atomic piles by breeding reactions.

breeder reaction A nuclear reaction in which isotopes that will not undergo spontaneous fission are changed into isotopes that will.

A **breeder reaction** is an induced reaction in which the product nuclei undergo fission reactions. The plutonium used in the Nagasaki bomb was made by bombarding uranium-238 with neutrons in the first atomic pile. Uranium-238 is the most abundant isotope of uranium found in nature, but it will not undergo fission, as will the much less abundant uranium-235. However, uranium-238 reacts as follows when placed in a neutron-rich environment.

$$^{238}_{92}\text{U} + ^{1}_{0}n \rightarrow ^{239}_{92}\text{U} \tag{10.17}$$

$$^{239}_{92}\text{U} \rightarrow ^{239}_{93}\text{Np} + ^{0}_{-1}\beta \tag{10.18}$$

$$^{239}_{93}\text{Np} \rightarrow ^{239}_{94}\text{Pu} + ^{0}_{-1}\beta \tag{10.19}$$

The plutonium-239 will undergo fission and, when collected into a supercritical mass, will cause an explosion. Reactions such as 10.17–10.19 are called breeding reactions because a fissionable material is "bred" from a nonfissionable material.

Curiosity about the seemingly unlimited energy supply of the stars (especially the sun) eventually led scientists to an understanding of a second nuclear process that releases vast amounts of energy. In 1920, Sir Arthur Eddington suggested that the energy of the stars is a by-product of a reaction in which hydrogen is changed into helium. This was followed in 1929 by the concept of a **thermonuclear reaction**, a nuclear reaction that is started by very high temperatures. According to this idea, at high temperatures the nuclei of very lightweight elements will combine to form nuclei of heavier elements, and in the process some mass is converted into energy. This idea was consistent with Eddington's suggestion that hydrogen is converted to helium in the sun and other stars. It was not until nuclear fission reactions became available that high enough temperatures could be achieved to test the theory of **nuclear fusion**. The theory proved to be correct, and today most nuclear weapons in the world are based on fusion reactions.

thermonuclear reaction A nuclear fusion reaction that requires a very high temperature to start.

The fusion reactions responsible for the energy output of the sun are thought to be

$$2{}^1_1\text{H} \rightarrow {}^2_1\text{H} + {}^0_1\beta \tag{10.20}$$

$$ {}^1_1\text{H} + {}^2_1\text{H} \rightarrow {}^3_2\text{He} + \gamma \tag{10.21}$$

$$ {}^3_2\text{He} + {}^3_2\text{He} \rightarrow {}^4_2\text{He} + 2{}^1_1\text{H} \tag{10.22}$$

nuclear fusion A process in which small nuclei combine or fuse to form larger nuclei.

The net overall reaction is

$$4{}^1_1\text{H} \rightarrow {}^4_2\text{He} + 2{}^0_1\beta + 2\gamma \tag{10.23}$$

which is what Eddington proposed; hydrogen (${}^1_1\text{H}$) is converted to helium (${}^4_2\text{He}$).

A goal of many of the early developers of nuclear energy was to harness it and make it a virtually unlimited source of useful energy. Today, that goal has been partially achieved. Fission reactors generate a small percentage of the electricity used in the world, and researchers are making progress in their attempts to harness fusion reactions.

The essential components of a fission-powered generating plant are shown in **Figure 10.13**. The use of nuclear power plants has created much controversy. The plants release large amounts of waste heat and thus cause thermal pollution of natural waters. In addition, radioactive wastes create disposal problems—where do you dump radioactive materials that will be a health hazard for thousands of years? Electricity consumers and producers have lost some of their confidence in the safety of nuclear power plants as a result of accidents such as the well-known incidents at Three-Mile Island (1979), Chernobyl (1986), and Fukushima (2011). Another problem is the relatively short supply of fissionable uranium-235. Breeder reactors could solve this problem, since they can convert abundant supplies of uranium-238 and thorium-232 into fissionable fuels, but the breeder reactor program in the United States is at a standstill. As a result of these factors, the future of fission power generation is uncertain at the present time.

Figure 10.13 Components of a nuclear electricity power plant.

Medications to Avoid on Test Day

There you were, beginning to take the chemistry exam that you had studied so hard for. If anyone was prepared, it was you. You attended lecture, took copious notes, studied the text, and spent hours memorizing the material. What more could a student do? You were feeling good about your preparation, and then as you read the questions, you realized that you couldn't remember the material you had studied. How could that be?

This scenario is every student's nightmare, and too frequently it becomes a reality, because memory is a complicated molecular process that involves the transfer of molecules from one place to another in the brain. Could you have taken something that would interfere with that transfer? There are numerous medications commonly taken by students that might diminish the ability to learn and remember. Some of these culprit medications are available over the counter without a prescription.

To avoid memory problems, be sure to follow good health practices like proper nutrition, and be sure to get an appropriate amount of nightly sleep. Next, have a pharmacist evaluate all medications/supplements you take. Sometimes, the chronic use of certain compounds will make learning and remembering more difficult. The most likely situation that causes acute memory impairment in students is the use of medications that block the action of acetylcholine. These substances are called anticholinergics.

Acetylcholine is the primary neurotransmitter involved with memory and learning. If the acetylcholine pathway is inhibited, you can't get all of the information stored in the brain out of the brain.

The most common drugs used by students that have this negative memory effect are antihistamines. These medications, are commonly taken to relieve cold, flu, allergy, and insomnia symptoms. The most commonly used antihistamine is diphenhydramine (popular brand name is Benadryl). The good news is that these medications do relieve some symptoms. The bad news is your memory might suffer. The reason for this is that some antihistamines not only block the histamine receptor but also block acetylcholine receptors involved in memory. Some good news is that new second- and third-generation antihistamines are becoming available. These compounds have been designed so that they do not block acetylcholine receptors, and thus have no effect on memory.

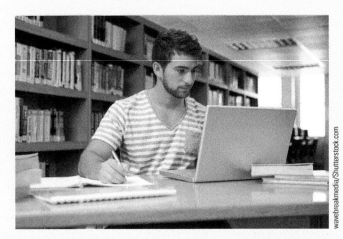

Some medications can diminish your ability to learn and remember.

In principle, fusion reactors can be used as controlled energy sources just as fission reactors are. Fusion reactors would have many advantages over fission reactors. The deuterium fuel (hydrogen-2) is readily available since it is a minor constituent of all water on Earth, no significant amounts of radioactive wastes would be produced during operation, and the possibility of a serious accident is much smaller. Unfortunately, the feasibility of fusion reactors has not yet been established, although scientists are getting closer through continuing research.

▶ Case Study Follow-up

A complete cystectomy is a surgical procedure that removes the urinary bladder (and the prostrate gland in men). Patients who undergo this procedure will dispose of urinary wastes through a urostomy, which is constructed from a section of small intestine into which the ureters are connected. This section of small intestine exits the abdomen and connects to a special collection apparatus that the patient empties manually when it is full. The procedure is permanent and bypasses the urinary bladder.

Smoking is the most important risk factor for developing bladder cancer. Smokers are three times more likely to develop bladder cancer than non-smokers. Tobacco contains carcinogens (cancer-causing agents) that enter the blood through the lungs, then concentrate in the urine. These concentrated toxins can cause cancer in the bladder, ureters, or kidneys even years after exposure. The best prevention for bladder cancer is to drink plenty of fluids and do not smoke.

Radioactive Nuclei Some nuclei are unstable and undergo radioactive decay. The common types of radiation emitted during decay processes are alpha, beta, and gamma, which can be characterized by mass and charge values.
Objective 1 (Section 10.1), Exercise 10.2

Equations for Nuclear Reactions Nuclear reactions can be represented by balanced equations in which the focus is on mass number and atomic number balance on each side. Symbols are used in nuclear equations that make it convenient to balance mass and atomic numbers.
Objective 2 (Section 10.2), Exercise 10.12

Isotope Half-Life Different radioisotopes generally decay at different rates, which are indicated by half-lives. One half-life is the time required for one-half of the unstable nuclei in a sample to undergo radioactive decay.
Objective 3 (Section 10.3), Exercise 10.16

The Health Effects of Radiation Radiation generates free radicals in tissue as it passes through. Radiation is hazardous even at low intensity if there is long-term exposure. Radiation sickness is caused by short-term intense radiation. Those working around radioactive sources can minimize exposure by using shielding or distance as a protection.
Objective 4 (Section 10.4), Exercise 10.22

Measurement Units for Radiation Two systems, physical and biological, are used to describe quantities of radiation. Physical units indicate the number of nuclei of radioactive material that decay per unit of time. Biological units are related to the damage caused by radiation in living tissue. The common physical units are the curie and its fractions and the becquerel. Biological units include the roentgen (for gamma and X-rays), the rad, the gray, and the rem.
Objective 5 (Section 10.5), Exercise 10.24

Medical Uses of Radioisotopes Radioisotopes behave chemically like nonradioisotopes of the same element and can be used diagnostically and therapeutically. Diagnostically, radioisotopes are used as tracers whose movement or localization in the body can be followed. Therapeutic radioisotopes localize in diseased areas of the body, where their radiation can destroy diseased tissue.
Objective 6 (Section 10.6), Exercise 10.30

Nonmedical Uses of Radioisotopes Nonmedical uses of radioisotopes include: (1) tracers in chemical reactions, (2) boundary markers between liquids in pipelines, (3) tracers to detect leaks, (4) metal-wear indicators, and (5) thickness indicators for foils and sheet metal. An especially interesting application is the determination of the ages of artifacts and rocks.
Objective 7 (Section 10.7), Exercise 10.36

Induced Nuclear Reactions The bombardment of nuclei with small high-energy particles causes the bombarded nuclei to (1) change into stable nuclei, (2) change into radioactive nuclei, (3) break into smaller pieces (fission), or (4) form larger nuclei (fusion). Most bombarding particles are accelerated by using cyclic or linear accelerators. Bombardment reactions have been used to produce elements that are not found in nature.
Objective 8 (Section 10.8), Exercise 10.38

Nuclear Energy During spontaneous nuclear fission reactions, heavy nuclei split when bombarded by neutrons and release large amounts of energy. This process was used to produce two atomic bombs whose use ended World War II. A second energy-releasing nuclear process, fusion, is the basis for today's hydrogen bombs. Nuclear fission is in limited use as a source of electrical power; however, this use is controversial. Nuclear fusion has not yet proved feasible as a controlled source of power, but research toward this end continues.
Objective 9 (Section 10.9), Exercise 10.48

Key Terms and Concepts

Acute radiation syndrome (10.4)
Alpha particle (10.1)
Becquerel (10.5)
Beta particle (10.1)
Biological unit of radiation (10.5)
Breeder reaction (10.9)
Chain reaction (10.9)
Cold spot (10.6)
Critical mass (10.9)
Critical reaction (10.9)
Curie (10.5)
Cyclotron (10.8)
Daughter nuclei (10.2)
Electron capture (10.2)

Expanding or branching chain
 reaction (10.9)
Gamma ray (10.1)
Geiger–Müller tube (10.5)
Gray (10.5)
Half-life (10.3)
Hot spot (10.6)
Inverse square law of radiation (10.4)
Linear accelerator (10.8)
Moderators (10.8)
Nuclear fission (10.9)
Nuclear fusion (10.9)
Physical unit of radiation (10.5)
Positron (10.2)

Rad (10.5)
Radical or free radical (10.4)
Radioactive dating (10.7)
Radioactive decay (10.1)
Radioactive nuclei (10.1)
Radioisotope (10.2)
Rem (10.5)
Roentgen (10.5)
Scintillation counter (10.5)
Supercritical mass (10.9)
Supercritical reaction (10.9)
Thermonuclear reaction (10.9)
Tracer (10.6)
Transuranium elements (10.8)

Key Equations

1. Symbol for isotopes (Section 10.2):	$^A_Z X$	
2. Conversion of a neutron to a proton and electron in the nucleus, leading to beta emission (Section 10.2):	$^1_0 n \rightarrow {}^1_1 p + {}^{\ 0}_{-1}\beta$	Equation 10.1
3. Variation of radiation intensity with distance (Section 10.4):	$\dfrac{I_x}{I_y} = \dfrac{d_y^2}{d_y^2}$	Equation 10.2
4. Fission of uranium-235 (Section 10.9): See also Equations 10.14, 10.15, and 10.16.	$^{235}_{92}U + {}^1_0 n \rightarrow {}^{135}_{53}I + {}^{97}_{39}Y + 4{}^1_0 n$	Equation 10.13
5. Hydrogen fusion reaction of the sun (Section 10.9):	$4{}^1_1 H \rightarrow {}^4_2 He + 2{}^{\ 0}_{1}\beta + 2\gamma$	Equation 10.23

Exercises

Even-numbered exercises are answered in Appendix B.

Blue-numbered exercises are more challenging.

Radioactive Nuclei (Section 10.1)

10.1 Define the term *radioactive*; then criticize the following statements:

 a. Beta rays are radioactive.

 b. Radon is a stable radioactive element.

10.2 Group the common nuclear radiations (Table 10.1) into the following categories:

 a. Those with a mass number of 0

 b. Those with a positive charge

 c. Those with a charge of 0

10.3 Group the common nuclear radiations (Table 10.1) into the following categories:

 a. Those with a negative charge

 b. Those with a mass number greater than 0

 c. Those that consist of particles

10.4 Characterize the following nuclear particles in terms of the fundamental particles—protons, neutrons, and electrons:

 a. A beta particle

 b. An alpha particle

 c. A positron

10.5 Discuss how the charge and mass of particles that comprise radiation influence the range or ability of the radiation to penetrate matter.

Equations for Nuclear Reactions (Section 10.2)

10.6 Summarize how the atomic number and mass number of daughter nuclei compare with the original nuclei after:

 a. An alpha particle is emitted.

 b. A beta particle is emitted.

 c. An electron is captured.

 d. A gamma ray is emitted.

 e. A positron is emitted.

10.7 Write appropriate symbols for the following particles using the $^A_Z X$ symbolism:

 a. A tin-117 nucleus

 b. A nucleus of the chromium (Cr) isotope containing 26 neutrons

 c. A nucleus of element number 20 that contains 24 neutrons

10.8 Write appropriate symbols for the following particles using the $^A_Z X$ symbolism:

 a. A nucleus of the element in period 5 and group VB(5) with a mass number of 96

 b. A nucleus of element number 37 with a mass number of 80

 c. A nucleus of the calcium (Ca) isotope that contains 18 neutrons

10.9 Complete the following equations, using appropriate notations and formulas:

 a. $^{10}_4 Be \rightarrow ? + {}^{10}_5 B$

 b. $^{210}_{83} Bi \rightarrow {}^4_2\alpha + ?$

 c. $^{15}_8 O \rightarrow ? + {}^{15}_7 N$

 d. $^{44}_{22} Ti + {}^{\ 0}_{-1}e \rightarrow ?$

 e. $^8_4 Be \rightarrow ? + {}^4_2 He$

 f. $^{46}_{23} V \rightarrow ? + {}^{46}_{22} Ti$

10.10 Complete the following equations, using appropriate notations and formulas:

 a. $^{204}_{82} Pb \rightarrow ? + {}^4_2\alpha$

 b. $^{84}_{35} Br \rightarrow ? + {}^{\ 0}_{-1}\beta$

 c. $? + {}^{\ 0}_{-1}e \rightarrow {}^{41}_{19} K$

 d. $^{149}_{62} Sm \rightarrow {}^{145}_{60} Nd + ?$

 e. $? \rightarrow {}^{34}_{15} P + {}^{\ 0}_{-1}\beta$

 f. $^{15}_8 O \rightarrow {}^0_1\beta + ?$

10.11 Write balanced equations to represent decay reactions of the following isotopes. The decay process or daughter isotope is given in parentheses.

 a. $^{121}_{50}Sn$ (beta emission)

 b. $^{55}_{26}Fe$ (electron capture)

 c. $^{22}_{11}Na$ (daughter = neon-22)

 d. $^{190}_{78}Pt$ (alpha emission)

 e. $^{67}_{28}Ni$ (beta emission)

 f. $^{67}_{31}Ga$ (daughter = Zinc-67)

10.12 Write balanced equations to represent decay reactions of the following isotopes. The decay process or daughter isotope is given in parentheses.

 a. $^{157}_{63}Eu$ (beta emission)

 b. $^{190}_{78}Pt$ (daughter = osmium-186)

 c. $^{138}_{62}Sm$ (electron capture)

 d. $^{188}_{80}Hg$ (daughter = Au-188)

 e. $^{234}_{90}Th$ (beta emission)

 f. $^{218}_{85}At$ (alpha emission)

Isotope Half-Life (Section 10.3)

10.13 What is meant by a half-life?

10.14 Describe half-life in terms of something familiar, such as a cake or cookies or your checking account.

10.15 An isotope of lead, $^{194}_{82}Pb$, has a half-life of 11 minutes. What fraction of the lead-194 atoms in a sample would remain after 44 minutes had elapsed?

10.16 Technetium-99 has a half-life of 6 hours. This isotope is used diagnostically to perform brain scans. A patient is given a 9.0-ng dose. How many nanograms will be present in the patient 24 hours later?

10.17 An archaeologist sometime in the future analyzes the iron used in an old building. The iron contains tiny amounts of nickel-63, with a half-life of 92 years. On the basis of the amount of nickel-63 and its decay products found, it is estimated that about 0.78% (1/128) of the original nickel-63 remains. If the building was constructed in 1980, in what year did the archaeologist make the discovery?

10.18 An archaeologist unearths the remains of a wooden box, analyzes for the carbon-14 content, and finds that about 93.75% of the carbon-14 initially present has decayed. Estimate the age of the box. The half-life of carbon-14 is 5600 years.

10.19 Germanium-66 decays by positron emission, with a half-life of 2.5 hours. What mass of germanium-66 remains in a sample after 10.0 hours if the sample originally weighed 50.0 mg?

10.20 A grain sample was found in a cave. The ratio of $^{14}_{6}C/^{12}_{6}C$ was 1/8 the value in a fresh grain sample. How old was the grain in the cave? The half-life of C-14 is 5600 years.

The Health Effects of Radiation (Section 10.4)

10.21 A source of radiation has an intensity of 130 units at a distance of 10.0 feet. How far away from the source would you have to be to reduce the intensity to 15 units?

10.22 Compare and contrast the general health effects of long-term exposure to low-level radiation and short-term exposure to intense radiation.

10.23 A radiologist found that in a 30-minute period the dose from a radioactive source was 75 units at a distance of 12.0 m from the source. What would the dose be in the same amount of time at a distance of 2.00 m from the source?

Measurement Units for Radiation (Section 10.5)

10.24 Explain the difference between physical and biological units of radiation and give an example of each type.

10.25 Explain why the rem is the best unit to use when evaluating the radiation received by a person working in an area where exposure to several types of radiation is possible.

10.26 An individual receives a short-term whole-body dose of 3.1 rads of beta radiation. How many roentgens of X-rays would represent the same health hazard?

10.27 Describe how scintillation counters and Geiger–Müller counters detect radiation.

10.28 One Ci corresponds to 3.7×10^{10} nuclear disintegrations per second. How many disintegrations per second would take place in a sample containing 4.6 μ Ci of radioisotope?

Medical Uses of Radioisotopes (Section 10.6)

10.29 Explain what a diagnostic tracer is and list the ideal characteristics one should have.

10.30 Describe the importance of hot and cold spots in diagnostic work using tracers.

10.31 List the ideal characteristics of a radioisotope that is to be administered internally for therapeutic use.

10.32 Chromium-51 is used medically to monitor kidney activity. Chromium-51 decays by electron capture. Write a balanced equation for the decay process and identify the daughter that is produced.

10.33 Gold-198 is a β emitter used to treat leukemia. It has a half-life of 2.7 days. The dosage is 1.0 mCi/kg body weight. How long would it take for a 70.-mCi dose to decay to an activity of 2.2 mCi?

Nonmedical Uses of Radioisotopes (Section 10.7)

10.34 A mixture of water (H_2O) and hydrogen peroxide (H_2O_2) will give off oxygen gas when solid manganese dioxide is added as a catalyst. Describe how you could use a tracer to determine if the oxygen comes from the water or the peroxide.

10.35 Suppose you are planning a TV commercial for motor oil. Describe how you would set it up to show that your oil prevents engine wear better than a competing brand.

10.36 Propose a method for measuring the volume of water in an irregularly shaped swimming pool. You have 1 gallon of water that contains a radioisotope with a long half-life, and a Geiger–Müller counter.

10.37 Explain why carbon-14, with a half-life of about 5600 years, is not a good radioisotope to use if you want to determine the age of a coal bed thought to be several million years old.

Induced Nuclear Reactions (Section 10.8)

10.38 Write a balanced equation to represent the synthesis of silicon-27 that takes place when magnesium-24 reacts with an accelerated alpha particle. A neutron is also produced.

10.39 Complete the following reactions by identifying the missing nuclear symbols.

 a. $^{9}_{4}Be + ? \rightarrow ^{12}_{6}C + ^{1}_{0}n$

 b. $^{27}_{13}Al + ? \rightarrow ^{30}_{15}P + ^{1}_{0}n$

 c. $^{27}_{13}Al + ^{2}_{1}H \rightarrow ? + ^{4}_{2}\alpha$

 d. $^{14}_{7}N + ^{4}_{2}\alpha \rightarrow ? + ^{1}_{1}H$

10.40 Write a balanced equation to represent the net breeding reaction that occurs when uranium-238 reacts with a neutron to form plutonium-239 and two beta particles.

10.41 Explain why neutrons cannot be accelerated in a cyclotron or linear accelerator.

10.42 Describe the role of a moderator in nuclear reactions involving neutrons.

10.43 To make indium-111 for diagnostic work, silver-109 is bombarded with α particles. What isotope forms when a silver-109 nucleus captures an α particle? Write an equation to represent the process. What two particles must be emitted by the isotope to form indium-111?

10.44 To make gallium-67 for diagnostic work, zinc-66 is bombarded with accelerated protons. Write a balanced equation to represent the process when a zinc-66 nucleus captures a single proton.

Nuclear Energy (Section 10.9)

10.45 Write one balanced equation that illustrates nuclear fission.

10.46 Write one balanced equation that illustrates nuclear fusion.

10.47 Explain why critical mass is an important concept in neutron-induced fission chain reactions.

10.48 Describe the difference between nuclear fission and nuclear fusion processes. Use nuclear equations to illustrate the difference.

10.49 Complete the following equations, which represent two additional ways uranium-235 can undergo nuclear fission.

 a. $^{235}_{92}U + ^{1}_{0}n \rightarrow ^{160}_{62}Sm + ? + 4^{1}_{0}n$

 b. $^{235}_{92}U + ^{1}_{0}n \rightarrow ^{87}_{35}Br + ? + 3^{1}_{0}n$

10.50 Plutonium-238 is used to power batteries for pacemakers. It decays by α emission. Write a balanced equation for the process.

Additional Exercises

10.51 Polonium-210 decays by alpha emission and has a half-life of 138 days. Suppose a sample of Po-210 contains 6.02×10^{20} Po-210 atoms and undergoes decay for 138 days. What was the mass in grams of the original sample? Assume all alpha particles formed by the decay escape, and the daughter nuclei remain in the sample as nonradioactive isotopes; then calculate the mass of the sample in grams after 138 days.

10.52 Propose an explanation for the fact that the fusion reactions occurring in the sun only take place at extremely high temperatures.

10.53 Explain how you would make a 10.0-mL sample of 0.100 M NaI that contained iodine in the form of the $^{123}_{53}I$ isotope.

10.54 Zinc-71 decays by beta emission, and has a half-life of 2.4 minutes. Suppose a 0.200-g sample of Zn-71 was put inside the tube of a Geiger–Müller counter that counted every beta particle emitted by the sample for 2.4 minutes. What would the average reading of the Geiger–Müller counter be in counts per minute during the 2.4 minutes? How many curies is this?

10.55 At one time, nuclear bombs were tested by exploding them aboveground. The fallout from such tests contained some Sr-90, a radioactive isotope of strontium. If Sr-90 gets into the food supply, it can become incorporated into the bones of humans and other animals. Explain how this can happen even though strontium is not a normal component of bones.

Chemistry for Thought

10.56 One (unrealized) goal of ancient alchemists was to change one element into another (such as lead to gold). Do such changes occur naturally? Explain your reasoning.

10.57 Refer to Figure 10.2. Some isotopes used as positron emitters in PET scans are fluorine-18, oxygen-15, and nitrogen-13. What element results in each case when a positron is emitted?

10.58 Consider the concept of half-life and decide if, in principle, a radioactive isotope ever completely disappears by radioactive decay. Explain your reasoning.

10.59 Do you think Earth is more or less radioactive than it was when it first formed? Explain your answer.

10.60 Nuclear wastes typically have to be stored for at least 20 half-lives before they are considered safe. This can be a time of hundreds or thousands of years for some isotopes. With that in mind, would you consider sending such wastes into outer space a responsible solution to the nuclear waste disposal problem? Explain your answer.

10.61 Read Chemistry Around Us 10.2 and write an equation to represent the radioactive decay of radon-222. Then, write an equation to represent the decay of the daughter produced by the radon decay. The daughter decays by alpha emission. Then, write an equation for the decay of the daughter produced by this second reaction, which decays by beta emission. In each of the three reactions, assume that only one particle in addition to the daughter is produced. What element is radon converted into by this series of three decays?

10.62 Uranium-238 is the most abundant naturally occurring isotope of uranium. It undergoes radioactive decay to form other isotopes. In the first three steps of this decay, uranium-238 is converted to thorium-234, which is converted to protactinium-234, which is converted to uranium-234. Assume that only one particle in addition to the daughter is produced in each step, and use equations for the decay processes to determine the type of radiation emitted during each step.

Allied Health Exam Connection

The following questions are from these sources:

- *Nursing School Entrance Exam* © 2005, Learning Express, LLC.

- *McGraw-Hill's Nursing School Entrance Exams* by Thomas A. Evangelist, Tamara B. Orr, and Judy Unrein © 2009, The McGraw-Hill Companies, Inc.

- *NSEE Nursing School Entrance Exams,* 3rd edition © 2009, Kaplan Publishing.

- *Cliffs Test Prep: Nursing School Entrance Exams* by Fred N. Grayson © 2004, Wiley Publishing, Inc.

- *Peterson's Master the Nursing School and Allied Health Entrance Exams,* 18th edition by Marion F. Gooding © 2008, Peterson's, a Nelnet Company.

10.63 What happens when a radioactive element decays?

 a. The nucleus gives off particles.

 b. The ribosomes give off particles.

 c. The mitochondria give off particles.

 d. The Golgi apparatus gives off particles.

10.64 Which of the following particles/rays is not a type of radioactive emanation?

 a. alpha particle

 b. beta particle

 c. gamma ray

 d. carbon-12

10.65 Which of the following forms of radiation contains the least energy?

 a. radio waves

 b. gamma rays

 c. ultraviolet radiation

 d. visible light

10.66 Which of the following is NOT a form of radioactive decay?

 a. electron capture

 b. beta emission

 c. alpha emission

 d. proton emission

10.67 The least penetrating radiation given off by a radioactive substance consists of:

 a. alpha particles.

 b. beta particles.

 c. gamma rays.

 d. X-rays.

10.68 Alpha, beta, and gamma radiation are the three major products of natural radioactive decay. Which of the following is (are) the most penetrating radiation?

 a. alpha rays

 b. beta rays

 c. gamma rays

 d. They are all equally penetrating.

10.69 The isotope $^{234}_{92}U$ decays by releasing an alpha particle. What is the resulting isotope?

 a. $^{234}_{93}Np$

 b. $^{233}_{91}Pa$

 c. $^{234}_{91}Pa$

 d. $^{230}_{90}Th$

10.70 Carbon dating involves the decay of a carbon-14 isotope by beta emission. Which of the following equations describes this decay?

 a. $^{14}_{6}C \rightarrow ^{13}_{5}B + ^{1}_{1}H$

 b. $^{14}_{6}C \rightarrow ^{14}_{7}N + ^{0}_{-1}\beta$

 c. $^{14}_{6}C \rightarrow ^{13}_{5}B + ^{1}_{0}n$

 d. $^{14}_{7}N + ^{1}_{0}n \rightarrow ^{14}_{6}C + ^{1}_{1}H$

10.71 What is the name of particle X in the following reaction?

$$^{232}_{90}Th \rightarrow ^{228}_{88}Ra + X$$

 a. deuterium

 b. gamma radiation

 c. beta particle

 d. alpha particle

10.72 If element A below gives off an alpha particle, what is the atomic number and mass of the resulting element B? $^{210}_{83}A$

 a. $^{210}_{81}B$

 b. $^{206}_{81}B$

 c. $^{206}_{83}B$

 d. $^{204}_{81}B$

10.73 If element B below gives off a beta particle and gamma rays, what is the resulting element? $^{238}_{92}B$

 a. $^{238}_{93}B$

 b. $^{234}_{90}B$

 c. $^{239}_{92}B$

 d. $^{239}_{91}B$

10.74 The loss of an alpha particle from the radioactive atom $^{228}_{88}X$ would leave:

 a. $^{224}_{86}Rn$

 b. $^{222}_{86}Rn$

 c. $^{224}_{88}Ra$

 d. $^{230}_{90}Th$

10.75 What is the missing product?

$$^{42}_{17}X \rightarrow ^{42}_{18}Y + ?$$

 a. $^{4}_{2}He$

 b. $^{0}_{0}\gamma$

 c. $^{0}_{1}e$

 d. $^{0}_{-1}\beta$

10.76 What is the missing product?

$$^{60}_{24}A \rightarrow {}^{60}_{24}B + ?$$

 a. 4_2He

 b. $^0_0\gamma$

 c. $^0_{-1}e$

 d. $^0_1\beta$

10.77 The half-life of a certain radioactive isotope is 5 days. How much of a 100-g sample of this radioactive isotope will remain after 10 days?

 a. 25

 b. 50

 c. 75

 d. 100

10.78 The radioactive isotope I-131 has a half-life of 8 days. What fraction of a sample of I-131 will remain after 24 days?

 a. $\dfrac{1}{4}$

 b. $\dfrac{1}{16}$

 c. $\dfrac{1}{8}$

 d. $\dfrac{1}{2}$

10.79 If 40 g of a radioactive substance naturally decays to 10 g after 16 days, what is the half-life of the substance?

 a. 40 days

 b. 8 days

 c. 16 days

 d. 10 days

10.80 A dating technique involves electron capture by the potassium-40 isotope according to the following equation: $^{40}_{19}K + {}^0_{-1}e \rightarrow {}^{40}_{18}Ar$. If the half-life is 1.2×10^9 years, how long does it take for only 10 g to remain of the original 40 g of potassium-40 in a rock sample?

 a. 1.2×10^9 years

 b. 0.6×10^9 years

 c. 2.4×10^9 years

 d. 1.8×10^9 years

10.81 If the half-life of a certain isotope is 5 years, what fraction of a sample of that isotope will remain after 15 years?

 a. $\dfrac{1}{5}$

 b. $\dfrac{1}{8}$

 c. $\dfrac{1}{15}$

 d. $\dfrac{1}{20}$

10.82 The half-life of a given element is 70 years. How long will it take 5.0 g of this element to be reduced to 1.25 g?

 a. 70 years

 b. 140 years

 c. 210 years

 d. 35 years

10.83 I^{131} has a half-life of eight days. A 100-milligram sample of this radioactive element would decay to what amount after eight days?

 a. 50 milligrams

 b. 40 milligrams

 c. 30 milligrams

 d. 20 milligrams

10.84 If a radioactive element with a half-life ($t^{1/2}$) of 100 years has 31.5 kg remaining after 400 years of decay, the amount in the original sample was close to:

 a. 2,500 kg

 b. 500 kg

 c. 50 kg

 d. 5,000 kg

10.85 Carbon-14 has a half-life of 5.73×10^3 years. If a sample contained 1 gram of C-14, the time required to decay to only 0.0625 g would be:

 a. 11.46×10^3 years.

 b. 5.73×10^3 years.

 c. 22.92×10^3 years.

 d. none of the above.

10.86 The time required for $\frac{1}{2}$ of the atoms in a sample of a radioactive element to disintegrate is known as the element's:

 a. decay period.

 b. lifetime.

 c. radioactive period.

 d. half-life.

11

Organic Compounds: Alkanes

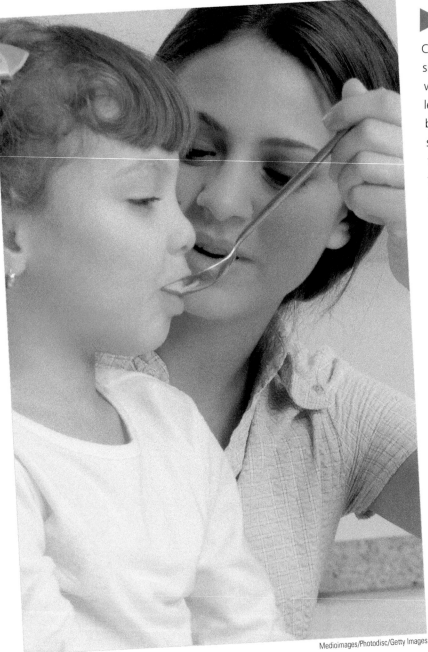

Medioimages/Photodisc/Getty Images

▶ Case Study

Christi was at the end of her rope—she felt as if she had tried everything. Her daughter Mellissa was 4 years old, would start kindergarten in less than a year, and still wasn't able to control bowel movements. Nearly every day, Mellissa soiled her pants. Christi tried rewarding her with candy and even resorted to spanking, and the condescending advice of her mother-in-law didn't help. Mellissa would urinate in the toilet, but she seemed too busy playing or too stubborn to use it for anything else. Finally, in desperation and embarrassment, Christi consulted a pediatrician. He quickly diagnosed Mellissa with encopresis due to constipation. The doctor assured Christi it was neither poor parenting on her part nor bad behavior from her daughter that was causing the problem, but a medical condition. In addition to dietary changes, the doctor prescribed laxatives and 3 tablespoons of mineral oil daily. Mellissa disliked the mineral oil.

▶ **What behavioral techniques could Christi use to encourage her daughter to cooperate with the treatment? Should she punish Mellissa if she has an "accident"?**

Follow-up to this Case Study appears at the end of the chapter before the *Concept Summary*.

Learning Objectives

When you have completed your study of this chapter, you should be able to:

1 Show that you understand the general importance of organic chemical compounds. **(Section 11.1)**

2 Be able to recognize the molecular formulas of organic and inorganic compounds. **(Section 11.1)**

3 Explain some general differences between inorganic and organic compounds. **(Section 11.2)**

4 Be able to use structural formulas to identify compounds that are isomers of each other. **(Section 11.3)**

5 Write condensed or expanded structural formulas for compounds. **(Section 11.4)**

6 Classify alkanes as normal or branched. **(Section 11.5)**

7 Use structural formulas to determine whether compounds are structural isomers. **(Section 11.6)**

8 Assign IUPAC names and draw structural formulas for alkanes. **(Section 11.7)**

9 Assign IUPAC names and draw structural formulas for cycloalkanes. **(Section 11.8)**

10 Name and draw structural formulas for geometric isomers of cycloalkanes. **(Section 11.9)**

11 Describe the key physical properties of alkanes. **(Section 11.10)**

12 Write alkane combustion reactions. **(Section 11.11)**

The word *organic* is used in several different contexts. Scientists of the 18th and 19th centuries studied compounds extracted from plants and animals and labeled them "organic" because they had been obtained from organized (living) systems. Organic fertilizer is organic in the original sense that it comes from a living organism. There is no universal definition of organic foods, but the term is generally taken to mean foods grown without the application of pesticides or synthetic fertilizers. When referring to organic chemistry, however, we mean the chemistry of carbon-containing compounds.

11.1 | Carbon: The Element of Organic Compounds

Learning Objectives

1. Show that you understand the general importance of organic chemical compounds.

2. Be able to recognize the molecular formulas of organic and inorganic compounds.

Early chemists thought organic compounds could be produced only through the action of a "vital force," a special force active only in living organisms. This idea was central to the study of organic chemistry until 1828, because up to that time, no one had been able to synthesize an organic compound from its elements or from naturally occurring minerals. In that year, Friedrich Wöhler, a German chemist, heated an inorganic salt called ammonium cyanate and produced urea. This compound, normally found in blood and urine, was unquestionably organic, and it had come from an inorganic source. The reaction is

$$\underset{\substack{\text{ammonium} \\ \text{cyanate}}}{NH_4NCO} \xrightarrow{\text{Heat}} \underset{\text{urea}}{H-\overset{\displaystyle H}{\overset{|}{N}}-\overset{\displaystyle O}{\overset{\|}{C}}-\overset{\displaystyle H}{\overset{|}{N}}-H} \qquad (11.1)$$

Wöhler's urea synthesis discredited the "vital force" theory, and his success prompted other chemists to attempt to synthesize organic compounds. Today, organic compounds are being synthesized in thousands of laboratories, and most of the synthetics have never been isolated from natural sources.

Organic compounds share one unique feature: They all contain carbon. Therefore, **organic chemistry** is defined as the study of carbon-containing compounds. There are a few exceptions to this definition; a small number of carbon compounds—such as CO, CO_2, carbonates, and cyanides—were studied before Wöhler's urea synthesis. These were classified as inorganic because they were obtained from nonliving systems, and even though they contain carbon, we still consider them to be a part of **inorganic chemistry**.

The importance of carbon compounds to life on Earth cannot be overemphasized. If all carbon compounds were removed from the Earth, its surface would be somewhat like the barren surface of the moon (see **Figure 11.1**). There would be no animals, plants, or any other form of life. If carbon-containing compounds were removed from the human body, all that would remain would be water, a very brittle skeleton, and a small residue of minerals. Many of the essential constituents of living matter—such as carbohydrates, fats, proteins, nucleic acids, enzymes, and hormones—are organic chemicals.

> **organic compound** A compound that contains the element carbon.
>
> **organic chemistry** The study of carbon-containing compounds.
>
> **inorganic chemistry** The study of the elements and all noncarbon compounds.

Figure 11.1 Organic chemistry makes a tremendous difference when comparing the physical makeup of the Earth and the moon.

NASA

The essential needs of daily human life are food, fuel, shelter, and clothing. The principal components of food (with the exception of water) are organic. The fuels we use (e.g., wood, coal, petroleum, and natural gas) are mixtures of organic compounds. Our homes typically incorporate wood construction, and our clothing, whether made of natural or synthetic fibers, is organic.

Besides the major essentials, many of the smaller everyday things often taken for granted are also derived from carbon and its compounds. Consider an ordinary pencil. The "lead" (actually graphite), the wood, the rubber eraser, and the paint on the surface are all either carbon or carbon compounds. The paper in this book, the ink on its pages, and the glue holding it all together are also made of carbon compounds.

11.2 | Organic and Inorganic Compounds Compared

Learning Objective

3. Explain some general differences between inorganic and organic compounds.

It is interesting that the subdivision of chemistry into its organic and inorganic branches results in one branch that deals with compounds composed mainly of one element and another branch that deals with compounds formed by the more than 100 remaining elements.

However, this classification seems more reasonable when we recognize that known organic compounds are much more numerous than inorganic compounds. An estimated 500,000 inorganic compounds have been identified, but more than 9 million organic compounds are known, and thousands of new ones are synthesized or isolated each year.

One of the reasons for the large number of organic compounds is the unique ability of carbon atoms to form stable covalent bonds with other carbon atoms and with atoms of other elements. The resulting covalently bonded molecules may contain as few as one or more than a million carbon atoms.

In contrast, inorganic compounds are often characterized by the presence of ionic bonding. Covalent bonding also may be present, but it is less common. These differences generally cause organic and inorganic compounds to differ physically (see **Figure 11.2**) and chemically, as shown in **Table 11.1**.

Figure 11.2 Many organic compounds, such as ski wax, have relatively low melting points. What does this fact reveal about the forces between organic molecules?

TABLE 11.1 Properties of Typical Organic and Inorganic Compounds

Property	Organic Compounds	Inorganic Compounds
Bonding within molecules	Covalent	Often ionic
Forces between molecules	Generally weak	Quite strong
Normal physical state	Gases, liquids, or low-melting-point solids	Usually high-melting-point solids
Flammability	Often flammable	Usually nonflammable
Solubility in water	Often low	Often high
Conductivity of water solutions	Nonconductor	Conductor
Rate of chemical reactions	Usually slow	Usually fast

✔ **LEARNING CHECK 11.1** Classify each of the following compounds as organic or inorganic:

a. NaCl d. NaOH
b. CH_4 e. CH_3OH
c. C_6H_6 f. $Mg(NO_3)_2$

✔ **LEARNING CHECK 11.2** Decide whether each of the following characteristics most likely describes an organic or inorganic compound:

a. Flammable b. Low boiling point c. Soluble in water

11.3 | Bonding Characteristics and Isomerism

Learning Objective

4. Be able to use structural formulas to identify compounds that are isomers of each other.

There are two major reasons for the astonishing number of organic compounds: the bonding characteristics of carbon atoms, and the isomerism of carbon-containing molecules. As a group IVA(14) element, a carbon atom has four valence electrons. Two of these outermost-shell electrons are in an *s* orbital, and two are in *p* orbitals (see **Section 3.4**):

$$\underset{2s}{\textcircled{\scriptsize ↓↑}} \qquad \underset{2p}{\textcircled{\scriptsize ↑}\;\textcircled{\scriptsize ↑}\;\textcircled{}}$$

With only two unpaired electrons, we might predict that carbon would form just two covalent bonds with other atoms. Yet, we know from the formula of methane (CH_4) that carbon forms four bonds.

Linus Pauling (1901–1994), winner of the Nobel Prize in chemistry (1954) and Nobel Peace Prize (1963), developed a useful model to explain the bonding characteristics of carbon. Pauling found that a mathematical mixing of the 2*s* and three 2*p* orbitals could produce four new, equivalent orbitals (see **Figure 11.3**). Each of these **hybrid orbitals** has the same energy and is designated sp^3. An sp^3 orbital has a two-lobed shape, similar to

hybrid orbital An orbital produced from the combination of two or more nonequivalent orbitals of an atom.

the shape of a *p* orbital but with different-sized lobes (see **Figure 11.4**). Each of the four *sp*³ hybrid orbitals contains a single unpaired electron available for covalent bond formation. Thus, carbon forms four bonds.

Figure 11.3 The mixing of 2*s* and three 2*p* orbitals of a carbon atom to generate four *sp*³ carbon orbitals, each with energy intermediate between 2*s* and 2*p* orbitals.

Each carbon–hydrogen bond in methane arises from an overlap of a C (*sp*³) and an H (1*s*) orbital. The sharing of two electrons in this overlap region creates a sigma (σ) bond. The four equivalent *sp*³ orbitals point toward the corners of a regular tetrahedron (see **Figure 11.5**).

Carbon atoms also have the ability to bond covalently to other carbon atoms to form chains and networks. This means that two carbon atoms can join by sharing two electrons to form a single covalent bond:

$$\cdot \dot{\overset{\cdot}{C}} \cdot + \cdot \dot{\overset{\cdot}{C}} \cdot \longrightarrow \cdot \dot{\overset{\cdot}{C}} : \dot{\overset{\cdot}{C}} \cdot \qquad \text{or} \qquad \cdot \dot{\overset{\cdot}{C}} - \dot{\overset{\cdot}{C}} \cdot \qquad (11.2)$$

A third carbon atom can join the end of this chain:

$$\cdot \dot{\overset{\cdot}{C}} - \dot{\overset{\cdot}{C}} \cdot + \cdot \dot{\overset{\cdot}{C}} \cdot \longrightarrow \cdot \dot{\overset{\cdot}{C}} - \dot{\overset{\cdot}{C}} - \dot{\overset{\cdot}{C}} \cdot \qquad (11.3)$$

This process can continue and form carbon chains of almost any length, such as

$$\cdot \dot{\overset{\cdot}{C}} - \dot{\overset{\cdot}{C}} - \dot{\overset{\cdot}{C}} - \dot{\overset{\cdot}{C}} - \dot{\overset{\cdot}{C}} - \dot{\overset{\cdot}{C}} - \dot{\overset{\cdot}{C}} - \dot{\overset{\cdot}{C}} - \dot{\overset{\cdot}{C}} \cdot$$

The electrons not involved in forming the chain can be shared with electrons of other carbon atoms (to form chain branches) or with electrons of other elements such as hydrogen, oxygen, or nitrogen. Carbon atoms may also share more than one pair of electrons to form multiple bonds:

$$\cdot \dot{\overset{\cdot}{C}} - \dot{\overset{\cdot}{C}} = \dot{\overset{\cdot}{C}} - \dot{\overset{\cdot}{C}} \cdot \qquad\qquad \cdot \dot{\overset{\cdot}{C}} - C \equiv C - \dot{\overset{\cdot}{C}} \cdot$$

<div align="center">

Chain with Chain with
double bond triple bond

</div>

In principle, there is no limit to the number of carbon atoms that can bond covalently. Thus, organic molecules range from the simple molecules such as methane (CH₄) to very complicated molecules containing over a million carbon atoms.

The variety of possible carbon atom arrangements is even more important than the size range of the resulting molecules. The carbon atoms in all but the very simplest organic molecules can bond in more than one arrangement, giving rise to different compounds with different structures and properties. This property, called **isomerism**, is characterized by compounds that have identical molecular formulas but different arrangements of atoms. One type of isomerism is characterized by compounds called **structural isomers**. Another type of isomerism is introduced in Section 11.9.

Example 11.1 Using Rules for Covalent Bonding

Use the usual rules for covalent bonding to show that a compound with the molecular formula C₂H₆O demonstrates the property of isomerism. Draw formulas for the isomers, showing all covalent bonds.

A 2*p* orbital

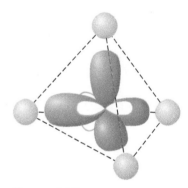

An *sp*³ hybrid orbital

Figure 11.4 A comparison of unhybridized *p* and *sp*³ hybridized orbital shapes. The atomic nucleus is at the junction of the lobes in each case.

Figure 11.5 Directional characteristics of *sp*³ hybrid orbitals of carbon and the formation of C—H bonds in methane (CH₄). The hybrid orbitals point toward the corners of a regular tetrahedron. Hydrogen 1*s* orbitals are illustrated in position to form bonds by overlap with the major lobes of the hybrid orbitals.

isomerism A property in which two or more compounds have the same molecular formula but different arrangements of atoms.

structural isomers Compounds that have the same molecular formula but in which the atoms bond in different patterns.

Solution

Carbon forms four covalent bonds by sharing its four valence-shell electrons. Similarly, oxygen should form two covalent bonds, and hydrogen a single bond. On the basis of these bonding relationships, two structural isomers are possible:

$$\underset{\text{ethyl alcohol}}{H-\overset{\overset{\displaystyle H}{|}}{\underset{\underset{\displaystyle H}{|}}{C}}-\overset{\overset{\displaystyle H}{|}}{\underset{\underset{\displaystyle H}{|}}{C}}-O-H} \qquad \underset{\text{dimethyl ether}}{H-\overset{\overset{\displaystyle H}{|}}{\underset{\underset{\displaystyle H}{|}}{C}}-O-\overset{\overset{\displaystyle H}{|}}{\underset{\underset{\displaystyle H}{|}}{C}}-H}$$

✔ **LEARNING CHECK 11.3** Which one of the structures below represents a structural isomer of

$$H-\overset{\overset{\displaystyle H}{|}}{\underset{\underset{\displaystyle H}{|}}{C}}-\overset{\overset{\displaystyle O}{\|}}{C}-\overset{\overset{\displaystyle H}{|}}{\underset{\underset{\displaystyle H}{|}}{C}}-H \quad ?$$

a. $H-\overset{\overset{\displaystyle H}{|}}{C}=\overset{\overset{\displaystyle H}{|}}{C}-\overset{\overset{\displaystyle O-H}{|}}{\underset{\underset{\displaystyle H}{|}}{C}}-H$

c. $H-\overset{\overset{\displaystyle H}{|}}{\underset{\underset{\displaystyle H}{|}}{C}}-\overset{\overset{\displaystyle H}{|}}{\underset{\underset{\displaystyle H}{|}}{C}}-\overset{\overset{\displaystyle O}{\|}}{C}-O-H$

b. $H-\overset{\overset{\displaystyle H}{|}}{C}=\overset{\overset{\displaystyle H}{|}}{C}-\overset{\overset{\displaystyle O}{\|}}{C}-H$

The two isomers of Example 11.1 are quite different. Ethyl alcohol (grain alcohol) is a liquid at room temperature, whereas dimethyl ether is a gas. As we've seen before, the structural differences exert a significant influence on properties. From this example, we can see that molecular formulas such as C_2H_6O provide much less information about a compound than do structural formulas. **Figure 11.6** shows ball-and-stick models of these two molecules.

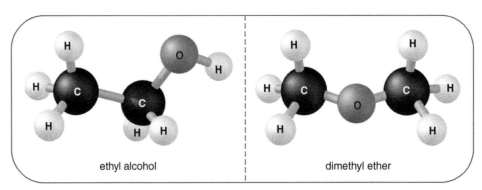

Figure 11.6 Ball-and-stick models of the isomers of C_2H_6O. Ethyl alcohol is a liquid at room temperature and completely soluble in water, whereas dimethyl ether is a gas at room temperature and only partially soluble in water.

As the number of carbon atoms in the molecular formula increases, the number of possible isomers increases dramatically. For example, 366,319 different isomers are possible for a molecular formula of $C_{20}H_{42}$. No one has prepared all these isomers or even drawn their structural formulas, but the number helps us understand why so many organic compounds have been either isolated from natural sources or synthesized.

11.4 | Functional Groups: The Organization of Organic Chemistry

Learning Objective

5. Write condensed or expanded structural formulas for compounds.

Because of the enormous number of possible compounds, the study of organic chemistry might appear to be hopelessly difficult. However, the arrangement of organic compounds into a relatively small number of classes can simplify the study a great deal. This organization is done on the basis of characteristic structural features called **functional groups**. For example, compounds with a carbon–carbon double bond

$$\overset{\diagdown}{\underset{\diagup}{}}C = C\overset{\diagup}{\underset{\diagdown}{}}$$

are classified as alkenes. The major classes and functional groups are given in **Table 11.2**. Notice that each functional group in Table 11.2 (except for alkanes) contains a multiple bond or at least one oxygen or nitrogen atom.

In Table 11.2, we have used both expanded and condensed structural formulas for the compounds. **Expanded structural formulas** show all covalent bonds, whereas **condensed structural formulas** show only specific bonds. You should become familiar with both types, but especially with condensed formulas because they will be used often.

> **functional group** A unique reactive combination of atoms that differentiates molecules of organic compounds of one class from those of another.

> **expanded structural formula** A structural molecular formula showing all the covalent bonds.

> **condensed structural formula** A structural molecular formula showing the general arrangement of atoms but without showing all the covalent bonds.

Example 11.2 Writing Condensed Structural Formulas

Write a condensed structural formula for each of the following compounds:

a.
$$\text{H} - \underset{\underset{\displaystyle H}{|}}{\overset{\overset{\displaystyle H}{|}}{C}} - \underset{\underset{\displaystyle H}{|}}{\overset{\overset{\displaystyle H}{|}}{C}} - \underset{\underset{\displaystyle H}{|}}{\overset{\overset{\displaystyle H}{|}}{C}} - \underset{\underset{\displaystyle H}{|}}{\overset{\overset{\displaystyle H}{|}}{C}} - \text{H}$$

b.
$$\text{H} - \underset{\underset{\displaystyle H}{|}}{\overset{\overset{\displaystyle H}{|}}{C}} \text{———} \underset{\underset{\displaystyle H}{|}}{\overset{\overset{\displaystyle H-C-H}{|}}{C}} \text{———} \underset{\underset{\displaystyle H}{|}}{\overset{\overset{\displaystyle H}{|}}{C}} - \underset{\underset{\displaystyle H}{|}}{\overset{\overset{\displaystyle H}{|}}{C}} - \underset{\underset{\displaystyle H}{|}}{\overset{\overset{\displaystyle H}{|}}{C}} - \text{H}$$

Solution

a. Usually the hydrogens belonging to a carbon are grouped to the right. Thus, the group

$$\text{H} - \underset{\underset{\displaystyle H}{|}}{\overset{\overset{\displaystyle H}{|}}{C}} -$$

condenses to CH_3—, and

$$- \underset{\underset{\displaystyle H}{|}}{\overset{\overset{\displaystyle H}{|}}{C}} -$$

condenses to —CH_2—. Thus, the formula condenses to

$$CH_3 - CH_2 - CH_2 - CH_3$$

Other acceptable condensations are

$$CH_3CH_2CH_2CH_3 \quad \text{and} \quad CH_3(CH_2)_2CH_3$$

Parentheses are used here to denote a series of two —CH_2— groups.

TABLE 11.2 Classes and Functional Groups of Organic Compounds

Class	Functional Group	Example of Expanded Structural Formula	Example of Condensed Structural Formula	Common Name
Alkane	None	$H-\overset{\displaystyle H}{\underset{\displaystyle H}{C}}-\overset{\displaystyle H}{\underset{\displaystyle H}{C}}-H$	CH_3CH_3	ethane
Alkene	$\diagdown C = C \diagup$	$\overset{H}{\diagdown} C = C \overset{H}{\diagup}$	$H_2C = CH_2$	ethylene
Alkyne	$-C \equiv C-$	$H-C \equiv C-H$	$HC \equiv CH$	acetylene
Aromatic	(ring structure)	(ring structure)	(hexagon ring)	benzene
Alcohol	$-\overset{\displaystyle \mid}{\underset{\displaystyle \mid}{C}}-O-H$	$H-\overset{H}{\underset{H}{C}}-\overset{H}{\underset{H}{C}}-O-H$	CH_3CH_2-OH	ethyl alcohol
Ether	$-\overset{\mid}{\underset{\mid}{C}}-O-\overset{\mid}{\underset{\mid}{C}}-$	$H-\overset{H}{\underset{H}{C}}-O-\overset{H}{\underset{H}{C}}-H$	CH_3-O-CH_3	dimethyl ether
Amine	$-\overset{H}{N}-H$	$H-\overset{H}{\underset{H}{C}}-\overset{H}{N}-H$	CH_3-NH_2	methylamine
Aldehyde	$-\overset{\displaystyle O}{\overset{\displaystyle \|}{C}}-H$	$H-\overset{H}{\underset{H}{C}}-\overset{O}{\overset{\|}{C}}-H$	$CH_3-\overset{O}{\overset{\|}{C}}-H$	acetaldehyde
Ketone	$-\overset{\mid}{\underset{\mid}{C}}-\overset{O}{\overset{\|}{C}}-\overset{\mid}{\underset{\mid}{C}}-$	$H-\overset{H}{\underset{H}{C}}-\overset{O}{\overset{\|}{C}}-\overset{H}{\underset{H}{C}}-H$	$CH_3-\overset{O}{\overset{\|}{C}}-CH_3$	acetone
Carboxylic acid	$-\overset{O}{\overset{\|}{C}}-O-H$	$H-\overset{H}{\underset{H}{C}}-\overset{O}{\overset{\|}{C}}-O-H$	$CH_3-\overset{O}{\overset{\|}{C}}-OH$	acetic acid
Ester	$-\overset{O}{\overset{\|}{C}}-O-\overset{\mid}{\underset{\mid}{C}}-$	$H-\overset{H}{\underset{H}{C}}-\overset{O}{\overset{\|}{C}}-O-\overset{H}{\underset{H}{C}}-H$	$CH_3-\overset{O}{\overset{\|}{C}}-O-CH_3$	methyl acetate
Amide	$-\overset{O}{\overset{\|}{C}}-\overset{H}{N}-H$	$H-\overset{H}{\underset{H}{C}}-\overset{O}{\overset{\|}{C}}-\overset{H}{N}-H$	$CH_3-\overset{O}{\overset{\|}{C}}-NH_2$	acetamide

b. The group

$$-\overset{\displaystyle |}{\underset{\displaystyle H}{C}}-$$

condenses to $-CH-$. The condensed formula is therefore

$$CH_3-\overset{\displaystyle \overset{CH_3}{|}}{CH}-CH_2-CH_2-CH_3 \qquad \text{or} \qquad CH_3\overset{\displaystyle \overset{CH_3}{|}}{CH}CH_2CH_2CH_3$$

or

$$(CH_3)_2CH(CH_2)_2CH_3$$

In the last form, the first set of parentheses indicates two identical CH_3 groups attached to the same carbon atom, and the second set of parentheses denotes two CH_2 groups.

✔ **LEARNING CHECK 11.4** Write a condensed structural formula for each of the following compounds. Retain the bonds to and within the functional groups.

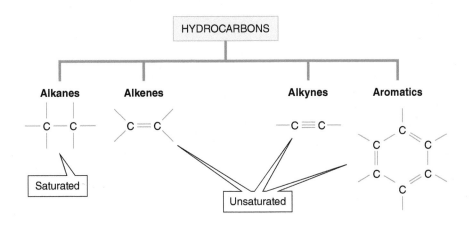

11.5 | Alkane Structures

Learning Objective

6. Classify alkanes as normal or branched.

Hydrocarbons, the simplest of all organic compounds, contain only two elements, carbon and hydrogen. **Saturated hydrocarbons** or **alkanes** are organic compounds in which carbon is bonded to four other atoms by single bonds; there are no double or triple bonds in the molecule. Unsaturated hydrocarbons, studied later, are called alkenes, alkynes, and aromatics and contain double bonds, triple bonds, or six-carbon rings, as shown in **Figure 11.7**.

hydrocarbon An organic compound that contains only carbon and hydrogen.

saturated hydrocarbon Another name for an alkane.

alkane A hydrocarbon that contains only single bonds.

Figure 11.7 Classification of hydrocarbons.

HYDROCARBONS

Alkanes — Alkenes — Alkynes — Aromatics

Saturated

Unsaturated

Q: Is organic food worth the price?

A: "Organic" is a designation used by the United States Department of Agriculture (USDA) to certify food that is produced without the use of antibiotics, hormones, synthetic chemicals or fertilizers, most pesticides, bioengineering, ionizing radiation, or sewage sludge. Organic agriculture is "an ecological production management system that promotes and enhances biodiversity, biological cycles and soil biological activity" according to the USDA. During the past decade, organic foods, including produce, meat, dairy, and even packaged foods, have undergone a huge growth in use. This growth has been tempered by the fact that they are still generally more expensive than conventionally produced food. Does the extra expense justify the higher price? Many believe that organic products are higher in nutrients, but a recent systematic review of studies performed over the past 50 years did not find a difference in nutrient quality between organic and conventionally grown produce. When it comes to pesticide and chemical exposure, the reality is that we are exposed to potentially harmful chemicals in many things, ranging from consumer products, air, drinking water, pollution, food, and even our homes. In fact, the levels of pesticides allowed in agricultural products in the United States are much more strictly regulated than what people can spray in their homes.

What is the clinical relevance of the exposure to pesticides from agricultural products? We do not know. The ability to detect very small amounts of pesticides has improved over the past several decades with advances in analytical chemistry, which now allow some labs to detect levels as low as 10 parts per trillion. However, improved detection methods do not necessarily reflect the potential health risk. Most pesticides are neurotoxicants, meaning that they damage the brain. For this reason, two of the most potentially vulnerable populations include pregnant women and children from birth to 6 years old. Preliminary studies suggest that the offspring of pregnant women subjected to above-average pesticide exposure from living in farmworking communities do show effects on several indicators of brain development. In addition, adult farm workers with very high exposure show an increased risk of neurodegenerative disease.

Until convincing evidence to the contrary becomes available, the benefits of eating fruits and vegetables—which are low in calories and loaded with vitamins, minerals, and disease-fighting phytonutrients—even if they are not organically grown, likely far outweigh the risk for most people. If you are pregnant or have children under the age of 6 and want to limit their pesticide exposure without limiting their intake of produce, the Environmental Working Group publishes an annual list of the top 12 foods that have been found to contain the most pesticides

Melina B. Jampolis, M.D., is a Physician Nutrition Specialist focusing on nutrition for weight loss and disease prevention and treatment.

and the 15 that contain the least. Choose organic when you can from the top 12, but the most important thing you can do is make sure to get plenty of variety when it comes to produce. This will minimize your exposure and maximize your nutrition and health. The least-contaminated produce are often those protected by an outer coating (such as a corn husk) or thick skin (pineapple, watermelon, kiwi, cantaloupe, avocado, grapefruit, mango), so you can use that as a rough guide to direct you to produce with lower levels of pesticides.

Outer coatings on produce help minimize the exposure to pesticides.

Most life processes are based on the reactions of functional groups. Since alkanes have no functional group, they are not abundant in the human body. However, most compounds in human cells contain parts consisting solely of carbon and hydrogen that behave very much like hydrocarbons. Thus, to understand the chemical properties of the more complex biomolecules, it is useful to have some understanding of the structure, physical properties, and chemical behavior of hydrocarbons.

Another important reason for becoming familiar with the characteristics of hydrocarbons is the crucial role they play in modern industrial society. We use naturally occurring hydrocarbons as primary sources of energy and as important sources of raw materials for the manufacture of plastics, synthetic fibers, drugs, and hundreds of other compounds used daily (see **Figure 11.8**).

Alkanes can be represented by the general formula C_nH_{2n+2}, where n is the number of carbon atoms in the molecule. The simplest alkane, methane, contains one carbon atom and therefore has the molecular formula CH_4. The carbon atom is at the center, and the four bonds of the carbon atom are directed toward the hydrogen atoms at the corners of a regular tetrahedron; each hydrogen atom is geometrically equivalent to the other three in the molecule (see **Figure 11.9**). A tetrahedral orientation of bonds with bond angles of 109.5° is typical for carbon atoms that form four single bonds. Methane is the primary compound in natural gas. Tremendous quantities of natural gas are consumed worldwide because methane is an efficient, clean-burning fuel. It is used to heat homes, cook food, and power factories.

Steve Gorton/Getty Images

Figure 11.8 This golf club has the strength and light weight of aluminum yet is made from graphite (carbon) fibers reinforced with plastic. What other sports equipment is made from graphite fibers?

Figure 11.9 Structural representations of methane, CH_4.

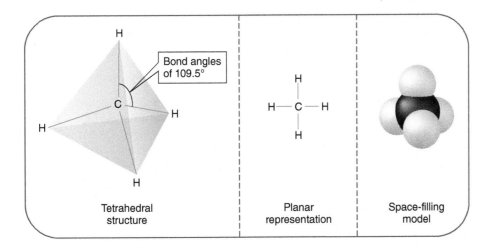

Tetrahedral structure Planar representation Space-filling model

The next alkane is ethane, which has the molecular formula C_2H_6 and the structural formula CH_3—CH_3. This molecule may be thought of as a methane molecule with one hydrogen removed and a —CH_3 put in its place. Again, the carbon bonds have a tetrahedral geometry as shown in **Figure 11.10**. Ethane is a minor component of natural gas.

Figure 11.10 Perspective models of the ethane molecule, CH_3CH_3.

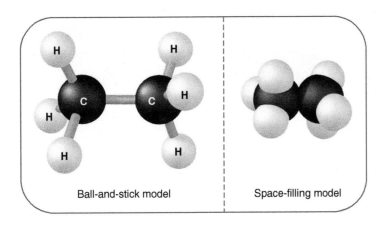

Ball-and-stick model Space-filling model

Propane, the third alkane, has the molecular formula of C_3H_8 and the structural formula CH_3—CH_2—CH_3. Again, we can produce this molecule by removing a hydrogen atom from the preceding compound (ethane) and substituting a —CH_3 in its place (see **Figure 11.11**). Since all six hydrogen atoms of ethane are equivalent, it makes no difference which one is replaced. Propane is used extensively as an industrial fuel, as well as for home heating and cooking (see **Figure 11.11c**).

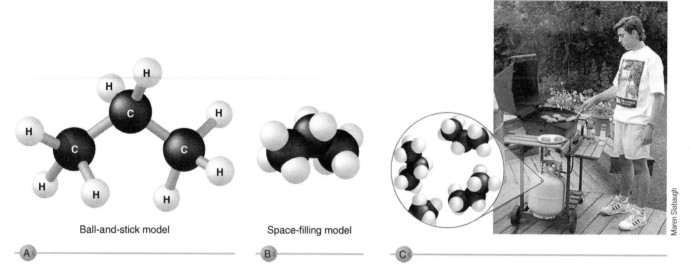

Ball-and-stick model

Space-filling model

A B C

Maren Slabaugh

Figure 11.11 Perspective models (Parts A and B) of the propane molecule $CH_3CH_2CH_3$. Propane is a common fuel for gas grills (Part C).

The fourth member of the series, butane, with molecular formula C_4H_{10}, can also be produced by removing a hydrogen atom (this time from propane) and adding a —CH_3. However, all the hydrogen atoms of propane are not geometrically equivalent, and more than one position is available for substitution. Replacing a hydrogen atom on one of the end carbons of propane with —CH_3 produces a butane molecule that has the structural formula CH_3—CH_2—CH_2—CH_3. If, however, substitution is made on the central carbon atom of propane, the butane produced is

$$CH_3-CH-CH_3$$
$$|$$
$$CH_3$$

Notice that both butanes have the same molecular formula, C_4H_{10}. These two possible butanes are structural isomers because they have the same molecular formulas, but they have different atom-to-atom bonding sequences. The straight-chain isomer is called a **normal alkane** and the other is a **branched alkane** (see **Figure 11.12**).

normal alkane Any alkane in which all the carbon atoms are aligned in a continuous chain.

branched alkane An alkane in which at least one carbon atom is not part of a continuous chain.

Figure 11.12 Space-filling models of the isomeric butanes.

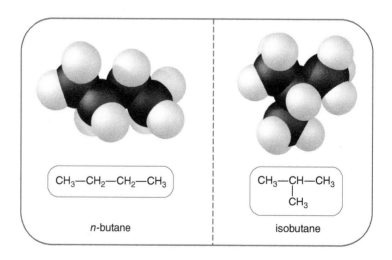

CH_3—CH_2—CH_2—CH_3

CH_3—CH—CH_3
$|$
CH_3

n-butane

isobutane

The number of possible structural isomers increases dramatically with the number of carbon atoms in an alkane, as shown in **Table 11.3**.

11.6 | Conformations of Alkanes

Learning Objective

7. Use structural formulas to determine whether compounds are structural isomers.

Remember that planar representations such as CH_3—CH_2—CH_3 or

$$\begin{array}{ccccc} & H & H & H & \\ & | & | & | & \\ H - & C - & C - & C - & H \\ & | & | & | & \\ & H & H & H & \end{array}$$

are given with no attempt to accurately portray correct bond angles or molecular geometries. Structural formulas are usually written horizontally simply because it is convenient. It is also important to know that actual organic molecules are in constant motion—twisting, turning, vibrating, and bending. Groups connected by a single bond are capable of rotating about that bond much like a wheel rotates around an axle (see **Figure 11.13**). As a result of such rotation about single bonds, a molecule can exist in many different orientations, called **conformations**. In a sample of butane containing billions of identical molecules, there are countless conformations present at any instant, and each conformation is rapidly changing into another. Two of the possible conformations of butane are shown in **Figure 11.14**. We must be sure to recognize that these different conformations do not represent different structural isomers. In each case, the four carbon atoms are bonded in a

	TABLE 11.3 Molecular Formulas and Possible Structural Isomers of Alkanes	
Molecular Formula	**Number of Possible Structural Isomers**	
C_4H_{10}	2	
C_5H_{12}	3	
C_6H_{14}	5	
$C_{10}H_{22}$	75	
$C_{20}H_{42}$	366,319	
$C_{30}H_{62}$	4,111,846,763	

conformations The different arrangements of atoms in space achieved by rotation about single bonds.

Figure 11.13 Rotation about single bonds (*n*-butane molecule).

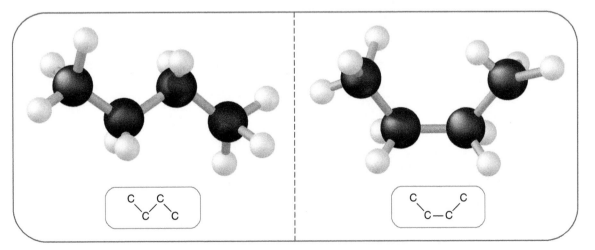

Figure 11.14 Perspective models and carbon skeletons of two conformations of *n*-butane.

continuous (unbranched) chain. Since the order of bonding is not changed, the conformations correspond to the same molecule. Two structures would be structural isomers only if bonds had to be broken and remade to convert one into the other.

Example 11.3 Identifying Structural Isomers

Which of the following pairs are structural isomers, and which are simply different representations of the same molecule?

a. $CH_3-CH_2-CH_2-CH_3$

$$\begin{array}{c} CH_3 \\ | \\ CH_2-CH_2 \\ | \\ CH_3 \end{array}$$

b. $\begin{array}{c} CH_3 \\ | \\ CH_3-CH-CH_2-CH_3 \end{array}$ $\begin{array}{c} CH_3 \\ | \\ CH_3-CH_2-CH-CH_3 \end{array}$

c. $\begin{array}{c} CH_3 \\ | \\ CH_3-CH-CH_2-CH_2-CH_3 \end{array}$ $\begin{array}{c} CH_3 \\ | \\ CH_3-CH_2-CH-CH_2-CH_3 \end{array}$

Solution

a. Same molecule: In both molecules, the four carbons are bonded in a continuous chain.

$$\overset{1}{CH_3}-\overset{2}{CH_2}-\overset{3}{CH_2}-\overset{4}{CH_3} \qquad \begin{array}{c} \overset{4}{CH_3} \\ | \\ \overset{2}{CH_2}-\underset{3}{CH_2} \\ | \\ \overset{1}{CH_3} \end{array}$$

b. Same molecule: In both molecules, there is a continuous chain of four carbons with a branch at position 2. The molecule has simply been turned around.

$$\begin{array}{c} CH_3 \\ | \\ \underset{1}{CH_3}-\underset{2}{CH}-\underset{3}{CH_2}-\underset{4}{CH_3} \end{array} \qquad \begin{array}{c} CH_3 \\ | \\ \underset{4}{CH_3}-\underset{3}{CH_2}-\underset{2}{CH}-\underset{1}{CH_3} \end{array}$$

c. Structural isomers: Both molecules have a continuous chain of five carbons, but the branch is located at different positions.

$$\begin{array}{c} CH_3 \\ | \\ \underset{1}{CH_3}-\underset{2}{CH}-\underset{3}{CH_2}-\underset{4}{CH_2}-\underset{5}{CH_3} \end{array} \qquad \begin{array}{c} CH_3 \\ | \\ \underset{1}{CH_3}-\underset{2}{CH_2}-\underset{3}{CH}-\underset{4}{CH_2}-\underset{5}{CH_3} \end{array}$$

✔ **LEARNING CHECK 11.6** Which of the following pairs represent structural isomers, and which are simply the same compound?

a. $\begin{array}{c} CH_3-CH_2-CH_2 \\ \qquad\qquad | \\ \qquad\quad CH_2-CH_3 \end{array}$ $\begin{array}{c} CH_3-CH_2 \\ \qquad | \\ \quad CH_2-CH_2 \\ \qquad\qquad | \\ \qquad\quad CH_3 \end{array}$

b. $\begin{array}{c} CH_3 \\ | \\ CH_3-CH-CH_2-CH_3 \end{array}$ $\begin{array}{c} CH_2-CH_3 \\ | \\ CH_3-CH-CH_3 \end{array}$

c. $\begin{array}{c} CH_3 \\ | \\ CH_3-CH_2-CH-CH_2-CH_3 \end{array}$ $\begin{array}{c} CH_3 \\ | \\ CH_3-CH_2-CH_2-CH-CH_3 \end{array}$

11.7 | Alkane Nomenclature

Learning Objective

8. Assign IUPAC names and draw structural formulas for alkanes.

When only a relatively few organic compounds were known, chemists gave them what are today called trivial or common names, such as methane, ethane, propane, and butane. The names for the larger alkanes were derived from the Greek prefixes that indicate the number of carbon atoms in the molecule. Thus, *pent*ane contains five carbons, *hex*ane has six, *hept*ane has seven, and so forth, as shown in **Table 11.4**.

TABLE 11.4 Names of Alkanes

Number of Carbon Atoms	Name	Molecular Formula	Structure of Normal Isomer
1	methane	CH_4	CH_4
2	ethane	C_2H_6	CH_3CH_3
3	propane	C_3H_8	$CH_3CH_2CH_3$
4	butane	C_4H_{10}	$CH_3CH_2CH_2CH_3$
5	pentane	C_5H_{12}	$CH_3CH_2CH_2CH_2CH_3$
6	hexane	C_6H_{14}	$CH_3CH_2CH_2CH_2CH_2CH_3$
7	heptane	C_7H_{16}	$CH_3CH_2CH_2CH_2CH_2CH_2CH_3$
8	octane	C_8H_{18}	$CH_3CH_2CH_2CH_2CH_2CH_2CH_2CH_3$
9	nonane	C_9H_{20}	$CH_3CH_2CH_2CH_2CH_2CH_2CH_2CH_2CH_3$
10	decane	$C_{10}H_{22}$	$CH_3CH_2CH_2CH_2CH_2CH_2CH_2CH_2CH_2CH_3$

As more compounds and isomers were discovered, however, it became increasingly difficult to devise unique names and much more difficult to commit them to memory. Obviously, a systematic method was needed. Such a method is now in use, but so are several common methods.

The names for the two isomeric butanes (*n*-butane and isobutane) illustrate the important features of the common nomenclature system used for alkanes. The stem *but-* indicates that four carbons are present in the molecule. The *-ane* ending signifies the alkane family. The prefix *n-* indicates that all carbons form an unbranched chain. The prefix *iso-* refers to compounds in which all carbons except one are in a continuous chain and in which that one carbon is branched from a next-to-the-end carbon, as shown:

$$CH_3CH_2CH_2CH_3$$
n-butane

$$\underset{\text{isobutane}}{CH_3\overset{\displaystyle CH_3}{\overset{|}{C}HCH_3}}$$

$$CH_3CH_2CH_2CH_2CH_3$$
n-pentane

$$\underset{\text{isopentane}}{CH_3\overset{\displaystyle CH_3}{\overset{|}{C}HCH_2CH_3}}$$

$$CH_3CH_2CH_2CH_2CH_2CH_3$$
n-hexane

$$\underset{\text{isohexane}}{CH_3\overset{\displaystyle CH_3}{\overset{|}{C}HCH_2CH_2CH_3}}$$

This common naming system has limitations. Pentane has three isomers, and hexane has five. The more complicated the compound, the greater the number of isomers, and the greater the number of special prefixes needed to name all the isomers. It would be extremely difficult and time-consuming to try to identify each of the 75 isomeric alkanes containing 10 carbon atoms by a unique prefix or name.

To devise a system of nomenclature that could be used for even the most complicated compounds, committees of chemists have met periodically since 1892. The system resulting from these meetings is called the IUPAC (International Union of Pure and Applied Chemistry) system. This system is much the same for all classes of organic compounds. The IUPAC name for an organic compound consists of three component parts:

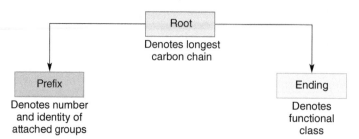

The *root* of the IUPAC name specifies the longest continuous chain of carbon atoms in the compound. The roots for the first 10 normal hydrocarbons are based on the names given in **Table 11.4**: C_1 *meth-*, C_2 *eth-*, C_3 *prop-*, C_4 *but-*, C_5 *pent-*, C_6 *hex-*, C_7 *hept-*, C_8 *oct-*, C_9 *non-*, C_{10} *dec-*.

The *ending* of an IUPAC name specifies the functional class or the major functional group of the compound. The ending *-ane* specifies an alkane. Each of the other functional classes has a characteristic ending; for example, *-ene* is the ending for alkenes, and the *-ol* ending designates alcohols.

Prefixes are used to specify the identity, number, and location of atoms or groups of atoms that are attached to the longest carbon chain. **Table 11.5** lists several common carbon-containing groups referred to as **alkyl groups**. Each alkyl group is a collection of atoms

alkyl group A group differing by one hydrogen from an alkane.

TABLE 11.5 Common Alkyl Groups			
Parent Alkane	Structure of Parent Alkane	Structure of Alkyl Group	Name of Alkyl Group
methane	CH_4	CH_3-	methyl
ethane	CH_3CH_3	CH_3CH_2-	ethyl
propane	$CH_3CH_2CH_3$	$CH_3CH_2CH_2-$	propyl
		CH_3CHCH_3	isopropyl
n-butane	$CH_3CH_2CH_2CH_3$	$CH_3CH_2CH_2CH_2-$	butyl
		$CH_3CH_2CHCH_3$	*sec*-butyl (secondary-butyl)[a]
isobutane	CH_3CHCH_3 with CH_3	CH_3CHCH_2- with CH_3	isobutyl
		CH_3CCH_3 with CH_3	*t*-butyl (tertiary-butyl)[a]

[a]For an explanation of *secondary* and *tertiary*, see Section 13.2.

that can be thought of as an alkane minus one hydrogen atom. Alkyl groups are named simply by dropping *-ane* from the name of the corresponding alkane and replacing it with *-yl*. For example, CH_3- is called a methyl group and CH_3-CH_2- an ethyl group:

CH_4	CH_3-CH_3	CH_3-	CH_3-CH_2-
methane	ethane	methyl group	ethyl group

Two different alkyl groups can be derived from propane, depending on which hydrogen is removed. Removal of a hydrogen from an end carbon results in a propyl group:

$CH_3-CH_2-CH_3$	$CH_3-CH_2-CH_2-$
propane	propyl group

Removal of a hydrogen from the center carbon results in an isopropyl group:

$$CH_3-CH_2-CH_3 \qquad CH_3-\overset{|}{C}H-CH_3$$
propane · · · · · · · · · · isopropyl group

An isopropyl group also can be represented by $(CH_3)_2CH-$. As shown in **Table 11.5**, four butyl groups can be derived from butane, two from the straight-chain, or normal, butane, and two from the branched-chain isobutane. A number of nonalkyl groups are also commonly used in naming organic compounds (see **Table 11.6**).

The following steps are useful when the IUPAC name of an alkane is written on the basis of its structural formula:

Step 1. *Name the longest chain.* The longest continuous carbon-atom chain is chosen as the basis for the name. The names are those given in **Table 11.4**.

TABLE 11.6 Common Nonalkyl Groups

Group	Name
—F	fluoro
—Cl	chloro
—Br	bromo
—I	iodo
—NO₂	nitro
—NH₂	amino

Example	Longest Chain	Comments
$CH_3-CH_2-\underset{\underset{CH_3}{\vert}}{C}H-CH_3$	$C-C-\underset{\underset{C}{\vert}}{C}-C$	This compound is a butane.
$CH_3-\underset{\underset{CH_3}{\vert}}{C}H-\underset{\overset{\vert}{CH_2}\,\vert\,CH_3}{C}H_2$	$C-\underset{\underset{C}{\vert}}{C}-\overset{\overset{C-C}{\vert}}{C}$	This compound is a pentane.
$CH_3-CH_2-\underset{\underset{CH_2}{\vert}\,\underset{CH_3-CH_2}{\vert}}{\overset{\overset{CH_2-CH_3}{\vert}}{C}}-CH_3$	$C-C-\overset{\overset{C-C}{\vert}}{\underset{\underset{C-C}{\vert}}{C}}-C$ $C-C-\underset{\underset{C-C}{\vert}}{\overset{\overset{C-C}{\vert}}{C}}-C$	Often there is more than one way to designate the longest chain. Either way, in this case, the compound is a hexane.

✔ **LEARNING CHECK 11.7** Identify the longest carbon chain in the following:

a. $\underset{CH_2-CH_2-CH_3}{\overset{CH_3}{\vert}}$

b. $\underset{CH_2-\underset{\underset{CH_3}{\vert}}{C}H-CH_2}{\overset{CH_3 \qquad CH_3}{\vert \qquad \vert}}$

c. $\underset{CH_3-\underset{\underset{CH_3}{\vert}}{C}H-CH_3}{\overset{CH_3-CH-CH_3}{\vert}}$

Step 2. *Number the longest chain.* The carbon atoms in the longest chain are numbered consecutively from the end that will give the lowest possible number to any carbon to which a group is attached.

$$\overset{4}{C}H_3 - \overset{3}{C}H_2 - \overset{2}{C}H - \overset{1}{C}H_3 \qquad \text{and not} \qquad \overset{1}{C}H_3 - \overset{2}{C}H_2 - \overset{3}{C}H - \overset{4}{C}H_3$$
$$\quad\quad\quad\quad | \quad\quad\quad\quad\quad\quad\quad\quad\quad\quad\quad\quad\quad | $$
$$\quad\quad\quad\quad CH_3 \quad\quad\quad\quad\quad\quad\quad\quad\quad\quad\quad CH_3$$

$$\overset{2}{C}H_2 - \overset{1}{C}H_3 \qquad\qquad\qquad \overset{4}{C}H_2 - \overset{5}{C}H_3$$
$$\quad \overset{3}{|} \qquad\qquad\qquad\qquad\qquad\qquad\qquad |$$
$$CH_3 - \overset{}{C}H - \overset{4}{C}H_2 \qquad \text{the chain may also} \qquad CH_3 - CH - \overset{2}{C}H_2$$
$$\quad\quad\quad \overset{5}{|} \qquad\qquad \text{be numbered} \qquad \overset{3}{\;} \quad \overset{1}{|}$$
$$\quad\quad\quad CH_3 \qquad\qquad\qquad\qquad\qquad\qquad\qquad CH_3$$

If two or more alkyl groups are attached to the longest chain and more than one numbering sequence is possible, the chain is numbered to get the lowest series of numbers. An easy way to follow this rule is to number from the end of the chain nearest a branch:

$$\overset{1}{C}H_3 - \overset{2}{C}H - \overset{3}{C}H_2 - \overset{4}{C}H - \overset{5}{C}H_2\overset{6}{C}H_3 \qquad \text{and not} \qquad \overset{6}{C}H_3 - \overset{5}{C}H - \overset{4}{C}H_2 - \overset{3}{C}H - \overset{2}{C}H_2 - \overset{1}{C}H_3$$
$$\quad\quad | \qquad\qquad | \qquad\qquad\qquad\qquad\qquad\qquad\qquad | \qquad\qquad\qquad |$$
$$\quad\quad CH_3 \qquad\quad CH_2 - CH_3 \qquad\qquad\qquad\qquad\qquad CH_3 \qquad\quad CH_2 - CH_3$$

Groups are located at positions 2 and 4 Groups are located at positions 3 and 5

$$\qquad\qquad\qquad\qquad\qquad\qquad CH_3 \qquad\qquad\qquad\qquad\qquad\qquad\qquad\qquad CH_3$$
$$\qquad\qquad\qquad\qquad\qquad\qquad | \qquad\qquad\qquad\qquad\qquad\qquad\qquad\qquad\qquad |$$
$$\overset{5}{C}H_3 - \overset{4}{C}H - \overset{3}{C}H_2 - \overset{2}{C} - \overset{1}{C}H_3 \qquad \text{and not} \qquad \overset{1}{C}H_3 - \overset{2}{C}H - \overset{3}{C}H_2 - \overset{4}{C} - \overset{5}{C}H_3$$
$$\quad\quad | \qquad\qquad | \qquad\qquad\qquad\qquad\qquad\qquad | \qquad\qquad\qquad\quad |$$
$$\quad\quad CH_3 \qquad\quad CH_3 \qquad\qquad\qquad\qquad\qquad CH_3 \qquad\qquad CH_3$$

Groups are located at Groups are located at
positions 2,2,4 positions 2,4,4

Here, a difference occurs with the second number in the sequence, so positions 2,2,4 is the lowest series and is used rather than positions 2,4,4.

> ✔ **LEARNING CHECK 11.8** Decide how to correctly number the longest chain in the following according to IUPAC rules:
>
> a. $CH_3 - CH - CH_2 - CH_2 - CH_3$
> $\qquad\quad |$
> $\qquad\quad CH_2$
> $\qquad\quad |$
> $\qquad\quad CH_3$
>
> CH_3
> |
> b. $CH_3 - CH_2 - CH_2 - CH_2 - C - CH_2 - CH - CH_3$
> | |
> CH_3 CH_3

Step 3. *Locate and name the attached alkyl groups.* Each group is located by the number of the carbon atom to which it is attached on the chain.

$$\overset{4}{C}H_3 - \overset{3}{C}H_2 - \overset{2}{C}H - \overset{1}{C}H_3 \qquad \text{The attached group is located on carbon 2 of the chain,}$$
$$\qquad\qquad\qquad | \qquad\qquad\qquad \text{and it is a methyl group.}$$
$$\qquad\qquad\qquad CH_3$$

$$\overset{1}{C}H_3 - \overset{2}{C}H - \overset{3}{C}H_2 - \overset{4}{C}H - \overset{5}{C}H_2 - \overset{6}{C}H_3 \qquad \text{The one-carbon group at position 2 is a}$$
$$\qquad\quad | \qquad\qquad | \qquad\qquad\qquad\qquad \text{methyl group. The two-carbon group at}$$
$$\qquad\quad CH_3 \qquad\quad CH_2 - CH_3 \qquad\qquad \text{position 4 is an ethyl group.}$$

✔ **LEARNING CHECK 11.9** Identify the alkyl groups attached to the horizontal line, which symbolizes a long carbon chain. Refer to Table 11.5 if necessary.

$$CH_3$$
$$|$$
$$CH - CH_3$$
$$|$$

CH_3	CH_2	CH — CH_3	CH_2 — CH_3
	CH_2	CH_2	
	CH_3	CH_3	

Step 4. *Combine the longest chain and the branches into the name.* The position and the name of the attached alkyl group are added to the name of the longest chain and written as one word:

$$\overset{4}{C}H_3 - \overset{3}{C}H_2 - \overset{2}{C}H - \overset{1}{C}H_3$$
$$|$$
$$CH_3$$

2-methylbutane

Additional steps are needed when more than one alkyl group is attached to the longest chain.

Step 5. *Indicate the number and position of attached alkyl groups.* If two or more of the same alkyl group occur as branches, the number of them is indicated by the prefixes *di-*, *tri-*, *tetra-*, *penta-*, etc., and the location of each is again indicated by a number. These position numbers, separated by commas, are put just before the name of the group, with hyphens before and after the numbers when necessary:

$$\overset{1}{C}H_3 - \overset{2}{C}H - \overset{3}{C}H_2 - \overset{4}{C}H - \overset{5}{C}H_3$$
with CH_3 above position 4 and CH_3 below position 2

2,4-dimethylpentane

$$\overset{1}{C}H_3 - \overset{2}{C}H_2 - \overset{3}{C} - \overset{4}{C}H_2 - \overset{5}{C}H_3$$
with CH_3 above and CH_3 below position 3

3,3-dimethylpentane

If two or more *different* alkyl groups are present, their names are alphabetized and added to the name of the basic alkane, again as one word. For purposes of alphabetizing, the prefixes *di-*, *tri-*, and so on are ignored, as are the italicized prefixes secondary *(sec)* and tertiary *(t)*. The prefix *iso-* is an exception and is used for alphabetizing:

$$\overset{1}{C}H_3 - \overset{2}{C}H - \overset{3}{C}H - \overset{4}{C}H_2 - \overset{5}{C}H - \overset{6}{C}H_2 - \overset{7}{C}H_2 - \overset{8}{C}H_3$$
with CH_3, CH_3 below positions 2, 3 and $CH - CH_3$ below position 5
$$|$$
$$CH_3$$

5-isopropyl-2,3-dimethyloctane

✔ LEARNING CHECK 11.10 Give the correct IUPAC name to each of the following:

a.
$$CH_3-CH_2-CH_2-CH_2-\overset{\overset{\displaystyle CH_3}{|}}{CH}-CH_3$$

b.
$$CH_3-CH_2-CH_2-\overset{\overset{\displaystyle CH_2-CH_3}{|}}{CH}-CH_2-\underset{\underset{\displaystyle CH_2-CH_3}{|}}{CH}-CH_3$$

c.
$$CH_3-CH_2-CH_2-\underset{\underset{\displaystyle \underset{\underset{\displaystyle CH_3}{|}}{CH-CH_3}}{|}}{CH}-\overset{\overset{\displaystyle CH_3}{|}}{CH}-\underset{\underset{\displaystyle CH_3}{|}}{CH}-CH_3$$

Naming compounds is a very important skill, as is the reverse process of using IUPAC nomenclature to specify a structural formula. The two processes are very similar. To obtain a formula from a name, determine the longest chain, number the chain, and add any attached groups.

Example 11.4 Drawing Condensed Structural Formulas

Draw a condensed structural formula for 3-ethyl-2-methylhexane.

Solution
Use the last part of the name to determine the longest chain. Draw a chain of six carbons. Then, number the carbon atoms.

$$\underset{1}{C}-\underset{2}{C}-\underset{3}{C}-\underset{4}{C}-\underset{5}{C}-\underset{6}{C}$$

Attach a methyl group at position 2 and an ethyl group at position 3.

$$\underset{1}{C}-\underset{2}{\overset{\overset{\displaystyle CH_3}{|}}{C}}-\underset{3}{\underset{\underset{\displaystyle CH_2CH_3}{|}}{C}}-\underset{4}{C}-\underset{5}{C}-\underset{6}{C}$$

Complete the structure by adding enough hydrogen atoms so that each carbon has four bonds.

$$\underset{1}{CH_3}-\underset{2}{\overset{\overset{\displaystyle CH_3}{|}}{CH}}-\underset{3}{\underset{\underset{\displaystyle CH_2CH_3}{|}}{CH}}-\underset{4}{CH_2}-\underset{5}{CH_2}-\underset{6}{CH_3}$$

✔ LEARNING CHECK 11.11 Draw a condensed structural formula for each of the following compounds:

a. 2,2,4-trimethylpentane
b. 3-isopropylhexane
c. 3-ethyl-2,4-dimethylheptane

11.8 | Cycloalkanes

Learning Objective

9. Assign IUPAC names and draw structural formulas for cycloalkanes.

From what we have said so far, the formula C_3H_6 cannot represent an alkane. Not enough hydrogens are present to allow each carbon to form four bonds, unless there are multiple bonds. For example, the structural formula $CH_3—CH=CH_2$ fits the molecular formula but cannot represent an alkane because of the double bond. The C_3H_6 formula does become acceptable for an alkane if the carbon atoms form a ring, or cyclic, structure rather than the open-chain structure shown:

Open chain Cyclic

The resulting saturated cyclic compound, called cyclopropane, has the structural formula

Alkanes containing rings of carbon atoms are called **cycloalkanes**. Like the other alkanes, cycloalkanes are not found in human cells. However, several important molecules in human cells do contain rings of five or six atoms, and the study of cycloalkanes will help you better understand the chemical behavior of these complex molecules.

According to IUPAC rules, cycloalkanes are named by placing the prefix *cyclo-* before the name of the corresponding alkane with the same number of carbon atoms. Chemists often abbreviate the structural formulas for cycloalkanes and draw them as geometric figures (triangles, squares, etc.) in which each corner represents a carbon atom. The hydrogens are omitted (see **Table 11.7**). It is important to remember that each carbon atom still possesses four bonds, and that hydrogen is assumed to be bonded to the carbon atoms unless something else is indicated. When substituted cycloalkanes (those with attached groups) are named, the position of a single attached group does not need to be specified

cycloalkane An alkane in which carbon atoms form a ring.

TABLE 11.7 Structural Formulas and Symbols for Common Cycloalkanes

Name	Structural Formula	Condensed Formula
cyclopropane		
cyclobutane		
cyclopentane		
cyclohexane		

in the name because all positions in the ring are equivalent. However, when two or more groups are attached, their positions of attachment are indicated by numbers, just as they were for alkanes. The ring numbering begins with the carbon attached to the first group alphabetically and proceeds around the ring in the direction that will give the lowest numbers for the locations of the other attached groups.

methylcyclopentane

1,2-dimethylcyclopentane
not 1,5-dimethylcyclopentane

1-chloro-3-methylcyclopentane
not 1-chloro-4-methylcyclopentane
not 3-chloro-1-methylcyclopentane

Example 11.5 Naming Cycloalkanes

Represent each of the following cycloalkanes by a geometric figure, and name each compound:

Solution

a. A pentagon represents a five-membered ring, which is called cyclopentane. This compound has a methyl group attached, so the name is methylcyclopentane. The position of a single alkyl group is not indicated by a number because the positions of all carbons in the ring are equivalent.

b. A hexagon represents a six-carbon ring, which is called cyclohexane. Two methyl groups are attached; thus we have a dimethylcyclohexane.

However, the positions of the two alkyl groups must be indicated. The ring is numbered beginning with a carbon to which a methyl group is attached, counting in the direction giving the lowest numbers. The correct name is 1,3-dimethylcyclohexane. Notice that a reverse numbering beginning at the same carbon would have given 1,5-dimethylcyclohexane. The number 3 in the correct name is lower than the 5 in the incorrect name.

11.9 | The Shape of Cycloalkanes

Learning Objective

10. Name and draw structural formulas for geometric isomers of cycloalkanes.

Recall from **Section 11.5** that a tetrahedral orientation of bonds with bond angles of 109.5° is characteristic of carbon atoms that form four single bonds. A tetrahedral arrangement is the most stable because it results in the least crowding of the atoms. In certain cycloalkanes, however, a tetrahedral arrangement for all carbon-to-carbon bonds is not possible. For example, the bond angles between adjacent carbon–carbon bonds in planar cyclopropane molecules must be 60° (see **Figure 11.15**). In cyclobutane, they are close to 90°. As a result, cyclopropane and cyclobutane rings are much less stable than compounds with bond angles of about 109°. Both cyclobutane and cyclopentane bend slightly from a planar structure to reduce the crowding of hydrogen atoms (Figure 11.15). In larger cycloalkanes, the bonds to carbon atoms can be tetrahedrally arranged only when the carbon atoms do not lie in the same plane. For example, cyclohexane can assume several nonplanar shapes. The chair and boat forms, where all the bond angles are 109.5°, are shown in Figure 11.15.

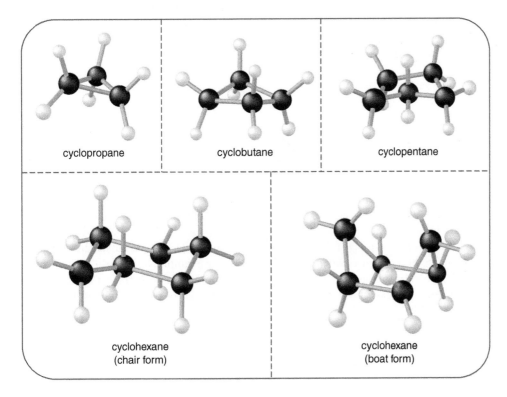

Figure 11.15 Ball-and-stick models for common cycloalkanes.

cyclopropane

cyclobutane

cyclopentane

cyclohexane (chair form)

cyclohexane (boat form)

CHEMISTRY AROUND US 11.1

Fracking Oil Wells

Natural gas and liquid petroleum are major sources of energy in the modern world. In its early history, liquid petroleum naturally found its way to the Earth's surface and formed puddles or ponds. It was collected and used for such things as a topical medicine, a protective coating, a waterproofing material, and a fuel for fires.

Today, liquid petroleum is used primarily as an energy source in the form of liquid fuels. As its use has increased, techniques have been developed to improve the efficiency of getting it out of the ground.

A modern oil well is created by drilling into a deposit and removing the liquid. However, the liquid is not found in underground pools. It is generally distributed in porous rock formations. In order to enhance the flow of liquid oil out of the porous rock, the rock formations are broken up or fractured. This is done by pumping liquid under high pressure into the porous rock. This causes the rock to further break up or fracture, hence the term "fracking."

Fracking is not without its critics. The liquid used to do the fracking contains gas, oil, water, and other contaminants that must be disposed of. In addition to these materials, there is also flammable methane, CH_4, in waste left from the fracking process. There is concern that this methane and other wastes from the fracking fluid will escape and contaminate groundwater. Because of the challenges associated with disposal of the fracking fluid, the fracking process has been banned or put on hold in some European countries.

Fracking procedures enable a greater recovery of oil and gas deposits.

The free rotation that can take place around C—C single bonds in alkanes (see **Section 11.6**) is not possible for the C—C bonds of cycloalkanes. The ring structure allows bending or puckering but prevents free rotation (see **Figure 11.16**). Any rotation of one carbon atom 180° relative to an adjacent carbon atom in a cycloalkane would require a single carbon–carbon bond to be broken somewhere in the ring. The breaking of such bonds would require a large amount of energy.

Figure 11.16 Rotation about C—C single bonds occurs in open-chain compounds but not within rings.

The lack of free rotation around C—C bonds in disubstituted cycloalkanes leads to an extremely important kind of isomerism called *stereoisomerism*. Two different compounds that have the same molecular formula and the same structural formula but different spatial arrangements of atoms are called **stereoisomers**. For example, consider a molecule of 1,2-dimethylcyclopentane. The cyclopentane ring is drawn in **Figure 11.17** as a planar pentagon with the heavy lines indicating that two of the carbons are in front as one views

stereoisomers Compounds with the same structural formula but different spatial arrangements of atoms.

Figure 11.17 Two geometric isomers of 1,2-dimethylcyclopentane.

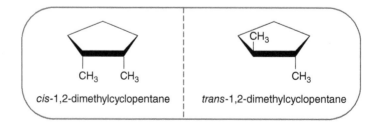

cis-1,2-dimethylcyclopentane trans-1,2-dimethylcyclopentane

the structure. The groups attached to the ring project above or below the plane of the ring. Two stereoisomers are possible: Either both groups may project in the same direction from the plane, or they may project in opposite directions from the plane of the ring. Since the methyl groups cannot rotate from one side of the ring to the other, molecules of the two compounds represented in Figure 11.17 are distinct.

These two compounds have physical and chemical properties that are quite different and therefore can be separated from each other. Stereoisomers of this type, in which the spatial arrangement or geometry of their groups is maintained by rings, are called **geometric isomers** or cis-trans isomers. The prefix *cis-* denotes the isomer in which both groups are on the same side of the ring, and *trans-* denotes the isomer in which they are on opposite sides. To exist as geometric isomers, a disubstituted cycloalkane must be bound to groups at two different carbons of the ring. For example, there are no geometric isomers of 1,1-dimethylcyclohexane:

geometric isomers Molecules with restricted rotation around C—C bonds that differ in the three-dimensional arrangements of their atoms in space and not in the order of linkage of atoms.

cis- On the same side (as applied to geometric isomers).

trans- On opposite sides (as applied to geometric isomers).

Example 11.6 Identifying Geometric Isomers

Name and draw structural formulas for all the isomers of dimethylcyclobutane. Indicate which ones are geometric isomers.

Solution

There are three possible locations for the two methyl groups: positions 1,1, positions 1,2, and positions 1,3.

 Geometric isomerism is not possible in this case with the two groups bound to the same carbon of the ring

1,1-dimethylcyclobutane

 Two groups on opposite sides of the planar ring

trans-1,2-dimethylcyclobutane

H₃C CH₃ Two groups on the same side of the ring

cis-1,2-dimethylcyclobutane

trans-1,3-dimethylcyclobutane *cis*-1,3-dimethylcyclobutane

✔ LEARNING CHECK 11.13

a. Identify each of the following cycloalkanes as a *cis*- or *trans*- compound:

b. Draw the structural formula for *cis*-1,2-dichlorocyclobutane.

Take Care of Dry Skin

Healthy, normal human skin is kept moist by natural body oils contained in sebum, a secretion of the skin's sebaceous glands. The oily sebum helps the epidermis, or outer layer of the skin, retain the 10–30% of water it normally contains. However, some people suffer from skin that is naturally dry, or dry because of aging or contact with materials like paint thinner that dissolve and remove the sebum. Such individuals may find some relief by using OTC products called *moisturizers.*

Two types of skin moisturizers are commonly available. One type, which behaves like natural sebum, contains nonpolar oily substances that form a barrier and prevent water from passing through and evaporating from the skin. These barrier-forming products often contain combinations of materials from various sources, including mineral oil and petroleum jelly from petroleum; vegetable oils such as apricot oil; sesame seed oil, palm kernel oil, olive oil, and safflower oil; and lanolin, an animal fat from sheep oil glands and wool. While they effectively keep water from leaving the skin, these products are somewhat messy to use and leave the skin feeling greasy.

A second, more popular type of moisturizer works by attracting water from the air and skin. These products form a water-rich layer that adheres to the skin without giving it a greasy feel. The substances that attract water are called *humectants*, and are compounds capable of forming hydrogen bonds with water. Some examples of humectants used in products of this type are glycerol, urea, lactic acid, and propylene glycol.

If you are shopping for a moisturizer, remember that the main characteristic you should look for in a product is the ability to form a barrier to prevent water evaporation, or the ability to act as a humectant. Some expensive products advertise that in addition to moisturizing, they also beautify the skin and even reverse aging because they contain proteins such as collagen and elastin, vitamins, hormones, or even DNA. It is unlikely that such substances can pass through the epidermis of the skin in sufficient amounts to provide the advertised benefits.

Alkane products marketed to soften and protect the skin.

11.10 | Physical Properties of Alkanes

Learning Objective

11. Describe the key physical properties of alkanes.

homologous series Compounds of the same functional class that differ by a —CH₂— group.

Since alkanes are composed of nonpolar carbon–carbon and carbon–hydrogen bonds, alkanes are nonpolar molecules. Alkanes have lower melting and boiling points than other organic compounds of comparable molecular weight (see **Table 11.8**). This is because their nonpolar molecules exert very weak attractions for each other. Alkanes are odorless compounds.

The normal, or straight-chain, alkanes make up what is called a **homologous series**. This term describes any series of compounds in which each member differs from a previous member only by having an additional —CH₂— unit. The physical and chemical properties of compounds making up a homologous series are usually closely related and vary in a systematic and predictable way. For example, the boiling points of normal alkanes increase smoothly as the length of the carbon chain increases (see **Figure 11.18**). This pattern results from increasing dispersion forces as molecular weight increases. At ordinary temperatures and pressures, normal alkanes with 1 to 4 carbon atoms are gases, those with 5 to 20 carbon atoms are liquids, and those with more than 20 carbon atoms are waxy solids.

Because they are nonpolar, alkanes and other hydrocarbons are insoluble in water, which is a highly polar solvent. They are also less dense than water and thus float on it.

These two properties of hydrocarbons are partly responsible for the well-known serious effects of oil spills from ships (see **Figure 11.19**).

TABLE 11.8 Physical Properties of Some Normal Alkanes

Number of Carbon Atoms	IUPAC Name	Condensed Structural Formula	Melting Point (°C)	Boiling Point (°C)	Density (g/mL)
1	methane	CH_4	−182.5	−164.0	0.55
2	ethane	CH_3CH_3	−183.2	−88.6	0.57
3	propane	$CH_3CH_2CH_3$	−189.7	−42.1	0.58
4	butane	$CH_3CH_2CH_2CH_3$	−133.4	−0.5	0.60
5	pentane	$CH_3CH_2CH_2CH_2CH_3$	−129.7	36.1	0.63
6	hexane	$CH_3CH_2CH_2CH_2CH_2CH_3$	−95.3	68.9	0.66
7	heptane	$CH_3CH_2CH_2CH_2CH_2CH_2CH_3$	−90.6	98.4	0.68
8	octane	$CH_3CH_2CH_2CH_2CH_2CH_2CH_2CH_3$	−56.8	125.7	0.70
9	nonane	$CH_3CH_2CH_2CH_2CH_2CH_2CH_2CH_2CH_3$	−53.5	150.8	0.72
10	decane	$CH_3CH_2CH_2CH_2CH_2CH_2CH_2CH_2CH_2CH_3$	−29.7	174.1	0.73

Figure 11.18 Normal alkane boiling points depend on chain length.

Simon Fraser/Science Photo Library/Science Source

Figure 11.19 Oil spills can have serious and long-lasting effects on the environment because of the insolubility of hydrocarbons in water.

hydrophobic Molecules or parts of molecules that repel (are insoluble in) water.

Liquid alkanes of higher molecular weight behave as *emollients* (skin softeners) when applied to the skin. An alkane mixture known as mineral oil is sometimes used to replace natural skin oils washed away by frequent bathing or swimming. Petroleum jelly (Vaseline is a well-known brand name) is a semisolid mixture of alkanes that is used as both an emollient and a protective film. Water and water solutions such as urine don't dissolve or penetrate the film, and the underlying skin is protected. Many cases of diaper rash have been prevented or treated this way.

The word **hydrophobic** (literally "water fearing") is often used to refer to molecules or parts of molecules that are insoluble in water. Many biomolecules, the large organic molecules associated with living organisms, contain nonpolar (hydrophobic) parts. Thus, such molecules are not water-soluble. Palmitic acid, for example, contains a large nonpolar hydrophobic portion and is insoluble in water.

Nonpolar portion

$$CH_3-CH_2-CH_2-CH_2-CH_2-CH_2-CH_2-CH_2-CH_2-CH_2-CH_2-CH_2-CH_2-CH_2-CH_2-\overset{\displaystyle O}{\overset{\displaystyle \|}{C}}-OH$$

palmitic acid

Reducing Your Carbon Footprint

Before something can be reduced, it must be defined. Your carbon footprint is defined as the total amount of greenhouse gases produced each year as a direct or indirect result of your activity. The value will be defined in terms of tons of carbon dioxide, CO_2, produced by your activities each year.

Some of your activities produce CO_2 directly. Examples of this include driving an automobile or heating your house by burning natural gas in a furnace. Other activities produce CO_2 indirectly. An example of this is when you purchase food or other produced goods in which CO_2 was released during their production.

The most common ways to reduce your CO_2 footprint can be summarized by the four *R's*: *Reduce*, *Reuse*, *Recycle*, and *Refuse*. For example, you can reduce your personal energy use by such things as turning off lights, unplugging electronics and appliances when they are not in use, and reducing the use of air conditioners in the summer and heating furnaces in the winter. An adjustment of just a degree or two in home temperature can make a significant difference in your personal carbon footprint. You can also switch to more efficient lightbulbs. LED bulbs use much less energy than incandescent bulbs.

The second R, Reuse, is demonstrated by people who take their own empty paper or cloth grocery bags with them when they go shopping. Another example of Reuse is the collection and reuse of lumber from a building demolition site.

Recycle, the third R, is catching on more and more in communities. Some communities provide special, recognizable, waste containers for recyclables that are picked up separately from regular refuse containers. Other communities have central locations with large bins for recyclables.

The fourth R, Refuse, is very personal. It requires a personal decision to act or refuse to act in specific ways. For example, you could refuse to drive your car for trips of less than a mile. You could ride a bicycle or walk instead. You could decide to fly less and drive your car instead.

LED bulbs use less energy than regular incandescent bulbs.

11.11 | **Alkane Reactions**

Learning Objective

12. Write alkane combustion reactions.

The alkanes are the least reactive of all organic compounds. In general, they do not react with strong acids (such as sulfuric acid), strong bases (such as sodium hydroxide), most oxidizing agents (such as potassium dichromate), or most reducing agents (such as sodium metal). This unreactive nature is reflected in the name *paraffins*, sometimes used to identify alkanes (from Latin words that mean "little affinity"). Paraffin wax, sometimes used to seal jars of homemade preserves, is a mixture of solid alkanes. Paraffin wax is also used in the preparation of milk cartons and wax paper. The inertness of the compounds in the wax makes it ideal for these uses. Alkanes do undergo reactions with halogens such as chlorine and bromine, but these are not important for our purposes.

Perhaps the most significant reaction of alkanes is the rapid oxidation called combustion. In the presence of ample oxygen, alkanes burn to form carbon dioxide and water, liberating large quantities of heat:

$$CH_4 + 2O_2 \rightarrow CO_2 + 2H_2O + 212.8 \text{ kcal/mol} \qquad (11.4)$$

It is this type of reaction that accounts for the wide use of hydrocarbons as fuels.

Natural gas contains methane (80–95%), some ethane, and small amounts of other hydrocarbons. Propane and butane are extracted from natural gas and sold in pressurized metal containers (bottled gas). In this form, they are used for heating and cooking in campers, trailers, boats, and rural homes.

Gasoline is a mixture of hydrocarbons (primarily alkanes) that contain 5 to 12 carbon atoms per molecule. Diesel fuel is a similar mixture, except the molecules contain 12 to 16 carbon atoms. The hot CO_2 and water vapor generated during combustion in an internal combustion engine have a much greater volume than the air and fuel mixture. It is this sudden increase in gaseous volume and pressure that pushes the pistons and delivers power to the crankshaft.

If there is not enough oxygen available, incomplete combustion of hydrocarbons occurs, and some carbon monoxide (CO) or even carbon (see **Figure 11.20**) may be produced (see **Reactions 11.5** and **11.6**):

$$2CH_4 + 3O_2 \rightarrow 2CO + 4H_2O \tag{11.5}$$

$$CH_4 + O_2 \rightarrow C + 2H_2O \tag{11.6}$$

These reactions are usually undesirable because CO is toxic, and carbon deposits hinder engine performance. Occasionally, however, incomplete combustion is deliberately caused; specific carbon blacks (particularly lampblack, a pigment for ink) are produced by the incomplete combustion of natural gas.

Michael C. Slabaugh

Figure 11.20 A luminous yellow flame from a laboratory burner produces a deposit of carbon when insufficient air (O_2) is mixed with the gaseous fuel.

▶ Case Study Follow-up

Mineral oil is a mixture of light alkanes in the C-15 to C-40 range. Because mineral oil is hydrophobic it has a slippery quality that many people find unpleasant. Mineral oil is frequently used to treat encopresis. Behavioral techniques have been successful in increasing compliance with medication regimens in both children and adults. It is important to set up a reliable routine so that a child knows what to expect. For example, medication should be given at the same time every day, in the same place, and in the same manner. Rewards should be immediate and of sufficient merit to encourage compliance.

Some children need significant rewards in order to be persuaded to take an unpleasant medicine. Never underestimate the power of parental praise as part of the reward! Many parents object to behavioral programs that use rewards because they equate them with bribery. However, they do work. Eventually, once the desired behavior is established, the reward system can be phased out or the parent may choose a different behavior to reward. Punishing a child for soiling is counterproductive. Transferring responsibility to the child (e.g., having even a young child help clean up the mess) is more effective.

Concept Summary

Carbon: The Element of Organic Compounds
Organic compounds contain carbon, and organic chemistry is the study of those compounds. Inorganic chemistry is the study of the elements and all noncarbon compounds.
Objective 1 (Section 10.1), Exercise 11.2

Carbon compounds are of tremendous everyday importance to life on the Earth and are the basis of all life processes.
Objective 2 (Section 10.1), Exercise 11.4

Organic and Inorganic Compounds Compared
The properties of organic and inorganic compounds often differ, largely as a result of bonding differences. Organic compounds contain primarily covalent bonds, whereas ionic bonding is more prevalent in inorganic compounds.
Objective 3 (Section 10.2), Exercise 11.8

Bonding Characteristics and Isomerism Large
numbers of organic compounds are possible because carbon

atoms link to form chains and networks. An additional reason for the existence of so many organic compounds is the phenomenon of isomerism. Isomers are compounds that have the same molecular formula but different arrangements of atoms.
Objective 4 (Section 10.3), Exercise 11.20

Functional Groups: The Organization of Organic
Chemistry All organic compounds are grouped into classes based on characteristic features called functional groups. Compounds with their functional groups are represented by two types of structural formulas. Expanded structural formulas show all covalent bonds, whereas condensed structural formulas show no covalent bonds or only selected bonds.
Objective 5 (Section 10.4), Exercise 11.24

Alkane Structures Alkanes are hydrocarbons that contain only single covalent bonds and can be represented by the formula C_nH_{2n+2}. Alkanes possess a three-dimensional geometry in which each carbon is surrounded by four bonds

directed to the corners of a tetrahedron. Methane, the simplest alkane, is an important fuel (natural gas) and a chemical feedstock for the preparation of other organic compounds. The number of structural isomers possible for an alkane increases dramatically with the number of carbon atoms present in the molecule. The straight-chain isomer is called a normal alkane; others are called branched isomers.

Objective 6 (Section 10.5), Exercise 11.28

Conformations of Alkanes Rotation about the single bonds between carbon atoms allows alkanes to exist in many different conformations. When an alkane is drawn using only two dimensions, the structure can be represented in a variety of ways as long as the order of bonding is not changed.

Objective 7 (Section 10.6), Exercise 11.30

Alkane Nomenclature Some simple alkanes are known by common names. More complex compounds are usually named using the IUPAC system. The characteristic IUPAC ending for alkanes is *-ane*.

Objective 8 (Section 10.7), Exercise 11.34

Cycloalkanes These are alkanes in which the carbon atoms form a ring. The prefix *cyclo-* is used in the names of these compounds to indicate their cyclic nature.

Objective 9 (Section 10.8), Exercise 11.44

The Shape of Cycloalkanes The carbon atom rings of cycloalkanes are usually shown as planar, although only cyclopropane is planar. Because rotation about the single bonds in the ring is restricted, certain disubstituted cycloalkanes can exist as geometric (*cis-trans*) isomers.

Objective 10 (Section 10.9), Exercise 11.54

Physical Properties of Alkanes The physical properties of alkanes are typical of all hydrocarbons: nonpolar, insoluble in water, less dense than water, and increasing melting and boiling points with increasing molecular weight.

Objective 11 (Section 10.10), Exercise 11.56

Alkane Reactions Alkanes are relatively unreactive and remain unchanged by most reagents. The reaction that is most significant is combustion.

Objective 12 (Section 10.11), Exercise 11.60

Key Terms and Concepts

Alkane (11.5)	Functional group (11.4)	Normal alkane (11.5)
Alkyl group (11.7)	Geometric isomers (11.9)	Organic chemistry (11.1)
Branched alkane (11.5)	Homologous series (11.10)	Organic compound (11.1)
cis- (11.9)	Hybrid orbital (11.3)	Saturated hydrocarbon (11.5)
Condensed structural formula (11.4)	Hydrocarbon (11.5)	Stereoisomers (11.9)
Conformations (11.6)	Hydrophobic (11.10)	Structural isomers (11.3)
Cycloalkane (11.8)	Inorganic chemistry (11.1)	*trans-* (11.9)
Expanded structural formula (11.4)	Isomerism (11.3)	

Key Equations

1.	Complete combustion of alkanes (Section 11.11):	Alkane $+ O_2 \rightarrow CO_2 + H_2O$
2.	Incomplete combustion of alkanes (Section 11.11):	Alkane $+ O_2 \rightarrow CO$ (or C) $+ H_2O$

Exercises

Even-numbered exercises are answered in Appendix B.

Blue-numbered exercises are more challenging.

Carbon: The Element of Organic Compounds (Section 11.1)

11.1 Why were the compounds of carbon originally called organic compounds?

11.2 Name at least six items you recognize to be composed of organic compounds.

11.3 Describe what Wöhler did that made the vital force theory highly questionable.

11.4 What is the unique structural feature shared by all organic compounds?

11.5 Classify each of the following compounds as organic or inorganic:

a. KBr

b. H_2O

c. H—C≡C—H

d. LiOH

e. CH_3—NH_2

Organic and Inorganic Compounds Compared (Section 11.2)

11.6 What kind of bond between atoms is most prevalent among organic compounds?

11.7 Are the majority of all compounds that are insoluble in water organic or inorganic? Why?

11.8 Indicate for each of the following characteristics whether it more likely describes an inorganic or an organic compound. Give one reason for your answer.

 a. This compound is a liquid that readily burns.

 b. A white solid upon heating is found to melt at 735°C.

 c. A liquid added to water floats on the surface and does not dissolve.

 d. This compound exists as a gas at room temperature and ignites easily.

 e. A solid substance melts at 65°C.

11.9 Devise a test, based on the general properties in Table 11.1, that you could use to quickly distinguish between the substances in each of the following pairs:

 a. Gasoline (liquid, organic) and water (liquid, inorganic)

 b. Naphthalene (solid, organic) and sodium chloride (solid, inorganic)

 c. Methane (gaseous, organic) and hydrogen chloride (gaseous, inorganic)

11.10 Explain why organic compounds are nonconductors of electricity.

11.11 Explain why the rate of chemical reactions is generally slow for organic compounds and usually fast for inorganic compounds.

Bonding Characteristics and Isomerism (Section 11.3)

11.12 Give two reasons for the existence of the tremendous number of organic compounds.

11.13 How many of carbon's electrons are unpaired and available for bonding according to an sp^3 hybridization model?

11.14 Describe what atomic orbitals overlap to produce a carbon–hydrogen bond in CH_4.

11.15 What molecular geometry exists when a central carbon atom bonds to four other atoms?

11.16 Compare the shapes of unhybridized p and hybridized sp^3 orbitals.

11.17 Use Example 11.1 and Tables 11.2 and 11.6 to determine the number of covalent bonds formed by atoms of the following elements: carbon, hydrogen, oxygen, nitrogen, and bromine.

11.18 Complete the following structures by adding hydrogen atoms where needed:

 a. C — C — C

 b. C — C = C

 c.
$$\begin{array}{c} O \\ \parallel \\ C - C \end{array}$$

 d. C — C — N

11.19 Complete the following structures by adding hydrogen atoms where needed.

 a.
$$\begin{array}{c} C \\ \vert \\ C - C - C \end{array}$$
 c. C — C ≡ C — C

 b.
$$\begin{array}{c} C \\ \vert \\ C - N - C \end{array}$$
 d. C — O — C

11.20 Which of the following pairs of compounds are structural isomers?

 a. CH_3 — CH = CH — CH_3 and CH_3 — CH_2 — CH_2 — CH_3

 b. CH_3 — CH_2 — CH_2 — CH_2 — CH_3 and
$$\begin{array}{c} CH_3 \\ \vert \\ CH_3 - C - CH_3 \\ \vert \\ CH_3 \end{array}$$

 c.
$$\begin{array}{c} CH_3 \\ \vert \\ CH_3 - CH_2 - CH - OH \end{array}$$
and
$$\begin{array}{c} O \\ \parallel \\ CH_3 - C - CH_2 - CH_3 \end{array}$$

 d.
$$\begin{array}{c} O \\ \parallel \\ CH_3 - CH_2 - C - H \end{array}$$
and
$$\begin{array}{c} O \\ \parallel \\ CH_3 - C - CH_3 \end{array}$$

 e. CH_3 — CH_2 — CH_2 — NH_2 and CH_3 — CH_2 — NH — CH_3

11.21 Group all the following compounds together that represent structural isomers of each other:

 a.
$$\begin{array}{c} O \\ \parallel \\ CH_3 - CH_2 - C - OH \end{array}$$

 b.
$$\begin{array}{c} O \\ \parallel \\ CH_3 - C - CH_2 - OH \end{array}$$

 c. CH_3 — CH_2 — CH_2 — OH

 d.
$$\begin{array}{c} O \\ \parallel \\ CH_3 - C - CH_2 - CH_3 \end{array}$$

 e.
$$\begin{array}{c} OH \quad O \\ \vert \quad\; \parallel \\ CH_3 - CH - C - H \end{array}$$

 f.
$$\begin{array}{c} O \\ \parallel \\ CH_3 - C - O - CH_3 \end{array}$$

 g. HO — CH_2 — CH_2 — CH_2 — OH

11.22 On the basis of the number of covalent bonds possible for each atom, determine which of the following structural formulas are correct. Explain what is wrong with the incorrect structures.

a.
$$
\begin{array}{ccc}
& H & H \\
& | & | \\
H- & C-H-C & -H \\
& | & | \\
& H & H
\end{array}
$$

b.
$$
\begin{array}{ccc}
& H & H \\
& | & | \\
H-N- & C-C & -H \\
| & | & | \\
H & H & H
\end{array}
$$

c.
$$
\begin{array}{ccc}
H & O & H \\
| & || & | \\
H-C\!\!-\!\!-\!\!-\!\!C\!\!-\!\!-\!\!-\!\!C-H \\
| & | & | \\
H & H-C-H & H \\
& | & \\
& H &
\end{array}
$$

d. $CH_3-C\!=\!C-CH_3$

e.
$$
\begin{array}{ccc}
& OH & O \\
& | & || \\
CH_3-CH-CH-C-H \\
& | & \\
& CH_3 &
\end{array}
$$

Functional Groups: The Organization of Organic Chemistry (Section 11.4)

11.23 Identify each of the following as a condensed structural formula, expanded structural formula, or molecular formula:

a. $CH_3CH_2CH_2-NH_2$

b.
$$
\begin{array}{cccc}
H & H & H & H \\
| & | & | & | \\
H-C-C-C-N-H \\
| & | & | \\
H & H & H
\end{array}
$$

c. $CH_3-CH_2-CH_2-NH_2$

d. C_3H_9N

11.24 Write a condensed structural formula for the following compounds:

a.
$$
\begin{array}{ccccccc}
H & H & H & & O & H & H \\
| & | & | & & || & | & | \\
H-C-C-C-C\!=\!C-C-C-H \\
| & | & | & | & | & | & | \\
H & H & H & H & H & H & H
\end{array}
$$

b.
$$
\begin{array}{ccc}
H & H & O \\
| & | & || \\
H-C-C-C-N-H \\
| & | & | \\
H & H & H
\end{array}
$$

11.25 Write a condensed structural formula for the following compounds:

a.
$$
\begin{array}{cccc}
& & H\;\;H & \\
& & \backslash\;/ & \\
H & & N & H\;\;H \\
| & & | & |\;\;| \\
H-C\!\!-\!\!-\!\!-\!\!-\!\!-\!\!C-C\!=\!C-H \\
| & & | & \\
H & & H-C-H & \\
& & | & \\
& & H &
\end{array}
$$

b.
$$
\begin{array}{ccccc}
O & H & & H & O \\
|| & | & & | & || \\
H-O-C-C-O-C-C-H \\
& | & & | \\
& H-C-H & & H \\
& | & \\
& H &
\end{array}
$$

11.26 Write an expanded structural formula for the following:

a.
$$
\begin{array}{c}
O \\
|| \\
H-C-CH_2-CH_2-NH_2
\end{array}
$$

b.
$$
\begin{array}{c}
CH_3-CH-O-CH_3 \\
| \\
CH_3
\end{array}
$$

Alkane Structures (Section 11.5)

11.27 The name of the normal alkane containing 10 carbon atoms is decane. What are the molecular and condensed structural formulas for decane?

11.28 Classify each of the following compounds as a normal alkane or a branched alkane:

a.
$$
\begin{array}{c}
CH_3-CH-CH_3 \\
| \\
CH_3
\end{array}
$$

b.
$$
\begin{array}{c}
CH_3-CH_2-CH_2 \\
| \\
CH_2-CH_3
\end{array}
$$

c.
$$
\begin{array}{c}
CH_3 \\
| \\
CH_2-CH_2 \\
| \\
CH_3
\end{array}
$$

d.
$$
\begin{array}{c}
CH_3-CH_2-CH_2 \\
| \\
CH_3-CH_2
\end{array}
$$

e.
$$
\begin{array}{cc}
CH_3 & CH_3 \\
| & | \\
CH_2-CH_2-CH \\
& | \\
& CH_3
\end{array}
$$

f.
$$
\begin{array}{c}
CH_3 \\
| \\
CH_3-CH_2-C-CH_3 \\
| \\
CH_3
\end{array}
$$

Conformations of Alkanes (Section 11.6)

11.29 Why are different conformations of an alkane not considered structural isomers?

11.30 Which of the following pairs represent structural isomers, and which are simply the same compound?

a. H_3C CH_2 and $CH_3CH_2CH_2CH_3$
 $CH_2\ CH_3$

b. $CH_3-CH_2-CH_2$ and $CH_3CH_2CH_2CH_2CH_2CH_3$
 $CH_3-CH_2-CH_2$

c. $CH_3-CH-CH_3$ and $CH_3CH_2CH_2CH_3$
 CH_3

d. $CH_3-CH-CH_2CH_3$ and $CH_3-CH_2-CH-CH_3$
 CH_3 CH_3

Alkane Nomenclature (Section 11.7)

11.31 For each of the following carbon skeletons, give the number of carbon atoms in the longest continuous chain:

a.
```
    C — C
    |
C — C — C — C
    |   |
    C   C — C
```

c.
```
    C — C — C
        |
C — C — C — C
        |
        C — C
```

b.
```
C — C — C — C — C
    |
    C — C — C — C
    |
    C — C — C
        |
        C — C
```

11.32 For each of the following carbon skeletons, give the number of carbon atoms in the longest continuous chain:

a.
```
    C   C
    |   |
    C   C
    |   |
C — C — C — C
```

c.
```
    C — C      C — C
        |      |
C — C — C — C — C
```

b.
```
    C — C
    |
C — C — C — C
```

11.33 Identify the following alkyl groups:

a. CH_3-CH-
 CH_3

b. $CH_3-CH_2-CH_2-CH_2-$

c. $CH_3-CH-CH_2-$
 CH_3

d. CH_3-

11.34 Give the correct IUPAC name for each of the following alkanes:

a.
CH_3-CH_2
 $|$
$CH_3-CH-CH_2-CH_3$

b.
$CH_3-CH-CH-CH_2-CH-CH_3$
 $|$ $|$ $|$
 CH_2 CH_3 CH_3
 $|$
 CH_3

c.
```
                CH3
                |
CH3 — CH — C — CH2 — CH3
       |   |
      CH3  CH2
           |
           CH3
```

d.
```
                CH3
                |
                CH2
                |
                CH2
                |
CH3 — CH2 — CH — CH2 — CH — CH3
                      |
        CH3 — CH2 — CH2 — CH
                          |
                          CH2 — CH3
```

e.
```
        CH2 — CH2 — CH2 — CH3
        |
CH3 — CH — CH2 — CH — CH2 — CH3
                 |
                 CH2 — CH3
```

11.35 Give the correct IUPAC name for each of the following alkanes:

a.
$CH_3 - CH_2 - CH - CH_3$
 $|$
 CH_3

b.
```
        CH3
        |
CH3 — C — CH3
        |
        CH3
```

c.
```
        CH2 — CH3
        |
CH3 — CH — CH — CH3
            |
            CH3
```

d.

$$CH_2 - CH - C - CH_2 - CH - CH_2 - CH_3$$

with substituents:
- top: CH_3 and CH_2 above C
- CH_2 connected to CH_3
- CH_3, CH_2 below
- CH_3
- CH_2 connected to $CH - CH_3$ and CH_3

e.

$$CH_3 - CH_2 - CH - CH - CH - CH_2 - CH - CH_3$$

with CH_3 groups on three carbons and a $CH_2 - CH_2 - CH_2 - CH_3$ chain

11.36 Draw a condensed structural formula for each of the following compounds:

a. 3-ethylpentane

b. 2,2-dimethylbutane

c. 4-ethyl-3,3-dimethyl-5-propyldecane

d. 5-*sec*-butyldecane

11.37 Draw a condensed structural formula for each of the following compounds:

a. 2,2,4-trimethylpentane

b. 4-isopropyloctane

c. 3,3-diethylhexane

d. 5-*t*-butyl-2-methylnonane

11.38 Draw the condensed structural formula for each of the three structural isomers of C_5H_{12}, and give the correct IUPAC names.

11.39 Isooctane is 2,2,4-trimethylpentane. Draw structural formulas for and name a branched heptane, hexane, pentane, and butane that are structural isomers of isooctane.

11.40 Draw structural formulas for the compounds and give correct IUPAC names for the five structural isomers of C_6H_{14}.

11.41 The following names are incorrect, according to IUPAC rules. Draw the structural formulas and tell why each name is incorrect. Write the correct name for each compound.

a. 2,2-methylbutane

b. 4-ethyl-5-methylheptane

c. 2-ethyl-1,5-dimethylhexane

11.42 The following names are incorrect, according to IUPAC rules. Draw the structural formulas and tell why each name is incorrect. Write the correct name for each compound.

a. 1,2-dimethylpropane

b. 3,4-dimethylpentane

c. 2-ethyl-4-methylpentane

d. 2-bromo-3-ethylbutane

Cycloalkanes (Section 11.8)

11.43 The general formula for alkanes is C_nH_{2n+2}. Write a general formula for cycloalkanes.

11.44 Write the correct IUPAC name for each of the following:

a.

b.

c.

d.

11.45 Write the correct IUPAC name for each of the following:

a.

b.

c.

d.

11.46 Draw the structural formulas corresponding to each of the following IUPAC names:

a. ethylcyclobutane

b. 1,1,2,5-tetramethylcyclohexane

c. 1-butyl-3-isopropylcyclopentane

11.47 Draw the structural formulas corresponding to each of the following IUPAC names:

a. 1-ethyl-1-propylcyclopentane

b. isopropylcyclobutane

c. 1,2-dimethylcyclopropane

11.48 Which of the following pairs of cycloalkanes represent structural isomers?

a.

b.

c.

d.

11.49 Draw structural formulas for the five structural isomers of C_5H_{10} that are cycloalkanes.

The Shape of Cycloalkanes (Section 11.9)

11.50 Why does cyclohexane assume a chair form rather than a planar hexagon?

11.51 Explain the difference between geometric and structural isomers.

11.52 Which of the following cycloalkanes could show geometric isomerism? For each that could, draw structural formulas, and name both the *cis*- and the *trans*- isomers.

a.

c.

b.

d.

11.53 Draw structural formulas for *cis*- and *trans*-1,2-dimethylcyclohexane.

11.54 Using the prefix *cis*- or *trans*-, name each of the following:

a.

c.

b.

d.

11.55 For each of the following molecular formulas, give the structural formulas requested. In most cases, there are several possible structures.

a. C_6H_{14}, a normal alkane

b. C_6H_{14}, a branched alkane

c. C_5H_{12}, a pair of conformations

d. C_5H_{12}, a pair of structural isomers

e. C_5H_{10}, a cyclic hydrocarbon

f. C_6H_{12}, two cycloalkane geometric isomers

g. C_6H_{12}, a cycloalkane that has no geometric isomers

Physical Properties of Alkanes (Section 11.10)

11.56 The compound decane is a straight-chain alkane. Predict the following:

a. Is decane a solid, liquid, or gas at room temperature?

b. Is it soluble in water?

c. Is it soluble in hexane?

d. Is it more or less dense than water?

11.57 Explain why alkanes of low molecular weight have lower melting and boiling points than water.

11.58 Suppose you have a sample of 2-methylhexane and a sample of 2-methylheptane. Which sample would you expect to have the higher melting point? Boiling point?

11.59 Identify (circle) the alkanelike portions of the following molecules:

a.
$$H_2N - \underset{\underset{\displaystyle CH_3}{\overset{\displaystyle |}{\underset{\displaystyle |}{CH_2}}}}{\overset{\displaystyle |}{CH}} - \overset{\overset{\displaystyle O}{\displaystyle \|}}{C} - OH$$

isoleucine (an amino acid)

b.
$$CH_3-CH_2-CH_2-CH_2-CH_2-CH_2-CH_2-CH_2$$
$$Na^+{}^-O - \overset{\overset{\displaystyle O}{\displaystyle \|}}{C} - CH_2 - CH_2 - CH_2 - CH_2 - CH_2 - CH_2 - CH_2$$

sodium palmitate (a soap)

c.
lipoic acid (a component of a coenzyme)

d.

menthol (a flavoring)

Alkane Reactions (Section 11.11)

11.60 Write a balanced equation to represent the complete combustion of each of the following:

a. Butane

b. $CH_3 - \underset{\underset{\displaystyle CH_3}{\overset{\displaystyle |}{\underset{\displaystyle |}{C}}}}{\overset{\overset{\displaystyle CH_3}{\displaystyle |}}{C}} - CH_3$

c.

11.61 Write a balanced equation to represent the complete combustion of each of the following:

a. Propane

b. CH₃CHCH₃
 |
 CH₃

c. (a square ring with a CH₃ group)

11.62 Write a balanced equation for the incomplete combustion of hexane, assuming the formation of carbon monoxide and water as the products.

11.63 Why is it dangerous to relight a furnace when a foul odor is present?

Additional Exercises

11.64 Using the concept of dispersion forces, explain why most cycloalkanes have higher boiling points than normal alkanes with the same number of carbon atoms.

11.65 Draw a Lewis electron dot formula for ethane (CH_3—CH_3). Explain why ethane molecules do not hydrogen-bond.

11.66 Your sometimes inept lab technician performed experiments to determine the vapor pressure of pentane, hexane, and heptane at 20°C. He gave you back the numbers of 113.9, 37.2, and 414.5 torr without identifying the compounds. Which vapor pressure value goes with which compound?

11.67 How many grams of water will be produced by the complete combustion of 10.0 g of methane (CH_4)?

11.68 How many liters of air at STP are needed to completely combust 1.00 g of methane (CH_4)? Air is composed of about 21% v/v oxygen (O_2).

Chemistry for Thought

11.69 Would you expect a molecule of urea produced in the body to have any different physical or chemical properties from a molecule of urea prepared in a laboratory?

11.70 Why might the study of organic compounds be important to someone interested in the health or life sciences?

11.71 Why do very few aqueous solutions of organic compounds conduct electricity?

11.72 The ski wax being examined in Figure 11.2 has a relatively low melting point. What does that fact reveal about the forces between molecules?

11.73 Charcoal briquettes sometimes burn with incomplete combustion when the air supply is limited. Why would it be hazardous to place a charcoal grill inside a home or a camper in an attempt to keep warm?

11.74 If carbon did not form hybridized orbitals, what would you expect to be the formula of the simplest compound of carbon and hydrogen?

11.75 What types of sports equipment are made from graphite fibers besides that shown in Figure 11.8?

11.76 A semi truck loaded with cyclohexane overturns during a rainstorm, spilling its contents over the road embankment. If the rain continues, what will be the fate of the cyclohexane?

11.77 On the way home from school, you drove through a construction zone, resulting in several tar deposits on the car's fender. What substances commonly found in the kitchen might help in removing the tar deposits?

11.78 Oil spills along coastal shores can be disastrous to the environment. What physical and chemical properties of alkanes contribute to the consequences of an oil spill?

11.79 Why might some farmers hesitate to grow and sell organic produce?

Allied Health Exam Connection

The following questions are from these sources:

• *Nursing School Entrance Exam* © 2005, Learning Express, LLC.

• *McGraw-Hill's Nursing School Entrance Exams* by Thomas A. Evangelist, Tamara B. Orr, and Judy Unrein © 2009, The McGraw-Hill Companies, Inc.

• *NSEE Nursing School Entrance Exams,* 3rd edition © 2009, Kaplan Publishing.

• *Cliffs Test Prep: Nursing School Entrance Exams* by Fred N. Grayson © 2004, Wiley Publishing, Inc.

• *Peterson's Master the Nursing School and Allied Health Entrance Exams,* 18th edition by Marion F. Gooding © 2008, Peterson's, a Nelnet Company.

11.80 Which of the following are examples of organic compounds?

 a. $CaCO_3$

 b. NH_3

 c. NaCl

 d. $C_6H_{12}O_6$

11.81 Use the generic formula for alkanes (C_nH_{2n+2}) to derive molecular and condensed structural formulas for:

 a. Propane, 3 carbon atoms

 b. Octane, 8 carbon atoms

 c. Butane, 4 carbon atoms

11.82 Organic compounds are the basis for life as we know it because:

 a. Carbon-to-carbon bonds are strong.

 b. Carbon can form long chains.

 c. Carbon chains can include other elements to give rise to different functional groups.

 d. All of the above are correct.

11.83 Which of the following is an example of an alkane?

 a. C_2H_6

 b. C_2H_4

 c. CH_3OH

 d. $C_6H_{12}O_6$

11.84 The name of the compound CH_3—CH_2—CH_2—CH_3 is most likely:

 a. cyclobutane

 b. butane

 c. butene

 d. butyne

11.85 Which of the following are the general products of a combustion reaction?

 a. $C(s)$, O_2, and H_2

 b. $C(s)$, H_2O, and O_2

 c. CO_2 and H_2

 d. CO_2 and H_2O

11.86 Which is mostly methane?

 a. oil

 b. natural gas

 c. coal waste

 d. propane

11.87 The deadly property of carbon monoxide, if inhaled, is due to its:

 a. high affinity for O_2.

 b. low affinity for hemoglobin.

 c. high affinity for hemoglobin.

 d. conversion to cyanide gas.

12

Unsaturated Hydrocarbons

▶ Case Study

Scott, a young man, worked as a sub-contractor at a copper mine. In spite of extensive safety training provided by the mining company, one of his co-workers failed to properly secure a load of steel pipe on a truck. Scott knelt to place a chock under a wheel of the truck, and a pipe fell, crushing his leg. Scott's leg was amputated above the knee. The company accepted responsibility for his injury and covered his medical costs. Thanks to technological advances in plastics and computer design, Scott was offered a choice of colors and designs for his prosthetic leg. His company refused to pay the additional cost of personalizing his prosthesis. His physician recommended that the bill be covered, considering the psychological impact of a man losing a leg in his youth.

> ▶ Is it a reasonable request for an insurance company to cover this additional cost? Is the ability of a patient to choose the design of the prosthesis important to self-esteem? How does self-esteem impact recovery?

Follow-up to this Case Study appears at the end of the chapter before the *Concept Summary*.

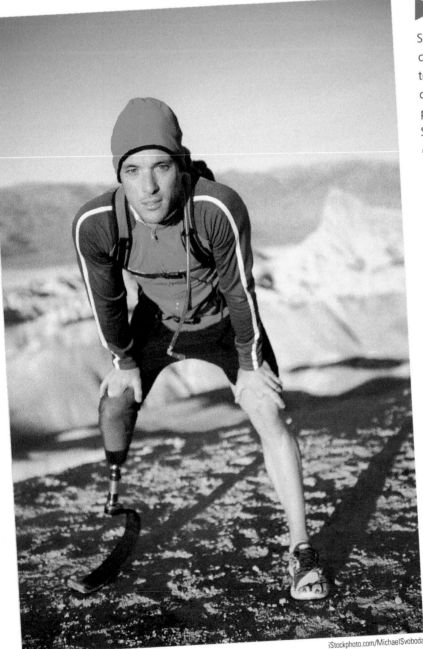
iStockphoto.com/MichaelSvoboda

Learning Objectives

When you have completed your study of this chapter, you should be able to:

1 Classify unsaturated hydrocarbons as alkenes, alkynes, or aromatics. **(Section 12.1)**

2 Write the IUPAC names of alkenes from their molecular structures. **(Section 12.1)**

3 Predict the existence of geometric (cis-trans) isomers from formulas of compounds. **(Section 12.2)**

4 Write the names and structural formulas for geometric isomers. **(Section 12.2)**

5 Write equations for addition reactions of alkenes, and use Markovnikov's rule to predict the major products of certain reactions. **(Section 12.3)**

6 Write equations for addition polymerization, and list uses for addition polymers. **(Section 12.4)**

7 Write the IUPAC names of alkynes from their molecular structures. **(Section 12.5)**

8 Classify organic compounds as aliphatic or aromatic. **(Section 12.6)**

9 Name and draw structural formulas for aromatic compounds. **(Section 12.7)**

10 Recognize uses for specific aromatic compounds. **(Section 12.8)**

An **unsaturated hydrocarbon** contains and belongs to one or more double or triple bonds between carbon atoms and belong to one of three classes: alkenes, alkynes, or aromatic hydrocarbons. An **alkene** contains one or more double bonds, an **alkyne** contains one or more triple bonds, and an **aromatic hydrocarbon** contains three double bonds alternating with three single bonds in a six-carbon ring. Ethylene (the simplest alkene), acetylene (the simplest alkyne), and benzene (the simplest aromatic) are represented by the following structural formulas:

unsaturated hydrocarbon A hydrocarbon containing one or more multiple bonds.

alkene A hydrocarbon containing one or more double bonds.

alkyne A hydrocarbon containing one or more triple bonds.

aromatic hydrocarbon Any organic compound that contains the characteristic benzene ring or similar feature.

ethylene
(an alkene)

acetylene
(an alkyne)

benzene
(an aromatic)

Alkenes and alkynes are called *unsaturated* because more hydrogen atoms can be added in somewhat the same sense that more solute can be added to an unsaturated solution. Benzene and other aromatic hydrocarbons also react to add hydrogen atoms; in general, however, they have chemical properties very different from those of alkenes and alkynes.

12.1 | The Nomenclature of Alkenes

Learning Objectives

1. Classify unsaturated hydrocarbons as alkenes, alkynes, or aromatics.
2. Write the IUPAC names of alkenes from their molecular structure.

The general formula for alkenes is C_nH_{2n} (the same as that for cycloalkanes). The simplest members are well known by their common names, ethylene and propylene:

$$CH_2 = CH_2 \qquad CH_3 - CH = CH_2$$
$$\text{ethylene, } C_2H_4 \qquad \text{propylene, } C_3H_6$$

Three structural isomers have the formula C_4H_8:

$$CH_3 - CH_2 - CH = CH_2 \qquad CH_3 - CH = CH - CH_3 \qquad \underset{\overset{|}{CH_3}}{CH_3 - C = CH_2}$$

The number of structural isomers increases rapidly as the number of carbons increases because, besides variations in chain length and branching, variations occur in the position of the double bond. IUPAC nomenclature is extremely useful in differentiating among these many alkene compounds.

The IUPAC rules for naming alkenes are similar to those used for the alkanes, with a few additions to indicate the presence and location of double bonds.

Step 1. Name the longest chain that contains the double bond. The characteristic name ending is *-ene.*

Step 2. Number the longest chain of carbon atoms so that the carbon atoms joined by the double bond have numbers as low as possible.

Step 3. Locate the double bond by the lower-numbered carbon atom bound by the double bond.

Step 4. Locate and name attached groups.

Step 5. Combine the names for the attached groups and the longest chain into the name.

Example 12.1 Naming Alkenes

Name the following alkenes:

a. $CH_3 - CH = CH - CH_3$

c. $\underset{\overset{|}{CH_3 - CH_2 - CH_2}}{\overset{\overset{\displaystyle CH_3 - CH_2}{\overset{|}{}}}{C = CH_2}}$

b. $\underset{\overset{|}{CH_3}}{CH_3 - CH - CH = CH_2}$

d.

Solution

a. The longest chain containing a double bond has four carbon atoms. The four-carbon alkane is butane. Thus, the compound is a butene:

$$\overset{1}{C}H_3 - \overset{2}{C}H = \overset{3}{C}H - \overset{4}{C}H_3$$

The chain can be numbered from either end because the double bond will be between carbons 2 and 3 either way. The position of the double bond is indicated by the lower-numbered carbon atom that is double-bonded, carbon 2 in this case. The name is 2-butene.

b. To give lower numbers to the carbons bound by the double bond, the chain is numbered from the right:

$$\overset{4}{C}H_3 - \overset{3}{C}H - \overset{2}{C}H = \overset{1}{C}H_2 \qquad \text{not} \qquad \overset{1}{C}H_3 - \overset{2}{C}H - \overset{3}{C}H = \overset{4}{C}H_2$$
$$\quad\quad\quad |\quad\quad\quad\quad\quad\quad\quad\quad\quad\quad\quad\quad |$$
$$\quad\quad\; CH_3 \quad\quad\quad\quad\quad\quad\quad\quad\quad\; CH_3$$

Thus, the compound is a 1-butene with an attached methyl group on carbon 3. Therefore, the name is 3-methyl-1-butene.

c. Care must be taken to select the longest chain containing the double bond. This compound is named as a pentene and not as a hexene because the double bond is not contained in the six-carbon chain:

$$CH_3 - CH_2 \qquad\qquad\qquad\qquad \overset{1}{C}H_3 - \overset{2}{C}H_2$$
$$\qquad\quad |\qquad\qquad\qquad\qquad\qquad\qquad\qquad |$$
$$\qquad \overset{2}{C} = \overset{1}{C}H_2 \qquad \text{not} \qquad \overset{3}{C} = CH_2$$
$$\qquad\quad |\qquad\qquad\qquad\qquad\qquad\qquad\qquad |$$
$$CH_3 - CH_2 - CH_2 \qquad\qquad CH_3 - CH_2 - CH_2$$
$$\;\;_5 \quad\quad\;\; _4 \quad\quad\; _3 \qquad\qquad\quad\;\; _6 \quad\quad\;\; _5 \quad\quad\; _4$$

The compound is a 1-pentene with an ethyl group at position 2. Therefore, the name is 2-ethyl-1-pentene.

d. In cyclic alkenes, the ring is numbered so as to give the lowest possible numbers to the double-bonded carbons (they become carbons 1 and 2). The numbering direction around the ring is chosen so that attached groups are located on the lowest-numbered carbon atoms possible. Thus, the name is 3-ethylcyclopentene:

Notice that it is not called 3-ethyl-1-cyclopentene because the double bond is always between carbons 1 and 2, and therefore its position need not be indicated.

✔ **LEARNING CHECK 12.1** Give the IUPAC name for each of the following:

a.
$$\overset{\displaystyle Br}{\overset{\displaystyle |}{CH_2}} - CH = CH_2$$

b.
$$CH_2 = C - CH_2 - CH_2 - CH_3$$
$$\qquad\quad |$$
$$\qquad\; CH_2$$
$$\qquad\quad |$$
$$\qquad\; CH_3$$

c.
$$CH_3 \quad CH_3$$

Some compounds contain more than one double bond per molecule. Molecules of this type are important components of natural and synthetic rubber and other useful materials. The nomenclature of these compounds is the same as for the alkenes with one double bond, except that the endings *-diene, -triene*, and the like are used to denote the number of double bonds. Also, the locations of all the multiple bonds must be indicated in all molecules, including those with rings:

$$\overset{1}{C}H_2 = \overset{2}{C}H - \overset{3}{C}H = \overset{4}{C}H_2$$
1,3-butadiene

1,3-cyclohexadiene

Example 12.2 More Naming Alkenes

Name the following compounds:

a. $CH_2=C-C=CH-CH=CH_2$ with CH_3 above the second carbon and CH_3 below the second carbon

b. A cyclohexadiene ring with CH_3 and CH_2CH_3 substituents

Solution

a. This compound is a methyl-substituted hexatriene. The chain is numbered from the end nearest the branch because the direction of numbering, again, makes no difference in locating the double bonds correctly. The name is 2,3-dimethyl-1,3,5-hexatriene:

$$\overset{1}{CH_2}=\overset{2}{C}-\overset{3}{C}=\overset{4}{CH}-\overset{5}{CH}=\overset{6}{CH_2} \quad \text{and not} \quad \overset{6}{CH_2}=\overset{5}{C}-\overset{4}{C}=\overset{3}{CH}-\overset{2}{CH}=\overset{1}{CH_2}$$

with CH_3 groups attached

 2,3-dimethyl-1,3,5-hexatriene 4,5-dimethyl-1,3,5-hexatriene

b. This compound is a substituted cyclohexadiene. The ring is numbered as shown. The name is 5-ethyl-1-methyl-1,3-cyclohexadiene:

 and not

5-ethyl-1-methyl-1,3-cyclohexadiene 6-ethyl-4-methyl-1,3-cyclohexadiene

✔ **LEARNING CHECK 12.2** Give the IUPAC name for each of the following:

a. $CH_2=CH-C=CH_2$ with CH_3 above the third carbon

b. $CH_2=C-CH=CH-CH_2-CH=CH-CH_3$ with CH_3 above the second carbon

c. Br attached to a cycloheptatriene ring

12.2 | The Geometry of Alkenes

Learning Objectives

3. Predict the existence of geometric (cis-trans) isomers from formulas of compounds.

4. Write the names and structural formulas for geometric isomers.

The hybridization of atomic orbitals discussed in Section 11.3 to explain the bonding characteristics of carbon atoms bonded to four other atoms can also be used to describe alkenes, compounds in which some carbon atoms are bonded to only three atoms. This hybridization involves mixing a $2s$ orbital and two $2p$ orbitals of a carbon atom to form three hybrid sp^2 orbitals (see **Figure 12.1**).

The three sp^2 hybrid orbitals lie in the same plane and are separated by angles of 120°. The unhybridized $2p$ orbital of the carbon atom is located perpendicular to the plane of the sp^2 hybrid orbitals (see **Figure 12.2**).

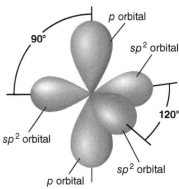

Figure 12.1 A representation of *sp*² hybridization of carbon. During hybridization, two of the 2*p* orbitals mix with the single 2*s* orbital to produce three *sp*² hybrid orbitals. One 2*p* orbital is not hybridized and remains unchanged.

Figure 12.2 The unhybridized *p* orbital is perpendicular to the plane of the three *sp*² hybridized orbitals.

The bonding between the carbon atoms of ethylene results partially from the overlap of one *sp*² hybrid orbital of each carbon to form a sigma (σ) bond. The second carbon–carbon bond is formed when the unhybridized 2*p* orbitals of the carbons overlap sideways to form what is called a pi (π) bond. The remaining two *sp*² hybrid orbitals of each carbon overlap with the 1*s* orbitals of the hydrogen atoms to form sigma bonds. Thus, each ethylene molecule contains five sigma bonds (one carbon–carbon and four carbon–hydrogen) and one pi bond (carbon–carbon), as shown in **Figure 12.3**.

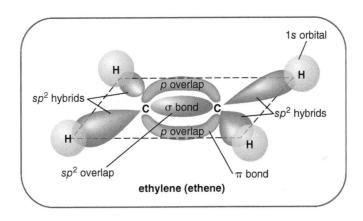

ethylene (ethene)

Figure 12.3 The bonding in ethylene is explained by combining two *sp*² hybridized carbon atoms. The C—H sigma bonds all lie in the same plane. The unhybridized *p* orbitals of the carbon atoms overlap above and below the molecular plane to form the pi bond.

Experimental data support this hybridization model. Ethylene has been found to be a planar molecule with bond angles close to 120° between the atoms (see **Figure 12.4**).

Figure 12.4 All six atoms of ethylene lie in the same plane.

In addition to geometry, alkenes also differ from open-chain alkanes in that the double bonds prevent the relatively free rotation that is characteristic of carbon atoms bonded by single bonds. As a result, alkenes can exhibit geometric isomerism, the same type of stereoisomerism seen earlier for the cycloalkanes (see **Section 11.9**). There are two geometric isomers of 2-butene:

CHEMISTRY AROUND US 12.1

Three-Dimensional Printers

We are all familiar with copiers that can make two-dimensional copies of a printed page. In these copiers, computer-controlled printer heads move horizontally, printing a one-dimensional line at a time. Eventually (and quite quickly) this creates many side-by-side lines which form the two-dimensional image or line of printed letters. This is somewhat like laying narrow, thin bricks end to end and side by side to form a narrow walk.

A three-dimensional printer works somewhat like a two-dimensional printer, with one big difference. After laying down one line like a two-dimensional printer, the three-dimensional printer lays a second line on top of the first line, giving it a thickness. In the brick walk analogy, this would be the equivalent of laying a second row of bricks on top of the first row to begin forming a wall.

The "ink" used in 3-D printers is a plastic with a specific melting point. The plastic comes wound on spools, and the 3-D printer heads heat the end of the plastic just enough to melt it so it can form a thin line. Multiple lines laid side by side form a layer. Layers laid on top of one another form an object with thickness. The computer controls these operations to give objects of the desired length, width, and thickness.

A 3-D printer is building a golf ball as a demonstration.

Once again, the prefix *cis*- is used for the isomer in which the two similar or identical groups are on the same side of the double bond and *trans*- for the one in which they are on opposite sides. The two isomers *cis*- and *trans*-2-butene represent distinctly different compounds with different physical properties (see **Table 12.1**).

TABLE 12.1 Physical Properties of a Pair of Geometric Isomers

Isomer	Melting Point (°C)	Boiling Point (°C)	Density (g/mL)
cis-2-butene	−139.9	3.7	0.62
trans-2-butene	−105.6	0.9	0.60

Not all double-bonded compounds show cis-trans stereoisomerism. Cis-trans stereoisomerism is found only in alkenes that have two different groups attached to each double-bonded carbon atom. In 2-butene, the two different groups are a methyl and a hydrogen for each double-bonded carbon:

two groups are different H H *two groups are different* *two groups are different* H CH₃ *two groups are different*

C=C

CH₃ CH₃

cis-2-butene

C=C

CH₃ H

trans-2-butene

If either double-bonded carbon is attached to identical groups, no cis-trans isomers are possible. Thus, there are no geometric isomers of ethene or propene:

H H *two groups are the same*

C=C

H H

ethene

CH₃ H *two groups are the same*

C=C

H H

propene

To see why this is so, let's try to draw geometric isomers of propene:

(A) (B)

Notice that structure (B) can be converted into (A) by just flipping it over. Thus, they are identical and not isomers.

Example 12.3 Drawing Geometric Isomers

Determine which of the following molecules can exhibit geometric isomerism, and draw structural formulas to illustrate your conclusions:

a. $Cl-CH=CH-Cl$
b. $CH_2=CH-Cl$
c. $Cl-CH=CH-CH_3$

Solution

a. Begin by drawing the carbon–carbon double bond with bond angles of 120° about each carbon atom:

Next, complete the structure and analyze each carbon of the double bond to see if it is attached to two different groups:

H and Cl are different groups *H and Cl are different groups*

In this case, each carbon is attached to two different groups and geometric isomers are possible:

cis trans

b. No geometric isomers are possible because one carbon contains two identical groups:

two identical groups

c. Geometric isomers are possible because there are two different groups on each carbon:

H and Cl are different *H and CH₃ are different*

cis-1-chloropropene *trans*-1-chloropropene

✔ **LEARNING CHECK 12.3** Determine which of the following can exhibit geometric isomerism, and draw structural formulas for the *cis* and *trans* isomers of those that can:

a. CH_2=C—CH_2—CH_3
 |
 CH_3

c. CH_3—C=CH—CH_3
 |
 CH_3

b. CH_3—CH—CH=CH—CH_3
 |
 CH_3

12.3 | Properties of Alkenes

Learning Objective

5. Write equations for addition reactions of alkenes, and use Markovnikov's rule to predict the major products of certain reactions.

12.3A Physical Properties

The alkenes are similar to the alkanes in physical properties—they are nonpolar compounds that are insoluble in water, soluble in nonpolar solvents, and less dense than water (see **Table 12.2**). Alkenes containing 4 carbon atoms or fewer are gases under ordinary conditions. Those containing 5 to 17 carbon atoms are liquids, and those with 18 or more carbon atoms are solids. Low molecular weight alkenes have somewhat unpleasant, gasoline-like odors.

TABLE 12.2 Physical Properties of Some Alkenes

IUPAC Name	Structural Formula	Boiling Point (°C)	Melting Point (°C)	Density (g/mL)
ethene	CH_2=CH_2	−104	−169	0.38
propene	CH_2=$CHCH_3$	−47	−185	0.52
1-butene	CH_2=$CHCH_2CH_3$	−6	−185	0.60
1-pentene	CH_2=$CHCH_2CH_2CH_3$	30	−138	0.64
1-hexene	CH_2=$CHCH_2CH_2CH_2CH_3$	63	−140	0.67
cyclohexene	⬡	83	−104	0.81

12.3B Chemical Properties

In contrast to alkanes, which are inert to almost all chemical reagents (see **Section 11.11**), alkenes are quite reactive chemically. Since the only difference between an alkane and an alkene is the double bond, it is not surprising to learn that most of the reactions of alkenes take place at the double bond. These reactions follow the pattern:

$$\begin{array}{c} & A & B \\ & | & | \\ \diagdown C = C \diagup + \quad A - B \longrightarrow -C - C- \\ \diagup \qquad \diagdown & | & | \end{array} \qquad (12.1)$$

addition reaction A reaction in which a compound adds to a multiple bond.

They are called **addition reactions** because a substance is added to the double bond. Addition reactions are characterized by two reactants that combine to form one product.

haloalkane or alkyl halide A derivative of an alkane in which one or more hydrogens are replaced by halogens.

Halogenation is one of the most common addition reactions. For example, when bromine, a halogen, is added to an alkene, the double bond reacts, and only a carbon–carbon single bond remains in the product. Without a double bond present, the product is referred to as a **haloalkane** or **alkyl halide**. Reaction 12.2 gives the general reaction, and Reaction 12.3 is a specific example, the bromination of 1-butene.

General reaction:

$$\overset{\diagdown}{\underset{\diagup}{C}} = \overset{\diagup}{\underset{\diagdown}{C}} \ + \ Br-Br \ \longrightarrow \ -\underset{\underset{Br}{|}}{C}-\underset{\underset{Br}{|}}{C}- \qquad (12.2)$$

Specific example:

$$CH_3-CH_2-CH=CH_2 \ + \ Br-Br \ \longrightarrow \ CH_3-CH_2-\underset{\underset{Br}{|}}{CH}-\underset{\underset{Br}{|}}{CH_2} \qquad (12.3)$$

 1-butene 1,2-dibromobutane

The addition of Br_2 to double bonds provides a simple laboratory test for unsaturation (see **Figure 12.5**). As the addition takes place, the characteristic red-brown color of the added bromine fades as it is used up, and the colorless dibromoalkane product forms.

Figure 12.5 The reaction of bromine with an unsaturated hydrocarbon.

1. The bromine vapor in the flask has a red brown color. The fat in bacon is partially unsaturated.

2. The bromine vapor loses its red brown color when it reacts with the unsaturated bacon fat.

The addition of halogens is also used to quantitatively determine the degree of unsaturation in vegetable oils, margarines, and shortenings (see **Section 18.4**). Chlorine reacts with alkenes to give dichloro products in an addition reaction similar to that of bromine. However, it is not used as a test for unsaturation because it is difficult to see the pale green color of chlorine in solution.

✔ **LEARNING CHECK 12.4** Write the structural formula for the product of each of the following reactions:

a. $CH_3-\underset{\underset{CH_3}{|}}{C}=CH_2+ Br_2 \longrightarrow$ b. ⬠ $+ \ Cl_2 \longrightarrow$

In the presence of an appropriate catalyst (such as platinum, palladium, or nickel), hydrogen adds to alkenes and converts them into the corresponding alkanes. This reaction, which is called **hydrogenation**, is illustrated in Reactions 12.4 and 12.5.

hydrogenation A reaction in which the addition of hydrogen takes place.

General reaction:

$$\overset{\diagdown}{\underset{\diagup}{C}} = \overset{\diagup}{\underset{\diagdown}{C}} \ + \ H-H \ \overset{Pt}{\longrightarrow} \ -\underset{\underset{H}{|}}{C}-\underset{\underset{H}{|}}{C}- \qquad (12.4)$$

 an alkane

Specific example: $CH_3CH = CHCH_3 + H - H \xrightarrow{Pt} CH_3CH - CHCH_3$ (12.5)

<center>H H</center>

<center>2-butene butane</center>

The hydrogenation of vegetable oils is a very important commercial process. Vegetable oils, such as soybean and cottonseed oil, are composed of long-chain organic molecules that contain many alkene bonds. The high degree of unsaturation characteristic of these oils gave rise to the term **polyunsaturated**. Upon hydrogenation, the melting point of the oils is raised, and the oils become low-melting-point solids. These products are used in the form of margarine and shortening.

polyunsaturated A term usually applied to molecules with several double bonds.

> ✔ **LEARNING CHECK 12.5** Write the structural formula for the product of each of the following reactions:
>
> a. $CH_3 - C = CH - CH_3 + H_2 \xrightarrow{Pt}$ b. CH_3
>
> CH_3
>
> $+ H_2 \xrightarrow{Pt}$

A number of acidic compounds, such as the hydrogen halides—HF, HCl, HBr, and HI—also add to alkenes to give the corresponding alkyl halide. The reaction with HCl is illustrated as follows:

General reaction: $\diagdown C = C \diagup + H - Cl \longrightarrow - \overset{|}{\underset{|}{C}} - \overset{|}{\underset{|}{C}} -$ (12.6)

<center>H Cl</center>

Specific example: $CH_3CH = CHCH_3 + H - Cl \longrightarrow CH_3CHCHCH_3$ (12.7)

<center>H Cl</center>

The addition reactions involving H_2, Cl_2, and Br_2 yield only one product because the same group (H and H or Br and Br) adds to each double-bonded carbon. However, with H—X, a different group adds to each carbon, and for certain alkenes, there are two possible products. For example, in the reaction of HBr with propene, two products might be expected: 1-bromopropane or 2-bromopropane,

<center>Br H H Br</center>

$CH_2 = CH - CH_3 + H - Br \longrightarrow CH_2 - CH - CH_3$ or $CH_2 - CH - CH_3$ (12.8)

<center>1-bromopropane 2-bromopropane</center>

It turns out that only one product, 2-bromopropane, is formed in significant amounts. This fact, first reported in 1869 by Russian chemist Vladimir Markovnikov, gave rise to a rule for predicting which product will be exclusively or predominantly formed. According to **Markovnikov's rule**, when a molecule of the form H—X adds to an alkene, the hydrogen becomes attached to a double-bonded carbon atom that is already bonded to more hydrogens. A phrase to help you remember this rule is "the rich get richer." Applying this rule to propene, we find:

Markovnikov's rule In the addition of H—X to an alkene, the hydrogen becomes attached to the carbon atom that is already bonded to more hydrogens.

<center><i>one hydrogen attached</i></center>

<center>$CH_2 = CH - CH_3$</center>

<center><i>two hydrogens three hydrogens but they</i></center>
<center><i>attached are not attached to the</i></center>
<center><i>double-bonded carbons</i></center>

Therefore, H attaches to the end carbon of the double bond, Br attaches to the second carbon, and 2-bromopropane is the major product.

Example 12.4 Using Markovnikov's Rule

Use Markovnikov's rule to predict the major product in the following reactions:

a.
$$CH_3-\underset{\underset{\displaystyle CH_3}{|}}{C}=CH_2+H-Cl$$

b.
$+ \ H-Cl \longrightarrow$

Solution

a. Analyze the C=C to see which carbon atom has more hydrogens attached:

$$CH_3-\underset{\underset{\displaystyle CH_3}{|}}{C}=CH_2$$

two hydrogens attached

no hydrogens attached

The H of H—Cl will attach to the position that has more hydrogens. Thus, 2-chloro-2-methylpropane is the major product:

$$CH_3-\underset{\underset{\displaystyle CH_3}{|}}{C}=CH_2$$

H attaches here to give

Cl attaches here

$$CH_3-\underset{\underset{\displaystyle Cl \quad H}{| \quad |}}{C}-CH_2$$

2-chloro-2-methylpropane

b. The challenge with a cyclic alkene is to remember that the hydrogens are not shown. Thus,

is the same as

As before, the H of H—Cl will attach to the double-bonded carbon that has more hydrogens:

Cl attaches here *H attaches here* \longrightarrow or

✔ **LEARNING CHECK 12.6** Use Markovnikov's rule to predict the major product in the following reactions:

a.
$$CH_3-CH=\underset{\underset{\displaystyle CH_3}{|}}{C}-CH_2-CH_3+HBr$$

b.
$+ \ HBr \longrightarrow$

STUDY SKILLS 12.1

Keeping a Reaction Card File

Remembering organic reactions for exams is challenging for most students. Because the number of reactions being studied increases rapidly over the semester, it is a good idea to develop a systematic way to organize them for easy and effective review.

One way to do this is to focus on the functional group concept. When an exam question asks you to complete a reaction by identifying the product, your first step should be to identify the functional group of the reactant. Usually, only the functional group portion of a molecule undergoes reaction. In addition, a particular functional group usually undergoes the same characteristic reactions regardless of the other features of the organic molecule to which it is bound. Thus, by remembering the behavior of a functional group under specific conditions, you can predict the reactions of many compounds, no matter how complex the structures look, as long as they contain the same functional group. For example, any structure containing a C=C will undergo reactions typical of alkenes. Other functional groups will be introduced in later chapters.

Keeping a reaction card file based on the functional group concept is a good way to organize reactions for review. Write the structures and names of the reactants on one side of an index card with an arrow showing any catalyst or special conditions. Write the product structure and name on the back of the card. We recommend that you do this for the general reaction (like those in the Key Reactions section at the end of most chapters) and for a specific example. Review your cards every day (this can even be done while waiting for a bus, etc.), and add to them as new reactions are studied. As an exam approaches, put aside the reactions you know well, and concentrate on the others in what should be a dwindling deck. This is an effective way to focus on learning what you don't know.

In the absence of a catalyst, water does not react with alkenes. But, if an acid catalyst such as sulfuric acid is added, water adds to carbon–carbon double bonds to give alcohols. In this reaction, which is called **hydration**, a water molecule is split in such a way that —H attaches to one carbon of the double bond, and —OH attaches to the other carbon. In Reactions 12.9–12.11, H_2O is written H—OH to emphasize the portions that add to the double bond. Notice that the addition follows Markovnikov's rule:

hydration The addition of water to a multiple bond.

$$\ce{C=C} + H-OH \xrightarrow{H_2SO_4} -\overset{|}{\underset{|}{C}}-\overset{|}{\underset{|}{C}}- \quad (12.9)$$
$$\overset{\quad H \quad OH}{}$$
an alcohol

$$CH_3CH=CHCH_3 + H-OH \xrightarrow{H_2SO_4} CH_3\overset{|}{\underset{|}{C}}H-\overset{|}{\underset{|}{C}}HCH_3 \quad (12.10)$$
2-butene
$$\overset{\qquad\quad H \quad OH}{}$$
2-butanol

$$CH_3CH=CH_2 + H-OH \xrightarrow{H_2SO_4} CH_3\overset{|}{\underset{|}{C}}H-\overset{|}{\underset{|}{C}}H_2 \quad (12.11)$$
propene
$$\overset{\qquad\quad OH \quad H}{}$$
2-propanol

The hydration of alkenes provides a convenient method for preparing alcohols on a large scale. The reaction is also important in living organisms, but the catalyst is an enzyme rather than sulfuric acid. For example, one of the steps in the body's utilization of carbohydrates for energy involves the hydration of fumaric acid, which is catalyzed by the enzyme fumarase:

$$HO-\overset{O}{\overset{||}{C}}-CH=CH-\overset{O}{\overset{||}{C}}-OH + H_2O \xrightarrow{fumarase} HO-\overset{O}{\overset{||}{C}}-\overset{H}{\underset{|}{C}}H-\overset{OH}{\underset{|}{C}}H-\overset{O}{\overset{||}{C}}-OH \quad (12.12)$$
fumaric acid
malic acid

✔ LEARNING CHECK 12.7 Draw structural formulas for the major organic product of each of the following reactions:

a. $CH_3CH_2CH_2CH=CH_2 + H_2O \xrightarrow{H_2SO_4}$

b. (pentagon) $+ H_2O \xrightarrow{H_2SO_4}$

12.4 | Addition Polymers

Learning Objective

6. Write equations for addition polymerization, and list uses for addition polymers.

Certain alkenes undergo a very important reaction in the presence of specific catalysts. In this reaction, alkene molecules undergo an addition reaction with one another. The double bonds of the reacting alkenes are converted to single bonds as hundreds or thousands of molecules bond and form long chains. For example, several ethylene molecules react as follows:

$$CH_2=CH_2 + CH_2=CH_2 + CH_2=CH_2 + CH_2=CH_2 \xrightarrow[\text{catalysts}]{\text{Heat, pressure,}} \quad (12.13)$$
$$\text{ethylene molecules}$$

$$-CH_2-CH_2-CH_2-CH_2-CH_2-CH_2-CH_2-CH_2-$$
$$\text{polyethylene}$$

The product is commonly called polyethylene even though there are no longer any double bonds present. The newly formed bonds in this long chain are shown in color. This type of reaction is called a **polymerization**, and the long-chain product made up of repeating units is a **polymer** (*poly* = many, *mer* = parts). The trade names of many polymers such as Orlon, Plexiglas, Lucite, and Teflon are familiar (see **Figure 12.6**). These products are referred to as **addition polymers** because of the addition reaction between double-bonded compounds that is used to produce them. The starting materials that make up the repeating units of polymers are called **monomers** (*mono* = one, *mer* = part). Quite often, common names are used for both the polymer and the monomer.

It is not possible to give an exact formula for a polymer produced by a polymerization reaction because the individual polymer molecules vary in size. We could represent polymerization reactions as in Reaction 12.13. However, since this type of reaction is inconvenient, we adopt a commonly used approach: The polymer is represented by a simple repeating unit based on the monomer. For polyethylene, the unit is $+CH_2-CH_2+$. The large number of units making up the polymer is denoted by *n*, a whole number that varies from several hundred to several thousand. The polymerization reaction of ethylene is then written as:

$$nCH_2=CH_2 \xrightarrow[\text{catalysts}]{\text{Heat, pressure,}} +CH_2-CH_2+_n \quad (12.14)$$
$$\text{ethylene} \qquad\qquad\qquad \text{polyethylene}$$

The lowercase *n* in Reaction 12.14 represents a large, unspecified number. From this reaction, we see that polyethylene is essentially a very long chain alkane. As a result, it has the chemical inertness of alkanes, a characteristic that makes polyethylene suitable for food storage containers, garbage bags, eating utensils, laboratory apparatus, and hospital equipment

George Frey/Getty Images

Figure 12.6 Gore-Tex is a thin, membranous material made by stretching Teflon fibers. Fabrics layered with Gore-Tex repel wind and rain but allow body perspiration to escape, making it an excellent fabric for sportswear.

polymerization A reaction that produces a polymer.

polymer A very large molecule made up of repeating units.

addition polymer A polymer formed by the linking together of many alkene molecules through addition reactions.

monomer The starting material that becomes the repeating units of polymers.

A Reaction Map for Alkenes

A diagram may help you visualize and remember the four common addition reactions in this section. In each case, the alkene double bond reacts, and an alkane or alkane derivative is produced. The specific reagent determines the outcome of the reaction.

(see **Figure 12.7**). Polymer characteristics are modified by using alkenes with different groups attached to either or both of the double-bonded carbons. For example, the polymerization of vinyl chloride gives the polymer poly (vinyl chloride), PVC (Reaction 12.15):

$$n\text{CH}_2 = \overset{\overset{\displaystyle \text{Cl}}{\displaystyle |}}{\text{CH}} \quad \xrightarrow[\text{catalysts}]{\text{Heat,}} \quad + \text{CH}_2 - \overset{\overset{\displaystyle \text{Cl}}{\displaystyle |}}{\text{CH}} +_n \qquad (12.15)$$

vinyl chloride poly (vinyl chloride)

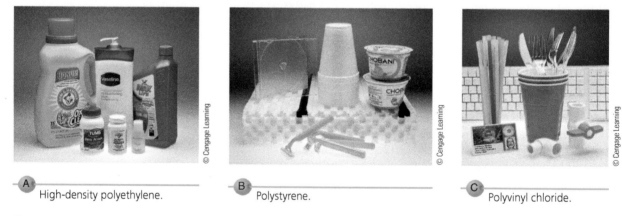

A ── High-density polyethylene.

B ── Polystyrene.

C ── Polyvinyl chloride.

Figure 12.7 Common polymer-based consumer products.

copolymer An addition polymer formed by the reaction of two different monomers.

The commercial product Saran Wrap is an example of a **copolymer**, which is a polymer made up of two different monomers (Reaction 12.16):

$$n\text{CH}_2 = \overset{\overset{\displaystyle \text{Cl}}{\displaystyle |}}{\underset{\underset{\displaystyle \text{Cl}}{\displaystyle |}}{\text{CH}}} + n\text{CH}_2 = \overset{\overset{\displaystyle \text{Cl}}{\displaystyle |}}{\underset{\underset{\displaystyle \text{Cl}}{\displaystyle |}}{\text{C}}} \xrightarrow{\text{Catalyst}} + \text{CH}_2 - \overset{\overset{\displaystyle \text{Cl}}{\displaystyle |}}{\underset{\underset{\displaystyle \text{Cl}}{\displaystyle |}}{\text{CH}}} - \text{CH}_2 - \overset{\overset{\displaystyle \text{Cl}}{\displaystyle |}}{\underset{\underset{\displaystyle \text{Cl}}{\displaystyle |}}{\text{C}}} +_n \qquad (12.16)$$

vinyl chloride vinylidene chloride Saran Wrap

A number of the more important addition polymers are shown in **Table 12.3**. As you can tell by looking at some of the typical uses, addition polymers have become nearly indispensable in modern life (see **Figure 12.8**).

TABLE 12.3 Common Addition Polymers

Chemical Name and Trade Name(s)	Monomer	Polymer	Typical Uses
polyethylene	$CH_2{=}CH_2$	$+CH_2{-}CH_2\,\overset{}{)_n}$	Bottles, plastic bags, film
polypropylene	$CH_2{=}CH$ $\quad\quad\mid$ $\quad\quad CH_3$	$+CH_2{-}CH\,\overset{}{)_n}$ $\quad\quad\quad\mid$ $\quad\quad\quad CH_3$	Carpet fiber, pipes, bottles, artificial turf
poly (vinyl chloride) (PVC)	$CH_2{=}CH$ $\quad\quad\mid$ $\quad\quad Cl$	$+CH_2{-}CH\,\overset{}{)_n}$ $\quad\quad\quad\mid$ $\quad\quad\quad Cl$	Synthetic leather, floor tiles, garden hoses, water pipe
polytetrafluoroethylene (Teflon)	$CF_2{=}CF_2$	$+CF_2{-}CF_2\,\overset{}{)_n}$	Pan coatings, plumbers' tape, heart valves, fabrics
poly (methyl methacrylate) (Lucite, Plexiglas)	$\quad\quad CH_3$ $\quad\quad\mid$ $CH_2{=}C$ $\quad\quad\mid$ $\quad\quad C{-}O{-}CH_3$ $\quad\quad\parallel$ $\quad\quad O$	$\quad\quad CH_3$ $\quad\quad\mid$ $+CH_2{-}C\,\overset{}{)_n}$ $\quad\quad\mid$ $\quad\quad C{-}O{-}CH_3$ $\quad\quad\parallel$ $\quad\quad O$	Airplane windows, paint, contact lenses, fiber optics
poly (vinyl acetate)	$CH_2{=}CH$ $\quad\quad\mid$ $\quad\quad O{-}C{-}CH_3$ $\quad\quad\quad\parallel$ $\quad\quad\quad O$	$+CH_2{-}CH\,\overset{}{)_n}$ $\quad\quad\quad\mid$ $\quad\quad\quad O{-}C{-}CH_3$ $\quad\quad\quad\quad\parallel$ $\quad\quad\quad\quad O$	Adhesives, latex paint, chewing gum
polyacrylonitrile (Orlon, Acrilan)	$CH_2{=}CH$ $\quad\quad\mid$ $\quad\quad CN$	$+CH_2{-}CH\,\overset{}{)_n}$ $\quad\quad\quad\mid$ $\quad\quad\quad CN$	Carpets, fabrics
polystyrene (Styrofoam)	$CH_2{=}CH$ $\quad\quad\mid$ $\quad\quad$⬡	$+CH_2{-}CH\,\overset{}{)_n}$ $\quad\quad\quad\mid$ $\quad\quad\quad$⬡	Food coolers, drinking cups, insulation

aceshot1/Shutterstock.com

A Automobile safety glass contains a sheet of polyvinyl polymer layered between two sheets of glass to prevent the formation of sharp fragments.

Ryan Jorgensen/Jorgo/Shutterstock.com

B The elasticity of bubble gum comes from a copolymer of styrene and 1,3-butadiene.

Figure 12.8 Two uses of addition polymers. What properties of addition polymers are exhibited in both of these products?

Polycarbonate—The Lucky Polymer

A **polymer** is a material that is composed of many identical units connected by chemical bonds. The word "polymer" literally means many (*poly*) parts (*mer*). In polycarbonates, the repeating parts have the following pattern:

$$\left(O-R-O-\overset{\displaystyle O}{\overset{\displaystyle \|}{C}} \right)_n$$

Polycarbonate was accidentally discovered in 1953 by a chemist named Daniel W. Fox who was working for the General Electric Company. At nearly the same time a chemist named Hermann Schnell who worked for the Bayer Company in Germany also discovered the same substance. General Electric and the Bayer Company worked together to commercially produce and market this unbreakable substance. It is called the lucky polymer because it is unbreakable, very adaptable, and versatile.

The polycarbonate polymer is a clear solid that is very strong. It weighs less than glass, resists heat, and is flexible. These superior characteristics allow polycarbonate to replace other materials in a wide variety of commercial products. It is now used as a headlight cover in vehicles, as a bulletproof material, and as a substitute for glass in windows. It is also used as food packaging in the form of bottles and other containers. Polycarbonates were even used in the helmets worn by Buzz Aldrin and other astronauts when they landed on the moon in 1969.

For music lovers, polycarbonate polymers have been very beneficial. The widely used music CDs are actually made of several layers. One of these layers is made of polycarbonate. It is this layer that contains the bumps and pits that constitute the code for the music. Other electronic applications of polycarbonate polymers are also used in making other electronic devices such as DVDs and cell phones.

Polycarbonate is an essential component of CDs.

✔ **LEARNING CHECK 12.8** Draw the structural formula of a portion of polypropylene containing four repeating units of the monomer propylene,

$$CH_2 = \overset{\displaystyle CH_3}{\overset{\displaystyle |}{CH}}$$

12.5 | Alkynes

Learning Objective

7. Write the IUPAC names of alkynes from their molecular structures.

The characteristic feature of alkynes is the presence of a triple bond between carbon atoms. Thus, alkynes are also unsaturated hydrocarbons. Only a few compounds containing the carbon–carbon triple bond are found in nature. The simplest and most important compound of this series is ethyne, more commonly called acetylene (C_2H_2):

$$H-C\equiv C-H$$
acetylene

Acetylene is used in torches for welding steel and in making plastics and synthetic fibers.

Once again, orbital hybridization provides an explanation for the bonding of the carbon atoms. Structurally, the hydrogen and carbon atoms of acetylene molecules lie in a straight line. This same linearity of the triple bond and the two atoms attached to the triple-bonded carbons is found in all alkynes. These characteristics are explained by mixing a $2s$ and a single $2p$ orbital of each carbon to form a pair of sp hybrid orbitals. Two of the $2p$ orbitals of each carbon are unhybridized (see **Figures 12.9** and **12.10**).

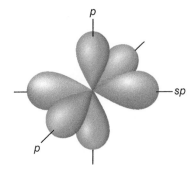

Figure 12.9 *sp* hybridization occurs when one of the 2*p* orbitals of carbon mixes with the 2*s* orbital. Two 2*p* orbitals remain unhybridized.

Figure 12.10 The unhybridized *p* orbitals are perpendicular to the two *sp* hybridized orbitals.

A carbon–carbon sigma bond in acetylene forms by the overlap of one *sp* hybrid orbital of each carbon. The other *sp* hybrid orbital of each carbon overlaps with a 1*s* orbital of a hydrogen to form a carbon–hydrogen sigma bond. The remaining pair of unhybridized *p* orbitals of each carbon overlap sideways to form a pair of pi bonds between the carbon atoms. Thus, each acetylene molecule contains three sigma bonds (two carbon–hydrogen and one carbon–carbon) and two pi bonds (both are carbon–carbon). This is shown in **Figure 12.11**.

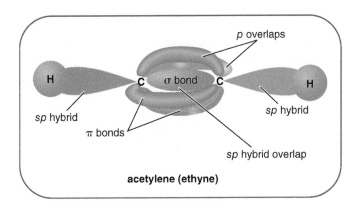

acetylene (ethyne)

Figure 12.11 The shape of acetylene is explained by sigma bonding between *sp* hybrid orbitals and pi bonding between unhybridized *p* orbitals.

Alkynes are named in exactly the same ways as alkenes, except the ending *-yne* is used:

$$\overset{4}{C}H_3\overset{3}{C}H_2-\overset{2}{C}\equiv\overset{1}{C}-H \qquad H-\overset{1}{C}\equiv\overset{2}{C}-\overset{3}{C}H-\overset{4}{C}H_3$$
$$\underset{CH_3}{\vert}$$

1-butyne 3-methyl-1-butyne

✔ **LEARNING CHECK 12.9** Give the IUPAC name for each of the following:

a. $CH_3-CH_2-C\equiv C-CH_3$

b. $CH_3-\underset{\underset{CH_3}{\vert}}{CH}-CH_2-C\equiv C-CH_3$

The physical properties of the alkynes are nearly the same as those of the corresponding alkenes and alkanes: They are insoluble in water, less dense than water, and have relatively low melting and boiling points. Alkynes also resemble alkenes in their addition reactions. The same substances (Br_2, H_2, HCl, etc.) that add to double bonds also add to triple bonds. The one significant difference is that alkynes consume twice as many moles of addition reagent as alkenes in addition reactions that go on to completion.

12.6 | Aromatic Compounds and the Benzene Structure

Learning Objective

8. Classify organic compounds as aliphatic or aromatic.

Some of the early researchers in organic chemistry became intrigued by fragrant oils that could be extracted from certain plants. Oil of wintergreen and the flavor component of the vanilla bean are examples. The compounds responsible for the aromas had similar chemical properties. As a result, they were grouped together and called aromatic compounds.

As more and more aromatic compounds were isolated and studied, chemists gradually realized that aromatics contained at least six carbon atoms, had low hydrogen-to-carbon ratios (relative to other organic hydrocarbons), and were related to benzene (C_6H_6). For example, toluene, an aromatic compound from the bark of the South American tolu tree, has the formula C_7H_8.

Chemists also learned that the term *aromatic* was not always accurate. Many compounds that belong to the class because of chemical properties and structures are not at all fragrant. Conversely, there are many fragrant compounds that do not have aromatic compound properties or structures. Today, the old class name is used but with a different meaning. Aromatic compounds are those that contain the characteristic benzene ring or its structural relatives. Compounds that do not contain this structure (non-aromatic compounds) are referred to as **aliphatic compounds**. Alkanes, alkenes, and alkynes are, therefore, aliphatic compounds.

> **aliphatic compound** Any organic compound that is not aromatic.

The molecular structure of benzene presented chemists with an intriguing puzzle after the compound was discovered in 1825 by Michael Faraday. The formula C_6H_6 indicated that the molecule was highly unsaturated. However, the compound did not show the typical reactivity of unsaturated hydrocarbons. Benzene underwent relatively few reactions, and these proceeded slowly and often required heat and catalysts. This was in marked contrast to alkenes, which reacted rapidly with many reagents, in some cases almost instantaneously. This apparent discrepancy between structure and reactivity plagued chemists until 1865, when Friedrich August Kekulé von Stradonitz (see **Figure 12.12**), a German chemist, suggested that the benzene molecule might be represented by a ring arrangement of carbon atoms with alternating single and double bonds between the carbon atoms:

He later suggested that the double bonds alternate in their positions between carbon atoms to give two equivalent structures:

Kekulé structures

Figure 12.12 Friedrich August Kekulé (1829–1896).

HOW REACTIONS OCCUR 12.1

The Hydration of Alkenes: An Addition Reaction

The mechanism for the hydration of alkenes is believed to begin when H^+ from the acid catalyst is attracted to the electrons of the carbon–carbon double bond. The H^+ becomes bonded to one of the carbons by a sharing of electrons:

Step 1.

$$-C=C- \; + H^+ \longrightarrow \; -\overset{\overset{\displaystyle H}{|}}{C}-\overset{+}{C}-$$

a carbocation

This process leaves the second carbon with only three bonds about it and thus a positive charge. Such ions, referred to as **carbocations**, are extremely reactive.

As soon as it forms, the positive carbocation attracts any species that has readily available nonbonding electrons, whether it is an anion or a neutral molecule. In the case of water, the oxygen atom has two unshared pairs of electrons:

$$\overset{\displaystyle \cdot\cdot\overset{\displaystyle \cdot\cdot}{O}}{\underset{\displaystyle H \qquad H}{}}$$

One pair of oxygen electrons forms a covalent bond with the carbocation:

Step 2.

$$-\overset{\overset{\displaystyle H}{|}}{C}-\overset{+}{C}- \; + \; H\overset{\displaystyle \cdot\cdot\overset{\displaystyle \cdot\cdot}{O}}{}H \longrightarrow \; -\overset{\overset{\displaystyle H}{|}}{C}-\overset{\overset{\displaystyle H\;\;\overset{+}{O}\overset{H}{\diagup}}{|}}{C}-$$

In the third step, H^+ is lost to produce the alcohol:

Step 3.

$$-\overset{\overset{\displaystyle H\;\;\overset{+}{O}\overset{H}{\diagup}}{|}}{C}-\overset{|}{C}- \longrightarrow \; -\overset{\overset{\displaystyle H}{|}}{C}-\overset{\overset{\displaystyle OH}{|}}{C}- \; + H^+$$

an alcohol

Notice that the acid catalyst, H^+, which initiated the reaction, is recovered unchanged in the final step of the mechanism.

By applying Step 2 of the mechanism for hydration, we can understand how HCl and HBr react with alkenes:

$$-\overset{\overset{\displaystyle H}{|}}{C}-\overset{+}{C}- \; + Br^- \longrightarrow \; -\overset{\overset{\displaystyle H}{|}}{C}-\overset{\overset{\displaystyle Br}{|}}{C}-$$

$$-\overset{\overset{\displaystyle H}{|}}{C}-\overset{+}{C}- \; + Cl^- \longrightarrow \; -\overset{\overset{\displaystyle H}{|}}{C}-\overset{\overset{\displaystyle Cl}{|}}{C}-$$

carbocation An ion of the form $-\overset{+}{C}-$

A modern interpretation of the benzene structure based on hybridization enables chemists to better understand and explain the chemical properties of benzene and other aromatic compounds. Each carbon atom in a benzene ring has three sp^2 hybrid orbitals and one unhybridized p orbital.

A single sigma bond between carbons of the benzene ring is formed by the overlap of two sp^2 orbitals, one from each of the double-bonded carbons. Because each carbon forms two single bonds, two of the sp^2 hybrid orbitals of each carbon are involved. The third sp^2 hybrid orbital of each carbon forms a single sigma bond with a hydrogen by overlapping with a $1s$ orbital of hydrogen. The unhybridized p orbitals of each carbon overlap sideways above and below the plane of the carbon ring to form two delocalized pi lobes that run completely around the ring in the shape of a double torus. The first torus represents the blended and overlapping orbital area of the six pi lobes that point up, and the torus underneath it represents the blending and overlapping orbital area of the six pi lobes that point down (see **Figure 12.13**).

Figure 12.13 A hybrid orbital view of the benzene structure.

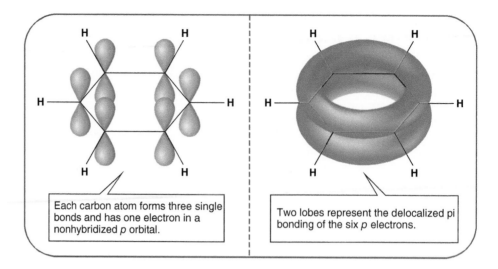

This interpretation leads to the conclusion that six bonding electrons move freely around the double ring structure formed by the overlapping delocalized pi lobes. Because of this, the benzene structure is often represented by the symbol:

with the circle representing the evenly distributed electrons in the pi lobes. All six carbon and six hydrogen atoms in benzene molecules lie in the same plane (see **Figure 12.13**). Therefore, substituted aromatic compounds do not exhibit cis-trans isomerism.

As you draw the structure of aromatic compounds, remember that only one hydrogen atom or group can be attached to a particular position on the benzene ring. For example, compounds (A) and (B) below exist, but (C) does not. Examination of the Kekulé structure of compound (C) shows that the carbon has five bonds attached to it in violation of the octet rule:

(A) exists (B) exists (C) does not exist Kekulé structure — *five bonds attached (not possible)*

12.7 | The Nomenclature of Benzene Derivatives

Learning Objective

9. Name and draw structural formulas for aromatic compounds.

The chemistry of the aromatic compounds developed somewhat haphazardly for many years before systematic nomenclature schemes were developed. Some of the common names used earlier have acquired historical respectability and are used today; some have even been incorporated into the modern systematic nomenclature system.

The following guidelines are all based on the IUPAC aromatic nomenclature system. They are not complete, and you will not be able to name all aromatic compounds by using them. However, you will be able to name and recognize those used in this book.

Guideline 1. When a single hydrogen of the benzene ring is replaced, the compound can be named as a derivative of benzene:

ethylbenzene	nitrobenzene	bromobenzene	chlorobenzene

Guideline 2. A number of benzene derivatives are known by common names that are also IUPAC-accepted and are used preferentially over other possibilities. Thus, toluene is favored over methylbenzene, and aniline is used rather than aminobenzene:

toluene	phenol	aniline	benzoic acid

Guideline 3. Compounds formed by replacing a hydrogen of benzene with a more complex hydrocarbon group can be named by designating the benzene ring as the group. This is shown as the benzene ring, C_6H_5—, which is also

called a **phenyl group**:

$$CH_3 - CH_2 - CH_2 - CH - CH_2 - CH_2 - CH_3$$

4-phenylheptane

1,1-diphenylcyclobutane

phenyl group A benzene ring with one hydrogen absent, C_6H_5—.

It's easy to confuse the words *phenyl* and *phenol*. The key to keeping them straight is the ending: *-ol* means an alcohol (C_6H_5—OH), and *-yl* means a group (C_6H_5—).

Guideline 4. When two groups are attached to a benzene ring, three isomeric structures are possible. They can be designated by the prefixes *ortho (o), meta (m),* and *para (p):*

o-dibromobenzene	*m*-dibromobenzene	*p*-dibromobenzene

Guideline 5. When two or more groups are attached to a benzene ring, their positions can be indicated by numbering the carbon atoms of the ring so as to obtain the lowest possible numbers for the attachment positions. Groups are arranged in

For many people, a well-tanned skin is considered to be very attractive. However, the methods used to achieve this stylish appearance sometimes have some serious health drawbacks.

The body protects itself from ultraviolet (UV) radiation by producing a skin pigment called melanin. The accumulation of melanin in the skin leads to the characteristic tan color and protects the skin from burning and other damage caused by UV radiation. Melanin takes 3 to 5 days to form after initial exposure to the sun. Melanin then helps the skin to gradually tan if the exposure time is kept within reasonable bounds.

Teens are particularly susceptible to the risk of overexposure because they are still experiencing tremendous growth at the cellular level. However, people run the risk of developing skin cancer if they are exposed to the sun or to artificial sunlight in tanning salons too often or for long periods of time.

Anyone using the facilities of a tanning salon should find out the recommended exposure times for each bed. At the beginning of developing a tan, an individual should not tan more than every other day. As the tan develops, it is wise to cut back tanning sessions to no more than twice a week. Many salon owners allow individuals to tan for greater periods of time than is safely recommended. Every tanning bed has a chart indicating what frequency and duration are approved by safety standards. If the beds in a salon do not have such charts, the salon is not following safety guidelines, and it would be a good idea to leave.

Exposure to UV rays is known to stimulate addictive behavior that results in increasing tanning frequency and duration. Although a tan is commonly thought to be beautiful, skin cancer certainly is not. Be a conscientious and perceptive consumer.

According to the Food and Drug Administration, the use of artificial tanning devices is not recommended for anyone.

alphabetical order. If there is a choice of identical sets of numbers, the group that comes first in alphabetical order is given the lower number. IUPAC-acceptable common names may be used:

m-bromochlorobenzene or 1-bromo-3-chlorobenzene

1,2,4-trichlorobenzene

3,5-dichlorobenzoic acid

Example 12.5 Naming Aromatic Compounds

Name each of the following aromatic compounds:

a. CH₃ / CH₂CH₃

b. CH₂CH₂CH₃

c. $C-OH$ with CH₃CH₂ and Cl

Controlled Substances

So what are *controlled substance,* anyway, and why do we have them? Before we answer those questions, let's briefly review the major legislation that pertains to products used as medicines. Prior to the 1900s there were no governmental regulations on foods or drugs. As a result, some products were contaminated and some not labeled accurately. Consequently, the U.S. Congress passed the Pure Food and Drug Act of 1906. It proved to be helpful, but opiates and cocaine were not regulated. A significant percentage of the population became addicted, and many deaths were attributed to the use of products that were "pure" and "labeled" correctly but still contained addicting materials. In 1914 the Harrison Act was passed. It regulated heroin and cocaine sales. During the next half century, several other drug laws were passed that dealt with efficacy and safety.

With the 1960s came a precipitous increase in the use of mind-altering drugs. Culturally, the use of these drugs became more acceptable, especially with young people who got involved in the "hippie" movement and "flower power." Also, members of the military who were deployed to foreign lands got exposure, access, and in some cases addictions to habituating compounds. Providing these drugs of abuse spawned a very lucrative underground empire in the United States that was responsible for an increase in crime and health care costs. The federal government joined the "war on drugs" by enacting legislation that included the creation of the Drug Enforcement Agency (DEA), which is responsible for suppressing illegal drug use and distribution. The federal government also passed the Controlled Substances Act (CSA). This law categorized dangerous drugs into five schedules (abbreviated C-I to C-V).

A controlled substance is generally defined as a drug or other chemical whose manufacture, possession, or use is regulated by the federal government. The five schedule definitions are:

Schedule I: The substance has a high potential for abuse. It has no currently accepted medical use in treatment in the United States. There is a lack of accepted safety procedures for use of the substance under medical supervision (e.g., LSD, Ecstasy).

Note: Tetrahydrocannabinol (THC, marijuana) is still considered a Schedule I drug by the DEA, even though some U.S. states have legalized marijuana for personal, recreational, or medical use.

Schedule II: The substance has a high potential for abuse. It has a currently accepted medical use for treatment in the United States, or a currently accepted medical use with severe restrictions. Abuse of the substance might lead to severe psychological or physical dependence (e.g., Percocet, Demerol, Ritalin).

Schedule III: The substance has a potential for abuse less than the compounds in Schedules I and II. The substance has a currently accepted medical use for treatment in the United States. Abuse of the substance might lead to moderate or low physical dependence or high psychological dependence (e.g., Tylenol with codeine used for pain, anabolic steroids).

Schedule IV: The substance has a low potential for abuse relative to the compounds in Schedule III. The substance has a currently accepted medical use for treatment in the United States. Abuse of the substance might lead to limited physical dependence or psychological dependence relative to the substances in Schedule III (e.g., lorazepam, Valium, drugs used to treat anxiety).

Schedule V: The substance has a low potential for abuse relative to the compounds in Schedule IV. The substance has a currently accepted medical use for treatment in the United States. Abuse of the substance might lead to limited physical dependence or psychological dependence relative to the compounds in Schedule IV (e.g., cough syrups with low doses of codeine).

All the compounds that are categorized into the five different schedules are classified as controlled substances. A substance being listed as a controlled substance does not mean that it shouldn't be used in clinical practice; it just helps clarify the risk/benefit ratio and the legal parameters associated with its use. Many of the compounds with the greatest clinical utility are classified as controlled substances. Whether you are the prescriber or the patient, use them wisely.

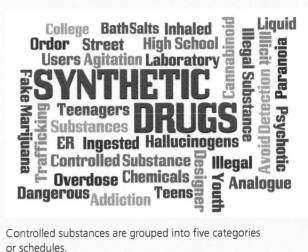

Controlled substances are grouped into five categories or schedules.

Solution

a. This compound is named as a substituted toluene. Both of the following are correct: 3-ethyltoluene and *m*-ethyltoluene. Note that in 3-ethyltoluene the methyl group, which is a part of the toluene structure, must be assigned to position number one.

b. This compound may be named as a substituted benzene or a substituted propane: propylbenzene, 1-phenylpropane.

c. When three groups are involved, the ring-numbering approach must be used: 3-chloro-5-ethylbenzoic acid.

✔ **LEARNING CHECK 12.10** Name the following aromatic compounds:

a.

c.

b.

d.

12.8 | Properties and Uses of Aromatic Compounds

Learning Objective

10. Recognize uses for specific aromatic compounds.

The physical properties of benzene and other aromatic hydrocarbons are similar to those of alkanes and alkenes. They are nonpolar and thus insoluble in water. This hydrophobic characteristic plays an important role in the chemistry of some proteins (see **Chapter 19**).

Aromatic rings are relatively stable chemically. Because of this, benzene often reacts in such a way that the aromatic ring remains intact. Thus, benzene does not undergo the addition reactions that are characteristic of alkenes and alkynes. The predominant type of reaction of aromatic molecules is substitution, in which one of the ring hydrogens is replaced by some other group. Such aromatic reactions are of lesser importance for our purposes and are not shown here.

All aromatic compounds discussed to this point contain a single benzene ring. There are also substances called **polycyclic aromatic compounds**, which contain two or more benzene rings sharing a common side, or "fused" together. The simplest of these compounds, naphthalene, is the original active ingredient in mothballs:

polycyclic aromatic compound
A derivative of benzene in which carbon atoms are shared between two or more benzene rings.

naphthalene

Graphene

Graphene is an allotropic form of pure carbon that was discovered in 1962, and then "rediscovered" in 2004. The two scientists who rediscovered it received Nobel Prizes for their "discovery."

At an atomic scale, graphene is made up of flat sheets of carbon atoms. Each sheet is made up of interconnected hexagons of carbon atoms. It can be considered as an indefinitely large aromatic molecule, the ultimate case of the group of flat polycyclic aromatic compounds (see **Section 12.8**). Graphene has numerous extraordinary properties. It is very flexible and is about 100 times stronger than steel of the same thickness. It conducts heat and electricity efficiently, and is nearly transparent to light.

Because of these characteristics, graphene has been used in a large number of commercial products. Its strength has led to its use in products such as tennis rackets. Its ability to conduct electricity has led to its use in inks designed for three-dimensional printing. Because of its diverse uses, the number of patents involving graphene has exploded in recent years.

In the last several years, the electrical properties of graphene have been somewhat of a focus. This has led to such products as graphene-based lightbulbs. Some mobile phone companies are even attempting to create flexible phones based on graphene.

Graphene forms a hexagonal lattice.

A number of more complex polycyclic aromatic compounds are known to be carcinogens—chemicals that cause cancer. Two of these compounds are:

a benzopyrene

a dibenzanthracene

These cancer-producing compounds are often formed as a result of heating organic materials to high temperatures. They are present in tobacco smoke (see **Figure 12.14**), automobile exhaust, and sometimes in burned or heavily browned food. Such compounds are believed to be at least partially responsible for the high incidence of lung and lip cancer among cigarette smokers. Those who smoke heavily face an increased risk of getting cancer. Chemists have identified more than 4000 compounds in cigarette smoke, including 43 known carcinogens. The Environmental Protection Agency (EPA) considers tobacco smoke a Class A carcinogen.

The major sources of aromatic compounds are petroleum and coal tar, a sticky, dark-colored material derived from coal. As with many classes of organic compounds, the simplest structures are the most important commercial materials.

Benzene and toluene are excellent laboratory and industrial solvents. In addition, they are the starting materials for the synthesis of hundreds of other valuable aromatic compounds that are intermediates in the manufacture of a wide variety of commercial products, including the important polymers Bakelite and polystyrene (see **Table 12.4**).

A number of aromatic compounds are important in another respect: They must be present in our diet for proper nutrition. Unlike plants, which have the ability to synthesize the

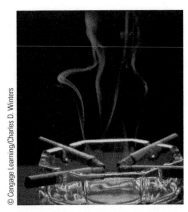

Figure 12.14 Cigarette smoke contains carcinogenic polycyclic aromatic compounds.

vitamin An organic nutrient that the body cannot produce in the small amounts needed for good health.

benzene ring from simpler materials, humans must obtain any necessary aromatic rings from their diet. This helps explain why certain amino acids, the building blocks of proteins, and some **vitamins** are dietary essentials (see **Table 12.4**).

TABLE 12.4 Some Important Aromatic Compounds

Name	Structural Formula	Use
benzene		Industrial solvent and raw material
toluene	CH_3	Industrial solvent and raw material
phenol	OH	Manufacture of Bakelite® and Formica®
aniline	NH_2	Manufacture of drugs and dyes
styrene	$CH = CH_2$	Preparation of polystyrene products
phenylalanine	$CH_2CH - \overset{\overset{O}{\|}}{C} - OH$, NH_2	An essential amino acid
riboflavin	$CH_2 - CH - CH - CH - CH_2$ with OH OH OH OH; CH_3, CH_3 ring system with N, N, O, NH, O	Vitamin B_2

▶ Case Study Follow-up

The Women's Health and Cancer Rights Act of 1998 requires insurance companies that cover mastectomies to cover all aspects of post-mastectomy reconstructive surgery. It could be argued that this surgery does nothing to prolong the cancer patient's life, but according to the act, it is considered a right for the woman to be restored to her former condition as much as possible. It is a matter of esthetic concern and self-esteem. In the same way, esthetic concerns are important to limb amputees. People who receive proper rehabilitation along with prosthetics are less likely to need expensive social welfare services. Prostheses help restore an amputee's dignity, productivity, and independence. The U.S. Armed Forces recognizes the value of prosthetics and allows service members with limb loss to choose a military or private prosthetist at no cost. There is a growing trend across the United States in which private insurance companies are significantly reducing prosthetic benefits or eliminating prosthetic coverage. Legislation is needed to increase parity between federal and private coverage. Self-assessment of physical appearance is important to self-esteem, which can influence levels of happiness, or, conversely, depression. Allowing amputees to be involved in the design of their prosthetics can be important to their recovery.

Concept Summary

The Nomenclature of Alkenes Compounds containing double or triple bonds between carbon atoms are said to be unsaturated. The alkenes contain double bonds, alkynes contain triple bonds, and aromatics contain a six-membered ring with three double bonds. In the IUPAC nomenclature system, alkene names end in *-ene*, and alkynes end in *-yne*.

Objective 1 (Section 12.1), Exercise 12.2
Objective 2 (Section 12.1), Exercise 12.4

The Geometry of Alkenes In alkenes, the double-bonded carbons and the four groups attached to these carbons lie in the same plane. Because rotation about the double bond is restricted, alkenes may exist as geometric, or cis-trans, isomers. This type of stereoisomerism is possible when each double-bonded carbon is attached to two different groups. IUPAC names of stereoisomers contain the prefixes *cis-* or *trans-*.

Objective 3 (Section 12.2), Exercise 12.18
Objective 4 (Section 12.2), Exercise 12.20

Properties of Alkenes The physical properties of alkenes are very similar to those of the alkanes. They are nonpolar, insoluble in water, less dense than water, and soluble in nonpolar solvents. Alkenes are quite reactive, and their characteristic reaction is addition to the double bond. Three important addition reactions are bromination (an example of halogenation) to give a dibrominated alkane, hydration to produce an alcohol, and the reaction with H—X to give an alkyl halide. The addition of H_2O and H—X are governed by Markovnikov's rule.

Objective 5 (Section 12.3), Exercise 12.26

Addition Polymers Addition polymers are formed from alkene monomers that undergo repeated addition reactions with each other. Many familiar and widely used materials, such as fibers and plastics, are addition polymers.

Objective 6 (Section 12.4), Exercise 12.36

Alkynes The alkynes contain triple bonds and possess a linear geometry of the two carbons and the two attached groups. Alkyne names end in *-yne*. The physical and chemical properties of alkynes are very similar to those of the alkenes.

Objective 7 (Section 12.5), Exercise 12.44

Aromatic Compounds and the Benzene Structure Benzene, the simplest aromatic compound, and other members of the aromatic class contain a six-membered ring with three double bonds. This aromatic ring is often drawn as a hexagon containing a circle, which represents the six electrons of the double bonds that move freely around the ring. All organic compounds that do not contain an aromatic ring are called aliphatic compounds.

Objective 8 (Section 12.6), Exercise 12.48

The Nomenclature of Benzene Derivatives Several acceptable IUPAC names are possible for many benzene compounds. Some IUPAC names are based on widely used common names such as toluene and aniline. Other compounds are named as derivatives of benzene or by designating the benzene ring as a phenyl group.

Objective 9 (Section 12.7), Exercises 12.52 and 12.54

Properties and Uses of Aromatic Compounds Aromatic hydrocarbons are nonpolar and have physical properties similar to those of the alkanes and alkenes. Benzene resists addition reactions typical of alkenes. Benzene and toluene are key industrial chemicals. Other important aromatics include phenol, aniline, and styrene.

Objective 10 (Section 12.8), Exercise 12.66

Key Terms and Concepts

Addition polymer (12.4)
Addition reaction (12.3)
Aliphatic compound (12.6)
Alkene (Introduction)
Alkyne (Introduction)
Aromatic hydrocarbon (Introduction)
Carbocation (12.6)

Copolymer (12.4)
Haloalkane or alkyl halide (12.3)
Hydration (12.3)
Hydrogenation (12.3)
Markovnikov's rule (12.3)
Monomer (12.4)
Phenyl group (12.7)

Polycyclic aromatic compound (12.8)
Polymer (12.4)
Polymerization (12.4)
Polyunsaturated (12.3)
Unsaturated hydrocarbon (Introduction)
Vitamin (12.8)

Key Reactions

1.	Halogenation of an alkene (Section 12.3):	$\overset{\diagdown}{\diagup}C=C\overset{\diagup}{\diagdown}$ + Br—Br \longrightarrow $-\overset{\mid}{\underset{\underset{Br}{\mid}}{C}}-\overset{\mid}{\underset{\underset{Br}{\mid}}{C}}-$	Reaction 12.2
2.	Hydrogenation of an alkene (Section 12.3):	$\overset{\diagdown}{\diagup}C=C\overset{\diagup}{\diagdown}$ + H—H $\xrightarrow{\text{Pt}}$ $-\overset{\mid}{\underset{\underset{H}{\mid}}{C}}-\overset{\mid}{\underset{\underset{H}{\mid}}{C}}-$	Reaction 12.4
3.	Addition of H—X to an alkene (Section 12.3):	$\overset{\diagdown}{\diagup}C=C\overset{\diagup}{\diagdown}$ + H—Cl \longrightarrow $-\overset{\mid}{\underset{\underset{H}{\mid}}{C}}-\overset{\mid}{\underset{\underset{Cl}{\mid}}{C}}-$	Reaction 12.6
4.	Hydration of an alkene (Section 12.3):	$\overset{\diagdown}{\diagup}C=C\overset{\diagup}{\diagdown}$ + H—OH $\xrightarrow{\text{H}_2\text{SO}_4}$ $-\overset{\mid}{\underset{\underset{H}{\mid}}{C}}-\overset{\mid}{\underset{\underset{OH}{\mid}}{C}}-$	Reaction 12.9
5.	Addition polymerization of an alkene (Section 12.4):	$n\text{CH}_2\text{=CH}_2$ $\xrightarrow[\text{catalysts}]{\text{Heat, pressure}}$ $-\!\!\left(\text{CH}_2-\text{CH}_2\right)\!\!_n$	Reaction 12.14

Exercises

Even-numbered exercises are answered in Appendix B.

Blue-numbered exercises are more challenging.

The Nomenclature of Alkenes (Section 12.1) and Alkynes (Section 12.5)

12.1 What is the definition of an *unsaturated hydrocarbon*?

12.2 Define the terms *alkene*, *alkyne*, and *aromatic hydrocarbon*.

12.3 Select those compounds that can be correctly called *unsaturated* and classify each one as an *alkene* or an *alkyne*:

a. $\text{CH}_3-\text{CH}_2-\text{CH}_3$

b. $\text{CH}_3\text{CH}=\text{CHCH}_3$

c. $\text{H}-\text{C}\equiv\text{C}-\underset{\underset{\text{CH}_3}{\mid}}{\text{CH}}-\text{CH}_3$

d. (cyclopentene structure)

e. (methylcyclohexane structure) CH₃

f. (cyclopentane with CH=CH₂ substituent)

g. $\underset{\text{CH}_2-\text{CH}_2}{\overset{\text{CH}=\text{CH}}{\mid\qquad\mid}}$

h. $\text{CH}_2=\text{CHCH}_2\text{CH}_3$

i. $\text{CH}_3\underset{\underset{\text{CH}_3}{\mid}}{\text{CH}}\text{CH}_3$

12.4 Give the IUPAC name for the following compounds:

a. $\text{CH}_3\text{CH}=\text{CHCH}_3$

b. $\text{CH}_3\text{CH}_2-\underset{\underset{\text{CH}_2\text{CH}_3}{\mid}}{\text{C}}=\text{CHCH}_3$

c. $\text{CH}_3-\text{C}\equiv\text{C}-\underset{\underset{\text{CH}_3}{\mid}}{\overset{\overset{\text{CH}_3}{\mid}}{\text{C}}}-\text{CH}_2\text{CH}_3$

d. (cyclopentene with CH₃ substituent)

e. $\text{CH}_3\underset{\underset{\text{Br}}{\mid}}{\text{CH}}\text{CH}_2-\text{C}\equiv\text{C}-\underset{\underset{\text{CH}_3}{\mid}}{\text{CH}}-\text{CH}_3$

f.

g. $\text{CH}_3\underset{\underset{\text{CH}_3}{\mid}}{\text{CH}}-\text{CH}=\text{CHCH}_2\text{CH}=\text{CH}_2$

12.5 Give the IUPAC name for the following compounds:

a. $CH_3CHCH=CHCH_2CH_3$
 |
 CH_3

b. $CH_3CH=CHCH=CHCHCH_3$
 |
 CH_3

c.

d. $CH_3-C{\equiv}C-CH_2CH_3$

e. $CH_3C-CH_2CH_2-CCH_3$
 || ||
 CH_2 CH_2

f.

g.

12.6 Draw structural formulas for the following compounds:

a. 3-ethyl-2-hexene

b. 3,4-dimethyl-1-pentene

c. 3-methyl-1,3-pentadiene

d. 2-isopropyl-4-methylcyclohexene

e. 1-butylcyclopropene

12.7 Draw structural formulas for the following compounds:

a. 4-methylcyclohexene

b. 2-ethyl-1,4-pentadiene

c. 4,4,5-trimethyl-2-heptyne

d. 4-isopropyl-3,3-dimethyl-1,5-octadiene

e. 2-methyl-1,3-cyclopentadiene

f. 3-sec-butyl-3-t-butyl-1-heptyne

12.8 A compound has the molecular formula C_5H_8. Draw a structural formula for a compound with this formula that would be classified as (a) an alkyne, (b) a diene, and (c) a cyclic alkene. Give the IUPAC name for each compound.

12.9 Draw structural formulas and give IUPAC names for the 13 alkene isomers of C_6H_{12}. Ignore geometric isomers and cyclic structures.

12.10 α-Farnesene is a constituent of the natural wax found on apples. Given that a 12-carbon chain is named as a dodecane, what is the IUPAC name of α-farnesene?

α-farnesene

12.11 Each of the following names is wrong. Give the structure and correct name for each compound.

a. 3-pentene

b. 3-methyl-2-butene

c. 2-ethyl-3-pentyne

12.12 Each of the following names is wrong. Give the structure and correct name for each compound.

a. 2-methyl-4-hexene

b. 3,5-heptadiene

c. 4-methylcyclobutene

The Geometry of Alkenes (Section 12.2)

12.13 What type of hybridized orbital is present on carbon atoms bonded by a double bond? How many of these hybrid orbitals are on each carbon atom?

12.14 What type of orbital overlaps to form a pi bond in an alkene? What symbol is used to represent a pi bond? How many electrons are in a pi bond?

12.15 Describe the geometry of the carbon–carbon double bond and the two atoms attached to each of the double-bonded carbon atoms.

12.16 Explain the difference between geometric and structural isomers of alkenes.

12.17 Draw structural formulas and give IUPAC names for all the isomeric pentenes (C_5H_{10}) that are:

a. Alkenes that do not show geometric isomerism. There are four compounds.

b. Alkenes that do show geometric isomerism. There is one cis and one trans compound.

12.18 Which of the following alkenes can exist as cis-trans isomers? Draw structural formulas and name the cis and trans isomers.

a. $CH_3CH_2CH_2CH_2CH=CH_2$

b. $CH_3CH_2CH=CHCH_2CH_3$

 CH_3
 |
c. $CH_3C=CHCH_2CH_3$

12.19 Which of the following alkenes can exist as cis-trans isomers? Draw structural formulas and name the cis and trans isomers.

a. $H_2C=CH-CH_3$

 Br
 |
b. $CH_3C=CHCH_3$

 Cl
 |
c. $HC=CHCH_3$

12.20 Draw structural formulas for the following:

a. cis-3-hexene

b. trans-3-heptene

12.21 Draw structural formulas for the following:

a. cis-4-methyl-2-pentene

b. trans-5-methyl-2-heptene

Properties of Alkenes (Section 12.3)

12.22 In what ways are the physical properties of alkenes similar to those of alkanes?

12.23 Which of the following reactions is an addition reaction?

a. $A_2 + C_3H_6 \rightarrow C_3H_6A_2$

b. $A_2 + C_6H_6 \rightarrow C_6H_5A + HA$

c. $HA + C_4H_8 \rightarrow C_4H_9A$

d. $3O_2 + C_2H_4 \rightarrow 2CO_2 + 2H_2O$

e. $C_7H_{16} \rightarrow C_7H_8 + 4H_2$

12.24 State Markovnikov's rule, and write a reaction that illustrates its application.

12.25 Complete the following reactions. Where more than one product is possible, show only the one expected according to Markovnikov's rule.

a. (cyclopentene with CH_3) $+ H_2 \xrightarrow{Pt}$

b. $CH_3CH_2CH{=}CH_2 + Cl_2 \longrightarrow$

c. $CH_3CH_2\overset{\overset{\displaystyle CH_3}{|}}{C}{=}CH_2 + H_2O \xrightarrow{H_2SO_4}$

d. $CH_3\overset{\overset{\displaystyle CH_3}{|}}{C}{=}CHCH_2CH_3 + HBr \longrightarrow$

12.26 Complete the following reactions. Where more than one product is possible, show only the one expected according to Markovnikov's rule.

a. (cyclohexane with $CH{=}CH{-}CH_3$) $+ Br_2 \longrightarrow$

b. (cyclohexene) $+ H_2O \xrightarrow{H_2SO_4}$

c. $CH_2{=}\overset{\overset{\displaystyle CH_3}{|}}{C}{-}CH{=}CH_2 + 2H_2 \xrightarrow{Pt}$

d. (cyclohexane with CH_2) $+ HCl \longrightarrow$

12.27 Draw the structural formula for the alkenes with molecular formula C_5H_{10} that will react to give the following products. Show all correct structures if more than one starting material will react as shown.

a. $C_5H_{10} + Br_2 \longrightarrow CH_3CH_2\overset{\overset{\displaystyle |}{}}{C}H\overset{\overset{\displaystyle |}{}}{C}HCH_3$
$\qquad\qquad\qquad\qquad\quad\;\; \underset{Br}{|}\;\underset{Br}{|}$

b. $C_5H_{10} + H_2 \xrightarrow{Pt} CH_3CH_2CH_2CH_2CH_3$

c. $C_5H_{10} + H_2O \xrightarrow{H_2SO_4} CH_3\overset{\overset{\displaystyle CH_3}{|}}{\underset{\underset{\displaystyle OH}{|}}{C}}CH_2CH_3$

d. $C_5H_{10} + HBr \longrightarrow CH_3\underset{\underset{\displaystyle Br}{|}}{C}HCH_2CH_2CH_3$

12.28 What reagents would you use to prepare each of the following from 3-hexene?

a. $CH_3CH_2\overset{\overset{\displaystyle Br}{|}}{C}H\overset{\overset{\displaystyle Br}{|}}{C}HCH_2CH_3$

b. $CH_3CH_2CH_2CH_2CH_2CH_3$

c. $CH_3CH_2CH_2\underset{\underset{\displaystyle Cl}{|}}{C}HCH_2CH_3$

d. $CH_3CH_2CH_2\underset{\underset{\displaystyle OH}{|}}{C}HCH_2CH_3$

12.29 What is an important commercial application of hydrogenation?

12.30 Cyclohexane and 2-hexene both have the molecular formula C_6H_{12}. Describe a simple chemical test that would distinguish one from the other.

12.31 Terpin hydrate is used medicinally as an expectorant for coughs. It is prepared by the following addition reaction. What is the structure of terpin hydrate?

(structure: cyclohexene ring with CH_3 at top and C bearing CH_3 and CH_2 at bottom) $+\; 2H_2O \xrightarrow{H_2SO_4}$ terpin hydrate

Addition Polymers (Section 12.4)

12.32 Explain what is meant by each of the following terms: *monomer, polymer, addition polymer,* and *copolymer.*

12.33 A section of polypropylene containing three units of monomer can be shown as:

$-CH_2-\overset{\overset{\displaystyle CH_3}{|}}{C}H-CH_2-\overset{\overset{\displaystyle CH_3}{|}}{C}H-CH_2-\overset{\overset{\displaystyle CH_3}{|}}{C}H-$

Draw structural formulas for comparable three-unit sections of:

a. Teflon

b. Orlon

c. Lucite

12.34 Identify a structural feature characteristic of all monomers listed in Table 12.3.

12.35 Rubber cement contains a polymer of 2-methylpropene (isobutylene) called polyisobutylene. Write an equation for the polymerization reaction.

12.36 Much of today's plumbing in newly built homes is made from a plastic called poly (vinyl chloride), or PVC. Using Table 12.3, write a reaction for the formation of poly (vinyl chloride).

12.37 Identify a major use for each of the following addition polymers:

a. Styrofoam

b. Acrilan

c. Plexiglas

d. PVC

e. polypropylene

Alkynes (Section 12.5)

12.38 What type of hybridized orbital is present on carbon atoms bonded by a triple bond? How many of these hybrid orbitals are on each carbon atom?

12.39 How many sigma bonds and how many pi bonds make up the triple bond of an alkyne?

12.40 Describe the geometry in an alkyne of the carbon–carbon triple bond and the two attached atoms.

12.41 Explain why geometric isomerism is not possible in alkynes.

12.42 Give the common name and major uses of the simplest alkyne.

12.43 Describe the physical and chemical properties of alkynes.

12.44 Write the structural formulas and IUPAC names for all the isomeric alkynes with the formula C_5H_8.

Aromatic Compounds and the Benzene Structure (Section 12.6)

12.45 What type of hybridized orbital is present on the carbon atoms of a benzene ring? How many sigma bonds are formed by each carbon atom in a benzene ring?

12.46 What type of orbital overlaps to form the pi bonding in a benzene ring?

12.47 What does the circle within the hexagon represent in the structural formula for benzene?

12.48 Define the terms *aromatic* and *aliphatic*.

12.49 Limonene, which is present in citrus peelings, has a very pleasant lemonlike fragrance. However, it is not classified as an aromatic compound. Explain.

limonene

12.50 A disubstituted cycloalkane such as (a) exhibits cis-trans isomerism, whereas a disubstituted benzene (b) does not. Explain.

a.

b.

The Nomenclature of Benzene Derivatives (Section 12.7)

12.51 Give an IUPAC name for each of the following hydrocarbons as a derivative of benzene:

a. b.

12.52 Give an IUPAC name for each of the following hydrocarbons as a derivative of benzene:

a. b.

12.53 Give an IUPAC name for the following as hydrocarbons with the benzene ring as a substituent:

a. b.

12.54 Give an IUPAC name for the following as hydrocarbons with the benzene ring as a substituent:

a. $CH_3CH_2CHCH{=}CH_2$

b. $CH_3CHCH_2CH_2CHCH_3$

12.55 Name the following compounds, using the prefixed abbreviations for *ortho*, *meta*, and *para* and assigning IUPAC-acceptable common names:

a.

b.

12.56 Name the following compounds, using the prefixed abbreviations for *ortho, meta,* and *para* and assigning IUPAC-acceptable common names:

a. b.

12.57 Name the following by numbering the benzene ring. IUPAC-acceptable common names may be used where appropriate:

a.

b.

12.58 Name the following by numbering the benzene ring. IUPAC-acceptable common names may be used where appropriate:

a. b.

12.59 Draw structural formulas for the following:

a. *p*-chloroaniline

b. *m*-ethylphenol

c. 2-phenyl-1-pentene

12.60 Write structural formulas for the following:

a. *o*-ethylphenol

b. *m*-chlorobenzoic acid

c. 3-methyl-3-phenylpentane

12.61 There are three chlorophenol derivatives. Draw their structures and give an IUPAC name for each one.

Properties and Uses of Aromatic Compounds (Section 12.8)

12.62 Describe the chief physical properties of aromatic hydrocarbons.

12.63 Why does benzene not readily undergo addition reactions characteristic of other unsaturated compounds?

12.64 Compare the chemical behavior of benzene and cyclohexane.

12.65 For each of the following uses, list an appropriate aromatic compound:

a. A solvent

b. A vitamin

c. An essential amino acid

d. Starting material for dyes

12.66 For each of the following uses, list an appropriate aromatic compound:

a. Used in the production of Formica®

b. A starting material for polystyrene

c. Used to manufacture drugs

d. A starting material for Bakelite®

Additional Exercises

12.67 In general, alkynes have slightly higher boiling points and densities than structurally equivalent alkanes. What interparticle force would this be attributable to?

12.68 In Reaction 12.14, heat, pressure, and catalysts are needed to convert ethylene gas to polyethylene. Explain the effects of each of the three conditions (heat, pressure, catalysts) in terms of factors that affect reaction rates.

12.69 Propene reacts with a diatomic molecule whose atoms have the electronic configuration of $1s^2 2s^2 2p^6 3s^2 3p^5$. Draw the structure of the product formed and give its IUPAC name.

12.70 Draw a generalized energy diagram for the following reaction. Is the reaction endothermic or exothermic?

$$\text{alkene} + \text{water} \xrightarrow{H_2SO_4} \text{alcohol} + 10 \text{ kcal/mol}$$

12.71 What will be the limiting reactant when 25.0 g of 2-butene reacts with 25.0 g of iodine to produce 2,3-diiodobutane? How many moles of product could be produced?

Chemistry for Thought

12.72 Napthalene is the simplest polycyclic aromatic compound:

Draw a Kekulé structure for this compound like that shown for benzene in Section 12.6.

12.73 Why does propene not exhibit geometric isomerism?

12.74 Limonene is present in the rind of lemons and oranges. Based on its structure (see Exercise 12.49), would you consider it to be a solid, liquid, or gas at room temperature?

12.75 If the average molecular weight of polyethylene is 5.0×10^4 u, how many ethylene monomers ($CH_2{=}CH_2$) are contained in a molecule of the polymer?

12.76 Reactions to synthesize the benzene ring of aromatic compounds do not occur within the human body, and yet many essential body components involve the benzene structure. How does the human body get its supply of aromatic compounds?

12.77 Answer the question in the caption to Figure 12.8 pertaining to addition polymers.

12.78 Why can't alkanes undergo addition polymerization?

12.79 Some polymers produce toxic fumes when they are burning. Which polymer in Table 12.3 produces hydrogen cyanide, HCN? Which produces hydrogen chloride, HCl?

12.80 "Super glue" is an addition polymer of the following monomer. Draw a structural formula for a three-unit section of super glue.

$$
\underset{\text{CH}_2}{}=\underset{\underset{\text{CN}}{|}}{\text{C}}-\text{O}-\underset{\overset{\text{O}}{\|}}{\text{C}}-\text{CH}_3
$$

12.81 One of the fragrant components in mint plants is menthene, a compound whose IUPAC name is 1-isopropyl-4-methylcyclohexene. Draw a structural formula for menthene.

Allied Health Exam Connection

The following questions are from these sources:

- *Nursing School Entrance Exam* © 2005, Learning Express, LLC.

- *McGraw-Hill's Nursing School Entrance Exams* by Thomas A. Evangelist, Tamara B. Orr, and Judy Unrein © 2009, The McGraw-Hill Companies, Inc.

- *NSEE Nursing School Entrance Exams,* 3rd edition © 2009, Kaplan Publishing.

- *Cliffs Test Prep: Nursing School Entrance Exams* by Fred N. Grayson © 2004, Wiley Publishing, Inc.

- *Peterson's Master the Nursing School and Allied Health Entrance Exams,* 18th edition by Marion F. Gooding © 2008, Peterson's, a Nelnet Company.

12.82 Which of the following are aromatic compounds?

 a. benzene

 b. ethyl alcohol

 c. methane

 d. phenol

12.83 Identify which of the following general formulas would be used to characterize (1) an alkane, (2) an alkene with one C=C bond, and (3) an alkyne with one C≡C bond.

 a. C_nH_{2n}

 b. C_nH_{2n+2}

 c. C_nH_{2n-2}

12.84 The compound CH_2=CH—CH_2—CH_2—CH_3 is an example of:

 a. a pentane.

 b. a hexene.

 c. an alkene.

 d. organic macromolecule.

12.85 The correct structural formula for ethyne is:

 a. H—C≡C—H

 b. H—C≡C=H

 c. H—C≡C—H

 d. H=C≡C=H

12.86 Protection of the skin from the harmful effects of ultraviolet light is provided by the pigment _____, which is produced by specialized cells within the stratum germinativum.

 a. carotene

 b. melanin

 c. keratin

 d. hemoglobin

13

Alcohols, Phenols, and Ethers

haveseen/Shutterstock.com

▶ Case Study

Ron was a 63-year-old retired Army sergeant. He went for a physical at his wife Sheila's urging. She was worried because Ron had become increasingly clumsy, including tripping and falling. He had difficulty hearing, and his memory seemed to be failing. Ron couldn't remember what day it was or what he had done two days ago, but he seemed to remember older memories just fine. Ron's father was in the last stages of Alzheimer's. Could Ron have the disease too? In addition to the routine exam and laboratory tests, the doctor asked some screening questions. She made it a habit to ask a particular question of all her adult patients: "On any one occasion during the past three months, have you had more than five alcoholic drinks?" Ron's affirmative reply caused the doctor to probe more deeply and to consider that alcoholism and the memory loss associated with Korsekoff's syndrome, a neurological disorder caused by the lack of vitamin B_1 in the brain, might be Ron's primary issue. By taking time to compassionately address his issues, the physician was able to convince Ron to check himself into treatment.

> ▶ How important is addiction screening to the routine physical? How should a health care professional prioritize a patient's complex set of symptoms?

Follow-up to this Case Study appears at the end of the chapter before the *Concept Summary*.

When you have completed your study of this chapter, you should be able to:

1 Name and draw structural formulas for alcohols and phenols. **(Section 13.1)**

2 Classify alcohols as primary, secondary, or tertiary on the basis of their structural formulas. **(Section 13.2)**

3 Discuss how hydrogen bonding influences the physical properties of alcohols. **(Section 13.3)**

4 Write equations for alcohol dehydration and oxidation reactions. **(Section 13.4)**

5 Recognize uses for specific alcohols. **(Section 13.5)**

6 Recognize uses for specific phenols. **(Section 13.6)**

7 Name and draw structural formulas for ethers. **(Section 13.7)**

8 Describe the key physical and chemical properties of ethers. **(Section 13.8)**

9 Write equations for a thiol reaction with heavy metal ions and the production of disulfides that results when thiols are oxidized. **(Section 13.9)**

10 Identify functional groups in polyfunctional compounds. **(Section 13.10)**

A lcohols, phenols, and ethers are important organic compounds that occur naturally and are produced synthetically in significant amounts. They are used in numerous industrial and pharmaceutical applications (see **Figure 13.1**). Ethanol, for example, is one of the simplest and best known of all organic substances; the fermentation method for producing it was known and used by ancient civilizations. Menthol, a 10-carbon alcohol obtained from peppermint oil, is widely used as a flavoring agent. Cholesterol, with its complicated multi-ring molecular structure, has been implicated in some forms of heart disease. The chemistry of all of these substances is similar because the alcohol functional group behaves essentially the same, regardless of the complexity of the molecule in which it is found.

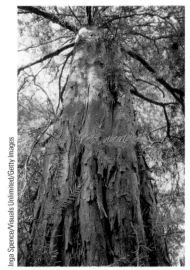

Figure 13.1 A promising anticancer drug, taxol, is isolated from the bark of the Pacific yew tree. Though taxol is a complex compound, what does its name suggest about its structure?

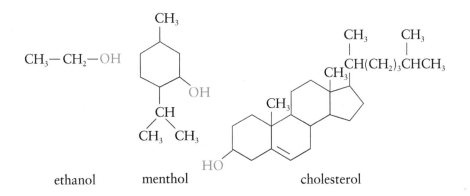

ethanol menthol cholesterol

Structurally, an **alcohol** is obtained from a hydrocarbon by replacing a hydrogen atom with a **hydroxy group** (—OH). A formula for an alcohol can be generalized as R—OH, where R— represents CH_3—, CH_3CH_2—, or any other singly bonded carbon–hydrogen group. Thus, R—OH can stand for CH_3—OH, CH_3CH_2—OH, and so on. If the replaced hydrogen was attached to an aromatic ring, the resulting compound is known as a **phenol**:

R—OH ⬡—OH

an alcohol a phenol

alcohol A compound in which an —OH group is connected to an aliphatic carbon atom.

hydroxy group The —OH functional group.

phenol A compound in which an —OH group is connected to a benzene ring. The parent compound is also called phenol.

Alcohols and phenols may also be considered to be derived from water by the replacement of one of its hydrogen atoms with an alkyl group or an aromatic ring:

$$H-OH \xrightarrow{\text{Replace H with R}} R-OH$$

water · an alcohol

$$H-OH \xrightarrow{\text{Replace H with}} \text{(benzene ring)}-OH$$

water · a phenol

ether A compound that contains a

$$-\overset{|}{\underset{|}{C}}-O-\overset{|}{\underset{|}{C}}-$$

functional group.

For most people, the word *ether* generates images of surgeons and operating rooms. Chemists define an **ether** as an organic compound in which two carbon atoms are bonded to an oxygen atom. This structure results when both hydrogen atoms of water are replaced by alkyl groups:

$$H-O-H \xrightarrow{\substack{\text{Replace both} \\ \text{H's with R}}} R-O-R'$$

water · an ether

R' indicates that the two R groups can be the same or different.

13.1 The Nomenclature of Alcohols and Phenols

Learning Objective

1. Name and draw structural formulas for alcohols and phenols.

The simpler alcohols are often known by common names, where the alkyl group name is followed by the word *alcohol:*

CH_3-OH	CH_3CH_2-OH	$CH_3CH_2CH_2-OH$	CH_3CHCH_3	
methyl alcohol	ethyl alcohol	propyl alcohol	$\overset{	}{OH}$

isopropyl alcohol

The IUPAC rules for naming alcohols that contain a single hydroxy group are as follows:

Step 1. Name the longest chain to which the hydroxy group is attached. The chain name is obtained by dropping the final *-e* from the name of the hydrocarbon that contains the same number of carbon atoms and adding the ending *-ol*.

Step 2. Number the longest chain to give the lower number to the carbon with the attached hydroxy group.

Step 3. Locate the position of the hydroxy group by the number of the carbon atom to which it is attached.

Step 4. Locate and name any branches attached to the chain.

Step 5. Combine the name and location for other groups, the hydroxy group location, and the longest chain into the final name.

Example 13.1 Naming Alcohols

Name the following alcohols according to the IUPAC system:

a. CH_3-OH

b. $CH_3CH_2CH_2CHCH_2CH_3$
 $\quad\quad\quad\quad\quad\overset{|}{CH_2}-OH$

c.

d.

Solution

a. The longest chain has one carbon atom; thus, the alcohol will be called methanol (methane − e + ol). Since the —OH group can be attached only to the single carbon atom, no location number for —OH is needed.

b. The longest chain containing the —OH group is numbered as follows:

$$\overset{5}{C}H_3\overset{4}{C}H_2\overset{3}{C}H_2\overset{2}{C}HCH_2CH_3$$
$$\overset{}{\underset{1}{C}H_2-OH}$$

Note that there is a longer chain (six carbon atoms), but the —OH group is not directly attached to it. The five-carbon chain makes the alcohol a pentanol. The —OH group is on carbon number 1; hence, the compound is a 1-pentanol. An ethyl group is attached to carbon number 2 of the chain, so the final name is 2-ethyl-1-pentanol.

c. The longest chain contains three carbon atoms, with the —OH group attached to carbon number 2. Hence, the compound is a 2-propanol. A benzene ring attached to carbon number 1 is named as a phenyl group. Therefore, the complete name is 1-phenyl-2-propanol. Note that the phenyl group is on carbon number 1 rather than number 3.

d. The —OH group is given preference, so the carbon to which it is attached is carbon number 1 of the ring. The —CH₃ group is located on carbon number 3 (since 3 is lower than 5, which would be obtained by counting the other way). The complete name is 3-methylcyclohexanol. Note that because the —OH group is always at the number-1 position in a ring, the position is not shown in the name.

✔ **LEARNING CHECK 13.1** Provide IUPAC names for the following alcohols:

a.

b.

c.

Compounds containing more than one hydroxy group are known. Alcohols containing two hydroxy groups are called *diols*. Those containing three —OH groups are called *triols*. The IUPAC nomenclature rules for these compounds are essentially the same as those for the single hydroxy alcohols, except that the ending *-diol* or *-triol* is attached to the name of the parent hydrocarbon.

Example 13.2 More Naming Alcohols

Name the following alcohols according the IUPAC system:

a. CH_2-CH_2
 | |
 OH OH

b. CH_3
 |
 $CH_2CHCH_2CH_2$
 | |
 OH OH

c. $CH_2-CH-CH_2$
 | | |
 OH OH OH

Solution

a. The longest chain has two carbons, making this an ethanediol. The —OH groups attached to carbons 1 and 2 give the final name 1,2-ethanediol. This substance is also called ethylene glycol.

b. Because the longest chain has four carbons, the compound is a butanediol. The —OH groups are attached to carbons 1 and 4, and a methyl group is attached to carbon 2. The name is 2-methyl-1,4-butanediol.

c. The longest chain contains three carbons, so the compound is a propanetriol. The —OH groups are attached to carbons 1, 2, and 3, making the name 1,2,3-propanetriol. Common names for this substance are glycerin and glycerol.

✔ **LEARNING CHECK 13.2** Give IUPAC names to the following diols:

a. OH OH
 | |
 $CH_2CH_2CH_2CH_2$

b.

Substituted phenols are usually named as derivatives of the parent compound phenol:

phenol 4-bromophenol 2,4,6-tribromophenol

✔ **LEARNING CHECK 13.3** Name this compound as a phenol:

13.2 | Classification of Alcohols

Learning Objective

2. Classify alcohols as primary, secondary, or tertiary on the basis of their structural formulas.

primary alcohol An alcohol in which the OH group is attached to CH_3 or to a carbon attached to one other carbon atom.

The characteristic chemistry of an alcohol sometimes depends on the groups bonded to the carbon atom that bears the hydroxy group. Alcohols are classified as primary, secondary, or tertiary on the basis of these attached groups (see **Table 13.1**). In a **primary alcohol**, the

hydroxy-bearing carbon atom is attached to one other carbon atom and two hydrogen atoms. The simplest alcohol, CH_3—OH, is also considered a primary alcohol. In **secondary alcohol**, the hydroxy-bearing carbon atom is attached to two other carbon atoms and one hydrogen atom. In a **tertiary alcohol**, the hydroxy-bearing carbon atom is attached to three other carbon atoms. The use of R, R′, and R″ in Table 13.1 signifies that each of those alkyl groups may be different.

TABLE 13.1 Primary, Secondary, and Tertiary Alcohols

	Primary	Secondary	Tertiary
General formula	$R-CH_2-OH$	$\begin{array}{c} R-CH-OH \\ \vert \\ R' \end{array}$	$\begin{array}{c} R \\ \vert \\ R'-C-OH \\ \vert \\ R'' \end{array}$
Specific example	$CH_3CH_2CH_2-OH$ 1-propanol (propyl alcohol)	$\begin{array}{c} CH_3CH-OH \\ \vert \\ CH_3 \end{array}$ 2-propanol (isopropyl alcohol)	$\begin{array}{c} CH_3 \\ \vert \\ CH_3-C-OH \\ \vert \\ CH_3 \end{array}$ 2-methyl-2-propanol (t-butyl alcohol)

✔ **LEARNING CHECK 13.4** Classify the following alcohols as primary, secondary, or tertiary:

a. $CH_3CH_2CH_2CH_2-OH$

b. [cyclohexane ring]—OH

c. $\begin{array}{c} OH \\ \vert \\ CH_3CH \\ \vert \\ CH_2CH_3 \end{array}$

13.3 | Physical Properties of Alcohols

Learning Objective

3. Discuss how hydrogen bonding influences the physical properties of alcohols.

The replacement of one hydrogen of water with an organic group does not cause all the waterlike properties to disappear. Thus, the lower molecular weight alcohols—methyl, ethyl, propyl, and isopropyl alcohols—are completely miscible with water. As the size of the alkyl group in an alcohol increases, the physical properties become less waterlike and more alkanelike. Long-chain alcohols are less soluble in water (see **Figure 13.2**) and more soluble in nonpolar solvents such as benzene, carbon tetrachloride, and ether. The solubility of alcohols in water depends on the number of carbon atoms per hydroxy group in the molecule. In general, one hydroxy group can solubilize three to four carbon atoms. The high solubility of the lower molecular weight alcohols in water can be attributed to hydrogen bonding between the alcohol and water molecules (see **Figure 13.3**).

Figure 13.2 The solubility of alcohols and linear alkanes in water.

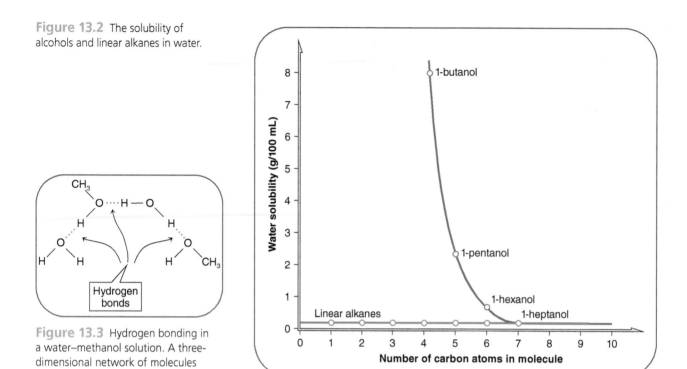

Figure 13.3 Hydrogen bonding in a water–methanol solution. A three-dimensional network of molecules and hydrogen bonds is formed.

In long-chain alcohols, the hydrophobic alkyl group does not form hydrogen bonds with water. Thus, in 1-heptanol, water molecules surround and form bonds with the —OH end but do not bond to the alkyl portion (see **Figure 13.4**). Because the alkyl portion is unsolvated, 1-heptanol is insoluble in water. Its solubility in water is comparable to that of heptane (seven carbon atoms), as shown in Figure 13.2.

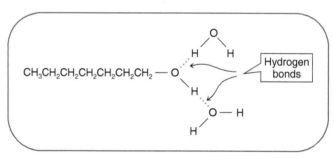

Figure 13.4 Water interacts only with the —OH group of 1-heptanol.

✔ **LEARNING CHECK 13.5** Arrange the following compounds in order of increasing solubility in water (least soluble first, most soluble last):

a. $CH_3CH_2CH_2$—OH
b. $CH_3CH_2CH_3$
c. $CH_3(CH_2)_3CH_2$—OH

Hydrogen bonding also causes alcohols to have much higher boiling points than most other compounds of similar molecular weight. In this case, the hydrogen bonding is

between alcohol molecules (see **Figure 13.5**). Ethanol (CH_3CH_2OH) boils at 78°C, whereas methyl ether (CH_3—O—CH_3) (with the same molecular weight) boils at −24°C. Propane ($CH_3CH_2CH_3$) has nearly the same molecular weight but boils at −42°C (see **Figure 13.6**).

Figure 13.5 Hydrogen bonding in pure ethanol.

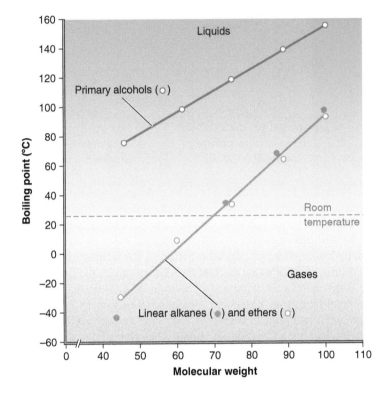

Figure 13.6 The boiling points of alcohols are higher than those of alkanes and ethers of similar molecular weights.

> ✔ **LEARNING CHECK 13.6** How can you explain the observation that methanol has a boiling point of 65°C, whereas propane, with a higher molecular weight, has a boiling point of −42°C?

13.4 | Reactions of Alcohols

Learning Objective

4. Write equations for alcohol dehydration and oxidation reactions.

Alcohols undergo many reactions, but we will consider only two important ones at this time—alcohol dehydration and alcohol oxidation. In an alcohol **dehydration reaction**, water is chemically removed from an alcohol. This reaction can occur in two different ways, depending on the reaction temperature. At 140°C, the main product is an ether, whereas at higher temperatures (180°C), alkenes are the predominant products.

dehydration reaction A reaction in which water is chemically removed from a compound.

13.4A Dehydration to Produce an Alkene

When the elements of a water molecule are removed from a single alcohol molecule, an alkene is produced. A reaction of this type, in which two or more covalent bonds are broken and a new multiple bond is formed, is called an **elimination reaction**. In this case, a water molecule is eliminated from an alcohol. Reaction 13.1 gives a general reaction, and Reactions 13.2 and 13.3 are two specific examples:

elimination reaction A reaction in which two or more covalent bonds are broken and a new multiple bond is formed.

$$\text{General reaction: } -\underset{\underset{H}{|}}{C}-\underset{\underset{OH}{|}}{C}- \xrightarrow[180°C]{H_2SO_4} \underset{/}{\overset{\backslash}{}}C=C\overset{/}{\underset{\backslash}{}} + H_2O \qquad (13.1)$$

alcohol alkene

$$\text{Specific example: } CH_3CH-\underset{\underset{H}{|}}{CH_2} \xrightarrow[180°C]{H_2SO_4} CH_3CH=CH_2 + H_2O \qquad (13.2)$$

$\quad\quad\quad\quad\quad\quad\quad\quad\;\; \underset{OH}{|}$

2-propanol propene

$$\text{Specific example: } CH_3CH_2\underset{\underset{OH}{|}}{CHCH_3} \xrightarrow[180°C]{H_2SO_4} CH_3CH=CHCH_3 + CH_3CH_2CH=CH_2 + H_2O \qquad (13.3)$$

2-butanol

2-butene
Principal product (90%)
(two carbon groups on double bond)

1-butene
Minor product (10%)
(one carbon group on double bond)

Notice that in Reaction 13.3 two alkenes can be produced, depending on which carbon next to the hydroxy-bearing carbon loses the hydrogen atom. The principal alkene produced in such cases will be the one in which the higher number of carbon groups is bonded to the double-bonded carbon atoms.

Alcohol dehydration is an important reaction in the human body, where enzymes (rather than sulfuric acid) function as catalysts. For example, the dehydration of citrate is part of the citric acid cycle discussed in Section 23.5. Note that the dehydration of citrate, 2-propanol (Reaction 13.2), and 2-butanol (Reaction 13.3) all involve exactly the same changes in bonding. In each reaction, an H and an —OH are removed from adjacent carbons, leaving a C=C double bond. The other molecular bonds in all three of those compounds do not change:

$$^-OOC-\underset{\underset{H}{|}}{\overset{\overset{H}{|}}{C}}-\underset{\underset{COO^-}{|}}{\overset{\overset{OH}{|}}{C}}-CH_2-COO^- \xrightarrow{Enzyme} \underset{^-OOC}{\overset{H}{\backslash}}C=C\underset{COO^-}{\overset{CH_2-COO^-}{/}} + H_2O \qquad (13.4)$$

citrate *cis*-aconitate

✔ **LEARNING CHECK 13.7** Predict the major products of the following reactions:

a.

$\quad\quad\quad\quad \underset{|}{OH}$

$CH_3CH_2CHCH_2CH_3 \xrightarrow[180°C]{H_2SO_4}$

b.

[cyclohexane ring with CH₃ and OH groups] $\xrightarrow[180°C]{H_2SO_4}$

HOW REACTIONS OCCUR 13.1

The Dehydration of an Alcohol

Step 1. A strong acid such as sulfuric acid initially reacts with an alcohol in much the same way that it reacts with water in solution. Ionization of the acid produces a proton (H^+) that is attracted to an oxygen atom of the alcohol:

$$CH_3CH_2-\overset{..}{\underset{..}{O}}: + H_2SO_4 \rightleftarrows CH_3CH_2-\overset{+}{\overset{H}{\underset{H}{O}}}: + HSO_4^-$$

protonated alcohol

Step 2. The creation of a positive charge on the oxygen atom weakens the carbon–oxygen bond, which allows the following equilibrium reaction to take place:

$$CH_3CH_2-\overset{+}{\overset{H}{\underset{H}{O}}} \rightleftarrows CH_3\overset{+}{C}H_2 + :\overset{..}{\underset{H}{O}}\overset{H}{}$$

a carbocation water
(one product)

Step 3. The ethyl carbocation is unstable because the carbon has only three bonds (six outer electrons) around it. To achieve an octet of electrons about this carbon, the carbocation donates an H^+ to a proton acceptor, such as HSO_4^-, and forms the double bond of the alkene product:

$$CH_2-\overset{+}{C}H_2 \longrightarrow CH_2=CH_2 + H_2SO_4$$
$$\underset{HSO_4^-}{\overset{|}{H}}$$

ethene regenerated
catalyst

13.4B Dehydration to Produce an Ether

Alcohol dehydration can also occur when the elements of water are removed from two alcohol molecules. The residual molecular fragments of the alcohols bond and form an ether. This ether-forming reaction is useful mainly with primary alcohols. Reaction 13.5 gives the general reaction, and the formation of diethyl ether (the anesthetic ether) is a specific example (Reaction 13.6):

General reaction:

$$R-O-H+H-O-R \xrightarrow[140°C]{H_2SO_4} R-O-R+ H_2O \qquad (13.5)$$

an alcohol an alcohol an ether

Specific example:

$$CH_3CH_2-OH + HO-CH_2CH_3 \xrightarrow[140°C]{H_2SO_4} CH_3CH_2-O-CH_2CH_3 + H_2O \qquad (13.6)$$

ethanol ethanol diethyl ether
(ethyl alcohol) (ethyl alcohol)

Thus, an alcohol can be dehydrated to produce either an alkene or an ether. The type of dehydration that takes place is controlled by the temperature, with sulfuric acid serving as a catalyst. Although a mixture of alkene and ether products often results, we will use these temperatures to indicate that the major dehydration product is an alkene (at 180°C) or an ether (at 140°C).

The reaction that produces ethers is an example of a dehydration synthesis in which two molecules are joined and a new functional group is generated by the removal of a water molecule. Dehydration synthesis is important in living organisms because it is part of the formation of carbohydrates, fats, proteins, and many other essential substances.

✔ **LEARNING CHECK 13.8** What catalyst and reaction temperature would you use to accomplish each of the following reactions?

a. $2CH_3CH_2CH_2-OH \rightarrow CH_3CH_2CH_2-O-CH_2CH_2CH_3+H_2O$

b. $CH_3CH-OH \rightarrow CH_3CH=CH_2+H_2O$
 $\underset{CH_3}{\overset{|}{}}$

13.4C Oxidation

An oxidation reaction occurs when a molecule gains oxygen atoms or loses hydrogen atoms. Under appropriate conditions, alcohols can be oxidized in a controlled way (not combusted) by removing two hydrogen atoms per molecule. A number of oxidizing agents can be used, including potassium dichromate ($K_2Cr_2O_7$) and potassium permanganate ($KMnO_4$). The symbol (O) is used to represent an oxidizing agent. Reaction 13.7 gives the general reaction:

$$-\overset{|}{\underset{|}{C}}-O-H + (O) \longrightarrow -\overset{O}{\overset{||}{C}}- + H_2O \qquad (13.7)$$

alcohol aldehyde
or ketone

The three classes of alcohols behave differently toward oxidizing agents.

Primary Alcohols. Reaction 13.8 gives the general reaction, and the formation of acetic acid is a specific example (Reaction 13.9):

General reaction:

$$R-\overset{OH}{\underset{H}{\overset{|}{\underset{|}{C}}}}-H + (O) \longrightarrow R-\overset{O}{\overset{||}{C}}-H + H_2O \qquad (13.8)$$

primary aldehyde
alcohol

$$(O) \xrightarrow{\text{further oxidation}} R-\overset{O}{\overset{||}{C}}-OH$$

carboxylic
acid

Specific example:

$$CH_3CH_2-OH + (O) \longrightarrow CH_3-\overset{O}{\overset{||}{C}}-H + H_2O \qquad (13.9)$$

ethanol acetaldehyde

$$(O) \xrightarrow{\text{further oxidation}} CH_3-\overset{O}{\overset{||}{C}}-OH$$

acetic acid

The immediate product of the oxidation of a primary alcohol is an aldehyde. However, aldehydes are readily oxidized by the same oxidizing agents (see **Section 14.3**) to give carboxylic acids. Therefore, the oxidation of a primary alcohol normally results in the corresponding carboxylic acid as the product. The oxidation of ethanol by $K_2Cr_2O_7$ is shown in **Figure 13.7**. Because aldehydes do not hydrogen bond (see **Section 14.2**), they have lower boiling points than the corresponding alcohol or carboxylic acid. This makes it possible to isolate the aldehyde product before it is oxidized by maintaining the reaction temperature high enough to boil the aldehyde out of the reaction mixture before it can react.

✔ **LEARNING CHECK 13.9** Draw the structural formulas of the first and second products of the following reaction:

$$\text{(benzene ring)}-CH_2-OH + (O) \longrightarrow$$

① The tube on the left contains orange $K_2Cr_2O_7$ and is next to the colorless ethanol.

② After mixing, the ethanol is oxidized, and chromium is reduced, forming a grayish-green precipitate.

Figure 13.7 Oxidation of ethanol by $K_2Cr_2O_7$. How might a color change like this one be used to measure alcohol concentration?

Secondary Alcohols. Reactions 13.10 and 13.11 give the general reaction and a specific example.

$$\underset{\substack{\text{secondary} \\ \text{alcohol}}}{R - \overset{\overset{\displaystyle OH}{|}}{\underset{\underset{\displaystyle H}{|}}{C}} - R' } + (O) \longrightarrow \underset{\text{ketone}}{R - \overset{\overset{\displaystyle O}{\|}}{C} - R'} + H_2O \qquad (13.10)$$

General reaction:

$$\text{Specific example:} \quad \underset{\text{2-propanol}}{CH_3 - \overset{\overset{\displaystyle OH}{|}}{\underset{\underset{\displaystyle H}{|}}{C}} - CH_3} + (O) \longrightarrow \underset{\text{acetone}}{CH_3 - \overset{\overset{\displaystyle O}{\|}}{C} - CH_3} + H_2O \quad (13.11)$$

Secondary alcohols are oxidized to ketones in exactly the same way primary alcohols are oxidized to aldehydes, and by the same oxidizing agents. However, unlike aldehydes, ketones resist further oxidation, so this reaction provides an excellent way to prepare ketones.

✔ **LEARNING CHECK 13.10** Complete the following oxidation reaction:

+ (O) ⟶

Tertiary Alcohols. A general reaction is given in Reaction 13.12:

$$\text{General reaction:} \quad R - \overset{\overset{\displaystyle OH}{|}}{\underset{\underset{\displaystyle R''}{|}}{C}} - R' + (O) \longrightarrow \text{no reaction} \quad (13.12)$$

Tertiary alcohols do not have any hydrogen on the —OH-bearing carbon and thus do not react with oxidizing agents.

✔ **LEARNING CHECK 13.11** Which of the following alcohols will react with an oxidizing agent?

a. b. CH₃ OH c.

13.4D Multistep Reactions

Alcohols are important organic compounds not only for their own commercial applications but because they can be converted by chemical reactions into a wide variety of other useful products. Most alcohols do not occur naturally in commercial quantities and therefore are prepared from alkenes.

It is very common in both laboratory and industrial processes for such preparations to involve a sequence of reactions. For example, the commercial production of vinyl plastic (PVC) begins with the alkene ethene:

$$CH_2=CH_2 \longrightarrow \underset{\text{vinyl chloride}}{\overset{\displaystyle \overset{\text{Cl}}{\underset{\displaystyle |}{}}}{CH_2=CH}} \longrightarrow \underset{\substack{\text{poly (vinyl chloride)}\\ \text{(PVC)}}}{\overset{\displaystyle \overset{\text{Cl}}{\underset{\displaystyle |}{}}}{+CH_2-CH+}}$$

ethene

Nature, too, employs multistep reactions to carry out processes essential to life. The glycolysis pathway used by the body to initiate the utilization of glucose for energy production involves a series of 10 reactions (see **Section 23.3**).

Example 13.3 Writing Reactions

The following conversion requires more than one step. Show the reaction, reagents, and intermediate structures necessary to carry out the synthesis.

Solution

The best way to initially solve a multistep synthesis is to work backward from the final product. Identify the functional group present in the product—in this case, a ketone. Think how to prepare a ketone—by oxidizing an alcohol.

Now, working backward again, think how to prepare an alcohol—from an alkene through hydration.

Adding these two steps together gives

✔ **LEARNING CHECK 13.12** Show the reactions necessary to carry out the following conversion.

$$CH_2=CH_2 \longrightarrow CH_3CH_2-O-CH_2CH_3$$

13.5 | Important Alcohols

Learning Objective

5. Recognize uses for specific alcohols.

The simplest alcohol, methanol or methyl alcohol, is a very important industrial chemical; more than 1 billion gallons are produced and used annually. It is sometimes known as wood alcohol because the principal source for many years was the distillation of wood.

Today, it is synthetically produced in large quantities by reacting hydrogen gas with carbon monoxide (Reaction 13.13):

$$CO + 2H_2 \xrightarrow[\text{heat, pressure}]{\text{Catalysts,}} CH_3-OH \qquad (13.13)$$

The industrial importance of methanol is due to its oxidation product, formaldehyde, which is a major starting material for the production of plastics. Methanol is used as a fuel in Indy-style racing cars (see **Figure 13.8**). It is also used as a fuel in products such as alcohol burners found in camping supplies.

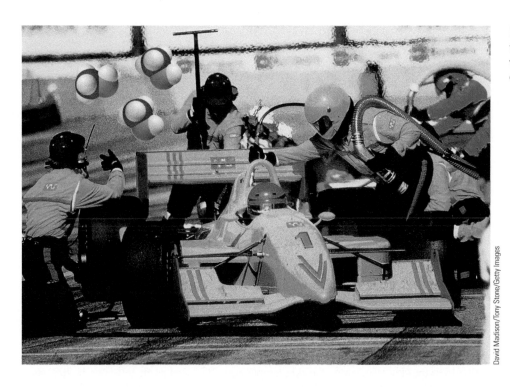

David Madison/Tony Stone/Getty Images

Figure 13.8 Racing cars are fueled by methanol. What would be the equation for the combustion of methanol?

Methanol is highly toxic and causes permanent blindness if taken internally. Deaths and injuries have resulted as a consequence of mistakenly substituting methanol for ethyl alcohol in beverages.

Ethanol, or ethyl alcohol, CH_3CH_2-OH, is probably the most familiar alcohol because it is the active ingredient in alcoholic beverages. It is also used in pharmaceuticals as a solvent (tinctures are ethanol solutions) and in aftershave lotions as an antiseptic and skin softener. It is an important industrial solvent as well. The gasohol now marketed at many locations around the United States is a mixture of ethanol and gasoline.

Most ethanol used industrially is produced on a large scale by the direct hydration of ethylene obtained from petroleum (Reaction 13.14):

$$H_2C=CH_2+H-OH \xrightarrow[\text{300°C}]{\text{70 atm}} \begin{array}{c} H_2C-CH_2 \\ | \quad | \\ H \quad OH \end{array} \qquad (13.14)$$

ethylene ethanol

Ethanol for beverages and increasingly for gasohol is produced by the yeast **fermentation** of carbohydrates such as sugars, starch, and cellulose. For example, a common method is the fermentation of glucose, grape sugar (Reaction 13.15):

$$C_6H_{12}O_6 \xrightarrow{\text{Yeast}} 2CH_3CH_2-OH+2CO_2 \qquad (13.15)$$

glucose ethanol

Ethanol

fermentation A reaction of sugars, starch, or cellulose to produce ethanol and carbon dioxide.

STUDY SKILLS 13.1

A Reaction Map for Alcohols

To solve a test question, using a stepwise approach is often helpful. First, decide what type of question is being asked. Is it a nomenclature, typical uses, physical properties, or a reaction question? Second, if you recognize it as, say, a reaction question, look for the functional group. Third, if there is an alcohol group in the starting material, identify the reagent and use the following diagram to predict the right products.

Figure 13.9 Isopropyl alcohol is used as an antiseptic.

Maren Hansen

Figure 13.10 Ethylene glycol ($HO—CH_2CH_2—OH$) is used as an automotive antifreeze.

2-propanol (isopropyl alcohol), the main component of rubbing alcohol, acts as an astringent on the skin. It causes tissues to contract, hardens the skin, and limits secretions. Its rapid evaporation from the skin also lowers the skin temperature. It is used as an antiseptic in 70% solutions (see **Figure 13.9**). Isopropyl alcohol is toxic and should not be taken internally.

1,2,3-propanetriol is known more commonly as glycerol or glycerin:

$$\begin{array}{ccc} OH & OH & OH \\ | & | & | \\ CH_2 & -CH- & CH_2 \end{array}$$

1,2,3-propanetriol (glycerol)

The hydrogen bonding resulting from the three hydroxy groups causes glycerol to be a syrupy, high-boiling liquid with an affinity for water. This property and its nontoxic nature make it valuable as a moistening agent in tobacco and many food products, such as candy and shredded coconut. Florists use glycerol to retain water that maintains the freshness of cut flowers. Glycerol is also used for its soothing qualities in shaving and toilet soaps and in many cosmetics. In Chapter 18 on lipids, we will see that glycerol forms part of the structure of body fats and oils.

Due to hydrogen bonding, 1,2-ethanediol (ethylene glycol) and 1,2-propanediol (propylene glycol) are both high-boiling liquids that are completely miscible with water.

$$\begin{array}{cc} OH & OH \\ | & | \\ CH_2 & -CH_2 \end{array} \qquad \begin{array}{ccc} OH & OH \\ | & | \\ CH_2 & -CH- & CH_3 \end{array}$$

1,2-ethanediol 1,2-propanediol
(ethylene glycol) (propylene glycol)

Their main uses are as automobile antifreeze (see **Figure 13.10**) and as a starting material for the manufacture of polyester fibers (Reaction 15.11).

CHEMISTRY AROUND US 13.1

Alcohol and Antidepressants Don't Mix

Individuals suffering from depression often demonstrate one or more of the following symptoms: extreme feelings of sadness or melancholy; feelings of inadequacy in facing life's challenges; inability to make decisions; physical inactivity; lack of appetite; inability to sleep; and frequent crying.

Antidepressant medications are designed to counteract these symptoms. However, when alcohol, itself a depressant, is combined with antidepressant medications, serious consequences can result. For example, some antidepressants cause sedation and drowsiness. Alcohol has similar effects, and the combination of the two could easily cause a person to fall asleep while driving. Initially, alcohol might seem to improve the mood of a depressed person, but its overall effect is often to increase the depression symptoms.

The combination of some types of antidepressants and alcohol influences coordination, judgment, and reaction time and other motor skills more than alcohol alone. When combined with certain types of alcoholic beverages and foods, antidepressants known as monoamine oxidase inhibitors (MAOIs) cause bodily reactions such as a dangerous increase in blood pressure.

It has been found that people suffering from depression have an increased risk of becoming addicted to alcohol or other addictive substances. Clearly, when these factors are all taken into consideration, it is not worth the risks for people being treated for depression to mix their antidepressant medications with alcohol.

When alcohol is combined with antidepressants, serious consequences can result.

Alcohols are very important biologically because the alcohol functional group occurs in a variety of compounds associated with biological systems. For example, sugars contain several hydroxy groups, and starch and cellulose contain thousands of hydroxy groups (see **Chapter 17**). The structures and uses (see **Figure 13.10**) of some common alcohols are summarized in **Table 13.2**.

TABLE 13.2 Examples of Alcohols

Name	Structural Formula	Typical Uses
methanol (methyl alcohol)	CH_3—OH	Solvent, making formaldehyde
ethanol (ethyl alcohol)	CH_3CH_2—OH	Solvent, alcoholic beverages
2-propanol (isopropyl alcohol)	CH_3CHCH_3 \| OH	Rubbing alcohol, solvent
1-butanol (butanol)	$CH_3CH_2CH_2CH_2$—OH	Solvent, hydraulic fluid
1,2-ethanediol (ethylene glycol)	HO—CH_2CH_2—OH	Automobile antifreeze, polyester fibers
1,2-propanediol (propylene glycol)	CH_3CH—CH_2 \| \| OH OH	Moisturizer in lotions and foods, automobile antifreeze
1,2,3-propanetriol (glycerin, glycerol)	CH_2—CH—CH_2 \| \| \| OH OH OH	Moisturizer in foods, tobacco, and cosmetics
menthol	CH_3CHCH_3 ... OH ... CH_3	Cough drops, shaving lotion, mentholated tobacco

✔ **LEARNING CHECK 13.13** What is an important use for each of the following alcohols?

a. methanol c. 2-propanol e. glycerol
b. ethanol d. ethylene glycol f. menthol

13.6 | Characteristics and Uses of Phenols

Learning Objective

6. Recognize uses for specific phenols.

Phenol, C_6H_5OH, is a colorless, low-melting-point solid with a medicinal odor. The addition of a small amount of water causes the solid to liquefy at room temperature. This liquid mixture is sometimes called carbolic acid. Unlike alcohols, phenols are weak acids (Reactions 13.16 and 13.17), which can damage skin by denaturing the proteins in the skin, and they must be handled with care.

$$\text{phenol (a weak acid)} + H_2O \longrightarrow \text{(phenoxide)} + H_3O^+ \qquad (13.16)$$

$$\text{phenol} + KOH \longrightarrow \text{(potassium phenoxide)} + H_2O \qquad (13.17)$$

In dilute solutions, phenol is used as an antiseptic and disinfectant. Joseph Lister, an English surgeon, introduced the use of phenol as a hospital antiseptic in the late 1800s. Before that time, antiseptics had not been used, and very few patients survived even minor surgery because of postoperative infections. Today, there are effective derivatives used as antiseptics instead of phenol that are less irritating to the skin. Two such compounds, 4-chloro-3,5-dimethylphenol and 4-hexylresorcinol, are shown here.

phenol
(Lister's original
disinfectant)

4-chloro-3,5-dimethylphenol
(nonirritating topical
antiseptic)

4-hexylresorcinol
(used in mouthwashes
and throat lozenges)

ASK A PHARMACIST 13.1

Marijuana: A Gateway Drug

According to the Federal Controlled Substances Act of 1970, marijuana is classified as a Schedule I controlled substance. Materials placed in this category are recognized as having a high potential for addiction and have no recognized medical uses. In spite of these characteristics, marijuana has been legalized by numerous state legislatures for medical and even recreational use. Many testimonials can be found on public media sites from both physicians and patients that praise the effectiveness of marijuana in treating a wide variety of medical conditions.

From a chemical perspective, marijuana plants contain over 100 chemical compounds that are called cannabinoids. The most well-known of these is a mind-altering substance called tetrahydrocannabinol that is commonly shortened to THC. Most of the pharmacological effects associated with using marijuana are caused by THC, although a user would be exposed to other cannabinoids that can bind to several different cannabinoid receptors in the nervous system.

Users take marijuana into the body by inhalation of smoke or vapor, by drinking brewed tea, or by eating after combining it with other foods. Smoking is the preferred ingestion route by users, because the THC is quickly absorbed by the lungs.

The short-term effects of THC use include feeling exhilarated (high), impaired memory, and decreased ability to think clearly and solve problems. Long-term use has been linked to several temporary mental conditions including hallucinations, paranoia, depression, and anxiety.

Marijuana is commonly believed to not be addictive in spite of the federal government's classification of it as an addictive substance. However, evidence gained from users indicates that it can be addictive. The potential for addiction increases if the substance is used by teenagers, and it is often the gateway drug to make other forms of drug abuse and addiction more likely.

Tetrahydrocannibinol, THC, is present in marijuana.

Jan Mika/Shutterstock.com

Two other derivatives of phenol, *o*-phenylphenol and 2-benzyl-4-chlorophenol, are ingredients in Lysol®, a disinfectant for walls and furniture in hospitals and homes:

o-phenylphenol

2-benzyl-4-chlorophenol

Other important phenols act as **antioxidants** by interfering with oxidizing reactions. These phenols are useful in protecting foods from spoilage and a variety of other materials from unwanted reactions. BHA and BHT are widely used as antioxidants in gasoline, lubricating oils, rubber, certain foods, and packaging materials for foods that might turn rancid (see **Figure 13.11**). In food or containers used to package food, the amount of these antioxidants is limited to 200 ppm (parts per million) or 0.02%, based on the fat or oil

antioxidant A substance that prevents another substance from being oxidized.

Take Advantage of Hand Sanitizers

Hospitals, schools, and businesses often provide alcohol-based hand sanitizers to the public. Wall-mounted dispensers of foams and gels or wipes designed to sanitize a shopping cart offer safety, but do you feel comfortable using them? Do they really make a difference for the better?

According to the Centers for Disease Control and Prevention (CDC), proper hand sanitizer use does offer significant safe and healthy benefits. Clean hands prevent the spread of disease and help keep the public healthy. When available, plain soap and water is the best choice for cleaning hands. Rubbing the hands together with soap mechanically removes grime and germs, and the water rinses them away. Washing for at least 20 seconds with soap and water has been shown to remove microbes, and is the preferred method to use for hand cleaning. However, soap and water might not always be available. That is when hand sanitizers can provide a useful alternative.

Some hand sanitizer labels claim to protect against colds and flu. Should they be believed? Research confirms that hand sanitizers do provide a level of protection against disease. A 2010 published study documented that workers who used hand sanitizers at least five times during the workday became sick two-thirds as often as workers who only washed their hands. An earlier study showed that families who regularly used hand sanitizers had 60% as many gastrointestinal infections as those who did not use hand sanitizers.

Still, some people wonder if hand sanitizers are safe and effective. The CDC recommends that only alcohol-based hand sanitizers with an alcohol content between 60–95% be used. Non-alcohol-based hand sanitizers might actually contribute to the spread of resistant bacteria because they only reduce the growth rate of germs rather than kill them outright.

Alcohol-based hand sanitizers can only perform well when used properly. Often, people do not use enough of the product. To ensure effectiveness, one should use at least a dime-sized amount of the sanitizer and rub the product all over the hands, including between the fingers and across the nails. Also, be careful not to wipe the product off before it dries.

If you use enough, the sanitizer should take at least 30 seconds to evaporate. If hands are visibly dirty, hand sanitizer might not effectively kill germs because the grime provides a shield from the alcohol. Sanitizers work best on visibly clean hands and in conjunction with soap and water hand washing.

Alcohol-based hand sanitizers kill germs effectively and help prevent the spread of disease, but there are some safety concerns. Adults should supervise children when they are using hand sanitizers to ensure that they do not ingest the product. Adults should make sure that the sanitizer is completely dry on their hands before children are allowed to eat or drink anything.

Also, since alcohol is flammable, hand sanitizers should be stored away from sources of heat. Some non-alcohol-based sanitizers contain triclosan. This is an antibacterial product that has been shown to interfere with hormone regulation in animals, promote the growth of antibiotic-resistant bacteria, and might be harmful to immune systems. Despite these few safety risks, research regarding alcohol-based hand sanitizers suggests that conscientious and proper use of these products is a positive step in preventing the spread of disease and protecting individual health by reducing infection.

The use of hand sanitizers helps prevent the spread of disease.

Figure 13.11 BHT is listed as a food ingredient that will "maintain freshness."

content of the food. Notice that industrial nomenclature for these compounds does not follow IUPAC rules.

BHA (butylated hydroxy anisole)
2-*t*-butyl-4-methoxyphenol
(IUPAC name)

BHT (butylated hydroxy toluene)
2, 6-di-*t*-butyl-4-methylphenol
(IUPAC name)

✔ **LEARNING CHECK 13.14** What is an important use for each of the following phenols?

a. BHA b. 4-hexylresorcinol

13.7 | Ethers

Learning Objective

7. Name and draw structural formulas for ethers.

Ethers were introduced at the start of this chapter as compounds whose formula is R—O—R′. Common names for ethers are obtained by first naming the two groups attached to the oxygen and then adding the word *ether.* If the groups are the same, it is appropriate to use the prefix *di-*, as in diethyl ether, but many people omit the *di-* and simply say "ethyl ether."

Dimethyl ether

CH_3—O—CH_2CH_3
ethyl methyl ether

isopropyl phenyl ether

CH_3—O—CH_3
dimethyl ether
(methyl ether)

CH_3CH_2—O—CH_2CH_3
diethyl ether
(ethyl ether)

✔ **LEARNING CHECK 13.15** Assign a common name to the following:

a. CH_3CH_2—O—⬡

b. CH_3CH—O—$CHCH_3$ (with CH_3 groups on each CH)

In the IUPAC naming system for ethers, the —O—R group is called an **alkoxy group**. The *-yl* ending of the smaller R group is replaced by *-oxy*. For example, —O—CH$_3$ is called a methoxy group, —O—CH$_2$CH$_3$ is called ethoxy, and so on:

$$CH_3—O—CH_3$$
methoxymethane

$$CH_3CH_2\overset{\displaystyle O—CH_2CH_3}{\overset{|}{C}HCH_3}$$
2-ethoxybutane

p-methoxybenzoic acid

✔ **LEARNING CHECK 13.16** Give IUPAC names for the following:

a.

b. $CH_3CH_2CHCH_2CH_3$

heterocyclic ring A ring in which
one or more atoms are an atom other
than carbon.

The ether linkage is also found in cyclic structures. A ring that contains elements in addition to carbon is called a **heterocyclic ring**. Two common structures with oxygen-containing heterocyclic rings are:

furan pyran

Many carbohydrates contain the fundamental ring systems of furan and pyran (see **Chapter 17**).

13.8 | Properties of Ethers

Learning Objective

8. Describe the key physical and chemical properties of ethers.

The oxygen atom of ethers can form hydrogen bonds with water (see **Figure 13.12A**), so ethers are slightly more soluble in water than hydrocarbons but less soluble than alcohols of comparable molecular weight. Ethers, however, cannot form hydrogen bonds with other ether molecules in the pure state (see **Figure 13.12B**), which results in low boiling points close to those of hydrocarbons with similar molecular weights.

Hydrogen
bond

No hydrogen
bond

A

B

Figure 13.12 Hydrogen bonding of dimethyl ether: **A** with water and **B** no hydrogen bonding in the pure state.

The chief chemical property of ethers is that, like the alkanes, they are inert and do not react with most reagents. It is this property that makes diethyl ether such a useful solvent. Like hydrocarbons, ethers are flammable, and diethyl ether is especially flammable because of its high volatility. Diethyl ether was the first general anesthetic used in surgery. For many years, it was the most important compound used as an anesthetic, but it has now been largely replaced by other compounds.

✔ **LEARNING CHECK 13.17**

a. Explain why an ether boils at a lower temperature than an alcohol of similar molecular weight.
b. What is the most significant chemical property of ethers?

13.9 | Thiols

Learning Objective

9. Write equations for a thiol reaction with heavy metal ions and the production of disulfides that results when thiols are oxidized.

Sulfur and oxygen belong to the same group of the periodic table, so it is not surprising that they form similar compounds. The sulfur counterparts (analogues) of alcohols contain an —SH group rather than an —OH group and are called **thiols** (an older name is mercaptans). The —SH is known as a **sulfhydryl group**.

thiol A compound containing an —SH group.

sulfhydryl group The —SH functional group.

$$R—OH \qquad R—SH$$
$$\text{an alcohol} \qquad \text{a thiol}$$

Perhaps the most distinguishing feature of thiols is their strong and disagreeable odors. Two different thiols and a disulfide are responsible for the odor associated with skunks (see **Figure 13.13**).

trans-2-butene-1-thiol 3-methyl-1-butanethiol methyl-1-(*trans*-2-butenyl)disulfide

Figure 13.13 Skunks use thiol chemistry as an effective defense mechanism. Thiols are more volatile than alcohols but less soluble in water. Why are these characteristics an advantage to skunks?

General Anesthetics

Anesthetics are compounds that induce a loss of sensation in a specific part (local anesthetic) or all (general anesthetic) of the body. A general anesthetic acts on the brain to produce unconsciousness as well as insensitivity to pain.

The word *ether* is often associated with general anesthetics because of the well-known use of ethyl ether for this purpose. Ethyl ether became generally used as an anesthetic for surgical operations in about 1850. Before that time, surgery was an agonizing procedure. The patient was strapped to a table and (if lucky) soon fainted from the pain. Divinyl ether (Vinethene) is an anesthetic that acts more rapidly and is less nauseating than ethyl ether:

$$CH_2{=}CH{-}O{-}CH{=}CH_2$$
divinyl ether

Another common ether anesthetic is enflurane:

$$H{-}\underset{H}{\overset{F}{\underset{|}{\overset{|}{C}}}}{-}O{-}\underset{H}{\overset{F}{\underset{|}{\overset{|}{C}}}}{-}\underset{Cl}{\overset{F}{\underset{|}{\overset{|}{C}}}}{-}H$$
enflurane

Numerous inhalation anesthetics are not ethers at all. Nitrous oxide, for example, is a simple inorganic compound with the formula N_2O. Also known as "laughing gas," it is used as a general anesthetic by some dentists because its effects wear off quickly. Halothane, currently a popular general anesthetic, is a simple halogen derivative of ethane. It is nonflammable, does not cause nausea or similar upsets, and its effects wear off quickly:

$$H{-}\underset{Cl}{\overset{Br}{\underset{|}{\overset{|}{C}}}}{-}\underset{F}{\overset{F}{\underset{|}{\overset{|}{C}}}}{-}F$$
halothane

In modern surgical practice, the use of a single anesthetic has become quite rare. Usually, a patient is given an injection of a strong sedative that causes unconsciousness. A general anesthetic is then administered to provide insensitivity to pain and maintain the patient in an unconscious condition. A muscle relaxant may also be used to produce complete relaxation, while minimizing the need for, and hazards associated with, deep anesthesia.

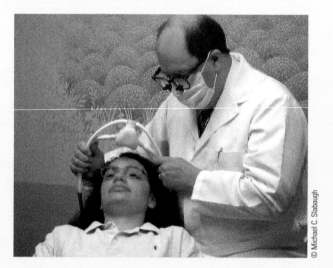
© Michael C. Slabaugh

A patient receives an anesthetic prior to dental surgery.

Some less-offensive odors (and flavors) are also characteristic of thiols and disulfides. Freshly chopped onions emit propanethiol, and 1-propene-3-thiol and 3,3-di-(1-propenyl) disulfide are partially responsible for the odor and flavor of garlic. It is amazing that the simple substitution of a sulfur atom for an oxygen atom can cause such a dramatic change in odor.

$$CH_3CH_2CH_2{-}SH \quad CH_2{=}CHCH_2 \quad CH_2{=}CHCH_2{-}S{-}S{-}CH_2CH{=}CH_2$$
$$\underset{SH}{\overset{|}{\phantom{CH_2{=}CHCH_2}}}$$

propanethiol 1-propene-3-thiol 3,3-di-(1-propenyl)disulfide

Gas companies make effective use of the odor of thiols. Natural gas (methane) has no odor, so the companies add a tiny amount of a thiol such as ethanethiol ($CH_3CH_2{-}SH$) to make it possible to detect leaking gas before a spark or match sets off an explosion.

Two reactions of the —SH group are important in the chemistry of proteins. When thiols are oxidized by agents such as O_2, a coupling reaction occurs (two molecules become one) to form a **disulfide**, a compound containing an —S—S— linkage:

disulfide A compound containing an —S—S— group.

General reaction:
$$2R-SH + (O) \rightarrow R-S-S-R + H_2O \qquad (13.18)$$
$$\text{a thiol} \qquad\qquad\quad \text{a disulfide}$$

Specific example:
$$2CH_3-SH + (O) \rightarrow CH_3-S-S-CH_3 + H_2O \qquad (13.19)$$

✔ **LEARNING CHECK 13.18** Complete the following reaction:

$$2CH_3CH-SH + (O) \rightarrow$$
$$\qquad\quad | $$
$$\qquad\quad CH_3$$

Disulfide linkages are important structural features of some proteins, especially those of hair (see **Section 19.6**). The coupling reaction can be reversed by treating disulfides with a reducing agent (H) such as H_2, which regenerates the thiol:

General reaction:
$$R-S-S-R + 2(H) \rightarrow 2R-SH \qquad (13.20)$$

Specific example:
$$CH_3-S-S-CH_3 + 2(H) \rightarrow 2CH_3-SH \qquad (13.21)$$

✔ **LEARNING CHECK 13.19** What products result from the reduction of the following disulfides?

a.
$$CH_3CH-S-S-CHCH_3 + 2(H) \rightarrow$$
$$\quad\; | \qquad\qquad |$$
$$\quad\; CH_3 \qquad\quad\; CH_3$$

b. $-S-S-$ $+ 2(H) \rightarrow$

The reaction of thiols with heavy metal ions, such as those of lead and mercury, to form insoluble compounds has adverse biological results. The metal ions (M^{2+}) react with sulfhydryl groups in the following way:

General reaction:
$$2R-SH + M^{2+} \rightarrow R-S-M-S-R + 2H^+ \qquad (13.22)$$

Specific example:
$$2CH_3-SH + Hg^{2+} \rightarrow CH_3-S-Hg-S-CH_3 + 2H^+ \qquad (13.23)$$

Many enzymes (catalysts for biological reactions— see **Section 20.7**) contain —SH groups. Exposure to heavy metal ions ties up these groups, through Reaction 13.22, rendering the enzyme ineffective. The loss of enzymes in this manner can lead to serious biological consequences.

✔ **LEARNING CHECK 13.20** Complete the following reaction:

$$2CH_3CH-SH + Pb^{2+} \rightarrow$$
$$\qquad\quad |$$
$$\qquad\quad CH_3$$

13.10 | Polyfunctional Compounds

Learning Objective

10. Identify functional groups in polyfunctional compounds.

polyfunctional compound
A compound with two or more functional groups.

So far our study of organic molecules has included several compounds with more than one functional group. Compounds of this type, such as dienes and diols, are said to be **polyfunctional compounds**. Substances with different functional groups are also referred to as polyfunctional. Compounds with multiple functional groups are very common in nature, and many play essential roles in life processes. Carbohydrate structures (see **Chapter 17**) such as glucose, typically involve a functional group at every carbon position.

$$HO-CH_2-\underset{|}{\overset{\overset{\displaystyle OH}{|}}{CH}}-\underset{|}{\overset{\overset{\displaystyle OH}{|}}{CH}}-\underset{|}{\overset{\overset{\displaystyle OH}{|}}{CH}}-\underset{|}{\overset{\overset{\displaystyle OH}{|}}{CH}}-\overset{\overset{\displaystyle O}{\|}}{C}-H$$
glucose

Even though some organic structures look quite complex, with multiple rings and long carbon chains, remember that it's the functional group or groups that determine the chemical properties of the compound. Thus, cholesterol, with two functional groups, should exhibit the reactions of both alkenes and alcohols.

cholesterol

cholesterol

Example 13.4 Identifying Functional Groups

Identify the functional groups in tetrahydrocannabinol, the active ingredient in marijuana.

tetrahydrocannabinol

Solution
To find functional groups, look for unique atoms or collections of atoms while ignoring ordinary rings, chains, and alkyl groups. Look for multiple bonds and atoms other than carbon. The tetrahydrocannabinol molecule has four functional groups: an ether, a phenolic —OH, the benzene ring, and a carbon–carbon double bond.

tetrahydrocannabinol

✔ LEARNING CHECK 13.21 Identify the functional groups in vitamin E.

vitamin E

Case Study Follow-up

Alcohol abuse is the overconsumption of ethanol, usually in the form of alcoholic beverages but can include non-food substances such as mouthwash or nighttime cold medication. Alcohol abuse is the third leading preventable cause of death in the United States. Research has shown that screening of patients by health care providers for possible alcohol abuse can promote significant changes in the behavior of heavy drinkers that can positively influence their own and public health. Nearly 40% of traffic fatalities are alcohol related. Helping even a small percentage of alcoholics make a change can have a huge impact.

Palliative care focuses on relieving and preventing the suffering of patients. This approach can be useful in prioritizing symptoms. Focusing on a patient's most distressing symptoms based on patient self-assessment and self-report is often the first priority. It is important to address the underlying causes—which might produce significant distress for the patient in the short term but will likely result in overall gains in the long run. Support for the patient in the form of alcohol and chemical abuse treatment centers is invaluable for successful recovery.

Concept Summary

The Nomenclature of Alcohols and Phenols Alcohols are aliphatic compounds that contain the hydroxy functional group (—OH). Aromatic compounds with an —OH group attached to the ring are called phenols. Several alcohols are well-known by common names. In the IUPAC system, the characteristic ending -ol is used to designate alcohols. Phenols are named as derivatives of the parent compound phenol.
Objective 1 (Section 13.1), Exercises 13.4 and 13.10

Classification of Alcohols Alcohols are classified on the basis of the number of carbon atoms bonded to the carbon that is attached to the —OH group. In primary alcohols, the OH-bonded carbon is bonded to one other carbon. In secondary alcohols, the OH-bonded carbon is bonded to two other carbons. In tertiary alcohols, the OH-bonded carbon is bonded to three other carbons.
Objective 2 (Section 13.2), Exercise 13.14

Physical Properties of Alcohols Hydrogen bonding can occur between alcohol molecules and between alcohol molecules and water. As a result, alcohols have higher boiling points than hydrocarbons of similar molecular weight, and low molecular weight alcohols are soluble in water.
Objective 3 (Section 13.3), Exercise 13.18

Reactions of Alcohols Two important reactions of alcohols are dehydration and oxidation. Alcohols may be dehydrated in two different ways depending on the reaction temperature. With H_2SO_4 at 180°C, an alkene is produced, whereas at 140°C an ether is the product. The oxidation products obtained from the alcohols depend on the class of alcohol being oxidized. A primary alcohol produces an aldehyde that is further oxidized to a carboxylic acid. A secondary alcohol produces a ketone. A tertiary alcohol does not react with oxidizing agents.
Objective 4 (Section 13.4), Exercises 13.22 and 13.26

Important Alcohols Some simple alcohols with commercial value are methanol (a solvent and fuel), ethanol (present in alcoholic beverages and gasohol), isopropyl alcohol (rubbing alcohol), ethylene glycol (antifreeze), and glycerol (a moistening agent).
Objective 5 (Section 13.5), Exercise 13.36

Characteristics and Uses of Phenols Phenols are weak acids that are widely used for their antiseptic and disinfectant properties. Some phenols are used as antioxidants in foods and a variety of other materials.
Objective 6 (Section 13.6), Exercise 13.38

Ethers Ethers contain an oxygen attached to two carbons as the characteristic functional group. In the IUPAC system, ethers are named as alkoxy derivatives of alkanes. The —O—R group is the alkoxy group that may have names such as methoxy (—O—CH₃) or ethoxy (—O—CH₂CH₃).

Objective 7 (Section 13.7), Exercise 13.42

Properties of Ethers Like the alkanes, ethers are very unreactive substances. Pure ethers cannot form hydrogen bonds. As a result, their boiling points are comparable to those of hydrocarbons and lower than those of alcohols of similar molecular weight. Ethers are much less polar than alcohols but are still slightly soluble in water due to the limited formation of hydrogen bonds.

Objective 8 (Section 13.8), Exercise 13.46

Thiols Thiols contain an —SH group, which often imparts a strong, disagreeable odor to the compound. Two reactions of thiols that are important in protein chemistry are their oxidation to produce a disulfide and their reaction with heavy metals such as mercury. Disulfides may be converted back to thiols by a reducing agent.

Objective 9 (Section 13.9), Exercise 13.52

Polyfunctional Compounds It is common for organic compounds to contain more than one functional group. Many of the compounds that serve essential roles in life processes are polyfunctional.

Objective 10 (Section 13.10), Exercise 13.56

Key Terms and Concepts

Alcohol (Introduction)
Alkoxy group (13.7)
Antioxidant (13.6)
Dehydration reaction (13.4)
Disulfide (13.9)
Elimination reaction (13.4)

Ether (Introduction)
Fermentation (13.5)
Heterocyclic ring (13.7)
Hydroxy group (Introduction)
Phenol (Introduction)
Polyfunctional compound (13.10)

Primary alcohol (13.2)
Secondary alcohol (13.2)
Sulfhydryl group (13.9)
Tertiary alcohol (13.2)
Thiol (13.9)

Key Reactions

1.	Dehydration of alcohols to give alkenes (Section 13.4):	Reaction 13.1
2.	Dehydration of alcohols to give ethers (Section 13.4): R—O—H + H—O—R →(H₂SO₄, 140°C) R—O—R + H₂O	Reaction 13.5
3.	Oxidation of a primary alcohol to give an aldehyde and then a carboxylic acid (Section 13.4):	Reaction 13.8

4.	Oxidation of a secondary alcohol to give a ketone (Section 13.4):	$\underset{\substack{\text{secondary}\\\text{alcohol}}}{R - \overset{\displaystyle OH}{\underset{\displaystyle H}{C}} - R' + (O)} \longrightarrow \underset{\substack{\text{ketone}}}{R - \overset{\displaystyle O}{C} - R' + H_2O}$	Reaction 13.10
5.	Attempted oxidation of a tertiary alcohol gives no reaction (Section 13.4):	$R - \overset{\displaystyle OH}{\underset{\displaystyle R''}{C}} - R' + (O) \longrightarrow \text{ no reaction}$	Reaction 13.12
6.	Oxidation of a thiol to give a disulfide (Section 13.9):	$2R - SH + (O) \longrightarrow \underset{\substack{\text{a disulfide}}}{R - S - S - R} + H_2O$ $\underset{\text{a thiol}}{}$	Reaction 13.18
7.	Reduction of a disulfide to give thiols (Section 13.9):	$R{-}S{-}S{-}R + 2(H) \rightarrow 2R{-}SH$	Reaction 13.20
8.	Reaction of thiols with heavy metals (Section 13.9):	$2R{-}SH + M^{2+} \rightarrow R{-}S{-}M{-}S{-}R + 2H^+$	Reaction 13.22

Exercises

Even-numbered exercises are answered in Appendix B.

Blue-numbered exercises are more challenging.

The Nomenclature of Alcohols and Phenols (Section 13.1)

13.1 Draw general formulas for an alcohol and phenol, showing the functional group.

13.2 Draw a general formula for an ether, emphasizing the functional group.

13.3 Assign IUPAC names to the following alcohols:

a. $CH_3CH_2CH_2 - OH$

b. with $CH_2CH_2CH_2 - OH$

c. $CH_3\overset{\displaystyle CH_3}{\underset{\displaystyle}{CH}}CHCH_2 - OH$ with Cl below second carbon

d. $\overset{\displaystyle OH}{CH_2CH_2CH_2CH_2 - OH}$

e. cyclopentane with Cl, OH, and CH$_2$CH$_3$ substituents

f. cyclohexane with OH, OH, and OH substituents

13.4 Assign IUPAC names to the following alcohols:

a. $CH_3\overset{\displaystyle OH}{\underset{\displaystyle}{CH}}CH_3$

b. $CH_3CH_2\overset{\displaystyle CH_2CH_3}{\underset{\displaystyle OH}{C}}CH_3$

c. $\overset{\displaystyle OH}{CH_2}CH_2\overset{\displaystyle CH_2CH_3}{CH}CH_2CH_2 - Br$

d.

CH$_3$CHCH$_3$

(cyclopropane with OH and CH$_3$ substituents)

e.

(cyclopentane with OH, CH$_3$, CH$_3$ substituents)

f.

OH

OH

(cyclopentane with two OH substituents)

g. CH$_3$CH$_2$CHCH$_2$CHCH$_2$—OH
 | |
 OH OH

13.5 Several important alcohols are well known by common names. Give a common name for each of the following:

a. CH$_3$—OH

b. CH$_3$CHCH$_3$
 |
 OH

c. CH$_3$CH$_2$—OH

d. CH$_2$—CH$_2$
 | |
 OH OH

e. CH$_2$—CH—CH$_2$
 | | |
 OH OH OH

13.6 Give each of the structures in Exercise 13.5 an IUPAC name.

13.7 Draw structural formulas for each of the following:

a. 2-methyl-1-butanol

b. 2-bromo-3-methyl-3-pentanol

c. 1-methylcyclopentanol

13.8 Draw structural formulas for each of the following:

a. 2-methyl-2-pentanol

b. 1,3-butanediol

c. 1-ethylcyclopentanol

13.9 Name each of the following as a derivative of phenol:

a.

CH$_3$
|
CHCH$_3$

OH

(benzene ring with isopropyl group and OH)

b.

OH

CH$_3$

CH$_3$

(benzene ring with OH and two CH$_3$ groups)

13.10 Name each of the following as a derivative of phenol:

a.

CH$_2$CH$_3$

OH

(benzene ring with ethyl and OH groups)

b.

OH

H$_3$C CH$_2$CH$_3$

CH$_3$

(benzene ring with methyl, ethyl, OH, and methyl groups)

13.11 Draw structural formulas for each of the following:

a. *m*-methylphenol b. 2,3-dichlorophenol

13.12 Draw structural formulas for each of the following:

a. *p*-chlorophenol b. 2,5-diisopropylphenol

Classification of Alcohols (Section 13.2)

13.13 What is the difference between a primary, secondary, and tertiary alcohol?

13.14 Classify the following alcohols as primary, secondary, or tertiary:

a.

CH$_3$
|
CH$_3$CH$_2$C — OH
|
CH$_3$

b. CH$_3$CH$_2$CH$_2$CH$_2$ — OH

c.

CH$_3$CH$_2$CH — OH
 |
 CH$_3$

13.15 Classify the following alcohols as primary, secondary, or tertiary:

a.

CH$_3$
|
CH$_3$CH$_2$C—OH
|
CH$_3$

b.

CH$_3$CH$_2$CH$_2$CH$_2$—OH

c.

OH

(cyclohexane with OH)

13.16 Draw structural formulas for the four aliphatic alcohols with the molecular formula $C_4H_{10}O$. Name each compound using the IUPAC system and classify it as a primary, secondary, or tertiary alcohol.

Physical Properties of Alcohols (Section 13.3)

13.17 Why are the boiling points of alcohols much higher than the boiling points of alkanes with similar molecular weights?

13.18 Arrange the compounds of each group in order of increasing boiling point.

a. ethanol, 1-propanol, methanol

b. butane, ethylene glycol, 1-propanol

13.19 Which member of each of the following pairs would you expect to be more soluble in water? Briefly explain your choices.

 a. butane or 2-butanol

 b. 2-propanol or 2-pentanol

 c. 2-butanol or 2,3-butanediol

13.20 Draw structural formulas for the following molecules and use a dotted line to show the formation of hydrogen bonds:

 a. one molecule of 1-butanol and one molecule of ethanol

 b. cyclohexanol and water

13.21 Explain why the use of glycerol (1,2,3-propanetriol) in lotions helps retain water and keep the skin moist.

Reactions of Alcohols (Section 13.4)

13.22 Draw the structures of the chief product formed when the following alcohols are dehydrated to alkenes:

 a. CH₃CHCH₂CH₃
 |
 OH

 b. CH₃CHCHCH₃
 |
 CH₃ (above), OH (below)

13.23 Draw the structures of the chief product formed when the following alcohols are dehydrated to alkenes:

 a. cyclohexane ring with OH and CH₃

 b. CH₃CCH₂CHCH₃ with CH₃, OH, CH₃ substituents

13.24 Draw the structures of the ethers that can be produced from the following alcohols:

 a. CH₃CH₂CH₂—OH

 b. phenol (benzene ring with OH)

 c. benzene ring with CH₂CH₂—OH

13.25 Draw the structures of the two organic compounds that can be obtained by oxidizing

 CH₃CHCH₂—OH
 |
 CH₃

13.26 Give the structure of an alcohol that could be used to prepare each of the following compounds:

 a. cyclopentanone

 b. CH₃CH₂CCH—CH₃ with C=O and CH₃ substituent

 c. CH₃CH₂CH₂—C—OH with C=O

13.27 Give the structure of an alcohol that could be used to prepare each of the following compounds:

 a. cyclopentanone with CH₃ substituent

 b. benzene ring with CH₂C—H and C=O

 c. CH₃CH₂—C—OH with C=O

13.28 What products would result from the following processes? Write an equation for each reaction.

 a. 2-Methyl-2-butanol is subjected to controlled oxidation.

 b. 1-Propanol is heated to 140°C in the presence of sulfuric acid.

 c. 3-Pentanol is subjected to controlled oxidation.

 d. 3-Pentanol is heated to 180°C in the presence of sulfuric acid.

 e. 1-Hexanol is subjected to an *excess* of oxidizing agent.

13.29 What products would result from the following processes? Write an equation for each reaction.

 a. 1-Butanol is heated to 140°C in the presence of sulfuric acid.

 b. 1-Butanol is subjected to an *excess* of oxidizing agent.

 c. 2-Pentanol is subjected to controlled oxidation.

 d. 2-Pentanol is heated to 180°C in the presence of sulfuric acid.

 e. 2-Methyl-2-pentanol is subjected to controlled oxidation.

13.30 Each of the following conversions requires more than one step, and some reactions studied in previous chapters may be needed. Show the reagents you would use and draw structural formulas for intermediate compounds formed in each conversion.

 a. CH₃CH₂CH=CH₂ → CH₃CH₂CCH₃ with C=O

b.

c. $CH_3CH_2CH_2CH_2 \overset{\overset{\displaystyle OH}{|}}{} \rightarrow CH_3CH_2\overset{\overset{\displaystyle O}{||}}{C}CH_3$

13.31 Each of the following conversions requires more than one step, and some reactions studied in previous chapters may be needed. Show the reagents you would use and draw structural formulas for intermediate compounds formed in each conversion.

a. $CH_2 = CH_2 \rightarrow CH_3CH_2-O-CH_2CH_3$

b.

c. $CH_3CH_2CH_2 \overset{\overset{\displaystyle OH}{|}}{} \rightarrow CH_3\overset{\overset{\displaystyle OH}{|}}{C}HCH_3$

d. $CH_3CH_2CH_2CH_2 \overset{\overset{\displaystyle OH}{|}}{} \rightarrow CH_3CH_2\overset{\overset{\displaystyle Cl}{|}}{C}HCH_3$

13.32 The three-carbon diol used in antifreeze is

$$CH_3 - \overset{\overset{\displaystyle OH}{|}}{C}H - \overset{\overset{\displaystyle OH}{|}}{C}H_2$$

It is nontoxic and is used as a moisturizing agent in foods. Oxidation of this substance within the liver produces pyruvic acid, which can be used by the body to supply energy. Give the structure of pyruvic acid.

Important Alcohols (Section 13.5)

13.33 Methanol is fairly volatile and evaporates quickly if spilled. Methanol is also absorbed quite readily through the skin. If, in the laboratory, methanol accidentally spilled on your clothing, why would it be a serious mistake to just let it evaporate?

13.34 Why is methanol such an important industrial chemical?

13.35 Suppose you are making some chocolate cordials (chocolate-coated candies with soft fruit filling). Why might you want to add a little glycerol to the filling?

13.36 Name an alcohol used in each of the following ways:

a. A moistening agent in many cosmetics

b. The solvent in solutions called tinctures

c. Automobile antifreeze

d. Rubbing alcohol

e. A flavoring in cough drops

f. Present in gasohol

Characteristics and Uses of Phenols (Section 13.6)

13.37 Name a phenol associated with each of the following:

a. A vitamin

b. A disinfectant cleaner

13.38 Name a phenol used in each of the following ways:

a. A disinfectant used for cleaning walls

b. An antiseptic found in some mouthwashes

c. An antioxidant used to prevent rancidity in foods

Ethers (Section 13.7)

13.39 Assign a common name to each of the following ethers:

a. $CH_3 - O - CH_2CH_3$

b.

c. $CH_3CH_2\overset{\overset{\displaystyle CH_3}{|}}{C}H - O - \overset{\overset{\displaystyle CH_3}{|}}{C}HCH_2CH_3$

13.40 Assign a common name to each of the following ethers:

a. $CH_3CH_2 - O - \overset{\underset{\displaystyle CH_3}{|}}{C}HCH_3$

c.

b. $CH_3 - O - CH_2CH_2CH_2CH_3$

13.41 Assign the IUPAC name to each of the following ethers. Name the smaller alkyl group as the alkoxy group.

a. $CH_3 - O - CH_2CH_3$

b.

c.

$CH_3\overset{\underset{\displaystyle CH_3}{|}}{C}H - O-$

d. $CH_3CH_2CH_2 - O - \overset{\overset{\displaystyle CH_3}{|}}{C}HCH_2CH_2CH_2CH_3$

13.42 Assign the IUPAC name to each of the following ethers. Name the smaller alkyl group as the alkoxy group.

a. $CH_3CH_2 - O - CH_2CH_2CH_3$

b. $CH_3CH_2 - O - \overset{\underset{\displaystyle CH_3}{|}}{C}HCH_3$

c. $CH_3CH_2 - O -$

d.

13.43 Draw structural formulas for the following:

a. ethyl methyl ether

b. butyl phenyl ether

c. 3-ethoxypentane

d. dipropyl ether

e. 1,3-dipropoxycyclohexane

13.44 Draw structural formulas for the following:

 a. methyl isopropyl ether

 b. phenyl propyl ether

 c. 2-methoxypentane

 d. 1,2-diethoxycyclopentane

 e. 2-phenoxy-2-butene

13.45 Give the IUPAC names and draw structural formulas for the six isomeric ethers of the molecular formula $C_5H_{12}O$.

Properties of Ethers (Section 13.8)

13.46 What is the chief chemical property of ethers?

13.47 Why is diethyl ether hazardous to use as an anesthetic or as a solvent in the laboratory?

13.48 Arrange the following compounds in order of decreasing solubility in water. Explain the basis for your decisions.

$CH_3CH_2CH_2$—OH $CH_3CH_2CH_2CH_3$ CH_3CH_2—O—CH_3

13.49 Arrange the compounds in Exercise 13.48 in order of decreasing boiling point. Explain your answer.

13.50 Draw structural formulas and use a dotted line to show hydrogen bonding between a molecule of diethyl ether and water.

Thiols (Section 13.9)

13.51 Complete the following reactions:

 a. $2CH_3CH_2 + Hg^{2+} \rightarrow$
 $|$
 SH

 b. $+ (O) \longrightarrow$

 c. CH_3—S—S—$CH_2CH_2CH_3 + 2(H) \rightarrow$

13.52 Complete the following reactions:

 a. $2CH_3CH_2CH_2$—SH + $Hg^{2+} \rightarrow$

 b. $2CH_3CH_2CH_2$—SH + $(O) \rightarrow$

 c. $+ 2(H) \longrightarrow$

13.53 Lipoic acid is required by many microorganisms for proper growth. As a disulfide, it functions in the living system by catalyzing certain oxidation reactions and is reduced in the process. Write the structure of the reduction product.

lipoic acid

13.54 Alcohols and thiols can both be oxidized in a controlled way. What are the differences in the products?

Polyfunctional Compounds (Section 13.10)

13.55 Identify the functional groups in vanillin, the active ingredient in vanilla flavoring.

13.56 Identify the functional groups in cinnamaldehyde, present in cinnamon flavoring.

Additional Exercises

13.57 alkene + water $\overset{H_2SO_4}{\rightleftharpoons}$ alcohol + heat

 In simplistic terms, it can be thought that an equilibrium exists in the hydration-dehydration reactions above. Use LeChâtelier's principle to explain why the dehydration reaction is favored at 180°C.

13.58 Thiols have lower boiling points and are less water soluble than the corresponding alcohol (e.g., CH_3CH_2—OH and CH_3CH_2—SH). Explain why.

13.59 Determine the oxidation numbers on the carbon atom in the following reaction.

13.60 Using the idea of aromaticity (the electronic structure of benzene), explain why phenol is a stronger acid than cyclohexanol.

13.61 Use general reactions to show how an alcohol (R—O—H) can behave as a weak Brønsted acid when it reacts with a strong Brønsted base (X^-). Next, show how an alcohol can behave as a weak Brønsted base in the presence of a strong Brønsted acid (HA).

Chemistry for Thought

13.62 Many aftershaves are 50% ethanol. What do you think is the purpose of the ethanol?

13.63 A mixture of ethanol and 1-propanol is heated to 140°C in the presence of sulfuric acid. Careful analysis reveals that three ethers are formed. Write formulas for these three products, name each, and explain why three form, rather than a single product.

13.64 The two-carbon alcohol in some antifreeze

is toxic if ingested. Reactions within the liver oxidize the diol to oxalic acid, which can form insoluble salts and damage the kidneys. Give the structure of oxalic acid.

13.65 Although ethanol is only mildly toxic to adults, it poses greater risks for fetuses. Why do you think this is the case?

13.66 The medicinal compound taxol is mentioned in Figure 13.1. What does the name indicate about its possible structure?

13.67 When diethyl ether is spilled on the skin, the skin takes on a dry appearance. Explain whether this effect is more likely due to a removal of water or a removal of natural skin oils.

13.68 Figure 13.8 points out that methanol is used as a fuel in racing cars. Write a balanced equation for the combustion of methanol.

13.69 Figure 13.13 focuses on the use of thiol chemistry by skunks. If thiols are more volatile than alcohols and less soluble in water, why do these properties represent an advantage to skunks?

13.70 Why do you think a dye is added to antifreeze before it is sold in stores?

13.71 Glycerol, or glycerin, is used as a moisturizer in foods such as candy and shredded coconut. What is the molecular explanation for its utility as a moisturizer?

Allied Health Exam Connection

The following questions are from these sources:

- *Nursing School Entrance Exam* © 2005, Learning Express, LLC.

- *McGraw-Hill's Nursing School Entrance Exams* by Thomas A. Evangelist, Tamara B. Orr, and Judy Unrein © 2009, The McGraw-Hill Companies, Inc.

- *NSEE Nursing School Entrance Exams,* 3rd edition © 2009, Kaplan Publishing.

- *Cliffs Test Prep: Nursing School Entrance Exams* by Fred N. Grayson © 2004, Wiley Publishing, Inc.

- *Peterson's Master the Nursing School and Allied Health Entrance Exams,* 18th edition by Marion F. Gooding © 2008, Peterson's, a Nelnet Company.

13.72 Which compound below contains the alcohol functional group?

 a. $C_2H_5OCH_3$

 b. C_3H_4

 c. CH_3CH_2CHO

 d. C_3H_7OH

13.73 The compound that has the greatest polarity is:

 a. $CH_3—CH_2—O—CH_2—CH_3$

 b. $CH_3—CH_2CH_2CH_2—CH_3$

 c. $CH_3—CH_2—CH_2—CH_2—CH_2—Cl$

 d. $CH_3—CH_2—CH_2—CH_2—CH_2—OH$

13.74 Alcoholic beverages contain:

 a. wood alcohol.

 b. isopropyl alcohol.

 c. glyceryl alcohol.

 d. ethyl alcohol.

13.75 Rubbing alcohol is:

 a. methyl alcohol.

 b. ethyl alcohol.

 c. phenol.

 d. isopropyl alcohol.

13.76 Which of the following compounds is an ether?

 a. CH_2CHO

 b. $CH_3—O—CH_3$

 c. $CH_3—COOH$

 d. $CH_3—CH_2OH$

14

Aldehydes and Ketones

Creations/Shutterstock.com

▶ Case Study

Chris was a go-getter, always giving 110%. However, when the economy tanked, he lost his engineering job. Through a temporary service, he secured two jobs that together paid the bills. Unfortunately, one of the jobs was a graveyard shift and the other was a day shift on weekends. To say the least, this created havoc with his sleep schedule. Chris coped with his lack of energy by eating high-calorie junk food, and in just a few months he had gained 20 lbs. Disgusted with himself, he changed his diet; he almost completely cut out carbohydrates and consumed mainly meat products. His wife was annoyed by his overboard approach to weight loss and complained that he had "bad breath" since he started dieting.

Chris explained that was a symptom of ketosis—a sign that his body was burning fat. Chris was happy with his progress until he found himself in the emergency room. Kidney stones had to be the worst pain he had ever experienced. Chris was told the stones were a result of ketosis.

▶ **How can Chris be encouraged to manage his health more gently and effectively? How do excess ketones affect the kidneys? What is a reasonable approach to weight loss?**

Follow-up to this Case Study appears at the end of the chapter before the *Concept Summary*.

Learning Objectives

When you have completed your study of this chapter, you should be able to:

1 Recognize the carbonyl group in compounds and classify the compounds as aldehydes or ketones. **(Section 14.1)**

2 Assign IUPAC names to aldehydes and ketones. **(Section 14.1)**

3 Compare the physical properties of aldehydes and ketones with those of compounds in other classes. **(Section 14.2)**

4 Write key reactions for aldehydes and ketones. **(Section 14.3)**

5 Give specific uses for aldehydes and ketones. **(Section 14.4)**

The carbonyl functional group is characteristic of both aldehydes and ketones. In a **carbonyl group**, a carbon is double bonded to an oxygen atom and single bonded to two other atoms:

$$
\begin{array}{c} O \\ \parallel \\ -C- \end{array}
$$

carbonyl group

carbonyl group

$$
\begin{array}{c} O \\ \parallel \\ \text{The } -C- \text{ group.} \end{array}
$$

Aldehydes and ketones occur widely in nature and play important roles in living organisms. For example, the carbonyl group is found in numerous carbohydrates, including glucose and fructose. Glucose, a major source of energy in living systems, is found combined with fructose in cane sugar:

glucose fructose

In this chapter, we will study the naming of aldehydes and ketones, as well as some of their important reactions. These concepts will be very useful when we study the chemistry of carbohydrates in Chapter 17.

14.1 | The Nomenclature of Aldehydes and Ketones

Learning Objectives

1. Recognize the carbonyl group in compounds and classify the compounds as aldehydes or ketones.
2. Assign IUPAC names to aldehydes and ketones.

When a carbonyl group is directly bonded to at least one hydrogen atom, the compound is an **aldehyde**. In **ketones**, two carbon atoms are directly bonded to the carbonyl group:

aldehyde

A compound that contains the

$$-\overset{\overset{\displaystyle O}{\|}}{C}-H \text{ group; the general}$$

formula is $R-\overset{\overset{\displaystyle O}{\|}}{C}-H.$

ketone

A compound that contains the

$$-\overset{|}{\underset{|}{C}}-\overset{\overset{\displaystyle O}{\|}}{C}-\overset{|}{\underset{|}{C}}- \text{ group;}$$

the general formula is

$$R-\overset{\overset{\displaystyle O}{\|}}{C}-R'.$$

$$-\overset{\overset{\displaystyle O}{\|}}{C}- \qquad -\overset{\overset{\displaystyle O}{\|}}{C}-H \qquad -\overset{|}{\underset{|}{C}}-\overset{\overset{\displaystyle O}{\|}}{C}-\overset{|}{\underset{|}{C}}-$$

carbonyl group aldehyde ketone

Structural models of the simplest aldehyde (formaldehyde) and the simplest ketone (acetone) are shown in **Figure 14.1**.

Figure 14.1 Ball-and-stick models of formaldehyde and acetone.

Several aldehydes and ketones, in addition to formaldehyde and acetone, are well known by common names (see **Table 14.1**). Notice that aldehyde and ketone common names are easily recognized by the endings *-aldehyde* and *-one*, respectively.

In the IUPAC system, aldehydes are named by selecting the longest chain that includes the carbonyl carbon. The *-e* of the alkane that has the same number of carbons is replaced by *-al*. Because the carbonyl carbon in aldehydes is always at the end of the chain, it is always designated as carbon number 1.

TABLE 14.1 Some Common Aldehydes and Ketones

Structural Formula	IUPAC Name	Common Name	Boiling Point (°C)
Aldehydes			
$H-\overset{\overset{\displaystyle O}{\|\|}}{C}-H$	methanal	formaldehyde	−21
$CH_3-\overset{\overset{\displaystyle O}{\|\|}}{C}-H$	ethanal	acetaldehyde	21
$CH_3CH_2-\overset{\overset{\displaystyle O}{\|\|}}{C}-H$	propanal	propionaldehyde	49
$CH_3CH_2CH_2-\overset{\overset{\displaystyle O}{\|\|}}{C}-H$	butanal	butyraldehyde	76
$\bigcirc\!\!-\overset{\overset{\displaystyle O}{\|\|}}{C}-H$	benzaldehyde	benzaldehyde	178
Ketones			
$CH_3-\overset{\overset{\displaystyle O}{\|\|}}{C}-CH_3$	propanone	acetone	56
$CH_3-\overset{\overset{\displaystyle O}{\|\|}}{C}-CH_2CH_3$	butanone	methyl ethyl ketone (MEK)	80
$\bigcirc\!\!=\!\!O$	cyclohexanone	cyclohexanone	156

Example 14.1 Naming Aldehydes

Name the following aldehydes according to the IUPAC system.

a. $CH_3CH_2CH_2-\overset{\overset{\displaystyle O}{\|\|}}{C}-H$

b. $CH_3\overset{\overset{\displaystyle Br}{\|}}{C}HCH_2-\overset{\overset{\displaystyle O}{\|\|}}{C}-H$

c. $CH_3CH_2\overset{\overset{\displaystyle \overset{\displaystyle O}{\|\|}}{\overset{\displaystyle C}{\|}}-H}{C}HCH_2CH_3$

Solution

a. This aldehyde contains four carbon atoms, and the name is butanal. The aldehyde group is always position number 1, so the number designating the aldehyde position is omitted from the name.

b. Carbon chain numbering begins with the carbonyl end of the molecule even though a lower number for the Br could be obtained by numbering from the left:

$$\underset{4}{CH_3}\underset{3}{CHCH_2}\underset{2}{\overset{\overset{\displaystyle Br}{|}}{}}\underset{1}{\overset{\overset{\displaystyle O}{\|}}{C}}-H$$

The compound is 3-bromobutanal.

c. The carbonyl group must be position number 1 in the chain even though a longer carbon chain (pentane) exists:

$$\overset{\overset{\displaystyle O}{\|}}{\underset{1}{C}}-H$$
$$\underset{4}{CH_3}\underset{3}{CH_2}\underset{2}{CH}CH_2CH_3$$

The name is 2-ethylbutanal.

✔ **LEARNING CHECK 14.1** Give IUPAC names to the following aldehydes:

a. $CH_3CH_2CH_2CH_2-\overset{\overset{\displaystyle O}{\|}}{C}-H$

b. $CH_3\overset{\overset{\displaystyle CH_3}{|}}{CH}CHCH_2-\overset{\overset{\displaystyle O}{\|}}{\underset{|}{C}}-H$ with CH_3 below

In naming ketones by the IUPAC system, the longest chain that includes the carbonyl group is given the name of the corresponding alkane, with the *-e* replaced by *-one.* The chain is numbered from the end closest to the carbonyl group, and the carbonyl group position is indicated by a number.

Example 14.2 Naming Ketones

Give IUPAC names to the following ketones:

a. $CH_3-\overset{\overset{\displaystyle O}{\|}}{C}-CH_3$

b. $CH_3-\overset{\overset{\displaystyle O}{\|}}{C}-\overset{\overset{\displaystyle }{|}}{CH}CH_2CH_3$ with CH_3 below

c. (cyclohexanone ring with $=O$ and CH_3 substituent)

Solution

a. There are three carbons in the chain, and the correct name is propanone. The number 2 is not used in this case because the carbonyl group in a ketone cannot be on the end of a chain. Therefore, it must be in the number 2 position.

b. The chain is numbered from the left to give a lower number for the carbonyl carbon:

$$\underset{1}{CH_3}-\underset{2}{\overset{\overset{\displaystyle O}{\|}}{C}}-\underset{3}{CH}\underset{4}{CH_2}\underset{5}{CH_3}$$
with CH_3 below

The complete name is 3-methyl-2-pentanone.

c. In cyclic ketones, the carbonyl carbon will always be at position 1, and so the number is omitted from the name:

The correct name is 2-methylcyclohexanone.

✔ **LEARNING CHECK 14.2** Give IUPAC names to the following ketones:

a.
$$
\underset{\substack{| \\ \text{Br}}}{\text{CH}_3\text{CH}}\text{CH}_2\underset{\substack{| \\ \text{Br}}}{\text{CH}}\text{CH}_2 - \underset{\substack{\| \\ \text{O}}}{\text{C}} - \text{CH}_3
$$

b.

14.2 | Physical Properties

Learning Objective

3. Compare the physical properties of aldehydes and ketones with those of compounds in other classes.

The physical properties of aldehydes and ketones can be explained by an examination of their structures. First, the lack of a hydrogen on the oxygen prevents the formation of hydrogen bonds between molecules:

no H attached to the oxygen

$$
\underset{\substack{\| \\ \text{R} - \text{C} - \text{H}}}{\text{O}} \qquad \underset{\substack{\| \\ \text{R} - \text{C} - \text{R}'}}{\text{O}}
$$

Therefore, boiling points of pure aldehydes and ketones are expected to be lower than those of alcohols with similar molecular weights. Remember, alcohols can form hydrogen bonds with one another. **Table 14.2** shows that the boiling points of propanal and acetone are 49°C and 56°C, respectively, whereas the alcohol of comparable molecular weight, 1-propanol, has a boiling point of 97°C.

TABLE 14.2 A Comparison of Physical Properties

Class	Example	Formula	Molecular weight	Boiling point (°C)	Water solubility
Alkane	butane	$CH_3CH_2CH_2CH_3$	58	0	Insoluble
Aldehyde	propanal	$CH_3CH_2 - \overset{\text{O}}{\underset{\|}{C}} - H$	58	49	Soluble
Ketone	propanone (acetone)	$CH_3 - \overset{\text{O}}{\underset{\|}{C}} - CH_3$	58	56	Soluble
Alcohol	1-propanol	$CH_3CH_2CH_2 - OH$	60	97	Soluble

Faking a Tan

Many people believe that a suntan makes one look healthy and attractive. Studies, however, indicate that this perception is far from the truth. According to these studies, sunbathing, especially when sunburn results, ages the skin prematurely and increases the risk of skin cancer. Cosmetic companies have developed a tanning alternative for those not willing to risk using the sun but who want to be "fashionably" tan.

Tanning lotions and creams that chemically darken the skin are now available. The active ingredient in these "bronzers" is dihydroxyacetone (DHA), a colorless compound classified by the Food and Drug Administration as a safe skin dye.

$$HO-CH_2-\overset{\overset{\displaystyle O}{\|}}{C}-CH_2-OH$$
dihydroxyacetone (DHA)

Within several hours of application, DHA produces a brown skin color by reacting with the outer layer of the skin, which consists of dead cells. Only the dead cells react with DHA, so the color gradually fades as the dead cells slough off and are replaced. This process generally leads to the fading of chemical tans within a few weeks. Another problem with chemical tans is uneven skin color. Areas of skin such as elbows and knees, which contain a thicker layer of dead cells, may absorb and react with more tanning lotion and become darker than other areas.

Perhaps the greatest problem with chemical tans is the false sense of security they might give. Some people with chemical tans think it is safe to go into the sun and get a deeper tan. This isn't true. Sunlight presents the same hazards to chemically tanned skin that it does to untanned skin.

Some DHA-containing products.

Also notice in Table 14.2 that the boiling points of the aldehyde and ketone are higher than that of the comparable alkane. This can be explained by differences in polarity. Whereas alkanes are nonpolar, the carbonyl group of aldehydes and ketones is polar due to the more electronegative oxygen atom:

$$-\overset{\overset{\displaystyle O^{\delta-}}{\|}}{\underset{\delta+}{C}}-$$

The attractive forces between molecules possessing such dipoles are not as strong as hydrogen bonds, but they do cause aldehydes and ketones to boil at higher temperatures than the nonpolar alkanes. The graph in **Figure 14.2** summarizes the boiling point comparison for several families of compounds.

Water is a polar solvent and would be expected to be effective in dissolving compounds containing polar carbonyl groups. Besides exhibiting dipolar attractions, the carbonyl oxygen also forms hydrogen bonds with water:

$$
\begin{array}{ccc}
& \text{O} & & & & \text{O} \\
\text{H} & & \text{H} & & \text{H} & & \text{H} \\
\\
\text{O} & & & & \text{O} \\
\| & & & & \| \\
\text{R}-\text{C}-\text{H} & & & & \text{R}-\text{C}-\text{R}' \\
& \textit{hydrogen bond} & & & \textit{hydrogen bond}
\end{array}
$$

As a result, low molecular weight aldehydes and ketones are water soluble (see **Table 14.2**).

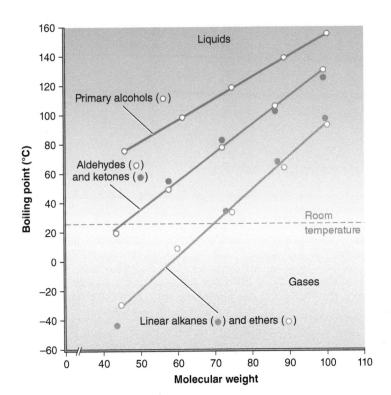

Figure 14.2 The boiling points of aldehydes and ketones are higher than those of alkanes and ethers but lower than those of alcohols of comparable molecular weights.

14.3 | Chemical Properties

Learning Objective

4. Write key reactions for aldehydes and ketones.

14.3A Oxidation

We learned in Section 13.4 that when aldehydes are prepared by oxidizing primary alcohols with $KMnO_4$ or $K_2Cr_2O_7$, the reaction may continue and produce carboxylic acids. Ketones are prepared similarly, but they are much less susceptible to further oxidation, especially by mild oxidizing agents:

$$\textbf{General reaction: } R-CH_2-OH \xrightarrow{(O)} \underset{\text{aldehyde}}{R-\overset{\displaystyle O}{\overset{\|}{C}}-H} \xrightarrow{(O)} \underset{\substack{\text{carboxylic}\\\text{acid}}}{R-\overset{\displaystyle O}{\overset{\|}{C}}-OH} \quad (14.1)$$

$$\underset{\substack{\text{primary}\\\text{alcohol}}}{}$$

$$\textbf{General reaction: } \underset{\substack{\text{secondary}\\\text{alcohol}}}{R-\overset{\displaystyle OH}{\overset{|}{C}H}-R'} \xrightarrow{(O)} \underset{\text{ketone}}{R-\overset{\displaystyle O}{\overset{\|}{C}}-R'} \xrightarrow{(O)} \text{no reaction} \quad (14.2)$$

The difference in reactivity toward oxidation is the chief reason why aldehydes and ketones are classified in separate families (see **Figure 14.3**).

Specific example:

$$\underset{\text{benzaldehyde}}{\text{C}_6\text{H}_5-\overset{\displaystyle O}{\overset{\|}{C}}-H} + (O) \longrightarrow \underset{\text{benzoic acid}}{\text{C}_6\text{H}_5-\overset{\displaystyle O}{\overset{\|}{C}}-OH} \quad (14.3)$$

① From left to right, three test tubes containing potassium dichromate ($K_2Cr_2O_7$), acetone, and benzaldehyde.

② After the addition of equal amounts of $K_2Cr_2O_7$ from the first test tube, the acetone remains unreacted, whereas the benzaldehyde is oxidized.

Figure 14.3 Attempted oxidation of acetone and oxidation of benzaldehyde. As the reaction proceeds, chromium is reduced, forming a grayish green precipitate.

Specific example:

$$CH_3 - \overset{\overset{\displaystyle O}{\|}}{C} - CH_3 + (O) \longrightarrow \text{no reaction} \qquad (14.4)$$

acetone

The oxidation of aldehydes occurs so readily that even atmospheric oxygen causes the reaction. A few drops of liquid benzaldehyde (Reaction 14.3) placed on a watch glass will begin to oxidize to solid benzoic acid within an hour. In the laboratory, it is not unusual to find that bottles of aldehydes that have been opened for a length of time contain considerable amounts of carboxylic acids.

> ✔ **LEARNING CHECK 14.3** Draw the structural formula for each product. Write "no reaction" if none occurs.
>
> a.
> $$CH_3CHCH_2 - \overset{\overset{\displaystyle O}{\|}}{C} - H + (O) \longrightarrow$$
> with CH_3 on the second carbon
>
> b. $+ (O) \longrightarrow$

The effectiveness of two common chemical reagents used to test for the presence of aldehydes—Tollens' reagent and Benedict's reagent—depends on the ease with which aldehydes are oxidized. In general, ketones fail to react with these same reagents.

Tollens' reagent is a mild oxidizing solution containing complexed Ag^+ ions. The oxidizing agent is the silver ion, which is reduced to metallic silver in the process and precipitates out (Reaction 14.5):

Tollens' reagent A mild oxidizing solution containing silver ions used to test for aldehydes.

$$\underset{\text{an aldehyde}}{R - \overset{\overset{\displaystyle O}{\|}}{C} - H} + \underset{\substack{\text{Tollens'} \\ \text{reagent}}}{2Ag(NH_3)_2^+} + 2OH^- \longrightarrow \underset{\substack{\text{a carboxylate} \\ \text{salt}}}{R - \overset{\overset{\displaystyle O}{\|}}{C} - O^-NH_4^+} + 3NH_3 + \underset{\text{silver}}{2Ag} + H_2O \qquad (14.5)$$

When the reaction is carried out in a clean, grease-free glass container, the metallic silver deposits on the glass and forms a mirror. This reaction is used commercially to silver some mirrors (see **Figure 14.4**).

Figure 14.4 This young woman can see herself because the silver coating on the back of the mirror reflects light.

Benedict's reagent is a mild oxidizing solution containing Cu^{2+} ions. When an aldehyde oxidation takes place in an alkaline solution containing Cu^{2+}, a red precipitate of Cu_2O is produced:

Benedict's reagent A mild oxidizing solution containing Cu^{2+} ions used to test for the presence of aldehydes.

$$\underset{\text{certain aldehydes}}{\overset{\overset{\displaystyle O}{\displaystyle \|}}{R-C-H}} + \underset{\substack{\text{Benedict's}\\\text{reagent}\\\text{(blue)}}}{2Cu^{2+}} + 5OH^- \longrightarrow \underset{\text{a carboxylate ion}}{\overset{\overset{\displaystyle O}{\displaystyle \|}}{R-C-O^-}} + \underset{\substack{\text{red}\\\text{precipitate}}}{Cu_2O} + 3H_2O \quad (14.6)$$

The presence of Cu^{2+} in Benedict's reagent causes this alkaline solution to have a bright blue color. A positive Benedict's test consists of the appearance of a colored (usually red) precipitate. All aldehydes give a positive Tollens' test, but only certain aldehydes and one type of easily oxidized ketone readily give a positive Benedict's test. The necessary structural features are shown.

$$\underset{\substack{A\\\text{an aldehyde with}\\\text{an adjacent}\\\text{carbonyl group}}}{\overset{\overset{\displaystyle O \quad O}{\displaystyle \| \quad \|}}{R-C-C-H}} \qquad \underset{\substack{B\\\text{an aldehyde with}\\\text{an adjacent}\\\text{alcohol group}}}{\overset{\overset{\displaystyle \qquad\quad O}{\displaystyle \qquad\quad \|}}{\underset{\overset{\displaystyle |}{\displaystyle OH}}{R-CH-C-H}}} \qquad \underset{\substack{C\\\text{a ketone with}\\\text{an adjacent}\\\text{alcohol group}}}{\overset{\overset{\displaystyle \qquad\quad O}{\displaystyle \qquad\quad \|}}{\underset{\overset{\displaystyle |}{\displaystyle OH}}{R-CH-C-R'}}}$$

The structural features (parts B or C) are found in a number of sugars, including glucose. Accordingly, Benedict's reagent reacts with glucose to produce a red precipitate of Cu_2O (see **Figure 14.5**).

14.3B The Addition of Hydrogen

In Section 12.3, we learned that addition reactions are common for compounds containing carbon–carbon double bonds. The same is true for the carbon–oxygen double bonds of carbonyl groups. An important example is the addition of H_2 in the presence of catalysts such as nickel or platinum:

$$\text{General reaction:} \quad \underset{\text{aldehyde}}{R-\overset{\overset{\displaystyle O}{\|}}{C}-H} + H_2 \xrightarrow{Pt} \underset{\text{primary alcohol}}{R-\overset{\overset{\displaystyle OH}{|}}{\underset{\underset{\displaystyle H}{|}}{C}}-H} \quad (14.7)$$

$$\text{General reaction:} \quad \underset{\text{ketone}}{R-\overset{\overset{\displaystyle O}{\|}}{C}-R'} + H_2 \xrightarrow{Pt} \underset{\text{secondary alcohol}}{R-\overset{\overset{\displaystyle OH}{|}}{\underset{\underset{\displaystyle H}{|}}{C}}-R'} \quad (14.8)$$

$$\text{Specific example:} \quad \underset{\text{propanal}}{CH_3CH_2-\overset{\overset{\displaystyle O}{\|}}{C}-H} + H_2 \xrightarrow{Pt} \underset{\text{1-propanol}}{CH_3CH_2CH_2-\overset{\overset{\displaystyle OH}{|}}{}} \quad (14.9)$$

$$\text{Specific example:} \quad \underset{\text{acetone}}{CH_3-\overset{\overset{\displaystyle O}{\|}}{C}-CH_3} + H_2 \xrightarrow{Pt} \underset{\text{2-propanol}}{CH_3-\overset{\overset{\displaystyle OH}{|}}{C}H-CH_3} \quad (14.10)$$

We see that the addition of hydrogen to an aldehyde produces a primary alcohol, whereas the addition to a ketone gives a secondary alcohol. Notice that these reactions are reductions that are essentially the reverse of the alcohol oxidations used to prepare aldehydes and ketones (Reactions 14.1 and 14.2).

✔ **LEARNING CHECK 14.4** Draw the structural formula for each product of the following reactions:

a.

$$CH_3CH_2-\overset{\overset{\displaystyle O}{\|}}{C}-CH_2CH_3 + H_2 \xrightarrow{Pt}$$

b.

$$\overset{\overset{\displaystyle O}{\|}}{C}-H$$ (benzaldehyde) $$+ H_2 \xrightarrow{Pt}$$

14.3C The Addition of Alcohols

Aldehydes react with alcohols to form **hemiacetals**, a functional group that is very important in carbohydrate chemistry (Chapter 17). Hemiacetals are usually not very stable and are difficult to isolate. With excess alcohol present and with an acid catalyst, however, a stable product called an acetal is formed:

$$\text{General reaction:} \quad \underset{\text{aldehyde}}{R-\overset{\overset{\displaystyle O}{\|}}{C}-H} + \underset{\text{alcohol}}{R'-OH} \rightleftharpoons \underset{\substack{\text{hemiacetal} \\ \text{intermediate}}}{R-\overset{\overset{\displaystyle OH}{|}}{\underset{\underset{\displaystyle OR'}{|}}{C}}-H} \underset{R'-OH}{\overset{H^+ \text{ and}}{\rightleftharpoons}} \underset{\text{acetal}}{R-\overset{\overset{\displaystyle OR'}{|}}{\underset{\underset{\displaystyle OR'}{|}}{C}}-H} + H_2O \quad (14.11)$$

① From left to right, three test tubes containing Benedict's reagent, 0.5% glucose solution, and 2.0% glucose solution.

② The addition of Benedict's reagent from the first tube produces colors (due to the red Cu_2O) that indicate the amount of glucose present.

Figure 14.5 The Benedict's test for glucose.

hemiacetal

A compound that contains the functional group $-\overset{\overset{\displaystyle OH}{|}}{\underset{\underset{\displaystyle OR}{|}}{C}}-H$

Specific example:

$$CH_3 - \overset{\overset{\displaystyle O}{\|}}{C} - H \ + \ CH_3 - OH \ \rightleftharpoons \ CH_3 - \overset{\overset{\displaystyle OH}{|}}{\underset{\underset{\displaystyle OCH_3}{|}}{C}} - H \ \overset{H^+ \ and}{\underset{CH_3 - OH}{\rightleftharpoons}} \ CH_3 - \overset{\overset{\displaystyle OCH_3}{|}}{\underset{\underset{\displaystyle OCH_3}{|}}{C}} - H \ + \ H_2O \quad (14.12)$$

acetaldehyde	methanol	hemiacetal intermediate	acetal

Hemiacetals contain an —OH group, hydrogen, and an —OR group on the same carbon. Thus, a hemiacetal appears to have alcohol and ether functional groups. **Acetals** contain a carbon that has a hydrogen and two —OR groups attached; they look like diethers:

$$CH_3 - \overset{\overset{\displaystyle OH}{|}}{\underset{\underset{\displaystyle OCH_3}{|}}{C}} - H \qquad\qquad CH_3 - \overset{\overset{\displaystyle OCH_3}{|}}{\underset{\underset{\displaystyle OCH_3}{|}}{C}} - H$$

hemiacetal carbon *acetal carbon*

acetal

A compound that contains the

$$- \overset{\overset{\displaystyle OR}{|}}{\underset{\underset{\displaystyle OR}{|}}{C}} - H \quad \text{arrangement of atoms.}$$

To better understand the changes that have occurred, let's look first at hemiacetal formation and then at acetal formation. The first molecule of alcohol adds across the double bond of the aldehyde in a manner similar to other addition reactions:

—H attaches here

$$CH_3 - \overset{\overset{\displaystyle O}{\|}}{C} - H \ + \ CH_3 - O - H \ \rightleftharpoons \ CH_3 - \overset{\overset{\displaystyle OH}{|}}{\underset{\underset{\displaystyle OCH_3}{|}}{C}} - H \qquad (14.13)$$

— OCH₃ attaches here

hemiacetal

The second step, in which the hemiacetal is converted to an acetal, is like earlier substitution reactions. In this case, an —OCH₃ substitutes for an —OH:

$$CH_3 - \overset{\overset{\displaystyle OH}{|}}{\underset{\underset{\displaystyle OCH_3}{|}}{C}} - H \ + \ CH_3 - O - H \ \overset{H^+}{\rightleftharpoons} \ CH_3 - \overset{\overset{\displaystyle OCH_3}{|}}{\underset{\underset{\displaystyle OCH_3}{|}}{C}} - H \ + \ H - OH \qquad (14.14)$$

hemiacetal acetal

In a very similar way, ketones can react with alcohols to form **hemiketals** and **ketals**:

hemiketal

A compound that contains the

$$- \overset{\overset{\displaystyle OH}{|}}{\underset{\underset{\displaystyle OR}{|}}{C}} - \overset{|}{C} - \overset{|}{C} - \text{ group.}$$

ketal

A compound that contains the

$$- \overset{\overset{\displaystyle OR}{|}}{\underset{\underset{\displaystyle OR}{|}}{C}} - \overset{|}{C} - \overset{|}{C} - \text{ group.}$$

General reaction:

hemiketal carbon *ketal carbon*

$$R - \overset{\overset{\displaystyle O}{\|}}{C} - R' \ + \ R'' - O - H \ \rightleftharpoons \ R - \overset{\overset{\displaystyle OH}{|}}{\underset{\underset{\displaystyle OR''}{|}}{C}} - R' \ \overset{H^+ \ and}{\underset{R'' - OH}{\rightleftharpoons}} \ R - \overset{\overset{\displaystyle OR''}{|}}{\underset{\underset{\displaystyle OR''}{|}}{C}} - R' \ + \ H_2O \quad (14.15)$$

ketone	alcohol	hemiketal intermediate	ketal

Specific example:

$$CH_3 - \overset{\overset{\displaystyle O}{\|}}{C} - CH_3 \; + \; CH_3CH_2OH \; \rightleftharpoons \; CH_3 - \underset{\underset{\displaystyle OCH_2CH_3}{|}}{\overset{\overset{\displaystyle OH}{|}}{C}} - CH_3 \; \underset{\xleftarrow{\hspace{1cm}}}{\overset{H^+ \text{ and}}{\overset{CH_3CH_2 - OH}{\xrightarrow{\hspace{1cm}}}}} \; CH_3 - \underset{\underset{\displaystyle OCH_2CH_3}{|}}{\overset{\overset{\displaystyle OCH_2CH_3}{|}}{C}} - CH_3 \; + \; H_2O \quad (14.16)$$

acetone ethanol hemiketal intermediate ketal

Example 14.3 Identifying Functional Groups

Identify each of the following as a hemiacetal, hemiketal, acetal, ketal, or none of the above:

a. $CH_3—O—CH_2CH_2—OH$

b. $CH_3 - O - \overset{\overset{\displaystyle OH}{|}}{C}HCH_3$

c. $CH_3CH_2 - O - \underset{\underset{\displaystyle CH_3}{|}}{\overset{\overset{\displaystyle OCH_3}{|}}{C}} - CH_3$

d.

Solution

a. Look first at any structure to see if there is a carbon with two oxygens attached. Hemiacetals, hemiketals, acetals, and ketals are all alike in that regard. This compound has two oxygens, but they are on different carbons; thus it is none of the possibilities.

b. After finding a carbon with two oxygens attached, next look to see if one is an —OH. If it is, that makes it a hemiacetal or hemiketal. The third step is to look at the carbon containing the two oxygens. If it is attached to an —H, the compound was originally an aldehyde and not a ketone. Thus, this compound is a hemiacetal:

$$CH_3 - O - \overset{\overset{\displaystyle OH}{|}}{C}HCH_3$$

OH ← *one group is* —OH

two oxygens and one hydrogen attached

c. There is a carbon with two oxygens attached and neither is an —OH. Thus, the compound is an acetal or ketal:

$$CH_3CH_2 - O - \underset{\underset{\displaystyle CH_3}{|}}{\overset{\overset{\displaystyle OCH_3}{|}}{C}} - CH_3$$

no hydrogens attached

No hydrogens are attached to the carbon bonded to two oxygens. The compound is a ketal.

d. Two oxygens are attached to the same carbon. One group is an —OH. No hydrogens are attached to the central carbon. The compound is a hemiketal.

✔ LEARNING CHECK 14.5 Draw the structural formula for the hemiacetal and hemiketal intermediates and for the acetal and ketal products of the following reactions. Label each structure as a hemiacetal, hemiketal, ketal, or acetal.

a.

$$\text{C}_6\text{H}_5\text{-C(=O)-H} + 2\text{CH}_3\text{CH}_2-\text{OH} \xrightarrow{\text{H}^+}$$

b.

$$\text{C}_6\text{H}_5\text{-C(=O)-CH}_3 + 2\text{CH}_3-\text{OH} \xrightarrow{\text{H}^+}$$

Some molecules contain both an —OH and a C=O group on different carbon atoms. In such cases, an intramolecular (within the molecule) reaction can occur, and a cyclic hemiacetal or hemiketal can be formed:

$$\text{open-chain molecule} \rightleftharpoons \text{cyclic structure} \quad (14.17)$$

open-chain molecule containing an aldehyde and alcohol group

cyclic structure containing a hemiacetal group

Cyclic hemiacetals and hemiketals are much more stable than the open-chain compounds discussed earlier. In Chapter 17, we'll see that sugars such as glucose and fructose exist predominantly in the form of cyclic hemiacetals and hemiketals.

HOW REACTIONS OCCUR 14.1

Hemiacetal Formation

Step 1. In the first step of the mechanism, the carbonyl oxygen is protonated:

$$R-\overset{\overset{\displaystyle H}{|}}{\underset{\overset{\displaystyle \|}{O^+}}{C}}-H \rightleftharpoons R-\overset{\overset{\displaystyle H}{|}}{\underset{\overset{\displaystyle +}{\ddot{O}}}{C}}-H$$

Step 2. The alcohol oxygen then bonds to the carbonyl carbon:

$$R-\overset{\overset{\displaystyle :\ddot{O}\text{H}}{|}}{\underset{\overset{\displaystyle +}{|}}{C}}-H \rightleftharpoons R-\overset{\overset{\displaystyle :\ddot{O}\text{H}}{|}}{\underset{\overset{\displaystyle O^+}{|}}{C}}-H$$

R'—Ö—H R' H

Step 3. In the final step, a proton is given off to form the hemiacetal and regenerate the acid catalyst.

$$R-\overset{\overset{\displaystyle :\ddot{O}\text{H}}{|}}{\underset{\overset{\displaystyle O^+}{|}}{C}}-H \rightleftharpoons R-\overset{\overset{\displaystyle :\ddot{O}\text{H}}{|}}{\underset{\overset{\displaystyle O:}{|}}{C}}-H + H^+$$

R' H R'

hemiacetal

In another reaction important in carbohydrate chemistry (see **Chapter 17**), cyclic acetals and ketals may form from the hemistructures. As before, all reactions are reversible:

$$\text{cyclic hemiacetal} + CH_3OH \rightleftharpoons \text{cyclic acetal} + H_2O \qquad (14.18)$$

cyclic hemiacetal cyclic acetal

Example 14.4 More Identifying Functional Groups

Identify each of the following as a cyclic hemiacetal, hemiketal, acetal, or ketal:

a.

b.

c.

d.

STUDY SKILLS 14.1

A Reaction Map for Aldehydes and Ketones

This reaction map is designed to help you master organic reactions. Whenever you are trying to complete an organic reaction, use these two basic steps: (1) Identify the functional group that is to react, and (2) Identify the reagent that is to react with the functional group. If the reacting functional group is an aldehyde or a ketone, find the reagent in the summary diagram, and use the diagram to predict the correct products.

Solution

Use the same criteria that were applied to open-chain compounds in Example 14.3. Compound a has two oxygens attached to the same carbon. One is an —OH (hemiacetal or hemiketal). The central carbon has no hydrogens. Thus, the compound is a hemiketal. These criteria give the following results for each structure:

a.

hemiketal carbon

b.

*ketal carbon
(—OH absent,
—H absent)*

c.

*acetal carbon
(—OH absent,
—H present)*

d.

*hemiacetal carbon
(—OH present,
—H present)*

We have shown each of the reactions in acetal and ketal formation as being reversible. Simple hemiacetals and hemiketals are characteristically unstable and easily revert to the starting materials. Acetals and ketals are stable structures, but the reactions may be reversed by using water and an acid catalyst:

$$
\text{General reaction:} \quad
\underset{\text{acetal}}{R-\overset{\overset{\displaystyle OR'}{|}}{\underset{\underset{\displaystyle OR'}{|}}{C}}-H} \; + \; H_2O \; \underset{}{\overset{H^+}{\rightleftharpoons}} \; \underset{\text{aldehyde}}{R-\overset{\overset{\displaystyle O}{\|}}{C}-H} \; + \; \underset{\text{alcohol}}{2R'-OH} \qquad (14.19)
$$

$$
\text{General reaction:} \quad
\underset{\text{ketal}}{R-\overset{\overset{\displaystyle OR''}{|}}{\underset{\underset{\displaystyle OR''}{|}}{C}}-R'} \; + \; H_2O \; \underset{}{\overset{H^+}{\rightleftharpoons}} \; \underset{\text{ketone}}{R-\overset{\overset{\displaystyle O}{\|}}{C}-R'} \; + \; \underset{\text{alcohol}}{2R''-OH} \qquad (14.20)
$$

$$
\text{Specific example:} \quad
CH_3-\overset{\overset{\displaystyle OCH_2CH_3}{|}}{\underset{\underset{\displaystyle OCH_2CH_3}{|}}{C}}-CH_3 \; + \; H_2O \; \overset{H^+}{\rightleftharpoons} \; CH_3-\overset{\overset{\displaystyle O}{\|}}{C}-CH_3 \; + \; 2CH_3CH_2-OH \qquad (14.21)
$$

This type of reaction, in which water causes a compound to split into component substances, is called **hydrolysis**. It will be an important reaction to remember as we discuss, in later chapters, the chemistry of biomolecules such as carbohydrates, lipids, and proteins.

hydrolysis Bond breakage by reaction with water.

✔ **LEARNING CHECK 14.6** Draw the structural formulas needed to complete the following hydrolysis reactions:

a. $CH_3CH_2CH_2-\overset{\overset{\displaystyle OCH_2CH_3}{|}}{\underset{\underset{\displaystyle OCH_2CH_3}{|}}{C}}-H + H_2O \overset{H^+}{\rightleftharpoons}$

b. (cyclopentane ring) $\overset{OCH_3}{\underset{OCH_3}{<}} + H_2O \overset{H^+}{\rightleftharpoons}$

Vanilloids: Hot Relief from Pain

The word *vanilloids* might bring to mind images of vanilla-flavored candies or ice cream. However, vanilloids have properties very different from those of vanilla, and their inclusion in foods would result in a hot and spicy flavor that might even be extremely irritating, rather than a smooth vanilla flavor.

The best-known vanilloids are capsaicin and resiniferatoxin (RTX). Capsaicin is the compound in various types of hot peppers that gives them their characteristic hot and spicy flavor. It is used in a variety of products ranging from personal-defense pepper sprays to squirrel-proof birdseed. RTX is a powerful irritant extracted from a cactuslike plant, *Euphorbia resinifera*. The milky latex obtained from the plant is referred to in medical literature dating as far back as the first century. In these writings, it was described as a nose and skin irritant and as a treatment for chronic pain. In powdered form, RTX causes sneezing, a characteristic used by some practical jokers.

Structurally, molecules of capsaicin and RTX (not shown) are similar to those of vanillin in that they both have a homovanillyl group in their structure.

problem when it was instilled into the urinary bladder, but the first application caused an intense burning sensation that was intolerable and unacceptable to patients. RTX fared better in the tests. It also relieved the condition, and caused only mild discomfort and an itching sensation with the first application.

Basic research has led to some important discoveries about these compounds. The gene for the capsaicin receptor has been cloned, making available the large quantities of the receptor needed to do further research and develop new drugs. For example, it has been found that the receptor responds not only to capsaicin but also to levels of heat that can damage tissue. It is suggested that possibly the reason hot peppers feel hot in the mouth is because they stimulate the same pain receptors that heat does.

Another important development is the discovery of a way to synthesize RTX in the laboratory. The availability of synthetic RTX may reduce the cost for the material and make possible further advances in the use of RTX in medicines.

capsaicin

vanillin

homovanillyl group

When either of the compounds is applied to the skin or mouth, a burning sensation results. This sensation is caused by the compounds bonding to receptors of small sensory neurons associated with the transmission of pain. Following repeated application of capsaicin or RTX, these neurons, or C sensory nerve fibers, become desensitized to the initial irritation or pain the compounds cause. This characteristic allows the compounds to be used topically to relieve chronic pain. Capsaicin is now used as an ingredient in a few OTC pain-relieving creams such as Capzasin-P, Heet, Menthacin, and Pain Doctor.

Both capsaicin and RTX have also been tested as treatments for a condition called *bladder hyperreflexia*, which is characterized by the urge to urinate even when there is very little urine in the bladder. Capsaicin effectively relieved the

Hot peppers are a source of capsaicin.

Example 14.5 Writing Reaction Products

Draw structural formulas for the products of the following reactions:

a. + CH$_3$CH$_2$OH $\overset{H^+}{\rightleftharpoons}$

b. + H$_2$O $\overset{H^+}{\rightleftharpoons}$

c. + H$_2$O $\overset{H^+}{\rightleftharpoons}$

Solution

a. The starting material is a cyclic hemiketal. Reaction with an alcohol produces a cyclic ketal by a substitution of —OCH$_2$CH$_3$ for the —OH:

+ CH$_3$CH$_2$O—H \rightleftharpoons + H—OH

b. The starting compound is an acetal. Hydrolysis with water will form an aldehyde and two alcohol groups:

c. The starting compound is a ketal. Hydrolysis with water will form a ketone and two alcohol groups:

✔ **LEARNING CHECK 14.7** Identify each of the following as cyclic hemiacetals, hemiketals, acetals, or ketals and show structural formulas for the hydrolysis products:

a.

b.

Get the Right Dose of Exercise

Experts agree that exercise is one of the best preventative "medicines" available. It increases energy, stamina, and one's sense of well-being. In the long term it also reduces the risk of premature death from cardiovascular disease. Put simply, it makes you feel better and live longer. We expect medicines to make us feel better when we are ill. But exercise acts as a powerful medicine to prevent illness. How do you know what the proper dose is? Do you need to exercise on a daily basis or will a weekly dose provide the desired health benefits? Just how little can you get away with and stay healthy?

Researchers arrive at the proper dose by examining health survey data that includes the exercise habits of several hundred thousand people. This information is compared to death records. The resulting recommendations vary. The U.S. government recommended in 2008 that Americans should get 30 minutes of moderate-intensity exercise each day. This daily dose has come under scrutiny in recent years. Thirty minutes a day is preferable to less exercise. However, researchers examining exercise rates and heart failure rates over 13 years found that those who exercised and followed governmental guidelines experienced only a modest reduction in heart disease. Those who doubled the recommended 30-minute dose experienced a 20% reduction, and those who exercised 2 hours per day experienced a 35% reduction.

As shown by the information above, even a little exercise produces significant benefits. People who exercise less than the recommended 30 minutes per day still reduce their risk of premature death by 20%. Those who complete 30 minutes of moderate exercise (such as walking) per day reduce their risk of premature death by 31%. Those who increase the daily dose of exercise to an hour or more per day are 39% less likely to die prematurely. For those who exercise more frequently, health benefits increase only modestly. At high rates of exercise (up to 10 times the recommended amount), health benefits are comparable to those achieved by people who merely meet the minimum requirements. In other words, many extra hours of exercise do not equate to huge gains in longevity. On the other hand, many times the recommended exercise level is not considered to be harmful. It is difficult to overdose on moderate exercise.

Intensity, as well as frequency, should be considered when calculating the ideal exercise dose. People who spend part of their daily exercise time in vigorous activity, rather than moderate activity alone (e.g., running instead of walking) reap additional health benefits. People who spent up to 30% of their exercise time in vigorous activity were 9% less likely to die prematurely. Those who spent more than 30% of their exercise time in strenuous activity gained 13% against early death.

Studies confirm that taking a daily dose of at least 20 to 30 minutes of moderate exercise is a key to enjoying a long and healthful life. Strengthening this daily dose with vigorous activity significantly improves your chances for health and longevity. Take your exercise, it's good for you!

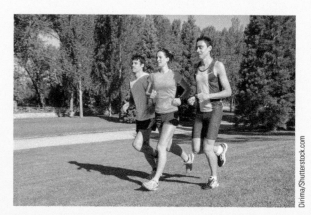

Vigorous activity is a part of the ideal exercise dose.

14.4 | Important Aldehydes and Ketones

Learning Objective

5. Give specific uses for aldehydes and ketones.

14.4A Formaldehyde: $H-\overset{\overset{\displaystyle O}{\|}}{C}-H$

As seen with other functional classes, the most important compound of the class is usually the simplest member. Formaldehyde, the simplest aldehyde, is a gas at room temperature, but it is often supplied and used in the form of a 37% aqueous solution called *formalin*. This solution kills microorganisms and is effective in sterilizing surgical instruments. It is also used to embalm cadavers. Formaldehyde is a key industrial chemical in the production of plastics such as Bakelite®, the first commercial polymer invented in 1901, and Formica®.

14.4B Acetone: $CH_3 - \overset{\overset{\textstyle O}{\|}}{C} - CH_3$

Judging from the quantity used, acetone is by far the most important of the ketones. Over 1 billion pounds are used annually in the United States. It is particularly useful as a solvent because it dissolves most organic compounds and yet it is miscible in water. It is widely used as a solvent for coatings, which range from nail polish to exterior enamel paints (see **Figure 14.6**).

Figure 14.6 Acetone is an excellent solvent for many organics, including nail polish. Why does acetone-based polish remover evaporate fairly quickly when used?

A number of naturally occurring aldehydes and ketones play important roles within living systems. The carbonyl functional group is fairly common in biological compounds. For example, progesterone and testosterone, female and male sex hormones, respectively, are both ketones.

Progesterone progesterone testosterone

Several naturally occurring aldehydes and ketones have very fragrant odors and are used in flavorings (see **Figure 14.7** and **Table 14.3**).

Figure 14.7 Vanilla ice cream owes its characteristic flavor to a special aldehyde.

TABLE 14.3 Some Fragrant Aldehydes and Ketones

Name	Structural Formula	Sources and Typical Uses
Vanillin		Vanilla bean; flavoring
Cinnamaldehyde		Oil of cinnamon; flavoring
Citral		Rinds of lemons, limes, and oranges; citrus flavoring
Biacetyl		Butter; flavoring for margarines
Camphor		Camphor tree; characteristic medicinal odor; an ingredient in some inhalants
Menthone		Mint plants; peppermint flavoring

► Case Study Follow-up

Ketones are partially broken-down fats. A buildup of ketones in blood can cause the body to produce high levels of uric acid, a risk factor for kidney stones. Uric acid kidney stones can be caused by ingestion of too much meat, as in this case study.

According to the National Institute of Diabetes and Digestive and Kidney Diseases (NIDDK), to achieve slow and steady weight loss a person should not only consider a change in diet but also a weight-loss program that encourages healthy eating behaviors and moderately intense physical activity. This should then be followed by a plan to keep the weight off. Fad diets do not provide this. Programs such as Weight Watchers offer tools to manage weight in a healthy way. Healthy weight loss is a slow process. Many people need the help of a support group to continue with their weight-loss program in the long term.

The Nomenclature of Aldehydes and Ketones

The functional groups characteristic of aldehydes and ketones are very similar; they both contain a carbonyl group:

$$
\begin{array}{c}
O \\
\parallel \\
-C-
\end{array}
$$

Aldehydes have a hydrogen attached to the carbonyl carbon,

$$
\begin{array}{c}
O \\
\parallel \\
-C-H
\end{array}
$$

whereas ketones have two carbons attached to the carbonyl carbon:

$$
\begin{array}{c}
O \\
|\quad \parallel \quad| \\
-C-C-C- \\
||
\end{array}
$$

Objective 1 (Section 14.1), Exercise 14.4

The IUPAC ending for aldehyde names is *-al*, whereas the IUPAC ending for ketones is *-one*.

Objective 2 (Section 14.1), Exercise 14.6

Physical Properties

Molecules of aldehydes and ketones cannot form hydrogen bonds with each other. As a result, they have lower boiling points than alcohols of similar molecular weight. Aldehydes and ketones have higher boiling points than alkanes of similar molecular weight because of the presence of the polar carbonyl group. The polarity of the carbonyl group and the fact that aldehydes and ketones can form hydrogen bonds with water explain why the low molecular weight compounds of these organic classes are water-soluble.

Objective 3 (Section 14.2), Exercise 14.16

Chemical Properties

Aldehydes and ketones are prepared by the oxidation of primary and secondary alcohols, respectively. Aldehydes can be further oxidized to carboxylic acids, but ketones resist oxidation. Thus, aldehydes are oxidized by Tollens' reagent (Ag^+) and Benedict's solution (Cu^{2+}), whereas ketones are not. A characteristic reaction of both aldehydes and ketones is the addition of hydrogen to the carbonyl double bond to form alcohols. In a reaction that is very important in sugar chemistry, an alcohol can add across the carbonyl group of an aldehyde to produce a hemiacetal. The substitution reaction of a second alcohol molecule with the hemiacetal produces an acetal. Ketones can undergo similar reactions to form hemiketals and ketals.

Objective 4 (Section 14.3), Exercise 14.42

Important Aldehydes and Ketones

Formaldehyde is a key industrial chemical in the production of plastics such as Bakelite® and Formica®. Acetone is an important solvent. Some naturally occurring aldehydes and ketones are important in living systems. Some function as sex hormones; others are used as flavorings.

Objective 5 (Section 14.4), Exercise 14.52

Key Terms and Concepts

Acetal (14.3)
Aldehyde (14.1)
Benedict's reagent (14.3)
Carbonyl group (Introduction)

Hemiacetal (14.3)
Hemiketal (14.3)
Hydrolysis (14.3)
Ketal (14.3)

Ketone (14.1)
Tollens' reagent (14.3)

Key Reactions

1.	Oxidation of an aldehyde to give a carboxylic acid (Section 14.3):	$$R-\overset{\displaystyle O}{\overset{\displaystyle \parallel}{C}}-H + (O) \longrightarrow R-\overset{\displaystyle O}{\overset{\displaystyle \parallel}{C}}-OH$$	Reaction 14.1
2.	Attempted oxidation of a ketone (Section 14.3):	$$R-\overset{\displaystyle O}{\overset{\displaystyle \parallel}{C}}-R' + (O) \longrightarrow \text{no reaction}$$	Reaction 14.2

(continued)

Key Reactions (continued)

3.	Hydrogenation of an aldehyde to give a primary alcohol (Section 14.3):	$R-\overset{\displaystyle O}{\overset{\displaystyle \|}{C}}-H + H_2 \xrightarrow{Pt} R-\overset{\displaystyle OH}{\underset{\displaystyle H}{\overset{\displaystyle \|}{C}}}-H$	Reaction 14.7
4.	Hydrogenation of a ketone to give a secondary alcohol (Section 14.3):	$R-\overset{\displaystyle O}{\overset{\displaystyle \|}{C}}-R' + H_2 \xrightarrow{Pt} R-\overset{\displaystyle OH}{\underset{\displaystyle H}{\overset{\displaystyle \|}{C}}}-R'$	Reaction 14.8
5.	Addition of an alcohol to an aldehyde to form a hemiacetal and then an acetal (Section 14.3):	$R-\overset{\displaystyle O}{\overset{\displaystyle \|}{C}}-H + R'-OH \rightleftarrows R-\overset{\displaystyle OH}{\underset{\displaystyle OR'}{C}}-H \underset{R'-OH}{\overset{H^+ \text{ and}}{\rightleftarrows}} R-\overset{\displaystyle OR'}{\underset{\displaystyle OR'}{C}}-H + H_2O$	Reaction 14.11
6.	Addition of an alcohol to a ketone to form a hemiketal and a ketal (Section 14.3):	$R-\overset{\displaystyle O}{\overset{\displaystyle \|}{C}}-R' + R''-O-H \rightleftarrows R-\overset{\displaystyle OH}{\underset{\displaystyle OR''}{C}}-R' \underset{R''-OH}{\overset{H^+ \text{ and}}{\rightleftarrows}} R-\overset{\displaystyle OR''}{\underset{\displaystyle OR''}{C}}-R' + H_2O$	Reaction 14.15
7.	Hydrolysis of an acetal to yield an aldehyde and two molecules of alcohol (Section 14.3):	$R-\overset{\displaystyle OR'}{\underset{\displaystyle OR'}{C}}-H + H_2O \overset{H^+}{\rightleftarrows} R-\overset{\displaystyle O}{\overset{\displaystyle \|}{C}}-H + 2R'-OH$	Reaction 14.19
8.	Hydrolysis of a ketal to yield a ketone and two molecules of alcohol (Section 14.3):	$R-\overset{\displaystyle OR''}{\underset{\displaystyle OR''}{C}}-R' + H_2O \overset{H^+}{\rightleftarrows} R-\overset{\displaystyle O}{\overset{\displaystyle \|}{C}}-R' + 2R''-OH$	Reaction 14.20

Exercises

Even-numbered exercises are answered in Appendix B.

Blue-numbered exercises are more challenging.

The Nomenclature of Aldehydes and Ketones (Section 14.1)

14.1 What is the structural difference between an aldehyde and a ketone?

14.2 Draw structural formulas for an aldehyde and a ketone that contain the fewest number of carbon atoms possible.

14.3 Identify each of the following compounds as an aldehyde, a ketone, or neither:

a. $CH_3CH_2\overset{\displaystyle O}{\overset{\displaystyle \|}{C}}-CH_2CH_3$

b. (cyclohexyl)$-\overset{C=O}{\underset{OH}{}}$

c. (six-membered lactone ring)

d. (cyclobutanone)

e. (benzyl)$-CH_2\overset{\displaystyle O}{\overset{\displaystyle \|}{C}}-H$

f. $NH_2-\overset{\displaystyle O}{\overset{\displaystyle \|}{C}}-NH_2$

14.4 Identify each of the following compounds as an aldehyde, a ketone, or neither:

a. CH₃CH₂ — C(=O) — H

b. CH₃ — C(=O) — O — CH₃

c. (cyclopentanone structure)

d. CH₃CH₂CH₂ — C(=O) — OH

e. CH₃CH₂ — C(=O) — CH₃

f. CH₃CH₂ — C(=O) — NH₂

14.5 Assign IUPAC names to the following aldehydes and ketones:

a. CH₃ — C(=O) — H

b. CH₃CHCH₂CH — C(=O) — H, with CH₃ and CH₃ substituents

c. CH₃CHCH₂ — C(=O) — CH₃, with CH₃ substituent

d. CH₃CH₂CHCH₂ — C(=O) — H, with phenyl substituent

e. (cyclopropanone with CH₃ substituent)

14.6 Assign IUPAC names to the following aldehydes and ketones:

a. CH₃CH₂ — C(=O) — H

b. CH₃CHCH₂ — C(=O) — H, with Br substituent

c. (benzene ring)—CH₂CH₂ — C(=O) — H

d. CH₃CHCH₂ — C(=O) — CH₃, with CH₃ substituent

e. (cyclopentanone with CHCH₃ and CH₃ substituent)

14.7 Draw structural formulas for each of the following compounds:
a. methanal
b. 3-ethyl-2-pentanone
c. 3-methylcyclohexanone
d. 2,4,6-trimethylheptanal

14.8 Draw structural formulas for each of the following compounds:
a. propanal
b. 3-methyl-2-butanone

c. 2,2-dimethylcyclopentanone
d. 3-bromo-4-phenylbutanal

14.9 Draw structural formulas and give IUPAC names for all the isomeric aldehydes and ketones that have the molecular formula $C_5H_{10}O$.

14.10 Draw structural formulas and give IUPAC names for all the isomeric aldehydes and ketones that have the molecular formula C_3H_6O.

14.11 Each of the following names is wrong. Give the structure and correct name for each compound.
a. 3-ethyl-2-butanone
b. 2-ethyl-propanal
c. 4-methyl-5-propyl-3-hexanone
d. 2-methyl-2-phenylethanal

14.12 Each of the following names is wrong. Give the structure and correct name for each compound.
a. 3-ethyl-2-methylbutanal
b. 2-methyl-4-butanone
c. 4,5-dibromocyclopentanone

Physical Properties (Section 14.2)

14.13 Why does hydrogen bonding *not* take place between molecules of aldehydes or ketones?

14.14 Most of the remaining water in washed laboratory glassware can be removed by rinsing the glassware with acetone (propanone). Explain how this process works (acetone is much more volatile than water).

14.15 The boiling point of propanal is 49°C, whereas the boiling point of propanol is 97°C. Explain this difference in boiling points.

14.16 Explain why propane boils at −42°C, whereas ethanal, which has the same molecular weight, boils at 20°C.

14.17 Use a dotted line to show hydrogen bonding between molecules in each of the following pairs:

a. CH₃ — C(=O) — CH₃ and O with two H (water)

b. (cyclopentane)—C(=O)—H and O with two H (water)

14.18 Use a dotted line to show hydrogen bonding between molecules in each of the following pairs:

a. (cyclopentane)—C(=O)—CH₃ and O with two H

b. (benzene ring)—C(=O)—H and O with two H

14.19 Arrange the following compounds in order of increasing boiling point:

a. O b. OH c. CH₃

14.20 The compounds menthone and menthol are fragrant substances present in an oil produced by mint plants:

menthone menthol

When pure, one of these pleasant-smelling compounds is a liquid at room temperature; the other is a solid. Identify the solid and the liquid and explain your reasoning.

Chemical Properties (Section 14.3)

14.21 Write an equation for the formation of the following compounds from the appropriate alcohol:

a. $CH_3CH_2CH_2-\overset{\overset{O}{\|}}{C}-H$

b. [structure: methylcyclohexanone with CH₃ and =O]

c. $CH_3-\overset{\overset{O}{\|}}{C}-CH_2CH_3$

14.22 Write an equation for the formation of the following compounds from the appropriate alcohol:

a. $CH_3-\overset{\overset{O}{\|}}{C}-CH_2\overset{\overset{CH_3}{|}}{CH}CH_3$

b. [cyclopentane with $\overset{\overset{O}{\|}}{C}-H$]

c. [cyclopentanone with CH₂CH₃]

14.23 Identify the following structures as hemiacetals, hemiketals, or neither:

a. $CH_3-O-\overset{\overset{CH_3}{|}}{\underset{\underset{CH_3}{|}}{C}}-OH$

b. $CH_3CH_2\overset{\overset{OH}{|}}{CH}-OCH_2CH_3$

c. [cyclopentane with OH and OCH₃]

d. [benzene ring with $O-\overset{\overset{OH}{|}}{CH}-CH_2CH_3$]

14.24 Identify the following structures as hemiacetals, hemiketals, or neither:

a. $CH_3-\overset{\overset{CH_3}{|}}{\underset{\underset{H}{|}}{C}}-OCH_3$

b. $CH_3CH_2\overset{\overset{OCH_3}{|}}{CH}-OH$

c. $CH_3-\overset{\overset{OH}{|}}{\underset{\underset{OH}{|}}{C}}-CH_2CH_3$

d. $CH_3\overset{\overset{OH}{|}}{\underset{\underset{CH_3}{|}}{C}}-OCH_2CH_3$

14.25 Label each of the following as acetals, ketals, or neither:

a. [cyclohexane with OCH₃ and OCH₂CH₃]

b. [cyclopentane with $O-\overset{\overset{CH_3}{|}}{\underset{\underset{CH_3}{|}}{C}}-OCH_3$]

c. $\overset{\overset{OCH_3}{|}}{\underset{\underset{OCH_2CH_3}{|}}{CH_2}}$

14.26 Label each of the following as acetals, ketals, or neither:

a. $CH_3-\overset{\overset{OCH_2CH_3}{|}}{\underset{\underset{OCH_2CH_3}{|}}{C}}-CH_3$

b. [two cyclopentane rings joined by $O-CHCH_3$ with O]

c. $CH_3CH_2-\overset{\overset{OH}{|}}{\underset{\underset{OCH_3}{|}}{C}}-CH_3$

14.27 Label each of the following structures as a cyclic hemiacetal, hemiketal, acetal, ketal, or none of these:

a. [pyran ring with OH] b. [pyran ring with OH] c. [pyran ring with OCH₃]

14.28 Label each of the following structures as a hemiacetal, hemiketal, acetal, ketal, or none of these:

a. [pyran ring with OCH₂CH₃ and CH₃]

b. [cyclopentane with CH₂CH₃ and OH]

c. [cyclopentane with CH—OH and OCH₃]

14.29 What two functional groups react to form the following?

 a. A hemiacetal

 b. An acetal

 c. A ketal

 d. A hemiketal

14.30 Hemiacetals are sometimes referred to as potential aldehydes. Explain.

14.31 Complete the following statements:

 a. Oxidation of a secondary alcohol produces _____.

 b. Oxidation of a primary alcohol produces an aldehyde that can be further oxidized to a _____.

 c. Hydrogenation of a ketone produces _____.

 d. Hydrogenation of an aldehyde produces _____.

 e. Hydrolysis of an acetal produces _____.

 f. Hydrolysis of a ketal produces _____.

14.32 What observation characterizes a positive Tollens' test?

14.33 Which of the following compounds will react with Tollens' reagent (Ag^+, NH_3, and H_2O)? For those that do react, draw a structural formula for the organic product.

a. $\underset{\underset{\displaystyle OH}{|}}{CH_2} - CH_2 - CH_2 - OCH_3$

b. $H - \overset{\overset{\displaystyle O}{||}}{C} - CH_2 - CH_2 - CH_3$

c.

d.

e.

14.34 Which of the following compounds will react with Tollens' reagent (Ag^+, NH_3, and H_2O)? For those that do react, draw a structural formula for the organic product.

a.

b.

c. $CH_3 - O - \underset{\underset{\displaystyle OH}{|}}{CH}CH_3$

d. $CH_3CH_2 - \overset{\overset{\displaystyle O}{||}}{C} - H$

e. $CH_3 - \overset{\overset{\displaystyle O}{||}}{C} - CH_3$

14.35 What is the formula of the precipitate that forms in a positive Benedict's test?

14.36 Not all aldehydes give a positive Benedict's test. Which of the following aldehydes do?

a. $CH_3CH_2 - \overset{\overset{\displaystyle O}{||}}{C} - H$

b.

c.

d. $CH_3\underset{\underset{\displaystyle OH}{|}}{CH} - \overset{\overset{\displaystyle O}{||}}{C} - H$

e. $CH_3 - O - CH_2 - \overset{\overset{\displaystyle O}{||}}{C} - H$

14.37 A stockroom assistant prepares three bottles, each containing one of the following compounds:

$$CH_3CH_2 - \overset{\overset{\displaystyle O}{||}}{C} - CH_3 \qquad CH_3CH_2CH_2\overset{\overset{\displaystyle O}{||}}{CH} \qquad CH_3CH_2\underset{\underset{\displaystyle OH}{|}}{CH}CH$$

However, the bottles are left unlabeled. Instead of discarding the samples, another assistant runs tests on the contents, using Tollens' and Benedict's reagents. Match the compound to the bottle, based on these results.

	Tollens' Reagent	**Benedict's Reagent**
Bottle A	A dark precipitate forms	A red-orange precipitate forms
Bottle B	No reaction	No reaction
Bottle C	A dark precipitate and slight mirror form	No reaction

14.38 Glucose, the sugar present within the blood, gives a positive Benedict's test. Circle the structural features that enable glucose to react.

$$\underset{\underset{\displaystyle OH}{|}}{CH_2} - \underset{\underset{\displaystyle OH}{|}}{CH} - \underset{\underset{\displaystyle OH}{|}}{CH} - \underset{\underset{\displaystyle OH}{|}}{CH} - \underset{\underset{\displaystyle OH}{|}}{CH} - \overset{\overset{\displaystyle O}{||}}{C} - H$$

14.39 Fructose, present with glucose in honey, reacts with Benedict's reagent. Circle the structural features that enable fructose to react.

$$\underset{\underset{\displaystyle OH}{|}}{CH_2} - \underset{\underset{\displaystyle OH}{|}}{CH} - \underset{\underset{\displaystyle OH}{|}}{CH} - \underset{\underset{\displaystyle OH}{|}}{CH} - \overset{\overset{\displaystyle O}{||}}{C} - \underset{\underset{\displaystyle OH}{|}}{CH_2}$$

14.40 Glucose can exist as a cyclic hemiacetal because one of the alcohol groups reacts with the aldehyde:

$$\underset{\underset{\displaystyle OH}{|}}{CH_2} - \underset{\underset{\displaystyle OH}{|}}{CH} - \underset{\underset{\displaystyle OH}{|}}{CH} - \underset{\underset{\displaystyle OH}{|}}{CH} - \underset{\underset{\displaystyle OH}{|}}{CH} - \overset{\overset{\displaystyle O}{||}}{C} - H \rightleftharpoons$$

Draw an arrow to the carbon of the cyclic hemiacetal that was originally part of the aldehyde group. Circle the alcohol group in the open-chain structure that became a part of the cyclic hemiacetal.

14.41 After an intramolecular reaction, fructose forms a hemiketal:

$$\underset{CH_2}{\overset{OH}{|}} - \underset{CH}{\overset{OH}{|}} - \underset{CH}{\overset{OH}{|}} - \underset{CH}{\overset{OH}{|}} - \underset{C}{\overset{O}{\|}} - \underset{CH_2}{\overset{OH}{|}} \rightleftharpoons$$

Draw an arrow to the carbon of the cyclic hemiketal that was originally part of the ketone group. Circle the alcohol group in the open-chain structure that became part of the cyclic hemiketal.

14.42 Complete the following equations. If no reaction occurs, write "no reaction."

a.

+ H_2 \xrightarrow{Pt}

b. $CH_3CH_2 - \underset{\underset{CH_3}{|}}{\overset{OH}{\underset{|}{C}}} - OCH_3 + CH_3CH_2CH_2 - OH \overset{H^+}{\rightleftharpoons}$

c. $CH_3CH_2 - \overset{O}{\overset{\|}{C}} - H + 2CH_3\overset{OH}{\overset{|}{C}}HCH_3 \overset{H^+}{\rightleftharpoons}$

d. $CH_2 - \overset{O}{\overset{\|}{C}} - H$
 $+ (O) \longrightarrow$

e. $CH_3CH_2\overset{CH_3}{\overset{|}{C}}H - \overset{O}{\overset{\|}{C}} - H + H_2 \xrightarrow{Pt}$

14.43 Complete the following equations. If no reaction occurs, write "no reaction."

a. $CH_3CH_2 - \overset{O}{\overset{\|}{C}} - CH_3 + H_2 \xrightarrow{Pt}$

b. $CH_3\overset{Br}{\overset{|}{C}}HCH - \overset{O}{\overset{\|}{C}} - H + (O) \longrightarrow$
 $\overset{|}{CH_3}$

c. $\overset{O}{\overset{\|}{C}} - H$
 $+ 2CH_3 - OH \overset{H^+}{\rightleftharpoons}$

d. $+ (O) \longrightarrow$

e. $+ CH_3CH_2 - OH \overset{H^+}{\rightleftharpoons}$

14.44 Describe the products that result when hydrogen (H_2) is added to alkenes, alkynes, aldehydes, and ketones. You might review Chapter 12 briefly.

14.45 Draw structural formulas for the products of the following hydrolysis reactions:

a. $CH_3\overset{OCH_2CH_3}{\overset{|}{C}}H - OCH_3 + H_2O \overset{H^+}{\rightleftharpoons}$

b. $\overset{OCH_3}{\underset{OCH_3}{<}} + H_2O \overset{H^+}{\rightleftharpoons}$

c.
 $CH_3 - \overset{O-}{\overset{|}{\underset{\underset{CH_3}{|}}{C}}} - O- $ $ + H_2O \overset{H^+}{\rightleftharpoons}$

d. $\overset{CH_2 - OCH_2CH_3}{\underset{OCH_2CH_3}{|}} + H_2O \overset{H^+}{\rightleftharpoons}$

14.46 Draw structural formulas for the products of the following hydrolysis reactions:

a. $\overset{OCH_2CH_3}{\underset{OCH_2CH_3}{<}} + H_2O \overset{H^+}{\rightleftharpoons}$

b. $CH_3CH_2 - \overset{OCH_3}{\overset{|}{\underset{\underset{CH_2CH_3}{|}}{C}}} - OCH_3 + H_2O \overset{H^+}{\rightleftharpoons}$

c. $CH_3CH_2CH_2O - CH_2 - OCH_2CH_2CH_3 + H_2O \overset{H^+}{\rightleftharpoons}$

d. $CH_3O - \overset{}{\underset{\underset{OCH_3}{|}}{C}}H - CH_2CH_3 + H_2O \overset{H^+}{\rightleftharpoons}$

14.47 The following compounds are cyclic acetals or ketals. Write structural formulas for the hydrolysis products.

a. $+ H_2O \overset{H^+}{\rightleftharpoons}$

b. $+ H_2O \overset{H^+}{\rightleftharpoons}$

14.48 The following compounds are cyclic acetals or ketals. Write structural formulas for the hydrolysis products.

a.

$$\text{(cyclic structure with O, CH}_3\text{, OCH}_2\text{CH}_3) + H_2O \rightleftharpoons \overset{H^+}{}$$

b.

$$\text{(cyclic structure with O, OCH}_2\text{CH}_2\text{CH}_3\text{, H)} + H_2O \rightleftharpoons \overset{H^+}{}$$

14.49 Write equations to show how the following conversions can be achieved. More than one reaction is required, and reactions from earlier chapters may be necessary.

a. $CH_3CH = CHCH_3 \longrightarrow CH_3 - \overset{\overset{\displaystyle O}{\|}}{C} - CH_2CH_3$

b. $CH_3CH_2CH_2 - OH \longrightarrow CH_3CH_2 - \overset{\overset{\displaystyle O}{\|}}{C} - OH$

14.50 Write equations to show how the following conversions can be achieved. More than one reaction is required, and reactions from earlier chapters may be necessary.

a. $CH_3CH_2CH_2 \xrightarrow{\overset{\displaystyle OH}{|}} CH_3CH_2 - \underset{\underset{\displaystyle OCH_2CH_3}{|}}{\overset{\overset{\displaystyle OCH_2CH_3}{|}}{CH}}$

b. $CH_3\overset{\overset{\displaystyle CH_3}{|}}{CH} - \overset{\overset{\displaystyle O}{\|}}{C} - CH_3 \longrightarrow CH_3\overset{\overset{\displaystyle CH_3}{|}}{C} = CH - CH_3$

Important Aldehydes and Ketones (Section 14.4)

14.51 Identify the most important aldehyde and ketone from Section 14.4 (on the basis of amount used), and list at least one characteristic for each that contributes to its usefulness.

14.52 Using Table 14.3, name an aldehyde or ketone used in the following ways:

a. Peppermint flavoring

b. Flavoring for margarines

c. Cinnamon flavoring

d. Vanilla flavoring

Additional Exercises

14.53 Arrange the following compounds starting with the carbon atom that is the most reduced and ending with the carbon that is the least reduced:

a. $R - CH_2 - OH$

b. $R - \overset{\overset{\displaystyle O}{\|}}{CH}$

c. $R - CH_3$

14.54 $CH_3 - \overset{\overset{\displaystyle O}{\|}}{C} - H \xrightarrow{(O)} CH_3 - \overset{\overset{\displaystyle O}{\|}}{C} - OH$

acetaldehyde acetic acid

You need to produce 500 g of acetic acid using the reaction above. The actual yield of the reaction is 79.6% of the theoretical yield. How many grams of acetaldehyde do you need to start with to produce 500 g of acetic acid?

14.55 $CH_3 - \overset{\overset{\displaystyle O}{\|}}{C} - CH_3 + H_2O \rightleftharpoons$

The addition of water to aldehydes and ketones occurs rapidly, although it is not thermodynamically favored. What would be the product for the reaction above? *Hint:* Think of the self-ionization of water and the polarity of the carbonyl group.

14.56 Draw the product that would result when 3-pentanone reacts with an excess of 1-propanol in the presence of an acid catalyst.

14.57 Formaldehyde levels above 0.10 mg/1000 L of ambient air can cause some humans to experience difficulty breathing. How many molecules of formaldehyde would be found in 1.0 L of air at this concentration?

Chemistry for Thought

14.58 In the IUPAC name for the following ketone, it is not common to use a number for the position of the carbonyl group. Why not?

$$CH_3CH_2 - \overset{\overset{\displaystyle O}{\|}}{C} - CH_3$$

14.59 Why can formaldehyde (CH_2O) be prepared in the form of a 37% solution in water, whereas decanal ($C_{10}H_{20}O$) cannot?

14.60 Other addition reactions of aldehydes occur. Water, for example, adds to the carbonyl of trichloroacetaldehyde to form chloral hydrate, a strong hypnotic and sedative known as "knock-out drops" or (when mixed with alcohol) a "Mickey Finn." Complete the reaction by drawing the structural formula of chloral hydrate:

$$Cl - \overset{\overset{\displaystyle Cl}{|}}{\underset{\underset{\displaystyle Cl}{|}}{C}} - \overset{\overset{\displaystyle O}{\|}}{C} - H + H - OH \longrightarrow$$

trichloroacetaldehyde chloral hydrate

14.61 In the labels of some consumer products, ketone components listed with an *-one* ending can be found. However, very few aldehyde-containing (*-al* ending) products are found. How can you explain this?

14.62 In Figure 14.6, it is noted that acetone is used as a solvent in fingernail polish remover. Why do you think fingernail polish remover evaporates fairly quickly when used?

14.63 Vanilla flavoring is either extracted from a tropical orchid or synthetically produced from wood pulp by-products. What differences in chemical structure would you expect in these two commercial products: vanilla extract and imitation vanilla extract?

14.64 The structure of fructose in Exercise 14.39 reveals a ketone group and several alcohol groups. If the alcohol groups are referred to as hydroxys, give an IUPAC name for fructose.

14.65 The use of acetone in laboratory experiments must be carefully monitored because of the highly flammable nature of acetone. Give an equation for the combustion of acetone.

14.66 Look ahead to Chapter 17 and locate the structure of maltose in Section 17.7. Draw the structure of maltose and circle the acetal group.

Allied Health Exam Connection

The following questions are from these sources:

- *Nursing School Entrance Exam* © 2005, Learning Express, LLC.

- *McGraw-Hill's Nursing School Entrance Exams* by Thomas A. Evangelist, Tamara B. Orr, and Judy Unrein © 2009, The McGraw-Hill Companies, Inc.

- *NSEE Nursing School Entrance Exams,* 3rd edition © 2009, Kaplan Publishing.

- *Cliffs Test Prep: Nursing School Entrance Exams* by Fred N. Grayson © 2004, Wiley Publishing, Inc.

- *Peterson's Master the Nursing School and Allied Health Entrance Exams,* 18th edition by Marion F. Gooding © 2008, Peterson's, a Nelnet Company.

14.67 Identify the functional group designated by each of the following:

 a. RCHO

 b. ROH

 c.
$$R-\overset{\displaystyle O}{\underset{\|}{C}}-R'$$

Wait, let me present c correctly:

 c.
$$R-\overset{}{C}-R' \quad \text{with} \quad \|O$$

$$R-C-R'$$
$$\|$$
$$O$$

14.68 Which of the following would be classified as a ketone?

 a.
$$H-\overset{H}{\underset{H}{C}}-\overset{H}{\underset{H}{C}}-\overset{H}{C}=O$$

 b.
$$H-\overset{H}{\underset{H}{C}}-\overset{O}{\underset{\|}{C}}-\overset{H}{\underset{H}{C}}-\overset{H}{\underset{H}{C}}-OH$$

 c.
$$H-\overset{H}{\underset{H}{C}}-\overset{O}{\underset{\|}{C}}-\overset{H}{\underset{H}{C}}-H$$

15

Carboxylic Acids and Esters

Triff/Shutterstock.com

▶ Case Study

Caitlin felt the familiar mid-back pain, nausea, and fever. This time, drinking copious quantities of cranberry juice and water had not cured her urinary tract infection. She faced an unavoidable trip to the doctor. Caitlin dreaded explaining that she couldn't swallow pills. Normally a confident manager in a successful company, Caitlin felt embarrassed by this problem. No matter what "trick" she tried, even including liquids with the pills, she always sputtered and gagged before she could get the pills down. How could she overcome this persistent problem?

When Caitlin talked with Dr. Sadiq, he explained that many patients suffer with this same problem. He kindly suggested she go to a "compounding pharmacist," a specially trained expert who understands medication and flavor chemistry, and how to mask unpleasant tastes. The pharmacist tried a cherry-pineapple-flavored liquid to take with

her prescribed pills. Because of this individualized care, Caitlin found that she could swallow her prescribed medications with little problem. Caitlin overcame her medication problems and was able to return to work.

▶ **What additional education do compounding pharmacists obtain? What patients require compounded medications? How important are compounded medications in veterinary practice?**

Follow-up to this Case Study appears at the end of the chapter before the *Concept Summary*.

Learning Objectives

When you have completed your study of this chapter, you should be able to:

1 Assign IUPAC names and draw structural formulas for carboxylic acids. **(Section 15.1)**

2 Explain how hydrogen bonding affects the physical properties of carboxylic acids. **(Section 15.2)**

3 Recognize and write key reactions of carboxylic acids. **(Section 15.3)**

4 Assign common and IUPAC names to carboxylic acid salts. **(Section 15.4)**

5 Describe uses for carboxylate salts. **(Section 15.4)**

6 Recognize and write key reactions for ester formation. **(Section 15.5)**

7 Assign common and IUPAC names to esters. **(Section 15.6)**

8 Recognize and write key reactions of esters. **(Section 15.7)**

9 Write reactions for the formation of phosphate esters. **(Section 15.8)**

The tart flavor of foods that taste sour is generally caused by the presence of one or more carboxylic acids. Vinegar contains acetic acid, lemons and other citrus fruits contain citric acid (see **Figure 15.1**), and the tart taste of apples is caused by malic acid.

The characteristic functional group of **carboxylic acids** is the **carboxyl group**:

$$\begin{matrix} & O \\ & \| \\ - C & - OH \end{matrix} \qquad -COOH \quad \text{or} \quad -CO_2H$$

carboxyl group abbreviated forms

Carboxyl-containing compounds are found abundantly in nature (see **Figure 15.2**). Examples are lactic acid, which is formed in muscle cells during vigorous exercise, and citric acid, which is an important intermediate in cellular energy production, as well as a flavoring agent in citrus fruits (see **Table 15.1**).

© Mark Slabaugh

Figure 15.1 Citrus fruits and juices are rich in citric acid. What flavor sensation is characteristic of acids?

carboxylic acid
An organic compound that contains the
$$\begin{matrix} & O \\ & \| \\ - C & - OH \end{matrix}$$
functional group.

carboxyl group
$$\begin{matrix} & O \\ & \| \\ The - C & - OH \end{matrix}$$
group.

© Mark Slabaugh

Figure 15.2 Rhubarb and spinach contain oxalic acid, a compound with two carboxyl groups. The nomenclature of dicarboxylic acids is not discussed here. What would you propose as the IUPAC ending for molecules with two carboxyl groups?

TABLE 15.1 Examples of Carboxylic Acids

Common Name	IUPAC Name	Structural Formula	Characteristics and Primary Uses
Formic acid	methanoic acid	$\overset{\displaystyle O}{\overset{\displaystyle \|}{H-C-OH}}$	Stinging agents of certain ants and nettles; used in food preservation
Acetic acid	ethanoic acid	$\overset{\displaystyle O}{\overset{\displaystyle \|}{CH_3-C-OH}}$	Active ingredient in vinegar; used in food preservation
Propionic acid	propanoic acid	$\overset{\displaystyle O}{\overset{\displaystyle \|}{CH_3CH_2-C-OH}}$	Salts used as mold inhibitors
Butyric acid	butanoic acid	$\overset{\displaystyle O}{\overset{\displaystyle \|}{CH_3(CH_2)_2-C-OH}}$	Odor-causing agent in rancid butter
Caproic acid	hexanoic acid	$\overset{\displaystyle O}{\overset{\displaystyle \|}{CH_3(CH_2)_4-C-OH}}$	Characteristic odor of Limburger cheese
Oxalic acid	ethanedioic acid	$HO-\overset{\displaystyle O}{\overset{\displaystyle \|}{C}}-\overset{\displaystyle O}{\overset{\displaystyle \|}{C}}-OH$	Present in leaves of some plants such as rhubarb and spinach; used as a cleaning agent for rust stains on fabric and porcelain
Citric acid	2-hydroxy-1,2,3-propane-tricarboxylic acid	$HO-\overset{O}{\overset{\|}{C}}-CH_2-\overset{OH}{\underset{\underset{O}{\overset{\|}{C-OH}}}{\overset{\|}{C}}}-CH_2-\overset{O}{\overset{\|}{C}}-OH$	Present in citrus fruits; used as a flavoring agent in foods; present in cells
Lactic acid	2-hydroxypropanoic acid	$CH_3\underset{OH}{\overset{\overset{\displaystyle O}{\overset{\displaystyle\|}{}}}{CH}}-C-OH$	Found in sour milk and sauerkraut; formed in muscles during exercise

Figure 15.3 Acetic acid is the active component in vinegar and a familiar laboratory weak acid.

fatty acid A long-chain carboxylic acid found in fats.

15.1 | The Nomenclature of Carboxylic Acids

Learning Objective

1. Assign IUPAC names and draw structural formulas for carboxylic acids.

Because of their abundance in nature, carboxylic acids were among the first organic compounds studied in detail. No systematic nomenclature system was then available, so the acids were usually named after some familiar source. Acetic acid,

$$\overset{\displaystyle O}{\overset{\displaystyle \|}{CH_3-C-OH}}$$

acetic acid

the sour constituent in vinegar (see **Figure 15.3**), is named after *acetum*, the Latin word for vinegar. Butyric acid occurs in rancid butter (the Latin word for butter is *butyrum*). Many of these common names are still widely used today (see **Table 15.1**). Carboxylic acids with long hydrocarbon chains, generally 12 to 20 carbon atoms, are called **fatty acids** (see **Chapter 18**) because they were first isolated from natural fats. These acids, such as stearic acid with 18 carbons, are known almost exclusively by their common names.

To name a carboxylic acid using the IUPAC system, the longest carbon chain including the carboxyl group is found and numbered. The numbering begins with the carboxyl

carbon, so in monocarboxylic acids it is always located at the beginning of the chain, and no number is needed to locate it. The final -e of the parent hydrocarbon is dropped, and the ending -oic is added, followed by the word acid. Groups attached to the chain are named and located as before. Aromatic acids are given names derived from the parent compound, benzoic acid.

Example 15.1 Naming Carboxylic Acids

Use IUPAC rules to name the following carboxylic acids:

a. $CH_3CH_2 - \overset{\overset{\displaystyle O}{\|}}{C} - OH$ b. $CH_3\overset{\overset{\displaystyle }{|}}{\underset{\underset{\displaystyle Br}{|}}{C}}HCH_2 - \overset{\overset{\displaystyle O}{\|}}{C} - OH$ c. $\overset{\overset{\displaystyle O}{\|}}{C} - OH$

Solution

a. The carbon chain contains three carbons. Thus, the name is based on the parent hydrocarbon, propane. The name is propanoic acid.
b. The chain contains four carbon atoms, and a bromine is attached to carbon number 3. The name is 3-bromobutanoic acid.
c. The aromatic acids are named as derivatives of benzoic acid. The carboxyl group is at position one, and the acid name is 3-methylbenzoic acid.

✔ **LEARNING CHECK 15.1** Give the IUPAC name to the following:

a. $\overset{\overset{\displaystyle O}{\|}}{C} - OH$ b. $\overset{\overset{\displaystyle O}{\|}}{C} - OH$
 $CH_3CH_2CHCH_3$

15.2 | Physical Properties of Carboxylic Acids

Learning Objective

2. Explain how hydrogen bonding affects the physical properties of carboxylic acids.

Carboxylic acids with low molecular weights are liquids at room temperature and have characteristically sharp or unpleasant odors (see **Table 15.2**). Butyric acid, for example, is a component of perspiration and is partially responsible for the odor of locker rooms and unwashed socks. As the molecular weight of carboxylic acids increases, so does the boiling point. The heavier acids (containing more than ten carbons) are waxlike solids. Stearic acid ($C_{18}H_{36}O_2$), for example, is mixed with paraffin to produce a wax used to make candles.

When boiling points of compounds with similar molecular weights are compared, the carboxylic acids have the highest boiling points of any organic compounds we have studied so far. See **Table 15.3** and **Figure A**.

For simple aliphatic compounds, the boiling points usually increase in this order: hydrocarbons and ethers, aldehydes and ketones, alcohols, and carboxylic acids. Carboxylic acids, like alcohols, can form intermolecular hydrogen bonds. It is this bonding that causes high boiling points. Acids have higher boiling points than alcohols with similar molecular

TABLE 15.2 Physical Properties of Some Carboxylic Acids

Common Name	Structural Formula	Boiling Point (°C)	Melting Point (°C)	Solubility (g/100 mL H₂O)
Formic acid	H—COOH	101	8	Infinite
Acetic acid	CH_3—COOH	118	17	Infinite
Propionic acid	CH_3—CH_2—COOH	141	−21	Infinite
Butyric acid	CH_3—$(CH_2)_2$—COOH	164	−5	Infinite
Valeric acid	CH_3—$(CH_2)_3$—COOH	186	−34	5
Caproic acid	CH_3—$(CH_2)_4$—COOH	205	−3	1
Caprylic acid	CH_3—$(CH_2)_6$—COOH	239	17	Insoluble
Capric acid	CH_3—$(CH_2)_8$—COOH	270	32	Insoluble
Lauric acid	CH_3—$(CH_2)_{10}$—COOH	299	44	Insoluble
Myristic acid	CH_3—$(CH_2)_{12}$—COOH	Decomposes	58	Insoluble
Palmitic acid	CH_3—$(CH_2)_{14}$—COOH	Decomposes	63	Insoluble
Stearic acid	CH_3—$(CH_2)_{16}$—COOH	Decomposes	71	Insoluble

TABLE 15.3 Boiling Points of Compounds with Similar Molecular Weight

Class	Compound	Molecular Weight	Boiling Point (°C)
Alkane	pentane	72	35
Ether	diethyl ether	74	35
Aldehyde	butanal	72	76
Alcohol	1-butanol	74	118
Carboxylic acid	propanoic acid	74	141

Figure 15.4 The boiling points of carboxylic acids are higher than those of all the other organic compounds studied thus far.

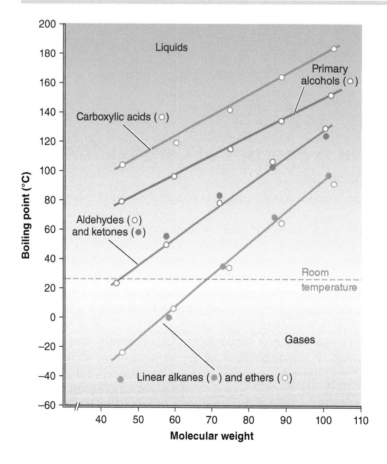

weights because hydrogen-bonded acids are held together by two hydrogen bonds rather than the one hydrogen bond that is characteristic of alcohols. A structure where two identical molecules are joined together is called a **dimer**:

dimer Two identical molecules bonded together.

$$R-C \underset{O-H \cdots O}{\overset{O \cdots H-O}{<}} C-R$$

Low molecular weight acids are very soluble in water (see **Table 15.2**) because of hydrogen bonds that form between the carboxyl group and water molecules (intermolecular hydrogen bonds). As the length of the nonpolar hydrocarbon portion (the R group) of the carboxylic acid increases, the water solubility decreases. Carboxylic acids containing eight or more carbon atoms are not appreciably soluble in water and are considered to be insoluble.

Organic compounds may be arranged according to increasing water solubility as follows: hydrocarbons, ethers, aldehydes and ketones, alcohols, and carboxylic acids. Because forces resulting from hydrogen bonding play important roles in determining both boiling points and water solubility, it is not surprising that the same order is generally observed for these two properties.

15.3 | The Acidity of Carboxylic Acids

Learning Objective

3. Recognize and write key reactions of carboxylic acids.

The most important chemical property of carboxylic acids is the acidic behavior implied by their name. The hydrogen attached to the oxygen of the carboxyl group gives carboxylic acids their acidic character. In water, this proton may leave the acid, converting it into a **carboxylate ion**. Reaction 15.1 gives the general reaction, and Reaction 15.2 gives a specific example:

carboxylate ion

The $R-C-O^-$ ion that results from the dissociation of a carboxylic acid.

General reaction:

$$\underset{\text{carboxylic acid}}{R-\overset{O}{\overset{\|}{C}}-O-H} + H_2O \rightleftharpoons \underset{\text{carboxylate ion}}{R-\overset{O}{\overset{\|}{C}}-O^-} + \underset{\substack{\text{hydronium} \\ \text{ion}}}{H_3O^+} \qquad (15.1)$$

Generally, carboxylic acids behave as weak acids. A 1-M solution of acetic acid, for example, is only about 0.5% dissociated.

Specific example:

$$\underset{\text{acetic acid}}{CH_3-\overset{O}{\overset{\|}{C}}-O-H} + H_2O \rightleftharpoons \underset{\text{acetate ion}}{CH_3-\overset{O}{\overset{\|}{C}}-O^-} + H_3O^+ \qquad (15.2)$$

It is important to realize that Reactions 15.1 and 15.2 are reversible. According to Le Châtelier's principle, the addition of H_3O^+ (low pH) should favor formation of the carboxylic acid, and removal of H_3O^+ by adding base (high pH) should favor formation of the carboxylate ion:

$$\underset{\substack{\text{carboxylic} \\ \text{acid}}}{R-\overset{O}{\overset{\|}{C}}-O-H} + H_2O \underset{\substack{\text{acidic conditions} \\ \text{(low pH)}}}{\overset{\substack{\text{basic conditions} \\ \text{(high pH)}}}{\rightleftharpoons}} \underset{\substack{\text{carboxylate} \\ \text{ion}}}{R-\overset{O}{\overset{\|}{C}}-O^-} + H_3O^+ \qquad (15.3)$$

Thus, the pH of a solution determines the form in which a carboxylic acid exists in the solution. At pH 7.4, the normal pH of body fluids, the carboxylate form predominates. For example, citric acid in body fluids is often referred to as citrate. This is an important point for later chapters in biochemistry.

> ✔ **LEARNING CHECK 15.2** Pyruvic acid, an important intermediate in the energy-conversion reactions in living organisms, is usually called pyruvate because it is commonly in the carboxylate form. Draw the structure of pyruvate.
>
> $$CH_3 - \overset{\overset{\displaystyle O}{\|}}{C} - \overset{\overset{\displaystyle O}{\|}}{C} - OH$$
>
> pyruvic acid

Although they are weak, carboxylic acids react readily with strong bases such as sodium hydroxide or potassium hydroxide to form salts. Reaction 15.4 gives the general reaction, and Reactions 15.5 and 15.6 are two specific examples:

General reaction:

$$R - \overset{\overset{\displaystyle O}{\|}}{C} - O - H + NaOH \longrightarrow R - \overset{\overset{\displaystyle O}{\|}}{C} - O^-Na^+ + H_2O \qquad (15.4)$$

carboxylic acid sodium carboxylate

Specific examples:

$$CH_3 - \overset{\overset{\displaystyle O}{\|}}{C} - O - H + NaOH \longrightarrow CH_3 - \overset{\overset{\displaystyle O}{\|}}{C} - O^-Na^+ + H_2O \qquad (15.5)$$

acetic acid sodium acetate

benzoic acid potassium benzoate

$$\qquad (15.6)$$

> ✔ **LEARNING CHECK 15.3** Write the structural formulas for the products of the following reaction:
>
> $$CH_3CH_2CH_2 - \overset{\overset{\displaystyle O}{\|}}{C} - OH + NaOH \longrightarrow$$

15.4 | Salts of Carboxylic Acids

Learning Objectives

4. Assign common and IUPAC names to carboxylic acid salts.
5. Describe uses for carboxylate salts.

15.4A Nomenclature

Both common names and IUPAC names are assigned to carboxylic acid salts by naming the metal first and changing the *-ic* ending of the acid name to *-ate*.

Example 15.2 Naming Salts

Give both a common name and an IUPAC name to the following salts:

$$\text{a. } CH_3-\overset{\overset{\textstyle O}{\|}}{C}-O^-Na^+ \qquad \text{b. } \underset{\text{(benzene ring)}}{\bigcirc}-\overset{\overset{\textstyle O}{\|}}{C}-O^-K^+$$

Solution

a. This salt is formed from $CH_3-\overset{\overset{\textstyle O}{\|}}{C}-OH$, which is called acetic acid (common name) or ethanoic acid (IUPAC name). Thus, the salt is sodium acetate (common name) or sodium ethanoate (IUPAC name).

b. This compound is the potassium salt of benzoic acid (both the common and IUPAC names). The name is potassium benzoate.

✔ **LEARNING CHECK 15.4** Give the IUPAC name for the following salts:

$$\text{a. } \underset{\underset{\textstyle CH_3}{|}}{\bigcirc}-\overset{\overset{\textstyle O}{\|}}{C}-O^-Li^+ \qquad \text{b. } CH_3CH_2CH_2-\overset{\overset{\textstyle O}{\|}}{C}-O^-Na^+$$

Carboxylic acid salts are solids at room temperature and are usually soluble in water because they are ionic. Even long-chain acids with an extensive nonpolar hydrocarbon portion can be solubilized by converting them into salts (Reaction 15.7):

$$CH_3(CH_2)_{16}-\overset{\overset{\textstyle O}{\|}}{C}-O-H + NaOH \longrightarrow CH_3(CH_2)_{16}-\overset{\overset{\textstyle O}{\|}}{C}-O^-Na^+ + H_2O \quad (15.7)$$

<div align="center">

stearic acid sodium stearate
(insoluble) (soluble)

</div>

15.4B Useful Carboxylic Acid Salts

A number of acid salts are important around the home. Sodium stearate $(CH_3(CH_2)_{16}COO^-Na^+)$ and other sodium and potassium salts of long-chain carboxylic acids are used as soaps (see **Figure 15.5**).

Calcium and sodium propanoate are used commercially as preservatives in bread, cakes, and cheese to prevent the growth of bacteria and molds (see **Figure 15.6**). The parent acid, CH_3CH_2COOH, occurs naturally in Swiss cheese. The labels for these bakery products often contain the common name propionate, rather than the IUPAC-acceptable name propanoate.

Figure 15.5 Soap products enable us to keep ourselves and our homes clean.

Figure 15.6 Propanoates extend the shelf life of bread, preventing the formation of mold.

Sodium benzoate, another common food preservative, occurs naturally in many foods, especially cranberries and prunes. It is used in bakery products, ketchup, carbonated beverages, and a host of other foods:

$$\text{sodium benzoate}$$

Zinc 10-undecylenate, the zinc salt of $CH_2 \!=\! CH(CH_2)_8COOH$, is commonly used to treat athlete's foot. One commercial product that contains zinc 10-undecylenate is Desenex®. A mixture of sodium citrate and citric acid

$$\underset{\text{citric acid}}{HO - \overset{\overset{\displaystyle O}{\|}}{C} - CH_2 - \overset{\overset{\displaystyle OH}{|}}{\underset{\underset{\displaystyle O}{\|}}{\underset{C - OH}{C}}} - CH_2 - \overset{\overset{\displaystyle O}{\|}}{C} - OH}$$

Citric acid

is widely used as a buffer to control pH. Products sold as foams or gels such as jelly, ice cream, candy, and whipped cream maintain their desirable characteristics only at certain pHs, which can be controlled by the citrate/citric acid buffer. This same buffer is used in medicines and in human blood that is used for transfusions. In blood, it also functions as an anticoagulant.

15.5 | Carboxylic Esters

Learning Objective

6. Recognize and write key reactions for ester formation.

A very important reaction occurs when carboxylic acids are heated with alcohols or phenols in the presence of an acid catalyst. A molecule of water splits out and a **carboxylic ester** is formed.

General reaction:

$$\underset{\substack{\text{carboxylic} \\ \text{acid}}}{R - \overset{\overset{\displaystyle O}{\|}}{C} - OH} + \underset{\substack{\text{alcohol} \\ \text{or phenol}}}{H - O - R'} \underset{\xrightarrow{\hspace{1cm}}}{\overset{H^+,\ heat}{\rightleftharpoons}} \underset{\substack{\text{carboxylic} \\ \text{ester}}}{R - \overset{\overset{\displaystyle O}{\|}}{C} - O - R'} + H_2O \qquad (15.8)$$

As before, R′ indicates that the two R groups can be the same or different. The process of ester formation is called **esterification**, and the carbonyl carbon–oxygen single bond of the ester group is called the **ester linkage**:

$$- \overset{\overset{\displaystyle O}{\|}}{C} \overset{\curvearrowright}{-} O - \overset{|}{\underset{|}{C}} - \qquad \textit{ester linkage}$$

ester group

The ester functional group is a key structural feature in fats, oils, and other lipids (see **Chapter 18**). Also widely found in fruits and flowers, many esters are very fragrant and represent some of nature's most pleasant odors (see **Figure 15.7**). Because of this characteristic, esters are commonly used as flavoring agents in foods and as scents in personal products.

carboxylic ester
A compound with the

$$- \overset{\overset{\displaystyle O}{\|}}{C} - OR \ \text{functional group.}$$

esterification The process of forming an ester.

ester linkage The carbonyl carbon–oxygen single bond of the ester group.

Figure 15.7 Esters are partially responsible for the fragrance of oranges, pears, bananas, pineapples, and strawberries.

Reactions 15.9 and 15.10 illustrate the formation of two widely used esters.

Specific examples:

$$CH_3CH_2CH_2-\overset{\overset{\displaystyle O}{\|}}{C}-OH + HO-CH_2CH_3 \underset{}{\overset{H^+}{\rightleftharpoons}} CH_3CH_2CH_2-\overset{\overset{\displaystyle O}{\|}}{C}-O-CH_2CH_3 + H_2O \quad (15.9)$$

butanoic acid ethyl alcohol ethyl butanoate
(strawberry flavoring)

salicylic acid methyl alcohol methyl salicylate (oil of wintergreen)

$$+ H_2O \quad (15.10)$$

Example 15.3 Showing Ester Formation

Give the structure of the ester formed in the following reactions:

a. $CH_3-\overset{\overset{\displaystyle O}{\|}}{C}-OH + CH_3-OH \overset{H^+,\,heat}{\rightleftharpoons}$

b. $CH_3CH_2CH_2-\overset{\overset{\displaystyle O}{\|}}{C}-OH + \left\langle\bigcirc\right\rangle-OH \overset{H^+,\,heat}{\rightleftharpoons}$

Solution

a. $CH_3-\overset{\overset{\displaystyle O}{\|}}{C}-OH + CH_3OH \overset{H^+,\,heat}{\rightleftharpoons} CH_3-\overset{\overset{\displaystyle O}{\|}}{C}-O-CH_3 + H_2O$

b. $CH_3CH_2CH_2-\overset{\overset{\displaystyle O}{\|}}{C}-OH + \left\langle\bigcirc\right\rangle-OH \overset{H^+,\,heat}{\rightleftharpoons} CH_3CH_2CH_2-\overset{\overset{\displaystyle O}{\|}}{C}-O-\left\langle\bigcirc\right\rangle + H_2O$

✔ **LEARNING CHECK 15.5** Give the structure of the products formed in the following reactions:

a. $CH_3CH_2-\overset{\overset{\displaystyle O}{\|}}{C}-OH$ and CH_3-OH

b. $CH_3-\overset{\overset{\displaystyle O}{\|}}{C}-OH$ and $\left\langle\bigcirc\right\rangle-OH$

c. $\left\langle\bigcirc\right\rangle-\overset{\overset{\displaystyle O}{\|}}{C}-OH$ and CH_3CH-OH
 $|$
 CH_3

Polyesters, which are found in many consumer products, are just what the name implies—polymers made by an esterification reaction. The process used is an example of **condensation polymerization**, in which monomers combine and form a polymer; a small molecule such as water is formed as well. This is in contrast to addition polymerization, in which all atoms of the alkene monomer are incorporated into the polymer (see **Section 12.4**). In many condensation polymerization reactions, each monomer has two functional groups so that the chain can grow from either end. A typical example is the reaction of terephthalic acid and ethylene glycol:

$$nHO-\overset{\overset{O}{\parallel}}{C}-\bigcirc-\overset{\overset{O}{\parallel}}{C}-OH + nHO-CH_2CH_2-OH \underset{\longleftarrow}{\overset{H^+,\text{ heat}}{\rightleftharpoons}} \qquad (15.11)$$

terephthalic acid ethylene glycol

$$+O-\overset{\overset{O}{\parallel}}{C}-\bigcirc-\overset{\overset{O}{\parallel}}{C}-O-CH_2CH_2+_n + nH_2O$$

polyethylene terephthalate (PET)

Molecules of product continue to react with available monomers until, finally, a polyester is formed with the general structure given above, where the n means that the unit in parentheses is repeated many times.

Over 3 billion pounds of the polyester PET are produced annually. Fibers are formed by melting the polymer and forcing the liquid through tiny holes in devices called spinnerettes. The resulting fibers are spun into thread or yarn and marketed under the trade names Dacron®, Fortrel®, and Terylene®, depending on the manufacturer. Typical uses include automobile tire cord and permanent-press clothing. In medicine, Dacron thread is used for sutures, and woven fabric is used to replace damaged or diseased sections of blood vessels and the esophagus. Dacron fibers are often blended with cotton to make a fabric that is more comfortable to wear on hot, humid days than those containing 100% polyester and that retains the latter's wrinkle resistance. Besides being forced through spinnerettes, PET melts may be forced through a narrow slit to produce thin sheets or films called Mylar® (see **Figure 15.8**). This form of polyester is used extensively in the manufacture of magnetic tapes for tape recorders.

Figure 15.8 Mylar, a polyester developed by E. I. DuPont deNemours Co., was used to cover the wings and pilot's compartment of the Gossamer Albatross, the first human-powered aircraft to fly across the English Channel.

condensation polymerization
The process by which monomers combine together with the simultaneous elimination of a small molecule.

Although many esters can be prepared by the method just outlined in Reaction 15.8, the reversible nature of the reaction does not always allow for good yields of product. Two compounds that are more reactive than carboxylic acids and that produce esters in good yield are **carboxylic acid chlorides** and **carboxylic acid anhydrides**:

$$
\begin{array}{c}
O \\
\parallel \\
R-C-Cl
\end{array}
$$

carboxylic acid chloride

$$
\begin{array}{c}
O \qquad O \\
\parallel \qquad \parallel \\
R-C-O-C-R
\end{array}
$$

carboxylic acid anhydride

carboxylic acid chloride
An organic compound that

$$
\begin{array}{c}
O \\
\parallel \\
contains\ the -C-Cl
\end{array}
$$

functional group.

carboxylic acid anhydride
An organic compound that

$$
\begin{array}{c}
O \qquad O \\
\parallel \qquad \parallel \\
contains\ the -C-O-C-
\end{array}
$$

functional group.

These substances react vigorously with alcohols in a nonreversible manner.

General reactions:

$$
\begin{array}{c}
O \\
\parallel \\
R-C-Cl + HO-R' \longrightarrow R-C-O-R' + HCl
\end{array}
\qquad (15.12)
$$

carboxylic acid chloride alcohol carboxylic ester

$$
\begin{array}{c}
O \qquad O \\
\parallel \qquad \parallel \\
R-C-O-C-R + HO-R' \longrightarrow R-C-O-R' + R-C-OH
\end{array}
\qquad (15.13)
$$

carboxylic acid anhydride alcohol carboxylic ester carboxylic acid

Reactions 15.14 and 15.15 illustrate two specific examples. In another example, the laboratory synthesis of aspirin (Reaction 15.18), using acetic anhydride, gives excellent yields of the ester.

Specific reactions:

$$
\begin{array}{c}
O \\
\parallel \\
CH_3-C-Cl + HO-CH_2CH_3 \longrightarrow CH_3-C-O-CH_2CH_3 + HCl
\end{array}
\qquad (15.14)
$$

acetyl chloride ethyl alcohol ethyl acetate

$$
\begin{array}{c}
O \qquad O \\
\parallel \qquad \parallel \\
CH_3-C-O-C-CH_3 +
\end{array}
$$

acetic anhydride phenol phenyl acetate acetic acid (15.15)

Example 15.4 Writing Structures for Ester Formation

Give the structure of the ester formed in the following reactions:

a.

$$C_6H_5-\overset{\displaystyle O}{\overset{\displaystyle \|}{C}}-Cl \quad + CH_3CH_2-OH \longrightarrow$$

b.

$$CH_3CH_2-\overset{\displaystyle O}{\overset{\displaystyle \|}{C}}-O-\overset{\displaystyle O}{\overset{\displaystyle \|}{C}}-CH_2CH_3 + CH_3CH_2-OH \longrightarrow$$

Solution

a. The products are

$$C_6H_5-\overset{\displaystyle O}{\overset{\displaystyle \|}{C}}-O-CH_2CH_3 + HCl$$

b. The products are

$$CH_3CH_2-\overset{\displaystyle O}{\overset{\displaystyle \|}{C}}-O-CH_2CH_3 + CH_3CH_2-\overset{\displaystyle O}{\overset{\displaystyle \|}{C}}-OH$$

The first product in each case is the ester.

✔ **LEARNING CHECK 15.6** Write equations to represent ester formation when the following pairs of compounds are reacted:

a.

$$C_6H_5-\overset{\displaystyle O}{\overset{\displaystyle \|}{C}}-O-\overset{\displaystyle O}{\overset{\displaystyle \|}{C}}-C_6H_5 \quad \text{and} \quad CH_3-OH$$

b.

$$CH_3CH_2-\overset{\displaystyle O}{\overset{\displaystyle \|}{C}}-Cl \quad \text{and} \quad \underset{\underset{\displaystyle CH_3}{\displaystyle |}}{CH_3CH}-OH$$

15.6 | The Nomenclature of Esters

Learning Objective

7. Assign common and IUPAC names to esters.

Esters are named by a common system as well as by IUPAC rules; the common names are used most often. It is helpful with both IUPAC and common names to think of the esters as being formed from a carboxylic acid, the

$$R-\overset{\displaystyle O}{\overset{\displaystyle \|}{C}}- \text{ part, and an alcohol, the } -OR' \text{ part: } \underbrace{R-\overset{\displaystyle O}{\overset{\displaystyle \|}{C}}}_{\text{acid part}}-\overset{\overset{\textit{alcohol part}}{\displaystyle \swarrow}}{OR'}$$

The first word of the name of an ester is the name of the alkyl or aromatic group (R′) contributed by the alcohol. The second word is the carboxylic acid name, with the *-ic acid* ending changed to *-ate*. This is similar to the method used for naming carboxylic acid salts. Thus, an ester of acetic acid becomes an acetate, one of butyric acid becomes a butyrate, one of lactic acid becomes a lactate, and so on.

Example 15.5 Naming Esters

Give common and IUPAC names for the following esters:

a.
$$CH_3 - \overset{\overset{\textstyle O}{\|}}{C} - O - CH_3$$

b.
$$CH_3CH_2CH_2 - \overset{\overset{\textstyle O}{\|}}{C} - O - \text{(phenyl)}$$

c.
$$\text{(phenyl)} - \overset{\overset{\textstyle O}{\|}}{C} - O - CH_2CH_3$$

Solution

a. The first word is methyl, which refers to the alcohol portion:

$$CH_3 - \overset{\overset{\textstyle O}{\|}}{C} - O - CH_3 \quad \nearrow methyl$$

$$\underbrace{}_{}$$
acetate (common name)
ethanoate (IUPAC name)

The second word is derived from the name of the two-carbon carboxylic acid (acetic acid or ethanoic acid). Thus, we have methyl acetate (common name) and methyl ethanoate (IUPAC name).

b.
$$CH_3CH_2CH_2 - \overset{\overset{\textstyle O}{\|}}{C} - O - \text{(phenyl)} \quad \nearrow phenyl$$

butyrate (common name)
butanoate (IUPAC name)

Thus, the two names are phenyl butyrate (common name) and phenyl butanoate (IUPAC name).

c.
$$\text{(phenyl)} - \overset{\overset{\textstyle O}{\|}}{C} - O - CH_2CH_3 \quad \nearrow ethyl$$

benzoate

Thus, we have ethyl benzoate (common name and IUPAC name).

✔ **LEARNING CHECK 15.7** Give both the common and the IUPAC names for each of the following esters:

a.
$$\underset{\text{}}{H - \overset{\overset{\textstyle O}{\|}}{C} - O - \underset{\underset{\textstyle CH_3}{|}}{CH}CH_3}$$

b.
(benzene ring)—$\overset{\overset{\textstyle O}{\|}}{C} - OCH_3$

15.7 | Reactions of Esters

Learning Objective

8. Recognize and write key reactions of esters.

The most important reaction of carboxylic esters, both in commercial processes and in the body, involves the breaking of the ester linkage. The process is called either ester hydrolysis or saponification, depending on the reaction conditions.

15.7A Ester Hydrolysis

Ester hydrolysis is the reaction of an ester with water to break an ester linkage and produce an alcohol and a carboxylic acid. This process is simply the reverse of the esterification reaction (Reaction 15.8). Strong acids are frequently used as catalysts to hydrolyze esters.

General reaction:

$$R - \overset{\overset{\textstyle O}{\|}}{C} - OR' + H-OH \overset{H^+}{\rightleftharpoons} R - \overset{\overset{\textstyle O}{\|}}{C} - OH + R'-OH \qquad (15.16)$$

ester carboxylic alcohol
 acid or phenol

Specific example:

$$CH_3 - \overset{\overset{\textstyle O}{\|}}{C} - O - CH_2CH_3 + H-OH \overset{H^+}{\rightleftharpoons} CH_3 - \overset{\overset{\textstyle O}{\|}}{C} - OH + CH_3CH_2 - OH \quad (15.17)$$

ethyl acetate acetic acid ethyl alcohol

Notice that when ester linkages break during hydrolysis, the elements of water (—OH and —H) are attached to the ester fragments to form the acid and alcohol (Reaction 15.16). The breaking of a bond and the attachment of the elements of water to the fragments are characteristic of all hydrolysis reactions.

The human body and other biological organisms are constantly forming and hydrolyzing a variety of esters. The catalysts used in these cellular processes are very efficient protein catalysts called enzymes (see **Chapter 20**). Animal fats and vegetable oils are esters. Their enzyme-catalyzed hydrolysis is an important process that takes place when they are digested.

Consider Low-Dose Aspirin

Aspirin, an ester of salicylic acid first prepared in 1893, is remarkably versatile. In the body it reduces fever but does not reduce normal body temperature. It also relieves a variety of simple pains such as headache, sprain, and toothache, and is an anti-inflammatory agent often used to treat arthritis.

use before surgery can present a danger because it increases the risk of severe bleeding after the operation.

Recent research reported in medical journals provides evidence of another important medical use of aspirin that is not related to its ability to inhibit blood clotting: Aspirin helps prevent colorectal cancers. This indication is very significant

$$\underset{\substack{\text{salicylic}\\\text{acid}}}{} + \underset{\substack{\text{acetic}\\\text{anhydride}}}{CH_3-C-O-C-CH_3} \longrightarrow \underset{\substack{\text{acetylsalicylic acid}\\\text{(aspirin)}}}{} + \underset{\text{acetic acid}}{CH_3-C-OH} \tag{15.18}$$

Americans consume an estimated 80 billion aspirin tablets each year. The *Physicians' Desk Reference* lists more than 50 OTC (over-the-counter) drugs in which aspirin is the principal ingredient. Yet, despite the fact that aspirin has been in routine use for nearly a century, scientific journals continue to publish reports about new uses for this old remedy.

In recent years, researchers have discovered one particular effect of aspirin in the body that helps explain its benefits as well as some of its side effects: Aspirin interferes with the clotting action of blood. When blood vessels are injured and bleed, cells in the blood called platelets accumulate at the injury site and clump together to form a plug that seals the opening in the blood vessel. Aspirin reduces the ability of the platelets to clump together, thereby reducing the tendency of blood to clot. In this same way, aspirin may help reduce or prevent clots from forming in arteries that are narrowed by diseases such as atherosclerosis. A heart attack results when a coronary artery that provides blood to the heart is blocked by a blood clot. One study shows that aspirin reduces the chances of having a second heart attack by more than 30%. Other studies suggest that aspirin use helps prevent a first heart attack in healthy individuals, and aspirin can also help prevent first or recurrent strokes of the type caused by blood clots that block the blood supply to the brain.

The same process that reduces blood clotting is also responsible for some of the undesirable side effects of using aspirin. It is known that aspirin use causes increased rates of gastrointestinal bleeding and hemorrhagic stroke (stroke caused by a ruptured blood vessel in the brain). Heavy aspirin

because colorectal cancer is the second leading cancer killer after lung cancer. Researchers theorize that aspirin exerts this effect by blocking the production of hormone-like prostaglandins, which in turn inhibits the growth of tumors.

The U.S. Preventative Services Task Force has recommended that adults over 50 take a daily low-dose aspirin to prevent both heart attacks and colorectal cancer. When such therapy is recommended, a low dose of 81 mg per day (one baby aspirin) is generally used. Because the long-term, regular use of aspirin poses risks of serious bleeding, the decision to begin such a program should be made only after consultation with a physician.

A variety of aspirin products is available.

✔ **LEARNING CHECK 15.8** Give the structure of the products formed in the following hydrolysis reaction:

$$
\underset{\text{O}}{\overset{\text{O}}{\underset{\|}{\text{C}}}} - \text{O} - \text{CH}_2\text{CH}_2\text{CH}_3 + \text{H}_2\text{O} \overset{\text{H}^+}{\rightleftharpoons}
$$

15.7B Saponification

saponification The basic cleavage of an ester linkage.

Like ester hydrolysis, **saponification** results in a breaking of the ester linkage and produces an alcohol and a carboxylic acid (see **Section 18.4** for more on saponification). However, unlike ester hydrolysis, saponification is done in solutions containing strong bases such as potassium or sodium hydroxide. The carboxylic acid is converted to a salt under these basic conditions.

General reaction:

$$
\underset{\text{ester}}{\text{R} - \overset{\overset{\text{O}}{\|}}{\text{C}} - \text{O} - \text{R}'} + \text{NaOH} \longrightarrow \underset{\substack{\text{carboxylic} \\ \text{acid salt}}}{\text{R} - \overset{\overset{\text{O}}{\|}}{\text{C}} - \text{O}^-\text{Na}^+} + \underset{\text{alcohol}}{\text{R}' - \text{OH}} \qquad (15.19)
$$

Specific example:

$$
\underset{\text{methyl benzoate}}{\overset{\overset{\text{O}}{\|}}{\text{C}} - \text{O} - \text{CH}_3} + \text{NaOH} \longrightarrow \underset{\text{sodium benzoate}}{\overset{\overset{\text{O}}{\|}}{\text{C}} - \text{O}^-\text{Na}^+} + \underset{\text{methyl alcohol}}{\text{CH}_3 - \text{OH}} \qquad (15.20)
$$

We will see in Chapter 18 that saponification of fats and oils is important in the production of soaps.

STUDY SKILLS 15.1

A Reaction Map for Carboxylic Acids

We hope you have become familiar with the reaction maps of Chapters 12, 13, and 14 and have found them useful in solving problems. Another reaction map is given below.

It includes the key reactions of this chapter and should be useful as you answer questions and prepare for exams.

Ester Saponification

The cleavage of an ester linkage under basic conditions is initiated by the attraction of the carbonyl carbon for hydroxide:

$$CH_3-\overset{\overset{O}{\parallel}}{C}-O-CH_3 \rightleftharpoons CH_3-\overset{\overset{:\ddot{O}:^-}{|}}{\underset{\underset{OH}{|}}{C}}-O-CH_3$$

As the —OH group bonds to the carbonyl carbon, the double bond must break (two electrons move to the carbonyl oxygen) so that carbon does not possess five bonds.

In the next step of the mechanism, the electrons reestablish the double bond. Once again, carbon will have five bonds if there is not a simultaneous bond breakage. In this case, it is the ester linkage that is cleaved:

$$CH_3-\overset{\overset{:\ddot{O}:^-}{|}}{\underset{\underset{OH}{|}}{C}}-O-CH_3 \rightleftharpoons CH_3-\overset{\overset{O}{\parallel}}{C}-OH + {}^-:\ddot{O}-CH_3$$

carboxylic acid strong base

A carboxylic acid is formed momentarily but it quickly donates its proton to the strongly basic anion:

$$CH_3-\overset{\overset{O}{\parallel}}{C}-OH + {}^-:\ddot{O}-CH_3 \longrightarrow$$

$$CH_3-\overset{\overset{O}{\parallel}}{C}-O^- + CH_3-OH$$

carboxylate product alcohol product

If the source of hydroxide for the reaction was sodium hydroxide, the carboxylate product would be coupled with the positive sodium ion to form the salt.

LEARNING CHECK 15.9 Give structural formulas for the products of the following saponification reactions.

a. $CH_3CH_2-\overset{\overset{O}{\parallel}}{C}-O-\underset{\underset{CH_3}{|}}{CH}CH_3 + NaOH \longrightarrow$

b. $\langle\!\!\bigcirc\!\!\rangle-\overset{\overset{O}{\parallel}}{C}-O-CH_2CH_3 + KOH \longrightarrow$

15.8 | Esters of Inorganic Acids

Learning Objective

9. Write reactions for the formation of phosphate esters.

We have defined a carboxylic ester as the product of a reaction between a carboxylic acid and an alcohol. Alcohols can also form **esters** by reacting with inorganic acids such as sulfuric, nitric, and phosphoric acids. The most important of these in biochemistry are the esters of phosphoric acid (H_3PO_4).

ester A compound in which the —OH of an acid is replaced by an —OR.

General reaction:

$$R-OH + HO-\overset{\overset{O}{\parallel}}{\underset{\underset{OH}{|}}{P}}-OH \rightleftharpoons R-O-\overset{\overset{O}{\parallel}}{\underset{\underset{OH}{|}}{P}}-OH + H_2O \qquad (15.21)$$

alcohol phosphoric acid phosphate ester

Specific example:

$$CH_3CH_2-OH \ + \ HO-\overset{\displaystyle \overset{O}{\|}}{\underset{\displaystyle \underset{OH}{|}}{P}}-OH \ \rightleftharpoons \ CH_3CH_2-O-\overset{\displaystyle \overset{O}{\|}}{\underset{\displaystyle \underset{OH}{|}}{P}}-OH \ + \ H_2O \quad (15.22)$$

<div align="center">
ethyl alcohol phosphoric ethyl

acid phosphate
</div>

Because phosphoric acid has three —OH groups, it can form mono-, di-, and triesters:

$$RO-\overset{\displaystyle \overset{O}{\|}}{\underset{\displaystyle \underset{OH}{|}}{P}}-OH \qquad RO-\overset{\displaystyle \overset{O}{\|}}{\underset{\displaystyle \underset{OH}{|}}{P}}-OR' \qquad RO-\overset{\displaystyle \overset{O}{\|}}{\underset{\displaystyle \underset{OR''}{|}}{P}}-OR'$$

<div align="center">
monoester, diester, triester,

one R group two R groups three R groups
</div>

Mono- and diesters are essential to life and represent some of the most important biological molecules. An example of an important monoester is glucose-6-phosphate, which represents the first intermediate compound formed when glucose is oxidized to supply energy for the body (see **Chapter 23**):

<div align="center">glucose-6-phosphate</div>

At body pH, the two —OH groups of the phosphoric acid are ionized, and the phosphate group has a charge of 2^-. Examples of important diesters include the phospholipids (see **Chapter 18**) and the nucleic acids (see **Chapter 21**).

As we study key compounds in the storage and transfer of chemical energy in living systems (see **Chapter 22**), we will encounter phosphate esters in which two or three phosphate groups are linked. These **phosphoric anhydrides** are referred to as diphosphates and triphosphates.

> **phosphoric anhydride** A compound that contains the
> $$-O-\overset{\displaystyle \overset{O}{\|}}{\underset{\displaystyle \underset{O^-}{|}}{P}}-O-\overset{\displaystyle \overset{O}{\|}}{\underset{\displaystyle \underset{O^-}{|}}{P}}-O- \text{ group.}$$

$$R-O-\overset{\displaystyle \overset{O}{\|}}{\underset{\displaystyle \underset{O^-}{|}}{P}}-O-\overset{\displaystyle \overset{O}{\|}}{\underset{\displaystyle \underset{O^-}{|}}{P}}-O^- \qquad\qquad R-O-\overset{\displaystyle \overset{O}{\|}}{\underset{\displaystyle \underset{O^-}{|}}{P}}-O-\overset{\displaystyle \overset{O}{\|}}{\underset{\displaystyle \underset{O^-}{|}}{P}}-O-\overset{\displaystyle \overset{O}{\|}}{\underset{\displaystyle \underset{O^-}{|}}{P}}-O^-$$

<div align="center">
a diphosphate ester a triphosphate ester
</div>

Adenosine diphosphate (ADP) and adenosine triphosphate (ATP) are the most important of these compounds. We will study their rather complex structures in more detail in

CHEMISTRY AROUND US 15.1
Nitroglycerin in Dynamite and in Medicine

Nitroglycerin is a nitrate ester resulting from the reaction of nitric acid and glycerol:

$$
\begin{array}{l}
CH_2-OH \\
| \\
CH-OH + 3HO-NO_2 \rightleftarrows \\
| \\
CH_2-OH
\end{array}
\qquad
\begin{array}{l}
CH_2-O-NO_2 \\
| \\
CH-O-NO_2 + 3H_2O \\
| \\
CH_2-O-NO_2
\end{array}
$$

glycerol nitric acid nitroglycerin

First made in 1846 by the Italian chemist Sobrero, nitroglycerin was discovered to be a powerful explosive with applications in both war and peace. However, as a shock-sensitive, unstable liquid, nitroglycerin proved to be extremely dangerous to manufacture and handle. The Swedish chemist Alfred Nobel later perfected its synthesis and devised a safe method for handling it. Nobel mixed nitroglycerin with a clay-like absorbent material. The resulting solid, called dynamite, is an explosive that is much less sensitive to shock than liquid nitroglycerin. Dynamite is still one of the most important explosives used for mining, digging tunnels, and blasting hills for road building.

Surprising as it may seem, nitroglycerin is also an effective medicine. It is used to treat patients with angina pectoris—sharp chest pains caused by an insufficient supply of oxygen to the heart muscle. Angina pectoris is usually found in patients with coronary artery diseases, such as arteriosclerosis (hardening of the arteries). During overexertion or excitement, the partially clogged coronary arteries prevent the heart from getting an adequate supply of oxygenated blood, and pain results. Nitroglycerin relaxes the cardiac muscle and causes a dilation of the arteries, thus increasing blood flow to the heart and relieving the chest pains.

Nitroglycerin (combined with other substances to render it nonexplosive) can be administered in small tablets, which are placed under the tongue during an attack of angina. The nitroglycerin is rapidly absorbed into the bloodstream and finds its way to the heart muscle within seconds. It can also be applied directly to the skin as a cream or absorbed through the skin from a transdermal patch attached to the skin. This last delivery system is a recent development that allows the nitroglycerin to be continuously absorbed for 24 hours.

A transdermal patch allows nitroglycerin to be absorbed through the skin.

Section 22.7. For now, note the ester linkages in each structure and the presence of the di- and triphosphate groups. Again, at the pH of body fluids, the —OH groups attached to the phosphorus lose H^+ and acquire negative charges.

ADP

ATP

Before the advent of manufactured drugs in the 1950s and '60s, pharmacists compounded most of the medications they sold. This involved preparing the medications from individual ingredients. This practice of compounding medications has experienced a resurgence. In addition to flavoring options, compounding pharmacists can adjust strengths, combinations, and delivery options according to physician requirements. Compounding pharmacists are especially helpful with hormone replacement therapies in which tiny dosage adjustments are critical. Flavored medications are also helpful in pediatric

medications, especially with chronically ill children who require multiple medications. Compounding pharmacists also work with dentists to create custom mouthwashes, and with veterinarians to create medications that pets are willing to take.

Compounding pharmacists rely on flavor chemists to develop flavors that work well with different medications and in different applications. Many flavor chemists consider their work an art form and work very creatively. Compounding pharmacists along with flavor chemists often make the difference in a patient's ability to follow a prescribed treatment regime.

Concept Summary

The Nomenclature of Carboxylic Acids The characteristic functional group of carboxylic acids is the carboxyl group

$$\begin{array}{c} O \\ \parallel \\ -C-OH \end{array}$$

Many of the simpler carboxylic acids are well known by common names. In the IUPAC system, the ending of *-oic acid* is used in the names of these compounds. Aromatic acids are named as derivatives of benzoic acid.
Objective 1 (Section 15.1), Exercise 15.6

Physical Properties of Carboxylic Acids At room temperature, low molecular weight carboxylic acids are liquids with distinctively sharp or unpleasant odors. High molecular weight, long-chain acids are waxlike solids. Carboxylic acids are quite effective in forming dimers, two molecules held together by hydrogen bonds. Thus, they have relatively high boiling points, and those with lower molecular weights are soluble in water.
Objective 2 (Section 15.2), Exercise 15.10

The Acidity of Carboxylic Acids Soluble carboxylic acids behave as weak acids; they dissociate only slightly in water to form an equilibrium mixture with the carboxylate ion. The equilibrium concentrations of the carboxylic acid and the carboxylate ion depend on pH. At low pH, the acid form predominates, and at pH 7.4 (the pH of cellular fluids) and above, the carboxylate ion predominates. Carboxylic acids react with bases to produce carboxylate salts and water.
Objective 3 (Section 15.3), Exercise 15.26

Salts of Carboxylic Acids The carboxylate salts are named by changing the *-ic* ending of the acid to *-ate.*
Objective 4 (Section 15.4), Exercise 15.28

The ionic nature of the salts makes them water soluble. A number of carboxylate salts are useful as food preservatives, soaps, and medicines.
Objective 5 (Section 15.4), Exercise 15.32

Carboxylic Esters Carboxylic acids, acid chlorides, and acid anhydrides react with alcohols to produce esters. Polyesters result from the reaction of dicarboxylic acids and diols. Polyesters are an example of condensation polymers; these are produced when monomers react to form a polymer plus a small molecule such as water. Many esters are very fragrant and represent some of nature's most pleasant odors. Because of this characteristic, esters are widely used as flavoring agents.
Objective 6 (Section 15.5), Exercise 15.36

The Nomenclature of Esters Both common and IUPAC names for esters are formed by first naming the alkyl group of the alcohol portion followed by the name of the acid portion in which the *-ic acid* ending has been changed to *-ate.*
Objective 7 (Section 15.6), Exercise 15.46

Reactions of Esters Esters can be converted back to carboxylic acids and alcohols under either acidic or basic conditions. Hydrolysis, the reaction with water in the presence of an acid, produces the carboxylic acid and alcohol. Saponification occurs in the presence of a base to produce the carboxylate salt and alcohol.
Objective 8 (Section 15.7), Exercise 15.54

Esters of Inorganic Acids Alcohols can also form esters by reacting with inorganic acids such as phosphoric acid. Phosphate esters represent some of the most important biological compounds.
Objective 9 (Section 15.8), Exercise 15.56

Key Terms and Concepts

Carboxylate ion (15.3)
Carboxyl group (Introduction)
Carboxylic acid (Introduction)
Carboxylic acid anhydride (15.5)
Carboxylic acid chloride (15.5)

Carboxylic ester (15.5)
Condensation polymerization (15.5)
Dimer (15.2)
Ester (15.8)
Esterification (15.5)

Ester linkage (15.5)
Fatty acid (15.1)
Phosphoric anhydride (15.8)
Saponification (15.7)

Key Reactions

1.	Dissociation of a carboxylic acid to give a carboxylate ion (Section 15.3):	**Reaction 15.1** $$R-\overset{\overset{\displaystyle O}{\|}}{C}-O-H + H_2O \rightleftarrows R-\overset{\overset{\displaystyle O}{\|}}{C}-O^- + H_3O^+$$
2.	Reaction of a carboxylic acid with base to produce a carboxylate salt plus water (Section 15.3):	**Reaction 15.4** $$R-\overset{\overset{\displaystyle O}{\|}}{C}-O-H + NaOH \longrightarrow R-\overset{\overset{\displaystyle O}{\|}}{C}-O^-Na^+ + H_2O$$
3.	Reaction of a carboxylic acid with an alcohol to produce an ester plus water (Section 15.5):	**Reaction 15.8** $$R-\overset{\overset{\displaystyle O}{\|}}{C}-OH + H-O-R' \overset{H^+, \text{heat}}{\rightleftarrows} R-\overset{\overset{\displaystyle O}{\|}}{C}-O-R' + H_2O$$
4.	Reaction of a carboxylic acid chloride with an alcohol to produce an ester plus hydrogen chloride (Section 15.5):	**Reaction 15.12** $$R-\overset{\overset{\displaystyle O}{\|}}{C}-Cl + HO-R' \longrightarrow R-\overset{\overset{\displaystyle O}{\|}}{C}-O-R' + HCl$$
5.	Reaction of a carboxylic acid anhydride with an alcohol to produce an ester plus a carboxylic acid (Section 15.5):	**Reaction 15.13** $$R-\overset{\overset{\displaystyle O}{\|}}{C}-O-\overset{\overset{\displaystyle O}{\|}}{C}-R + HO-R' \longrightarrow R-\overset{\overset{\displaystyle O}{\|}}{C}-O-R' + R-\overset{\overset{\displaystyle O}{\|}}{C}-OH$$
6.	Ester hydrolysis to produce a carboxylic acid and alcohol (Section 15.7):	**Reaction 15.16** $$R-\overset{\overset{\displaystyle O}{\|}}{C}-OR' + H-OH \overset{H^+}{\rightleftarrows} R-\overset{\overset{\displaystyle O}{\|}}{C}-OH + R'-OH$$
7.	Ester saponification to give a carboxylate salt and alcohol (Section 15.7):	**Reaction 15.19** $$R-\overset{\overset{\displaystyle O}{\|}}{C}-O-R' + NaOH \longrightarrow R-\overset{\overset{\displaystyle O}{\|}}{C}-O^-Na^+ + R'-OH$$
8.	Phosphate ester formation (Section 15.8):	**Reaction 15.21** $$R-OH + HO-\underset{\underset{\displaystyle OH}{\|}}{\overset{\overset{\displaystyle O}{\|}}{P}}-OH \rightleftarrows R-O-\underset{\underset{\displaystyle OH}{\|}}{\overset{\overset{\displaystyle O}{\|}}{P}}-OH + H_2O$$

Exercises

The Nomenclature of Carboxylic Acids (Section 15.1)

15.1 What is the structure of the carboxylic acid functional group? How does it differ from the structure of an alcohol and from that of an aldehyde or a ketone?

15.2 What structural features are characteristic of fatty acids? Why are fatty acids given that name?

15.3 What compound is responsible for the sour or tart taste of Italian salad dressing (vinegar and oil)?

15.4 What carboxylic acid is present in sour milk and sauerkraut?

15.5 Write the correct IUPAC name for each of the following:

a. $CH_3CH_2CH_2 - \overset{\overset{\displaystyle O}{\|}}{C} - OH$

b. $CH_3 - O - CH_2CH_2 - \overset{\overset{\displaystyle O}{\|}}{C} - OH$

c. $CH_3\overset{\overset{\displaystyle CH_3}{|}}{C}HCHCH_2 - \overset{\overset{\displaystyle O}{\|}}{C} - OH$ (with Br below)

d. benzene ring with $\overset{\overset{\displaystyle O}{\|}}{C} - OH$ and CH_2CH_3

e. $CH_3CHCH_2CH_2 - Br$ (with $\overset{\overset{\displaystyle O}{\|}}{C} - OH$ above)

15.6 Write the correct IUPAC name for each of the following:

a. $CH_3CH_2CH_2CH_2 - \overset{\overset{\displaystyle O}{\|}}{C} - OH$

b. $CH_3CH_2\overset{\overset{\displaystyle }{|}}{C}HCH_2 - \overset{\overset{\displaystyle O}{\|}}{C} - OH$ (with CH_3 below)

c. $CH_3CH_2CHCH_2CH_2 - \overset{\overset{\displaystyle O}{\|}}{C} - OH$ (with benzene ring below)

d. benzene ring with $\overset{\overset{\displaystyle O}{\|}}{C} - OH$ and CH_3 and CH_3

e. $CH_3CHCH_2CHCH_2 - \overset{\overset{\displaystyle O}{\|}}{C} - OH$ (with CH_3 and CH_2CH_3 below)

15.7 Write a structural formula for each of the following:

a. hexanoic acid

b. 4-bromo-3-methylpentanoic acid

c. *o*-ethylbenzoic acid

15.8 Write a structural formula for each of the following:

a. pentanoic acid

b. 2-bromo-3-methylhexanoic acid

c. 4-propylbenzoic acid

Physical Properties of Carboxylic Acids (Section 15.2)

15.9 Of the classes of organic compounds studied so far, which have particularly unpleasant odors?

15.10 Which compound in each of the following pairs would you expect to have the higher boiling point? Explain your answer.

a. acetic acid or 1-propanol

b. propanoic acid or butanone

c. acetic acid or butyric acid

15.11 List the following compounds in order of increasing boiling point:

a. pentanal

b. hexane

c. 1-pentanol

d. butanoic acid

15.12 Draw the structure of the dimer formed when two molecules of propanoic acid hydrogen bond with each other.

15.13 The two isomers propanoic acid and methyl acetate are both liquids. One boils at 141°C, the other at 57°C.

$CH_3CH_2 - \overset{\overset{\displaystyle O}{\|}}{C} - OH$ $CH_3 - \overset{\overset{\displaystyle O}{\|}}{C} - OCH_3$

propanoic acid methyl acetate

a. Predict which compound boils at 57°C and which at 141°C. Explain your reasoning.

b. Which of these two compounds would you predict to be more soluble in water? Explain.

15.14 Caproic acid, a six-carbon acid, has a solubility in water of 1 g/100 mL of water (Table 15.2). Which part of the structure of caproic acid is responsible for its solubility in water, and which part prevents greater solubility?

15.15 Why are acetic acid, sodium acetate, and sodium caprate all soluble in water, whereas capric acid, a 10-carbon fatty acid, is not?

15.16 List the following compounds in order of increasing water solubility:

a. ethoxyethane

b. propanoic acid

c. pentane

d. 1-butanol

15.17 List the following compounds in order of increasing water solubility:

a. hexane **c.** 3-pentanone

b. 2-pentanol **d.** pentanoic acid

The Acidity of Carboxylic Acids (Section 15.3)

15.18 Draw the structural formula for the carboxylate ion of propanoic acid.

15.19 What is the most important chemical property of carboxylic acids?

15.20 As we discuss the cellular importance of lactic acid in a later chapter, we will refer to this compound as lactate. Explain why.

$$CH_3 - \underset{\underset{OH}{|}}{CH} - \underset{\underset{O}{\parallel}}{C} - OH$$

lactic acid

15.21 Write an equation to illustrate the equilibrium that is present when propanoic acid is dissolved in water. What structure predominates when OH^- is added to raise the pH to 12? What structure predominates as acid is added to lower the pH to 2?

15.22 Draw a structural formula that shows which form of butyric acid would predominate at a pH of 2.

15.23 Complete each of the following reactions:

a. $CH_3(CH_2)_7 - \underset{\underset{}{\overset{O}{\parallel}}}{C} - OH + NaOH \longrightarrow$

b.
$$\underset{\text{(benzene ring)}}{\overset{CH_3}{\underset{}{}}} \overset{O}{\underset{\parallel}{C}} - OH$$
$$+ KOH \longrightarrow$$

15.24 Complete each of the following reactions:

a. $H - \underset{\underset{}{\overset{O}{\parallel}}}{C} - OH + NaOH \longrightarrow$

b. $CH_3CH_2CH_2 - \underset{\underset{}{\overset{O}{\parallel}}}{C} - OH + KOH \longrightarrow$

15.25 Write a balanced equation for the reaction of butanoic acid with each of the following:

a. KOH

b. H_2O

c. NaOH

15.26 Write a balanced equation for the reaction of acetic acid with each of the following:

a. NaOH

b. KOH

c. $Ca(OH)_2$

Salts of Carboxylic Acids (Section 15.4)

15.27 Give the IUPAC name for each of the following:

a. $CH_3CH_2 - \underset{\underset{}{\overset{O}{\parallel}}}{C} - O^- Na^+$

b. $CH_3\underset{\underset{CH_3}{|}}{CH} - \underset{\underset{}{\overset{O}{\parallel}}}{C} - O^- K^+$

c.
$$\underset{\text{(benzene ring with }NO_2\text{)}}{} - \underset{\underset{}{\overset{O}{\parallel}}}{C} - O^- Na^+$$

15.28 Give the IUPAC name for each of the following:

a. $CH_3\underset{\underset{Br}{|}}{CH}CH_2 - \underset{\underset{}{\overset{O}{\parallel}}}{C} - O^- Na^+$

b. $(H - \underset{\underset{}{\overset{O}{\parallel}}}{C} - O^-)_2 Ca^{2+}$

c.
$$\underset{\text{(benzene ring)}}{} O - CH_2CH_2 - \underset{\underset{}{\overset{O}{\parallel}}}{C} - O^- K^+$$

15.29 Draw structural formulas for the following:

a. sodium formate

b. calcium benzoate

c. sodium 2-methylbutanoate

15.30 Draw structural formulas for the following:

a. potassium ethanoate

b. sodium *m*-methylbenzoate

c. sodium 2-methylbutanoate

15.31 Name each of the following:

a. The sodium salt of valeric acid

b. The magnesium salt of lactic acid

c. The potassium salt of citric acid

15.32 Give the name of a carboxylic acid or carboxylate salt used in each of the following ways:

a. As a soap

b. As a general food preservative used to pickle vegetables

c. As a preservative used in soft drinks

d. As a treatment for athlete's foot

e. As a mold inhibitor used in bread

f. As a food additive noted for its pH buffering ability

Carboxylic Esters (Section 15.5)

15.33 Which of the following compounds are esters?

a. $\text{C}_6\text{H}_5\text{—CH}_2\text{—O—CH}_2\text{—OH}$

b. $\text{CH}_3\text{—}\overset{\displaystyle O}{\overset{\displaystyle \|}{\text{C}}}\text{—O—CH}_2\text{CH}_3$

c. $\text{CH}_3\text{CH}_2\text{—CH—OCH}_3$
 $\overset{|}{\text{OCH}_3}$

d. $\text{C}_6\text{H}_5\text{—O—}\overset{\displaystyle O}{\overset{\displaystyle \|}{\text{C}}}\text{—CH}_3$

e. $\text{CH}_3\text{—O—CH}_2\text{—}\overset{\displaystyle O}{\overset{\displaystyle \|}{\text{C}}}\text{—CH}_2\text{—OH}$

f. $\text{C}_6\text{H}_5\text{—}\overset{\displaystyle O}{\overset{\displaystyle \|}{\text{C}}}\text{—O—CH}_2\text{CH}_3$

15.34 For each ester in Exercise 15.33, draw a circle around the portion that came from the acid and use an arrow to point out the *ester linkage*.

15.35 Complete the following reactions:

a. $\text{C}_6\text{H}_5\text{—}\overset{\displaystyle O}{\overset{\displaystyle \|}{\text{C}}}\text{—OH} + \text{CH}_3\text{CH—OH} \;\overset{\text{H}^+,\ \text{heat}}{\rightleftarrows}$
 $\overset{|}{\text{CH}_3}$

b. $\text{C}_6\text{H}_5\text{—}\overset{\displaystyle O}{\overset{\displaystyle \|}{\text{C}}}\text{—Cl} + \text{CH}_3\text{CH—OH} \longrightarrow$
 $\overset{|}{\text{CH}_3}$

c. $\text{C}_6\text{H}_5\text{—}\overset{\displaystyle O}{\overset{\displaystyle \|}{\text{C}}}\text{—O—}\overset{\displaystyle O}{\overset{\displaystyle \|}{\text{C}}}\text{—C}_6\text{H}_5 + \text{CH}_3\text{—OH} \longrightarrow$

15.36 Complete the following reactions:

a. $\text{CH}_3\text{—}\overset{\displaystyle O}{\overset{\displaystyle \|}{\text{C}}}\text{—O—}\overset{\displaystyle O}{\overset{\displaystyle \|}{\text{C}}}\text{—CH}_3 + \text{CH}_3\text{—OH} \longrightarrow$

b. $\text{CH}_3\text{CH—}\overset{\displaystyle O}{\overset{\displaystyle \|}{\text{C}}}\text{—OH} + \text{C}_6\text{H}_5\text{—CH}_2\text{—OH} \;\overset{\text{H}^+,\ \text{heat}}{\rightleftarrows}$
 $\overset{|}{\text{CH}_3}$

c. $\text{CH}_3\text{CH—}\overset{\displaystyle O}{\overset{\displaystyle \|}{\text{C}}}\text{—Cl} +$ $\text{C}_6\text{H}_4(\text{OH}) \longrightarrow$
 $\overset{|}{\text{CH}_3}$

15.37 Using the alcohol $\text{CH}_3\text{CH}_2\text{—OH}$, show three different starting compounds that may be used to synthesize the ester:

$$\text{CH}_3\text{CH}_2\text{CH}_2\text{—}\overset{\displaystyle O}{\overset{\displaystyle \|}{\text{C}}}\text{—O—CH}_2\text{CH}_3$$

15.38 Give the structures of the ester that forms when propanoic acid is reacted with the following:

a. methyl alcohol

b. phenol

c. 2-methyl-1-propanol

15.39 The structures of two esters used as artificial flavorings are given below. Write the structure of the acid and the alcohol from which each ester could be synthesized.

a. Pineapple flavoring,

$$\text{CH}_3\text{CH}_2\text{CH}_2\text{—}\overset{\displaystyle O}{\overset{\displaystyle \|}{\text{C}}}\text{—O—CH}_2\text{CH}_2\text{CH}_2\text{CH}_3$$

b. Apple flavoring,

$$\text{CH}_3\text{CH}_2\text{CH}_2\text{—}\overset{\displaystyle O}{\overset{\displaystyle \|}{\text{C}}}\text{—O—CH}_3$$

15.40 Heroin is formed by reacting morphine with 2 mol of

$$\text{CH}_3\text{—}\overset{\displaystyle O}{\overset{\displaystyle \|}{\text{C}}}\text{—Cl}$$

to form the diester. Show the structure of heroin.

morphine

15.41 How do the acids and alcohols involved in polyester formation differ from those that commonly form simple esters?

15.42 Draw the structure of the repeating monomer unit in the polyester formed in the condensation of oxalic acid and 1,3-propanediol.

$$\text{HO—}\overset{\displaystyle O}{\overset{\displaystyle \|}{\text{C}}}\text{—}\overset{\displaystyle O}{\overset{\displaystyle \|}{\text{C}}}\text{—OH} + \text{CH}_2\text{—CH}_2\text{—CH}_2 \;\rightleftarrows$$
 $\quad\quad\quad\quad\overset{|}{\text{OH}}\quad\quad\quad\overset{|}{\text{OH}}$

The Nomenclature of Esters (Section 15.6)

15.43 Assign common names to the following esters. Refer to Table 15.1 for the common names of the acids.

a. $H—\overset{\overset{\displaystyle O}{\|}}{C}—O—CH_2CH_3$

b. $CH_3(CH_2)_4—\overset{\overset{\displaystyle O}{\|}}{C}—O—\underset{\underset{\displaystyle CH_3}{|}}{C}HCH_3$

15.44 Assign common names to the following esters. Refer to Table 15.1 for the common names of the acids.

a. $CH_3\underset{\underset{\displaystyle OH}{|}}{C}H—\overset{\overset{\displaystyle O}{\|}}{C}—O—CH_2CH_3$

b. $CH_3CH_2CH_2—\overset{\overset{\displaystyle O}{\|}}{C}—O—CH_2CH_2CH_3$

15.45 Give the IUPAC name for each of the following:

a. $CH_3CH_2CH_2—\overset{\overset{\displaystyle O}{\|}}{C}—O—CH_2CH_3$

b. (benzene ring with) $\overset{\overset{\displaystyle O}{\|}}{C}—O—\underset{\underset{\displaystyle CH_3}{|}}{C}HCH_3$ and CH_3 substituent

15.46 Give the IUPAC name for each of the following:

a. $CH_3\underset{\underset{\displaystyle CH_3}{|}}{C}H—\overset{\overset{\displaystyle O}{\|}}{C}—OCH_3$

b. (benzene ring with) $\overset{\overset{\displaystyle O}{\|}}{C}—O—CH_2CH_3$ and Cl, Cl substituents

15.47 Assign IUPAC names to the simple esters produced by a reaction between ethanoic acid and the following:

a. ethanol

b. 1-propanol

c. 1-butanol

15.48 Assign common names to the simple esters produced by a reaction between methanol and the following:

a. propionic acid

b. butyric acid

c. lactic acid

15.49 Draw structural formulas for the following:

a. methyl ethanoate

b. propyl 2-bromobenzoate

c. ethyl 3,4-dimethylpentanoate

15.50 Draw structural formulas for the following:

a. phenyl formate

b. methyl 4-nitrobenzoate

c. ethyl 2-chloropropanoate

Reactions of Esters (Section 15.7)

15.51 Define and compare the terms *ester hydrolysis* and *saponification*.

15.52 Write equations for the hydrolysis and saponification of ethyl acetate:

$$CH_3—\overset{\overset{\displaystyle O}{\|}}{C}—OCH_2CH_3$$

15.53 Complete the following reactions:

a. (benzene ring)$—\overset{\overset{\displaystyle O}{\|}}{C}—O—CH_2CH_3 + NaOH \longrightarrow$

b. $CH_3(CH_2)_4—\overset{\overset{\displaystyle O}{\|}}{C}—O—\underset{\underset{\displaystyle CH_3}{|}}{C}HCH_3 + H_2O \overset{H^+}{\rightleftharpoons}$

15.54 Complete the following reactions:

a. $CH_3(CH_2)_{16}—\overset{\overset{\displaystyle O}{\|}}{C}—O—CH_2CH_3 + NaOH \longrightarrow$

b. $CH_3\underset{\underset{\displaystyle CH_3}{|}}{C}H—\overset{\overset{\displaystyle O}{\|}}{C}—O—$(benzene ring)$+ H_2O \overset{H^+}{\rightleftharpoons}$

Esters of Inorganic Acids (Section 15.8)

15.55 Glyceraldehyde-3-phosphate, an important compound in the cellular oxidation of carbohydrates, is an ester of phosphoric acid and glyceraldehyde.

$$\overset{\displaystyle O\diagdown\diagup H}{\underset{\displaystyle C}{}}$$
$$H—\underset{|}{C}—OH \quad \overset{\displaystyle O}{}$$
$$CH_2—O—\overset{\overset{\displaystyle O}{\|}}{\underset{\underset{\displaystyle O^-}{|}}{P}}—O^-$$

a. Give the structure of glyceraldehyde.

b. Why is glyceraldehyde-3-phosphate shown with a 2^- charge?

15.56 Dihydroxyacetone reacts with phosphoric acid to form the monoester called dihydroxyacetone phosphate. Complete the reaction for its formation.

$$\underset{\underset{\displaystyle OH}{|}}{C}H_2—\overset{\overset{\displaystyle O}{\|}}{C}—\underset{\underset{\displaystyle OH}{|}}{C}H_2 + HO—\overset{\overset{\displaystyle O}{\|}}{\underset{\underset{\displaystyle OH}{|}}{P}}—OH \longrightarrow$$
dihydroxyacetone

15.57 Give the structure for the following:

 a. monoethyl phosphate (with a 2^- charge)

 b. monoethyl diphosphate (with a 3^- charge)

 c. monoethyl triphosphate (with a 4^- charge)

Additional Exercises

15.58 Explain how a sodium bicarbonate ($NaHCO_3$) solution can be used to distinguish between a carboxylic acid solution and an alcohol solution.

15.59 The following reaction requires several steps. Show the reagents you would use and draw structural formulas for intermediate compounds formed in each step.

$$H_2C = CH_2 \longrightarrow CH_3\overset{\overset{\text{O}}{\|}}{C} - O - CH_2CH_3$$

15.60 Draw the structural formula and give the IUPAC name for the ester that results when 3-ethylhexanoic acid reacts with phenol.

15.61 How many mL of a 0.100 M NaOH solution would be needed to titrate 0.156 g of butanoic acid to a neutral pH endpoint?

15.62 Identify two things that can be done to the following equilibrium to achieve 100% conversion of butanoic acid to ethyl butanoate.

$$CH_3CH_2CH_2\overset{\overset{\text{O}}{\|}}{C} - OH + CH_3CH_2OH \rightleftharpoons$$

$$CH_3CH_2CH_2\overset{\overset{\text{O}}{\|}}{C} - O - CH_2CH_3 + H_2O$$

Chemistry for Thought

15.63 Citric acid is often added to carbonated beverages as a flavoring. Why is citric acid viewed as a "safe" food additive?

15.64 Ester formation (Reaction 15.8) and ester hydrolysis (Reaction 15.16) are exactly the same reaction, only written in reverse. What determines which direction the reaction proceeds and what actually forms?

15.65 Write a mechanism for the following reaction. Show the final location of the oxygen atom in the second color.

$$CH_3CH_2 - \overset{\overset{\text{O}}{\|}}{C} - O - CH_3 + NaOH \longrightarrow$$

15.66 Answer the question in Figure 15.1. What flavor sensation is characteristic of acids?

15.67 Oxalic acid, the substance mentioned in Figure 15.2, has two carboxyl groups. If you were to give this compound an IUPAC name, what ending would be appropriate?

15.68 Why is it safe for us to consume foods like vinegar that contain acetic acids?

15.69 A Dacron patch is sometimes used in heart surgery. Why do polyesters like the Dacron patch resist hydrolysis reactions that would result in deterioration of the patch?

15.70 Citrus fruits such as grapefruit and oranges may be kept for several weeks without refrigeration. Offer some possible reasons why the spoilage rate is not higher.

15.71 In Section 18.3, locate the structure of a triglyceride. Draw its structure and circle the ester linkages.

15.72 Section 21.2 shows a tetranucleotide structure. Draw a portion of the structure and circle the portion representing the phosphate diester.

Allied Health Exam Connection

The following questions are from these sources:

- *Nursing School Entrance Exam* © 2005, Learning Express, LLC.

- *McGraw-Hill's Nursing School Entrance Exams* by Thomas A. Evangelist, Tamara B. Orr, and Judy Unrein © 2009, The McGraw-Hill Companies, Inc.

- *NSEE Nursing School Entrance Exams*, 3rd edition © 2009, Kaplan Publishing.

- *Cliffs Test Prep: Nursing School Entrance Exams* by Fred N. Grayson © 2004, Wiley Publishing, Inc.

- *Peterson's Master the Nursing School and Allied Health Entrance Exams,* 18th edition by Marion F. Gooding © 2008, Peterson's, a Nelnet Company.

15.73 Identify the functional group designated by each of the following:

 a. ROR

 b. RCOOH

 c. R

 d. RCHO

 e. $R - \overset{\overset{\text{O}}{\|}}{C} - R'$

15.74 Rank the following compounds from the one containing the greatest polarity to the one with the least polarity. Also, rank the compounds in order of increasing boiling points.

 a. $CH_3CH_2 - OH$ **c.** $CH_3CH_2CH_3$

 b. $CH_3 - O - CH_3$ **d.** CH_3COOH

15.75 Fats belong to the class of organic compounds represented by the general formula, RCOOR′, where R and R′ represent hydrocarbon groups; therefore, fats are:

 a. ethers.

 b. soaps.

 c. esters.

 d. lipases.

16 Amines and Amides

▶ Case Study

Koji was a rambunctious 6-year-old. His mother, Masumi, and her friend went to the park and let their children play together. Koji and his friend Ben were exploring the edge of the park, their "magical forest." They gathered sticks and built a fort—such fun! That night, however, Koji broke out in an itchy rash. An Internet search led Masumi to conclude it was poison ivy. Calamine lotion and antihistamines were recommended. She rushed to the store, where she bought supplies to treat the rash. She gave Koji a generic allergy medication containing diphenhydramine then slathered him with the pink lotion. The next morning she had trouble waking him, and he went right back to sleep. The doctor she called instructed her to bring him to the hospital along with bottles of any medications he had been given. Masumi was told that Koji had been overdosed with antihistamine. Not only had Masumi given him too much oral medication for his weight, she had not noticed that the lotion she gave him contained diphenhydramine as well as calamine. Koji was treated and expected to recover from the frightening ordeal.

▶ **How can the public be better educated about using remedies that contain mixtures of common medications? Are mixed medications a good idea? What are the symptoms of antihistamine overdose?**

Follow-up to this Case Study appears at the end of the chapter before the *Concept Summary*.

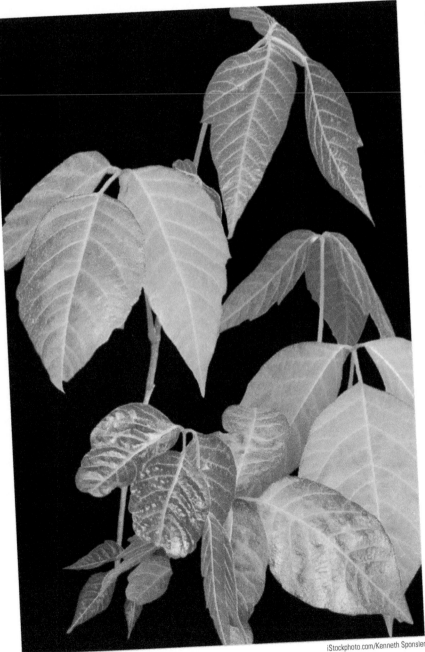

iStockphoto.com/Kenneth Sponsler

When you have completed your study of this chapter, you should be able to:

1 Given structural formulas, classify amines as primary, secondary, or tertiary. **(Section 16.1)**

2 Assign common and IUPAC names to simple amines. **(Section 16.2)**

3 Discuss how hydrogen bonding influences the physical properties of amines. **(Section 16.3)**

4 Recognize and write key reactions for amines. **(Section 16.4)**

5 Name amines used as neurotransmitters. **(Section 16.5)**

6 Give uses for specific biological amines. **(Section 16.6)**

7 Assign IUPAC names for amides. **(Section 16.7)**

8 Show the formation of hydrogen bonds with amides. **(Section 16.8)**

9 Give the products of acidic and basic hydrolysis of amides. **(Section 16.9)**

The effectiveness of a wide variety of important medicines depends either entirely or partly on the presence of a nitrogen-containing group in their molecules. Nitrogen-containing functional groups are found in more medications than any other type of functional group. In this chapter, we will study two important classes of organic nitrogen compounds, the amines and the amides.

Both amines and amides are abundant in nature, where they play important roles in the chemistry of life. Our study of these two functional classes will help prepare us for later chapters dealing with amino acids, proteins, and nucleic acids, and provide a basis for understanding the structures and chemistry of a number of medicines.

16.1 | Classification of Amines

Learning Objective

1. Given structural formulas, classify amines as primary, secondary, or tertiary.

Amines are organic derivatives of ammonia, NH_3, in which one or more of the hydrogens are replaced by an aromatic or alkyl group (R). Like alcohols, amines are classified as primary, secondary, or tertiary on the basis of molecular structure. However, the classification is done differently. When classifying amines, the number of groups (R) that have replaced hydrogens in the NH_3 is counted. The nitrogen atom in a **primary amine** is bonded to one R group. The nitrogen in **secondary amines** is bonded to two R groups, and that of **tertiary amines** is bonded to three R groups. These amine subclasses are summarized in **Table 16.1**.

amine An organic compound derived by replacing one or more of the hydrogen atoms of ammonia with alkyl or aromatic groups, as in RNH_2, R_2NH, and R_3N.

primary amine An amine having one alkyl or aromatic group bonded to nitrogen, as in $R—NH_2$.

secondary amine An amine having two alkyl or aromatic groups bonded to nitrogen, as in R_2NH.

tertiary amine An amine having three alkyl or aromatic groups bonded to nitrogen, as in R_3N.

✔ **LEARNING CHECK 16.1** Classify each of the following amines as primary, secondary, or tertiary:

a. $CH_3 - \overset{\displaystyle H}{\underset{\displaystyle |}{N}} - CH_2CH_3$

b. ⬡$-N-H$

c. ⬡$-\overset{\displaystyle}{\underset{\displaystyle |}{N}}-CH_3$
　　　　　　CH_3

TABLE 16.1 Subclasses of Amines

Subclass	General Formula	Example
Primary (1°)	R—N—H $\quad\ \|$ \quad H	CH_3—N—H $\qquad\ \|$ \qquad H
Secondary (2°)	R—N—R′ $\quad\ \|$ \quad H	CH_3—N—CH_3 $\qquad\ \|$ \qquad H
Tertiary (3°)	R—N—R′ $\quad\ \|$ \quad R″	CH_3—N—CH_3 $\qquad\ \|$ $\qquad CH_3$

16.2 | The Nomenclature of Amines

Learning Objective

2. Assign common and IUPAC names to simple amines.

Several approaches are used in naming amines. Common names are used extensively for those with low molecular weights. These are named by alphabetically linking the names of the alkyl or aromatic groups bonded to the nitrogen and attaching the suffix -amine. The name is written as one word, and the prefixes di- and tri- are used when identical alkyl groups are present. Some examples are

CH_3—NH_2
methylamine

CH_3—NH—CH_3
dimethylamine

CH_3CH_2—NH—CH_3
ethylmethylamine

✔ **LEARNING CHECK 16.2** Assign a common name to the following amines:

a. $CH_3CH_2CH_2$—NH_2 b. ⬡—NH—CH_3 c. CH_3—N—CH_3
$\qquad\qquad\qquad\qquad\qquad\qquad\qquad\qquad\qquad\qquad\qquad\ \ |$
$\qquad\qquad\qquad\qquad\qquad\qquad\qquad\qquad\qquad\qquad\ CH_3$

An IUPAC approach for naming amines is similar to that for alcohols. The root name is determined by the longest continuous chain of carbon atoms. The -e ending in the alkane name is changed to –amine. The position of the amino group on the chain is shown by a number. Numbers are also used to locate other substituents on the carbon chain. An italic capital N is used as a prefix for a substituent on nitrogen.

$CH_3CH_2CH_2CHCH_3$
$\qquad\qquad\quad |$
$\qquad\qquad\ NH_2$
2-pentanamine

$\qquad CH_3\quad\ NH_2$
$\qquad\ |\qquad\ |$
$CH_3CHCH_2CH_2$
3-methyl-1-butanamine

$\qquad\qquad\ NHCH_3$
$\qquad\qquad\quad |$
$CH_3CH_2CHCH_3$
N-methyl-2-butanamine

✔ **LEARNING CHECK 16.3** Give an IUPAC name for the following amines:

a. $CH_3CH_2CH_2CH_2$ — NH_2 b. $CH_3CH_2CH_2CH_2$
$\qquad\qquad\qquad\qquad\qquad\qquad\qquad\qquad\qquad\quad |$
$\qquad\qquad\qquad\qquad\qquad\qquad\qquad\qquad\ NHCH_3$

Aromatic amines are usually given names (both common and IUPAC) based on aniline, the simplest aromatic amine. They are named as substituted anilines, and an italic capital N is used to indicate that an alkyl group is attached to the nitrogen and not to the ring.

ASK AN EXPERT 16.1

Q: **Does caffeine help with weight loss?**

Melina B. Jampolis, M.D., is a Physician Nutrition Specialist focusing on nutrition for weight loss and disease prevention and treatment.

A: Caffeine, a synthetic and naturally occurring alkaloid, is found in numerous plant products including kola nuts, cacao bean, yerba mate, guarana berries, and the two most common dietary sources, roasted coffee beans and tea leaves. It is also included in many energy drinks and weight-reducing supplements due to its ability to stimulate the sympathetic nervous system, thereby boosting metabolism and increasing lipolysis, the breakdown of fat. Caffeine is also a well-established ergogenic aid that enhances athletic performance and decreases fatigue by increasing mental alertness and improving the utilization of fat as fuel during exercise.

Although numerous studies have found that caffeine does increase daily caloric expenditure, the effect is modest in terms of weight loss. One study showed that after consuming 300 mg of caffeine (the amount in three regular cups of coffee), overweight subjects burned 79 more calories during the following 24 hours. When you consider that most weight-loss programs recommend a 500-calorie deficit per day in order to lose one pound of fat per week, it is clear that eating a lower calorie diet and exercising regularly are far more effective in terms of weight loss than caffeine consumption. In addition, most of the coffee drinks, bottled teas, and energy drinks that Americans consume contain far more than 79 calories, so their metabolism-boosting effect is completely offset by the additional calories ingested.

In terms of weight loss, the research is somewhat more promising when caffeine is combined with other bioactive ingredients. In particular, research looking at the combination of green tea catechins (GTCs), polyphenols found in unfermented tea leaves, and caffeine has shown positive results in terms of weight loss, especially metabolically toxic visceral adiposity, commonly referred to as "belly fat." Similar to caffeine, GTCs stimulate the central nervous system, so they might have a synergistic effect in terms of boosting metabolism and improving fat oxidation. In addition, GTC might decrease appetite and nutrient absorption, which might contribute further to their role in weight loss.

Epigallocatechin gallate (EGCG) is the most abundant catechin in green tea and is often found in extract form in weight-loss supplements. A recent study found a modest but significant reduction in body mass index, body weight, and waist circumference when GTCs were combined with caffeine, but not when they were consumed alone. Most of the research on GTCs has been done with green tea extracts because the higher doses required for weight loss would be very difficult to obtain from simply drinking green tea. The optimal dose of GTC plus caffeine for weight loss has not been established, and a supplement might not be as effective if you are already consuming several cups of coffee per day.

So, if your goal is weight loss, caffeine alone probably won't help very much. If a little caffeine motivates you to exercise—or gives you a pre-workout boost, allowing you to exercise more effectively—there is no harm in drinking a cup of coffee or green tea (without a lot of sugar or cream) before workouts. In terms of supplements, the modest effect (often achieved at high financial cost) and the variability in quality and amount of active ingredient make them difficult to recommend for most people trying to lose weight. If you are watching what you eat and exercising regularly, or if you have hit a weight-loss plateau, they might be worth a try, but the effects will probably be modest at best. And if caffeine in any form interferes with your sleep, this could have a negative impact on weight loss because sleep deprivation has been shown in numerous studies to correlate with weight gain.

Tea is a common source of caffeine.

If two groups are attached to the nitrogen, an italic capital *N* is used for each one. Some examples are

aniline *N*-methylaniline 2-ethyl-*N*-methylaniline *N*,*N*-dimethylaniline

Amines and Amides **519**

✔ **LEARNING CHECK 16.4** Name the following amines as derivatives of aniline:

a. [structure: benzene ring attached to NH—CH₂CH₃]

c. [structure: benzene ring with NH₂, CH₃, and CH₂CH₃ substituents]

b. [structure: benzene ring attached to N(CH₃)—CH₂CH₃]

16.3 | Physical Properties of Amines

Learning Objective

3. Discuss how hydrogen bonding influences the physical properties of amines.

The simple, low molecular weight amines are gases at room temperature (see **Table 16.2**). Heavier, more complex compounds are liquids or solids.

TABLE 16.2 Properties of Some Amines

Name	Structure	Melting Point (°C)	Boiling Point (°C)
methylamine	CH_3-NH_2	☒94	☒6
ethylamine	$CH_3CH_2-NH_2$	☒81	17
dimethylamine	$CH_3-NH-CH_3$	☒93	7
diethylamine	$CH_3CH_2-NH-CH_2CH_3$	☒48	56
trimethylamine	CH_3-N-CH_3 with CH_3	☒117	3
triethylamine	$CH_3CH_2-N-CH_2CH_3$ with CH_2CH_3	☒114	89

Like alcohols, primary and secondary amines form hydrogen bonds among themselves:

$$CH_3-\underset{\underset{H}{|}}{N}-H \cdots \underset{\underset{H}{|}}{\overset{\overset{H}{|}}{N}}-CH_3$$

Because nitrogen is less electronegative than oxygen, the hydrogen bonds formed by amines are weaker than those formed by alcohols. For this reason, the boiling points of primary and secondary amines are usually somewhat lower than the boiling points of alcohols with similar molecular weights. Tertiary amines cannot form hydrogen bonds among themselves (because the nitrogen has no attached hydrogen atom), and their boiling points are similar to alkanes that have about the same molecular weights.

Amines with fewer than six carbon atoms are generally soluble in water as a result of hydrogen bond formation between amine functional groups and water molecules. This is true for tertiary as well as primary and secondary amines (see **Figure 16.1**).

Figure 16.1 Hydrogen bonding
between amines in water.

1° amine 2° amine 3° amine

Low molecular weight amines have a sharp, penetrating odor similar to that of ammonia. Somewhat larger amines have an unpleasant odor reminiscent of decaying fish. In fact, some amines are partially responsible for the odor of decaying animal tissue. Two such compounds have especially descriptive common names—putrescine and cadaverine:

$$H_2N—CH_2CH_2CH_2CH_2—NH_2$$
putrescine
(1,4-diaminobutane)

$$H_2N—CH_2CH_2CH_2CH_2CH_2—NH_2$$
cadaverine
(1,5-diaminopentane)

✔ **LEARNING CHECK 16.5** Draw structural formulas to show how each of the following amines forms hydrogen bonds with water:

a. $CH_3—\underset{\underset{H}{|}}{N}—H$ b. $CH_3CH_2—\underset{\underset{CH_3}{|}}{N}—CH_3$

16.4 | Chemical Properties of Amines

Learning Objective

4. Recognize and write key reactions for amines.

Amines undergo many chemical reactions, but we will limit our study to the two most important properties of amines: basicity and amide formation.

16.4A Basicity and Amine Salts

The single most distinguishing feature of amines is their behavior as weak bases. They are the most common organic bases. We know from inorganic chemistry that ammonia behaves as a Brønsted base by accepting protons, and in the process becomes a conjugate acid. The reaction of NH_3 with HCl gas illustrates this point (Reaction 16.1):

$$\underset{\text{ammonia}}{H—\overset{\overset{H}{|}}{\underset{\underset{H}{|}}{N}}:} + H—Cl \longrightarrow \left[H—\overset{\overset{H}{|}}{\underset{\underset{H}{|}}{N}}—H \right]^+ \underset{\underset{\text{ion}}{\text{ammonium}}}{} + \underset{\underset{\text{ion}}{\text{chloride}}}{Cl^-} \qquad (16.1)$$

We also remember from inorganic chemistry that Brønsted bases such as ammonia react with water to liberate OH^- ions (Reaction 16.2):

$$\underset{\text{ammonia}}{H—\overset{\overset{H}{|}}{\underset{\underset{H}{|}}{N}}:} + \underset{\text{water}}{H—OH} \rightleftharpoons \left[H—\overset{\overset{H}{|}}{\underset{\underset{H}{|}}{N}}—H \right]^+ \underset{\underset{\text{ion}}{\text{ammonium}}}{} + \underset{\underset{\text{ion}}{\text{hydroxide}}}{OH^-} \qquad (16.2)$$

Because the amines are derivatives of ammonia, they react in similar ways. In water, they produce OH^- ions:

General reaction:	$R-NH_2 + H_2O \rightleftarrows R-NH_3^+ + OH^-$	(16.3)
Specific reaction:	$CH_3-NH_2 + H_2O \rightleftarrows CH_3-NH_3^+ + OH^-$	(16.4)
	methylamine \qquad methylammonium ion	

Example 16.1 Completing Reactions

Complete the following reaction:

$$CH_3-NH-CH_3 + H_2O \rightleftarrows$$

Solution

Primary, secondary, and tertiary amines are all basic and exhibit the same reaction in water.

$$CH_3-NH-CH_3 + H_2O \rightleftarrows CH_3-\overset{+}{N}H_2-CH_3 + OH^-$$

✔ **LEARNING CHECK 16.6** Complete the following reactions:

a. $CH_3 - \underset{\underset{CH_3}{|}}{N} - CH_2CH_3 + H_2O \rightleftarrows$

b. ⬡$-NH_2 + H_2O \rightleftarrows$

All amines behave as weak bases and form salts when they react with acids such as HCl:

General reaction:	$R-NH_2 + HCl \rightarrow R-NH_3^+Cl^-$	(16.5)
	amine \quad acid \quad amine salt	
Specific example:	$CH_3-NH_2 + HCl \rightarrow CH_3-NH_3^+Cl^-$	(16.6)
	methylamine $\;$ acid $\;$ methylammonium chloride	

Other acids, such as sulfuric, nitric, phosphoric, and carboxylic acids (Reaction 16.7), also react with amines to form salts:

$$CH_3CH_2-NH_2 + CH_3COOH \rightarrow CH_3CH_2-NH_3^+CH_3COO^- \qquad (16.7)$$
$$\text{ethylamine} \qquad \text{acetic acid} \qquad \text{ethylammonium acetate}$$

✔ **LEARNING CHECK 16.7** Complete the following reactions:

a. ⬡$-NH - CH_3 + HCl \longrightarrow$

b. ⬠$^{NH_2} + CH_3CH_2COOH \longrightarrow$

As we have seen, amine salts may be named by changing "amine" to "ammonium" and adding the name of the negative ion derived from the acid. Some more examples are

$$CH_3 - \underset{\underset{CH_2CH_3}{|}}{\overset{\overset{CH_3}{|}}{N}} - H^+Br^- \qquad\qquad (CH_3CH_2)_3NH^+CH_3COO^-$$

ethyldimethylammonium bromide \qquad triethylammonium acetate

Amine salts have physical properties characteristic of other ionic compounds. They are white crystalline solids with high melting points. Because they are ionic, amine salts are more soluble in water than the parent amines. For this reason, amine drugs are often given in the form of salts because they will dissolve in body fluids.

morphine
(water insoluble)

morphine sulfate
(water soluble)

(16.8)

Amine salts are easily converted back to amines by adding a strong base:

$$R—NH_3^+Cl^- + NaOH \rightarrow R—NH_2 + H_2O + NaCl \qquad (16.9)$$

The form in which amines occur in solutions is pH dependent, just as is the case for carboxylic acids (see **Section 15.3**). In solutions of high H^+ (low pH), the salt form of amines predominates. In solutions of high OH^- (high pH), the amine form is favored:

$$R—NH_2 \underset{OH^-}{\overset{H^+}{\rightleftarrows}} R—NH_3^+ \qquad (16.10)$$

amine amine salt
(high pH) (low pH)

The amine cations in the salts we have discussed to this point have one, two, or three alkyl groups attached to the nitrogen:

$$R—\overset{+}{N}H_3 \qquad R—\overset{+}{N}H_2—R' \qquad R—\overset{+}{N}H—R'$$
$$\qquad\qquad\qquad\qquad\qquad\qquad\qquad\qquad | $$
$$\qquad\qquad\qquad\qquad\qquad\qquad\qquad\qquad R''$$

primary amine
cation

secondary amine
cation

tertiary amine
cation

It is also possible to have amine cations in which four alkyl groups or benzene rings are attached to the nitrogen. An example is:

$$\left[CH_3CH_2 - \overset{\overset{\displaystyle CH_2CH_3}{|}}{\underset{\underset{\displaystyle CH_3}{|}}{N}} - CH_2CH_3 \right]^+ \quad Cl^-$$

triethylmethylammonium chloride

These salts are called **quaternary ammonium salts** and are named the same way as other amine salts. Unlike other amine salts, quaternary salts contain no hydrogen attached to the nitrogen that can be removed by adding a base. Thus, quaternary salts are present in solution in only one form, which is independent of the pH of the solution.

quaternary ammonium salt
An ionic compound containing a positively charged ion in which four alkyl or aromatic groups are bonded to nitrogen, as in R_4N^+.

A Wake-Up Call for Treating Insomnia

Insomnia is defined differently by people who claim to have it. Clinically, it is considered to be a persistent disorder that makes it difficult to fall or stay asleep despite being in an environment conducive to sleep. Without adequate sleep a person typically feels unrefreshed, fatigued, and lacking in energy. This leads to a decrease in mental and physical performance as well as adverse health consequences that influence the quality of life and longevity. Conditions such as diabetes, obesity, heart disease, and high blood pressure are a few disorders that have been associated with insomnia.

As with treating any health disorder, the underlying cause or causes must be considered, and not just the symptoms treated. Numerous causes of insomnia have been identified, including stress, anxiety, depression, and some diseases. Even changes in home or work environments can contribute to an inability to sleep. Social and personal habits such as inconsistent sleep scheduling or participation in stimulating activities or eating near bedtime have also been found to contribute to insomnia. The ingestion of stimulants such as caffeine or medications that alter sleep patterns can also contribute to the problem. The nighttime urination associated with prostrate problems in older men and the night sweats associated with menopause in older women are examples of age-related conditions that frequently disrupt sleep.

The first line of treatment focuses on management of any underlying causes. If that doesn't correct insomnia, there are several therapy options. Since chemistry students are interested in chemical solutions to problems, some chemical options will be considered. These will include substances available by prescription only. These substances will be indicated by Rx following the name. Substances available over the counter without a prescription will be indicated by OTC following the name. The list of substances follows:

Melatonin (OTC): This compound is a hormone secreted by the pineal gland. It regulates our day/night sleep cycle. This compound is considered to be safe for short-term use, with good effectiveness in older patients.

Sedating herbs (OTC): As the category title implies, these are herbs such as valerian, kava kava, and chamomile. All are considered to be safe, but they carry the same concerns as any sedating agent.

Rozerem (Rx): This compound is a melatonin receptor competitor. It is approved for short-term use to initiate sleep. It is not a controlled substance.

Benzodiazepines (Rx): Dalmane, Restoril, Valium, etc. Many compounds are in this class. They are all approved for short-term use. All are Schedule IV controlled substances.

Non-benzodiazepine hypnotics (Rx): Ambien, Zolpidem, and many others. They are approved for short-term use. There is little rebound insomnia when stopped. There is a lower risk of dependence than benzodiazepines. All are Schedule IV controlled substances.

Belsomra (Suvorexant) (Rx): It is approved for sleep onset and maintenance. There is little or no rebound. It is a Schedule IV controlled substance.

Over-the-counter sleep aids: Sominex (diphenhydramines), Unisom (doxylamine), and other brands. These contain sedating antihistamines. They are often associated with daytime tiredness, cognitive deficits, and confusion.

There are many medications that induce and maintain sleep. Some are available over the counter, some are available only by prescription, and some are controlled substances. These therapies are not equally effective or safe. Great care should be used in their selection and continued use. These therapies can be very helpful when used correctly, but there is a risk of adverse effects and habit formation. The wake-up call for treating insomnia is that it is not easily treated by just taking a pill. It is a serious problem. Professional help should be obtained and their directions followed.

Insomnia is fairly common and has many causes.

Choline is an important quaternary ammonium ion that is a component of certain lipids (see **Section 18.6**). Acetylcholine is a compound involved in the transmission of nerve impulses from one cell to another (see **Section 16.5**):

$$\left[HO-CH_2CH_2-\overset{\overset{\displaystyle CH_3}{|}}{\underset{\underset{\displaystyle CH_3}{|}}{N}}-CH_3 \right]^+ \quad \left[CH_3-\overset{\overset{\displaystyle O}{\|}}{C}-O-CH_2CH_2-\overset{\overset{\displaystyle CH_3}{|}}{\underset{\underset{\displaystyle CH_3}{|}}{N}}-CH_3 \right]^+$$

<div align="center">choline cation acetylcholine cation</div>

Some quaternary ammonium salts are important because they have disinfectant properties. For example, benzalkonium chloride (Zephiran®) is a well-known antiseptic compound that kills many pathogenic (disease-causing) bacteria and fungi on contact:

$$\left[\bigcirc\!\!\!-CH_2-\overset{\overset{\displaystyle CH_3}{|}}{\underset{\underset{\displaystyle CH_3}{|}}{N}}-R \right]^+ \quad Cl^-$$

<div align="center">Zephiran chloride
(R represents a long alkyl chain)</div>

Its detergent action destroys the membranes that coat and protect the microorganisms. Zephiran chloride is recommended as a disinfectant solution for skin and hands prior to surgery and for the sterile storage of instruments. The trade names of some other anti-infectives that contain quaternary ammonium salts are Phemerol®, Bactine®, and Ceepryn®.

16.4B Amide Formation

Amines react with acid chlorides or acid anhydrides to form amides. The characteristic functional group of **amides** is a carbonyl group attached to nitrogen. The single bond linking the carbonyl carbon and nitrogen atoms in the group is called an **amide linkage**:

$$\overset{\overset{\displaystyle O}{\|}}{-C}\overset{\text{— amide linkage}}{-}\overset{}{\underset{|}{N}}-$$

<div align="center">amide functional group</div>

amide
An organic compound having the functional group

$$-\overset{\overset{\displaystyle O}{\|}}{C}-\underset{|}{N}-.$$

amide linkage The carbonyl carbon–nitrogen single bond of the amide group.

Amides may be thought of as containing an ammonia or amine portion and a portion derived from a carboxylic acid:

$$-\overset{\overset{\displaystyle O}{\|}}{C}-\underset{|}{N}-$$

<div align="center">from a carboxylic acid
from ammonia or an amine</div>

However, we've seen that the reaction of an amine with a carboxylic acid normally produces a salt and not an amide. For this reason, the preparation of amides requires the more reactive acid chlorides or acid anhydrides.

Both primary and secondary amines can be used to prepare amides in this manner, a method similar to the formation of esters (see **Section 15.5**). Tertiary amines, however, do not form amides because they lack a hydrogen atom on the nitrogen.

General reactions:

$$\underset{\text{acid chloride}}{R-\overset{\overset{\displaystyle O}{\|}}{C}-Cl} \;+\; \underset{\text{amine}}{R'-\underset{\underset{\displaystyle H}{|}}{N}-H} \;\longrightarrow\; \underset{\text{amide}}{R-\overset{\overset{\displaystyle O}{\|}}{C}-\underset{\underset{\displaystyle H}{|}}{N}-R'} \;+\; HCl \qquad (16.11)$$

$$\underset{\text{acid anhydride}}{R-\overset{\overset{\displaystyle O}{\|}}{C}-O-\overset{\overset{\displaystyle O}{\|}}{C}-R} \;+\; \underset{\text{amine}}{R'-\underset{\underset{\displaystyle H}{|}}{N}-H} \;\longrightarrow\; \underset{\text{amide}}{R-\overset{\overset{\displaystyle O}{\|}}{C}-\underset{\underset{\displaystyle H}{|}}{N}-R'} \;+\; \underset{\substack{\text{carboxylic}\\\text{acid}}}{R-\overset{\overset{\displaystyle O}{\|}}{C}-OH} \quad (16.12)$$

Specific examples:

$$\underset{\text{acetic anhydride}}{CH_3-\overset{\overset{\displaystyle O}{\|}}{C}-O-\overset{\overset{\displaystyle O}{\|}}{C}-CH_3} \;+\; \underset{\text{ammonia}}{H-\underset{\underset{\displaystyle H}{|}}{N}-H} \;\longrightarrow\; \underset{\text{acetamide}}{CH_3-\overset{\overset{\displaystyle O}{\|}}{C}-\underset{\underset{\displaystyle H}{|}}{N}-H} \;+\; \underset{\text{acetic acid}}{CH_3-\overset{\overset{\displaystyle O}{\|}}{C}-OH} \quad (16.13)$$

$$\underset{\substack{\text{benzoyl chloride}}}{\text{(ring)}-\overset{\overset{\displaystyle O}{\|}}{C}-Cl} \;+\; \underset{\substack{\text{methylamine}\\\text{(a 1° amine)}}}{H-\underset{\underset{\displaystyle H}{|}}{N}-CH_3} \;\longrightarrow\; \underset{\substack{\text{N-methylbenzamide}}}{\text{(ring)}-\overset{\overset{\displaystyle O}{\|}}{C}-NH-CH_3} \;+\; HCl \quad (16.14)$$

$$\underset{\substack{\text{benzoyl chloride}}}{\text{(ring)}-\overset{\overset{\displaystyle O}{\|}}{C}-Cl} \;+\; \underset{\substack{\text{dimethylamine}\\\text{(a 2° amine)}}}{H-\underset{\underset{\displaystyle CH_3}{|}}{N}-CH_3} \;\longrightarrow\; \underset{\substack{\text{N,N-dimethylbenzamide}}}{\text{(ring)}-\overset{\overset{\displaystyle O}{\|}}{C}-\underset{\underset{\displaystyle CH_3}{|}}{N}-CH_3} \;+\; HCl \quad (16.15)$$

✔ **LEARNING CHECK 16.8** Complete the following reactions:

a. $CH_3-\overset{\overset{\displaystyle O}{\|}}{C}-O-\overset{\overset{\displaystyle O}{\|}}{C}-CH_3 \;+\; CH_3CH_2CH_2-NH_2 \;\longrightarrow$

b. $CH_3CH_2-\overset{\overset{\displaystyle O}{\|}}{C}-Cl \;+\; NH_3 \;\longrightarrow$

Reaction of diacid chlorides with diamines produces polyamides that, like polyesters, are condensation polymers. The repeating units in polyamides are held together by amide linkage. An example is the formation of nylon, shown in Reaction 16.16 and **Figure 16.2**.

$$n\text{Cl}-\overset{\overset{\text{O}}{\|}}{\text{C}}-(\text{CH}_2)_4-\overset{\overset{\text{O}}{\|}}{\text{C}}-\text{Cl} \;+\; n\text{H}_2\text{N}-(\text{CH}_2)_6-\text{NH}_2 \longrightarrow$$

adipoyl chloride 　　　 hexamethylenediamine

$$---\overset{\overset{\text{O}}{\|}}{\text{C}}-(\text{CH}_2)_4-\overset{\overset{\text{O}}{\|}}{\text{C}}\overset{amide\ linkages}{-\text{NH}-(\text{CH}_2)_6-\text{NH}}-\overset{\overset{\text{O}}{\|}}{\text{C}}-(\text{CH}_2)_4-\overset{\overset{\text{O}}{\|}}{\text{C}}\overset{amide\ linkage}{-\text{NH}-(\text{CH}_2)_6-\text{NH}}--- \;+\; n\text{HCl}$$

polyamide nylon

(16.16)

Three billion pounds of nylon and related polyamides are produced annually. About 60% of that total goes to make the nylon fiber used in home furnishings such as carpet. The remainder is used largely as a textile fiber in clothing (shirts, dresses, stockings, underwear, etc.) and as tire cord. Minor but important uses include fasteners, rope, parachutes, paintbrushes, and electrical parts (see **Figure 16.3**). In medicine, nylon is used in specialized tubing. Nylon sutures were the first synthetic sutures and are still used.

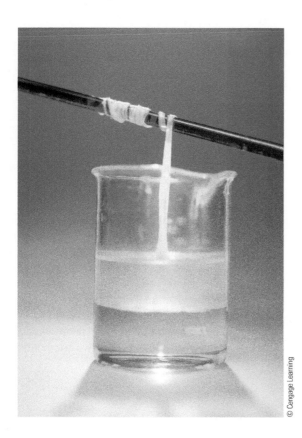

Figure 16.2 The preparation of nylon. The reactants occupy separate immiscible layers. Hexamethylenediamine is dissolved in water (bottom layer), and adipoyl chloride is dissolved in hexane (top layer). The two compounds react at the interface between the two layers to form a film of nylon, which is lifted by the rod.

© Cengage Learning

Silk and wool are natural polyamides that we classify as proteins (see **Chapter 19** for a detailed discussion). For now, note the repeating units and amide linkages characteristic of proteins as shown in the following portion of a polyamide protein chain:

$$---\text{NH}-\underset{\underset{\text{R}}{|}}{\text{CH}}-\overset{\overset{\text{O}}{\|}}{\text{C}}-\text{NH}-\underset{\underset{\text{R}'}{|}}{\text{CH}}-\overset{\overset{\text{O}}{\|}}{\text{C}}-\text{NH}-\underset{\underset{\text{R}''}{|}}{\text{CH}}-\overset{\overset{\text{O}}{\|}}{\text{C}}-\text{NH}-\underset{\underset{\text{R}'''}{|}}{\text{CH}}-\overset{\overset{\text{O}}{\|}}{\text{C}}---$$

amide linkages

Aspirin Substitutes

Even though aspirin is an excellent and useful drug it occasionally produces adverse effects that cannot be tolerated by some people. These effects include gastrointestinal bleeding and such allergic reactions as skin rashes and asthmatic attacks. Children who are feverish with influenza or chickenpox should not be given aspirin because of a strong correlation with Reye's syndrome, a devastating illness characterized by liver and brain dysfunction.

The amide *acetaminophen,* marketed under trade names such as Tylenol®, Excedrin Aspirin Free®, Panadol®, and Anacin-3®, is an effective aspirin substitute available for use in such situations. Acetaminophen does not irritate the intestinal tract and yet has comparable analgesic (pain-relieving) and antipyretic (fever-reducing) effects. Acetaminophen is available in a stable liquid form suitable for administration to children and other patients who have difficulty taking tablets or capsules. Unlike aspirin, acetaminophen has virtually no anti-inflammatory effect and is therefore a poor substitute for aspirin in the treatment of inflammatory disorders such as rheumatoid arthritis. The overuse of acetaminophen has been linked to liver and kidney damage.

acetaminophen

Ibuprofen, a carboxylic acid rather than an amide, is a drug with analgesic, antipyretic, and anti-inflammatory properties that was available for decades only as a prescription drug under the trade name Motrin®.

ibuprofen

In 1984, it became available in an over-the-counter (OTC) form under trade names such as Advil®, Ibuprin®, Mediprin®, and Motrin IB®. Ibuprofen is thought to be somewhat superior to aspirin as an anti-inflammatory drug, but it has about the same effectiveness as aspirin in relieving mild pain and reducing fever.

In 1994, the FDA approved another compound for sale as an over-the-counter pain reliever. *Naproxen,* marketed under the trade names Aleve® and Anaprox®, exerts its effects for a longer time in the body (8–12 hours per dose) than ibuprofen (4–6 hours per dose) or aspirin and acetaminophen (4 hours per dose). However, the chances for causing slight intestinal bleeding and stomach upset are greater with naproxen than with ibuprofen, and naproxen is not recommended for use by children under age 12.

naproxen

Ketoprofen, a prescription pain reliever, has the advantage of requiring very small doses when compared to other pain-relieving products. A dose of ketoprofen is 12.5 mg, compared to doses of 200–500 mg for the other products. The tiny, easy-to-swallow tablet is a characteristic emphasized by the manufacturers in advertising. The side effects of ketoprofen are similar to those of aspirin and ibuprofen, so ketoprofen should not be used by people known to have problems with either of these other two compounds.

ketoprofen

Figure 16.3 Parachutes made of nylon fabric.

16.5 | Amines as Neurotransmitters

Learning Objective

5. Name amines used as neurotransmitters.

Neurotransmitters, the chemical messengers of the nervous system, carry nerve impulses from one nerve cell (neuron) to another. A neuron consists of a bulbous body called the soma (or cell body) attached to a long stemlike projection called an axon (see **Figure 16.4**). Numerous short extensions called dendrites are attached to the large bulbous end of the soma, and filaments called synaptic terminals are attached to the end of the axon. At low magnification, the synaptic terminals of one neuron's axon appear to be attached to the dendrites

neurotransmitter A substance that acts as a chemical bridge in nerve impulse transmission between nerve cells.

Figure 16.4 A nerve cell and the transmission of a nerve signal.

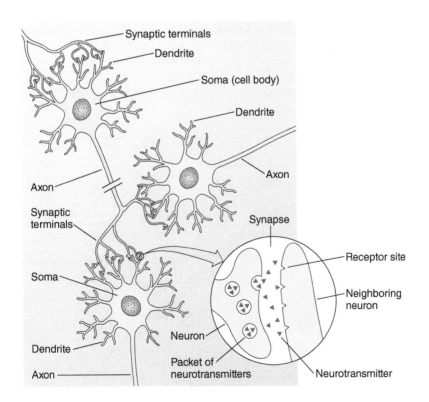

of an adjacent neuron. However, upon further magnification, a small gap called a synapse is visible between the synaptic terminals of the axon and the dendrites of the next neuron.

Molecules of a neurotransmitter are stored in small pockets in the axon near the synapse. These pockets release neurotransmitter molecules into the synapse when an electrical current flows from the soma, along the axon to the synaptic terminals. The electrical current is generated by the exchange of positive and negative ions across membranes. The released neurotransmitter molecules diffuse across the synapse and bind to receptors on the dendrites of the next neuron. Once the neurotransmitter binds to the receptor, the message has been delivered. The receiving cell then sends an electrical current down its own axon until a synapse is reached, and a neurotransmitter delivers the message across the synapse to the next neuron. In the central nervous system—the brain and spinal cord—the most important neurotransmitters are acetylcholine and three other amines: norepinephrine, dopamine, and serotonin:

norepinephrine

dopamine

serotonin

acetylcholine cation

Neurotransmitters are not only chemical messengers for the nervous system, they may also be partly responsible for our moods. A simplified biochemical theory of mental illness is based on two amines found in the brain. The first is norepinephrine (NE). When an excess of norepinephrine is formed in the brain, the result is a feeling of elation. Extreme excesses of NE can even induce a manic state, while low NE levels may be a cause of depression.

At least six different receptors in the body can be activated by norepinephrine and related compounds. A class of drugs called beta blockers reduces the stimulant action of epinephrine and norepinephrine on various kinds of cells. Some beta blockers are used to treat cardiac arrhythmias, angina, and hypertension (high blood pressure). They function by slightly decreasing the force of each heartbeat. Depression is sometimes a side effect of using such drugs.

In a complex, multistep process, the amino acid tyrosine is used in the body to synthesize norepinephrine:

tyrosine

dopa

dopamine

norepinephrine

A Reaction Map for Amines

We have discussed four different reactions of amines. When trying to decide what happens in a reaction, remember to focus on the functional groups. After identifying an amine as a starting material, next decide what other reagent is involved and which of the four pathways is appropriate.

Each step in the process is catalyzed by at least one enzyme, and each of the intermediates formed in the synthesis has physiological activity. Dopa is used as a treatment for Parkinson's disease, and dopamine has been used to treat low blood pressure. Tyrosine is an essential amino acid that must be obtained from the diet, so what we eat may make a direct contribution to our mental health.

The second amine often associated with the biochemical theory of mental illness is serotonin. Like norepinephrine, it also functions as a neurotransmitter. Serotonin is produced in the body from the amino acid tryptophan:

tryptophan serotonin

Serotonin

Serotonin has been found to influence sleeping, the regulation of body temperature, and sensory perception, but its exact role in mental illness is not yet clear. Unusually low levels of 5-hydroxyindoleacetic acid, a product of serotonin utilization, are characteristically found in the spinal fluid of victims of violent suicide. Drugs that mimic serotonin are sometimes used to treat depression, anxiety, and obsessive-compulsive disorder. Serotonin blockers are used to treat migraine headaches and relieve the nausea that accompanies cancer chemotherapy. A better understanding of the biochemistry of the brain may lead to better medications for treating various forms of mental illness.

16.6 | Other Biologically Important Amines

Learning Objective

6. Give uses for specific biological amines.

16.6A Epinephrine

Epinephrine, or adrenaline, has a molecular structure similar to that of norepinephrine. The difference is that epinephrine contains an *N*-methyl group:

epinephrine
(adrenaline)

Epinephrine is more important as a hormone than as a neurotransmitter. It is synthesized in the adrenal gland and acts to increase the blood level of glucose, a key source of energy for body processes. The release of epinephrine—usually in response to pain, anger, or fear—provides glucose for a sudden burst of energy. For this reason, epinephrine has been called the fight-or-flight hormone.

Epinephrine also raises blood pressure. It does this by increasing the rate and force of heart contractions and by constricting the peripheral blood vessels. Injectable local anesthetics usually contain epinephrine because it constricts blood vessels in the vicinity of the injection. This prevents the blood from rapidly "washing" the anesthetic away and prolongs the anesthetic effect in the target tissue. Epinephrine is also used to reduce hemorrhage, treat asthma attacks, and combat anaphylactic shock (collapse of the circulatory system).

16.6B Amphetamines

Amphetamine (also known as Benzedrine®) is a powerful nervous system stimulant with an amine structure similar to that of epinephrine:

Amphetamine

amphetamine
(Benzedrine)

epinephrine
(adrenaline)

Notice that both compounds contain an aromatic ring attached to a substituted ethylamine. That structural arrangement, sometimes referred to as a phenethylamine, is apparently an essential feature of several physiologically active compounds. These structural similarities lead to similar behaviors in the body: Both compounds raise the glucose level in the blood and increase pulse rate and blood pressure. Other closely related phenethylamine compounds, such as Methedrine®, also act as powerful nervous system stimulants. Compounds of this type can be thought of as being derived from amphetamine and as a result are sometimes collectively referred to as **amphetamines**:

amphetamines A class of drugs structurally similar to epinephrine, used to stimulate the central nervous system.

N-methylamphetamine
(Methedrine, or "speed")

Amphetamines are used both legally (as prescribed by a physician) and illegally to elevate mood or reduce fatigue. In the drug culture, amphetamine tablets go by such names as bennies, pep pills, reds, red devils, speed, dexies, and uppers. Some amphetamines (known as STP, speed, and mescaline) cause hallucinations. The abuse of such drugs results in severe detrimental effects on both the body and the mind. They are addictive and concentrate in the brain and nervous system. Abusers often experience long periods of sleeplessness, weight loss, and paranoia. Abusers of amphetamines often compound their problems by taking other drugs in order to prevent the "crash," or deep depression, brought on by discontinuation of the drug.

16.6C Alkaloids

Plants are the sources of some of the most powerful drugs known. Numerous primitive tribes in the world possess knowledge about the physiological effects that result from eating or chewing the leaves, roots, flowers, or bark of certain plants. The effects vary from plant species to plant species. Some cure diseases; others, such as opium, are addictive drugs. Still others are deadly poisons, such as the leaves of the belladonna plant and the hemlock herb. The substances responsible for these marked physiological effects are often nitrogen-containing compounds called **alkaloids**. The name, which means alkali-like, reflects the fact that these amine compounds are weakly basic.

alkaloids A class of nitrogen-containing organic compounds obtained from plants.

The molecular structures of alkaloids vary from simple to complex. Two common and relatively simple alkaloids are nicotine, which is found in tobacco, and caffeine, which is present in coffee and cola drinks:

nicotine

Nicotine

caffeine

Caffeine

Many people absorb nicotine into their bodies as a result of smoking or chewing tobacco. In small doses, nicotine behaves like a stimulant and is not especially harmful. However, it is habit-forming, and habitual tobacco users are exposed to other harmful substances such as tars, carbon monoxide, and polycyclic carcinogens.

Besides coffee and cola drinks, other sources of caffeine are tea, chocolate, and cocoa. Caffeine is a mild stimulant of the respiratory and central nervous systems, the reason for its well-known side effects of nervousness and insomnia. These characteristics, together with its behavior as a mild diuretic, account for the use of caffeine in a wide variety of products, including pain relievers, cold remedies, diet pills, and "stay-awake" pills (No-Doz®). Because caffeine is considered to be a drug, pregnant women should be prudent about how much caffeine they consume. Like most other drugs, caffeine enters the bloodstream, crosses the placental barrier, and reaches the fetus.

A number of alkaloids are used medicinally. Quinine, for example, is used to treat malaria. Atropine is used as a preoperative drug to relax muscles and reduce the secretion of saliva in surgical patients, and to dilate the pupil of the eye in patients undergoing eye examinations.

quinine

atropine

Try a Little Chocolate

Chocolate has a history that goes back beyond 1492 when Columbus arrived on the American continent. Chocolate in various forms is prepared from the seeds of the cacao plant. The seeds are roasted, husked, and then ground into a powder. The Mayan word "cacao" comes from two Mayan words that together mean "bitter juice." The Mayans were able to convert this bitter material into a form that we, today, call chocolate. In this form it was reserved for Mayan and Aztec royalty, and other high-ranking individuals.

Today, chocolate in various forms is commonplace. Americans celebrate major holidays with delicious gifts of this once rare food. Many of us eat it daily, and some refer to themselves as "chocoholics." Like other rich foods, chocolate is regarded by many as an indulgence or guilty pleasure. Because of its high sugar and fat content, chocolate has been associated with health problems such as obesity, acne, diabetes, heart disease, and tooth decay. But scientists are now discovering health benefits, and offering some justification for including chocolate in a healthy diet.

Some recent studies correlate chocolate consumption with reduced cardiovascular disease, lower blood pressure, lower cholesterol levels, and improved cognition and memory. A Canadian study with a sample size of nearly 45,000 people concluded that chocolate eaters are 22% less likely to suffer a stroke, and for those who did experience a stroke, chocolate eaters were 46% less likely to die from the stroke. Some Harvard Medical School research indicates that drinking 2 cups of hot chocolate per day helps preserve brain health and prevent memory decline in older people.

Most of the world's chocolate is grown in a narrow area of the world located 10 degrees on either side of the equator (mostly in West Africa). The exported chocolate from this area reaches nearly every corner of the world in multiple forms. The word "chocolate" can refer to anything from white chocolate to milk chocolate. However, the chocolate that provides the most health benefits is *dark* chocolate. Dark chocolate is rich in phenolic compounds called flavonoids. Flavonoids are also found in fruits, vegetables, tea, red wine, and beer. The flavonoids in chocolate are powerful antioxidants that protect cells against oxidation damage. Darker chocolate has a higher flavonoid content than lighter varieties. So, if you are seeking health benefits from chocolate, the darker the chocolate, the better. Because the flavor of dark chocolate is so strong, pairing it with fruit or nuts softens the taste and adds other nutrients.

The bottom line is that high-quality dark chocolate that is not loaded with sugar and fat can provide impressive health benefits. The recommended "dose" of chocolate is between 1.5 and 3 ounces per day. In addition to improving your chances for a long and healthy life, chocolate provides an enjoyable highlight to any day.

Dark chocolate can be part of a healthy diet.

Opium, the dried juice of the poppy plant, has been used for centuries as a painkilling drug. It contains numerous alkaloids, including morphine and its methyl ether, codeine. Both are central nervous system depressants, with codeine being much weaker. They exert a soothing effect on the body when administered in sufficient amounts. These properties make them useful as painkillers. Morphine is especially useful in this regard, being one of the most effective painkillers known. Codeine is used in some cough syrups to depress the action of the cough center in the brain. The major drawback to using these drugs is that they are all addictive. Heroin, a derivative of morphine, is one of the more destructive hard drugs used illegally in our society. A substantial number of addicts commit crimes to support their habits.

morphine

codeine

heroin

Morphine

16.7 | The Nomenclature of Amides

Learning Objective

7. Assign IUPAC names for amides.

Simple amides are named after the corresponding carboxylic acids by changing the *-ic* ending (common names) or the *-oic* ending (IUPAC names) of the acid to *-amide*. Common names are used more often than IUPAC names. Some examples of the assignment of such names follow:

$$\underset{\text{O}}{\overset{\text{O}}{\underset{\|}{\text{H}-\text{C}-\text{OH}}}} \longrightarrow \underset{\text{O}}{\overset{\text{O}}{\underset{\|}{\text{H}-\text{C}-\text{NH}_2}}}$$

Common name	formic acid	formamide
IUPAC name	methanoic acid	methanamide

$$\overset{\text{O}}{\underset{\|}{\text{CH}_3-\text{C}-\text{OH}}} \longrightarrow \overset{\text{O}}{\underset{\|}{\text{CH}_3-\text{C}-\text{NH}_2}}$$

Common name	acetic acid	acetamide
IUPAC name	ethanoic acid	ethanamide

$$\overset{\text{O}}{\underset{\|}{\text{C}-\text{OH}}} \longrightarrow \overset{\text{O}}{\underset{\|}{\text{C}-\text{NH}_2}}$$

Common name	benzoic acid	benzamide
IUPAC name	benzoic acid	benzamide

✔ LEARNING CHECK 16.9

a. Give the IUPAC name for the following simple amide:

$$\overset{\text{O}}{\underset{\|}{\text{CH}_3\text{CH}_2\text{CH}_2-\text{C}-\text{NH}_2}}$$

b. What is the IUPAC name for the simple amide derived from hexanoic acid?

If alkyl groups are attached to the nitrogen atom of amides (substituted amides), the name of the alkyl group precedes the name of the amide, and an italic capital *N* is used to indicate that the alkyl group is bonded to the nitrogen.

See the following examples.

$$H - \overset{\overset{\displaystyle O}{\|}}{C} - NH - CH_3 \qquad CH_3 - \overset{\overset{\displaystyle O}{\|}}{C} - \underset{\underset{\displaystyle CH_3}{|}}{N} - CH_3$$

Common name N-methylformamide N,N-dimethylacetamide N-ethyl-N-methylbenzamide
IUPAC name N-methylmethanamide N,N-dimethylethanamide N-ethyl-N-methylbenzamide

✔ **LEARNING CHECK 16.10** Give either a common name or the IUPAC name for each of the following amides:

a. $CH_3 - \overset{\overset{\displaystyle O}{\|}}{C} - \underset{\underset{\displaystyle CH_2CH_3}{|}}{N} - CH_2CH_3$

b. $CH_3CH_2CH_2 - \overset{\overset{\displaystyle O}{\|}}{C} - NH - CH_2\overset{\overset{\displaystyle CH_3}{|}}{C}HCH_3$

c. (benzene ring)$- \overset{\overset{\displaystyle O}{\|}}{C} - \underset{\underset{\displaystyle CH_2CH_3}{|}}{N} -$(benzene ring)

d. (benzene ring with CH₃)$- \overset{\overset{\displaystyle O}{\|}}{C} - NH - CH_3$

16.8 | Physical Properties of Amides

Learning Objective

8. Show the formation of hydrogen bonds with amides.

Formamide is a liquid at room temperature. It is the only unsubstituted amide that is a liquid; all others are solids.

Unsubstituted amides can form a complex network of intermolecular hydrogen bonds (see **Figure 16.5**). It is this characteristic that causes melting points of these substances to be so high.

Figure 16.5 Intermolecular hydrogen bonding in an unsubstituted amide.

The substitution of alkyl or aromatic groups for hydrogen atoms on the nitrogen reduces the number of intermolecular hydrogen bonds that can form and causes the melting points to decrease. Thus, disubstituted amides often have lower melting and boiling points than monosubstituted and unsubstituted amides. We'll see in Chapter 19 that hydrogen bonding between amide groups is important in maintaining the shape of protein molecules.

Amides are rather water soluble, especially those containing fewer than six carbon atoms. This solubility results from the ability of amides to form hydrogen bonds with water. Even disubstituted amides do this because of the presence of the carbonyl oxygen (see **Figure 16.6**).

Figure 16.6 Hydrogen bonding between water.

① an unsubstituted amide ② a disubstituted amide

✔ **LEARNING CHECK 16.11** Show how the amide below can form the following:

a. Intermolecular hydrogen bonds with other amide molecules
b. Hydrogen bonds with water molecules

$$CH_3 - \overset{\overset{\text{O}}{\|}}{C} - \overset{\overset{}{\underset{\underset{\text{H}}{|}}{N}}}{} - CH_3$$

16.9 | **Chemical Properties of Amides**

Learning Objective

9. Give the products of acidic and basic hydrolysis of amides.

16.9A Neutrality

Because basicity is the most important property of amines, it is only natural to wonder if amides are also basic. The answer is no. Although they are formed from carboxylic acids and basic amines, amides are neither basic nor acidic; they are neutral. The carbonyl group bonded to the nitrogen has destroyed the basicity of the original amine, and the nitrogen of the amine has replaced the acidic —OH of the carboxylic acid.

16.9B Amide Hydrolysis

The most important reaction of amides is hydrolysis. This reaction corresponds to the reverse of amide formation; the amide is cleaved into two compounds, a carboxylic acid and an amine or ammonia:

$$R - \overset{\overset{\text{O}}{\|}}{C} - NH - R' + H_2O \xrightarrow[\text{Heat}]{\substack{\text{Acid} \\ \text{or} \\ \text{base}}} R - \overset{\overset{\text{O}}{\|}}{C} - OH + R' - NH_2 \quad (16.17)$$

cleavage here

amide carboxylic acid amine

As with hydrolysis of carboxylic esters, amide hydrolysis requires the presence of a strong acid or base, and the nature of the final products depends on which of these reactants is used. Notice in the following examples that under acidic conditions, the salt of the amine is produced along with the carboxylic acid, and under basic conditions, the salt of the carboxylic acid is formed along with the amine. Also, notice that heat is required for the hydrolysis of amides.

General reactions:

$$R-\underset{\substack{\| \\ O}}{C}-NH-R' + H_2O + HCl \xrightarrow{\text{Heat}} R-\underset{\substack{\| \\ O}}{C}-OH + R'-NH_3^+Cl^- \quad (16.18)$$

\qquad amide $\qquad\qquad\qquad\qquad\qquad\qquad$ carboxylic acid \qquad amine salt

$$R-\underset{\substack{\| \\ O}}{C}-NH-R' + NaOH \xrightarrow{\text{Heat}} R-\underset{\substack{\| \\ O}}{C}-O^-Na^+ + R'-NH_2 \quad (16.19)$$

\qquad amide $\qquad\qquad\qquad\qquad\qquad$ carboxylate salt \qquad amine

Specific examples:

$$CH_3-\underset{\substack{\| \\ O}}{C}-NH_2 + H_2O + HCl \xrightarrow{\text{Heat}} CH_3-\underset{\substack{\| \\ O}}{C}-OH + NH_4^+Cl^- \quad (16.20)$$

\qquad acetamide $\qquad\qquad\qquad\qquad\qquad$ acetic acid \qquad ammonium chloride

$$CH_3-\underset{\substack{\| \\ O}}{C}-NH-CH_3 + H_2O + HCl \xrightarrow{\text{Heat}} CH_3-\underset{\substack{\| \\ O}}{C}-OH + CH_3-NH_3^+Cl^- \quad (16.21)$$

N-methylacetamide $\qquad\qquad\qquad\qquad$ acetic acid \qquad methylammonium chloride

$$CH_3-\underset{\substack{\| \\ O}}{C}-NH-CH_3 + NaOH \xrightarrow{\text{Heat}} CH_3-\underset{\substack{\| \\ O}}{C}-O^-Na^+ + CH_3-NH_2 \quad (16.22)$$

N-methylacetamide $\qquad\qquad\qquad\qquad$ sodium acetate \qquad methylamine

Because the form in which carboxylic acids and amines occur in solution depends on the solution pH, one of the hydrolysis products must always be in the form of a salt.

✔ **LEARNING CHECK 16.12** Complete the following hydrolysis reactions:

a. (benzene ring)—$\underset{\substack{\| \\ O}}{C}$—NH—CH$_2CH_3$ + H$_2$O + HCl $\xrightarrow{\text{Heat}}$

b. (benzene ring)—$\underset{\substack{\| \\ O}}{C}$—NH—CH$_2CH_3$ + NaOH $\xrightarrow{\text{Heat}}$

Amide hydrolysis, a very important reaction in biochemistry, is a central reaction in the digestion of proteins and the breakdown of proteins within cells. However, most amide hydrolysis in the body is catalyzed by enzymes rather than by strong acids or bases.

Because of their physiological properties, a number of amides play valuable roles in medicine. **Table 16.3** lists some of these important compounds.

TABLE 16.3 Some Important Amides in Medicine

Structure	Generic Name (Trade Name)	Use
	thiopental (Pentothal®)	Intravenous anesthesia
	amobarbital (Amytal®)	Treatment of insomnia
	diazepam (Valium®)	Tranquilizer
	ampicillin (Polycillin®)	Antibiotic

▶ Case Study Follow-up

In addition to helpful information on its website, the U.S. Food and Drug Administration has an educational pamphlet entitled *Medicines in My Home*. It can be viewed online or printed for distribution in doctor's offices, clinics, or schools. Many schools conduct education during the last week of October to help kids stay drug free, known as Red Ribbon Week. Educating children about over-the-counter (OTC) medications could easily be incorporated into such existing programs. Clinics that provide educational materials about OTC medications also provide a valuable public service.

Over-the-counter remedies that are mixtures of more than one medication can be convenient for the consumer. However, they significantly increase the choices available and can add

to confusion. Since the chance of adverse reactions increases with the number of medications taken, combined medications can be problematic. Educating the public about the effects of medications and their common and chemical names is of utmost importance.

Unfortunately, antihistamine overdose is common. The list of symptoms include agitation, blurred vision, coma, confusion, convulsions, delirium, diarrhea, drowsiness, dry mouth, fever, flushing, inability to urinate, incoherence, lack of sweat, nausea, rapid heart rate, stomach pain, and unsteadiness. The patient may be treated with activated charcoal, oxygen, IV fluids, and gastric lavage. If the patient survives the first 24 hours after an antihistamine overdose, fatalities are rare.

Concept Summary

Classification of Amines Amines are organic derivatives of ammonia in which one or more of the ammonia hydrogens are replaced by alkyl or aromatic groups. Amines are classified as primary, secondary, or tertiary, depending on the number of groups (one, two, or three) attached to the nitrogen.
Objective 1 (Section 16.1), Exercise 16.4

The Nomenclature of Amines Common names are given to simple amines by adding the ending *-amine* to the names of the alkyl groups attached to the nitrogen. In IUPAC names, the *-e* ending in the alkane name is changed to *-amine*. Aromatic amines are named as derivatives of aniline.
Objective 2 (Section 16.2), Exercises 16.8 and 16.10

Physical Properties of Amines Primary and secondary amines have boiling points slightly lower than those of corresponding alcohols. Tertiary amines have boiling points similar to those of alkanes. Low molecular weight amines are water soluble.
Objective 3 (Section 16.3), Exercises 16.16 and 16.18

Chemical Properties of Amines Amines are weak bases. They react with water to liberate hydroxide ions, and they react with acids to form salts. Amines react with acid chlorides and acid anhydrides to form amides.
Objective 4 (Section 16.4), Exercise 16.26

Amines as Neurotransmitters Neurotransmitters are the chemical messengers of the nervous system. They carry nerve impulses from one nerve cell (neuron) to another. The most important neurotransmitters are acetylcholine and three other amines: norepinephrine, dopamine, and serotonin.
Objective 5 (Section 16.5), Exercise 16.38

Other Biologically Important Amines Four neurotransmitters are acetylcholine, norepinephrine, dopamine, and serotonin. Epinephrine is also known as the fight-or-flight hormone. The amphetamines have structures similar to that of epinephrine. Alkaloids are nitrogen-containing compounds isolated from plants. They exhibit a variety of physiological effects on the body. Examples of alkaloids include nicotine, caffeine, quinine, atropine, morphine, and codeine.
Objective 6 (Section 16.6), Exercises 16.40 and 16.44

The Nomenclature of Amides Amides are named by changing the *-ic acid* or *-oic acid* ending of the carboxylic acid portion of the compound to *-amide*. Groups attached to the nitrogen of the amide are denoted by an italic capital N that precedes the name of the attached group.
Objective 7 (Section 16.7), Exercise 16.46

Physical Properties of Amides Low molecular weight amides are water soluble due to the formation of hydrogen bonds. The melting and boiling points of unsubstituted amides are higher than those of comparable substituted amides due to intermolecular hydrogen bonding.
Objective 8 (Section 16.8), Exercise 16.50

Chemical Properties of Amides Amides undergo hydrolysis in acidic conditions to yield a carboxylic acid and an amine salt. Hydrolysis under basic conditions produces a carboxylate salt and an amine.
Objective 9 (Section 16.9), Exercise 16.52

Key Terms and Concepts

Alkaloids (16.6)
Amide (16.4)
Amide linkage (16.4)
Amine (16.1)

Amphetamines (16.6)
Neurotransmitter (16.5)
Primary amine (16.1)
Quaternary ammonium salt (16.4)

Secondary amine (16.1)
Tertiary amine (16.1)

Key Reactions

1.	Reaction of amines with water (Section 16.4):	$R-NH_2 + H_2O \rightleftarrows R-NH_3^+ + OH^-$	Reaction 16.3
2.	Reaction of amines with acids (Section 16.4):	$R-NH_2 + HCl \rightarrow R-NH_3^+ + Cl^-$	Reaction 16.5

3.	Conversion of amine salts back to amines (Section 16.4):	$R—NH_3^+Cl^- + NaOH \rightarrow R—NH_2 + H_2O + NaCl$	Reaction 16.9
4.	Reaction of amines with acid chlorides to form amides (Section 16.4):	$R—\overset{\overset{\displaystyle O}{\|\|}}{C}—Cl + R'—\underset{\underset{\displaystyle H}{\|}}{N}—H \longrightarrow R—\overset{\overset{\displaystyle O}{\|\|}}{C}—\underset{\underset{\displaystyle H}{\|}}{N}—R' + HCl$	Reaction 16.11
5.	Reaction of amines with acid anhydrides to form amides (Section 16.4):	$R—\overset{\overset{\displaystyle O}{\|\|}}{C}—O—\overset{\overset{\displaystyle O}{\|\|}}{C}—R + R'—\underset{\underset{\displaystyle H}{\|}}{N}—H \rightarrow R—\overset{\overset{\displaystyle O}{\|\|}}{C}—\underset{\underset{\displaystyle H}{\|}}{N}—R' + R—\overset{\overset{\displaystyle O}{\|\|}}{C}—OH$	Reaction 16.12
6.	Acid hydrolysis of amides (Section 16.9):	$R—\overset{\overset{\displaystyle O}{\|\|}}{C}—NH—R' + H_2O + HCl \xrightarrow{\text{Heat}} R—\overset{\overset{\displaystyle O}{\|\|}}{C}—OH + R'—NH_3^+Cl^-$	Reaction 16.18
7.	Basic hydrolysis of amides (Section 16.9):	$R—\overset{\overset{\displaystyle O}{\|\|}}{C}—NH—R' + NaOH \xrightarrow{\text{Heat}} R—\overset{\overset{\displaystyle O}{\|\|}}{C}—O^-Na^+ + R'—NH_2$	Reaction 16.19

Exercises

Even-numbered exercises are answered in Appendix B.

Blue-numbered exercises are more challenging.

Classification of Amines (Section 16.1)

16.1 What is the difference among primary, secondary, and tertiary amines in terms of bonding to the nitrogen atom?

16.2 Give a general formula for a primary amine, a secondary amine, and a tertiary amine.

16.3 Classify each of the following as a primary, secondary, or tertiary amine:

a. $CH_3CH_2—\underset{\underset{\displaystyle CH_3}{\|}}{N}—CH_2CH_3$

b. $CH_3—\underset{\underset{\displaystyle CH_3}{\|}}{\overset{\overset{\displaystyle CH_3}{\|}}{C}}—NH_2$

c. a benzene ring with $NH—CH_3$

d. a pyrrolidine ring with $N—CH_3$

16.4 Classify each of the following as a primary, secondary, or tertiary amine:

a. $CH_3CH_2CH_2$ with $NH—CH_2CH_3$

b. a benzene ring with $\underset{\displaystyle}{\overset{\overset{\displaystyle CH_3}{\|}}{N}}—CH_3$

c. a cyclobutane ring with NH_2

d. $CH_3CH—NH—CHCH_3$ with CH_3 and CH_3 substituents

16.5 Draw structural formulas for all amines that have the molecular formula $C_4H_{11}N$. Label each one as a primary, secondary, or tertiary amine.

16.6 Draw structural formulas for the four amines that have the molecular formula C_3H_9N. Label each one as a primary, secondary, or tertiary amine.

The Nomenclature of Amines (Section 16.2)

16.7 Give each of the following amines a common name by adding the ending *-amine* to alkyl group names:

a. $CH_3\underset{\underset{\displaystyle NH_2}{\|}}{CH}CH_3$

b. a benzene ring with $NH—CH_2CH_3$

c. $CH_3CH_2CH_2—\underset{\underset{\displaystyle CH_2CH_3}{\|}}{N}—CH_3$

16.8 Give each of the following amines a common name by adding the ending *-amine* to alkyl group names:

a.

$$\text{C}_6\text{H}_5-\text{NH}-\text{CH}_3$$ (drawn as benzene ring) NH—CH₃

b. $CH_3CH_2-NH-CH_2CH_3$

c. CH_3CHCH_3 with NH—CH₂CH₃ substituent

16.9 Give each of the following amines an IUPAC name:

a. CH_3CHCH_3 with NH_2

b. CH_3CCH_3 with NH_2 and CH_3

c. $CH_3CH_2CH_2$ with NH—CH₂CH₃

16.10 Give each of the following amines an IUPAC name:

a. cyclohexane with NH_2

b. cyclohexane with NH—CH₂CH₃

c. $CH_3CH_2CH-NH_2$ with CH_3

16.11 Name the following aromatic amines as derivatives of aniline:

a. benzene ring with NH—CH₂CH₂CH₃

b. benzene ring with N—CH₂CH₃ and CH₃

16.12 Name the following aromatic amines as derivatives of aniline:

a. benzene ring with NH—CHCH₃ and CH₃

b. benzene ring with N—CH₂CH₃ and CH₂CH₃

16.13 Name each of the following amines by one of the methods used in Exercises 16.7, 16.9, and 16.11:

a. $CH_3-\underset{\underset{CH_3}{|}}{\overset{\overset{CH_3}{|}}{C}}-NH_2$

b. $CH_3CH_2-N-CH_2CH_3$ (with benzene ring below N)

c. benzene ring with NH_2 and Cl

d. $CH_3CHCH_2CHCH_3$ with CH_3 and NH_2

16.14 Draw the structural formula for each of the following amines:

a. 3-ethyl-2-pentanamine

b. *m*-ethylaniline

c. *N,N*-diphenylaniline

16.15 Draw the structural formula for each of the following amines:

a. 2,3-dimethyl-1-butanamine

b. *p*-propylaniline

c. *N,N*-dimethylaniline

Physical Properties of Amines (Section 16.3)

16.16 Explain why all classes of low molecular weight amines are water soluble.

16.17 Why are the boiling points of amines lower than those of corresponding alcohols?

16.18 Why are the boiling points of tertiary amines lower than those of corresponding primary and secondary amines?

16.19 Draw diagrams similar to Figure 16.1 to illustrate hydrogen bonding between the following compounds:

a. $CH_3CH_2-\overset{\overset{\text{H}}{|}}{N}-CH_3$ and $CH_3CH_2-\overset{\overset{\text{H}}{|}}{N}-CH_3$

b. pyrrolidine ring with N—H and H₂O

16.20 Draw diagrams similar to Figure 16.1 to illustrate hydrogen bonding between the following compounds:

a. $CH_3CH_2-NH-CH_3$ and H₂O

b. pyrrolidine ring with N—H and pyrrolidine ring with N—H

16.21 Arrange the following compounds in order of increasing boiling point:

a. $CH_3CH_2CH_2-NH_2$

b. $CH_3-\underset{\underset{CH_3}{|}}{N}-CH_3$

c. $CH_3CH_2CH_2-OH$

16.22 Arrange the following compounds in order of increasing boiling point:

a.

CH_3 ... NH_2 ... CH_3 (benzene ring with CH_3, NH_2, and CH_3 substituents)

b.

CH_3
$N-CH_3$ (benzene ring with $N(CH_3)_2$ group)

c.

$NH-CH_3$... CH_3 (benzene ring with $NH-CH_3$ and CH_3 substituents)

Chemical Properties of Amines (Section 16.4)

16.23 Explain why CH_3-NH_2 is a Brønsted base.

16.24 When diethylamine is dissolved in water, the solution becomes basic. Write an equation to account for this observation.

16.25 Complete the following equations. If no reaction occurs, write "no reaction."

a.

(cyclopentane ring)—$\overset{\overset{\displaystyle O}{\|}}{C}-Cl$ $+ NH_3 \longrightarrow$

b.

(benzene ring)—$\underset{\underset{\displaystyle CH_3}{|}}{N}-CH_3 + CH_3\overset{\overset{\displaystyle O}{\|}}{C}-OH \longrightarrow$

c. $CH_3CH_2-\underset{\underset{\displaystyle CH_3}{|}}{N}-CH_3 + CH_3-\overset{\overset{\displaystyle O}{\|}}{C}-Cl \longrightarrow$

d. (pyrrolidine ring)$N-H + HCl \longrightarrow$

e. (pyrrolidine ring)$N-H + H_2O \rightleftharpoons$

f. $CH_3CH_2-\underset{\underset{\displaystyle CH_3}{|}}{NH_2}{}^+Cl^- + NaOH \longrightarrow$

16.26 Complete the following equations. If no reaction occurs, write "no reaction."

a. $CH_3CH_2CH_2CH_2-NH_2 + HCl \rightarrow$

b.

(cyclopentane ring)$\overset{NH_2}{} + H_2O \rightleftharpoons$

c.

(cyclopentane ring)$\overset{NH_2}{} + CH_3\overset{\overset{\displaystyle O}{\|}}{C}-OH \longrightarrow$

d. $CH_3\underset{\underset{\displaystyle}{}}{\overset{\overset{\displaystyle NH_3{}^+Cl^-}{|}}{C}}HCH_3 + NaOH \longrightarrow$

e. $CH_3CH-\overset{\overset{\displaystyle O}{\|}}{C}-Cl + CH_3-NH_2 \longrightarrow$
 $\underset{\displaystyle CH_3}{|}$

f.

(benzene ring)—$\overset{\overset{\displaystyle O}{\|}}{C}-O-\overset{\overset{\displaystyle O}{\|}}{C}$—(benzene ring) $+ NH_3 \longrightarrow$

16.27 How does the structure of a quaternary ammonium salt differ from the structure of a salt of a tertiary amine?

16.28 Why are amine drugs commonly administered in the form of their salts?

16.29 Write an equation to show how the drug dextromethorphan, a cough suppressant, could be made more water soluble.

(structure of dextromethorphan with CH_3 on N, and CH_3O substituent)

dextromethorphan

16.30 Write equations for two different methods of synthesizing the following amide:

$CH_3-\overset{\overset{\displaystyle O}{\|}}{C}-\underset{\underset{\displaystyle CH_3}{|}}{N}-CH_3$

16.31 Write structures for the chemicals needed to carry out the following conversions:

a.

(benzene ring)—$NH_2 \longrightarrow$ (benzene ring)—$NH_3{}^+ CH_3-\overset{\overset{\displaystyle O}{\|}}{C}-O^-$

b. $CH_3-NH_2 \longrightarrow$ (benzene ring)—$\overset{\overset{\displaystyle O}{\|}}{C}-NH-CH_3$

c. $CH_3\underset{\underset{\displaystyle CH_3}{|}}{CH}-NH_2 \rightarrow CH_3\underset{\underset{\displaystyle CH_3}{|}}{CH}-NH_3{}^+ + OH^-$

16.32 Write structures for the chemicals needed to carry out the following conversions:

a. $\text{C}_6\text{H}_5-\text{NH}_3{}^+\text{Cl}^- \longrightarrow \text{C}_6\text{H}_5-\text{NH}_2$

b. $\text{CH}_3\text{CH}_2-\overset{\displaystyle O}{\overset{\|}{\text{C}}}-\text{Cl} \longrightarrow \text{CH}_3\text{CH}_2-\overset{\displaystyle O}{\overset{\|}{\text{C}}}-\text{NH}_2$

c. $\text{CH}_3-\underset{\underset{\displaystyle \text{CH}_3}{|}}{\text{NH}} \longrightarrow \text{CH}_3-\underset{\underset{\displaystyle \text{CH}_3}{|}}{\text{NH}_2{}^+\text{Cl}^-}$

Amines as Neurotransmitters (Section 16.5)

16.33 Describe the general structure of a neuron.

16.34 What term is used to describe the gap between one neuron and the next?

16.35 Name the two amino acids that are starting materials for the synthesis of neurotransmitters.

16.36 Name the two amines often associated with the biochemical theory of mental illness.

16.37 What role do neurotransmitters play in nerve impulse transmission?

16.38 List four neurotransmitters important in the central nervous system.

Other Biologically Important Amines (Section 16.6)

16.39 Why is epinephrine called the fight-or-flight hormone?

16.40 Describe one clinical use of epinephrine.

16.41 What are the physiological effects of amphetamines on the body?

16.42 What is the source of alkaloids?

16.43 Why are alkaloids weakly basic?

16.44 Give the name of an alkaloid for the following:

 a. Found in cola drinks

 b. Used to reduce saliva flow during surgery

 c. Present in tobacco

 d. A cough suppressant

 e. Used to treat malaria

 f. An effective painkiller

The Nomenclature of Amides (Section 16.7)

16.45 Assign IUPAC names to the following amides:

a. $\text{CH}_3\text{CH}_2\underset{\underset{\displaystyle \text{CH}_2\text{CH}_3}{|}}{\text{CH}}-\overset{\displaystyle O}{\overset{\|}{\text{C}}}-\text{NH}_2$

b. $\text{CH}_3\text{CH}_2\text{CH}_2-\overset{\displaystyle O}{\overset{\|}{\text{C}}}-\text{NH}-\text{C}_6\text{H}_5$

c. $\text{C}_6\text{H}_5-\overset{\displaystyle O}{\overset{\|}{\text{C}}}-\text{NH}-\underset{\underset{\displaystyle \text{CH}_3}{|}}{\text{CHCH}_3}$

d. $\text{CH}_3\text{CH}_2-\overset{\displaystyle O}{\overset{\|}{\text{C}}}-\underset{\underset{\displaystyle \text{CH}_3}{|}}{\text{N}}-\text{CH}_3$

16.46 Assign IUPAC names to the following amides:

a. $\text{CH}_3\text{CH}_2\text{CH}_2-\overset{\displaystyle O}{\overset{\|}{\text{C}}}-\text{NH}-\text{CH}_3$

b. $\text{CH}_3-\overset{\displaystyle O}{\overset{\|}{\text{C}}}-\underset{\underset{\displaystyle \text{CH}_3}{|}}{\text{N}}-\text{CH}_3$

c. $\text{(}o\text{-CH}_3\text{C}_6\text{H}_4\text{)}-\overset{\displaystyle O}{\overset{\|}{\text{C}}}-\text{NH}_2$

d. $\text{CH}_3\text{CH}_2\underset{\underset{\displaystyle \text{CH}_3}{|}}{\text{CH}}-\overset{\displaystyle O}{\overset{\|}{\text{C}}}-\text{NH}_2$

16.47 Draw structural formulas for the following amides:

 a. benzamide

 b. *N*-methylethanamide

 c. *N*-methyl-3-phenylbutanamide

16.48 Draw structural formulas for the following amides:

 a. butanamide

 b. *N*-ethylbenzamide

 c. *N,N*-dimethylpropanamide

Physical Properties of Amides (Section 16.8)

16.49 Draw diagrams similar to those in Figure 16.1 to illustrate hydrogen bonding between the following molecules:

a. $\text{CH}_3-\overset{\displaystyle O}{\overset{\|}{\text{C}}}-\underset{\underset{\displaystyle \text{CH}_3}{|}}{\text{N}}-\text{CH}_3$ and H_2O

b. $\text{CH}_3-\overset{\displaystyle O}{\overset{\|}{\text{C}}}-\text{NH}_2$ and $\text{CH}_3-\overset{\displaystyle O}{\overset{\|}{\text{C}}}-\text{NH}_2$

16.50 Draw diagrams similar to those in Figure 16.1 to illustrate hydrogen bonding between the following molecules:

a. $\text{(cyclopentyl)}-\overset{\displaystyle O}{\overset{\|}{\text{C}}}-\text{NH}_2$ and H_2O

b. $\text{CH}_3\underset{\underset{\displaystyle \text{CH}_3}{|}}{\text{CH}}-\overset{\displaystyle O}{\overset{\|}{\text{C}}}-\text{NH}_2$ and $\text{CH}_3\underset{\underset{\displaystyle \text{CH}_3}{|}}{\text{CH}}-\overset{\displaystyle O}{\overset{\|}{\text{C}}}-\text{NH}_2$

16.51 Explain why the boiling points of disubstituted amides are often lower than those of unsubstituted amides.

Chemical Properties of Amides (Section 16.9)

16.52 Complete the following reactions:

a.
$$CH_3 - \underset{\underset{O}{\|}}{C} - NH - CH_2CH_3 + NaOH \xrightarrow{\text{Heat}}$$

b.
$$CH_3CH_2 - \underset{\underset{O}{\|}}{C} - NH_2 + H_2O + HCl \xrightarrow{\text{Heat}}$$

16.53 Complete the following reactions:

a.

b.

16.54 One of the most successful mosquito repellents has the following structure and name. What are the products of the basic hydrolysis of *N,N*-diethyl-*m*-toluamide?

N,N-diethyl-*m*-toluamide

16.55 What are the products of the acid hydrolysis of the local anesthetic lidocaine?

lidocaine (Xylocaine)

Additional Exercises

16.56 The following molecular formulas represent saturated amine compounds or unsaturated amine compounds that contain C=C bonds. Determine whether each compound is saturated and, if not, how many C=C bonds are present.

a. $C_4H_{11}N$

b. $C_5H_{12}N_2$

c. C_4H_7N

16.57 16.50 mL of 0.100 M HCl was needed to titrate 0.216 g of an unknown amine to a neutral pH endpoint. What is the molecular weight of the amine? Assume the compound only contains one nitrogen atom and there are no other reactive groups present.

16.58 What products would result from the hydrolysis with NaOH of nylon 66 (polymer shown here)?

16.59 Finely ground plant material can be treated with a base to aid in the extraction of alkaloids present in the tissue. Explain why.

16.60 After a nerve impulse occurs, excess acetylcholine cation molecules in the synapse are hydrolyzed in about 10^{-6} seconds by the enzyme cholinesterase. What products are formed? *Hint:* Identify the functional group that can undergo hydrolysis (upon the addition of water).

acetylcholine cation

Chemistry for Thought

16.61 Why might it be more practical for chemical suppliers to ship amines as amine salts?

16.62 Give structural formulas for the four-carbon diamine and the four-carbon diacid chloride that would react to form nylon.

16.63 Why might it be dangerous on a camping trip to prepare a salad using an unknown plant?

16.64 One of the active ingredients in Triaminic® cold syrup is listed as chlorpheniramine maleate. What does the word *maleate* indicate about the structure of this active ingredient?

16.65 Write reactions to show how the following conversion can be achieved. An amine reagent is required, and reactions from earlier chapters are necessary.

$$CH_3CH_2 - OH \longrightarrow CH_3 - NH_3{}^+ CH_3 - \underset{\underset{O}{\|}}{C} - O^-$$

16.66 Several OTC products are available that contain amine salts. Sometimes the active ingredients are given common names ending in hydrochloride. What acid do you think is used to prepare an amine hydrochloride salt?

16.67 In Section 21.1, several components of nucleic acids, DNA and RNA, are identified as "bases." What molecular feature of these substances might explain why they are referred to as bases?

16.68 The hydrolysis of amides requires rather vigorous conditions of strong acid or base and heat. What technique might be employed to achieve hydrolysis under milder conditions?

16.69 Alkaloids can be extracted from plant tissues using dilute hydrochloric acid. Why does the use of acid enhance their water solubility?

Allied Health Exam Connection

16.70 Define neurotransmitter.

16.71 Give general chemical formulas for a primary, a secondary, and a tertiary amine.

16.72 The stimulant in coffee is:

 a. tannic acid

 b. theobromine

 c. theophylline

 d. caffeine

16.73 What are the most likely products of a reaction between $CH_3CH_2NH_2$ and H_2O?

 a. $CH_3CH_2NH_3^+ + OH^-$

 b. $(CH_3)_2NH_2^+ + OH^-$

 c. $CH_3CH_2NH_2OH^- + H^+$

 d. $(CH_2)NH_2^- + H_3O^+$

17 Carbohydrates

© fredredhat/Shutterstock.com

▶ Case Study

Flu season had just begun, and a wave of infection was making its way through the daycare center. Rayla was 3 years old, and this was her first year of preschool. With only a 24-hour incubation period, the virus spread fast. It hit Rayla hard. She was vomiting, with diarrhea and a fever. Her mother tried to get her to eat, but Rayla couldn't keep anything down. The doctor recommended clear liquids for the first 24 hours, then the BRAT diet (bananas, rice, applesauce, and toast), followed by yogurt and a normal diet as soon as Rayla could tolerate it. The doctor explained that dehydration was the first concern and that it was also important to introduce carbohydrates for energy, and then add protein and fat for nutrition. The key was the careful introduction of foods as rapidly as Rayla could tolerate them, but not so quickly as to trigger more stomach

upset. Rayla's mother was surprised that she was not supposed to give her daughter food for 24 hours, but the treatment plan really helped. Rayla was back at preschool in a few days.

▶ **What foods qualify as clear liquids? How much juice should be given to a sick child? When are flu symptoms dangerous?**

Follow-up to this Case Study appears at the end of the chapter before the *Concept Summary*.

Learning Objectives

When you have completed your study of this chapter, you should be able to:

1 Describe the four major functions of carbohydrates in living organisms. **(Section 17.1)**

2 Classify carbohydrates as monosaccharides, disaccharides, or polysaccharides. **(Section 17.1)**

3 Identify molecules possessing chiral carbon atoms. **(Section 17.2)**

4 Use Fischer projections to represent D and L compounds. **(Section 17.3)**

5 Classify monosaccharides as aldoses or ketoses, and classify them according to the number of carbon atoms they contain. **(Section 17.4)**

6 Write reactions for monosaccharide oxidation and glycoside formation. **(Section 17.5)**

7 Describe uses for important monosaccharides. **(Section 17.6)**

8 Draw the structures and list sources and uses for important disaccharides. **(Section 17.7)**

9 Write reactions for the hydrolysis of disaccharides. **(Section 17.7)**

10 Describe the structures and list sources and uses for important polysaccharides. **(Section 17.8)**

arbohydrates are compounds of tremendous biological and commercial importance. Widely distributed in nature, they include such familiar substances as cellulose, table sugar, and starch (see **Figure 17.1**). Carbohydrates have four important functions in living organisms:

• To provide energy through their oxidation.
• To supply carbon for the synthesis of cell components.
• To serve as a stored form of chemical energy.
• To form a part of the structural elements of some cells and tissues.

With carbohydrates, we begin our study of **biomolecules**, substances closely associated with living organisms. All life processes, as far as we know, involve carbohydrates and other biomolecules, including lipids, proteins, and nucleic acids. Biomolecules are organic because they are compounds of carbon, but they are also a starting point for **biochemistry**, the chemistry of living organisms. Thus, these chapters on biomolecules represent an overlap of two areas of chemistry: organic and biochemistry.

biomolecule A general term referring to organic compounds essential to life.

biochemistry A study of the compounds and processes associated with living organisms.

17.1 | Classes of Carbohydrates

Learning Objectives

1. Describe the four major functions of carbohydrates in living organisms.
2. Classify carbohydrates as monosaccharides, disaccharides, or polysaccharides.

The name *carbohydrate* comes from an early observation that heating these compounds produced water and a black residue of carbon, which was incorrectly interpreted to mean that they were hydrates of carbon.

Figure 17.1 Starches like pasta, bread, and rice are examples of carbohydrates.

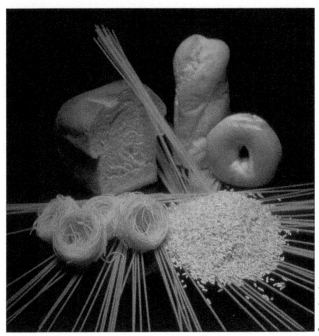

The most striking chemical characteristic of carbohydrates is the large number of functional groups they have. For example, ribose has a functional group on every carbon atom of its five-carbon chain:

$$
\begin{array}{c}
\overset{H}{}\diagdown \overset{O}{}\diagup \\
C \\
| \\
H - C - OH \\
| \\
H - C - OH \\
| \\
H - C - OH \\
| \\
CH_2OH
\end{array}
$$

ribose

carbohydrate A polyhydroxy aldehyde or ketone, or substance that yields such compounds on hydrolysis.

monosaccharide A simple carbohydrate most commonly consisting of three to six carbon atoms.

disaccharide A carbohydrate formed by the combination of two monosaccharide units.

polysaccharide A carbohydrate formed by the combination of many monosaccharide units.

In addition to —OH groups, most carbohydrates have as part of their molecular structure a carbonyl group—either an aldehyde (sometimes shown as —CHO) or a ketone functional group. This is the origin of the modern definition: **Carbohydrates** are polyhydroxy aldehydes or ketones, or substances that yield such compounds on hydrolysis.

Carbohydrates can be classified according to the size of the molecules. **Monosaccharides** consist of a single polyhydroxy aldehyde or ketone unit. **Disaccharides** are carbohydrates composed of two monosaccharide units linked together chemically. Oligosaccharides (less common and of minor importance) contain from three to ten units. **Polysaccharides** consist of very long chains of linked monosaccharide units (see **Figure 17.2**).

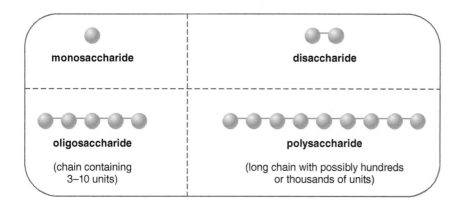

Figure 17.2 Carbohydrate classification.

monosaccharide

disaccharide

oligosaccharide

(chain containing 3–10 units)

polysaccharide

(long chain with possibly hundreds or thousands of units)

17.2 | The Stereochemistry of Carbohydrates

Learning Objective

3. Identify molecules possessing chiral carbon atoms.

Glyceraldehyde, the simplest carbohydrate, exhibits a fascinating stereoisomerism shown by many organic compounds but especially characteristic of carbohydrates and many other natural substances. Glyceraldehyde can exist in two isomeric forms that are mirror images of each other (see **Figure 17.3**). Mirror-image isomers, called **enantiomers**, have the same molecular and structural formulas but different spatial arrangements of their atoms. No amount of rotation can convert one of these structures into the other. The molecules can be compared to right and left hands, which are also mirror images of each other (see **Figure 17.4**). Test this yourself. Put your right hand up to a mirror, and verify that its mirror image is the same as the direct image of your left hand.

enantiomers Stereoisomers that are mirror images.

Figure 17.3 Enantiomers of glyceraldehyde. The designations L and D are discussed in Section 17.3.

Mirror Right hand

Figure 17.4 The reflection of a right hand is a left hand.

Compare your right hand with the right hand of a friend. Note that they can be overlapped or superimposed (see **Figure 17.5**). Now try to rotate or twist your right and left hands so that one is superimposed on the other and turned the exact same way. It cannot be done. Even if you place your hands palm-to-palm so the profiles superimpose, the nails are on the wrong sides. The same thing happens when you try to put a left-hand glove on your right hand.

When an object cannot be superimposed on its mirror image, the object is said to be **chiral**. Thus, a hand, a glove, and a shoe are chiral objects; they cannot be superimposed on their mirror images. However, a drinking glass, a sphere, and a cube are achiral objects because they can be superimposed on their mirror images.

chiral A descriptive term for compounds or objects that cannot be superimposed on their mirror image.

Figure 17.5 Two right hands can be superimposed; a right hand and a left hand cannot.

Many molecules in addition to glyceraldehyde are chiral, but fortunately it is not necessary to try to superimpose them on their mirror images in order to determine whether or not they are chiral. An organic molecule that contains a carbon atom attached to four different groups will be chiral and thus will have two non-superimposable mirror images. A carbon atom with four different groups attached is called a **chiral carbon**.

chiral carbon A carbon atom with four different groups attached.

The center carbon of glyceraldehyde

$$\boxed{CHO}$$
$$\boxed{H} - C - \boxed{OH}$$
$$\boxed{CH_2OH}$$

is chiral because it is attached to CHO, H, OH, and CH_2OH groups. Like your hands, the two enantiomers of glyceraldehyde are nonsuperimposable mirror images. No matter how hard we try, there is no way, short of breaking bonds, to superimpose the four groups so that they coincide.

Because the presence of a single chiral carbon atom gives rise to stereoisomerism, it is important to be able to recognize one when you see it. It is really not difficult. If a carbon atom is connected to four different groups, no matter how slight the differences, it is a chiral carbon. If any two groups are identical, it is not a chiral carbon.

Example 17.1 Identifying Chiral Carbons

Which of the following compounds contain a chiral carbon?

a. CH_3CHCH_3
 |
 OH

b. $CH_3CHCH_2CH_3$
 |
 OH

c. $CH_3CCH_2CH_3$
 ‖
 O

d.
 OH
 |
 CH — CH₂
 / \
 CH₂ O
 \ /
 CH₂ — CH₂

Solution

It may help to draw the structures in a more complete form.

a. The central carbon atom has one H, one OH, but two CH_3 groups attached. It is not chiral. Neither of the other two carbons has four different groups.

$$\begin{array}{c} H \\ | \\ CH_3 - C - CH_3 \\ | \\ OH \end{array}$$

b. The central carbon atom has one H, one OH, one CH_3, and one CH_2CH_3 group attached. Both CH_3 and CH_2CH_3 are bonded to the central carbon atom through a carbon atom. However, when analyzing for chiral carbons, we look at the entire collection of atoms in a group and not just at the four atoms directly bonded. Thus, CH_3 and CH_2CH_3 groups are different (as are H and OH), so the central carbon (identified by an asterisk) is chiral. None of the other carbons are chiral.

$$
\begin{array}{c}
H \\
| \\
CH_3 - C^* - CH_2CH_3 \\
| \\
OH
\end{array}
$$

c. The carbonyl carbon is attached to only three groups, so it is not chiral, even though the three groups are different. None of the other carbons have four different groups attached.

$$
\begin{array}{c}
CH_3 - C - CH_2 - CH_3 \\
\| \\
O
\end{array}
$$

d. The rules are no different for carbons in rings. There is a chiral carbon marked with an asterisk. H and OH are two of the groups. As we proceed around the ring, $-CH_2-O-$ (the third group) is encountered in the clockwise direction and $-CH_2-CH_2-$ (the fourth group) in the counterclockwise direction.

$$
\begin{array}{c}
OH \\
| \\
^*CH - CH_2 \\
\diagup \qquad \diagdown \\
CH_2 \qquad\quad O \\
\diagdown \qquad \diagup \\
CH_2 - CH_2
\end{array}
$$

✔ **LEARNING CHECK 17.1** Which of the carbon atoms shown in color is chiral?

a.
$$
\begin{array}{c}
CH_2OH \\
| \\
CH - OH \\
| \\
CH_2OH
\end{array}
$$

b.
$$
\begin{array}{c}
CHO \\
| \\
CH - OH \\
| \\
CH - OH \\
| \\
CH_2OH
\end{array}
$$

c.
$$
\begin{array}{c}
CH_2OH \\
| \\
C = O \\
| \\
CH - OH \\
| \\
CH_2OH
\end{array}
$$

d.
$$
\begin{array}{c}
CH_2OH \\
| \\
CH - O \\
\diagup \qquad \diagdown \\
CH_2 \qquad\quad CH_2 \\
\diagdown \qquad \diagup \\
CH_2 - CH_2
\end{array}
$$

Organic molecules (and especially carbohydrates) may contain more than one chiral carbon. Erythrose, for example, has two chiral carbon atoms:

$$
\begin{array}{c}
OH \quad\ OH \quad OH \\
| \qquad\ | \qquad\ | \\
CH_2 - \underset{*}{CH} - \underset{*}{CH} - CHO
\end{array}
$$

erythrose

Example 17.2 More Identifying Chiral Carbons

Identify the chiral carbons in glucose:

$$
\begin{array}{c}
\overset{H}{} \diagdown \quad \diagup \overset{O}{} \\
\overset{1}{C} \\
| \\
H-\overset{2}{C}-OH \\
| \\
HO-\overset{3}{C}-H \\
| \\
H-\overset{4}{C}-OH \\
| \\
H-\overset{5}{C}-OH \\
| \\
\overset{6}{CH_2OH}
\end{array}
$$

Solution

Carbon number 1 is not chiral, because it has only three groups attached to it:

$$
\begin{array}{c}
\boxed{H} \diagdown \quad \diagup \boxed{O} \\
\overset{1}{C} \\
| \\
\boxed{
\begin{array}{c}
H-C-OH \\
| \\
HO-C-H \\
| \\
H-C-OH \\
| \\
H-C-OH \\
| \\
CH_2OH
\end{array}
}
\end{array}
$$

Carbon atoms 2, 3, 4, and 5 are chiral because they each have four different groups attached:

$$
\begin{array}{c}
\boxed{
\begin{array}{c}
H \diagdown \quad \diagup O \\
C
\end{array}
} \\
| \\
\boxed{H}-\overset{2}{C}-\boxed{OH} \\
\boxed{
\begin{array}{c}
HO-C-H \\
| \\
H-C-OH \\
| \\
H-C-OH \\
| \\
CH_2OH
\end{array}
}
\end{array}
\quad
\begin{array}{c}
H \diagdown \quad \diagup O \\
C \\
| \\
H-C-OH \\
\boxed{HO}-\overset{3}{C}-\boxed{H} \\
\boxed{
\begin{array}{c}
H-C-OH \\
| \\
H-C-OH \\
| \\
CH_2OH
\end{array}
}
\end{array}
\quad
\begin{array}{c}
H \diagdown \quad \diagup O \\
C \\
| \\
H-C-OH \\
| \\
HO-C-H \\
\boxed{H}-\overset{4}{C}-\boxed{OH} \\
\boxed{
\begin{array}{c}
H-C-OH \\
| \\
CH_2OH
\end{array}
}
\end{array}
\quad
\begin{array}{c}
H \diagdown \quad \diagup O \\
C \\
| \\
\boxed{
\begin{array}{c}
H-C-OH \\
| \\
HO-C-H \\
| \\
H-C-OH
\end{array}
} \\
\boxed{H}-\overset{5}{C}-\boxed{OH} \\
\boxed{CH_2OH}
\end{array}
$$

Carbon number 6 is not chiral because two of the four attached groups are the same.

✔ LEARNING CHECK 17.2 Identify the chiral carbon atoms in the following:

a.
$$CH_2OH$$
$$|$$
$$C=O$$
$$|$$
$$HO-C-H$$
$$|$$
$$H-C-OH$$
$$|$$
$$CH_2OH$$

b.
$$CHO$$
$$|$$
$$CH-OH$$
$$|$$
$$CH-OH$$
$$|$$
$$CH-OH$$
$$|$$
$$CH_2OH$$

c. HOCH₂ ... OH
(structure with H, H, OH, OH)

When a molecule contains more than one chiral carbon, the possibility exists for two arrangements of attached groups at each chiral carbon atom. Thus, nearly all molecules having two chiral carbon atoms exist as two pairs of enantiomers, for a total of four stereoisomers. In a similar manner, when there are three chiral carbons present, there are eight stereoisomers (four pairs of enantiomers). The general formula is

$$\text{Maximum number of possible stereoisomers} = 2^n$$

where n is the number of chiral carbon atoms.

Example 17.3 Determining Number of Stereoisomers

How many stereoisomers are possible for a molecule that contains four chiral carbons such as glucose?

Solution
Stereoisomers $= 2^n$

$$n = 4 \text{ chiral carbons}$$

$$2^4 = 2 \times 2 \times 2 \times 2 = 16$$

✔ LEARNING CHECK 17.3 How many stereoisomers are possible for the following?

$$\begin{array}{cccc} OH & OH & OH & OH \\ | & | & | & | \\ CH_2-CH-CH-CH-CHO \end{array}$$

17.3 | Fischer Projections

Learning Objective

4. Use Fischer projections to represent D and L compounds.

It is time-consuming to draw molecules in the three-dimensional shapes shown for the two enantiomers of glyceraldehyde in Figure 17.3, but there is a way to represent these mirror images in two dimensions. Emil Fischer, a German chemist known for pioneering work in carbohydrate chemistry, introduced a method late in the 19th century. His two-dimensional structures, called **Fischer projections**, are illustrated in **Figure 17.6** for glyceraldehyde.

In Fischer projections, the chiral carbon is represented by the intersection of two lines. If the compound is a carbohydrate, the carbonyl group is placed at or near the top. The molecule is also positioned so that the two bonds coming toward you out of the plane of the paper are drawn horizontally in the Fischer projection. The two bonds projecting away from you into the plane of the paper are drawn vertically. Thus, when you see a Fischer projection, you must realize that the molecule has a three-dimensional shape with horizontal bonds coming toward you and vertical bonds going away from you.

Fischer projection A method of depicting three-dimensional shapes for chiral molecules.

Figure 17.6 Ball-and-stick models (top) and Fischer projections (bottom) of the two enantiomers of glyceraldehyde.

Fischer designated the two enantiomers of glyceraldehyde as ʟ and ᴅ compounds. Using this system, we can represent the isomers of similar compounds by the direction of bonds about the chiral carbon. A small capital ʟ is used to indicate that an —OH group (or another functional group) is on the left of the chiral carbon when the carbonyl is at the top. A small capital ᴅ means the —OH is on the right of the chiral carbon.

Example 17.4 Drawing Fischer Projections

Draw Fischer projections for the ᴅ and ʟ forms of the following:

a.
$$\begin{array}{c} \text{OH} \\ | \\ \text{CH}_3-\text{CH}-\text{COOH} \\ \text{lactic acid} \end{array}$$

b.
$$\begin{array}{c} \text{NH}_2 \\ | \\ \text{CH}_3-\text{CH}-\text{COOH} \\ \text{alanine} \end{array}$$

Solution

a. The second carbon of lactic acid is chiral with four different groups attached: H, OH, CH₃, and COOH. The chiral carbon is placed at the intersection of two lines. Lactic acid is not an aldehyde, but it does contain a carbon–oxygen double bond (in the carboxyl group). This carboxyl group is placed at the top of the vertical line. The direction of the —OH groups on the chiral carbons determines the ᴅ and ʟ notations:

$$\begin{array}{ccc} \text{COOH} & & \text{COOH} \\ \text{HO}\!-\!\!\!-\!\!\!-\!\!\!|\!\!\!-\!\!\!-\!\!\!-\!\text{H} & & \text{H}\!-\!\!\!-\!\!\!-\!\!\!|\!\!\!-\!\!\!-\!\!\!-\!\text{OH} \\ \text{CH}_3 & & \text{CH}_3 \end{array}$$

Mirror

ʟ-lactic acid ᴅ-lactic acid

b. Similarly, the amino acid alanine has a chiral carbon and a carboxyl group. The direction of the —NH_2 group determines D and L notations:

$$
\begin{array}{ccc}
\text{COOH} & \Big| & \text{COOH} \\
H_2N \!-\!\!\!\!\!-\!\! H & \Big| & H \!-\!\!\!\!\!-\!\! NH_2 \\
\text{CH}_3 & \Big| & \text{CH}_3
\end{array}
$$

Mirror

L-alanine D-alanine

✔ **LEARNING CHECK 17.4** Draw Fischer projections for the D and L forms of the following:

$$
\begin{array}{cc}
\text{CH}_3 & \text{NH}_2 \\
| & | \\
\end{array}
$$
$$\text{CH}_3\text{CH}-\text{CH}-\text{COOH}$$

The existence of more than one chiral carbon in a carbohydrate molecule could lead to confusion of D and L designations. This is avoided by focusing on the hydroxy group attached to the chiral carbon farthest from the carbonyl group. By convention, the D family of compounds is that in which this hydroxy group projects to the right when the carbonyl is "up." In the L family of compounds, it projects to the left:

D-erythrose L-erythrose D-glucose L-glucose

Notice that D and L compounds are mirror images of each other (enantiomers).

✔ **LEARNING CHECK 17.5** Identify each structure as D or L:

a.
$$
\begin{array}{c}
\text{CHO} \\
HO \!-\!\!\!\!\!-\!\! H \\
HO \!-\!\!\!\!\!-\!\! H \\
HO \!-\!\!\!\!\!-\!\! H \\
\text{CH}_2\text{OH}
\end{array}
$$

b.
$$
\begin{array}{c}
\text{CH}_2\text{OH} \\
\text{C}=\text{O} \\
HO \!-\!\!\!\!\!-\!\! H \\
H \!-\!\!\!\!\!-\!\! OH \\
H \!-\!\!\!\!\!-\!\! OH \\
\text{CH}_2\text{OH}
\end{array}
$$

c.
$$
\begin{array}{c}
\text{CHO} \\
H \!-\!\!\!\!\!-\!\! OH \\
HO \!-\!\!\!\!\!-\!\! H \\
\text{CH}_2\text{OH}
\end{array}
$$

The physical properties of the two compounds that make up a pair of D and L isomers (enantiomers) are generally the same. The only exception is the way solutions of the two compounds affect polarized light that is passed through them. Light is polarized by passing it through a special polarizing lens, such as those found in polarized sunglasses. When polarized light is passed through a solution of one enantiomer, the plane of polarization of the light is rotated to either the right or the left when viewed by looking toward the source of the light (see **Figure 17.7**). The enantiomer that rotates it to the left is called the **levorotatory** (to the left) or (−) enantiomer. The one that rotates it to the right is the **dextrorotatory** (to the right) or (+) enantiomer.

Figure 17.7 The rotation of plane-polarized light by the solution of an enantiomer.

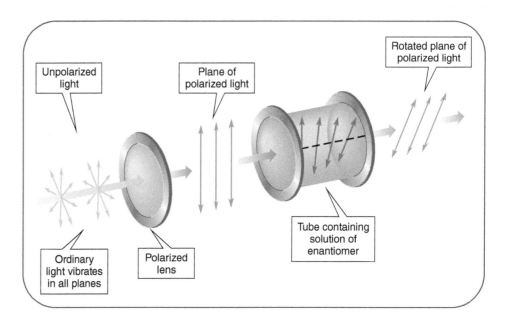

The D and L designations do not represent the words *dextrorotatory* and *levorotatory* but only the spatial relationships of a Fischer projection. Thus, for example, some D compounds rotate polarized light to the right (dextrorotatory), and some rotate it to the left (levorotatory). The spatial relationships D and L and the rotation of polarized light are entirely different concepts and should not be confused with each other.

The property of rotating the plane of polarized light is called optical activity, and molecules with this property are said to be **optically active**. Measurements of optical activity are useful for differentiating between enantiomers.

Why is it so important to be able to describe stereoisomerism and recognize it in molecules? The reason stems from the fact that living organisms, both plant and animal, consist largely of chiral substances. Most of the compounds in natural systems—including biomolecules such as carbohydrates and amino acids—are chiral. Although these molecules can in principle exist as a mixture of stereoisomers, quite often only one stereoisomer is found in nature.

In some instances, two enantiomers are found in nature, but they are not found together in the same biological system. For example, lactic acid occurs naturally in both forms. The L-lactic acid is found in living muscle, whereas the D-lactic acid is present in sour milk. When we realize that only one enantiomer is found in a given biological system, it is not too surprising to find that the system can usually use or assimilate only one enantiomer. Humans utilize the D-isomers of monosaccharides and are unable to metabolize the L-isomers. The D form of glucose tastes sweet, is nutritious, and is an important component

of our diets. The L-isomer, on the other hand, is tasteless, and the body cannot use it. Yeast can ferment only D-glucose to produce alcohol, and most animals are able to utilize only L-amino acids in the synthesis of proteins.

17.4 | Monosaccharides

Learning Objective

5. Classify monosaccharides as aldoses or ketoses, and classify them according to the number of carbon atoms they contain.

The simplest carbohydrates are the monosaccharides, consisting of a single polyhydroxy aldehyde or ketone unit. Monosaccharides are further classified according to the number of carbon atoms they contain (see **Table 17.1**). Thus, simple sugars containing three, four, five, and six carbon atoms are called, respectively, trioses, tetroses, pentoses, and hexoses. The presence of an aldehyde group in a monosaccharide is indicated by the prefix *aldo-*. Similarly, a ketone group is denoted by the prefix *keto-*. Thus, glucose is an aldohexose, and ribulose is a ketopentose:

TABLE 17.1 Monosaccharide Classification Based on the Number of Carbons in Their Chains

Number of Carbon Atoms	Sugar Class
3	Triose
4	Tetrose
5	Pentose
6	Hexose

```
        CHO                    CH₂OH
         |                      |
   H — C — OH              C = O
         |                      |
  HO — C — H             H — C — OH
         |                      |
   H — C — OH             H — C — OH
         |                      |
   H — C — OH                CH₂OH
         |
       CH₂OH

    glucose, an            ribulose, a
    aldohexose             ketopentose
```

✔ **LEARNING CHECK 17.6** Classify each of the following monosaccharides by combining the aldehyde–ketone designation with terminology indicating the number of carbon atoms:

```
a.      CHO          b.     CH₂OH        c.      CHO
         |                   |                    |
   H — C — OH           C = O           HO — C — H
         |                   |                    |
  HO — C — H           H — C — OH        H — C — OH
         |                   |                    |
   H — C — OH             CH₂OH             CH₂OH
         |
       CH₂OH
```

Most monosaccharides are aldoses, and almost all natural monosaccharides belong to the D series. The family of D aldoses is shown in **Figure 17.8**. D-glyceraldehyde, the smallest monosaccharide with a chiral carbon, is the standard on which the whole series is based. Notice that the bottom chiral carbon in each compound is directed to the right. The 2^n formula tells us there must be 2 trioses, 4 tetroses, 8 pentoses, and 16 hexoses. Half of those are the D compounds shown in Figure 17.8. The other half (not shown) are the enantiomers or L compounds.

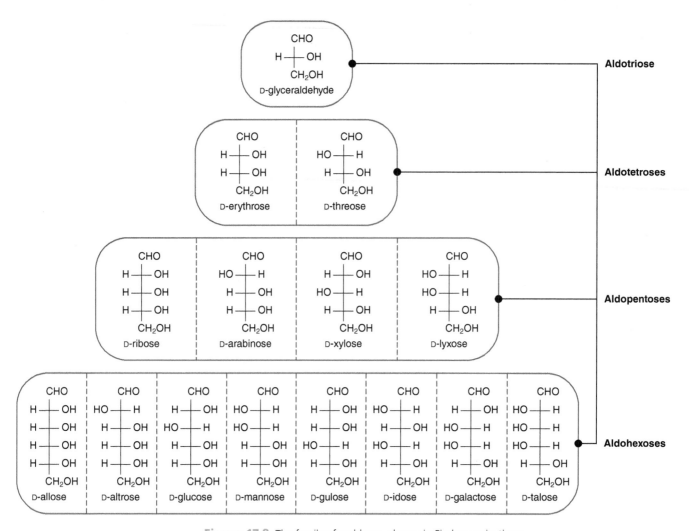

Figure 17.8 The family of D aldoses, shown in Fischer projections.

17.5 | Properties of Monosaccharides

Learning Objective

6. Write reactions for monosaccharide oxidation and glycoside formation.

17.5A Physical Properties

Most monosaccharides (and disaccharides) are called sugars because they taste sweet. The degree of sweetness varies, as shown in **Table 17.2**. Fructose, the sweetest of the common sugars, is about 73% sweeter than sucrose, ordinary table sugar.

All carbohydrates are solids at room temperature, and because of the many —OH groups present, monosaccharide carbohydrates are extremely soluble in water. The —OH groups form numerous hydrogen bonds with the surrounding water molecules. In the body, this solubility allows carbohydrates to be transported rapidly in the circulatory system.

17.5B Chemical Properties, Cyclic Forms

So far we have represented the monosaccharides as open-chain polyhydroxy aldehydes and ketones. These representations are useful in discussing the structural features and stereochemistry of monosaccharides. However, we learned in **Section 14.3** that aldehydes and ketones react with alcohols to form hemiacetals and hemiketals. When these functional groups are present in the same molecule, we noted that the result of their reaction is a

TABLE 17.2 The Relative Sweetness of Sugars (Sucrose = 1.00)

Sugar	Relative Sweetness	Type
Lactose	0.16	Disaccharide
Galactose	0.22	Monosaccharide
Maltose	0.32	Disaccharide
Xylose	0.40	Monosaccharide
Glucose	0.74	Monosaccharide
Sucrose	1.00	Disaccharide
Invert sugar	1.30	Mixture of glucose and fructose
Fructose	1.73	Monosaccharide

stable cyclic structure. You should not be surprised then to find that all monosaccharides with at least five carbon atoms exist predominantly as cyclic hemiacetals and hemiketals.

To help depict the cyclization of glucose in Reaction 17.1, the open-chain structure has been bent around to position the functional groups in closer proximity. Notice the numbering of the carbon atoms that begins at the end of the chain, giving the lowest number to the carbonyl group carbon:

(17.1)

β-D-glucose α-D-glucose

In the reaction, the alcohol group on carbon 5 adds to the aldehyde group on carbon 1. The result is a **pyranose ring**, a six-membered ring containing an oxygen atom. The attached groups have been drawn above or below the plane of the ring. This kind of drawing, called a **Haworth structure**, is used extensively in carbohydrate chemistry. Note that as the reaction occurs, a new chiral carbon is produced at position 1. Thus, two stereoisomers are possible: one with the —OH group pointing down (the α form), and the other with the —OH group pointing up (the β form). The C-1 carbon is called an **anomeric carbon**,

pyranose ring A six-membered sugar ring system containing an oxygen atom.

Haworth structure A method of depicting three-dimensional carbohydrate structures.

anomeric carbon An acetal, ketal, hemiacetal, or hemiketal carbon atom giving rise to two stereoisomers.

Carbohydrates **561**

anomers Stereoisomers that differ in the three-dimensional arrangement of groups at the carbon of an acetal, ketal, hemiacetal, or hemiketal group.

and the α and β forms are called **anomers**. Convenient condensed structures for the cyclic compounds omit the carbon atoms in the ring and the hydrogen atoms attached to the ring carbons, as shown in Reaction 17.2:

$$(17.2)$$

α-D-glucose
(36%)

D-glucose
(0.02%)

β-D-glucose
(64%)

Thus, there are three forms of D-glucose: an open-chain structure and two ring forms. Because the reaction to form a cyclic hemiacetal is reversible, the three isomers of D-glucose are interconvertible. Studies indicate that the equilibrium distribution in water solutions is approximately 36% α-glucose, 0.02% open-chain form, and 64% β-glucose.

Other monosaccharides also form cyclic structures. However, some form five-membered rings, called **furanose rings**, rather than the six-membered kind. An example is D-fructose.

furanose ring A five-membered sugar ring system containing an oxygen atom.

The five-membered ring cyclization occurs like the ring formation in glucose. An alcohol adds across the carbonyl double bond (in this case, a ketone). The orientation of the —OH group at position 2 determines whether fructose is in the α or β form:

D-fructose

$$(17.3)$$

α-D-fructose

β-D-fructose

In drawing cyclic Haworth structures of monosaccharides, certain rules should be followed so that all our structures will be consistent. First we must draw the ring with its oxygen to the back:

furanose ring

pyranose ring

We must also put the anomeric carbon on the right side of the ring. Then we envision the ring as planar with groups pointing up or down. The terminal —CH_2OH group (position 6) is always shown above the ring for D-monosaccharides.

Example 17.5 Drawing Haworth Structures

The aldohexose D-galactose exists predominantly in cyclic forms. Given the structure below, draw the Haworth structure for the anomer. Label the new compound as α or β.

Solution

Draw the pyranose ring with the oxygen atom to the back. Number the ring starting at the right side. Position number 1 will be the anomeric carbon. Place the —OH group at position 1 in the up direction (β form) so that it is the anomer of the given compound. Place groups at the other positions exactly as they are in the given compound. Remember that anomers differ only in the position of the OH attached to the anomeric carbon.

β-D-galactose

✔ **LEARNING CHECK 17.7** Draw the Haworth structure for the anomer of D-ribose. Label the new compound as α or β.

D-ribose

17.5C Oxidation

We learned in **Section 14.3** that aldehydes and those ketones that contain an adjacent —OH group are readily oxidized by an alkaline solution of Cu^{2+} (Benedict's reagent). Open-chain forms of monosaccharides exist as aldehydes or hydroxyketones and are readily oxidized. As the oxidation proceeds, the cyclic forms in equilibrium are converted to open-chain forms (see Le Châtelier's principle, **Section 8.8**) and also react.

Sugars that can be oxidized by weak oxidizing agents are called **reducing sugars**. Thus, all monosaccharides are reducing sugars.

reducing sugar A sugar that can be oxidized by Cu^{2+} solutions.

General reaction:

$$\text{Reducing sugar} \quad + \quad Cu^{2+} \quad \rightarrow \quad \text{oxidized compound} \quad + \quad Cu_2O \qquad (17.4)$$

(complex)
deep blue solution

red-orange
precipitate

Sugar-Free Foods and Diabetes

In the past, ice cream, cookies, and other sweets were considered to be off-limits to people with diabetes mellitus. However, advances in food science and a better understanding of the relationship between different foods and blood glucose levels have changed the rules, and sweet foods are now acceptable as part of the meal plan for people with diabetes. Diabetes is characterized by an impairment of the body's ability to utilize blood glucose.

Among the food science developments helpful to diabetics have been the discovery and commercial distribution of several synthetic noncarbohydrate sweetening agents. Three that have been approved by the American Diabetes Association are saccharin, acesulfame-K, and aspartame:

saccharin

aspartame

acesulfame-K

When compared with an equal weight of sucrose (table sugar), saccharin is about 450 times sweeter, acesulfame-K is about 200 times sweeter, and aspartame is about 180 times sweeter. Saccharin and acesulfame-K are noncaloric,

whereas aspartame contributes very few calories when used in the quantities necessary to provide sweetness. Thus, these substances are useful dietary substitutes for sucrose or other calorie-laden sweeteners for people who wish to control calorie intake or for those who must avoid sugar, such as diabetics.

Until the approval of aspartame (trade name Nutrasweet®) in 1981, people with diabetes shopped for "dietetic" foods primarily in a special section of the supermarket. Aspartame vastly expanded the range of acceptable foods. Studies have verified its general safety when used in moderate amounts. The one exception is that it should not be used by individuals suffering from phenylketonuria (PKU), an inherited inability to metabolize phenylalanine properly. This condition is easily detected in newborn babies and infants, so routine screening is done shortly after birth.

In 1988, after six years of review, the FDA approved acesulfame-K under the trade name Sunette®. Coca-Cola® is using this artificial sweetener in a soft drink called Coca-Cola Zero®. Unlike aspartame, acesulfame-K is heat-stable and can survive the high temperatures of cooking processes.

Other sweeteners that have expanded the food options for diabetics are known as sugar alcohols. These are carbohydrate derivatives, such as sorbitol, in which the carbon–oxygen double bond of the aldehyde or ketone has been converted to an alcohol:

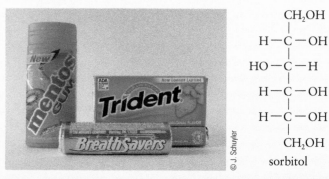

sorbitol

Sugar alcohols are incompletely absorbed during digestion and consequently contribute fewer calories than other carbohydrates. Sugar alcohols are found naturally in fruits. They are also produced commercially from carbohydrates for use in sugar-free candies, cookies, and chewing gum.

Specific example:

$$
\begin{array}{ccc}
\begin{array}{c}
\text{CHO} \\
\text{H}-\overset{|}{\text{C}}-\text{OH} \\
\text{HO}-\overset{|}{\text{C}}-\text{H} \\
\text{H}-\overset{|}{\text{C}}-\text{OH} \\
\text{H}-\overset{|}{\text{C}}-\text{OH} \\
\text{CH}_2\text{OH} \\
\text{D-glucose}
\end{array}
& + \; \text{Cu}^{2+} \longrightarrow &
\begin{array}{c}
\text{COOH} \\
\text{H}-\overset{|}{\text{C}}-\text{OH} \\
\text{HO}-\overset{|}{\text{C}}-\text{H} \\
\text{H}-\overset{|}{\text{C}}-\text{OH} \\
\text{H}-\overset{|}{\text{C}}-\text{OH} \\
\text{CH}_2\text{OH} \\
\text{D-gluconic} \\
\text{acid}
\end{array}
\; + \; \text{Cu}_2\text{O}
\end{array}
\tag{17.5}
$$

Benedict's reagent is deep blue in color. As the open-chain form of a monosaccharide is oxidized, Cu^{2+} is reduced and precipitated as Cu_2O, a red-orange solid.

When we study energy transformations in cells (see **Chapters 22 and 23**), we will encounter many reactions that involve the stepwise oxidation of carbohydrates to CO_2 and H_2O. The controlled oxidation of carbohydrates in the body serves as an important source of heat and other forms of energy.

17.5D Phosphate Esters

The hydroxy groups of monosaccharides can behave as alcohols and react with acids to form esters. Esters formed from phosphoric acid and various monosaccharides are found in all cells, and some serve as important intermediates in carbohydrate metabolism. The structures of two representative phosphate esters are shown below:

glucose 6-phosphate fructose 6-phosphate

17.5E Glycoside Formation

In **Section 14.3**, we learned that hemiacetals and hemiketals can react with alcohols in acid solutions to yield acetals and ketals, respectively. Thus, cyclic monosaccharides (hemiacetals and hemiketals) readily react with alcohols in the presence of acid to form acetals and ketals. The general name for these carbohydrate products is **glycosides**. The reaction of α-D-glucose with methanol is shown in Reaction 17.6.

glycoside Another name for a carbohydrate containing an acetal or ketal group.

α-D-glucose methyl α-D-glucopyranoside methyl β-D-glucopyranoside

glycosides

$$\tag{17.6}$$

The hemiacetal in glucose is located at position 1, and the acetal thus forms at that same position. All the other —OH groups in glucose are ordinary alcohol groups and do not react under these conditions. As the glycoside reaction takes place, a new bond is established between the pyranose ring and the —OCH₃ groups. The new group may point up or down from the ring. Thus, a mixture of α- and β-glycosides is formed. The new carbon–oxygen–carbon linkage that joins the components of the glycoside is called a **glycosidic linkage**.

Although carbohydrate cyclic hemiacetals and hemiketals are in equilibrium with open-chain forms of the monosaccharides, the glycosides (acetals and ketals) are much more stable and do not exhibit open-chain forms. Therefore, glycosides of monosaccharides are not reducing sugars.

As we'll see later in this chapter, both disaccharides and polysaccharides are examples of glycosides in which monosaccharide units are joined together by acetal (glycosidic) linkages.

✔ **LEARNING CHECK 17.8** Two glycosides are shown below. Circle any acetal or ketal groups and use an arrow to identify the glycosidic linkages.

a.

CH_2OH

HO

O

OH

OCH_2CH_3

OH

b.

$HOCH_2$

O

OCH_3

CH_2OH

HO OH

17.6 | Important Monosaccharides

Learning Objective

7. Describe uses for important monosaccharides.

17.6A Ribose and Deoxyribose

Two pentoses, ribose and deoxyribose, are extremely important because they are used in the synthesis of nucleic acids (DNA and RNA), substances essential in protein synthesis and the transfer of genetic material (see **Chapter 21**). Ribose (along with phosphate groups) forms the long chains that make up ribonucleic acid (RNA). Deoxyribose (along with phosphate groups) forms the long chains of deoxyribonucleic acid (DNA). Deoxyribose differs from ribose in that the OH group on carbon 2 has been replaced by a hydrogen atom (deoxy form).

$HOCH_2$ O OH

1

2

OH OH

β-D-ribose

$HOCH_2$ O OH

1

2

two hydrogens at this position

OH

β-D-deoxyribose

17.6B Glucose

Of the monosaccharides, the hexose glucose is the most important nutritionally and the most abundant in nature. Glucose is present in honey and fruits such as grapes, figs, and dates. Ripe grapes, for example, contain 20–30% glucose. Glucose is sometimes called dextrose. Glucose is also known as blood sugar because it is the sugar transported by the blood to body tissues to satisfy energy requirements (see **Figure 17.9**). Other sugars absorbed into the body must be converted to glucose by the liver. Glucose is commonly used as a sweetener in confections and other foods, including some baby foods.

Figure 17.9 A glucose solution prepared for intravenous use.

JinYoung Lee/Shutterstock.com

β-D-glucose

17.6C Galactose

Galactose is a hexose with a structure very similar to that of glucose. The only difference between the two is the orientation of the hydroxy group attached to carbon 4. Like glucose, galactose can exist in α, β, or open-chain forms; the β form is shown below:

β-D-galactose

Galactose is synthesized in the mammary glands, where it is incorporated into lactose, the sugar found in milk. It is also a component of substances present in nerve tissue.

17.6D Fructose

Fructose (Section 17.5) is the most important ketohexose. It is also known as levulose and, because of its presence in many fruits, fruit sugar. It is present in honey in a 1:1 ratio with glucose and is abundant in corn syrup. This sweetest of the common sugars is important as a food sweetener because less fructose is needed than other sugars to achieve the same degree of sweetness.

17.7 | Disaccharides

Learning Objectives

8. Draw the structures and list sources and uses for important disaccharides.
9. Write reactions for the hydrolysis of disaccharides.

Disaccharides are sugars composed of two monosaccharide units linked together by the acetal or ketal linkages described earlier. They can be hydrolyzed to yield their monosaccharide building blocks by boiling with dilute acid or reacting them with appropriate enzymes. Nutritionally, the most important members of this group are maltose, lactose, and sucrose.

17.7A Maltose

Maltose, also called malt sugar, contains two glucose units joined by a glycosidic linkage between carbon 1 of the first glucose unit and carbon 4 of the second unit (Reaction 17.7). The configuration of the linkage on carbon 1 in the glycosidic linkage between glucose units is α, and the linkage is symbolized α(1 → 4) to indicate this fact.

(17.7)

The chapters on organic chemistry (see **Chapters 11–16**) were organized around the functional group concept. Each chapter dealt with a particular functional group or related groups. The nomenclature, properties, and uses of the compounds were discussed. The focus is different in this and the next four chapters (lipids, proteins, enzymes, and nucleic acids). Each of these chapters is devoted to a particular class of biomolecules—substances closely associated with life. Reactions and nomenclature receive much less emphasis; more time is spent describing structures and how they contribute to the processes of living organisms. Another key difference between these and earlier chapters is the reduced emphasis on mastering specific skills, such as solving numerical problems, balancing equations, and naming compounds. More attention should now be given to recognizing structures and understanding key terms and processes. It is a good time to increase your efforts to highlight important concepts in both the text and your lecture notes.

Notice that maltose contains both an acetal carbon (left glucose unit, position 1) and a hemiacetal carbon (right glucose unit, position 1). The —OH group at the hemiacetal position can point either up or down (the β is shown). The presence of a hemiacetal group also means that the right glucose ring can open up to expose an aldehyde group. Thus, maltose is a reducing sugar (Reaction 17.8).

hemiacetal group

aldehyde group

maltose

(17.8)

Maltose, which is found in germinating grain, is formed during the digestion (hydrolysis) of starch to glucose. Its name is derived from the fact that during the germinating (or malting) of barley, starch is hydrolyzed, and the disaccharide is formed. On hydrolysis, maltose forms two molecules of D-glucose:

$$\text{maltose} + \text{H}_2\text{O} \xrightarrow{\text{H}^+} \text{D-glucose} + \text{D-glucose} \qquad (17.9)$$

17.7B Lactose

Lactose, or milk sugar, constitutes 5% of cow's milk and 7% of human milk by weight (see **Figure 17.10**). Pure lactose is obtained from whey, the watery by-product of cheese production. Lactose is composed of one molecule of D-galactose and one of D-glucose. The linkage between the two sugar units is β(1 → 4):

Figure 17.10 Cow's milk is about 5% lactose. What two products form when the lactose is hydrolyzed in the child's digestive system?

hemiacetal of α-D-glucose unit

β-D-galactose unit

lactose

The presence of a hemiacetal group in the glucose unit makes lactose a reducing sugar.

Put Fiber into Snacks and Meals

Fiber is a type of carbohydrate found only in food derived from plants. It is not found in any meat or dairy products. While not considered to be a "nutrient," fiber still plays a vital role in maintaining good digestive and cardiovascular health, as well as weight loss and weight maintenance. Research indicates that eating a diet high in fiber also reduces the risk of developing cancer. Fiber, along with adequate fluid intake, plays a cleansing role in the body. The recommended daily fiber intake is 21–25 grams for women and 30–38 grams for men. Nutrition labels on foods list fiber content. This is usually given in grams per serving, or in percentage of daily recommended amounts per serving. Most foods that contain fiber offer small amounts (2–5 grams) per serving, so including fiber throughout the day in snacks as well as meals is important in order to reach a goal of 25–30 grams per day.

Knowing which foods contain fiber assists in making healthy choices and reaching daily fiber intake goals. Since many people eat fruits as a snack, fruits are a good place to start when counting daily fiber intake. Fresh raspberries contain 8 grams of fiber per cup, which is more than 30% of the daily fiber intake recommendation. Apples, oranges, and bananas each contain 3 to 5 grams per serving. Vegetable snacks that contribute fiber, such as raw broccoli and carrots, offer convenience and portability. Other vegetables might not make suitable raw snacks, but can add significant fiber to the diet when added to daily meals. Good vegetable choices are cooked peas, beans, corn, and artichokes.

Grains and grain products are significant sources of fiber. Many choices are found in the bread and cereal aisles of grocery stores. The range of fiber content in these products varies widely. White bread and cereals based on white rice contain little or no significant amounts of fiber. In contrast, a typical piece of whole wheat bread contains 2 to 4 grams of fiber. Some high-fiber breads can contain as much as 9 grams of fiber per slice. One peanut butter sandwich made using high-fiber bread can satisfy the fiber recommendations for an entire day. Choose cereals with "bran" or "fiber" in the name, and check the fiber content on the nutrition label. Whole grains provide the easiest way to add fiber to a diet. Cereal eaten as a snack is an easy way to boost fiber intake. A small container of shredded wheat squares, or granola either in a snack bar, sprinkled on yogurt, or simply eaten plain, can provide between 5 and 13 grams of fiber per 100 g serving, depending on brand.

A great way to boost fiber intake is to sprinkle high-fiber foods on normally eaten food. Oat bran, wheat bran, ground flax seeds, or berries can easily be added to sweet or savory foods such as oatmeal, pancakes, soups, baked potatoes, casseroles, or salads. Any meal can become a high-fiber meal with a sprinkling of bran.

It is important to increase fiber intake slowly, and to drink plenty of water to avoid digestive upset. Foods high in fiber are more satisfying, and provide a feeling of fullness for a longer time. High-fiber diets are often associated with successful weight-loss programs in addition to other health benefits. Conscientious shopping, by noting the fiber content in purchased food, can be one of the simplest and best things you can do to maintain and possibly improve your health.

Fruits are a good source of fiber.

17.7C Sucrose

The disaccharide sucrose, common household sugar, is extremely abundant in the plant world. It occurs in many fruits, in the nectar of flowers (see **Figure 17.11**), and in the juices of many plants, especially sugar cane and sugar beets, the commercial sources. Sucrose contains two monosaccharides, glucose and fructose, joined together by a linkage that is α from carbon 1 of glucose and β from carbon 2 of fructose:

Figure 17.11 Hummingbirds depend on the sucrose and other carbohydrates of nectar for their energy.

CH₂OH

O

OH

HO

OH ⎯⎯1

α-D-glucose unit

O

HOCH₂ O

HO ⎯⎯2

CH₂OH

β-D-fructose unit

OH

sucrose

Neither ring in the sucrose molecule contains a hemiacetal or hemiketal group that is necessary for ring opening. This is true because the anomeric positions in both glucose and fructose are part of the glycosidic linkage that connects the rings to each other. Thus, in contrast to maltose and lactose, both rings of sucrose are locked in the cyclic form, and sucrose is therefore not a reducing sugar (see **Figure 17.12**).

© Spencer L. Seager

© Spencer L. Seager

1 From left to right, four test tubes containing Benedict's reagent, 2% maltose solution, 2% sucrose solution, and 2% lactose solution.

2 Both maltose and lactose are reducing sugars. The sucrose has remained unreacted.

Figure 17.12 Benedict's test on disaccharides.

invert sugar A mixture of equal amounts of glucose and fructose.

© Christina Slabaugh

Figure 17.13 Sucrose is hydrolyzed by bees in making honey. Why is honey sweeter than the starting sucrose?

The hydrolysis of sucrose (see **Figure 17.13**) produces a 1:1 mixture of D-glucose and D-fructose called **invert sugar** (Reaction 17.10):

$$\text{sucrose} + \text{H}_2\text{O} \xrightarrow{\text{H}^+} \text{D-glucose} + \text{D-fructose} \tag{17.10}$$

This mixture is sweeter than the original sucrose (see **Table 17.2**).

When sucrose is cooked with acid-containing fruits or berries, partial hydrolysis takes place, and some invert sugar is formed. The jams and jellies prepared in this manner are actually sweeter than the pure sugar originally put in them. **Table 17.3** summarizes some features of the three disaccharides we have discussed.

TABLE 17.3 Some Important Disaccharides

Name	Monosaccharide Constituents	Glycoside Linkage	Source
Maltose	Two glucose units	$\alpha(1 \rightarrow 4)$	Hydrolysis of starch
Lactose	Galactose and glucose	$\beta(1 \rightarrow 4)$	Mammalian milk
Sucrose	Glucose and fructose	$\alpha\text{-}1 \rightarrow \beta\text{-}2$	Sugar cane and sugar beet juices

17.8 | Polysaccharides

Learning Objective

10. Describe the structures and list sources and uses for important polysaccharides.

Just as two sugar units are linked together to form a dissacharide, additional units can be added to form higher saccharides. Nature does just this in forming polysaccharides, which are condensation polymers containing thousands of units. Because of their size, polysaccharides are not water soluble, but their many hydroxy groups become hydrated individually when exposed to water, and some polysaccharides form thick colloidal dispersions when heated in water. Thus, the polysaccharide known as starch can be used as a thickener in sauces, gravies, pie fillings, and other food preparations. As shown in **Table 17.4**, the properties of polysaccharides differ markedly from those of monosaccharides and disaccharides.

TABLE 17.4 Properties of Polysaccharides Compared with Those of Monosaccharides and Disaccharides

Property	Monosaccharides and Disaccharides	Polysaccharides
Molecular weight	Low	Very high
Taste	Sweet	Tasteless
Solubility in water	Soluble	Insoluble or form colloidal dispersions
Size of particles	Pass through a membrane	Do not pass through a membrane
Test with Cu^{2+} for reducing sugars	Positive (except for sucrose)	Negative

17.8A Starch

Starch is a polymer consisting entirely of D-glucose units. It is the major storage form of D-glucose in plants. Two starch fractions, amylose (10–20%) and amylopectin (80–90%), can usually be isolated from plants. Amylose is made up of long unbranched chains of glucose units connected by $\alpha(1 \rightarrow 4)$ linkages (see **Figure 17.14**). This is the same type of linkage found in maltose. The long chain, often containing between 1000 and 2000 glucose units, is flexible enough to allow the molecules to twist into the shape of a helix (see **Figure 17.15A**).

$\alpha(1 \rightarrow 4)$ linkage

Figure 17.14 The structure of amylose.

An important test for the presence of starch is the reaction that occurs between iodine, I_2, and the coiled form of amylose. The product of the reaction is deep blue in color (see **Figure 17.16**) and is thought to consist of the amylose helix filled with iodine molecules. This same iodine reaction is also widely used to monitor the hydrolysis of starch. The color gradually fades and finally disappears, as starch is hydrolyzed by either acid or enzymes to form dextrins (smaller polysaccharides), then maltose, and finally glucose. The disappearance of the blue iodine color is thought to be the result of the breakdown of the starch helix.

© Mark Slabaugh

Figure 17.16 A dark bluish-black color is the characteristic result when a solution of iodine encounters the starch in potatoes. What other foods would you expect to give a positive starch test?

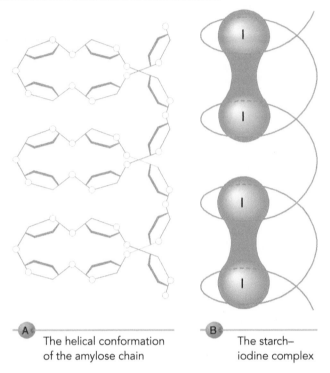

A: The helical conformation of the amylose chain

B: The starch–iodine complex

Figure 17.15 The molecular conformation of starch and the starch–iodine complex.

Amylopectin, the second component of starch, is not a straight-chain molecule like amylose but contains random branches. The branching point is the $\alpha(1 \rightarrow 6)$ glycosidic linkage (see **Figure 17.17**). There are usually 24 to 30 D-glucose units, all connected by $\alpha(1 \rightarrow 4)$ linkages, between each branch point of amylopectin. These branch points give amylopectin the appearance of a bushy molecule. Amylopectin contains as many as 10^5 glucose units in one gigantic molecule.

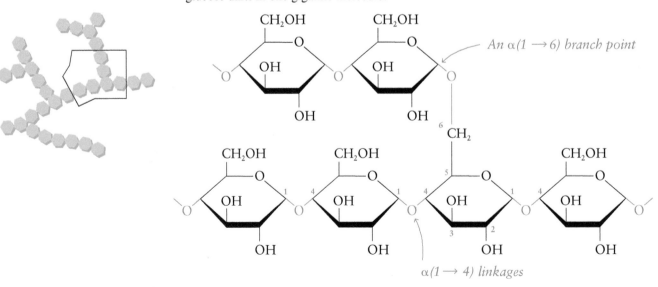

An $\alpha(1 \rightarrow 6)$ branch point

$\alpha(1 \rightarrow 4)$ linkages

Figure 17.17 The partial structure of an amylopectin molecule. Glycogen has a similar structure.

17.8B Glycogen

The polysaccharide glycogen is sometimes called animal starch because it is a storage carbohydrate for animals that is analogous to the starch of plants. It is especially abundant in the liver and muscles, where excess glucose taken in by an animal is stored for future use. On hydrolysis, glycogen forms D-glucose, which helps maintain the normal blood sugar level and provide the muscles with energy. Structurally, glycogen is very similar to amylopectin, containing both $\alpha(1 \rightarrow 4)$ and $\alpha(1 \rightarrow 6)$ linkages between glucose units (see **Figure 17.18**). The main difference between amylopectin and glycogen is that glycogen is even more highly branched; there are only 8 to 12 D-glucose units between branch points.

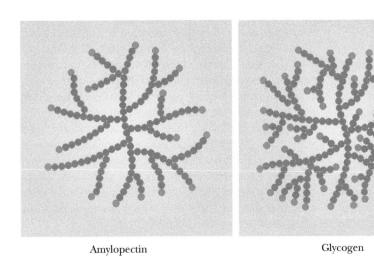

Amylopectin Glycogen

Figure 17.18 A simplified representation of the branched polysaccharide glycogen (branches every 8–12 glucose units). Amylopectin is much less densely branched (branches every 24–30 glucose units). Each small circle represents a single glucose unit.

17.8C Cellulose

Cellulose is the most important structural polysaccharide and is the single most abundant organic compound on Earth (see **Figure 17.19**). It is the material in plant cell walls that provides strength and rigidity. Wood is about 50% cellulose. Like amylose, cellulose is a linear polymer consisting of D-glucose units joined by the $1 \rightarrow 4$ linkage. It may contain from 300 to 3000 glucose units in one molecule. The main structural difference

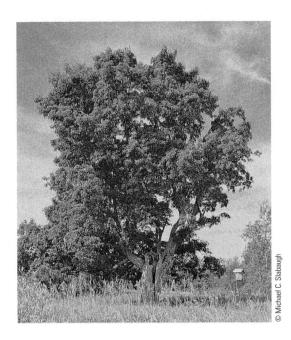

Figure 17.19 Cellulose and the plant kingdom are responsible for much of the world's natural beauty.

Q: Is high-fructose corn syrup worse for your health than table sugar?

Melina B. Jampolis, M.D., is a Physician Nutrition Specialist focusing on nutrition for weight loss and disease prevention and treatment.

A: High-fructose corn syrup (HFCS) is a low-cost liquid sweetener that was first introduced in the United States in the late 1960s. It is made from a mixture of cornstarch and water by chemically reacting the mixture using enzymes as catalysts. The reaction converts the mixture into a liquid that contains water, glucose, and fructose in various ratios depending on the reaction conditions and the enzymes used. There are two common types of HFCS. HFCS-55, found mostly in soft drinks, contains 55% fructose and 42% glucose (the remaining 3% is made up of starch-like glucose polymers). HFCS-42, found in processed foods including ice cream and baked goods, contains 42% fructose and 53% glucose (the remaining 5% is made up of glucose polymers). Sucrose, commonly known as table sugar, contains 50% glucose and 50% fructose. The only major chemical difference between sucrose and HFCS is the presence of a chemical bond in sucrose linking together glucose and fructose to form a disaccharide. This bond is easily broken by enzymes located in the small intestine. In HFCS, glucose and fructose exist in free-monosaccharide form. Also unlike sucrose, HFCS is stable in an acidic environment (such as soft drinks), has a lower freezing point, and occurs as a syrup rather than a granular solid that has to be dissolved in water before use. These characteristics, in addition to its low cost and widespread availability due to government corn subsidies, make it very appealing to the food and beverage industry. Its use grew exponentially for the 35 years after its introduction, but in the mid-2000s, leading nutrition experts began to suspect that the rapid rise in obesity in the United States during the same time period might be connected to our increased consumption of HFCS.

However, this hypothesis has not been clinically substantiated and is the subject of ongoing discussion and continued research. Most experts now believe that due to the similarities in the chemical structure of HFCS and sucrose, HFCS does not uniquely contribute to the obesity epidemic. They argue that it is the overall high sugar, fat, and calorie intake from the average U.S. diet that is responsible. Nutrition experts agree that fructose consumption from both HFCS and sucrose is particularly damaging when it occurs in high doses. Because fructose is absorbed via passive rather than active transport, it is rapidly taken up by the liver and bypasses a key regulatory step in glycolysis. This can lead to increased fat production in the liver. High fructose intake, particularly from sugar or HFCS-sweetened beverages, has been associated with significant increases in obesity (particularly abdominal obesity), fatty liver disease, and metabolic syndrome, a condition that often leads to type 2 diabetes.

Until research uncovers more convincing evidence pointing to a uniquely detrimental effect of HFCS compared to sucrose, the best approach is to limit overall added sugar consumption, particularly in the form of sugar-sweetened beverages including sodas, fruit drinks, "energy" drinks, and sugar-filled water. As a category these account for the greatest percentage of added sugar intake in the United States.

Our body does not process HFCS the same way it processes regular table sugar. Some scientists believe that HFCS may contribute to obesity by disrupting normal metabolic function.

Mountain Dew contains high amounts of high-fructose corn syrup.

iStockphoto.com/Stephen Krow

between amylose and cellulose is that all the $1 \rightarrow 4$ glycosidic linkages in cellulose are β instead of α (see **Figure 17.20**). This seemingly small difference causes tremendous differences between amylose and cellulose. The shapes of the molecules are quite different. The $α(1 \rightarrow 4)$-linked amylose tends to form loose spiral structures (see **Figure 17.15**), whereas $β(1 \rightarrow 4)$-linked cellulose tends to form extended straight chains. These chains become aligned side by side to form well-organized, water-insoluble fibers in which the hydroxy groups form numerous hydrogen bonds with the neighboring chains. These parallel chains of cellulose confer rigidity and strength to the molecules.

Figure 17.20 The structure of cellulose.

Second, although starch with its α linkages is readily digestible, humans and other animals lack the necessary enzymes to hydrolyze the β linkages of cellulose. Thus, cellulose passes unchanged through the digestive tract and does not contribute to the caloric value of food. However, it still serves a useful purpose in digestion. Cellulose, a common constituent of dietary fiber, is the roughage that provides bulk, stimulates contraction of the intestines, and aids the passage of food through the digestive system.

Animals that use cellulose as a food do so only with the help of bacteria that possess the necessary enzymes for breakdown. Herbivores such as cows, sheep, and horses are animals in this category. Each has a colony of such bacteria somewhere in the digestive system and uses the simple carbohydrates resulting from the bacterial hydrolysis of cellulose. Fortunately, the soil also contains organisms with appropriate enzymes. Otherwise, debris from dead plants would accumulate rather than disappear through biodegradation.

▶ Case Study Follow-up

Liquids such as water, sports drinks, broth, and plain gelatin, which are easily digested and leave no residue in the intestinal tract, are considered to be clear liquids. A clear liquid diet cannot supply adequate nutrition and should not be continued for more than a few days. Juice that contains no pulp such as apple, grape, or cranberry is considered to be a clear liquid. Any liquid, whether juice or water, should be given to a sick child only in small amounts so as not to trigger nausea. Popsicles are a good alternative to liquids because they take longer to ingest.

"Stomach flu" is the common term used for symptoms that typically include nausea, vomiting, diarrhea, and fever. It is not the same as influenza virus infection, which affects the respiratory system. What most people are referring to when they use the term "stomach flu" is viral gastroenteritis. Signs of severe gastroenteritis requiring medical attention in infants include fever of over 102°F, no wet diapers for 6 hours, sunken fontanel, crying without tears, dry mouth, bloody stools, vomiting that lasts more than several hours, and lethargy. For children, the warning signs are similar: dehydration, fever, excessive vomiting, and bloody stools. In adults, the fever can be as high as 104°F, and vomiting continuing for more than two days is cause for concern.

Classes of Carbohydrates Carbohydrates are poly-hydroxy aldehydes or ketones, or substances that yield such compounds on hydrolysis. Carbohydrates are used as energy sources, biosynthetic intermediates, energy storage, and structural elements in organisms.

Objective 1 (Section 17.1), Exercise 17.2

Carbohydrates can exist either as single units (monosaccharides) or joined together in molecules ranging from two units (disaccharides) to hundreds of units (polysaccharides).

Objective 2 (Section 17.1), Exercise 17.4

The Stereochemistry of Carbohydrates Carbohydrates, along with many other natural substances, exhibit a type of isomerism in which two isomers are mirror images of each other. The two isomers are called enantiomers and contain a chiral carbon atom. When a molecule has more than one chiral carbon, the maximum number of stereoisomers possible is 2^n, where n is the number of chiral carbons.

Objective 3 (Section 17.2), Exercise 17.8

Fischer Projections A useful way of depicting the structure of chiral molecules employs crossed lines (Fischer projections) to represent chiral carbon atoms. The prefixes D- and L- are used to distinguish between enantiomers. Signs indicating the rotation of plane-polarized light to the right (+) or to the left (−) may also be used to designate enantiomers.

Objective 4 (Section 17.3), Exercise 17.12

Monosaccharides Monosaccharides that contain an aldehyde group are called aldoses, whereas those containing a ketone group are ketoses. Monosaccharides are also classified by the number of carbon atoms as trioses, tetroses, etc. Most natural monosaccharides belong to the D family.

Objective 5 (Section 17.4), Exercise 17.22

Properties of Monosaccharides Monosaccharides are sweet-tasting solids that are very soluble in water. Noncarbohydrate low-calorie sweeteners such as aspartame have been developed as sugar substitutes. Pentoses and hexoses form cyclic hemiacetals or hemiketals whose structures can be represented by Haworth structures. Two isomers referred to as anomers (the α and β forms) are produced in the cyclization reaction. All monosaccharides are oxidized by Benedict's reagent and are called reducing sugars. Monosaccharides can react with alcohols to produce acetals or ketals that are called glycosides.

Objective 6 (Section 17.5), Exercise 17.34

Important Monosaccharides Ribose and deoxyribose are important as components of nucleic acids. The hexoses glucose, galactose, and fructose are the most important nutritionally and the most abundant in nature. Glucose, also known as blood sugar, is transported within the bloodstream to body tissues, where it supplies energy.

Objective 7 (Section 17.6), Exercise 17.37

Disaccharides Glycosidic linkages join monosaccharide units together to form disaccharides. Three important disaccharides are maltose (two glucose units α(1 → 4)-linked), lactose (a galactose linked to glucose by a β(1 → 4) glycosidic linkage), and sucrose (α-glucose joined to β-fructose).

Objective 8, (Section 17.7), Exercise 17.44

Hydrolysis of disaccharides produces monosaccharides.

Objective 9 (Section 17.7), Exercise 17.52

Polysaccharides Cellulose, starch, and glycogen are three important polysaccharides. Starch is the major storage form of glucose in plants, whereas glycogen is the storage form of glucose in animals. Cellulose is the structural material of plants.

Objective 10 (Section 17.8), Exercise 17.54

Key Terms and Concepts

Anomeric carbon (17.5)	Disaccharide (17.1)	Levorotatory (17.3)
Anomers (17.5)	Enantiomers (17.2)	Monosaccharide (17.1)
Biochemistry (Introduction)	Fischer projection (17.3)	Optically active molecule (17.3)
Biomolecule (Introduction)	Furanose ring (17.5)	Polysaccharide (17.1)
Carbohydrate (17.1)	Glycoside (17.5)	Pyranose ring (17.5)
Chiral (17.2)	Glycosidic linkage (17.5)	Reducing sugar (17.5)
Chiral carbon (17.2)	Haworth structure (17.5)	
Dextrorotatory (17.3)	Invert sugar (17.7)	

Key Reactions

1.	Oxidation of a sugar (Section 17.5):	reducing sugar + Cu^{2+} → oxidized compound + Cu$_2$O	Reaction 17.4
2.	Glycoside formation (Section 17.5):	monosaccharide + alcohol $\xrightarrow{\text{H}^+}$ acetals or ketals	
3.	Hydrolysis of disaccharides (Section 17.7):	disaccharide + H$_2$O $\xrightarrow[\text{enzymes}]{\text{H}^+ \text{ or}}$ two monosaccharides	

Exercises

Even-numbered exercises are answered in Appendix B.

Blue-numbered exercises are more challenging.

Classes of Carbohydrates (Section 17.1)

17.1 What are the four important roles of carbohydrates in living organisms?

17.2 Describe whether each of the following substances serves primarily as an energy source, a form of stored energy, or a structural material (some serve as more than one):

a. cellulose **c.** glycogen

b. sucrose, table sugar **d.** starch

17.3 What are the structural differences among monosaccharides, disaccharides, and polysaccharides?

17.4 Match the terms *carbohydrate, monosaccharide, disaccharide,* and *polysaccharide* to each of the following (more than one term may fit):

a. table sugar

b.

c. starch

d. fructose

e. cellulose

f.

g. glycogen

h. amylose

17.5 Define *carbohydrate* in terms of the functional groups present.

The Stereochemistry of Carbohydrates (Section 17.2)

17.6 Why are carbon atoms 1 and 3 of glyceraldehyde not considered chiral?

17.7 Locate the chiral carbon in amphetamine and identify the four different groups attached to it.

17.8 Which of the following molecules can have enantiomers? Identify any chiral carbon atoms.

a. $CH_3CH_2 - CH - CH_2CH_3$
 |
 OH

b. $CH_3CH_2 - CH - \overset{\overset{O}{\|}}{C} - CH_3$
 |
 OH

c.

17.9 Which of the following molecules can have enantiomers? Identify any chiral carbon atoms.

a.

b.

c.

Fischer Projections (Section 17.3)

17.10 Explain what the following Fischer projection denotes about the three-dimensional structure of the compound:

17.11 Identify each of the following as a D or an L form and draw the structural formula of the enantiomer:

a.

b.

17.12 Identify each of the following as a D or an L form and draw the structural formula of the enantiomer:

a.
```
        CHO
   HO ──┼── H
   HO ──┼── H
    H ──┼── OH
       CH₂OH
```

b.
```
       CH₂OH
        C = O
    H ──┼── OH
   HO ──┼── H
   HO ──┼── H
       CH₂OH
```

17.13 Draw Fischer projections for both the D and L isomers of the following:

a. 2,3-dihydroxypropanoic acid

b. 2-chloro-3-hydroxypropanoic acid

17.14 Draw Fischer projections for both the D and L isomers of the following amino acids:

```
Alanine,    H₂N─CH─COOH
a.                │
                 CH₃
```

```
Leucine,    H₂N─CH─COOH
b.                │
                 CH₂
                  │
                 CH
               ╱     ╲
            CH₃      CH₃
```

17.15 How many chiral carbon atoms are there in each of the following sugars? How many stereoisomers exist for each compound?

```
      OH         O        OH
       │         ║         │
a. CH₂─CH─C─CH─CH₂
           │        │
          OH      OH
```

```
      O   OH  OH  OH  OH
      ║    │    │    │    │
b. H─C─CH─CH─CH─CH─CH₂─OH
```

17.16 How many chiral carbon atoms are there in each of the following sugars? How many stereoisomers exist for each compound?

```
a. OH    OH   OH    O
    │     │     │     ║
   CH₂─CH─CH─C─CH₂─OH
```

```
b. OH    OH   OH   OH
    │     │     │     │
   CH₂─CH─CH─CH─CHO
```

17.17 How many aldoheptoses are possible? How many are the D form and how many are the L form?

17.18 How many aldopentoses are possible? How many are the D form and how many are the L form?

17.19 Why is the study of chiral molecules important in biochemistry?

17.20 What physical property is characteristic of optically active molecules?

Monosaccharides (Section 17.4)

17.21 Classify each of the following monosaccharides as an aldo- or keto-triose, tetrose, pentose, or hexose:

a.
```
        CHO
    H ──┼── OH
    H ──┼── OH
   HO ──┼── H
   HO ──┼── H
       CH₂OH
```

b.
```
       CH₂OH
        C = O
    H ──┼── OH
    H ──┼── OH
       CH₂OH
```

17.22 Classify each of the following monosaccharides as an aldo- or keto-triose, tetrose, pentose, or hexose:

a.
```
       CH₂OH
        C = O
    H ──┼── OH
   HO ──┼── H
    H ──┼── OH
       CH₂OH
```

b.
```
        CHO
   HO ──┼── H
    H ──┼── OH
   HO ──┼── H
       CH₂OH
```

17.23 Draw Fischer projections of any aldotetrose and any ketopentose.

Properties of Monosaccharides (Section 17.5)

17.24 Explain why certain carbohydrates are called sugars.

17.25 Explain why monosaccharides are soluble in water.

17.26 Using a Fischer projection of a glucose molecule, identify the site(s) on the molecule where water can hydrogen-bond.

17.27 Identify each of the following as an α or a β form and draw the structural formula of the other anomer:

17.28 Identify each of the following as an α or a β form and draw the structural formula of the other anomer:

a.
```
        CH₂OH
   HO ╱──── O
      │
      OH
         ╲──── OH
```

b.
```
   HOCH₂  O   OH
      ╲────╱
       OH
```

17.29 The structure of talose differs from galactose only in the direction of the ─OH group at position 2. Draw and label Haworth structures for α- and β-talose.

17.30 The structure of mannose differs from glucose only in the direction of the ─OH group at position 2. Draw and label Haworth structures for α- and β-mannose.

17.31 What is the difference between pyranose and furanose rings?

17.32 An unknown sugar failed to react with a solution of Cu^{2+}. Classify the compound as a reducing or nonreducing sugar.

17.33 Explain how the cyclic compound β-D-galactose can react with Cu^{2+} and be classified as a reducing sugar.

17.34 Complete the following reactions:

a.

b.

CH₂OH structure + CH₃—OH $\xrightarrow{H^+}$

c.

HOCH₂ structure + CH₃CH₂CH₂—OH $\xrightarrow{H^+}$

17.35 Use an arrow to identify the glycosidic linkage in each of the following:

a.

CH₂OH structure with OCH₂CH₂CH₃

b.

HOCH₂ structure with OCH₂CH₃

Important Monosaccharides (Section 17.6)

17.36 What monosaccharides are used in the synthesis of nucleic acids?

17.37 Explain why D-glucose can be injected directly into the bloodstream to serve as an energy source.

17.38 Give two other names for D-glucose.

17.39 Which of the following are ketohexoses?

a. glucose **b.** fructose **c.** galactose

17.40 How do the hexoses glucose and galactose differ structurally?

17.41 Give the natural sources for each of the following:

a. glucose **b.** fructose **c.** galactose

17.42 Explain why fructose can be used as a low-calorie sweetener.

Disaccharides (Section 17.7)

17.43 What one monosaccharide is a component of all three of the disaccharides sucrose, maltose, and lactose?

17.44 Identify a disaccharide that fits each of the following:

a. The most common household sugar

b. Formed during the digestion of starch

c. An ingredient of human milk

d. Found in germinating grain

e. Hydrolyzes when cooked with acidic foods to give invert sugar

f. Found in high concentrations in sugar cane

17.45 What type of linkage is broken when disaccharides are hydrolyzed to monosaccharides?

17.46 Explain the process of how the hemiacetal group of a lactose molecule is able to react with Benedict's reagent.

17.47 Why is invert sugar sweeter than sucrose?

17.48 Sucrose and honey are commonly used sweeteners. Suppose you had a sweet-tasting water solution that contained either honey or sucrose. How would you chemically determine which sweetener was present?

17.49 Draw Haworth projection formulas of the disaccharides maltose and sucrose. Label the hemiacetal, hemiketal, acetal, or ketal carbons.

17.50 Using structural characteristics, show why

a. lactose is a reducing sugar.

b. sucrose is not a reducing sugar.

17.51 What type of linkage, for example, α(1 → 4), is present in the following?

a. lactose **b.** maltose **c.** sucrose

17.52 The disaccharide melibiose is present in some plant juices.

a. What two monosaccharides are formed on hydrolysis of melibiose?

b. Is melibiose a reducing sugar? Explain.

c. Describe the glycosidic linkage between the two monosaccharide units.

Polysaccharides (Section 17.8)

17.53 Match the following structural characteristics to the polysaccharides amylose, amylopectin, glycogen, and cellulose (a characteristic may fit more than one):

a. Contains both α(1 → 4) and α(1 → 6) glycosidic linkages

b. Is composed of glucose monosaccharide units

c. Contains acetal linkages between monosaccharide units

d. Is composed of highly branched molecular chains

e. Is composed of unbranched molecular chains

f. Contains only $\alpha(1 \rightarrow 4)$ glycosidic linkages

g. Contains only $\beta(1 \rightarrow 4)$ glycosidic linkages

17.54 Name a polysaccharide that fits each of the following:

a. The unbranched polysaccharide in starch

b. A polysaccharide widely used as a textile fiber

c. The most abundant polysaccharide in starch

d. The primary constituent of paper

e. A storage form of carbohydrates in animals

17.55 Polysaccharides are abundant in celery, and yet celery is a good snack for people on a diet. Explain why.

Additional Exercises

17.56 There are two solutions:

(1) a 10% (w/v) water–glucose solution

(2) a 10% (w/v) water–maltose solution

Can you differentiate between the two solutions based on boiling points? Explain your answer.

17.57 Why does one mole of D-glucose only react with one mole of methanol (instead of two) to form an acetal?

17.58 If 100 g of sucrose is completely hydrolyzed to invert sugar, how many grams of glucose will be present?

17.59 Hexanal and 1-hexanol are liquids at room temperature, but glucose (a hexose) is a solid. Give two reasons that may explain this phenomenon.

17.60 Look at Figure 7.14 on page 233 "Osmosis through carrot membranes." Suppose maltose was used instead of molasses in the demonstration and the solution level rose 5 cm above the carrot in the tube. Next, maltase, an enzyme that hydro-

lyzes the glycosidic linkage, was added to the solution. What would happen to the solution level in the tube? Explain your reasoning.

Chemistry for Thought

17.61 Suppose human intestinal bacteria were genetically altered so they could hydrolyze the β linkages of cellulose. Would the results be beneficial? Explain.

17.62 A sample of starch is found to have a molecular weight of 2.80×10^5 u. How many glucose units are present in a molecule of the starch?

17.63 The open-chain form of glucose constitutes only a small fraction of any glucose sample. Yet, when Cu^{2+} is used to oxidize the open-chain form, nearly all the glucose in a sample reacts. Use Le Châtelier's principle to explain this observation.

17.64 Aspartame (Nutrasweet) contains calories and yet is used in diet drinks. Explain how a drink can contain aspartame as a sweetener and yet be low in calories.

17.65 From Figure 17.10, what two products form when the lactose in cow's milk is hydrolyzed in a child's digestive system?

17.66 Answer the question raised in Figure 17.16. What foods in addition to potatoes would you expect to give a positive starch test?

17.67 Amylopectin is a component of starch, yet it does not give a positive iodine test (turn bluish-black) for the presence of starch when it is tested in its pure form. Why?

17.68 Amylose is a straight-chain glucose polymer similar to cellulose. What would happen to the longevity of paper manufactured with amylose instead of cellulose?

17.69 Raffinose, a trisaccharide found in some plants, contains three monosaccharide components: galactose, glucose, and fructose. Disregarding the types of linkages, how many possible structures are there for raffinose?

Allied Health Exam Connection

The following questions are from these sources:

- *Nursing School Entrance Exam* © 2005, Learning Express, LLC.

- *McGraw-Hill's Nursing School Entrance Exams* by Thomas A. Evangelist, Tamara B. Orr, and Judy Unrein © 2009, The McGraw-Hill Companies, Inc.

- *NSEE Nursing School Entrance Exams,* 3rd edition © 2009, Kaplan Publishing.

- *Cliffs Test Prep: Nursing School Entrance Exams* by Fred N. Grayson © 2004, Wiley Publishing, Inc.

- *Peterson's Master the Nursing School and Allied Health Entrance Exams,* 18th edition by Marion F. Gooding © 2008, Peterson's, a Nelnet Company.

17.70 The molecular formula for glucose is $C_6H_{12}O_6$, and the molecular formula for fructose is $C_6H_{12}O_6$. Which of the following describe glucose and fructose?

a. stereoisomers c. hexoses

b. isomers d. anomers

17.71 Which of the following are composed of long chains of glucose molecules?

a. cellulose

b. starch

c. cholesterol

d. glycogen

17.72 Identify each of the following as a monosaccharide or a disaccharide. If it is a disaccharide, identify the individual monosaccharide components.

a. dextrose

b. fructose

c. sucrose

d. maltose

e. lactose

f. ribose

17.73 What is the general type of reaction used to decompose sucrose to glucose + fructose?

17.74 Glucose is a reducing sugar, which if boiled in Benedict's reagent produces an orange to brick-red color. What chemical species is being reduced during the reaction?

17.75 Which polysaccharide is a branched polymer of glucose found in the liver and muscle cells?

a. amylase

b. cellulose

c. glycogen

d. amylopectin

18

Lipids

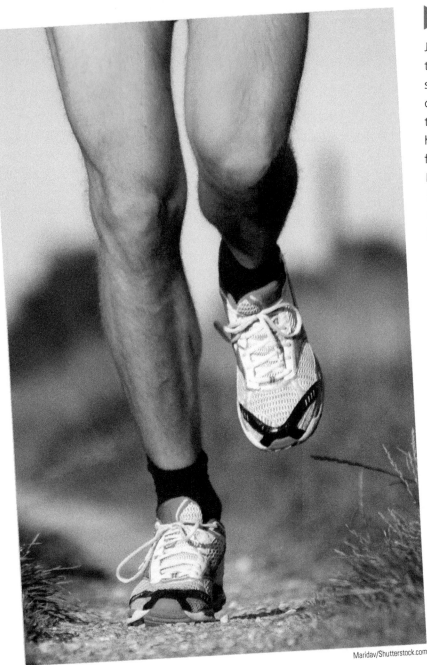

Maridav/Shutterstock.com

▶ Case Study

Jamal ran cross-country for his high school track team 18 years ago and had regularly placed at statewide competitions. Now, family and work concerns made keeping fit difficult. Competition had been a powerful motivating factor for him in his teens, and Jamal thought training for a marathon might get him back on track. He set a goal of running a marathon for his 35th birthday. His wife encouraged him to get a physical first. The doctor ordered a fasting lipoprotein profile. Jamal's results were LDL 169 mg/dL, HDL 50 mg/dL, total cholesterol 189 mg/dL, and triglycerides 155 mg/dL. These levels of total cholesterol and triglycerides are considered elevated for Jamal's age. The doctor explained that exercise and a healthy diet would work together to lower his total cholesterol and increase his high-density lipoprotein (HDL). Now he had even more motivation to hit the track!

▶ **What is the difference between "good" and "bad" cholesterol? What role do triglycerides play in heart health?**

Follow-up to this Case Study appears at the end of the chapter before the *Concept Summary*.

When you have completed your study of this chapter, you should be able to:

1 Classify lipids as saponifiable or nonsaponifiable and list five major functions of lipids. **(Section 18.1)**

2 Describe four general characteristics of fatty acids. **(Section 18.2)**

3 Draw structural formulas of triglycerides given the formulas of the component parts. **(Section 18.3)**

4 Describe the structural similarities and differences of fats and oils. **(Section 18.3)**

5 Write key reactions for fats and oils. **(Section 18.4)**

6 Compare the structures of fats and waxes. **(Section 18.5)**

7 Draw structural formulas and describe uses for phospho-glycerides. **(Section 18.6)**

8 Draw structural formulas and describe uses for sphingolipids. **(Section 18.7)**

9 Describe the major features of cell membrane structure. **(Section 18.8)**

10 Identify the structural characteristic typical of steroids and list important groups of steroids in the body. **(Section 18.9)**

11 Name the major categories of steroid hormones. **(Section 18.10)**

12 Describe the biological importance and therapeutic uses of the prostaglandins. **(Section 18.11)**

The group of compounds called lipids is made up of substances with widely different compositions and structures. Unlike carbohydrates, which are defined in terms of structure, lipids are defined in terms of a physical property—solubility. Lipids are biological molecules that are insoluble in water but soluble in nonpolar solvents. Lipids are the waxy, greasy, or oily compounds found in plants and animals (see **Figure 18.1** and **Figure 18.2**). Lipids repel water, a useful characteristic of protective wax coatings found on some plants. Fats and oils are energy-rich and have relatively low densities. These properties account for their use as storage forms of energy in plants and animals. Still other lipids are used as structural components, especially in the formation of cellular membranes.

Figure 18.1 Waxes form a protective coating on these leaves.

Figure 18.2 Thick layers of fat help insulate penguins against low temperatures.

18.1 | Classification of Lipids

Learning Objective

1. Classify lipids as saponifiable or nonsaponifiable and list the major functions of lipids.

lipid A biological compound that is soluble only in nonpolar solvents.

simple lipid An ester-containing lipid with just two types of components: an alcohol and one or more fatty acids.

complex lipid An ester-containing lipid with more than two types of components: an alcohol, fatty acids—plus others.

Lipids are grouped into two main classes: saponifiable lipids and nonsaponifiable lipids. Saponification refers to the process in which esters are hydrolyzed under basic conditions (see **Section 15.7**). Triglycerides, waxes, phospholipids, and sphingolipids are esters and thus all belong to the first class. Nonsaponifiable lipids are not esters and cannot be hydrolyzed. Steroids and prostaglandins belong to this class. **Figure 18.3** summarizes the classification of different types of lipids. Notice from the figure that saponifiable lipids are further classified into categories of simple and complex, based on the number of components in their structure. **Simple lipids** contain just two types of components (fatty acids and an alcohol), whereas **complex lipids** contain more than two (fatty acids, an alcohol, plus other components).

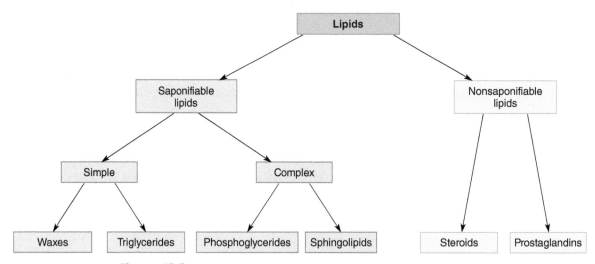

Figure 18.3 The major types of lipids.

18.2 | Fatty Acids

Learning Objective

2. Describe four general characteristics of fatty acids.

Prior to discussing the chemical structures and properties of the saponifiable lipids, it is useful to first describe the chemistry of fatty acids, the fundamental building blocks of many lipids. Fatty acids (defined in **Section 15.1**) are long-chain carboxylic acids, as shown in **Figure 18.4**. It is the long nonpolar tails of fatty acids that are responsible for most of the fatty or oily characteristics of fats. The carboxyl group, or polar head of fatty acids, is very hydrophilic under conditions of physiological pH, where it exists as the carboxylate anion —COO^-.

micelle A spherical cluster of molecules in which the polar portions of the molecules are on the surface and the nonpolar portions are located in the interior.

In aqueous solution, the ions of fatty acids associate with one another in a rather unique way. The ions form spherical clusters, called **micelles**, in which the nonpolar chains extend toward the interior of the structure away from water, and the polar carboxylate groups face outward in contact with the water. Some micelles are large on a molecular scale and contain hundreds or even thousands of fatty acid molecules. The nonpolar chains in a micelle are held together by weak dispersion forces. The structure of a micelle in a radial cross section is shown in **Figure 18.5**. Micelle formation and structure are important in a number of biological functions, such as the transport of insoluble lipids in the blood (see **Section 24.1**).

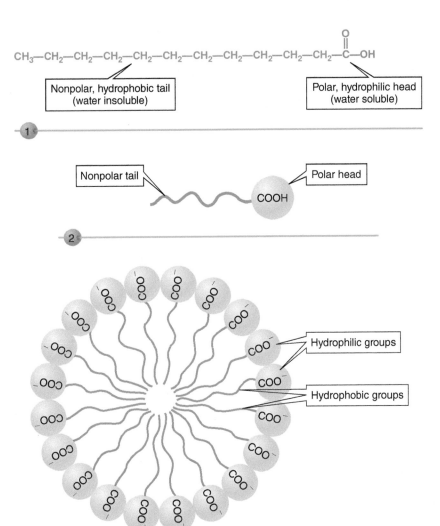

Figure 18.4 The molecular structure of fatty acids: (part A) lauric acid and (part B) a simplified diagram of a fatty acid with a nonpolar tail and a polar head.

Figure 18.5 A cross section of a fatty acid micelle.

The fatty acids found in natural lipids have several characteristics in common:

1. They are usually straight-chain carboxylic acids (no branching).
2. The sizes of most common fatty acids range from 10 to 20 carbons.
3. Fatty acids usually have an even number of carbon atoms (including the carboxyl group carbon).
4. Fatty acids can be saturated (containing no double bonds between carbons) or unsaturated (containing one or more double bonds between carbons). Apart from the carboxyl group and double bonds, there are usually no other functional groups present.

Table 18.1 lists some important fatty acids, along with their formulas and melting points. Unsaturated fatty acids usually contain double bonds in the cis configuration.

$$\underset{\text{long chain}}{\overset{\text{H}}{\diagdown}}C=\underset{\text{long chain}}{\overset{\text{H}}{\diagup}}$$

This configuration creates a long characteristic bend, or kink, in the fatty acid chain that is not found in saturated fatty acids.

These kinks prevent unsaturated fatty acid chains from packing together as closely as do the chains of saturated acids (see **Figure 18.6**). As a result, the intermolecular forces are weaker, and unsaturated fatty acids have lower melting points and are usually liquids at room temperature. For example, the melting point of stearic acid (18 carbons) is 71°C, whereas that of oleic acid (18 carbons with one cis double bond) is 13°C. The melting point of linoleic acid (18 carbons and two double bonds) is even lower (−5°C).

TABLE 18.1 Some Important Fatty Acids

Compound Type and Number of Carbons	Name	Formula	Melting Point (°C)	Common Sources
Saturated				
14	Myristic acid	$CH_3(CH_2)_{12}$—COOH	54	Butterfat, coconut oil, nutmeg oil
16	Palmitic acid	$CH_3(CH_2)_{14}$—COOH	63	Lard, beef fat, butterfat, cottonseed oil
18	Stearic acid	$CH_3(CH_2)_{16}$—COOH	70	Lard, beef fat, butterfat, cottonseed oil
20	Arachidic acid	$CH_3(CH_2)_{18}$—COOH	76	Peanut oil
Monounsaturated				
16	Palmitoleic acid	$CH_3(CH_2)_5CH{=}CH(CH_2)_7$—COOH	−1	Cod liver oil, butterfat
18	Oleic acid	$CH_3(CH_2)_7CH{=}CH(CH_2)_7$—COOH	13	Lard, beef fat, olive oil, peanut oil
Polyunsaturated				
18	Linoleic acid[a]	$CH_3(CH_2)_4(CH{=}CHCH_2)_2(CH_2)_6$—COOH	−5	Cottonseed oil, soybean oil, corn oil, linseed oil
18	Linolenic acid[b]	$CH_3CH_2(CH{=}CHCH_2)_3(CH_2)_6$—COOH	−11	Linseed oil, corn oil
20	Arachidonic acid[a]	$CH_3(CH_2)_4(CH{=}CHCH_2)_4(CH_2)_2$—COOH	−50	Corn oil, linseed oil, animal tissues
20	Eicosapentaenoic acid[b]	$CH_3CH_2(CH{=}CHCH_2)_5(CH_2)_2$—COOH	−54	Fish oil, seafoods
22	Docosahexaenoic acid[b]	$CH_3CH_2(CH{=}CHCH_2)_6CH_2$—COOH	−44	Fish oil, seafoods

[a]Omega-6 fatty acid.

[b]Omega-3 fatty acid.

Figure 18.6 Space-filling models of fatty acids. Unsaturated fatty acids do not pack as tightly as saturated acids, and their melting points are lower than those of saturated acids.

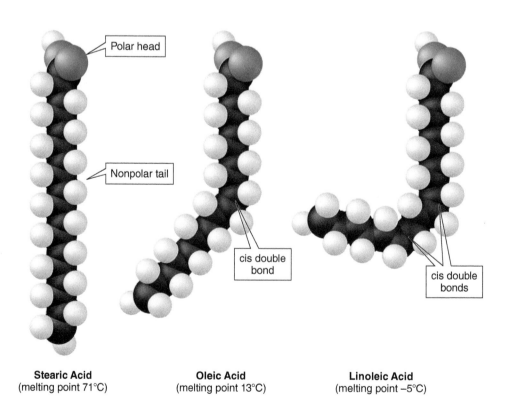

Stearic Acid
(melting point 71°C)

Oleic Acid
(melting point 13°C)

Linoleic Acid
(melting point −5°C)

Chain length also affects the melting point, as illustrated by the fact that palmitic acid (16 carbons) melts at 7°C lower than stearic acid (18 carbons). The presence of double bonds and the length of fatty acid chains in membrane lipids partly explain the fluidity of biological membranes, an important feature discussed in **Section 18.8**.

The human body can synthesize all except two of the fatty acids it needs. These two, linoleic acid and linolenic acid, are polyunsaturated fatty acids that contain 18 carbon atoms (see **Table 18.1**). Because they are not synthesized within the body and must be obtained from the diet, they are called **essential fatty acids**. Both are widely distributed in plant and fish oils. In the body, both acids are used to produce hormonelike substances that regulate a wide range of functions and characteristics, including blood pressure, blood clotting, blood lipid levels, the immune response, and the inflammation response to injury and infection.

essential fatty acid A fatty acid needed by the body but not synthesized within the body.

Linolenic acid is called an omega-3 fatty acid, which means that the endmost double bond is three carbons from the methyl end of the chain:

$$CH_3-CH_2-(CH{=}CHCH_2)_3-(CH_2)_6-COOH$$
linolenic acid

In linoleic acid, an omega-6 fatty acid, the endmost double bond is located six carbons from the methyl end of the chain. Both linoleic and linolenic acids can be converted to other omega-3 and omega-6 fatty acids.

Omega-3 fatty acids have been a topic of interest in medicine since 1985. In that year, researchers reported the results of a study involving natives of Greenland. These individuals have a very low death rate from heart disease, even though their diet is very high in fat. Studies led researchers to conclude that the abundance of fish in the diet of the Greenland natives was involved. Continuing studies led to a possible involvement of the omega-3 fatty acids in the oil of the fish, which may decrease serum cholesterol and triglyceride levels.

A recent five-year study on the benefits of omega-3 fatty acids involved 18,000 patients with unhealthy cholesterol levels. At the end of the study, those patients receiving omega-3 fatty acids had superior cardiovascular function, and nonfatal coronary events were significantly reduced.

18.3 | The Structure of Fats and Oils

Learning Objectives

3. Draw structural formulas of triglycerides given the formulas of the component parts.
4. Describe the structural similarities and differences of fats and oils.

Animal fats and vegetable oils are the most widely occurring lipids. Chemically, fats and oils are both esters, and as we learned in **Chapter 15**, esters consist of an alcohol portion and an acid portion. In fats and oils, the alcohol portion is always derived from glycerol, and the acid portion is furnished by fatty acids. Because glycerol has three —OH groups, a single molecule of glycerol can be attached to three different acid molecules. An example is the esterification reaction of stearic acid and glycerol:

The following diagram may help you remember the components of a triglyceride.

$$
\begin{array}{ccc}
CH_2-OH \quad HO-\overset{\displaystyle O}{\overset{\|}{C}}-C_{17}H_{35} & CH_2-O-\overset{\displaystyle O}{\overset{\|}{C}}-C_{17}H_{35} & \\
| \qquad\qquad\qquad O & | \qquad\qquad O & \\
CH-OH + HO-\overset{\|}{C}-C_{17}H_{35} \rightarrow & CH-O-\overset{\|}{C}-C_{17}H_{35} + 3H_2O & \text{(18.1)}\\
| \qquad\qquad\qquad O & | \qquad\qquad O & \\
CH_2-OH \quad HO-\overset{\|}{C}-C_{17}H_{35} & CH_2-O-\overset{\|}{C}-C_{17}H_{35} &
\end{array}
$$

glycerol (1 molecule) stearic acid (3 molecules) glyceryl tristearate (a triglyceride — 1 molecule)

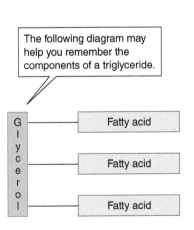

triglyceride or triacylglycerol
A triester of glycerol in which all three alcohol groups are esterified.

The resulting esters are called **triglycerides** or **triacylglycerols**.

The fatty acid components in naturally occurring triglyceride molecules are rarely identical (as in the case of glyceryl tristearate). In addition, natural triglycerides (fats and oils) are usually mixtures of different triglyceride molecules. Butterfat, for example, contains at least 14 different fatty acid components.

✔ **LEARNING CHECK 18.1** Write one structure (several possibilities exist) for a triglyceride derived from stearic acid, oleic acid, and palmitic acid.

fat A triglyceride that is a solid at room temperature.

oil A triglyceride that is a liquid at room temperature.

With some exceptions, triglycerides that come from animals other than fish are solids at room temperature and are called **fats**. Triglycerides from plants or fish are liquids at room temperature and are usually referred to as **oils**. Although both fats and oils are triglycerides, they differ structurally in one important respect—the degree of unsaturation. As shown in **Figure 18.7**, animal fats contain primarily triglycerides of long-chain saturated fatty acids (higher melting points). In contrast, vegetable oils, such as corn oil and sunflower oil, consist of triglycerides containing unsaturated fatty acids (lower melting points) (see **Figure 18.8**). Thus, we see that the structural and physical properties of fatty acids determine the properties of the triglycerides derived from them.

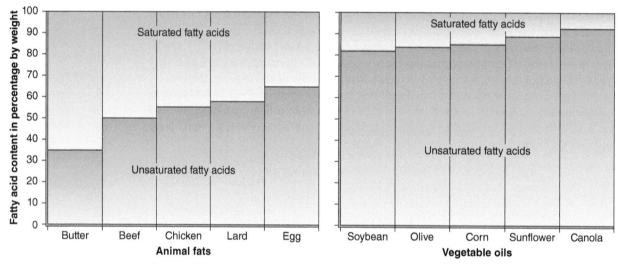

Figure 18.7 A comparison of saturated and unsaturated fatty acids in some foods.

Figure 18.8 Sunflowers are an important crop in some regions of the country because the seeds are a good source of unsaturated oils. Why do you think seeds are rich in oils?

Excessive fat in the diet is recognized as one risk factor influencing the development of chronic disease. The main concern about excessive dietary fat, especially saturated fat, centers on its role in raising blood cholesterol levels (see **Section 18.9**). High blood cholesterol is a recognized risk factor in the development of coronary heart disease, the leading cause of death in Americans every year.

18.4 Chemical Properties of Fats and Oils

Learning Objective

5. Write key reactions for fats and oils.

The chemical properties of triglycerides are typical of esters and alkenes because these are the two functional groups present.

18.4A Hydrolysis

Hydrolysis is one of the most important reactions of fats and oils, just as it was for carbohydrates. The treatment of fats or oils with water and an acid catalyst causes them to hydrolyze to form glycerol and fatty acids. This reaction is simply the reverse of ester formation:

$$
\begin{array}{l}
CH_2-O-\overset{\overset{\textstyle O}{\|}}{C}-(CH_2)_{14}CH_3 \\[2pt]
CH-O-\overset{\overset{\textstyle O}{\|}}{C}-(CH_2)_7CH=CH(CH_2)_7CH_3 \qquad + \ 3H_2O \ \xrightarrow[\text{or}\atop\text{lipase}]{H^+} \\[2pt]
CH_2-O-\overset{\overset{\textstyle O}{\|}}{C}-(CH_2)_6(CH_2CH=CH)_2(CH_2)_4CH_3
\end{array}
$$

(18.2)

$$
\begin{array}{l}
CH_2-OH \\
CH-OH \qquad + \\
CH_2-OH
\end{array}
\quad
\left\{
\begin{array}{l}
CH_3(CH_2)_{14}-\overset{\overset{\textstyle O}{\|}}{C}-OH \\
\qquad\text{palmitic acid} \\[4pt]
CH_3(CH_2)_7CH=CH(CH_2)_7-\overset{\overset{\textstyle O}{\|}}{C}-OH \\
\qquad\text{oleic acid} \\[4pt]
CH_3(CH_2)_4(CH=CHCH_2)_2(CH_2)_6-\overset{\overset{\textstyle O}{\|}}{C}-OH \\
\qquad\text{linoleic acid}
\end{array}
\right.
$$

glycerol

Enzymes (lipases) of the digestive system also catalyze the hydrolysis process. This reaction represents the only important change that takes place in fats and oils during digestion. The breakdown of cellular fat deposits to supply energy also begins with the lipase-catalyzed hydrolysis reaction.

✔ **LEARNING CHECK 18.2** Write the structures of the hydrolysis products of the following reaction:

$$
\begin{array}{l}
CH_2-O-\overset{\overset{\textstyle O}{\|}}{C}-(CH_2)_7CH=CH(CH_2)_7CH_3 \\[2pt]
CH-O-\overset{\overset{\textstyle O}{\|}}{C}-(CH_2)_{14}CH_3 \qquad\qquad\qquad + 3H_2O \xrightarrow{H^+} \\[2pt]
CH_2-O-\overset{\overset{\textstyle O}{\|}}{C}-(CH_2)_{16}CH_3
\end{array}
$$

18.4B Saponification

soap A salt of a fatty acid often
used as a cleaning agent.

When triglycerides are reacted with a strong base, the process of saponification (soap making) occurs. In this commercially important reaction, the products are glycerol and the salts of fatty acids, which are also called **soaps** (Reaction 18.3).

$$
\begin{array}{l}
CH_2-O-\overset{\displaystyle O}{\overset{\|}{C}}-(CH_2)_{14}CH_3 \\
CH-O-\overset{\displaystyle O}{\overset{\|}{C}}-(CH_2)_7CH=CH(CH_2)_7CH_3 \ +\ 3NaOH\ \longrightarrow \\
\hspace{12em}\text{strong base} \\
CH_2-O-\overset{\displaystyle O}{\overset{\|}{C}}-(CH_2)_6(CH_2CH=CH)_2(CH_2)_4CH_3
\end{array}
\tag{18.3}
$$

$$
\begin{array}{l}
CH_2-OH \\
CH-OH \\
CH_2-OH \\
\text{glycerol}
\end{array}
\ +\
\left\{
\begin{array}{l}
CH_3(CH_2)_{14}-\overset{\displaystyle O}{\overset{\|}{C}}-O^-Na^+ \\
\text{sodium palmitate} \\
CH_3(CH_2)_7CH=CH(CH_2)_7-\overset{\displaystyle O}{\overset{\|}{C}}-O^-Na^+ \\
\text{sodium oleate} \\
CH_3(CH_2)_4(CH=CHCH_2)_2(CH_2)_6-\overset{\displaystyle O}{\overset{\|}{C}}-O^-Na^+ \\
\text{sodium linoleate}
\end{array}
\right.
$$

$$\textit{soaps}$$

The salts obtained from saponification depend on the base used. Sodium salts, known as hard salts, are found in most cake soap used in the home. Potassium salts, known as soft soaps, are used in some shaving creams and liquid soap preparations.

In traditional soap making, animal fat is the source of triglycerides, and lye (crude NaOH) or an aqueous extract of wood ashes is the source of the base. The importance of soap making was amply proved after the soap maker's art was lost with the fall of the Roman Empire. The soapless centuries between A.D. 500 and 1500 were notorious for the devastating plagues that nearly depopulated an unsanitary western Europe.

✔ **LEARNING CHECK 18.3** Write structures for the products formed when the following triglyceride is saponified using NaOH:

$$
\begin{array}{l}
CH_2-O-\overset{\displaystyle O}{\overset{\|}{C}}-(CH_2)_{16}CH_3 \\
CH-O-\overset{\displaystyle O}{\overset{\|}{C}}-(CH_2)_7CH=CH(CH_2)_7CH_3 \\
CH_2-O-\overset{\displaystyle O}{\overset{\|}{C}}-(CH_2)_7CH=CH(CH_2)_7CH_3
\end{array}
$$

18.4C Hydrogenation

In **Chapter 12**, we learned that double bonds can be reduced to single bonds by treatment with hydrogen (H_2) in the presence of a catalyst. An important commercial reaction of

590 Chapter 18

fats and oils uses this same hydrogenation process, in which some of the fatty acid double bonds are converted to single bonds. The result is a decrease in the degree of unsaturation and a corresponding increase in the melting point of the fat or oil (see **Figure 18.6**). The peanut oil in many popular brands of peanut butter has been partially hydrogenated to convert the oil to a semisolid that does not separate. Hydrogenation is most often used in the production of semisolid cooking shortenings (such as Crisco®, Spry®, and Fluffo®) or margarines from liquid vegetable oils (see **Figure 18.9**).

Partially hydrogenated vegetable oils were developed in part to help displace highly saturated animal fats used in frying, baking, and spreads. It is important not to complete the reaction and totally saturate all the double bonds. If this is done, the product is hard and waxy—not the smooth, creamy product desired by consumers. Equation 18.4 gives an example in which some of the double bonds react, showing one of the possible products.

Figure 18.9 Hydrogenation of vegetable oils is used to prepare both shortenings and margarines. How do oils used in automobiles differ chemically from these vegetable oils?

$$
\begin{array}{c}
CH_2-O-\overset{\overset{\displaystyle O}{\|}}{C}-(CH_2)_{14}CH_3 \\
| \qquad \overset{\displaystyle O}{\|} \\
CH-O-C-(CH_2)_7CH=CH(CH_2)_7CH_3 \;+\; 2H_2 \\
| \qquad \overset{\displaystyle O}{\|} \\
CH_2-O-C-(CH_2)_6(CH_2CH=CH)_2(CH_2)_4CH_3
\end{array}
\;\xrightarrow{\text{Ni}}\;
\begin{array}{c}
CH_2-O-\overset{\overset{\displaystyle O}{\|}}{C}-(CH_2)_{14}CH_3 \\
| \qquad \overset{\displaystyle O}{\|} \\
CH-O-C-(CH_2)_7CH=CH(CH_2)_7CH_3 \\
| \qquad \overset{\displaystyle O}{\|} \\
CH_2-O-C-(CH_2)_6(CH_2CH_2CH_2)_2(CH_2)_4CH_3
\end{array}
\qquad (18.4)
$$

✔ **LEARNING CHECK 18.4** If an oil with the following structure is completely hydrogenated, what is the structure of the product?

$$
\begin{array}{c}
CH_2-O-\overset{\overset{\displaystyle O}{\|}}{C}-(CH_2)_7CH=CH(CH_2)_7CH_3 \\
| \qquad \overset{\displaystyle O}{\|} \\
CH-O-C-(CH_2)_7CH=CHCH_2CH=CHCH_2CH=CHCH_2CH_3 \\
| \qquad \overset{\displaystyle O}{\|} \\
CH_2-O-C-(CH_2)_7CH=CHCH_2CH=CH(CH_2)_4CH_3
\end{array}
$$

During hydrogenation, some fatty acid molecules in the common cis configuration (see **Section 18.2**) are isomerized into trans fatty acids. The main sources of trans fatty acids in the diet are stick margarine, shortening, and high-fat baked goods. Recent clinical studies have shown that blood cholesterol levels may be raised by the consumption of trans fatty acids. Current dietary advice is to reduce the consumption of saturated fatty acids and products that contain significant amounts of trans fatty acids. In 2015, the Food and Drug Administration (FDA) took action to reduce the use of partially hydrogenated oils, the major source of trans fatty acids (also called trans fats) in the food supply. This action is expected to reduce coronary heart disease and prevent thousands of fatal heart attacks each year in the United States.

18.5 | Waxes

Learning Objective

6. Compare the structures of fats and waxes.

wax An ester of a long-chain fatty acid and a long-chain alcohol.

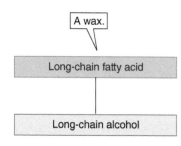

Waxes represent a second group of simple lipids that, like fats and oils, are esters of fatty acids. However, the alcohol portion of waxes is derived from long-chain alcohols (12–32 carbons) rather than glycerol. Beeswax, for example, contains a wax with the following structure:

$$CH_3(CH_2)_{14} - \overset{\overset{\displaystyle O}{\|}}{C} - O - (CH_2)_{29}CH_3$$

$$\underbrace{\phantom{CH_3(CH_2)_{14}C}}_{\substack{\text{palmitic acid} \\ \text{portion}}} \quad \underbrace{\phantom{(CH_2)_{29}CH_3}}_{\substack{\text{long-chain alcohol} \\ \text{portion}}}$$

Waxes are water insoluble and not as easily hydrolyzed as fats and oils; consequently, they often occur in nature as protective coatings on feathers, fur, skin, leaves, and fruits. Sebum, a secretion of the sebaceous glands of the skin, contains many different waxes. It keeps the skin soft and prevents dehydration. Waxes are used commercially to make cosmetics, candles, ointments, and protective polishes.

18.6 | Phosphoglycerides

Learning Objective

7. Draw structural formulas and describe uses for phosphoglycerides.

phosphoglyceride A complex lipid containing glycerol, fatty acids, phosphoric acid, and an aminoalcohol component.

phospholipid A phosphorus-containing lipid.

Phosphoglycerides are complex lipids that serve as the major components of cell membranes. Phosphoglycerides and related compounds are also referred to as **phospholipids**. In these compounds, one of the —OH groups of glycerol is joined by an ester linkage to phosphoric acid, which in turn is linked to another alcohol (usually an aminoalcohol).

The other two —OH groups of glycerol are linked to fatty acids, resulting in the following general structure:

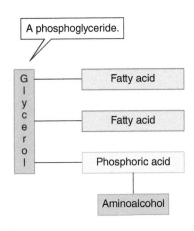

A phosphoglyceride.

$$
\text{glycerol}
\begin{cases}
\overset{}{\underset{}{CH_2}} - O - \overset{\displaystyle O}{\overset{\displaystyle \|}{C}} - R \quad \textit{fatty acid} \\[2mm]
CH - O - \overset{\displaystyle O}{\overset{\displaystyle \|}{C}} - R' \quad \textit{fatty acid} \\[2mm]
CH_2 - O - \overset{\displaystyle O}{\overset{\displaystyle \|}{\underset{\displaystyle \underset{\displaystyle O^-}{|}}{P}}} - OR'' \quad \textit{phosphate and an aminoalcohol}
\end{cases}
$$

a phosphoglyceride

The most abundant phosphoglycerides have one of the alcohols choline, ethanolamine, or serine attached to the phosphate group. These aminoalcohols are shown below in charged forms:

$$HO - CH_2CH_2 - \overset{+}{N}(CH_3)_3 \qquad HO - CH_2CH_2 - \overset{+}{N}H_3 \qquad HO - \underset{\underset{COO^-}{|}}{CH_2CH} - \overset{+}{N}H_3$$

choline
(a quaternary
ammonium cation)

ethanolamine
(cation form)

serine
(two ionic groups
present)

A typical phosphoglyceride is phosphatidylcholine, which is commonly called **lecithin**:

lecithin A phosphoglyceride containing choline.

$$
\begin{aligned}
&^1CH_2 - O - \overset{\displaystyle O}{\overset{\displaystyle \|}{C}} - (CH_2)_{16}CH_3 \\[2mm]
&^2CH - O - \overset{\displaystyle O}{\overset{\displaystyle \|}{C}} - (CH_2)_7CH = CHCH_2CH = CH(CH_2)_4CH_3 \\[2mm]
&^3CH_2 - O - \overset{\displaystyle O}{\overset{\displaystyle \|}{\underset{\displaystyle \underset{\displaystyle O^-}{|}}{P}}} - O - CH_2CH_2 - \overset{+}{N}(CH_3)_3
\end{aligned}
$$

phosphatidylcholine

All phosphoglycerides that contain choline are classified as lecithins. Because different fatty acids may be bonded at positions 1 and 2 of the glycerol, a number of different lecithins are possible. Note that the lecithin shown contains a negatively charged phosphate group and a positively charged quaternary nitrogen. These charges make that end of the molecule strongly hydrophilic, whereas the rest of the molecule is hydrophobic. This structure of lecithins enables them to function as very important structural components of most cell membranes (see **Section 18.8**). It also allows lecithins to function as emulsifying and micelle-forming agents. Such micelles play an important role in the transport of lipids in the blood (see **Chapter 24**). Lecithin (phosphatidylcholine) is commercially extracted from soybeans for use as an emulsifying agent in food. It gives a smooth texture to such products as margarine and chocolate candies (see **Figure 18.10**).

Phosphoglycerides in which the alcohol is ethanolamine or serine, rather than choline, are called **cephalins**. They are found in most cell membranes and are particularly abundant

cephalin A phosphoglyceride containing ethanolamine or serine.

in brain tissue. Cephalins are also found in blood platelets, where they play an important role in the blood-clotting process. A typical cephalin is represented as:

$$CH_2-O-\overset{\displaystyle O}{\overset{\displaystyle \|}{C}}-(CH_2)_{14}CH_3$$

$$CH-O-\overset{\displaystyle O}{\overset{\displaystyle \|}{C}}-(CH_2)_7CH=CH(CH_2)_7CH_3$$

$$CH_2-O-\overset{\displaystyle O}{\underset{\displaystyle O^-}{\overset{\displaystyle \|}{P}}}-O-CH_2\overset{\displaystyle +}{CH}-\overset{+}{NH_3}$$

Note the presence of the negatively charged phosphate and carboxyl groups and positively charged nitrogen. This characteristic allows cephalins, like the lecithins, to function as emulsifiers.

✔ **LEARNING CHECK 18.5** Draw a typical structure for a cephalin containing the cation of ethanolamine:

$$HO-CH_2CH_2-\overset{+}{N}H_3$$

18.7 | Sphingolipids

Learning Objective

8. Draw structural formulas and describe uses for sphingolipids.

sphingolipid A complex lipid containing the aminoalcohol sphingosine.

Sphingolipids, a second type of complex lipid found in cell membranes, do not contain glycerol. Instead, they contain sphingosine, a long-chain unsaturated aminoalcohol. This substance and a typical sphingolipid, called a sphingomyelin, are represented below:

$$CH_3(CH_2)_{12}CH=CH-CH-OH$$
$$\underset{\displaystyle CH-NH_2}{|}$$
$$\underset{\displaystyle CH_2OH}{|}$$

sphingosine

$$CH_3(CH_2)_{12}CH = CH - CH - OH$$

fatty acid

$$CH - NH - \overset{\overset{\displaystyle O}{\|}}{C} - (CH_2)_7CH = CH(CH_2)_7CH_3$$

$$CH_2 - O - \overset{\overset{\displaystyle O}{\|}}{\underset{\underset{\displaystyle O^-}{|}}{P}} - O - CH_2CH_2\overset{+}{N}(CH_3)_3$$

phosphate *choline*

sphingomyelin

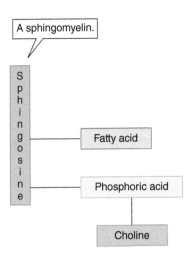

A sphingomyelin.

Sphingosine — Fatty acid

Sphingosine — Phosphoric acid — Choline

In sphingomyelin, choline is attached to sphingosine through a phosphate group. A single fatty acid is also attached to the sphingosine, but an amide linkage is involved instead of an ester linkage. A number of sphingomyelins are known; they differ only in the fatty acid component. Large amounts of sphingomyelins are found in brain and nerve tissue and in the protective myelin sheath that surrounds nerves.

Glycolipids are another type of sphingolipid. Unlike sphingomyelins, however, these complex lipids contain carbohydrates (usually monosaccharides such as glucose or galactose). Glycolipids are often called cerebrosides because of their abundance in brain tissue. The structure of a typical cerebroside is

glycolipid A complex lipid containing a sphingosine, a fatty acid, and a carbohydrate.

$$CH_3(CH_2)_{12}CH = CH - CH - OH$$

fatty acid

$$CH - NH - \overset{\overset{\displaystyle O}{\|}}{C} - (CH_2)_7CH = CH(CH_2)_7CH_3$$

CH_2

CH_2OH

D-galactose

HO — O — O

OH

OH

OH

a cerebroside

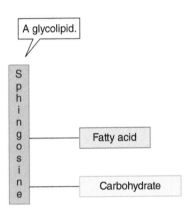

A glycolipid.

Sphingosine — Fatty acid

Sphingosine — Carbohydrate

Notice that, unlike the sphingomyelins, cerebrosides do not contain phosphate.

Several human diseases are known to result from an abnormal accumulation of sphingomyelins and glycolipids in the body (see **Table 18.2**). Research has shown that each of these diseases is the result of an inherited absence of an enzyme needed to break down these complex lipids.

TABLE 18.2 Diseases Originating from Abnormal Metabolism and Accumulation of Glycolipids and Sphingomyelins

Disease	Organ(s) Affected	Type of Lipid Accumulated
Tay-Sachs	Brain	Glycolipid (ganglioside G_{M2})
Gaucher's	Spleen, liver	Cerebrosides containing glucose
Niemann-Pick	Several, particularly liver and spleen	Sphingomyelins

18.8 | Biological Membranes

Learning Objective

9. Describe the major features of cell membrane structure.

18.8A Cell Structure

prokaryotic cell A simple unicellular organism that contains no nucleus and no membrane-enclosed organelles.

eukaryotic cell A cell containing membrane-enclosed organelles, particularly a nucleus.

organelle A specialized structure within a cell that performs a specific function.

Two cell types are found in living organisms: prokaryotic and eukaryotic. **Prokaryotic cells**, comprising bacteria and cyanobacteria (blue-green algae), are smaller and less complex than eukaryotic cells. The more complex **eukaryotic cells** make up the tissues of all other organisms, including humans. Both types of cells are essentially tiny membrane-enclosed units of fluid containing the chemicals involved in life processes. In addition, eukaryotic cells contain small bodies called **organelles**, which are suspended in the cellular fluid, or cytoplasm. Organelles are the sites of many specialized functions in eukaryotes. The most prominent organelles and their functions are listed in **Table 18.3**. Membranes perform two vital functions in living organisms: The external cell membrane acts as a selective barrier between the living cell and its environment, and internal membranes surround some organelles, creating cellular compartments that have separate organization and functions.

TABLE 18.3 The Functions of Some Cellular Organelles

Organelle	Function
Endoplasmic reticulum	Synthesis of proteins, lipids, and other substances
Lysosome	Digestion of substances taken into cells
Microfilaments and microtubules	Cellular movements
Mitochondrion	Cellular respiration and energy production
Nucleus	Contains hereditary material (DNA), which directs protein synthesis
Plastids	Contains plant pigments such as chlorophyll (photosynthesis)
Ribosome	Protein synthesis

18.8B Membrane Structure

fluid-mosaic model A model of membrane structure in which proteins are embedded in a flexible lipid bilayer.

lipid bilayer A structure found in membranes, consisting of two sheets of lipid molecules arranged so that the hydrophobic portions are facing each other.

Most cell membranes contain about 60% lipid and 40% protein. Phosphoglycerides (such as lecithin and cephalin), sphingomyelin, and cholesterol are the predominant types of lipids found in most membranes. Precisely how the lipids and proteins are organized to form membranes has been the subject of a great deal of research. A widely accepted model called the **fluid-mosaic model** is diagrammed in **Figure 18.11**. The lipids are organized in a **lipid bilayer** in which the hydrophobic chains extend toward the inside of the bilayer, and the hydrophilic groups (the phosphate groups and other polar groups) are oriented toward the outside, where they come in contact with water (see **Figure 18.12**). Like the micelle structure discussed in **Section 18.2**, a lipid bilayer is a very stable arrangement where "like" associates with "like." The hydrophobic tails of the lipids are protected from water, and the hydrophilic heads are in a position to interact with water. When a lipid bilayer is broken and the interior hydrocarbon tails are exposed to water, the resulting repulsion causes the bilayer to re-form, and the break seals spontaneously.

Figure 18.11 The fluid-mosaic model of membrane structure. Phosphoglycerides are the chief lipid component. They are arranged in a bilayer. Proteins float like icebergs in a sea of lipid.

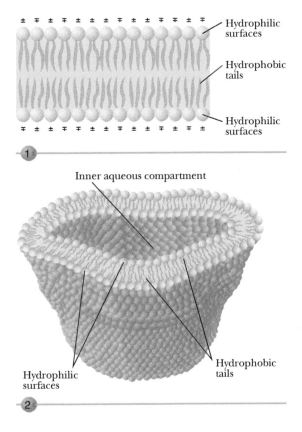

Figure 18.12 Lipid bilayers. (part a) Portion of a bilayer. Circles represent the charged polar portion of phosphoglycerides. The nonpolar hydrocarbon "tails" are within the interior of the bilayer. (part b) Cutaway view of a lipid bilayer vesicle, showing an inner aqueous compartment.

Membrane lipids usually contain unsaturated fatty acid chains that fit into bilayers more loosely than do saturated fatty acids (see **Section 18.2**). This increases the flexibility or fluidity of the membrane. Some of the proteins in the membrane float in the lipid bilayer like icebergs in the sea, whereas others extend completely through the bilayer. The lipid molecules are free to move laterally within the bilayer like dancers on a crowded dance floor—hence the term "fluid mosaic."

18.9 | Steroids

Learning Objective

10. Identify the structural characteristic typical of steroids and list important groups of steroids in the body.

Steroids exhibit the distinguishing feature of other lipids: They are soluble in nonpolar solvents. Structurally, however, they are completely different from the lipids already discussed. **Steroids** have as their basic structure a set of three six-membered rings and a single five-membered ring fused together:

steroid A compound containing four rings fused in a particular pattern.

steroid ring system

18.9A Cholesterol

The most abundant steroid in the human body, and the most important, is cholesterol:

cholesterol

Cholesterol is an essential component of cell membranes and is a precursor for other important steroids, including bile salts, male and female sex hormones, vitamin D, and the adrenocorticoid hormones (see **Section 18.10**). Cholesterol for use in these important functions is synthesized in the liver. Additional amounts of cholesterol are present in the foods we eat.

On the other side of the coin, cholesterol has received much attention because of a correlation between its levels in the blood and the disease known as atherosclerosis, or hardening of the arteries (see **Section 24.1**). Although our knowledge of the role played by cholesterol in atherosclerosis is incomplete, it is now considered advisable to reduce the amount of cholesterol in the foods we eat. In addition, reducing the amount of saturated fatty acids in the diet appears to lower cholesterol production by the body.

18.9B Bile Salts

Bile is a yellowish-brown or green liver secretion that is stored and concentrated in the gallbladder. The entry of partially digested fatty food into the small intestine causes the gallbladder to contract and empty bile into the intestine. Bile does not catalyze any digestive

Biofuels Move into the Kitchen

In the early part of this century when George W. Bush was the U.S. president, he wanted to provide research funding for biofuels, which are fuels derived from organic and renewable products like corn, soybean oil, algae, and other plant products. During Bush's presidency, some people imagined that biofuels would dramatically change what we used as fuel for transportation vehicles. The Obama administration, which followed the Bush administration, continued funding biofuel research.

Ethanol (ethyl alcohol), a biofuel made from corn, has been mixed with gasoline and used as a fuel for transportation in states that have an abundance of corn. But overall transportation fuel options in the United States have not changed as much as the government and researchers hoped.

Much of the funding given to biofuel companies to develop, change, and test fuels used for transportation has been used instead to create and test the technology needed to create non-fuel products. Over the years, the price of imported oil has decreased significantly primarily because of decreases in the amount of oil imported by the United States. Because less oil is imported and most transportation has become more fuel efficient, biofuel companies have made adjustments in order to keep up. Some biofuel companies that received funds from the government have turned their focus completely away from transportation and toward other household products including food, lotions, cleaning products, and perfumes.

Biofuels, particularly those made from algae, are beginning to appear in kitchens rather than in vehicles. Algae- and yeast-based oils are now being marketed as healthy alternatives to other foods. These oils can be used in place of butter, cooking oil, and egg yolks. Some companies are using algae-based oils as starting materials for producing nutritional supplements such as healthy fatty acids called omega-3. Other companies are researching and producing protein supplements based on algae and yeast.

For now, it seems that biofuels have been put on the back burner with regard to their use as transportation fuels. In fact, in a summit on climate change held in Paris at the end of 2015, there was little mention of the future potential of transportation biofuels. As a result of funding and consumer demands, interest in biofuels seems to have turned from only transportation fuel uses toward everyday household uses, including food items.

Pond algae being harvested.

reactions, but it is important in lipid digestion. Fats are more difficult to hydrolyze than starch and proteins because they are not soluble in water. In the watery medium of the digestive system, fats tend to form large globules with a limited surface area exposed to attack by the digestive juices. The chief constituents of bile are the bile salts and two waste components: cholesterol and bile pigments. One of the principal bile salts is sodium glycocholate:

sodium glycocholate

Q: How significantly can diet really lower cholesterol?

Melina B. Jampolis, M.D., is a Physician Nutrition Specialist focusing on nutrition for weight loss and disease prevention and treatment.

A: Hippocrates said, "Let food be thy medicine and medicine be thy food." Unfortunately, the majority of Americans, and their physicians, do not follow this advice. According to a 2010 report from the National Center for Health Statistics, one in four Americans over the age of 45 takes a cholesterol-reducing statin medication—which can have detrimental side effects—rather than maintaining healthy cholesterol levels through diet. Many mistakenly believe that diet is simply not as effective as medication in terms of cholesterol reduction, but a series of recent studies suggests otherwise. Although no single food is as effective as the powerful statin drugs, research investigating a dietary portfolio of cholesterol-lowering foods has shown very promising results, with far fewer side effects and at far less cost.

One study, published in a major nutrition journal in 2005, involved 34 people with high cholesterol. They were divided into three groups, then randomly placed into three different treatment scenarios. They followed each treatment for one month, during which all food and medication was provided and monitored. The control treatment involved placing the subjects on a very low saturated fat diet. The second treatment combined a very low saturated fat diet with a 20-mg dose of a popular statin medication. The third treatment combined a very low saturated fat diet with a special diet containing specific "doses" of cholesterol-lowering foods that included the following:

1. *Plant sterols (1.0 gram for every 1,000 calories consumed)* Plant sterols are plant-based compounds that compete with cholesterol for absorption in the gut, thereby lowering blood cholesterol levels. They are found naturally in plants in much smaller doses, so supplementation or fortification is required for higher doses. This study used margarine enhanced with plant sterols.

2. *Soy-protein foods, including soy milk and soy burgers (21.4 grams per 1,000 calories)*

3. *Almonds (14 grams per 1,000 calories)*

4. *Viscous fiber (10 grams per 1,000 calories)* Viscous fiber is a type of dietary fiber that forms a gel in solution with water. The dietary sources used in this study included oats, barley, psyllium, okra, and eggplant. Other good sources of soluble fiber not included in this study include beans, legumes, citrus fruit, and apples.

After one month of following the control diet, participants lowered their LDL (low-density lipoprotein, or "bad" cholesterol) by an average of 8.5%. After one month of taking the statin medication, the second group lowered their LDL by an average of 33.3%, and after one month of incorporating the special diet, subjects in the third group showed an average LDL reduction of 29.6%. One fourth of the participants in the third group actually achieved their lowest LDL levels on record after following the special diet for one month.

The following year, the researchers tested the special diet of the third group in the real world. They prescribed the same special diet, this time without providing food, to 66 people with high cholesterol for one year. Overall, the subjects showed an average of 13% reduction in LDL. One third achieved a greater than 20% reduction in LDL, and success was directly related to how closely the participants

Products containing plant sterols.

Soy milk is made from soybeans.

followed the diet. The researchers went on to show that adding strawberries, an excellent source of antioxidants, helped further protect the heart by decreasing oxidative damage due to LDL cholesterol, a major risk factor in the progression of heart disease.

It seems as though Hippocrates had the right idea. By incorporating these heart-healthy foods into your diet regularly, along with exercise, maintaining a healthy weight, and not smoking, you can significantly lower cholesterol and improve your heart health.

Almonds may help lower cholesterol levels.

Beans and certain grains are sources of viscous fiber.

The bile salts act much like soaps; they emulsify the lipids and break the large globules into many smaller droplets. After such action, the total amount of lipids is still the same, but a much larger surface area is available for the hydrolysis reactions. Some researchers feel that bile salts also remove fatty coatings from particles of other types of food and thus aid in their digestion.

A second function of bile salts is to emulsify cholesterol found in the bile. Excess body cholesterol is concentrated in bile and is passed into the small intestine for excretion. If cholesterol levels become too high in the bile, or if the concentration of bile salts is too low, the cholesterol precipitates and forms gallstones (see **Figure 18.13**). The most common gallstones contain about 80% cholesterol and are colored by entrapped bile pigments.

The passage of a gallstone from the gallbladder down the common bile duct to the intestine causes excruciating pain. Sometimes one or more stones become lodged in the duct and prevent bile from passing into the duodenum, and fats can no longer be digested normally. A person afflicted with this condition feels great pain and becomes quite nauseated and ill. Bile pigments are absorbed into the blood, the skin takes on a yellow coloration, and the stool becomes gray-colored because of lack of excreted bile pigments. If the condition becomes serious, both the gallbladder and the stones can be surgically removed.

Figure 18.13 Three gallstones (green) in a gallbladder (red).

18.10 | **Steroid Hormones**

Learning Objective

11. Name the major categories of steroid hormones.

A number of steroids in the body serve important roles as **hormones**. The two major categories of steroid hormones are the adrenocorticoid hormones and the sex hormones.

18.10A Adrenocorticoid Hormones

Adrenal glands are small mounds of tissue located at the top of each kidney. The outer layer of the gland, the adrenal cortex, produces a number of potent steroid hormones, the

hormone A chemical messenger secreted by specific glands and carried by the blood to a target tissue, where it triggers a particular response.

adrenocorticoids. They are classified into two groups according to function: Mineralocorticoids regulate the concentration of ions (mainly Na^+) in body fluids, and glucocorticoids enhance carbohydrate metabolism.

Cortisol, the major glucocorticoid, functions to increase the glucose and glycogen concentrations in the body. This is accomplished by the conversion of lactate and amino acids from body proteins into glucose and glycogen. The reactions take place in the liver under the influence of cortisol.

cortisol

Cortisol and its ketone derivative, cortisone, exert powerful anti-inflammatory effects in the body. These, or similar synthetic derivatives such as prednisolone, are used to treat inflammatory diseases such as rheumatoid arthritis and bronchial asthma.

cortisone

prednisolone

By far the most important mineralocorticoid is aldosterone, which influences the absorption of Na^+ and Cl^- in kidney tubules. An increase in the level of the hormone results in a corresponding increase in absorption of Na^+ and Cl^-. Because the concentration of Na^+ in tissues influences the retention of water, aldosterone is involved in water balance in the body.

aldosterone

18.10B Sex Hormones

The testes and ovaries produce steroids that function as sex hormones. Secondary sex characteristics that appear at puberty—including a deep voice, beard, and increased muscular development in males and a higher voice, increased breast size, and lack of facial hair in females—develop under the influence of these sex hormones.

The testes in males perform two functions: One is to produce sperm; the other is to produce male sex hormones (androgens), the most important of which is testosterone.

Consider Olive Oil

When individuals decide to go on a weight-loss diet, they often focus primarily on the caloric content of food. When this is done, highly caloric lipids—fats and oils—are usually among the first foods to be eliminated from the diet. While this makes sense from a calorie-counting perspective, it might not when other factors are considered, such as the health benefits of some eliminated lipids.

Olive oil is a good example of a lipid that provides numerous health benefits when included in the diet (see **Chemistry Tips for Living Well 4.1**). The main type of lipid found in all types of olive oil is monounsaturated fatty acids, or MUFAs, with the 18-carbon oleic acid being the predominant component (see **Table 18.1** for its formula). MUFAs are considered to be healthy lipids. There is evidence that dietary MUFAs help lower total cholesterol levels in the body, and especially the levels of undesirable low-density lipoproteins (LDLs). Some research results show that MUFAs may help maintain normal blood clotting and normal blood pressure and may aid in the control of blood sugar levels.

Olive oil is also known as a good source of potent antioxidants called polyphenols, which are thought to help slow the development of atherosclerosis, or narrowing of the arteries. There is even some evidence suggesting that olive oil in the diet may provide protection against age-related dementia, Alzheimer's disease, and some types of cancer, such as breast cancer.

In spite of all this good news, the high calorie content of olive oil should not be overlooked. It is important to pay attention to serving sizes and to remember to consume olive oil in place of and not in addition to other high-calorie foods. If this is not done, the beneficial health gains might be nullified by unhealthy weight gain.

Numerous health benefits are associated with olive oil.

iStockphoto.com/Valentyn Volkov

Testosterone promotes the normal growth of the male genital organs and aids in the development of secondary sex characteristics.

Testosterone testosterone methandrostenolone

Growth-promoting (anabolic) steroids, including the male hormone testosterone and its synthetic derivatives such as methandrostenolone (Dianabol), are among the most widely used drugs banned by sports-governing organizations because they are alleged to be dangerous and to confer an unfair advantage to users. These drugs, used by both male and female athletes, promote muscular development without excessive masculinization. Their use is particularly prevalent in sports where strength and muscle mass are advantageous, such as weight lifting, track-and-field events like the shot put and hammer throw, and body building.

The side effects of anabolic steroids range from acne to deadly liver tumors. Effects on the male reproductive system include testicular atrophy, a decrease in sperm count, and, occasionally, temporary infertility.

The primary female sex hormones are the estrogens estradiol and estrone, and progesterone. Estrogens and progesterone are important in the reproductive process. Estrogens are involved in egg (ovum) development in the ovaries, and progesterone causes changes in the wall of the uterus to prepare it to accept a fertilized egg and maintain the resulting pregnancy.

estradiol estrone progesterone

18.11 | Prostaglandins

Learning Objective

12. Describe the biological importance and therapeutic uses of the prostaglandins.

This group of compounds was given its name because the first prostaglandins were identified from the secretions of the male prostate gland. Recent research has identified as many as 20 prostaglandins in a variety of tissues within both males and females.

prostaglandin A substance derived from unsaturated fatty acids, with hormonelike effects on a number of body tissues.

Prostaglandins are cyclic compounds synthesized in the body from the 20-carbon unsaturated fatty acid arachidonic acid. Prostaglandins are designated by codes that refer to the ring substituents and the number of side-chain double bonds. For example, prostaglandin E_2 (PGE_2) has a carbonyl group on the ring and a side chain containing two double bonds:

Prostaglandins are similar to hormones in the sense that they are intimately involved in a host of body processes. Clinically, it has been found that prostaglandins are involved in almost every phase of the reproductive process; they can act to regulate menstruation, prevent conception, and induce uterine contractions during childbirth. Certain prostaglandins stimulate blood clotting. They also lead to inflammation and fever. It has been found that aspirin inhibits prostaglandin production. This explains in part why aspirin is such a powerful drug in the treatment of inflammatory diseases such as arthritis.

From a medical viewpoint, prostaglandins have enormous therapeutic potential. For example, PGE_2 and PGF_2 induce labor and are used for therapeutic abortion in early pregnancy. PGE_2 in aerosol form is used to treat asthma; it opens up the bronchial tubes by relaxing the surrounding muscles. Other prostaglandins inhibit gastric secretions and are used in treating peptic ulcers. Many researchers believe that when prostaglandins are fully understood, they will be found useful for treating a much wider variety of ailments.

High-density lipoproteins are considered "good choles-terol" and low-density lipoproteins are considered "bad" cholesterol. Some remember the difference by thinking, "H is for health and L is for lethal." HDL levels should be above 40 mg/dL and LDL levels should be below 130 mg/dL for a total cholesterol below 200 mg/dL. Elevated LDL cholesterol is the primary target of cholesterol-lowering therapies (nicotinic acid, bile sequestering, or statin medications).

If more calories are consumed than are used, the body stores the excess in the form of triglycerides. Triglycerides are stored in fat cells and converted to energy between meals. Eating refined foods containing sugar or white flour can contribute to elevated levels of triglycerides in the blood. High triglycerides levels (over 150 mg/dL) may contribute to atherosclerosis (hardening of the arteries). Lifestyle choices that contribute to lowering total cholesterol levels, such as a diet low in saturated fat and adequate exercise, tend to lower triglyceride levels as well.

Concept Summary

Classification of Lipids Lipids are a family of naturally occurring compounds grouped together on the basis of their relative insolubility in water and solubility in nonpolar solvents. Lipids are energy-rich compounds that are used as waxy coatings, energy storage compounds, and as structural components by organisms. Lipids are classified as saponifiable (ester-containing) or nonsaponifiable. Saponifiable lipids are further classified as simple or complex, depending on the number of structural components.

Objective 1 (Section 18.1), Exercises 18.2 and 18.4

Fatty Acids A fatty acid consists of a long nonpolar chain of carbon atoms, with a polar carboxylic acid group at one end. Most natural fatty acids contain an even number of carbon atoms. They may be saturated, unsaturated, or polyunsaturated (containing two or more double bonds).

Objective 2 (Section 18.2), Exercise 18.6

The Structure of Fats and Oils Triglycerides or triacylglycerols in the form of fats and oils are the most abundant lipids. Fats and oils are simple lipids that are esters of glycerol and fatty acids.

Objective 3 (Section 18.3), Exercise 18.14

The difference between fats and oils is the melting point, which is essentially a function of the fatty acids in the compound.

Objective 4 (Section 18.3), Exercise 18.12

Chemical Properties of Fats and Oils Fats and oils can be hydrolyzed in the presence of acid to produce glycerol and fatty acids. When the hydrolysis reaction is carried out in the presence of a strong base, salts of the fatty acids (soaps) are produced. During hydrogenation, some multiple bonds of unsaturated fatty acids contained in fats or oils are reacted with hydrogen and converted to single bonds.

Objective 5 (Section 18.4), Exercise 18.18

Waxes Waxes are simple lipids composed of a fatty acid esterified with a long-chain alcohol. Waxes are insoluble in water and serve as protective coatings in nature.

Objective 6 (Section 18.5), Exercise 18.24

Phosphoglycerides Phosphoglycerides consist of glycerol esterified to two fatty acids and phosphoric acid. The phosphoric acid is further esterified to choline (in the lecithins) and to ethanolamine or serine (in the cephalins). The phosphoglycerides are particularly important in membrane formation.

Objective 7 (Section 18.6), Exercises 18.28 and 18.30

Sphingolipids These complex lipids contain a backbone of sphingosine rather than glycerol and only one fatty acid component. They are abundant in brain and nerve tissue.

Objective 8 (Section 18.7), Exercise 18.34

Biological Membranes Membranes surround tissue cells as a selective barrier, and they encase the organelles found in eukaryotic cells. Membranes contain both proteins and lipids. According to the fluid-mosaic model, the lipids are arranged in a bilayer fashion with the hydrophobic portions on the inside of the bilayer. Proteins float in the bilayer.

Objective 9 (Section 18.8), Exercise 18.42

Steroids Steroids are compounds that have four rings fused together in a specific way. The most abundant steroid in humans is cholesterol, which serves as a starting material for other important steroids such as bile salts, adrenocorticoid hormones, and sex hormones.

Objective 10 (Section 18.9), Exercises 18.44 and 18.46

Steroid Hormones Hormones are chemical messengers, synthesized by specific glands, that affect various target tissues in the body. The adrenal cortex produces a number of

steroid hormones that regulate carbohydrate utilization (the glucocorticoids) and electrolyte balance (the mineralocorticoids). The testes and ovaries produce steroid hormones that determine secondary sex characteristics and regulate the reproductive cycle in females.

Objective 11 (Section 18.10), Exercise 18.50

Prostaglandins These compounds are synthesized from the 20-carbon fatty acid arachiodonic acid. They exert many hormonelike effects on the body and are used therapeutically to induce labor, treat asthma, and control gastric secretions.

Objective 12 (Section 18.11), Exercise 18.58

Key Terms and Concepts

Cephalin (18.6)
Complex lipid (18.1)
Essential fatty acid (18.2)
Eukaryotic cell (18.8)
Fat (18.3)
Fluid-mosaic model (18.8)
Glycolipid (18.7)
Hormone (18.10)

Lecithin (18.6)
Lipid (18.1)
Lipid bilayer (18.8)
Micelle (18.2)
Oil (18.3)
Organelle (18.8)
Phosphoglyceride (18.6)
Phospholipid (18.6)

Prokaryotic cell (18.8)
Prostaglandin (18.11)
Simple lipid (18.1)
Soap (18.4)
Sphingolipid (18.7)
Steroid (18.9)
Triglyceride or triacylglycerol (18.3)
Wax (18.5)

Key Reactions

1. Hydrolysis of a triglyceride to glycerol and fatty acids—general reaction (Section 18.4):

$$
\begin{array}{c}
CH_2-O-\overset{\overset{O}{\|}}{C}-R \\[2pt]
CH-O-\overset{\overset{O}{\|}}{C}-R + 3H_2O \xrightarrow[\text{lipase}]{\overset{H^+}{\text{or}}} \\[2pt]
CH_2-O-\overset{\overset{O}{\|}}{C}-R
\end{array}
\qquad
\begin{array}{c}
CH_2-OH \\[2pt]
CH-OH + 3R-\overset{\overset{O}{\|}}{C}-OH \\[2pt]
CH_2-OH
\end{array}
$$

2. Saponification of a triglyceride to glycerol and fatty acid salts—general reaction (Section 18.4):

$$
\begin{array}{c}
CH_2-O-\overset{\overset{O}{\|}}{C}-R \\[2pt]
CH-O-\overset{\overset{O}{\|}}{C}-R + 3NaOH \longrightarrow \\[2pt]
CH_2-O-\overset{\overset{O}{\|}}{C}-R
\end{array}
\qquad
\begin{array}{c}
CH_2-OH \\[2pt]
CH-OH + 3R-\overset{\overset{O}{\|}}{C}-O^-Na^+ \\[2pt]
CH_2-OH
\end{array}
$$

3. Hydrogenation of a triglyceride—general reaction (Section 18.4):

$$
\begin{array}{c}
CH_2-O-\overset{\overset{O}{\|}}{C}-R \\[2pt]
CH-O-\overset{\overset{O}{\|}}{C}-(CH_2)_7CH=CH(CH_2)_7CH_3 + H_2 \xrightarrow{Ni} \\[2pt]
CH_2-O-\overset{\overset{O}{\|}}{C}-R
\end{array}
\qquad
\begin{array}{c}
CH_2-O-\overset{\overset{O}{\|}}{C}-R \\[2pt]
CH-O-\overset{\overset{O}{\|}}{C}-(CH_2)_{16}CH_3 \\[2pt]
CH_2-O-\overset{\overset{O}{\|}}{C}-R
\end{array}
$$

Exercises

Even-numbered exercises are answered in Appendix B.

Blue-numbered exercises are more challenging.

Introduction and Classification of Lipids (Section 18.1)

18.1 What is the basis for deciding if a substance is a lipid?

18.2 List two major functions of lipids in the human body.

18.3 What functional group is common to all saponifiable lipids?

18.4 Classify the following as saponifiable or nonsaponifiable lipids:

 a. A steroid **d.** A phosphoglyceride

 b. A wax **e.** A glycolipid

 c. A triglyceride **f.** A prostaglandin

Fatty Acids (Section 18.2)

18.5 Draw the structure of a typical saturated fatty acid. Label the polar and nonpolar portions of the molecule. Which portion is hydrophobic? Hydrophilic?

18.6 Describe four structural characteristics exhibited by most fatty acids.

18.7 Describe the structure of a micelle formed by the association of fatty acid molecules in water. What forces hold the micelle together?

18.8 Name two essential fatty acids, and explain why they are called essential.

18.9 Indicate whether each of the following fatty acids is saturated or unsaturated. Which of them are solids and which are liquids at room temperature?

 a. $CH_3(CH_2)_{14}COOH$

 b. $CH_3(CH_2)_4CH = CHCH_2CH = CH(CH_2)_7COOH$

 c. $CH_3(C_{14}H_{24})COOH$

 d. $CH_3(C_{10}H_{20})COOH$

18.10 Explain why the melting points of unsaturated fatty acids are lower than those of saturated fatty acids.

18.11 What structural feature of a fatty acid is responsible for the name omega-3 fatty acid?

The Structure of Fats and Oils (Section 18.3)

18.12 How are fats and oils structurally similar? How are they different?

18.13 From Figure 18.7, arrange the following substances in order of increasing percentage of unsaturated fatty acids: chicken fat, beef fat, corn oil, butter, and sunflower oil.

18.14 Draw the structure of a triglyceride that contains one myristic acid, one palmitoleic acid, and one linoleic acid. Identify the ester bonds.

18.15 From what general source do triglycerides tend to have more saturated fatty acids? More unsaturated fatty acids?

18.16 The percentage of fatty acid composition of two triglycerides is reported below. Predict which triglyceride has the lower melting point.

	Palmitic Acid	Stearic Acid	Oleic Acid	Linoleic Acid
Triglyceride A	20.4	28.8	38.6	10.2
Triglyceride B	9.6	7.2	27.5	55.7

18.17 Why is the amount of saturated fat in the diet a health concern?

Chemical Properties of Fats and Oils (Section 18.4)

18.18 What process is used to prepare a number of useful products such as margarines and cooking shortenings from vegetable oils?

18.19 Write equations for the following reactions using the oil shown below:

$$CH_2 - O - \overset{\displaystyle O}{\overset{\|}{C}} - (CH_2)_7 CH = CH(CH_2)_7 CH_3$$

$$CH - O - \overset{\displaystyle O}{\overset{\|}{C}} - (CH_2)_7 CH = CH(CH_2)_7 CH_3$$

$$CH_2 - O - \overset{\displaystyle O}{\overset{\|}{C}} - (CH_2)_{16} CH_3$$

 a. Lipase-catalyzed hydrolysis

 b. Saponification

 c. Hydrogenation

18.20 Why is the hydrogenation of vegetable oils of great commercial importance?

18.21 In general terms, name the products of the reactions below:

 a. Acid hydrolysis of a fat

 b. Acid hydrolysis of an oil

 c. Saponification of a fat

 d. Saponification of an oil

18.22 Write reactions to show how each of the following products might be prepared from a typical triglyceride present in the source given. Use Table 18.1 as an aid.

 a. Glycerol from beef fat

 b. Stearic acid from beef fat

 c. A margarine from corn oil

 d. Soaps from lard

Waxes (Section 18.5)

18.23 Draw the structure of a wax formed from oleic acid and cetyl alcohol $(CH_3(CH_2)_{14}CH_2 — OH)$.

18.24 Like fats, waxes are esters of long-chain fatty acids. What structural difference exists between them that warrants placing waxes in a separate category?

18.25 Draw the structure of a wax formed from stearic acid and 1-dodecanol $(CH_3(CH_2)_{10}CH_2 — OH)$

18.26 What role do waxes play in nature?

Phosphoglycerides (Section 18.6)

18.27 How do phosphoglycerides differ structurally from triglycerides?

18.28 Draw the general block diagram structure of a phosphoglyceride.

18.29 Draw the structure of a phosphoglyceride containing serine.

$$HO-CH_2-CH-\overset{+}{N}H_3$$
$$\underset{COO^-}{|}$$

Is it a lecithin or a cephalin?

18.30 Describe two biological roles served by the lecithins.

18.31 Draw the structure of a lecithin. What structural features make lecithin an important commercial emulsifying agent in certain food products?

18.32 What is the structural difference between a lecithin and a cephalin?

18.33 Where are cephalins found in the human body?

Sphingolipids (Section 18.7)

18.34 Draw the general block diagram structure of a sphingolipid.

18.35 List two structural differences between sphingolipids and phosphoglycerides.

18.36 List three diseases caused by abnormal metabolism and accumulation of sphingolipids.

18.37 Describe the structural similarities and differences between the sphingomyelins and the glycolipids.

18.38 Give another name for glycolipids. In what tissues are they found?

Biological Membranes (Section 18.8)

18.39 Where would you find membranes in a prokaryotic cell? Where would you find membranes in a eukaryotic cell? What functions do the membranes serve?

18.40 What three classes of lipids are found in membranes?

18.41 How does the polarity of the phosphoglycerides contribute to their function of forming cell membranes?

18.42 Describe the major features of the fluid-mosaic model of cell membrane structure.

Steroids (Section 18.9)

18.43 Why is it suggested that some people restrict cholesterol intake in their diet?

18.44 Draw the characteristic chemical structure that applies to all steroid molecules.

18.45 Explain how bile salts aid in the digestion of lipids.

18.46 List three important groups of compounds the body synthesizes from cholesterol. Give the location in bodily tissues where cholesterol is an essential component.

18.47 What symptoms might indicate the presence of gallstones in the gallbladder?

18.48 What is the major component in gallstones?

Steroid Hormones (Section 18.10)

18.49 What is a hormone? What are the two major categories of steroid hormones?

18.50 Name the two groups of adrenocorticoid hormones, give a specific example of each group, and explain the function of those compounds in the body.

18.51 How are testosterone and progesterone structurally similar? How are they different?

18.52 Name the primary male sex hormone and the three principal female sex hormones.

18.53 Why do athletes use anabolic steroids? What side effects are associated with their use?

18.54 What role do the estrogens and progesterone serve in preparation for pregnancy?

Prostaglandins (Section 18.11)

18.55 What structural features characterize the prostaglandins?

18.56 What compound serves as a starting material for prostaglandin synthesis?

18.57 What body processes appear to be regulated in part by prostaglandins?

18.58 Name three therapeutic uses of prostaglandins.

Additional Exercises

18.59 Unsaturated fatty acids are susceptible to oxidation, which causes rancidity in food products. Suggest a way to decrease the amount of oxidative rancidity occurring in a food product.

18.60 A red-brown solution of bromine (Br_2) was added to a lipid, and the characteristic red-brown bromine color disappeared. What can be deduced about the lipid structure from this information?

18.61 Which of the following waxes would have the lowest melting point? Explain your answer.

$$\text{a.} \quad CH_3(CH_2)_{14}-\overset{\overset{\textstyle O}{\|}}{C}-O-(CH_2)_{29}CH_3$$

$$\text{b.} \quad CH_3(CH_2)_7CH=CH(CH_2)_5-\overset{\overset{\textstyle O}{\|}}{C}-O-(CH_2)_{29}CH_3$$

18.62 Suggest a reason why complex lipids, such as phosphoglycerides and sphingomyelin, are more predominant in cell membranes than simple lipids.

18.63 Many esters are pleasantly fragrant. Examples include esters that are partially responsible for the smell of oranges and strawberries. What is the main difference between a fragrant ester and a wax? Explain why some esters are fragrant and waxes are not especially aromatic.

Chemistry for Thought

18.64 Gasoline is soluble in nonpolar solvents. Why is gasoline not classified as a lipid?

18.65 The structure of cellular membranes is such that ruptures are closed naturally. Describe the molecular forces that cause the closing to occur.

18.66 In Figure 18.7, the five vegetable oils listed are derived from seeds. Why do you think seeds are rich in oils?

18.67 Answer the question asked in Figure 18.9. How do oils used in automobiles differ chemically from vegetable oils?

18.68 Why doesn't honey dissolve beeswax?

18.69 In Figure 18.5 (where the nonpolar tails are together), why is the structure more stable than a structure in which the polar heads are together?

18.70 Why are there so many structurally different kinds of lipids?

18.71 In what ways are the structural features of lecithin similar to those of a soap?

18.72 When a doughnut is placed on a napkin, the napkin will often absorb a liquid and appear "moist." What is most likely the nature of the liquid?

Allied Health Exam Connection

The following questions are from these sources:

- *Nursing School Entrance Exam* © 2005, Learning Express, LLC.

- *McGraw-Hill's Nursing School Entrance Exams* by Thomas A. Evangelist, Tamara B. Orr, and Judy Unrein © 2009, The McGraw-Hill Companies, Inc.

- *NSEE Nursing School Entrance Exams,* 3rd edition © 2009, Kaplan Publishing.

- *Cliffs Test Prep: Nursing School Entrance Exams* by Fred N. Grayson © 2004, Wiley Publishing, Inc.

- *Peterson's Master the Nursing School and Allied Health Entrance Exams,* 18th edition by Marion F. Gooding © 2008, Peterson's, a Nelnet Company.

18.73 Fats belong to the class of organic compounds represented by the general formula, RCOOR′, where R and R′ represent hydrocarbon groups. What is the name of the functional group present in fats? What functional group is common to all saponifiable lipids?

18.74 Identify each of the following characteristics as describing an unsaturated fatty acid or a saturated fatty acid:
 a. Contains more hydrogen atoms
 b. Is more healthy
 c. More plentiful in plant sources
 d. Is usually a solid at room temperature

18.75 Identify which sex hormones (testosterone, estrogens, or progesterone) are produced in:
 a. The ovaries
 b. The testes

18.76 Substance X passes through a cell membrane easily. Substance X would best be described as:
 a. Hydrophilic or hydrophobic.
 b. Polar or nonpolar.
 c. A lipid or a protein.

18.77 In diseases of the gallbladder, which of the following nutrients is limited in its digestibility? Explain your choice(s).
 a. Starches
 b. Proteins
 c. Fats

18.78 Steroids are classified as:
 a. carbohydrates.
 b. nucleic acids.
 c. lipids.
 d. proteins.

18.79 Accumulation of cholesterol leads to the hardening of the arteries. This is called:
 a. vasoconstriction.
 b. venipuncture.
 c. atherosclerosis.
 d. hypertension.

18.80 Cholesterol, in spite of its bad reputation, is an essential component of:
 a. microtubules.
 b. the cell membrane.
 c. ribosomes.
 d. cytosol.

18.81 Bile is manufactured in the:
 a. duodenum.
 b. liver.
 c. gall bladder.
 d. pancreas.

18.82 The basic structure of cell membranes is a:
 a. protein bilayer.
 b. protein-impregnated phospholipid bilayer.
 c. carbohydrate bilayer.
 d. phospholipid bilayer.

18.83 The mineralocorticoid _____ is a product of the adrenal cortex.
 a. epinephrine
 b. norepinephrine
 c. ACTH
 d. aldosterone

18.84 Which cell type has a nucleus?
 a. bikaryotic
 b. prokaryotic
 c. eukaryotic
 d. postkaryotic

18.85 The cells of a human body are:
 a. bikaryotic cells.
 b. prokaryotic cells.
 c. unkaryotic cells.
 d. eukaryotic cells.

19

Proteins

Shawn Hempel/Shutterstock.com

▶ Case Study

Laura always wore her hair long and loose. Even when summer weather caused the sweat to trickle down her face as she worked in the garden, she refused to put her hair in a ponytail or use a headband. Laura just couldn't show her ears. "Peanut ear," the children called her. Years of teasing had crippled Laura's self-image. Missing an ear seemed like such a little thing. Why couldn't she just get over it? *Microtia*, the birth defect responsible for her condition, not only affected Laura's hearing, it made her feel unattractive.

Laura's mom called her to come and watch the news on TV. According to a news report, researchers were able to grow outer ears from a collagen scaffold printed with a 3-D printer. Collagen is a fibrous protein found in cartilage, skin, and other connective tissue in the body. The printed ears not only looked realistic, they were actually alive. Laura's heart skipped a beat as the newscaster said that human trials would begin next year. Laura was determined to volunteer for the trials and study.

► What is microtia? How is ear reconstruction compatible with bioprinting? How far into the future should we expect to see more complex organs produced by manufacturing?

Follow-up to this Case Study appears at the end of the chapter before the *Concept Summary*.

Learning Objectives

When you have completed your study of this chapter, you should be able to:

1 Identify the characteristic parts of alpha-amino acids. **(Section 19.1)**

2 Draw structural formulas to illustrate the various ionic forms assumed by amino acids. **(Section 19.2)**

3 Write reactions to represent the formation of peptides and proteins.**(Section 19.3)**

4 Describe uses for important peptides. **(Section 19.4)**

5 Describe proteins in terms of the following characteristics: size, function, classification as fibrous or globular, and classification as simple or conjugated. **(Section 19.5)**

6 Explain what is meant by the primary structure of proteins. **(Section 19.6)**

7 Describe the role of hydrogen bonding in the secondary structure of proteins. **(Section 19.7)**

8 Describe the role of side-chain interactions in the tertiary structure of proteins. **(Section 19.8)**

9 Explain what is meant by the quaternary structure of proteins. **(Section 19.9)**

10 Describe the conditions that can cause proteins to hydrolyze or become denatured. **(Section 19.10)**

The name *protein*, coined more than 100 years ago, is derived from the Greek *proteios*, which means "of first importance," and is appropriate for these most important of all biological compounds. Proteins are indispensable components of all living things, where they play crucial roles in all biological processes. In this chapter, we begin with a discussion of the structures and chemistry of the common amino acids, then progress to show how combinations of amino acids form large molecules called peptides and larger molecules called proteins.

19.1 | The Amino Acids

Learning Objective

1. Identify the characteristic parts of alpha-amino acids.

19.1A Structure

All proteins, regardless of their origin or biological function, are structurally similar. They are polymers consisting of chains of amino acids chemically bound to each other. As the name implies, an amino acid is an organic compound that contains both an amino group and a carboxylate group in ionic form (see **Section 19.2**). Hundreds of different amino acids, both synthetic and naturally occurring, are known, but only 20 are commonly found in natural proteins. The amino acids in proteins are called **alpha (α)-amino acids** because the amino group is attached to the α carbon (the carbon next to the carboxylate group), as shown in **Figure 19.1**. Each amino acid has a characteristic side chain, or R group, that imparts chemical individuality to the molecule. These R side chains contain different structural features, such as aromatic rings, —OH groups, —NH$_3^+$ groups, and —COO$^-$ groups. This variety in side chains causes differences in the properties of the individual amino acids and the proteins containing different combinations of them.

alpha-amino acid An organic compound containing both an amino group and a carboxylate group, with the amino group attached to the carbon next to the carboxylate group.

$$H_2\overset{+}{N}-CH-\overset{\overset{\textstyle O}{\|}}{C}-O^-$$
$$\begin{array}{cc} | & | \\ CH_2 & CH_2 \\ \diagdown & \diagup \\ & CH_2 \end{array}$$

The structure of proline differs slightly from the general formula because the amino group and R group are part of a ring.

Figure 19.1 The general
structure of α-amino acids shown
in ionic form.

The polarity of the R groups is an important characteristic and is the basis for classifying amino acids into the four groups shown in **Table 19.1**: neutral and nonpolar side chains, neutral but polar side chains, basic side chains, and acidic side chains. Note that acidic side chains contain a carboxylate group and that basic side chains contain an additional amino group. The three-letter and one-letter abbreviations that are used to represent amino acids are given after the names:

cysteine (Cys) C

glutamine (Gln) Q

asparagine (Asn) N

19.1B Stereochemistry

A glance at the general formula for the 20 amino acids

shows that with the exception of glycine, in which the R group is —H, there are four different groups attached to the α carbon. Thus, 19 of the 20 amino acids contain a chiral α carbon atom and can exist as two enantiomers; these are the L and D forms:

an L-amino acid

a D-amino acid

As we saw when we studied carbohydrates, one of the two possible forms usually predominates in natural systems. With few exceptions, the amino acids found in living systems exist in the L form.

NEUTRAL, NONPOLAR SIDE CHAINS

$\overset{+}{H_3N}-CH-COO^-$ with H	$\overset{+}{H_3N}-CH-COO^-$ with CH_3	$\overset{+}{H_3N}-CH-COO^-$ with CH and CH_3 CH_3	$\overset{+}{H_3N}-CH-COO^-$ with CH_2, CH, CH_3 CH_3
Glycine (Gly) G	**Alanine (Ala) A**	**Valine (Val) V**	**Leucine (Leu) L**
$\overset{+}{H_3N}-CH-COO^-$ with $CH-CH_3$, CH_2, CH_3	$\overset{+}{H_3N}-CH-COO^-$ with CH_2 and benzene ring	$\overset{+}{H_2N}-CH-COO^-$ with CH_2 CH_2, CH_2 (ring)	$\overset{+}{H_3N}-CH-COO^-$ with CH_2, CH_2, $S-CH_3$
Isoleucine (Ile) I	**Phenylalanine (Phe) F**	**Proline (Pro) P**	**Methionine (Met) M**

NEUTRAL, POLAR SIDE CHAINS

$\overset{+}{H_3N}-CH-COO^-$ with CH_2, OH	$\overset{+}{H_3N}-CH-COO^-$ with $CH-CH_3$, OH	$\overset{+}{H_3N}-CH-COO^-$ with CH_2, benzene ring, OH	$\overset{+}{H_3N}-CH-COO^-$ with CH_2 and indole ring (N, H)
Serine (Ser) S	**Threonine (Thr) T**	**Tyrosine (Tyr) Y**	**Tryptophan (Trp) W**
$\overset{+}{H_3N}-CH-COO^-$ with CH_2, SH	$\overset{+}{H_3N}-CH-COO^-$ with CH_2, $C=O$, NH_2	$\overset{+}{H_3N}-CH-COO^-$ with CH_2, CH_2, $C=O$, NH_2	
Cysteine (Cys) C	**Asparagine (Asn) N**	**Glutamine (Gln) Q**	

BASIC, POLAR SIDE CHAINS			ACIDIC, POLAR SIDE CHAINS	
$\overset{+}{H_3N}-CH-COO^-$ with CH_2 and imidazole ring ($\overset{+}{HN}$, NH)	$\overset{+}{H_3N}-CH-COO^-$ with CH_2, CH_2, CH_2, CH_2, $\overset{+}{NH_3}$	$\overset{+}{H_3N}-CH-COO^-$ with CH_2, CH_2, CH_2, NH, $C=\overset{+}{NH_2}$, NH_2	$\overset{+}{H_3N}-CH-COO^-$ with CH_2, COO^-	$\overset{+}{H_3N}-CH-COO^-$ with CH_2, CH_2, COO^-
Histidine (His) H	**Lysine (Lys) K**	**Arginine (Arg) R**	**Aspartate (Asp) D**	**Glutamate (Glu) E**

19.2 | Zwitterions

Learning Objective

2. Draw structural formulas to illustrate the various ionic forms assumed by amino acids.

zwitterion A dipolar ion that carries both a positive and a negative charge as a result of an internal acid–base reaction in an amino acid molecule.

The structural formulas used to this point have shown amino acids in an ionized form. The fact that amino acids are white crystalline solids with relatively high melting points and high water solubilities suggests that they exist in an ionic form. Studies of amino acids confirm that this is true both in the solid state and in solution. The presence of a carboxyl group and a basic amino group in the same molecule makes possible the transfer of a hydrogen ion in a kind of internal acid–base reaction (Reaction 19.1). The product of this reaction is a dipolar ion called a **zwitterion**.

$$H_2N-\underset{\underset{R}{|}}{CH}-\overset{\overset{O}{||}}{C}-OH \rightarrow \overset{+}{H_3}N-\underset{\underset{R}{|}}{CH}-\overset{\overset{O}{||}}{C}-O^- \qquad (19.1)$$

nonionized form zwitterion
(does not exist) (present in solids
and solutions)

The structure of an amino acid in solution varies with the pH of the solution. For example, if the pH of a solution is lowered by adding a source of H_3O^+, such as hydrochloric acid, the carboxylate group ($-COO^-$) of the zwitterion can pick up a proton to form $-COOH$ (Reaction 19.2):

$$\overset{+}{H_3}N-\underset{\underset{R}{|}}{CH}-\overset{\overset{O}{||}}{C}-O^- + H_3O^+ \rightarrow \overset{+}{H_3}N-\underset{\underset{R}{|}}{CH}-\overset{\overset{O}{||}}{C}-OH + H_2O \qquad (19.2)$$

zwitterion
(0 net charge) (positive net charge)

The zwitterion form has a net charge of 0, but the form in acid solution has a net positive charge.

When the pH of the solution is increased by adding OH^-, the $-NH_3^+$ of the zwitterion can lose a proton, and the zwitterion is converted into a negatively charged form (Reaction 19.3):

$$\overset{+}{H_3}N-\underset{\underset{R}{|}}{CH}-\overset{\overset{O}{||}}{C}-O^- + OH^- \rightarrow H_2N-\underset{\underset{R}{|}}{CH}-\overset{\overset{O}{||}}{C}-O^- + H_2O \qquad (19.3)$$

zwitterion
(0 net charge) (negative net charge)

Changes in pH also affect acidic and basic side chains and those proteins in which they occur. Thus, all amino acids exist in solution in ionic forms, but the actual form (and charge) is determined by the solution pH.

Because amino acids assume a positively charged form in acidic solutions and a negatively charged form in basic solutions, it seems reasonable to assume that at some pH between the acidic and basic extremes the amino acid will have no net charge (it will be a zwitterion). This assumption is correct, and the pH at which the zwitterion forms is called the **isoelectric point**. Each amino acid has a unique and characteristic isoelectric point; those with neutral R groups are all near a pH of 6 (that of glycine, e.g., is at pH 6.06), those with basic R groups have higher values (that of lysine is at pH 9.47), and

isoelectric point The characteristic solution pH at which an amino acid has a net charge of 0.

the two acidic amino acids have lower values (that of aspartate is at pH 2.98). Proteins, which are composed of amino acids, also have characteristic isoelectric points.

The ability of amino acids and some proteins to react with both H_3O^+ and OH^-, as shown by Reactions 19.2 and 19.3, allows solutions of amino acids or proteins to behave as buffers. The buffering action of blood proteins is one of their most important functions.

Example 19.1 Drawing Zwitterion Structures

Draw the structure of the amino acid leucine

$$\overset{+}{H_3}N-CH-\overset{\overset{\textstyle O}{\|}}{C}-O^-$$
$$|$$
$$CH_2$$
$$|$$
$$CH$$
$$\diagup \diagdown$$
$$CH_3 \quad CH_3$$

a. in acidic solution at a pH below the isoelectric point;
b. in basic solution at a pH above the isoelectric point.

Solution

a. In acidic solution, the carboxylate group of the zwitterion picks up a proton.

$$\overset{+}{H_3}N-CH-\overset{\overset{\textstyle O}{\|}}{C}-O^- + H_3O^+ \rightarrow \overset{+}{H_3}N-CH-\overset{\overset{\textstyle O}{\|}}{C}-OH + H_2O$$

with side chains:
$$CH_2 \quad CH_2$$
$$CH \quad CH$$
$$CH_3 \ CH_3 \quad CH_3 \ CH_3$$

b. In basic solution, the $\overset{+}{H_3}N-$ group of the zwitterion loses a proton.

$$\overset{+}{H_3}N-CH-\overset{\overset{\textstyle O}{\|}}{C}-O^- + OH^- \rightarrow H_2N-CH-\overset{\overset{\textstyle O}{\|}}{C}-O^- + H_2O$$

with side chains:
$$CH_2 \quad CH_2$$
$$CH \quad CH$$
$$CH_3 \ CH_3 \quad CH_3 \ CH_3$$

✔ **LEARNING CHECK 19.1** Draw the structure of the amino acid serine

$$\overset{+}{H_3}N-CH-\overset{\overset{\textstyle O}{\|}}{C}-O^-$$
$$|$$
$$CH_2-OH$$

a. in acidic solution at a pH below the isoelectric point;
b. in basic solution at a pH above the isoelectric point.

19.3 | Reactions of Amino Acids

Learning Objective

3. Write reactions to represent the formation of peptides and proteins.

Amino acids can undergo reactions characteristic of any functional group in the molecule. This includes reactions of the carboxylate group, the amino group, and any other functional groups attached to the R side chain. However, two reactions are of special interest because of their influence in determining protein structure.

19.3A Oxidation of Cysteine

Cysteine, the only sulfhydryl (—SH)-containing amino acid of the 20 commonly found in proteins, has a chemical property not shared by the other 19. We learned in **Section 13.9** that a sulfhydryl group can easily be oxidized to form a disulfide bond (—S—S—). Thus, two cysteine molecules react readily to form a disulfide compound called cystine. The disulfide, in turn, is easily converted to —SH groups by the action of reducing agents:

As will be seen, this reaction is important in establishing the structure of some proteins.

19.3B Peptide Formation

In **Section 16.4** we noted that amides could be thought of as being derived from a carboxylic acid and an amine (Reaction 19.5), although in actual practice the acid chlorides or acid anhydrides are generally used.

In the same hypothetical way, it is possible to envision the carboxylate group of one amino acid reacting with the amino group of a second amino acid. For example, glycine could combine with alanine:

Q: **Can a higher-protein diet help me lose weight?**

A: There is extensive debate among experts over which type of diet is the most effective for weight loss. Protein has several potential benefits due to its effect on hunger, metabolism, and body composition. Of the three macronutrients—protein, carbohydrates, and fat—protein has the greatest ability to satisfy hunger, which is critical for dieters. Dietary protein functions by triggering a greater release of the powerful appetite-suppressing hormone peptide YY (PYY), which communicates with the brain to decrease hunger. Studies have found that eating larger amounts of protein, especially at breakfast, helps improve hunger control throughout the day.

Protein can also boost metabolism slightly, which is particularly important during dieting, when metabolism can drop significantly. The total metabolic rate, or the number of calories you burn each day, includes three components: resting metabolic rate (RMR, the calories your body, including all of your organs, needs to function at rest), physical activity (the component over which you have the most control), and the thermic effect of feeding (the calories burned as a result of the digestion and processing of foods that you consume). Although the thermic effect of feeding represents only about 10% of your daily caloric expenditure, research shows that protein boosts the thermic effect of eating significantly

> Melina B. Jampolis, M.D., is a Physician Nutrition Specialist focusing on nutrition for weight loss and disease prevention and treatment.

more than carbohydrates and fat do. This enables you to burn more calories by consuming slightly more protein with each meal. The effect is fairly small, however, and can easily be negated by overconsumption of calories. So, it is important to eat a slightly higher percentage of protein without increasing total calories.

Protein may also benefit metabolism through its impact on body composition. Research has shown that protein in the diet can help maintain lean body mass during weight loss, particularly when combined with resistance training. If and when weight is regained, one study showed that by eating just 3% more protein, the regained weight was more likely to be muscle than fat. Lean body mass, which includes muscle, bone, and water, is one of the only components of RMR that you can control to some extent. Therefore, preserving muscle may help minimize the drop in metabolism both during and after weight loss, which can improve both weight loss and weight maintenance. In addition, research suggests that consuming a higher percentage of protein may help increase abdominal fat loss, particularly in women who tend to carry fat around their midsection rather than their hips and thighs. Higher-protein diets may also help lower triglycerides more effectively in women with excess belly fat.

Finally, eating slightly more protein during weight loss may help preserve bone mineral content, which is particularly important in women as they get older and their risk of osteoporosis increases. If you have kidney problems, however, consuming higher amounts of protein is not recommended because it can worsen kidney disease.

The bottom line is that calories count the most when it comes to losing weight. However, consuming a slightly higher percentage of protein (25–30%) while consuming fewer calories may make weight loss slightly easier and more sustainable. For optimal health, make sure to consume mainly low-fat and plant-based protein.

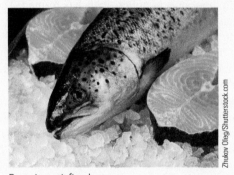

Protein satisfies hunger.

Zhukov Oleg/Shutterstock.com

Dietary protein may help increase abdominal fat loss.

FXQuadro/Shutterstock.com

Dietary protein can help maintain lean body mass.

Andresr/Shutterstock.com

Compounds of this type made up of two amino acids are called **dipeptides**, and the amide linkage that holds them together is called a **peptide linkage** or **peptide bond**.

It is important to realize that a different dipeptide could also form by linking glycine and alanine the other way (alanine on the left in this case):

$$
\overset{+}{H_3N}-\underset{\underset{CH_3}{|}}{CH}-\overset{\overset{O}{||}}{C}-O^- + \overset{+}{H_3N}-CH_2-\overset{\overset{O}{||}}{C}-O^- \longrightarrow \overset{+}{H_3N}-\underset{\underset{CH_3}{|}}{CH}-\overset{\overset{O}{||}}{C}-NH-CH_2-\overset{\overset{O}{||}}{C}-O^- + H_2O \qquad (19.7)
$$

peptide linkage

alanine glycine alanylglycine

Glycylalanine and alanylglycine are structural isomers of each other, and each has a unique set of properties. The reactions to form these and other dipeptides are more complex than Reactions 19.6 and 19.7 imply. In nature, these processes are carried out according to genetic information using a mechanism we shall examine in **Section 21.8**.

The presence of amino and carboxylate groups on the ends of dipeptides allows for the attachment of a third amino acid to form a tripeptide. For example, valine could be attached to alanylglycine:

two peptide linkages

$$
\overset{+}{H_3N}-\underset{\underset{CH_3}{|}}{CH}-\overset{\overset{O}{||}}{C}-NH-CH_2-\overset{\overset{O}{||}}{C}-NH-\underset{\underset{\underset{CH_3}{|}}{CH-CH_3}}{CH}-\overset{\overset{O}{||}}{C}-O^- + H_2O \qquad (19.8)
$$

alanylglycylvaline, a tripeptide

More amino acids can react in the same way to form a tetrapeptide, a pentapeptide, and so on, until a chain of hundreds or even thousands of amino acids has formed. The compounds with the shortest chains are often simply called **peptides**, those with longer chains are called **polypeptides**, and those with still longer chains are called **proteins**. Chemists differ about where to draw the lines in the use of these names, but generally polypeptide chains with more than 50 amino acids are called proteins. However, the terms *protein* and *polypeptide* are often used interchangeably.

Amino acids that have been incorporated into chains are called **amino acid residues**. The amino acid residue with an unreacted or free amino group at the end of the chain is designated the **N-terminal residue**, and, according to convention, it is written on the left end of the peptide chain. Similarly, the residue with a free carboxylate group at the other end of the chain is the **C-terminal residue**; it is written on the right end of the chain. Thus, alanine is the N-terminal residue and valine is the C-terminal residue in the tripeptide alanylglycylvaline:

alanine, the N-terminal residue

$$
\overset{+}{H_3N}-\underset{\underset{CH_3}{|}}{CH}-\overset{\overset{O}{||}}{C}-NH-CH_2-\overset{\overset{O}{||}}{C}-NH-\underset{\underset{\underset{CH_3}{|}}{CH-CH_3}}{CH}-\overset{\overset{O}{||}}{C}-O^-
$$

valine, the C-terminal residue

Peptides are named by starting at the N-terminal end of the chain and listing the amino acid residues in order from left to right. Structural formulas and even full names for large peptides become very unwieldy and time-consuming to write. This is simplified by representing peptide and protein structures in terms of three-letter abbreviations for the amino acid residues with dashes to show peptide linkages. Thus, the structure of glycylalanine

is represented by Gly-Ala. The isomeric dipeptide, alanylglycine, is Ala-Gly. In this case, alanine is the N-terminal residue, and glycine is the C-terminal residue.

The tripeptide alanylglycylvaline can be represented by Ala-Gly-Val. In addition to this sequence, five other arrangements of these three components are possible, and each one represents an isomeric tripeptide with different properties. These sequences are Ala-Val-Gly, Val-Ala-Gly, Val-Gly-Ala, Gly-Val-Ala, and Gly-Ala-Val.

Insulin, with 51 amino acids, has 1.55×10^{66} sequences possible! From these possibilities, the body reliably produces only one, which illustrates the remarkable precision of the life process. From the simplest bacterium to the human brain cell, only the amino acid sequences needed to form essential cell peptides are produced.

> ✔ **LEARNING CHECK 19.2** Use **Table 19.1** to draw the full structure of the following tetrapeptide. Label the N-terminal and C-terminal residues.
>
> <div align="center">Phe-Cys-Ser-Ile</div>

19.4 | Important Peptides

Learning Objective

4. Describe uses for important peptides.

More than 200 peptides have been isolated and identified as essential to the proper functioning of the human body. The identity and sequence of amino acid residues in these peptides is extremely important to proper physiological function. Two well-known examples are vasopressin and oxytocin, which are hormones released by the pituitary gland. Each hormone is a nonapeptide (nine amino acid residues) with six of the residues held in the form of a loop by a disulfide bond (see **Figure 19.2**). The disulfide bond in these hormones is formed by the oxidation of cysteine residues in the first and sixth positions (counting from the N-terminal end) of these peptide chains. In peptides and proteins, disulfide bonds such as these that draw a single peptide chain into a loop or hold two peptide chains together are called **disulfide bridges**.

disulfide bridge A bond produced by the oxidation of —SH groups on two cysteine residues. The bond loops or holds two peptide chains together.

Figure 19.2 The structures of vasopressin and oxytocin. Differences are in amino acid residues 3 and 8. The normal carboxylate groups of the C-terminal residues (Gly) have been replaced by amide groups.

Figure 19.2 shows that the structures of vasopressin and oxytocin differ only in the third and eighth amino acid residues of their peptide chains. The result of these variations is a significant difference in their biological functions. Vasopressin is known as an antidiuretic hormone (ADH) because of its action in reducing the volume of urine formed, thus conserving the body's water. It also raises blood pressure. Oxytocin causes the smooth muscles of the uterus to contract and is often administered to induce labor. Oxytocin also acts on the smooth muscles of lactating mammary glands to stimulate milk ejection.

Go for the Good Grains

Eating whole grains is important for good health. Dietary whole grains contribute to good heart health, lower cholesterol, decrease in obesity, and good gastrointestinal health. Dieticians recommend that at least half the grains in your diet should come from whole grains.

What exactly is a grain? Grains are grass seeds. Wheat and corn dominate American grain consumption. In other countries rice or millet constitute the main grain consumption. Whether we talk of rice, cereal, flour, cornmeal, masa, hominy, grits, oatmeal, flax seed, or quinoa, we are talking about grains. Breads, cereal, granola, and pasta are good examples of foods made from grains. Grains make up a large portion, if not the largest part, of most diets.

The difference between whole grains and those that are not considered "whole" is the refining process. Refining or milling removes the bran and germ of the seeds making up the grain. The "germ" of a grain is the part of the seed that germinates or sprouts when the seed is planted. The majority of vitamins and minerals such as selenium, potassium, and magnesium are located in this part of the grain seed, as well as the healthful fat or oil. Removing the germ from grain increases the shelf life because it removes the oil that can become rancid. Whole grains should be kept fresh by refrigeration or storing them in a cool, dry place. Bran, the outer covering of the seed, provides fiber in the diet. Dietary fiber is essential to maintaining good gastrointestinal and cardiac health.

Many refined grains are "enriched," as in "enriched white flour." This means that some of the nutrients, such as B vitamins, removed by the milling process, are added back to the product. Many enriched products are also "fortified." Fortified grains are refined grains with nutrients added to the product that do not naturally occur in the grain, such as folic acid and iron. Most breakfast cereals are enriched and fortified. Some countries, like the United States, require refined grain products to be enriched, in order to prevent malnutrition and diseases such as rickets and kwashiorkor disease.

Unrefined grains are usually darker in color than their refined counterparts. White bread, white rice, white flour tortillas, and white pasta are examples of refined foods. Whole-grain options are whole-wheat bread, brown rice, whole-wheat tortillas, and whole-wheat pasta. Switching to whole-grain options of the foods you already consume is a simple way to significantly improve your nutrition. Be careful to check the food labels for the word "whole" to insure that you are getting whole grains, not just a brown-colored product. Not all whole grains are colored brown. Rolled oats and popcorn are both examples of white whole grains. There are also varieties of white whole wheat. Bread made from white whole wheat has similar nutritional value to regular whole-wheat bread.

Overall, whole-grain products are much better for your health than products made from refined grains. However, most Americans receive the majority of a nutrient called folic acid from fortified cereals to which folic acid has been added. If you don't eat breakfast cereals and all of the grains in your diet are whole, you might be lacking in folic acid, and you might need to add a folic acid supplement to your diet. This is especially true if you are a woman who might become pregnant. Folic acid is important in preventing birth defects such as spina bifida. Others can get adequate folic acid from foods such as fruits, vegetables, and legumes.

Choosing dietary whole grains aids your health in multiple ways. To increase your intake of whole grains, you can try adding barley and brown rice to soups, adding buckwheat to pancakes or waffles, adding flax seed or cooked quinoa to salads, or eating millet instead of white rice. Whole grains are more satisfying than their refined counterparts and, therefore, can play an important role in weight control. They can also add variety to meals and snacks.

Whole grains are an important part of a healthy diet.

Another important peptide hormone is adrenocorticotropic hormone (ACTH). It is synthesized by the pituitary gland and consists of a single polypeptide chain of 39 amino acid residues with no disulfide bridges (see **Figure 19.3**). The major function of ACTH is to regulate the production of steroid hormones in the cortex of the adrenal gland. **Table 19.2** lists these and some other important peptide or protein hormones.

Figure 19.3 The amino acid sequence of ACTH.

TABLE 19.2 **Examples of Peptide or Protein Hormones**

Name	Origin	Action
Adrenocorticotropic hormone (ACTH)	Pituitary	Stimulates production of adrenal hormones
Angiotensin II	Blood plasma	Causes blood vessels to constrict
Follicle-stimulating hormone (FSH)	Pituitary	Stimulates sperm production and follicle maturation
Gastrin	Stomach	Stimulates stomach to secrete acid
Glucagon	Pancreas	Stimulates glycogen metabolism in liver
Human growth hormone (HGH)	Pituitary	Stimulates growth and cell reproduction
Insulin	Pancreas	Controls metabolism of carbohydrates
Oxytocin	Pituitary	Stimulates contraction of uterus and other smooth muscles
Prolactin	Pituitary	Stimulates lactation
Somatostatin	Hypothalamus	Inhibits production of HGH
Vasopressin	Pituitary	Decreases volume of urine excreted

19.5 | Characteristics of Proteins

Learning Objective

5. Describe proteins in terms of the following characteristics: size, function, classification as fibrous or globular, and classification as simple or conjugated.

19.5A Size

Proteins are extremely large natural polymers of amino acids with molecular weights that vary from about 6000 to several million u. The immense size of proteins (in the molecular sense) can be appreciated by comparing glucose with hemoglobin, a relatively small protein. Glucose has a molecular weight of 180 u, whereas the molecular weight of hemoglobin is 65,000 u (see **Table 19.3**). The molecular formula of glucose is $C_6H_{12}O_6$ and that of hemoglobin is $C_{2952}H_{4664}O_{832}N_{812}S_8Fe_4$.

Protein molecules are too large to pass through cell membranes and are contained inside the normal cells where they were formed. However, if cells become damaged by disease or trauma, the protein contents can leak out. Thus, persistent excessive amounts of proteins in the urine are indicative of damaged kidney cells. A routine urinalysis usually includes a test for protein. Similarly, a heart attack can be confirmed by the presence in the blood of certain proteins (enzymes) that are normally confined to cells in heart tissue (see **Section 20.9**).

19.5B Acid–Base Properties

The 20 different R groups of amino acids are important factors in determining the physical and chemical properties of proteins that contain the amino acids. Acid–base behavior is

TABLE 19.3 Molecular Weights of Some Common Proteins

Protein	Molecular Weight (u)	Number of Amino Acid Residues
Insulin	6,000	51
Cytochrome c	16,000	104
Growth hormone	49,000	191
Hemoglobin	65,000	574
Hexokinase	96,000	730
Gamma globulin	176,000	1320
Myosin	800,000	6100

one of the most important of these properties. Proteins, like amino acids, take the form of zwitterions. The R groups of glutamate and aspartate contain —COO^- groups and provide H^+ ions, whereas the R groups of arginine and histidine are basic (see **Table 19.1**). Thus, proteins, like amino acids, have characteristic isoelectric points and can behave as buffers in solutions.

The tendency for large molecules such as proteins to remain in solution or form a stable colloidal dispersion depends to a large extent on the repulsive forces acting between molecules with like charges on their surfaces. When protein molecules are at a pH where they take forms that have a net positive or negative charge, the presence of these like charges causes the molecules to repel one another and remain dispersed. These repulsive forces are smallest at the isoelectric point, when the net molecular charges are essentially zero. Thus, protein molecules tend to clump together and precipitate from solutions in which the pH is equal to or close to the isoelectric point.

19.5C Protein Function

A large part of the importance of proteins results from the crucial roles they serve in all biological processes. The various functions of proteins are discussed below and summarized in **Table 19.4**.

1. **Catalytic function:** Nearly all the reactions that take place in living organisms are catalyzed by proteins functioning as enzymes. Without these catalytic proteins, biological reactions would take place too slowly to support life. Enzyme-catalyzed reactions range from relatively simple processes to the very complex, such as the duplication of hereditary material in cell nuclei. Enzymes are discussed in more detail in the next chapter.

2. **Structural function:** We saw in **Section 17.8** that the main structural material for plants is cellulose. In the animal kingdom, structural materials other than the inorganic components of the skeleton are composed of protein. Collagen, a fiberlike protein, is responsible for the mechanical strength of skin and bone. Keratin, the chief constituent of hair, skin, and fingernails, is another example.

3. **Storage function:** Some proteins provide a way to store small molecules or ions. Ovalbumin, for example, is a stored form of amino acids that is used by embryos developing in bird eggs. Casein, a milk protein, and gliadin, in wheat seeds, are also stored forms of protein intended to nourish animals and plants, respectively. Ferritin, found in all cells of the body, attaches to iron ions and forms a storage complex in humans and other animals.

4. **Protective function:** **Antibodies** are tremendously important proteins that protect the body from disease. These highly specific proteins combine with and help destroy viruses, bacteria, and other foreign substances that get into the blood or tissue of the body. Blood clotting, another protective process, is carried out by the proteins thrombin and fibrinogen. Without this process, even small wounds would result in life-threatening bleeding.

antibody A substance that helps protect the body from invasion by viruses, bacteria, and other foreign substances.

TABLE 19.4 Biological Functions of Proteins

Function	Examples	Occurrence or Role
Catalysis	lactate dehydrogenase	Oxidizes lactic acid
	cytochrome c	Transfers electrons
	DNA polymerase	Replicates and repairs DNA
Structure	viral-coat proteins	Sheath around nucleic acid of viruses
	glycoproteins	Cell coats and walls
	α-keratin	Skin, hair, feathers, nails, and hooves
	β-keratin	Silk of cocoons and spiderwebs
	collagen	Fibrous connective tissue
	elastin	Elastic connective tissue
Storage	ovalbumin	Egg-white protein
	casein	A milk protein
	ferritin	Stores iron in the spleen
	gliadin	Stores amino acids in wheat
	zein	Stores amino acids in corn
Protection	antibodies	Form complexes with foreign proteins
	fibrinogen	Involved in blood clotting
	thrombin	Involved in blood clotting
Regulation	insulin	Regulates glucose metabolism
	growth hormone	Stimulates growth of bone
Nerve impulse transmission	rhodopsin	Involved in vision
	acetylcholine receptor protein	Impulse transmission in nerve cells
Movement	myosin	Thick filaments in muscle fiber
	actin	Thin filaments in muscle fiber
	dynein	Movement of cilia and flagella
Transport	hemoglobin	Transports O_2 in blood
	myoglobin	Transports O_2 in muscle cells
	serum albumin	Transports fatty acids in blood
	transferrin	Transports iron in blood
	ceruloplasmin	Transports copper in blood

5. **Regulatory function:** Numerous body processes are regulated by hormones, many of which are proteins. Examples are growth hormone, which regulates the growth rate of young animals, and thyrotropin, which stimulates the activity of the thyroid gland.

6. **Nerve impulse transmission:** Some proteins behave as receptors of small molecules that pass between gaps (synapses) separating nerve cells. In this way, they transmit nerve impulses from one nerve to another. Rhodopsin, a protein found in the rod cells of the retina of the eye, functions this way in the vision process.

7. **Movement function:** Every time we climb stairs, push a button, or blink an eye, we use muscles that have proteins as their major components. The proteins actin and myosin are particularly important in processes involving movement. They are long-filament proteins that slide along each other during muscle contraction.

8. **Transport function:** Numerous small molecules and ions are transported effectively through the body only after binding to proteins. For example, fatty acids are carried between fat (adipose) tissue and other tissues or organs by serum albumin, a blood protein. Hemoglobin, a well-known example, carries oxygen from the lungs to other body tissues, and transferrin is a carrier of iron in blood plasma.

Proteins perform other functions, but these are among the most important. It is easy to see that properly functioning cells of living organisms must contain many proteins. It has

Alzheimer's Disease

Alzheimer's disease is a progressive, degenerative disease that attacks the brain, causing impaired thinking, behavior, and memory. As the disease progresses, nerve cells in the brain degenerate, and the size of the brain actually decreases. About 4 million Americans suffer from this malady; it afflicts 1 in 10 people over age 65, and nearly half of all people over age 85. The disease, first identified by Alois Alzheimer in 1906, is best described by the words of one of its victims: "I have lost myself."

Some warning signs of Alzheimer's disease include difficulty in performing familiar tasks, language problems, disorientation related to time or place, poor or decreased quality of judgment, memory loss affecting job skills, problems with abstract thinking, misplacing things or putting them in inappropriate places, passivity and loss of initiative, and changes in mood, personality, or behavior. If several of these symptoms are present in an older person, a medical evaluation for Alzheimer's disease is in order.

The causes of the disease are under debate. Some researchers feel it is caused by the formation in the brain of fibrils (strands) and plaque deposits consisting largely of amyloid beta-protein. Others think the formation of tangles in a microtubule-associated protein called Tau is the culprit. Still others think both of these explanations are correct, and that the formation of abnormal amyloid beta strands causes the abnormal killer forms of the Tau protein to form. This position is supported by a recent study of the brains of aging rhesus monkeys. This study showed that amyloid beta-protein caused the same damage to the monkey brains that was found in the brains of Alzheimer's victims. In addition, the amyloid beta-protein caused changes in the Tau protein of the monkey brains that made it capable of forming tangles.

Current research is focused on preventing or reversing the process of changing normal amyloid protein into the deformed amyloid protein found in the brains of Alzheimer's patients.

The α-helices of normal proteins are changed to β-pleated sheets in the abnormal proteins characteristic of the disease. In a recent promising study, an experimental peptide containing five amino acids was shown to mimic a region of amyloid beta-protein that regulates folding. However, the experimental peptide is so constructed that it prevents the other amino acids of amyloid beta-protein from folding into the β-pleated sheet conformation necessary to form amyloid beta-protein fibrils. In cell culture studies, the experimental peptide inhibited the formation of amyloid fibrils, broke down existing fibrils, and prevented the death of neurons. This type of promising discovery may eventually lead to a drug that can stop the progress—or even reverse the effects—of Alzheimer's disease.

An MRI showing brain atrophy in Alzheimer's disease.

James Cavallini/Science Source

been estimated that a typical human cell contains 9000 different proteins, and that a human body contains about 100,000 different proteins.

19.5D Classification

Proteins can be classified by the functions just discussed. They can also be classified into two major types, fibrous and globular, on the basis of their structural shape. **Fibrous proteins** are made up of long rod-shaped or stringlike molecules that intertwine with one another and form strong fibers. They are water insoluble and are usually found as major components of connective tissue, elastic tissue, hair, and skin. Examples are collagen, elastin, and keratin (see **Figure 19.4**).

Globular proteins are more spherical and either dissolve in water or form stable suspensions in water. They are not found in structural tissue but are transport proteins, or

fibrous protein A protein made up of long rod-shaped or stringlike molecules that intertwine to form fibers.

globular protein A spherical protein that usually forms stable suspensions in water or dissolves in water.

proteins that may be moved easily through the body by the circulatory system. Examples of these are hemoglobin and transferrin.

Another classification scheme for proteins is based on composition—proteins are either simple or conjugated. **Simple proteins** are those that contain only amino acid residues. **Conjugated proteins** contain amino acid residues and other organic or inorganic components. The non–amino acid parts of conjugated proteins are called **prosthetic groups**. The type of prosthetic group present is often used to further classify conjugated proteins. For example, proteins containing lipids, carbohydrates, or metal ions are called lipoproteins, glycoproteins, and metalloproteins, respectively. Thus, the iron-containing proteins hemoglobin and myoglobin are metalloproteins. **Table 19.5** lists the common classes of conjugated proteins.

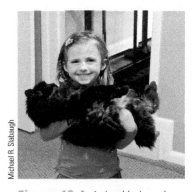

Michael R. Slabaugh

Figure 19.4 Animal hair and spiderwebs are constructed of the fibrous proteins α-keratin and β-keratin, respectively. What properties of fibrous proteins make hair, fur, and spiderwebs useful?

simple protein A protein made up entirely of amino acid residues.

conjugated protein A protein made up of amino acid residues and other organic or inorganic components.

prosthetic group The non–amino acid parts of conjugated proteins.

TABLE 19.5 Conjugated Proteins

Class	Prosthetic Groups	Examples
Nucleoproteins	nucleic acids (DNA, RNA)	Viruses
Lipoproteins	lipids	Fibrin in blood, serum lipoproteins
Glycoproteins	carbohydrates	Gamma globulin in blood, mucin in saliva
Phosphoproteins	phosphate groups	Casein in milk
Hemoproteins	heme	Hemoglobin, myoglobin, cytochromes
Metalloproteins	iron	Ferritin, hemoglobin
	zinc	Alcohol dehydrogenase

19.5E Protein Structure

As you might expect, the structures of extremely large molecules, such as proteins, are much more complex than those of simple organic compounds. Many protein molecules consist of a chain of amino acids twisted and folded into a complex three-dimensional structure. This structural complexity imparts unique features to proteins that allow them to function in the diverse ways required by biological systems. To better understand protein function, we must look at four levels of organization in their structure. These levels are referred to as primary, secondary, tertiary, and quaternary.

19.6 | The Primary Structure of Proteins

Learning Objective

6. Explain what is meant by the primary structure of proteins.

Every protein, be it hemoglobin, keratin, or insulin, has a similar backbone of carbon and nitrogen atoms held together by peptide bonds:

$$-NH-CH-\underset{\underset{R}{|}}{\overset{\overset{O}{\parallel}}{C}}-NH-CH-\underset{\underset{R'}{|}}{\overset{\overset{O}{\parallel}}{C}}-NH-CH-\underset{\underset{R''}{|}}{\overset{\overset{O}{\parallel}}{C}}-NH-CH-\underset{\underset{R'''}{|}}{\overset{\overset{O}{\parallel}}{C}}-NH-CH-\underset{\underset{R''''}{|}}{\overset{\overset{O}{\parallel}}{C}}-$$

protein backbone (in color)

The difference in proteins is the variation in the length of the backbone and the sequence of the side chains (R groups) that are attached to the backbone. The order in which amino acid residues are linked together in a protein is called the **primary structure**.

primary protein structure The linear sequence of amino acid residues in a protein chain.

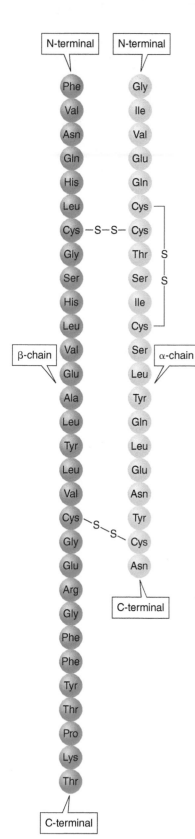

Figure 19.5 The amino acid sequence (primary structure) of human insulin.

Each different protein in a biological organism has a unique sequence of amino acid residues. It is this sequence that causes a protein chain to fold and curl into the distinctive shape that, in turn, enables the protein to function properly. As we will see in **Section 19.10**, a protein molecule that loses its characteristic three-dimensional shape cannot function.

Biochemists have devised techniques for finding the order in which the residues are linked together in protein chains. A few of the proteins whose primary structures are known are listed in **Table 19.6**.

TABLE 19.6 Some Proteins Whose Sequence of Amino Acids Is Known

Protein	Number of Amino Acid Residues
Insulin	51
Hemoglobin	
α-chain	141
β-chain	146
Myoglobin	153
Human growth hormone	191
Trypsinogen	229
Carboxypeptidase A	307
Gamma globulin	1320

Figure 19.5 shows the primary structure of human insulin. Notice that the individual molecules of this hormone consist of two chains having a total of 51 amino acid residues. In the molecule, two disulfide bridges hold the chains together, and there is one disulfide linkage within a chain. Insulin plays an essential role in regulating the use of glucose by cells. Diabetes mellitus can result when the body does not produce enough insulin or when cells do not respond to the insulin that is produced.

As we saw in the case of vasopressin and oxytocin (see **Section 19.4**), small changes in amino acid sequence can cause profound differences in the functioning of proteins. For example, a minor change in sequence in the blood protein hemoglobin causes the fatal sickle-cell disease.

19.7 | The Secondary Structure of Proteins

Learning Objective

7. Describe the role of hydrogen bonding in the secondary structure of proteins.

If the only structural characteristics of proteins were their amino acid sequences (primary structures), all protein molecules would consist of long chains arranged in random fashion. However, protein chains fold and become aligned in such a way that certain orderly patterns result. These orderly patterns, referred to as **secondary structures**, result from hydrogen bonding and include the α-helix (alpha-helix) and the β-pleated (beta-pleated) sheet.

19.7A The α-Helix

In 1951, American chemists Linus Pauling and Robert Corey suggested that proteins could exist in the shape of an α-**helix**, a form in which a single protein chain twists so that it resembles a coiled helical spring (see **Figure 19.6**). The chain is held in the helical shape

Carbon
Oxygen
Nitrogen
R group
Hydrogen

Primary structure

Secondary structure

by numerous intramolecular hydrogen bonds between carbonyl oxygens and amide hydrogens in adjacent turns of the helical backbone:

$$-\text{C}=\text{O} \cdots \text{H}-\text{N}-$$

The carbonyl group of each amino acid residue is hydrogen-bonded to the amide hydrogen of the amino acid that is four residues away in the chain; thus, all amide groups in the helix are hydrogen-bonded (see **Figure 19.6**). The protein backbone forms the coil, and the side chains (R groups) extend outward from the coil.

Six years after Pauling and Corey proposed the α-helix, its presence in proteins was detected. Since that time it has been found that the amount of α-helical content is quite variable in proteins. In some it is the major structural component, whereas in others a random coil predominates and there is little or no α-helix coiling.

In the proteins α-keratin (found in hair), myosin (found in muscle), epidermin (found in skin), and fibrin (found in blood clots), two or more helices interact (supracoiling) to form a cable (see **Figure 19.7**). These cables make up bundles of fibers that lend strength to the tissue in which they are found.

secondary protein structure The arrangement of protein chains into patterns as a result of hydrogen bonds between amide groups of amino acid residues in the chain. The common secondary structures are the α-helix and the β-pleated sheet.

α-helix The helical structure in proteins that is maintained by hydrogen bonds.

Figure 19.7 The supracoiling of three α-helices in the keratins of hair and wool.

19.7B The β-Pleated Sheet

A less common type of secondary structure is found in some proteins. This type, called a β-**pleated sheet**, results when several protein chains lie side by side and are held in position by hydrogen bonds between the carbonyl oxygens of one chain and the amide hydrogens of an adjacent chain (see **Figure 19.8**).

β-pleated sheet A secondary protein structure in which protein chains are aligned side by side in a sheet-like array held together by hydrogen bonds.

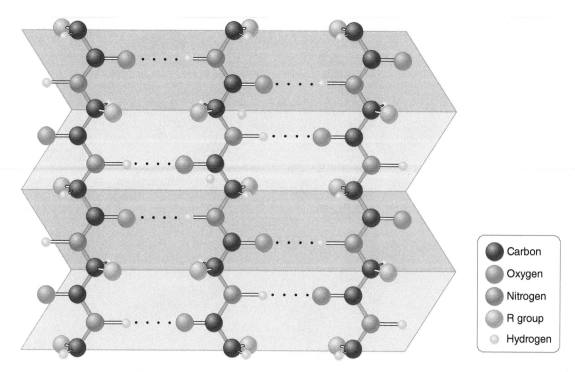

Figure 19.8 The β-pleated sheet. The four dots show hydrogen bonds between adjacent protein chains.

The β-pleated sheet is found extensively only in the protein of silk. However, a number of proteins in which a single polypeptide chain forms sections of a β-pleated sheet structure by folding back on itself are known. **Figure 19.9** depicts a protein chain exhibiting the β-pleated sheet, the α-helix, and other structural conformations. Note the position of the hydrogen bonds in the two secondary structures.

Figure 19.9 A segment of a protein showing areas of α-helical, β-pleated sheet, and coil molecular structure.

A Milk Primer

There are many different available options for milk today. There is coconut milk, almond milk, and soy milk in addition to regular dairy milk. Even dairy milk has numerous available options such as whole milk, 2%, 1%, and skim. It can be hard to decide what will be the best milk for you. To make this decision, it is important to know the differences between the kinds of milk available.

Whole milk contains 3.5% fat by weight of the milk. This means that whole milk has about 8.7 grams of saturated fat in every 8-ounce serving. In addition to the fat, whole milk also contains conjugated linoleic acid (CLA). CLA can reduce the risk of heart disease and cancer, but this benefit is reduced somewhat by the large amount of fat in whole milk. Reduced-fat milk or 2% has 122 calories in an 8-ounce serving. It has 5 grams of fat, but only 3 of the 5 grams are saturated fat. Low-fat milk or 1% has only 102 calories and 2 grams of fat per serving. Skim milk, or nonfat milk, contains 83 calories and little or no fat per 8-ounce serving. Skim milk has an advantage of a little more calcium per serving than other milks. Information about the fat content of different types of milk is important because dietary saturated fats are known to increase cholesterol levels in the body and increase the risk of heart disease.

There are now milks on the market that are "omega-3-fortified." This means that the milk has 32–50 milligrams of omega-3 supplement in each 8-ounce serving. The source of the omega-3 supplement is fish or algae oil that has been added to the milk. Omega-3 is recognized as a good fat, but the amount found in these milk products is tiny when compared to the recommended daily omega-3 intake of 500 milligrams. The recommended daily intake of omega-3 supplement can be satisfied by including two servings of fish per week in the diet.

There are even milks available that are labeled "shelf stable." This means that they have a relatively long shelf life as indicated by their expiration dates. These types of milk are made differently than others. Regular milk is made using a high temperature, short time (HTST) pasteurization process that destroys 99.9% of the bacteria in the milk. Shelf-stable milks are pasteurized using an ultra-high temperature (UHT) process. The UHT process destroys more than the 99.9% of bacteria that is destroyed by the HTST process. This gives the milk a shelf-life of one month in shelf-stable boxes that are designed to keep oxygen and light out. The UHT process does slightly alter the milk proteins, and causes the milk to taste slightly cooked. The UHT process can also somewhat reduce the amount of vitamin B in the milk.

The word "primer" included in the title to this feature correctly indicates that the included material is a basic introduction (a primer) to what is known about the chemistry of the important beverage we call milk.

There are a number of options in the purchase of milk.

19.8 | **The Tertiary Structure of Proteins**

Learning Objective

8. Describe the role of side-chain interactions in the tertiary structure of proteins.

Tertiary structure, the next higher level of complexity in protein structure, refers to the bending and folding of the protein into a specific three-dimensional shape. This bending and folding may seem rather disorganized, but nevertheless it results in a favored arrangement for a given protein.

The tertiary structure results from interactions between the R side chains of the amino acid residues. These R-group interactions are of four types:

1. **Disulfide bridges:** As in the structure of insulin (see **Figure 19.5**), a disulfide linkage can form between two cysteine residues that are close to each other in the same chain or between cysteine residues in different chains. The existence and location of these

tertiary protein structure A specific three-dimensional shape of a protein resulting from interactions between R groups of the amino acid residues in the protein.

disulfide bridges is a part of the tertiary structure because these interactions hold the protein chain in a loop or some other three-dimensional shape.

2. **Salt bridges:** These interactions are a result of ionic bonds that form between the ionized side chain of an acidic amino acid (—COO⁻) and the side chain of a basic amino acid (—NH₃⁺).

3. **Hydrogen bonds:** Hydrogen bonds can form between a variety of side chains, especially those that possess the following functional groups:

$$\text{—OH} \qquad \text{—NH}_2 \qquad \overset{\displaystyle \overset{O}{\|}}{\text{—C—NH}_2}$$

One possible interaction, between the —OH groups of two serine residues, is shown in **Figure 19.10**. We saw in **Section 19.7** that hydrogen bonding also determines the secondary structure of proteins. The distinction is that the tertiary hydrogen bonding occurs between R groups, whereas in secondary structures, the hydrogen bonding is between backbone —C═O and —N groups.

Figure 19.10 R-group interactions leading to tertiary protein structure.

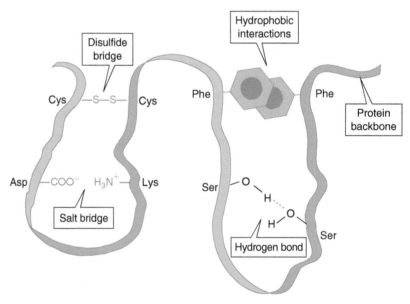

4. **Hydrophobic interactions:** These result when nonpolar groups are either attracted to one another or forced together by their mutual repulsion of aqueous solvent. Interactions of this type are common between R groups such as the nonpolar phenyl rings of phenylalanine residues (**Figure 19.10**). This is illustrated by the compact structure of globular proteins in aqueous solutions (see **Figure 19.11**). Much like in the

Figure 19.11 A globular protein with a hydrophobic region on the inside and polar groups on the outside extending into the aqueous surroundings.

fatty acid micelles in **Section 18.2**, the globular shape results because polar groups are pointed outward toward the aqueous solvent, and nonpolar groups are pointed inward, away from the water molecules. This type of interaction is weaker than the other three, but it usually acts over large surface areas so that the net effect is an interaction strong enough to stabilize the tertiary structure.

The interactions leading to tertiary protein structures are summarized in **Figure 19.10**.

✔ **LEARNING CHECK 19.3** What kind of R-group interaction might be expected if the following side chains were in close proximity?

a. $-\overset{\overset{\displaystyle CH_3}{|}}{CH}CH_2CH_3$ and $-\overset{\overset{\displaystyle CH_3}{|}}{CH}CH_2CH_3$

b. $-\overset{\overset{\displaystyle O}{||}}{C}-NH_2$ and ⬡—OH

c. $-CH_2-\overset{\overset{\displaystyle O}{||}}{C}-O^-$ and $-CH_2CH_2CH_2CH_2-NH_3^+$

19.9 | The Quaternary Structure of Proteins

Learning Objective

9. Explain what is meant by the quaternary structure of proteins.

Many functional proteins contain two or more polypeptide chains held together by forces such as ionic attractions, disulfide bridges, hydrogen bonds, and hydrophobic forces. Each of the polypeptide chains, called **subunits**, has its own primary, secondary, and tertiary structure. The arrangement of subunits to form the larger protein is called the **quaternary structure** of the protein.

subunit A polypeptide chain having primary, secondary, and tertiary structural features that is a part of a larger protein.

quaternary protein structure The arrangement of subunits that form a larger protein.

STUDY SKILLS 19.1

Visualizing Protein Structure

A visual aid to help you "see" protein structure and remember the difference between primary, secondary, and tertiary structure is the cord of a telephone receiver. The chainlike primary structure is represented by the long straight cord.

The coiling of the cord into a helical arrangement represents the secondary structure, and the tangled arrangement the cord adopts after the receiver has been used several times is the tertiary structure.

Primary structure

Secondary structure

Tertiary structure

Hemoglobin is a well-known example of a protein exhibiting quaternary structure. Hemoglobin is made of four chains (subunits): two identical chains (called alpha) containing 141 amino acid residues each and two other identical chains (beta) containing 146 residues each. Each of these four hemoglobin subunits contains a heme group (a planar ring structure centered around an iron atom, shown in **Figure 19.12B**) located in crevices near the exterior of the molecule. The hemoglobin molecule is nearly spherical, with the four subunits held together rather tightly by hydrophobic forces. A schematic drawing of the quaternary structure of hemoglobin is shown in **Figure 19.12A**.

A Hemoglobin exhibits quaternary structure with two α-chains and two β-chains. The purple disks are heme groups.

B A heme group.

Figure 19.12 The structure of hemoglobin.

ASK A PHARMACIST 19.1

Who Really Needs Gluten-Free Food?

The National Institutes of Health estimates that 1 in 133 people, or 0.75% of the U.S. population, suffers from a digestive disorder called "celiac disease." These individuals cannot tolerate in their diet a group of substances collectively called "gluten." Their digestive systems become very upset when they ingest gluten-containing foods.

Celiac disease is a genetic disorder that affects the small intestines. The word "celiac" comes from the Greek word for "belly-ache," which is the most frequent symptom of the disease. In patients with the disease, gluten causes an immune response that inflames and damages the interior of the small intestines. Chemically, gluten is a mixture of proteins found in grains such as wheat. The predominant proteins in gluten are gliadin, glutenin, globulin, and albumin.

Unfortunately, gluten-free diets and products are becoming something of a fad in society. Gluten-free products are being marketed heavily, and supported by celebrity endorsements. This has led to the perception that a gluten-free diet is especially healthy. Unless you suffer from celiac disease,

Individuals with celiac disease cannot tolerate gluten.

this is not true. If a medical history review and symptoms suggest a likelihood of celiac disease, an individual should seek advice from a medical professional before beginning a gluten-free diet.

19.10 | Protein Hydrolysis and Denaturation

Learning Objective

10. Describe the conditions that can cause proteins to hydrolyze or become denatured.

19.10A Hydrolysis

We learned in **Section 16.8** that amides can be hydrolyzed by an aqueous acid or base. The amide linkages of peptides and proteins also show this characteristic. As a result, the heating of a peptide or protein in the presence of an acid or base causes it to break into smaller peptides, or even amino acids, depending on the hydrolysis time, temperature, and pH.

$$\text{protein} + H_2O \xrightarrow[\text{OH}^-]{H^+ \text{ or}} \text{smaller peptides} \xrightarrow[\text{OH}^-]{H^+ \text{ or}} \text{amino acids} \qquad (19.9)$$

The process of protein digestion (see **Section 22.6**) involves hydrolysis reactions that are catalyzed by enzymes in the digestive tract. In **Chapter 24**, we'll also see that the hydrolysis of cellular proteins to amino acids is an ongoing process as the body resynthesizes needed molecules and tissues.

19.10B Denaturation

Proteins are maintained in their natural three-dimensional conformation, or **native state**, by stable secondary and tertiary structures and through the aggregation of subunits in quaternary structures. If proteins are exposed to physical or chemical conditions, such as extreme temperatures or pH values that disrupt these stabilizing structures, the folded native structure breaks down and the protein takes on a random, disorganized conformation (see **Figure 19.13**). This process, called **denaturation**, causes the protein to become

native state The natural three-dimensional conformation of a functional protein.

denaturation The process by which a protein loses its characteristic native structure and function.

Proteins **633**

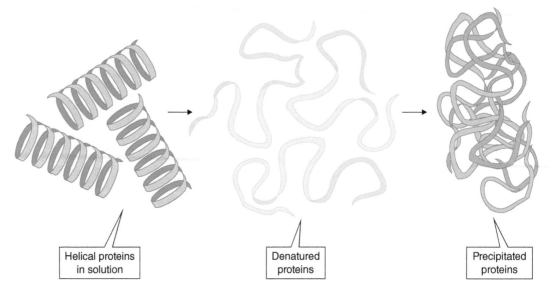

Helical proteins in solution

Denatured proteins

Precipitated proteins

Figure 19.13 Protein denaturation and coagulation of the proteins.

© Christina Slabaugh

Figure 19.14 Denaturing the protein in eggs.

inactive, and it may precipitate because hydrophobic residues are likely to be exposed to the aqueous surroundings. The change of egg white from a clear, jellylike material to a white solid when heated is an example of this process (see **Figure 19.14**). Denaturation also explains why most proteins are biologically active only over a narrow temperature range, typically 0°C–40°C. Temperatures higher than those normal for a living organism can cause intramolecular vibrations to become so intense that hydrogen bonds or other stabilizing forces are disrupted and denaturation results.

The correlation between denaturation and loss of biological activity is clearly illustrated by the common methods used to kill microorganisms and deactivate their toxins. Most bacterial toxins, which are proteins, lose their ability to cause disease when they are heated. For example, the toxic protein excreted by *Clostridium botulinum* is responsible for botulism, an often fatal form of food poisoning. This dangerous substance loses its toxic properties after being heated at 100°C for a few minutes. Heating also deactivates the toxins of the bacteria that cause diphtheria and tetanus. Heat denaturation is used to prepare vaccines against these particular diseases. The deactivated (denatured) toxin can no longer cause the symptoms of the disease to appear, but it can stimulate the body to produce substances that induce immunity to the functional toxins. It is well known that high temperatures have been used for many years in hospitals and laboratories to sterilize surgical instruments and glassware that must be germ-free. The high temperatures denature and deactivate proteins needed within the germ cells for survival.

Similar changes in protein conformation resulting in denaturation can be brought about by heavy-metal ions such as Hg^{2+}, Ag^+, and Pb^{2+} that interact with —SH groups and carboxylate (—COO^-) groups. Organic materials containing mercury (mercurochrome and merthiolate) were once commonly used as antiseptics in the home to treat minor skin abrasions because they denature proteins in bacteria. Mercury, silver, and lead salts are also poisonous to humans for the same reason: They deactivate critical proteins we need to live. Large doses of raw egg white and milk are often given to victims of heavy-metal poisoning. The proteins in the egg and milk bind to the metal ions and form a precipitate. The metal-containing precipitate is removed from the stomach by pumping or is ejected by induced vomiting. If the precipitated protein is not removed, it will be digested, and the toxic metal ions will be released again.

Some of the various substances and conditions that will denature proteins are summarized in **Table 19.7**.

TABLE 19.7 Substances and Conditions That Denature Proteins

Substance or Condition	Effect on Proteins
Heat and ultraviolet light	Disrupt hydrogen bonds and ionic attractions by making molecules vibrate too violently; produce coagulation, as in cooking an egg
Organic solvents (ethanol and others miscible with water)	Disrupt hydrogen bonds in proteins and probably form new ones with the proteins
Strong acids or bases	Disrupt hydrogen bonds and ionic attractions; prolonged exposure results in hydrolysis of protein
Detergents	Disrupt hydrogen bonds, hydrophobic interactions, and ionic attractions
Heavy-metal ions (Hg^{2+}, Ag^+, Pb^{2+})	Form bonds to thiol groups and precipitate proteins as insoluble heavy-metal salts

▶ Case Study Follow-up

Bioprinting has the potential to lead to great improvements in bioengineering. Using computers and 3-D printers, researchers have created beating hearts and functional lungs that have been successfully implanted into research animals. Most bioprinting techniques consist of printing collagen scaffolds that are later "seeded" with living cells. Collagen, the main structural protein in animal connective tissue, provides an ideal structure on which to grow new tissues.

Collagen is the most abundant protein in the human body, and functions as the "glue" that holds the body together. In addition to printing a collagen scaffold from a 3-D printer, collagen scaffolds are obtained by washing away cells from an organ such as a pig kidney or heart, then seeding the resulting protein structure with heart or kidney cells from another animal or human. Researchers believe that entire organs for transplant will be manufactured using this method in the future. Less complex structures such as skin or bone tissue used for grafting are already available. Tracheal tissue transplants are now produced by growing a patient's own tissue on collagen scaffolds obtained from cadavers. This type of transplant does not require any anti-rejection medication, because the transplanted tissue is from the patient. It is estimated that within a matter of decades, organs grown on either printed, animal, or cadaver scaffolds might be common.

Concept Summary

The Amino Acids All proteins are polymers of amino acids, which are bifunctional organic compounds that contain both an amino group and a carboxylate group. Differences in the R groups of amino acids cause differences in the properties of amino acids and proteins.
Objective 1 (Section 19.1), Exercise 19.2

Zwitterions The presence of both amino groups and carboxyl groups in amino acids makes it possible for amino acids to exist in several ionic forms, including the form of a zwitterion. The zwitterion is a dipolar form in which the net charge on the ion is zero.
Objective 2 (Section 19.2), Exercise 19.12

Reactions of Amino Acids Amino acids can undergo reactions characteristic of any functional group in the molecule. Two important reactions are the reaction of two cysteine molecules to form a disulfide, and the reaction of amino groups and carboxylate groups of different molecules to form peptide (amide) linkages.
Objective 3 (Section 19.3), Exercise 19.16

Important Peptides More than 200 peptides have been shown to be essential for the proper functioning of the human body. Hormones are among the peptides for which functions have been identified.
Objective 4 (Section 19.4), Exercise 19.22

Characteristics of Proteins Proteins are large polymers of amino acids. Acidic and basic properties of proteins are determined by the acidic or basic character of the R groups of the amino acids constituting the protein. Proteins perform numerous important functions in the body. Proteins are classified structurally as fibrous or globular.

They are classified on the basis of composition as simple or conjugated.

Objective 5 (Section 19.5), Exercises 19.30 and 19.32

The Primary Structure of Proteins The primary structure of a protein is the sequence of amino acids in the polymeric chain. This gives all proteins an identical backbone of carbon and nitrogen atoms held together by peptide linkages. The difference in proteins is the sequence of R groups attached to the backbone.

Objective 6 (Section 19.6), Exercise 19.34

The Secondary Structure of Proteins Protein chains are held in characteristic shapes called secondary structures by hydrogen bonds. Two specific structures that have been identified are the α-helix and the β-pleated sheet.

Objective 7 (Section 19.7), Exercise 19.38

The Tertiary Structure of Proteins A third level of complexity in protein structure results from interactions between the R groups of protein chains. These interactions include disulfide bridges, salt bridges, hydrogen bonds, and hydrophobic attractions.

Objective 8 (Section 19.8), Exercise 19.42

The Quaternary Structure of Proteins Some functional proteins consist of two or more polypeptide chains held together by forces such as ionic attractions, disulfide bridges, hydrogen bonds, and hydrophobic forces. The arrangement of these polypeptides to form the functional protein is called the quaternary structure of the protein.

Objective 9 (Section 19.9), Exercise 19.46

Protein Hydrolysis and Denaturation The peptide (amide) linkages of peptides and proteins can be hydrolyzed under appropriate conditions. This destroys the primary structure and produces smaller peptides or amino acids. The characteristic secondary, tertiary, and quaternary structures of proteins can also be disrupted by certain physical or chemical conditions such as extreme temperatures or pH values. The disruption of these structures is called denaturation and causes the protein to become nonfunctional and, in some cases, to precipitate.

Objective 10 (Section 19.10), Exercise 19.50

Key Terms and Concepts

Alpha-amino acid (19.1)
α-helix (19.7)
Amino acid residue (19.3)
Antibody (19.5)
β-pleated sheet (19.7)
Conjugated protein (19.5)
C-terminal residue (19.3)
Denaturation (19.10)
Dipeptide (19.3)

Disulfide bridge (19.4)
Fibrous protein (19.5)
Globular protein (19.5)
Isoelectric point (19.2)
Native state (19.10)
N-terminal residue (19.3)
Peptide (19.3)
Peptide linkage (peptide bond) (19.3)
Polypeptide (19.3)

Primary protein structure (19.6)
Prosthetic group (19.5)
Protein (19.3)
Quaternary protein structure (19.9)
Secondary protein structure (19.7)
Simple protein (19.5)
Subunit (19.9)
Tertiary protein structure (19.8)
Zwitterion (19.2)

Key Reactions

1. Formation of a zwitterion (Section 19.2):	$$H_2N-\underset{\underset{R}{\mid}}{CH}-\overset{\overset{O}{\mid\mid}}{C}-OH \rightarrow \overset{+}{H_3}N-\underset{\underset{R}{\mid}}{CH}-\overset{\overset{O}{\mid\mid}}{C}-O^-$$	Reaction 19.1
2. Conversion of a zwitterion to a cation in an acidic solution (Section 19.2):	$$\overset{+}{H_3}N-\underset{\underset{R}{\mid}}{CH}-\overset{\overset{O}{\mid\mid}}{C}-O^- + H_3O^+ \rightarrow \overset{+}{H_3}N-\underset{\underset{R}{\mid}}{CH}-\overset{\overset{O}{\mid\mid}}{C}-OH + H_2O$$	Reaction 19.2

3.	Conversion of a zwitterion to an anion in a basic solution (Section 19.2):	$\overset{+}{H_3N}-CH(R)-\overset{O}{\overset{\|}{C}}-O^- + OH^- \rightarrow H_2N-CH(R)-\overset{O}{\overset{\|}{C}}-O^- + H_2O$	Reaction 19.3
4.	Oxidation of cysteine to cystine (Section 19.3):	$\overset{+}{H_3N}-CH(CH_2-SH)-\overset{O}{\overset{\|}{C}}-O^-$ and $\overset{+}{H_3N}-CH(CH_2-SH)-\overset{O}{\overset{\|}{C}}-O^-$ $+ (O) \xrightarrow{\text{Oxidation}}$ cystine (S—S) $+ H_2O$	Reaction 19.4

5. Formation of a peptide linkage—general reaction (Section 19.3):

$$\overset{+}{H_3N}-CH(R)-\overset{O}{\overset{\|}{C}}-O^- + \overset{+}{H_3N}-CH(R')-\overset{O}{\overset{\|}{C}}-O^- \rightarrow \overset{+}{H_3N}-CH(R)-\overset{O}{\overset{\|}{C}}-NH-CH(R')-\overset{O}{\overset{\|}{C}}-O^- + H_2O$$

Reaction 19.6

6.	Hydrolysis of proteins in acid or base (Section 19.10):	$\text{protein} + H_2O \xrightarrow[\text{OH}^-]{\text{H}^+ \text{ or}} \text{smaller peptides} \xrightarrow[\text{OH}^-]{\text{H}^+ \text{ or}} \text{amino acids}$	Reaction 19.9

Exercises

Even-numbered exercises are answered in Appendix B.

Blue-numbered exercises are more challenging.

The Amino Acids (Section 19.1)

19.1 Draw the structure of hexanoic acid. Label the alpha carbon.

19.2 What functional groups are found in all amino acids?

19.3 Identify the R group of the side chain in the following amino acids that results in the side-chain classification indicated in parentheses (see Table 19.1):

a. tyrosine (neutral, polar)

b. glutamate (acidic, polar)

c. methionine (neutral, nonpolar)

d. histidine (basic, polar)

e. cysteine (neutral, polar)

f. valine (neutral, nonpolar)

19.4 Draw structural formulas for the following amino acids, identify the chiral carbon atom in each one, and circle the four different groups attached to the chiral carbon.

a. threonine

b. aspartate

c. serine

d. phenylalanine

19.5 Draw structural formulas for the following amino acids, identify the chiral carbon atom in each one, and circle the four different groups attached to the chiral carbon.

a. valine

b. glutamate

c. asparagine

d. cysteine

19.6 Isoleucine contains two chiral carbon atoms. Draw the structural formula for isoleucine twice. In the first, identify one of the chiral carbons and circle the four groups attached to it. In the second, identify the other chiral carbon and circle the four groups attached to it.

19.7 Draw Fischer projections representing the D and L forms of the following:

 a. aspartate

 b. phenylalanine

19.8 Draw Fischer projections representing the D and L forms of the following:

 a. serine

 b. valine

Zwitterions (Section 19.2)

19.9 What is meant by the term *zwitterion*?

19.10 What characteristics indicate that amino acids exist as zwitterions?

19.11 What is meant by the term *isoelectric point*?

19.12 Write structural formulas to show the form the following amino acids would have in a solution with a pH higher than the amino acid isoelectric point:

 a. cysteine

 b. alanine

19.13 Write structural formulas to show the form the following amino acids would have in a solution with a pH lower than the amino acid isoelectric point:

 a. valine

 b. threonine

19.14 Write ionic equations to show how serine acts as a buffer against the following added ions:

 a. OH^-

 b. H_3O^+

Reactions of Amino Acids (Section 19.3)

19.15 Write two reactions to represent the formation of the two dipeptides that form when leucine and threonine react.

19.16 Write a complete structural formula and an abbreviated formula for the tripeptide formed from aspartate, cysteine, and valine in which the C-terminal residue is cysteine and the N-terminal residue is valine.

19.17 Write a complete structural formula and an abbreviated formula for the tripeptide formed from tyrosine, serine, and phenylalanine in which the C-terminal residue is serine and the N-terminal residue is phenylalanine.

19.18 Write abbreviated formulas for the six isomeric tripeptides of alanine, phenylalanine, and arginine.

19.19 Write abbreviated formulas for the six isomeric tripeptides of asparagine, glutamate, and proline.

19.20 How many tripeptide isomers that contain one residue each of leucine, threonine, and tyrosine are possible?

19.21 How many tripeptide isomers that contain one glycine residue and two alanine residues are possible?

Important Peptides (Section 19.4)

19.22 What special role does the amino acid cysteine have in the peptides vasopressin and oxytocin?

19.23 What is the difference between a peptide and a protein?

Characteristics of Proteins (Section 19.5)

19.24 Explain why the presence of certain proteins in body fluids such as urine or blood can indicate that cellular damage has occurred in the body.

19.25 Write equations to show how a protein containing both lysine and aspartate could act as a buffer against the following added ions:

 a. OH^-

 b. H_3O^+

19.26 Explain why a protein is least soluble in an aqueous medium that has a pH equal to the isoelectric point of the protein.

19.27 Wool is a protein used as a textile fiber. Classify the protein of wool into a category of Table 19.4.

19.28 List the eight principal functions of proteins.

19.29 Classify each of the following proteins into one of the eight functional categories of proteins:

 a. insulin

 b. collagen

 c. hemoglobin

 d. cytochrome c

 e. α-keratin

 f. serum albumin

19.30 For each of the following two proteins listed in Table 19.4, predict whether it is more likely to be a globular or a fibrous protein. Explain your reasoning.

 a. collagen

 b. lactate dehydrogenase

19.31 For each of the following two proteins listed in Table 19.4, predict whether it is more likely to be a globular or a fibrous protein. Explain your reasoning.

 a. elastin

 b. ceruloplasmin

19.32 Differentiate between simple and conjugated proteins.

19.33 Give one example of a conjugated protein that contains the following prosthetic group:

 a. iron

 b. lipid

 c. phosphate group

 d. carbohydrate

 e. heme

The Primary Structure of Proteins (Section 19.6)

19.34 Describe what is meant by the term *primary structure of proteins*.

19.35 What type of bonding is present to account for the primary structure of proteins?

19.36 Write the structure for a protein backbone. Make the backbone long enough to attach four R groups symbolizing amino acid side chains.

The Secondary Structure of Proteins (Section 19.7)

19.37 Describe the differences between alpha and beta secondary structures of proteins.

19.38 What type of bonding between amino acid residues is most important in holding a protein or polypeptide in a specific secondary configuration?

19.39 Describe what is meant by "supracoiling" in proteins.

The Tertiary Structure of Proteins (Section 19.8)

19.40 How do hydrogen bonds involved in tertiary protein structures differ from those involved in secondary structures?

19.41 Which amino acids have side-chain groups that can form salt bridges?

19.42 Refer to Table 19.1 and list the type of side-chain interaction expected between the side chains of the following pairs of amino acid residues.

 a. tyrosine and glutamine

 b. aspartate and lysine

 c. leucine and isoleucine

 d. phenylalanine and valine

19.43 Which amino acids have side-chain groups that can participate in hydrogen bonding?

19.44 A globular protein in aqueous surroundings contains the following amino acid residues: phenylalanine, methionine, glutamate, lysine, and alanine. Which amino acid side chains would be directed toward the inside of the protein and which would be directed toward the aqueous surroundings?

The Quaternary Structure of Proteins (Section 19.9)

19.45 What is meant by the term *quaternary protein structure*?

19.46 What types of forces give rise to quaternary structure?

19.47 Describe the quaternary protein structure of hemoglobin.

19.48 What is meant by the term *subunit*?

Protein Hydrolysis and Denaturation (Section 19.10)

19.49 What is meant by the term *denaturation*?

19.50 Suppose a sample of a protein is completely hydrolyzed and another sample of the same protein is denatured. Compare the final products of each process.

19.51 As fish is cooked, the tissue changes from a soft consistency to a firm one. A similar change takes place when fish is pickled by soaking it in vinegar (acetic acid). In fact, some people prepare fish for eating by pickling instead of cooking. Propose an explanation of why the two processes give somewhat similar results.

19.52 In what way is the protein in a raw egg the same as that in a cooked egg?

19.53 In addition to emulsifying greasy materials, what other useful function might be served by a detergent used to wash dishes?

19.54 Once cooked, egg whites remain in a solid form. However, egg whites that are beaten to form meringue will partially change back to a jellylike form if allowed to stand for a while. Explain these behaviors using the concept of reversible protein denaturation.

19.55 Lead is a toxic material. What form of the lead is actually the harmful substance?

19.56 Explain how egg white can serve as an emergency antidote for heavy-metal poisoning.

Additional Exercises

19.57 A protein has a molecular weight of about 12,000 u. This protein contains 6.4% (w/w) sulfur in the form of cystine (the cysteine disulfide). How many cystine units are present in the protein?

19.58 Which amino acids could be referred to as derivatives of butanoic acid?

19.59 A Tyndall effect is observed as light is scattered as it passes through a small polypeptide–water mixture with a pH of 6.7. What can be deduced about the polypeptide's isoelectric point from this information?

19.60 What is the conjugate acid for the following zwitterion? What is the conjugate base for it?

$$\overset{\displaystyle H}{\underset{\displaystyle R}{H_3\overset{+}{N}-\overset{|}{\underset{|}{C}}-COO^-}}$$

19.61 The K_a values recorded for alanine are 5.0×10^{-3} and 2.0×10^{-10}. Why are two values recorded? What functional groups do the values correspond to? How many K_a values would be recorded for glutamate?

Chemistry for Thought

19.62 Some researchers feel that the "high" experienced by runners may be due to the brain's synthesis of peptides. Explain why this theory might be a reasonable explanation.

19.63 Suggest a reason why individuals with the sickle-cell trait might sometimes encounter breathing problems on high-altitude flights.

19.64 In Section 19.2, you learned that amino acids like alanine are crystalline solids with high melting points. However, the ethyl ester of alanine, which has a free NH_2 group, melts at only 87°C. Explain why the ethyl ester melts 200 degrees below the melting point of alanine.

19.65 Why do health care workers wipe a patient's skin with solutions of rubbing alcohol before giving injections?

19.66 Would you expect plasma proteins to be fibrous or globular? Explain.

19.67 Why must the protein drug insulin be given by injection rather than taken by mouth?

19.68 In Figure 19.4, it is noted that animal hair is made of fibrous proteins. What properties of fibrous proteins make hair, fur, and spiderwebs useful?

19.69 Why is it necessary that protein molecules be enormous?

19.70 Name two elements that are found in proteins but are not present in fats, oils, or carbohydrates.

Allied Health Exam Connection

The following questions are from these sources:

- *Nursing School Entrance Exam* © 2005, Learning Express, LLC.
- *McGraw-Hill's Nursing School Entrance Exams* by Thomas A. Evangelist, Tamara B. Orr, and Judy Unrein © 2009, The McGraw-Hill Companies, Inc.
- *NSEE Nursing School Entrance Exams,* 3rd edition © 2009, Kaplan Publishing.
- *Cliffs Test Prep: Nursing School Entrance Exams* by Fred N. Grayson © 2004, Wiley Publishing, Inc.
- *Peterson's Master the Nursing School and Allied Health Entrance Exams,* 18th edition by Marion F. Gooding © 2008, Peterson's, a Nelnet Company.

19.71 _____ stimulates the smooth muscle of the uterine wall during the labor and delivery process. After delivery, it promotes the ejection of milk.

19.72 What functional groups are found in all amino acids? How many different amino acids are found in naturally occurring proteins?

19.73 Which of the following are true concerning the chemical bond that forms between the carboxyl (RCOOH) group of one amino acid and the amino (RC—NH$_2$) group of another?

 a. The bond is called a peptide bond.

 b. It is formed by inserting a water molecule between them.

 c. It is formed by a dehydration reaction.

 d. A polypeptide has more of these bonds than a protein.

19.74 Rank the following components of hemoglobin in decreasing amounts as would normally be found in a hemoglobin molecule:

 a. Iron atoms

 b. Amino acids

 c. Heme groups

19.75 Describe the quaternary protein structure of hemoglobin.

19.76 Complete degradation of a protein into individual amino acids involves (choose all that are correct):

 a. Removal of a water molecule from between two amino acids

 b. Addition of a water molecule between two amino acids

 c. A hydrolysis reaction

 d. The breaking of peptide linkage

19.77 Which is NOT a characteristic of proteins?

 a. They contain genetic information.

 b. They can act as hormones.

 c. They can catalyze chemical reactions.

 d. They act in cell membrane trafficking.

19.78 Which of the following describes the primary structure of proteins?

 a. The collective shape assumed by all of the chains in a protein containing multiple chains.

 b. The folding of an individual protein molecule.

 c. The regular repeated shape of the protein molecule's backbone.

 d. The sequence of amino acids bonded together by peptide bonds.

20 Enzymes

Doug Struthers/Getty Images

▶ Case Study

Myocardial infarction sounded like a word from a foreign language or even a joke—Bob wasn't sure which. How could he, a 34-year-old construction worker, possibly be having a heart attack? When he complained of shoulder pain and nausea at work, his supervisor insisted on taking him to the ER. He had thought his boss was overreacting, but now Bob realized the supervisor's caution saved his life. At the emergency room, an EKG and tests for the enzyme creatine phosphokinase and the protein troponin confirmed that his symptoms were indeed caused by a heart attack. The doctor explained that Bob's genetic inheritance, his poor diet, and his smoking habit formed a deadly combination. In fact, for many men his age, death is the first "symptom" of a heart attack. Immediate treatment saved his life. If Bob wanted to live to be an old man, he would need to change his diet and stop smoking.

▶ **What resources are available to help people like Bob stop smoking? How do heart attack symptoms differ for men and women? What role do diet and genetics play in cholesterol levels?**

Follow-up to this Case Study appears at the end of the chapter before the *Concept Summary*.

When you have completed your study of this chapter, you should be able to:

1 Describe the general characteristics of enzymes and explain why enzymes are vital in body chemistry. (Section 20.1)

2 Determine the function and/or substrate of an enzyme on the basis of its name. (Section 20.2)

3 Identify the general function of cofactors. (Section 20.3)

4 Use the lock-and-key theory to explain specificity in enzyme action. (Section 20.4)

5 List two ways of describing enzyme activity. (Section 20.5)

6 Identify the factors that affect enzyme activity. (Section 20.6)

7 Compare the mechanisms of competitive and noncompetitive enzyme inhibition. (Section 20.7)

8 Describe the three methods of cellular control over enzyme activity. (Section 20.8)

9 Discuss the importance of measuring enzyme levels in the diagnosis of disease. (Section 20.9)

The catalytic behavior of proteins acting as enzymes is one of the most important functions performed by cellular proteins. Without catalysts, most cellular reactions would take place much too slowly to support life (see **Figure 20.1**). With the exception of a few recently discovered RNA molecules (discussed in **Chapter 21**) that catalyze their own reactions, all enzymes are globular proteins. Enzymes are the most efficient of all known catalysts; some can increase reaction rates by 10^{20} times that of uncatalyzed reactions.

20.1 | General Characteristics of Enzymes

Learning Objective

1. Describe the general characteristics of enzymes and explain why enzymes are vital in body chemistry.

Figure 20.1 One miracle of life is that hundreds of chemical reactions take place simultaneously in each living cell; the activity of enzymes in the cells makes it all possible.

Enzymes are well-suited to their essential roles in living organisms in three major ways: They have enormous catalytic power, they are highly specific in the reactions they catalyze, and their activity as catalysts can be regulated.

20.1A Catalytic Efficiency

As we know, catalysts increase the rate of chemical reactions but are not used up in the process. Although a catalyst actually participates in a chemical reaction, it is not permanently modified and may be used again and again. Enzymes are true catalysts that speed chemical reactions by lowering activation energies (see **Figure 20.2**) and allowing reactions to achieve equilibrium more rapidly.

Many of the important enzyme-catalyzed reactions are similar to the reactions we studied in the chapters on organic chemistry: ester hydrolysis, alcohol oxidation, and so on. However, laboratory conditions cannot match what happens when these reactions are carried out in the body; enzymes cause such reactions to proceed under mild pH and temperature conditions. In addition, enzyme catalysis within the body can accomplish in seconds reactions that ordinarily take weeks or even months under laboratory conditions.

The influence of enzymes on the rates of reactions essential to life is truly amazing. A good example is the need to move carbon dioxide, a waste product of cellular respiration, out of the body. Before this can be accomplished, the carbon dioxide must be combined with water to form carbonic acid:

enzyme A biomolecule that catalyzes chemical reactions.

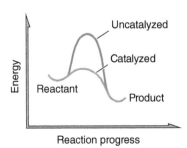

Figure 20.2 An energy diagram for a reaction when uncatalyzed and catalyzed.

$$CO_2 + H_2O \xrightarrow{\text{carbonic anhydrase}} H_2CO_3 \qquad (20.1)$$

In the absence of the appropriate enzyme, carbonic anhydrase, the formation of carbonic acid takes place much too slowly to support the required exchange of carbon dioxide between the blood and the lungs. But in the presence of carbonic anhydrase, this vital reaction proceeds rapidly. Each molecule of the enzyme can catalyze the formation of carbonic acid at the rate of 36 million molecules per minute.

20.1B Specificity

Enzyme specificity is a second characteristic that is important in life processes. Unlike other catalysts, enzymes are often quite specific in the type of reaction they catalyze and even the particular substance that will be involved in the reaction. For example, strong acids catalyze the hydrolysis of any amide, the dehydration of any alcohol, and a variety of other processes. However, the enzyme urease catalyzes only the hydrolysis of a single amide, urea (Reaction 20.2 and **Figure 20.3**):

$$H_2N - \overset{\overset{\displaystyle O}{\|}}{C} - NH_2 + H_2O \xrightarrow{\text{urease}} CO_2 + 2NH_3 \qquad (20.2)$$

1 The enzyme urease is added to a urea solution containing the indicator phenolphthalein.

2 As urea is hydrolyzed, forming ammonia, the indicator turns pink. How else might the presence of ammonia be detected?

Figure 20.3 The hydrolysis of urea catalyzed by urease.

absolute specificity The characteristic of an enzyme that it acts on one and only one substance.

relative specificity The characteristic of an enzyme that it acts on several structurally related substances.

stereochemical specificity The characteristic of an enzyme that it is able to distinguish between stereoisomers.

This is an example of **absolute specificity**, the catalyzing of the reaction of one and only one substance. Other enzymes display **relative specificity** by catalyzing the reaction of structurally related substances. For example, the lipases hydrolyze lipids, proteases split up proteins, and phosphatases hydrolyze phosphate esters. **Stereochemical specificity** extends even to enantiomers: D-amino acid oxidase will not catalyze the reactions of L-amino acids.

20.1C Regulation

A third significant property of enzymes is that their catalytic behavior can be regulated. Even though each living cell contains thousands of different molecules that could react with each other in an almost unlimited number of ways, only a relatively small number of these possible reactions take place because of the enzymes present. The cell controls the rates of these reactions and the amount of any given product formed by regulating the action of the enzymes.

20.2 | Enzyme Nomenclature and Classification

Learning Objective

2. Determine the function and/or substrate of an enzyme on the basis of its name.

Some of the earliest discovered enzymes were given names ending with *-in* to indicate their protein composition. For example, three of the digestive enzymes that catalyze protein hydrolysis are named pepsin, trypsin, and chymotrypsin.

The large number of enzymes now known has made it desirable to adopt a systematic nomenclature system known as the Enzyme Commission (EC) system. Enzymes are grouped into six major classes on the basis of the reaction catalyzed (see **Table 20.1**).

TABLE 20.1 The EC Classification of Enzymes

Group Name	Type of Reaction Catalyzed
Oxidoreductases	Oxidation–reduction reactions
Transferases	Transfer of functional groups
Hydrolases	Hydrolysis reactions
Lyases	Addition to double bonds or the reverse of that reaction
Isomerases	Isomerization reactions
Ligases	Formation of bonds with ATP cleavage[a]

[a]ATP is discussed in Chapter 22.

In the EC system, each enzyme has an unambiguous (and often long) systematic name that specifies the **substrate** (substance acted on), the functional group acted on, and the type of reaction catalyzed. All EC names end in *-ase*. The hydrolysis of urea provides a typical example:

substrate The substance that undergoes a chemical change catalyzed by an enzyme.

$$H_2N - \overset{\overset{\textstyle O}{\|}}{C} - NH_2 + H_2O \xrightarrow{\text{enzyme}} CO_2 + 2NH_3 \qquad (20.3)$$

EC name: urea amidohydrolase
Substrate: urea
Functional group: amide
Type of reaction: hydrolysis

Enzymes are also assigned common names, which are usually shorter and more convenient. Common names are derived by adding *-ase* to the name of the substrate or to a combination of the substrate name and type of reaction. For example, urea amidohydrolase is assigned the common name urease:

Substrate: urea
Common name: urea + ase = urease

The enzyme name *alcohol dehydrogenase* is an example of a common name derived from both the name of the substrate and the type of reaction:

Substrate: alcohol (ethyl alcohol)
Reaction type: dehydrogenation (removal of hydrogen)
Common name: alcohol dehydrogenation + ase = alcohol dehydrogenase

Throughout this text, we will refer to most enzymes by their accepted common names.

Cut Back on Processed Meat

Dietary guidelines continue to change in accordance with research-based evidence. A panel of experts convened by the World Health Organization (WHO) and the International Agency for Research on Cancer (IARC) concluded in December of 2015 that eating processed meats such as hot dogs, lunch meat, salami, ham, pepperoni, and bacon increased the risk of developing colon cancer. This was disturbing news for many Americans. Bacon had been used in numerous creative ways during the last few years. The results were such concoctions as maple-bacon donuts and bacon-topped chocolate caramel sundaes. Pepperoni pizza and hot dogs were remembered as favorite childhood foods. Americans still rely on these processed foods to make camping outings, Fourth of July celebrations, and birthday parties special.

So, exactly how significant is the danger? The panel concluded that there was sufficient evidence to include processed meats in its "Group 1 category," meaning that they could cause cancer. Other substances in this category include alcohol, asbestos, and tobacco smoke. However, all these items do not have the same level of risk. Of about 900 substances or factors studied by the panel, only 120 have been assigned Group 1 category status. Other categories range from "possibly carcinogenic" to "probably not carcinogenic." Other materials or conditions evaluated by the study group include coffee, sunlight, and night shift work. In these studies only one factor has been assigned a "probably not carcinogenic" category. Within categories, there is a huge range of risk, from barely any at all, to very risky.

The IARC panel defined processed meats as those "transformed through salting, curing, fermentation, smoking, or other processes to enhance flavor or improve preservation." Based on largely observational studies, the panel added processed meats to the Group 1 category, which also includes substances such as mineral oil, radiation, and diesel exhaust. The panel cautioned that not all substances are associated with cancer to the same extent. In addition to processed meats, the panel recommended against consumption of red meat, which is defined as muscle meat such as beef, veal, pork, and lamb.

The risk of eating processed and red meat comes not only from the food itself, but also from the cooking method used. Cooking at high temperatures or in direct contact with a flame (as in roasting a hot dog over a fire or grilling a steak) can produce certain substances that are known to be carcinogenic. For example, high-temperature meat cooking produces heterocyclic amines (HCA) and polycyclic aromatic hydrocarbons (PAH). Both of these substances have been shown to cause cancer in animals.

How much processed or red meat poses a health risk? Or, how much can one person consume safely? The new report indicates that eating 50 grams of processed meat per day increases the chances of developing colorectal cancer by 18%. To be clear, this means that the risk of developing cancer increases from an existing lifetime risk of 5% for those who do not consume these products to a lifetime risk of 6% for those that do. Fifty grams of processed meat is equivalent to one of the following: 1 hot dog, 5 slices of salami, 2 slices of Canadian bacon, or 2 slices of ham. If you are eating processed meat every day, this 50 grams is too much. Turkey versions of processed meats are no safer than their red meat counterparts because the danger comes from the materials used in the processing and cooking methods associated with these foods more than from the meat itself.

Does this mean no more of your favorite party foods? Experts, not involved in the report, suggest that people should moderate their intake of processed meats and maintain that the increased risk of cancer from eating processed meats is relatively small. In other words, a commonsense approach is advisable. For example, eating processed meat doesn't come close to the risk of developing cancer from cigarette smoking. But, just like sugar or fried food consumption, moderation is the key. The majority of a healthful diet should consist of fresh fruits and vegetables with fish, poultry, and legumes being the main sources of protein. Processed meats and red meat should be a rare treat, not a daily indulgence.

Hot dogs, lunch meat, and salami are typical processed meats.

a. maltase
b. peptidase
c. glucose 6-phosphate isomerase

20.3 | Enzyme Cofactors

Learning Objective

3. Identify the general function of cofactors.

In **Section 19.5**, we learned that proteins may be simple (containing only amino acid residues) or conjugated with a prosthetic group present. Many enzymes are simple proteins, whereas many others function only in the presence of specific nonprotein molecules or metal ions. If these nonprotein components are tightly bound to and form an integral part of the enzyme structure, they are true prosthetic groups. Often, however, a nonprotein component is only weakly bound to the enzyme and is easily separated from the protein structure. This type of nonprotein component is referred to as a **cofactor**. When the cofactor is an organic substance, it is called a **coenzyme**. The cofactor may also be an inorganic ion (usually a metal ion). The protein portion of enzymes requiring a cofactor is called the **apoenzyme**. Thus, the combination of an apoenzyme and a cofactor produces an active enzyme:

$$\text{apoenzyme} + \text{cofactor (coenzyme or inorganic ion)} \rightarrow \text{active enzyme} \qquad (20.4)$$

Typical inorganic ions are metal ions such as Mg^{2+}, Zn^{2+}, and Fe^{2+}. For example, the enzyme carbonic anhydrase functions only when Zn^{2+} is present, and rennin needs Ca^{2+} in order to curdle milk. Numerous other metal ions are essential for proper enzyme function in humans; hence, they are required for good health.

Like metal ions, the small organic molecules that act as coenzymes bind reversibly to an enzyme and are essential for its activity. An interesting feature of coenzymes is that many of them are formed in the body from vitamins (see **Table 20.2**), which explains why it is necessary to have certain vitamins in the diet for good health. For example, the coenzyme nicotinamide adenine dinucleotide (NAD^+), which is a necessary part of some

cofactor A nonprotein molecule or ion required by an enzyme for catalytic activity.

coenzyme An organic molecule required by an enzyme for catalytic activity.

apoenzyme A catalytically inactive protein formed by removal of the cofactor from an active enzyme.

TABLE 20.2 Vitamins and Their Coenzyme Forms

Vitamin	Coenzyme Form	Function
biotin	biocytin	Carboxyl group removal or transfer
folacin	tetrahydrofolic acid	One-carbon group transfer
lipoic acid	lipoamide	Acyl group transfer
niacin	nicotinamide adenine dinucleotide (NAD^+)	Hydrogen transfer
	nicotinamide adenine dinucleotide phosphate ($NADP^+$)	Hydrogen transfer
pantothenic acid	coenzyme A (CoA)	Acyl group carrier
pyridoxal, pyridoxamine, pyridoxine (B_6 group)	pyridoxal phosphate	Amino group transfer
riboflavin	flavin mononucleotide (FMN) flavin adenine dinucleotide (FAD)	Hydrogen transfer Hydrogen transfer
thiamin (B_1)	thiamin pyrophosphate (TPP)	Aldehyde group transfer
vitamin B_{12}	coenzyme B_{12}	Shift of hydrogen atoms between adjacent carbon atoms; methyl group transfer

Treatment Options for the Common Cold

Wouldn't it be nice if the most common cause of misery and the most often stated reason for absenteeism from work and school could be prevented? But it can't, at least not yet. With over 100 different viruses that rapidly mutate as the cause for the common cold, it is not surprising that a preventive vaccine does not yet exist as it does for several other viral infections. Fortunately, a cold is usually not a serious infection that leads to significant complications, but during the time you are suffering the symptoms, it might feel serious. With so many different viruses causing colds, the cold signs and symptoms might vary dramatically. Most sufferers complain of multiple symptoms that might include various combinations of the following: stuffy or running nose, body aches, mild headache, itchy and scratchy eyes, cough, mild fatigue, and sore throat. It is no wonder that some sort of treatment and relief is desired. The first concern is how do you know your problem is just a cold? If a fever higher than 104°F in adults is present, or severe symptoms continue for more than a week, the advice of a health care provider should be obtained.

Luckily, most common cold symptoms persist for no more than a few days. However, because the symptoms disrupt common activities and responsibilities so much, remedies are looked for. Unluckily, no cure for the common cold is known. Antibiotics should not be used because the cold is caused by a virus and antibiotics are not effective against virus infections. The best strategy is to use medications that will decrease the severity of the symptoms. The following are remedies to treat specific symptoms.

Pain Relievers/Fever Reducers: Aspirin, ibuprofen, and acetaminophen are all over-the-counter medications that provide relief from pain and fever. Each one has advantages and disadvantages. Aspirin and ibuprofen are more likely to cause stomach upset. Aspirin should not be used in children and teenagers with flu symptoms because of the increased risk of "Reye's syndrome" (encephalopathy). Acetaminophen can cause liver damage if not dosed correctly. Otherwise, they have similar effectiveness with regards to reducing a fever and relieving the "aches and pains" associated with a cold.

Antihistamines: Antihistamines relieve runny noses and itchy eyes. For maximum effectiveness they should be administered at the first signs of these symptoms. Some antihistamines cause drowsiness. Antihistamines are much better at preventing symptoms rather than reversing them.

Cough Syrups: Seek professional guidance if dosing a child who is less than 4 years old. Dextromethorphan is an over-the-counter, non-narcotic cough reliever that is often helpful.

Oral Decongestants: Oral decongestants (pseudoephedrine or phenylephrine) open nasal passages effectively. But they have side effects of central nervous system stimulation that make it difficult to fall or stay asleep, so these medications should not be taken near bedtime.

Additional Remedies/Suggestions:

Wash your hands: This prevents contamination and the spread of virus particles. Cough into your elbow or use a tissue. Keep your children out of day care when sick children are evident. Select a facility that has good, visible hygiene practices and requires sick children to stay home.

Use a humidifier: These moisturize airways to help keep the virus from attaching to the respiratory tract cells and might diminish congestion and cough.

Throat lozenges/gargling: This soothes a sore throat. Lozenges can also contain a cough suppressant and/or a local anesthetic.

Stay home or keep children home, and get rest: The rest will help the patient, and prevent spreading the virus.

Increase fluid intake: This does not mean caffeinated beverages or alcohol, which might actually contribute to dehydration. This does mean to drink plenty of water to replace lost fluids.

Vitamin C: This does not cure the common cold as many claim. But, if taken at the onset of symptoms, it might shorten the duration slightly.

Echinacea: Studies dealing with the use of this herb are ongoing. Consistent data showing effectiveness are lacking. For most people it is probably not harmful.

Zinc compounds: Studies are inconclusive. Nasal sprays containing zinc compounds should not be used, as this might cause irreversible loss of the sense of smell.

Cold sufferers have multiple symptoms.

Andrey_Popov/Shutterstock.com

enzyme-catalyzed oxidation–reduction reactions, is formed from the vitamin precursor nicotinamide. Reaction 20.5 shows the participation of NAD^+ in the oxidation of lactate by the enzyme lactate dehydrogenase (LDH). Like other cofactors, NAD^+ is written separately from the enzyme so that the change in its structure may be shown and to emphasize that the enzyme and cofactor are easily separated.

$$\underset{\text{lactate}}{CH_3-\overset{\overset{\displaystyle OH}{|}}{CH}-COO^-} + NAD^+ \underset{\text{dehydrogenase}}{\overset{\text{lactate}}{\rightleftharpoons}} \underset{\text{pyruvate}}{CH_3-\overset{\overset{\displaystyle O}{||}}{C}-COO^-} + NADH + H^+ \quad (20.5)$$

We can see from the reaction that the coenzyme NAD^+ is essential because it is the actual oxidizing agent. NAD^+ is typical of coenzymes that often aid in the transfer of chemical groups from one compound to another. In Reaction 20.5, NAD^+ accepts hydrogen from lactate and will transfer it to other compounds in subsequent reactions. In Chapter 22, we will discuss in greater detail several coenzymes that act as shuttle systems in the exchange of chemical substances among various biochemical pathways.

20.4 | The Mechanism of Enzyme Action

Learning Objective

4. Use the lock-and-key theory to explain specificity in enzyme action.

Although enzymes differ widely in structure and specificity, a general theory has been proposed to account for their catalytic behavior. Because enzyme molecules are large compared with the molecules whose reactions they catalyze (substrate molecules), it has been proposed that substrate and enzyme molecules come in contact and interact over only a small region of the enzyme surface. This region of interaction is called the **active site**.

The binding of a substrate molecule to the active site of an enzyme may occur through hydrophobic attraction, hydrogen bonding, and/or ionic bonding. The complex formed when a substrate and an enzyme bond is called the enzyme–substrate (ES) complex. Once this complex is formed, the conversion of substrate (S) to product (P) may take place:

active site The location on an enzyme where a substrate is bound and catalysis occurs.

General reaction:

$$\underset{\text{enzyme}}{E} + \underset{\text{substrate}}{S} \rightleftharpoons \underset{\substack{\text{enzyme–}\\\text{substrate}\\\text{complex}}}{ES} \longrightarrow \underset{\text{enzyme}}{E} + \underset{\text{product}}{P} \quad (20.6)$$

Specific example:

$$\underset{\text{enzyme}}{\text{sucrase}} + \underset{\text{substrate}}{\text{sucrose}} \rightleftharpoons \left\{ \begin{array}{c} \text{sucrase–}\\ \text{sucrose}\\ \text{complex} \end{array} \right\} \longrightarrow \underset{\text{enzyme}}{\text{sucrase}} + \underset{\text{products}}{\text{glucose} + \text{fructose}} \quad (20.7)$$

The chemical transformation of the substrate occurs at the active site, usually aided by enzyme functional groups that participate directly in the making and breaking of chemical bonds. After chemical conversion has occurred, the product is released from the active site, and the enzyme is free for another round of catalysis.

An extension of the theory of ES complex formation is used to explain the high specificity of enzyme activity. According to the **lock-and-key theory**, enzyme surfaces will accommodate only those substrates having specific shapes and sizes. Thus, only specific substrates "fit" a given enzyme and can form complexes with it (see **Figure 20.4A**).

A limitation of the lock-and-key theory is the implication that enzyme conformations are fixed or rigid. However, research results suggest that the active sites of some enzymes are not rigid. This is taken into account in a modification of the lock-and-key theory known as the **induced-fit theory**, which proposes that enzymes have somewhat flexible conformations

lock-and-key theory A theory of enzyme specificity proposing that a substrate has a shape fitting that of the enzyme's active site, as a key fits a lock.

induced-fit theory A theory of enzyme action proposing that the conformation of an enzyme changes to accommodate an incoming substrate.

A In the **lock-and-key model**, the rigid enzyme and substrate have matching shapes.

B In the **induced-fit model**, the flexible enzyme changes shape to match the substrate.

Figure 20.4 Models representing enzyme action.

that may adapt to incoming substrates. The active site has a shape that becomes complementary to that of the substrate only after the substrate is bound (see **Figure 20.4B**).

20.5 | Enzyme Activity

Learning Objective

5. List two ways of describing enzyme activity.

enzyme activity The rate at which an enzyme catalyzes a reaction.

turnover number The number of molecules of substrate acted on by one molecule of enzyme per minute.

Enzyme activity refers in general to the catalytic ability of an enzyme to increase the rate of a reaction. The amazing rate (36 million molecules per minute) at which one molecule of carbonic anhydrase converts carbon dioxide to carbonic acid was mentioned earlier. This rate, called the **turnover number**, is one of the highest known for enzyme systems. More common turnover numbers for enzymes are closer to 10^3/min, or 1000 reactions per minute. Nevertheless, even such low numbers dramatize the speed with which a small number of enzyme molecules can transform a large number of substrate molecules. **Table 20.3** gives the turnover numbers of several enzymes.

TABLE 20.3 Examples of Enzyme Turnover Numbers

Enzyme	Turnover Number (per minute)	Reaction Catalyzed
carbonic anhydrase	36,000,000	$CO_2 + H_2O \rightleftharpoons H_2CO_3$
catalase	5,600,000	$2H_2O_2 \rightleftharpoons 2H_2O + O_2$
cholinesterase	1,500,000	Hydrolysis of acetylcholine
chymotrypsin	6,000	Hydrolysis of specific peptide bonds
DNA polymerase I	900	Addition of nucleotide monomers to DNA chains
lactate dehydrogenase	60,000	pyruvate + NADH + H$^+$ \rightleftharpoons lactate + NAD$^+$
penicillinase	120,000	Hydrolysis of penicillin

Experiments that measure enzyme activity are called *enzyme assays*. Assays for blood enzymes are routinely performed in clinical laboratories. Such assays are often done by determining how fast the characteristic color of a product forms or the color of a substrate decreases. Reactions in which protons (H$^+$) are produced or used up can be followed by measuring how fast the pH of the reacting mixture changes with time.

Because some clinical assays are done many times a day, the procedures for running the assays have been automated and computerized. The determined enzyme activity levels are usually reported in terms of **enzyme international units (IU)**, which define enzyme activity as the amount of enzyme that will convert a specified amount of substrate to a product within a certain time.

One standard enzyme international unit (1 IU) is the quantity of enzyme that catalyzes the conversion of 1 micromole (μmol) of substrate per minute under specified reaction conditions. Thus, unlike the turnover number, which is a constant characteristic for one molecule of a particular enzyme, international units measure how much enzyme is present. For example, an enzyme preparation with an activity corresponding to 40 IU contains a concentration of enzyme 40 times greater than the standard. This is a useful way to measure enzyme activity because the level of enzyme activity compared with normal activity is significant in the diagnosis of many diseases (see **Section 20.9**).

> ✔ **LEARNING CHECK 20.2** Differentiate between the terms *turnover number* and *enzyme international unit*.

enzyme international unit (IU) A quantity of enzyme that catalyzes the conversion of 1 μmol of substrate per minute under specified conditions.

20.6 | Factors Affecting Enzyme Activity

Learning Objective

6. Identify the factors that affect enzyme activity.

Several factors affect the rate of enzyme-catalyzed reactions. The most important factors are enzyme concentration, substrate concentration, temperature, and pH. In this section, we'll look at each of these factors in some detail. In **Section 20.7**, we'll consider another very important factor, the presence of enzyme inhibitors.

20.6A Enzyme Concentration

In an enzyme-catalyzed reaction, the concentration of enzyme is normally very low compared with the concentration of substrate. When the enzyme concentration is increased, the concentration of ES also increases in compliance with reaction rate theory:

$$E + S \rightleftarrows ES$$
increased [E] gives more [ES]

Thus, the availability of more enzyme molecules to catalyze a reaction leads to the formation of more ES and a higher reaction rate. The effect of enzyme concentration on the rate of a reaction is shown in **Figure 20.5**. As the figure indicates, the rate of a reaction is directly proportional to the concentration of the enzyme—that is, if the enzyme concentration is doubled, the rate of conversion of substrate to product is also doubled.

Figure 20.5 The dependence of reaction rate (initial velocity) on enzyme concentration.

20.6B Substrate Concentration

As shown in **Figure 20.6**, the concentration of substrate significantly influences the rate of an enzyme-catalyzed reaction. Initially, the rate is responsive to increases in substrate concentration. However, at a certain concentration, the rate levels out and remains constant. This maximum rate (symbolized by V_{max}) occurs because the enzyme is saturated with substrate and cannot work any faster under the conditions imposed.

20.6C Temperature

Enzyme-catalyzed reactions, like all chemical reactions, have rates that increase with temperature (see **Figure 20.7**). However, because enzymes are proteins, there is a temperature limit beyond which the enzyme becomes vulnerable to denaturation. Thus, every enzyme-catalyzed reaction has an **optimum temperature**, usually in the range of 25°C–40°C. Above or below that value, the reaction rate will be lower. This effect is illustrated in Figure 20.7.

optimum temperature The temperature at which enzyme activity is highest.

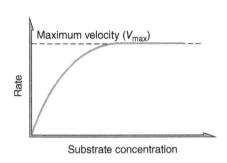

Rate

Maximum velocity (V_{max})

Substrate concentration

Figure 20.6 The dependence of reaction rate (initial velocity) on substrate concentration.

Rate

Optimum temperature

Increasing temperature results in denaturation

Temperature
Enzyme-catalyzed reaction

Figure 20.7 The effect of temperature on enzyme-catalyzed reaction rates.

Rate

Optimum pH value

Increasing pH

Figure 20.8 A typical plot of the effect of pH on reaction rate.

© Mark Slabaugh

Figure 20.9 Pickles are prepared under acidic conditions. Why do they resist spoilage?

optimum pH The pH at which enzyme activity is highest.

20.6D The Effect of pH

The graph of enzyme activity as a function of pH is somewhat similar to the behavior as a function of temperature (see **Figure 20.8**). Notice in Figure 20.8 that an enzyme is most effective in a narrow pH range and is less active at pH values lower or higher than this optimum. This variation in enzyme activity with changing pH may be due to the influence of pH on acidic and basic side chains within the active site. In addition, most enzymes are denatured by pH extremes. The resistance of pickled foods (see **Figure 20.9**) to spoilage is due to the enzymes of microorganisms being less active under acidic conditions.

Many enzymes have an **optimum pH** near 7, the pH of most biological fluids. However, the optimum pH of a few is considerably higher or lower than 7. For example, pepsin, a digestive enzyme of the stomach, shows maximum activity at a pH of about 1.5, the pH of gastric fluids. **Table 20.4** lists optimum pH values for several enzymes.

TABLE 20.4 Examples of Optimum pH for Enzyme Activity		
Enzyme	**Source**	**Optimum pH**
pepsin	Gastric mucosa	1.5
β-glucosidase	Almond	4.5
sucrase	Intestine	6.2
urease	Soybean	6.8
catalase	Liver	7.3
succinate dehydrogenase	Beef heart	7.6
arginase	Beef liver	9.0
alkaline phosphatase	Bone	9.5

✔ **LEARNING CHECK 20.3** Indicate how each of the following affects the rate of an enzyme-catalyzed reaction:

a. Increase in enzyme concentration
b. Increase in substrate concentration
c. Increase in temperature
d. Increase in pH

Learning Objective

7. Compare the mechanisms of competitive and noncompetitive enzyme inhibition.

An **enzyme inhibitor** is any substance that can decrease the rate of an enzyme-catalyzed reaction. An understanding of enzyme inhibition is important for several reasons. First, the characteristic function of many poisons and some medicines is to inhibit one or more enzymes and to decrease the rates of the reactions they catalyze. Second, some substances normally found in cells inhibit specific enzyme-catalyzed reactions and thereby provide a means for the internal regulation of cellular metabolism.

Enzyme inhibitors are classified into two categories, reversible and irreversible, on the basis of how they behave at the molecular level.

enzyme inhibitor A substance that decreases the activity of an enzyme.

20.7A Irreversible Inhibition

An irreversible inhibitor forms a covalent bond with a specific functional group of the enzyme and as a result renders the enzyme inactive. A number of very deadly poisons act as irreversible inhibitors.

The cyanide ion (CN^-) is an example of an irreversible enzyme inhibitor. It is extremely toxic and acts very rapidly. The cyanide ion interferes with the operation of an iron-containing enzyme called cytochrome oxidase. The ability of cells to use oxygen depends on the action of cytochrome oxidase. When the cyanide ion reacts with the iron of this enzyme, it forms a very stable complex (see **Reaction 20.8**), and the enzyme can no longer function properly. As a result, cell respiration stops, causing death in a matter of minutes.

$$Cyt\!-\!Fe^{3+} \;+\; CN^- \;\rightarrow\; Cyt\!-\!Fe\!-\!CN^{2+} \qquad (20.8)$$

cytochrome cyanide stable complex
oxidase ion

Any antidote for cyanide poisoning must be administered quickly. One antidote is sodium thiosulfate (the "hypo" form is used in the developing of photographic film). This substance converts the cyanide ion to a thiocyanate ion that does not bind to the iron of cytochrome oxidase.

$$CN^- \;+\; S_2O_3^{2-} \;\rightarrow\; SCN^- \;+\; SO_3^{2-} \qquad (20.9)$$

cyanide thiosulfate thiocyanate sulfite

The toxicity of heavy metals such as mercury and lead is due to their ability to render the protein part of enzymes ineffective. These metals act by combining with the —SH groups found on many enzymes (see **Reaction 20.10**). In addition, as we learned in **Chapter 19**, metals can cause nonspecific protein denaturation. Both mercury and lead poisoning can cause permanent neurological damage.

$$\text{active enzyme} \;+\; Hg^{2+} \longrightarrow \text{inactive enzyme} \;+\; 2H^+ \qquad (20.10)$$

active enzyme inactive enzyme

Heavy-metal poisoning is treated by administering chelating agents—substances that combine with the metal ions and hold them very tightly. One effective antidote is the chelating agent ethylenediaminetetraacetic acid, or EDTA, which will chelate all heavy metals except mercury.

The calcium salt of EDTA is administered intravenously. In the body, calcium ions of the salt are displaced by heavy-metal ions, such as lead, that bind to the chelate more tightly. The lead–EDTA complex is soluble in body fluids and is excreted in the urine.

$$CaEDTA^{2-} + Pb^{2+} \rightarrow PbEDTA^{2-} + Ca^{2+} \qquad (20.11)$$

Not all enzyme inhibitors act as poisons toward the body; some, in fact, are useful therapeutic agents. Sulfa drugs and the group of compounds known as penicillins are

Enzyme Discovery Heats Up

Enzymes provide the same advantages to a commercial process as they do to a living organism. They are highly specific and extremely fast-acting catalysts. The use of enzymes in commercial industrial processes is developing rapidly and has a potentially bright future because of the existence of enzymes in microorganisms that thrive in extreme environments. Such enzymes have been called **extremozymes** because of the extreme conditions under which they function. These conditions include temperatures greater than 100°C in hot springs and deep-sea thermal vents, or below 0°C in Antarctic waters. Some enzymes function under immense pressure on the ocean floor, whereas others are unaffected by extremes of pH and salt concentrations in solution.

One goal of current research is to obtain useful enzymes from extremophiles, the microbes that function in these extreme environments. The process used by one company to obtain commercially useful enzymes consists of the following steps:

1. Biomass samples are obtained from the extreme environment.
2. DNA is extracted from the biomass and purified.
3. Fragments of the purified DNA are cloned to produce large libraries of genes.
4. The cloned genes are screened to find any that are coded to produce enzymes.
5. The DNA sequence is determined for any genes that are coded to produce enzymes.
6. The DNA sequence for enzyme-coded genes is cloned to produce large amounts of the gene, which in turn is inserted into organisms and used to produce large amounts of the enzyme.
7. The enzyme is optimized by random DNA mutations.

The last step is particularly interesting. An organism's DNA is randomly mutated over several generations. Offspring from each generation are selected that produce a specific enzyme with activity greater than or different from that of the enzyme from the previous generation. This technique, known as *mutagenesis*, is used to alter enzymes so they will perform new tasks or do old ones better.

Because extremozymes function under conditions that are so far removed from normal, research into their behavior is intense.

One of the research goals is to develop ways to use enzyme catalysis in industrial processes that require extreme temperatures or pH. One example is the current use of heat-stable DNA polymerase in the DNA replication technique called the polymerase chain reaction (see **Section 21.3**). Another research goal is to develop enzyme-catalyzed industrial processes that don't generate large amounts of undesirable waste material. For example, an esterase that catalyzes the hydrolysis of the antibiotic intermediate *p*-nitrobenzyl ester was developed. That hydrolysis reaction was usually done using catalytic zinc and organic solvents in a process that generated a lot of industrial waste. Some other applications being considered are the use of hydrolases at high temperature in the food-processing industry and the use of heat-stable enzymes in the production of corn syrup.

Science Source

Extremozymes enable microorganisms to survive in the harsh environment of deep-sea thermal vents.

extremozyme A nickname for certain enzymes isolated from microorganisms that thrive in extreme environments.

antibiotic A substance produced by one microorganism that kills or inhibits the growth of other microorganisms.

two well-known families of **antibiotics** that inhibit specific enzymes essential to the life processes of bacteria. The first penicillin was discovered in 1928 by Alexander Fleming. He noticed that the bacteria on a culture plate did not grow in the vicinity of a mold that had contaminated the culture; the mold was producing penicillin. The general penicillin structure is shown in **Table 20.5**, along with four widely used forms. Penicillins interfere with transpeptidase, an enzyme that is important in bacterial cell wall construction. The inability to form strong cell walls prevents the bacteria from surviving.

TABLE 20.5 Four Widely Used Penicillins

$$R-\overset{\overset{\text{O}}{\|}}{C}-NH \quad\quad S \quad CH_3$$

General penicillin structure

Name	Year Marketed	Side Chain (R—)	Comments
Penicillin G	1943	⬡—CH₂—	Usually given by injection
Penicillin V	1953	⬡—O—CH₂—	An oral penicillin; resistant to stomach hydrolysis
Methicillin	1960	⬡ (OCH₃ top, OCH₃ bottom)	Given by injection
Ampicillin	1961	⬡—CH— (NH₂)	Given by injection or taken orally; effective against a broad spectrum of organisms

20.7B Reversible Inhibition

A reversible inhibitor (in contrast to one that is irreversible) reversibly binds to an enzyme. An equilibrium is established; therefore, the inhibitor can be removed from the enzyme by shifting the equilibrium:

$$E + I \rightleftarrows EI$$

There are two types of reversible inhibitors: competitive and noncompetitive.

A **competitive inhibitor** binds to the active site of an enzyme and thus "competes" with substrate molecules for the active site. Competitive inhibitors often have molecular structures that are similar to the normal substrate of the enzyme. The competitive inhibition of succinate dehydrogenase by malonate is a classic example. Succinate dehydrogenase catalyzes the oxidation of the substrate succinate to form fumarate by transferring two hydrogens to the coenzyme FAD:

competitive inhibitor An inhibitor that competes with substrate for binding at the active site of the enzyme.

$$
\begin{array}{c}
\text{COO}^- \\
| \\
\text{CH}_2 \\
| \\
\text{CH}_2 \\
| \\
\text{COO}^-
\end{array}
\; + \; \text{FAD}
\;\xrightarrow[\text{dehydrogenase}]{\text{succinate}}\;
\begin{array}{c}
\text{COO}^- \\
| \\
\text{CH} \\
\| \\
\text{HC} \\
| \\
\text{COO}^-
\end{array}
\; + \; \text{FADH}_2
\quad\quad (20.12)
$$

succinate fumarate

No Milk Please

Lactose intolerance is a condition that causes a person to be unable to digest lactose, the natural sugar found in milk and dairy products. Thirty million Americans live with this condition today. Lactose is a disaccharide like sucrose and maltose. Each disaccharide is made up of two monosaccharides. For example, sucrose is made of glucose and fructose, maltose is made of two units of glucose, and lactose is made of glucose and galactose. In order to digest the different disaccharides, they must first be broken down into their constituent monosaccharides.

Lactase is a catalytic enzyme in our bodies that catalyzes the breakdown of the disaccharide lactose into the monosaccharides glucose and galactose. When people are lactose intolerant, they do not produce enough lactase enzyme to catalyze the total breakdown of all the lactose that has been ingested. As a result, some lactose passes through the small intestine and enters the large intestine where it creates digestive problems. In order to overcome the inability of the body to create enough lactase enzymes, lactose intolerant people can take lactase supplements before they eat or drink milk or dairy products. Some milk and dairy products contain the enzyme as an additive. The milk product called "Lactaid" is milk that contains added lactase enzyme to help those who are lactose intolerant.

It is also common for people with lactose intolerance to be able to eat yogurt that contains live cultures without experiencing the unpleasant symptoms that other milk and dairy products cause. This is because yogurt contains probiotics or live bacteria that provide health benefits. The live bacteria in yogurt have the ability to produce the lactase enzyme that assists in breaking down the lactose in the yogurt. Probiotics can also be taken in pill form. When the bacteria in the pill reach the small intestine, they begin to reproduce. This provides continuing benefits from the helpful bacteria.

When lactose intolerant people drink milk without adding lactase, the lactose disaccharide eventually makes its way into the large intestine and begins to cause pain and other discomfort. The bacteria in the large intestine begin to break down the lactose. When they do this, they also release gas, causing uncomfortable bloating and cramps. When lactose enters the large intestine it osmotically draws in water and causes diarrhea. The lactose also undergoes fermentation in the large intestine, and produces lactic acid, which can be detected in the stool. That is one way to test for lactose intolerance in a person. Lactose intolerance is often passed on genetically, and unfortunately, there is no known cure for the condition.

Drinking milk can be painful for individuals who are lactose intolerant.

Malonate, having a structure similar to succinate, competes for the active site of succinate dehydrogenase and thus inhibits the enzyme:

$$
\begin{array}{cc}
\text{COO}^- & \text{COO}^- \\
| & | \\
\text{CH}_2 & \text{CH}_2 \\
| & | \\
\text{CH}_2 & \text{COO}^- \\
| & \\
\text{COO}^- & \\
\text{succinate} & \text{malonate}
\end{array}
$$

The action of sulfa drugs on bacteria is another example of competitive enzyme inhibition. Folic acid, a substance needed for growth by some disease-causing bacteria, is normally synthesized within the bacteria by a chemical process that requires p-aminobenzoic acid.

Because sulfanilamide, the first sulfa drug, resembles *p*-aminobenzoic acid and competes with it for the active site of the bacterial enzyme involved, it can prevent bacterial growth (see **Figure 20.10**). This is possible because the enzyme binds readily to either of these molecules. Therefore, the introduction of large quantities of sulfanilamide into a patient's body causes most of the active sites to be bound to the wrong (from the bacterial viewpoint) substrate. Thus, the synthesis of folic acid is stopped or slowed, and the bacteria are prevented from multiplying. Meanwhile, the normal body defenses work to destroy them.

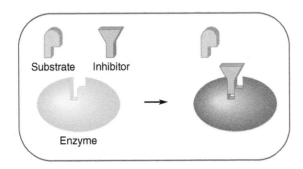

Figure 20.10 Structural relationships of sulfanilamide, *p*-aminobenzoic acid, and folic acid.

Human beings also require folic acid, but they get it from their diet. Consequently, sulfa drugs exert no toxic effect on humans. However, one danger of sulfa drugs, or of any antibiotic, is that excessive amounts can destroy important intestinal bacteria that perform many symbiotic, life-sustaining functions. Although sulfa drugs have largely been replaced by antibiotics such as the penicillins and tetracyclines, some are still used in treating certain bacterial infections, such as those occurring in the urinary tract.

The nature of competitive inhibition is represented in **Figure 20.11**. There is competition between the substrate and the inhibitor for the active site. Once the inhibitor combines with the enzyme, the active site is blocked, preventing further catalytic action.

Figure 20.11 The behavior of competitive inhibitors. The active site is occupied by the inhibitor, and the substrate is prevented from bonding.

Competitive inhibition can be reversed by increasing the concentration of substrate and letting Le Châtelier's principle operate. This is illustrated by the following equilibria that would exist in a solution containing enzyme (E), substrate (S), and competitive inhibitor (I):

$$\text{equilibrium 1:} \quad \text{E} + \text{S} \xrightleftharpoons{\text{Increasing [S]}} \text{ES}$$

$$\text{equilibrium 2:} \quad \text{E} + \text{I} \underset{\text{Decreasing [E]}}{\rightleftharpoons} \text{EI}$$

If the concentration of substrate is increased, then according to Le Châtelier's principle, equilibrium 1 should shift to the right. As it does, the concentration of enzyme decreases, which causes equilibrium 2 to shift to the left. The overall effect is that the concentration of ES has increased while the concentration of EI has decreased. Thus, the competition between inhibitor and substrate is won by whichever molecular species is in greater concentration.

A **noncompetitive inhibitor** bears no resemblance to the normal enzyme substrate and binds reversibly to the surface of an enzyme at a site other than the catalytically active site (see **Figure 20.12**). The interaction between the enzyme and the noncompetitive inhibitor causes the three-dimensional shape of the enzyme and its active site to change. The enzyme no longer binds the normal substrate, or the substrate is improperly bound in a way that prevents the catalytic groups of the active site from participating in catalyzing the reaction.

noncompetitive inhibitor An inhibitor that binds to the enzyme at a location other than the active site.

Figure 20.12 The behavior of noncompetitive inhibitors. The inhibitor bonds to the enzyme at a site other than the active site and changes the shape of the active site.

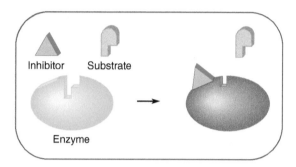

Unlike competitive inhibition, noncompetitive inhibition cannot be reversed by the addition of more substrate because additional substrate has no effect on the enzyme-bound inhibitor (it can't displace the inhibitor because it can't bond to the site occupied by the inhibitor).

> ✔ **LEARNING CHECK 20.4** Compare a noncompetitive and a competitive inhibitor with regard to the following:
>
> a. Resemblance to substrate
> b. Binding site on the enzyme
> c. The effect of increasing substrate concentration

20.8 | The Regulation of Enzyme Activity

Learning Objective

8. Describe the three methods of cellular control over enzyme activity.

Enzymes work together in an organized yet complicated way to facilitate all the biochemical reactions needed by a living organism. For an organism to respond to changing conditions and cellular needs, very sensitive controls over enzyme activity are required. We will discuss three mechanisms by which this is accomplished: activation of zymogens, allosteric regulation, and genetic control.

20.8A The Activation of Zymogens

zymogen or proenzyme The inactive precursor of an enzyme.

A common mechanism for regulating enzyme activity is the synthesis of enzymes in the form of inactive precursors called **zymogens** or **proenzymes.** Some enzymes in their active form would degrade the internal structures of the cells and, ultimately, the cells themselves. Such enzymes are synthesized and stored as zymogens. When the active enzyme is needed, the stored zymogen is released from storage and activated at the location of

the reaction. Activation usually involves the cleavage of one or more peptide bonds of the zymogen. The digestive enzymes pepsin, trypsin, and chymotrypsin and several of the enzymes involved in blood clotting are produced and activated this way. Trypsinogen, the inactive form of trypsin, is synthesized in the pancreas and activated in the small intestine by the hydrolysis of a single peptide bond. The products are the active enzyme (trypsin) and a hexapeptide:

$$\text{trypsinogen} + H_2O \xrightarrow{\text{enteropeptidase}} \text{trypsin} + \text{hexapeptide} \qquad (20.13)$$

Several examples of zymogens are listed in **Table 20.6**.

TABLE 20.6 Examples of Zymogens

Zymogen	Active Enzyme	Function
chymotrypsinogen	chymotrypsin	Digestion of proteins
pepsinogen	pepsin	Digestion of proteins
procarboxypeptidase	carboxypeptidase	Digestion of proteins
proelastase	elastase	Digestion of proteins
prothrombin	thrombin	Blood clotting
trypsinogen	trypsin	Digestion of proteins

20.8B Allosteric Regulation

A second method of enzyme regulation involves the combination of the enzyme with some other compound such that the three-dimensional conformation of the enzyme is altered and its catalytic activity is changed. Compounds that alter enzymes this way are called **modulators** of the activity; they may increase the activity (activators) or decrease the activity (inhibitors). The noncompetitive inhibitors we studied in Section 20.7 are examples of modulators. Enzymes that have a quaternary protein structure with distinctive binding sites for modulators are referred to as **allosteric enzymes**. These variable-rate enzymes are often located at key control points in cellular processes. In general, **activators** act as signals for increased production, whereas inhibitors are used to stop the formation of excess products.

An excellent example of allosteric regulation—the control of an allosteric enzyme—is the five-step synthesis of the amino acid isoleucine (see Figure 20.13). Threonine deaminase, the enzyme that catalyzes the first step in the conversion of threonine to isoleucine, is subject to inhibition by the final product, isoleucine. The structures of isoleucine and threonine are quite different, so isoleucine is not a competitive inhibitor.

modulator A substance that binds to an enzyme at a location other than the active site and alters the catalytic activity.

allosteric enzyme An enzyme with a quaternary structure whose activity is changed by the binding of modulators.

activator A substance that binds to an allosteric enzyme and increases its activity.

Figure 20.13 The allosteric regulation of threonine deaminase by isoleucine, an example of feedback inhibition.

Also, the site to which isoleucine binds to the enzyme is different from the enzyme active site that binds to threonine. This second site, called the allosteric site, specifically recognizes isoleucine, whose presence there induces a change in the conformation of the enzyme such that threonine binds poorly to the active site. Thus, isoleucine exerts an inhibiting effect on the enzyme activity. As a result, the reaction slows as the concentration of isoleucine increases, and no excess isoleucine is produced. When the concentration of isoleucine falls to a low enough level, the enzyme becomes more active, and more isoleucine is synthesized. This type of allosteric regulation in which the enzyme that catalyzes the first step of a series of reactions is inhibited by the final product is called **feedback inhibition.** The control of enzyme activity by allosteric regulation of key enzymes is of immense benefit to an organism because it allows the concentration of cellular products to be maintained within very narrow limits.

feedback inhibition A process in which the end product of a sequence of enzyme-catalyzed reactions inhibits an earlier step in the process.

20.8C Genetic Control

One way to increase production from an enzyme-catalyzed reaction, given a sufficient supply of substrate, is for a cell to increase the number of enzyme molecules present. The synthesis of all proteins, including enzymes, is under genetic control by nucleic acids (see **Chapter 21**). An example of the genetic control of enzyme activity involves **enzyme induction**, the synthesis of enzymes in response to a temporary need of the cell.

enzyme induction The synthesis of an enzyme in response to a cellular need.

STUDY SKILLS 20.1

A Summary Chart of Enzyme Inhibitors

At this point in your study of enzymes, a summary chart is a useful tool for representing the various types of enzyme inhibitors in a compact, easy-to-review way.

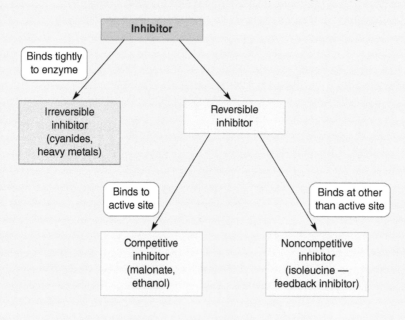

The first demonstrated example of enzyme induction came from studies conducted in the 1950s on the bacterium *Escherichia coli.* β-galactosidase, an enzyme required for the utilization of lactose, catalyzes the hydrolysis of lactose to D-galactose and D-glucose.

β-lactose β-galactoside bond β-D-galactose β-D-glucose

In the absence of lactose in the growth medium, there were fewer than 10 β-galactosidase molecules per cell of *E. coli*. However, if *E. coli* are added to a lactose-containing medium, within minutes the bacterium begins to produce thousands of molecules of β-galactosidase. This is an example of enzyme induction. If lactose is removed from the growth medium, the production of β-galactosidase again falls back to a low level.

The biological significance of the genetic control of enzyme activity is that it allows an organism to adapt to environmental changes. The coupling of genetic control and allosteric regulation of enzyme activity puts the cellular processes in the body under constant control.

✔ **LEARNING CHECK 20.5** Explain how each of the following processes is involved in the regulation of enzyme activity:

a. Activation of zymogens
b. Allosteric regulation
c. Genetic control

20.9 | Medical Application of Enzymes

Learning Objective

9. Discuss the importance of measuring enzyme levels in the diagnosis of disease.

20.9A Enzymes in Clinical Diagnosis

Certain enzymes are normally found almost exclusively inside tissue cells and are released into the blood and other biological fluids only when these cells are damaged or destroyed. Because of the normal breakdown and replacement of tissue cells that go on constantly, the blood serum contains these enzymes but at very low concentrations—often a million times lower than the concentration inside cells. However, blood serum levels of cellular enzymes increase significantly when excessive cell injury or destruction occurs, or when cells grow rapidly as a result of cancer. Changes in blood serum concentrations of specific enzymes can be used clinically to detect cell damage or uncontrolled growth, and even to suggest the site of the damage or cancer. Also, the extent of cell damage can often be estimated by the magnitude of the serum concentration increase above normal levels. For these reasons, the measurement of enzyme concentrations in blood serum and other biological fluids has become a major diagnostic tool, particularly in diagnosing diseases of the heart, liver, pancreas, prostate, and bones. In fact, certain enzyme determinations are performed so often that they have become routine procedures in the clinical chemistry laboratory. **Table 20.7** lists some enzymes used in medical diagnosis.

20.9B Isoenzymes

Lactate dehydrogenase (LDH) and creatine kinase (CK), two of the enzymes listed in **Table 20.7**, are particularly useful in clinical diagnosis because each occurs in multiple forms called **isoenzymes**. Although all forms of a particular isoenzyme catalyze the same reaction, their molecular structures are slightly different and their locations within the body tissues may vary.

isoenzyme A slightly different form of the same enzyme produced by different tissues.

TABLE 20.7 Diagnostically Useful Assays of Blood Serum Enzymes

Enzyme	Pathological Condition
acid phosphatase	Prostate cancer
alanine transaminase (ALT)	Hepatitis
alkaline phosphatase (ALP)	Liver or bone disease
amylase	Diseases of the pancreas
aspartate transaminase (AST)	Heart attack or hepatitis
creatine kinase (CK)	Heart attack
lactate dehydrogenase (LDH)	Heart attack, liver damage
lipase	Acute pancreatitis
lysozyme	Monocytic leukemia

Probably the most fully studied group of isoenzymes is that of lactate dehydrogenase (LDH). The quaternary structure of the enzyme consists of four subunits that are of two different types. One is labeled H because it is the predominant subunit present in the LDH enzyme found in heart muscle cells. The other, labeled M, predominates in other muscle cells. There are five possible ways to combine these four subunits to form the enzyme (see **Figure 20.14**). Each combination has slightly different properties, which allows them to be separated and identified by electrophoresis.

Figure 20.14 Isomeric forms (isoenzymes) of lactate dehydrogenase.

Table 20.8 shows the distribution of the five LDH isoenzymes in several human tissues. Notice that each type of tissue has a distinct pattern of isoenzyme percentages. Liver and skeletal muscle, for example, contain particularly high percentages of LDH_5. Also notice the differences in LDH_1 and LDH_2 levels in heart and skeletal muscles.

TABLE 20.8 Tissue Distribution of LDH Isoenzymes

Tissue	LDH_1 (%)	LDH_2 (%)	LDH_3 (%)	LDH_4 (%)	LDH_5 (%)
Brain	23	34	30	10	3
Heart	50	36	9	3	2
Kidney	28	34	21	11	6
Liver	4	6	17	16	57
Lung	10	20	30	25	15
Serum	28	41	19	7	5
Skeletal muscle	5	5	10	22	58

Because of the difference in tissue distribution of LDH isoenzymes, serum levels of LDH are used in the diagnosis of a wide range of diseases: anemias involving the rupture of red blood cells, acute liver diseases, congestive heart failure, and muscular diseases such as muscular dystrophy. This wide range of diseases makes an LDH assay a good initial diagnostic test. Elevated levels of LDH_1 and LDH_2, for example, indicate myocardial infarction, whereas elevated LDH_5 indicates possible liver damage. If LDH activity is elevated, then this and other isoenzyme assays can be used to pinpoint the location and type of disease more accurately.

▶ Case Study Follow-up

Middle-aged men are the stereotypical models of heart disease, but many younger men and women experience heart attack. Men typically report a feeling of pressure and crushing chest pain just prior to a heart attack. Women most frequently experience unusual fatigue and sleep disturbance, as well as shortness of breath and anxiety, preceding a heart attack. Women as well as men might experience jaw, arm, back, or chest pain; sweating; and nausea. Women typically postpone seeking treatment for up to two weeks after experiencing symptoms. Cardiac events are frequently misdiagnosed in women and younger men.

Diet, genetic inheritance, or both may cause hypercholesterolemia (high cholesterol). Familial hypercholesterolemia is caused by mutations in several genes. These mutations cause a decrease in the number of low-density lipoprotein receptors (LDLR), the main carrier of cholesterol in the blood. The condition is more prevalent in Afrikaner, Finnish, French Canadian, and Lebanese populations. However, the majority of hypercholesterolemia is caused by lifestyle choices—diet, exercise, and smoking.

Concept Summary

General Characteristics of Enzymes Enzymes are highly efficient protein catalysts involved in almost every biological reaction. They are often quite specific in terms of the substance acted on and the type of reaction catalyzed.
Objective 1 (Section 20.1), Exercise 20.2

Enzyme Nomenclature and Classification Enzymes are grouped into six major classes on the basis of the type of reaction catalyzed. Common names for enzymes often end in -ase and are based on the substrate and/or the type of reaction catalyzed.
Objective 2 (Section 20.2), Exercise 20.12

Enzyme Cofactors Cofactors are nonprotein molecules required for an enzyme to be active. Cofactors are either organic (coenzymes) or inorganic ions.
Objective 3 (Section 20.3), Exercise 20.14

The Mechanism of Enzyme Action The behavior of enzymes is explained by a theory in which the formation of an enzyme–substrate complex is assumed to occur. The specificity of enzymes is explained by the lock-and-key theory and the induced-fit theory.
Objective 4 (Section 20.4), Exercise 20.20

Enzyme Activity The catalytic ability of enzymes is described by turnover number and enzyme international units. Experiments that measure enzyme activity are referred to as enzyme assays.
Objective 5 (Section 20.5), Exercise 20.26

Factors Affecting Enzyme Activity The catalytic activity of enzymes is influenced by numerous factors. The most important are substrate concentration, enzyme concentration, temperature, and pH.
Objective 6 (Section 20.6), Exercise 20.28

Enzyme Inhibition Chemical substances called inhibitors decrease the rates of enzyme-catalyzed reactions. Irreversible inhibitors render enzymes permanently inactive and include several very toxic substances such as the cyanide ion and heavy-metal ions. Reversible inhibitors are of two types: competitive and noncompetitive.
Objective 7 (Section 20.7), Exercise 20.34

The Regulation of Enzyme Activity Three mechanisms of cellular control over enzyme activity exist. One method involves the synthesis of enzyme precursors called zymogens, which are activated when needed by the cell. The second mechanism relies on the binding of small molecules (modulators), which increase or decrease enzyme activity. Genetic control of enzyme synthesis, the third method, regulates the amount of enzyme available.
Objective 8 (Section 20.8), Exercise 20.38

Medical Application of Enzymes Numerous enzymes have become useful as aids in diagnostic medicine. The presence of specific enzymes in body fluids such as blood has been related to certain pathological conditions.
Objective 9 (Section 20.9), Exercise 20.46

Key Terms and Concepts

Absolute specificity (20.1)	Enzyme activity (20.5)	Noncompetitive inhibitor (20.7)
Activator (20.8)	Enzyme induction (20.8)	Optimum pH (20.6)
Active site (20.4)	Enzyme inhibitor (20.7)	Optimum temperature (20.6)
Allosteric enzyme (20.8)	Enzyme international unit (IU) (20.5)	Relative specificity (20.1)
Antibiotic (20.7)	Extremozyme (20.7)	Stereochemical specificity (20.1)
Apoenzyme (20.3)	Feedback inhibition (20.8)	Substrate (20.2)
Coenzyme (20.3)	Induced-fit theory (20.4)	Turnover number (20.5)
Cofactor (20.3)	Isoenzyme (20.9)	Zymogen or proenzyme (20.8)
Competitive inhibitor (20.7)	Lock-and-key theory (20.4)	
Enzyme (20.1)	Modulator (20.8)	

Key Reactions

1.	Formation of an active enzyme (Section 20.3):	apoenzyme + cofactor (coenzyme or inorganic ion) \rightarrow active enzyme	Reaction 20.4
2.	Mechanism of enzyme action (Section 20.4):	$E + S \rightleftarrows ES \rightarrow E + P$	Reaction 20.6

Exercises

Even-numbered exercises are answered in Appendix B.

Blue-numbered exercises are more challenging.

General Characteristics of Enzymes (Section 20.1)

20.1 What is the role of enzymes in the body?

20.2 List two ways that enzyme catalysis of a reaction is superior to normal laboratory conditions.

20.3 What is the relationship between an enzyme and the energy of activation for a reaction?

20.4 Why are so many different enzymes needed?

20.5 Define what is meant by the term *enzyme specificity*.

20.6 Define what is meant by the term *absolute specificity*.

20.7 List three ways in which enzymes are particularly well-suited for their essential roles in living organisms.

Enzyme Nomenclature and Classification (Section 20.2)

20.8 What name is given to the nomenclature system for enzymes?

20.9 What is the relationship between urea and urease? Between maltose and maltase?

20.10 Match the following enzymes and substrates:

Enzyme	Substrate
a. sucrase	maltose
b. amylase	amylose
c. lactase	sucrose
d. maltase	arginine
e. arginase	lactose

20.11 Match the following general enzyme names and reactions catalyzed:

Enzyme	Reaction Catalyzed
a. decarboxylase	formation of ester linkages
b. phosphatase	removal of carboxyl groups from compounds
c. peptidase	hydrolysis of peptide linkages
d. esterase	hydrolysis of phosphate ester linkages

20.12 Because one substrate may undergo a number of reactions, it is often convenient to use an enzyme nomenclature system that includes both the substrate name (or general type) and the type of reaction catalyzed. Identify the substrate and type of reaction for the following enzyme names:

a. succinate dehydrogenase

b. L-amino acid reductase

c. cytochrome oxidase

d. glucose-6-phosphate isomerase

Enzyme Cofactors (Section 20.3)

20.13 Some enzymes consist of protein plus another component. Which of the terms *cofactor* or *coenzyme* correctly describes each of the following nonprotein components?

a. an inorganic ion

b. a nonspecific component

c. an organic material

d. nicotinic acid,

$$\underset{\substack{\\ \text{N}}}{\overset{\substack{\text{O} \\ \| \\ \text{C} - \text{OH}}}{\bigcirc}}$$

20.14 What are the relationships among the terms *cofactor*, *active enzyme*, and *apoenzyme*?

20.15 State the relationship between vitamins and enzyme activity.

20.16 List some typical inorganic ions that serve as cofactors.

20.17 What vitamins are required for FAD and NAD$^+$ formation?

The Mechanism of Enzyme Action (Section 20.4)

20.18 Explain what is meant by the following equation:
$$\text{E} + \text{S} \rightleftarrows \text{ES} \rightarrow \text{E} + \text{P}$$

20.19 In what way are the substrate and active site of an enzyme related?

20.20 How is enzyme specificity explained by the lock-and-key theory?

20.21 Compare the lock-and-key theory with the induced-fit theory.

20.22 An enzyme can catalyze reactions involving propanoic acid, butanoic acid, and pentanoic acid. Would the lock-and-key theory or the induced-fit theory best explain this enzyme's mechanism of action? Why?

Enzyme Activity (Section 20.5)

20.23 List two ways of describing enzyme activity.

20.24 What observations may be used in experiments to determine enzyme activity?

20.25 Define the term *turnover number.*

20.26 What is an enzyme international unit? Why is the international unit a useful method of expressing enzyme activity in medical diagnoses?

Factors Affecting Enzyme Activity (Section 20.6)

20.27 Use graphs to illustrate enzyme activity as a function of the following:

 a. substrate concentration

 b. enzyme concentration

 c. pH

 d. temperature

20.28 Write a single sentence to summarize the information of each graph in Exercise 20.27.

20.29 What happens to the rate of an enzyme-catalyzed reaction as substrate concentration is raised beyond the saturation point?

20.30 How might V_{max} for an enzyme be determined?

20.31 How would you expect hypothermia to affect enzyme activity in the body?

20.32 When handling or storing solutions of enzymes, the pH is usually kept near 7.0. Explain why.

Enzyme Inhibition (Section 20.7)

20.33 Distinguish between irreversible and reversible enzyme inhibition.

20.34 Distinguish between competitive and noncompetitive enzyme inhibition.

20.35 How does each of the following irreversible inhibitors interact with enzymes?

 a. cyanide

 b. heavy-metal ions

20.36 List an antidote for each of the two poisons in Exercise 20.35 and describe how each functions.

The Regulation of Enzyme Activity (Section 20.8)

20.37 Why is the regulation of enzyme activity necessary?

20.38 List three mechanisms for the control of enzyme activity.

20.39 Describe the importance of zymogens in the body. Give an example of an enzyme that has a zymogen.

20.40 The clotting of blood occurs by a series of zymogen activations. Why are the enzymes that catalyze blood clotting produced as zymogens?

20.41 What type of enzyme possesses a binding site for modulators?

20.42 Name and contrast the two types of modulators.

20.43 Explain how feedback enzyme inhibition works and why it is advantageous for the cell.

20.44 What is meant by genetic control of enzyme activity and enzyme induction?

Medical Application of Enzymes (Section 20.9)

20.45 Why are enzyme assays of blood serum useful in a clinical diagnosis?

20.46 How is each of the following enzyme assays useful in diagnostic medicine?

 a. CK

 b. ALP

 c. amylase

20.47 What are isoenzymes? List two examples of isoenzymes.

20.48 Why is an LDH assay a good initial diagnostic test?

20.49 What specific enzyme assays are particularly useful in diagnosing liver damage or disease?

Additional Exercises

20.50 Carboxypeptidase is a proteolytic enzyme that cleaves the C-terminal amino acid residue from a polypeptide. The amino acid arginine is found in the active site of carboxypeptidase and is responsible for holding the C-terminal end of the polypeptide in place so the cleavage of the peptide bond can occur. What type of interaction occurs between the C-terminal end of the polypeptide and the arginine side chain?

20.51 An enzyme preparation that decarboxylates butanoic acid has an activity of 50 IU. How many grams of butanoic acid

will this enzyme preparation decarboxylate in 10 minutes? Assume an unlimited supply of substrate.

20.52 Propanoic acid chloride was found to react with the side chain of a serine residue on an enzyme to cause irreversible inhibition of that enzyme. Suggest a mechanism of how this occurs.

$$CH_3CH_2\overset{\overset{\displaystyle O}{\|}}{C}-Cl$$

20.53 1.0×10^{12} enzyme molecules were added to 1.0 L of substrate solution with an initial pH of 7.0. After 1.0 minute, the solution had a pH of 6.0. Assume an unlimited supply of substrate and any change in hydrogen ion concentration was due to the following reaction.

$$\text{Substrate} \xrightarrow{\text{enzyme}} \text{Product} + H^+$$

 a. What was the change in H^+ concentration?

 b. How many molecules of substrate were reacted to cause this change in H^+ concentration?

 c. What is the turnover number of this enzyme?

20.54 The following reaction has come to equilibrium and it is found that there are 100 times as many molecules of A as there are of B in the solution.

$$A \rightleftharpoons B$$

An enzyme that catalyzes this reaction is then added to the solution. In which direction will the equilibrium shift?

Chemistry for Thought

20.55 Explain how the pasteurization of milk utilizes one of the factors that influence enzyme activity.

20.56 Explain how food preservation by freezing utilizes one of the factors that influence enzyme activity.

20.57 Some pottery glazes contain lead compounds. Why would it be unwise to use such pottery to store or serve food?

20.58 Urease can catalyze the hydrolysis of urea, but not the hydrolysis of methyl urea. Why?

20.59 The active ingredient in meat tenderizer is the enzyme papain, a protease. Explain how treating meat with this material before cooking would make it more tender.

20.60 Why are enzymes that are used for laboratory or clinical work stored in refrigerators?

20.61 Describe the differences between graphs showing temperature versus reaction rate for an enzyme-catalyzed reaction and an uncatalyzed reaction.

20.62 Answer the question associated with Figure 20.3. How else might ammonia be detected?

20.63 In Figure 20.9, it is noted that pickles resist spoilage. Why is that true for this acidic food?

20.64 One of the steps in preparing frozen corn for the grocery market involves "blanching" (briefly placing the corn in boiling water). Why is blanching necessary?

20.65 What experiment would you conduct to determine if an inhibitor is competitive or noncompetitive?

20.66 Why might an enzyme be selected over an inorganic catalyst for a certain industrial process?

20.67 Enzymes are enormous molecules, and yet the active site is relatively small. Why is a huge molecule necessary for enzymes?

Allied Health Exam Connection

The following questions are from these sources:

• *Nursing School Entrance Exam* © 2005, Learning Express, LLC.

• *McGraw-Hill's Nursing School Entrance Exams* by Thomas A. Evangelist, Tamara B. Orr, and Judy Unrein © 2009, The McGraw-Hill Companies, Inc.

• *NSEE Nursing School Entrance Exams,* 3rd edition © 2009, Kaplan Publishing.

• *Cliffs Test Prep: Nursing School Entrance Exams* by Fred N. Grayson © 2004, Wiley Publishing, Inc.

• *Peterson's Master the Nursing School and Allied Health Entrance Exams,* 18th edition by Marion F. Gooding © 2008, Peterson's, a Nelnet Company.

20.68 Which of the following is not a characteristic of enzymes?

 a. They are macromolecules.

 b. They act on substances.

 c. They are phospholipids.

 d. They initiate and decelerate chemical reactions.

 e. They act as catalysts.

20.69 The human body has an average pH of about 7 and a temperature of about $37°C$. Use graphs to illustrate enzyme activity in the human body as a function of the following:

 a. Substrate concentration

 b. Enzyme concentration

 c. pH (include pH optimum value)

 d. Temperature (include temperature optimum value)

20.70 Look at Table 20.4 and identify the proteolytic enzyme that begins protein digestion in the stomach. Does this name use an EC name ending?

20.71 Saliva contains mucus, water, and _____, which partially digests polysaccharides. Fill in the blank with the correct response from below:

 a. lipase

 b. amylase

 c. trypsin

 d. insulin

20.72 Most human enzymes function best in the temperature range of:

a. 0°C–10°C

b. 10°C–25°C

c. 25°C–40°C

d. 40°C–55°C

20.73 The site on an enzyme molecule that does the catalytic work is called the:

a. binding site.

b. allosteric site.

c. lock.

d. active site.

20.74 The process by which an enzyme acts on the substrate can be described by the:

a. lock-and-key model.

b. enzyme-and-substrate model.

c. enzyme folding model.

d. catalytic model.

20.75 In the multienzyme sequence shown below, molecules of D are able to fit to the enzyme E_1 and prevent the conversion of A to B. What is this action of E called?

$$A \xrightarrow{E_1} B \xrightarrow{E_2} C \xrightarrow{E_3} D$$

a. effector inhibition

b. allosteric inhibition

c. feedback inhibition

d. competitive inhibition by nonproduct

21

Nucleic Acids and Protein Synthesis

esolla/Shutterstock.com

▶ Case Study

"Thin" and "sickly" described Paulina, yet she managed to lead a full life. She played the drums in a band and taught the second grade at an elementary school. Friends and co-workers worried about her health. They wondered if she suffered from anorexia. Paulina explained that certain foods upset her digestive system. She also struggled with frequent respiratory infections. Paulina thought part of her health problems resulted from her exposure to many children at work.

Then, her doctor diagnosed Paulina with lupus. The diagnosis seemed largely correct, but not quite right according to what she read on the Internet. Paulina changed doctors when she contracted pneumonia for the third time that year. Her new doctor, acting on a hunch, performed a skin test and confirmed that Paulina had cystic fibrosis (CF). Since CF is usually diagnosed in children, other doctors didn't think to test for it. Now, Paulina receives proper treatment. She has gained 15 pounds and feels better than ever.

▶ How does Paulina's case differ from a typical case of cystic fibrosis (CF)? How does a skin test diagnose CF? Why are most diagnoses made in childhood? What treatment is available?

Follow-up to this Case Study appears at the end of the chapter before the *Concept Summary*.

Learning Objectives

When you have completed your study of this chapter, you should be able to:

1 Identify the components of nucleotides and correctly classify the sugars and bases. **(Section 21.1)**

2 Describe the structure of DNA. **(Section 21.2)**

3 Outline the process of DNA replication. **(Section 21.3)**

4 Contrast the structures of DNA and RNA and list the function of the three types of cellular RNA. **(Section 21.4)**

5 Describe what is meant by the terms *transcription* and *translation*. **(Section 21.5)**

6 Describe the process by which RNA is synthesized in cells. **(Section 21.6)**

7 Explain how the genetic code functions in the flow of genetic information. **(Section 21.7)**

8 Outline the process by which proteins are synthesized in cells. **(Section 21.8)**

9 Describe how genetic mutations occur and how they influence organisms. **(Section 21.9)**

10 Describe the technology used to produce recombinant DNA. **(Section 21.10)**

O ne of the most remarkable properties of living cells is their ability to produce nearly exact replicas of themselves through hundreds of generations. Such a process requires that certain types of information be passed unchanged from one generation to the next. The transfer of necessary genetic information to new cells is accomplished by means of biomolecules called **nucleic acids**. The high molecular weight compounds represent coded information, much as words represent information in a book. It is the nearly infinite variety of possible structures that enables nucleic acids to represent the huge amount of information that must be transmitted sexually or asexually to reproduce a living organism. This information controls the inherited characteristics of the individuals in the new generation and determines the life processes as well (see **Figure 21.1**).

nucleic acid A biomolecule involved in the transfer of genetic information from existing cells to new cells.

Figure 21.1 Nucleic acids passed from the parents to offspring determine the inherited characteristics.

© Michael C. Slabaugh

21.1 | Components of Nucleic Acids

Learning Objective

1. Identify the components of nucleotides and correctly classify the sugars and bases.

ribonucleic acid (RNA) A nucleic acid found mainly in the cytoplasm of cells.

deoxyribonucleic acid (DNA) A nucleic acid found primarily in the nuclei of cells.

nucleotide The repeating structural unit or monomer of polymeric nucleic acids.

Nucleic acids are classified into two categories: **ribonucleic acid (RNA)**, found mainly in the cytoplasm of living cells, and **deoxyribonucleic acid (DNA)**, found primarily in the nuclei of cells. Both DNA and RNA are polymers, consisting of long, linear molecules. The repeating structural units, or monomers, of the nucleic acids are called **nucleotides**. Nucleotides, however, are composed of three simpler components: a heterocyclic base, a sugar, and a phosphate (see **Figure 21.2**). These components will be discussed individually.

Figure 21.2 The composition of nucleic acids.

```
                    ┌─────────────┐
                    │ Nucleic acid│
                    └─────────────┘
                           │
                      (Composed of)
                           │
                    ┌─────────────┐
                    │ Nucleotides │
                    └─────────────┘
                 ╱         │         ╲
           (Composed of)
         ╱             │               ╲
┌──────────────┐ ┌──────────────┐ ┌───────────┐
│ Pyrimidines  │ │  Ribose or   │ │ Phosphate │
│ and purines  │ │ 2-deoxyribose│ │           │
└──────────────┘ └──────────────┘ └───────────┘
```

Each of the five bases commonly found in nucleic acids are heterocyclic compounds that can be classified as either a pyrimidine or a purine, the parent compounds from which the bases are derived:

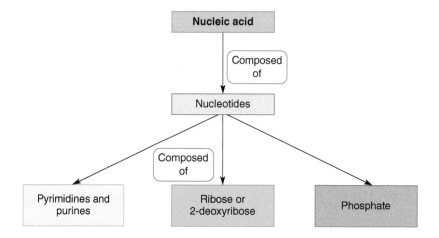

pyrimidine purine

The three pyrimidine bases are uracil, thymine, and cytosine, usually abbreviated U, T, and C. Adenine (A) and guanine (G) are the two purine bases. Adenine, guanine, and cytosine are found in both DNA and RNA, but uracil is ordinarily found only in RNA, and thymine only in DNA. Structural formulas of the five bases are given in **Figure 21.3**.

The sugar component of RNA is D-ribose, as the name ribonucleic acid implies. In deoxyribonucleic acid (DNA), the sugar is D-deoxyribose. Both sugars occur in the β-configuration.

D-ribose D-deoxyribose

β-configuration

no — OH group

Figure 21.3 The major bases of nucleic acids.

Phosphate, the third component of nucleotides, is derived from phosphoric acid (H_3PO_4), which under cellular pH conditions exists in ionic form:

The formation of a nucleotide from these three components is represented in **Reaction 21.1**:

$$+ 2H_2O \qquad (21.1)$$

In a nucleotide, the base is always attached to the 1′ position of the sugar, and the phosphate is generally located at the 5′ position. The carbon atoms in the sugar are designated with a number followed by a prime to distinguish them from the atoms in the bases.

Because the nucleotide of **Reaction 21.1** contains ribose as the sugar, the nucleotide must be one found in RNA. The general structure of a nucleotide can be represented as

$$
^-O-\overset{\overset{\displaystyle O}{\|}}{\underset{\underset{\displaystyle O^-}{|}}{P}}-O-CH_2 \quad \text{base}
$$

OH

— OH present in RNA

or

| phosphate | sugar |

base

21.2 | The Structure of DNA

Learning Objective

2. Describe the structure of DNA.

DNA molecules, among the largest molecules known, contain between 1 and 100 million nucleotide units. The nucleotides are joined together in nucleic acids by phosphate groups that connect the 5′ carbon of one nucleotide to the 3′ carbon of the next. These linkages are referred to as *phosphodiester bonds*. The result is a chain of alternating phosphate and sugar units to which the bases are attached. The sugar–phosphate chain is referred to as the **nucleic acid backbone**, and it is constant throughout the entire DNA molecule (see **Figure 21.4**). Thus, one DNA molecule differs from another only in the sequence, or order of attachment, of the bases along the backbone. Just as the amino acid sequence of a protein determines the primary structure of the protein, the order of the bases provides the primary structure of DNA.

nucleic acid backbone The sugar–phosphate chain that is common to all nucleic acids.

Figure 21.4 The nucleic acid backbone.

Figure 21.5 shows the structure of a tetranucleotide that represents a partial structural formula of a DNA molecule. The end of the polynucleotide segment of DNA in Figure 21.5 that has no nucleotide attached to the 5′ CH_2 is called the 5′ end of the segment. The other end of the chain is the 3′ end. By convention, the sequence of bases along the backbone is read from the 5′ end to the 3′ end. Because the backbone structure along the chain does not vary, a polynucleotide structure may conveniently be abbreviated by giving only the sequence of the bases along the backbone. For example, ACGT represents the tetranucleotide in Figure 21.5.

✔ **LEARNING CHECK 21.1** Draw the structural formula for a trinucleotide portion of DNA with the sequence CTG. Point out the 5′ and 3′ ends of the molecule. Indicate with arrows the phosphodiester linkages. Enclose the nucleotide corresponding to T in a box with dotted lines.

21.2A The Secondary Structure of DNA

The secondary structure of DNA was proposed in 1953 by American molecular biologist James D. Watson and English biologist Francis H. C. Crick (see **Figure 21.6**). This was perhaps the greatest discovery of modern biology, and it earned its discoverers the Nobel Prize in Physiology or Medicine in 1962.

Figure 21.5 The structure of ACGT, a tetranucleotide segment of DNA. The nucleic acid backbone is in color.

Figure 21.6 J. D. Watson (left) and F. H. C. Crick (right), working in the Cavendish Laboratory at Cambridge, England, built scale models of the DNA double helix based on the X-ray crystallography data of M. H. F. Wilkins and Rosalind Franklin.

A. Barrington Brown/Science Source

The model of DNA established by Watson and Crick was based on key contributions of other researchers, including the analysis of the base composition of DNA. The analysis of DNA from many different forms of life revealed an interesting pattern. The relative amounts of each base often varied from one organism to another, but in all DNA the percentages of adenine and thymine were always equal to each other as were the percentages of guanine and cytosine. For example, human DNA contains 20% guanine, 20% cytosine, 30% adenine, and 30% thymine. Watson and Crick concluded from these and other types of data that DNA is composed of two strands entwined around each other in a double helix, as shown in **Figure 21.7**.

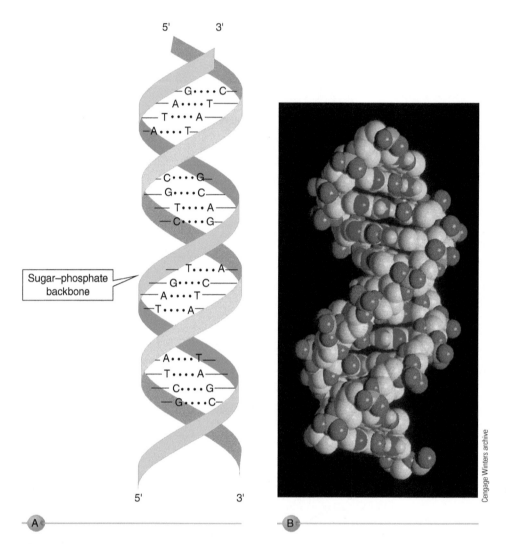

Figure 21.7 The double helix of DNA: (**A**) a schematic drawing and (**B**) a three-dimensional molecular model. Both representations show the bases pointing toward the center of the helix away from the sugar–phosphate backbone.

The two intertwined polynucleotide chains of the DNA double helix run in opposite (antiparallel) directions. Thus, each end of the double helix contains the 5′ end of one chain and the 3′ end of the other. The sugar–phosphate backbone is on the outside of the helix, and the bases point inward. The unique feature of the Watson and Crick structure is the way in which the chains are held together to form the double helix. They theorized that the DNA structure is stabilized by hydrogen bonding between the bases that extend inward from the sugar–phosphate backbone (see **Figure 21.8**). The spacing in the interior of the

double helix is such that adenine always hydrogen bonds to thymine, and guanine always hydrogen bonds to cytosine. Thus, in double-helical DNA, wherever there is an adenine on one strand of the helix, there must be a thymine on the other strand, and similarly for guanine and cytosine. The two DNA strands with these matched sequences are said to be **complementary** to each other.

complementary DNA strands Two strands of DNA in a double-helical form such that adenine and guanine of one strand are matched and hydrogen bonded to thymine and cytosine, respectively, of the second strand.

Figure 21.8 Base pairing. Hydrogen bonding between complementary base pairs holds the two strands of a DNA molecule together.

Example 21.1 Writing Base Sequences

One strand of a DNA molecule has the base sequence CCATTG. What is the base sequence for the complementary strand?

Solution

Three things must be remembered and used to solve this problem: (1) The base sequence of a DNA strand is always written from the 5′ end to the 3′ end; (2) adenine (A) is always paired with its complementary base thymine (T), and guanine (G) is always paired with its complement cytosine (C); and (3) in double-stranded DNA, the two strands run in opposite directions, so that the 5′ end of one strand is associated with the 3′ end of the other strand. These three ideas lead to the following:

Original strand:	5′ end C–C–A–T–T–G 3′ end
Complementary strand:	3′ end G–G–T–A–A–C 5′ end

When we follow convention and write the sequence of the complementary strand in the 5′ to 3′ direction, it becomes

$$5′ \ C–A–A–T–G–G \ 3′$$

✔ **LEARNING CHECK 21.2** Write the complementary base sequence for the DNA strand TTACG.

21.3 | DNA Replication

Learning Objective

3. Outline the process of DNA replication.

chromosome A tightly packed bundle of DNA and protein that is involved in cell division.

DNA is responsible for one of the most essential functions of living organisms, the storage and transmission of hereditary information. A human cell normally contains 46 structural units called **chromosomes** (see **Figure 21.9**). Each chromosome contains one molecule of DNA coiled tightly about a group of small, basic proteins called histones. Individual sections

Figure 21.9 Chromosomes. (**A**) The 46 chromosomes of a human cell. (**B**) Each chromosome is a protein-coated strand of multicoiled DNA.

Chromosome

© L. Lisco and D. W. Fawcett/Visuals Unlimited

A

DNA double helix

Histones

B

of DNA molecules make up **genes**, the fundamental units of heredity. In **Section 21.5**, you'll learn that each gene directs the synthesis of a specific protein. The number of genes contained in the structural unit of an organism varies with the type of organism. For example, a virus, the smallest structure known to carry genetic information, is thought to contain from a few to several hundred genes. A bacterial cell, such as *Escherichia coli (E. coli)*, contains about 1000 genes, whereas a human cell contains approximately 25,000. One of the elegant features of the Watson–Crick three-dimensional structure for DNA was that it provided the first satisfactory explanation for the transmission of hereditary information—it explained how DNA is duplicated for the next generation.

The process by which an exact copy of DNA is produced is called **replication**. It occurs when two strands of DNA separate and each serves as a template (pattern) for the construction of its own complement. The process generates new DNA double-stranded molecules that are exact replicas of the original DNA molecule. A schematic diagram of the replication process is presented in **Figure 21.10**. Notice that the two daughter DNA molecules have exactly the same base sequences as the original parent DNA. Notice, too, that each daughter contains one strand of the parent and one new strand that is complementary to the parent strand. This type of replication is called **semiconservative replication**.

gene An individual section of a chromosomal DNA molecule that is the fundamental unit of heredity.

replication The process by which an exact copy of a DNA molecule is produced.

semiconservative replication A replication process that produces DNA molecules containing one strand from the parent and a new strand that is complementary to the strand from the parent.

Figure 21.10 A schematic diagram of the replication of DNA.

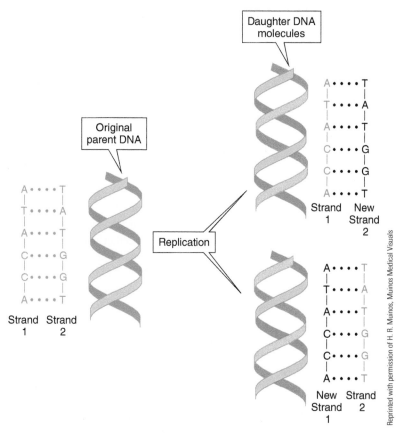

The process of replication can be viewed as occurring in three steps.

Step 1. Unwinding of the double helix: Replication begins when the enzyme helicase catalyzes the separation and unwinding of the nucleic acid strands at a specific point along the DNA helix. In this process, hydrogen bonds between complementary base pairs are broken, and the bases that were formerly in the center of the helix are exposed. The point where this unwinding takes place is called a **replication fork** (see **Figure 21.11**).

Step 2. Synthesis of DNA segments: DNA replication takes place along both nucleic acid strands separated in Step 1. The process proceeds from the 3′ end toward the 5′ end of the exposed strand (the template). Because the two strands are antiparallel, the synthesis of new nucleic acid strands proceeds toward the replication fork on

replication fork A point where the double helix of a DNA molecule unwinds during replication.

Figure 21.11 The replication of DNA. Both new strands are growing in the 5' to 3' direction.

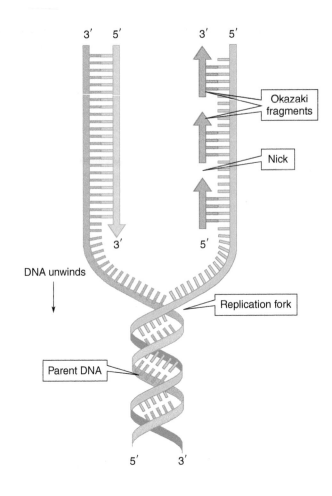

3' 5' 3' 5'

Okazaki fragments

Nick

3' 5'

DNA unwinds

Replication fork

Parent DNA

5' 3'

Okazaki fragment A DNA fragment produced during replication as a result of strand growth in a direction away from the replication fork.

one exposed strand and away from the replication fork on the other strand (see **Figure 21.11**). New (daughter) DNA strands form as nucleotides, complementary to those on the exposed strands, and are linked together under the influence of the enzyme DNA polymerase. Notice that the daughter chain grows from the 5' end toward the 3' end as the process moves from the 3' end of the exposed (template) strand toward the 5' end. As the synthesis proceeds, a second replication fork is created when a new section of DNA unwinds. The daughter strand that was growing toward the first replication fork continues growing smoothly toward the new fork. However, the other daughter strand was growing away from the first fork. A new segment of this strand begins growing from the new fork but is not initially bound to the segment that grew from the first fork. Thus, as the parent DNA progressively unwinds, this daughter strand is synthesized as a series of fragments that are bound together in Step 3. The gaps or breaks between segments in this daughter strand are called nicks, and the DNA fragments separated by the nicks are called **Okazaki fragments** after their discoverer, Reiji Okazaki (Figure 21.11).

Step 3. Closing the nicks: One daughter DNA strand is synthesized without any nicks, but the Okazaki fragments of the other strand must be joined together. An enzyme called DNA ligase catalyzes this final step of DNA replication. The result is two DNA double-helical molecules that are identical.

Observations with an electron microscope show that the replication of DNA molecules in eukaryotic cells occurs simultaneously at many points along the original DNA molecule. These replication zones blend together as the process of replication continues. This is necessary if long molecules are to be replicated rapidly. For example, it is estimated that the replication of the largest chromosome in *Drosophila* (the fruit fly) would take more than 16 days if there were only one origin for replication. Research results indicate that the actual replication is accomplished in less than 3 minutes because

it takes place simultaneously at more than 6000 replication forks per DNA molecule. **Figure 21.12** shows a schematic diagram in which replication is proceeding at several points in the DNA chain.

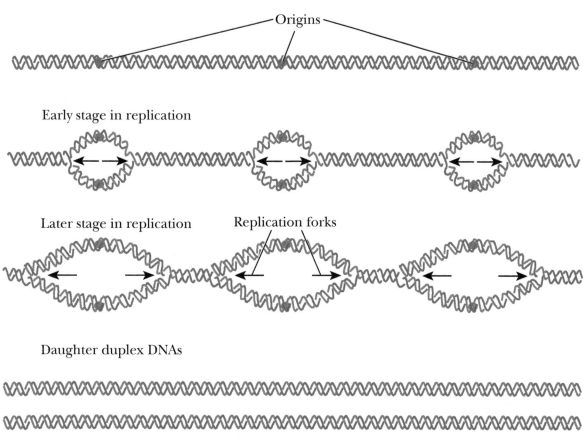

Figure 21.12 A schematic representation of eukaryotic chromosome replication, showing several replication forks. From Campbell, *Biochemistry*, 8E. © 2015 Cengage Learning.

A knowledge of the DNA replication process led scientist Kary Mullis in 1983 to a revolutionary laboratory technique called the *polymerase chain reaction (PCR)*. The PCR technique mimics the natural process of replication, in which the DNA double helix unwinds. As the two strands separate, DNA polymerase makes a copy using each strand as a template. To perform PCR, a small quantity of the target DNA is added to a test tube along with a buffered solution containing DNA polymerase; the cofactor $MgCl_2$, which is needed by the polymerase to function; the four nucleotides; the four building blocks; and primers. The primers are short polynucleotide segments that will bind to the separated DNA strands and serve as starting points for new chain growth. The PCR mixture is taken through three-step replication cycles:

1. Heat (94°C–96°C) is used for one to several minutes to unravel (denature) the DNA into single strands.
2. The tube is cooled to 50°C–65°C for one to several minutes, during which the primers hydrogen-bond to the separated strands of target DNA.
3. The tube is heated to 72°C for one to several minutes, during which time the DNA polymerase synthesizes new strands.

Each cycle doubles the amount of DNA. Following 30 such cycles, a theoretical amplification factor of 1 billion is attained.

Almost overnight, PCR became a standard research technique for detecting all manner of mutations associated with genetic disease (see **Section 21.9**). PCR can also be used to detect the presence of unwanted DNA, as in the case of a bacterial or viral infection.

Conventional tests that involve the culture of microorganisms or the use of antibodies can take weeks to perform. PCR offers a fast and simple alternative. The ability of PCR to utilize degraded DNA samples and sometimes the DNA from single cells is of great interest to forensic scientists. PCR has also permitted DNA to be amplified from some unusual sources, such as extinct mammals, Egyptian mummies, and ancient insects trapped in amber.

21.4 | Ribonucleic Acid (RNA)

Learning Objective

4. Contrast the structures of DNA and RNA and list the function of the three types of cellular RNA.

RNA, like DNA, is a long, unbranched polymer consisting of nucleotides joined by $3' \rightarrow 5'$ phosphodiester bonds. The number of nucleotides in an RNA molecule ranges from as few as 73 to many thousands. The primary structure of RNA differs from that of DNA in two ways. As we learned in **Section 21.1**, the sugar unit in RNA is ribose rather than deoxyribose. The other difference is that RNA contains the base uracil (U) instead of thymine (T).

The secondary structure of RNA is also different from that of DNA. RNA molecules are single-stranded, except in some viruses. Consequently, an RNA molecule need not contain complementary base ratios of 1:1. However, RNA molecules do contain regions of double-helical structure where they form loops (see **Figure 21.13**). In these regions, adenine (A) pairs with uracil (U), and guanine (G) pairs with cytosine (C). The proportion of helical regions varies over a wide range depending on the kind of RNA studied, but a value of 50% is typical.

Figure 21.13 A portion of RNA that has folded back on itself and formed a double-helical region.

RNA is distributed throughout cells; it is present in the nucleus, the cytoplasm, and in mitochondria (see **Table 18.3**; **Section 22.7**). Cells contain three kinds of RNA: messenger RNA (mRNA), ribosomal RNA (rRNA), and transfer RNA (tRNA). Each of these kinds of RNA performs an important function in protein synthesis.

21.4A Messenger RNA

messenger RNA (mRNA) RNA that carries genetic information from the DNA in the cell nucleus to the site of protein synthesis in the cytoplasm.

Messenger RNA (mRNA) functions as a carrier of genetic information from the DNA of the cell nucleus directly to the cytoplasm, where protein synthesis takes place. The bases of mRNA are in a sequence that is complementary to the base sequence of one of the strands of the nuclear DNA. In contrast to DNA, which remains intact and unchanged throughout the life of the cell, mRNA has a short lifetime—usually less than an hour. It is synthesized as needed and then rapidly degraded to the constituent nucleotides.

21.4B Ribosomal RNA

ribosomal RNA (rRNA) RNA that constitutes about 65% of the material in ribosomes, the sites of protein synthesis.

ribosome A subcellular particle that serves as the site of protein synthesis in all organisms.

Ribosomal RNA (rRNA) constitutes 80–85% of the total RNA of the cell. It is located in the cytoplasm in organelles called **ribosomes**, which are about 65% rRNA and 35% protein. The function of ribosomes as the sites of protein synthesis is discussed in Section 21.8.

The Clone Wars

Any gardener who has used a cutting from one plant to grow an identical plant has created a clone of the original plant, since clones are organisms that contain identical DNA. Based on that definition, identical twins are essentially clones of each other. The first successful laboratory cloning of an animal involved a tadpole, and occurred in 1952. It took more than 40 years of research before a mammal, Dolly the sheep, was cloned from an adult cell rather than an embryonic cell. Since that success in 1996, numerous other successful mammalian clonings have been achieved in the laboratory.

Research into cloning methods and procedures continues today and has proved to be a sensitive ethical and societal issue. A repeatedly asked question is: Why do it? Supporters of the practice give numerous answers to the question, including the need to save rare or endangered animal species, and the need to produce more quality food for an ever-increasing world population.

The cloning procedure involves three basic steps. First, a donor cell is obtained from an animal that is to be cloned. Next, an egg cell is obtained from an animal that will act as the surrogate mother for the cloned animal. The DNA-containing nucleus of the egg cell is removed and replaced with the nucleus (and the DNA) from the donor cell of the animal being cloned. Finally, the modified egg cell is reimplanted in the surrogate mother, and the embryo, which now contains DNA identical to that of the donor animal, develops. The resulting cloned animal is an exact genetic copy of the donor animal and has all the properties, characteristics, etc., of the donor animal.

This cloning technique has already been used to produce cloned copies of two endangered species of cows: the guar, an ox-like animal from India, and the banteng, a cow from Southeast Asia. The San Diego Zoo maintains a "frozen zoo" that contains preserved frozen cells of animals that can be used as future sources of DNA to reproduce or expand the numbers of selected animals.

The high cost of cloning has not prevented some enterprising individuals from commercializing the process. For example, a California company will clone your favorite pet for a nominal fee of $50,000. Surprisingly (or not for pet lovers), the company has had some customers interested in paying for their services.

The obvious extension of this technique to the cloning of humans is the basis for strong opposition from some individuals and organizations. They take the position that no one should play God and create animals, especially humans, even if it is scientifically possible. Cloning supporters point out that rather than creating humans, cloning could help humans have better-quality lives. They say that cloning is not an end in itself but can lead to scientific advances that will benefit human beings. One potential use in this category is to genetically engineer and then mass produce animals that can produce and secrete specific human proteins in their blood or milk. Potentially, such transgenic animals could produce therapeutic proteins for treating or studying human diseases. Another potential application is the use of cloning techniques to produce animal cells and organs that would not trigger the immune system of humans to reject such cells and organs used as transplants. Thus, an animal pancreas with such characteristics could be transplanted into a human suffering from diabetes and cure the disease.

It appears likely that disagreements about the appropriateness of conducting cloning research will continue. If cool heads prevail, a reasonable compromise will be found, and a level of research that provides minimal upset to opponents and maximum benefits to humanity in general will continue.

The first successfully cloned Afghan hound sits with his genetic father.

21.4C Transfer RNA

Transfer RNA (tRNA) molecules deliver amino acids, the building blocks of proteins, to the site of protein synthesis. Cells contain at least one specific type of tRNA for each of the 20 common amino acids found in proteins. These tRNA molecules are the smallest of all the nucleic acids, containing 73–93 nucleotides per chain.

A comparison of RNA characteristics for the three different forms is summarized in **Table 21.1**.

transfer RNA (tRNA) RNA that delivers individual amino acid molecules to the site of protein synthesis.

TABLE 21.1 Different Forms of RNA Molecules in *Escherichia coli (E. coli)*			
Class of RNA	% in Cells	Number of RNA Subtypes	Number of Nucleotides
Ribosomal RNA (rRNA)	80	3	120
			1700
			3700
Transfer RNA (tRNA)	15	46	73–93
Messenger RNA (mRNA)	5	Many	75–3000

Because of their small size, a number of tRNAs have been studied extensively. **Figure 21.14** shows a representation of the secondary structure of a typical tRNA:

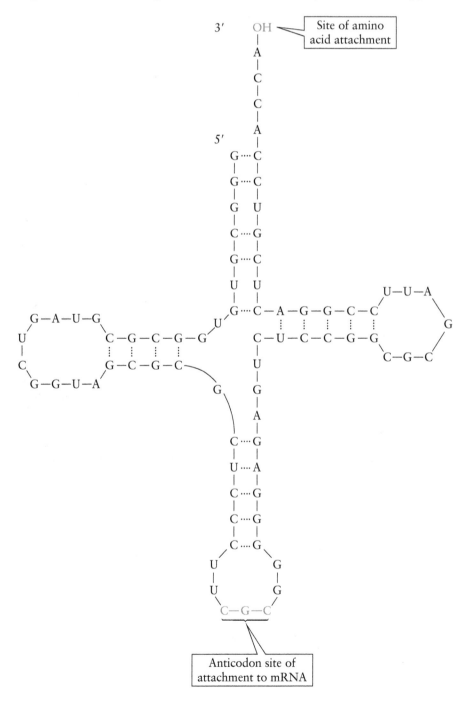

Figure 21.14 The typical tRNA cloverleaf structure.

This tRNA molecule, like all others, has regions where there is hydrogen bonding between complementary bases, and regions (loops) where there is no hydrogen bonding.

Two regions of tRNA molecules have important functions during protein synthesis. One of these regions, designated the **anticodon**, enables the tRNA to bind to mRNA during protein synthesis. The second important site is the 3′ end of the molecule, which binds to an amino acid and transports it to the site of protein synthesis. Each amino acid is joined to the 3′ end of its specific tRNA by an ester bond that forms between the carboxyl group of the amino acid and the 3′ hydroxy group of ribose. The reaction is catalyzed by an enzyme that matches tRNA molecules to their proper amino acids. These enzymes, which are very specific for both the structure of the amino acid and the tRNA, rarely cause a bond to form between an amino acid and the wrong tRNA. When a tRNA molecule is attached to its specific amino acid, it is said to be "activated" because it is ready to participate in protein synthesis. **Figure 21.15** shows the general structure of an activated tRNA and a schematic representation that will be used in **Section 21.8** to describe protein synthesis.

21.5 | The Flow of Genetic Information

Learning Objective

5. Describe what is meant by the terms *transcription* and *translation*.

You learned in **Section 21.3** that DNA is the storehouse of genetic information in the cell and that the stored information can be passed on to new cells as DNA undergoes replication. In this section, you will discover how the information stored in DNA is expressed within the cell.

This process of expression is so well-established that it is called the **central dogma of molecular biology**. The accepted dogma, or principle, says that genetic information contained in DNA molecules is transferred to RNA molecules. The transferred information of RNA molecules is then expressed in the structure of synthesized proteins. In other words, the genetic information in DNA (genes) directs the synthesis of certain proteins. In fact, there is a specific DNA gene for every protein in the body. DNA does not direct the synthesis of carbohydrates, lipids, or the other nonprotein substances essential for life. However, these other materials are manufactured by the cell through reactions made possible by enzymes (proteins) produced under the direction of DNA. Thus, in this respect, the information stored in DNA really does determine every characteristic of the living organism.

Two steps are involved in the flow of genetic information: transcription and translation. In higher organisms (eukaryotes), the DNA containing the stored information is located in the nucleus of the cell, and protein synthesis occurs in the cytoplasm. Thus, the stored information must first be carried out of the nucleus. This is accomplished by **transcription**,

Nucleic Acids and Protein Synthesis **683**

or transferring the necessary information from a DNA molecule onto a molecule of messenger RNA. The appropriately named messenger RNA carries the information (the message) from the nucleus to the site of protein synthesis in the cellular cytoplasm.

In the second step, the mRNA serves as a template on which amino acids are assembled in the proper sequence necessary to produce the specified protein. This takes place when the code or message carried by mRNA is translated into an amino acid sequence by tRNA. There is an exact word-to-word **translation** from mRNA to tRNA. Thus, each word in the mRNA language has a corresponding word in the amino acid language of tRNA. This communicative relationship between mRNA nucleotides and amino acids is called the *genetic code* (see **Section 21.7**). **Figure 21.16** summarizes the mechanisms for the flow of genetic information in the cell.

translation The conversion of the code carried by messenger RNA into an amino acid sequence of a protein.

Figure 21.16 The flow of genetic information in the cell.

21.6 | Transcription: RNA Synthesis

Learning Objective

6. Describe the process by which RNA is synthesized in cells.

An enzyme called RNA polymerase catalyzes the synthesis of RNA. During the first step of the process, the DNA double helix begins to unwind at a point near the gene that is to be transcribed. Since the end product will be single-stranded, only one strand of the DNA molecule is transcribed. Ribonucleotides are linked together along the unwound DNA strand in a sequence determined by complementary base pairing of the DNA strand bases and ribonucleotide bases.

The DNA strand always has one sequence of bases recognized by RNA polymerase as the initiation or starting point. Starting at this point, the enzyme catalyzes the synthesis of mRNA in the 5′ to 3′ direction until it reaches another sequence of bases that designates the termination point. Because the complementary chains of RNA and DNA run in opposite directions, the enzyme must move along the DNA template in the 3′ to 5′ direction (see **Figure 21.17**). Once the mRNA molecule has been synthesized, it moves away from the DNA template, which then rewinds to form the original double-helical structure.

Figure 21.17 The synthesis of mRNA.

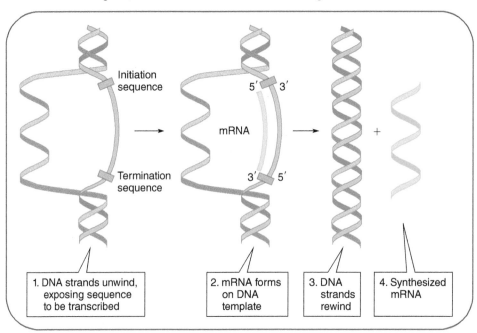

Transfer RNA and ribosomal RNA are synthesized in the same way, with DNA serving as a template.

Example 21.2 Writing RNA Sequences

Write the sequence for the mRNA that could be synthesized using the following DNA base sequence as a template:

$$5' \qquad G–C–A–A–C–T–T–G \qquad 3'$$

Solution

RNA synthesis begins at the 3' end of the DNA template and proceeds toward the 5' end. The complementary RNA strand is formed from the bases C, G, A, and U. Uracil (U) is the complement of adenine (A) on the DNA template.

direction of strand
→

DNA template: 5' G–C–A–A–C–T–T–G 3'
New mRNA: 3' C–G–U–U–G–A–A–C 5'

←
direction of strand

Writing the sequence of the new mRNA in the 5' to 3' direction, it becomes:

$$5' \qquad C–A–A–G–U–U–G–C \qquad 3'$$

LEARNING CHECK 21.3 Write the sequence for the mRNA that could be synthesized on the following DNA template:

$$5' \qquad A–T–T–A–G–C–C–G \qquad 3'$$

Although the general process of transcription operates universally, there are differences in detail between the processes in prokaryotes and eukaryotes. The genes found in prokaryotic cells exist as a continuous segment of a DNA molecule. Transcription of this gene segment produces mRNA that undergoes translation into a protein almost immediately because there is no nuclear membrane in a prokaryotic cell separating DNA from the cytoplasm. In 1977, however, it was discovered that the genes of eukaryotic cells are segments of DNA that are "interrupted" by segments that do not code for amino acids. These DNA segments that carry no amino acid code are called **introns**, and the coded DNA segments are called **exons**.

When transcription occurs in the nuclei of eukaryotic cells, both introns and exons are transcribed. This produces what is called **heterogeneous nuclear RNA (hnRNA)**. This long molecule of hnRNA then undergoes a series of enzyme-catalyzed reactions that cut and splice the hnRNA to produce mRNA (see **Figure 21.18**). The mRNA resulting from this process contains only the sequence of bases that actually codes for protein synthesis. Although the function of introns in eukaryotic DNA is not yet understood, it is an area of active research.

intron A segment of a eukaryotic DNA molecule that carries no codes for amino acids.

exon A segment of a eukaryotic DNA molecule that is coded for amino acids.

heterogeneous nuclear RNA (hnRNA) RNA produced when both introns and exons of eukaryotic cellular DNA are transcribed.

Figure 21.18 Segments of hnRNA formed from introns of eukaryotic cells are removed by special enzymes to produce mRNA.

Is There a DNA Checkup in Your Future?

The human genome consists of the genetic material contained in the human body in the form of genes and control factors for the genes. These genes and control factors, the fundamental units of heredity, are made of DNA and are organized into the chromosomes found in each cell. The genes carry the codes, or directions, for the synthesis of every type of protein produced by the body.

Scientists around the world have been working for many years on the "Human Genome Project," with the goal of identifying and determining the sequence of the approximately 3 billion letters of genetic code contained in the 20,000 genes of the human body. In June of 2000, scientists announced that the sequence of the complete human genome had been compiled. Today, that same work of reading a human genome—which required thousands of scientists much more than 10 years to accomplish at a cost of $2.7 billion—can be done in one lab in a matter of hours at a cost of less than $1,000.

What is the benefit of knowing the genome of a human? In 2000, only a handful of genes had been identified that were known to influence the development of common diseases such as heart disease, diabetes, and some cancers. Today, as a result of research in the Human Genome Project, the number of such predictive genes has increased to 600–700 and involves more than 100 diseases.

While it is probably a number of years in the future, one clinical expectation from the Human Genome Project is that physicians will one day be able to prescribe personal genome readings for patients. The results of such DNA checkups might allow physicians to precisely estimate individual risks for a significant number of common diseases and then provide advice on lifestyle changes, or even prescribe preventive medications.

Continuing efforts are underway to reduce the costs associated with DNA-sequencing equipment and procedures to bring them in line with the costs of relatively routine diagnostic procedures. So, it appears probable that DNA checkups are likely to be in your future, especially if you are relatively young.

A researcher prepares a sample for DNA sequencing.

Cultura Limited/Superstock

21.7 | The Genetic Code

Learning Objective

7. Explain how the genetic code functions in the flow of genetic information.

21.7A Discovery of the Genetic Code

By 1961, it was clear that the sequence of bases in DNA serves to direct the synthesis of mRNA, and that the sequence of bases in mRNA corresponds to the order of amino acids in a particular protein. However, the genetic code, the exact relationship between mRNA sequences and amino acids, was unknown. At least 20 mRNA "words" are needed to represent uniquely each of the 20 amino acids found in proteins. If the mRNA words consisted of a single letter represented by a base (A, C, G, or U), only four amino acids

could be uniquely represented. Thus, it was proposed that it is not one mRNA base but a combination of bases that codes for each amino acid. For example, if a code word consists of a sequence of two mRNA bases, there are $4^2 = 16$ possible combinations, so 16 amino acids could be represented uniquely. This is a more extensive code, but it still contains too few words to do the job for 20 amino acids. If the code consists of a sequence of three bases, there are $4^3 = 64$ possible combinations, more than enough to specify uniquely each amino acid in the primary sequence of a protein. Research has confirmed that nature does indeed use three-letter code words (a triplet code) to store and express genetic information. Each sequence of three nucleotide bases that represents code words on mRNA molecules is called a **codon**.

After the discovery of three-letter codons, researchers were anxious to answer the next question: Which triplets of bases (codons) code for which amino acids? In 1961, Marshall Nirenberg and his co-workers at the National Institutes of Health (NIH) in Bethesda, Maryland, attempted to break the code in a very ingenious way. They made a synthetic molecule of mRNA consisting of uracil bases only. Thus, this mRNA contained only one codon, the triplet UUU. They incubated this synthetic mRNA with ribosomes, amino acids, tRNAs, and the appropriate enzymes for protein synthesis. The exciting result of this experiment was that a polypeptide that consisted only of phenylalanine was synthesized. Thus, the first word of the genetic code had been deciphered; UUU phenylalanine.

A series of similar experiments by Nirenberg and other researchers followed, and by 1967 the entire genetic code had been broken. The complete code is shown in **Table 21.2**.

codon A sequence of three nucleotide bases that represents a code word on mRNA molecules.

TABLE 21.2 The Genetic Code: mRNA Codons for Each of the 20 Amino Acids

Amino Acid	Codons	Number of Codons
alanine	GCA, GCC, GCG, GCU	4
arginine	AGA, AGG, CGA, CGC, CGG, CGU	6
asparagine	AAC, AAU	2
aspartic acid	GAC, GAU	2
cysteine	UGC, UGU	2
glutamic acid	GAA, GAG	2
glutamine	CAA, CAG	2
glycine	GGA, GGC, GGG, GGU	4
histidine	CAC, CAU	2
isoleucine	AUA, AUC, AUU	3
leucine	CUA, CUC, CUG, CUU, UUA, UUG	6
lysine	AAA, AAG	2
methionine, initiation	AUG	1
phenylalanine	UUC, UUU	2
proline	CCA, CCC, CCG, CCU	4
serine	UCA, UCC, UCG, UCU, AGC, AGU	6
threonine	ACA, ACC, ACG, ACU	4
tryptophan	UGG	1
tyrosine	UAC, UAU	2
valine	GUA, GUC, GUG, GUU	4
Stop signals	UAG, UAA, UGA	3
Total number of codons		64

Three very important processes discussed in this chapter are replication, transcription, and translation. New terminology is always easier to learn if we can make associations with something we already know. See if these ideas help. To replicate means to copy or to repeat. Associate the word *replicate* with *repeat*.

For the processes of transcription and translation, think of an analogy with the English language. If you were to transcribe your lecture notes from a recording, you would rewrite them in the same language. In the same way, transcription of DNA to mRNA is a rewriting of the information using the same nucleic acid language. If you translated your lecture notes, you would convert them to another language, such as German. Translation of mRNA to an amino acid sequence is converting information from a nucleic acid language to a language of proteins.

21.7B Characteristics of the Genetic Code

The first important characteristic of the genetic code is that it applies almost universally. With very minor exceptions, the same amino acid is represented by the same three-base codon (or codons) in every organism.

Another interesting feature of the genetic code is apparent from **Table 21.2**. Most of the amino acids are represented by more than one codon, a condition known as *degeneracy*. Only methionine and tryptophan are represented by single codons. Leucine, serine, and arginine are the most degenerate, with each one represented by six codons. The remaining 15 amino acids are each coded for by at least two codons. Even though most amino acids are represented by more than one codon, it is significant that the reverse is not true—no single codon represents more than one amino acid. Each three-base codon represents one and only one amino acid.

Only 61 of the possible 64 base triplets represent amino acids. The remaining three (UAA, UAG, and UGA) are signals for chain terminations. They tell the protein-synthesizing process when the primary structure of the synthesized protein is complete and it is time to stop adding amino acids to the chain. These three codons are indicated in Table 21.2 as stop signals.

The presence of stop signals in the code implies that start signals must also exist. There is only one initiation (start) codon, AUG, which also is the codon for the amino acid methionine. AUG functions as an initiation codon only when it occurs as the first codon of a sequence. When this happens, protein synthesis begins at that point (see **Section 21.8**).

These general features of the genetic code are summarized in **Table 21.3**.

TABLE 21.3 General Characteristics of the Genetic Code

Characteristic	Example
Codons are three-letter words.	GCA = alanine
The code is degenerate.	GCA, GCC, GCG, GCU all represent alanine
The code is precise.	GCC represents only alanine
Chain initiation is coded.	AUG
Chain termination is coded.	UAA, UAG, and UGA
The code is almost universal.	GCA = alanine, perhaps, in all organisms

21.8 | Translation and Protein Synthesis

Learning Objective

8. Outline the process by which proteins are synthesized in cells.

To this point, you have become familiar with the molecules that participate in protein synthesis and the genetic code, the language that directs the synthesis. We now investigate the actual process by which polypeptide chains are assembled. There are three major stages in protein synthesis (see **Figure 21.19**): initiation of the polypeptide chain, elongation of the chain, and termination of the completed polypeptide chain.

21.8A Initiation of the Polypeptide Chain

The first amino acid that starts the process of protein synthesis in prokaryotic (bacterial) cells is a derivative of methionine. This compound, *N*-formylmethionine, initiates the growing polypeptide chain as the N-terminal amino acid. The fact that most proteins do not have *N*-formylmethionine as the N-terminal amino acid indicates that when protein synthesis is completed, the *N*-formylmethionine is cleaved from the finished protein.

Figure 21.19 The stages of protein synthesis. From Campbell, *Biochemistry*, 8E. © 2015 Cengage Learning.

N-formylmethionine
(fMet)

A ribosome comprises two subunits, a large subunit and a small subunit. The initiation process begins when mRNA is aligned on the surface of a small ribosomal subunit in such a way that the initiating codon, AUG, occupies a specific site on the ribosome called the P site (peptidyl site). When the AUG codon is used this way to initiate synthesis of a polypeptide chain in prokaryotic cells, it represents *N*-formylmethionine (fMet) instead of methionine. When AUG is located anywhere else on the mRNA, it simply represents methionine. For the eukaryotic cells of humans, AUG always specifies methionine even when it is the initiating codon. Next, a tRNA molecule with its attached fMet binds to the codon through hydrogen bonds. The resulting complex binds to the large ribosomal subunit to form a unit called an *initiation complex* (see **Figure 21.20**).

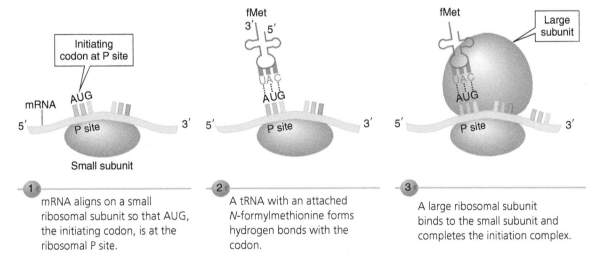

1. mRNA aligns on a small ribosomal subunit so that AUG, the initiating codon, is at the ribosomal P site.

2. A tRNA with an attached *N*-formylmethionine forms hydrogen bonds with the codon.

3. A large ribosomal subunit binds to the small subunit and completes the initiation complex.

Figure 21.20 Initiation complex formation.

Stem Cell Research

"**Stem cell** research" is a phrase many individuals are familiar with, but only understand in a superficial way. This ethically charged and very controversial research has great potential to provide treatments and possibly cures for serious conditions such as diabetes, heart disease, and Parkinson's disease. Despite this potential to benefit human society, numerous countries have limited and even banned specific practices associated with this research on the basis of ethical concerns.

Stem cells possess the unique capability to develop into many different types of cells found in the body. When a stem cell divides, each new cell has the ability to either remain as a stem cell or transform into another type of specialized cell such as a muscle cell, a red blood cell, or a brain cell. Because of this ability, stem cells can serve as a repair system for the body and theoretically divide without limit to replenish other cells that are lost through injury, disease, or normal wear and tear.

There are two types of stem cells: embryonic stem cells and adult stem cells. Human embryonic stem cells are thought to have much greater developmental potential than adult stem cells because they are *pluripotent*, which means they are able to develop into any type of cell. Adult stem cells are *multipotent*, meaning they can become cells corresponding to the type of tissue in which they reside.

After many years of research, it is now possible to isolate stem cells from human embryos and get them to multiply in the laboratory. These are called human embryonic stem cells. The embryos used in these studies were created by in vitro fertilization processes in order to treat infertility. When the embryos were no longer needed, they were donated for research with the informed consent of the donor.

The existence of two types of stem cells has prompted some critics of the embryonic cell research to ask why adult stem cells are not used instead of embryonic stem cells. Supporters of the research point out that embryonic stem cells have a much greater versatility because they can become any type of cell, not just the same type of cell as the cells of the tissue in which they are found. Embryonic stem cells could be used to treat a damaged organ, a broken bone, or an injured spinal cord. In some research, embryonic stem cells have been grown into heart muscle cells that clump together in a laboratory dish and pulse in unison. And when those heart cells were injected into mice and pigs with heart disease, they filled in for the injured or dead cells and accelerated the animal's recovery. Similar studies have suggested that embryonic stem cells have the potential to provide similar cures for conditions such as diabetes and spinal cord injury. In our society where waiting lists for donated organs are too long to help many patients, stem cell research could hold the answer.

A researcher prepares stem cells for culture.

stem cell An unspecialized cell that has the ability to replicate and differentiate, giving rise to a specialized cell.

21.8B Elongation of the Chain

A second site, called the A site (aminoacyl site), is located on the mRNA–ribosome complex next to the P site. The A site is where an incoming tRNA carrying the next amino acid will bond. Each of the tRNA molecules representing the 20 amino acids can try to fit the A site, but only the one with the correct anticodon that is complementary to the next codon on the mRNA will fit properly.

Once at the A site, the second amino acid (in this case, phenylalanine) is linked to *N*-formylmethionine by a peptide bond whose formation is catalyzed by the enzyme peptidyl transferase (see **Figure 21.21**). After the peptide bond forms, the tRNA bound to the P site is "empty," and the growing polypeptide chain is now attached to the tRNA bound to the A site.

In the next phase of elongation, the whole ribosome moves one codon along the mRNA toward the 3′ end. As the ribosome moves, the empty tRNA is released from the P site, and tRNA attached to the peptide chain moves from the A site and takes its

Figure 21.21 The amino acid at the P site bonds through a peptide bond to the amino acid at the A site.

place on the P site. This movement of the ribosome along the mRNA is called *translocation* and makes the A site available to receive the next tRNA with the proper anticodon. The amino acid carried by this tRNA bonds to the peptide chain, and the elongation process is repeated. This occurs over and over until the entire polypeptide chain is synthesized. The elongation process is represented in **Figure 21.22** for the synthesis of the tripeptide fMet—Phe—Val.

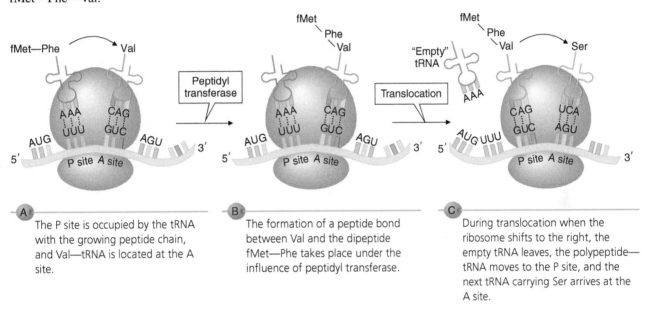

A The P site is occupied by the tRNA with the growing peptide chain, and Val—tRNA is located at the A site.

B The formation of a peptide bond between Val and the dipeptide fMet—Phe takes place under the influence of peptidyl transferase.

C During translocation when the ribosome shifts to the right, the empty tRNA leaves, the polypeptide—tRNA moves to the P site, and the next tRNA carrying Ser arrives at the A site.

Figure 21.22 Polypeptide chain elongation.

21.8C The Termination of Polypeptide Synthesis

The chain elongation process continues and polypeptide synthesis continues until the ribosome complex reaches a stop codon (UAA, UAG, or UGA) on the mRNA. At that point, a specific protein known as a termination factor binds to the stop codon and catalyzes the hydrolysis of the completed polypeptide chain from the final tRNA. The "empty" ribosome dissociates and can then bind to another strand of mRNA to once again begin the process of protein synthesis.

Several ribosomes can move along a single strand of mRNA one after another (see **Figure 21.23**). Thus, several identical polypeptide chains can be synthesized almost simultaneously from a single mRNA molecule. This markedly increases the efficiency of utilization of the mRNA. Such complexes of several ribosomes and mRNA are called **polyribosomes** or **polysomes**. Growing polypeptide chains extend from the ribosomes into the cellular cytoplasm and spontaneously fold to assume characteristic three-dimensional secondary and tertiary configurations.

polyribosome or polysome A complex of mRNA and several ribosomes.

Figure 21.23 A polyribosome, several ribosomes proceeding simultaneously along mRNA.

> ✔ **LEARNING CHECK 21.4** Write the primary structure of the polypeptide produced during translation of the following mRNA sequence:
>
> 5′ AUG–CAC–CAU–GUA–UUG–UGU–UAG 3′

21.9 | Mutations

Learning Objective

9. Describe how genetic mutations occur and how they influence organisms.

In **Section 21.3**, we pointed out that the base-pairing mechanism provides a nearly perfect way to copy a DNA molecule during replication. However, not even the copying mechanism involved in DNA replication is totally error free. It is estimated that, on average, one in every 10^{10} bases of DNA (i.e., 1 in 10 billion) is incorrect. Any change resulting in an incorrect sequence of bases on DNA is called a **mutation**. The faithful transcription of mutated DNA leads to an incorrect base sequence in mRNA. This can lead to an incorrect amino acid sequence for a protein, or possibly the failure of a protein to be synthesized at all.

Some mutations occur naturally during DNA replication; others can be induced by environmental factors such as ionizing radiation (X-rays, ultraviolet light, gamma rays, etc.). Repeated exposure to X-rays greatly increases the rate of mutation. Thus, patients are given X-rays only when necessary, and technicians who administer X-rays remain behind protective barriers. A large number of chemicals (e.g., nitrous acid and dimethyl sulfate) can also induce mutations by reacting with DNA. Such chemicals are called **mutagens**.

Some mutations might be beneficial to an organism by making it more capable of surviving in its environment. Mutations may also occur without affecting an organism. Conversely, mutations may be lethal or produce genetic diseases whenever an important protein (or enzyme) is not correctly synthesized. Sickle-cell disease, phenylketonuria (PKU), hemophilia, muscular dystrophy, and many other disorders are results of such mutations that have become permanently incorporated into the genetic makeup of certain individuals.

mutation Any change resulting in an incorrect base sequence on DNA.

mutagen A chemical that induces mutations by reacting with DNA.

21.10 | Recombinant DNA

Learning Objective

10. Describe the technology used to produce recombinant DNA.

Remarkable technology is available that allows segments of DNA from one organism to be introduced into the genetic material of another organism. The resulting new DNA (containing the foreign segment) is referred to as **recombinant DNA**. The application of

recombinant DNA DNA of an organism that contains genetic material from another organism.

this technology, commonly called *genetic engineering*, has produced major advances in human health care and holds the promise for a future of exciting advances in biology, agriculture (see **Figure 21.24**), and many other areas of study.

An early success of genetic engineering was the introduction of the gene for human insulin into the DNA of the common bacterium *Escherichia (E.) coli*. The bacterium then transcribed and translated the information carried by the gene and produced the protein hormone. By culturing such *E. coli*, it has become possible to produce and market large quantities of human insulin. The availability of human insulin is very important for diabetics allergic to the insulin traditionally used, which is isolated from the pancreatic tissue of slaughtered pigs or cattle. **Table 21.4** lists several other medically important materials that have been produced through genetic engineering.

Figure 21.24 This tomato, a beneficiary of genetic engineering, has an increased shelf life.

TABLE 21.4 **Some Substances Produced by Genetic Engineering**	
Substance	**Use**
Human insulin	Treatment of diabetes mellitus
Human growth hormone	Treatment of dwarfism
Interferon	Fights viral infections
Hepatitis B vaccine	Protection against hepatitis
Malaria vaccine	Protection against malaria
Tissue plasminogen activator	Promotes the dissolving of blood clots

21.10A Enzymes Used in Genetic Engineering

The discovery of restriction enzymes in the 1960s and 1970s made genetic engineering possible. **Restriction enzymes**, which are found in a wide variety of bacterial cells, catalyze the cleaving of DNA molecules. These enzymes are normally part of a mechanism that protects certain bacteria from invasion by foreign DNA, such as that of a virus. In these bacteria, some of the bases in their DNA have methyl groups attached. For example, the following compounds are found:

restriction enzyme A protective enzyme found in some bacteria that catalyzes the cleaving of all but a few specific types of DNA.

1-methylguanine 5-methylcytosine

The methylated DNA of these bacteria is left untouched by the restriction enzymes, but foreign DNA that lacks the methylated bases undergoes rapid cleavage of both strands and thus becomes nonfunctional. Because there is a "restriction" on the type of DNA allowed in the bacterial cell, the protective enzymes are called restriction enzymes.

Restriction enzymes act at sites on DNA called palindromes. In language, a palindrome is any word or phrase that reads the same in either direction, such as "radar" or "Madam, I'm Adam." For double-stranded DNA, a palindrome is a section in which the two strands have the same sequence but run in opposite directions. Examples of DNA palindromes follow; the arrows indicate the points of attack by restriction enzymes:

$$
\begin{array}{ccccccc}
& & \downarrow & & & & \\
5' & C-C-G-C-G-G & 3' & \quad & 5' & G-G-A-T-C-C & 3' \\
& \vdots \vdots \vdots \vdots \vdots \vdots & & & & \vdots \vdots \vdots \vdots \vdots \vdots & \\
3' & G-G-C-G-C-C & 5' & \quad & 3' & C-C-T-A-G-G & 5' \\
& \uparrow & & & & \uparrow &
\end{array}
$$

DNA and the Crime Scene

DNA, commonly referred to as the blueprint of life, passes along hereditary information to offspring and directs all life processes. DNA also plays a key role in forensic science, made popular by TV shows such as *CSI: Crime Scene Investigation*. However, DNA fingerprinting isn't just a gimmick used in Hollywood scripts; it is a valuable tool in solving real-world crimes.

DNA profiling—also called DNA typing or DNA fingerprinting—is possible because most of the DNA molecule is common to all humans, but small parts of it differ from person to person. In fact, 99.9% of human DNA is identical, but 0.1% differs from person to person. Small as this difference seems, it is equivalent to 3 million bases in an individual DNA. Therefore, with the exception of identical twins, the DNA of every human is unique. Unlike a conventional fingerprint, which occurs only on the fingertips, a DNA fingerprint is identical for every cell of each tissue and organ of the person's body. Consequently, DNA fingerprinting has become the method of choice for identifying and distinguishing among individual human beings.

Law enforcement laboratories around the world routinely use DNA fingerprints to link suspects to crime-scene biological evidence such as blood, semen, or hair. DNA fingerprints used to establish paternity in custody and child-support litigations provide nearly perfect accuracy in parental identification. The U.S. armed services have switched from dog tags and dental records to DNA fingerprint records as a means to identify all personnel, including those killed in combat.

The use of DNA by the U.S. armed services has even made it possible to identify soldiers from past wars. In 1997, DNA collected from bones of American servicemen killed during the Vietnam War helped the Army identify the bodies. It wasn't the only method used in the identification process, but it provided conclusive evidence.

The technique of DNA fingerprinting involves cleaving DNA molecules at specific points using restriction enzymes. The mixture of DNA fragments is then separated into specific patterns by a process called gel electrophoresis. In this process, the cleaved DNA is put into a gel through which electricity is passed. The current organizes the pieces of DNA in the gel by size. The resulting patterns are made visible by attaching radioactive reagents to the DNA fragments and exposing X-ray film to the results of the electrophoresis. The film shows a specific pattern of dark bands when it is developed.

The ability to obtain DNA fingerprints from exceedingly small tissue samples has become possible by using the polymerase chain reaction technique (PCR) (see **Section 21.3**).

The PCR technique allows billions of copies of a DNA sample to be prepared in a few hours. PCR enables researchers to study extremely faint and tiny traces of DNA obtained from biological samples, such as specks of dried blood, strands of hair, chips of bone, or even from fingerprints. PCR amplification of the DNA in a single sperm cell has been used to link suspects to rape victims. A single cell from saliva on the stamp of a letter bomb could provide enough DNA to help identify the person who licked the stamp.

These few examples illustrate the tremendous identification potential of the PCR technique coupled with the uniqueness of each individual's DNA. It is getting more and more difficult to hide.

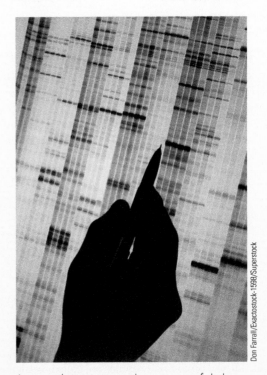

A researcher compares the patterns of dark bands produced by DNA fragments.

At least 100 bacterial restriction enzymes are known, and each catalyzes DNA cleavage in a specific and predictable way. These enzymes are the tools used to take DNA apart and reduce it to fragments of known size and nucleotide sequence.

Another set of enzymes important in genetic engineering, called DNA ligases, has been known since 1967. These enzymes normally function to connect DNA fragments during replication, and they are used in genetic engineering to put together pieces of DNA produced by restriction enzymes.

21.10B Plasmids

The introduction of a new DNA segment (gene) into a bacterial cell requires the assistance of a DNA carrier called a **vector**. The vector is often a circular piece of double-stranded DNA called a **plasmid**. Plasmids range in size from 2000 to several hundred thousand nucleotides and are found in the cytoplasm of bacterial cells. Plasmids function as accessories to chromosomes by carrying genes for the inactivation of antibiotics and the production of toxins. They have the unusual ability to replicate independently of chromosomal DNA. A typical bacterial cell contains about 20 plasmids and one or two chromosomes.

vector A carrier of foreign DNA into a cell.

plasmid Circular, double-stranded DNA found in the cytoplasm of bacterial cells.

21.10C The Formation of Recombinant DNA

The recombinant DNA technique begins with the isolation of a plasmid from a bacterium. A restriction enzyme is added to the plasmid, which is cleaved at a specific site:

~G–G–A–T–C–C~ restriction → ~G G–A–T–C–C~
~C–C–T–A–G–G~ enzyme ~C–C–T–A–G G~

sticky ends

Because a plasmid is circular, cleaving it this way produces a double-stranded chain with two ends (see **Figure 21.25**). These are called "sticky ends" because each has one strand in which several bases are unpaired and ready to pair up with a complementary section if available.

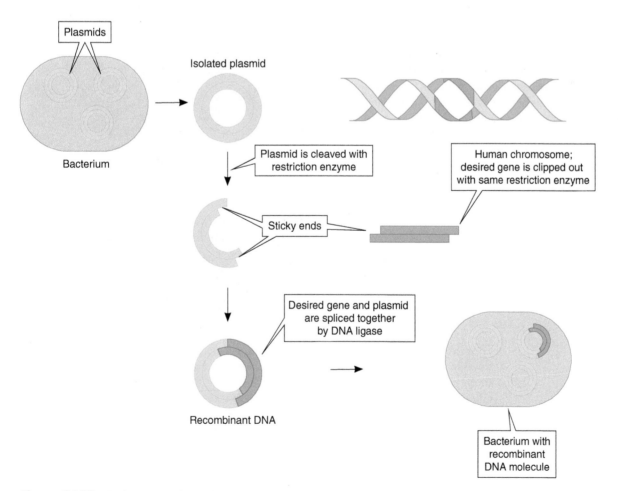

Figure 21.25 The formation of recombinant DNA.

Reduce Your Chances for Developing Cancer

If you could take a pill to reduce your chances of developing cancer by one third, would you take it? Statistically, one in three Americans will develop some form of cancer in their lifetime. You can increase your odds of being among the two-thirds who do not develop cancer by eating the proper foods.

Antioxidant phytochemicals present in fruits and vegetables (see **Chemistry Tips for Living Well 5.1**, page 156) form the basis of some popular cancer-preventing diets. People who regularly consume three to five servings of fruits and vegetables daily have been shown to have a lower risk of developing numerous types of cancer. People who eat even more fruits and vegetables have a correspondingly lower risk. Because of this correlation, some experts recommend up to nine servings of fruits and vegetables per day.

In addition to consuming a variety of plant foods (including whole grains), cancer prevention has been found to depend on other diet and lifestyle choices. Choosing a diet low in saturated fats is one key to cancer prevention. Diets low in saturated fats limit the consumption of red meat and dairy products. Replacing meat and dairy products with plant-based sources of protein such as nuts, seeds, dried beans, and soy products has been shown to help prevent cancer. Choosing fish in place of meat also reduces the risk of developing cancer.

In addition to saturated fats from meat and dairy products, certain types of meat preparation are linked to increased cancer risk. Grilling meat produces carcinogenic materials if the meat is charred or covered in an oily glaze produced from dripping fat flare-ups. Experts suggest that charred portions be removed and any oily coating be rinsed off before grilled meat is eaten.

Since tobacco smoke and tobacco products are related to many forms of cancer, not smoking or using any tobacco products is key to preventing cancer. Excessive consumption of alcohol (more than one drink per day for women or two per day for men) has been linked to mouth, throat, breast, and liver cancers. Excess body fat, especially midsection body fat, has been correlated with cancer development. Maintaining a healthful weight and exercising five or more times per week is considered to be a cancer prevention process.

In summary, cancer prevention is not a matter of taking the correct handful of chemicals; it consists of making healthful choices each day. This includes eating a plant-based, whole-grain-rich, low saturated fat diet. The healthful diet combined with exercise, healthy weight maintenance, alcohol use moderation, and no tobacco use place the odds in your favor for avoiding cancer.

Diets rich in fruits and vegetables offer protection.

The next step is to provide the sticky ends with complementary sections for pairing. A human chromosome is cleaved into several fragments using the same restriction enzyme. Because the same enzyme is used, the human DNA is cleaved such that the same sticky ends result:

$$G-A-T-C-C \sim\sim\sim\sim G$$
$$\text{human DNA segment}$$
$$G \sim\sim\sim\sim C-C-T-A-G$$

To splice the human gene into the plasmid, the two are brought together under conditions suitable for the formation of hydrogen bonds between the complementary bases of the sticky ends. The breaks in the strands are then joined by using DNA ligase, and the plasmid once again becomes a circular piece of double-stranded DNA that now incorporates both human and bacterial DNA (see **Figure 21.25**); recombinant DNA has been formed.

Bacterial cells are bathed in a solution containing recombinant DNA plasmids, which diffuse into the cells. When the bacteria reproduce, they replicate all the genes, including the recombinant DNA plasmid. Bacteria multiply quickly, and soon a large number of bacteria, all containing the modified plasmid, are manufacturing the new protein as directed by the recombinant DNA.

The use of this technology makes possible (in principle) the large-scale production of virtually any polypeptide or protein. By remembering the many essential roles these substances play in the body, it is easy to predict that genetic engineering will make tremendous contributions to improving health care in the future.

▶ Case Study Follow-up

Cystic fibrosis (CF) is a life-threatening inherited disease. In those who suffer from the disease, faulty genes disrupt the normal function of epithelial cells. Epithelial cells line the passageways of the liver, pancreas, digestive tract, reproductive system, and lungs. Cystic fibrosis causes a thick, sticky mucus to form in these passageways. The mucus traps bacteria in the lungs, causing chronic respiratory infections including bronchitis and pneumonia. In the pancreas, the mucus interferes with the delivery of pancreatic digestive enzymes, resulting in malnutrition. The thick mucus can also interfere with fertility. Before the development of antibiotics, youngsters with cystic fibrosis rarely lived long enough to attend elementary school. Now, with improved therapies, the life expectancy is 41 years.

Because so many genes cause cystic fibrosis, there is a wide range of severity among those afflicted with the disease. Most cystic fibrosis patients are diagnosed by age 2, using a chloride skin test. An indication of cystic fibrosis is very salty skin. People with cystic fibrosis can have up to five times the amount of sodium chloride in their perspiration as those without the disease.

Treatments for cystic fibrosis include vibrating vests that break up the thick mucus in the lungs, and pancreatic enzymes given as dietary supplements. In 2015, the Food and Drug Administration (FDA) approved a new drug that inhibits the production of defective proteins that are characteristic of the disease. This inhibitor might significantly influence the life span of CF patients.

Concept Summary

Components of Nucleic Acids Nucleic acids are classified into two categories: ribonucleic acids (RNA) and deoxyribonucleic acids (DNA). Both types are polymers made up of monomers called nucleotides. All nucleotides are composed of a pyrimidine or purine base, a sugar, and phosphate. The sugar component of RNA is ribose, and that of DNA is deoxyribose. The bases adenine, guanine, and cytosine are found in all nucleic acids. Uracil is found only in RNA, and thymine is found only in DNA.
Objective 1 (Section 21.1), Exercises 21.2 and 21.4

The Structure of DNA The nucleotides of DNA are joined by linkages between phosphate groups and sugars. The resulting sugar–phosphate backbone is the same for all DNA molecules, but the order of attached bases along the backbone varies. This order of nucleotides with attached bases is the primary structure of nucleic acids. The secondary structure is a double-stranded helix held together by hydrogen bonds between complementary base pairs on the strands.
Objective 2 (Section 21.2), Exercise 21.10

DNA Replication The replication of DNA occurs when a double strand of DNA unwinds at specific points. The exposed bases match up with complementary bases of nucleotides. The nucleotides bind together to form two new strands that are complementary to the strands that separated. Thus, the two new DNA molecules each contain one old strand and one new strand.
Objective 3 (Section 21.3), Exercise 21.20

Ribonucleic Acid (RNA) Three forms of ribonucleic acid are found in cells: messenger RNA, ribosomal RNA, and transfer RNA. Each form serves an important function during protein synthesis. All RNA molecules are single stranded, but some contain loops or folds.
Objective 4 (Section 21.4), Exercises 21.26 and 21.28

The Flow of Genetic Information The flow of genetic information occurs in two steps called transcription and translation. In transcription, information stored in DNA molecules is passed to molecules of messenger RNA. In translation, the messenger RNA serves as a template that directs the assembly of amino acids into proteins.
Objective 5 (Section 21.5), Exercise 21.32

Transcription: RNA Synthesis The various RNAs are synthesized in much the same way as DNA is replicated. Nucleotides with complementary bases align themselves against one strand of a partially unwound DNA segment that contains the genetic information that is to be transcribed. The aligned nucleotides bond together to form the RNA. In eukaryotic cells, the produced RNA is heterogeneous and is cut and spliced after being synthesized to produce the functional RNA.
Objective 6 (Section 21.6), Exercise 21.34

The Genetic Code The genetic code is a series of three-letter words that represent the amino acids of proteins as well as start and stop signals for protein synthesis. The letters of the words are the bases found on mRNA, and the

words on mRNA are called codons. The genetic code is the same for all organisms and is degenerate for most amino acids.

Objective 7 (Section 21.7), Exercise 21.38

Translation and Protein Synthesis The translation step in the flow of genetic information results in the synthesis of proteins. The synthesis takes place when properly coded mRNA forms a complex with the component of a ribosome. Transfer RNA molecules carrying amino acids align themselves along the mRNA in an order representing the correct primary structure of the protein. The order is determined by the matching of complementary codons on the mRNA to anticodons on the tRNA. The amino acids sequentially bond together to form the protein, which then spontaneously forms characteristic secondary and tertiary structures.

Objective 8 (Section 21.8), Exercise 21.44

Mutations Any change that results in an incorrect sequence of bases on DNA is called a mutation. Some mutations occur naturally during DNA replication, whereas others are induced by environmental factors. Some mutations are beneficial to organisms; others may be lethal or result in genetic diseases.

Objective 9 (Section 21.9), Exercise 21.48

Recombinant DNA The discovery and application of restriction enzymes and DNA ligases have resulted in a technology called genetic engineering. The primary activities of genetic engineers are the isolation of genes (DNA) that code for specific useful proteins and the introduction of these genes into the DNA of bacteria. The new (recombinant) DNA in the rapidly reproducing bacteria mediates production of the useful protein, which is then isolated for use.

Objective 10 (Section 21.10), Exercise 21.52

Key Terms and Concepts

Anticodon (21.4)
Central dogma of molecular biology (21.5)
Chromosome (21.3)
Codon (21.7)
Complementary DNA strands (21.2)
Deoxyribonucleic acid (DNA) (21.1)
Exon (21.6)
Gene (21.3)
Heterogeneous nuclear RNA (hnRNA) (21.6)
Intron (21.6)

Messenger RNA (mRNA) (21.4)
Mutagen (21.9)
Mutation (21.9)
Nucleic acid (Introduction)
Nucleic acid backbone (21.2)
Nucleotide (21.1)
Okazaki fragment (21.3)
Plasmid (21.10)
Polyribosome (polysome) (21.8)
Recombinant DNA (21.10)
Replication (21.3)

Replication fork (21.3)
Restriction enzyme (21.10)
Ribonucleic acid (RNA) (21.1)
Ribosomal RNA (rRNA) (21.4)
Ribosome (21.4)
Semiconservative replication (21.3)
Stem cell (21.8)
Transcription (21.5)
Transfer RNA (tRNA) (21.4)
Translation (21.5)
Vector (21.10)

Exercises

Even-numbered exercises are answered in Appendix B.

Blue-numbered exercises are more challenging.

Components of Nucleic Acids (Section 21.1)

21.1 What is the principal location of DNA within the eukaryotic cell?

21.2 Which pentose sugar is present in DNA? In RNA?

21.3 Name the three components of nucleotides.

21.4 Indicate whether each of the following is a pyrimidine or a purine:

 a. guanine

 b. thymine

 c. uracil

 d. cytosine

 e. adenine

21.5 Which bases are found in DNA? In RNA?

21.6 Write the structural formula for the nucleotide thymidine 5′-monophosphate. The base component is thymine.

The Structure of DNA (Section 21.2)

21.7 In what way might two DNA molecules that contain the same number of nucleotides differ?

21.8 Identify the 3′ and 5′ ends of the DNA segment AGTCAT.

21.9 What data obtained from the chemical analysis of DNA supported the idea of complementary base pairing in DNA proposed by Watson and Crick?

21.10 Describe the secondary structure of DNA as proposed by Watson and Crick.

21.11 Describe the role of hydrogen bonding in the secondary structure of DNA.

21.12 How many total hydrogen bonds would exist between the following strands of DNA and their complementary strands?

 a. CAGTAG

 b. TTGACA

21.13 How many total hydrogen bonds would exist between the following strands of DNA and their complementary strands?

 a. GCATGC

 b. TATGGC

21.14 A strand of DNA has the base sequence ATGCATC. Write the base sequence for the complementary strand.

21.15 A strand of DNA has the base sequence GATTCA. Write the base sequence for the complementary strand.

DNA Replication (Section 21.3)

21.16 What is a chromosome? How many chromosomes are in a human cell? What is the approximate number of genes in a human cell?

21.17 What is meant by the term *semiconservative replication*?

21.18 What is a replication fork?

21.19 Describe the function of the enzyme helicase in the replication of DNA.

21.20 List the steps involved in DNA replication.

21.21 What enzymes are involved in DNA replication?

21.22 In what direction is a new DNA strand formed?

21.23 Explain how the synthesis of a DNA daughter strand growing toward a replication fork differs from the synthesis of a daughter strand growing away from a replication fork.

21.24 The base sequence ACGTCT represents a portion of a single strand of DNA. Draw the complete double-stranded molecule for this portion of the strand and use the representation to illustrate the replication of the DNA strand. Be sure to clearly identify the nucleotide bases involved, the new strands formed, and the daughter DNA molecules.

21.25 Explain the origin of Okazaki fragments.

Ribonucleic Acid (RNA) (Section 21.4)

21.26 How does the sugar–phosphate backbone of RNA differ from the backbone of DNA?

21.27 Compare the secondary structures of RNA and DNA.

21.28 Briefly describe the characteristics and functions of the three types of cellular RNA.

21.29 Must the ratio of guanine to cytosine be 1:1 in RNA? Explain.

21.30 What are the two important regions of a tRNA molecule?

The Flow of Genetic Information (Section 21.5)

21.31 What is the central dogma of molecular biology?

21.32 In the flow of genetic information, what is meant by the terms *transcription* and *translation*?

Transcription: RNA Synthesis (Section 21.6)

21.33 Briefly describe the synthesis of mRNA in the transcription process.

21.34 Write the base sequence for the mRNA that would be formed during transcription from the DNA strand with the base sequence GCCATATTG.

21.35 Describe the differences in the transcription and translation processes as they occur in prokaryotic cells and eukaryotic cells.

21.36 What is the relationship among exons, introns, and hnRNA?

The Genetic Code (Section 21.7)

21.37 What is a codon?

21.38 For each of the following mRNA codons, give the tRNA anticodon and use Table 21.2 to determine the amino acid being coded for by the codon.

 a. UAU

 b. CAU

 c. UCA

 d. UCU

21.39 Describe the experiment that allowed researchers to first identify the codon for a specific amino acid.

21.40 Which of the following statements about the genetic code are true and which are false? Correct each false statement.

 a. Each codon is composed of four bases.

 b. Some amino acids are represented by more than one codon.

 c. All codons represent an amino acid.

 d. Each living species is thought to have its own unique genetic code.

 e. The codon AUG at the beginning of a sequence is a signal for protein synthesis to begin at that codon.

 f. It is not known if the code contains stop signals for protein synthesis.

Translation and Protein Synthesis (Section 21.8)

21.41 The β-chain of hemoglobin is a protein that contains 146 amino acid residues. What minimum number of nucleotides must be present in a strand of mRNA that is coded for this protein?

21.42 What is a polysome?

21.43 Beginning with DNA, describe in simple terms (no specific codons, etc.) how proteins are coded and synthesized.

21.44 List the three major stages in protein synthesis.

21.45 Does protein synthesis begin with the N-terminal or with the C-terminal amino acid?

21.46 What is meant by the terms *A site* and *P site*?

21.47 Beginning with DNA, describe specifically the coding and synthesis of the following tetrapeptide that represents the first four amino acid residues of the hormone oxytocin: Gly–Leu–Pro–Cys. Be sure to include processes such as formation of mRNA (use correct codons, etc.), attachment of mRNA to a ribosome, attachment of tRNA to mRNA-ribosome complex, and so on.

Mutations (Section 21.9)

21.48 What is a genetic mutation?

21.49 Briefly explain how genetic mutations can do the following:

 a. Harm an organism

 b. Help an organism

21.50 What is the result of a genetic mutation that causes the mRNA sequence GCC to be replaced by CCC?

Recombinant DNA (Section 21.10)

21.51 Explain the function and importance of restriction enzymes and DNA ligase in the formation of recombinant DNA.

21.52 Explain how recombinant DNA differs from normal DNA.

21.53 Explain what plasmids are and how they are used to get bacteria to synthesize a new protein that they normally do not synthesize.

21.54 List three substances likely to be produced on a large scale by genetic engineering, and give an important use for each.

Additional Exercises

21.55 Explain how heating a double-stranded DNA fragment to 94°C–96°C when performing PCR causes the DNA to unravel into single strands.

21.56 Review Figure 21.5 and suggest a reaction that would be used to excise introns from hnRNA. Also suggest a reaction that would be used to join the exon segments together.

21.57 A segment of DNA has an original code of –ACA–. This segment was mutated to –AAA–. After the mutation occurred, the protein that was coded for was no longer able to maintain its tertiary structure. Explain why.

21.58 A species of bacteria can use sucrose as its sole source of carbon energy. The bacterial enzyme Q is used to hydrolyze sucrose into its constituent monosaccharides. In the absence of sucrose, there are fewer than 10 enzyme Q molecules in a bacterial cell. In the presence of sucrose there are more than several thousand enzyme molecules in a single cell. Enzyme Q does not exist as a zymogen.

 a. What are the constituent monosaccharides of sucrose?

 b. What is the specific type of enzyme regulation that controls enzyme Q?

 c. Suggest a mechanism of how this type of enzyme regulation occurs.

21.59 During DNA replication, about 1000 nucleotides are added per minute per molecule of DNA polymerase. The genetic material in a single human cell consists of about 3 billion nucleotides.

 a. How many years would it take one DNA polymerase enzyme to replicate the genetic material in one cell?

 b. How many DNA polymerase molecules would be needed to replicate a cell's genetic material in 10 minutes (assuming equidistance between replication forks)?

Chemistry for Thought

21.60 Genetic engineering shows great promise for the future but has been controversial at times. Discuss with some classmates the pros and cons of genetic engineering. List two benefits and two concerns that come from your discussion.

21.61 Two samples of DNA are compared, and one has a greater percentage of guanine-cytosine base pairs. How should that greater percentage affect the attractive forces holding the double helix together?

21.62 If DNA replication were not semiconservative, what might be another possible structure for the daughter DNA molecules?

21.63 The genetic code contains three stop signals and 61 codons that code for 20 amino acids. From the standpoint of mutations, why are we fortunate that the genetic code does not have only the required 20 amino acid codons and 44 stop signals?

21.64 Azidothymidine (AZT), a drug used to fight HIV, is believed to act as an enzyme inhibitor. What type of enzyme inhibition is most likely caused by this drug?

21.65 If DNA specifies only the primary structure of a protein, how does the correct three-dimensional protein structure develop?

21.66 When were the first experiments carried out that produced genetically modified plants?

21.67 How does the DNA content determine what reactions occur within an organism?

21.68 What would be the ramifications if DNA were single stranded?

Allied Health Exam Connection

The following questions are from these sources:

- *Nursing School Entrance Exam* © 2005, Learning Express, LLC.

- *McGraw-Hill's Nursing School Entrance Exams* by Thomas A. Evangelist, Tamara B. Orr, and Judy Unrein © 2009, The McGraw-Hill Companies, Inc.

- *NSEE Nursing School Entrance Exams,* 3rd edition © 2009, Kaplan Publishing.

- *Cliffs Test Prep: Nursing School Entrance Exams* by Fred N. Grayson © 2004, Wiley Publishing, Inc.

- *Peterson's Master the Nursing School and Allied Health Entrance Exams,* 18th edition by Marion F. Gooding © 2008, Peterson's, a Nelnet Company.

21.69 Put the following terms in the correct order of information transfer according to the central dogma of molecular biology:

RNA, DNA, protein

21.70 Which of the following are components of a nucleotide in a DNA molecule?

a. Sugar

b. A phospholipid

c. A nitrogen base

21.71 Explain the function and importance of each of the following in the formation of recombinant DNA:

a. Plasmid

b. Restriction enzyme

c. DNA ligase

21.72 The genes that encode for eukaryotic protein sequences are passed from one generation to the next via:

a. other proteins.

b. rRNA.

c. tRNA.

d. DNA.

21.73 Which base is found in DNA but not in RNA?

a. thymine

b. cytosine

c. guanine

d. adenine

21.74 Which of the following best describes tRNA?

a. It is the site of rRNA synthesis.

b. It is the product of transcription, and encodes for translated proteins.

c. It binds specific amino acids and carries them to the ribosomes during protein synthesis.

d. It contains the genome and is the site at which genes are used to make mRNA.

21.75 Which of the following is the site of protein synthesis within a eukaryotic cell?

a. the ribosomes

b. the nucleus

c. the mitochondria

d. the Golgi apparatus

21.76 In messenger RNA, a codon contains how many nucleotides?

a. one

b. two

c. three

d. four

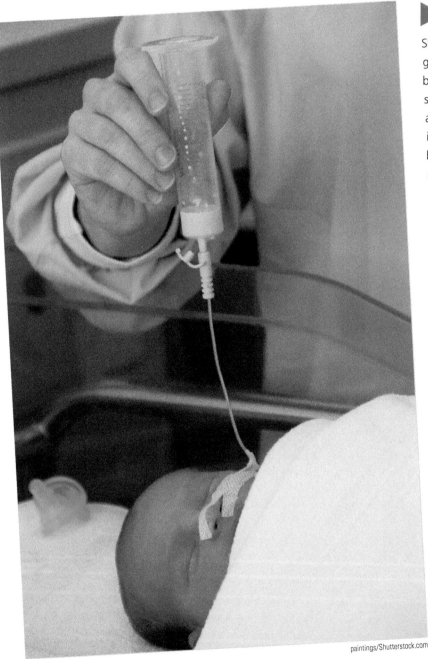

22

Nutrition and Energy for Life

▶ Case Study

Shani was shocked and disappointed. She had gone into labor at 32 weeks' gestation, and her baby had been delivered by C-section. She had so wanted a natural birth! Now, just a few hours after the surgery, a lactation consultant was telling Shani that the best thing she could do for her baby was to breastfeed him. The consultant also said that because he was premature, her newborn would not be able to nurse for at least five weeks. However, they could feed her milk to him through a naso-gastric tube. The lactation consultant explained that breast milk contains the perfect balance of water, fat, protein, and antibodies, and that it changes over time to match the baby's changing needs. Shani should expect to pump eight times a day and one to two times at night. She was exhausted by the surgery but was determined to do the best for her baby. Even though she couldn't take her baby home, she could do something important to help him.

▶ **Why is breast milk a superior food for any baby, but especially a premature one? To what age of the child does the World Health Organization and the American Pediatric Association recommend breastfeeding?**

Follow-up to this Case Study appears at the end of the chapter before the *Concept Summary*.

paintings/Shutterstock.com

When you have completed your study of this chapter, you should be able to:

1 Describe the difference between macronutrients and micronutrients in terms of amounts required and their functions in the body. **(Section 22.1)**

2 Describe the primary functions in the body of each macronutrient. **(Section 22.2)**

3 Distinguish between and classify vitamins as water-soluble or fat-soluble on the basis of name and behavior in the body. **(Section 22.3)**

4 List a primary function in the body for each major mineral. **(Section 22.4)**

5 Describe the major steps in the flow of energy in the biosphere. **(Section 22.5)**

6 Differentiate among metabolism, anabolism, and catabolism. **(Section 22.6)**

7 Outline the three stages in the extraction of energy from food. **(Section 22.6)**

8 Explain how ATP plays a central role in the production and use of cellular energy. **(Section 22.7)**

9 Explain the role of coenzymes in the common catabolic pathway. **(Section 22.8)**

The science of **nutrition** is an applied field that focuses on the study of food, water, and other nutrients and the ways in which living organisms utilize them. Nutrition scientists use the principles of nutrition to obtain answers to questions that have a great deal of practical significance. For example, they might try to determine the proper components of a sound diet, the best way to maintain proper body weight, or the foods and other nutrients needed by people with specific illnesses or injuries.

The fuels of the human body are the sugars, lipids, and proteins derived from food (see **Figure 22.1**). The reactions that release energy from these substances are among the body's most important biochemical processes. In this chapter, we will study an introduction to nutrition and how cells extract energy from food.

nutrition An applied science that studies food, water, and other nutrients and the ways living organisms use them.

Figure 22.1 Carbohydrates, lipids, and proteins from food supply the energy for all of our activities.

22.1 | Nutritional Requirements

Learning Objective

1. Describe the difference between macronutrients and micronutrients in terms of amounts required and their functions in the body.

A human body must be supplied with appropriate nutrients if it is to function properly and remain healthy. Some nutrients are required in relatively large amounts (gram quantities daily) because they provide energy and materials required to repair damaged tissue or build new tissue. These **macronutrients** are the carbohydrates, lipids, and proteins contained in food. **Micronutrients** are required by the body in only small amounts (milligrams or micrograms daily). The small quantities needed suggest correctly that at least some micronutrients are utilized in enzymes. Micronutrients are classified as either vitamins or minerals.

In addition to macro- and micronutrients, the human body must receive appropriate amounts of water and fiber. The importance of water becomes obvious when we learn that 45%–75% of the human body mass is water. Fiber is an indigestible plant material composed primarily of cellulose. Fiber makes no contribution to the body in the form of macro- or micronutrients, but it prevents or relieves constipation by absorbing water and softening the stool for easier elimination.

macronutrient A substance needed by the body in relatively large amounts.

micronutrient A substance needed by the body only in small amounts.

A number of countries of the world have established nutritional guidelines for their citizens in an attempt to improve and maintain good national health. In the United States, the Nutrition Labeling and Education Act of 1990 brought sweeping changes to the regulations that define what is required on a food label. The official guidelines are called **Reference Daily Intakes (RDIs)** for proteins and 19 vitamins and minerals, and **Daily Reference Values (DRVs)** for other nutrients of public health importance. For simplicity, all reference values on food labels are referred to as **Daily Values (DVs)** (see **Figure 22.2**). In 2016, the Food and Drug Administration (FDA) finalized the new Nutrition Facts label for packaged foods. The new label will make it easier for consumers to make better informed food choices. The Food and Drug Administration (FDA) decided to use 2000 calories as a standard for energy intake in calculating the DRVs.

Figure 22.2 An example of a food label.

A 2000-calorie diet is considered about right for many adults. Recall from **Chapter 1** that the nutritional calorie is equal to 1 kilocalorie (kcal) of energy. The guidelines are reviewed and revised about every five years. The U.S. Department of Agriculture (USDA) has issued the food guide MyPlate, designed to replace the old MyPyramid posters with new recommendations (see **Figure 22.3**).

Figure 22.3 The food guide MyPlate developed by the U.S. Department of Agriculture. ChooseMyPlate.gov will help you select foods and amounts that are right for you.

22.2 | The Macronutrients

Learning Objective

2. Describe the primary functions in the body of each macronutrient.

22.2A Carbohydrates

Carbohydrates are ideal energy sources for most body functions and also provide useful materials for the synthesis of cell and tissue components. These facts, plus the relatively low cost and ready availability of carbohydrates, have led to their worldwide use as the main dietary source of energy.

Despite their importance as a dominant source of energy, many people consider foods rich in carbohydrates to be inferior, at least in part because of their reputation for being fattening. It is now recognized that this reputation is generally not deserved. Most of the excess calories associated with eating carbohydrates are actually due to the high-calorie foods eaten *with* the carbohydrates—for example, potatoes and bread are often eaten with butter, a high-energy lipid.

Dietary carbohydrates are often classified as simple or complex. **Simple carbohydrates** are the sugars we classified earlier as monosaccharides and disaccharides (see **Chapter 17**). **Complex carbohydrates** consist essentially of the polysaccharides amylose and amylopectin, which are collectively called starch. Cellulose, another polysaccharide, is also a complex carbohydrate; however, because it cannot be digested by humans, it serves a nonnutritive role as fiber.

Numerous nutritional studies include recommendations about dietary carbohydrates in their reports. One conclusion common to most of these studies is that a typical American diet does not include enough complex carbohydrates. These studies recommend that about 58% of daily calories should come from carbohydrate food. Currently, only about 46% of the typical diet comes from carbohydrates, and too much of that total comes from simple carbohydrates (see **Figure 22.4**). The 2015–2020 Dietary Guidelines issued by the U.S. Department of Health and Human Services (HHS) and the U.S. Department of Agriculture (USDA) recommends the consumption of less than 10% of calories per day from added sugars.

22.2B Lipids

Lipids, or fats, like carbohydrates, have a somewhat negative dietary image. In part, this is because the word *fat* has several meanings, and its relationship to health is sometimes

simple carbohydrates Monosaccharides and disaccharides, commonly called sugars.

complex carbohydrates The polysaccharides amylose and amylopectin, collectively called starch.

Figure 22.4 (**A**) The composition of a typical American diet; (**B**) the composition of a healthful diet with lower fat intake and reduced levels of fat.

Figure 22.5 Oil-fried foods are rich in triglycerides.

satiety A state of satisfaction or fullness.

misunderstood. In our modern society, some people think of fat as something to be avoided at all costs, regardless of body size.

As we saw in **Chapter 18**, a number of different substances are classified as lipids. However, about 95% of the lipids in foods and in our bodies are triglycerides, in which three fatty acid molecules are bonded to a molecule of glycerol by ester linkages. The fatty acids in triglycerides can be either saturated or unsaturated. Generally, triglycerides containing a high percentage of unsaturated fatty acids are liquids at room temperature and are called oils (see **Figure 22.5**). A higher concentration of saturated fatty acids causes triglyceride melting points to increase, and they are solids (fats) at room temperature.

Lipids are important dietary constituents for a number of reasons: They are a concentrated source of energy and provide more than twice the energy of an equal mass of carbohydrate; they contain some fat-soluble vitamins and help carry them through the body; and some fatty acids needed by the body cannot be synthesized and must come from the diet—they are the essential fatty acids.

In addition to metabolic needs, dietary lipids are desirable for other reasons. Lipids improve the texture of foods, and absorb and retain flavors, thus making foods more palatable. Lipids are digested more slowly than other foods and therefore prolong **satiety**, a feeling of satisfaction and fullness after a meal.

The essential fatty acids are the polyunsaturated linoleic and linolenic acids. Generally, vegetable oils are good sources of unsaturated fatty acids (see **Figure 22.6**). Some oils that are especially rich in linoleic acid come from corn, cottonseed, soybean, and wheat germ.

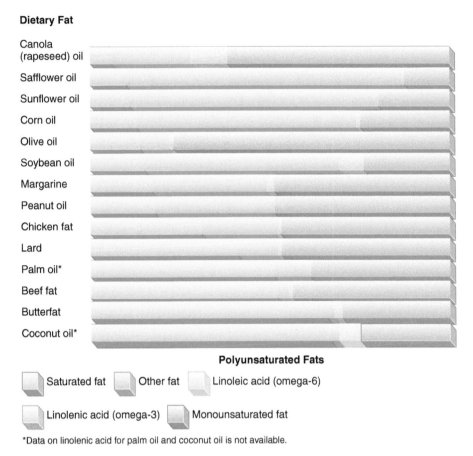

Dietary Fat

Canola (rapeseed) oil
Safflower oil
Sunflower oil
Corn oil
Olive oil
Soybean oil
Margarine
Peanut oil
Chicken fat
Lard
Palm oil*
Beef fat
Butterfat
Coconut oil*

Polyunsaturated Fats

Saturated fat Other fat Linoleic acid (omega-6)

Linolenic acid (omega-3) Monounsaturated fat

*Data on linolenic acid for palm oil and coconut oil is not available.

Figure 22.6 The fatty acid composition of common food fats.

Infants especially need linoleic acid for good health and growth; human breast milk contains a higher percentage of it than does cow's milk.

We still do not have a complete understanding of the relationship between dietary lipids and health. A moderate amount of fat is needed in everyone's diet, but many people consume much more than required. Research results indicate a correlation between the consumption of too much fat, and fat of the wrong type (saturated fatty acids), and two of the greatest health problems: obesity and cardiovascular disease. Because of these results, there is a recommended reduction in the percentage of calories obtained from fats— from the national average of almost 42% to no more than 30%. The 2015–2020 Dietary Guidelines issued by the U.S. Department of Health and Human Services (HHS) and the U.S. Department of Agriculture (USDA) recommends the consumption of less than 10% of calories per day from saturated fats.

22.2C Proteins

Proteins are the only macronutrients for which a Reference Daily Intake (RDI) has been established. The RDI varies for different groups of people, as shown in **Table 22.1**, which shows that nursing mothers have a greater need for protein than do other adults. This reflects one of the important uses of protein in the body, the production of new tissue as the body grows. In addition, proteins are needed for the maintenance and repair of cells and for the production of enzymes, hormones, and other important nitrogen-containing compounds of the body. Proteins can also be used to supply energy; they provide about 4 calories/gram, the same as carbohydrates. It is recommended that 12% of the calories obtained from food be obtained from dietary proteins (see **Figure 22.4**).

As we learned in **Chapter 19**, proteins are natural polymers of amino acids joined by amide (or peptide) linkages. On digestion, proteins are broken down to individual amino

TABLE 22.1 The RDI for Protein

Group	RDI (g)
Pregnant women	60
Nursing mothers	65
Infants under age 1	14
Children age 1 to 4	16
Adults	50

essential amino acid An amino acid that cannot be synthesized within the body at a rate adequate to meet metabolic needs.

complete protein Protein in food that contains all essential amino acids in the proportions needed by the body.

acids that are absorbed into the body's amino acid pool and used for the purposes mentioned earlier. The **essential amino acids** listed in **Table 22.2** must be obtained from the diet because they cannot be synthesized in the body in sufficient amounts to satisfy the body's needs.

The minimum quantity of each essential amino acid needed per day by an individual will vary, depending on the use in the body. For example, results of one study showed that young men need a daily average of only 7 g (0.034 mol) of tryptophan but 31 g (0.21 mol) of methionine. According to this study, the body of an average young man requires about seven times as many moles of methionine as it does of tryptophan to carry out the processes involving amino acids.

Proteins in foods are classified as **complete proteins** if they contain all the essential amino acids in the proper proportions needed by the body. Protein foods that don't meet this requirement are classified as incomplete. Several protein-containing foods are shown in **Figure 22.7**.

TABLE 22.2 **The Essential Amino Acids**
Histidine
Isoleucine
Leucine
Lysine
Methionine
Phenylalanine
Threonine
Tryptophan
Valine

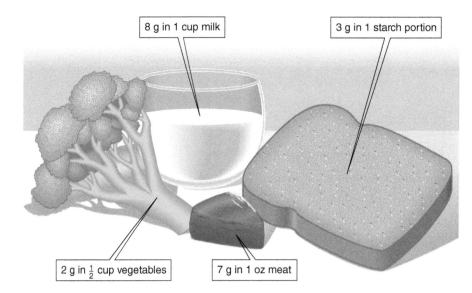

8 g in 1 cup milk

3 g in 1 starch portion

2 g in ½ cup vegetables

7 g in 1 oz meat

Figure 22.7 The protein content of several foods.

22.3 | Micronutrients I: Vitamins

Learning Objective

3. Distinguish between and classify vitamins as water-soluble or fat-soluble on the basis of name and behavior in the body.

Vitamins are organic micronutrients that the body cannot produce in amounts needed for good health. The highly polar nature of some vitamins renders them water-soluble. Nine water-soluble vitamins have been identified. Some are designated by names, some by letters, and some by letters and numbers. **Table 22.3** presents the names nutritionists use (the "correct" names), along with other common names you will likely encounter in books and on product labels. With the exception of vitamin C, all water-soluble vitamins have been shown to function as coenzymes.

Some vitamins (A, D, E, and K) have very nonpolar molecular structures and therefore dissolve only in nonpolar solvents. In the body, the nonpolar solvents are the lipids we have classified as fats, so these vitamins are called fat-soluble. The fat-soluble vitamins have diverse functions in the body, and they act somewhat like hormones (Table 22.3).

TABLE 22.3 Vitamin Sources, Functions, and Deficiency Conditions

Vitamin	Dietary Sources	Functions	Deficiency Conditions
Water-soluble			
B$_1$ (thiamin)	Bread, beans, nuts, milk, peas, pork, rice bran	Coenzyme in decarboxylation reactions	Beriberi: nausea, severe exhaustion, paralysis
B$_2$ (riboflavin)	Milk, meat, eggs, dark green vegetables, bread, beans, peas	Forms the coenzymes FMN and FAD, which are hydrogen transporters	Dermatitis (skin problems)
Niacin	Meat, whole grains, poultry, fish	Forms the coenzyme NAD$^+$, which is a hydride transporter	Pellagra: weak muscles, no appetite, diarrhea, dermatitis
B$_6$ (pyridoxine)	Meat, whole grains, poultry, fish	Coenzyme form carries amino and carboxyl groups	Dermatitis, nervous disorders
B$_{12}$ (cobalamin)	Meat, fish, eggs, milk	Coenzyme in amino acid metabolism	Rare except in vegetarians; pernicious anemia
Folic acid	Leafy green vegetables, peas, beans	Coenzyme in methyl group transfers	Anemia
Pantothenic acid	All plants and animals, nuts; whole-grain cereals	Part of coenzyme A, acyl group carrier	Anemia
Biotin	Found widely; egg yolk, liver, yeast, nuts	Coenzyme form used in fatty acid synthesis	Dermatitis, muscle weakness
C (ascorbic acid)	Citrus fruits, tomatoes, green pepper, strawberries, leafy green vegetables	Synthesis of collagen for connective tissue	Scurvy: tender tissues; weak, bleeding gums; swollen joints
Fat-soluble			
A (retinol)	Eggs, butter, cheese, dark green and deep orange vegetables	Synthesis of visual pigments	Inflamed eye membranes, night blindness, scaliness of skin
D (calciferol)	Fish-liver oils, fortified milk	Regulation of calcium and phosphorus metabolism	Rickets (malformation of the bones)
E (tocopherol)	Whole-grain cereals, margarine, vegetable oil	Prevention of oxidation of vitamin A and fatty acids	Breakage of red blood cells
K	Cabbage, potatoes, peas, leafy green vegetables	Synthesis of blood-clotting substances	Blood-clotting disorders

Care must be taken to avoid overdoses of the fat-soluble vitamins. Toxic effects are known to occur, especially with vitamin A, when excess amounts of these vitamins accumulate in body tissue. Excesses of water-soluble vitamins are excreted readily through the kidneys and are not normally a problem.

CHEMISTRY AROUND US 22.1

The 10 Most Dangerous Foods to Eat While Driving

We don't often think of it, but every time we drive a car, we are maneuvering a two-ton piece of machinery at speeds up to 80 mph! And yet, as if that weren't enough, some drivers eat and drink as they do it! Exxon conducted a survey in 2001 and found that 70% of auto drivers eat while they are driving; 85% drink coffee, juice, or soda; and a few confessed they would use a microwave in the car if they could! The National Highway Traffic Safety Administration tracked 6.3 million auto crashes in one year, and attributed 25% to food and drink distractions.

With the hurried pace of society, it is no wonder many of us try to eat and drive at the same time. However, some foods and drinks have disaster written all over them when combined with driving. Liquids are especially problematic as they spill easily. Most drivers do not take time to pull over and clean up a mess, but instead continue to drive and simultaneously clean the car and/or themselves. According to studies, the following are the top 10 food and drink culprits that contribute to causing accidents and collisions:

1. Coffee: It is usually way too hot and burns the mouth. The coffee also finds a way out of the cup and onto a clean white blouse.

iStockphoto.com/Andyd

2. Hot Soup: It has the same problem as coffee because most people drink it.

iStockphoto.com/Viktor Pravdica

3. Tacos: Notorious for coming apart at the first bite, most tacos end up scattered all over the seat, creating a salad bar effect.

iStockphoto.com/DNY59

4. Chili: The spoon rarely makes it to the mouth without losing some of the food.

iStockphoto.com/DNY59

5. Hamburgers: Our hands get greasy, and the mustard and ketchup squeeze out the other end, making steering almost impossible!

iStockphoto.com/Kelly Cline

6. Barbecued food: It tastes good, but that slippery, messy sauce will end up everywhere!

iStockphoto.com/Ryerson Clark

7. Fried chicken: The grease ends up on your hands, face, seat, and clothes. The steering wheel usually becomes greasy, which makes it nearly impossible to steer.

iStockphoto.com/Robert Byron

8. Jelly or cream-filled doughnuts: No one has ever eaten these types of doughnuts without creating some sort of mess. Why would you want to try to eat them while driving a car?

iStockphoto.com/tonyoquias

9. Soft drinks: They are subject to spills, and the carbonation can also trigger that moment when the nose gets all tingly and distracts you from your driving.

iStockphoto.com/Vasko Miokovic

10. Chocolate: It tastes yummy but ends up all over the hands and everything you touch! If the car is warm, good luck trying to keep the chocolate from melting!

iStockphoto.com/Picsfive

Select a Heart-Healthful Diet

People make food choices in the context of cultural norms and lifelong habits. Regional foods and family practices become part of our lives from birth. Families pass down food traditions to each generation. These food choices become deeply ingrained habits. If your food-choice inheritance is not healthful, it can be difficult to change your dietary habits.

Making changes requires education. Knowing what foods to choose, and why, allows you to make healthful choices. Since heart health is the cornerstone of longevity, shaping your diet to meet heart-health goals is a good place to start. Here are some guidelines for fostering a healthy heart:

1. Be aware of portion size.

Portion size is a fundamental part of a healthful diet. In America, portion size has increased over time to unhealthful proportions. Most restaurant portions are made too large in an attempt to convince customers that they are getting "more for their money." However, eating all that food ensures that it is going to "waist."

Recommended daily serving sizes can be found at *choosemyplate.gov*. Once you know the correct portion size, a good strategy when dining out is to ask for a take-home container at the beginning of the meal and remove the extra food from your plate before you start eating, so you won't be tempted to overeat.

2. Eat a plant-based diet.

Research shows that diets low in calories, low in fats, high in fiber, and high in phytochemicals, such as lycopene, promote heart health. Dieticians recommend building your meals around fruits and vegetables, including whole grains. Now, rather than a plate of steak with a side of potato and a vegetable, the image of a healthful meal consists of a large salad including a variety of vegetables and whole grains with a small portion of fish or chicken, or better yet, with nuts or legumes for protein. Meat plays a minimal role in these healthful diets, while the whole grains contribute soluble and insoluble fiber that help reduce cholesterol levels.

3. Choose healthful proteins.

Muscle meats—such as pork, beef, lamb, and veal, as well as processed meats—are not healthful additions to a heart healthful diet. Choose lean meats such as fish and skinless chicken or turkey. Choose low-fat varieties of dairy products, including no-fat yogurt, skim milk, low-fat cheeses, and ice milk rather than ice cream. Legumes such as dried beans, peas, and lentils offer excellent meat alternatives and provide additional fiber, which is an important part of a heart-healthful diet.

4. Limit unhealthful fats.

Saturated and trans fats are known to contribute to coronary artery disease; therefore limiting these unhealthful fats is an important part of a heart-healthful diet. Saturated fats such as butter, shortening, and margarine are solids at room temperature, and should be minimized in the diet. To further avoid saturated fats, trim all visible fat from meats and choose only lean meats. Replace saturated fats with monounsaturated or polyunsaturated fats such as olive oil, canola oil, fish oils, avocados, nuts, and seeds. A good compromise is using butter made spreadable by adding canola oil.

5. Reduce sodium intake.

High sodium intake contributes to high blood pressure, which is an important risk factor in heart disease. Healthy adults should not exceed an intake of 2300 mg of sodium per day. This is the amount of sodium in about a half teaspoon of table salt. This is very little salt, and it is much less than most Americans consume. People over the age of 50, African Americans, and anyone with diabetes or kidney disease should limit their sodium intake even more, to not exceed 1500 mg per day. Omitting salt from home recipes is a good start, but in order to reach those recommended goals, most processed foods, especially canned soups and stews, should be avoided. Also, be aware that sea salt contains no less sodium than regular table salt.

6. Allow an occasional treat.

Since dietary habits are significant parts of our behavior, changing too much too soon can undermine your success. Life without cheese or ice cream might seem too difficult; some people feel a healthful diet is just not worth the effort involved in giving up their favorite foods. Make changes slowly, but always be moving toward more healthful choices. As a treat, allow for an occasional splurge, such as a small piece of cake at a party. This will help you stay on track by avoiding an all-or-nothing mentality.

7. Plan ahead.

Planning is key to successful healthful eating. Prosperous American communities are filled with attractive but unhealthful eating choices. In order to avoid unhealthful food, you must plan ahead. Keep washed and cut vegetables available for quick snacks. Place fruit in a bowl on the counter to remind you to eat it. Cook from scratch when possible, and make enough to provide for the next few days. Freeze leftovers for a quick dinner.

Planning your own healthful meals overcomes the temptation to grab a high-calorie, sodium- and fat-laden alternative. Never shop when you are hungry; eat a piece of fruit or a handful of nuts before you go to the grocery store just to keep you on track. Avoid impulse buying by shopping from a list. Think about your heart today, and plan your meals for tomorrow.

Strive to eat heart-healthful foods every day, so you can reap the benefits of a long and healthy life.

Allow yourself an occasional treat.

22.4 | Micronutrients II: Minerals

Learning Objective

4. List a primary function in the body for each major mineral.

mineral A metal or nonmetal used in the body in the form of ions or compounds.

major mineral A mineral found in the body in quantities greater than 5 g.

trace mineral A mineral found in the body in quantities smaller than 5 g.

The word **mineral** is generally used to describe inorganic substances, often as if they were in the elemental form. Thus, we might hear that a food is a good source of phosphorus, when in fact the food does not contain elemental phosphorus, a caustic nonmetal, but probably contains phosphate ions such as HPO_4^{2-} and $H_2PO_4^-$ or, more likely, organic substances that contain phosphorus, such as nucleic acids and phospholipids.

In the body, the elements classified as minerals are never used in elemental form, but rather in the form of ions or compounds. They are classified as **major minerals** and **trace minerals** on the basis of the amount present in the body, as shown in **Figure 22.8**. In general, the major minerals are required in larger daily amounts than are the trace minerals. For example, the RDIs for calcium and phosphorus, the major minerals found in the body in largest amounts, are 800 mg. The RDIs for iron and iodine, two of the trace minerals, are 10–18 mg and 0.12 mg, respectively. By way of comparison, the RDIs for vitamin C and niacin are 55–60 mg and 18 mg, respectively.

Figure 22.8 Minerals in a 60-kg person. The major minerals are those present in amounts larger than 5 g (a teaspoon); only four of the numerous trace minerals are shown. Only calcium and phosphorus appear in amounts larger than a pound (454 g).

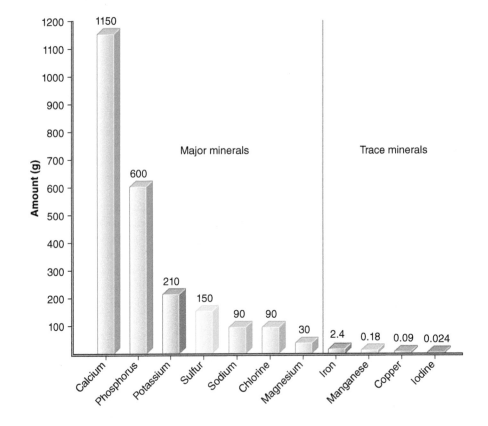

As shown in **Table 22.4**, the functions of minerals are consistent with their classification as major or trace and with the amount required daily in the diet. For example, compounds of some major minerals (Ca and P) are the primary inorganic structural components of bones and teeth. Other major minerals (Na, K, Cl, and Mg) form principal ions that are distributed throughout the body's various fluids. Some trace minerals are components of vitamins (Co), enzymes (Zn and Se), hormones (I), or specialized proteins (Fe and Cu). Thus, we see that even though trace minerals are required in small quantities, their involvement in critical enzymes, hormones, and the like makes them equally as important for good health as the major minerals.

TABLE 22.4 Major and Trace Mineral Sources, Functions, and Deficiency Conditions

Mineral	Dietary Sources	Functions	Deficiency Conditions
Major minerals			
Calcium	Dairy foods, dark green vegetables	Bone and teeth formation, blood clotting, nerve impulse transmission	Stunted growth, rickets, weak and brittle bones
Chlorine	Table salt, seafood, meat	HCl in gastric juice, acid–base balance	Muscle cramps, apathy, reduced appetite
Magnesium	Whole-grain cereals, meat, nuts, milk, legumes	Activation of enzymes, protein synthesis	Inhibited growth, weakness, spasms
Phosphorus	Milk, cheese, meat, fish, grains, legumes, nuts	Enzyme component, acid–base balance, bone and tooth formation	Weakness, calcium loss, weak bones
Potassium	Meat, milk, many fruits, cereals, legumes	Acid–base and water balance, nerve function	Muscle weakness, paralysis
Sodium	Most foods except fruit	Acid–base and water balance, nerve function	Muscle cramps, apathy, reduced appetite
Sulfur	Protein foods	Component of proteins	Deficiencies are very rare
Trace minerals			
Arsenic[a]	Many foods	Growth and reproduction	Poor growth and reproduction
Cobalt	Meat, liver, dairy foods	Component of vitamin B_{12}	Pernicious anemia (vitamin-deficiency symptom)
Copper	Drinking water, liver, grains, legumes, nuts	Component of numerous enzymes, hemoglobin formation	Anemia, fragility of arteries
Chromium	Fats, vegetable oils, grains, meat	Enhances insulin action	Reduced ability to metabolize glucose
Fluorine	Drinking water, seafood, onions, spinach	Maintenance of bones and teeth	Higher frequency of tooth decay
Iodine	Iodized salt, fish, dairy products	Component of thyroid hormones	Hypothyroidism, goiter
Iron	Liver, lean meat, whole grains, dark green vegetables	Component of enzymes and hemoglobin	Anemia
Manganese	Grains, beet greens, legumes, fruit	Component of enzymes	Deficiencies are rare
Molybdenum	Legumes, cereals, organ meats, dark green vegetables	Component of enzymes	Deficiencies are rare
Nickel	Many foods	Needed for health of numerous tissues	Organ damage, deficiencies are rare
Selenium	Grains, meat, poultry, milk	Component of enzymes	Deficiencies are rare
Silicon[a]	Many foods	Bone calcification	Poor bone development, deficiencies are rare
Tin[a]	Many foods	Needed for growth	Poor growth
Vanadium[a]	Many foods	Growth, bone development, reproduction	Poor growth, bone development, and reproduction
Zinc	Milk, liver, shellfish, wheat bran	Component of numerous enzymes	Poor growth, lack of sexual maturation, loss of appetite, abnormal glucose tolerance

[a]Need and deficiency symptoms determined for animals, probable in humans, but not yet recognized.

22.5 | The Flow of Energy in the Biosphere

Learning Objective

5. Describe the major steps in the flow of energy in the biosphere.

The sun is the ultimate source of energy used in all biological processes. The sun's enormous energy output is the result of the fusion of hydrogen atoms into helium atoms. The net reaction for this process is:

$$4H \xrightarrow{\text{nuclear fusion}} He + energy \qquad (22.1)$$

A portion of the liberated energy reaches Earth as sunlight and is absorbed by chlorophyll pigments in plants. In the plants, the energy drives the reactions of photosynthesis that convert carbon dioxide and water first into glucose, and then into starch, triglycerides, and other storage forms of energy. The net reaction for glucose production is:

$$6CO_2 + 6H_2O + energy \xrightarrow{photosynthesis} C_6H_{12}O_6 + 6O_2 \qquad (22.2)$$
$$glucose$$

All animals obtain energy either directly or indirectly from these energy stores in plants.

cellular respiration The entire process involved in the use of oxygen by cells.

During **cellular respiration**, both plants and animals combine these energy-rich compounds with oxygen from the air, producing carbon dioxide and water, and releasing energy. Cellular respiration, which should not be confused with pulmonary respiration (breathing), is represented as:

$$glucose\ and\ other\ storage\ forms\ of\ energy + O_2 \xrightarrow{respiration} CO_2 + H_2O + energy\ released \qquad (22.3)$$

A portion of the energy released during respiration is captured within the cells in the form of adenosine triphosphate (ATP), a chemical carrier of energy, from which energy can be obtained and used directly for the performance of biological work. The remainder of the energy from respiration is liberated as heat. Steps in the flow of energy in the biosphere are summarized in **Figure 22.9**.

Figure 22.9 Energy flow in the biosphere.

Notice in Figure 22.9 that atmospheric carbon dioxide is the initial source of carbon for carbohydrates. Another important feature of Figure 22.9 is that overall, the respiration process is the reverse of the photosynthetic process. Thus, some of Earth's carbon compounds are repeatedly recycled by living organisms (see **Figure 22.10**).

Figure 22.10 A simplified carbon cycle.

22.6 | Metabolism and an Overview of Energy Production

Learning Objectives

6. Differentiate among metabolism, anabolism, and catabolism.
7. Outline the three stages in the extraction of energy from food.

22.6A Metabolism

Living cells are very active chemically, with thousands of different reactions occurring at the same time. The sum total of all the chemical reactions involved in maintaining a living cell is called **metabolism**. The reactions of metabolism are divided into two categories, catabolism and anabolism. **Catabolism** consists of all reactions that lead to the breakdown of biomolecules. **Anabolism** includes all reactions that lead to the synthesis of biomolecules. In general, energy is released during catabolism and required during anabolism.

A characteristic of all living organisms is a high degree of order in their structures and chemical processes. Even the simplest organisms exhibit an enormous array of chemical reactions that are organized into orderly, well-regulated sequences known as **metabolic pathways**. Each metabolic pathway consists of a series of consecutive chemical reactions or steps that convert a starting material into an end product. For example, two pathways that will be covered in **Chapter 23** are the citric acid cycle and the electron transport chain.

Fortunately for those who study the biochemistry of living systems, a great many similarities exist among the major metabolic pathways found in all life forms. These similarities enable scientists to study the metabolism of simple organisms and use the results to help explain the corresponding metabolic pathways in more complex organisms, including humans. This text concentrates on human biochemistry, but you should realize that much of what is said about human metabolism can be applied equally well to the metabolism of most other organisms.

metabolism The sum of all reactions occurring in an organism.

catabolism All reactions involved in the breakdown of biomolecules.

anabolism All reactions involved in the synthesis of biomolecules.

metabolic pathway A sequence of reactions used to produce one product or accomplish one process.

22.6B The Catabolism of Food

The three stages involved in the catabolic extraction of energy from food are illustrated in **Figure 22.11**.

Figure 22.11 The three stages in the extraction of energy from food.

Sport Supplements: Where Is My Edge?

It is not uncommon for people participating in competitive physical activities to seek some kind of performance advantage, or at least to have reasons to believe that they are performing at their best. This is true not only in competitive activities, but in life itself. The Olympic mantra of *Citius, Altius, Fortius* (Faster, Higher, Stronger), prompts some of us to search for dietary supplements to help in achieving that ideal. You adhere to your training and nutrition regimen, and you are a disciple of good health behaviors. Is that it? Are you as good as you can be? You wonder, are there dietary supplements available that might allow me to perform better? You would think so, at least in terms of the countless products available, the many promotions and testimonials found in the media, and the numerous "you should try this" comments that occur in the gym. This discussion contains a brief review of some of the most common dietary supplements used to improve performance.

Protein is the most common nutrient used to improve performance. The building blocks of proteins are amino acids. Proteins from different sources will contain different mixtures of amino acids. Before protein supplements became available, a high-protein diet was used to increase protein intake. But that often meant an increase in caloric intake beyond what was wanted. Today, there are good sources of protein available as dietary supplements that are usually ingested in the form of a "protein shake" (protein powder mixed with a liquid). These shakes are usually ingested after a workout to decrease muscle breakdown and encourage muscle growth. The timing of the dose is important, and it is considered to be ideal to ingest it within 30 minutes of completing an exercise workout. Each type of protein has its own advantages and disadvantages, as shown in the following list:

Whey Protein: This "quick-acting" milk protein is quickly absorbed. It can be taken at least 30 minutes before a workout and immediately after a workout. Timing is important for maximum benefit. This type of protein satisfies the appetite in addition to enhancing muscle building.

Casein Protein: This long-acting milk protein is digested slowly. The onset of action is slow, but the duration of its effect is the longest. It is best taken at least one hour after working out or at bedtime.

Soy Protein: This intermediate-acting vegetable protein can be taken at least 45 minutes before a workout and immediately after a workout. This is a good choice for individuals with an allergy or sensitivity to milk proteins. It is thought that its high arginine content might improve vascular health by increasing nitric oxide production.

Amino Acids: Some protein supplements are fortified by the addition of particular amino acids. The added amino acids are available individually or in different proprietary combinations.

Branched Chain Amino Acids: Supplementing foods with the branched chain amino acids leucine, isoleucine, and valine is particularly useful because they can be used by muscles as fuel, and consequently are easily depleted. Other advantages of these supplements is they increase muscle growth and decrease muscle soreness and recovery time following a workout.

Glutamine: This is the most prevalent amino acid in muscle tissue. It helps replace back muscle that is damaged by exercise. It shortens the muscle recovery time following exercise.

Creatine: Creatine is normally found in meat and fish and is produced in the kidneys and liver. Creatine is stored mainly in muscles. Oral consumption increases the concentration in muscle tissue, which aids in the phosphorylation process of adenosine diphosphate (ADP) to produce adenosine triphosphate (ATP), which is the energy source for cells. Creatine appears to provide modest benefits for intense, repetitive exercise lasting less than 30 seconds. It does not appear to provide significant benefits for endurance aerobic exercise.

Vasodilators: The amino acid arginine is frequently used as a pre-workout supplement that increases nitric oxide production. Nitric oxide is secreted by endothelial cells and causes the dilation of blood vessels, which increases the rate of blood flow. It is unclear if using a supplement that increases nitric oxide production provides any benefits to healthy individuals.

Caffeine: Almost all pre-workout supplements contain caffeine, which is a central nervous system stimulant. Depending on dose levels and tolerance, caffeine is a mild stimulant that provides a boost in energy and alertness, and modest increases in the rate of metabolism.

Omega-3 Fatty Acids: The typical American diet includes significantly more omega-6 fatty acids than omega-3 fatty acids. That ratio leads to a greater tendency for tissue inflammation. Supplementing the diet with omega-3 fatty acids can decrease the likelihood of tissue inflammation. The most beneficial of the omega-3 fatty acids in the body are eicosapentaenoic acid (abbreviated EPA) and docosahexaenoic acid (abbreviated DHA). Fish oils are the usual dietary source of omega-3 fatty acids, but read the labels of fish oil products carefully, because the majority of the fatty acids might not be the beneficial EPA and DHA. Select a

fish oil product that contains primarily EPA and DHA. The benefits of using products containing EPA and DHA have been found to include decreased muscle soreness, improved insulin sensitivity, and increased blood flow. There is evidence supporting the use of omega-3 supplements; however, some data are contradictory. The studies have often been done with athletes as the subjects. This raises the question about how the study results should translate to people with different levels of fitness, age, and gender. Potential users of omega-3 supplements should objectively assess the results of using the supplements and follow the recommended dosing guidelines on product labels.

Protein shakes are a typical method of supplementing dietary protein.

Stage I is digestion, in which large, complex molecules are chemically broken into relatively small, simple ones (see **Figure 22.12**). The most common reaction in digestion is hydrolysis, in which proteins are converted to amino acids, carbohydrates are converted to monosaccharides (primarily glucose, galactose, and fructose), and fats are converted to fatty acids and glycerol. The smaller molecules are then absorbed into the body through the lining of the small intestine.

$$\text{proteins} \xrightarrow{\text{hydrolysis}} \text{amino acids}$$

$$\text{disaccharides and polysaccharides} \xrightarrow{\text{hydrolysis}} \text{monosaccharides}$$

$$\text{fats and oils} \xrightarrow{\text{hydrolysis}} \text{fatty acids} + \text{glycerol}$$

Figure 22.12 The digestion of food involves enzyme-catalyzed hydrolysis reactions.

In Stage II, the small molecules from digestion are degraded to even simpler units, primarily the two-carbon acetyl portion of acetyl coenzyme A (acetyl CoA):

$$\underbrace{CH_3 - \overset{\overset{\textstyle O}{\|}}{C}}_{\text{acetyl}} - \underbrace{S - CoA}_{\text{coenzyme A}}$$

Some energy is released in the second stage, but much more is produced during the oxidation of the acetyl units of acetyl CoA in the third stage. Stage III consists of the citric acid cycle followed by electron transport chain and oxidative phosphorylation. Because the reactions of Stage III are the same regardless of the type of food being degraded, Stage III is sometimes referred to as the **common catabolic pathway**. Energy released during Stage III appears in the form of energy-rich molecules of ATP. The whole purpose of the catabolic pathway is to convert the chemical energy in foods to molecules of ATP. The common catabolic pathway will be covered in Chapter 23. **Chapters 23–25** discuss the ways in which carbohydrates, lipids, and proteins provide molecules that are degraded by reactions of the common catabolic pathway.

common catabolic pathway The reactions of the citric acid cycle plus those of the electron transport chain and oxidative phosphorylation.

22.7 | ATP: The Primary Energy Carrier

Learning Objective

8. Explain how ATP plays a central role in the production and use of cellular energy.

Certain bonds in ATP store the energy released during the oxidation of carbohydrates, lipids, and proteins. ATP molecules act as energy carriers and deliver the energy to the parts of the cell where energy is needed to power muscle contraction, biosynthesis, and other cellular work.

22.7A The Structure and Hydrolysis of ATP

The structure of the ATP molecule is relatively simple (see **Figure 22.13**). It consists of a heterocyclic base, adenine, bonded to the sugar ribose. Taken together, this portion of the molecule is called adenosine. The triphosphate portion is bonded to the ribose to give adenosine triphosphate, or ATP. At the cell pH of 7.4, all the protons of the triphosphate group are ionized, giving the ATP molecule a charge of -4. In the cell, ATP is complexed with Mg^{2+} in a 1:1 ratio. Thus, the 1:1 complex exhibits a net charge of -2. Two other triphosphates—GTP, containing the base guanine, and UTP, containing the base uracil—are important in carbohydrate metabolism (see **Chapter 23**).

Figure 22.13 The structure of ATP.

The triphosphate end of ATP is the part of the molecule that is important in the transfer of biochemical energy. The key reaction in this energy delivery system is the transfer of a phosphoryl group, $-PO_3^{2-}$, from ATP to another molecule. For example, during the hydrolysis of ATP in water, a phosphoryl group is transferred from ATP to water. The products of this hydrolysis are adenosine diphosphate (ADP) and a phosphate ion, often referred to as an inorganic phosphate, P_i, or simply as *phosphate*.

The transfer of a phosphoryl group from ATP to water is accompanied by a release of energy. This energy is called free energy and is represented by the symbol ΔG. When energy is released by a reaction or process, ΔG will have a negative value. When energy is absorbed, ΔG will have a positive value. When ΔG is measured under standard conditions (temperature, concentration, etc.), it is represented by $\Delta G°$, and a superscript prime ($\Delta G°'$) is added if $\Delta G°$ is measured at body conditions. The liberated free energy is available for use by the cell to carry out processes requiring an input of energy. This energy-releasing hydrolysis reaction (see **Figure 22.14**) is represented by the following equation:

$$\text{ATP} + \text{H}_2\text{O} \rightarrow \text{ADP} + \text{P}_i + \text{H}^+ \quad \Delta G°' = -7.3 \text{ kcal/mol} \quad (22.5)$$

Figure 22.14 Schematic representation of the lowering of energy. From Campbell, *Biochemistry*, 8E. © 2015 Cengage Learning.

A. A ball rolls down a hill, releasing potential energy.

B. ATP is hydrolyzed to produce ADP and phosphate ion, releasing energy. The release of energy when a ball rolls down a hill is analogous to the release of energy in a chemical reaction.

Many other phosphate-containing compounds found in biological systems also liberate energy on hydrolysis. Several are listed in **Table 22.5**, along with values for their free standard energy of hydrolysis.

Compounds that liberate a great amount of free energy on hydrolysis are called **high-energy compounds**. Remember, a negative sign for $\Delta G°'$ means the energy is liberated when the reaction takes place. Thus, ATP ($\Delta G°' = -7.3$ kcal/mol) is considered a high-energy compound, but not the only one found in cells, as you can see from Table 22.5.

high-energy compound A substance that on hydrolysis liberates a great amount of free energy.

TABLE 22.5 Free Energy Changes for the Hydrolysis of Some Phosphate Compounds

Reaction			$\Delta G°'$ (kcal/mol)
Phosphoenolpyruvate + H_2O	\rightarrow	pyruvate + phosphate	−14.8
Carbamoyl phosphate + H_2O	\rightarrow	ammonia + CO_2 + phosphate	−12.3
1,3-bisphosphoglycerate + H_2O	\rightarrow	3-phosphoglycerate + phosphate	−11.8
Phosphocreatine + H_2O	\rightarrow	creatine + phosphate	−10.3
ATP + H_2O	\rightarrow	ADP + phosphate	−7.3
ADP + H_2O	\rightarrow	AMP + phosphate	−7.3
Glucose-1-phosphate + H_2O	\rightarrow	glucose + phosphate	−5.0
AMP + H_2O	\rightarrow	adenosine + phosphate	−3.4
Glucose-6-phosphate + H_2O	\rightarrow	glucose + phosphate	−3.3
Glycerol-3-phosphate + H_2O	\rightarrow	glycerol + phosphate	−2.2

The hydrolysis of ATP to ADP is the principal energy-releasing reaction for ATP. However, some other hydrolysis reactions also play important roles in energy transfer. An example is the hydrolysis of ATP to adenosine monophosphate (AMP) and pyrophosphate (PP_i):

$$\text{adenosine}-O-\underset{\underset{O^-}{|}}{\overset{\overset{O}{\|}}{P}}-O-\underset{\underset{O^-}{|}}{\overset{\overset{O}{\|}}{P}}-O-\underset{\underset{O^-}{|}}{\overset{\overset{O}{\|}}{P}}-O^- + H_2O \longrightarrow \text{adenosine}-O-\underset{\underset{O^-}{|}}{\overset{\overset{O}{\|}}{P}}-O^- + {}^-O-\underset{\underset{O^-}{|}}{\overset{\overset{O}{\|}}{P}}-O-\underset{\underset{O^-}{|}}{\overset{\overset{O}{\|}}{P}}-O^- + 2H^+ \tag{22.6}$$

<div align="center">ATP AMP PP_i
adenosine pyrophosphate
monophosphate</div>

The reaction may also be written as:

$$\text{ATP} + H_2O \rightarrow \text{AMP} + PP_i + 2H^+ \qquad \Delta G^{\circ\prime} = -10.9 \text{ kcal/mol} \tag{22.7}$$

STUDY SKILLS 22.1

Bioprocesses

The last four chapters of the text complete our study of the molecular basis of life. This chapter and the next three build on your knowledge of the biomolecules (carbohydrates, lipids, proteins, and nucleic acids) and examine the processes that support living organisms. Your attention should now be directed toward gaining an understanding of these processes, where they occur in the cell, their purpose, and, in some cases, key summary reactions.

A number of detailed pathways such as glycolysis, the fatty acid spiral, and the urea cycle will be presented. Don't let the numerous reactions in these pathways intimidate you. Although each reaction could be looked at individually, we have tried to help you focus your attention by emphasizing the overall importance of each pathway by identifying its purpose, cellular location, starting material, products, energy requirements, and regulation.

When the hydrolysis of ATP to AMP occurs during metabolism, it is usually followed by the immediate hydrolysis of the pyrophosphate, which releases even more free energy:

$$^-O-\underset{\underset{O^-}{|}}{\overset{\overset{O}{\|}}{P}}-O-\underset{\underset{O^-}{|}}{\overset{\overset{O}{\|}}{P}}-O^- + H_2O \longrightarrow 2HO-\underset{\underset{O^-}{|}}{\overset{\overset{O}{\|}}{P}}-O^- \quad \Delta G^{\circ\prime} = -4.6 \text{ kcal/mol} \tag{22.8}$$

<div align="center">PP_i P_i</div>

The hydrolyses of ATP and related compounds are summarized in **Table 22.6**.

TABLE 22.6 Hydrolyses of Some ATP-Related Compounds

Reaction			$\Delta G^{\circ\prime}$ (kcal/mol)
$\text{ATP} + H_2O$	\rightarrow	$\text{ADP} + P_i + H^+$	-7.3
$\text{ATP} + H_2O$	\rightarrow	$\text{AMP} + PP_i + 2H^+$	-10.9
$PP_i + H_2O$	\rightarrow	$2P_i$	-4.6
$\text{ADP} + H_2O$	\rightarrow	$\text{AMP} + P_i + H^+$	-7.3

In the United States, obesity is now considered to be a national problem. Many adults are overweight, obese, or even morbidly obese. Even more concerning is the increase in adolescent obesity. The concern with obesity is not an issue of vanity, but rather is an issue of health. Obesity is directly or indirectly linked to a wide array of serious negative health conditions that impact not only the health of individuals, but also the health of our society as a whole.

Part of the problem is the demand for quick and satisfying foods to fit into our hectic lives. It is easy to grab burgers, fries, and soda for the family at a fast-food restaurant as we rush from school and the office to soccer games and ballet. But the tide is changing. People want healthier choices. As a result, fast-food restaurants and prepackaged food industries are making changes. But are these changes in our convenience foods really a significant improvement? Or do they make us believe we are eating better when, in reality, we are simply trading one bad choice for another, but are unaware of it?

If we educate ourselves on the meaning of the new marketing language being used, we might avoid falling for misleading calorie-content claims for foods. The following legal definitions will help you make more informed decisions about your foods.

- "Calorie free" — items contain less than 5 calories **per serving**
- "Low calorie" — items contain less than 40 calories **per serving**

- "Reduced" or "fewer calories" — items contain 25% **fewer calories than the original product**
- "Light" — items contain 50% less fat OR 1/3 the calories of the original product

The above terminology can be tricky and confusing, especially when you need to break out a calculator and a scale to determine how many servings you are consuming and the total calories. Keep in mind that there are very few truly calorie-free items that we consume, aside from pure unadulterated water. For more information or if you need assistance wading through the nutrition labels of foods, check out the following website: choosemyplate.gov

Fast-food restaurants are beginning to offer more healthful choices.

iStockphoto.com/Jaimie D. Travis

22.7B The ATP–ADP Cycle

In biological systems, ATP functions as an immediate donor of free energy rather than as a storage form of free energy. In a typical cell, an ATP molecule is hydrolyzed within one minute following its formation. Thus, the turnover rate of ATP is very high. For example, a resting human body hydrolyzes ATP at the rate of about 40 kg every 24 hours. During strenuous exertion, this rate of ATP utilization may be as high as 0.5 kg per minute. Motion, biosynthesis, and other forms of cellular work can occur only if ATP is continuously regenerated from ADP. **Figure 22.15** depicts the central role served by the ATP–ADP cycle in linking energy production with energy utilization.

Figure 22.15 The ATP–ADP cycle. ATP, which supplies the energy for cellular work, is continuously regenerated from ADP during the oxidation of fuel molecules.

22.7C Mitochondria

Enzymes that catalyze the formation of ATP, as well as the other processes in the common catabolic pathway, are all located within the cell in organelles called **mitochondria**

mitochondrion A cellular organelle where reactions of the common catabolic pathway occur.

Nutrition and Energy for Life **721**

(see **Figure 22.16**). Mitochondria are often called cellular "power stations" because they are the sites for most ATP synthesis in the cells.

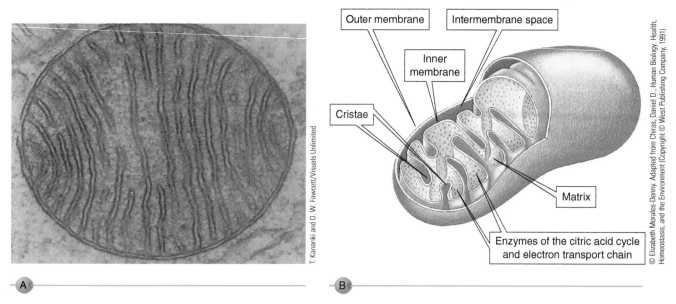

Figure 22.16 The mitochondrion: (**A**) a photomicrograph; (**B**) a schematic drawing, cut away to reveal the internal organization.

A mitochondrion contains both inner and outer membranes. As a result of extensive folding, the inner membrane has a surface area that is many times that of the outer membrane. The folds of the inner membrane are called *cristae*, and the gel-filled space that surrounds them is called the *matrix*. The enzymes for ATP synthesis (electron transport and oxidative phosphorylation) are located on the matrix side of the cristae. The enzymes for the citric acid cycle are found within the matrix, attached or near to the surface of the inner membrane. Thus, all the enzymes involved in the common catabolic pathway are in close proximity.

22.8 | Important Coenzymes in the Common Catabolic Pathway

Learning Objective

9. Explain the role of coenzymes in the common catabolic pathway.

As discussed in **Section 20.3**, coenzymes often participate in enzyme-catalyzed reactions by acting as shuttle systems in the transfer of chemical groups from one compound to another. For example, a coenzyme may accept hydrogen from one compound and later donate hydrogen to another compound. We also pointed out that many important coenzymes are formed from vitamins. The three substances discussed in this section exhibit both of these coenzyme characteristics.

22.8A Coenzyme A

Coenzyme A is a central compound in metabolism. It is a part of acetyl coenzyme A (acetyl CoA), the substance formed from all foods as they pass through Stage II of catabolism (see **Figure 22.17**).

Figure 22.17 Stage II of the oxidation of foods. Acetyl CoA is formed as the pathways of catabolism converge.

$$\left.\begin{array}{l} \text{Fatty acids and glycerol} \\ \text{Glucose and other sugars} \\ \text{Amino acids} \end{array}\right\} \longrightarrow \underset{\substack{\text{acetyl coenzyme A} \\ \text{(acetyl CoA)}}}{CH_3 - \overset{\overset{\displaystyle O}{\|}}{C} - S - CoA}$$

Coenzyme A is derived from the B vitamin pantothenic acid (B₅) (see **Figure 22.18**). It also contains two other components, a phosphate derivative of ADP and b-mercapto-ethylamine. A key feature in the structure of coenzyme A is the presence of the sulfhydryl group (—SH), which is the reactive portion of the molecule. For this reason, coenzyme A is often abbreviated as CoA—SH.

Figure 22.18 The structure of coenzyme A. The sulfhydryl group (—SH) is the reactive site.

The letter *A* is included in the coenzyme name to signify its participation in the transfer of acetyl groups. It is now known that coenzyme A transfers acyl groups in general rather than just the acetyl group. This is demonstrated in fatty acid oxidations (see **Section 24.4**), fatty acid synthesis (see **Section 24.7**), and other biological reactions. Acyl groups are linked to coenzyme A through the sulfur atom in a **thioester bond**.

thioester bond
The carbon–sulfur bond of the

$$\underset{\text{functional group.}}{-\overset{\overset{\displaystyle O}{\|}}{C}-S-}$$

22.8B The Coenzymes NAD⁺ and FAD

The majority of energy for ATP synthesis is released when the oxygen we breathe is reduced. During that process, oxygen accepts electrons and hydrogen ions, and water is produced. The electrons come from the oxidation of fuel molecules, such as glucose, but the electrons are not transferred directly to the oxygen from the fuel molecules or their breakdown products. Instead, these substrates first transfer the electrons to special coenzyme carriers. The reduced forms of these coenzymes then transfer the electrons to oxygen through the reactions of the electron transport chain. ATP is formed from ADP and P_i as a result of this flow of electrons.

The first of these coenzymes, nicotinamide adenine dinucleotide (NAD⁺), is a derivative of ADP and the vitamin nicotinamide (see **Figure 22.19**). The reactive site of NAD⁺ is located in the nicotinamide portion of the molecule. During the oxidation of a substrate, the nicotinamide ring of NAD⁺ accepts two electrons and one proton, which forms the reduced coenzyme NADH.

Figure 22.19 The structure of nicotinamide adenine dinucleotide (NAD^+). The reactive site is in the nicotinamide part of the molecule.

$$+ 2e^- + H^+ \rightleftharpoons \tag{22.9}$$

NAD$^+$
(oxidized form)

NADH
(reduced form)

A typical cellular reaction in which NAD^+ serves as an electron acceptor is the oxidation of an alcohol:

General reaction:

$$R - \overset{\overset{\displaystyle OH}{|}}{\underset{\underset{\displaystyle H}{|}}{C}} - H + NAD^+ \rightarrow R - \overset{\overset{\displaystyle O}{\|}}{C} - H + NADH + H^+ \tag{22.10}$$

alcohol aldehyde

Specific example:

$$CH_3CH_2 - OH + NAD^+ \xrightarrow[\text{dehydrogenase}]{\text{alcohol}} CH_3 - \overset{\overset{\displaystyle O}{\|}}{C} - H + NADH + H^+ \tag{22.11}$$

ethanol acetaldehyde

In this reaction, one hydrogen atom of the alcohol substrate is directly transferred to NAD^+, whereas the other appears in solution as H^+. Both electrons lost by the alcohol have been transferred to the nicotinamide ring in NADH.

Biochemical reactions in which coenzymes participate are often written in a more concise way. For example, Reaction 22.10 may be written:

$$
\begin{array}{ccc}
\overset{\displaystyle OH}{\underset{\displaystyle H}{R-C-H}} & \xrightarrow{\hspace{2cm}} & R-\overset{\displaystyle O}{\overset{\|}{C}}-H \\
& NAD^+ \quad NADH + H^+ &
\end{array}
\tag{22.12}
$$

This format allows for the separate emphasis of the oxidation reaction and the involvement of a coenzyme.

✔ **LEARNING CHECK 22.1** Write the following reaction in the normal form of a chemical equation:

$$
\begin{array}{ccc}
\overset{\displaystyle OH}{\underset{\displaystyle CH_2-COO^-}{CH-COO^-}} & \xrightarrow{\hspace{2cm}} & \overset{\displaystyle O}{\overset{\|}{\underset{\displaystyle CH_2-COO^-}{C-COO^-}}} \\
& NAD^+ \quad NADH + H^+ &
\end{array}
$$

Flavin adenine dinucleotide (FAD) is the other major electron carrier in the oxidation of fuel molecules. This coenzyme is a derivative of ADP and the vitamin riboflavin. The reactive site is located within the riboflavin ring system (see **Figure 22.20**).

Figure 22.20 The structure of flavin adenine dinucleotide (FAD). The reactive portion (shown in blue) is in the riboflavin ring system.

ASK AN EXPERT 6.1

Q: **Is it better to take a fiber supplement or to eat fiber-fortified foods?**

Melina B. Jampolis, M.D., is a Physician Nutrition Specialist focusing on nutrition for weight loss and disease prevention and treatment.

A: Added fiber is popping up in everything these days, including yogurt, ice cream, nutrition bars, cookies, juice, and even water. The health benefits of fiber are numerous and include promoting bowel health and regularity, cholesterol and blood sugar reduction, and increasing satiety, which can help with weight control. The recommended daily intake of fiber ranges between 25 and 38 grams per day, yet most Americans consume only about 12 grams per day. So, fiber-fortified foods or supplements could help us narrow the gap—right?

One problem with this idea is that added fibers, such as those found in many supplements or used to fortify foods, might not have the same health benefits as naturally occurring, or intact, fiber. Added fiber generally comes in the form of isolated fibers, which include inulin, polydextrose, and maltrodextrin. These fibers, also known as functional fiber, can be extracted from foods or chemically synthesized. Naturally occurring fiber-rich foods like whole grains, beans, fruits, nuts, seeds, and vegetables contain a variety of fiber types, often in a combination of water-soluble and insoluble fiber. These foods also contain vitamins, minerals, and phytonutrients, but isolated fibers offer only one type of fiber without the accompanying nutrients. In addition, they are often added to less nutrient-rich foods, so the potential health benefits, which have not been thoroughly investigated, are likely to be significantly less.

Isolated fibers do not appear to have the same benefit as intact soluble fiber in terms of cholesterol and blood sugar reduction. This is because they are not viscous, meaning they do not form a gel when added to water. Polydextrose, a synthetic indigestible glucose polymer, acts as a prebiotic, stimulating the growth of healthy probiotic bacteria in the colon. But in high doses it can have a laxative or gas-producing effect. Inulin, a soluble, indigestible fructose polymer derived from chicory root, is also a prebiotic, thereby promoting bowel health. Inulin might also help with satiety, and some studies suggest it boosts the absorption of minerals, including calcium and magnesium. In higher doses, however, it too can cause gas, bloating, and diarrhea.

Fiber supplements vary in the types of fiber that they contain. Psyllium, a naturally occurring soluble fiber, has been studied extensively, and research suggests a benefit in terms of constipation prevention, cholesterol reduction, blood sugar control, and weight loss, particularly when combined with a healthy diet. It also seems to help those suffering from irritable bowel syndrome (IBS), a common condition associated with bowel irregularity, bloating, and abdominal pain.

There is no question that obtaining fiber directly from regular foods is optimal, but there are some cases in which a supplement or fortified food might be helpful. However, more research is needed to determine the health benefits of isolated fibers. If they are added to less-healthy products like frozen desserts, fruit drinks, baked goods, or sugary cereals, do not be fooled into believing that they balance out the negative aspect of these foods. And simply adding a fiber supplement to an unhealthy diet is much less beneficial than changing the diet to include more healthful foods.

Naturally occurring fiber-rich foods include whole grains, nuts, and seeds.

Fiber supplements may have certain health benefits.

As in the reduction of NAD^+, substrates of enzymes that use FAD as the coenzyme give up two electrons to the coenzyme. However, unlike NAD^+, the FAD coenzyme accepts both of the hydrogen atoms lost by the substrate. Thus, the abbreviation for the reduced form of FAD is $FADH_2$:

$$+ 2H^+ + 2e^- \rightleftharpoons \tag{22.13}$$

FAD
(oxidized form)

$FADH_2$
(reduced form)

The substrates for oxidation reactions in which FAD is the coenzyme differ from those involving NAD^+. FAD is often involved in oxidation reactions in which a $—CH_2—CH_2—$ portion of the substrate is oxidized to a double bond:

General reaction:

$$R—\underset{\underset{H}{|}}{\overset{\overset{H}{|}}{C}}—\underset{\underset{H}{|}}{\overset{\overset{H}{|}}{C}}—R + FAD \longrightarrow R—CH{=}CH—R + FADH_2 \qquad (22.14)$$

saturated (reduced) → unsaturated (oxidized)

Specific example:

$$CH_3CH_2CH_2—\overset{\overset{O}{\|}}{C}—S—CoA \longrightarrow CH_3CH{=}CH—\overset{\overset{O}{\|}}{C}—SCoA \quad (22.15)$$

butyryl CoA

FAD FADH_2

✔ **LEARNING CHECK 22.2** Complete the following reactions:

a. $^-OOC—CH_2—CH_2—COO^- + FAD \longrightarrow$

b. $CH_3—\underset{\underset{}{\overset{\overset{OH}{|}}{CH}}}—COO^- + NAD^+ \longrightarrow$

▶ Case Study Follow-up

The American Association of Pediatrics recommends exclusive breastfeeding during the first six months, then for at least one additional year, and as long thereafter as mutually desired by mother and child. "Human milk is species-specific, and all substitute feeding preparations differ markedly from it, making human milk uniquely superior for infant feeding. In addition, human milk–fed premature infants receive significant benefits with respect to parasite protection and improved developmental outcomes compared with formula-fed premature infants. Breastfeeding decreases the incidence of a wide range of diseases, is associated with lower rates of sudden infant death, and is associated with enhanced cognitive development."

The World Health Organization recommends exclusive breastfeeding for the first six months and then continued breastfeeding along with other foods for two years or more. Breastfeeding should be "on demand" as often as the baby wants it, day and night.

Concept Summary

Nutritional Requirements Every human body requires certain amounts of various macro- and micronutrients, water, and fiber to function properly. The amounts needed vary with a number of factors such as body size, age, and sex. Various countries have established nutritional guidelines in attempts to maintain good health for their citizens. In the United States, the Reference Daily Intakes (RDIs) and Daily Reference Values (DRVs) are designed to represent appropriate nutrient intake for 95% of the population.
Objective 1 (Section 22.1), Exercise 22.2

The Macronutrients Macronutrients are substances required by the body in relatively large amounts. They are used by the body to provide energy and the materials necessary to form new or replacement tissue. The macronutrients are carbohydrates, lipids, and proteins. An RDI has been established only for proteins, but other groups have recommended the amounts of carbohydrates, lipids, and proteins that should be included in the diet.
Objective 2 (Section 22.2), Exercise 22.4

Micronutrients I: Vitamins Vitamins are organic micronutrients that the body cannot produce in the amounts needed for good health. A number of vitamins have a high water solubility resulting from the highly polar nature of their molecules. All but one of the water-soluble vitamins are known to function as coenzymes in the body and are involved in many important metabolic processes. Fat-soluble vitamins have nonpolar molecular structures. As a result, they are insoluble in water but soluble in fat or other nonpolar solvents. Fat-soluble vitamins act somewhat like hormones in the body. Fat-soluble vitamins do not dissolve in water-based body fluids, so they are not excreted through the kidneys. Amounts in excess of bodily requirements are stored in body fat. Thus, it is much easier to produce toxic effects by overdosing with fat-soluble vitamins than with water-soluble vitamins.

Objective 3 (Section 22.3), Exercise 22.10

Micronutrients II: Minerals Minerals are inorganic substances needed by the body. Those present in the body in amounts equal to or greater than 5 g are called major minerals. Those present in smaller amounts are called trace minerals. Minerals perform many useful functions in the body.

Objective 4 (Section 22.4), Exercise 22.16

The Flow of Energy in the Biosphere Energy for all living processes comes from the sun. A portion of the sun's energy is trapped in the process of photosynthesis, which produces carbohydrates. We derive energy from the foods we eat: carbohydrates, lipids, and proteins. As these substances are oxidized, molecules of ATP that supply the energy needed for cellular processes are formed.

Objective 5 (Section 22.5), Exercise 22.24

Metabolism and an Overview of Energy Production Metabolism, the sum of all cellular reactions, involves the breakdown (catabolism) and synthesis (anabolism) of molecules.

Objective 6 (Section 22.6), Exercise 22.26

There are three stages in the catabolism of foods to provide energy. In Stage I, digestion converts foods into smaller molecules. In Stage II, these smaller molecules are converted into two-carbon acetyl units that combine with coenzyme A, forming acetyl CoA. Stage III, which is called the common catabolic pathway, consists of the citric acid cycle followed by the electron transport chain and oxidative phosphorylation. The main function of catabolism is to produce ATP molecules.

Objective 7 (Section 22.6), Exercise 22.28

ATP: The Primary Energy Carrier Molecular ATP is the link between energy production and energy use in cells. ATP is called a high-energy compound because of the large amount of free energy liberated on hydrolysis. Cells are able to harness the free energy liberated by ATP to carry out cellular work. ATP is consumed immediately following its formation and is quickly regenerated from ADP as fuel molecules are oxidized. Mitochondria play a key role in energy production because they house the enzymes for both the citric acid cycle and the electron transport chain.

Objective 8 (Section 22.7), Exercise 22.34

Important Coenzymes in the Common Catabolic Pathway Three very important coenzymes are involved in catabolism: coenzyme A, NAD^+, and FAD. Coenzyme A binds to the two-carbon fragments produced in Stage II to form acetyl CoA, the direct fuel of the citric acid cycle. NAD^+ and FAD are the oxidizing agents that participate in the oxidation reactions of the citric acid cycle. They transport hydrogen atoms and electrons from the citric acid cycle to the electron transport chain.

Objective 9 (Section 22.8), Exercise 22.46

Key Terms and Concepts

Anabolism (22.6)
Catabolism (22.6)
Cellular respiration (22.5)
Common catabolic pathway (22.6)
Complete protein (22.2)
Complex carbohydrates (22.2)
Daily Reference Values (DRVs) (22.1)
Daily Values (DVs) (22.1)

Essential amino acid (22.2)
High-energy compound (22.7)
Macronutrient (22.1)
Major mineral (22.4)
Metabolic pathway (22.6)
Metabolism (22.6)
Micronutrient (22.1)
Mineral (22.4)

Mitochondrion (22.7)
Nutrition (Introduction)
Reference Daily Intakes (RDIs) (22.1)
Satiety (22.2)
Simple carbohydrates (22.2)
Thioester bond (22.8)
Trace mineral (22.4)

Key Reactions

1.	Photosynthesis (Section 22.5):	$6CO_2 + 6H_2O + energy \rightarrow C_6H_{12}O_6 \rightarrow 6O_2$	Reaction 22.2
2.	Hydrolysis of ATP (Section 22.7):	$ATP + H_2O \rightarrow ADP + P_i + H^+$	Reaction 22.5

Even-numbered exercises answered in Appendix B.

3. Oxidation by NAD$^+$ (Section 22.8):	$$R-\underset{\underset{\displaystyle H}{\displaystyle	}}{\overset{\overset{\displaystyle OH}{\displaystyle	}}{C}}-H + NAD^+ \longrightarrow R-\overset{\overset{\displaystyle O}{\displaystyle \|}}{C}-H + NADH + H^+$$	Reaction 22.10		
5. Oxidation by FAD (Section 22.8):	$$R-\underset{\underset{\displaystyle H}{\displaystyle	}}{\overset{\overset{\displaystyle H}{\displaystyle	}}{C}}-\underset{\underset{\displaystyle H}{\displaystyle	}}{\overset{\overset{\displaystyle H}{\displaystyle	}}{C}}-R + FAD \longrightarrow R-CH=CH-R + FADH_2$$	Reaction 22.14

Exercises

Even-numbered exercises are answered in Appendix B.

Blue-numbered exercises are more challenging.

Nutritional Requirements (Section 22.1)

22.1 What is the principal component of dietary fiber?

22.2 What is the primary difference between a macronutrient and a micronutrient?

22.3 Explain the importance of sufficient fiber in the diet.

The Macronutrients (Section 22.2)

22.4 List two general functions in the body for each of the macronutrients.

22.5 For each of the following macronutrients, list a single food item that would be a good source:

 a. carbohydrate (simple) **c.** lipid

 b. carbohydrate (complex) **d.** protein

22.6 List the types of macronutrients found in each of the following food items. List the nutrients in approximate decreasing order of abundance (most abundant first, etc.).

 a. Potato chips

 b. Buttered toast

 c. Plain toast with jam

 d. Cheese sandwich

 e. A lean steak

 f. A fried egg

22.7 Why is linoleic acid called an essential fatty acid for humans?

22.8 Use Figure 22.2 and list the Daily Values (DV) for fat, saturated fat, total carbohydrate, and fiber based on a 2000-calorie diet.

22.9 Use the Daily Values (DVs) and the calories per gram conversions given in Figure 22.2 to calculate the number of calories in a 2000-calorie diet that should be from fat, saturated fat, and total carbohydrate.

Micronutrients I: Vitamins (Section 22.3)

22.10 Identify each of the following vitamins as water-soluble or fat-soluble.

 a. tocopherol **c.** folic acid

 b. niacin **d.** retinol

22.11 What general function is served by eight of the water-soluble vitamins?

22.12 Why is there more concern about large doses of fat-soluble vitamins than of water-soluble vitamins?

22.13 What fat-soluble vitamin deficiency is associated with the following?

 a. Night blindness

 b. Blood clotting

 c. Calcium and phosphorus use in forming bones and teeth

 d. Preventing oxidation of fatty acids

22.14 What water-soluble vitamin deficiency is associated with the following?

 a. scurvy

 b. beriberi

 c. pernicious anemia

 d. pellagra

Micronutrients II: Minerals (Section 22.4)

22.15 What is the difference between a vitamin and a mineral?

22.16 List at least one specific function for each major mineral of the body.

22.17 What determines whether a mineral is classified as a major or trace mineral?

22.18 What are the general functions of trace minerals in the body?

22.19 How can the ingestion of a small lump of elemental sodium metal be toxic when sodium is a required major mineral?

The Flow of Energy in the Biosphere (Section 22.5)

22.20 What element serves as a fuel for the fusion process occurring on the sun?

22.21 What is the source of energy for life on Earth?

22.22 Write a net equation for the photosynthetic process.

22.23 Explain how Earth's supply of carbon compounds may be recycled by the processes of photosynthesis and respiration.

22.24 Describe in general terms the steps in the flow of energy from the sun to molecules of ATP.

Metabolism and an Overview of Energy Production (Section 22.6)

22.25 What is a metabolic pathway?

22.26 Classify catabolism and anabolism as synthetic or breakdown processes.

22.27 Outline the three stages in the extraction of energy from food.

22.28 Which stage of energy production is concerned primarily with the following?

 a. Formation of ATP

 b. Digestion of fuel molecules

 c. Consumption of O_2

 d. Generation of acetyl CoA

22.29 Explain why Stage III of energy production is referred to as the common catabolic pathway.

22.30 In terms of energy production, what is the main purpose of the catabolic pathway?

22.31 Would the digestion of protein to amino acids that are then used to build muscle in the body be considered a catabolic pathway? Why or why not?

ATP: The Primary Energy Carrier (Section 22.7)

22.32 Is ATP involved in anabolic processes or catabolic processes?

22.33 What do the symbols ATP, ADP, and AMP represent?

22.34 Explain why ATP is referred to as the primary energy carrier.

22.35 Using the partial structure shown below, write the structure of ATP, ADP, and AMP:

$$-\overset{\overset{\displaystyle O}{\|}}{\underset{\underset{\displaystyle O^-}{|}}{P}}-O-\text{adenosine}$$

22.36 What do the symbols P_i and PP_i represent?

22.37 Using symbolic formulas such as ADP and PP_i, write equations for the hydrolysis of ATP to ADP and the hydrolysis of ATP to AMP.

22.38 Using Table 22.5, explain why phosphoenolpyruvate is called a high-energy compound but glycerol-3-phosphate is not.

22.39 What does the symbol $\Delta G^{\circ\prime}$ represent? How does the sign for $\Delta G^{\circ\prime}$ indicate whether a reaction is exergonic or endergonic?

22.40 Which portion of the ATP molecule is particularly responsible for it being described as a high-energy compound?

22.41 Why is it more accurate to refer to ATP as a carrier or donor of free energy rather than as a storage form of free energy?

22.42 What is the role of mitochondria in the use of energy by living organisms?

22.43 Describe the importance and function of the ATP–ADP cycle.

22.44 Describe the structure of a mitochondrion and identify the location of enzymes important in energy production.

Important Coenzymes in the Common Catabolic Pathway (Section 22.8)

22.45 What coenzymes are produced by the body from the following vitamins?

 a. riboflavin **b.** pantothenic acid **c.** nicotinamide

22.46 What roles do the following coenzymes serve in the catabolic oxidation of foods?

 a. coenzyme A **b.** FAD **c.** NAD^+

22.47 Write the abbreviations for the oxidized and reduced forms of:

 a. flavin adenine dinucleotide

 b. nicotinamide adenine dinucleotide

22.48 Which compound is oxidized in the following reaction? Which compound is reduced?

$$^-OOC-CH_2-CH_2-COO^- + FAD \longrightarrow$$
$$^-OOC-CH=CH-COO^- + FADH_2$$

22.49 Which compound is oxidized in the following reaction? Which compound is reduced?

$$CH_3CH_2-OH + NAD^+ \longrightarrow CH_3\overset{\overset{\displaystyle O}{\|}}{C}-H + NADH + H^+$$

22.50 Write balanced reactions for the following oxidations using NAD^+ or FAD:

 a. $HO-CH_2-COOH \rightarrow O=CH-COOH$

 b. $HO-CH_2CH_2-OH \rightarrow HO-CH=CH-OH$

22.51 Write balanced reactions for the following oxidations using NAD^+:

 a. $CH_3-\overset{\overset{\displaystyle OH}{|}}{CH}-COO^- \rightarrow CH_3-\overset{\overset{\displaystyle O}{\|}}{C}-COO^-$

 b. $CH_3-OH \rightarrow H-\overset{\overset{\displaystyle O}{\|}}{C}-H$

22.52 Draw abbreviated structural formulas to show the reactive portion of FAD and the reduced compound $FADH_2$.

Additional Exercises

22.53 Based on the solubility characteristics of vitamin B_1, would you expect to find it in the interior of an enzyme or on the exterior surface of the enzyme when it acts as a coenzyme?

22.54 Balance the following equation. How many electrons will be transferred to NADH?

$$CH_4 + O_2 + ____ NAD^+ \rightarrow CO_2 + ____ NADH + ____ H^+$$

22.55 Draw an energy diagram that depicts the two hydrolysis reactions of ATP below.

$$ATP \rightarrow ADP + P_i \rightarrow AMP + 2P_i$$

22.56 What would be the main process in the human body that would be interrupted by a deficiency of an essential amino acid? What specific part of that process would be affected?

22.57 Average 18-year-olds should get 1300 mg of calcium in their diet daily. Use Figure 22.8 and calculate the percentage of the daily intake that represents the total amount of calcium found in a 60-kg person.

Chemistry for Thought

22.58 At one time, a system of MDR (minimum daily requirement) values for nutrients was used. Why do you think Daily Values (DVs) might be more informative and useful than MDRs?

22.59 Briefly explain why you think it would be undesirable or beneficial (decide which position to defend in each case) to follow a diet that is (a) extremely low in carbohydrates, (b) extremely low in lipids, or (c) extremely low in proteins.

22.60 Doughnuts, even those without frosting, are high in triglycerides. Explain.

22.61 One of the oils listed in Figure 22.6 is palm oil, which for several years was used in fast-food restaurants to fry potatoes. Why was this use considered an unhealthy practice?

22.62 In Figure 22.7, it is observed that vegetables are a source of protein. We most often think of meats, grains, and dairy products as sources of protein, but not vegetables. Use Table 19.4 to list the types of protein that might be found in vegetables.

22.63 Vitamin tablets and oranges are both plentiful sources of vitamin C. Why might an orange be better for you than taking a vitamin C tablet?

22.64 A cold cereal manufactured from corn is deficient in the amino acid lysine. What other foods in a breakfast meal might supply lysine?

22.65 The body has a storage capacity for excess carbohydrates and fats but not for excess proteins. What consequences does this fact have for a daily diet?

22.66 Which macronutrient is a major source of dietary sulfur?

Allied Health Exam Connection

The following questions are from these sources:

- *Nursing School Entrance Exam* © 2005, Learning Express, LLC.

- *McGraw-Hill's Nursing School Entrance Exams* by Thomas A. Evangelist, Tamara B. Orr, and Judy Unrein © 2009, The McGraw-Hill Companies, Inc.

- *NSEE Nursing School Entrance Exams,* 3rd edition © 2009, Kaplan Publishing.

- *Cliffs Test Prep: Nursing School Entrance Exams* by Fred N. Grayson © 2004, Wiley Publishing, Inc.

- *Peterson's Master the Nursing School and Allied Health Entrance Exams,* 18th edition by Marion F. Gooding © 2008, Peterson's, a Nelnet Company.

22.67 Identify which vitamin(s) is indicated by the following:

 a. The fat-soluble vitamins

 b. Thiamin

 c. A vitamin that helps prevent rickets

 d. A vitamin that helps clot blood

 e. A vitamin useful in combatting pernicious anemia

22.68 Answer the following:

 a. Which mineral is the most abundant in the human body?

 b. Which mineral is necessary for the proper functioning of the thyroid gland?

 c. Which mineral is necessary for the proper formation of hemoglobin?

 d. What is the difference between a vitamin and a mineral?

22.69 Which of the following are not a requirement for photosynthesis?

 a. oxygen **c.** sunlight

 b. carbon dioxide **d.** chlorophyll

22.70 With which organelle is the synthesis of ATP associated?

22.71 Use Figure 22.2 to determine which macronutrient would produce the greatest amount of energy from 1 gram of the macronutrient.

22.72 Outline the three stages in the extraction of energy from food. What type of chemical reaction occurs most often during digestion?

22.73 Amino acids that cannot be manufactured by the body are called:

 a. essential amino acids **c.** basic amino acids

 b. synthetic amino acids **d.** dependent amino acids

22.74 More than 90% of dietary fat is in the form of:

 a. triglycerides **c.** cholesterol

 b. phospholipids **d.** lipase

22.75 Water-soluble vitamins include:

 a. vitamin A **c.** vitamin D

 b. vitamin C **d.** vitamin K

22.76 The chemical reaction that supplies immediate energy for muscular contractions can be summarized as:

 a. $ATP \rightarrow ADP + P$

 b. lactic acid $\rightarrow CO_2 + H_2O$

 c. lactic acid \rightarrow glycogen

 d. glycogen \rightarrow ATP

22.77 Which process does the following equation represent?

 $6CO_2 + \text{light energy} + 6H_2O \rightarrow C_6H_{12}O_6 + 6O_2$

 a. glycolysis

 b. fermentation

 c. photosynthesis

 d. cellular respiration

23 Carbohydrate Metabolism

iStockphoto.com/Mark Hatfield

▶ Case Study

Natalie's husband is an F-16 pilot for the U.S. Air Force. They moved from a military base in humid San Antonio, Texas, to Tucson, Arizona. Natalie thought she was having difficulty adjusting to the dry climate. She was drinking large quantities of water and could not quench her thirst. She had also lost a few pounds. Could it be dehydration? Eventually she felt so ill that she visited the emergency room. The ER physician immediately suspected diabetic ketoacidosis and ordered blood and urine tests. The results supported a diagnosis of type 1 diabetes mellitus. Natalie now has to check her sugar levels before every meal and several times during the day, and she has an insulin pump. She tries to keep her blood sugar in the range of 90 mg/dL to 130 mg/dL before meals and below 180 mg/dL two hours after meals.

▶ What should she do if her levels dip or go too high? What type of community support might be available to Natalie? How might Natalie's husband need support, since her diabetes diagnosis impacts his life as well?

Follow-up to this Case Study appears at the end of the chapter before the *Concept Summary*.

Learning Objectives

When you have completed your study of this chapter, you should be able to:

1. Identify the products of carbohydrate digestion. **(Section 23.1)**
2. Explain the importance to the body of maintaining proper blood sugar levels. **(Section 23.2)**
3. List the starting material and products of the glycolysis pathway. **(Section 23.3)**
4. Describe how the glycolysis pathway is regulated in response to cellular needs. **(Section 23.3)**
5. Name the three fates of pyruvate. **(Section 23.4)**
6. Identify the two major functions of the citric acid cycle. **(Section 23.5)**
7. Describe how the citric acid cycle is regulated in response to cellular energy needs. **(Section 23.5)**

8. Explain the function of the electron transport chain and describe how electrons move down the chain. **(Section 23.6)**
9. List the major features of the chemiosmotic hypothesis. **(Section 23.7)**
10. Calculate the amount of ATP produced by the complete oxidation of a mole of glucose. **(Section 23.8)**
11. Explain the importance of the processes of glycogenesis and glycogenolysis. **(Section 23.9)**
12. Describe gluconeogenesis and the operation of the Cori cycle. **(Section 23.10)**
13. Describe how hormones regulate carbohydrate metabolism. **(Section 23.11)**

Carbohydrates play major roles in cell metabolism. The central substance in carbohydrate metabolism is glucose; it is the key food molecule for most organisms. As the major carbohydrate metabolic pathways are discussed, you will see that each one involves either the use or the synthesis of glucose. You will also see how the chemical degradation of glucose provides energy in the cell, how glucose is stored, how the blood glucose level is regulated, and how diabetes results from changes in blood glucose regulation mechanisms.

23.1 | The Digestion of Carbohydrates

Learning Objective

1. Identify the products of carbohydrate digestion.

The major function of dietary carbohydrates is to serve as an energy source. In a typical American diet, carbohydrates meet about 45%–55% of the daily energy needs. Fats and proteins furnish the remaining energy. During carbohydrate digestion, di- and polysaccharides are hydrolyzed to monosaccharides, primarily glucose, fructose, and galactose:

$$\text{polysaccharides} + H_2O \xrightarrow{\text{digestion}} \text{glucose} \qquad (23.1)$$

$$\text{sucrose} + H_2O \xrightarrow{\text{digestion}} \text{glucose} + \text{fructose} \qquad (23.2)$$

$$\text{lactose} + H_2O \xrightarrow{\text{digestion}} \text{glucose} + \text{galactose} \qquad (23.3)$$

$$\text{maltose} + H_2O \xrightarrow{\text{digestion}} \text{glucose} \qquad (23.4)$$

Figure 23.1 Blood glucose levels.

blood sugar level The amount of glucose present in blood, normally expressed as milligrams per 100 mL of blood.

hypoglycemia A lower-than-normal blood sugar level.

hyperglycemia A higher-than-normal blood sugar level.

renal threshold The blood glucose level at which glucose begins to be excreted in the urine.

glucosuria A condition in which elevated blood sugar levels result in the excretion of glucose in the urine.

glycolysis A series of reactions by which glucose is oxidized to pyruvate.

After digestion is completed, glucose, fructose, and galactose are absorbed into the bloodstream through the lining of the small intestine and transported to the liver. In the liver, fructose and galactose are rapidly converted to glucose or to compounds that are metabolized by the same pathway as glucose.

23.2 | Blood Glucose

Learning Objective

2. Explain the importance to the body of maintaining proper blood sugar levels.

Glucose is by far the most plentiful monosaccharide found in blood, and the term *blood sugar* usually refers to glucose. In adults, the normal **blood sugar level** measured after a fast of 8–12 hours is in the range of 70–110 mg/100 mL. (In clinical reports, such levels are often expressed in units of milligrams per deciliter, mg/dL.) The blood sugar level reaches a maximum of approximately 140–160 mg/100 mL about 1 hour after a carbohydrate-containing meal. It returns to normal after 2–2½ hours.

If a blood sugar level is below the normal fasting level, a condition called **hypoglycemia** exists. Because glucose is the only nutrient normally used by the brain for energy, mild hypoglycemia leads to dizziness and fainting as brain cells are deprived of energy. Severe hypoglycemia can cause convulsions and coma. When the blood glucose concentration is above normal, the condition is referred to as **hyperglycemia**. If blood glucose levels exceed approximately 180 mg/100 mL, the sugar is not completely reabsorbed by the kidneys, and glucose is excreted in the urine (see **Figure 23.1**). The blood glucose level at which this occurs is called the **renal threshold**, and the condition in which glucose appears in the urine is called **glucosuria**. Prolonged hyperglycemia at a glucosuric level must be considered serious because it indicates that something is wrong with the body's normal ability to control the blood sugar level.

The liver is the key organ involved in regulating the blood glucose level. The liver responds to the increase in blood glucose that follows a meal by removing glucose from the bloodstream. The removed glucose is primarily converted to glycogen or triglycerides for storage. Similarly, when blood glucose levels are low, the liver responds by converting stored glycogen to glucose and by synthesizing new glucose from noncarbohydrate substances.

23.3 | Glycolysis

Learning Objectives

3. List the starting material and products of the glycolysis pathway.
4. Describe how the glycolysis pathway is regulated in response to cellular needs.

Glycolysis is an important part of the catabolic process by which the body gets energy from carbohydrates. It consists of a series of 10 reactions, with the net result being the conversion of a glucose molecule into two molecules of pyruvate. All enzymes for glycolysis are present in the cellular cytoplasm. The 10 reactions of glycolysis catalyzed by these enzymes are represented in **Figure 23.2**. The addition of all these reactions gives a net reaction for glycolysis:

$$\text{glucose} + 2\text{P}_i + 2\text{ADP} + 2\text{NAD}^+ \rightarrow$$
$$2 \text{ pyruvate} + 2\text{ATP} + 2\text{NADH} + 4\text{H}^+ + 2\text{H}_2\text{O} \qquad (23.5)$$

This equation indicates a net gain of 2 mol of ATP for every mole of glucose converted to pyruvate. In addition, 2 mol of the coenzyme NADH is formed when NAD^+ serves as an oxidizing agent in glycolysis. Fructose and galactose from digested food are both converted into intermediates that enter into the glycolysis pathway.

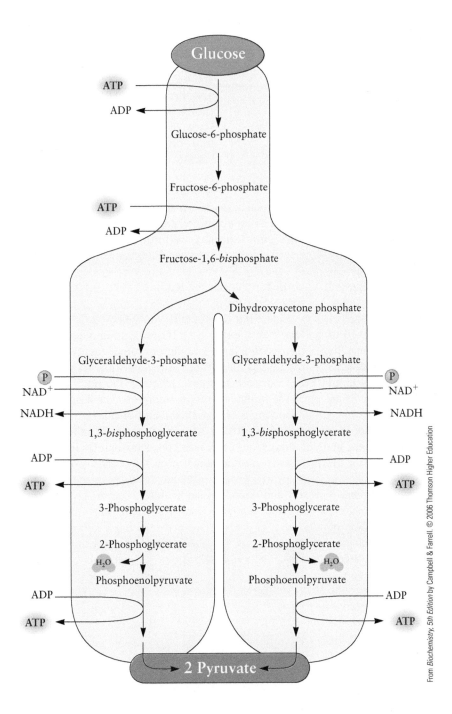

From *Biochemistry, 5th Edition* by Campbell & Farrell. © 2006 Thomson Higher Education

Fructose enters glycolysis as dihydroxyacetone phosphate and glyceraldehyde-3-phosphate, whereas galactose is metabolized to glucose and enters in the form of glucose-6-phosphate.

23.3A The Regulation of Glycolysis

The glycolysis pathway, like all metabolic pathways, is controlled constantly. The precise regulation of the concentrations of pathway intermediates is efficient and makes possible the rapid adjustment of the output level of pathway products as the need arises.

Q: How can we avoid energy crashes?

Melina B. Jampolis, M.D., is a Physician Nutrition Specialist focusing on nutrition for weight loss and disease prevention and treatment.

A: An "energy crash" typically occurs one to three hours after a meal, usually late morning, after breakfast, and early afternoon, after lunch. The term *energy crash* is not scientific, but is generally used to indicate a rapid drop in blood sugar that is accompanied by symptoms including headaches, hunger, decreased ability to concentrate, and even mood changes. The reason for the rapid decline in blood sugar is usually related to the nutrient content of the previous meal. For example, meals containing large amounts of rapidly digested carbohydrates can cause a rapid increase in blood sugar. The blood sugar increase leads to a surge in the release of insulin, a hormone that transfers sugar from the blood to cells of the brain or muscles for use as energy, or to fat cells for storage. Thus, a quick spike in blood sugar, leading to an exaggerated release of insulin, can cause a fairly rapid drop in blood sugar, which in turn causes symptoms of an energy crash.

The measure of how quickly and significantly a specific food increases blood sugar levels is known as the glycemic index (GI) of the food. The GI is obtained by comparing the total increase in blood sugar during the two hours following the consumption of a sample of the food that contains 50 grams of carbohydrate. This increase is compared to the increase caused by a control that generally consists of the consumption of 50 grams of glucose, which has a GI of 100. Foods with generally high GI values include refined grains, white potatoes, white rice, white bread, sugary cereals, candy, baked goods, crackers, pretzels, fruit drinks, and sodas. The GI indicates the ability of carbohydrates to increase blood sugar levels, but in order to understand the true effect of a food on blood sugar, the amount of carbohydrates present in the food is equally important. The glycemic load of a food includes both the ability of carbohydrates in the food to increase blood sugar levels and the amount of carbohydrates present in the food. It is calculated by multiplying together the glycemic index and the amount of carbohydrates in a typical serving of the food.

The glycemic index of a meal can be decreased in several ways. Consuming lower GI carbohydrates like whole grains, oats, whole-grain cereals and breads, brown rice, and beans instead of refined grains, sugary cereals, white rice, and white potatoes is key. Opting for fiber-rich whole fruit instead of juice is helpful, and limiting the intake of sugary beverages, snacks, baked goods, and candy is important. Finally, combining healthy fat or lean protein with carbohydrates at each meal also helps lower the GI of the meal. By lowering the GI of meals, blood sugar levels remain more constant throughout the day, helping you avoid energy crashes. Lowering the GI of your diet may also decrease your risk of type 2 diabetes, heart disease, and age-related macular degeneration, the leading cause of blindness in older Americans.

One final thought: If you consume a high-GI breakfast, like a bagel with fat-free cream cheese and orange juice *with* a large amount of coffee, energy crashes can be particularly significant one to two hours after breakfast. Caffeine reaches peak levels in the body approximately 30–45 minutes after consumption. The drop in blood sugar following a high-GI meal combined with the drop in the stimulant effects of caffeine can enhance symptoms of lethargy and decreased concentration. By limiting yourself to moderate amounts of caffeine and lowering the GI of your diet, you can stay alert and energized throughout the day and improve your health.

Sugary foods have high GI values.

The glycolysis pathway is regulated by three enzymes: hexokinase, phosphofructokinase, and pyruvate kinase. The regulation of the pathway is shown in **Figure 23.3**.

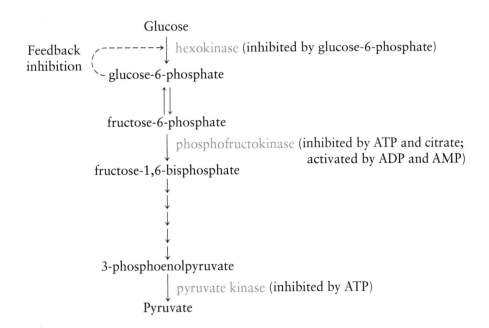

Glucose

Feedback inhibition ------→ hexokinase (inhibited by glucose-6-phosphate)

glucose-6-phosphate

fructose-6-phosphate

phosphofructokinase (inhibited by ATP and citrate; activated by ADP and AMP)

fructose-1,6-bisphosphate

3-phosphoenolpyruvate

pyruvate kinase (inhibited by ATP)

Pyruvate

Figure 23.3 The regulation of the glycolysis pathway. Glucose-6-phosphate is a noncompetitive inhibitor of hexokinase. Phosphofructokinase and pyruvate kinase are also allosteric enzymes subject to control.

Hexokinase catalyzes the conversion of glucose to glucose-6-phosphate and thus initiates the glycolysis pathway. The enzyme is inhibited by a high concentration of glucose-6-phosphate, the product of the reaction it catalyzes. Thus, the phosphorylation of glucose is controlled by feedback inhibition (see **Section 20.8**).

A second and very important control point for glycolysis shown in Figure 23.3 is the step catalyzed by phosphofructokinase. Once glucose is converted to fructose-1, 6-bisphosphate, its carbon skeleton must continue through the glycolytic pathway. The phosphorylation of fructose-6-phosphate is a committed step that will occur as long as the enzyme is active. As an allosteric enzyme (Section 20.8), phosphofructokinase is inhibited by high concentrations of ATP and citrate, and it is activated by high concentrations of ADP and AMP.

The third control point is the last step of glycolysis, the conversion of phosphoenolpyruvate to pyruvate catalyzed by pyruvate kinase. This enzyme, too, is an allosteric enzyme that is inhibited by higher concentrations of ATP.

To understand how these controls interact, remember that when the glycolysis pathway is operating at a maximum rate, so also are the citric acid cycle and the electron transport chain. As a result of their activity, large quantities of ATP are produced. The entire pathway continues to operate at a high rate, as long as the produced ATP is used up by processes within the cell. However, if ATP use decreases, the concentration of ATP increases, and more diffuses to the binding sites of the allosteric enzymes sensitive to ATP. Thus, the activity of these enzymes decreases. When phosphofructokinase is inhibited in this way, the entire glycolysis pathway slows or stops. As a result, the concentration of glucose-6-phosphate increases, thus inhibiting hexokinase and the initial step of the glycolysis pathway.

Conversely, when ATP concentrations are low, the concentrations of AMP and ADP are high. The AMP and ADP function as activators (Section 20.8) that influence phosphofructokinase and thereby speed up the entire glycolysis pathway.

✔ **LEARNING CHECK 23.1** Explain how each of the following enzymes helps regulate the glycolysis pathway:

a. hexokinase
b. phosphofructokinase
c. pyruvate kinase

23.4 | The Fates of Pyruvate

Learning Objective

5. Name the three fates of pyruvate.

The sequence of reactions that converts glucose to pyruvate is very similar in all organisms and in all kinds of cells. However, the fate of the pyruvate as it is used to generate energy is variable. Three different fates of pyruvate are considered here.

As glucose is oxidized to pyruvate in glycolysis, NAD^+ is reduced to NADH. The need for a continuous supply of NAD^+ for glycolysis is a key to understanding the fates of pyruvate.

$$\text{glucose} + 2NAD^+ \longrightarrow 2 \text{ pyruvate} + 2NADH + 2H^+ \quad (23.6)$$
$$2ADP + 2P_i \qquad 2ATP$$

In each case, pyruvate is metabolized in ways that regenerate NAD^+ so that glycolysis can continue.

23.4A Oxidation to Acetyl CoA

aerobic In the presence of oxygen.

The mitochondrial membrane is permeable to pyruvate formed by glycolysis in the cytoplasm, and under **aerobic** conditions (a plentiful oxygen supply), pyruvate is converted inside the mitochondria to acetyl CoA.

$$\underset{\text{pyruvate}}{CH_3 - \overset{\overset{\displaystyle O}{\|}}{C} - COO^-} + CoA - S - H + NAD^+ \xrightarrow[\text{complex}]{\substack{\text{pyruvate} \\ \text{dehydrogenase}}} \underset{\text{acetyl CoA}}{CH_3 - \overset{\overset{\displaystyle O}{\|}}{C} - S - CoA} + NADH + CO_2 \quad (23.7)$$

Acetyl CoA formed in this reaction can enter the citric acid cycle on its way to complete oxidation to CO_2. The NAD^+ needed for the conversion of pyruvate to acetyl CoA (Reaction 23.7) and for the operation of the glycolysis pathway (Step 6) is regenerated when NADH transfers its electrons to O_2 in the electron transport chain.

Not all acetyl CoA generated metabolically enters the citric acid cycle. Some also serve as a starting material for fatty acid biosynthesis (Section 24.7). The fatty acid components of triglycerides, for example, are synthesized using acetyl CoA. Thus, as most people are aware, the intake of excess dietary carbohydrates can lead to an increase in body fat storage.

23.4B Reduction to Lactate

anaerobic In the absence of oxygen.

Under **anaerobic** conditions, such as those that accompany strenuous or long-term muscle activity, the cellular supply of oxygen is not adequate for the reoxidation of NADH to NAD^+. Under such conditions, the cells begin reducing pyruvate to lactate as a means of regenerating NAD^+:

$$\underset{\text{pyruvate}}{CH_3 - \overset{\overset{\displaystyle O}{\|}}{C} - COO^-} + NADH + H^+ \underset{\xleftarrow{\hspace{1cm}}}{\overset{\text{lactate}}{\underset{\text{dehydrogenase}}{\xrightarrow{\hspace{1cm}}}}} \underset{\text{lactate}}{CH_3 - \overset{\overset{\displaystyle OH}{|}}{CH} - COO^-} + NAD^+ \quad (23.8)$$

When Reaction 23.8 is added to the net reaction of glycolysis, an overall reaction for **lactate fermentation** results:

lactate fermentation The production of lactate from glucose.

$$\underset{\text{glucose}}{C_6H_{12}O_6} + 2ADP + 2P_i \xrightarrow[\text{fermentation}]{\text{lactate}} \underset{\text{lactate}}{2CH_3 - \overset{\overset{\displaystyle OH}{|}}{CH} - COO^-} + 2ATP + 2H_2O \quad (23.9)$$

Notice that the NAD^+ produced when pyruvate is reduced (Reaction 23.8) is used up and does not appear in lactate fermentation (Reaction 23.9). Although much more energy is

available from the complete oxidation of pyruvate under aerobic conditions, the two ATPs produced from lactate fermentation are sufficient to sustain the life of anaerobic microorganisms. In human metabolism, those two molecules of ATP also play a critical role by furnishing energy when cellular supplies of oxygen are insufficient for ATP production through the complete oxidation of pyruvate.

23.4C Reduction to Ethanol

Several organisms, including yeast, have developed an alternative pathway for the regeneration of NAD^+ under anaerobic conditions. The process, called **alcoholic fermentation**, involves the decarboxylation of pyruvate to form acetaldehyde:

$$CH_3 - \overset{\overset{\text{O}}{\|}}{C} - COO^- + H^+ \xrightarrow[\text{decarboxylase}]{\text{pyruvate}} CH_3 - \overset{\overset{\text{O}}{\|}}{C} - H + CO_2 \qquad (23.10)$$
$$\text{pyruvate} \qquad\qquad\qquad\qquad \text{acetaldehyde}$$

The carbon dioxide produced in this reaction causes beer to foam and provides carbonation for naturally fermented wines and champagnes (see **Figure 23.4**).

The second step in alcoholic fermentation is the reduction of acetaldehyde to ethanol by NADH:

$$CH_3 - \overset{\overset{\text{O}}{\|}}{C} - H + NADH + H^+ \underset{\text{}}{\overset{\text{alcohol}}{\underset{\text{dehydrogenase}}{\rightleftharpoons}}} CH_3 - \overset{\overset{\text{OH}}{|}}{\underset{\overset{|}{\text{H}}}{C}} - H + NAD^+ \qquad (23.11)$$
$$\text{acetaldehyde} \qquad\qquad\qquad\qquad\qquad \text{ethanol}$$

Figure 23.4 Bread, wine, and cheese are all produced through fermentation processes.

The NAD^+ regenerated by this reaction may then be used in the glycolysis pathway.

An overall reaction for alcoholic fermentation is obtained by combining the reaction for the conversion of pyruvate to ethanol and the reactions of glycolysis. Notice that because NAD^+ is used up in glycolysis but regenerated in ethanol formation, it does not appear in the overall equation:

$$C_6H_{12}O_6 + 2ADP + 2P_i \xrightarrow[\text{fermentation}]{\text{alcoholic}} 2CH_3CH_2 - OH + 2CO_2 + 2ATP \qquad (23.12)$$
$$\text{glucose} \qquad\qquad\qquad\qquad\qquad \text{ethanol}$$

Figure 23.5 summarizes the various pathways pyruvate can follow as it is used to regenerate NAD^+.

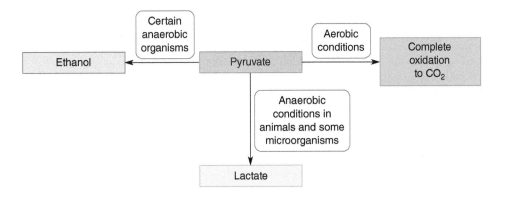

Figure 23.5 The fates of pyruvate.

✔ **LEARNING CHECK 23.2** Why is it important that each biochemical fate of pyruvate involve the production of NAD^+?

23.5 | The Citric Acid Cycle

Learning Objectives

6. Identify the two major functions of the citric acid cycle.
7. Describe how the citric acid cycle is regulated in response to cellular energy needs.

citric acid cycle A series of reactions in which acetyl CoA is oxidized to carbon dioxide and reduced forms of coenzymes FAD and NAD$^+$ are produced.

Stage III in the oxidation of fuel molecules (see **Section 22.6**) begins when the two-carbon acetyl units (of acetyl CoA) enter the **citric acid cycle**. This process is called the citric acid cycle because one of the key intermediates is citric acid. However, it is also called the tricarboxylic acid cycle (TCA) in reference to the three carboxylic acid groups in citric acid, and the Krebs cycle in honor of Sir Hans A. Krebs, who deduced its reaction sequence in 1937 (see **Figure 23.6**).

The citric acid cycle is the principal process for generating the reduced coenzymes NADH and FADH$_2$, which are necessary for the reduction of oxygen and ATP synthesis in the electron transport chain. Another important function of the citric acid cycle as a source of intermediates for biosynthesis is discussed in **Section 24.11**.

23.5A Reactions of the Citric Acid Cycle

The reactions of the citric acid cycle occur within the matrix of the mitochondrion. **Figure 23.7** summarizes the details of the eight reactions that make up the citric acid cycle. Of particular importance are those reactions in which molecules of NADH and FADH$_2$ are produced. Notice that a two-carbon unit (the acetyl group) enters the cycle and that two carbon atoms are liberated in the form of two CO$_2$ molecules, one each in Reactions 3 and 4. Thus, this series of reactions both begins and ends with a C$_4$ compound, oxaloacetate. This is why the pathway is called a cycle. In each trip around the cycle, the starting material is regenerated and the reactions can proceed again as long as more acetyl CoA is available as fuel.

Figure 23.6 Sir Hans Adolf Krebs (1900–1981), German-born British biochemist, received the 1953 Nobel Prize in physiology or medicine.

23.5B A Summary of the Citric Acid Cycle

The individual reactions of the cycle can be added to give an overall net equation:

$$\text{acetyl CoA} + 3\text{NAD}^+ + \text{FAD} + \text{GDP} + \text{P}_i + 2\text{H}_2\text{O} \rightarrow$$
$$2\text{CO}_2 + \text{CoA—S—H} + 3\text{NADH} + 2\text{H}^+ + \text{FADH}_2 + \text{GTP} \qquad (23.13)$$

Some important features of the citric acid cycle are the following:

1. Acetyl CoA, available from the breakdown of carbohydrates, lipids, and amino acids, is the fuel of the citric acid cycle.
2. The operation of the cycle requires a supply of the oxidizing agents NAD$^+$ and FAD. The cycle is dependent on reactions of the electron transport chain to supply the necessary NAD$^+$ and FAD (Section 23.6). Because oxygen is the final acceptor of electrons in the electron transport chain, the continued operation of the citric acid cycle depends ultimately on an adequate supply of oxygen.
3. Two carbon atoms enter the cycle as an acetyl unit, and two carbon atoms leave the cycle as two molecules of CO$_2$. However, the carbon atoms leaving the cycle correspond to carbon atoms that entered in the previous cycle; there is a one-cycle delay between the entry of two carbon atoms as an acetyl unit and their release as CO$_2$.
4. In each complete cycle, four oxidation–reduction reactions produce three molecules of NADH and one molecule of FADH$_2$.
5. One molecule of the high-energy phosphate compound guanosine triphosphate (GTP) (see **Section 22.7**) is generated.

23.5C Regulation of the Citric Acid Cycle

The rate at which the citric acid cycle operates is precisely adjusted to meet cellular needs for ATP. One important control point is the very first step in the cycle, the synthesis of citrate. The catalyst for this reaction, citrate synthetase, is an allosteric enzyme that is

Bettmann/Getty Images

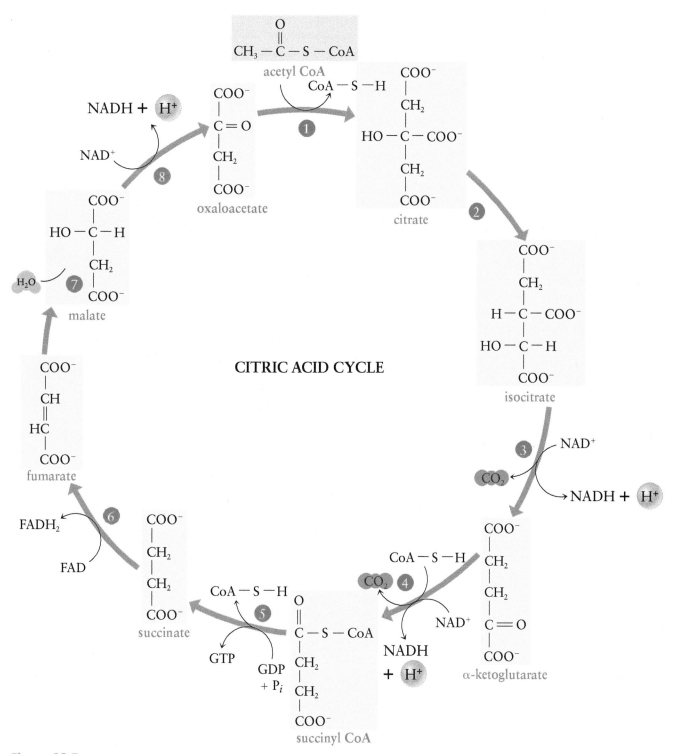

Figure 23.7 The citric acid cycle.

inhibited by ATP and NADH and activated by ADP. When cellular needs for ATP are met, ATP interacts with citrate synthetase and reduces its affinity for acetyl CoA. Thus, ATP acts as a modulator (inhibitor) of citrate synthetase and inhibits the entry of acetyl CoA into the citric acid cycle. Similarly, NADH acts as an inhibitor and signals the citric acid cycle to slow the production of NADH and $FADH_2$. On the other hand, when ATP levels are low, ADP levels are usually high. ADP, an activator of citrate synthetase, stimulates the entry of acetyl CoA into the citric acid cycle and thus boosts the production of NADH and $FADH_2$.

Lactate Accumulation

During vigorous exercise, the oxygen supply to muscles can become too low to support increased aerobic conversion of pyruvate to acetyl CoA. Under such conditions, a shift occurs from aerobic pyruvate oxidation to anaerobic production of lactate as the means for producing ATP. The resulting accumulation of lactate causes the muscle pain and cramps that accompany physical exhaustion. The buildup of lactate also causes a slight decrease in blood pH, which triggers an increase in the rate and depth of breathing, a response that provides more oxygen to the cells. The combination of pain from lactate accumulation and the maximum rate of breathing represents a limit to physical activities. Regular training and exercise increase this limit by improving the efficiency with which oxygen is delivered to the cells. The muscle cells of a well-conditioned athlete can maintain the aerobic oxidation of pyruvate for a much longer period of activity than can those of a nonathlete.

The shift to anaerobic lactate formation also occurs in the heart muscle during an episode of cardiac arrest, in which the oxygen-carrying blood supply to the heart is cut off by the blockage of an artery leading to the heart muscles. Aerobic oxidation of pyruvate and its ATP production are stopped. Glycolysis proceeds at an accelerated rate to meet the energy needs, and lactate accumulates. The heart muscle contracts, producing a cramp, and stops beating. Massaging the heart muscle relieves such cramps, just as it does for skeletal muscles. Thus, it is sometimes possible to start the heart beating by massaging it.

Premature infants, whose lungs are not fully developed, are given pure oxygen to breathe in order to prevent a shift to anaerobic lactate production. In addition, they are given bicarbonate solutions intravenously to combat the increasing acidity of the blood due to lactate buildup.

Vigorous exercise causes a buildup of lactate in tissues.

Isocitrate dehydrogenase (Reaction 3), a second controlling enzyme, is an allosteric enzyme that is activated by ADP and inhibited by NADH in very much the same way as is citrate synthetase. The third control point (Reaction 4) is catalyzed by the α-ketoglutarate dehydrogenase complex. This group of enzymes is inhibited by succinyl CoA and NADH, products of the reaction it catalyzes, and also by ATP. These control points are indicated in **Figure 23.8**. In short, the entry of acetyl CoA into the citric acid cycle and the rate at which the cycle operates are reduced when cellular ATP levels are high. When ATP supplies are low (and ADP levels are high), the cycle is stimulated.

Figure 23.8 Control points in the citric acid cycle.

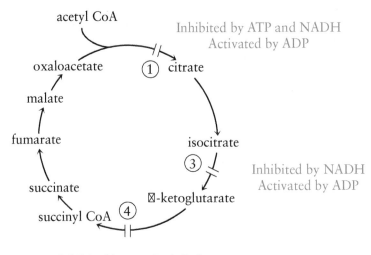

23.6 | The Electron Transport Chain

Learning Objective

8. Explain the function of the electron transport chain and describe how electrons move down the chain.

The reduced coenzymes NADH and FADH$_2$ are end products of the citric acid cycle. In the final stage of food oxidation, the hydrogen ions and electrons carried by these coenzymes combine with oxygen and form water.

$$4H^+ + 4e^- + O_2 \rightarrow 2H_2O \qquad (23.14)$$

This simple energy-releasing process does not take place in a single step. Instead, it occurs as a series of steps that involve a number of enzymes and cofactors located within the inner membrane of the mitochondria. Electrons from the reduced coenzymes are passed from one electron carrier to another within the membrane in assembly-line fashion and are finally combined with molecular oxygen, the final electron acceptor. The series of reactions involved in the process is called the **electron transport chain**. As illustrated in **Figure 23.9**, there are several intermediate electron carriers between NADH and molecular oxygen. The electron carriers are lined up in order of increasing affinity for electrons.

electron transport chain A series of reactions in which protons and electrons from the oxidation of foods are used to reduce molecular oxygen to water.

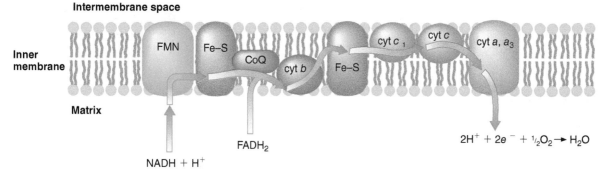

Figure 23.9 The electron transport chain. The heavy arrows show the path of the electrons as they pass from one carrier to the next.

The first electron carrier in the electron transport chain is an enzyme that contains a tightly bound coenzyme. The coenzyme has a structure similar to FAD. The enzyme formed by the combination of this coenzyme with a protein is called flavin mononucleotide (FMN). Two electrons and one H$^+$ ion from NADH plus another H$^+$ ion from a mitochondrion pass to FMN, then to an iron–sulfur (Fe—S) protein, and then to coenzyme Q (CoQ). CoQ is also the entry point into the electron transport chain for the two electrons and two H$^+$ ions from FADH$_2$. As NADH and FADH$_2$ release their hydrogen atoms and electrons, NAD$^+$ and FAD are regenerated for reuse in the citric acid cycle.

Four of the five remaining electron carriers are **cytochromes** (abbreviated cyt), which are structurally related proteins that contain an iron group. CoQ passes along the two electrons to two molecules of cytochrome b, and the two H$^+$ ions are given up to the mitochondrion. The electrons are passed along the chain, and in the final step, an oxygen atom accepts the electrons and combines with two H$^+$ ions to form water.

cytochrome An iron-containing enzyme located in the electron transport chain.

23.7 | Oxidative Phosphorylation

Learning Objective

9. List the major features of the chemiosmotic hypothesis.

As electrons are transported along the electron transport chain, a significant amount of free energy is released (52.6 kcal/mol). Some of this energy is conserved by the

synthesis of ATP from ADP and P_i. Because this synthesis of ATP (a phosphorylation reaction) is linked to the oxidation of NADH and FADH$_2$, it is termed **oxidative phosphorylation**. It takes place at three different locations of the electron transport chain (see **Figure 23.10**).

Figure 23.10 ATP is synthesized at three sites within the electron transport chain. Energy decreases at other locations result in the release of heat energy.

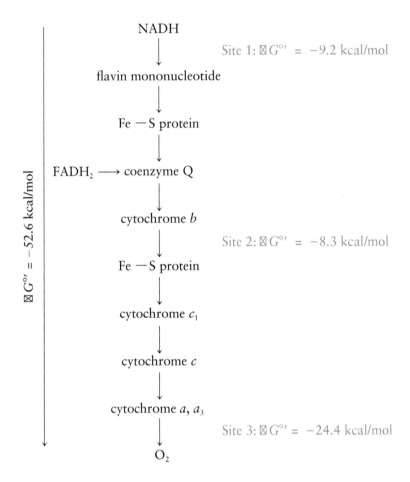

For each mole of NADH oxidized in the electron transport chain, 2.5 mol of ATP is formed. The free energy required for the phosphorylation of ADP is 7.3 kcal/mol, or 18 kcal for 2.5 mol of ATP. Because the oxidation of NADH liberates 52.6 kcal/mol, coupling the oxidation and phosphorylation reactions conserves 18 kcal/52.6 kcal, or approximately 34% of the energy released.

The electron donor FADH$_2$ enters the electron transport chain later than does NADH. Thus, for every FADH$_2$ oxidized to FAD, only 1.5 ATP molecules are formed during oxidative phosphorylation, and approximately 25% of the released free energy is conserved as ATP.

Now we can calculate the energy yield for the entire catabolic pathway (citric acid cycle, electron transport chain, and oxidative phosphorylation combined). According to **Section 23.5**, every acetyl CoA entering the citric acid cycle produces three NADH and one FADH$_2$ plus one GTP, which is equivalent in energy to one ATP. Thus, 10 ATP molecules are formed per molecule of acetyl CoA catabolized.

3 NADH ultimately produce	7.5 ATP
1 FADH$_2$ ultimately produces	1.5 ATP
1 GTP is equivalent to	1.0 ATP
	10.0 ATP total produced

The complex mechanism by which the cell couples the oxidations of the electron transport chain and the synthesis of ATP is believed to involve a flow of protons. This explanation, called the **chemiosmotic hypothesis**, proposes that the flow of electrons through

the electron transport chain causes protons to be "pumped" from the matrix across the inner membrane and into the space between the inner and outer mitochondrial membranes (see **Figure 23.11**). This creates a difference in proton concentration across the inner

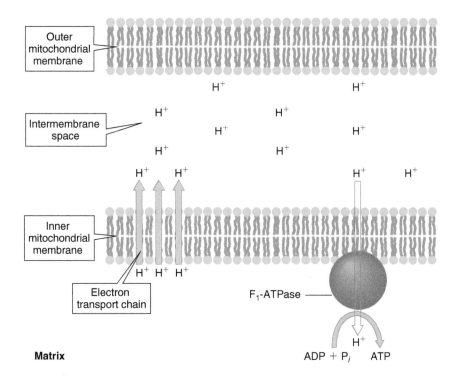

Figure 23.11 The chemiosmotic hypothesis. Electron transport pumps protons (H^+) out of the matrix. The formation of ATP accompanies the flow of protons through F_1-ATPase back into the matrix.

mitochondrial membrane, along with an electrical potential difference. As a result of the concentration and potential differences, protons flow back through the membrane using a channel formed by the enzyme F_1-ATPase. The flow of protons through this enzyme is believed to drive the phosphorylation reaction.

$$ADP + P_i \xrightarrow[\text{F_1-ATPase}]{\text{H^+ flow through}} ATP \qquad (23.15)$$

✔ **LEARNING CHECK 23.3** How many molecules of ATP are produced from the complete catabolism of the following?

a. Two molecules of $FADH_2$
b. Two molecules of NADH
c. Two molecules of acetyl CoA

23.8 | The Complete Oxidation of Glucose

Learning Objective

10. Calculate the amount of ATP produced by the complete oxidation of a mole of glucose.

The information given to this point about glycolysis, the citric acid cycle, and the electron transport chain almost makes it possible to calculate the ATP yield from the total oxidation of 1 mol of glucose. However, one more factor must be considered. NADH produced in the cytoplasm during glycolysis does not pass through the mitochondrial membrane to the site of the electron transport chain. Brain and muscle cells employ a transport mechanism

that passes electrons from the cytoplasmic NADH through the membrane to FAD molecules inside the mitochondria, forming $FADH_2$. In **Section 23.7**, we pointed out that one molecule of $FADH_2$ gives 1.5 molecules of ATP using the electron transport chain. Thus, with this transport mechanism, one molecule of cytoplasmic NADH generates only 1.5 molecules of ATP, whereas 2.5 molecules of ATP come from each molecule of mitochondrial NADH. A more efficient shuttle mechanism is found in liver, heart, and kidney cells, where one cytoplasmic NADH results in one mitochondrial NADH. Thus, 2.5 molecules of ATP are formed in this case.

This production of ATP from the complete aerobic catabolism of 1 mol of glucose in the liver is summarized in **Table 23.1**. Notice from the table that of the 32 mol of ATP generated, the great majority (28 mol) is formed as a result of oxidative phosphorylation (last category in Table 23.1). In **Section 23.4**, we learned that only 2 mol of ATP is produced per mole of glucose by lactate fermentation and by alcoholic fermentation. Thus, in terms of ATP production from fuels, the complete aerobic oxidation of glucose is 16 times more efficient than either lactate or alcoholic fermentation.

TABLE 23.1 The ATP Yield from the Complete Oxidation of One mol of Glucose

Reaction	Comments	Yield of ATP (moles)
Cytoplasmic reactions		
Glycolysis		
glucose → glucose-6-phosphate	1 mol consumes 1 ATP	−1
glucose-6-phosphate → fructose-1,6-bisphosphate	1 mol consumes 1 ATP	−1
glyceraldehyde-3-phosphate → 1,3-bisphosphoglycerate	2 mol produces 2 mol of cytoplasmic NADH	
1,3-bisphosphoglycerate → 3-phosphoglycerate	2 mol produces 2 ATP	+2
phosphoenolpyruvate → pyruvate	2 mol produces 2 ATP	+2
Mitochondrial reactions		
Oxidation of pyruvate		
pyruvate → acetyl CoA + CO_2	2 mol produces 2 NADH	
Citric acid cycle		
isocitrate → α-ketoglutarate + CO_2	2 mol produces 2 NADH	
α-ketoglutarate → succinyl CoA + CO_2	2 mol produces 2 NADH	
succinyl CoA → succinate	2 mol produces 2 GTP	+2
succinate → fumarate	2 mol produces 2 $FADH_2$	
malate → oxaloacetate	2 mol produces 2 NADH	
Electron transport chain and oxidative phosphorylation		
2 cytoplasmic NADH formed in glycolysis	each yields 2.5 ATP	+5[a]
2 NADH formed in the oxidation of pyruvate	each yields 2.5 ATP	+5
2 $FADH_2$ formed in the citric acid cycle	each yields 1.5 ATP	+3
6 NADH formed in the citric acid cycle	each yields 2.5 ATP	+15
	Net yield of ATP	+32

[a]In muscle and brain cells, 3 ATP are produced in this step and only 30 ATP overall.

Another useful way to look at the efficiency of the complete oxidation of glucose is to compare the amount of free energy available in glucose and the amount of free energy stored in synthesized ATP.

Glucose oxidation:

$$C_6H_{12}O_6 + 6O_2 \rightarrow 6CO_2 + H_2O \qquad \Delta G^{\circ\prime} = -686 \text{ kcal/mol} \qquad (23.16)$$

ATP synthesis:

$$32ADP + 32P_i \rightarrow 32ATP + 32H_2O \qquad \Delta G^{\circ\prime} = +234 \text{ kcal/mol} \qquad (23.17)$$

Overall reaction:

$$C_6H_{12}O_6 + 6O_2 + 32ADP + 32P_i \rightarrow 6CO_2 + 32ATP + 38H_2O \qquad (23.18)$$

$$\text{Total: } \Delta G^{\circ\prime} = -452 \text{ kcal/mol}$$

Thus, glucose oxidation liberates 686 kcal/mol, whereas the synthesis of 32 mol of ATP stores 234 kcal/mol. The efficiency of energy stored is

$$\frac{\text{energy stored}}{\text{energy available}} \times 100 = \frac{234 \text{ kcal/mol}}{686 \text{ kcal/mol}} \times 100 = 34.1\%$$

It is remarkable that living cells can capture 34% of the released free energy and make it available to do biochemical work. By contrast, an automobile engine makes available only about 20%–30% of the energy actually released by burning gasoline.

23.9 | Glycogen Metabolism

Learning Objective

11. Explain the importance of the processes of glycogenesis and glycogenolysis.

23.9A Glycogen Synthesis

Glucose consumed in excess of immediate body requirements is converted to glycogen (see **Section 17.8**), which is stored primarily in the liver and muscle tissue. The liver of an average adult can store about 110 g of glycogen, and the muscles can store about 245 g. The process by which glucose is converted to glycogen is called **glycogenesis**. Glycogenesis can occur in all cells, but it is an especially important function of liver and muscle cells. The net result of this anabolic process is the bonding of glucose units to a growing glycogen chain, with the hydrolysis of a high-energy nucleotide, UTP (uridine triphosphate) (see **Section 22.7**), providing the energy. The enzyme glycogen synthase forms the new $\alpha(1 \rightarrow 4)$ linkages between each glucose unit and lengthens the chain. The $\alpha(1 \rightarrow 6)$ branch points on the chain are inserted by a branching enzyme called amylo (1,4-1,6)-trans-glycosylase.

glycogenesis The synthesis of glycogen from glucose.

The overall process of glycogenesis is summarized by Reaction 23.19:

$$(\text{glucose})_n \xrightarrow{\hspace{3cm}} (\text{glucose})_{n+1} = \text{glycogen} \qquad (23.19)$$
$$\text{UTP} \qquad \text{UDP} + P_i$$

23.9B Glycogen Breakdown

The breakdown of glycogen back to glucose is called **glycogenolysis**. Although major amounts of glycogen are stored in both muscle tissue and the liver, glycogenolysis can occur in the liver (and kidney and intestinal cells) but not in muscle tissue because one essential enzyme (glucose-6-phosphatase) is lacking. The first step in glycogen breakdown is the cleavage of $\alpha(1 \rightarrow 4)$ linkages, catalyzed by glycogen phosphorylase. In this reaction, glucose units are released from the glycogen chain as glucose-1-phosphate:

glycogenolysis The breakdown of glycogen to glucose.

$$(\text{glucose})_n + P_i \rightarrow (\text{glucose})_{n-1} + \text{glucose-1-phosphate} \qquad (23.20)$$
$$\text{glycogen} \qquad \text{glycogen with one} \\ \text{fewer glucose unit}$$

The second step in glycogen breakdown is the cleavage of the $\alpha(1 \rightarrow 6)$ linkages by hydrolysis, which is carried out by the debranching enzyme. When the branch points are eliminated, the phosphorylase can then continue to act on the rest of the chain.

In the debranching reaction, phosphoglucomutase catalyzes the isomerization of glucose-1-phosphate to glucose-6-phosphate:

$$\text{glucose-1-phosphate} \rightleftharpoons \text{glucose-6-phosphate} \qquad (23.21)$$

The final step in glycogen breakdown is the hydrolysis of glucose-6-phosphate to produce free glucose. The enzyme for this reaction, glucose-6-phosphatase, is the one found only in liver, kidney, and intestinal cells:

$$\text{glucose-6-phosphate} + H_2O \rightarrow \text{glucose} + P_i \qquad (23.22)$$

Without glucose-6-phosphatase, muscle cells cannot form free glucose from glycogen. However, they can carry out the first two steps of glycogenolysis to produce glucose-6-phosphate. This form of glucose is the first intermediate in the glycolysis pathway and can still be used to produce energy even if the reaction does not proceed to the formation of D-glucose. In this way, muscle cells utilize glycogen only for energy production. An important function of the liver is to maintain a relatively constant level of blood glucose, necessitating that this organ have a broader biochemical function than muscle. Thus, it has the capacity to degrade glycogen all the way to glucose, which is released into the blood during muscular activity and between meals.

✔ **LEARNING CHECK 23.4** What enzyme enables the liver to form glucose from glycogen? Why can't muscles carry out the same conversion?

23.10 | Gluconeogenesis

Learning Objective

12. Describe gluconeogenesis and the operation of the Cori cycle.

The supply of glucose in the form of liver and muscle glycogen can be depleted by about 12–18 hours of fasting. These stores of glucose are also depleted in a short time as a result of heavy work or strenuous exercise (see **Figure 23.12**). Nerve tissue, including the brain, would be deprived of glucose under such conditions if the only source was glycogen. Fortunately, a metabolic pathway exists that overcomes this problem.

Glucose can be synthesized from noncarbohydrate materials in a process called **gluconeogenesis**. When carbohydrate intake is low and when glycogen stores are depleted, the carbon skeletons of lactate, glycerol (derived from the hydrolysis of fats), and certain amino acids are used to synthesize pyruvate, which is then converted to glucose:

$$\text{(lactate, certain amino acids, glycerol)} \rightarrow \text{pyruvate} \rightarrow \text{glucose} \qquad (23.23)$$

About 90% of gluconeogenesis takes place in the liver. Very little occurs in the kidneys, brain, skeletal muscle, or heart, even though these organs have a high demand for glucose. Thus, gluconeogenesis taking place in the liver helps maintain the blood glucose level so that tissues needing glucose can extract it from the blood.

Gluconeogenesis involving lactate is especially important. As might be expected, this process follows a separate and different pathway from glycolysis. During active exercise, lactate levels increase in muscle tissue, and the compound diffuses out of the tissue into the blood. It is transported to the liver, where lactate dehydrogenase, the enzyme that catalyzes lactate formation in muscle, converts it back to pyruvate:

$$\underset{\text{lactate}}{CH_3-\overset{\overset{\displaystyle OH}{|}}{CH}-COO^-} + NAD^+ \rightleftharpoons \underset{\text{pyruvate}}{CH_3-\overset{\overset{\displaystyle O}{||}}{C}-COO^-} + NADH + H^+ \qquad (23.24)$$

The pyruvate is then converted to glucose by the gluconeogenesis pathway, and the glucose enters the blood. In this way, the liver increases a low blood glucose level and makes glucose available to muscle. This cyclic process, involving the transport of lactate from muscle to the liver, the resynthesis of glucose by gluconeogenesis, and the return of glucose to muscle tissue, is called the **Cori cycle** (see **Figure 23.13**).

Figure 23.12 Gluconeogenesis provides a source of glucose when glycogen supplies are exhausted after vigorous exercise.

gluconeogenesis The synthesis of glucose from noncarbohydrate molecules.

Cori cycle The process in which glucose is converted to lactate in muscle tissue, lactate is reconverted to glucose in the liver, and glucose is returned to the muscle.

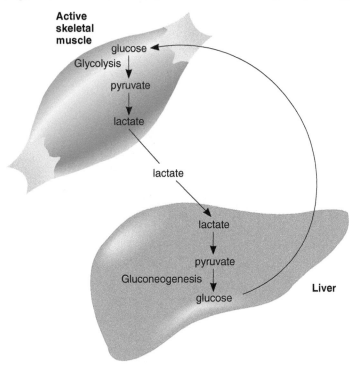

Figure 23.13 The Cori cycle, showing the relationship between glycolysis (in the muscle) and gluconeogenesis (in the liver).

What Is the Best Weight-Loss Strategy?

On the basis of the amount of advertising presented on TV and in other media, it would be easy to conclude that losing weight is a favorite pastime of many people. Special diets, numerous exercise devices, food supplements, and even meditation techniques are advertised with promises of near-miraculous weight-loss results.

The correct approach to weight loss is relatively easy to understand, but often quite challenging to put into practice. The human body must have a source of energy in order to function properly. The normal source of energy is digested food that is converted into chemical energy through a variety of biochemical processes. The produced energy is used to do work such as the movement of body parts and the synthesis of new tissue used for growth and the repair of damaged tissue. In the event that enough food is ingested to provide more energy than the body needs at that time, the energy is not wasted. It is converted into adipose tissue (fat) and stored in the body for later use as needed. Thus, the easily understood but challenging-to-use approach to weight loss can be represented by the following three conditions:

1. **Constant body weight:** Energy into body (food) = Energy used by body
2. **Weight gain:** Energy into body is greater than energy used by body
3. **Weight loss:** Energy into body is less than energy used by body

The challenge for those trying to lose weight is to achieve condition number three.

Once again, numerous approaches are widely advertised. We hear of the grapefruit diet, the high-protein diet, the low-carbohydrate diet, the twist-and-turn exercise device, and the portable doorway gym. In addition, each of the diets and devices is accompanied by testimonials from individuals in great physical condition. It can be hard to resist these well-designed "miracle" approaches, but the real truth about weight loss was given earlier in condition three. This condition can be achieved in several ways: decrease energy intake (food) into the body, increase energy used by the body (exercise), or both.

Nutritional studies indicate that for most people it is easier to decrease energy intake (food calories) than it is to use up the equivalent number of calories by exercising. For example, one pound (0.45 kg) of body fat represents 3500 calories of energy. If a person reduced their total food calorie intake by 500 calories per day and did not change their exercise habits, the result would be the loss of 1 pound of body fat per week (52 pounds per year!). In order to use an extra 500 calories per day by exercising, a 140-pound person would have to walk 6.5 miles in about 1.5 hours each day in addition to their normal daily physical activities.

It is not surprising that studies indicate it is more difficult for most people to lose weight by increased exercising than by dieting. Exercising usually requires a significant change in habits, not to mention the requirement to set aside the needed time on a regular basis. Calorie-counting, on the other hand, is relatively easy to do because of such things as the information given on most packaged food labels about serving sizes and the number of calories per serving. With such available information, it is not too hard to keep a record of current daily caloric intake for a couple of weeks, then plan on how to reduce it by the desired amount. The same studies show that calorie counting and reduced caloric intake coupled with increased physical activity results in the greatest long-term (permanent) weight loss. This is consistent with condition number 3 because both the energy into the body and the energy used by the body are each being changed in the direction that will lead to weight loss.

Exercise

+

Good nutrition

= Weight Loss

Figure 23.14 shows the relationship of gluconeogenesis to the other major pathways of glucose metabolism.

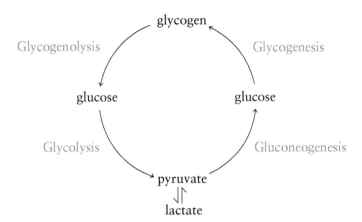

Figure 23.14 A summary of the major pathways in glucose metabolism. Note that the breakdown processes end in *-lysis*, whereas the synthesis pathways end in *-genesis*.

23.11 | The Hormonal Control of Carbohydrate Metabolism

Learning Objective

13. Describe how hormones regulate carbohydrate metabolism.

It is important that metabolic pathways be responsive to cellular conditions so that energy is not wasted in producing unneeded materials. Besides the regulation of enzymes at key control points (see **Section 23.3**), the body also uses three important regulatory hormones: epinephrine, glucagon, and insulin.

In **Section 23.2**, we pointed out that when carbohydrates are consumed, the blood glucose level increases. This increase signals the β-cells of the pancreas to release a small amount of the hormone insulin into the bloodstream. The hormone enhances the absorption of glucose from the blood into cells of active tissue such as skeletal and heart muscles (see **Section 19.6**). In addition, insulin increases the rate of synthesis of glycogen, fatty acids, and proteins and stimulates glycolysis. Each of these activities uses up glucose. As a result of insulin released into the bloodstream, the blood glucose level begins to decrease within 1 hour after carbohydrates have been ingested and returns to normal within about 3 hours.

Glucagon, a second polypeptide hormone, is synthesized and secreted by the α-cells of the pancreas. Glucagon activates the breakdown of glycogen (glycogenolysis) in the liver and thus counteracts the effect of insulin by raising blood glucose levels. **Figure 23.15** emphasizes the fact that insulin and glucagon work in opposition to each other and that the blood sugar level at any given time depends, in part, on the biochemical balance between these two hormones. Blood levels of glucose are also influenced to some degree by growth hormones and adrenal cortex steroids.

Figure 23.15 The biochemical balance between two opposing hormones, insulin and glucagon.

Choose Complex Carbohydrates

Humans are omnivorous; we have become adapted to eating fruits, vegetables, whole grains, and protein in the form of beans, legumes, or meat. In the past, people used to expend significant amounts of energy obtaining and preparing food. With much of the work to obtain food removed, and the increasing availability of processed foods, it is very easy to consume excessive amounts of calories and sugars. This excess overtaxes the pancreas, and is stored as body fat. The resulting obesity is linked to many health problems, so it is important to develop healthful dietary habits to avoid it and reduce the risk of chronic disease.

With this goal in mind, all food and beverage choices matter. To increase the percentage of complex carbohydrates in your diet as recommended in **Section 22.2**, focus on foods that are often described as "starchy." Examples of foods that are rich in dietary starches include potatoes (white, red, or sweet), green vegetables, beans, lentils, peas, and grains, or foods made from these materials such as breads or pastas. Starchy foods that are eaten in their natural state (whole grains and unprocessed vegetables) are considered to be healthful foods. Processing removes many of the nutrients and fiber, so choose carbohydrates that are not only complex, but are also unprocessed.

Because complex carbohydrates are made of simple sugars, the body has to expend energy and do work to break them down into simple sugars. The digestive system in a healthy body is able to do this quite well. Choosing foods that require the various parts of the digestive system to function is a practice that adds to good bodily health. This means that, where possible, foods that contain complex carbohydrates (starches) should be chosen rather than foods that contain simple sugars. Examples of doing this include eating whole fruit instead of drinking fruit juice; eating whole-grain bread and pasta rather than white bread and pasta. When choosing carbohydrates as food, consider how much work your body will have to do in order to convert that food into energy. Give your body something beneficial to do; choose complex carbohydrates.

Several food sources of complex carbohydrates.

The third hormone affecting glucose metabolism is epinephrine, synthesized by the adrenal gland (see **Section 16.6**). Epinephrine stimulates glycogen breakdown in muscles and, to a smaller extent, in the liver. This glycogenolysis reaction—usually a response to pain, anger, or fear—provides energy for a sudden burst of muscular activity. **Table 23.2** summarizes the hormonal control of glycogen.

TABLE 23.2 The Hormonal Control of Glycogen

Hormone	Source	Effect on Glycogen	Impact on Blood Glucose
Insulin	β-cells of pancreas	Increases formation	Lowers blood glucose levels
Glucagon	α-cells of pancreas	Activates breakdown of liver glycogen	Raises blood glucose levels
Epinephrine	Adrenal medulla	Stimulates breakdown	Raises blood glucose levels

▶ Case Study Follow-up

The Juvenile Diabetes Research Foundation (JDRF) is an excellent resource for the newly diagnosed adult with type 1 diabetes. The JDRF has extensive information available online and free publications such as the "Adult Type 1 Toolkit (Newly Diagnosed)." It covers the procedures to managing diabetes and the common emotional reactions that accompany the diagnosis. The Toolkit outlines steps to educating friends and family concerning diabetes and suggests conducting a "fire drill," or rehearsal, for dealing with the possibility of discovering a person with diabetes in a hypoglycemic coma. The American Diabetes Association (ADA) also has excellent

resources for people with diabetes. Support groups meeting locally or online are also available.

The steps for treating diabetes are complex and can be frustrating. Diabetes can affect a person's ability to fight infection; have an adverse effect on vision; and, for women, complicate pregnancy. Diabetes is considered a coronary heart disease (CHD) risk factor equal to smoking. For couples, all of these complications may have a negative impact on their relationship. Adequate emotional support for both partners is important to managing the disease and maintaining a positive relationship.

Concept Summary

The Digestion of Carbohydrates Glucose is the key food molecule for most organisms, and it is the central substance in carbohydrate metabolism. During digestion, carbohydrates are hydrolyzed to the monosaccharides glucose, fructose, and galactose, which are absorbed into the bloodstream through the lining of the small intestine.

Objective 1 (Section 23.1), Exercise 23.2

Blood Glucose Glucose is the most abundant carbohydrate in the blood. The liver regulates blood glucose levels so that a sufficient concentration is always available to meet the body's energy needs. A lower-than-normal blood glucose level is referred to as hypoglycemia; a higher-than-normal blood glucose level is termed hyperglycemia.

Objective 2 (Section 23.2), Exercise 23.6

Glycolysis Glycolysis, a series of 10 reactions that occurs in the cytoplasm, is a process in which one glucose molecule is converted into two molecules of pyruvate. Two molecules of ATP and two molecules of NADH are produced in the process.

Objective 3 (Section 23.3), Exercise 23.10

Both galactose and fructose are converted into intermediates that enter the glycolysis pathway. The glycolysis pathway is controlled by the regulation of three key enzymes in the reaction sequence.

Objective 4 (Section 23.3), Exercise 23.14

The Fates of Pyruvate Pyruvate, the product of glycolysis, has three different fates. Under aerobic conditions, pyruvate enters mitochondria and is converted to acetyl CoA. The acetyl CoA enters the citric acid cycle. Products of the citric acid cycle enter the electron transport chain, where oxidation is completed and ATP is synthesized. A second fate of pyruvate is conversion to lactate. This takes place in anaerobic

microorganisms and in humans when glycolysis occurs faster than the oxygen-dependent citric acid cycle and the electron transport chain can operate. Some microorganisms convert pyruvate to ethanol, the third fate of pyruvate. In each of these three processes, NAD^+ is regenerated so that glycolysis can continue.

Objective 5 (Section 23.4), Exercise 23.20

The Citric Acid Cycle The citric acid cycle generates NADH and $FADH_2$, which are necessary for ATP synthesis. It also produces intermediates used for biosynthesis.

Objective 6 (Section 23.5), Exercise 23.24

This pathway consists of eight reactions that process incoming molecules of acetyl CoA. The carbon atoms leave the cycle in the form of molecules of CO_2. The hydrogen atoms and electrons leave the cycle in the form of reduced coenzymes NADH and $FADH_2$. The cycle is regulated by three allosteric enzymes in response to cellular levels of ATP. One acetyl CoA molecule entering the citric acid cycle produces three molecules of NADH, one of $FADH_2$, and one of GTP.

Objective 7 (Section 23.5), Exercise 23.32

The Electron Transport Chain This pathway involves a series of reactions that pass electrons from NADH and $FADH_2$ to molecular oxygen. Each carrier in the series has an increasing affinity for electrons. Four of the carriers, which are referred to as cytochromes, contain iron, which accepts and then transfers the electrons. As NADH and $FADH_2$ release their hydrogen atoms and electrons, NAD^+ and FAD are regenerated for return to the citric acid cycle.

Objective 8 (Section 23.6), Exercises 23.34 and 23.36

Oxidative Phosphorylation At three sites within the electron transport chain, the decrease in free energy is sufficient to convert ADP to ATP. One molecule of NADH

Carbohydrate Metabolism **753**

produces 2.5 molecules of ATP. $FADH_2$, which enters the chain one step later, produces 1.5 molecules of ATP. According to the chemiosmotic hypothesis, the synthesis of ATP takes place because of a flow of protons across the inner mitochondrial membrane.

Objective 9 (Section 23.7), Exercise 23.46

The Complete Oxidation of Glucose The complete oxidation of 1 mol of glucose produces 32 (or 30) mol of ATP. In contrast, lactate and alcoholic fermentation produce only 2 mol of ATP.

Objective 10 (Section 23.8), Exercise 23.48

Glycogen Metabolism When blood glucose levels are high, excess glucose is converted into glycogen by the process of glycogenesis. The glycogen is stored within the liver and muscle tissue. Glycogenolysis, the breakdown of glycogen into glucose, occurs when muscles need energy and when the liver is restoring a low blood sugar level to normal.

Objective 11 (Section 23.9), Exercise 23.50

Gluconeogenesis Lactate, certain amino acids, and glycerol can be converted into glucose. This process, called gluconeogenesis, takes place in the liver when glycogen supplies are being depleted and when carbohydrate intake is low. Gluconeogenesis from lactate is especially important during periods of high muscle activity; the liver converts excess lactate from the muscles into glucose, which is then cycled back to the muscles. This process is called the Cori cycle.

Objective 12 (Section 23.10), Exercise 23.58

The Hormonal Control of Carbohydrate Metabolism Three hormones exert major interactive effects on carbohydrate metabolism: insulin, glucagon, and epinephrine. Insulin promotes the uptake and utilization of glucose by the cells. Thus, insulin lowers blood sugar levels. Glucagon stimulates the conversion of glycogen to glucose and thus raises blood glucose levels. Epinephrine is released in response to anger, fear, and excitement. It also stimulates the release of glucose from glycogen and raises blood glucose levels.

Objective 13 (Section 23.11), Exercise 23.62

Key Terms and Concepts

Aerobic (23.4)	Cytochrome (23.6)	Glycolysis (23.3)
Alcoholic fermentation (23.4)	Electron transport chain (23.6)	Hyperglycemia (23.2)
Anaerobic (23.4)	Gluconeogenesis (23.10)	Hypoglycemia (23.2)
Blood sugar level (23.2)	Glucosuria (23.2)	Lactate fermentation (23.4)
Chemiosmotic hypothesis (23.7)	Glycogenesis (23.9)	Oxidative phosphorylation (23.7)
Citric acid cycle (23.5)	Glycogenolysis (23.9)	Renal threshold (23.2)
Cori cycle (23.10)		

Key Reactions

1. Glycolysis (Section 23.3): $glucose + 2P_i + 2ADP + 2NAD^+ \longrightarrow 2\ pyruvate + 2ATP + 2NADH + 4H^+ + 2H_2O$	Reaction 23.5
2. Oxidation of pyruvate to acetyl CoA (Section 23.4): $CH_3 - \overset{\overset{O}{\|}}{C} - COO^- + CoA - S - H + NAD^+ \longrightarrow CH_3 - \overset{\overset{O}{\|}}{C} - S - CoA + NADH + CO_2$	Reaction 23.7
3. Reduction of pyruvate to lactate (Section 23.4): $CH_3 - \overset{\overset{O}{\|}}{C} - COO^- + NADH + H^+ \rightleftarrows CH_3 - \overset{\overset{OH}{\|}}{CH} - COO^- + NAD^+$	Reaction 23.8
4. Net reaction of lactate fermentation (Section 23.4): $C_6H_{12}O_6 + 2ADP + 2P_i \longrightarrow 2CH_3 - \overset{\overset{OH}{\|}}{CH} - COO^- + 2ATP + 2H_2O$	Reaction 23.9

5. Reduction of pyruvate to ethanol (Section 23.4):	$$CH_3-\overset{\overset{\displaystyle O}{\|}}{C}-COO^- + 2H^+ + NADH \longrightarrow CH_3CH_2-OH + NAD^+ + CO_2$$	Reactions 23.10 and 23.11
6. Net reaction of alcoholic fermentation (Section 23.4):	$$C_6H_{12}O_6 + 2ADP + 2P_i \longrightarrow 2CH_3CH_2-OH + 2CO_2 + 2ATP$$	Reaction 23.12

7. The citric acid cycle (Section 23.5):

$$\text{acetyl CoA} + 3NAD^+ + FAD + GDP + P_i + 2H_2O \longrightarrow 2CO_2 + CoA-S-H + 3NADH + 2H^+ + FADH_2 + GTP$$

Reaction 23.13

8. The electron transport chain (Section 23.6):	$$4H^+ + 4e^- + O_2 \longrightarrow 2H_2O$$	Reaction 23.14
9. Oxidative phosphorylation (Section 23.7):	$$ADP + P_i \longrightarrow ATP$$	Reaction 23.15
10. Complete oxidation of glucose (Section 23.8):	$$C_6H_{12}O_6 + 6O_2 + 32ADP + 32P_i \longrightarrow 6CO_2 + 32ATP + 38H_2O$$	Reaction 23.18
11. Glycogenesis (Section 23.9):	glucose \longrightarrow glycogen	Reaction 23.19
12. Glycogenolysis (Section 23.9):	glycogen \longrightarrow glucose	Reactions 23.20, 23.21, and 23.22
13. Gluconeogenesis (Section 23.10):	$$\left.\begin{matrix} \text{lactate,} \\ \text{certain amino acids,} \\ \text{glycerol} \end{matrix}\right\} \longrightarrow \text{pyruvate} \longrightarrow \text{glucose}$$	Reaction 23.23

Exercises

Even-numbered exercises are answered in Appendix B.

Blue-numbered exercises are more challenging.

The Digestion of Carbohydrates (Section 23.1)

23.1 Why is glucose considered the pivotal compound in carbohydrate metabolism?

23.2 Name the products produced by the digestion of

 a. starch **c.** sucrose

 b. lactose **d.** maltose

23.3 What type of reaction is involved as carbohydrates undergo digestion?

Blood Glucose (Section 23.2)

23.4 Describe what is meant by the terms *blood sugar level* and *normal fasting level.*

23.5 What range of concentrations for glucose in blood is considered a normal fasting level?

23.6 How do each of the following terms relate to blood sugar level?

 a. Hypoglycemia

 b. Hyperglycemia

 c. Renal threshold

 d. Glucosuria

23.7 Explain how the synthesis and degradation of glycogen in the liver help regulate blood sugar levels.

23.8 Explain why severe hypoglycemia can be very serious.

Glycolysis (Section 23.3)

23.9 What is the main purpose of glycolysis?

23.10 What is the starting material for glycolysis? The product?

23.11 In what part of the cell does glycolysis take place?

23.12 How many steps in glycolysis require ATP? How many steps produce ATP? What is the net production of ATP from one glucose molecule undergoing glycolysis?

23.13 Which coenzyme serves as an oxidizing agent in glycolysis?

23.14 What is the third control point in glycolysis? How is it responsive to cellular needs?

23.15 Explain how fructose and galactose can enter the glycolysis pathway.

23.16 Explain how a high concentration of glucose-6-phosphate causes a decrease in the rate of glycolysis.

23.17 Why is the conversion of fructose-6-phosphate to fructose-1, 6-bisphosphate sometimes called the "committed step" of glycolysis?

The Fates of Pyruvate (Section 23.4)

23.18 Distinguish between the terms *aerobic* and *anaerobic*.

23.19 Explain how lactate formation allows glycolysis to continue under anaerobic conditions.

23.20 Compare the fate of pyruvate in the body under (a) aerobic conditions and (b) anaerobic conditions.

23.21 Write a summary equation for the conversion of glucose to:

 a. lactate

 b. pyruvate

 c. ethanol

23.22 What is the fate of the pyruvate molecule's carboxylate group during alcoholic fermentation? Explain how ethanol formation from pyruvate allows yeast to survive with limited oxygen supplies.

The Citric Acid Cycle (Section 23.5)

23.23 What are two other names for the citric acid cycle?

23.24 What is the primary function of the citric acid cycle in ATP production? What other vital role is served by the citric acid cycle?

23.25 Why is it appropriate to refer to the sequence of reactions in the citric acid cycle as a cycle?

23.26 Describe the citric acid cycle by identifying the following:

 a. The fuel needed by the cycle

 b. The form in which carbon atoms leave the cycle

 c. The form in which hydrogen atoms and electrons leave the cycle

23.27 The citric acid cycle is located in close cellular proximity to the electron transport chain. Why?

23.28 How many molecules of each of the following are produced by one run through the citric acid cycle?

 a. NADH

 b. $FADH_2$

 c. GTP

 d. CO_2

23.29 Write reactions for the steps in the citric acid cycle that involve oxidation by NAD^+.

23.30 Write a reaction for the step in the citric acid cycle that involves oxidation by FAD.

23.31 Write a reaction for the step in the citric acid cycle that involves the phosphorylation of GDP. What is the structural difference between ATP and GTP? Explain why GTP can function as a high-energy compound.

23.32 The reactions in the citric acid cycle involve enzymes.

 a. Identify the enzymes at the three control points.

 b. Each control-point enzyme is inhibited by NADH and ATP. Explain how this enables the cell to be responsive to energy needs.

 c. Two of the control-point enzymes are activated by ADP. Explain the benefit of this regulation to the cell.

The Electron Transport Chain (Section 23.6)

23.33 What is the primary function of the electron transport chain?

23.34 Which coenzymes bring electrons to the electron transport chain?

23.35 List the participants of the electron transport chain in the order in which they accept electrons.

23.36 What is the role of the cytochromes in the electron transport chain?

23.37 What is the role of oxygen in the functioning of the electron transport chain? Write a reaction that summarizes the fate of the oxygen.

Oxidative Phosphorylation (Section 23.7)

23.38 The third stage in the oxidation of foods to provide energy involves oxidative phosphorylation. What is oxidized? What is phosphorylated?

23.39 How many ATP molecules are synthesized when NADH passes electrons to molecular oxygen?

23.40 How many ATP molecules are synthesized when $FADH_2$ passes electrons to molecular oxygen?

23.41 Explain what is meant by the statement: "Oxidation of NADH in the electron transport chain results in 34% energy conservation."

23.42 How many ATP molecules can be formed by the oxidation of 10 acetyl CoA molecules in the common catabolic pathway?

23.43 Identify the steps in the electron transport chain where ATP is synthesized.

23.44 The change in free energy for the complete oxidation of glucose to CO_2 and H_2O is -686 kcal per mole of glucose. If all this free energy could be used to drive the synthesis of ATP, how many moles of ATP could be formed from 1 mol of glucose?

23.45 What is the function of the protein F_1-ATPase?

23.46 According to the chemiosmotic theory, ATP synthesis results from the flow of what particles? Across what membrane does this flow occur?

The Complete Oxidation of Glucose (Section 23.8)

23.47 Explain why NADH produced in muscles during glycolysis ultimately gives rise to just 1.5 molecules of ATP.

23.48 A total of 32 mol of ATP can be produced by the complete oxidation of 1 mol of glucose in the liver. Indicate the number of these moles of ATP produced by:

 a. glycolysis

 b. the citric acid cycle

 c. the electron transport chain and oxidative phosphorylation

23.49 The complete biochemical oxidation of glucose to ATP is only 34% efficient in capturing the total amount of energy released. The remaining 66% is released as what form of energy?

Glycogen Metabolism (Section 23.9)

23.50 Compare the terms *glycogenesis* and *glycogenolysis*.

23.51 Fructose and galactose from digested food enter the glycolysis pathway. Name the intermediates of the pathway which serve as entry points for these substances.

23.52 What high-energy compound is involved in the conversion of glucose to glycogen?

23.53 Why is it important that the liver store glycogen? How does this aid the controlling of blood sugar levels?

23.54 Why can liver glycogen, but not muscle glycogen, be used to raise blood sugar levels?

23.55 How does glycogen enter the glycolysis pathway?

Gluconeogenesis (Section 23.10)

23.56 What organ serves as the principal site for gluconeogenesis?

23.57 A friend has been fasting for three days. Where does the energy come from to keep his brain functioning?

23.58 What are the sources of compounds that undergo gluconeogenesis?

23.59 Explain the operation of the Cori cycle and its importance to muscles.

23.60 Would the Cori cycle operate if the body's muscles were completely oxidizing the glucose supplied? Why or why not?

The Hormonal Control of Carbohydrate Metabolism (Section 23.11)

23.61 What cells produce insulin and glucagon in the body?

23.62 Describe the influence of glucagon and insulin on the following:

 a. Blood sugar level

 b. Glycogen formation

23.63 Why is it more important that epinephrine stimulate the breakdown of glycogen in the muscles rather than elevating blood sugar levels?

Additional Exercises

23.64 You eat a dextrose tablet that contains 10.0 g of glucose for an energy boost. How many molecules of ATP can be produced with the complete oxidation of this glucose by muscle cells?

23.65 As an athlete's cells begin to go anaerobic, what happens to the rate of pyruvate oxidation by the citric acid cycle? What happens to the rate of glycolysis? Explain how each of these rate changes occurs and which enzymes are affected.

23.66 Lactate dehydrogenase catalyzes the following equilibrium reaction in both liver and muscle cells:

$$\text{pyruvate} + \text{NADH} + \text{H}^+ \rightleftharpoons \text{lactate} + \text{NAD}^+$$

In terms of equilibrium, explain why lactate must be removed from muscle cells under anaerobic conditions and why lactate does not accumulate in the liver under these conditions. (*Hint*: Think of the Cori cycle.)

23.67 Identify all chiral carbon atoms in citrate and isocitrate.

citrate isocitrate

23.68 The pyruvate dehydrogenase enzyme complex becomes inactive when a specific serine residue in the enzyme complex is phosphorylated by ATP. Phosphorylation is the covalent addition of a phosphate group. Show how this reaction occurs and name the type of bond formed.

Chemistry for Thought

23.69 The citric acid cycle does not utilize oxygen, yet it cannot function without oxygen. Why?

23.70 Malonate ($^-\text{OOC}-\text{CH}_2-\text{COO}^-$) is an extremely effective competitive inhibitor of the citric acid cycle. Which enzyme do you think malonate inhibits? Explain.

23.71 A friend started to make wine by adding yeast to fresh grape juice and placing the mixture in a sealed bottle. Several days later the bottle exploded. Explain.

23.72 Two students are debating whether 30 or 32 molecules of ATP can be formed by the catabolism of one molecule of glucose. Who is correct?

23.73 Explain why monitoring blood lactate levels might be a useful technique to gauge the amount of conditioning in an Olympic runner.

23.74 You may have experienced a rush of epinephrine during a recent exam or athletic event. Are there other physiological responses to epinephrine in addition to a higher blood sugar level?

23.75 In Figure 23.4, it is noted that making bread involves a fermentation process. Why, then, is the alcohol content of bread zero?

23.76 A friend has the habit of eating four or five candy bars per day as snacks. What might be the metabolic consequences of such a practice?

23.77 List some pros and cons of a low-carbohydrate diet.

23.78 Cyanide is believed to act as an inhibitor of the electron transport chain. Why is such an inhibitor so toxic?

23.79 Drinking milk always seems to make you slightly nauseous. What might be an explanation for this?

Allied Health Exam Connection

The following questions are from these sources:

- *Nursing School Entrance Exam* © 2005, Learning Express, LLC.

- *McGraw-Hill's Nursing School Entrance Exams* by Thomas A. Evangelist, Tamara B. Orr, and Judy Unrein © 2009, The McGraw-Hill Companies, Inc.

- *NSEE Nursing School Entrance Exams*, 3rd edition © 2009, Kaplan Publishing.

- *Cliffs Test Prep: Nursing School Entrance Exams* by Fred N. Grayson © 2004, Wiley Publishing, Inc.

- *Peterson's Master the Nursing School and Allied Health Entrance Exams*, 18th edition by Marion F. Gooding © 2008, Peterson's, a Nelnet Company.

23.80 In order to be absorbed into the blood, complex carbohydrates must first be hydrolyzed into _____ units, primarily _____.

23.81 Which of the following are true statements about glycolysis?

 a. It requires 1 ATP molecule to activate a molecule of glucose.

 b. It occurs in the mitochondria.

 c. It does not require oxygen.

23.82 Compare the fate of pyruvate (a) in the body under aerobic conditions, (b) in the body under anaerobic conditions, and (c) in alcoholic fermentative microbes under anaerobic conditions.

23.83 Which of the following are true statements about the citric acid cycle?

 a. It produces H_2O, CO_2, and NADH.

 b. It occurs in the mitochondria.

 c. It is the first stage of aerobic cellular respiration.

 d. It is also called the Krebs cycle.

23.84 What is the final electron acceptor of the electron transport system? Write the reaction for the last step of the electron transport chain.

23.85 Most soluble food substances enter the bloodstream through the:

 a. small intestine. **c.** capillaries in the stomach.

 b. duodenum. **d.** hepatic vein.

23.86 Glycogenesis occurs primarily in the:

 a. blood cells and spleen.

 b. pancreas and gallbladder.

 c. small intestines and stomach.

 d. liver and muscles.

23.87. Insulin is produced in the:

 a. pituitary gland.

 b. thymus.

 c. pancreas.

 d. pineal gland.

23.88 Which of the following describes the function of insulin in the body?

 a. It decreases glucose uptake by cells.

 b. It stimulates the release of glucagon.

 c. It increases serum levels of glucose.

 d. It increases glucose conversion to glycogen.

23.89 For the aerobic respiration pathway, electron transport systems are located in the:

 a. cytoplasm.

 b. Golgi bodies.

 c. lysosomes.

 d. mitochondrion.

23.90 During strenuous exercise, a buildup of what substance may cause muscle cramps?

 a. lactic acid

 b. lactose

 c. adrenaline

 d. serotonin

24

Lipid and Amino Acid Metabolism

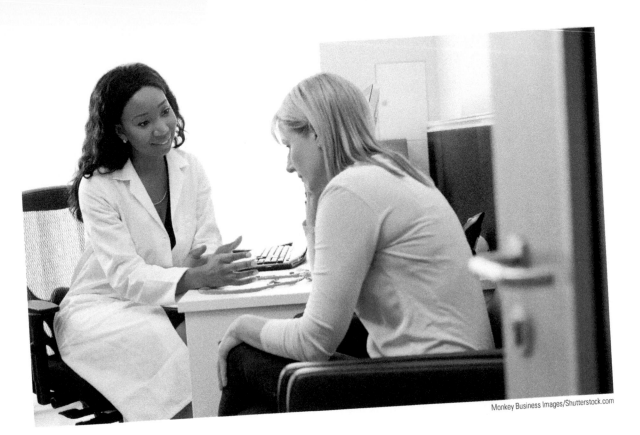

Monkey Business Images/Shutterstock.com

▶ Case Study

Fish Oil and Antidepressants: Laura brought her teenaged daughter, Rebecca, to the psychiatrist to obtain an expert opinion on medications. Rebecca had been visiting therapists because of her mood swings and difficult behaviors since she was 9 years old. Now, at age 15, she began engaging in self-harming behaviors of binge eating and inflicting small cuts on her arms. Laura felt desperate to get her daughter some help. Their family doctor prescribed a common antidepressant and therapy, but Rebecca's symptoms got worse. Laura then took Rebecca to a psychiatrist who recommended a different medication that was taken along with fish oil supplements. He gave Laura a handout containing publications that showed connections between fish oil supplements and improved response to antidepressant medications. Laura was anxious for Rebecca to try it, and it worked! After 3 weeks, Rebecca's mood improved enough that she could work with her therapist to learn healthier behaviors and care for herself. Now, she enjoys school, sports, and friends.

▶ How common is depression in teenagers? How might fish oil improve a patient's response to antidepressants? Can you get enough fish oil from eating fish, or should you take supplements?

Follow-up to this Case Study appears at the end of the chapter before the *Concept Summary*.

Learning Objectives

When you have completed your study of this chapter, you should be able to:

1 Describe the digestion, absorption, and distribution of lipids in the body. **(Section 24.1)**
2 Explain what happens during fat mobilization. **(Section 24.2)**
3 Identify the metabolic pathway by which glycerol is catabolized. **(Section 24.3)**
4 Outline the steps of β-oxidation for fatty acids. **(Section 24.4)**
5 Determine the amount of ATP produced by the complete catabolism of a fatty acid. **(Section 24.5)**
6 Name the three ketone bodies and list the conditions that cause their overproduction. **(Section 24.6)**

7 Describe the pathway for fatty acid synthesis. **(Section 24.7)**
8 Describe the source and function of the body's amino acid pool. **(Section 24.8)**
9 Write equations for transamination and deamination reactions. **(Section 24.9)**
10 Explain the overall results of the urea cycle. **(Section 24.9)**
11 Describe how amino acids can be used for energy production, the synthesis of triglycerides, and gluconeogenesis. **(Section 24.10)**
12 State the relationship between intermediates of carbohydrate metabolism and the synthesis of nonessential amino acids. **(Section 24.11)**

Triglycerides (fats and oils) are important dietary sources of energy. Fat also functions as the major form of energy storage in humans. Fat, with a caloric value of 9 calories/g, contains more than twice the energy per gram of glycogen or starch. Thus, fat is a more concentrated and efficient storage form of energy than carbohydrates. Because it is water-insoluble, fat can be stored in much larger quantities in our bodies. For example, the limited carbohydrate reserves in the body are depleted after about 1 day without food, but stored fat can provide needed calories (but not other nutrients) in an individual for 30 to 40 days. Amino acids perform three vital functions in the human body: They are the building blocks for the synthesis of proteins, they provide carbon and nitrogen atoms for the synthesis of other biomolecules, and they can be sources of energy.

In this chapter, we discuss the storage, degradation, and synthesis of lipids. We also study the degradation and synthesis of amino acids and the relationships between amino acid metabolism and the metabolism of carbohydrates and lipids.

24.1 | Blood Lipids

Learning Objective

1. Describe the digestion, absorption, and distribution of lipids in the body.

During digestion, triglycerides are hydrolyzed to glycerol, fatty acids, and monoglycerides (one fatty acid attached to glycerol). Phosphoglycerides are also hydrolyzed to their component substances. These smaller molecules, along with cholesterol, are absorbed into cells of the intestinal mucosa, where resynthesis of the triglycerides

and phosphoglycerides occurs. For transport within the aqueous environment of the lymph and blood, these insoluble lipids are complexed with proteins to form lipoprotein aggregates called **chylomicrons**. The chylomicrons (which contain triglycerides, phosphoglycerides, and cholesterol) pass into the lymph system and then into the bloodstream. Chylomicrons are modified by the liver into smaller lipoprotein particles, the form in which most lipids are transported to various parts of the body by the bloodstream (see **Figure 24.1**).

Figure 24.1 The digestion and absorption of triglycerides.

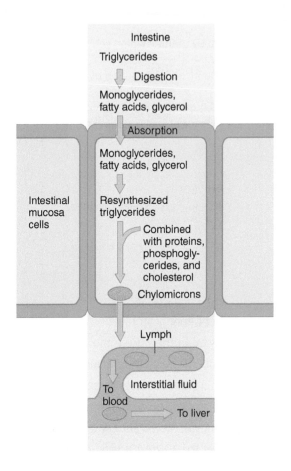

To some extent, the behavior of blood lipids parallels that of blood sugar. The concentration of both substances increases following a meal and returns to normal as a result of storage in fat depots and of oxidation to provide energy. As shown in **Figure 24.2**, the concentration of plasma lipids usually begins to rise within 2 hours after a meal, reaches a peak in 4 to 6 hours, and then drops rather rapidly to a normal level.

One general method of classifying lipoproteins is by density. Because lipids are less dense than proteins, lipoprotein particles with a higher proportion of lipids are less dense than those with a lower proportion. Chylomicrons are the least dense because they have the highest ratio of lipid to protein. **Figure 24.3** is a schematic model of a low-density lipoprotein (LDL). The compositions of the four classes of lipoprotein are compared in **Figure 24.4**. We see from Figure 24.4 that the protein content of chylomicrons is significantly lower than that of other plasma lipoproteins, and that this low percentage results in a density of the particle that is less than 0.95 g/mL. Notice that the so-called high-density lipoproteins (HDL) have a protein mass that is 45% to 50% of the total mass of the particles. The densities of HDL particles have correspondingly high values, ranging from 1.06 to 1.21 g/mL.

Figure 24.2 The rise in blood lipids following a meal containing fat.

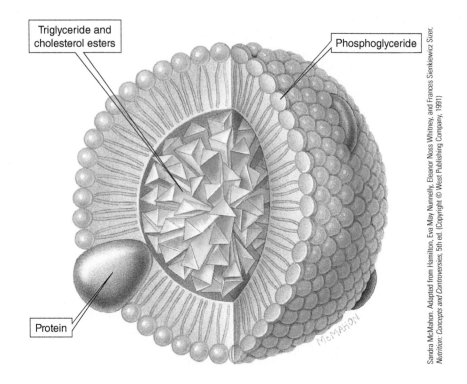

Figure 24.3 A schematic model of low-density lipoprotein (LDL).

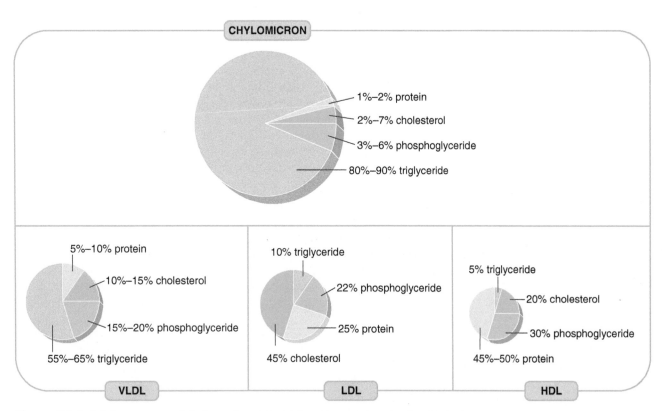

Figure 24.4 Relative amounts of cholesterol, protein, triglyceride, and phosphoglyceride in the four classes of lipoproteins.

Q: Are certain foods better for the brain?

A: When it comes to eating better for your brain, there are several foods and nutrients that may improve mood and memory and prevent age-related declines in cognitive function. Most research points to the so-called Mediterranean diet—one that is high in healthy fats, legumes, nuts, cereal, fish, fruits, and vegetables—as being the best for the brain. This dietary pattern, which is also lower in saturated fat and meat and contains moderate amounts of alcohol and dairy products, has several benefits when it comes to memory, cognitive function, and mood.

Research shows that higher levels of HDL cholesterol, commonly referred to as "good" cholesterol, may play a role in the preservation of memory. Many of the key components of the Mediterranean diet—including replacing saturated fat with unsaturated fat, eating plenty of soluble fiber found in legumes and cereal, and moderate alcohol consumption—may also help support healthy HDL levels.

Melina B. Jampolis, M.D., is a Physician Nutrition Specialist focusing on nutrition for weight loss and disease prevention and treatment.

Nuts and seeds may also be particularly beneficial for the brain. Almonds and sunflower seeds are good sources of monounsaturated fat and vitamin E, an antioxidant that helps protect neuronal (brain cell) membranes. Neuronal membranes are lined with fat. Healthy dietary fats, like monounsaturated fat, may improve the binding of neurotransmitters to these membranes. Neurotransmitters allow brain cells to communicate with each other, and their improved binding might improve mood. Animal studies indicate that walnuts—rich in polyphenols, polyunsaturated fatty acids, and antioxidants—provide a potent brain boost.

The omega-3 fatty acid DHA (docosahexaenoic acid) is also an important component of neuronal membranes. DHA plays a key role in brain structure and function and may offer protection against age-related cognitive decline. Since our bodies do not produce DHA, it is considered an essential fatty acid and must be obtained from our diet. Fatty fish—including salmon, tuna, mackerel, herring, and sardines—are excellent sources of DHA and should be consumed at least twice per week. Research also suggests that higher intakes of fish may be linked to lower levels of depression.

Berries—rich in fiber, vitamin C, antioxidants, and polyphenols—also appear to play an important role in brain health by improving brain signaling and protecting the brain from oxidative stress. Blueberries seem to be especially useful in terms of memory. Curcumin, another potent polyphenol and the active ingredient in the spice turmeric, also plays an active role in brain health. This might partially explain the lower incidence of Alzheimer's disease in Northern India, where turmeric, as a curry ingredient, is widely consumed.

Almonds are a good source of vitamin E.

DHA is present in salmon.

The spice turmeric contains a polyphenol.

Foods with whole grains are important.

Staying hydrated is essential.

It is also important to eat adequate amounts of healthy carbohydrates for optimal brain function. The brain does not store glucose, and studies suggest that low-carbohydrate diets might impair cognitive function. Low-carbohydrate diets might also negatively affect mood over time. Try to consume mainly healthy carbs—such as whole grains, fruits, and vegetables—to prevent large fluctuations in blood sugar, which can cause fatigue and thereby impair brain function.

Finally, staying well hydrated plays a part in optimal brain function. A study of college students found that even a mild 1 to 2% dehydration led to fatigue and confusion. Exercise is also key. Moderate aerobic activity, such as walking 20 to 30 minutes per day, helps to significantly improve cognitive function and performance.

So, while food may not necessarily be able to boost your IQ, eating a healthy, nutrient-rich Mediterranean-type diet, staying well hydrated, and getting regular exercise can help keep your brain fit and healthy.

Aerobic activity helps keep your brain fit.

24.2 | Fat Mobilization

Learning Objective

2. Explain what happens during fat mobilization.

Carbohydrates from dietary sources and glycogen catabolism are the compounds utilized preferentially for energy production by some tissues. The brain and active skeletal muscles are examples of such tissues. But as we have seen, the body's stores of glycogen are depleted after only a few hours of fasting. When this happens, fatty acids stored in triglycerides are called on as energy sources. Even when glycogen supplies are adequate, many body cells, such as those of resting muscle and liver, utilize energy derived from the breakdown of fatty acids. This helps conserve the body's glycogen stores and glucose for use by brain cells and red blood cells. Red blood cells cannot oxidize fatty acids because red blood cells have no mitochondria, the site of fatty acid oxidation. Brain cells are bathed in cerebrospinal fluid and do not obtain nutrients directly from the blood because of the blood-brain barrier. Glucose and many other substances can cross the barrier into the cerebrospinal fluid, but fatty acids cannot.

When body cells need fatty acids for energy, the endocrine system is stimulated to produce several hormones, including epinephrine, which interact with **adipose tissue**. Earlier we saw that epinephrine stimulates the conversion of glycogen to glucose. In the fat cells of adipose tissue, epinephrine stimulates the hydrolysis of triglycerides to fatty acids and glycerol, which enter the bloodstream. This process is called **fat mobilization**. In the blood, mobilized fatty acids form a lipoprotein with the plasma protein called serum albumin. In this form, the fatty acids are transported to the tissue cells that need them. The glycerol produced by the hydrolysis of the triglyceride is water-soluble, so it dissolves in the blood and is also transported to cells that need it.

adipose tissue A kind of connective tissue where triglycerides are stored.

fat mobilization The hydrolysis of stored triglycerides, followed by the entry of fatty acids and glycerol into the bloodstream.

24.3 | Glycerol Metabolism

Learning Objective

3. Identify the metabolic pathway by which glycerol is catabolized.

The glycerol hydrolyzed from fats can provide energy to cells. Glycerol is converted in the cytoplasm of the cell to dihydroxyacetone phosphate in two steps:

$$
\begin{array}{ccccc}
H_2C-OH & & H_2C-OH & & H_2C-OH \\
| & & | & & | \\
HC-OH & \longrightarrow & HC-OH & \longrightarrow & C=O & \longrightarrow & \text{Glycolysis} \\
| & & | & & | \\
H_2C-OH & & H_2C-OPO_3{}^{2-} & & H_2C-OPO_3{}^{2-} & & (24.1)
\end{array}
$$

$$\text{ATP} \quad \text{ADP} \qquad \text{NAD}^+ \quad \text{NADH} + \text{H}^+$$

glycerol glycerol- dihydroxyacetone
3-phosphate phosphate

Dihydroxyacetone phosphate is one of the chemical intermediates of glycolysis (see **Section 23.3**). Thus, glycerol, by entering glycolysis, can be converted to pyruvate and contribute to cellular energy production. The pyruvate formed from glycerol can also be converted to glucose through gluconeogenesis (see **Section 23.10**).

24.4 | The Oxidation of Fatty Acids

Learning Objective

4. Outline the steps of β-oxidation for fatty acids.

Fatty acids that enter tissue cells cannot be oxidized to produce energy until they pass through the mitochondrial membrane. This cannot occur until the fatty acid is converted into fatty acyl CoA by reaction with coenzyme A. Fatty acyl refers to the

$$
\begin{array}{c}
O \\
\|\\
R-C-
\end{array}
\text{ portion of the molecule.}
$$

$$
\begin{array}{ccc}
O & & O \\
\|\\
R-C-OH + HS-CoA & \longrightarrow & R-C-S-CoA + H_2O \qquad (24.2) \\
\text{fatty acid} & & \text{fatty acyl CoA}
\end{array}
$$

$$\text{ATP} \quad \text{AMP} + \text{PP}_i$$

This reaction, which is catalyzed by acyl CoA synthetase, is referred to as activation of the fatty acid because the fatty acyl CoA is a high-energy compound. The energy needed for its synthesis is provided by the hydrolysis of ATP to AMP and PP_i and the subsequent hydrolysis of PP_i to $2P_i$. The process of forming fatty acyl CoA molecules demonstrates how the energy stored in the phosphate bonds of ATP can be used to drive chemical processes that normally could not occur. The cost to the cell of activating one molecule of fatty acid is the loss of two high-energy bonds in an ATP molecule.

The fatty acyl CoA molecules that enter mitochondria are then degraded in a catabolic process called **β-oxidation**. During β-oxidation, the second (or beta) carbon down the chain from the carbonyl of the fatty acyl CoA molecule is oxidized to a ketone:

β-oxidation process A pathway in which fatty acids are broken down into molecules of acetyl CoA.

$$R \overset{\beta}{\underset{H}{\overset{H}{\underset{|}{\overset{|}{C}}}}} - \overset{\alpha}{CH_2} - \overset{O}{\overset{||}{C}} - S - CoA \xrightarrow{\text{Oxidation}} R - \overset{O}{\overset{||}{C}} - CH_2 - \overset{O}{\overset{||}{C}} - S - CoA \qquad (24.3)$$

fatty acyl CoA

Four reactions are involved in oxidizing the β-carbon to a ketone. We will use the fatty acyl CoA formed from one 18-carbon acid, stearic acid, as an example of the reactions. We will also keep track of any NADH or FADH$_2$ produced in the reactions because these molecules can enter the respiration process (electron transport chain). The reactions involved in β-oxidation are shown in **Figure 24.5**.

In the final step of β-oxidation (Step 4), the chain is broken between the α- and β-carbons by reaction with coenzyme A. As a result, a new fatty acyl CoA is formed with a chain shortened by two carbon atoms, and a unit of acetyl CoA is released:

$$CH_3(CH_2)_{14} - \overset{O}{\overset{||}{C}} - CH_2 - \overset{O}{\overset{||}{C}} - S - CoA + H - S - CoA \rightarrow CH_3(CH_2)_{14} - \overset{O}{\overset{||}{C}} - S - CoA + CH_3 - \overset{O}{\overset{||}{C}} - S - CoA \qquad (24.4)$$

β-carbon α-carbon new fatty acyl CoA acetyl CoA
18 carbons 16 carbons 2 carbons

The new fatty acyl compound enters the β-oxidation process at Step 1, and the sequence is repeated until the fatty acyl CoA is completely degraded to acetyl CoA (see **Figure 24.5**). The fatty acyl CoA molecule that starts each run through the β-oxidation process is two carbons shorter than the one going through the previous run. For this reason, the β-oxidation pathway for fatty acid degradation to acetyl CoA is often called the fatty acid spiral (rather than cycle).

Every run through the spiral produces one molecule each of acetyl CoA, NADH, and FADH$_2$ until the fatty acyl CoA is only four carbons long:

$$R - CH_2CH_2 - \overset{O}{\overset{||}{C}} - S - CoA + NAD^+ + FAD + H_2O + CoA - SH \longrightarrow$$

fatty acyl CoA
entering the spiral

$$\qquad (24.5)$$

$$R - \overset{O}{\overset{||}{C}} - S - CoA + CH_3\overset{O}{\overset{||}{C}} - S - CoA + NADH + H^+ + FADH_2$$

fatty acyl CoA acetyl CoA
2 carbons shorter

Lipid and Amino Acid Metabolism **767**

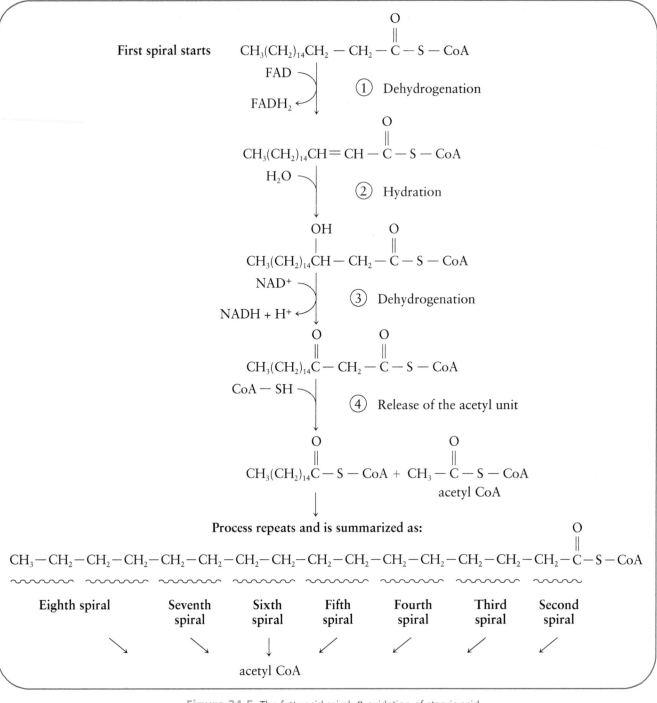

Figure 24.5 The fatty acid spiral: β-oxidation of stearic acid.

In the last spiral, the four-carbon chain of butyryl CoA passes through the β-oxidation sequence, and it produces one molecule of $FADH_2$, one molecule of NADH, and two molecules of acetyl CoA:

$$CH_3-CH_2-CH_2-\overset{\overset{\displaystyle O}{\|}}{C}-S-CoA + NAD^+ + H_2O + FAD + CoA-SH \longrightarrow$$

butyryl CoA

$$2CH_3-\overset{\overset{\displaystyle O}{\|}}{C}-S-CoA + NADH + H^+ + FADH_2$$

2 acetyl CoA molecules

(24.6)

In other words, the complete conversion of a fatty acyl CoA to two-carbon fragments of acetyl CoA always produces one more molecule of acetyl CoA than of FADH$_2$ or NADH. Thus, the breakdown of stearic acid requires eight passes through the β-oxidation sequence and produces 9 molecules of acetyl CoA, but only 8 molecules of FADH$_2$ and 8 molecules of NADH.

Net reaction:

$$CH_3(CH_2)_{16}\overset{\overset{O}{\|}}{C}-S-CoA + 8FAD + 8NAD^+ + 8H_2O + 8CoA-SH \longrightarrow$$

$$9CH_3\overset{\overset{O}{\|}}{C}-S-CoA + 8FADH_2 + 8NADH + 8H^+ \qquad (24.7)$$

✔ **LEARNING CHECK 24.1** Capric acid (10 carbon atoms) is converted to acetyl CoA through β-oxidation. What are the yields of the following?

a. acetyl CoA b. FADH$_2$ c. NADH

24.5 | The Energy from Fatty Acids

Learning Objective

5. Determine the amount of ATP produced by the complete catabolism of a fatty acid.

To compare the energy yield from fatty acids with that from glucose, let's continue our discussion with stearic acid, a typical and abundant fatty acid.

We start with the activation of stearic acid by reacting it with coenzyme A to form stearoyl CoA. The energy needed to cause this reaction to occur comes from the hydrolysis of ATP to AMP and PP$_i$ and the subsequent hydrolysis of PP$_i$ to 2P$_i$. This is equivalent to hydrolyzing two molecules of ATP to ADP.

As a stearoyl CoA molecule (18 carbons) passes through the β-oxidation spiral, 9 acetyl CoA, 8 FADH$_2$, and 8 NADH molecules are produced. Acetyl CoA produced in the fatty acid spiral can enter the citric acid cycle (followed by the electron transport chain), where each molecule of acetyl CoA results in the production of 10 ATP molecules. In addition, when the FADH$_2$ and NADH molecules enter the electron transport chain, each FADH$_2$ yields 1.5 ATP molecules, and each NADH yields 2.5 molecules. The calculations are summarized in **Table 24.1**, which shows a total of 120 molecules of ATP formed from the 18-carbon fatty acid.

TABLE 24.1 Energy Produced by the Oxidation of Stearic Acid

Oxidation Product or Step	ATP/Unit	Total ATP
Activation step		-2
9 acetyl CoA	10.0	90
8 FADH$_2$	1.5	12
8 (NADH + H$^+$)	2.5	20
Total ATP from the 18-carbon fatty acid		120

It is instructive to compare the energy yield from the complete oxidation of fatty acids with that obtained from an equivalent amount of glucose, because both are important constituents of the diet. In **Section 23.8**, we saw that the oxidation of a single glucose molecule

produces 32 ATP molecules. The complete oxidation of three 6-carbon glucose molecules (18 carbons) yields 3 × 32, or 96, ATP molecules. Thus, on the basis of equal numbers of carbons, lipids are nearly 25% more efficient than carbohydrates as energy-storage systems.

On a mass basis, the difference is even more striking. One mole of stearic acid weighs 284 g, and, as we have seen, it yields 120 mol of ATP on complete degradation. Three moles of glucose weigh 540 g and yield 96 mol of ATP. On this basis, 284 g of glucose would generate 50 mol of ATP. On an equal-mass basis, lipids contain more than twice the energy of carbohydrates. The reason for this difference is that fatty acids are a more reduced form of fuel, with a number of —CH₂— groups as plentiful sources of H atoms for the energy-yielding oxidation process. Glucose, in contrast, is already partially oxidized, with an oxygen atom located on each carbon atom.

> ✔ **LEARNING CHECK 24.2** Calculate the number of molecules of ATP produced by the complete oxidation of palmitic acid (16 carbon atoms) to carbon dioxide and water.

24.6 | Ketone Bodies

Learning Objective

6. Name the three ketone bodies and list the conditions that cause their overproduction.

It is generally agreed by nutrition scientists that a diet balanced in carbohydrates and fats as energy sources is best for good health. When such a balanced diet is followed, most of the acetyl CoA produced by fatty acid metabolism is processed through the citric acid cycle.

Under certain conditions, the balance between carbohydrate and fatty acid metabolism is lost. During a fast, for example, fatty acids from stored fats become the body's primary energy source. With a minimum amount of cellular glucose available, the level of glycolysis decreases, and a reduced amount of oxaloacetate is synthesized. In addition, oxaloacetate is used for gluconeogenesis to a greater-than-normal extent as the cells react to make their own glucose. The lack of oxaloacetate reduces the activity of the citric acid cycle. As a result of these events, more acetyl CoA is produced by fatty acid oxidation than can be processed through the citric acid cycle. As the concentration of acetyl CoA builds up, the excess is converted within the liver to three substances called **ketone bodies**— acetoacetate, β-hydroxybutyrate, and acetone.

ketone bodies Three compounds— acetoacetate, β-hydroxybutyrate, and acetone—formed from acetyl CoA.

$$
\underset{\text{acetoacetate}}{CH_3 - \overset{\overset{\displaystyle O}{\|}}{C} - CH_2 - \overset{\overset{\displaystyle O}{\|}}{C} - O^-} \qquad
\underset{\substack{\beta\text{-hydroxybutyrate} \\ \text{(not a ketone)}}}{CH_3 - \overset{\overset{\displaystyle OH}{|}}{CH} - CH_2 - \overset{\overset{\displaystyle O}{\|}}{C} - O^-} \qquad
\underset{\text{acetone}}{CH_3 - \overset{\overset{\displaystyle O}{\|}}{C} - CH_3}
$$

The ketone bodies are carried by the blood to body tissues—mainly the brain, heart, and skeletal muscle—where they may be oxidized to meet energy needs (see **Figure 24.6**). You can see from their structural formulas that β-hydroxybutyrate does not actually contain a ketone group, but it is nevertheless included in the designation because it is always produced along with the other two. Under normal conditions, the concentration of ketone bodies in the blood averages a very low 0.5 mg/100 mL.

Diabetes mellitus, like a long fast or starvation, also produces an imbalance in carbohydrate and lipid metabolism. Even though blood glucose reaches hyperglycemic levels, a deficiency of insulin prevents the glucose from entering tissue cells in sufficient amounts to meet cellular energy needs. The resulting increase in fatty acid metabolism leads to excessive production of acetyl CoA and a substantial increase in the level of ketone bodies in the blood of poorly controlled diabetes. A concentration higher than about 20 mg/100 mL of blood is called **ketonemia** ("ketones in the blood"). At a level of about 70 mg/100 mL of blood, the renal threshold for ketone bodies is exceeded, and ketone bodies are excreted in the urine. This condition is called **ketonuria** ("ketones in the urine"). A routine check

ketonemia An elevated level of ketone bodies in the blood.

ketonuria The presence of ketone bodies in the urine.

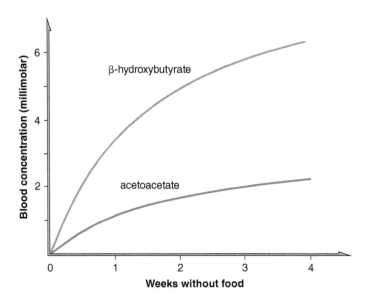

Figure 24.6 The effect of starvation on blood levels of β-hydroxybutyrate and acetoacetate.

of urine for ketone bodies is useful in the diagnosis of diabetes. Occasionally, the concentration of acetone in the blood reaches levels that cause it to be expelled through the lungs, and the odor of acetone can be detected on the breath. When ketonemia, ketonuria, and **acetone breath** exist simultaneously, the condition is called **ketosis**.

Two of the ketone bodies are acids, and their accumulation in the blood as ketosis worsens results in a condition of **acidosis** called **ketoacidosis** because it is caused by ketone bodies in the blood. If ketoacidosis is not controlled, the person becomes severely dehydrated because the kidneys excrete excessive amounts of water in response to low blood pH. Prolonged ketoacidosis leads to general debilitation, coma, and even death.

Patients suffering from diabetes-related ketosis are usually given insulin as a first step in their treatment. The insulin restores normal glucose metabolism and reduces the rate of formation of ketone bodies. If the patient is also suffering severe dehydration, the return to fluid and acid–base balance is sped up by the intravenous administration of solutions containing sodium bicarbonate. Excessive alcohol consumption can also cause ketone bodies to be produced with acetone on the breath.

acetone breath A condition in which acetone can be detected in the breath.

ketosis A condition in which ketonemia, ketonuria, and acetone breath exist together.

acidosis A low blood pH.

ketoacidosis A low blood pH due to elevated levels of ketone bodies.

STUDY SKILLS **24.1**

Key Numbers for ATP Calculations

The calculation of the energy yield in terms of ATP molecules from the metabolism of a fatty acid is a useful exercise because it may help you better understand this metabolic process. The three key substances to account for in such calculations are acetyl CoA, NADH, and $FADH_2$. Two carbon atoms must be transported by coenzyme A in order to synthesize acetyl CoA; thus, the number of acetyl CoA molecules formed is determined by the following equation:

$$\text{acetyl CoA} = \frac{\text{number of fatty acid carbons}}{2}$$

For example, a 10-carbon fatty acid would produce 10/2, or 5, acetyl CoA molecules.

Every trip through the fatty acid spiral produces one $NADH + H^+$ molecule and one $FADH_2$ molecule. Thus,

number of trips = molecules NADH = molecules $FADH_2$

Furthermore, the number of trips equals one less than the number of acetyl CoA molecules produced (because the last trip of the spiral produces 2 molecules of acetyl CoA). Thus, for the 10-carbon fatty acid above, 5 acetyl CoA molecules are produced during four trips through the spiral. Also, the four trips produce $4\ NADH + H^+$ molecules and $4\ FADH_2$ molecules.

Finally, remember the multipliers: Every acetyl CoA gives rise to 10 ATPs, every NADH produces 2.5 ATPs, and every $FADH_2$ gives 1.5 ATPs. The summary for the 10-carbon fatty acid is:

(5 acetyl CoA) × 10	= 50 ATP
($4\ NADH + H^+$) × 2.5	= 10 ATP
($4\ FADH_2$) × 1.5	= 6 ATP
Activation step	= −2 ATP
Total ATP	= 64 ATP

24.7 | Fatty Acid Synthesis

Learning Objective

7. Describe the pathway for fatty acid synthesis.

When organisms (including humans) take in more nutrients than are needed for energy requirements, the excesses are not excreted; they are converted first to fatty acids and then to body fat. Most of the conversion reactions take place in the liver, adipose tissue, and mammary glands. The mammary glands become especially active in the process during lactation.

As is true in the opposing processes of glycolysis and gluconeogenesis, the pathways for the opposing processes of fatty acid degradation and fatty acid synthesis are not simply the reverse of each other. The processes are separate from each other, utilize different enzyme systems, and are even located in different cellular compartments. Degradation by the β-oxidation pathway takes place in cellular mitochondria, whereas biosynthesis of fatty acids occurs in the cytoplasm.

However, one aspect of fatty acid synthesis and fatty acid degradation is the same. Both processes require units of two carbon atoms as a result of involving acetyl CoA. Thus, fatty acid chains are built up two carbons at a time during synthesis and broken down two carbons at a time during β-oxidation.

Acetyl CoA is generated inside mitochondria and so must be transported into the cytoplasm if it is to be used to synthesize fatty acids. This transport is done by first reacting acetyl CoA with oxaloacetate (the first step of the citric acid cycle). The product is citrate:

$$\text{acetyl CoA} + \text{oxaloacetate} + H_2O \longrightarrow \text{citrate} + \text{CoA—SH} \qquad (24.8)$$

Mitochondrial membranes contain a citrate transport system that enables excess citrate to move out of the mitochondria into the cytoplasm. Once in the cytoplasm, the citrate reacts to regenerate acetyl CoA and oxaloacetate.

Fatty acid synthesis occurs by way of a rather complex series of reactions catalyzed by a multienzyme complex called the *fatty acid synthetase system*. This system is made up of six enzymes and an additional protein, acyl carrier protein (ACP), to which all intermediates are attached.

The summary equation for the synthesis of palmitic acid from acetyl CoA demonstrates that a great deal of energy, 7 ATP + 14 NADPH (a phosphate derivative of NADH), is required:

$$8CH_3 \overset{\overset{\displaystyle O}{\displaystyle \|}}{-} C - S - CoA + 14NADPH + 13H^+ + 7ATP \longrightarrow$$

acetyl CoA

$$(24.9)$$

$$CH_3(CH_2)_{14}COO^- + 8CoA - SH + 6H_2O + 14NADP^+ + 7ADP + 7P_i$$

palmitate

This large input of energy is stored in the synthesized fatty acids and is one of the reasons it is so difficult to lose excess weight due to fat. After synthesis, the fatty acids are incorporated into triglycerides and stored in the form of fat in adipose tissues.

The liver is the most important organ involved in fatty acid and triglyceride synthesis. It is able to modify body fats by lengthening or shortening and saturating or unsaturating the fatty acid chains. The only fatty acids that cannot be synthesized by the body are those that are polyunsaturated. However, linoleic acid (two double bonds) and linolenic acid (three double bonds) from the diet can be converted to other polyunsaturated fatty acids. This utilization of linoleic and linolenic acids as sources of other polyunsaturated fatty acids is the basis for classifying them as essential fatty acids for humans (see **Section 18.2**).

The human body can convert glucose to fatty acids, but it cannot convert fatty acids to glucose. Our cells contain no enzyme that can catalyze the conversion of acetyl CoA to pyruvate, a compound required for gluconeogenesis (see **Figure 23.13**). However, plants and some bacteria do possess such enzymes and convert fats to carbohydrates as part of their normal metabolism.

24.8 | Amino Acid Metabolism

Learning Objective

8. Describe the source and function of the body's amino acid pool.

In terms of amount used, the most important function of amino acids is to provide building blocks for the synthesis of proteins in the body. It is estimated that about 75% of amino acid utilization in a normal, healthy adult is for this function. The maintenance of body proteins must occur constantly in the body because tissue proteins break down regularly from normal wear and tear, as well as from diseases and injuries. The amino acids used in this maintenance come from proteins that are eaten and hydrolyzed during digestion, from the body's own degraded tissue, and from the synthesis in the liver of certain amino acids. The amino acids from these three sources constitute what is called the **amino acid pool** of the body. This cellular supply of amino acids is constantly being restocked to allow the synthesis of new proteins and other necessary metabolic processes to take place as needed.

amino acid pool The total supply of amino acids in the body.

The dynamic process in which body proteins are continuously hydrolyzed and resynthesized is called **protein turnover**. The turnover rate, or life expectancy, of body proteins is a measure of how fast the proteins are broken down and resynthesized. The turnover rate is usually expressed as a half-life. The use of radioactive amino acids has enabled researchers to estimate half-lives by measuring the exchange rate between body proteins and the amino acid pool. For example, the half-life of liver proteins is about 10 days. This means that over a 10-day period, half the proteins in the liver are hydrolyzed to amino acids and replaced by equivalent proteins. Plasma proteins also have a half-life of about 10 days, hemoglobin about 120 days, and muscle protein about 180 days. The half-life of collagen, a protein of connective tissue, is considerably longer—some estimates are as high as 1000 days. Other proteins, particularly enzyme and polypeptide hormones, have much shorter half-lives of only a few minutes. Once it is released from the pancreas, insulin has a half-life estimated to be only 7 to 10 minutes.

protein turnover The continuing process in which body proteins are hydrolyzed and resynthesized.

Because of the high turnover rate of proteins in the body, manufactured proteins do not have enough time to become a permanent part of the body structure or function. The frequent turnover of proteins is advantageous, for it allows the body to continually renew important molecules and to respond quickly to its own changing needs.

In addition to tissue protein synthesis, there is a constant draw on the amino acid pool for the synthesis of other nitrogen-containing biomolecules. These molecules include the purine and pyrimidine bases of nucleic acids; heme structures for hemoglobin and myoglobin; choline and ethanolamine, which are the building blocks of phospholipids; and neurotransmitters such as acetylcholine and dopamine. Like proteins, these compounds are also constantly being degraded and replaced. **Table 24.2** lists some of the important nitrogen-containing compounds that are synthesized from amino acids.

TABLE 24.2 Compounds Derived from Amino Acids

Amino Acid	Product	Function
Tyrosine	dopamine	Neurotransmitter
	norepinephrine	Neurotransmitter, hormone
	epinephrine	Hormone
	thyroxine (T3 & T4)	Hormone
	melanin	Skin pigmentation
Tryptophan	serotonin	Neurotransmitter
Histidine	histamine	Involved in allergic reactions
Serine	ethanolamine	Required in cephalin synthesis
Cysteine	taurine	A compound of bile salts

Unlike carbohydrates and fatty acids, amino acids in excess of immediate body requirements cannot be stored for later use. They are degraded and, depending on the organism, their nitrogen atoms are converted to either ammonium ions, urea, or uric acid and excreted. Their carbon skeletons are converted to pyruvate, acetyl CoA, or one of the intermediates of the citric acid cycle and used for energy production, the synthesis of glucose through gluconeogenesis (see **Section 23.10**), or conversion to triglycerides. The various pathways involved in amino acid metabolism are summarized in **Figure 24.7**.

Figure 24.7 The major metabolic pathways of amino acids.

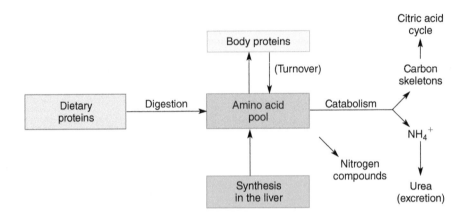

24.9 | Amino Acid Catabolism: The Fate of the Nitrogen Atoms

Learning Objectives

9. Write equations for transamination and deamination reactions.

10. Explain the overall results of the urea cycle.

As previously noted, excess amino acids cannot be stored in the body but are catabolized for energy production, and the nitrogen is either excreted or used to synthesize other nitrogen-containing compounds. In this section, we will study the chemical pathways involved with the nitrogen. The chemical fate of the carbon skeleton of the amino acids is discussed in **Section 24.10**.

In essence, there are three stages in nitrogen catabolism (see **Figure 24.8**). These processes (which occur in the liver) are transamination, deamination, and urea formation.

Figure 24.8 An overview of the catabolic fate of nitrogen atoms in amino acids.

transaminase An enzyme that catalyzes the transfer of an amino group.

transamination The enzyme-catalyzed transfer of an amino group to a keto acid.

24.9A Stage 1: Transamination

In the tissues, amino groups freely move from one amino acid to another under the catalytic influence of enzymes called aminotransferases, or more commonly, **transaminases**. A key reaction for amino acids undergoing catabolism is a **transamination** involving the

transfer of amino groups to α-ketoglutarate. The carbon skeleton of the amino acid remains behind and is transformed into a new α-keto acid:

General reaction:

$$
\begin{array}{c}
\overset{\displaystyle NH_3{}^+}{\underset{\displaystyle |}{}} \\
R-CH-COO^- + {}^-OOC-CH_2CH_2-\overset{O}{\overset{||}{C}}-COO^- \quad \underset{\rightleftharpoons}{\text{transaminase}}
\end{array}
$$

a donor α-ketoglutarate
α-amino acid an α-keto acid

(24.10)

$$
R-\overset{O}{\overset{||}{C}}-COO^- + {}^-OOC-CH_2CH_2-\overset{NH_3{}^+}{\overset{|}{C}H}-COO^-
$$

a new glutamate
α-keto acid the new amino acid

Specific example:

$$
CH_3-\overset{NH_3{}^+}{\overset{|}{C}H}-COO^- + {}^-OOC-CH_2CH_2-\overset{O}{\overset{||}{C}}-COO^- \quad \underset{\rightleftharpoons}{\overset{\text{alanine}}{\text{transaminase}}}
$$

alanine α-ketoglutarate

(24.11)

$$
CH_3-\overset{O}{\overset{||}{C}}-COO^- + {}^-OOC-CH_2CH_2-\overset{NH_3{}^+}{\overset{|}{C}H}-COO^-
$$

pyruvate glutamate

Notice that in the reaction, the $-NH_3{}^+$ group trades places with a carbonyl oxygen.

A second important example of transamination is the production of aspartate, which is used in Stage 3, urea formation. In this case, the amino group is transferred to oxaloacetate:

Specific example:

$$
\begin{array}{c}
CH_3-\underset{\underset{\displaystyle CH_3}{\displaystyle |}}{CH}-\overset{NH_3{}^+}{\overset{|}{C}H}-COO^- + {}^-OOC-CH_2-\overset{O}{\overset{||}{C}}-COO^- \rightleftharpoons
\end{array}
$$

oxaloacetate

valine

(24.12)

$$
CH_3-\underset{\underset{\displaystyle CH_3}{\displaystyle |}}{CH}-\overset{O}{\overset{||}{C}}-COO^- + {}^-OOC-CH_2-\overset{NH_3{}^+}{\overset{|}{C}H}-COO^-
$$

aspartate

The net effect of the transamination reactions in the catabolism of amino acids is to use nitrogen from a variety of amino acids to form glutamate and aspartate. In **Section 24.11**, we'll see that transamination is also an important method in the biosynthesis of amino acids.

✔ LEARNING CHECK 24.3 Complete the following transamination reaction:

$$
HO-\bigcirc\!\!\!\!\!\!\!-CH_2-\overset{NH_3{}^+}{\overset{|}{C}H}-COO^- + {}^-OOC-CH_2CH_2-\overset{O}{\overset{||}{C}}-COO^- \rightleftharpoons
$$

tyrosine

Pick the Right Fats

Fat, the most vilified macronutrient, is an essential part of a healthful diet. In fact, you can't live without it. However, fat exists in several forms, and some of them are not beneficial to good health. Specifically, trans fats and saturated fats are the ones to avoid. These types to avoid occur naturally in meats and dairy products. Some are manufactured from oils and are used to make commercial baked goods. Saturated fats such as shortening, coconut oil, butter, and stick margarine are solids at room temperature. Trans fats begin as liquids at room temperature, but are converted to solids at room temperature by a process called hydrogenation. These trans fats are widely used to make commercial baked goods such as cookies and crackers. The characteristic of being solids at low temperatures gives these fats the potential to obstruct arteries in the human body.

Food scientists developed hydrogenated oils with trans fats in response to the fact that the use of saturated animal fats like lard and beef tallow in baked goods was associated with coronary heart disease. The animal fats were replaced in baked goods with hydrogenated oils. The use of hydrogenated oils gave foods an appealing crispy texture and increased the shelf life of baked goods. However, research began to show that trans fats caused more health issues than the saturated fats they replaced. Currently, food scientists and government policy makers no longer consider trans fats to be desirable food ingredients.

At the current time, monounsaturated and polyunsaturated fats in contrast to saturated fats are considered to be healthful. They remain in liquid form at room temperature. Examples of these healthful fats are olive, peanut, sunflower, and canola oil. Studies indicate that unsaturated dietary fats help maintain healthful cholesterol levels, promote heart health, and reduce the chances of developing diabetes. In addition to helping protect your health, they add flavor and satisfaction to the foods that contain them.

Since the leading sources of saturated fat in the typical American diet are meat and dairy products, choosing low-fat dairy products like skim milk and frozen yogurt in place of whole milk and ice cream, or choosing plant sources of protein and fats such as nuts, seeds, olives, and avocados results in a healthier diet that is lower in saturated fats.

In addition, these plant-based foods contain a variety of vitamins, antioxidants, minerals, fiber, and protein that add to your overall good health. Eating fatty fish like tuna or mackerel twice a week could be one of the best dietary decisions you can make. These fish contain heart-healthy omega-3 fatty acids.

A major factor in misunderstanding dietary fat is thinking that fat-free foods are better for you than foods that contain fat. Recent research shows that replacing fats with the refined carbohydrates found in most "fat-free" snacks is worse than eating the fat-containing foods the fad foods replace. Because the fat-free foods are less satisfying than their fat-containing counterparts, people who eat them tend to consume more calories than those who eat the regular fat-containing foods. Quite often, people who go on fat-free diets ultimately gain weight rather than losing it.

To be clear, fats should be included in a balanced diet, but in moderation. They are calorie-dense macronutrients that have a powerful impact on the overall diet. Not only should fats occupy the smallest category of the diet, but they should be chosen carefully. When you do choose to consume fats, make certain they are good for you. Choose monounsaturated oils, polyunsaturated plant-based fats, or heart-healthy omega-3 fats from fish or low-fat dairy products. A little fat goes a long way, so choose wisely.

Several food sources of healthy fats.

24.9B Stage 2: Deamination

This phase of amino acid catabolism uses the glutamate produced in Stage 1. The enzyme glutamate dehydrogenase catalyzes the removal of the amino group as an ammonium ion and regenerates α-ketoglutarate, which can again participate in the first stage (transamination). This reaction is the principal source of NH_4^+ in humans. Because the deamination results in the oxidation of glutamate, it is called **oxidative deamination**:

oxidative deamination An oxidation process resulting in the removal of an amino group.

$$^-OOC-CH_2CH_2-\underset{\underset{\text{glutamate}}{|}}{\overset{\overset{+}{NH_3}}{\underset{}{CH}}}-COO^- + NAD^+ + H_2O \rightleftharpoons$$

$$(24.13)$$

$$NH_4^+ + {}^-OOC-CH_2CH_2-\overset{\overset{O}{\|}}{C}-COO^- + NADH + H^+$$
$$\alpha\text{-ketoglutarate}$$

The NADH produced in this second stage enters the electron transport chain and eventually produces 2.5 ATP molecules. Other amino acids in addition to glutamate may also be catabolized by oxidative deamination. The reactions are catalyzed by enzymes in the liver called amino acid oxidases. In each case, an α-keto acid and NH_4^+ are the products.

24.9C Stage 3: Urea Formation

The ammonium ions released by the glutamate dehydrogenase reaction are toxic to the body and must be prevented from accumulating. In the **urea cycle**, which occurs only in the liver, ammonium ions are converted to urea (see **Figure 24.9**). This compound is

urea cycle A metabolic pathway in which ammonium ions are converted to urea.

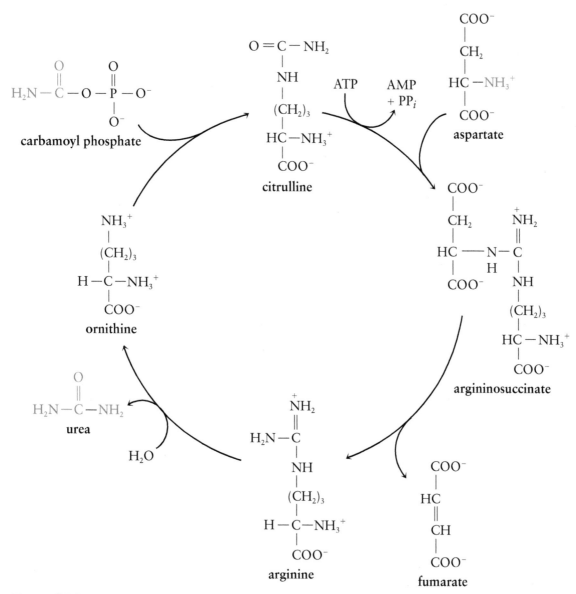

Figure 24.9 The urea cycle. Nitrogen atoms enter the cycle from carbamoyl phosphate and aspartate.

less toxic than ammonium ions and can be allowed to concentrate until it is convenient to excrete it in urine.

The urea cycle processes the ammonium ions in the form of carbamoyl phosphate, the fuel for the urea cycle. This compound is synthesized in the mitochondria from ammonium ions and bicarbonate ions.

$$NH_4^+ + HCO_3^- + 2ATP + H_2O \xrightarrow{\substack{\text{carbamoyl} \\ \text{phosphate} \\ \text{synthetase}}} \underset{\text{carbamoyl phosphate}}{H_2N-\overset{\displaystyle O}{\overset{\|}{C}}-O-\overset{\displaystyle O}{\underset{\displaystyle O^-}{\overset{\|}{P}}}-O^- + 2ADP + P_i + 3H^+} \quad (24.14)$$

As we see, the energy for the synthesis of one molecule of carbamoyl phosphate is derived from two molecules of ATP. **Figure 24.9** summarizes the reactions of the urea cycle. The net reaction for carbamoyl phosphate formation and the urea cycle is

$$NH_4^+ + HCO_3^- + 3ATP + 2H_2O + \underset{\text{aspartate}}{\overset{\displaystyle COO^-}{\underset{\displaystyle COO^-}{\overset{\displaystyle |}{\underset{\displaystyle |}{\overset{\displaystyle CH_2}{\underset{\displaystyle CH-NH_3^+}{\overset{\displaystyle |}{\overset{\displaystyle |}{}}}}}}}} \longrightarrow \quad (24.15)$$

$$\underset{\text{urea}}{H_2N-\overset{\displaystyle O}{\overset{\|}{C}}-NH_2} + \underset{\text{fumarate}}{\overset{\displaystyle H}{\underset{\displaystyle {}^-OOC}{\overset{\diagdown}{\diagup}}}C=C\overset{\displaystyle COO^-}{\underset{\displaystyle H}{\overset{\diagup}{\diagdown}}}} + 2ADP + 2P_i + AMP + PP_i$$

After urea is formed, it diffuses out of liver cells into the blood, the kidneys filter it out, and it is excreted in the urine. Normal urine from an adult usually contains about 25 to 30 g of urea daily, although the exact amount varies with the protein content of the diet. The direct excretion of NH_4^+ accounts for a small but important amount of total urinary nitrogen. Ammonium ions can be excreted along with acidic ions, a mechanism that helps the kidneys control the acid–base balance of body fluids.

✔ LEARNING CHECK 24.4

a. What is the primary function of the urea cycle?
b. What is the fuel for the urea cycle?

24.10 | Amino Acid Catabolism: The Fate of the Carbon Skeleton

Learning Objective

11. Describe how amino acids can be used for energy production, the synthesis of triglycerides, and gluconeogenesis.

After removal of the amino group of an amino acid is accomplished by transamination or oxidative deamination (see **Section 24.9**), the remaining carbon skeleton undergoes catabolism and is converted into one of several products. We will not study the various pathways followed, but we will focus on the final forms the skeleton assumes and the metabolic fate of these forms.

Phenylketonuria (PKU)

At least 35 different hereditary changes in amino acid metabolism have been identified. One of the more well known is an inherited disease associated with abnormal aromatic amino acid metabolism. In *phenylketonuria (PKU)*, there is a lack of the enzyme phenylalanine hydroxylase. As a result, phenylalanine cannot be converted to tyrosine, leading to the accumulation of phenylalanine and its metabolites (phenylpyruvate and phenylacetate) in the tissues and blood:

phenylpyruvate

phenylacetate

Abnormally high levels of phenylpyruvate and phenylacetate damage developing brain cells, causing severe mental retardation. It is estimated that 1 out of every 20,000 newborns is afflicted with PKU. A federal regulation requires that all infants be tested for this disease, which can be detected very easily by analyzing blood or urine samples for phenylalanine or phenylpyruvic acid.

The disease is treated by restricting the amount of phenylalanine in the diet of the infant starting 2 to 3 weeks after birth. The diet is also supplemented with tyrosine because it is not being formed in the PKU patient. Strict dietary management is not as critical after the age of 6 because the brain has matured by this time and is not as susceptible to the toxic effects of phenylalanine metabolites.

Infants with PKU have an excellent chance for a normal healthy life, if the disease is detected and treated early. Here a nurse is taking blood samples from an infant's foot to be used in a simple chemical test for PKU.

After the amino group is gone, the skeletons of all 20 amino acids are degraded into either pyruvate, acetyl CoA, acetoacetyl CoA (which is degraded to acetyl CoA), or various substances that are intermediates in the citric acid cycle. As shown in **Figure 24.10**,

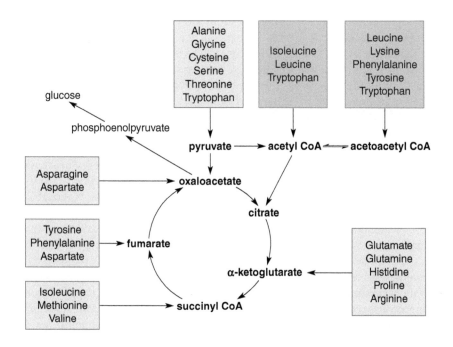

Figure 24.10 Fates of the carbon atoms of amino acids. Glucogenic amino acids are shaded purple, and ketogenic amino acids are shaded green.

CHEMISTRY AROUND US **24.2**

Phenylalanine and Diet Foods

In human society, we often want to indulge ourselves but we don't want to suffer any negative consequences of the indulgence. A good example of this is the desire to satisfy our craving for a sweet taste in foods, but not wanting to suffer the results of excessive caloric intake. As a result, a large industry has developed based on the production and distribution of low-calorie or no-calorie food sweeteners (see Chemistry Around Us 17.1). One popular product of this industry is a dipeptide of the two amino acids aspartic acid and phenylalanine. The product, called *aspartame,* has a sweet taste about 180 times that of ordinary table sugar (sucrose). When aspartame is hydrolyzed during digestion, the two amino acids are released.

Most foods that contain aspartame carry a warning on the label stating that they contain phenylalanine, although the phenylalanine is only released during the digestion of the food. The label warning is, for the most part, a protection for the food manufacturer against the somewhat remote possibility that the user of the food product suffers from a genetic metabolic disorder called phenylketonuria, or PKU (see Chemistry Around Us 24.1). The label warning also complies with federal regulations that require any food product containing aspartame to carry the warning.

People who have PKU suffer serious complications from ingesting phenylalanine, such as mental retardation, brain damage, seizures, and other problems. People who don't have PKU do not have to worry about ingesting phenylalanine or aspartame, with a certain few exceptions. It is known that aspartame or phenylalanine-rich foods can cause a rapid increase in brain levels of phenylalanine. So, aspartame and other products containing phenylalanine should be used cautiously by individuals with the following characteristics:

- Those who are taking medications such as monoamine oxidase inhibitors, neuroleptics, or medications containing levodopa
- Those who have a sleep disorder
- Those who have an anxiety disorder or other mental health disorder

Aspartame is present in many low-calorie drinks.

all these degraded forms of the carbon skeletons are a part of or can enter the citric acid cycle, and thus may be very important in the production of energy.

Even though the degraded carbon skeletons can be used to produce energy via the citric acid cycle, not all are used this way because cellular needs might dictate other uses. For example, once a carbon skeleton is catabolized to pyruvate, there are two possible uses for it: the production of energy or the synthesis of glucose through gluconeogenesis (see **Section 23.10**). The skeleton of an amino acid that is degraded into pyruvate or intermediates of the citric acid cycle can be used to make glucose and is classified as a **glucogenic amino acid**. Amino acids with skeletons that are degraded into acetyl CoA or acetoacetyl CoA cannot be converted into glucose but can, in addition to energy production, be used to make ketone bodies and fatty acids. Such amino acids are classified as **ketogenic**. **Figure 24.10** shows this catabolic classification of the 20 amino acids. Notice that some amino acids appear in more than one location in the diagram.

Glucogenic amino acids play a special role in the metabolic changes that occur during starvation. To maintain the blood glucose level, the body first uses the glycogen stored in the liver. After 12 to 18 hours without food, the glycogen supplies are exhausted. Glucose is then synthesized from glucogenic amino acids through the process of gluconeogenesis. Hydrolysis of proteins from body tissues, particularly from muscle, supplies the amino acids for gluconeogenesis. Because each protein in the body has an important function, the body can meet its energy needs for only a limited time by sacrificing proteins.

glucogenic amino acid An amino acid whose carbon skeleton can be converted metabolically to an intermediate used in the synthesis of glucose.

ketogenic amino acid An amino acid whose carbon skeleton can be converted metabolically to acetyl CoA or acetoacetyl CoA.

24.11 | Amino Acid Biosynthesis

Learning Objective

12. State the relationship between intermediates of carbohydrate metabolism and the synthesis of nonessential amino acids.

As we have already seen, the liver is highly active biochemically. In biosynthesis, it is responsible for producing most of the amino acids the body can synthesize. Amino acids that can be synthesized in the amounts needed by the body are called **nonessential amino acids** (see **Table 24.3**). As explained in **Chapter 22**, essential amino acids either cannot be made or cannot be made in large enough amounts to meet bodily needs and must be included in the diet.

nonessential amino acid An amino acid that can be synthesized within the body in adequate amounts.

TABLE 24.3 Essential and Nonessential Amino Acids

Essential	Nonessential	Essential	Nonessential
Histidine	Alanine	Threonine	Glutamine
Isoleucine	Arginine	Tryptophan	Glycine
Leucine	Asparagine	Valine	Proline
Lysine	Aspartate		Serine
Methionine	Cysteine		Tyrosine
Phenylalanine	Glutamate		

The key starting materials for the synthesis of 10 nonessential amino acids are intermediates of the glycolysis pathway and the citric acid cycle (see **Figure 24.11**). Tyrosine, the only nonessential amino acid with an aromatic side chain, is produced from the essential amino acid phenylalanine in a conversion that requires a single oxidation step catalyzed by the enzyme phenylalanine hydroxylase:

$$\underset{\text{phenylalanine}}{\boxed{\bigcirc}-CH_2-\underset{\underset{NH_3^+}{|}}{CH}-COO^-} + O_2 + NADPH + H^+ \xrightarrow{\text{phenylalanine hydroxylase}} \underset{\text{tyrosine}}{HO-\boxed{\bigcirc}-CH_2-\underset{\underset{NH_3^+}{|}}{CH}-COO^-} + NADP^+ + H_2O \qquad (24.16)$$

Figure 24.11 A summary of amino acid biosynthesis.

Three nonessential amino acids (glutamate, alanine, and aspartate) are synthesized from α-keto acids via reactions catalyzed by transaminases. For example, alanine is produced from pyruvate and glutamate (which furnishes the amino group):

$$\underset{\text{pyruvate}}{CH_3-\underset{\underset{O}{\|}}{C}-COO^-} + \text{glutamate} \underset{\text{transaminase}}{\overset{\text{glutamic pyruvic}}{\rightleftharpoons}} \underset{\text{alanine}}{CH_3-\underset{\underset{NH_3^+}{|}}{CH}-COO^-} + \text{α-ketoglutarate} \qquad (24.17)$$

Thus, we see that transaminases perform two vital functions in amino acid metabolism. By taking part in the biosynthesis of nonessential amino acids, they provide a means to help readjust the relative proportions of amino acids to meet the particular needs of the body. This is a vital function because most of our diets do not contain amino acids in the exact proportions the body requires. Also, as we noted in Section 24.9, transamination reactions allow the nitrogen atoms of all amino acids to be transferred to α-keto acids to form glutamate and aspartate when disposal of nitrogen is necessary.

Two other nonessential amino acids, asparagine and glutamine, are formed from aspartate and glutamate by reaction of the side-chain carboxylate groups with ammonium ions:

$$^-O-\overset{\overset{\displaystyle O}{\|}}{C}-CH_2-\overset{\overset{\displaystyle NH_3^+}{|}}{CH}-COO^- + NH_4^+ \xrightarrow[\hspace{1cm}]{} H_2N-\overset{\overset{\displaystyle O}{\|}}{C}-CH_2-\overset{\overset{\displaystyle NH_3^+}{|}}{CH}-COO^- \quad (24.18)$$
$$\text{aspartate} \qquad\qquad\qquad\qquad\qquad\qquad \text{asparagine}$$
$$\text{ATP} \qquad \text{ADP} + P_i$$

$$^-O-\overset{\overset{\displaystyle O}{\|}}{C}-CH_2-CH_2-\overset{\overset{\displaystyle NH_3^+}{|}}{CH}-COO^- + NH_4^+ \xrightarrow[\hspace{1cm}]{} H_2N-\overset{\overset{\displaystyle O}{\|}}{C}-CH_2CH_2-\overset{\overset{\displaystyle NH_3^+}{|}}{CH}-COO^- \quad (24.19)$$
$$\text{glutamate} \qquad\qquad\qquad\qquad\qquad\qquad \text{glutamine}$$
$$\text{ATP} \qquad \text{ADP} + P_i$$

So far, we have shown how 6 of the 11 nonessential amino acids are synthesized. The syntheses of the remaining 5 are more complex, and we will not consider them here.

▶ Case Study Follow-up

Depression is defined as periods of two weeks or longer during which a person has a depressed mood or loss of interest in pleasurable activities and interactions with others. According to a National Institute of Mental Health report in 2014, depression affects 11.4% of the U.S. population aged 12 to 17 (see reference 1 below). Forty-two percent of the people who suffer from depression do not respond well to antidepressant medications, but people who eat fish regularly or take fish oil supplements report an improved response to antidepressant medications (see reference 2 below). The main fatty acids in the human diet are omega-3 and omega-6 fatty acids. Foods that contain omega-3 fatty acids are fish and certain plant and nut oils. Fish oil contains two omega-3 fatty acids, docosahexaenoic acid (DHA) and eicosapentaenoic acid (EPA). (See **Ask a Pharmacist 22.1**, page 716 for

more discussion of DHA and EPA). These omega-3 fatty acids have been found to be important for proper brain function and are thought to support heart health by lowering cholesterol levels. There is some evidence that omega-3 fatty acids lower triglyceride levels in the body and reduce the risk of heart attacks and strokes in patients with heart disorders.

References:

1. National Institute of Mental Health. 2014. *Major Depression Among Adolescents*. Retrieved from http://www.nimh.nih .gov/health/statistics/prevalence/major-depression-among -adolescents.shtml

2. McNamee, David. 2014. "Could Eating More Fish Make Antidepressants Work Better?" *Medical News Today*. Retrieved from http://www.medicalnewstoday.com/articles/284102.php

Concept Summary

Blood Lipids Fatty acids, glycerol, and monoglycerides that result from lipid digestion are absorbed into the lymph, recombined into triglycerides, and deposited into the bloodstream as lipoprotein aggregates.

Objective 1 (Section 24.1), Exercises 24.2 and 24.4

Fat Mobilization When stored fats are needed for energy, hydrolysis reactions liberate the fatty acids and glycerol. These component substances of fat then enter the bloodstream and travel to tissues, where they are utilized.

Objective 2 (Section 24.2), Exercise 24.8

Glycerol Metabolism The glycerol available from fat mobilization is first phosphorylated and then oxidized to dihydroxyacetone phosphate, an intermediate of glycosis. By having an entry point into the glycolysis pathway, glycerol can ultimately be converted into glucose or oxidized to CO_2 and H_2O.

Objective 3 (Section 24.3), Exercise 24.12

The Oxidation of Fatty Acids The catabolism of fatty acids begins in the cytoplasm, where they are activated by combining with CoA—SH. After being transported into mitochondria, the degradation of fatty acids occurs by the β-oxidation pathway. The β-oxidation pathway produces reduced coenzymes ($FADH_2$ and NADH) and cleaves the fatty acid chain into two-carbon fragments bound to coenzyme A (acetyl CoA).

Objective 4 (Section 24.4), Exercise 24.22

The Energy from Fatty Acids The energy available from a fatty acid for ATP synthesis can be calculated by determining the amount of acetyl CoA, $FADH_2$, and NADH produced. On an equal-mass basis, fatty acids contain more than twice the energy of glucose.

Objective 5 (Section 24.5), Exercise 24.26

Ketone Bodies Acetoacetate, β-hydroxybutyrate, and acetone are known as ketone bodies. They are synthesized in the liver from acetyl CoA. During starvation, in unchecked diabetes, and after excessive alcohol consumption, the level of ketone bodies becomes very high, leading to several conditions: ketonemia, ketonuria, acetone breath, and ketosis.

Objective 6 (Section 24.6), Exercise 24.28

Fatty Acid Synthesis This process occurs in the cytoplasm and uses acetyl CoA as a starting material. Two-carbon fragments are added to growing fatty acid molecules as energy is supplied by ATP and NADPH.

Objective 7 (Section 24.7), Exercise 24.34

Amino Acid Metabolism Amino acids and proteins are not stored within the body. Amino acids absorbed from digested food or synthesized in the body become a part of a temporary supply or pool that may be used in metabolic processes. The turnover, or life expectancy, of proteins is usually expressed in half-lives. Protein half-lives range from several minutes to hundreds of days. Some amino acids in the body are converted into relatively simple but vital nitrogen-containing compounds. The catabolism of amino acids produces intermediates used for energy production, synthesis of glucose, or formation of triglycerides.

Objective 8 (Section 24.8), Exercise 24.38

Amino Acid Catabolism: The Fate of the Nitrogen Atoms As an amino acid undergoes catabolism, the amino group may be transferred to an α-keto acid through a process called transamination. In the process of deamination, an amino acid is converted into a keto acid and ammonia.

Objective 9 (Section 24.9), Exercise 24.44

The urea cycle converts toxic ammonium ions into urea for excretion.

Objective 10 (Section 24.9), Exercise 24.54

Amino Acid Catabolism: The Fate of the Carbon Skeleton Amino acids are classified as glucogenic or ketogenic based on their catabolic products. Glucogenic amino acids are degraded into pyruvate or intermediates of the citric acid cycle and can be used for glucose synthesis. Ketogenic amino acids are degraded into acetoacetyl CoA and acetyl CoA for energy production and triglyceride synthesis.

Objective 11 (Section 24.10), Exercise 24.58

Amino Acid Biosynthesis The body can synthesize amino acids, known as nonessential amino acids. The key starting materials are intermediates of the glycolysis pathway and the citric acid cycle, and phenylalanine in the case of tyrosine.

Objective 12 (Section 24.11), Exercise 24.64

Key Terms and Concepts

Acetone breath (24.6)
Acidosis (24.6)
Adipose tissue (24.2)
Amino acid pool (24.8)
β-oxidation process (24.4)
Chylomicron (24.1)
Fat mobilization (24.2)

Glucogenic amino acid (24.10)
Ketoacidosis (24.6)
Ketogenic amino acid (24.10)
Ketone bodies (24.6)
Ketonemia (24.6)
Ketonuria (24.6)
Ketosis (24.6)

Nonessential amino acid (24.11)
Oxidative deamination (24.9)
Protein turnover (24.8)
Transaminase (24.9)
Transamination (24.9)
Urea cycle (24.9)

Key Reactions

1.	Digestion of triglycerides (Section 24.1):	triglyceride $+ H_2O \xrightarrow{\text{lipase}}$ fatty acids $+$ glycerol $+$ monoglycerides	
2.	Fat mobilization (Section 24.2):	triglyceride $+ 3H_2O \xrightarrow{\text{lipase}}$ 3 fatty acids $+$ glycerol	

3. Activation of fatty acids (Section 24.4):

$$R-\overset{\overset{\displaystyle O}{\|}}{C}-OH + HS-CoA \longrightarrow R-\overset{\overset{\displaystyle O}{\|}}{C}-S-CoA + H_2O$$

fatty acid $\qquad\qquad\qquad$ fatty acyl CoA

$$ATP \quad AMP + PP_i$$

Reaction 24.2

4. Oxidation of fatty acids (Section 24.4):

Reaction 24.5

$$R-CH_2CH_2-\overset{\overset{\displaystyle O}{\|}}{C}-S-CoA + NAD^+ + FAD + H_2O + CoA-SH \longrightarrow$$

$$R-\overset{\overset{\displaystyle O}{\|}}{C}-S-CoA + CH_3\overset{\overset{\displaystyle O}{\|}}{C}-S-CoA + NADH + H^+ + FADH_2$$

5. Fatty acid synthesis (Section 24.7):

Reaction 24.9

$$8CH_3-\overset{\overset{\displaystyle O}{\|}}{C}-S-CoA + 14NADPH + 13H^+ + 7ATP \longrightarrow$$

$$CH_3(CH_2)_{14}COO^- + 8CoA-SH + 6H_2O + 14NADP^+ + 7ADP + 7P_i$$

6. Transamination (Section 24.9):

Reaction 24.10

$$R-\overset{\overset{\displaystyle NH_3^+}{|}}{CH}-COO^- + {}^-OOC-CH_2CH_2-\overset{\overset{\displaystyle O}{\|}}{C}-COO^- \rightleftharpoons R-\overset{\overset{\displaystyle O}{\|}}{C}-COO^- + {}^-OOC-CH_2CH_2-\overset{\overset{\displaystyle NH_3^+}{|}}{CH}-COO^-$$

7. Oxidative deamination (Section 24.9):

Reaction 24.13

$${}^-OOC-CH_2CH_2-\overset{\overset{\displaystyle NH_3^+}{|}}{CH}-COO^- + NAD^+ + H_2O \rightleftharpoons NH_4^+ + {}^-OOC-CH_2CH_2-\overset{\overset{\displaystyle O}{\|}}{C}-COO^- + NADH + H^+$$

8. Urea formation (Section 24.9):

Reaction 24.15

$$NH_4^+ + HCO_3^- + 3ATP + 2H_2O + \underset{\substack{|\\CH_2\\|\\ \overset{\displaystyle CH-NH_3^+}{|}\\COO^-}}{COO^-} \longrightarrow H_2N-\overset{\overset{\displaystyle O}{\|}}{C}-NH_2 + \underset{{}^-OOC}{\overset{H}{}}C=C\overset{COO^-}{\underset{H}{}} + 2ADP + 2P_i + AMP + PP_i$$

aspartate $\qquad\qquad$ urea \qquad fumarate

Exercises

Even-numbered exercises are answered in Appendix B.

Blue-numbered exercises are more challenging.

Blood Lipids (Section 24.1)

24.1 How does the caloric value of fat compare with the caloric value of glycogen and starch?

24.2 What are the products of triglyceride digestion?

24.3 What happens to the products of triglyceride digestion as they pass into the cells of the intestinal mucosa?

24.4 What mechanism exists in the body for transporting insoluble lipids?

24.5 Describe what happens to the level of blood lipids after a meal rich in fat.

24.6 List the major types of lipoproteins.

Fat Mobilization (Section 24.2)

24.7 Describe the relationship between epinephrine and fat mobilization.

24.8 Describe what is meant by the term *fat mobilization*.

24.9 Explain why fatty acids cannot serve as fuel for the brain.

24.10 Which cells utilize fatty acids over glucose to satisfy their energy needs?

Glycerol Metabolism (Section 24.3)

24.11 At what point does glycerol enter the glycolysis pathway?

24.12 Describe the two fates of glycerol after it has been converted to an intermediate of glycolysis.

The Oxidation of Fatty Acids (Section 24.4)

24.13 Where in the cell does fatty acid catabolism take place?

24.14 How is the fatty acid prepared for catabolism? Where in the cell does fatty acid activation take place?

24.15 Why is the oxidation of fatty acids referred to as β-oxidation?

24.16 What hydrogen-transporting coenzymes play important roles in the oxidation of fatty acids?

24.17 Label the α- and β-positions in the following acids:

a. $CH_3CH_2CH_2CH_2CH_2 \overset{\overset{\displaystyle O}{\|}}{-C} - OH$

b. $CH_3(CH_2)_6 \overset{\overset{\displaystyle OH}{|}}{CH} - CH_2 \overset{\overset{\displaystyle O}{\|}}{-C} - OH$

24.18 Write an equation for the activation of palmitic acid (16 carbons).

24.19 Write formulas for the products that result when activated palmitic acid is taken through the following:

a. One turn of the fatty acid spiral

b. Two turns of the fatty acid spiral

24.20 Why is the oxidation of fatty acids referred to as the fatty acid spiral rather than the fatty acid cycle?

24.21 How many molecules of acetyl CoA are produced in the catabolism of a molecule of stearic acid, an 18-carbon

carboxylic acid? How many times does the spiral reaction sequence occur?

24.22 Complete the following equations for one turn of the fatty acid spiral:

a. $CH_3CH_2CH_2CH_2CH_2 \overset{\overset{\displaystyle O}{\|}}{-C} - S - CoA + FAD \longrightarrow$

_____ + _____

b. _____ + $H_2O \rightarrow$ _____

c. _____ + $NAD^+ \rightarrow$ ___ + ___ + ___

d. _____ + CoA—SH \rightarrow ___ + ___

24.23 Outline the four reactions that occur in the fatty acid spiral.

The Energy from Fatty Acids (Section 24.5)

24.24 Determine the ATP yield per molecule of acetyl CoA that enters the citric acid cycle.

24.25 What is the energy cost in ATP molecules to activate a fatty acid?

24.26 Calculate the number of ATP molecules that can be produced from the complete oxidation of a 10-carbon fatty acid to carbon dioxide and water.

24.27 Compare the energy available (in ATP molecules) from a six-carbon fatty acid to the energy available from a molecule of glucose.

Ketone Bodies (Section 24.6)

24.28 List the three substances known as ketone bodies.

24.29 Explain how excessive ketone bodies may form in the following:

a. During starvation

b. In patients with diabetes mellitus

24.30 Where are ketone bodies formed? What tissues utilize ketone bodies to meet energy needs?

24.31 Differentiate between (a) ketonemia, (b) ketonuria, (c) acetone breath, and (d) ketosis.

24.32 Why is ketosis frequently accompanied by acidosis?

Fatty Acid Synthesis (Section 24.7)

24.33 Where within a liver cell does fatty acid synthesis occur?

24.34 What substance supplies the carbons for fatty acid synthesis?

24.35 What substances furnish the energy for fatty acid synthesis?

24.36 Why can the liver convert glucose to fatty acids but cannot convert fatty acids to glucose?

Amino Acid Metabolism (Section 24.8)

24.37 List three vital functions served by amino acids in the body.

24.38 What name is given to the cellular supply of amino acids?

24.39 What are the sources of amino acids in the pool?

24.40 What term is used to denote the breakdown and resynthesis of body proteins?

24.41 What percentage of amino acids is used to synthesize proteins?

24.42 Name four important biomolecules other than proteins that are synthesized from amino acids.

24.43 Compare the half-life of plasma proteins with the half-life of enzymes.

Amino Acid Catabolism: The Fate of the Nitrogen Atoms (Section 24.9)

24.44 Match the terms *transamination* and *deamination* to the following:

 a. An amino group is removed from an amino acid and donated to a keto acid.

 b. An ammonium ion is produced.

 c. New amino acids are synthesized from other amino acids.

 d. A keto acid is produced from an amino acid.

24.45 Complete the following transamination reaction and name the amino acid that is produced:

$$\text{C}_6\text{H}_5-\text{CH}_2-\overset{\overset{\displaystyle O}{\|}}{\text{C}}-\text{COO}^- + \text{H}_3\overset{+}{\text{N}}-\text{CH}-\text{COO}^- \xrightarrow{\text{transaminase}}$$

with side chain $-\text{CH}_2-\text{CH}_2-\text{COO}^-$

24.46 Write an equation to illustrate transamination between leucine and pyruvate.

24.47 Complete the following oxidative deamination reaction:

$$\text{H}_3\overset{+}{\text{N}}-\text{CH}-\text{COO}^- + \text{NAD}^+ + \text{H}_2\text{O} \xrightarrow{\substack{\text{amino acid} \\ \text{oxidase}}}$$

with side chain $-\text{CH}_2-\overset{\overset{\displaystyle}{}}{\text{C}}=\text{O}$ bearing $-\text{NH}_2$

24.48 Write an equation to illustrate the oxidative deamination of glycine.

24.49 By writing two reactions, the first a transamination and the second an oxidative deamination, show how the amino group of alanine can be removed as NH_4^+.

24.50 In what organ of the body is urea synthesized?

24.51 What are the sources of the carbon atom and two nitrogen atoms in urea?

24.52 Name the compound that enters the urea cycle by combining with ornithine.

24.53 How much energy (in number of ATP molecules) is expended to synthesize a molecule of urea?

24.54 Write a summary equation of the formation of urea.

Amino Acid Catabolism: The Fate of the Carbon Skeleton (Section 24.10)

24.55 Differentiate between glucogenic and ketogenic amino acids.

24.56 Is any amino acid both glucogenic and ketogenic?

24.57 Identify the five locations at which the carbon skeletons of amino acids enter the citric acid cycle.

24.58 Based on the following conversions, classify each amino acid as glucogenic or ketogenic.

 a. aspartate → oxaloacetate

 b. leucine → acetyl CoA

 c. tyrosine → acetoacetyl CoA

24.59 What special role is played by glucogenic amino acids during fasting or starvation?

Amino Acid Biosynthesis (Section 24.11)

24.60 Differentiate between essential and nonessential amino acids.

24.61 What essential amino acid makes tyrosine nonessential?

24.62 List two vital functions performed by transaminases.

24.63 What amino acids are synthesized from glutamate?

24.64 List two general sources of intermediates for the biosynthesis of amino acids.

Additional Exercises

24.65 Trace glycerol metabolism after it enters the glycolytic pathway as dihydroxyacetone phosphate and give an overall net reaction (including ATP, NAD$^+$, etc.) for the conversion of glycerol to pyruvate.

24.66 A change in the concentration of enzymes that catalyze fatty acid synthesis has been shown to be a major mechanism for controlling synthesis rate. Give two regulatory means of changing enzyme concentrations.

24.67 A ketone body has a molecular weight of 58.1 g/mol and a concentration of 0.050 mg/100 mL in a patient's blood. Which ketone body is it, and how many molecules of it would be found in 1.0 mL of the patient's blood?

24.68 A 7.48-mg sample of a blood lipid was found to contain 1.87 mg of protein. Use Figure 24.4 to classify this lipid.

24.69 The following reaction proceeded in an aqueous solution as the pH was monitored. After a while, the pH quit changing even though there was excess urea present in the solution.

$$\underset{\text{urea}}{\text{NH}_2-\overset{\overset{\displaystyle O}{\|}}{\text{C}}-\text{NH}_2} + \text{H}_2\text{O} \xrightarrow{\text{urease}} 2\text{NH}_{3(aq)} + \text{CO}_{2(g)}$$

Was the initial pH when the reaction began acidic, neutral, or basic? How did the pH change over time? Why did the pH quit changing?

Chemistry for Thought

24.70 Cellular reactions for the biosynthesis and catabolism of amino acids are not simply the reversals of the same metabolic pathways. Why is this an advantage for the cell?

24.71 Why do you think the half-lives of enzymes are comparatively short?

24.72 Why do you think the half-life of collagen is comparatively long?

24.73 How might HDL levels suggest a partial explanation for the observation that women as a group have fewer heart attacks than do men as a group?

24.74 A student calculates the net yield of ATP molecules from a 12-carbon fatty acid to be 80. What did the student probably overlook?

24.75 If researchers discovered a method of reducing blood cholesterol to extremely low levels, why might it be undesirable to do so?

24.76 Could there be a connection between high sugar intake and high blood levels of triglycerides? Explain.

Allied Health Exam Connection

The following questions are from these sources:

- *Nursing School Entrance Exam* © 2005, Learning Express, LLC.
- *McGraw-Hill's Nursing School Entrance Exams* by Thomas A. Evangelist, Tamara B. Orr, and Judy Unrein © 2009, The McGraw-Hill Companies, Inc.
- *NSEE Nursing School Entrance Exams,* 3rd edition © 2009, Kaplan Publishing.
- *Cliffs Test Prep: Nursing School Entrance Exams* by Fred N. Grayson © 2004, Wiley Publishing, Inc.
- *Peterson's Master the Nursing School and Allied Health Entrance Exams,* 18th edition by Marion F. Gooding © 2008, Peterson's, a Nelnet Company.

24.77 Which of the following processes must occur before amino acids can be metabolized to release energy?
 a. Fermentation
 b. Deamination
 c. Dehydrogenation

24.78 When in excess, which element is eliminated from the human body through the formation of urea? In what organ of the body is urea synthesized?

24.79 What substances are produced by the digestion of proteins?

24.80 What is the name for amino acids that cannot be manufactured by the body?

24.81 The clinic nurse is evaluating a man wearing a diabetic medic alert band, who appears confused, with hot, dry, flushed skin. His respirations are deep and fast, and he says he is nauseated. His breath smells fruity. What diabetes-related medical condition is this man exhibiting?

25

Body Fluids

Timothey Kosachev/Shutterstock.com

▶ **Case Study**

Wesley had studied blood types in biology class. He understood that he had O negative blood and was a universal donor. Recently, his cousin Trina had been in an auto accident involving a drunk-driver, and Wesley was astounded to find that she had received 50 units of blood. For the first time, the need for donated blood became a personal issue, but Wesley was ashamed to say he had never donated blood. There were blood drives at his work twice a year, but he hadn't participated because, to be honest, he was a little scared. He didn't like doctor appointments or medical procedures of any kind, but now he realized his cousin was alive only because of the generosity of other people. He felt that if he donated, it would be fighting back against the drunk-driver who hurt Trina. He was determined to make the effort to donate before the week was over.

▶ **How many units of blood are required each year in the United States? Why do cancer patients require transfusions? To what extent can artificial blood substitutes be used?**

Follow-up to this Case Study appears at the end of the chapter before the *Concept Summary*.

When you have completed your study of this chapter, you should be able to:

1 Compare the chemical compositions of plasma, interstitial fluid, and intracellular fluid. **(Section 25.1)**

2 Explain how oxygen and carbon dioxide are transported within the bloodstream. **(Section 25.2)**

3 Explain how materials move from the blood into the body cells and from the body cells into the blood. **(Section 25.3)**

4 List the normal and abnormal constituents of urine. **(Section 25.4)**

5 Discuss how proper fluid and electrolyte balance is maintained in the body. **(Section 25.5)**

6 Explain how acid–base balance is maintained in the body. **(Section 25.6)**

7 Explain how buffers work to control blood pH. **(Section 25.7)**

8 Describe respiratory control of blood pH. **(Section 25.8)**

9 Describe urinary control of blood pH. **(Section 25.9)**

10 List the causes of acidosis and alkalosis. **(Section 25.10)**

To carry out its complex function, a living cell requires a steady supply of reactants, such as nutrients and oxygen. A cell also requires a reliable system for removing the resulting waste products—carbon dioxide and water, for example. Both of these conditions are met by the simple process of diffusion. Reactants can diffuse into the cell through the thin cell membrane, and the products of cell function can diffuse out of the cell and be absorbed by the surrounding liquid.

However, a complex organism, which consists of many closely packed cells, cannot depend solely on diffusion. Some circulating system must bring necessary materials to the cells and carry away the wastes. Otherwise, the liquid surrounding the cells would soon be depleted of reactants and saturated with wastes.

25.1 | A Comparison of Body Fluids
Learning Objective

1. Compare the chemical compositions of plasma, interstitial fluid, and intracellular fluid.

The average adult body contains about 42 L of fluids, which accounts for two-thirds of the total body weight. Most of these fluids are compartmentalized in three regions of the body: the interior of cells, tissue spaces between cells (including the lymph vessels), and the blood vessels. The majority, about 28 L, is located inside the cells and is called **intracellular fluid**. All body fluids not located inside the cells are collectively known as **extracellular fluids**. Thus, fluids in two of the fluid compartments are extracellular: the **interstitial fluid**, which fills the space between tissue cells and moves in lymph vessels, and the fluid of the bloodstream—**plasma**. Interstitial fluid constitutes about 25% (10.5 L) of the total body fluid, and the plasma (nearly 3.5 L) makes up about 8% of the total. Other body fluids that occur in lesser amounts are urine, digestive juices, and cerebrospinal fluid. The distribution of the various body fluids is illustrated in **Figure 25.1**. Extracellular fluid provides a relatively constant environment for the cells and transports substances to and from the cells. Intracellular fluid, on the other hand, is the medium in which the vital life-maintaining reactions take place.

Chemically, the two extracellular fluids, plasma and interstitial fluid, are nearly identical. The only difference is that plasma contains an appreciable amount of protein not found in interstitial fluid. Intracellular fluid is chemically different from extracellular fluid.

intracellular fluid Body fluid located inside cells.

extracellular fluid Body fluid located outside cells.

interstitial fluid The fluid surrounding individual tissue cells.

plasma The liquid portion of whole blood.

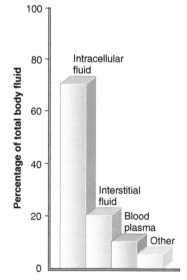

Figure 25.1 The distribution of body fluids.

Intracellular fluid contains larger amounts of protein and a much different ionic population (see **Figure 25.2**). From Figure 25.2, we can see that:

1. The principal cation of plasma and interstitial fluid is Na^+. The principal cation of intracellular fluid is K^+ (actually, about 98% of the body's potassium is found inside cells).
2. The principal anion of extracellular fluid is chloride, Cl^-, whereas phosphate (mainly HPO_4^{2-}) is the main anion inside cells.
3. Intracellular fluid contains more than four times as much protein as plasma, and there is remarkably little protein in interstitial fluid.

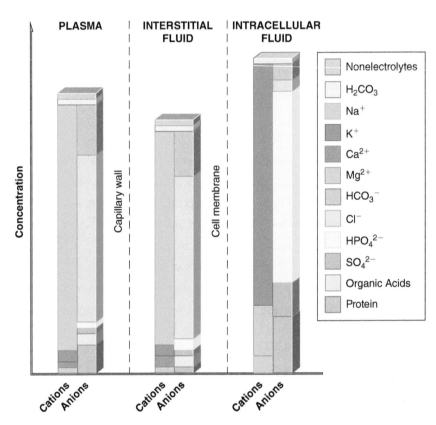

Figure 25.2 The relative chemical composition of the major body fluids.

Figure 25.3 Red blood cells passing through a narrow capillary.

Professors P. M. Motta and S. Correr/Science Photo Library/Science Source

oxyhemoglobin Hemoglobin combined with oxygen.

deoxyhemoglobin (hemoglobin) Nonoxygenated hemoglobin.

25.2 | Oxygen and Carbon Dioxide Transport

Learning Objective

2. Explain how oxygen and carbon dioxide are transported within the bloodstream.

One of the most important functions of blood is to carry oxygen from the lungs to the tissues. An average adult at rest requires about 350 mL of oxygen per minute. The limited solubility of oxygen in plasma allows only about 2% of this total to be dissolved and transported in solution. About 98% of the needed oxygen is carried by red blood cells (see **Figure 25.3**) in the form of an oxygen-hemoglobin combination called **oxyhemoglobin**. The nonoxygenated hemoglobin is called **deoxyhemoglobin**, or simply **hemoglobin**. Normal human blood contains about 15 g of hemoglobin per 100 mL.

Because of its physiological importance, hemoglobin has been studied extensively. It is known to be a conjugated protein containing four chains (the globin portion of the molecule) and four heme groups. Each heme group consists of an Fe^{2+} ion in the center of a nitrogenous heterocyclic structure (see **Figure 25.4**). The iron in oxyhemoglobin is bonded at six positions, as shown in Figure 25.4. The four nitrogens are part of the heme group, and the iron is directly attached to the histidine, which is a part of the globin molecule.

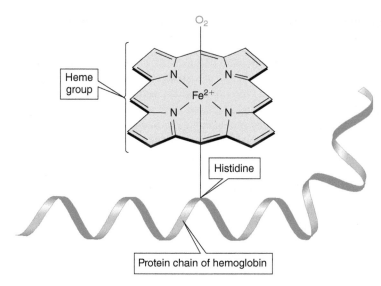

Figure 25.4 The iron in oxyhemoglobin.

The sixth bond is between iron and oxygen. Because each hemoglobin molecule contains four heme groups, it can bind a total of four O_2 molecules. When one of the four heme groups binds an oxygen molecule, the affinity of the other heme groups for oxygen is increased. The reversible binding that takes place between hemoglobin (HHb) and molecular oxygen is represented like this:

$$HHb + O_2 \rightleftharpoons HbO_2^- + H^+ \qquad (25.1)$$
$$\text{hemoglobin} \quad \text{oxyhemoglobin}$$

When CO_2 is present, hemoglobin can combine with it to form **carbaminohemoglobin**. This reversible reaction is favored by conditions found in body tissues:

$$HHb + CO_2 \rightleftharpoons HHbCO_2 \qquad (25.2)$$
$$\text{carbaminohemoglobin}$$

carbaminohemoglobin Hemoglobin combined with carbon dioxide.

About 25% of the total CO_2 in the body is carried from body tissues to the lungs in the form of carbaminohemoglobin. Another 5% dissolves in the plasma and travels in solution; the remaining 70% is transported in the form of bicarbonate ions (HCO_3^-).

The transport of oxygen to tissues begins when inhaled air fills the alveoli of the lungs. These clusters of small sacs are surrounded by many tiny blood capillaries, and it is here that oxygen is absorbed into the blood. **Figure 25.5** represents the processes that take place as oxygen is absorbed and carbon dioxide is released. The main steps (circled numbers in the figure) are:

1. Oxygen partial pressure is higher in the alveoli than in the red blood cells. As a result, oxygen diffuses from the alveoli through the capillary wall and into the red blood cells.
2. Inside the red blood cells, oxygen reacts with hemoglobin to form oxyhemoglobin and H^+ ions. Oxyhemoglobin formation is favored (according to Le Châtelier's principle) by a high partial pressure of oxygen, a slight cooling of the blood, and an increase in pH (decrease in H^+ ions)—conditions encountered by the blood as it passes through the capillaries of the lungs.
3. Bicarbonate ions (HCO_3^-) diffuse from the plasma into the red blood cells.
4. The bicarbonate ions are replaced in the plasma by chloride ions that diffuse out of the blood cells. This step, called the **chloride shift**, maintains charge balance and osmotic pressure relationships between the plasma and red blood cells.

chloride shift The movement of chloride ions from one fluid to another to maintain charge neutrality.

Performance-Enhancing Drugs

Could anyone be convinced to consume a substance that was known to improve athletic performance even though there might be a risk associated with its use? In the past, a reporter for *Sports Illustrated* magazine interviewed a group of elite Olympic athletes. One of the questions asked was: "If you were given a performance-enhancing substance, and you would not be caught, and you would win your event, would you take it?" Ninety-eight percent of the interviewed athletes answered yes. When the question was changed to: "If you were given a performance-enhancing substance, and you would not be caught, and you would win all of your competitions for 5 years, then die, would you take it?" More than 50% still said yes. The lure to win is so great that some of these athletes would forfeit their lives just to "win." Winning at all costs is a common goal for some athletes and their coaches or sponsors. Scholarships, awards, recognition, contracts, and millions of dollars are at stake.

Performance-enhancing drugs, abbreviated PED, are substances used to improve any form of physical activity performance in humans. The practice of using them is commonly called "doping," and it is not limited to enhancing physical performance. Some students use cognitive performance enhancers to improve their performance in class. The practice is called "academic doping." Some military personnel are known to have used PEDs to enhance their combat performance. Truck drivers are known to have used them to stay awake while driving. In the athletic community, the use of PEDs is often regulated by drug testing conducted by the respective sport's governing bodies, such as the International Olympic Committee and the National Collegiate Athletic Association. However, players unions and collective bargaining agreements of some professional sports associations have prevented extensive testing.

Information found on the Internet can strongly influence an individual's decision about whether or not to use PEDs. The classification of a substance as a PED is not exactly clear-cut, but lists of substances banned by various organizations are readily available for review. A violation of these regulations could disqualify an athlete from competition. The most commonly used drug classifications are given in the following list.

Anabolic Steroids: These are used to increase muscle mass and strength. They also reduce the recovery time from workouts, and allow the training to be harder and longer. They exert effects similar to the naturally produced steroid called testosterone. Because anabolic steroids are included on the banned substances list, an entire industry has developed that is devoted to the creation of illicit anabolic steroids that will not give positive results in the testing process. These so called "designer steroids" have not yet been tested adequately and pose an unknown risk to users. Even when legal anabolic steroids are used by athletes for unapproved uses, the dose used is usually larger than FDA guidelines. The risks of high-dose steroid use by men include infertility, impotence, baldness, and decreases in testicular size and function. The risks of high-dose steroid use by women include infrequent or absent menstrual cycles, deepening of the voice, increases in body hair, and baldness. Both men and women have experienced severe acne, liver function changes, and increased blood pressure.

Human Growth Hormone: This peptide hormone is used for its anabolic effects of increases in muscle mass and performance that are similar to the anabolic steroid group. The results of using this hormone are not always beneficial. Some adverse effects of long-term use include enlargement of the head and extremities, joint pain, enlargement of the heart, diabetes, blood pressure increase, and carpel tunnel syndrome.

Amphetamines/Stimulants: These are used to increase energy, alertness, and aggressiveness, and to decrease fatigue, improve endurance, and suppress appetite. Their effects range from mild, as found in uncontrolled substances like caffeine and guarana that are commonly used in energy drinks and pre-workout products, to the very powerful, as found in amphetamines and cocaine. The predictable risks associated with this group of PEDs include nervousness, irritability, tremors, heart rhythm changes, increases in blood pressure, stroke, weight loss, tolerance, and addiction.

Creatine: This is an amino acid found in meat and fish. It is also produced in the liver and kidneys. It appears to provide modest benefits for repetitive exercise lasting less than 30 seconds. It does not appear to provide significant benefits for endurance aerobic exercise.

Erythropoietin: A hormone secreted by the kidneys that stimulates the production of red blood cells and hemoglobin, which increases the movement of oxygen to muscle tissue. It is commonly used by endurance athletes. Risks associated with its use include heart attack and pulmonary blood clots. A number of infamous Tour de France winners have been stripped of their titles for abusing this hormone, which is banned in cycling.

Diuretics: As the name indicates, these drugs stimulate the process of passing urine. In this process, body water and electrolytes are lost from the body. The loss of water weight might allow an athlete to compete in a lighter weight category, or a jockey to reach a weight for an important horse race. However, the accompanying electrolyte loss can upset the balance of other chemicals and systems of the body. Occasionally, athletes will try to mask other banned substances by taking a diuretic in an attempt to increase urine flow and reduce the banned substance to a level below the detection's threshold of recognition.

Analgesics: Obviously, masking pain will allow an athlete to "play through the pain." Pain is a normal protection mechanism of the body. Inhibiting this defense might lead to worse tissue damage.

A Final Word about PEDs: The use of PEDs has the potential to drastically alter the human body and its functions. Some of these alterations come with the potential for improved athletic performance, but not without some risk, including even death. Some of the complications resulting from PED use are reversible and some are not. PED users should decide whether the temporary glory, improved appearance, or performance are worth the long-term effects. Choose wisely.

Performance-enhancing drugs are known as PEDs.

Joe Belanger/Shutterstock.com

Figure 25.5 At the lungs, the carbon dioxide of red blood cells is exchanged for oxygen.

5. Protons released by hemoglobin as it is oxygenated (Step 2) react with bicarbonate ions to form carbonic acid.
6. An enzyme in the red blood cells, carbonic anhydrase, promotes the breakdown of carbonic acid to water and carbon dioxide. The reaction is reversible, but the low pressure of carbon dioxide in the lungs favors the formation of carbon dioxide and water (Le Châtelier's principle, again).
7. Carbaminohemoglobin ($HHbCO_2$), formed from hemoglobin and CO_2 at the cellular level, breaks apart to yield hemoglobin and carbon dioxide.
8. Because the carbon dioxide pressure is high in the red blood cells but low in the alveoli, carbon dioxide molecules diffuse out of the red blood cells and into the lung. The carbon dioxide and some of the water formed in Step 6 are expelled in the exhaled air.

Select the Right Pre-Exercise Foods

Are you ready to sweat? Fluid intake before a workout is a precaution that should be followed. But should you eat before a workout? If the answer is yes, what type of food should be eaten? Opinions on this question are plentiful and widely shared thanks to the Internet. You can find information about such topics as carbohydrate loading, protein shakes, protein bars, and energy gels. But how do you know which to believe?

Evidence, based on research, provides some answers to questions about sensible pre-workout fuel. During a workout, in addition to plenty of water, the body needs energy that is delivered in a steady stream without any significant spikes or letdowns. This can be accomplished by supplying your body with a combination of simple and complex carbohydrates, a little protein, and a very small amount of beneficial fat. The following are examples of pre-workout meals and snacks that satisfy these guidelines.

- *Greek yogurt with fruit and some sensible granola (containing oats, nuts, and seeds, but no candy).* This meal provides protein, simple sugars, and complex carbohydrates to fuel you for a significant part of the day.
- *Whole-wheat toast with peanut butter and a sliced banana.* This meal contains a combination of simple and complex carbohydrates, with protein. The banana provides a healthy dose of potassium to replace this electrolyte lost in the sweat.
- *Oatmeal cereal with milk and fruit.* The liquid in the milk and fruit helps with hydration, and the fiber-rich whole-grain cereal lasts through the entire workout. The sugars in the fruit provide an energy boost at the start of the

workout. Add a little milk, and the protein will last to the end of the workout.
- *Smoothie made of yogurt, fruit, and 10 to 20 grams of protein powder.* This portable meal is easy on the stomach and adds to your hydration.
- *Whole-wheat toast with a sliced apple and low-fat cheese.* This is a good alternative to a sweet choice and provides an essential carbohydrate, sugar, and protein combination that your exercising body needs.
- *Breakfast burrito consisting of a scrambled egg in a whole-wheat tortilla with salsa and low-fat cheese.* This is another non-sweet choice that can fuel your workout very well. The goal in pre-exercise eating is to obtain the necessary energy without overloading your digestive system or upsetting your stomach. Keep the portions reasonable. You want to eat just enough to get you through the workout; then, finish with a protein-rich snack or small meal to ensure a smooth post-workout recovery.

The right pre-exercise foods can help sustain a vigorous workout.

Oxygenated red blood cells are carried by the bloodstream into the capillaries surrounding the cells of active tissues. **Figure 25.6** shows the steps involved in delivering oxygen to the tissues and accepting carbon dioxide for transport to the lungs. These steps are essentially the reverse of those at the lungs:

1. Tissue cells use oxygen to produce energy; carbon dioxide and water are formed as byproducts. The concentration of carbon dioxide is higher in the tissue cells than in the red blood cells, so carbon dioxide diffuses from the tissue cells into the interstitial fluid and then into the red blood cells.
2. Inside the red blood cells, carbon dioxide rapidly reacts with water in the presence of carbonic anhydrase to produce carbonic acid.
3. Carbonic acid dissociates to give hydrogen ions and bicarbonate ions.
4. The bicarbonate ions diffuse into the plasma. Most carbon dioxide in the plasma is transported in the bicarbonate ion form, moving from tissue cells to the lungs.
5. The chloride shift again takes place (this time in the opposite direction) to maintain electrolyte balance.
6. A smaller amount (25%) of carbon dioxide from Step 1 reacts with hemoglobin to form carbaminohemoglobin ($HHbCO_2$) for transport to the lungs.

Figure 25.6 At the tissue cells, oxygen of red blood cells is exchanged for carbon dioxide.

7. The increase in concentration of hydrogen ions inside the red blood cells promotes a reaction with oxyhemoglobin, which releases oxygen.

8. The free oxygen diffuses out of the red blood cells through the plasma, capillary membrane, interstitial fluid, and into the tissue cells.

✔ **LEARNING CHECK 25.1** Decide whether each of the following red blood cell reactions occurs at the lungs or at the site of the tissue cells:

a. $HHbCO_2 \rightleftarrows HHb + CO_2$

b. $H_2O + CO_2 \rightleftarrows H_2CO_3$

c. $H^+ + HCO_3^- \rightleftarrows H_2CO_3$

d. $HbO_2^- + H^+ \rightleftarrows HHb + O_2$

25.3 | Chemical Transport to the Cells

Learning Objective

3. Explain how materials move from the blood into the body cells and from the body cells into the blood.

For substances to be chemically transported in the body, they must become part of the moving bloodstream. They may dissolve in the water-based plasma (as do sugars, amino acids, ions, and gases to some extent), they may become chemically bonded to cellular components (as do oxygen and carbon dioxide with the hemoglobin of blood cells), or they may form a suspension in the plasma of the blood (as do lipids).

Materials transported by the bloodstream to the cells must move through the capillary wall into the interstitial fluid and through the cell membrane into the cell. The reverse must take place when waste products move from inside tissue cells into the bloodstream. During these processes, capillary walls behave as selectively permeable membranes or filters, allowing water containing dissolved nutrients—including oxygen—to pass in one direction and water containing dissolved wastes to pass in the other direction.

The movement of water and dissolved materials through the capillary walls is governed by two factors. They are the pressure of blood against the capillary walls and the differences in protein concentration on each side of the capillary walls.

The concentration of protein in plasma is much higher than the protein concentration of the interstitial fluid outside the blood vessels. This concentration difference results in an osmotic pressure of about 18 torr, which tends to move fluid into the bloodstream. However, the pumping action of the heart puts the blood under pressure (blood pressure) that is greater at the arterial end of a capillary (about 32 torr) than at the venous end (about 12 torr). Because the blood pressure at the arterial end is higher than the osmotic pressure, the tendency is for a net flow to occur from the capillary into the interstitial fluid (see **Figure 25.7**). The fluid that leaves the capillary contains the dissolved nutrients, oxygen, hormones, and vitamins needed by the tissue cells.

Figure 25.7 The flow of fluid between capillaries and interstitial spaces.

TABLE 25.1 Major Dissolved Solids in Normal Urine	
Constituent[a]	**Grams per 24 Hours**[b]
Amino acids	0.80
Urea	25.0
Creatinine	1.5
Uric acid	0.7
H^+	pH 4.5–8.0
Na^+	3.0
K^+	1.7
NH_4^+	0.8
Ca^{2+}	0.2
Mg^{2+}	0.15
Cl^-	6.3
SO_4^{2-}	1.4
$H_2PO_4^-$	1.2
HCO_3^-	0–3
Other compounds	2–3

[a]The 24-hour volume and composition of urine vary widely, depending on the fluid intake and diet.

[b]These data are for an average 24-hour specimen with a total volume of 1400 mL.

As the blood moves through the capillary, its pressure decreases, until at the venous end, it becomes lower than the osmotic pressure, and a net flow of fluid takes place from the interstitial fluid into the capillary. The incoming fluid contains the metabolic waste products, such as carbon dioxide and excess water. This mechanism allows the exchange of nutrients and wastes and maintains fluid balance between the blood and interstitial space.

25.4 | The Constituents of Urine

Learning Objective

4. List the normal and abnormal constituents of urine.

Urine contains about 96% water and 4% dissolved organic and inorganic waste products. Approximately 40 to 50 g of dissolved solids are contained in the daily urine output of an adult (see **Table 25.1**). The urine of a healthy person has a pH of 4.5 to 8.0, with 6.6 being a reasonable average for a person on an ordinary diet. Fruits and vegetables tend to make the urine alkaline, whereas high-protein foods tend to make the urine acidic.

The composition of urine is an excellent indicator of the state of health. This accounts for the importance of urinalysis as part of a physical checkup or a diagnosis. Normal constituents are considered abnormal when they are eliminated in excess amounts. **Table 25.2** lists some conditions that lead to abnormally high urinary levels of normal constituents; some other substances (such as pus or blood), which are considered abnormal in *any* measurable amount, are also listed.

A urine specimen is often collected from a patient entering a hospital and is checked with a paper test strip that contains bands of different reagents that will react with the different abnormal constituents (see **Figure 25.8**). This one simple test quickly checks for indications of pathological conditions.

TABLE 25.2 Some Abnormal Urine Constituents

Abnormal Constituent	Condition	Causes
Glucose (in large amounts)	Glucosuria	Diabetes mellitus, renal diabetes, alimentary glycosuria
Protein	Proteinuria or albuminuria	Kidney damage, nephritis, bladder infection
Ketone bodies	Ketonuria	Diabetes mellitus, starvation, high-fat diets
Pus (leukocytes)		Kidney or bladder infection
Hemoglobin	Hemoglobinuria	Excessive hemolysis of red blood cells
Red blood cells	Hematuria	Hemorrhage in the urinary tract
Bile pigments (in large amounts)	Jaundice	Blockage of bile duct, hepatitis, cirrhosis

Figure 25.8 Urinalysis test strips. A urine sample is tested for several components simultaneously by dipping a test strip into a sample and comparing the reagent areas to a color chart.

25.5 | Fluid and Electrolyte Balance

Learning Objective

5. Discuss how proper fluid and electrolyte balance is maintained in the body.

The body is 45% to 75% water (by weight), the percentage depending on a number of factors, including age, sex, and especially the amount of body fat. In general, the percentage of water increases as fat content decreases.

Fluid balance within the body is maintained not only by a balance in the total amount but by a normal and stable distribution of fluid in the body's three fluid-containing areas. Thus, the amounts of fluid inside the cells, in the interstitial spaces, and in the blood vessels must remain relatively constant if a fluid balance is to exist.

Fluid and electrolyte balance are interdependent. If one deviates from normal, so does the other. In modern medical practice, a great deal of attention is given to these matters. Many of today's hospital patients receive some form of fluid and electrolyte balance therapy.

An important and obvious principle of fluid balance is that fluid output and intake must be equal. Water from food and drink is absorbed into the body from the digestive tract. Besides this well-known source, water is also derived within the cells from food metabolism. This by-product water enters the bloodstream along with the water that is absorbed from the digestive tract.

Water intake is regulated in part by the thirst mechanism (see **Figure 25.9**). When the body loses large amounts of water, salivary secretions decrease, and a dry feeling develops in the mouth. This and other sensations are recognized as thirst, and water is drunk to relieve the condition. The fluid intake compensates for the fluid lost, and balance is reestablished.

Water normally leaves the body through the kidneys (urine), lungs (water vapor in expired air), skin (by diffusion and perspiration), and intestines (feces) (see **Table 25.3**).

TABLE 25.3 Typical Normal Values for Water Entry and Exit

Intake (mL)	
Ingested liquids	1500
Water in foods	700
Water formed by catabolism	200
Total	2400

Output (mL)	
Kidneys (urine)	1400
Lungs (water in expired air)	350
Skin	
By diffusion	350
By perspiration	100
Intestines (in feces)	200
Total	2400

Figure 25.9 The thirst mechanism.

Normal body-water volume → Fluid loss → Decreased body-water volume → Decreased saliva secretion → Dry mouth, thirst → Fluid intake → Normal body-water volume

Pulse Oximetry

Often when we think of a hospital, we think of rooms full of people lying in high-tech beds with tubes leading into their veins carrying antibiotics or other fluids, and debilitated people wearing oxygen masks to assist with their breathing. To complete the picture, we might imagine a small clamp-like device attached to the finger of all the patients. This device glows with a subdued red light and has wires leading to an important-looking monitor. But what exactly is this device? What is its purpose? How does it work? Is it safe?

The object in question is called a pulse oximeter. It measures both the pulse rate and the oxygen saturation level of the blood. This information is used to monitor the breathing of a patient. Monitoring breathing is one method used to determine if a patient's condition is stable, improving, or worsening. Generally, a young healthy individual will have an oxygen saturation level of 95% to 99%. Naturally, this percentage of oxygen saturation varies with age, aerobic fitness, and the altitude at which the person lives and at which the measurement is being made. Because of these characteristics, a single pulse oximetry reading is not very informative. It is much better to make multiple readings over time as a means of monitoring change over that time.

A pulse oximeter determines the quality of breathing by measuring the level of oxygen saturation of the blood. The color of blood changes depending on the amount of oxygen that is attached to each molecule of hemoglobin, i.e., the oxygen saturation of the blood. Highly oxygenated blood is bright red in color, while deoxygenated blood is a deep red color. These changes in color are detected by passing a beam of red light and a beam of infrared light through an extremity such as a finger, toe, or an ear lobe. The amount of light absorbed by the blood changes as the color of the blood changes. Thus, the readings of the pulse oximeter give an indication of the extent of oxygen saturation of the blood, which gives an indication of how well a patient is breathing.

A pulse oximeter alone poses few dangers for the patient to whom it is attached. In some cases, particularly older models, a pulse oximeter can cause burns or pressure ulcers. However, greater hazards involving the use of a pulse oximeter come when it is the only means of monitoring the health of a patient. If the readings from the pulse oximeter look good, a health care provider might have a false sense of security and overlook other indicators of the patient's overall condition. Furthermore, a pulse oximeter requires a strong signal to operate correctly. Once again, a false sense of security might be obtained if the instrument appears to have a strong signal but is in fact displaying inaccurate values.

Technology can aid the medical profession. But there is no substitute for personal care from an attentive, knowledgeable medical professional.

A pulse oximeter measures the pulse rate and oxygen concentration of the blood.

Abnormally high fluid losses, and possibly dehydration, can be caused by hyperventilation, excessive sweating, vomiting, or diarrhea.

Fluid balance in the body is maintained or restored primarily by variations in urine output. Normally, this amounts to about 1400 mL/day, but the total excretion fluctuates according to water intake. The primary factor that controls urine production is the rate of water reabsorption from the renal tubules in the kidneys. This rate is regulated chiefly by the pituitary hormone vasopressin and by the adrenal cortex hormone aldosterone.

Vasopressin, also called antidiuretic hormone (ADH), exerts its action by affecting the permeability of the renal tubules to water. In the absence of vasopressin, less water is reabsorbed into the bloodstream, and a large volume of light yellow, dilute urine results. Patients with the rare disease diabetes insipidus do not produce vasopressin; their urine is extremely voluminous and dilute. In the presence of vasopressin, the tubules become more permeable to water that is drawn into the capillaries, thus increasing blood volume. The reabsorption of water from the tubules produces a low volume of dark yellow, concentrated urine. (Diabetes insipidus is very different from diabetes mellitus, mentioned elsewhere in the text.)

With the help of vasopressin, a healthy person can vary the intake of water widely and yet preserve a stable overall concentration of substances in the blood. But sometimes the

water level of the body gets dangerously low, from not drinking enough water or excessive water loss caused by diarrhea or excessive perspiration. Under such circumstances, the adrenal gland secretes the steroid hormone aldosterone. At the kidneys, aldosterone stimulates the reabsorption of sodium ions (electrolytes) from the renal tubes into the blood. Chloride ions follow the sodium ions to maintain electrical neutrality, and water follows the sodium chloride. Thus, aldosterone secretion conserves both salt and water in the body. When the fluid level of the body returns to normal, the amount of aldosterone secreted decreases.

alkalosis An abnormally high blood pH.

25.6 | Acid–Base Balance

Learning Objective

6. Explain how acid–base balance is maintained in the body.

Blood pH normally remains within the rather narrow range of 7.35 to 7.45. A decrease in pH, called acidosis, or an increase in pH, called **alkalosis**, is serious and requires prompt attention. Death can result when the pH value falls below 6.8 or rises above 7.8 (see **Figure 25.10**). Acidosis, an abnormally low blood pH, was introduced in **Section 24.6**.

Large amounts of acids and smaller amounts of bases normally enter the blood. Some mechanisms must neutralize or eliminate these substances if the blood pH is to remain constant. In practice, a constant pH is maintained by the interactive operation of three systems: buffer, respiratory, and urinary. Only when all three parts of this complex mechanism function properly can the acid–base balance be maintained.

Figure 25.10 Human circulatory and respiratory processes operate only within a very narrow range of blood pH.

25.7 | Buffer Control of Blood pH

Learning Objective

7. Explain how buffers work to control blood pH.

A buffer system prevents marked changes in the pH of a solution when an acid or base is added. Three major buffer systems of the blood are the bicarbonate buffer, the phosphate buffer, and the plasma proteins. The most important of these is the bicarbonate buffer system, consisting of a mixture of bicarbonate ions (HCO_3^-) and carbonic acid (H_2CO_3).

Buffer systems work by neutralizing H^+ or OH^- ions that come from the dissociation of acids or bases added to a system. As an example of buffers in action, consider the fate of lactic acid (HLac) entering the bloodstream. When lactic acid dissolves in the blood, it dissociates to produce H^+ and lactate ions (Lac^-). The liberated H^+ ions drive the pH of the blood down. The bicarbonate ions of the bicarbonate buffer system react with the excess H^+ and return the pH to the normal range:

$$HLac \rightleftharpoons H^+ + Lac^- \tag{25.3}$$

$$H^+ + HCO_3^- \rightleftharpoons H_2CO_3 \tag{25.4}$$

Reaction 25.4 is a net ionic reaction for the buffering process. When a total reaction is written, we see that the effect of the buffering is to replace lactic acid in the blood with carbonic acid, which is a weaker acid and so releases fewer H^+ ions to the blood than does lactic acid:

$$\underset{\substack{\text{stronger} \\ \text{acid}}}{HLac} + HCO_3^- \longrightarrow \underset{\substack{\text{weaker} \\ \text{acid}}}{H_2CO_3} + Lac^- \tag{25.5}$$

This replacement of a stronger acid with a weaker acid is how biochemists often describe buffering reactions.

Although the bicarbonate system is not the strongest buffer system in the body (the most powerful and plentiful one utilizes the proteins of plasma and cells), it is very important because the concentrations of both bicarbonate and carbonic acid are regulated by the kidneys and by the respiratory system. Without the respiratory and urinary processes, the capacity of buffers would eventually be exceeded.

25.8 | Respiratory Control of Blood pH

Learning Objective

8. Describe respiratory control of blood pH.

The respiratory system helps control the acidity of blood by regulating the elimination of carbon dioxide and water molecules. These molecules, exhaled in every breath, come from carbonic acid as follows (see **Figure 25.5**):

$$H_2CO_3 \rightleftharpoons H_2O + CO_2 \qquad (25.6)$$
$$\text{carbonic acid}$$

The more CO_2 and H_2O that are exhaled, the more carbonic acid is removed from the blood, thus elevating the blood pH to a more alkaline level.

The respiratory mechanism for controlling blood pH begins in the brain with respiratory center neurons that are sensitive to blood CO_2 levels and pH. A significant increase in the CO_2 of arterial blood, or a decrease below about 7.38 of arterial blood pH, causes the breathing to increase both in rate and depth, resulting in **hyperventilation**. This increased ventilation eliminates more carbon dioxide, reduces carbonic acid and hydrogen-ion concentrations, and increases the blood pH back toward the normal level (see **Figure 25.11**).

hyperventilation Rapid, deep breathing.

Figure 25.11 Respiratory control of blood pH.

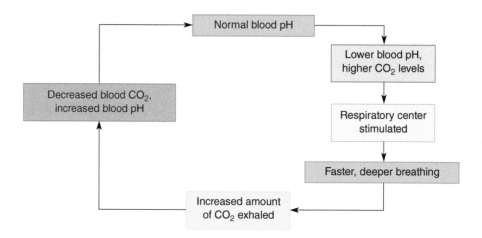

hypoventilation Slow, shallow breathing.

In the opposite situation, an increase in blood pH above normal causes **hypoventilation**, a reduced rate of respiration. Less CO_2 is exhaled, and the higher concentration of carbonic acid remaining in the blood lowers the pH back to normal.

25.9 | Urinary Control of Blood pH

Learning Objective

9. Describe urinary control of blood pH.

Because they can excrete varying amounts of acid and base, the kidneys, like the lungs, play a vital role in pH control. For example, with acidic blood, the excretion of hydrogen ions by the kidneys decreases the urine pH and simultaneously increases the blood pH back toward normal.

Figure 25.12 traces the reactions involved in the excretion of hydrogen ions by the kidneys:

1. Carbon dioxide diffuses from the blood capillaries into the kidney distal tubule cells.
2. Catalyzed by carbonic anhydrase, water and carbon dioxide react to give carbonic acid.
3. The carbonic acid ionizes to give hydrogen ions and bicarbonate ions. The hydrogen ions diffuse into the developing urine.
4. For every hydrogen ion entering the urine, a sodium ion passes into the tubule cells.
5. The sodium ions and bicarbonate ions enter the bloodstream capillaries.

Figure 25.12 Urinary control of blood pH in the kidney.

The net result of these reactions is the conversion of CO_2 to HCO_3^- within the blood. Both the decrease in CO_2 and the increase in HCO_3^- tend to increase blood pH levels back to normal. The developing urine has picked up the hydrogen ions, which now react with buffering ions such as phosphate that are present in urine:

$$H^+ + HPO_4^{2-} \rightarrow H_2PO_4^- \tag{25.7}$$

The presence of the phosphate buffer system $H_2PO_4^-/HPO_4^{2-}$ usually keeps urine from going much below pH 6, but too great an excess of protons from carbonic acid can exceed the buffer capacity of the system and result in quite acidic urine.

25.10 | Acidosis and Alkalosis

Learning Objective

10. List the causes of acidosis and alkalosis.

When blood pH is normal and a state of acid–base balance exists, components of the bicarbonate buffer are present in plasma in a ratio of 20 parts of bicarbonate (HCO_3^-) to 1 part of carbonic acid (H_2CO_3). When respiratory and urinary systems work together to maintain the blood at pH 7.4, they do so primarily by maintaining the HCO_3^-/H_2CO_3 ratio at 20:1. Any change in the ratio produces a change in pH. An increase in the ratio of bicarbonate to carbonic acid causes the pH to increase (alkalosis); and a decrease in the ratio, which is much more common, causes the pH to decrease (acidosis) (see **Figure 25.13**).

Figure 25.13 Blood pH depends on the relative concentrations of carbonic acid and bicarbonate ion.

pH 7.35 pH 7.45

Acidosis Normal Alkalosis

pH 6.80 pH 7.80

Death Death

1 part carbonic acid H_2CO_3

20 parts bicarbonate HCO_3^-

The normal ratio of 1 part H_2CO_3 to 20 parts HCO_3^- indicating that the body is in acid–base balance

Two adjectives describe the general cause or origin of pH imbalances in body fluids: metabolic and respiratory. Respiratory acid–base imbalances (respiratory acidosis and respiratory alkalosis) result from abnormal breathing patterns. Metabolic acidosis and metabolic alkalosis are caused by factors other than abnormal breathing.

25.10A Respiratory Alkalosis

respiratory alkalosis A condition of alkalosis caused by hyperventilation.

Respiratory alkalosis is caused by hyperventilating—breathing rapidly and deeply. Hysteria, anxiety, or prolonged crying may result in hyperventilation in which too much carbon dioxide is exhaled, which disturbs the blood's normal carbon dioxide/carbonic acid ratio. According to Le Châtelier's principle, Reaction 25.6 (written again below as part of Reaction 25.8) shifts to the right as it attempts to restore the equilibrium:

$$\underset{\underset{\text{restore equilibrium}}{\underset{\text{direction of shift to}}{\longrightarrow}}}{H_2CO_3 \rightleftharpoons H_2O + CO_2\uparrow} \qquad \overset{\textit{loss of } CO_2}{}$$

(25.8)

The equilibrium shift reduces the amount of carbonic acid present and rapidly increases the HCO_3^-/H_2CO_3 ratio to a value greater than 20:1. As a result, blood pH rises to 7.54 or higher within a few minutes. A corresponding loss of HCO_3^- also occurs, but it is not large enough to maintain the HCO_3^-/H_2CO_3 ratio at 20:1. The treatment of respiratory alkalosis involves rebreathing one's own exhaled air by breathing into a paper bag (*not* a plastic bag), administering carbon dioxide, and treating the underlying causes of the hyperventilation.

25.10B Respiratory Acidosis

respiratory acidosis A condition of acidosis caused by hypoventilation.

Respiratory acidosis is sometimes the result of slow or shallow breathing (hypoventilation) caused by an overdose of narcotics or barbiturates. Anesthetists need to be particularly aware of respiratory acidosis, because most inhaled anesthetics depress respiration rates. Lung diseases, such as emphysema and pneumonia, or an object lodged in the windpipe can also result in hypoventilation. In respiratory acidosis, too little carbon dioxide is exhaled, and its concentration within the blood increases. The carbon dioxide/carbonic acid equilibrium shifts to the left to restore the equilibrium:

$$\underset{\underset{\text{of equilibrium}}{\underset{\text{direction of shift}}{\longleftarrow}}}{H_2CO_3 \rightleftharpoons H_2O + CO_2} \qquad \overset{\textit{increased } CO_2}{}$$

(25.9)

As a result of the shift, the concentration of carbonic acid in the blood increases, the HCO_3^-/H_2CO_3 ratio decreases to less than 20:1, and respiratory acidosis sets in. The treatment of respiratory acidosis involves identifying the underlying causes and possibly the intravenous administration of isotonic sodium bicarbonate solution or hemodialysis.

25.10C Metabolic Acidosis

Various body processes—collectively called metabolism—produce acidic substances that liberate H^+ in solution. The diffusion of these substances into the bloodstream causes a shift in the carbonic acid–bicarbonate equilibrium:

$$\underset{\underset{\substack{\text{direction of shift to}\\\text{restore equilibrium}}}{\longleftarrow}}{H_2CO_3 \rightleftharpoons H^+ + HCO_3^-} \qquad \overset{\textit{increased } H^+}{} \tag{25.10}$$

The shift decreases the concentration of bicarbonate ions and increases the concentration of carbonic acid. As a result, the HCO_3^-/H_2CO_3 ratio decreases, as does the blood pH. **Metabolic acidosis** is often a serious problem in uncontrolled diabetes mellitus, and it also occurs temporarily after heavy exercise. Other causes include severe diarrhea and aspirin overdose.

Symptoms of metabolic acidosis, besides low blood pH and high levels of H_2CO_3, include the following: hyperventilation as the respiratory center tries to get excess CO_2 out of the lungs, increased urine formation as the kidneys try to remove excess acids, thirst (to replace urination water loss), drowsiness, headache, and disorientation. The treatment again depends on the cause; it may involve insulin therapy, intravenous bicarbonate, or hemodialysis. Note carefully that hyperventilation can be a symptom of metabolic acidosis and a cause of respiratory alkalosis, two different conditions requiring very different treatments.

metabolic acidosis A condition of acidosis resulting from causes other than hypoventilation.

25.10D Metabolic Alkalosis

In **metabolic alkalosis**, the body has lost acid in some form, perhaps from prolonged vomiting of the acidic stomach contents, or there has been an ingestion of alkaline substances. Excessive use of sodium bicarbonate (baking soda), for example, a common home remedy for upset stomach, increases the concentration of bicarbonate ions in the blood.

A decrease in acid concentration within the body (from vomiting) or an increase in HCO_3^- levels (from excessive $NaHCO_3$) causes the HCO_3^-/H_2CO_3 ratio to rise above 20:1, and measurable alkalosis is observed. The respiratory center responds to the higher blood pH by slowing the breathing (hypoventilation). Other symptoms may include numbness, tingling, and headache. **Table 25.4** is a summary of causes of acidosis and alkalosis.

metabolic alkalosis A condition of alkalosis resulting from causes other than hyperventilation.

TABLE 25.4 Some Causes of Blood Acid–Base Imbalance

Condition	Causes
Respiratory	
Acidosis: $CO_2 \uparrow pH \downarrow$	Hypoventilation, blockage of diffusion within lungs, respiratory center depressants
Alkalosis: $CO_2 \downarrow pH \uparrow$	Hyperventilation, excitement, trauma
Metabolic	
Acidosis: $H^+ \uparrow pH \downarrow$	Kidney failure, prolonged diarrhea, ketone bodies from diabetes mellitus
Alkalosis: $H^+ \downarrow pH \uparrow$	Kidney disease, prolonged vomiting, excessive intake of baking soda

Case Study Follow-up

Around 30 million blood components are transfused each year in the United States. The average red blood cell transfusion is approximately 3 pints, resulting in a need of 38,000 blood donations a day. Blood cannot be manufactured and therefore must come from donors. Keeping up with the demand for acceptable (disease-free) blood donations is not guaranteed, and the practice of donating is declining among young people. As a result, the cost of blood is steadily rising.

Blood is a complex tissue, and making a synthetic substitute that performs all of its functions is not currently possible. The U.S. FDA approves no blood substitute; however, some are used in South Africa and Russia. "Pharming" blood from stem cells gathered from discarded umbilical cord blood is showing some potential. Perfluorocarbon (PFC) or hemoglobin-based oxygen carriers (HBOC) blood substitutes are promising, but still experimental. Many require human donated blood as a starting material, and others use bovine hemoglobin.

Cancer patients may require blood transfusions for several reasons, one of which is the cancer itself, if it affects the ability of the body to manufacture blood, or if it causes internal bleeding, as in the case of digestive system cancers. Cancer surgeries are often major operations requiring transfusion. Most chemotherapy and radiation treatments negatively impact the patient's ability to produce blood.

Concept Summary

A Comparison of Body Fluids Most fluids of the body are located in three regions of the body: inside tissue cells (intracellular fluid), outside tissue cells (extracellular fluid), and in blood vessels. Volumetrically, most of the extracellular fluids (outside tissue cells) are of two types—blood and interstitial fluid, the fluid that occupies spaces between cells and moves in the lymph vessels. Other extracellular body fluids that occur in smaller amounts are urine, digestive juices, and cerebrospinal fluid. Chemically, blood (plasma) and interstitial fluid (including lymph) are similar. Intracellular fluid is chemically different from extracellular fluid.

Objective 1 (Section 25.1), Exercise 25.2

Oxygen and Carbon Dioxide Transport Nearly all oxygen that is transported from the lungs to tissue cells is transported by the blood in the form of oxyhemoglobin. Most carbon dioxide transported from the cells to the lungs is transported by the blood in the form of bicarbonate ions dissolved in the plasma. Smaller amounts also travel as carbaminohemoglobin and as CO_2 gas dissolved in plasma. The formation of the two hemoglobin compounds and the dissolving of various materials in the plasma are influenced by equilibria related to concentration differences of materials in the blood as it passes through capillaries in the lungs or near tissue cells.

Objective 2 (Section 25.2), Exercise 25.10

Chemical Transport to the Cells Substances transported by the blood to tissue cells move from the blood, through capillary walls, into the interstitial fluid, then through cell membranes, and finally into the cytoplasm of the cells. Waste products of the cells move in the opposite direction. The movement of fluid through capillary walls is governed by the blood pressure against the capillary walls and by osmotic pressure differences that arise from protein concentration differences between blood and interstitial fluid.

Objective 3 (Section 25.3), Exercise 25.14

The Constituents of Urine Normal urine is about 96% water and 4% dissolved organic and inorganic wastes. Because urine composition is an excellent indicator of a person's health, urinalysis is a routine procedure in physical checkups.

Objective 4 (Section 25.4), Exercise 25.18

Fluid and Electrolyte Balance Fluid balance in the body involves a balance between fluid intake and output, and a proper distribution of fluid between the three fluid areas. Fluid and electrolyte balance are interdependent. Water enters the body through food and drink and leaves in the form of urine, water vapor in exhaled air, perspiration, and feces. The amount of fluid in the body is maintained or restored primarily by variations in urine output, which is regulated chiefly by the hormones vasopressin and aldosterone.

Objective 5 (Section 25.5), Exercise 25.22

Acid–Base Balance Blood pH is normally in the range of 7.35 to 7.45. Three systems cooperate to maintain the pH in this narrow range. Acidosis and alkalosis exist when the blood pH becomes lower or higher, respectively, than these values.

Objective 6 (Section 25.6), Exercise 25.26

Buffer Control of Blood pH The blood contains three major buffer systems: the bicarbonate system, the phosphate system, and the plasma protein system. The most important is the bicarbonate system, which consists of a mixture of bicarbonate ions and carbonic acid.

Objective 7 (Section 25.7), Exercise 25.28

Respiratory Control of Blood pH The respiratory system influences blood pH by regulating the elimination of carbon dioxide and water. Because carbon dioxide and water are produced during the decomposition of carbonic acid, the greater the elimination of carbon dioxide and water from the body, the greater is the elimination of carbonic acid from the blood.

Objective 8 (Section 25.8), Exercise 25.30

Urinary Control of Blood pH The kidneys influence blood pH by varying the amount of hydrogen ions excreted in the urine. When the blood is too acidic, carbonic acid is converted to bicarbonate ions and hydrogen ions. The hydrogen ions are excreted, and the bicarbonate ions enter the blood.

Objective 9 (Section 25.9), Exercise 25.34

Acidosis and Alkalosis Acidosis and alkalosis can result from metabolic or respiratory causes. Respiratory acidosis and alkalosis are the result of abnormal breathing patterns. Alkalosis is caused by hyperventilation that accompanies hysteria, anxiety, or prolonged crying. Acidosis can result from hypoventilation brought on by such factors as an overdose of depressant narcotics or anesthesia. Metabolic acidosis can result from the metabolic production of acids in the body or the elimination of alkaline materials, as during severe diarrhea. Metabolic alkalosis results from the loss of acid in some form such as the prolonged vomiting of acidic stomach contents or the ingestion of alkaline materials.

Objective 10 (Section 25.10), Exercise 25.38

Key Terms and Concepts

Alkalosis (25.6)
Carbaminohemoglobin (25.2)
Chloride shift (25.2)
Deoxyhemoglobin (hemoglobin) (25.2)
Extracellular fluid (25.1)

Hyperventilation (25.8)
Hypoventilation (25.8)
Interstitial fluid (25.1)
Intracellular fluid (25.1)
Metabolic acidosis (25.10)

Metabolic alkalosis (25.10)
Oxyhemoglobin (25.2)
Plasma (25.1)
Respiratory acidosis (25.10)
Respiratory alkalosis (25.10)

Key Reactions

1.	Formation of oxyhemoglobin (Section 25.2):	$HHb + O_2 \rightleftarrows HbO_2^- + H^+$	Reaction 25.1
2.	Formation of carbaminohemoglobin (Section 25.2):	$HHb + CO_2 \rightleftarrows HHbCO_2$	Reaction 25.2
3.	Decomposition of carbonic acid (Section 25.8):	$H_2CO_3 \rightleftarrows H_2O + CO_2$	Reaction 25.6

Exercises

Even-numbered exercises are answered in Appendix B.

Blue-numbered exercises are more challenging.

A Comparison of Body Fluids (Section 25.1)

25.1 Where is most fluid in the body located? What is it called?

25.2 Which two of the following have nearly the same chemical composition: plasma, intracellular fluid, interstitial fluid?

25.3 In what way does the composition of the two similar fluids in Exercise 25.2 differ?

25.4 Which of the fluids of Exercise 25.2 contains the highest concentration of the following?

- **a.** protein
- **b.** K^+
- **c.** Na^+
- **d.** HPO_4^{2-}
- **e.** HCO_3^-
- **f.** Mg^{2+}
- **g.** Cl^-

25.5 What is the principal cation found in the following?

- **a.** blood plasma
- **b.** interstitial fluid
- **c.** intracellular fluid

25.6 What is the principal anion in the following?

- **a.** blood plasma
- **b.** interstitial fluid
- **c.** intracellular fluid

25.7 What is the principal type of organic compound found in intracellular fluid?

Oxygen and Carbon Dioxide Transport (Section 25.2)

25.8 Write a reversible chemical equation that shows the reaction of hemoglobin (Hb) with:

- **a.** oxygen
- **b.** carbon dioxide

25.9 Hemoglobin exists in several forms—oxyhemoglobin, deoxyhemoglobin, and carbaminohemoglobin. Match each of the following to the type of hemoglobin involved:

- **a.** Formed when CO_2 combines with hemoglobin
- **b.** The oxygen-transporting form of hemoglobin
- **c.** The form of hemoglobin designated as HHb

25.10 What percentages of the total O_2 and CO_2 transported by the blood are in the following forms?

 a. O_2 as dissolved gas in plasma

 b. O_2 as oxyhemoglobin

 c. CO_2 as carbaminohemoglobin

 d. CO_2 as dissolved gas in plasma

 e. CO_2 as bicarbonate ions

25.11 What term is given to the reversible flow of chloride ions across the red blood cell membrane during the oxygen and carbon dioxide transport process?

25.12 Why are large capillary surface areas critical to O_2 and CO_2 transport processes? What driving force causes O_2 to move from lung alveoli into the bloodstream and CO_2 to move from the bloodstream into lung alveoli?

25.13 Write the equilibrium equation representing the reaction between oxygen and hemoglobin. In which direction will the equilibrium shift when the following occurs?

 a. The pH is increased.

 b. The concentration (pressure) of O_2 is increased.

 c. The concentration of CO_2 is increased (see Figure 25.5).

 d. The blood is in a capillary near actively metabolizing tissue cells.

Chemical Transport to the Cells (Section 25.3)

25.14 Which factor, blood pressure or osmotic pressure, has the greater influence on the direction of fluid movement through capillary walls at (a) the venous end of a capillary and (b) the arterial end of a capillary?

25.15 What types of compounds leave the capillaries and enter the interstitial fluid? What types of compounds leave the interstitial fluid and enter the capillaries to be transported away?

The Constituents of Urine (Section 25.4)

25.16 What organic material is normally excreted in the largest amount in urine?

25.17 What cation and anion are normally excreted in the largest amounts in urine?

25.18 List the abnormal constituents of urine.

25.19 Under what circumstances would the presence of a normal constituent of urine be considered abnormal?

Fluid and Electrolyte Balance (Section 25.5)

25.20 List four routes by which water leaves the body.

25.21 What are three sources of water that contribute to the fluid intake of a human body?

25.22 What is the primary way fluid balance is maintained in the body?

25.23 Describe how vasopressin and aldosterone function to maintain fluid levels in the blood.

25.24 Trace the events of the regulation of fluid balance by the thirst mechanism.

Acid–Base Balance (Section 25.6)

25.25 What is the normal pH range of human blood? Outside what pH range will death occur?

25.26 List three systems that cooperate to maintain blood pH in an appropriate narrow range. Which one of these provides the most rapid response to changes in blood pH?

25.27 What is meant by the terms *acidosis* and *alkalosis*?

Buffer Control of Blood pH (Section 25.7)

25.28 List three major buffer systems of the blood.

25.29 Hemoglobin (HHb) is a weak acid that forms the Hb^- anion. Write equations to show how the HHb/Hb^- buffer would react to added H^+ and added OH^-.

Respiratory Control of Blood pH (Section 25.8)

25.30 Describe the respiratory mechanism for controlling blood pH.

25.31 Hyperventilation is a symptom of what type of acid–base imbalance in the body?

25.32 Hypoventilation is a symptom of what type of acid–base imbalance in the body?

Urinary Control of Blood pH (Section 25.9)

25.33 Write a short summary description of the role of the kidneys and urine in the control of blood pH.

25.34 What happens to CO_2 in the blood when the kidneys function to control blood pH?

25.35 What buffer system in the urine usually keeps the urine pH from dropping lower than 6?

25.36 What ionic shift maintains electron charge balance when the kidneys function to increase the pH of blood?

Acidosis and Alkalosis (Section 25.10)

25.37 What is the normal ratio of bicarbonate to carbonic acid in the blood?

25.38 What is the difference between the causes of respiratory acid–base imbalances and metabolic acid–base imbalances in the body?

25.39 What type of acid–base imbalance in the blood is caused by hypoventilation? By hyperventilation?

25.40 What type of acid–base imbalance might develop in an individual who has taken an overdose of a depressant narcotic? In an individual with uncontrolled diabetes mellitus?

25.41 Explain how prolonged vomiting can lead to alkalosis.

Additional Exercises

25.42 An average adult excretes about 25.0 g of urea in his or her urine daily. Assuming both nitrogen atoms of the urea molecule come from the deamination of amino acids (one N atom/amino acid), how many amino acid molecules were deaminated to produce this amount of urea?

$$\underset{\text{urea}}{H_2N - \overset{\displaystyle O}{\overset{\|}{C}} - NH_2}$$

25.43 A person inhales O_2 and exhales CO_2. What is the ultimate use for the O_2 in the cell? Is the O_2 that we inhale incorporated into the CO_2 we exhale? What is the source of the CO_2 we exhale?

25.44 When the body is dehydrated, aldosterone causes the active readsorption of Na^+ ions from the renal tubules into the blood, which causes water to "follow" the Na^+ ions into the blood. Explain what causes this flow of water to occur.

25.45 The body uses the phosphate buffer system shown in the following reaction (a).

$$H^+ + HPO_4^{2-} \rightleftarrows H_2PO_4^- \text{ (a)}$$

$$H^+ + H_2PO_4^- \rightleftarrows H_3PO_4 \text{ (b)}$$

Explain why reaction (b) is not used in the human body as a buffer system. (*Hint*: What are the pKa values for the dissociations of the individual hydrogen ions?)

25.46 Explain how heavy exercise can cause metabolic acidosis.

Chemistry for Thought

25.47 Explain why a hysterical individual might develop alkalosis.

25.48 Explain why a person suffering from metabolic acidosis might develop a thirst.

25.49 Ordinary water may be the most effective "sports drink." Explain why this might be true.

25.50 How does the concentration of hemoglobin of someone living in the Himalayas compare with that of someone living at a lower altitude?

25.51 What happens if the amount of H+ ions in the urine exceeds urine's buffering capacity?

25.52 Aerobic exercise causes your cells to produce CO_2 and H_2O. Why can you dehydrate during aerobic exercise even if you don't perspire?

25.53 What types of food can make urine acidic? What types of food will make it basic?

25.54 Explain how uncontrolled diarrhea can lead to acidosis.

Allied Health Exam Connection

The following questions are from these sources:

- *Nursing School Entrance Exam* © 2005, Learning Express, LLC.

- *McGraw-Hill's Nursing School Entrance Exams* by Thomas A. Evangelist, Tamara B. Orr, and Judy Unrein © 2009, The McGraw-Hill Companies, Inc.

- *NSEE Nursing School Entrance Exams,* 3rd edition © 2009, Kaplan Publishing.

- *Cliffs Test Prep: Nursing School Entrance Exams* by Fred N. Grayson © 2004, Wiley Publishing, Inc.

- *Peterson's Master the Nursing School and Allied Health Entrance Exams,* 18th edition by Marion F. Gooding © 2008, Peterson's, a Nelnet Company.

25.55 Rank the following in order from least important to most important in removing CO_2 from the body: carbaminohemoglobin, dissolving it in plasma, bicarbonate.

25.56 Movement of respiratory gases into and out of blood is primarily dependent upon the presence of a _____ gradient.

25.57 The most rapid method of pH regulation involves the blood's _____ system.

25.58 Which of the following are normal constituents of urine?

a. creatinine **e.** sugar

b. urea **f.** hormones

c. water **g.** blood

d. ammonia

25.59 Uncontrolled diabetes mellitus results in _____ acidosis.

25.60 The liquid portion of the blood is called:

a. serum. **c.** plasma.

b. gamma globulin. **d.** lymph.

25.61 The cation in the highest concentration within extracellular fluid that plays a major role in water distribution is:

a. sodium. **c.** calcium.

b. potassium. **d.** magnesium.

25.62 Water reabsorption in the collecting duct of the kidneys is controlled by _____ from the posterior pituitary.

a. oxytocin **c.** epinephrine

b. ADH **d.** aldosterone

25.63 The hormone most responsible for the renal regulation of sodium is:

a. thyroxine. **c.** glucagon.

b. insulin. **d.** aldosterone.

25.64 The presence of protein in the urine is called:

a. polyuria. **c.** albuminuria.

b. anuria. **d.** hematuria.

25.65 Hyperventilation resulting from hysteria may cause:

a. respiratory acidosis. **c.** metabolic acidosis.

b. respiratory alkalosis. **d.** metabolic alkalosis.

The International System of Measurements

The International System of Units (SI units) was established in 1960 by the International Bureau of Weights and Measures. The system was established in an attempt to streamline the metric system, which included certain traditional units that had historical origins but that were not logically related to other metric units. The International System established fundamental units to represent seven basic physical quantities. These quantities and the fundamental units used to express them are given in Table A.1.

TABLE A.1 Fundamental SI Units

Physical Quantity	SI Unit	Abbreviation
Length	meter	m
Mass	kilogram	kg
Temperature	kelvin	K
Amount of substance	mole	mol
Electrical current	ampere	A
Time	second	s
Luminous intensity	candela	cd

All other SI units are derived from the seven fundamental units. Prefixes are used to indicate multiples or fractions of the fundamental units (Table 1.2). In some cases, it has been found convenient to give derived units specific names. For example, the derived unit for force is $kg\ m/s^2$, which has been given the name newton (abbreviated N) in honor of Sir Isaac Newton (1642–1727). Some examples of derived units are given in Table A.2.

TABLE A.2 Examples of Derived SI Units

Physical Quantity	Definition in Fundamental Units	Specific Name	Abbreviation
Volume	m^3 (see NOTE below)	—	—
Force	$kg\ m/s^2$	newton	N
Energy	$kg\ m^2/s^2 = N\ m$	joule	J
Power	$kg\ m^2/s^3 = J/s$	watt	W
Pressure	$kg/m\ s^2 = N/m^2$	pascal	Pa
Electrical charge	$A\ s$	coulomb	C
Electrical potential	$kg\ m^2/A\ s^3 = W/A$	volt	V
Electrical resistance	$kg\ m^2/A^2\ s^3 = V/A$	ohm	Ω
Frequency	$1/s$	hertz	Hz

NOTE: The liter (L), a popular volume unit of the metric system, has been redefined in terms of SI units as $1\ dm^3$ or $1/1000\ m^3$. It and its fractions (mL, µL, etc.) are still widely used to express volume.

APPENDIX B

Answers to Even-Numbered End-of-Chapter Exercises

Chapter 1

1.2 Mass is a measurement of the amount of matter in an object. Weight is a measurement of the gravitational force acting on an object.

1.4 **a.** The distance the ball could be thrown on the moon would be greater. Although the throwing force would be the same, the smaller gravitational force pulling the ball downward would allow the ball to travel farther before hitting the surface of the moon.

 b. The rolling distance would be the same in both locations because gravity does not influence rolling motion on a smooth surface.

1.6 The gravitational force is greater the closer an object is to the center of Earth. Thus, a person would weigh more at the poles (closer to the center) and less at the equator.

1.8 **a.** Physical: Both pieces of the stick are still made of the same substance.

 b. Chemical: The smoke is a new substance.

 c. Physical: Each smaller piece is still the substance salt.

 d. Chemical: The change in color indicates a change in the substances present.

1.10 **a.** Physical: The state of a substance (solid, liquid, or gas) can be changed without changing the identity of the substance.

 b. Chemical: The word *reacts* indicates a chemical change.

 c. Physical: Freezing is a physical change, so the temperature at which it occurs is a physical property.

 d. Chemical: Because rusting is a chemical change (rust is a different substance from the metal), the ability to rust or not to rust is a chemical property.

 e. Physical: Color is a physical property because it can be observed without trying to change the substance into another substance.

1.12 **a.** Yes: The change in melting point indicates a different substance has been formed. Also, the evolved gas is a different substance.

 b. No: It must be different because its melting point is different from that of succinic acid.

 c. The succinic acid molecules must be larger because the atoms of the succinic acid were divided between the molecules of the new solid and the molecules of the evolved gas.

 d. Heteroatomic: Because the succinic acid changed into two different substances, molecules of succinic acid must contain at least two different kinds of atoms.

1.14 Heteroatomic: Because only a single substance is formed, its molecules must contain at least one atom from each of the two reacting substances.

1.16 Heteroatomic: The molecules of the products hydrogen and oxygen contain hydrogen atoms and oxygen atoms, respectively. The two different kinds of atoms must have come from the water.

1.18 **a.** A is a compound because compounds are made up of heteroatomic molecules.

 b. D is an element because elements are made up of homoatomic molecules.

 c. E is a compound because it can be changed into the simpler substances G and J. The simpler substances G and J cannot be classified without conducting further tests on them to see if they can be changed into other, still simpler substances.

1.20 **a.** R cannot be classified as an element or a compound. Even though none of the tests changed it into a simpler substance, all possible tests were not done on it, and some test might exist that would change it.

 b. T can be classified as a compound because it was changed into simpler substances.

 c. The remaining solid cannot be classified because no tests were done to see if it could be changed into simpler substances.

1.22 **a.** Homogeneous

 b. Homogeneous

 c. Heterogeneous

 d. Heterogeneous

 e. Homogeneous

 f. Heterogeneous

 g. Homogeneous

1.24 **a.** Pure substance

 b. Solution

 e. Solution

 g. Solution

1.26 In a modern society, quantities must be expressed with precision. Goods, services, time, and so on must be clearly expressed as quantities in order for trade, purchases, salaries, and the like to be carried out or determined.

1.28 The stone would be useful for expressing relatively large quantities such as the weights of people, cows, or bags of grain. It is likely that originally "one stone" was equal to the weight of some field stone that had been accepted as a standard.

1.30 **a.** Metric

 b. Metric

 c. Nonmetric

d. Metric

e. Nonmetric

f. Metric

1.32 **c.** Meters

e. Liters

1.34 **a.** 1 million μL

b. 75 thousand watts

c. 15 million hertz

d. 200×10^{-12} meters

1.36 0.240 L, 240 cm^3

1.38 8.8 lb

1.40 **a.** About 3 in.

b. 65 K

c. About 2 kg

1.42 **a.** 0.001 kg, or 1 g

b. About 68 fl oz

c. 333 mg, or about 300 mg

1.44 97.0°F in the morning and 99.0°F at bedtime

1.46 **a.** Incorrect: No zero should precede the first number in the nonexponential number. 2.7×10^{-3} is correct.

b. Correct

c. Incorrect: The decimal is not in the standard position in the nonexponential number. 7.19×10^{-5} is correct.

d. Incorrect: No nonexponential number is given. 1×10^3 is correct.

e. Incorrect: The decimal is not in the standard position in the nonexponential number. 4.05×10^{-4} is correct.

f. Incorrect: No exponent is used, and the decimal is not in the standard position in the nonexponential number. 1.19×10^{-1} is correct.

1.48 **a.** 1.4×10^4

b. 3.65×10^2

c. 2.04×10^{-3}

d. 4.618×10^2

e. 1.00×10^{-3}

f. 9.11×10^2

1.50 1.86×10^5 mi/s; 1.100×10^9 km/h

1.52 0.000 000 000 000 000 000 000 105

1.54 **a.** 9.0×10^{-5}

b. 1.4×10^7

c. 7.6×10^{-2}

d. 1.9×10^1

e. 2.6×10^{12}

1.56 **a.** $(1.44 \times 10^2)(8.76 \times 10^{-2}) = 1.26 \times 10^1$

b. $(7.51 \times 10^2)(1.06 \times 10^2) = 7.96 \times 10^4$

c. $(4.22 \times 10^{-2})(1.19 \times 10^{-3}) = 5.02 \times 10^{-5}$

d. $(1.28000 \times 10^5)(3.16 \times 10^{-5}) = 4.04 \times 10^0$

1.58 **a.** 2.6×10^{-5}

b. 2.2×10^2

c. 6.4×10^{-4}

d. 7.25×10^{-4}

e. 3.1×10^{-2}

1.60 **a.** 1.7×10^{-1}

b. 1.0×10^{-9}

c. 2.6×10^6

d. 2.3×10^0

e. 1.5×10^7

1.62 **a.** 0.01 cm

b. 0.01 mm

c. 0.1°

d. 0.1 lb/in.2

1.64 **a.** 6.00 mL

b. 37.0°C

c. 9.00 s

d. 15.5°

1.66 **a.** Measured: 5.06 lb; exact: 16 potatoes; 0.316 lb/potato

b. Measured: the individual percentages; exact: the 5 players; 71.34%/player. Note that the sum of the five percentages has four significant figures, so the calculated average must have four.

1.68 **a.** 3

b. 3

c. 4

d. 2

e. 4

f. 5

1.70 **a.** 5.2

b. 0.104

c. 0.518

d. 1.0×10^2

e. 2.52×10^{-18}

1.72 **a.** 6.2

b. 230

c. 0.589

d. 0.58

e. 27.75

f. 21.64

1.74 **a.** 4.60×10^{-3}

b. 2.208

c. 2.65

d. −13

e. 3

f. 0.81

1.76 **a.** Black rectangle: Area $= 124.8$ cm^2; perimeter $= 44.80$ cm

Red rectangle: Area $= 48.9$ cm^2; perimeter $= 45.24$ cm

Green rectangle: Area $= 8.11$ cm^2; perimeter $= 11.46$ cm

Orange rectangle: Area $= 9.0$ cm^2; perimeter $= 27.80$ cm

b. Black rectangle: Area $= 0.01248$ m^2; perimeter $= 0.4480$ m

Red rectangle: Area $= 0.00489$ m^2; perimeter $= 0.4524$ m

Green rectangle: Area $= 0.000811$ m^2; perimeter $= 0.1146$ m

Orange rectangle: Area $= 0.00090$ m^2; perimeter $= 0.2780$ m

c. No

1.78 **a.** $\dfrac{0.015 \text{ grain}}{1 \text{ mg}}$

b. $\dfrac{0.0338 \text{ fl oz}}{1 \text{ mL}}$

c. $\dfrac{1 \text{ L}}{1.057 \text{ qt}}$

d. $\dfrac{1 \text{ m}}{1.094 \text{ yd}}$

1.80 The factor is $\dfrac{1 \text{ km}}{0.621 \text{ mi}}$. The race is about 42 km long.

1.82 1.06 cups or 1 cup

1.84 39.6 lb; your baggage is not overweight.

1.86 1.31 g/L

1.88 $\$25.73/\$467.80 \times 100 = 5.500\%$

1.90 71%

1.92 IgG $= 75.09\%$; IgA $= 16.23\%$; IgM $= 7.58\%$; IgD $= 1.10\%$; IgE $= 0.008\%$

1.94 **a.** 0.792 g/mL

b. 1.03 g/mL

c. 1.98 g/L

d. 8.90 g/cm^3

1.96 Volume $= 63.0$ cm^3; density $= 11.4$ g/cm^3

1.98 380. mL

1.100 **a.** 4.5×10^6 mm

b. 6.0×10^3 g

c. 9.86×10^{12} km

d. 1.91×10^2 mg

e. 5.0×10^{-6} mg

1.102 54 days

1.104 9.5×10^2 mg

1.116 b

1.118 b

1.120 d

1.122 b

1.124 d

1.126 c

1.128 a

1.130 b

1.132 c

1.134 c

1.136 b

Chapter 2

2.2 **a.**

b.

c.

d.

2.4 **a.** H_2O

b. H_2O_2

c. H_2SO_4

d. C_2H_6O

2.6 **a.** One sulfur (S), two oxygen (O)

b. Four carbon (C), ten hydrogen (H)

c. One hydrogen (H), one chlorine (Cl), two oxygen (O)

d. One boron (B), three fluorine (F)

2.8 **a.** The two hydrogen atoms should be denoted by a subscript: H_2S

b. The letter L of chlorine should be lowercase: $HClO_2$

c. The coefficient 2 should not be used to indicate the number of atoms in the molecule. Subscripts should be used instead: H_2N_4

d. The numbers in the formula should be written as subscripts: C_2H_6

2.10 **a.** Charge $= 4+$; mass $= 9$ u

b. Charge $= 9+$; mass $= 19$ u

c. Charge $= 20+$; mass $= 43$ u

d. Charge $= 47+$; mass $= 107$ u

2.12 **a.** 4

b. 9

c. 20

d. 47

2.14 **a.** 14 electrons, 14 protons

b. 50 electrons, 50 protons

c. 74 electrons, 74 protons

2.16 **a.** 16 protons, 18 neutrons, 16 electrons

b. 40 protons, 51 neutrons, 40 electrons

c. 54 protons, 77 neutrons, 54 electrons

2.18 a. $^{28}_{14}Si$

b. $^{40}_{18}Ar$

c. $^{88}_{38}Sr$

2.20 a. Atomic number = 4; mass number = 9; $^{9}_{4}Be$

b. Atomic number = 9; mass number = 19; $^{19}_{9}F$

c. Atomic number = 20; mass number = 43; $^{43}_{20}Ca$

d. Atomic number = 47; mass number = 107; $^{107}_{47}Ag$

2.22 a. $^{37}_{17}Cl$

b. $^{65}_{29}Cu$

c. $^{66}_{30}Zn$

2.24 Atomic weights rounded to nearest whole number give He = 4 u and C = 12 u. Three He total 12 u, which will balance one C, which also totals 12 u.

2.26 Ca, calcium

2.28 N, nitrogen

2.30 a. 46.01 u

b. 17.03 u

c. 180.16 u

d. 48.00 u

e. 62.07 u

2.32 Ethylene, C_2H_4

2.34 $y = 3$

2.36 a. 14 neutrons

b. 27.0 u

2.38 Calculated value = 10.81 u

Periodic table value = 10.81 u

2.40 Calculated value = 63.55 u

Periodic table value = 63.55 u

2.42 1.90 g F

2.44 a. 1 mol P atoms = 6.02×10^{23} P atoms

6.02×10^{23} P atoms = 30.97 g P

1 mol P atoms = 30.97 g P

b. 1 mol Al atoms = 6.02×10^{23} Al atoms

6.02×10^{23} Al atoms = 26.98 g Al

1 mol Al atoms = 26.98 g Al

c. 1 mol Kr atoms = 6.02×10^{23} Kr atoms

6.02×10^{23} Kr atoms = 83.80 g Kr

1 mol Kr atoms = 83.80 g Kr

2.46 a. 6.14×10^{-23} g

b. 44.5 g

c. 20.95 g

2.48 Molecular weight BF_3 = 67.8 u

Molecular weight H_2S = 34.1 u

0.68 g BF_3

2.50 a. 1. 2 $C_4H_{10}O$ molecules contain 8 C atoms, 20 H atoms, and 2 O atoms.

2. 10 $C_4H_{10}O$ molecules contain 40 C atoms, 100 H atoms, and 10 O atoms.

3. 100 $C_4H_{10}O$ molecules contain 400 C atoms, 1000 H atoms, and 100 O atoms.

4. 6.02×10^{23} $C_4H_{10}O$ molecules contain 24.08×10^{23} C atoms, 60.2×10^{23} H atoms, and 6.02×10^{23} O atoms.

5. 1 mol of $C_4H_{10}O$ molecules contains 4 mol of C atoms, 10 mol of H atoms, and 1 mol of O atoms.

6. 74.12 g $C_4H_{10}O$ contain 48.04 g of C, 10.08 g of H, and 16.00 g of O.

b. 1. 2 $C_2H_3O_2F$ molecules contain 4 C atoms, 6 H atoms, 4 O atoms, and 2 F atoms.

2. 10 $C_2H_3O_2F$ molecules contain 20 C atoms, 30 H atoms, 20 O atoms, and 10 F atoms.

3. 100 $C_2H_3O_2F$ molecules contain 200 C atoms, 300 H atoms, 200 O atoms, and 100 F atoms.

4. 6.02×10^{23} $C_2H_3O_2F$ molecules contain 12.04×10^{23} C atoms, 18.06×10^{23} H atoms, 12.04×10^{23} O atoms, and 6.02×10^{23} F atoms.

5. 1 mol of $C_2H_3O_2F$ molecules contains 2 mol of C atoms, 3 mol of H atoms, 2 mol of O atoms, and 1 mol of F atoms.

6. 78.04 g of $C_2H_3O_2F$ contain 24.02 g of C, 3.02 g of H, 32.00 g of O, and 19.00 g of F.

c. 1. 2 C_6H_7N molecules contain 12 C atoms, 14 H atoms, and 2 N atoms.

2. 10 C_6H_7N molecules contain 60 C atoms, 70 H atoms, and 10 N atoms.

3. 100 C_6H_7N molecules contain 600 C atoms, 700 H atoms, and 100 N atoms.

4. 6.02×10^{23} C_6H_7N molecules contain 36.12×10^{23} C atoms, 42.14×10^{23} H atoms, and 6.02×10^{23} N atoms.

5. 1 mol of C_6H_7N molecules contains 6 mol of C atoms, 7 mol of H atoms, and 1 mol of N atoms.

6. 93.13 g of C_6H_7N contain 72.06 g of C, 7.06 g of H, and 14.01 g of N.

2.52 a. 5.0 mol hydrogen atoms

b. 3.0×10^{23} carbon atoms

c. 14.1 g hydrogen

2.54 34.6 g C_2H_6O

2.56 In CH_4, mass % H = 25.3%.

In C_2H_6, mass % H = 20.1%.

2.58 Statements 4–6 are:

4. 6.02×10^{23} $C_6H_{12}O_6$ molecules contain 36.1×10^{23} C atoms, 72.2×10^{23} H atoms, and 36.1×10^{23} O atoms.

5. 1 mol of $C_6H_{12}O_6$ molecules contains 6 mol of C atoms, 12 mol of H atoms, and 6 mol of O atoms.

6. 180.16 g of $C_6H_{12}O_6$ contain 72.06 g of C, 12.10 g of H, and 96.00 g of O.

a. 23.18 g of O

b. 18.0 mol of H atoms

c. 4.50×10^{23} C atoms

2.60 Magnetite = 72.36% Fe; hematite = 69.94% Fe. Magnetite has the higher % Fe by mass.

2.62 U-238 would have a greater density because each U-238 nucleus would contain 3 more neutrons than a U-235 nucleus and thus have a greater mass. However, the size (volume) of each atom (measured from the nucleus to the outer electrons) would be the same, so one cm^3 of each isotope would contain the same number of atoms.

2.64 2.32×10^{-23} g

2.66 The density would increase tremendously because many more of the smaller atoms (measured from the nucleus to the outer electrons) would fit into a cm^3. Each of the smaller atoms would have the same mass as the larger atoms, so the density (mass/volume) would be much greater.

2.76 b

2.78 a

2.80 b

2.82 b

2.84 a

2.86 d

2.88 b

2.90 c

2.92 a

2.94 b

Chapter 3

3.2 **a.** Group VIIIB (9), period 4
 b. Group IVA (14), period 6
 c. Group VA (15), period 4
 d. Group IIA (2), period 6

3.4 **a.** Kr, krypton
 b. Sn, tin
 c. Mo, molybdenum
 d. Nd, neodymium

3.6 **a.** 4
 b. 10
 c. 18

3.8 **a.** Period
 b. Period
 c. Group
 d. Group

3.10 protons

3.12 **a.** 2
 b. 6
 c. 8

3.14 Four orbitals; one 2s, and three 2p.

3.16 Seven orbitals in the subshell can contain a maximum of 14 electrons.

3.18 **a.** 8
 b. 5
 c. 4
 d. 4

3.20 Cesium, Cs, has chemical properties most like sodium, Na. Both elements have one valence-shell electron.

3.22 An ore deposit that contained copper might also be expected to contain gold and silver because all three elements belong to group IB (11) of the periodic table and thus have similar chemical properties.

3.24 **a.** $1s^2 2s^2 2p^6 3s^2 3p^6 4s^2 3d^{10} 4p^6 5s^1$; one electron is unpaired.
 b. $1s^2 2s^2 2p^6 3s^2 3p^2$; two electrons are unpaired.
 c. $1s^2 2s^2 2p^6 3s^2 3p^6 4s^2 3d^2$; two electrons are unpaired.
 d. $1s^2 2s^2 2p^6 3s^2 3p^6$; no electrons are unpaired.

3.26 **a.** $1s^2 2s^2 2p^6 3s^2$; six s electrons.
 b. $1s^2 2s^2 2p^3$; three unpaired electrons.
 c. $1s^2 2s^2 2p^6 3s^2 3p^1$; four subshells are filled.

3.28 **a.** C, carbon
 b. Na, sodium
 c. Si, silicon, and S, sulfur
 d. Nb, niobium
 e. V, vanadium, and Co, cobalt

3.30 **a.** $[Ar]4s^2 3d^{10} 4p^3$
 b. $[Ar]4s^2 3d^5$
 c. $[Ne]3s^2 3p^2$
 d. $[Kr]5s^2 4d^{10} 5p^5$

3.32 Na: $[Ne]3s^1$
Mg: $[Ne]3s^2$
Al: $[Ne]3s^2 3p^1$
Si: $[Ne]3s^2 3p^2$
P: $[Ne]3s^2 3p^3$
S: $[Ne]3s^2 3p^4$
Cl: $[Ne]3s^2 3p^5$
Ar: $[Ne]3s^2 3p^6$

3.34 **a.** p area
 b. d area
 c. f area
 d. s area

3.36 **a.** Transition
 b. Representative
 c. Inner-transition
 d. Noble gas
 e. Representative

3.38 **a.** Nonmetal
 b. Metal
 c. Metalloid

d. Metalloid

e. Metal

3.40 a. K

b. Bi

c. Sr

d. Sn

3.42 a. Ga

b. Sb

c. C

d. Te

3.44 a. K

b. Sn

c. Mg

d. Li

3.46 The chemical properties would be the same because chemical properties depend on the number of valence electrons in the atom. The number of valence electrons in isotopes of the same element is identical.

3.48 8.0 mg

3.50 The element has an atomic weight of 40.0, and so is calcium (Ca). It is a representative element and will conduct electricity because it is a metal.

3.58 d

3.60 d

3.62 a

3.64 b

3.66 b

3.68 b

3.70 d

3.72 d

3.74 a

3.76 b

Chapter 4

4.2 a. $:\overset{..}{\underset{..}{I}}\cdot$

b. $\overset{.}{Sr}\cdot$

c. $\cdot\overset{.}{Sn}\cdot$

d. $:\overset{.}{\underset{..}{S}}\cdot$

4.4 a. $[Kr]\,5s^2 4d^{10} 5p^2$

b. $[Xe]\,6s^1$

c. $[Kr]\,5s^2 4d^{10} 5p^1$

d. $[Kr]\,5s^2$

4.6 a. $\cdot\overset{.}{Sn}\cdot$

b. $Cs\cdot$

c. $\overset{.}{In}\cdot$

d. $\overset{.}{Sr}\cdot$

4.8 a. $\overset{.}{E}\cdot$

b. $\cdot\overset{.}{\underset{..}{E}}:$

4.10 a. Add 4 or lose 14

b. Add 31 or lose 1

c. Add 5 or lose 13

d. Add 16 or lose 2

4.12 a. Lose 1: $Cs \rightarrow Cs^+ + e^-$

b. Add 2: $O + 2e^- \rightarrow O^{2-}$

c. Add 3: $N + 3e^- \rightarrow N^{3-}$

d. Add 1: $I + e^- \rightarrow I^-$

4.14 a. Se^{2-}

b. Rb^{1+}

c. Al^{3+}

4.16 a. Sulfur, S

b. Aluminum, Al

c. Sodium, Na

d. Chlorine, Cl

4.18 a. Helium, He

b. Xenon, Xe

c. Argon, Ar

d. Krypton, Kr

4.20 a. $Ca \rightarrow Ca^{2+} + 2e^-$; $Cl + e^- \rightarrow Cl^-$ Formula is $CaCl_2$.

b. $Li \rightarrow Li^+ + e^-$; $Br + e^- \rightarrow Br^-$ Formula is LiBr.

c. $Mg \rightarrow Mg^{2+} + 2e^-$; $S + 2e^- \rightarrow S^{2-}$ Formula is MgS.

4.22 a. BaSe

b. Ba_3P_2

c. BaI_2

d. Ba_3As_2

4.24 a. Binary

b. Binary

c. Not binary

d. Binary

e. Not binary

4.26 a. Lithium ion

b. Magnesium ion

c. Barium ion

d. Cesium ion

4.28 a. Bromide ion

b. Oxide ion

c. Phosphide ion

d. Fluoride ion

4.30 **a.** Strontium sulfide

b. Calcium fluoride

c. Barium chloride

d. Lithium oxide

e. Magnesium oxide

4.32 **a.** Lead (II) oxide and lead (IV) oxide

b. Copper (I) chloride and copper (II) chloride

c. Gold (I) sulfide and gold (III) sulfide

d. Cobalt (II) oxide and cobalt (III) oxide

4.34 **a.** Plumbous oxide and plumbic oxide

b. Cuprous chloride and cupric chloride

c. Aurous sulfide and auric sulfide

d. Cobaltous oxide and cobaltic oxide

4.36 **a.** Hg_2O

b. PbO

c. PtI_4

d. Cu_3N

e. CoS

4.38 **a.** 102.9 u

b. 78.1 u

c. 159.2 u

d. 34.8 u

4.40 **a.** Na^+ and Br^-

b. Ca^{2+} and F^-

c. Cu^+ and S^{2-}

d. Li^+ and N^{3-}

4.42 **a.** 22.99 g of Na^+ and 79.90 g of Br^-

b. 40.08 g of Ca^{2+} and 38.00 g of F^-

c. 127.10 g of Cu^+ and 32.06 g of S^{2-}

d. 20.82 g of Li^+ and 14.01 g of N^{3-}

4.44 **a.** 6.02×10^{23} Na^+ ions and 6.02×10^{23} Br^- ions

b. 6.02×10^{23} Ca^{2+} ions and 12.04×10^{23} F^- ions

c. 12.04×10^{23} Cu^+ ions and 6.02×10^{23} S^{2-} ions

d. 18.06×10^{23} Li^+ ions and 6.02×10^{23} N^{3-} ions

4.46

4.48

a. $H:\ddot{C}:H$ or $H-C-H$

b. $:\ddot{O}::C::\ddot{O}:$ or $O=C=O$

c. $H:\ddot{Se}:H$ or $H-Se-H$

d. $H:\ddot{N}:H$ or $H-N-H$

4.50 **a.**

b.

c.

4.52 **a.** $H:\ddot{S}:H$ Bent

b. $:\ddot{Cl}:\ddot{P}:\ddot{Cl}:$ Triangular pyramid with P at the top

c. $:\ddot{F}:\ddot{O}:\ddot{F}:$ Bent

d. Tetrahedral with Sn in center

4.54 **a.** $[:\ddot{O}:\ddot{N}::\ddot{O}:]^-$ Bent

b. $[:\ddot{O}:\ddot{Cl}:\ddot{O}:]^-$ Triangular pyramid with Cl at the top

c. $[:\ddot{O}:C::\ddot{O}:]^{2-}$ Flat triangle with C in the center

d. $[H:\ddot{O}:H]^+$ Triangular pyramid with O at the top

4.56 **a.**

b.

c. Nonpolarized

4.58 **a.** Polar covalent

b. Polar covalent

c. Nonpolar covalent

d. Polar covalent

e. Ionic

4.60 **a.** Polar

b. Polar

c. Nonpolar

4.62 **a.** Polar covalent

b. Polar covalent

c. Nonpolar covalent

4.64 **a.** Polar

b. Polar

c. Nonpolar

$$I^{\delta-}$$
$$Al^{\delta 3+}$$
$$_{\delta-}I \quad I^{\delta-}$$

4.66 a. Phosphorus trichloride

 b. Dinitrogen pentoxide

 c. Carbon tetrachloride

 d. Boron trifluoride

 e. Carbon disulfide

4.68 a. ClO_2

 b. N_2O

 c. SO_2

 d. CCl_4

4.70 a. $Ca(ClO)_2$, calcium hypochlorite

 b. $CsNO_2$, cesium nitrite

 c. $MgSO_3$, magnesium sulfite

 d. $K_2Cr_2O_7$, potassium dichromate

4.72 a. $Ba(OH)_2$

 b. $MgSO_3$

 c. $CaCO_3$

 d. $(NH_4)_2SO_4$

 e. $LiHCO_3$

4.74 a. M_2SO_3

 b. $MC_2H_3O_2$

 c. MCr_2O_7

 d. MPO_4

 e. $M(NO_3)_3$

4.76 The stronger the attractive forces between molecules, the higher the melting and boiling points. Thus, we can conclude that the forces attracting dimethyl ether molecules to one another are weaker than the forces attracting ethyl alcohol molecules to one another. Because both compounds are covalently bonded, it is likely that the attractive forces are polar forces, and we might conclude that ethyl alcohol is more polar than dimethyl ether.

4.78 The predominant forces that exist between molecules of noble gases are dispersion forces because noble gas molecules are nonpolar. Dispersion forces increase with increasing mass of the particles involved, so the boiling points of the noble gases would increase in the order of He, Ne, Ar, Kr, Xe, and Rn.

4.80 The relatively high melting point indicates that it is unlikely that weak dispersion forces are the predominant forces present. Dipolar forces are also quite weak and therefore are also unlikely to be the forces present.

4.82 NH_3 H—N—H····N—H

H (top), H (bottom), H (bottom)

4.84 8.06 g

4.86 N_2O, covalent, dinitrogen monoxide

4.96 d

4.98 c

4.100 d

4.102 b

4.104 c

4.106 c

4.108 c

4.110 a

4.112 d

4.114 a

4.116 b

Chapter 5

5.2

	Reactants	Products
a.	H_2, Cl_2	HCl
b.	$KClO_3$	KCl, O_2
c.	magnesium oxide, carbon	magnesium, carbon monoxide
d.	ethane, oxygen	carbon dioxide, water

5.4 a. Not consistent: The number of O atoms is not the same on both sides.

 b. Consistent

 c. Not consistent: The masses of reactants and products are not equal.

 d. Consistent

5.6

	Left side	Right side	Classification
a.	1Ag, 1Cu, 2N, 6O	1Ag, 1Cu, 1N, 3O	Unbalanced
b.	4N, 8O	4N, 8O	Balanced
c.	1Mg, 2O	2Mg, 2O	Unbalanced
d.	4H, 1S, 6O, 1Ca	4H, 1S, 6O, 1Ca	Balanced

5.8 a. $2KClO_3(s) \rightarrow 2KCl(s) + 3O_2(g)$

 b. $2C_2H_6(g) + 7O_2(g) \rightarrow 4CO_2(g) + 6H_2O(\ell)$

 c. $2N_2(g) + 5O_2(g) \rightarrow 2N_2O_5(g)$

 d. $MgCl_2(s) + H_2O(g) \rightarrow MgO(s) + 2HCl(g)$

 e. $CaH_2(s) + 2H_2O(\ell) \rightarrow Ca(OH)_2(s) + 2H_2(g)$

 f. $2Al(s) + Fe_2O_3(s) \rightarrow Al_2O_3(s) + 2Fe(s)$

 g. $2Al(s) + 3Br_2(g) \rightarrow 2AlBr_3(s)$

 h. $Hg_2(NO_3)_2(aq) + 2NaCl(aq) \rightarrow$
 $Hg_2Cl_2(s) + 2NaNO_3(aq)$

5.10 a. +1

 b. +3

 c. 0

 d. +1

 e. +3

 f. +5

5.12 a. Cr $(+6)$

b. S $(+2)$

c. N $(+5)$

d. P $(+5)$

e. Cl $(+7)$

f. Cl $(+3)$

5.14 a. Oxidized

b. Neither

c. Reduced

d. Oxidized

e. Oxidized

5.16 a. $2Cu(s) + O_2(g) \rightarrow 2CuO(s)$
$\ 0 0 +2\ -2$
Oxidizing agent: O_2

Reducing agent: Cu

b. $Cl_2(g) + 2KI(aq) \rightarrow 2KCl(aq) + I_2(aq)$
$\ 0 +1\ -1 +1\ -1 0$
Oxidizing agent: Cl_2

Reducing agent: KI (I)

c. $3MnO_2(s) + 4Al(s) \rightarrow 2Al_2O_3(s) + 3Mn(s)$
$+4\ -2 0 +3\ -2 0$
Oxidizing agent: MnO_2 (Mn)

Reducing agent: Al

d. $2H^+(aq) + 3SO_3^{2-}(aq) + 2NO_3^-(aq) \rightarrow$
$+1 +4\ -2 +5\ -2$
$2NO(g) + H_2O(\ell) + 3SO_4^{2-}(aq)$
$+2\ -2 +1\ -2 +6\ -2$
Oxidizing agent: NO_3^- (N)

Reducing agent: SO_3^{2-} (S)

e. $Mg(s) + 2HCl(aq) \rightarrow MgCl_2(aq) + H_2(g)$
$0 +1\ -1 +2\ -1 0$
Oxidizing agent: HCl (H^+)

Reducing agent: Mg

f. $4NO_2(g) + O_2(g) \rightarrow 2N_2O_5(g)$
$+4\ -2 0 +5\ -2$
Oxidizing agent: O_2

Reducing agent: NO_2 (N)

5.18 Oxidizing agent: NaOH (H)

Reducing agent: Al

5.20 a. Nonredox, decomposition

b. Redox, single replacement

c. Nonredox, double replacement

d. Nonredox, combination

e. Redox, combination

f. Redox, combination

5.22 Nonredox, decomposition

5.24 Redox

5.26 Redox

5.28 a. Li^+, NO_3^-

b. Na^+, HPO_4^{2-}

c. Ca^{2+}, ClO_3^-

d. K^+, OH^-

e. Mg^{2+}, Br^-

f. NH_4^+, SO_4^{2-}

5.30 a. Total ionic: $SO_2(aq) + H_2O(\ell) \rightarrow 2H^+(aq) + SO_3^{2-}(aq)$

Spectator ions: none

Net ionic: $SO_2(aq) + H_2O(\ell) \rightarrow 2H^+(aq) + SO_3^{2-}(aq)$

b. Total ionic: $Cu^{2+}(aq) + SO_4^{2-}(aq) + Zn(s) \rightarrow$
$\phantom{Total ionic: Cu^{2+}(aq)}Cu(s) + Zn^{2+}(aq) + SO_4^{2-}(aq)$

Spectator ions: SO_4^{2-}

Net ionic: $Cu^{2+}(aq) + Zn(s) \rightarrow Cu(s) + Zn^{2+}(aq)$

c. Total ionic:

$2K^+(aq) + 2Br^-(aq) + 4H^+(aq) + 2SO_4^{2-}(aq) \rightarrow$

$Br_2(aq) + SO_2(aq) + 2K^+(aq) + SO_4^{2-}(aq) + 2H_2O(\ell)$

Spectator ions: K^+, one SO_4^{2-}

Net ionic: $2Br^-(aq) + 4H^+(aq) + SO_4^{2-}(aq) \rightarrow$
$Br_2(aq) + SO_2(aq) + 2H_2O(\ell)$

d. Total ionic:
$Ag^+(aq) + NO_3^-(aq) + Na^+(aq) + OH^-(aq) \rightarrow$
$AgOH(s) + Na^+(aq) + NO_3^-(aq)$

Spectator ions: Na^+, NO_3^-

Net ionic: $Ag^+(aq) + OH^-(aq) \rightarrow AgOH(s)$

e. Total ionic: $BaCO_3(s) + 2H^+(aq) + 2NO_3^-(aq) \rightarrow$
$Ba^{2+}(aq) + 2NO_3^-(aq) + CO_2(g) + H_2O(\ell)$

Spectator ions: NO_3^-

Net ionic:

$BaCO_3(s) + 2H^+(aq) \rightarrow Ba^{2+}(aq) + CO_2(g) + H_2O(\ell)$

f. Total ionic: $N_2O_5(aq) + H_2O(\ell) \rightarrow 2H^+(aq) + 2NO_3^-(aq)$

Spectator ions: none

Net ionic: $N_2O_5(aq) + H_2O(\ell) \rightarrow 2H^+(aq) + 2NO_3^-(aq)$

5.32 a. $H^+(aq) + NO_3^-(aq) + K^+(aq) + OH^-(aq) \rightarrow$
$K^+(aq) + NO_3^-(aq) + H_2O(\ell)$

Spectator ions: K^+, NO_3^-

Net ionic: $H^+(aq) + OH^-(aq) \rightarrow H_2O(\ell)$

b. $3H^+(aq) + PO_4^{3-}(aq) + 3NH_4^+(aq) + 3OH^-(aq) \rightarrow$
$\phantom{3H^+(aq)+PO_4^{3-}(aq)+}3NH_4^+(aq) + PO_4^{3-}(aq) + 3H_2O(\ell)$

Spectator ions: NH_4^+, PO_4^{3-}

Net ionic: $3H^+(aq) + 3OH^-(aq) \rightarrow 3H_2O(\ell)$

c. $H^+(aq) + I^-(aq) + Na^+(aq) + OH^-(aq) \rightarrow$
$Na^+(aq) + I^-(aq) + H_2O(\ell)$

Spectator ions: Na^+, I^-

Net ionic: $H^+(aq) + OH^-(aq) \rightarrow H_2O(\ell)$

In every net ionic reaction, H^+ reacts with OH^- to form H_2O

5.34 Exothermic: Heat is a product of the reaction.

5.36 The insulation around the ice slows its rate of melting by preventing it from absorbing heat. However, this prevents the ice from performing its cooling function in the cooler. For the

cooler to work properly, heat must move from the warmer food to the cooler ice, where it is absorbed as the ice melts.

5.38 **a.** $S(s) + O_2(g) \rightarrow SO_2(g)$

 2. $1 \text{ mol } S + 1 \text{ mol } O_2 \rightarrow 1 \text{mol } SO_2$

 3. 6.02×10^{23} S atoms $+ 6.02 \times 10^{23}$ O_2 molecules \rightarrow
 6.02×10^{23} SO_2 molecules

 4. $32.06 \text{ g } S + 32.00 \text{ g } O_2 \rightarrow 64.06 \text{ g } SO_2$

 b. $Sr(s) + 2H_2O(\ell) \rightarrow Sr(OH)_2(s) + H_2(g)$

 2. $1 \text{ mol } Sr + 2 \text{ mol } H_2O \rightarrow 1 \text{ mol } Sr(OH)_2 + 1 \text{ mol } H_2$

 3. 6.02×10^{23} Sr atoms $+ 12.04 \times 10^{23}$ H_2O
 molecules $\rightarrow 6.02 \times 10^{23}$ $Sr(OH)_2$ molecules
 $+ 6.02 \times 10^{23}$ H_2 molecules

 4. $87.62 \text{ g } Sr + 36.03 \text{ g } H_2O \rightarrow 121.64 \text{ g } Sr(OH)_2 +$
 $2.02 \text{ g } H_2$

 c. $2H_2S(g) + 3O_2(g) \rightarrow 2H_2O(g) + 2SO_2(g)$

 2. $2 \text{ mol } H_2S + 3 \text{ mol } O_2 \rightarrow 2 \text{ mol } H_2O + 2 \text{ mol } SO_2$

 3. 12.04×10^{23} H_2S molecules $+ 18.06 \times 10^{23}$
 O_2 molecules $\rightarrow 12.04 \times 10^{23}$ H_2O molecules $+$
 12.04×10^{23} SO_2 molecules

 4. $68.15 \text{ g } H_2S + 96.00 \text{ g } O_2 \rightarrow$
 $36.03 \text{ g } H_2O + 128.12 \text{ g } SO_2$

 d. $4NH_3(g) + 5O_2(g) \rightarrow 4NO(g) + 6H_2O(g)$

 2. $4 \text{ mol } NH_3 + 5 \text{ mol } O_2 \rightarrow 4 \text{ mol } NO + 6 \text{ mol } H_2O$

 3. 24.08×10^{23} NH_3 molecules $+ 30.10 \times 10^{23}$ O_2
 molecules $\rightarrow 24.08 \times 10^{23}$ NO molecules $+$
 36.12×10^{23} H_2O molecules

 4. $68.14 \text{ g } NH_3 + 160.00 \text{ g } O_2 \rightarrow$
 $120.04 \text{ g } NO + 108.10 \text{ g } H_2O$

 e. $CaO(s) + 3C(s) \rightarrow CaC_2(s) + CO(g)$

 2. $1 \text{ mol } CaO + 3 \text{ mol } C \rightarrow 1 \text{ mol } CaC_2 + 1 \text{ mol } CO$

 3. 6.02×10^{23} CaO molecules $+ 18.06 \times 10^{23}$ C
 atoms $\rightarrow 6.02 \times 10^{23}$ CaC_2 molecules $+$
 6.02×10^{23} CO molecules

 4. $56.08 \text{ g } CaO + 36.03 \text{ g } C \rightarrow$
 $64.10 \text{ g } CaC_2 + 28.01 \text{ g } CO$

5.40 $2SO_2(g) + O_2(g) \rightarrow 2SO_3(g)$

 2. $2 \text{ mol } SO_2 + 1 \text{ mol } O_2 \rightarrow 2 \text{ mol } SO_3$

 3. 12.0×10^{23} SO_2 molecules $+ 6.02 \times 10^{23}$ O_2
 molecules $\rightarrow 12.0 \times 10^{23}$ SO_3 molecules

 4. $128 \text{ g } SO_2 + 32.0 \text{ g } O_2 \rightarrow 160 \text{ g } SO_3$

Factors:

$$\frac{12.0 \times 10^{23} \text{ SO}_2 \text{ molecules}}{6.02 \times 10^{23} \text{ O}_2 \text{ molecules}}; \quad \frac{12.0 \times 10^{23} \text{ SO}_3 \text{ molecules}}{12.0 \times 10^{23} \text{ SO}_2 \text{ molecules}};$$

$$\frac{32.0 \text{ g O}_2}{128 \text{ g SO}_2}; \quad \frac{128 \text{ g SO}_2}{160 \text{ g SO}_3}; \quad \frac{32.0 \text{ g O}_2}{160 \text{ g SO}_3}; \quad \frac{128 \text{ g SO}_2}{32.0 \text{ g O}_2};$$

$$\frac{2 \text{ mol SO}_2}{1 \text{ mol O}_2}; \quad \frac{2 \text{ mol SO}_3}{2 \text{ mol SO}_2}; \quad \frac{1 \text{ mol O}_2}{2 \text{ mol SO}_3}; \quad \frac{1 \text{ mol O}_2}{2 \text{ mol SO}_2}; \quad \frac{2 \text{ mol SO}_3}{32.0 \text{ g O}_2};$$

$$\frac{128 \text{ g SO}_2}{2 \text{ mol SO}_2}; \quad \frac{2 \text{ mol SO}_2}{32.0 \text{ g O}_2}; \quad \frac{2 \text{ mol SO}_3}{1 \text{ mol O}_2}; \quad \frac{160 \text{ g SO}_3}{32.0 \text{ g O}_2}; \quad \frac{32.0 \text{ g O}_2}{1 \text{ mol O}_2}$$

This list does not include all possible factors.

5.42 892 g

5.44 445 g

5.46 495 g

5.48 1.02×10^3 g

5.50 256 g

5.52 **a.** O_2 is the limiting reactant.

 b. 71.9 g NO_2

5.54 144 g

5.56 72.96%

5.58 72.5%

5.60 89.0%

5.62 $BaCl_2(aq) + Na_2SO_4(aq) \rightarrow 2NaCl(aq) + BaSO_4(s)$

 Net ionic: $Ba^{2+}(aq) + SO_4^{2-}(aq) \rightarrow BaSO_4(s)$

5.64 7.502 g

5.74 d

5.76 b

5.78 a

5.80 c

5.82 b

5.84 a

5.86 d

5.88 c

5.90 c

5.92 b

5.94 a

5.96 b

Chapter 6

6.2 **a.** 121 mL

 b. 158 mL

 c. 67.9 mL

6.4 2 mL

6.6 **a.** The density will decrease because density is equal to mass/volume, and the volume of carbon dioxide will increase on heating while the mass remains unchanged.

 b. 1.67 g/L

6.8 As the ball travels upward and slows, more and more of its kinetic energy is converted into potential energy. At the maximum height, when the ball stops before starting to fall, all the energy is in the form of potential energy. As the ball falls faster and faster, more and more of the potential energy is converted back into kinetic energy. If there were no energy losses from air resistance, the ball would gain back as much kinetic energy as it had when it was initially thrown.

6.10 Each type of molecule has the same kinetic energy of 3.18×10^{10} u cm²/s².

6.12 **a.** Particles (molecules) in a liquid move freely and therefore assume the container shape, but the cohesive forces

between the particles prevent them from separating completely from each other and filling the container.

b. There is very little space between the particles (molecules) of liquids and solids. As a result, increased pressure cannot push the particles closer together and will have little influence on their volumes.

c. Gas molecules move around essentially unrestricted, and on average, equal numbers collide with each of the walls of a container in a given amount of time. The collisions cause the pressure on the walls, so the pressure on all the walls is the same.

6.14 Heat must be added to a substance to convert it from the solid to the liquid state, and then from the liquid to the gaseous state. If the added heat causes no temperature difference between the two states, as when a solid melts to a liquid at the melting point, then the added heat increases the potential energy of the particles of the substance. Thus, at the melting point, the particles of both solid and liquid forms of a substance will have the same kinetic energy because they are at the same temperature. However, particles of the liquid will have more potential energy, corresponding to the energy that had to be added to change the solid to a liquid. Similar arguments apply for a comparison of the liquid and gaseous states of a substance.

6.16 **a.** Gaseous

b. Solid

c. Gaseous

d. Gaseous

6.18 **a.** 0.957 atm

b. 727 torr

c. 14.1 psi

d. 0.966 bar

6.20 **a.** 14.3 atm

b. 14.4 bar

c. 1.09×10^4 mmHg

d. 427 in. Hg

6.22 **a.** 337 K

b. $-259°C$

c. 4 K

6.24 Column A: $P_f = 1.50$ atm; Column B: $V_f = 4.49$ L; Column C: $T_f = 221$ K

6.26 138 mL

6.28 3.7 atm

6.30 31.6 L

6.32 684 mL

6.34 4.6 L

6.36 2.5 L

6.38 100. atm

6.40 1.4×10^4 ft^3

6.42 884 mL

6.44 5.00 g/L

6.46 **a.** 0.0130 mol

b. 7.65 atm

c. 8.35 L

6.48 3.19 atm

6.50 26.3 g

6.52 2.7 mol

6.54 46.0 g/mol

6.56 44.1 g/mol; the gas is most likely CO_2

6.58 370 torr

6.60 O_2 mass/H_2 mass = 16 according to Graham's law. From the periodic table, O_2 mass/H_2 mass = 32 u/2 u = 16. The agreement is very good.

6.62 The helium-filled balloon would appear to go flat first because helium molecules diffuse through the rubber balloon faster than do nitrogen molecules. The rates of diffusion according to Graham's law would be He rate/N_2 rate = $\sqrt{28.0 \text{ u}/4.00 \text{ u}}$ = 2.6. Thus, He would be expected to escape through the balloon 2.6 times as fast as nitrogen.

6.64 **a.** Exothermic

b. Exothermic

c. Endothermic

6.66 A change in state refers to a process in which a substance changes from one of the three states of matter—solid, liquid, or gas—to another of the three states.

6.68 Liquid methylene chloride evaporates very rapidly, and when it evaporates, it absorbs heat. Thus, the liquid acted to cool and temporarily anesthetize any tissue on which it was sprayed.

6.70 The boiling points of the two liquids are different. Thus, a measurement of the temperature of each boiling liquid would differentiate between them. The ethylene glycol would have the higher boiling point of the two.

6.72 The temperature of boiling water is influenced by the atmospheric pressure on the water. On top of Mount Everest the atmospheric pressure is quite low, so water boils at a temperature significantly lower than the normal boiling point of 100°C. The temperature of the burning campfire would not be influenced by the altitude. Thus, the potato would cook faster if it were put into the campfire.

6.74 Since sand does not sublime, heat the mixture and lead the hot iodine vapor out of the hot chamber and to a cool surface. The pure iodine vapor will condense to a solid on the cool surface, and can be scraped off and collected.

6.76 **a.** 3.6×10^2 cal

b. 7×10^3 cal

c. 2.4×10^3 cal

6.78 **a.** 4.07×10^7 cal

b. 7.07×10^7 cal

c. 5.71×10^7 cal

6.80 7.72×10^4 cal

6.82 1.78×10^{-3} g/mL

6.84 1.16 L The reaction is a redox combination reaction.

6.86 The air of the atmosphere contains about 21 volume percent oxygen at any altitude. However, the pressure of the atmosphere at high altitudes is less than the pressure at lower altitudes. If the air at a higher altitude is at the same temperature as at a lower altitude, and since $P = nRT/V$, the same volume (V) of a lung full of air at a higher altitude and lower pressure (P) contains fewer moles of gas (n), and so fewer moles of oxygen. So, from the point of view of the number of molecules of oxygen, the higher altitude air has less oxygen, but from a percentage point of view it still contains 21% oxygen.

6.96 a

6.98 c

6.100 d

6.102 a

6.104 c

6.106 b

6.108 d

6.110 d

6.112 a

6.114 a

6.116 b

6.118 c

6.120 d

6.122 d

6.124 b

Chapter 7

7.2 **a.** Solvent: water; solute: sodium hypochlorite

 b. Solvent: isopropyl alcohol; solute: water

 c. Solvent: water; solute: hydrogen peroxide

 d. Solvent: SD alcohol; solutes: water, glycerin, fragrance, menthol, benzophenone, coloring

7.4 **a.** Not a solution because it is not homogeneous. It is cloudy rather than clear.

 b. Solution; solvent: water; solutes: all dissolved components.

 c. Not a solution because it is not homogeneous. Bits of pulp can be seen.

 d. Solution if prepared carefully with a tea bag or good strainer; solvent: water; solutes: all dissolved components.

 e. Not a solution because it is not homogeneous.

7.6 **a.** The oil and vinegar are immiscible.

 b. The alcohol is soluble in the water.

 c. The tar is soluble in the chloroform.

7.8 **a.** Unsaturated

 b. Supersaturated

 c. Saturated

7.10 The solution could be carefully cooled to a temperature lower than room temperature. Another method would be to allow some of the solvent to evaporate.

7.12 **a.** Soluble

 b. Insoluble

 c. Slightly soluble

 d. Very soluble

 e. Slightly soluble

7.14 Add pure water to the mixture. The $BaCl_2$ is soluble and will dissolve in the water. The liquid that contains the dissolved $BaCl_2$ can then be poured off carefully, leaving behind the insoluble solid $BaSO_4$.

7.16 **a.** Soluble in benzene

 b. Soluble in benzene

 c. Soluble in water

 d. Soluble in benzene

7.18 The structure indicates that Freon-114 is a nonpolar material. Grease is also nonpolar, as indicated by its insolubility in water. Thus, grease would dissolve readily in Freon-114.

7.20 **a.** 0.364 M

 b. 0.860 M

 c. 1.75 M

7.22 **a.** 0.500 M

 b. 0.200 M

 c. 0.26 M

7.24 **a.** 0.376 mol

 b. 0.0750 mol

 c. 571 mL

7.26 **a.** 0.820 g

 b. 0.130 L

 c. 0.516 L

7.28 **a.** 5.0%

 b. 5.0%

 c. 5.0%

7.30 **a.** 7.41%

 b. 15.3%

 c. 54.5%

 d. 44.1%

7.32 **a.** 11.1%

 b. 10.7%

 c. 7.48%

7.34 **a.** 7.5%

 b. 7.5%

 c. 3.0%

 d. 43.0%

7.36 0.025 L or 25 mL

7.38 **a.** 5.00%

 b. 5.00%

 c. 8.77%

7.40 About 40%

7.42 **a.** Put 0.0300 mol (4.26 g) of solid Na_2SO_4 into a 200 mL volumetric flask and add enough pure water to fill the flask to the mark after the solid dissolves.

b. Put 0.0625 mol (11.8 g) of solid $Zn(NO_3)_2$ into a 250 mL volumetric flask and add enough pure water to fill the flask to the mark after the solid dissolves.

c. Mix together 3.38 g of solid NaCl and 147 g (147 mL) of pure water. Allow the NaCl to dissolve.

d. Put 0.94 g of solid KCl into a 125 mL volumetric flask and add enough pure water to fill the flask to the mark after the solid dissolves.

7.44 %(w/w) = 25.0; %(w/v) = 23.8

7.46 **a.** 32.3 g

b. 0.700 mol

c. 31.3 mL

d. 2.10 g

7.48 **a.** Add 1.67 L of 18.0 M H_2SO_4 solution to enough water to give 5.00 L of solution. The 18.0 M acid should be added to water, and not the water to the acid. One way to make the desired solution is to add the 1.67 L of 18.0 M acid to about 3 L of pure water. Stir the resulting solution, then add enough additional water to bring the total volume to 5.00 L.

b. Put 41.7 mL of 3.00 M $CaCl_2$ solution into a 250 mL volumetric flask and add enough pure water to fill the flask to the mark when the contents are well mixed.

c. Put 30.0 mL of a 10.0% solution into a calibrated volumetric container (such as a volumetric flask or large volumetric cylinder) and add enough pure water to bring the total volume of well-mixed solution to 200 mL.

d. Put 100 mL of a 50.0% solution into a calibrated volumetric container and add enough pure water to bring the total volume of well-mixed solution to 500 mL.

7.50 **a.** 0.00650 M

b. 0.0488 M

c. 0.0195 M

d. 0.0139 M

7.52 16.6 g

7.54 20.0 mL

7.56 50.7 mL

7.58 42.3 mL

7.60 1.3 g

7.62 Because ice cream is essentially a solution of substances dissolved in water, the freezing point of the solution is lower than that of pure water. The temperature of a mixture of ice and pure water is the same as the freezing point of pure water. Thus, a mixture of ice and water has a temperature higher than the freezing point of the ice cream solution. A mixture of ice, salt, and water has a temperature (freezing point) lower than that of pure water. If enough salt is added to the ice and water mixture, the temperature can be made lower than the freezing point of the ice cream solution, and it will freeze.

7.64 **a.** Boiling point = 101.6°C; freezing point = −5.58°C

b. Boiling point = 100.78°C; freezing point = −2.79°C

c. Boiling point = 102.3°C; freezing point = −8.37°C

d. Boiling point = 103.1°C; freezing point = −11.2°C

7.66 **a.** Boiling point = 100.26°C; freezing point = 20.93°C

b. Boiling point = 100.39°C; freezing point = 21.40°C

c. Boiling point = 103.35°C; freezing point = 212.0°C

7.68 **a.** Osmolarity = 0.15

b. Osmolarity = 0.45

c. Osmolarity = 1.19

7.70 $\pi = 4.65 \times 10^3$ torr = 4.65×10^3 mmHg = 6.12 atm

7.72 $\pi = 1.12 \times 10^4$ torr = 1.12×10^4 mmHg = 14.7 atm

7.74 $\pi = 5.88 \times 10^4$ torr = 5.88×10^4 mmHg = 77.4 atm

7.76 $\pi = 3.5 \times 10^3$ torr = 3.5×10^3 mmHg = 4.6 atm

7.78 $\pi = 9.95 \times 10^3$ torr = 9.95×10^3 mmHg = 13.1 atm

7.80 Water will flow from the 5.00% solution into the 10.0% solution. The 10.0% solution will become diluted as osmosis takes place.

7.82 Much of the soiling material to be removed by washing consists of fats and oils, which are nonpolar and will not dissolve in water. The soaps or detergents act as emulsifying agents, which allow the nonpolar materials to be suspended in water and removed from the material being washed.

7.84 **a.** Na^+ and Cl^-

b. Urea molecules

c. K^+, Cl^- and glucose molecules

7.86 1.07 g/mL

7.88 10.2 mL

7.98 a

7.100 d

7.102 d

7.104 b

7.106 c

7.108 b

7.110 a

7.112 c

7.114 a

7.116 a

7.118 b

7.120 c

7.122 c

Chapter 8

8.2 **a.** Nonspontaneous: The rocket engines of the shuttle must operate continuously to push the shuttle into an orbit.

b. Spontaneous: Once ignited, the fuel continues to burn without additional input of energy.

c. Nonspontaneous: Heat must be supplied continuously to keep the water boiling.

d. Nonspontaneous: The only way to raise the temperature of a substance is to supply heat.

e. Nonspontaneous: To make your room orderly, someone must provide some work to move the clothing, etc., around. The work done represents an input of energy to the room.

8.4 **a.** Exergonic: The person doing the pushing is giving up energy to the automobile.

b. Endergonic: The ice must gain energy in the form of heat in order to melt.

c. Exergonic: The surroundings lose heat to the ice in order for the ice to melt.

d. Exergonic: In order to condense, the steam must give up some energy in the form of heat.

e. Endergonic: The heat given up by the steam as it condenses is accepted by the surroundings.

8.6 **a.** Both the energy and the entropy of the water decrease when water freezes. The process is spontaneous if the temperature of the surroundings is lower than 0°C.

b. The energy of the water decreases while the entropy stays essentially constant. The process is spontaneous.

c. The energy of the perfume increases slightly as energy is absorbed during evaporation. The entropy of the perfume increases a great deal as the molecules evaporate from the liquid and disperse throughout the room. The process is spontaneous.

8.8 **a.** Entropy is highest at the end of the play because the players are in a more mixed-up state as compared to just before the play began.

b. The answer is the 10% copper/gold alloy or the 2% alloy, depending on the atoms discussed. A lower concentration of copper atoms allows more disorder among the copper atoms, but more order among the gold atoms. A higher concentration of copper allows less disorder for the copper atoms, but increases the disorder of the gold atoms.

c. Entropy is highest with the purse on the ground and the contents scattered.

d. The mixture characteristic of coins in a piggy bank would have the highest entropy of the choices.

e. The mixture of loose pearls in a box would have the highest entropy of the choices.

8.10 The answers are subjective and depend on the meaning given to the terms *very slow*, *slow*, and *fast*.

a. Very slow

b. Slow

c. Very slow

d. Slow

e. Fast

8.12 **a.** Any of the following could be measured at different times: the weight of the block, the weight or volume of the water produced by the melting block, and the physical dimensions of the melting block.

b. Either of the following could be measured at different times: the distance a weighted object penetrates into a sample, and the force required to break a specific-sized piece.

c. Either of the following could be measured at different times: the length of the unburned candle and the weight of the candle.

8.14 **a.** Rate = 0.0330 M/min

b. Rate = 0.0194 M/min

8.16 Rate = 3.96×10^{-3} M/min

8.18 5.66×10^{-4} moles N_2. Rate = 1.51×10^{-6} M/sec

8.20 **a.** Assumption #1 is being violated. According to assumption #1, decreasing the concentrations of reactants would decrease the number of collisions and so decrease the rate at which the molecules would react.

b. Assumption #2 is being violated. According to assumption #2, molecules must collide with a certain minimum energy in order to react. At a lower temperature, fewer molecules would have the necessary minimum energy, so the reaction should go slower, not faster.

c. Assumption #1 is being violated. According to assumption #1, increasing the concentration of either of the reactants should increase the number of collisions and so increase the rate of the reaction.

8.22 An increase in temperature increases the speeds of the molecules involved in a reaction. The increased speed acts to increase the reaction rate in two ways. First, it causes more collisions to occur between molecules in a given amount of time. Second, it causes the energy of the colliding molecules to be greater so that more of the collisions will involve molecules that have at least the minimum energy needed for a reaction to take place.

8.24 **a.**

b.

8.26 **a.**

b.

AT 150° C

Both reactions are exothermic and the products have 500 kJ/mol less energy than the reactants, so both reactions give up 500 kJ/mol. The reaction that takes place at room temperature has a smaller activation energy than the reaction that must be heated to 150°C.

8.28 **a.** Slow to rapid, depending on the state of division of the solid. If large lumps of CaO were exposed to gaseous HCl, the reaction would be slow, since only the CaO on the surface of the lumps would be available. If the CaO were divided into dust-size particles and blown into the gaseous HCl, the reaction would go quite rapidly.

b. Slow. This reaction involves only solids, so only the material at the surface where two solid particles come into contact can react. As in part a, dividing the solids into a fine powder before mixing would speed the reaction significantly.

c. Won't react. The particles of both reactants carry a negative electrical charge, so the particles will repel each other, and collisions between them are highly unlikely.

d. Rapid. The particles of both reactants are in solution, where collisions occur frequently. Also, even though one of the particles is electrically charged, the other is not, so the charge will not diminish or enhance the chances for collisions between reactants.

8.30 Heat the reaction mixture, add a catalyst, or increase the concentration of one or more of the reactants.

8.32 About 0.93 hours or 56 minutes

8.34 Catalysts provide reaction pathways with lower activation energy. They may do this by reacting to form an intermediate structure that yields products when it breaks up or by providing a surface on which the reactants react.

8.36 **a.** Equilibrium would be indicated when the intensity of the violet color of the reaction mixture remained constant with time.

b. Equilibrium would be indicated when the amount of undissolved sugar remained constant with time.

c. Equilibrium would be indicated when either the intensity of the red-brown color of the reaction mixture remained constant with time or the total pressure of the reaction mixture remained constant with time.

8.38 The concentrations of H_2 and Br_2 would decrease, and the concentration of HBr would increase. The intensity of the red-brown color of the mixture would stop decreasing and would become constant when equilibrium was established.

8.40 **a.** $K = \dfrac{[CO_2]^2}{[CO]^2[O_2]}$

b. $K = \dfrac{[NO_2]^2}{[N_2O_4]}$

c. $K = \dfrac{[CO_2]^4[H_2O]^6}{[C_2H_6]^2[O_2]^7}$

d. $K = \dfrac{[NO]^2[Cl_2]}{[NOCl]^2}$

e. $K = \dfrac{[O_2][ClO_2]^4}{[Cl_2O_5]^2}$

8.42 **a.** $K = \dfrac{[Fe(CN)_6^{3-}]}{[Fe^{3+}][CN^-]^6}$

b. $K = \dfrac{[Ag(NH_3)_2^+]}{[Ag^+][NH_3]^2}$

c. $K = \dfrac{[AuCl_4^-]}{[Au^{3+}][Cl^-]^4}$

8.44 **a.** $CH_4 + 2O_2 \rightleftarrows CO_2 + 2H_2O$

b. $3H_2 + CO \rightleftarrows CH_4 + H_2O$

c. $2O_3 \rightleftarrows 3O_2$

d. $4NO_2 + 6H_2O \rightleftarrows 4NH_3 + 7O_2$

8.46 $K = 0.47$

8.48 $K = 0.099$

8.50 **a.** The concentration of products will be larger than the concentration of reactants.

b. The concentration of products will be larger than the concentration of reactants.

c. The concentration of products will be smaller than the concentration of reactants.

d. The concentration of products will be smaller than the concentration of reactants.

8.52 **a.** Shift left

b. Shift right

c. Shift left

8.54 **a.** Shift right: The color will become less blue and more purple.

b. Shift right: The amount of solid $PbCl_2$ will increase.

c. Shift right: The intensity of the violet color of the mixture will decrease.

d. Shift right: The intensity of the violet color of the mixture will decrease.

e. Shift left: The intensity of the brown color of the mixture will increase.

8.56 **a.** The amount of solid LiOH will increase, the amount of solid $LiHCO_3$ will decrease, and the concentration of CO_2 gas will be decreased.

b. The amount of solid $NaHCO_3$ will increase, the amount of solid Na_2O will decrease, and the concentration of CO_2 gas and H_2O gas will decrease.

c. The amount of solid $CaCO_3$ will increase, the amount of solid CaO will decrease, and the concentration of CO_2 gas will decrease.

8.58
a. Shift right
b. Shift left
c. Shift right
d. Shift left
e. No shift will occur
f. Shift left

8.60 Apply pressure to the balloon and decrease its volume. This would increase the concentrations of the gaseous reactants A and B inside the balloon and speed up the reaction.

8.62
a. Criterion 2
b. Criterion 3
c. Criterion 2
d. Criterion 3
e. Criterion 3

8.74 a

8.76 a

8.78 c

8.80 b

8.82 a

8.84 a

8.86 a

8.88 a

8.90 a

8.92 c

8.94 d

Chapter 9

9.2
a. $HBrO_2(aq) \rightarrow H^+(aq) + BrO_2^-(aq)$
b. $HS^-(aq) \rightarrow H^+(aq) + S^{2-}(aq)$
c. $HBr(aq) \rightarrow H^+(aq) + Br^-(aq)$
d. $HC_2H_3O_2(aq) \rightarrow H^+(aq) + C_2H_3O_2^-(aq)$

9.4
a. Not an Arrhenius base
b. Arrhenius base: $RbOH(aq) \rightarrow Rb^+(aq) + OH^-(aq)$
c. Not an Arrhenius base
d. Arrhenius base: $Ba(OH)_2(aq) \rightarrow Ba^{2+}(aq) + 2OH^-(aq)$

9.6
a. Acids: $HC_2O_4^-$ and H_3O^+
Bases: H_2O and $C_2O_4^{2-}$
b. Acids: HNO_2 and H_3O^+
Bases: H_2O and NO_2^-
c. Acids: H_2O and HPO_4^{2-}
Bases: PO_4^{3-} and OH^-
d. Acids: H_2SO_3 and H_3O^+
Bases: H_2O and HSO_3^-
e. Acids: H_2O and HF
Bases: F^- and OH^-

9.8

	Acid	Conjugate base	Base	Conjugate acid
a.	$HC_2O_4^-$	$C_2O_4^{2-}$	H_2O	H_3O^+
b.	HNO_2	NO_2^-	H_2O	H_3O^+
c.	H_2O	OH^-	PO_4^{3-}	HPO_4^{2-}
d.	H_2SO_3	HSO_3^-	H_2O	H_3O^+
e.	H_2O	OH^-	F^-	HF

9.10
a. $HF + H_2O \rightleftarrows H_3O^+ + F^-$
b. $HClO_3 + H_2O \rightleftarrows H_3O^+ + ClO_3^-$
c. $HClO + H_2O \rightleftarrows H_3O^+ + ClO^-$
d. $HS^- + H_2O \rightleftarrows H_3O^+ + S^{2-}$

9.12
a. SO_4^{2-}
b. CH_3NH_2
c. ClO_4^-
d. NH_3
e. Cl^-

9.14
a. H_2CO_3
b. HS^-
c. H_2S
d. $H_2C_2O_4$
e. $H_2N_2O_2$

9.16 The substance added to complete each equation appears in blue.
a. $H_2AsO_4^-(aq) + NH_3(aq) \rightarrow NH_4^+(aq) + HAsO_4^{2-}(aq)$
b. $C_6H_5NH_2(aq) + H_2O(\ell) \rightarrow C_6H_5NH_3^+(aq) + OH^-(aq)$
c. $S^{2-}(aq) + H_2O(\ell) \rightarrow HS^-(aq) + OH^-(aq)$
d. $(CH_3)_2NH(aq) + HBr(aq) \rightarrow (CH_3)_2NH_2^+(aq) + Br^-(aq)$
e. $CH_3NH_2(aq) + HCl(aq) \rightarrow CH_3NH_3^+(aq) + Cl^-(aq)$

9.18
a. $H_3O^+(aq) + NH_2^-(aq) \rightarrow H_2O(\ell) + NH_3(aq)$
b. $H_2PO_4^-(aq) + NH_3(aq) \rightarrow HPO_4^{2-}(aq) + NH_4^+(aq)$
c. $HS_2O_3^-(aq) + OCl^-(aq) \rightarrow S_2O_3^{2-}(aq) + HOCl(aq)$
d. $H_2O(\ell) + ClO_4^-(aq) \rightarrow OH^-(aq) + HClO_4(aq)$
e. $H_2O(\ell) + NH_3(aq) \rightarrow OH^-(aq) + NH_4^+(aq)$

9.20 hydrocyanic acid

9.22
a. Hydrotelluric acid
b. Hypochlorous acid
c. Sulfurous acid
d. Nitrous acid

9.24 succinic acid

9.26 $HMnO_4$

9.28
a. 2.3×10^{-13} M
b. 7.7×10^{-11} M
c. 1.1×10^{-12} M
d. 1.3×10^{-5} M
e. 3.0×10^{-13} M

9.30 **a.** 1.4×10^{-10} M

b. 1.4×10^{-13} M

c. 2.0×10^{-15} M

d. 5.9×10^{-12} M

e. 1.1×10^{-6} M

9.32 In Exercise 9.28: (a) acidic, (b) acidic, (c) acidic, (d) basic, (e) acidic

In Exercise 9.30: (a) basic, (b) basic, (c) basic, (d) basic, (e) acidic

9.34 **a.** Acidic

b. Basic

c. Acidic

d. Basic

9.36 **a.** 8.39, basic

b. 10.97, basic

c. 7.50, basic

d. 1.64, acidic

e. 4.71, acidic

9.38 **a.** 2.66, acidic

b. 11.41, basic

c. 5.12, acidic

d. 10.40, basic

e. 4.94, acidic

9.40 **a.** 5.4×10^{-10} M

b. 2.8×10^{-3} M

c. 3.8×10^{-6} M

9.42 **a.** $[H^+] = 1.1 \times 10^{-4}$ M; $[OH^-] = 8.9 \times 10^{-11}$ M

b. $[H^+] = 1.0 \times 10^{-4}$ M; $[OH^-] = 1.0 \times 10^{-10}$ M

c. $[H^+] = 1.4 \times 10^{-12}$ M; $[OH^-] = 7.2 \times 10^{-3}$ M

9.44 **a.** 8.9×10^{-9} M, basic

b. 1.2×10^{-4} M, acidic

c. 4.2×10^{-8} M, basic

d. 4.0×10^{-8} M, basic

e. 5.9×10^{-7} M, acidic

9.46 **a.** 1.2×10^{-3} M, acidic

b. 7.8×10^{-5} M, acidic

c. 4.8×10^{-3} M, acidic

d. 8.5×10^{-4} M, acidic

9.48 **a.** Add 1.0 L of dilute (6 M) HNO_3 to 1.0 L of pure water and mix well.

b. Add 50 mL of concentrated (15 M) NH_3 to 450 mL of pure water and mix well.

c. Add 83 mL of concentrated (12 M) HCl to 4.9 L of pure water and mix well.

9.50 **a.** $H_2SO_4(aq) + 2H_2O(\ell) \rightarrow 2H_3O^+(aq) + SO_4^{2-}(aq)$

b. $H_2SO_4(aq) + CaO(s) \rightarrow CaSO_4(aq) + H_2O(\ell)$

c. $H_2SO_4(aq) + Mg(OH)_2(s) \rightarrow MgSO_4(aq) + 2H_2O(\ell)$

d. $H_2SO_4(aq) + CuCO_3(s) \rightarrow CuSO_4(aq) + CO_2(g) + H_2O(\ell)$

e. $H_2SO_4(aq) + 2KHCO_3(s) \rightarrow K_2SO_4(aq) + 2CO_2(g) + 2H_2O(\ell)$

f. $H_2SO_4(aq) + Mg(s) \rightarrow MgSO_4(aq) + H_2(g)$

9.52 **a.** Total ionic: $2H^+(aq) + SO_4^{2-}(aq) + 2H_2O(\ell) \rightarrow 2H_3O^+(aq) + SO_4^{2-}(aq)$

Net ionic: $2H^+(aq) + 2H_2O(\ell) \rightarrow 2H_3O^+(aq)$

Simplified net ionic: $H^+(aq) + H_2O(\ell) \rightarrow H_3O^+(aq)$

b. Total ionic: $2H^+(aq) + SO_4^{2-}(aq) + CaO(s) \rightarrow Ca^{2+}(aq) + SO_4^{2-}(aq) + H_2O(\ell)$

Net ionic: $2H^+(aq) + CaO(s) \rightarrow Ca^{2+}(aq) + H_2O(\ell)$

c. Total ionic: $2H^+(aq) + SO_4^{2-}(aq) + Mg(OH)_2(s) \rightarrow Mg^{2+}(aq) + SO_4^{2-}(aq) + 2H_2O(\ell)$

Net ionic: $2H^+(aq) + Mg(OH)_2(s) \rightarrow Mg^{2+}(aq) + 2H_2O(\ell)$

d. Total ionic: $2H^+(aq) + SO_4^{2-}(aq) + CuCO_3(s) \rightarrow Cu^{2+}(aq) + SO_4^{2-}(aq) + CO_2(g) + H_2O(\ell)$

Net ionic: $2H^+(aq) + CuCO_3(s) \rightarrow Cu^{2+}(aq) + CO_2(g) + H_2O(\ell)$

e. Total ionic: $2H^+(aq) + SO_4^{2-}(aq) + 2KHCO_3(s) \rightarrow 2K^+(aq) + SO_4^{2-}(aq) + 2CO_2(g) + 2H_2O(\ell)$

Net ionic: $2H^+(aq) + 2KHCO_3(s) \rightarrow 2K^+(aq) + 2CO_2(g) + 2H_2O(\ell)$

Simplified net ionic: $H^+(aq) + KHCO_3(s) \rightarrow K^+(aq) + CO_2(g) + H_2O(\ell)$

f. Total ionic: $2H^+(aq) + SO_4^{2-}(aq) + Mg(s) \rightarrow Mg^{2+}(aq) + SO_4^{2-}(aq) + H_2(g)$

Net ionic: $2H^+(aq) + Mg(s) \rightarrow Mg^{2+}(aq) + H_2(g)$

9.54 $2HCl(aq) + SrO(s) \rightarrow SrCl_2(aq) + H_2O(\ell)$

$2HCl(aq) + Sr(OH)_2(s) \rightarrow SrCl_2(aq) + 2H_2O(\ell)$

$2HCl(aq) + SrCO_3(s) \rightarrow SrCl_2(aq) + H_2O(\ell) + CO_2(g)$

$2HCl(aq) + Sr(HCO_3)_2(s) \rightarrow SrCl_2(aq) + 2H_2O(\ell) + 2CO_2(g)$

$2HCl(aq) + Sr(s) \rightarrow SrCl_2(aq) + H_2(g)$

9.56 **a.** Molecular: $Sn(s) + H_2SO_3(aq) \rightarrow SnSO_3(aq) + H_2(g)$

Total ionic: $Sn(s) + 2H^+(aq) + SO_3^{2-}(aq) \rightarrow Sn^{2+}(aq) + SO_3^{2-}(aq) + H_2(g)$

Net ionic: $Sn(s) + 2H^+(aq) \rightarrow Sn^{2+}(aq) + H_2(g)$

b. Molecular: $3Mg(s) + 2H_3PO_4(aq) \rightarrow Mg_3(PO_4)_2(s) + 3H_2(g)$

Total ionic: $3Mg(s) + 6H^+(aq) + 2PO_4^{3-}(aq) \rightarrow Mg_3(PO_4)_2(s) + 3H_2(g)$

Net ionic: $3Mg(s) + 6H^+(aq) + 2PO_4^{3-}(aq) \rightarrow Mg_3(PO_4)_2(s) + 3H_2(g)$

c. Molecular: $Ca(s) + 2HBr(aq) \rightarrow CaBr_2(aq) + H_2(g)$

Total ionic: $Ca(s) + 2H^+(aq) + 2Br^-(aq) \rightarrow Ca^{2+}(aq) + 2Br^-(aq) + H_2(g)$

Net ionic: $Ca(s) + 2H^+(aq) \rightarrow Ca^{2+}(aq) + H_2(g)$

9.58 **a.** Molecular: $3\,RbOH(aq) + H_3PO_4(aq) \rightarrow$
$$Rb_3PO_4(aq) + 3H_2O(\ell)$$

Total ionic: $3Rb^+(aq) + 3OH^-(aq) + 3H^+(aq) +$
$PO_4^{3-}(aq) \rightarrow 3Rb^+(aq) + PO_4^{3-}(aq) + 3H_2O(\ell)$

Net ionic: $3OH^-(aq) + 3H^+(aq) \rightarrow 3H_2O(\ell)$

Simplified net ionic: $OH^-(aq) + H^+(aq) \rightarrow H_2O(\ell)$

b. Molecular: $2RbOH(aq) + H_2C_2O_4(aq) \rightarrow$
$$Rb_2C_2O_4(aq) + 2H_2O(\ell)$$

Total ionic: $2Rb^+(aq) + 2OH^-(aq) + 2H^+(aq) +$
$C_2O_4^{2-}(aq) \rightarrow 2Rb^+(aq) + C_2O_4^{2-}(aq) + 2H_2O(\ell)$

Net ionic: $2OH^-(aq) + 2H^+(aq) \rightarrow 2H_2O(\ell)$

Simplified net ionic: $OH^-(aq) + H^+(aq) \rightarrow H_2O(\ell)$

c. Molecular: $RbOH(aq) + HC_2H_3O_2(aq) \rightarrow$
$$RbC_2H_3O_2(aq) + H_2O(\ell)$$

Total ionic: $Rb^+(aq) + OH^-(aq) + H^+(aq) +$
$C_2H_3O_2^-(aq) \rightarrow Rb^+(aq) + C_2H_3O_2^-(aq) + H_2O(\ell)$

Net ionic: $OH^-(aq) + H^+(aq) \rightarrow H_2O(\ell)$

9.60 **a.** Molecular: $2KOH(aq) + H_3PO_4(aq) \rightarrow K_2HPO_4(aq) +$
$$2H_2O(\ell)$$

Total ionic:
$2K^+(aq) + 2OH^-(aq) + 2H^+(aq) + HPO_4^{2-}(aq) \rightarrow$
$$2K^+(aq) + HPO_4^{2-}(aq) + 2H_2O(\ell)$$

Net ionic: $2OH^-(aq) + 2H^+(aq) \rightarrow 2H_2O(\ell)$

Simplified net ionic: $OH^-(aq) + H^+(aq) \rightarrow H_2O(\ell)$

b. Molecular: $3KOH(aq) + H_3PO_4 \rightarrow K_3PO_4(aq) +$
$$3H_2O(\ell)$$

Total ionic:
$3K^+(aq) + 3OH^-(aq) + 3H^+(aq) + PO_4^{3-}(aq) \rightarrow$
$$3K^+(aq) + PO_4^{3-}(aq) + 3H_2O(\ell)$$

Net ionic: $3OH^-(aq) + 3H^+(aq) \rightarrow 3H_2O(\ell)$

Simplified net ionic: $OH^-(aq) + H^+(aq) \rightarrow H_2O(\ell)$

c. Molecular: $KOH(aq) + H_2C_2O_4(aq) \rightarrow$
$$KHC_2O_4(aq) + H_2O(\ell)$$

Total ionic:
$K^+(aq) + OH^-(aq) + H^+(aq) + HC_2O_4^-(aq) \rightarrow$
$$K^+(aq) + HC_2O_4^-(aq) + H_2O(\ell)$$

Net ionic: $OH^-(aq) + H^+(aq) \rightarrow H_2O(\ell)$

9.62 **a.** Cation: NH_4^+; anion: NO_3^-

b. Cation: Ca^{2+}; anion: Cl^-

c. Cation: Mg^{2+}; anion: HCO_3^-

d. Cation: K^+; anion: $C_2H_3O_2^-$

c. Cation: Li^+; anion: HSO_3^-

9.64 **a.** Acid: HNO_3; base: NH_3

b. Acid: HCl; base: $Ca(OH)_2$

c. Acid H_2CO_3; base: $Mg(OH)_2$

d. Acid: $HC_2H_3O_2$; base: KOH

e. Acid: H_2SO_3; base: $LiOH$

9.66 **a.** 1.3×10^2 g; the product would be $MgSO_4$

b. 1.8×10^2 g; the product would be $Na_2B_4O_7$

9.68 **a.** Acid: HNO_3; solid: $KHCO_3$

b. Acid: HCl; solid: Zn

c. Acid: HBr; solid Li_2O

9.70 **a.** $HNO_3(aq) + KHCO_3(s) \rightarrow KNO_3(aq) + H_2O(\ell) + CO_2(g)$

b. $2HCl(aq) + Zn(s) \rightarrow ZnCl_2(aq) + H_2(g)$

c. $2HBr(aq) + Li_2O(s) \rightarrow 2LiBr(aq) + H_2O(\ell)$

9.72 **a.** 1/2 mol

b. 1/2 mol

c. 1/3 mol

9.74 **a.** 0.44 eq; 4.4×10^2 meq

b. 0.45 eq; 4.5×10^2 meq

c. 6.24×10^{-2} eq; 62.4 meq

9.76 **a.** 3.50×10^{-2} eq; 35.0 meq

b. 4.01×10^{-2} eq; 40.1 meq

c. 1.35×10^{-1} eq; 135 meq

d. 4.17×10^{-2} eq; 41.7 meq

9.78 9.98×10^{-3} mol, 1.74 g

9.80 B, A, C, D

9.82 **a.** Acid order is B, A, C, D.

b. Conjugate base order is D, C, A, B.

9.84 **a.** $HSe^-(aq) \rightleftharpoons H^+(aq) + Se^{2-}(aq)$
$$K_a = \frac{[H^+][Se^{2-}]}{[HSe^-]}$$

b. $H_2BO_3^-(aq) \rightleftharpoons H^+(aq) + HBO_3^{2-}(aq)$
$$K_a = \frac{[H^+][HBO_3^{2-}]}{[H_2BO_3^-]}$$

c. $HBO_3^{2-}(aq) \rightleftharpoons H^+(aq) + BO_3^{3-}(aq)$
$$K_a = \frac{[H^+][BO_3^{3-}]}{[HBO_3^{2-}]}$$

d. $HAsO_4^{2-}(aq) \rightleftharpoons H^+(aq) + AsO_4^{3-}(aq)$
$$K_a = \frac{[H^+][AsO_4^{3-}]}{[HAsO_4^{2-}]}$$

e. $HClO(aq) \rightleftharpoons H^+(aq) + ClO^-(aq)$
$$K_a = \frac{[H^+][ClO^-]}{[HClO]}$$

9.86 According to the definitions given in the chapter, the person should be given the 20% acetic acid solution. If the other solution were the one desired, the term *dilute* should have been used rather than *weak*.

9.88 Titrations are done in order to analyze acids or bases. The amount of acid or base contained in a titrated sample can be determined from the volume of titrant added and the concentration of the titrant.

9.90 **a.** The endpoint and equivalence point will be the same if the chosen indicator changes color at the same pH as the pH of the salt solution produced by the reaction of the acid and base involved in the titration.

b. The endpoint and equivalence point will not be the same if the chosen indicator changes color at a pH that is different from the pH of the salt solution produced by the reaction of the acid and base involved in the titration.

9.92 a. 0.100 mol

b. 0.225 mol

9.94 a. $NaOH(aq) + HC_2O_2Cl_3(aq) \rightarrow NaC_2O_2Cl_3(aq) + H_2O(\ell)$

b. $2NaOH(aq) + H_2S_2O_6(aq) \rightarrow Na_2S_2O_6(aq) + 2H_2O(\ell)$

c. $4NaOH(aq) + H_4P_2O_6(aq) \rightarrow Na_4P_2O_6(aq) + 4H_2O(\ell)$

9.96 a. $2HCl(aq) + Zn(OH)_2(aq) \rightarrow ZnCl_2(aq) + 2H_2O(\ell)$

b. $3HCl(aq) + Tl(OH)_3(aq) \rightarrow TlCl_3(aq) + 3H_2O(\ell)$

c. $HCl(aq) + CsOH(aq) \rightarrow CsCl(aq) + H_2O(\ell)$

9.98 0.1660 M

9.100 a. 33.3 mL

b. 41.7 mL

c. 100 mL

d. 120 mL

e. 85.0 mL

f. 83.3 mL

9.102 a. 3.50 M

b. 0.891 M

c. 12.4 M

9.104 122.0 g/mol

9.106 The ions produced when Na_3PO_4 dissolves in water are Na^+ and PO_4^{3-}. The Na^+ ion is the conjugate acid of the strong base NaOH and therefore is a very weak Brønsted acid. The PO_4^{3-} ion is the conjugate base of the weak acid HPO_4^{2-} and therefore will act as a base in a hydrolysis reaction with water and will contribute to the pH of the solution. The reaction is:

$$PO_4^{3-}(aq) + H_2O(\ell) \rightarrow HPO_4^{2-}(aq) + OH^-(aq)$$

Thus, we see that the hydrolysis reaction contributes OH^- ions to the solution, causing it to have a pH greater than 7.

9.108 a. pH greater than 7. The ions produced when the salt dissolves in water are Na^+ and OCl^-. The Na^+ is the conjugate acid of a strong base (NaOH). It will be a weak acid and will not undergo a significant hydrolysis reaction. The OCl^- ion is the conjugate base of a weak acid (HOCl) and will therefore undergo a significant hydrolysis reaction that will contribute to the pH of the solution. The reaction is:

$$OCl^-(aq) + H_2O(\ell) \rightarrow HOCl(aq) + OH^-$$

The OH^- ions produced by this reaction will cause the pH of the solution to be greater than 7.

b. pH greater than 7. The explanation is the same as in part a. The hydrolysis reaction of the CHO_2^- ion is:

$$CHO_2^-(aq) + H_2O(\ell) \rightarrow HCHO_2(aq) + OH^-(aq)$$

c. pH equal to 7. The ions produced when the salt dissolves in water are K^+ and NO_3^-. The K^+ is the conjugate acid of the strong base KOH, and the NO_3^- is the conjugate base of the strong acid HNO_3. Since both ions come from

strong sources, they are both weak in their respective behaviors as a conjugate acid or base. As a result, neither of them will undergo a significant hydrolysis reaction that will influence the solution pH.

d. pH greater than 7. The explanation is the same as in part a. The hydrolysis reaction of the PO_4^{3-} ion is:

$$PO_4^{3-}(aq) + H_2O(\ell) \rightarrow HPO_4^{2-}(aq) + OH^-(aq)$$

9.110 Different salts will hydrolyze to different extents in water solution and produce solutions of different pH. In titrations, the pH at the equivalence point is the pH characteristic of a solution of the salt produced by the reaction of the acid and base involved in the titration. Different acids and bases will produce different salts and possibly different solution pH values at the equivalence point. Thus, an indicator must be chosen that will change color at a pH as near as possible to the pH of the salt solution resulting from the titration.

9.112 The buffering anions present in the solution of the two salts are HPO_4^{2-} and $H_2PO_4^-$. When acid is added to the solution, the buffering reaction (like Equation 9.48) is:

$$HPO_4^{2-}(aq) + H^+(aq) \rightleftharpoons H_2PO_4^-(aq)$$

When base is added, the buffering reaction (like Equation 9.49) is:

$$H_2PO_4^-(aq) + OH^-(aq) \rightleftharpoons HPO_4^{2-}(aq) + H_2O(\ell)$$

9.114 $HCO_3^-(aq) + H^+(aq) \rightleftharpoons H_2CO_3(aq)$

9.116 a. pH = 3.85

b. pH = 3.85

c. The difference is that the buffer of part b would have a greater buffer capacity; that is, it could buffer against larger added amounts of acid or base.

9.118 a. pH = 4.54

b. pH = 7.81

c. pH = 6.12

9.120 $[HPO_4^{2-}]/[H_2PO_4^-] = 2.8$

9.122 $Cl^-(aq) + H_3O^+(aq) \rightarrow HCl(aq) + H_2O(\ell)$

$$\left[:\ddot{\underset{..}{Cl}}: \right]^- + \left[H:\underset{H}{\overset{..}{\underset{..}{O}}}:H \right]^+ \rightarrow H:\ddot{\underset{..}{Cl}}: + H:\underset{H}{\overset{..}{O}:}$$

base acid acid base

A base donates a pair of electrons to form a covalent bond, while an acid accepts a pair.

9.124 pH = 6.63

9.138 d

9.140 b

9.142 b

9.144 d

9.146 d

9.148 b

9.150 b

9.152 a

9.154 c

9.156 c

10.2 **a.** Beta, gamma, and positron

 b. Alpha and positron

 c. Gamma and neutron

10.4 **a.** A beta particle is identical with an electron.

 b. An alpha particle is made up of two protons and two neutrons.

 c. A positron is a positively charged electron.

10.6 **a.** The mass number is reduced by 4, and the atomic number is reduced by 2.

 b. The mass number is unchanged, and the atomic number is increased by 1.

 c. The mass number is unchanged, and the atomic number is reduced by 1.

 d. The mass number and atomic number are both unchanged.

 e. The mass number is unchanged, and the atomic number is reduced by 1.

10.8 **a.** $^{96}_{41}\text{Nb}$

 b. $^{80}_{37}\text{Rb}$

 b. $^{38}_{20}\text{Ca}$

10.10 The quantity added to complete each equation appears in blue

 a. $^{204}_{82}\text{Pb} \rightarrow {}^{200}_{80}\text{Hg} + {}^{4}_{2}\alpha$

 b. $^{84}_{35}\text{Br} \rightarrow {}^{84}_{36}\text{Kr} + {}^{0}_{-1}\beta$

 c. $^{41}_{20}\text{Ca} + {}^{0}_{-1}\text{e} \rightarrow {}^{41}_{19}\text{K}$

 d. $^{149}_{62}\text{Sm} \rightarrow {}^{145}_{60}\text{Nd} + {}^{4}_{2}\alpha$

 e. $^{34}_{14}\text{Si} + {}^{34}_{15}\text{P} \rightarrow {}^{0}_{-1}\beta$

 f. $^{15}_{8}\text{O} \rightarrow {}^{0}_{-1}\beta + {}^{15}_{7}\text{N}$

10.12 **a.** $^{157}_{63}\text{Eu} \rightarrow {}^{0}_{-1}\beta + {}^{157}_{64}\text{Gd}$

 b. $^{190}_{78}\text{Pt} \rightarrow {}^{4}_{2}\alpha + {}^{186}_{76}\text{Os}$

 c. $^{138}_{62}\text{Sm} + {}^{0}_{-1}e \rightarrow {}^{138}_{61}\text{Pm}$

 d. $^{188}_{80}\text{Hg} \rightarrow {}^{188}_{79}\text{Au} + {}^{0}_{1}\beta$

 e. $^{234}_{90}\text{Th} \rightarrow {}^{0}_{-1}\beta + {}^{234}_{91}\text{Pa}$

 f. $^{218}_{85}\text{At} \rightarrow {}^{4}_{2}\alpha + {}^{214}_{83}\text{Bi}$

10.14 The half-life of a dozen cookies would be the time it takes for the number to be reduced to 6. The half-life of a $500 checking account would be the time it takes to reduce the amount in the account to $250.

10.16 0.56 ng remains

10.18 2.24×10^4 years old

10.20 1.68×10^4 years old

10.22 Long-term exposure to low-level radiation is more likely to cause changes to genetic material in cells. Such changes might result in genetic mutations, cancer, or other serious consequences. Short-term exposure to intense radiation destroys tissue rapidly and causes radiation sickness, which can lead to death, depending on the dose of radiation received.

10.24 A physical unit of radiation, such as a curie, indicates the activity of a source of radiation in terms of the number of nuclear decays that occur per minute. A biological unit of radiation, such as a rem, indicates the damage caused by radiation in living tissue.

10.26 3.2 R of X-rays

10.28 1.7×10^5 disintegrations/s

10.30 Hot spots are concentrations of radioisotope in tissue. Cold spots are areas of tissue that reject or keep out radioactive tracers.

10.32 $^{51}_{24}\text{Cr} + {}^{0}_{-1}\text{e} \rightarrow {}^{51}_{23}\text{V}$. The daughter is vanadium-51.

10.34 Use a radioactive isotope of oxygen to make water by burning hydrogen in it. Mix the radioactive water with hydrogen peroxide. Add the catalyst and collect the oxygen gas given off. If the gas is radioactive, it came from the water; if it is not radioactive, it came from the hydrogen peroxide.

10.36 Use a Geiger-Müller tube to measure the radioactivity of a gallon of water that contains the radioisotope. Pour that water into the pool and allow time for it to become completely and uniformly mixed with the water in the pool. Then, remove a gallon of the pool water and measure its radioactivity. For example, if the pool water has a radioactivity one ten-thousandth that of the original tracer-containing water, it means the original gallon of water has been diluted by a factor of 10,000, and therefore the pool contains 10,000 gal of water. Thus, the pool's volume is equal to the factor by which the radioactivity has been reduced.

10.38 $^{24}_{12}\text{Mg} + {}^{4}_{2}\alpha \rightarrow {}^{1}_{0}\text{n} + {}^{27}_{14}\text{Si}$

10.40 $^{238}_{92}\text{U} + {}^{1}_{0}\text{n} \rightarrow {}^{239}_{94}\text{Pu} + 2\,{}^{0}_{-1}\beta$

10.42 A moderator slows neutrons so that their chances of being captured and interacting with a target nucleus are greater.

10.44 $^{66}_{30}\text{Zn} + {}^{1}_{1}p \rightarrow {}^{67}_{31}\text{Ga}$

10.46 $2\,{}^{1}_{1}\text{H} \rightarrow {}^{2}_{1}\text{H} + {}^{0}_{1}\beta$

10.48 Nuclear fission is a process in which heavy nuclei split into smaller nuclei when they are hit by a neutron. Energy is released by the process. An example is:

$$^{235}_{92}\text{U} + {}^{1}_{0}\text{n} \rightarrow {}^{139}_{54}\text{Xe} + {}^{95}_{38}\text{Sr} + 2\,{}^{1}_{0}\text{n}$$

When nuclear fusion occurs, small nuclei fuse together and form heavier nuclei. Energy is released by the process. An example is: $\quad {}^{1}_{1}\text{H} + {}^{2}_{1}\text{H} \rightarrow {}^{3}_{2}\text{He} + \gamma$

10.50 $^{238}_{94}\text{Pu} \rightarrow {}^{4}_{2}\alpha + {}^{234}_{92}\text{U}$

10.52 The positive nuclei of the reacting hydrogen atoms naturally repel each other. In order to overcome this repulsion and get close enough together to fuse, the nuclei must have very high speeds (kinetic energy). The necessary high kinetic energy can only be obtained by the nuclei at very high temperatures. Remember, temperature is a measurement of the average kinetic energy of particles in matter.

10.54 3.5×10^{20} counts/minute $= 1.6 \times 10^8$ Ci

10.64 d

10.66 d

10.68 c

10.70 b

10.72 b

10.74 a

10.76 b

10.78 c

10.80 c

10.82 b

10.84 b

10.86 d

Chapter 11

11.2 Hair, skin, clothing, paper, plastic, carpet

11.4 They all contain carbon.

11.6 Covalent

11.8 **a.** Organic; most organic compounds are flammable.

b. Inorganic; a high-melting solid is typical of ionic compounds.

c. Organic; many organic compounds are insoluble in water.

d. Organic; most organic compounds can undergo combustion.

e. Organic; a low-melting solid is characteristic of covalent materials, which are usually organic.

11.10 Organic compounds do not readily accept electrons to transport them, thus making the flow of electrical current impossible.

11.12 The ability of carbon to bond to itself repeatedly; isomerism.

11.14 The s orbital of the hydrogen overlaps with the sp^3 orbital of carbon.

11.16 A p orbital and an sp^3 orbital both have a two-lobed shape. A p orbital has equal-sized lobes, whereas an sp^3 orbital has different-sized lobes.

11.18

11.20 b, d, e

11.22 **a.** Incorrect; H cannot have two bonds.

b. Correct

c. Incorrect; the carbon with the C=O bond has five bonds.

d. Incorrect; both carbons of the C=C double bond have three bonds.

e. Correct

11.24

a. $CH_3CH_2CH_2CH{=}CH-\overset{\displaystyle O}{\overset{\displaystyle \|}{C}}-CH_2CH_3$

b. $CH_3CH_2-\overset{\displaystyle O}{\overset{\displaystyle \|}{C}}-NH_2$

11.26 **a.**

b.

11.28 **a.** Branched

b. Normal

c. Normal

d. Normal

e. Branched

f. Branched

11.30 a. Same

b. Same

c. Isomers

d. Same

11.32 **a.** Six

b. Five

c. Seven

11.34 **a.** 3-methylpentane

b. 2,4,5-trimethylheptane

c. 3-ethyl-2,3-dimethylpentane

d. 4,7-diethyl-5-methyldecane

e. 3-ethyl-5-methylnonane

11.36

a. $CH_3-CH_2-CH-CH_2-CH_3$
 $\qquad\qquad\quad |$
 $\qquad\qquad\;CH_2CH_3$

b. $CH_3-\underset{\underset{\displaystyle CH_3}{|}}{\overset{\overset{\displaystyle CH_3}{|}}{C}}-CH_2-CH_3$

c. $CH_3-CH_2-\underset{\underset{\displaystyle CH_3}{|}}{\overset{\overset{\displaystyle CH_3}{|}}{C}}-\underset{\underset{\displaystyle CH_2CH_3}{|}}{\overset{\overset{\displaystyle CH_2CH_2CH_3}{|}}{CH}}-CH-CH_2-CH_2-CH_2-CH_2-CH_3$

d. $CH_3-CH_2-CH_2-CH_2-CH-CH_2-CH_2-CH_2-CH_2-CH_3$
 $\qquad\qquad\qquad\qquad\qquad\quad|$
 $\qquad\qquad\qquad\qquad CH_3CHCH_2CH_3$

11.38

$CH_3CH_2CH_2CH_2CH_3$ $CH_3\underset{\underset{\displaystyle CH_3}{}}{\overset{\overset{\displaystyle CH_3}{|}}{CH}}CH_2CH_3$

pentane 2-methylbutane

$CH_3\underset{\underset{\displaystyle CH_3}{|}}{\overset{\overset{\displaystyle CH_3}{|}}{C}}CH_3$

2,2-dimethylpropane

11.40 CH₃CH₂CH₂CH₂CH₂CH₃ → $CH_3CH_2CH_2CH_2CH_2CH_3$ hexane

$$CH_3\overset{\displaystyle CH_3}{\underset{\displaystyle |}{C}}HCH_2CH_2CH_3$$ 2-methylpentane

$$CH_3CH_2\overset{\displaystyle CH_3}{\underset{\displaystyle |}{C}}HCH_2CH_3$$ 3-methylpentane

$$CH_3\overset{\displaystyle CH_3}{\underset{\displaystyle \underset{\displaystyle CH_3}{|}}{C}}CH_2CH_3$$ 2,2-dimethylbutane

$$CH_3\overset{\displaystyle CH_3}{\underset{\displaystyle \underset{\displaystyle CH_3}{|}}{C}}HCHCH_3$$ 2,3-dimethylbutane

11.42 a. $CH_2\overset{\displaystyle CH_3}{\underset{\displaystyle \underset{\displaystyle CH_3}{|}}{C}}HCH_3$

The longest carbon chain is four carbon atoms rather than three. The correct name is 2-methylbutane.

b. $CH_3CH_2\overset{\displaystyle CH_3}{\underset{\displaystyle \underset{\displaystyle CH_3}{|}}{C}}HCHCH_3$

The chain should be numbered from the right to give 2,3-dimethylpentane.

c. $CH_3\overset{\displaystyle CH_2CH_3}{\underset{\displaystyle |}{C}}HCH_2\overset{\displaystyle}{\underset{\displaystyle CH_3}{C}}HCH_3$

The longest carbon chain is six carbon atoms and should be numbered from the right to give 2,4-dimethylhexane.

d. $CH_3\overset{\displaystyle Br}{\underset{\displaystyle |}{C}}HCHCH_3$ with CH_2CH_3

The longest carbon chain is five carbon atoms rather than four. The correct name is 2-bromo-3-methylpentane.

11.44 a. cyclopentane

 b. 1,2-dimethylcyclobutane

 c. 1,1-dimethylcyclohexane

 d. 1,2,3-trimethylcyclobutane

11.46 a.

b.

c. CH₂CH₂CH₂CH₃

11.48 b, c, and d

11.50 So that the carbon atoms with attached hydrogens can assume a tetrahedral shape.

11.52

a. none

b. none

c. *trans*-1,2-dimethylcyclobutane *cis*-1,2-dimethylcyclobutane

d. *trans*-1-isopropyl-2-methylcyclopropane *cis*-1-isopropyl-2-methylcyclopropane

11.54 a. *trans*-1-ethyl-2-methylcyclopropane

 b. *cis*-1-bromo-2-chlorocyclopentane

 c. *trans*-1-methyl-2-propylcyclobutane

 d. *trans*-1,3-dimethylcyclohexane

11.56 a. Liquid

 b. No

 c. Yes

 d. Less dense

11.58 2-methylheptane

11.60 a. $2C_4H_{10} + 13O_2 \rightarrow 8CO_2 + 10H_2O$

 b. $C_5H_{12} + 8O_2 \rightarrow 5CO_2 + 6H_2O$

 c. $C_4H_8 + 6O_2 \rightarrow 4CO_2 + 4H_2O$

11.62 $2C_6H_{14} + 13O_2 \rightarrow 12CO + 14H_2O$

11.64 Because of the rigid structure of cycloalkanes, stronger dispersion forces can develop between molecules. This increase of intermolecular attraction causes an increase in boiling point.

11.66 pentane: 414.5 torr; hexane: 113.9 torr; heptane: 37.2 torr

11.68 13 L of air

11.80 d

11.82 d

11.84 b

11.86 b

Chapter 12

12.2 An alkene contains one or more C=C bonds.

An alkyne contains one or more C≡C bonds.

An aromatic hydrocarbon contains the characteristic benzene ring or similar feature.

12.4 **a.** 2-butene

b. 3-ethyl-2-pentene

c. 4,4-dimethyl-2-hexyne

d. 4-methylcyclopentene

e. 6-bromo-2-methyl-3-heptyne

f. 1-ethyl-2,3-dimethylcyclopropene

g. 6-methyl-1,4-heptadiene

12.6

a.
$$CH_3-CH=\overset{\overset{\displaystyle CH_2CH_3}{|}}{C}-CH_2-CH_2-CH_3$$

b.
$$CH_2=CH-\overset{\overset{\displaystyle CH_3}{|}}{CH}-\overset{\overset{\displaystyle CH_3}{|}}{CH}-CH_3$$

c.
$$CH_2=CH-\overset{\overset{\displaystyle CH_3}{|}}{C}=CH-CH_3$$

d.

e.
$CH_2CH_2CH_2CH_3$

12.8 Several possibilities exist including these:

$CH\equiv CCH_2CH_2CH_3$ 1-pentyne

$CH_2=CHCH=CHCH_3$ 1,3-pentadiene

cyclopentene

12.10 3,7,11-trimethyl-1,3,6,10-dodecatetraene

12.12

a.
$$CH_3\overset{\overset{\displaystyle CH_3}{|}}{CH}-CH_2CH=CHCH_3$$
5-methyl-2-hexene

b. $CH_3CH_2CH=CHCH=CHCH_3$

2,4-heptadiene

c.

3-methylcyclobutene

12.14 The overlap of two *p* orbitals forms a π (pi) bond containing two electrons.

12.16 Structural isomers have a different order of linkage of atoms. Geometric isomers have the same order of linkage of atoms but different three-dimensional arrangements of their atoms in space.

12.18 **a.** and **c.** cannot have cis- and trans-isomers.

b.

cis-3-hexene

trans-3-hexene

12.20 **a.**

b.

12.22 Both alkanes and alkenes are nonpolar compounds that are insoluble in water, less dense than water, and soluble in non-polar solvents.

12.24 When H—X adds to an alkene, the hydrogen becomes attached to the carbon atom that is already bonded to more hydrogens.

$$CH_2=CH-CH_3 + HCl \longrightarrow CH_3\overset{\overset{\displaystyle Cl}{|}}{C}HCH_3$$

12.26

a.

b.

c.
$$CH_3\overset{\overset{\displaystyle CH_3}{|}}{C}HCH_2CH_3$$

d.

12.28 **a.** Br_2

b. H_2 with Pt catalyst

c. HCl

d. H_2O with H_2SO_4 catalyst

12.30 The addition of Br_2 to these samples will cause the cyclohexane solution to become light orange with unreacted Br_2, whereas the 2-hexene will react with the Br_2 and remain colorless.

12.32 A monomer is a starting material used in the preparation of polymers, long-chain molecules made up of many repeating units.

Addition polymers are long-chain molecules prepared from alkene monomers through numerous addition reactions. A copolymer is prepared from two monomer starting materials.

12.34 A carbon–carbon double bond

12.36 $nCH_2\!=\!CH \xrightarrow{\text{catalyst}} \ \begin{smallmatrix} \end{smallmatrix}\!\!+\!\!CH_2\!-\!CH\!+_{\overline{n}}$
 $\quad\ \ |$ $|$
 $\quad\ \ Cl$ Cl

12.38 *sp*; two

12.40 Linear

12.42 Acetylene is a fuel used by welding torches and in the synthesis of monomers for plastics and fibers.

12.44 H—C≡C—CH₂CH₂CH₃
1-pentyne

CH₃—C≡C—CH₂CH₃
2-pentyne

 CH₃
 |
CH₃CH—C≡C—H
3-methyl-1-butyne

12.46 Unhydridized *p* orbitals

12.48 Aromatic refers to compounds containing the benzene ring. Aliphatic substances do not contain a benzene ring.

12.50 In cyclopentane, attached groups are located above and below the plane of the ring. In benzene, attached groups are located in the same plane as the ring.

12.52 a. 1,3,5-trimethylbenzene
 b. 1,4-diethylbenzene

12.54 a. 3-phenyl-1-pentene
 b. 2,5-diphenylhexane

12.56 a. *m*-bromophenol
 b. *p*-ethylaniline

12.58 a. 2,6-dibromobenzoic acid
 b. 3-chloro-5-ethyltoluene

12.60

a. [benzene ring with OH and CH₂CH₃ substituents]

b. [benzene ring with HO—C=O and Cl substituents]

c. [structure: CH₃CH₂CCH₂CH₃ with CH₃ above and benzene ring below]

12.62 Nonpolar and insoluble in water

12.64 Cyclohexene readily undergoes addition reactions. Benzene resists addition reactions and favors substitution reactions. Both undergo combustion.

12.66 a. Phenol
 b. Styrene
 c. Aniline
 d. Phenol

12.68 Heat increases the kinetic energy of the ethylene molecules and increases the number of molecules having the required activation energy.

Pressure on a gas increases the temperature and concentration of the gas.

Catalysts lower the activation energy requirements.

12.70

12.82 a and d

12.84 c

12.86 b

Chapter 13

13.2 R—O—R

13.4 a. 2-propanol
 b. 3-methyl-3-pentanol
 c. 5-bromo-3-ethyl-1-pentanol
 d. 2-isopropyl-1-methylcyclopropanol
 e. 2,2-dimethylcyclopentanol
 f. 1,2-cyclopentanediol
 g. 1,2,4-hexanetriol

13.6 a. Methanol
 b. 2-propanol
 c. Ethanol
 d. 1,2-ethanediol
 e. 1,2,3-propanetriol

13.8

a. OH
 |
CH₃CCH₂CH₂CH₃
 |
 CH₃

b. OH OH
 | |
 CH₂CH₂CHCH₃

c. [cyclopentane ring with OH and CH₂CH₃ substituents]

13.10 a. 2-ethylphenol or *o*-ethylphenol

b. 2-ethyl-4,6-dimethylphenol

13.12

a.
4-chlorophenol structure (OH on benzene, Cl para)

b.
phenol with CHCH₃/CH₃ and CH₃CH/CH₃ substituents

13.14 a. Tertiary

b. Primary

c. Secondary

13.16 $CH_3CH_2CH_2CH_2-OH$ 1-butanol primary

$CH_3\overset{\underset{|}{OH}}{C}HCH_2CH_3$ 2-butanol secondary

$CH_3\overset{\underset{|}{CH_3}}{C}HCH_2-OH$ 2-methyl-1-propanol primary

$CH_3\overset{\underset{|}{CH_3}}{\underset{|}{C}}CH_3$ with OH on top 2-methyl-2-propanol tertiary

13.18 a. methanol, ethanol, 1-propanol

b. butane, 1-propanol, ethylene glycol

13.20 a.
$CH_3CH_2CH_2CH_2-O$ with H, hydrogen-bonded to $O-CH_2CH_3$

b.
cyclohexanol hydrogen bonded to water

13.22 a. $CH_3CH=CHCH_3$

b. $CH_3CH=\overset{\underset{|}{CH_3}}{C}CH_3$

13.24 a. $CH_3CH_2CH_2-O-CH_2CH_2CH_3$

b.
diphenyl ether

c.
$CH_2CH_2-O-CH_2CH_2$ connecting two benzene rings

13.26 a.
cyclopentanol (OH on cyclopentane)

b. $CH_3CH_2\overset{\underset{|}{OH}}{C}HCHCH_3$ with CH₃

c. $CH_3CH_2CH_2CH_2$ with OH

13.28 a. $CH_3\overset{\underset{|}{OH}}{\underset{|}{C}}CH_2CH_3 + (O) \longrightarrow$ no reaction, with CH_3

b. $2CH_3CH_2CH_2-OH \xrightarrow[140°C]{H_2SO_4}$
$CH_3CH_2CH_2-O-CH_2CH_2CH_3 + H_2O$

c. $CH_3CH_2\overset{\underset{|}{OH}}{C}HCH_2CH_3 + (O) \longrightarrow$
$CH_3CH_2\overset{\overset{O}{||}}{C}CH_2CH_3 + H_2O$

d. $CH_3CH_2\overset{\underset{|}{OH}}{C}HCH_2CH_3 \xrightarrow[180°C]{H_2SO_4}$
$CH_3CH=CHCH_2CH_3 + H_2O$

e. $CH_3CH_2CH_2CH_2CH_2CH_2-OH + 2(O) \longrightarrow$
$CH_3CH_2CH_2CH_2CH_2-\overset{\overset{O}{||}}{C}-OH + H_2O$

13.30

a. $CH_3CH_2CH=CH_2 + H_2O \xrightarrow{H_2SO_4} CH_3CH_2\overset{\underset{|}{OH}}{C}HCH_3$

$CH_3CH_2\overset{\underset{|}{OH}}{C}HCH_3 + (O) \rightarrow CH_3CH_2\overset{\overset{O}{||}}{C}CH_3 + H_2O$

b.
1,2-methylcyclopentanol $\xrightarrow[180°C]{H_2SO_4}$ 1-methylcyclopentene $+ H_2O$

1-methylcyclopentene $+ H_2O \xrightarrow{H_2SO_4}$ 1-methylcyclopentanol

c. $CH_3CH_2CH_2CH_2\overset{\underset{|}{OH}}{} \xrightarrow[180°C]{H_2SO_4} CH_3CH_2CH=CH_2 + H_2O$

$CH_3CH_2CH=CH_2 + H_2O \xrightarrow{H_2SO_4} CH_3CH_2\overset{\underset{|}{OH}}{C}HCH_3$

$CH_3CH_2\overset{\underset{|}{OH}}{C}HCH_3 + (O) \rightarrow CH_3CH_2\overset{\overset{O}{||}}{C}CH_3 + H_2O$

13.32 $CH_3\overset{\overset{O}{||}}{C}-\overset{\overset{O}{||}}{C}-OH$

Answers to Even-Numbered End-of-Chapter Exercises **B-25**

13.34 Methanol is used as an automobile fuel and in the preparation of formaldehyde, a starting material for certain plastics.

13.36 a. Glycerol

 b. Ethanol

 c. Ethylene glycol

 d. Isopropyl alcohol

 e. Menthol

 f. Ethanol

13.38 a. *o*-phenylphenol or 2-benzyl-4-chlorophenol

 b. 4-hexylresorcinol or phenol

 c. BHA or BHT

13.40 a. Ethyl isopropyl ether

 b. Butyl methyl ether

 c. Diphenyl ether

13.42 a. 1-ethoxypropane

 b. 2-ethoxypropane

 c. Ethoxybenzene

 d. 1,2-dimethoxycyclopentane

13.44

13.46 Inertness

13.48 $CH_3CH_2CH_2$—OH, CH_3CH_2—O—CH_3, $CH_3CH_2CH_2CH_3$

This order is based on the decreasing ability to form hydrogen bonds with water.

13.50

13.52 a. $CH_3CH_2CH_2$—S—Hg—S—$CH_2CH_2CH_3$ + 2H$^+$

 b. $CH_3CH_2CH_2$—S—S—$CH_2CH_2CH_3$ + H_2O

 c.

13.54 Alcohols are oxidized to aldehydes and ketones, whereas thiols are oxidized to disulfides.

13.56 aromatic, alkene, aldehyde

13.58 The electronegativity difference between S and H is less than between O and H. This means the S—H bond is less polar, leading to a decreased ability to hydrogen bond. The decrease in hydrogen bonding ability would decrease a thiol's boiling point and water solubility in comparison to the alcohol.

13.60 The negative charge that forms on the oxygen atom when phenol acts as an acid can be delocalized around the ring structure of benzene.

13.72 d

13.74 d

13.76 b

Chapter 14

14.2

Simplest aldehyde:

Simplest ketone:

14.4 a. Aldehyde

 b. Neither

 c. Ketone

 d. Neither

 e. Ketone

 f. Neither

14.6 a. Propanal

 b. 3-bromobutanal

 c. 3-phenylpropanal

 d. 4-methyl-2-pentanone

 e. 2-isopropycyclopentanone

14.8

d.

$$\text{CH}_2\text{CHCH}_2 - \overset{\displaystyle O}{\overset{\displaystyle \|}{\text{C}}} - \text{H}$$

(with Br on the CHCH$_2$ carbon and a phenyl ring attached)

14.10

$$\text{CH}_3\text{CH}_2\overset{\displaystyle O}{\overset{\displaystyle \|}{\text{C}}} - \text{H} \quad \text{and} \quad \text{CH}_3 - \overset{\displaystyle O}{\overset{\displaystyle \|}{\text{C}}} - \text{CH}_3$$

14.12

a.

$$\text{H} - \overset{\displaystyle O}{\overset{\displaystyle \|}{\text{C}}} - \overset{\displaystyle \text{CH}_2\text{CH}_3}{\underset{\displaystyle \text{CH}_3}{\text{CHCHCH}_3}}$$

The longest chain contains five carbon atoms. The correct name is 2,3-dimethylpentanal.

b. The compound is an aldehyde. Number from the carbonyl group.

$$\text{CH}_3\text{CHCH}_2 - \overset{\displaystyle O}{\overset{\displaystyle \|}{\text{C}}} - \text{H} \quad (\text{CH}_3 \text{ on the CHCH}_2 \text{ carbon})$$

The correct name is 3-methylbutanal.

c.

(cyclopentanone ring with Br at position 2 and Br at position 3)

Number the ring so that the bromines have lower numbers. The correct name is 2,3-dibromocyclopentanone.

14.14 Acetone dissolves remaining traces of water, and together they are discarded. The remaining small amounts of acetone evaporate.

14.16 Propane is nonpolar. Ethanal is a polar molecule and exhibits greater interparticle forces.

14.18

a.

(cyclopentane ring with C—CH$_3$ carbonyl; O···H—O—H hydrogen bond shown)

b.

(benzene ring with C—H carbonyl; O···H—O—H hydrogen bond shown)

14.20 The alcohol will have the higher melting point because of hydrogen bonding. Menthol is a solid, and menthone is a liquid.

14.22

a.

$$\text{CH}_3\text{CHCH}_2\text{CHCH}_3 + (\text{O}) \longrightarrow$$

(with OH and CH$_3$ substituents)

$$\text{CH}_3 - \overset{\displaystyle O}{\overset{\displaystyle \|}{\text{C}}} - \text{CH}_2\text{CHCH}_3 + \text{H}_2\text{O} \quad (\text{CH}_3 \text{ on the CHCH}_3 \text{ carbon})$$

b.

(cyclopentane ring with CH$_2$OH) $+ (\text{O}) \longrightarrow$ (cyclopentane ring with C—H carbonyl) $+ \text{H}_2\text{O}$

c.

(cyclopentane ring with OH and CH$_2$CH$_3$) $+ (\text{O}) \longrightarrow$ (cyclopentanone ring with CH$_2$CH$_3$) $+ \text{H}_2\text{O}$

14.24 a. Neither

b. Hemiacetal

c. Neither

d. Hemiketal

14.26 a. ketal

b. Acetal

c. Neither

14.28 a. Ketal

b. None of these

c. Hemiacetal

14.30 They are unstable, splitting into an alcohol and aldehyde.

14.32 The formation of a silver precipitate or mirror.

14.34

b.

(benzene ring with)

$$\overset{\displaystyle O}{\overset{\displaystyle \|}{\text{C}}} - \text{O}^-\text{NH}_4^+$$

d.

$$\text{CH}_3\text{CH}_2 - \overset{\displaystyle O}{\overset{\displaystyle \|}{\text{C}}} - \text{O}^-\text{NH}_4^+$$

14.36 a. No

b. No

c. Yes

d. Yes

e. No

14.38

$$\underset{\displaystyle \text{CH}_2}{\overset{\displaystyle \text{OH}}{|}} - \underset{\displaystyle \text{CH}}{\overset{\displaystyle \text{OH}}{|}} - \underset{\displaystyle \text{CH}}{\overset{\displaystyle \text{OH}}{|}} - \underset{\displaystyle \text{CH}}{\overset{\displaystyle \text{OH}}{|}} - \left(\underset{\displaystyle \text{CH}}{\overset{\displaystyle \text{OH}}{|}} - \overset{\displaystyle O}{\overset{\displaystyle \|}{\text{C}}} - \text{H}\right)$$

14.40

$$\underset{\displaystyle \text{CH}_2}{\overset{\displaystyle \text{OH}}{|}} - \underset{\displaystyle \text{CH}}{\overset{\displaystyle (\text{OH})}{|}} - \underset{\displaystyle \text{CH}}{\overset{\displaystyle \text{OH}}{|}} - \underset{\displaystyle \text{CH}}{\overset{\displaystyle \text{OH}}{|}} - \underset{\displaystyle \text{CH}}{\overset{\displaystyle \text{OH}}{|}} - \overset{\displaystyle O}{\overset{\displaystyle \|}{\text{C}}} - \text{H} \rightleftharpoons$$

(cyclic sugar structure with CH$_2$—OH, O, OH, H, OH, HO, OH, H substituents around the ring)

14.42

a.

OH
|
CH
(two cyclohexyl groups attached to CH)

b.
$$CH_3CH_2 - \overset{\overset{\displaystyle OCH_2CH_2CH_3}{|}}{\underset{\underset{\displaystyle CH_3}{|}}{C}} - OCH_3 \quad + H_2O$$

c.
$$CH_3CH_2 - \overset{\overset{\displaystyle O-CHCH_3 \, (CH_3)}{|}}{\underset{\underset{\displaystyle O-CHCH_3 \, (CH_3)}{|}}{C}} - H \quad + H_2O$$

d.

$$CH_2 - \overset{\overset{\displaystyle O}{\|}}{C} - OH$$

e.
$$CH_3CH_2\overset{\overset{\displaystyle CH_3}{|}}{CH}CH_2 - OH$$

14.44 Alkenes give alkanes.

Alkynes give alkenes and alkanes.

Aldehydes give primary alcohols.

Ketones give secondary alcohols.

14.46 a.

$+ 2CH_3CH_2OH$

b. $CH_3CH_2 - \overset{\overset{\displaystyle O}{\|}}{C} - CH_2CH_3 + 2CH_3OH$

c. $H - \overset{\overset{\displaystyle O}{\|}}{C} - H + 2CH_3CH_2CH_2OH$

d. $CH_3CH_2\overset{\overset{\displaystyle O}{\|}}{C} - H + 2CH_3OH$

14.48 a. $\overset{\overset{\displaystyle OH}{|}}{CH_2} \quad \overset{\overset{\displaystyle O}{\|}}{C} - CH_3 + HO - CH_2CH_3$

b.

$CH_2 - OH$
$C = O + HO - CH_2CH_2CH_3$
H

14.50

a. $CH_3CH_2CH_2\overset{\overset{\displaystyle OH}{|}}{} + (O) \longrightarrow CH_3CH_2 - \overset{\overset{\displaystyle O}{\|}}{C} - H$

$CH_3CH_2 - \overset{\overset{\displaystyle O}{\|}}{C} - H + 2CH_3CH_2 - OH \underset{}{\overset{H^+}{\rightleftharpoons}}$

$CH_3CH_2 - \overset{\overset{\displaystyle OCH_2CH_3}{|}}{\underset{\underset{\displaystyle OCH_2CH_3}{|}}{CH}} \quad + H_2O$

b. $CH_3\overset{\overset{\displaystyle CH_3}{|}}{CH} - \overset{\overset{\displaystyle O}{\|}}{C} - CH_3 + H_2 \overset{Pt}{\longrightarrow} CH_3\overset{\overset{\displaystyle CH_3}{|}}{CH} - \overset{\overset{\displaystyle OH}{|}}{CH} - CH_3$

$CH_3\overset{\overset{\displaystyle CH_3}{|}}{CH} - \overset{\overset{\displaystyle OH}{|}}{CH} - CH_3 \overset{H_2SO_4}{\underset{180°}{\longrightarrow}} CH_3\overset{\overset{\displaystyle CH_3}{|}}{C} = CH - CH_3 + H_2O$

14.52 a. Menthone

b. Biacetyl

c. Cinnamaldehyde

d. Vanillin

14.54 461 g

14.56

$CH_3CH_2\overset{\overset{\displaystyle O}{\|}}{C}CH_2CH_3 + 2CH_3CH_2CH_2 - OH \overset{H^+}{\longrightarrow}$

$CH_3CH_2\overset{\overset{\displaystyle O-CH_2CH_2CH_3}{|}}{\underset{\underset{\displaystyle O-CH_2CH_2CH_3}{|}}{C}}CH_2CH_3 \quad + H_2O$

14.68 b and c

Chapter 15

15.2 Carboxylic acids with more than ten carbon atoms are called fatty acids. They were originally produced by the saponification of fats.

15.4 lactic acid

15.6 a. Pentanoic acid

b. 3-methylpentanoic acid

c. 4-phenylhexanoic acid

d. 2,4-dimethylbenzoic acid

e. 3-ethyl-5-methylhexanoic acid

15.8

a. $CH_3CH_2CH_2CH_2 - \overset{\overset{\displaystyle O}{\|}}{C} - OH$

b. $CH_3CH_2CH_2\overset{\overset{\displaystyle Br}{|}}{\underset{\underset{\displaystyle CH_3}{|}}{CH}}CH - \overset{\overset{\displaystyle O}{\|}}{C} - OH$

c.

$$\underset{\text{CH}_2\text{CH}_2\text{CH}_3}{\underbrace{}} \overset{\text{O}}{\underset{\parallel}{\text{C}}}-\text{OH}$$

15.10 a. Acetic acid because it forms dimers

b. Propanoic acid because it forms dimers

c. Butyric acid because it has a higher molecular weight

15.12

$$\text{CH}_3\text{CH}_2-\overset{\text{O}}{\underset{\parallel}{\text{C}}}-\text{OH} \cdots \cdots \text{O} \qquad \text{O} \cdots \cdots \text{HO}-\overset{}{\underset{\parallel}{\text{C}}}-\text{CH}_2\text{CH}_3$$

15.14 The carboxyl group is hydrophilic; the carbon chain is hydrophobic.

15.16 pentane < ethoxyethane < 1-butanol < propanoic acid

15.18

$$\text{CH}_3\text{CH}_2-\overset{\text{O}}{\underset{\parallel}{\text{C}}}-\text{O}^-$$

15.20 Under cellular pH conditions, lactic acid ionizes to produce the anion lactate.

15.22

$$\text{CH}_3\text{CH}_2\text{CH}_2\overset{\text{O}}{\underset{\parallel}{\text{C}}}-\text{OH}$$

15.24

a. $\text{H}-\overset{\text{O}}{\underset{\parallel}{\text{C}}}-\text{O}^-\text{Na}^+ + \text{H}_2\text{O}$

b. $\text{CH}_3\text{CH}_2\text{CH}_2 -\overset{\text{O}}{\underset{\parallel}{\text{C}}}-\text{O}^-\text{K}^+ + \text{H}_2\text{O}$

15.26

a. $\text{CH}_3-\overset{\text{O}}{\underset{\parallel}{\text{C}}}-\text{OH} + \text{NaOH} \longrightarrow$

$$\text{CH}_3-\overset{\text{O}}{\underset{\parallel}{\text{C}}}-\text{O}^-\text{Na}^+ + \text{H}_2\text{O}$$

b. $\text{CH}_3-\overset{\text{O}}{\underset{\parallel}{\text{C}}}-\text{OH} + \text{KOH} \longrightarrow$

$$\text{CH}_3-\overset{\text{O}}{\underset{\parallel}{\text{C}}}-\text{O}^-\text{K}^+ + \text{H}_2\text{O}$$

c. $2\text{CH}_3-\overset{\text{O}}{\underset{\parallel}{\text{C}}}-\text{OH} + \text{Ca(OH)}_2 \longrightarrow$

$$(\text{CH}_3-\overset{\text{O}}{\underset{\parallel}{\text{C}}}-\text{O}^-)_2\text{Ca}^{2+} + 2\text{H}_2\text{O}$$

15.28 a. Sodium 3-bromobutanoate

b. Calcium methanoate

c. Potassium 3-phenoxypropanoate

15.30

a. $\text{CH}_3-\overset{\text{O}}{\underset{\parallel}{\text{C}}}-\text{O}^-\text{K}^+$

b.

$$\underset{\text{CH}_3}{\underbrace{}} \overset{\text{O}}{\underset{\parallel}{\text{C}}}-\text{O}^-\text{Na}^+$$

c. $\text{CH}_3\text{CH}_2\overset{\text{CH}_3}{\underset{|}{\text{CH}}}-\overset{\text{O}}{\underset{\parallel}{\text{C}}}-\text{O}^-\text{Na}^+$

15.32 a. Sodium stearate

b. Acetic acid

c. Sodium benzoate

d. Zinc 10-undecylenate

e. Sodium propanoate or calcium propanoate

f. Citric acid/sodium citrate

15.34 b.

$$\boxed{\text{CH}_3-\overset{\text{O}}{\underset{\parallel}{\text{C}}}} \underset{\uparrow}{\Big|}-\text{O}-\text{CH}_2\text{CH}_3$$

d.

$$\underset{}{\bigcirc}-\text{O}\underset{\uparrow}{\Big|}\boxed{\overset{\text{O}}{\underset{\parallel}{\text{C}}}-\text{CH}_3}$$

f.

$$\boxed{\underset{}{\bigcirc}-\overset{\text{O}}{\underset{\parallel}{\text{C}}}}\underset{\uparrow}{\Big|}-\text{O}-\text{CH}_2\text{CH}_3$$

15.36

a. $\text{CH}_3-\overset{\text{O}}{\underset{\parallel}{\text{C}}}-\text{O}-\text{CH}_3 + \text{CH}_3-\overset{\text{O}}{\underset{\parallel}{\text{C}}}-\text{OH}$

b. $\text{CH}_3\overset{}{\underset{\underset{\text{CH}_3}{|}}{\text{CH}}}-\overset{\text{O}}{\underset{\parallel}{\text{C}}}-\text{O}-\text{CH}_2-\bigcirc \quad + \text{ H}_2\text{O}$

c. $\text{CH}_3\overset{}{\underset{\underset{\text{CH}_3}{|}}{\text{CH}}}-\overset{\text{O}}{\underset{\parallel}{\text{C}}}-\text{O}-\bigcirc \quad + \text{ HCl}$

15.38

a. $\text{CH}_3\text{CH}_2-\overset{\text{O}}{\underset{\parallel}{\text{C}}}-\text{O}-\text{CH}_3$

b. $\text{CH}_3\text{CH}_2-\overset{\text{O}}{\underset{\parallel}{\text{C}}}-\text{O}-\bigcirc$

c. $\text{CH}_3\text{CH}_2-\overset{\text{O}}{\underset{\parallel}{\text{C}}}-\text{O}-\text{CH}_2\overset{\text{CH}_3}{\underset{|}{\text{CH}}}\text{CH}_3$

15.40

15.42

$$-(O-\overset{\overset{\displaystyle O}{\|}}{C}-\overset{\overset{\displaystyle O}{\|}}{C}-O-CH_2CH_2CH_2)_n-$$

15.44 a. Ethyl lactate

 b. Propyl butyrate

15.46 a. Methyl 2-methylpropanoate

 b. Ethyl 3,5-dichlorobenzoate

15.48 a. Methyl propionate

 b. Methyl butyrate

 c. Methyl lactate

15.50

 a. $H-\overset{\overset{\displaystyle O}{\|}}{C}-O-$⬡

 b. $\overset{\overset{\displaystyle O}{\|}}{C}-O-CH_3$ (on benzene ring with NO_2)

 c. $CH_3CH-\overset{\overset{\displaystyle O}{\|}}{C}-O-CH_2CH_3$, with Cl on the CH

15.52

$$CH_3-\overset{\overset{\displaystyle O}{\|}}{C}-OCH_2CH_3+H_2O \xrightarrow{H^+}$$

$$CH_3-\overset{\overset{\displaystyle O}{\|}}{C}-OH + CH_3CH_2-OH$$

$$CH_3-\overset{\overset{\displaystyle O}{\|}}{C}-OCH_2CH_3+NaOH \longrightarrow$$

$$CH_3-\overset{\overset{\displaystyle O}{\|}}{C}-O^-Na^+ + CH_3CH_2-OH$$

15.54

 a. $CH_3(CH_2)_{16}-\overset{\overset{\displaystyle O}{\|}}{C}-O^-Na^+ + CH_3CH_2-OH$

 b. $CH_3\overset{\overset{\displaystyle CH_3}{|}}{C}H-\overset{\overset{\displaystyle O}{\|}}{C}-OH + HO-$⬡

15.56

$$CH_2-\overset{\overset{\displaystyle O}{\|}}{\underset{\underset{\displaystyle }{}}{C}}-CH_2-O-\overset{\overset{\displaystyle O}{\|}}{\underset{\underset{\displaystyle OH}{|}}{P}}-OH + H_2O$$

(with OH on first CH$_2$)

15.58 When NaHCO$_3$ reacts with an acid, it forms CO$_2$ gas, water, and a salt:

$$RCO_2H + NaHCO_3 \rightarrow RCO_2^-Na^+ + CO_2 + H_2O$$

There would be no reaction with an alcohol. Therefore, the presence of effervescence ("bubbling") indicates an acid.

15.60

$$CH_3CH_2CH_2\overset{\overset{\displaystyle }{}}{\underset{\underset{\displaystyle CH_3CH_2}{|}}{C}}HCH_2\overset{\overset{\displaystyle O}{\|}}{C}-O-⬡$$

phenyl 3-ethylhexanoate

15.62 1. Use an excessive quantity of ethanol.

 2. Remove the water as it is being formed.

15.74 Polarity: d > a > b > c

Boiling Points: c < b < a < d

Chapter 16

16.2 R—NH$_2$ (primary), R$_2$NH (secondary), and R$_3$N (tertiary).

16.4 a. Secondary

 b. Tertiary

 c. Primary

 d. Secondary

16.6

 (1) $CH_3-CH_2-CH_2-NH_2$ primary

 (2) $CH_3-\overset{\overset{\displaystyle }{}}{\underset{\underset{\displaystyle CH_3}{|}}{C}}H-NH_2$ primary

 (3) $CH_3-CH_2-NH-CH_3$ secondary

 (4) $CH_3-\overset{\overset{\displaystyle }{}}{\underset{\underset{\displaystyle CH_3}{|}}{N}}-CH_3$ tertiary

16.8 a. Methylphenylamine

 b. Diethylamine

 c. Ethylisopropylamine

16.10 a. Cyclohexanamine

 b. *N*-ethylcyclohexanamine

 c. 2-butanamine

16.12 a. *N*-isopropylaniline

 b. *N,N*-diethylaniline

16.14

 a. $CH_3\overset{\overset{\displaystyle NH_2}{|}}{C}HCHCH_2CH_3$, with CH_2CH_3 branch

 b. benzene ring with NH_2 and CH_2CH_3 substituents

c.

16.16 Amines form hydrogen bonds with water.

16.18 Tertiary amines do not form hydrogen bonds among themselves.

16.20

a. $CH_3CH_2 - \overset{\overset{\displaystyle H}{|}}{\underset{\vdots}{N}} - CH_3$

 $\underset{H}{\overset{\displaystyle \vdots}{}} \overset{\displaystyle O}{} \underset{H}{}$

b. $\underset{\overset{}{\bigcirc}}{\overset{\overset{\displaystyle H}{|}}{N}} \cdots \overset{\displaystyle H}{\underset{\overset{}{\bigcirc}}{N}}$

16.22 b < c < a

16.24 $CH_3CH_2 - NH - CH_2CH_3 + H_2O \rightleftharpoons$

 $CH_3CH_2 - \overset{\overset{\displaystyle H}{|}}{\underset{\underset{\displaystyle H}{|}}{N^+}} - CH_2CH_3 + OH^-$

16.26 a. $CH_3CH_2CH_2CH_2 - NH_3{}^+Cl^-$

b. ⬠$- NH_3{}^+ + OH^-$

c. ⬠$- NH_3{}^+ \; CH_3\overset{\overset{\displaystyle O}{||}}{C} - O^-$

d. $CH_3\overset{\overset{\displaystyle NH_2}{|}}{C}HCH_3 + NaCl + H_2O$

e. $CH_3\overset{\underset{\displaystyle CH_3}{|}}{C}H - \overset{\overset{\displaystyle O}{||}}{C} - NH - CH_3 + HCl$

f. ⬡$- \overset{\overset{\displaystyle O}{||}}{C} - NH_2 \; + \;$ ⬡$- \overset{\overset{\displaystyle O}{||}}{C} - OH$

16.28 The amine salts are much more soluble in water (or the aqueous fluids in the body).

16.30

(1) $CH_3 - \overset{\overset{\displaystyle O}{||}}{C} - Cl + CH_3 - NH - CH_3 \longrightarrow$

 $CH_3 - \overset{\overset{\displaystyle O}{||}}{C} - \overset{\underset{\displaystyle CH_3}{|}}{N} - CH_3 + HCl$

(2) $CH_3 - \overset{\overset{\displaystyle O}{||}}{C} - O - \overset{\overset{\displaystyle O}{||}}{C} - CH_3 + CH_3 - NH - CH_3 \longrightarrow$

 $CH_3 - \overset{\overset{\displaystyle O}{||}}{C} - \overset{\underset{\displaystyle CH_3}{|}}{N} - CH_3 + HO - \overset{\overset{\displaystyle O}{||}}{C} - CH_3$

16.32 a. NaOH

b. NH_3

c. HCl

16.34 Synapse

16.36 Norepinephrine and serotonin

16.38 Norepinephrine, serotonin, acetylcholine, and dopamine

16.40 Epinephrine is used clinically in local anesthetics to constrict blood vessels in the area of injection.

16.42 Plants

16.44 a. Caffeine

b. Atropine

c. Nicotine

d. Codeine

e. Quinine

f. Morphine

16.46 a. N-methylbutanamide

b. N,N-dimethylethanamide

c. 2-methylbenzamide or o-methylbenzamide

d. 2-methylbutanamide

16.48

a. $CH_3CH_2CH_2 - \overset{\overset{\displaystyle O}{||}}{C} - NH_2$

b. ⬡$- \overset{\overset{\displaystyle O}{||}}{C} - NH - CH_2CH_3$

c. $CH_3CH_2 - \overset{\overset{\displaystyle O}{||}}{C} - \overset{\overset{\displaystyle CH_3}{|}}{N} - CH_3$

16.50

a. ⬠$- \overset{\overset{\displaystyle O}{||}}{C} - \overset{\underset{\displaystyle H}{|}}{N} - H \cdots \overset{\overset{}{}}{\underset{\displaystyle H}{O}} \overset{\displaystyle H}{}$

b. $CH_3\overset{\underset{\displaystyle CH_3}{|}}{C}H - \overset{\overset{\displaystyle O}{||}}{C} - \overset{\underset{\displaystyle H}{|}}{N} - H \cdots O = C - NH_2$ with $\underset{\displaystyle CH_3}{\overset{\displaystyle CHCH_3}{|}}$

16.52

a. $CH_3 - \overset{\overset{\displaystyle O}{||}}{C} - O^-Na^+ + CH_3CH_2 - NH_2$

b. $CH_3CH_2 - \overset{\overset{\displaystyle O}{||}}{C} - OH + NH_4{}^+Cl^-$

16.54

a benzoate structure with CH$_3$ substituent: $C-O^-Na^+ + CH_3CH_2-NH-CH_2CH_3$

16.56 a. Saturated

 b. One C=C bond

 c. Two C=C bonds

16.58 Na$^+$ $^-$O$_2$C(CH$_2$)$_4$CO$_2^-$ Na$^+$ and NH$_2$(CH$_2$)$_6$NH$_2$

16.60

CH_3C-OH and $HO-CH_2CH_2-N(CH_3)_3^+$

acetic acid choline

16.70 Neurotransmitter is a substance that acts as a chemical bridge in nerve impulse transmission between nerve cells.

16.72 d

Chapter 17

17.2 a. A structural material in plants

 b. An energy source in our diet

 c. A stored form of energy in animals

 d. A stored form of energy in plants

17.4 a. Carbohydrate, disaccharide

 b. Carbohydrate, monosaccharide

 c. Carbohydrate, polysaccharide

 d. Carbohydrate, monosaccharide

 e. Carbohydrate, polysaccharide

 f. Carbohydrate, disaccharide

 g. Carbohydrate, polysaccharide

 h. Carbohydrate, polysaccharide

17.6 They don't have four different groups attached; C number 1 has only three groups; C number 3 has four groups, but two are hydrogens.

17.8 a. No chiral carbon atoms

 b. CH$_3$CH$_2$—CH*—C(=O)—CH$_3$ with OH on the CH. Yes, can have enantiomers. * denotes the chiral carbon

 c. cyclopentane—C*—CH$_2$CH$_3$ with OH and CH$_3$. Yes, can have enantiomers. * denotes the chiral carbon

17.10 The chiral carbon is at the intersection of the lines. The H and OH are above the plane of the paper, and the CHO and CH$_2$CH$_3$ are below the plane.

17.12 a. The D form is given; the L enantiomer is:

CHO
H—OH
H—OH
HO—H
CH$_2$OH

 b. The L form is given; the D enantiomer is:

CH$_2$OH
C=O
HO—H
H—OH
H—OH
CH$_2$OH

17.14 a.

COOH COOH
H$_2$N—H H—NH$_2$
CH$_3$ CH$_3$
L-alanine D-alanine

 b.

COOH COOH
H$_2$N—H H—NH$_2$
CH$_2$CH(CH$_3$)$_2$ CH$_2$CH(CH$_3$)$_2$
L-leucine D-leucine

17.16 a. Two chiral carbon atoms; four stereoisomers

 b. Three chiral carbon atoms; eight stereoisomers

17.18 8 aldopentoses; 4 are in the D form, and 4 are in the L form.

17.20 Optically active molecules rotate the plane of polarized light.

17.22 a. Ketohexose

 b. Aldopentose

17.24 Some carbohydrates have a sweet taste; these are sugars.

17.26 * denotes sites where the hydrogens of a water molecule can hydrogen-bond.

CHO*
H—*OH
HO*—H
H—*OH
H—*OH
CH$_2$*OH

17.28 a. The α form is given in the text; the β form is:

b. The β form is given in the text; the α form is:

17.30

α-mannose β-mannose

17.32 Nonreducing sugar

17.34 a.

b.

c.

17.36 Ribose is found is RNA and deoxyribose is found in DNA.

17.38 Dextrose; blood sugar

17.40 The hexoses glucose and galactose differ in the positions of H and OH on carbon number 4.

17.42 Fructose can be used as a low-calorie sweetener because the caloric content per gram is about the same as for sucrose. However, since fructose is so much sweeter, less is needed to provide the same degree of sweetness.

17.44 a. Sucrose

 b. Maltose

 c. Lactose

 d. Maltose

 e. Sucrose

 f. Sucrose

17.46 The glucose ring opens at the hemiacetal group, exposing an aldehyde group that then reacts with the Cu^{2+} of the Benedict's reagent.

17.48 Perform a Benedict's test. Sucrose is not a reducing sugar, but honey contains reducing sugars.

17.50 The right ring in lactose has a hemiacetal group that can open to form the aldehyde and alcohol; hence, it is a reducing sugar. Sucrose has no hemiacetal bonds, so it can't reduce Cu^{2+}.

17.52 a. Galactose and glucose

 b. Yes; there is a hemiacetal on the glucose ring.

 c. $\alpha(1 \rightarrow 6)$

17.54 a. Amylose

 b. Cellulose

 c. Amylopectin

 d. Cellulose

 e. Glycogen

17.56 The boiling point of the glucose solution will be higher because the molarity of the solution is higher.

17.58 Glucose and fructose are formed in equal amounts (52.6 g). The extra mass comes from the water needed for hydrolysis.

17.60 As hydrolysis occurred, the molarity of the solution would increase, causing an increase in osmotic pressure, and the solution level in the tube would rise.

17.70 b and c

17.72 a. Monosaccharide

 b. Monosaccharide

 c. Disaccharide: glucose and fructose

 d. Disaccharide: two glucose units

 e. Disaccharide: glucose and galactose

 f. Monosaccharide

17.74 Cu^{2+}

Chapter 18

18.2 Energy storage, structural components, some hormones

18.4 a. Nonsaponifiable

 b. Saponifiable

 c. Saponifiable

 d. Saponifiable

 e. Saponifiable

 f. Nonsaponifiable

18.6 1. They have a straight-chain carbon segment with a carboxylic acid group.

2. The straight-chain carbon segment is 10–20 carbons long.

3. There is an even number of total C atoms.

4. The carbon chain may be saturated or unsaturated.

18.8 Linoleic acid and linolenic acid are essential fatty acids because they are needed by the body but are not synthesized within the body.

18.10 Unsaturated fatty acids have a bent structure that prevents the molecules from packing as tightly as saturated fatty acids.

18.12 Fats and oils are both triglycerides (esters). Fats are solid at room temperature; oils are liquid at room temperature. Fats tend to have more saturated fatty acids, whereas oils tend to have more unsaturated fatty acids.

18.14

$$CH_2-O-\overset{\overset{O}{\|}}{C}-(CH_2)_{12}CH_3$$
$$CH-O-\overset{\overset{O}{\|}}{C}-(CH_2)_7CH=CH(CH_2)_5CH_3$$
$$CH_2-O-\overset{\overset{O}{\|}}{C}-(CH_2)_6(CH_2CH=CH)_2(CH_2)_4CH_3$$

······ ester linkages

18.16 Triglyceride B

18.18 Hydrogenation

18.20 A number of commercially useful margarines and cooking shortenings are prepared by the hydrogenation of vegetable oils.

18.22 a.

$$CH_2-O-\overset{\overset{O}{\|}}{C}-(CH_2)_{16}CH_3$$
$$CH-O-\overset{\overset{O}{\|}}{C}-(CH_2)_{16}CH_3 + 3H_2O \xrightarrow{H^+}$$
$$CH_2-O-\overset{\overset{O}{\|}}{C}-(CH_2)_{16}CH_3$$

$$\begin{array}{l} CH_2-OH \\ CH-OH + 3HO-\overset{\overset{O}{\|}}{C}-(CH_2)_{16}CH_3 \\ CH_2-OH \qquad\quad \text{stearic acid} \\ \text{glycerol} \end{array}$$

b. Same reaction as in part a.

c.

$$CH_2-O-\overset{\overset{O}{\|}}{C}-(CH_2)_6(CH_2-CH=CH)_2(CH_2)_4CH_3$$
$$CH-O-\overset{\overset{O}{\|}}{C}-(CH_2)_6(CH_2-CH=CH)_3CH_2CH_3 + 5H_2 \xrightarrow{Ni}$$
$$CH_2-O-\overset{\overset{O}{\|}}{C}-(CH_2)_6(CH_2-CH=CH)_2(CH_2)_4CH_3$$

$$CH_2-O-\overset{\overset{O}{\|}}{C}-(CH_2)_{16}-CH_3$$
$$CH-O-\overset{\overset{O}{\|}}{C}-(CH_2)_{16}CH_3$$
$$CH_2-O-\overset{\overset{O}{\|}}{C}-(CH_2)_6(CH_2-CH=CH)_2(CH_2)_4CH_3$$

d.

$$CH_2-O-\overset{\overset{O}{\|}}{C}-(CH_2)_{16}CH_3$$
$$CH-O-\overset{\overset{O}{\|}}{C}-(CH_2)_{16}CH_3 + 3NaOH \longrightarrow$$
$$CH_2-O-\overset{\overset{O}{\|}}{C}-(CH_2)_{16}CH_3$$

$$\begin{array}{l} CH_2-OH \\ CH-OH + 3Na^{+}{}^{-}O-\overset{\overset{O}{\|}}{C}-(CH_2)_{16}CH_3 \\ CH_2-OH \qquad\qquad\qquad \text{soap} \\ \text{glycerol} \end{array}$$

18.24 Fats are a triester between glycerol and fatty acids. Waxes are a monoester between a long-chain alcohol and a fatty acid.

18.26 In nature, waxes often occur as protective coatings.

18.28
g — fatty acid
l
y
c — fatty acid
e
r
o — phosphoric acid
l
 alcohol

18.30 Lecithins serve as components in cell membranes and aid in lipid transport.

18.32 They differ in the identity of the amino alcohol portion. Lecithins contain choline, and cephalins contain ethanolamine or serine.

18.34
s
p
h
i
n — fatty acid
g
o
s
i — phosphoric acid
n
e choline

18.36 Tay-Sachs, Gaucher's, and Niemann-Pick

18.38 Cerebrosides, brain tissue

18.40 Phosphoglycerides, sphingomyelin, and cholesterol

18.42 The model is composed of a flexible bilayer structure with protein molecules floating in the bilayer. The lipid molecules are free to move in a lateral direction.

18.44 The fused ring system:

18.46 Bile salts, sex hormones, and adrenocorticoid hormones. Cell membranes.

18.48 Cholesterol

18.50 The two groups are glucocorticoids and mineralocorticoids. Cortisol, a glucocorticoid, acts to increase the glucose and glycogen concentrations in the body. Aldosterone, a mineralocorticoid, regulates the retention of Na^+ and Cl^- during urine formation.

18.52 Testosterone; estradiol, estrone, and progesterone

18.54 Estrogens are involved in egg development; progesterone causes changes in the wall of the uterus to prepare it to accept a fertilized egg and maintain the resulting pregnancy.

18.56 Arachidonic acid

18.58 Induction of labor, treatment of asthma, treatment of peptic ulcers

18.60 The lipid is unsaturated (see Figure 12.5).

18.62 The phosphate and amino alcohol groups found in complex lipids are much more hydrophilic and capable of interacting with water than simple lipids.

18.74 a. Saturated

b. Unsaturated

c. Unsaturated

d. Saturated

18.76 c

18.78 c

18.80 b

18.82 b

18.84 c

Chapter 19

19.2 The amine group (—NH_2) and the carboxylic acid group (—COOH), or the corresponding protonated —NH_3^+ group and unprotonated —COO^- group.

19.4 a. $H_3\overset{+}{N}$—C—COO^-, with H above C and CH—CH_3 / OH below.

b. $H_3\overset{+}{N}$—C—COO^-, with H above C and CH_2 / COO^- below.

c. $H_3\overset{+}{N}$—C—COO^-, with H above C and CH_2 / OH below.

d. $H_3\overset{+}{N}$—C—COO^-, with H above C and CH_2—(phenyl ring) below.

19.6

^+H_3N—$\overset{*}{C}$—COO^-, with H above $\overset{*}{C}$ and CH—CH_3 / CH_2 / CH_3 below

and ^+H_3N—CH—COO^-, with H—$\overset{*}{C}$—CH_3 and CH_2 / CH_3 below

19.8 a.

$$H_3\overset{+}{N}\text{——H} \qquad H\text{——}\overset{+}{N}H_3$$

with COO^- on top and CH_2OH on bottom

L-serine D-serine

b.

$$H_3\overset{+}{N}\text{——H} \qquad H\text{——}\overset{+}{N}H_3$$

with COO^- on top and CH—CH_3 / CH_3 on bottom

L-valine D-valine

19.10 They are crystalline solids with relatively high melting points and high water solubilities.

19.12

a. H_2N—CH—COO^-, with CH_2 / SH below

b. H_2N—CH—COO^-, with CH_3 below

19.14 a.

$$H_3\overset{+}{N}-CH-COO^- + OH^- \longrightarrow H_2N-CH-COO^- + H_2O$$

with side chain CH_2-OH on both.

b.

$$H_3\overset{+}{N}-CH-COO^- + H_3O^+ \longrightarrow H_3\overset{+}{N}-CH-COOH + H_2O$$

with side chain CH_2-OH on both.

19.16

$$H_3\overset{+}{N}-CH-\overset{\overset{\displaystyle O}{\|}}{C}-NH-CH-\overset{\overset{\displaystyle O}{\|}}{C}-NH-CH-\overset{\overset{\displaystyle O}{\|}}{C}-O^-$$

side chains: $CHCH_3 / CH_3$; CH_2 / COO^- ; CH_2 / SH

Val-Asp-Cys

19.18 Ala-Phe-Arg, Phe-Arg-Ala, Arg-Ala-Phe

Ala-Arg-Phe, Phe-Ala-Arg, Arg-Phe-Ala

19.20 Six

19.22 Each peptide contains a disulfide bridge in which two cysteines have been linked.

19.24 Normally, proteins do not pass through membranes, and the appearance of an abnormal protein in blood or urine indicates cellular damage, which has released those substances.

19.26 When a protein is at a pH corresponding to the isoelectric point, the net charge on the protein is 0. The protein is not as soluble with a 0 charge as when it possesses a net negative or positive charge.

19.28 Catalysis, structural, storage, protection, regulatory, nerve impulse transmission, motion, transport

19.30 a. Fibrous; collagen is a structural protein.

b. Globular; lactate dehydrogenase is an enzyme and needs to be water soluble.

19.32 Simple proteins contain only amino acid residues, whereas conjugated proteins contain additional components.

19.34 The sequence of the amino acid residues in the protein.

19.36

$$-NH-CH-\overset{\overset{\displaystyle O}{\|}}{C}-NH-CH-\overset{\overset{\displaystyle O}{\|}}{C}-NH-CH-\overset{\overset{\displaystyle O}{\|}}{C}-NH-CH-\overset{\overset{\displaystyle O}{\|}}{C}-$$

side chains: R, R', R'', R'''

19.38 Hydrogen bonding between the amide hydrogen and the carboxyl oxygen along the backbone.

19.40 In secondary structures, the hydrogen bonding is between the atoms in the backbone; in tertiary structures, the hydrogen bonding is between atoms in the side chains.

19.42 a. Hydrogen bonding

b. Salt bridges

c. Hydrophobic interactions

d. Hydrophobic interactions

19.44 The nonpolar R groups in phenylalanine, alanine, and methionine will point inward; the polar R groups in glutamate and lysine will point toward the water.

19.46 Ionic attractions, disulfide bridges, hydrogen bonds, and hydrophobic forces.

19.48 A subunit is a polypeptide chain that is part of a larger protein.

19.50 The products of hydrolysis are the separate, individual amino acid molecules. The product of denaturation has the same primary protein backbone with different secondary, tertiary, and quaternary structures.

19.52 The primary structure of the protein is the same.

19.54 Cooking egg white is an irreversible denaturation. Whipping air into them is a reversible denaturation.

19.56 The heavy metal ions denature the protein in the egg white by interacting with the —SH groups and the carboxylate (—COO⁻) groups. Vomiting should then be induced to remove the denatured proteins so they are not digested, releasing the metal ions.

19.58 valine, methionine, and threonine

19.60

Conjugate acid: $H_3\overset{+}{N}-\overset{\overset{\displaystyle H}{|}}{\underset{\underset{\displaystyle R}{|}}{C}}-COOH$

Conjugate base: $H_2N-\overset{\overset{\displaystyle H}{|}}{\underset{\underset{\displaystyle R}{|}}{C}}-COO^-$

19.72 Amino and carboxylate groups; 20

19.74 b > c,a

19.76 b, c, and d

19.78 d

Chapter 20

20.2 Enzymes are highly specific in the reactions they catalyze, and their activity as catalysts can be regulated.

20.4 Each one is specific for a single type of reaction and often is specific for a single substrate.

20.6 Enzymes that are specific for a single reactant exhibit absolute specificity.

20.8 Enzyme Commission (EC) system

20.10 a. Sucrose

b. Amylose

c. Lactose

d. Maltose

e. Arginine

20.12 a. Catalyzes the dehydrogenation (oxidation) of succinate

b. Catalyzes the reduction of L-amino acids

c. Catalyzes the oxidation of cytochrome

d. Catalyzes the isomerization of glucose-6-phosphate

20.14 An enzyme is a catalytically active structure made up of a protein portion called an apoenzyme and a nonprotein portion called a cofactor.

20.16 Mg^{2+}, Zn^{2+}, Fe^{2+}, Ca^{2+}

20.18 E = enzymes, S = substrate, ES = enzyme–substrate complex, and P = product of reaction. Thus $E + S \rightleftarrows ES \rightarrow E + P$ represents the reaction of an enzyme with a substrate molecule to form an enzyme–substrate complex, which then gives the reaction product and regenerates the enzyme.

20.20 The shape of the enzyme active site allows only specific substrates having the correct shape to bind and be acted upon.

20.22 Induced-fit theory, because the enzyme catalyzes reactions of several related compounds.

20.24 Any observation that allows the rate of product formation or reactant depletion to be measured will allow enzyme activity to be determined.

20.26 One international unit (IU) is the amount of enzyme that will convert 1 micromol (μmol) of substrate to product in 1 minute. The international unit is useful in medical diagnosis because it represents a standard quantity of enzyme (or enzyme activity level) against which the level present in a patient can be compared.

20.28 a. The enzyme activity (rate of reaction) increases with increased substrate concentration until all the enzyme is utilized, at which point no further increase in rate occurs.

b. The activity (rate) increases with an increase in enzyme concentration.

c. There is an optimum pH for the reaction, so at a pH above or below the optimum value, the rate will decrease.

d. The activity (rate) increases up to a maximum as the temperature increases, and then decreases with further temperature increase.

20.30 Prepare a series of reactions with constant enzyme amount and increasing substrate amounts. Plot reaction rate versus substrate concentration. A constant rate at higher substrate concentrations is V_{max}.

20.32 Enzymes could be denatured at pH values far from the optimum pH.

20.34 In competitive inhibition another substance, an inhibitor, competes with the substrate in binding to the active site of the enzyme. In noncompetitive inhibition the inhibitor binds to another site on the enzyme, and in doing so alters the shape of the enzyme so that it can no longer bind to the substrate, essentially prohibiting a reaction.

20.36 a. Cyanide poisoning is treated with $Na_2S_2O_3$ (sodium thiosulfate), which oxidizes CN^- to SCN^-. The thiocyanate does not bind to the iron in the enzyme.

b. Heavy metal poisoning is treated by administering chelating agents such as EDTA. The chelating agents bond to the metal ions to form water-soluble complexes that are excreted from the body in the urine.

20.38 1. Zymogens or proenzymes

2. Allosteric regulation

3. Genetic control over the synthesis of enzymes

20.40 So the blood doesn't clot as it circulates, but only when clotting is needed to stop bleeding.

20.42 Modulators may increase activity (activators) or decrease the activity (inhibitors).

20.44 Genetic control of enzyme activity means that a cell synthesizes an enzyme when conditions temporarily require it. Such a controlled synthesis is called enzyme induction.

20.46 a. CK assays are used in diagnosing a heart attack.

b. ALP assays are useful in diagnosing liver or bone disease.

c. Amylase assays are useful in diagnosing diseases of the pancreas.

20.48 An LDH assay is a good initial diagnostic test because the isoenzyme forms of LDH are so widespread in the human system. Thus, the locations of diseases can be pinpointed.

20.50 An ionic bond is formed:

sample substrate arginine side chain

20.52 A covalent ester linkage is formed between the acid chloride and the —OH of the serine:

20.54 There will be no shift in the equilibrium. Enzymes are catalysts and cannot change the position of equilibrium, only the speed of which an equilibrium is reached.

20.68 c

20.70 Pepsin; no

20.72 c

20.74 a

Chapter 21

21.2 Deoxyribose in DNA, ribose in RNA

21.4 a. Purine

b. Pyrimidine

c. Pyrimidine

d. Pyrimidine

e. Purine

21.6

21.8 By convention, in the DNA segment AGTCAT, the 5′ end is the A and the 3′ end is the T.

21.10 Each DNA has two complementary nucleic acid strands arranged in a double helix with the bases protruding into the interior of the helix and the 5′ end of one strand next to the 3′ end of the other. The bases on one strand form hydrogen bonds with the bases on the other.

21.12 a. 15

b. 14

21.14 GATGCAT

21.16 A chromosome is a tightly packed bundle of DNA and protein that is involved in cell division. A human cell normally contains 46 chromosomes with about 25,000 genes.

21.18 The point in the double helix where "unwinding" begins.

21.20 1. The double helix unwinds.

2. The new DNA segments are synthesized.

3. The "nicks" in the new strand are closed.

21.22 The daughter strand forms from its 5′ end toward the 3′ end.

21.24 a. ACGTCT

TGCAGA

b.

Old strand	5′ ACGTCT 3′	daughter
New strand	3′ TGCAGA 5′	DNA molecule
New strand	5′ ACGTCT 3′	daughter
Old strand	3′ TGCAGA 5′	DNA molecule

21.26 The sugar in the RNA backbone is ribose, and that in the DNA backbone is deoxyribose.

21.28 1. mRNA functions as a messenger. It consists of a base sequence that is complementary to a DNA strand. It is formed in the nucleus and then moves into the cytoplasm where protein synthesis occurs. It carries the genetic code to control protein synthesis.

2. Ribosomal RNA functions as the site of protein synthesis in the ribosome.

3. tRNA delivers an amino acid to the site of protein synthesis.

21.30 The two important regions of a tRNA molecule are the anticodon, which binds the tRNA to mRNA during protein synthesis, and the 3′ end, which binds to an amino acid by an ester linkage.

21.32 Transcription is the transferring of genetic DNA information to messenger RNA. Translation is the expression of messenger RNA information in the form of an amino acid sequence in a protein.

21.34 5′ CAAUAUGGC 3′

21.36 In the DNA of eukaryotic cells, segments called introns carry no amino acid coding, whereas exons are coded segments. Both introns and exons are copied (transcribed) and produce hnRNA. A series of enzyme-catalyzed reactions cut out the introns and splice the remaining segments to produce mRNA.

21.38 a. 5′ AUA 3′, tyrosine

b. 5′ AUG 3′, histidine

c. 5′ UGA 3′, serine

d. 5′ AGA 3′, serine

21.40 a. False; should be three bases.

b. True

c. False; 61 of 64 codons represent amino acids.

d. False; each living species is thought to use the same genetic code.

e. True

f. False; stop signals for protein synthesis are represented by three codons: UAA, UAG, and UGA.

21.42 A complex of mRNA and several ribosomes; also called polyribosome.

21.44 Initiation, elongation, termination

21.46 The A site (aminoacyl site) is where an incoming tRNA carrying the next amino acid will bond. The P site (peptidyl site) is where the tRNA carrying the peptide chain is located.

21.48 A change in the DNA of a cell wherein a wrong base sequence is produced on the DNA.

21.50 The protein synthesized would contain a proline rather than an alanine residue at that site in the protein.

21.52 Recombinant DNA is DNA of one organism that contains segments of DNA from another organism.

21.54 Human insulin will become widely available for treating diabetes. Human growth hormone will be used for treating pituitary disorders. Recombivax Hb, a vaccine for hepatitis B, will be marketed.

21.56 Hydrolysis of the phosphodiester linkage would break the nucleic acid backbone and allow the removal of intron segments. Esterification (ester formation) of the $5′—PO_4^{2-}$ group of one exon segment to the $3′—OH$ group of another exon segment would be used to join exons.

21.58 a. Glucose and fructose

b. Enzyme induction

c. Somehow the presence of sucrose activates the bacterial cell's DNA to produce more copies of mRNA that codes for enzyme Q.

21.70 a and c

21.72 d

21.74 c

21.76 c

Chapter 22

22.2 Macronutrients are needed in large quantities per day (1 gram or more). Micronutrients are needed in small quantities (milligrams or less).

22.4

Carbohydrates	Energy source
	Provide materials for cell and tissue synthesis
Proteins	Provide amino acids for cell repair and maintenance
	Provide ingredients for the synthesis of enzymes and hormones
Lipids	Energy source; solvent for fat-soluble vitamins
	Source of essential fatty acids

22.6 a. Carbohydrates, lipids, proteins

b. Carbohydrates, lipids, proteins

c. Carbohydrates, proteins

d. Carbohydrates, lipids, proteins

e. Proteins, lipids

f. Lipids, proteins, carbohydrates

22.8 Fat, 65 g; saturated fat, 20 g; total carbohydrate, 300 g; fiber, 25 g

22.10 a. Fat soluble

b. Water soluble

c. Water soluble

d. Fat soluble

22.12 Water-soluble vitamins can be excreted in water-based fluids much more readily than fat-soluble vitamins. Thus, large doses of fat-soluble vitamins accumulate in the body.

22.14 a. Vitamin C

b. Vitamin B_1

c. Vitamin B_{12}

d. Niacin

22.16 Ca: component of bones and teeth; regulation of muscle activity

P: component of bones and teeth, nucleic acids, phospholipids, and enzymes

K: charge balance in cells; regulates heartbeat

S: component of the amino acid cysteine; charge balance in cells (as SO_4^{2-})

Na: charge balance in cells

Cl: charge balance in cells

Mg: component of enzymes; regulation of muscle activity

22.18 The general functions of trace minerals in the body are to serve as components of vitamins (Co), enzymes (Zn, Se), hormones (I), or specialized proteins (Fe, Cu).

22.20 Hydrogen

22.22 $6CO_2 + 6H_2O + energy \rightarrow C_6H_{12}O_6 + 6O_2$

22.24 Energy from the sun powers the process of photosynthesis in plants. As the carbohydrates of plants are converted to CO_2 and H_2O, the released energy is trapped in molecules of ATP.

22.26 Catabolism includes all of the breakdown processes. Anabolism includes all of the synthesis processes.

22.28 a. Stage III

b. Stage I

c. Stage III

d. Stage II

22.30 The production of ATP.

22.32 ATP is involved in both.

22.34 ATP molecules store the energy released during the oxidation of food and deliver the energy to the parts of the cell where energy is needed.

22.36 P_i is the symbol for the phosphate ion. PP_i represents the pyrophosphate ion.

22.38 The hydrolysis of phosphoenolpyruvate liberates 14.8 kcal/mol, while the hydrolysis of glycerol-3-phosphate liberates only 2.2 kcal/mol.

22.40 The triphosphate portion.

22.42 Mitochondria are the sites for most ATP synthesis in the cells.

22.44 Mitochondria are cigar-shaped organelles with both an outer and an inner membrane. The inner membrane has many folds and contains the enzymes for the electron transport chain and oxidative phosphorylation. The enzymes for the citric acid cycle are attached to or near to the surface of the inner membrane.

22.46 a. Coenzyme A is part of acetyl coenzyme A, which is formed from all foods as they pass through the common metabolic pathway. Acetyl coenzyme A is the fuel for the citric acid cycle.

b. and c. FAD and NAD^+ carry hydrogen atoms from the citric acid cycle to the electron transport chain.

22.48 $^-OOC - CH_2CH_2 - COO^-$ is oxidized. FAD is reduced.

22.50 a. $HO-CH_2-COOH + NAD^+ \rightarrow$
$$O{=}CH-COOH + NADH + H^+$$

b. $HO-CH_2CH_2-OH + FAD \rightarrow$
$$HO-CH{=}CH-OH + FADH_2$$

22.52

FAD FADH₂

22.54 $CH_4 + O_2 + 2\,NAD^+ \rightarrow CO_2 + 2\,NADH + 2\,H^+$

A total of four electrons are transferred to the 2NADH molecules.

22.56 Protein synthesis, translation

22.68 a. Calcium

b. Iodine

c. Iron

d. A vitamin is organic; minerals are inorganic

22.70 Mitochondria

22.72 Stage 1: Food is digested. Stage 2: Acetyl CoA is formed. Stage 3: The citric acid cycle and the electron transport chain complete the conversion of food to ATP. Hydrolysis is the most common reaction of digestion.

22.74 a

22.76 a

Chapter 23

23.2 a. Glucose

b. Glucose and galactose

c. Glucose and fructose

d. Glucose

23.4 Blood sugar level is the concentration of glucose in the blood. The normal fasting level is the blood glucose concentration after a fast of 8–12 hours.

23.6 **a.** A blood sugar level below the normal range.

b. A blood sugar level above the normal fasting level.

c. A blood sugar level at which glucose is not reabsorbed by the kidneys.

d. An elevated blood sugar level resulting in the excretion of glucose in the urine.

23.8 If the blood sugar is too low, severe hypoglycemia can cause convulsions and coma.

23.10 Glucose gives pyruvate.

23.12 Steps 1 and 3 require ATP, Steps 7 and 10 produce ATP. A net of 2 ATP molecules are produced for each glucose molecule.

23.14 The third control point is in the last step of glycolysis. The enzyme for this step, pyruvate kinase, is inhibited by a high ATP concentration. If the cell does not need ATP, the ATP concentration will inhibit glycolysis and slow the production of ATP. As the need for ATP increases, the inhibition of the enzyme no longer occurs and glycolysis proceeds to make more ATP.

23.16 Glucose-6-phosphate is a feedback inhibitor of the enzyme hexokinase and causes a decrease in the enzymatic turnover rate (activity) of that enzyme.

23.18 Aerobic denotes the presence of oxygen, whereas anaerobic means without oxygen.

23.20 Under aerobic conditions, the pyruvate is converted to acetyl CoA. Under anaerobic conditions, the pyruvate is converted to lactate.

23.22 It is released as CO_2. Ethanol formation allows the yeast to regenerate NAD^+ in the absence of oxygen and still generate sufficient ATP to sustain life.

23.24 The citric acid cycle is the principal process for generating NADH and $FADH_2$, which are necessary for ATP synthesis. The citric acid cycle is also important as a source of intermediates in the biosynthesis of amino acids.

23.26 **a.** acetyl CoA

b. CO_2

c. NADH and $FADH_2$

23.28 **a.** Three

b. One

c. One

d. Two

23.30

$$\begin{array}{c} COO^- \\ | \\ CH_2 \\ | \\ CH_2 \\ | \\ COO^- \end{array} + FAD \longrightarrow \begin{array}{c} H \quad COO^- \\ \diagdown \ \diagup \\ C \\ \| \\ C \\ \diagup \ \diagdown \\ {}^-OOC \quad H \end{array} + FADH_2$$

23.32 **a.** Step 1, citrate synthetase

Step 3, isocitrate dehydrogenase

Step 4, α-ketoglutarate dehydrogenase

b. When supplies of NADH and ATP are abundant, the cell has sufficient energy for its needs, so the operation of the citric acid cycle is inhibited.

c. As the cell uses ATP and makes ADP, the enzymes controlling the cycle are activated to make more ATP.

23.34 NADH and $FADH_2$

23.36 Cytochromes pass electrons along the electron transport chain to oxygen, which combines with H^+ ions to form water.

23.38 NADH and $FADH_2$ are oxidized to NAD^+ and FAD, respectively. ADP is phosphorylated to ATP.

23.40 1.5 molecules of ATP are formed during a sequence of reactions.

23.42 100 ATP molecules

23.44 94 mol of ATP

23.46 ATP synthesis results from the flow of protons (H^+) across the inner mitochondrial membranes.

23.48 **a.** 2 **b.** 2 **c.** 28

23.50 Glycogenesis is the synthesis of glycogen, while glycogenolysis is the breakdown of glycogen.

23.52 UTP

23.54 Liver cells have the essential enzyme glucose-6-phosphatase, which is lacking in muscle tissue.

23.56 The liver

23.58 Lactate (formed from pyruvate generated in the muscles), glycerol (derived from the hydrolysis of fats), and certain amino acids (from the amino acid pool).

23.60 There would be no lactate produced, so the Cori cycle wouldn't operate.

23.62 **a.** Insulin promotes the utilization of blood glucose, causing the level to decrease. Glucagon activates the breakdown of glycogen, thereby raising blood glucose levels.

b. Insulin promotes glycogen formation, while glucagon stimulates the breakdown.

23.64 1.00×10^{24} ATP molecules

23.66 The removal of lactate from muscle cells shifts the equilibrium to the right, ensuring the regeneration of NAD^+ and the production of pyruvate by glycolysis. In the liver, pyruvate is used to make glucose, thus decreasing the pyruvate concentration, which shifts the equilibrium to the left and prevents the accumulation of lactate.

23.68

$$\begin{array}{c} \boxed{\begin{array}{c} e \\ n \\ z \\ y \\ m \\ e \end{array}} - CH_2 - OH \end{array} \xrightarrow[\text{ATP} \quad \text{ADP}]{} \begin{array}{c} \boxed{\begin{array}{c} e \\ n \\ z \\ y \\ m \\ e \end{array}} - CH_2 - O - \overset{\displaystyle O}{\underset{\displaystyle |}{\overset{\displaystyle \|}{P}}} - O^- \\ O^- \end{array}$$

An ester linkage is formed.

23.80 monosaccharide, glucose

23.82 **a.** Acetyl CoA

b. Lactate

c. Ethanol

23.84 Oxygen; $2H^+ + 2e^- + 1/2\ O_2 \rightarrow H_2O$

23.86 d

23.88 d

23.90 a

Chapter 24

24.2 Glycerol, fatty acids, and monoglycerides

24.4 Chylomicrons are lipoprotein aggregates found in the lymph and the bloodstream.

24.6 Chylomicrons, very low-density lipoproteins, low-density lipoproteins, high-density lipoproteins

24.8 Fat mobilization is the process of hydrolyzing triglycerides to produce fatty acids and glycerol, both of which can be used for energy in muscle cells.

24.10 Resting muscle and liver cells

24.12 Glycerol can be converted to pyruvate and contribute to cellular energy production or be converted to glucose.

24.14 A fatty acid is first converted to a fatty acyl CoA, a high-energy compound. This occurs in the cytoplasm of the cell.

24.16 FAD and NAD^+

24.18

$$CH_3(CH_2)_{14}-\overset{\overset{\displaystyle O}{\|}}{C}-OH + HS-CoA \longrightarrow$$

ATP AMP + 2P$_i$

$$CH_3(CH_2)_{14}-\overset{\overset{\displaystyle O}{\|}}{C}-SCoA + H_2O$$

24.20 The word *cycle* implies that the product is identical to the starting material, which is not the case. The fatty acyl product of a trip through the sequence has two fewer carbon atoms than the starting material.

24.22 a.

$$CH_3CH_2CH_2CH=CH-\overset{\overset{\displaystyle O}{\|}}{C}-S-CoA + FADH_2$$

b.

$$CH_3CH_2CH_2CH=CH-\overset{\overset{\displaystyle O}{\|}}{C}-S-CoA + H_2O \longrightarrow$$

$$CH_3CH_2CH_2\overset{\overset{\displaystyle OH}{|}}{C}HCH_2-\overset{\overset{\displaystyle O}{\|}}{C}-S-CoA$$

c.

$$CH_3CH_2CH_2\overset{\overset{\displaystyle OH}{|}}{C}HCH_2-\overset{\overset{\displaystyle O}{\|}}{C}-S-CoA + NAD^+ \longrightarrow$$

$$CH_3CH_2CH_2\overset{\overset{\displaystyle O}{\|}}{C}CH_2-\overset{\overset{\displaystyle O}{\|}}{C}-S-CoA + NADH + H^+$$

d.

$$CH_3CH_2CH_2\overset{\overset{\displaystyle O}{\|}}{C}CH_2-\overset{\overset{\displaystyle O}{\|}}{C}-S-CoA + CoA-SH \longrightarrow$$

$$CH_3CH_2CH_2-\overset{\overset{\displaystyle O}{\|}}{C}-S-CoA + CH_3-\overset{\overset{\displaystyle O}{\|}}{C}-S-CoA$$

24.24 10 ATP

24.26 64 ATP molecules

24.28 Acetoacetate, β-hydroxybutyrate, and acetone.

24.30 The liver; the brain, heart, and skeletal muscles

24.32 Two of the ketone bodies are carboxylic acids, which lower blood pH.

24.34 Fatty acids are built two carbons at a time and broken down two carbons at a time.

24.36 The required enzymes are missing.

24.38 Amino acid pool

24.40 Protein turnover

24.42 Purines and pyrimidines, heme, choline, ethanolamine

24.44 a. Transamination

b. Deamination

c. Transamination

d. Transamination and deamination

24.46

$$CH_3-\overset{\overset{\displaystyle O}{\|}}{C}-COO^- + \overset{+}{H_3}N-CH-COO^- \longrightarrow$$

$$\begin{array}{c}|\\CH_2\\|\\CH-CH_3\\|\\CH_3\end{array}$$

pyruvate leucine

$$CH_3-\overset{\overset{\displaystyle NH_3^+}{|}}{C}H-COO^- + \overset{\overset{\displaystyle O}{\|}}{C}-COO^-$$

$$\begin{array}{c}|\\CH_2\\|\\CH-CH_3\\|\\CH_3\end{array}$$

24.48

$$\overset{+}{H_3}N-CH_2-\overset{\overset{\displaystyle O}{\|}}{C}-O^- + NAD^+ + H_2O \longrightarrow$$

$$H-\overset{\overset{\displaystyle O}{\|}}{C}-\overset{\overset{\displaystyle O}{\|}}{C}-O^- + NADH + NH_4^+ + H^+$$

24.50 The liver

24.52 Carbamoyl phosphate

24.54 $NH_4^+ + HCO_3^- + 3ATP + 2H_2O + COO^- \longrightarrow$

$$\begin{array}{c}|\\CH_2\\|\\CH-NH_3^+\\|\\COO^-\end{array}$$

aspartate

$$H_2N-\overset{\overset{\displaystyle O}{\|}}{C}-NH_2 + {}^-OOC-CH=CH-COO^-$$

urea fumarate

$$+ 2ADP + 2P_i + AMP + PP_i$$

24.56 Yes; isoleucine, tryptophan, phenylalanine, and tyrosine

24.58 a. Glucogenic

b. Ketogenic

c. Ketogenic

24.60 Amino acids that can be synthesized by the body are called nonessential. Those amino acids that must be obtained from the diet because the body cannot produce them are called essential amino acids.

24.62 Transaminases are used in the biosynthesis of amino acids and in the transfer of nitrogen atoms from amino acids when the disposal of nitrogen is necessary.

24.64 The glycolysis pathway and the citric acid cycle

24.66 Activation of zymogens and the production of new enzymes through induction

24.68 It is an LDL.

24.78 Nitrogen, liver

24.80 essential amino acids

Chapter 25

25.2 Plasma and interstitial fluid

25.4 **a.** Intracellular fluid
 b. Intracellular fluid
 c. Plasma and interstitial fluid
 d. Intracellular fluid
 e. Plasma and interstitial fluid
 f. Intracellular fluid
 g. Plasma and interstitial fluid

25.6 **a.** Cl^-
 b. Cl^-
 c. HPO_4^{2-}

25.8 **a.** $HHb + O_2 \rightleftarrows HbO_2^- + H^+$
 b. $HHb + CO_2 \rightleftarrows HHbCO_2$

25.10 **a.** 2% **b.** 98% **c.** 25% **d.** 5% **e.** 70%

25.12 A larger surface area increases the amount of O_2 and CO_2 that can diffuse through the surface. The partial pressure gradient is the force that causes O_2 and CO_2 movement.

25.14 **a.** Osmotic pressure
 b. Blood pressure

25.16 Urea

25.18 Glucose (in large amounts), protein, ketone bodies, pus, hemoglobin, red blood cells, bile pigments (in large amounts)

25.20 Kidneys (urine), lungs (water vapor), skin (perspiration), and intestines (feces)

25.22 Variations in urine output

25.24 As fluid level decreases, saliva output decreases, giving the "thirst" feeling. Drinking water raises fluid level.

25.26 Blood buffers, urinary system, and respiratory system. The blood buffer system is the fastest to react.

25.28 Bicarbonate, phosphate, and protein

25.30 Exhaled H_2O and CO_2 are derived from H_2CO_3. The more CO_2 and H_2O that are exhaled, the more carbonic acid is removed from the blood, thus elevating the blood pH.

25.32 Respiratory alkalosis

25.34 CO_2 diffuses into the distal tubule cells to form carbonic acid, which then ionizes to H^+ and HCO_3^-. Thus, CO_2 concentration is lowered in the blood.

25.36 As H^+ enters the urine, Na^+ is reabsorbed into the blood from the urine.

25.38 Respiratory acid–base imbalances are the result of abnormal breathing patterns. Acid–base imbalances caused by any other factor are metabolic acid–base imbalances.

25.40 The depressant narcotic often causes a lower breathing rate, causing respiratory acidosis. Diabetes mellitus can cause metabolic acidosis.

25.42 5.01×10^{23} amino acid molecules

25.44 The increased concentration of Na^+ (and Cl^-) ions in the blood increases the osmotic pressure gradient and water flows from the region of low solute concentration to the region of high solute concentration.

25.46 Heavy exercise would cause anaerobic conditions to occur in some cells, which would then release lactate from glycolysis into the bloodstream.

25.56 Partial pressure

25.58 a, b, c, and d

25.60 c

25.62 b

25.64 c

Solutions to Learning Checks

Chapter 1

1.1 A change is chemical if new substances are formed and physical if no new substances are formed.

 a. Chemical: The changes in taste and odor indicate that new substances form.

 b. Physical: The handkerchief doesn't change, and the evaporated water is still water.

 c. Chemical: The changes in color and taste indicate that new substances form.

 d. Chemical: The gases and smoke released indicate that new substances form.

 e. Physical: The air is still air, as indicated by appearance, odor, and so on.

 f. Physical: On being condensed, the steam forms water, which is the same substance present before the change.

1.2 **a.** Triatomic and heteroatomic: The three atoms of two kinds (oxygen and hydrogen) make it both triatomic (three atoms) and heteroatomic (two or more kinds of atoms).

 b. Triatomic and homoatomic: The three atoms make it triatomic, but all three atoms are the same (oxygen).

 c. Polyatomic and heteroatomic: The total of five atoms makes it polyatomic, and the two different kinds of atoms (carbon and hydrogen) make it heteroatomic.

1.3 Molecules of elements can be diatomic, triatomic, or polyatomic, but all atoms will be identical and will be represented by the same color. Molecules of compounds must be diatomic, triatomic, or polyatomic (more than one atom), and the atoms will not all be identical.

 a. Compound: polyatomic and heteroatomic

 b. Element: diatomic and homoatomic

 c. Element: polyatomic and homoatomic

 d. Compound: polyatomic and heteroatomic

1.4 The product would have to be a compound because it would be polyatomic (more than one atom because two substances were used) and heteroatomic (different atoms from the two different substances).

1.5 Substitute the given radius and π values into the equation:

$A = \pi r^2 = (3.14)(3.5 \text{ cm})^2 = (3.14)(12.25 \text{ cm}^2) = 38 \text{ cm}^2$

1.6 The volume of a rectangular object is equal to the product of the three sides.

$V = (30.0 \text{ cm})(20.0 \text{ cm})(15.0 \text{ cm})$

$= 9000 \text{ cm}^3, \text{ or } (9.00 \times 10^3 \text{ cm}^3)$

1 cm³ = 1 mL, so the volume is 9.00×10^3 mL. A liter is equal to 1000 mL, so 9000 mL = 9 L. The correct answers are 9.00×10^3 mL and 9.00 L.

1.7 Because 1 kg = 1000 g,

$$0.819 \text{ kg} \times \frac{1000 \text{ g}}{1 \text{ kg}} = 819 \text{ g}$$

1.8 Use Equation 1.4 and the fact that 77°F = 25°C, as shown in Example 1.7.

$K = °C + 273 = 25°C + 273 = 298 \text{ K}$

1.9 **a.** Scientific notation is used. In nonscientific notation, the number is written 588.

 b. Scientific notation is not used. In scientific notation, the number is written 4.39×10^{-4}.

 c. Scientific notation is used. In nonscientific notation, the number is written 0.0003915.

 d. Scientific notation is not used. In scientific notation, the number is written 9.870×10^3.

 e. Scientific notation is not used. In scientific notation, the number is written 3.677×10^1.

 f. Scientific notation is not used. In scientific notation, the number is written 1.02×10^{-1}.

1.10 The two important things to note are the placement of the decimal in the standard position and the correct value of the exponent.

 a. Written incorrectly: The decimal is not in standard position. Move the decimal 1 place to the left and increase the exponent by 1 to account for moving the decimal: 6.25×10^5.

 b. Written incorrectly: The decimal is not in the standard position, and no exponent is used. Move the decimal 3 places to the right and use −3 for the exponent to account for moving the decimal: 9.8×10^{-3}.

 c. Written incorrectly: The decimal is not in the standard position. Move the decimal 3 places to the right and reduce the exponent by 3 to account for moving the decimal: 4.1×10^{-6}.

 d. Written correctly.

1.11 Perform multiplications and divisions separately for the nonexponential and exponential terms, then combine the results into the final answer.

 a. $(2.4 \times 10^3)(1.5 \times 10^4) = (2.4 \times 1.5)(10^3 \times 10^4) = (3.6)(10^{3+4}) = 3.6 \times 10^7$

 b. $(3.5 \times 10^2)(2.0 \times 10^{-3}) = (3.5 \times 2.0)(10^2 \times 10^{-3}) = (7.0)(10^{2+(-3)}) = 7.0 \times 10^{-1}$

 c. $\dfrac{6.3 \times 10^5}{2.1 \times 10^3} = \left(\dfrac{6.3}{2.1}\right)\left(\dfrac{10^5}{10^3}\right) = (3.0)(10^{5-3})$

$= 3.0 \times 10^2$

d. $\dfrac{4.4 \times 10^{-2}}{8.8 \times 10^{-3}} = \left(\dfrac{4.4}{8.8}\right)\left(\dfrac{10^{-2}}{10^{-3}}\right) = (.50)(10^{-2-(-3)})$

$\qquad\qquad = .50 \times 10^1 = 5.0 \times 10^0,\ \text{or } 5.0$

1.12 The primary challenge is to interpret correctly the significance of zeros. Leading zeros to the left of nonzero numbers are not significant. Trailing zeros are significant only when a decimal is present; all zeros between non-zero digits are significant.

 a. 2, trailing zero is not significant.

 b. 4

 c. 3, leading zeros are not significant.

 d. 2

 e. 1, leading zeros are not significant.

 f. 4, trailing zero is significant.

1.13 **a.** 3 significant figures; original decimal position is 2 places to the right of standard position, so exponent is $+2$. 1.01×10^2 m.

 b. 2 significant figures; original decimal position is 3 places to the right of standard position, so exponent is $+3$. 1.2×10^3 g.

 c. 3 significant figures; original decimal position is 3 places to the left of standard position, so exponent is -3. 2.30×10^{-3} kg.

 d. 4 significant figures; original decimal position is 3 places to the right of standard position, so exponent is $+3$. 1.296×10^{3}°C.

 e. 4 significant figures; original decimal position is 1 place to the right of standard position, so exponent is $+1$. 2.165×10^1 mL.

 f. 2 significant figures; original decimal position is 2 places to the left of standard position, so exponent is -2. 1.5×10^{-2} km.

1.14 **a.** Answer will be rounded to 2 significant figures to match 0.0019.

$$(0.0019)(21.39) = 0.04064 = 4.1 \times 10^{-2}$$

 b. Answer will be rounded to 2 significant figures to match 4.1.

$$\dfrac{8.321}{4.1} = 2.0295 = 2.0$$

 c. Answer will be rounded to 3 significant figures to match 0.0911 and 3.22.

$$\dfrac{(0.0911)(3.22)}{(1.379)} = 0.21272 = 0.213,\ \text{or } 2.13 \times 10^{-1}$$

1.15 In each case, the sum or difference is rounded to have the same number of places to the right of the decimal as the least number of places in the terms added or subtracted.

 a. Answer will be rounded to have one place to the right of the decimal to match 3.2.

$$8.01 + 3.2 = 11.21 = 11.2$$

 b. Answer will be rounded to have no places to the right of the decimal to match 3000.

$$3000. + 20.3 + 0.009 = 3020.309 = 3020$$

c. Answer will be rounded to have two places to the right of the decimal to match both 4.33 and 3.12.

$$4.33 - 3.12 = 1.21$$

d. Answer will be rounded to have two places to the right of the decimal to match 2.42.

$$6.023 - 2.42 = 3.603 = 3.60$$

1.16 Two factors result from the relationship 1 g = 1000 mg. The factors are:

$$\dfrac{1\text{ g}}{1000\text{ mg}} \quad \text{and} \quad \dfrac{1000\text{ mg}}{1\text{ g}}$$

The first factor is used because it cancels the mg unit and generates the g unit:

Step 1. 1.1 mg

Step 2. 1.1 mg = g

Step 3. $1.1 \text{ m\cancel{g}} \times \dfrac{1\text{ g}}{1000 \text{ m\cancel{g}}}$ = g

Step 4. $\dfrac{(1.1)(1\text{ g})}{(1000)} = 0.0011\text{ g} = 1.1 \times 10^{-3}\text{ g}$

The number of significant figures matches the number in 1.1 because the 1 and 1000 are exact numbers.

1.17 Two factors, 1 km/1000 m and 0.621 mi/1 km, are used to convert meters to miles, and two factors, 60 s/1 min and 60 min/1 h, are used to convert seconds in the denominator to hours.

$$10.0\,\dfrac{\cancel{\text{m}}}{\cancel{\text{s}}} \times \dfrac{1\text{ k\cancel{m}}}{1000\,\cancel{\text{m}}} \times \dfrac{0.621\text{ mi}}{1\text{ k\cancel{m}}} \times \dfrac{60\,\cancel{\text{s}}}{1\,\cancel{\text{min}}}$$

$$\times\ \dfrac{60\,\cancel{\text{min}}}{1\text{ h}} = 22.356\text{ mi/h}$$

Rounding to 3 significant figures gives 22.4 mi/h.

1.18 **a.** $\% = \dfrac{\text{part}}{\text{total}} \times 100;\ \% = \dfrac{\$988}{\$1200} \times 100 = 82.3\%$

 b. $\% = \dfrac{\text{part}}{\text{total}} \times 100$

$$90.4\% = \dfrac{\text{part}}{83} \times 100$$

$$\dfrac{(90.4\%)(83)}{100} = \text{part}$$

$$\qquad\qquad = \text{no. voting to not take final}$$

$$\qquad\qquad = 75.0$$

Number voting to take exam = 83 − 75 = 8.

1.19 The value of the density gives two factors that may be used to solve problems of this type:

$$\dfrac{2.7\text{ g}}{1.0\text{ cm}^3} \quad \text{and} \quad \dfrac{1.0\text{ cm}^3}{2.7\text{ g}}$$

 a. The sample volume is 60.0 cm³, and we wish to use a factor that will convert this to grams. The first factor given above will work.

$$60.0\text{ c\cancel{m}}^3 \times \dfrac{2.7\text{ g}}{1.0\text{ c\cancel{m}}^3} = 162\text{ g (calculator answer)}$$

$$= 1.6 \times 10^2\text{ g (properly rounded answer)}$$

b. The sample mass is 98.5 g, and we wish to convert this to cm^3. The second factor given above will work.

$$98.5 \text{ g} \times \frac{1.0 \text{ cm}^3}{2.7 \text{ g}} = 36.48 \text{ cm}^3 \text{ (calculator answer)}$$

$$= 3.6 \times 10^1 \text{ cm}^3 \text{ (properly rounded answer)}$$

1.20 **a.** The sample mass is equal to the difference between the mass of the container with the sample inside and the mass of the empty container:

$$m = 64.93 \text{ g} - 51.22 \text{ g} = 13.71 \text{ g}$$

The density of the sample is equal to the sample mass divided by the sample volume:

$$d = \frac{m}{V} = \frac{13.71 \text{ g}}{10.00 \text{ mL}}$$

$$= 1.371 \text{ g/mL (properly rounded answer)}$$

b. The identity can be determined by calculating the density of the sample from the data and comparing the density with the known densities of the two possible metals. The volume of the sample is equal to the difference between the cylinder readings with the sample present and with the sample absent:

$$V = 25.2 \text{ mL} - 21.2 \text{ mL} = 4.0 \text{ mL}$$

According to Table 1.3, 1 mL = 1 cm^3, so the sample volume is equal to 4.0 cm^3. The density of the sample is equal to the sample mass divided by the sample volume:

$$d = \frac{m}{V} = \frac{35.66 \text{ g}}{4.0 \text{ cm}^3}$$

$$= 8.9 \text{ g/cm}^3 \text{ (properly rounded answer)}$$

A comparison with the known densities of nickel and chromium allows the metal to be identified as nickel.

Chapter 2

2.1 The number of atoms in each molecule is represented by a subscript in the formula.

a. H_3PO_4 b. SO_3 c. $C_6H_{12}O_6$

2.2 **a.** The atomic number, Z, equals the number of protons: $Z = 4$. The mass number, A, equals the sum of the number of protons and the number of neutrons: $A = 4 + 5 = 9$. According to the periodic table, the element with an atomic number of 4 is beryllium, with the symbol Be. The isotope symbol is 9_4Be.

b. According to the periodic table, chlorine has the symbol Cl and an atomic number, Z, of 17. The mass number, 37, is equal to the sum of the number of protons and neutrons: $A = \#p + \#n$. Therefore, the number of neutrons is equal to the mass number, A, minus the number of protons: $\#n = A - 17 = 37 - 17 = 20$.

c. According to the periodic table, the element with the symbol Si is silicon, which has an atomic number of 14. Therefore, the atom contains 14 protons. Since $A = \#p + \#n$, we see that $\#n = A - \#p = 28 - 14 = 14$.

2.3 **a.** Nitrogen atoms have a periodic table weight of 14.01 u, which rounds to 14 u. Iron atoms have a periodic table weight of 55.85 u, which rounds to 56 u. Thus, four nitrogen atoms (4 × 14 u = 56 u) will balance one iron atom.

b. Carbon atoms have a periodic table weight of 12.01 u, which rounds to 12 u. Helium atoms have a periodic table weight of 4.00 u, which rounds to 4 u. Thus, two carbon atoms (2 × 12 u = 24 u) will balance six helium atoms (6 × 4 u = 24 u).

c. Calcium atoms have a periodic table weight of 40.08 u, which rounds to 40 u. Argon atoms have a periodic table weight of 39.95 u, which rounds to 40 u. Thus, two calcium atoms (2 × 40 u = 80 u) will balance two argon atoms (2 × 40 u = 80 u).

2.4 Molecular weights are obtained by adding the atomic weights of the atoms in molecules. We have used four significant figures in atomic weights.

a. H_2SO_4; molecular weight = 2(at. wt. H) + (at. wt. S) + 4(at. wt. O)

$$MW = 2(1.008 \text{ u}) + 32.06 \text{ u} + 4(16.00 \text{ u})$$

$$= 2.016 \text{ u} + 32.06 \text{ u} + 64.00 \text{ u}$$

$$= 98.08 \text{ u}$$

b. C_3H_8O; molecular weight = 3(at. wt. C) + 8(at. wt. H) + (at. wt. O)

$$MW = 3(12.01 \text{ u}) + 8(1.008 \text{ u}) + 16.00 \text{ u}$$

$$= 36.03 \text{ u} + 8.064 \text{ u} + 16.00 \text{ u}$$

$$= 60.09 \text{ u}$$

2.5 **a.** Because naturally occuring fluorine consists of only a single isotope, the equation for calculating atomic weight becomes:

$$\text{At. wt.} = \frac{(\% \text{ fluorine-19})(\text{mass fluorine-19})}{100}$$

$$= \frac{(100)(19.00 \text{ u})}{100} = 19.00 \text{ u}$$

The calculated value is the same as the periodic table value.

b.

$$\text{At. wt.} = \frac{(\% \text{ Mg-24})(\text{mass Mg-24})}{100}$$

$$+ \frac{(\% \text{ Mg-25})(\text{mass Mg-25}) + (\% \text{ Mg-26})(\text{mass Mg-26})}{100}$$

$$= \frac{(78.70)(23.99 \text{ u}) + (10.13)(24.99 \text{ u})}{100}$$

$$\frac{+ (11.17)(25.98 \text{ u})}{100}$$

$$= 24.31 \text{ u}$$

Once again, the calculated value is the same as the periodic table value.

2.6 The mass ratio is obtained by dividing the mass of the magnesium sample by the mass of the carbon sample:

$$\frac{\text{Mg mass}}{\text{C mass}} = \frac{13.66 \text{ g}}{6.748 \text{ g}} = 2.024$$

The number of atoms in each sample is calculated using the factors obtained earlier from the mass of one atom of each element:

$$13.66 \text{ g Mg} \times \frac{1 \text{ Mg atom}}{4.037 \times 10^{-23} \text{ g Mg}}$$

$$= 3.384 \times 10^{23} \text{ Mg atoms}$$

$$6.748 \text{ g } \cancel{C} \times \frac{1 \text{ C atom}}{1.994 \times 10^{-23} \text{ g } \cancel{C}} = 3.384 \times 10^{23} \text{ C atoms}$$

2.7 The known quantity is one O atom, and the unit of the unknown is grams of O. The factor comes from the relationship 6.02×10^{23} O atoms = 16.00 g O. This relationship provides two factors:

$$\frac{6.02 \times 10^{23} \text{ O atoms}}{16.00 \text{ g O}} \quad \text{and} \quad \frac{16.00 \text{ g O}}{6.02 \times 10^{23} \text{ O atoms}}$$

The second factor is the one used:

$$(1 \cancel{\text{O atom}})\left(\frac{16.00 \text{ g O}}{6.02 \times 10^{23} \cancel{\text{O atoms}}}\right) = 2.66 \times 10^{-23} \text{ g O}$$

The ratio of this mass to the mass of a carbon atom is:

$$\frac{2.66 \times 10^{-23} \text{ g O}}{1.994 \times 10^{-23} \text{ g C}} = 1.33 \text{ (properly rounded answer)}$$

The ratio of the atomic weights of oxygen and carbon is:

$$\frac{16.00 \text{ u}}{12.01 \text{ u}} = 1.33 \text{ (rounded to 3 significant figures)}$$

The two ratios are the same to 3 significant figures.

2.8 MW = 1(At. wt. C) + 1(At. wt. O) = 1(12.0 u) + 1(16.0 u) = 28.0 u

Application of the mole concept to CO molecules gives the following relationships:

1 mol CO molecules = 6.02×10^{23} CO molecules = 28.0 g CO

The calculation of the mass of one CO molecule is done as follows: The known quantity is 1 CO molecule, and the unit of the unknown is g CO. The factor comes from the relationship 6.02×10^{23} CO molecules = 28.0 g CO.

$$(1 \cancel{\text{CO molecule}})\left(\frac{28.0 \text{ g CO}}{6.02 \times 10^{23} \cancel{\text{CO molecules}}}\right)$$
$$= 4.65 \times 10^{-23} \text{ g CO}$$

Comparing the mass of a CO molecule to the mass of a CO_2 molecule gives:

$$\frac{\text{mass of 1 CO molecule}}{\text{mass of 1 } CO_2 \text{ molecule}} = \frac{4.65 \times 10^{-23} \text{ g}}{7.31 \times 10^{-23} \text{ g}} = 0.636$$

Comparing the molecular weight of CO to the molecular weight of CO_2 gives:

$$\frac{\text{MW CO}}{\text{MW } CO_2} = \frac{28.0 \text{ u}}{44.0 \text{ u}} = 0.636$$

It is seen that the ratio of the actual masses of the molecules is identical to the ratio of their relative masses (their molecular weights).

2.9 The relationships between moles of atoms in 1 mol of molecules is given by the subscripts of the atoms. Thus, 1 mol of glucose molecules would contain 6 mol of C atoms, 12 mol of H atoms, and 6 mol of O atoms. One-half mol of glucose molecules would contain half as many moles of each atom, or 3 mol of C atoms, 6 mol of H atoms, and 3 mol of O atoms.

2.10 $\% = \dfrac{\text{part}}{\text{total}} \times 100$

In each compound, the part will be the mass of carbon associated with some mass of compound (the total). The mass relationships are easily obtained by assuming a sample size equal to 1.00 mol of each compound.

CO_2: 1.00 mol CO_2 molecules = 1.00 mol C atoms + 2.00 mol O atoms or, using atomic weights,

44.01 g CO_2 = 12.01 g C + 32.00 g O

$$\% \text{ C} = \frac{(12.01 \text{ g C}) \times 100}{(44.01 \text{ g } CO_2)} = 27.29\% \text{ C}$$

CO: 1.00 mol CO molecules = 1.00 mol C atoms + 1.00 mol O atoms

28.01 g CO = 12.01 g C + 16.00 g O

$$\% \text{ C} = \frac{(12.01 \text{ g C})}{(28.01 \text{ g CO})} \times 100 = 42.88\% \text{ C}$$

Chapter 3

3.1 Each element is found at the intersection of the given group (vertical columns) and period (horizontal rows).

 a. Ge **b.** Re

3.2 **a.** Period 1 has 2 elements, H and He.

 b. Group IIB(12) has 4 elements, Zn, Cd, Hg, and Cn.

3.3 **a.** The maximum number of electrons that can be found in any orbital, including a $4p$ orbital, is 2.

 b. All d subshells, including the $5d$, contain 5 orbitals with a capacity of 2 electrons each. Thus, the $5d$ subshell can contain a maximum of 10 electrons.

 c. Shell number 1 contains only a single s orbital, the $1s$ orbital, and so can contain a maximum of 2 electrons.

3.4 The valence shell is the last shell with electrons. According to Table 3.2, group VA(15) elements all have 5 electrons in the valence shell. Group VIA(16) elements have 6, group VIIA(17) have 7, and the noble gases VIIIA(18) have 8.

3.5 **a.** Strontium (Sr) is in group IIA(2) and so contains 2 electrons in its valence shell.

 b. Once again, the roman numeral of the group heading gives the number of valence-shell electrons. The element is Ge, and because it is in group IVA(14), it has 4 electrons in the valence shell.

 c. The fifteenth element in period 4 is As, which is in group VA(15) and therefore has 5 electrons in the valence shell.

3.6 The subshell filling order is obtained from Figure 3.7 or Figure 3.8. All subshells except the last one are filled to their capacities as follows: s subshells—filled with 2, p subshells—filled with 6, d subshells—filled with 10, and f subshells—filled with 14. The last subshell has only enough electrons added to bring the total number of electrons to the desired value.

 a. Element 9 requires 9 electrons: $1s^2 2s^2 2p^5$; 1 unpaired

 b. Mg is element 12 and so needs 12 electrons:

$$1s^2 2s^2 2p^6 3s^2; \text{ 0 unpaired}$$

 c. The element in group VIA(16) and period 3 is S, which is element 16. Therefore, it needs 16 electrons:

$$1s^2 2s^2 2p^6 3s^2 3p^4; \text{ 2 unpaired}$$

 d. An atom with 23 protons needs 23 electrons:

$$1s^2 2s^2 2p^6 3s^2 3p^6 4s^2 3d^3; \text{ 3 unpaired}$$

3.7 **a.** From Learning Check 3.6, the electronic configuration of element number 9 is $1s^2 2s^2 2p^5$. The $1s^2$ electrons represent the configuration of helium. So, we can write the configuration as $[He]2s^2 2p^5$.

b. The configuration of Mg is $1s^2 2s^2 2p^6 3s^2$. The first 10 electrons represent the configuration of neon: $[Ne]3s^2$.

c. The configuration of S is $1s^2 2s^2 2p^6 3s^2 3p^4$. Once again, the first 10 electrons represent the configuration of neon: $[Ne]3s^2 3p^4$.

d. The 23 electrons are represented by $1s^2 2s^2 2p^6 3s^2 3p^6 4s^2 3d^3$. The first 18 electrons represent the configuration of argon: $[Ar]4s^2 3d^3$.

3.8 **a.** On the basis of the location of each element in Figure 3.9, the distinguishing electrons are of the following types: element 38(Sr): s; element 47(Ag): d; element 50(Sn): p; element 86(Rn): p.

b. Their classifications based on Figure 3.10 are element 38(Sr): representative element; element 47(Ag): transition element; element 50(Sn): representative element; element 86(Rn): noble gas.

3.9 The periodic table and Figure 3.12 are used.

a. Xe: nonmetal

b. As: metalloid

c. Hg: metal

d. Ba: metal

e. Th: metal

3.10 **a.** The slowest reaction involves Li, and the fastest involves K.

Li
Na
K

b. The rate of the reaction increases coming down the group, and according to Table 3.3, the ionization energy decreases coming down the group.

c. The decrease in ionization energy coming down the group means less energy is required to remove electrons from atoms of the elements located farther down the group. The reactions in Figure 3.17 involve the removal of an electron from atoms of the metals, so the easier it is to remove electrons (less energy required), the faster the reaction should go. This trend in speed is what is observed.

Chapter 4

4.1 The subshell filling order is obtained from Figure 3.7 or Figure 3.8.

F; 9 electrons: $1s^2 2s^2 2p^5$

K; 19 electrons: $1s^2 2s^2 2p^6 3s^2 3p^6 4s^1$

The configuration of F can be changed to that of He ($1s^2$) by giving up 7 electrons or to that of Ne ($1s^2 2s^2 2p^6$) by accepting 1 electron. The configuration of K can be changed to that of Ar ($1s^2 2s^2 2p^6 3s^2 3p^6$) by giving up 1 electron or to that of Kr ($1s^2 2s^2 2p^6 3s^2 3p^6 4s^2 3d^{10} 4p^6$) by accepting 17 electrons.

4.2 **a.** Element number 9 is fluorine, which is in group VIIA(17) and so has 7 valence-shell electrons. The Lewis structure is $:\!\overset{\cdot}{\underset{\cdot}{F}}\!\cdot$.

b. Magnesium is in group IIA(2) and so has 2 valence-shell electrons. The Lewis structure is $\overset{\cdot}{Mg}\cdot$.

c. Sulfur is in group VIA(16) and so has 6 valence-shell electrons. The Lewis structure is $:\!\overset{\cdot}{\underset{\cdot}{S}}\!\cdot$.

d. Krypton is a noble gas and so has 8 valence-shell electrons. The Lewis structure is $:\!\overset{\cdot\cdot}{\underset{\cdot\cdot}{Kr}}\!:$.

4.3 **a.** Lithium contains 3 electrons, with one of them classified as a valence-shell electron: $[He]2s^1$ and Li·.

b. Bromine contains 35 electrons, with 7 of them classified as valence-shell electrons: $[Ar]4s^2 3d^{10} 4p^5$ and $:\!\overset{\cdot}{\underset{\cdot}{Br}}\!\cdot$.

c. Strontium contains 38 electrons, with 2 classified as valence-shell electrons: $[Kr]5s^2$ and $\overset{\cdot}{Sr}\cdot$.

d. Sulfur contains 16 electrons, with 6 classified as valence-shell electrons: $[Ne]3s^2 3p^4$ and $:\!\overset{\cdot}{\underset{\cdot}{S}}\!\cdot$.

4.4 The change that will actually take place is the one that involves the fewest electrons. Thus, F will gain 1 electron rather than give up 7, and K will give up 1 electron rather than gain 17:

$$F + 1e^- \rightarrow F^-$$
$$K \rightarrow K^+ + 1e^-$$

4.5 **a.** Element number 34 is a nonmetal. It is in group VIA(16) and will, when treated using the rules for nonmetals, accept $8 - 6$ or 2 electrons during ionic bond formation:

$$Se + 2e^- \rightarrow Se^{2-}$$

b. Rb, a metal in group IA(1), will lose 1 electron during ionic bond formation:

$$Rb \rightarrow Rb^+ + 1e^-$$

c. Element 18 is a nonmetallic noble gas. It will not react to form ionic bonds.

d. In, a metal in group IIIA(13), will lose 3 electrons during ionic bond formation:

$$In \rightarrow In^{3+} + 3e^-$$

4.6 **a.** Mg, a metal in group IIA(2), will lose 2 electrons:

$$Mg \rightarrow Mg^{2+} + 2e^-$$

O, a nonmetal in group VIA(16), will gain 2 electrons:

$$O + 2e^- \rightarrow O^{2-}$$

The positive and negative ions are combined in the lowest numbers possible to give the compound formula. The combining requirement is that the total number of positive charges from the positive ion must equal the total number of negative charges from the negative ion. In the case of Mg^{2+} and O^{2-}, one of each ion satisfies the requirement, and the formula is MgO.

b. K, a group IA(1) metal, will lose 1 electron:

$$K \rightarrow K^+ + 1e^-$$

S, a group VIA(16) nonmetal, will gain 2 electrons:

$$S + 2e^- \rightarrow S^{2-}$$

In the formula, $2K^+$ will be needed for each S^{2-}. Thus, the compound formula is K_2S.

c. Ca, a group IIA(2) metal, will lose 2 electrons:

$$Ca \rightarrow Ca^{2+} + 2e^-$$

Br, a group VIIA(17) nonmetal, will gain 1 electron:

$$Br + 1e^- \rightarrow Br^-$$

In the formula, $2Br^-$ will combine with each Ca^{2+}: $CaBr_2$

4.7 **a.** magnesium oxide

 b. potassium sulfide

 c. calcium bromide

4.8 **a.** Charge balance requires that $2Br^-$ combine with each Co^{2+}: $CoBr_2$. Names are cobalt(II) bromide and cobaltous bromide.

 b. Charge balance requires that $3Br^-$ combine with each Co^{3+}: $CoBr_3$. Names are cobalt(III) bromide and cobaltic bromide.

4.9 **a.** For H_2S, the molecular weight is the sum of the atomic weights of the atoms in the formula:

$$MW = (2)(\text{at. wt. H}) + (1)(\text{at. wt. S})$$
$$= (2)(1.01\ u) + (1)(32.1\ u)$$
$$MW = 34.1\ u$$

For CaO, the formula weight is also equal to the sum of the atomic weights of the atoms in the formula:

$$FW = (1)(\text{at. wt. Ca}) + (1)(\text{at. wt. O})$$
$$= (1)(40.1\ u) + (1)(16.0\ u)$$
$$FW = 56.1\ u$$

 b. According to Section 2.6, 1.00 mol of a molecular compound has a mass in grams equal to the molecular weight of the compound. Thus, 1.00 mol H_2S = 34.1 g H_2S. For ionic compounds, 1.00 mol of compound has a mass in grams equal to the formula weight of the compound. Thus, 1.00 mol CaO = 56.1 g CaO.

 c. In Section 2.6, we learned that 1.00 mol of a molecular compound contains Avogadro's number of molecules. Thus, 1.00 mol H_2S = 6.02×10^{23} molecules of H_2S. In the case of ionic compounds, 1.00 mol of compound contains Avogadro's number of formula units. That is, 1.00 mol of calcium oxide contains Avogadro's number of CaO units, where each unit represents one Ca^{2+} ion and one O^{2-} ion. Thus:

1.00 mol CaO = 6.02×10^{23} CaO units or

1.00 mol CaO = 6.02×10^{23} Ca^{2+} ions

 + 6.02×10^{23} O^{2-} ions

4.10 **a.** N is in group VA(15) and therefore has 5 valence-shell electrons. Two N atoms will provide 10 total electrons to satisfy the octets. At least 2 electrons are needed to bond the atoms. This leaves only 8 electrons to satisfy the remaining octet requirements of both atoms. This deficiency of electrons means some multiple bonds will be needed between the atoms:

$$:\!\dot{N}\cdot + \cdot\dot{N}\!: \longrightarrow\ :N:::N:\quad \text{or} \quad N\equiv N$$

 b. Br is in group VIIA(17) and therefore has 7 valence-shell electrons. Two atoms will provide 14 electrons, and the bonding is readily shown:

$$:\!\ddot{Br}\cdot + \cdot\ddot{Br}\!: \longrightarrow\ :\!\ddot{Br}\!:\!\ddot{Br}\!:\quad \text{or} \quad Br - Br$$

4.11 **a.** Each C contributes 4 valence-shell electrons, and each H contributes 1. The total available is 8:

H:C:H or H—C—H (with H above and below C)

 b. Each H contributes 1 electron, the C contributes 4, and the O contributes 6. The total of 12 electrons cannot satisfy all octets unless two pair are shared between the C and O atoms:

H:C::Ö: or H—C=O (with H below C)

 c. Each H contributes 1 electron, each O contributes 6, and the N contributes 5. The 24 total electrons can satisfy all octets only if two pair are shared between the N and one of the O atoms:

H:Ö:N::Ö: or H—O—N=O (with :Ö: / O below N)

4.12 **a.** The P provides 5 electrons, each O provides 6, and the 3− charge indicates an additional 3 electrons. The total number of electrons is therefore 32, which is enough to satisfy all octets without any multiple bonds:

$$\left[\ddot{O} : \ddot{O} : \ddot{P} : \ddot{O} : \ddot{O} \right]^{3-}$$

 b. The S provides 6 electrons, each O provides 6, and the 2− charge indicates an additional 2 electrons. The total number of electrons is therefore 26, which is enough to satisfy all octets without any multiple bonds:

$$\left[\ddot{O} : \ddot{S} : \ddot{O} : \ddot{O} \right]^{2-}$$

 c. The N provides 5 electrons, each H provides 1, and the + charge indicates that 1 must be subtracted. The total number of electrons is therefore 8, which is enough to satisfy the octet of N, and each H requirement of 2 electrons without any multiple bonds:

$$\left[\begin{array}{c} H \\ H:N:H \\ H \end{array} \right]^{+}$$

4.13 **a.** The 3 electron pairs around the B atom will arrange themselves in a flat triangle around the B. Therefore, the molecule will be flat (planar) with the F atoms forming a triangle around the B.

 b. The Lewis structure is:

$$:\ddot{Cl}: \\ :\ddot{Cl}:C:\ddot{Cl}: \\ :\ddot{Cl}:$$

The 4 electron pairs around the C atom will arrange themselves in a tetrahedral orientation. Thus, the molecule will be tetrahedral with C in the center and a Cl atom at each of the corners of the tetrahedron around the C.

 c. The Lewis structure is:

$$:\ddot{O}:\ddot{S}::\ddot{O}:$$

The 3 electron pairs around the S will occupy the corners of a triangle around the S. This will give rise to a bent molecule

$$O - \ddot{S} \diagdown\kern-1em\diagup O$$

d. The Lewis structure is:

$$:\ddot{S}::C::\ddot{S}:$$

Each double bond on the central C atom is treated like a single pair of electrons when VSEPR theory is used. Thus, the double bonds are treated like 2 pairs and will arrange themselves to be on opposite sides of the C atom. A linear molecule results: $S=C=S$.

4.14 a. The Lewis structure from Learning Check 4.12 is:

$$\left[\begin{array}{c} :\ddot{O}: \\ :\ddot{O}:\overset{..}{P}:\ddot{O}: \\ :\ddot{O}: \end{array} \right]^{3-}$$

We see that phosphorus is the central atom, and it has four electron pairs around it. The four pairs will be located at the corners of a tetrahedron with the P in the middle. The four O atoms will occupy the corners of the tetrahedron, giving the same shape as that of the SO_4^{2-} ion of Example 4.14a.

b. The Lewis structure from Learning Check 4.12 is:

$$\left[\begin{array}{c} :\ddot{O}: \\ :\ddot{O}:\overset{..}{S}:\ddot{O}: \end{array} \right]^{2-}$$

We see that sulfur is the central atom, and it has four electron pairs around it. The four pairs will be located at the corners of a tetrahedron with the S in the middle. The shape of the ion is determined by the location of the oxygen atoms and the S atom. The O atoms will occupy three of the corners of the tetrahedron, and one pair of electrons will occupy the fourth. Thus, the shape of the ion will be a triangular-based pyramid with the O atoms at the three corners of the base and the S atom at the top.

c. The Lewis structure from Learning Check 4.12 is:

$$\left[\begin{array}{c} H \\ \overset{..}{H:N:H} \\ H \end{array} \right]^{+}$$

We see that nitrogen is the central atom, and it has four electron pairs around it. The four pairs will be located at the corners of a tetrahedron with the N in the middle. The four H atoms will thus be located at the corners of the tetrahedron with the N in the middle, giving a tetrahedral-shaped ion like the SO_4^{2-} ion of Example 4.14a.

4.15 a. No polarization because bound atoms are identical.

b. Br is located higher in the group and so is more electronegative than I:

$$\overset{\delta^+ \quad \delta^-}{I - Br}$$
$$\longrightarrow$$

c. Br is farther toward the right in the periodic table and so is more electronegative than H:

$$\overset{\delta^+ \quad \delta^-}{H - Br}$$
$$\longrightarrow$$

4.16 The values of ΔEN are used.

a. $\Delta EN = 4.0 - 0.8 = 3.2$. Bond is classified as ionic.

b. $\Delta EN = 3.5 - 3.0 = 0.5$. Bond is classified as polar covalent.

c. $\Delta EN = 3.0 - 1.5 = 1.5$. Bond is classified as polar covalent.

d. $\Delta EN = 3.5 - 0.8 = 2.7$. Bond is classified as ionic.

4.17 a. sulfur trioxide

b. boron trifluoride

c. disulfur heptoxide

d. carbon tetrachloride

4.18 a. Ca is a group IIA(2) metal and so forms Ca^{2+} ions. Formula: $CaHPO_4$; name: calcium hydrogen phosphate.

b. Mg is a group IIA(2) metal and so forms Mg^{2+} ions. Formula: $Mg_3(PO_4)_2$; name: magnesium phosphate.

c. K is a group IA(1) metal and so forms K^+ ions. Formula: $KMnO_4$; name: potassium permanganate.

d. The ammonium ion, NH_4^+, acts as the positive ion in this compound. Formula: $(NH_4)_2Cr_2O_7$; name: ammonium dichromate.

4.19 Assuming that dispersion forces are significant in establishing the melting and boiling points of these elements leads to the conclusion that, in each pair, the element with the higher atomic weight would have the higher melting and boiling points.

a. Se **b.** Sb **c.** Ne

Chapter 5

5.1 Both subscripts and coefficients are used.

N_2 : 2 atoms—on the basis of subscript.

$3H_2$: 6 atoms—on the basis of subscript and coefficient.

$2NH_3$: 2 N atoms (on the basis of coefficient) and 6 H atoms (on the basis of subscript and coefficient).

5.2 $2SO_2 + O_2 \rightarrow 2SO_3$

5.3 a. Oxygen is -2 (Rule 5).

Sulfur is $+6$ (Rule 6).

b. Calcium is $+2$ (Rule 3).

Oxygen is -2 (Rule 5).

Chlorine is $+5$ (Rule 6).

c. Oxygen is -2 (Rule 5).

Chlorine is $+7$ (Rule 7).

5.4 a. $Zn = 0$ (Rule 1), $H^+ = +1$ (Rule 2), $Zn^{2+} = +2$ (Rule 2), $H_2 = 0$ (Rule 1).

Zn is oxidized, so it is the reducing agent.

H^+ is reduced, so it is the oxidizing agent.

b. In KI: $K = +1$ (Rule 3), $I = -1$ (Rule 6).

$Cl_2 = 0$ (Rule 1).

In KCl: $K = +1$ (Rule 3), $Cl = -1$ (Rule 6).

$I_2 = 0$ (Rule 1).

I in KI is oxidized, so KI is the reducing agent.

Cl_2 is reduced, so Cl_2 is the oxidizing agent.

c. In IO_3^-: $O = -2$ (Rule 5), $I = +5$ (Rule 7).

In HSO_3^-: $H = +1$ (Rule 4), $O = -2$ (Rule 5), $S = +4$ (Rule 7).

$I^- = -1$ (Rule 2).

In HSO_4^-: $H = +1$ (Rule 4), $O = -2$ (Rule 5), $S = +6$ (Rule 7).

S in HSO_3^- is oxidized, so HSO_3^- is the reducing agent.

I in IO_3^- is reduced, so IO_3^- is the oxidizing agent.

5.5 **a.** The O.N. of H changes from $+1$ to 0, and the O.N. of I changes from -1 to 0. Reaction is redox. Because one substance changes into two substances, the reaction is a decomposition.

b. The O.N. of H does not change. The O.N. of O changes from -1 (in a peroxide) to -2 (in water) and 0 (in O_2). This is an example in which the same element is both oxidized and reduced. The reaction is redox. Because one substance changes into two substances, the reaction is a decomposition.

c. No oxidation numbers change. The reaction is nonredox. This is a double-replacement (metathesis) reaction.

d. The O.N. of P changes from 0 to $+5$, and the O.N. of O changes from 0 to -2. The reaction is redox. Because two substances combine to form a single substance, the reaction is a combination.

e. The O.N. of Na does not change. The O.N. of I changes from -1 to 0, and the O.N. of Cl changes from 0 to -1. The reaction is redox. Because the Cl simply replaces the I in the compound, this is a single-replacement reaction.

5.6 **a.** Total ionic: $2Na^+(aq) + 2I^-(aq) + Cl_2(aq) \rightarrow 2Na^+(aq) + 2Cl^-(aq) + I_2(aq)$

Na^+ is a spectator ion.

Net ionic: $2I^-(aq) + Cl_2(aq) \rightarrow 2Cl^-(aq) + I_2(aq)$

b. Total ionic: $Ca^{2+}(aq) + 2Cl^-(aq) + 2Na^+(aq) + CO_3^{2-}(aq) \rightarrow 2Na^+(aq) + 2Cl^-(aq) + CaCO_3(s)$

Na^+ and Cl^- are spectator ions.

Net ionic: $Ca^{2+}(aq) + CO_3^{2-}(aq) \rightarrow CaCO_3(s)$

c. Total ionic: $Ba^{2+}(aq) + 2OH^-(aq) + 2H^+(aq) + SO_4^{2-}(aq) \rightarrow 2H_2O(\ell) + BaSO_4(s)$

There are no spectator ions, so the net ionic reaction is the same as the total ionic reaction.

5.7 **a.** 2. $1\text{ mol }N_2(g) + 3\text{ mol }H_2(g) \longrightarrow 2\text{ mol }NH_3(g)$

3. $6.02 \times 10^{23}\ N_2$ molecules $+ 3(6.02 \times 10^{23})\ H_2$ molecules $\longrightarrow 2(6.02 \times 10^{23})\ NH_3$ molecules

4. $28.0\text{ g }N_2 + 3(2.00)\text{g }H_2 \longrightarrow 2(17.0)\text{g }NH_3$

b. **Step 1.** $2.11\text{ mol }H_2$

Step 2. $2.11\text{ mol }H_2 \qquad\qquad = \text{mol }NH_3$

Step 3. $2.11\text{ mol }H_2 \times \dfrac{2\text{ mol }NH_3}{3\text{ mol }H_2} = \text{mol }NH_3$

Step 4. $2.11 \times \dfrac{2\text{ mol }NH_3}{3} = 1.406\text{ mol }NH_3 = 1.41\text{ mol }NH_3$

The answer was rounded to 3 significant figures to match the three in 2.11. The 2 and 3 are exact counting numbers.

c. **Step 1.** $9.47\text{ g }H_2$

Step 2. $9.47\text{ g }H_2 \qquad\qquad = \text{g }N_2$

Step 3. $9.47\text{ g }H_2 \times \dfrac{28.0\text{ g }N_2}{6.00\text{ g }H_2} = \text{g }N_2$

Step 4. $9.47 \times \dfrac{28.0\text{ g }N_2}{6.00} = 44.19\text{ g }N_2 = 44.2\text{ g }N_2$

The answer was rounded to 3 significant figures to match the three in 9.47, 28.0 g N_2, and 6.00.

5.8 The mass of NH_3 possible from each reactant is calculated. The reactant giving the smallest mass of NH_3 is the limiting reactant, and the mass of NH_3 calculated for the limiting reactant is the amount that will be produced:

$2.00\text{ mol }H_2 \times \dfrac{34.0\ \text{ g }NH_3}{3\text{ mol }H_2} = 22.7\text{ g }NH_3$

$15.5\text{ g }N_2 \times \dfrac{34.0\text{ g }NH_3}{28.0\text{ g }N_2} = 18.8\text{ g }NH_3$

The results show the limiting reactant to be the 15.5 g of N_2, and the amount of NH_3 produced will be 18.8 g.

5.9 **a.** $\%\text{ yield} = \dfrac{\text{actual yield}}{\text{theoretical yield}} \times 100$

$= \dfrac{17.43\text{ g}}{21.34\text{ g}} \times 100 = 81.68\%$

b. The theoretical yield must be calculated.

$510.\text{ g }CaCO_3 \times \dfrac{56.08\text{ g }CaO}{100.1\text{ g }CaCO_3} = 285.722\text{ g }CaO = 286\text{ g }CaO$

The % yield is then calculated using this theoretical yield:

$\%\text{ yield} = \dfrac{\text{actual yield}}{\text{theoretical yield}} \times 100$

$= \dfrac{235\text{ g}}{286\text{ g}} \times 100 = 82.2\%$

Chapter 6

6.1 **a.** $d = \dfrac{\text{mass}}{\text{volume}}$

copper: $d = \dfrac{114.2\text{ g}}{12.8\text{ mL}} = 8.92\text{ g/mL}$

glycerin: $d = \dfrac{63.0\text{ g}}{50.0\text{ mL}} = 1.26\text{ g/mL}$

helium: $d = \dfrac{0.286\text{ g}}{1500\text{ mL}} = 1.91 \times 10^{-4}\text{ g/mL}$

b. Copper: Doubling the pressure will not influence the volume, mass, or density of a solid.

Glycerin: Doubling the pressure will not significantly influence the volume and will not influence the mass or density of a liquid.

Helium: The increased pressure will cause the volume to decrease. The mass will not be changed, so the density will increase.

6.2 $d = \dfrac{m}{V}$, so $m = Vd$

$m = (1200 \text{ mL})(1.18 \times 10^{-3} \text{ g/mL})$

$= 1.42 \text{ g}$

6.3 $KE = \frac{1}{2}mv^2 = \frac{1}{2}(3.00 \text{ g})(10.0 \text{ cm/s})^2 = 1.50 \times 10^2 \text{ g cm}^2/\text{s}^2$

$KE = \frac{1}{2}mv^2 = \frac{1}{2}(3.00 \text{ g})(20.0 \text{ cm/s})^2 = 6.00 \times 10^2 \text{ g cm}^2/\text{s}^2$

6.4 The factors used come from the information in Table 6.3.

a. $670 \text{ torr} \times \dfrac{1 \text{ atm}}{760 \text{ torr}} = 0.882 \text{ atm}$

b. $670 \text{ torr} \times \dfrac{14.7 \text{ psi}}{760 \text{ torr}} = 13.0 \text{ psi}$

6.5 **a.** $K = {}^\circ C + 273 = 27 + 273 = 300 \text{ K}$

b. $K = {}^\circ C + 273 = 0. + 273 = 273 \text{ K}$

c. ${}^\circ C = K - 273 = 0.00 - 273 = -273 {}^\circ C$

d. ${}^\circ C = K - 273 = 100. - 273 = -173 {}^\circ C$

6.6 **a.** Equation 6.8 is used. Temperatures must be converted to kelvins:

$P_i = 1.90 \text{ atm}$ $P_f = 1.00 \text{ atm}$

$V_i = 10.0 \text{ L}$ $V_f = ?$

$T_i = 30.{}^\circ C = 303 \text{ K}$ $T_f = -10.2{}^\circ C = 262.8 \text{ K}$

$$\dfrac{P_i V_i}{T_i} = \dfrac{P_f V_f}{T_f}$$

$$\dfrac{(1.90 \text{ atm})(10.0 \text{ L})}{303 \text{ K}} = \dfrac{(1.00 \text{ atm})(V_f)}{262.8 \text{ K}}$$

Solve for V_f:

$$V_f = \dfrac{(1.90 \text{ atm})(10.0 \text{ L})(262.8 \text{ K})}{(1.00 \text{ atm})(303 \text{ K})} = 16.5 \text{ L}$$

b. $P_i = 800. \text{ torr}$ $P_f = 900. \text{ torr}$

$V_i = 5.00 \times 10^2 \text{ mL}$ $V_f = 250. \text{ mL}$

$T_i = 300. \text{ K}$ $T_f = ?$

$$\dfrac{P_i V_i}{T_i} = \dfrac{P_f V_f}{T_f}$$

$$\dfrac{(800. \text{ torr})(5.00 \times 10^2 \text{ mL})}{300. \text{ K}} = \dfrac{(900. \text{ torr})(250. \text{ mL})}{T_f}$$

Invert both sides and solve for T_f:

$$T_f = \dfrac{(300. \text{ K})(900. \text{ torr})(250. \text{ mL})}{(800. \text{ torr})(5.00 \times 10^2 \text{ mL})} = 169 \text{ K}$$

$${}^\circ C = K - 273 = 169 \text{ K} - 273 = -104{}^\circ C$$

6.7 To use the ideal gas law, all quantities must have units to match those of R. The only unit that needs to be changed is the temperature:

$K = {}^\circ C + 273 = 30.{}^\circ C + 273 = 303 \text{ K}$

$PV = nRT$

Solve for P:

$$P = \dfrac{nRT}{V}$$

$$= \dfrac{(2.15 \text{ mol})\left(0.0821 \dfrac{\text{L atm}}{\text{mol K}}\right)(303 \text{ K})}{(12.6 \text{ L})}$$

$= 4.24 \text{ atm}$

6.8 The ideal gas law in the form $PV = mRT/MW$ is solved for MW:

$$MW = \dfrac{mRT}{PV}$$

The temperature must be converted to kelvins:

$$K = 27{}^\circ C + 273 = 300 \text{ K}$$

$$MW = \dfrac{(3.35 \text{ g})\left(0.0821 \dfrac{\text{L atm}}{\text{mol K}}\right)(300 \text{ K})}{(1.21 \text{ atm})(2.00 \text{ L})}$$

$= 34.1 \text{ g/mol}$

This matches the molecular weight of H_2S.

6.9 $P_{\text{total}} = P_{\text{He}} + P_{\text{N}_2} + P_{\text{O}_2}$

The partial pressures of N_2 and O_2 are changed to torr before adding. The factors used come from information in Table 6.3.

P_{N_2}: $0.200 \text{ atm} \times \dfrac{760 \text{ torr}}{1 \text{ atm}} = 152 \text{ torr}$

P_{O_2}: $7.35 \text{ psi} \times \dfrac{760 \text{ torr}}{14.7 \text{ psi}} = 380 \text{ torr}$

$P_{\text{total}} = P_{\text{He}} + P_{\text{N}_2} + P_{\text{O}_2} = 310 \text{ torr} + 152 \text{ torr} + 380 \text{ torr}$
$= 842 \text{ torr}$

6.10 The lighter (least massive) helium molecules will diffuse faster than the neon molecules. The relationship between the diffusion rates is given by Graham's law.

$$\dfrac{\text{Rate of He}}{\text{Rate of Ne}} = \sqrt{\dfrac{\text{mass of Ne}}{\text{mass of He}}} = \sqrt{\dfrac{20.18}{4.003}}$$

$$= \sqrt{5.041} = 2.25$$

Thus, rate of He = 2.25 rate of Ne. So, helium is seen to diffuse at a rate more than twice that of neon.

6.11 **a.** Evaporation is endothermic.

b. Freezing is exothermic.

c. Melting is endothermic.

6.12 The lower the forces between particles, the higher will be the vapor pressure.

a. Both substances form hydrogen bonds, but dispersion forces are stronger in propyl alcohol. Thus, methyl alcohol has the higher vapor pressure.

b. The only forces acting between the molecules in each case are dispersion forces. These are stronger between the heavier nitrogen molecules, so helium has the higher vapor pressure.

c. Molecules of HF form strong hydrogen bonds with one another, whereas the only forces between neon molecules are weak dispersion forces. Thus, neon has the higher vapor pressure.

6.13 Heat absorbed = (mass)(specific heat)(temp. change). The specific heat of helium is obtained from Table 6.8.

Heat absorbed = $(1000 \text{ g})(1.25 \text{ cal/g °C})(700°C - 25°C)$

$$= 8.44 \times 10^5 \text{ cal}$$

6.14 Heat absorbed = (mass)(heat of vaporization). The heat of vaporization for water was given in the text.

Heat absorbed = $(5000 \text{ g})(540 \text{ cal/g}) = 2.70 \times 10^6 \text{ cal}$

Chapter 7

7.1 **a.** Solvent is gold; solutes are copper, zinc, and nickel.

b. Solvent is nitrogen; solutes are oxygen, argon, and the other gases.

7.2 At 80°C, a saturated solution of KNO_3 contains about 181 g of solute/100 g of H_2O, and a saturated solution of NaBr contains about 118 g of solute per 100 g H_2O. At 50°C both solutions are still saturated, but the KNO_3 solution contains only about 102 g of solute per 100 g H_2O, while the NaBr solution contains about 117 g of solute per 100 g H_2O. Thus, the solubility of KNO_3 is seen to be much more temperature dependent than the solubility of NaBr.

7.3 **a.** The solubility of nitrates and lack of solubility of sulfates and carbonates of barium makes barium nitrate, $Ba(NO_3)_2$, the salt to use. More of it would dissolve and would release more Ba^{2+} ions.

b. Like dissolves like, so oil will dissolve best in oily solvents such as light mineral oil or gasoline. Light mineral oil is the solvent used because it leaves a residue that is similar to the natural oily secretions found on the bird feathers. Gasoline would be hazardous to use (flammable) and leaves no such useful residue.

7.4 **a.** The data may be substituted directly into Equation 7.5:

$$M = \frac{1.25 \text{ mol solute}}{2.50 \text{ L solution}} = 0.500 \frac{\text{mol solute}}{\text{L solution}}$$

The solution is 0.500 molar, or 0.500 M.

b. In this problem, the volume of solution must be converted to liters before the data are substituted into Equation 7.5:

$$(225 \text{ mL solution})\left(\frac{1 \text{ L}}{1000 \text{ mL}}\right) = 0.225 \text{ L solution}$$

$$M = \frac{0.486 \text{ mol solute}}{0.225 \text{ L solution}} = 2.16 \frac{\text{mol solute}}{\text{L solution}}$$

The solution is 2.16 molar, or 2.16 M.

c. In this problem, the volume of solution must be converted to liters and the mass of solute must be converted to moles before the data can be substituted into Equation 7.5:

$$(100 \text{ mL solution})\left(\frac{1 \text{ L}}{1000 \text{ mL}}\right) = 0.100 \text{ L solution}$$

$$(2.60 \text{ g NaCl})\left(\frac{1 \text{ mol NaCl}}{58.44 \text{ g NaCl}}\right) = 0.0445 \text{ mol NaCl}$$

In the last calculation, the factor $\frac{1 \text{ mol NaCl}}{58.44 \text{ g NaCl}}$ comes from the calculated formula weight of 58.44 u for NaCl. The data are now substituted into Equation 7.5:

$$M = \frac{0.0445 \text{ mol NaCl}}{0.100 \text{ L solution}} = 0.445 \frac{\text{mol NaCl}}{\text{L solution}}$$

The solution is 0.445 molar, or 0.445 M.

7.5 **a.** To calculate %(w/w), the mass of solute contained in a specific mass of solution is needed. The mass of solute is 0.900 g, and the mass of solution is equal to the solvent mass (100 g) plus the solute mass (0.900 g):

$$\%(w/w) = \frac{\text{solute mass}}{\text{solution mass}} \times 100 = \frac{0.900 \text{ g}}{100.9 \text{ g}} \times 100$$

$$= 0.892 \%(w/w)$$

To calculate %(w/v), the number of grams of solute must be known along with the number of milliliters of solution. The mass of solute is 0.900 g, and the solution volume is 100 mL:

$$\%(w/v) = \frac{\text{g of solute}}{\text{mL of solution}} \times 100 = \frac{0.900 \text{ g}}{100 \text{ mL}} \times 100$$

$$= 0.900 \%(w/v)$$

b. The given quantity is 30 mL of beverage, and the desired quantity is milliliters of pure alcohol. The %(v/v) provides the necessary factor:

$$30 \text{ mL beverage} \times \frac{45 \text{ mL alcohol}}{100 \text{ mL beverage}} = 14 \text{ mL alcohol}$$

7.6 **a.** $M = \dfrac{\text{mol of solute}}{\text{liters of solution}}$

$$\text{mol of solute} = (M)(\text{liters of solution})$$
$$= (1.00 \text{ M})(0.500 \text{ L})$$
$$= 0.500 \text{ mol of solute}$$

The solute is $MgCl_2$, which has a formula weight of 95.2 u. The formula weight provides the factor in the following calculation:

$$0.500 \text{ mol MgCl}_2 \times \frac{95.2 \text{ g MgCl}_2}{1 \text{ mol MgCl}_2} = 47.6 \text{ g MgCl}_2$$

The solution is prepared by putting 47.6 g of $MgCl_2$ into a 500-mL flask and adding pure water to the mark, making certain all the solid solute dissolves and the resulting solution is well mixed.

b. $\%(w/v) = \dfrac{\text{g of solute}}{\text{mL of solution}} \times 100$

$$\text{g of solute} = \frac{\%(w/v)(\text{mL of solution})}{100}$$

$$= \frac{(12.0\%)(100 \text{ mL})}{100} = 12.0 \text{ g}$$

The solution is prepared by putting 12.0 g of $MgCl_2$ into a 100-mL flask and adding pure water to the mark, making certain all the solid solute dissolves and the resulting solution is well mixed.

c. $\%(v/v) = \dfrac{\text{solute volume}}{\text{solution volume}} \times 100$

$$\text{solute volume} = \frac{\%(v/v)(\text{solution volume})}{100}$$

$$= \frac{(20.0\%)(1.00 \text{ L})}{100} = 0.200 \text{ L}$$

$$= 200 \text{ mL}$$

The solution is prepared by putting 200 mL of ethylene glycol into a 1.00-L flask and adding pure water to the mark, making certain the two liquids are completely mixed to form the final solution.

7.7 Equation 7.9 is used to calculate the volume of 6.00 M NaOH solution needed.

$$(C_c)(V_c) = (C_d)(V_d)$$

$$(6.00\ \text{M})(V_c) = (0.250\ \text{M})(500\ \text{mL})$$

$$V_c = \frac{(0.250\ \text{M})(500\ \text{mL})}{(6.00\ \text{M})} = 20.8\ \text{mL}$$

The solution is prepared by putting 20.8 mL of 6.00 M NaOH solution into a 500-mL flask, adding pure water to the mark, and making certain the resulting solution is well mixed.

7.8 a. Because $CaCl_2$ is a strong electrolyte, it dissociates completely:

$$CaCl_2 \rightarrow Ca^{2+} + 2Cl^-. \text{ Thus, } n = 3.$$

$$\Delta T_b = nK_b M = (3)(0.52°C/\text{M})(0.100\ \text{M}) = 0.16°C$$

Because the boiling point of a solution is higher than the boiling point of the pure solvent, ΔT_b is added to the boiling point of the pure solvent.

Solution B.P. = 100.00°C + 0.16°C = 100.16°C

$$\Delta T_f = nK_f M = (3)(1.86°C/\text{M})(0.100\ \text{M}) = 0.558°C$$

Because the freezing point of a solution is lower than the freezing point of the pure solvent, ΔT_f is subtracted from the freezing point of the solvent.

Solution F.P. = 0.00°C − 0.558°C = −0.558°C = −0.56°C

$$\pi = n\text{MRT}$$

$$= (3)(0.100\ \text{mol/L})(62.4\ \text{L torr/K mol})(300\ \text{K})$$

$$= 5.62 \times 10^3\ \text{torr} = 7.39\ \text{atm}$$

b. Because ethylene glycol does not dissociate in solution, $n = 1$.

$$\Delta T_b = nK_b M = (1)(0.52°C/\text{M})(0.100\ \text{M}) = 0.052°C$$

Solution B.P. = 100.00°C + 0.052°C = 100.05°C

$$\Delta T_f = nK_f M = (1)(1.86°C/\text{M})(0.100\ \text{M}) = 0.186°C$$

Solution F.P. = 0.00°C − 0.186°C = −0.186°C

$$= -0.19°C$$

$$\pi = n\text{MRT}$$

$$= (1)(0.100\ \text{mol/L})(62.4\ \text{L torr/K mol})(300\ \text{K})$$

$$= 1.87 \times 10^3\ \text{torr} = 2.46\ \text{atm}$$

Chapter 8

8.1 Because only Ce^{4+} and Fe^{2+} are mixed, the initial concentration of Ce^{3+} is 0.

$$\text{Rate} = \frac{C_t - C_0}{\Delta t} = \frac{1.50 \times 10^{-5}\ \text{mol/L} - 0\ \text{mol/L}}{75.0\ \text{s}}$$

$$= 2.00 \times 10^{-7}\ \frac{\text{mol/L}}{\text{s}}$$

8.2 a. $K = \dfrac{[NO_2]^2}{[N_2O_4]}$

b. $K = \dfrac{[N_2O][NO_2]}{[NO]^3}$

8.3 a. $K = \dfrac{[IBr]^2}{[Br_2][I_2]}$

$$= \frac{(1.96 \times 10^{-2})^2}{(1.50 \times 10^{-1})(5.00 \times 10^{-2})}$$

$$= 5.12 \times 10^{-2}$$

The small value of K indicates the equilibrium position is toward the left.

b. $K = \dfrac{[NH_3]^2}{[N_2][H_2]^3}$

$$= \frac{(9.00)^2}{(9.23 \times 10^{-3})(2.77 \times 10^{-2})^3}$$

$$= 4.13 \times 10^8$$

The large K value indicates the equilibrium position is toward the right.

8.4 a. Heat + $NH_4NO_3(s) \rightleftarrows NH_4^+(aq) + NO_3^-(aq)$

The equilibrium will shift to the right when heat is added in an attempt to use up the added heat. This shift corresponds to dissolving more of the solid at the higher temperature. The solubility is higher at a higher temperature.

b. $2SO_2(g) + O_2(g) \rightleftarrows 2SO_3(g)$

If O_2 were removed from an equilibrium mixture, the equilibrium would shift left in an attempt to replenish the removed O_2. This shift would reduce the amount of SO_3 present at equilibrium, which means the concentration of SO_3 would be lower than it was in the initial equilibrium.

Chapter 9

9.1 a. $HC_2H_3O_2 + H_2O \rightleftarrows H_3O^+ + C_2H_3O_2^-$
acid base acid base

conjugate pair
conjugate pair

b. $NO_2^- + H_2O \rightleftarrows OH^- + HNO_2$
base acid base acid

conjugate pair
conjugate pair

c. $HS^- + H_2O \rightleftarrows H_3O^+ + S^{2-}$
acid base acid base

conjugate pair
conjugate pair

9.2 a. The anhydrous compound is called hydrogen iodide. The name of the water solution is obtained by dropping *hydrogen* from the anhydrous compound name and adding the prefix *hydro-* to the stem *iod*. The *-ide* suffix on the *iod* stem is replaced by the suffix *-ic* to give the name *hydroiodic*. The word *acid* is added, giving the final name *hydroiodic acid* for the water solution.

b. The anhydrous compound is called hydrogen bromide. The name of the water solution is obtained the same way it was for HI(aq) in part a, using the stem *brom* in place of *iod*. The resulting name for the water solution is *hydrobromic acid*.

9.3 The removal of the two H^+ ions leaves behind the CO_3^{2-} polyatomic ion. From Table 4.6, this ion is the carbonate ion. According to Rules 1–4, the *-ate* suffix is replaced by the *-ic* suffix to give the name *carbonic acid*.

9.4 **a.** Solution is basic because $[OH^-]$ is greater than 1.0×10^{-7} mol/L:

$$[H_3O^+] = \frac{1.0 \times 10^{-14}(mol/L)^2}{[OH^-]} = \frac{1.0 \times 10^{-14}(mol/L)^2}{1.0 \times 10^{-5} \text{ mol/L}}$$

$$= 1.0 \times 10^{-9} \text{ mol/L}$$

b. Solution is basic because $[H_3O^+]$ is less than 1.0×10^{-7} mol/L:

$$[OH^-] = \frac{1.0 \times 10^{-14}(mol/L)^2}{[H_3O^+]} = \frac{1.0 \times 10^{-14}(mol/L)^2}{1.0 \times 10^{-9} \text{ mol/L}}$$

$$= 1.0 \times 10^{-5} \text{ mol/L}$$

This solution has the same $[H_3O^+]$ and $[OH^-]$ as the solution in part a.

c. Solution is acidic because $[H_3O^+]$ is greater than 1.0×10^{-7} mol/L:

$$[OH^-] = \frac{1.0 \times 10^{-14}(mol/L)^2}{[H_3O^+]} = \frac{1.0 \times 10^{-14}(mol/L)^2}{1.0 \times 10^{-2} \text{ mol/L}}$$

$$= 1.0 \times 10^{-12} \text{ mol/L}$$

9.5 **a.** $[H^+] = 1 \times 10^{-14}$ mol/L. pH is the negative of the exponent used to express $[H^+]$, so pH $= -(-14) = 14.0$.

b. $[OH^-] = 1.0$ mol/L. The value of $[H^+]$ is calculated:

$$[H^+] = \frac{1 \times 10^{-14}(mol/L)^2}{[OH^-]} = \frac{1 \times 10^{-14}(mol/L)^2}{1.0 \text{ mol/L}}$$

$$= 1 \times 10^{-14} \text{ mol/L}$$

pH is the negative of the exponent used to express $[H^+]$, so pH $= -(-14) = 14.0$.

c. $[OH^-] = 1 \times 10^{-8}$ mol/L. The value of $[H^+]$ is calculated:

$$[H^+] = \frac{1 \times 10^{-14}(mol/L)^2}{[OH^-]} = \frac{1 \times 10^{-14}(mol/L)^2}{1 \times 10^{-8} \text{ mol/L}}$$

$$= 1 \times 10^{-6} \text{ mol/L}$$

pH is the negative of the exponent used to express $[H^+]$, so pH $= -(-6) = 6.0$.

9.6 **a.** pH $= 10.0$, and

$$[H^+] = 1 \times 10^{-pH} = 1 \times 10^{-10} \text{ mol/L}$$

$$[OH^-] = \frac{K_w}{[H^+]} = \frac{1 \times 10^{-14} (mol/L)^2}{1 \times 10^{-10} \text{ mol/L}} = 1 \times 10^{-4} \text{ mol/L}$$

b. pH $= 4.0$, and $[H^+] = 1 \times 10^{-pH} = 1 \times 10^{-4}$

$$[OH^-] = \frac{K_w}{[H^+]} = \frac{1 \times 10^{-14} (mol/L)^2}{1 \times 10^{-4} \text{ mol/L}} = 1 \times 10^{-10} \text{ mol/L}$$

c. pH $= 5.0$, and $[H^+] = 1 \times 10^{-pH} = 1 \times 10^{-5}$

$$[OH^-] = \frac{K_w}{[H^+]} = \frac{1 \times 10^{-14} (mol/L)^2}{1 \times 10^{-5} \text{ mol/L}} = 1 \times 10^{-9} \text{ mol/L}$$

9.7 **a.** The calculator answer is 4.3768. The correctly rounded value is 4.38.

b. The calculator answer is 8.0915. The correctly rounded value is 8.09.

9.8 **a.** The calculator answer is 0.0017783, or 1.7783×10^{-3}. The correctly rounded value should have two significant figures to match the two figures to the right of the decimal in the pH value. The answer is 1.8×10^{-3} mol/L.

b. The calculator answer is 4.6774×10^{-9}. The correctly rounded answer is 4.7×10^{-9} mol/L.

9.9 According to Table 9.4, concentrated NH_3 stock solution is 15 M. Equation 7.9 is used to determine the amount of this solution needed.

$$C_c V_c = C_d V_d$$

$$(15 \text{ M})(V_c) = (3.0 \text{ M})(500 \text{ mL})$$

$$V_c = \frac{(3.0 \text{ M})(500 \text{ mL})}{(15 \text{ M})}$$

$$= 100 \text{ mL}$$

Because the resulting solution molarity is given using only two significant figures, the volumes can be measured using graduated cylinders. Measure 100 mL of stock (15 M) aqueous ammonia, put it into a container and add 400 mL of distilled water. Mix the resulting solution well.

9.10 $2HCl(aq) + Sr(HCO_3)_2(s) \rightarrow$

$$SrCl_2(aq) + 2CO_2(g) + 2H_2O(\ell)$$

Total Ionic: $2H^+(aq) + 2Cl^-(aq) + Sr(HCO_3)_2(s) \rightarrow$

$$Sr^{2+}(aq) + 2Cl^-(aq) + 2CO_2(g) + 2H_2O(\ell)$$

Cl^- is a spectator ion.

Net Ionic: $2H^+(aq) + Sr(HCO_3)_2(s) \rightarrow$

$$Sr^{2+}(aq) + 2CO_2(g) + 2H_2O(\ell)$$

9.11 **a.** Molecular: $Ca(s) + 2H_2O(\ell) \rightarrow H_2(g) + Ca(OH)_2(s)$

Total Ionic: $Ca(s) + 2H_2O(\ell) \rightarrow H_2(g) + Ca(OH)_2(s)$

Net Ionic: $Ca(s) + 2H_2O(\ell) \rightarrow H_2(g) + Ca(OH)_2(s)$

All three equations are the same because none of the reactants or products form ions in solution.

b. Molecular: $Mg(s) + H_2SO_4(aq) \rightarrow H_2(g) + MgSO_4(aq)$

Total Ionic: $Mg(s) + 2H^+(aq) + SO_4^{2-}(aq) \rightarrow$

$$H_2(g) + Mg^{2+}(aq) + SO_4^{2-}(aq)$$

The SO_4^{2-} is a spectator ion.

Net Ionic: $Mg(s) + 2H^+(aq) \rightarrow H_2(g) + Mg^{2+}(aq)$

9.12 **a.** Molecular: $HNO_3(aq) + NaOH(aq) \rightarrow$

$$H_2O(\ell) + NaNO_3(aq)$$

Total Ionic: $H^+(aq) + NO_3^-(aq) + Na^+(aq) + OH^-(aq) \rightarrow$

$$H_2O(\ell) + Na^+(aq) + NO_3^-(aq)$$

Na^+ and NO_3^- are spectator ions.

Net Ionic: $H^+(aq) + OH^-(aq) \rightarrow H_2O(\ell)$

b. Molecular: $H_2SO_4(aq) + 2KOH(aq) \rightarrow$

$$2H_2O(\ell) + K_2SO_4(aq)$$

Total Ionic:

$2H^+(aq) + SO_4^{2-}(aq) + 2K^+(aq) + 2OH^-(aq) \rightarrow$

$$2H_2O(\ell) + 2K^+(aq) + SO_4^{2-}(aq)$$

K^+ and SO_4^{2-} are spectator ions.

Net Ionic: $2H^+(aq) + 2OH^-(aq) \rightarrow 2H_2O(\ell)$

9.13 $6HCl(aq) + 2Al(s) \rightarrow 2AlCl_3(aq) + 3H_2(g)$

$6HCl(aq) + Al_2O_3(s) \rightarrow 2AlCl_3(aq) + 3H_2O(\ell)$

$3HCl(aq) + Al(OH)_3(s) \rightarrow AlCl_3(aq) + 3H_2O(\ell)$

$6HCl(aq) + Al_2(CO_3)_3(s) \rightarrow$

$$2AlCl_3(aq) + 3H_2O(\ell) + 3CO_2(g)$$

$3HCl(aq) + Al(HCO_3)_3(s) \rightarrow$

$$AlCl_3(aq) + 3H_2O(\ell) + 3CO_2(g)$$

9.14 **a.** The dissociation reaction is $NaCl(aq) \rightarrow Na^+(aq) + Cl^-(aq)$. Thus, 1 mol of NaCl produces 1 mol of Na^+ or 1 mol of positive charges. Thus, 1 mol NaCl = 1 eq NaCl. Therefore,

$$0.10 \text{ mol NaCl} \times \frac{1 \text{ eq NaCl}}{1 \text{ mol NaCl}} = 0.10 \text{ eq NaCl}$$

Also, because 1 eq = 1000 meq,

$$0.10 \text{ eq} \times \frac{1000 \text{ meq}}{1 \text{ eq}} = 1.0 \times 10^2 \text{ meq}$$

b. The dissociation reaction is $Mg(NO_3)_2(aq) \rightarrow Mg^{2+}(aq) + 2NO_3^-(aq)$. Thus, 1 mol of $Mg(NO_3)_2$ produces 1 mol of Mg^{2+} or 2 mol of positive charges. Thus, 1 mol $Mg(NO_3)_2$ = 2 eq $Mg(NO_3)_2$. Therefore,

$$0.10 \text{ mol Mg(NO}_3)_2 \times \frac{2 \text{ eq Mg(NO}_3)_2}{1 \text{ mol Mg(NO}_3)_2}$$
$$= 0.20 \text{ eq Mg(NO}_3)_2$$

Also, because 1 eq = 1000 meq,

$$0.20 \text{ eq} \times \frac{1000 \text{ meq}}{1 \text{ eq}} = 2.0 \times 10^2 \text{ meq}$$

9.15 The Cl^- ion has a single charge, so we may write:

1.00 mol NaCl = 1.00 mol Cl^- = 1.00 eq Cl^- = 1.00 eq NaCl

As in Example 9.11, the pattern for the calculation is liters solution $A \rightarrow$ eq A, and the pathway is liters Cl^- solution \rightarrow eq Cl^-. The conversion factor comes from the concentration of 0.103 eq/L given in the problem:

$$(0.250 \text{ L Cl}^- \text{ solution})\left(\frac{0.103 \text{ eq Cl}^-}{1.00 \text{ L solution}}\right) = 0.0258 \text{ eq Cl}^-$$

The equivalents of Cl^- are converted to the quantities asked for by using factors from the relationships given above and the formula weight for NaCl, 58.44 u:

$$(0.0258 \text{ eq Cl}^-)\left(\frac{1.00 \text{ eq NaCl}}{1.00 \text{ eq Cl}^-}\right) = 0.0258 \text{ eq NaCl}$$

$$(0.0258 \text{ eq NaCl})\left(\frac{1.00 \text{ mol NaCl}}{1.00 \text{ eq NaCl}}\right) = 0.0258 \text{ mol NaCl}$$

$$(0.0258 \text{ mol NaCl})\left(\frac{58.44 \text{ g NaCl}}{1.00 \text{ mol NaCl}}\right) = 1.51 \text{ g NaCl}$$

9.16 **a.** $HPO_4^{2-} \rightleftarrows H^+ + PO_4^{3-}$

$$K_a = \frac{[H^+][PO_4^{3-}]}{[HPO_4^{2-}]}$$

b. $HNO_2 \rightleftarrows H^+ + NO_2^-$

$$K_a = \frac{[H^+][NO_2^-]}{[HNO_2]}$$

c. $HF \rightleftarrows H^+ + F^-$

$$K_a = \frac{[H^+][F^-]}{[HF]}$$

9.17 **a.** In a series of similar oxyacids, acid strength increases with increases in the number of oxygens: $HClO_3, HClO_2, HClO$.

b. The anion from the weakest acid will be the strongest base. HNO_3 is a stronger acid than HNO_2, so NO_2^- is a stronger base than NO_3^-.

c. The anion of an acid behaves as a weaker acid than the acid from which it came. The anion actually behaves as a base in this case, so $HC_2H_3O_2$ is a stronger acid than $C_2H_3O_2^-$.

9.18 The pattern is liters solution $A \rightarrow$ mol B, and the pathway is liters NaOH solution \rightarrow mol NaOH \rightarrow mol H_3PO_4. In combined form, the steps in the factor-unit method are:

Step 1. 0.0141 L NaOH solution

Step 2. 0.0141 L NaOH solution = mol H_3PO_4

Step 3. $0.0141 \text{ L NaOH solution} \times \dfrac{0.250 \text{ mol NaOH}}{1 \text{ L NaOH solution}}$

$$\times \frac{1 \text{ mol H}_3PO_4}{3 \text{ mol NaOH}} = \text{mol H}_3PO_4$$

Step 4. $(0.0141)\left(\dfrac{0.250}{1}\right)\left(\dfrac{1 \text{ mol H}_3PO_4}{3}\right)$

$$= 0.00118 \text{ mol H}_3PO_4$$

The molarity of the H_3PO_4 solution is calculated:

$$\begin{aligned} M &= \frac{\text{mol H}_3PO_4}{\text{L H}_3PO_4 \text{ solution}} \\ &= \frac{0.00118 \text{ mol H}_3PO_4}{0.0250 \text{ L H}_3PO_4 \text{ solution}} \\ &= \frac{0.0472 \text{ mol H}_3PO_4}{\text{L H}_3PO_4 \text{ solution}} = 0.0472 \text{ M} \end{aligned}$$

9.19 **a.** The ions resulting from dissociation are Na^+ and NO_3^-. Na^+ is the cation of a strong base, and NO_3^- is the anion of a strong acid. Neither ion will hydrolyze significantly in water, so the solution pH will be about 7.

b. The ions resulting from dissociation are Na^+ and NO_2^-. Na^+ is the cation of a strong base and will not hydrolyze significantly. NO_2^- is the anion of a weak acid and will behave as a base in the hydrolysis reaction. The solution will therefore be basic, and the pH will be higher than 7.

c. The ions resulting from dissociation are K^+ and BO_3^{3-}. K^+ is the cation of a strong base and will not hydrolyze significantly. BO_3^{3-} is the anion of a weak acid and will behave as a base in the hydrolysis reaction. The solution will therefore be basic, and the pH will be higher than 7.

9.20 **a.** Because the acid (HCOOH) and conjugate base (HCOO$^-$) concentrations are the same, the pH will equal pK_a for the acid. This is shown by the following calculation:

$$pH = pK_a + \log \frac{[B^-]}{[HB]}$$

$$= 3.74 + \log \frac{[HCOO^-]}{[HCOOH]}$$

$$= 3.74 + \log \frac{0.22 \text{ mol/L}}{0.22 \text{ mol/L}}$$

$$= 3.74 + \log 1$$

$$= 3.74 + 0 = 3.74$$

b. The acid is H_2SO_3, and the conjugate base is HSO_3^- (produced when the $NaHSO_3$ salt dissociates).

$$pH = pK_a + \log \frac{[B^-]}{[HB]}$$

$$= 1.82 + \log \frac{(0.10 \text{ mol/L})}{(0.25 \text{ mol/L})}$$

$$= 1.82 + \log 0.40$$

$$= 1.82 - 0.40$$

$$= 1.42$$

c. The acid is H_2CO_3 and the conjugate base is HCO_3^-. The desired pH is 7.40. Substitution into Equation 9.54 gives:

$$pH = pK_a + \log \frac{[HCO_3^-]}{[H_2CO_3]}$$

$$7.40 = 6.37 + \log \frac{[HCO_3^-]}{[H_2CO_3]}$$

$$7.40 - 6.37 = \log \frac{[HCO_3^-]}{[H_2CO_3]}$$

$$1.03 = \log \frac{[HCO_3^-]}{[H_2CO_3]}$$

We now know the log of the desired ratio. Use Step 3 in Table 9.3 to get the antilog of 1.03. The correctly rounded value is 11. Thus, the concentration of HCO_3^- must be 11 times the concentration of H_2CO_3 in the blood to maintain the desired pH.

Chapter 10

10.1 **a.** $^{131}_{53}I$ **b.** $^{0}_{-1}\beta$ **c.** $^{4}_{4}X$ **d.** $^{0}_{-4}X$

10.2 $^{234}_{90}Th \rightarrow \, ^{0}_{-1}\beta + \, ^{234}_{91}Pa$

In general: $^{A}_{Z}X \rightarrow \, ^{0}_{-1}\beta + \, ^{A}_{Z+1}Y$

When a β particle is emitted, the daughter has the same mass number and an atomic number higher by 1 than that of the decaying nucleus.

10.3 $^{A}_{Z}X \rightarrow \, ^{0}_{1}\beta + \, ^{A}_{Z-1}Y$

The atomic number of the daughter is 1 less than that of the decaying nucleus.

10.4 $^{A}_{Z}X + \, ^{0}_{-1}e \rightarrow \, ^{A}_{Z-1}Y$

The atomic number of the daughter is 1 less than that of the decaying nucleus.

10.5 **a.** $^{50}_{25}Mn \rightarrow \, ^{0}_{1}\beta + \, ^{50}_{24}Cr$

 b. $^{54}_{25}Mn + \, ^{0}_{-1}e \rightarrow \, ^{54}_{24}Cr$

 c. $^{56}_{25}Mn \rightarrow \, ^{0}_{-1}\beta + \, ^{56}_{26}Fe$

 d. $^{224}_{88}Ra \rightarrow \, ^{4}_{2}\alpha + \, ^{220}_{86}Rn$

10.6 During the first half-life, the original 200.00 mg will be reduced to 100.00 mg. This will be reduced to 50.00 mg during the second half-life and to 25.00 mg during the third. Thus, 3 half-lives or 3(7.7 min) = 23.1 min will be required.

10.7 The elapsed time of 79.8 hours is 6 half-lives (79.8/13.3 = 6). The fraction remaining is 1/2 × 1/2 × 1/2 × 1/2 × 1/2 × 1/2 = 1/64 of the original diagnostic dose.

10.8 Equation 10.2 is used: $\dfrac{I_x}{I_y} = \dfrac{d_y^2}{d_x^2}$

$$\frac{10.0 \text{ units}}{I_5} = \frac{(5.0 \text{ ft})^2}{(25 \text{ ft})^2}$$

$$I_5 = \frac{(10.0 \text{ units})(25 \text{ ft})^2}{(5.0 \text{ ft})^2} = 250 \text{ units}$$

10.9 $^{55}_{25}Mn + \, ^{1}_{1}p \rightarrow \, ^{1}_{0}n + \, ^{55}_{26}Fe$

The product is iron-55.

Chapter 11

11.1 Look for the presence of carbon atoms.

 a. inorganic **d.** inorganic

 b. organic **e.** organic

 c. organic **f.** inorganic

11.2 Refer to Table 11.1 for help.

 a. organic

 b. organic

 c. inorganic

11.3 Compound (a) is a structural isomer because it has the same molecular formula but a different structural formula.

11.4 **a.**
$$CH_3 - \overset{\displaystyle OH}{\overset{|}{CH}} - CH_2 - CH_2 - CH_3$$

 b.
$$CH_3 - \overset{}{\underset{\underset{\textstyle CH_3}{|}}{CH}} - CH_2 - \overset{\overset{\textstyle O}{\|}}{C} - OH$$

11.5 **a.** The number of hydrogen atoms is twice the carbon atoms plus two: C_8H_{18}.

 b. $CH_3-CH_2-CH_2-CH_2-CH_2-CH_2-CH_2-CH_3$

11.6 **a.** Same molecule: In both molecules, the five carbons are bonded in a continuous chain.

 b. Same molecule: In both molecules, there is a continuous chain of four carbons with a branch at position 2.

$$\underset{1}{CH_3} - \overset{\overset{\textstyle CH_3}{|}}{\underset{2}{CH}} - \underset{3}{CH_2} - \underset{4}{CH_3} \qquad \underset{1}{CH_3} - \overset{\overset{3}{\overset{\textstyle CH_2}{|}} - \overset{4}{CH_3}}{\underset{2}{CH}} - \underset{}{CH_3}$$

 c. Structural isomers: Both molecules have a continuous chain of five carbons, but the branch is located at different positions.

$$\overset{\overset{1}{CH_3}}{\underset{}{|}}$$
$$\underset{2}{CH_2} - \underset{3}{CH_2} - \underset{4}{CH_3} \qquad \text{a butane}$$

11.7 **a.** $\underset{2}{CH_2} - \underset{3}{CH_2} - \underset{4}{CH_3}$ a butane

 b.
$$\underset{1}{CH_3} \qquad\qquad \overset{5}{CH_3}$$
$$\underset{2}{CH_2} - \underset{3}{CH} - \underset{4}{CH_2}$$
$$\qquad\quad \underset{}{\underset{\textstyle CH_3}{|}}$$
a pentane

 c.
$$\underset{1}{CH_3} - \overset{}{\underset{2}{CH}} - CH_3 \qquad \text{or}$$
$$CH_3 - \underset{3}{CH} - \underset{4}{CH_3}$$

$$\underset{3}{CH_3} - \overset{4}{\underset{}{CH}} - CH_3$$
$$\underset{1}{CH_3} - \underset{2}{CH} - CH_3 \qquad \text{a butane}$$

11.8 **a.**
$$\underset{3}{CH_3} - \overset{4}{CH} - \underset{5}{CH_2} - \underset{6}{CH_2} - CH_3$$
$$\qquad\quad \underset{2}{\underset{\textstyle CH_2}{|}}$$
$$\qquad\quad \underset{1}{\underset{\textstyle CH_3}{|}}$$

A CH$_3$ group is located at position 3. If the chain had been numbered beginning at the right, the CH$_3$ group would have been at position 4.

b. Numbering from the left, groups are located at positions 5, 5, 7. From the right, the groups are at positions 2, 4, 4. The first difference occurs with the first number, so numbering from the right (2, 4, 4) is correct.

11.9 Proceeding from the left, the groups are methyl, propyl, isopropyl, *sec*-butyl, and ethyl.

11.10 a. The chain is numbered from the right to give 2-methylhexane.

b.
$$CH_3 \overset{8}{-} CH_2 \overset{7}{-} CH_2 \overset{6}{-} \underset{\underset{\overset{|}{CH_2 \overset{2}{-} CH_3}}{\overset{|}{}}}{\overset{3}{CH}} \overset{5}{-} CH_2 \overset{4}{-} \underset{\overset{|}{CH_2 \overset{1}{-} CH_3}}{\overset{|}{}} \overset{3}{CH} \overset{}{-} CH_3$$

The chain is numbered from the right to give 5-ethyl-3-methyloctane.

c. The chain is numbered from the right to give 4-isopropyl-2, 3-dimethylheptane.

11.11

a.
$$CH_3 - \underset{\underset{CH_3}{|}}{\overset{\overset{CH_3}{|}}{C}} - CH_2 - \underset{\overset{CH_3}{|}}{CH} - CH_3$$

b.
$$CH_3 - CH_2 - \underset{\overset{|}{\underset{\overset{|}{CH_3}}{CH - CH_3}}}{CH} - CH_2 - CH_2 - CH_3$$

c.
$$CH_3 - CH - \underset{\overset{|}{CH_3}}{CH} - \underset{\overset{|}{CH_3}}{CH} - \underset{\overset{|}{CH_2 - CH_3}}{CH} - CH_2 - CH_2 - CH_3$$

11.12 a. 1,4-dimethylcyclohexane

b. The correct name is ethylcyclopropane. When only one group is attached to a ring, the position is not designated.

c. The name 1-ethyl-2-methylcyclopentane is correct, whereas 2-ethyl-1-methylcyclopentane is incorrect because the ring numbering begins with the carbon attached to the first group alphabetically.

11.13 a. (1) *Trans* because the two Br's are on opposite sides of the ring.

(2) *Cis*; both Cl's are on the same side.

(3) *Cis*; the two groups are on the same side.

b. In showing geometric isomers of ring compounds, it helps to draw the ring in perspective:

Cl Cl

Chapter 12

12.1 In each of these alkenes, the double-bonded carbons occur at positions 1 and 2.

a. 3-bromo-1-propene

b. 2-ethyl-1-pentene

c. 3,4-dimethylcyclohexene

12.2 a. The chain is correctly numbered from the right to give 2-methyl-1,3-butadiene.

b. The chain is correctly numbered from the left to give 2-methyl-1,3,6-octatriene.

c. 7-bromo-1,3,5-cycloheptatriene

12.3 a. This structure does not exhibit geometric isomerism because there are two H's attached to the carbon at position 1.

b.
$$\underset{CH_3 - \underset{\overset{|}{CH_3}}{CH}}{\overset{H}{\diagdown}}C = C \overset{H}{\diagup}\underset{CH_3}{} \qquad \underset{CH_3 - \underset{\overset{|}{CH_3}}{CH}}{\overset{H}{\diagdown}}C = C \overset{CH_3}{\diagup}\underset{H}{}$$

cis *trans*

c. This structure does not exhibit geometric isomerism because there are two methyl groups attached to the left double-bonded carbon.

12.4

a.
$$CH_3 - \underset{\overset{|}{CH_3}}{\overset{\overset{Br}{|}}{C}} \overset{Br}{\underset{}{\overset{|}{}}} - CH_2$$

b. (cyclopentane ring with Cl at top right and Cl at bottom)

12.5 a.
$$CH_3 - \underset{\overset{|}{CH_3}}{CH} - CH_2 - CH_3$$

b. (cyclopentane with CH$_3$ and H; = ; cyclopentane with CH$_3$)

12.6 a. The major product will be the one where H attaches to the CH carbon:
$$CH_3 - CH_2 - \underset{\underset{Br}{\overset{|}{}}}{\overset{\overset{CH_3}{|}}{C}} - CH_2 - CH_3$$

b. Position 1 on the ring has an attached hydrogen, whereas position 2 does not have any attached hydrogens:
(cyclohexene ring with CH$_3$, CH$_2$CH$_3$ at position 2, and H at position 1)

Thus, H attaches at position 1 to give:

12.7 a. Markovnikov's rule predicts that H will attach at position 1 to give:

b.

12.8 The double bond becomes a single bond:

12.9 Each chain is correctly numbered from the right.

a. 2-pentyne **b.** 5-methyl-2-hexyne

12.10 a. Numbers or the term *meta* may be used:

1,3-diethylbenzene or *m*-diethylbenzene

b. The compound must be named as a derivative of cyclopentane. The correct name is 1-chloro-3-phenylcyclopentane.

c. Numbers must be used when there are three groups: 1,2,3-tribromobenzene.

d. If the compound is named as a derivative of benzoic acid, then the methyl group is at position 2. The name is 2-methylbenzoic acid.

Chapter 13

13.1

a. 2-methyl-1-propanol

b. 2,5-dimethyl-4-heptanol

c. 2-methylcyclopentanol

13.2 a. 1,4-butanediol **b.** 1,2-cyclohexanediol

13.3 Because the compound is named as a phenol, the OH group is at position 1. The name is 2-ethylphenol or *o*-ethylphenol.

13.4 In each case, count the number of carbon atoms attached to the hydroxy-bearing carbon.

a. primary **b.** secondary **c.** secondary

13.5 b, c, a

13.6 Methanol molecules form hydrogen bonds, whereas molecules of propane do not.

13.7 a. There are two ways of removing water from this structure, but they both produce the same product:

$$CH_3CH=CHCH_2CH_3$$

b. Dehydration will occur in the direction that gives the highest number of carbon groups attached to the double-bonded carbons:

13.8 a. Ether formation is favored by H_2SO_4 at 140°C.

b. Alkene formation predominates when H_2SO_4 is used at 180°C.

13.9

first product second product

13.10

13.11 The alcohols in a and c react because they are secondary and primary, respectively. Alcohol b is tertiary and does not react.

13.12 Working backward, you could prepare an ether product from an alcohol. Thus, the starting alkene must first be converted to an alcohol.

$$CH_2=CH_2 + H_2O \xrightarrow{H_2SO_4} CH_3CH_2-OH$$

$$2CH_3CH_2-OH \xrightarrow[140°C]{H_2SO_4} CH_3CH_2-O-CH_2CH_3 + H_2O$$

13.13 Use Table 13.2 to answer these questions:

a. solvent

b. alcoholic beverages

c. rubbing alcohol

d. antifreeze

e. food moisturizer

f. cough drops

13.14 a. antioxidant in gasoline

b. throat lozenges

13.15 a. ethyl phenyl ether

b. diisopropyl ether

13.16 a. 1,2-dimethoxycyclopentane

b. 3-isopropoxypentane

13.17 a. Ether molecules do not form hydrogen bonds, whereas alcohol molecules do form hydrogen bonds.

b. They are quite unreactive.

13.18 Oxidation of a thiol produces a disulfide:

$$CH_3CH-S-S-CHCH_3 + H_2O$$
$$\quad | \qquad\qquad\quad |$$
$$\quad CH_3 \qquad\qquad CH_3$$

13.19 a. $2CH_3CH-SH$
$$\qquad\quad |$$
$$\qquad\quad CH_3$$

b.

13.20 $CH_3CH-S-Pb-S-CHCH_3 + 2H^+$
$$\qquad\quad | \qquad\qquad\qquad |$$
$$\qquad\quad CH_3 \qquad\qquad\quad CH_3$$

13.21 Ether, phenolic OH, and benzene ring

Chapter 14

14.1 a. The aldehyde contains five carbon atoms, and the name is pentanal.

b. The aldehyde carbon is position 1. The name is 3, 4-dimethylpentanal.

14.2 a. The chain is numbered from the right. The name is 4, 6-dibromo-2-heptanone.

b. The carbonyl carbon atom is position 1. The name is 2, 5-dimethylcyclopentanone.

14.3 a.
$$\qquad\quad CH_3 \qquad\quad O$$
$$\qquad\quad | \qquad\qquad\ \ ||$$
$$CH_3CHCH_2-C-OH$$

b. This is a ketone, and it does not undergo oxidation.

14.4 a.
$$\qquad\quad OH$$
$$\qquad\quad |$$
$$CH_3CH_2CHCH_2CH_3$$

b.

14.5 a.

hemiacetal acetal

b.

hemiketal ketal

14.6 a. $CH_3CH_2CH_2-\overset{\displaystyle O}{\overset{||}{C}}-H + 2CH_3CH_2-OH$

b.

$+ 2CH_3-OH$

14.7 a. This compound is not a hemiacetal or hemiketal because there is no OH group. The presence of the ring hydrogen makes it an acetal.

Hydrolysis breaks these bonds

b. An OH group is missing, so this compound is an acetal or ketal. Because the carbon with two attached oxygens does not have a hydrogen present, the compound is a ketal.

Hydrolysis breaks these bonds

Chapter 15

15.1 a. The longest chain beginning with the carboxyl group has four carbon atoms. The name is 2-methylbutanoic acid.

b. The carboxyl group is at position 1. The name is 3, 4-dibromobenzoic acid.

15.2 $CH_3-\overset{\displaystyle O}{\overset{||}{C}}-\overset{\displaystyle O}{\overset{||}{C}}-O^-$

15.3 $CH_3CH_2CH_2-\overset{\displaystyle O}{\overset{||}{C}}-O^-Na^+ + H_2O$

15.4 a. This is the lithium salt of 2-methylbenzoic acid. The name is lithium 2-methylbenzoate.

b. This is the sodium salt of butanoic acid. The name is sodium butanoate.

15.5 Each of these reactions requires H^+ as a catalyst. The products are as follows:

a. $CH_3CH_2-\overset{\displaystyle O}{\overset{||}{C}}-OCH_3 + H_2O$

b. $CH_3-\overset{\displaystyle O}{\overset{||}{C}}-O-$ $+ H_2O$

c.

15.6 a.

b.

$$CH_3CH_2-\overset{\overset{\displaystyle O}{\|}}{C}-Cl + CH_3\overset{\overset{\displaystyle}{|}}{\underset{\overset{\displaystyle}{CH_3}}{CH}}-OH \longrightarrow$$

$$CH_3CH_2-\overset{\overset{\displaystyle O}{\|}}{C}-O-\overset{\overset{\displaystyle}{|}}{\underset{\overset{\displaystyle}{CH_3}}{CHCH_3}} + HCl$$

15.7 a. The common name for the acid used to prepare this ester is formic acid. The alkyl group attached to oxygen is isopropyl. Thus, the common name is isopropyl formate. The IUPAC name for formic acid is methanoic acid. Thus, the IUPAC name for the ester is isopropyl methanoate.

b. The common and IUPAC names are identical when the carboxylic acid is benzoic acid. The name of the ester is methyl benzoate.

15.8

$$\text{(benzene ring)}-\overset{\overset{\displaystyle O}{\|}}{C}-OH \quad + HO-CH_2CH_2CH_3$$

15.9 a. $CH_3CH_2-\overset{\overset{\displaystyle O}{\|}}{C}-O^-Na^+ + HO-\overset{\overset{\displaystyle}{|}}{\underset{\overset{\displaystyle}{CH_3}}{CHCH_3}}$

b. $\text{(benzene ring)}-\overset{\overset{\displaystyle O}{\|}}{C}-O^-K^+ + HO-CH_2CH_3$

Chapter 16

16.1 In each case, count the number of carbon atoms directly attached to the nitrogen.

 a. secondary **b.** secondary **c.** tertiary

16.2 Identify the groups attached to the nitrogen.

 a. propylamine

 b. methylphenylamine

 c. trimethylamine

16.3 a. 1-butanamine

 b. *N*-methyl-1-butanamine

16.4 a. *N*-ethylaniline

 b. *N*-ethyl-*N*-methylaniline

 c. 3-ethyl-2-methylaniline

16.5 a.

b.

$$CH_3CH_2-\overset{\overset{\textstyle H \cdots\cdots\overset{\textstyle O}{\diagup\diagdown}\cdots H}{|}}{\underset{\overset{\displaystyle}{CH_3}}{N}}-CH_3$$

16.6 a.

$$CH_3-\overset{\overset{\displaystyle H}{\overset{|}{\underset{+}{N}}}}{\underset{\overset{\displaystyle}{CH_3}}{}}-CH_2CH_3 + OH^-$$

 b. $\text{(benzene ring)}-NH_3{}^+ + OH^-$

16.7 a. $\text{(benzene ring)}-\overset{\overset{\displaystyle}{\underset{+}{N}H_2}}{\underset{\overset{\displaystyle}{CH_3}}{}}{}^+Cl^-$

 b. $\text{(cyclopentane ring)}-NH_3{}^+CH_3CH_2COO^-$

16.8

 a. $CH_3-\overset{\overset{\displaystyle O}{\|}}{C}-NH-CH_2CH_2CH_3 + CH_3-\overset{\overset{\displaystyle O}{\|}}{C}-OH$

 b. $CH_3CH_2-\overset{\overset{\displaystyle O}{\|}}{C}-NH_2 + HCl$

16.9 a. This amide is derived from butanoic acid. The name is butanamide.

 b. hexanamide

16.10 a. This amide is a derivative of the two-carbon acid acetic acid (common name) or ethanoic acid (IUPAC name). The amide is named *N,N*-diethylacetamide (common name) or *N,N*-diethylethanamide (IUPAC name).

 b. This amide is a derivative of butyric acid (common name) or butanoic acid (IUPAC name). The amide is named *N*-isobutylbutyramide (common name) or *N*-isobutylbutanamide (IUPAC name).

 c. The acid portion of this amide came from benzoic acid. The amide is named *N*-ethyl-*N*-phenylbenzamide.

 d. The acid portion of this amide came from 2-methylbenzoic acid. The amide is named 2,*N*-dimethylbenzamide.

16.11 a.

$$CH_3-\overset{\overset{\displaystyle O}{\|}}{C}-\overset{\overset{\displaystyle}{\underset{\overset{\displaystyle}{H}}{N}}}{}-CH_3$$

$$\vdots$$

$$CH_3-\overset{\overset{\displaystyle O}{\|}}{C}-\overset{\overset{\displaystyle}{\underset{\overset{\displaystyle}{H}}{N}}}{}-CH_3$$

b.

16.12 a. Under acidic conditions, amide hydrolysis produces a carboxylic acid and the salt of an amine.

$$
\begin{array}{c}
\text{O} \\
\parallel \\
\text{C}-\text{OH}
\end{array}
$$

+ CH₃CH₂ — NH₃⁺Cl⁻

b. Under basic conditions, amide hydrolysis produces an amine and the salt of a carboxylic acid.

$$
\begin{array}{c}
\text{O} \\
\parallel \\
\text{C}-\text{O}^-\text{Na}^+
\end{array}
$$

+ CH₃CH₂ — NH₂

Chapter 17

17.1 In each case, look for the presence of four different groups attached to the colored atom.

a. The carbon is not chiral because of the two attached CH₂OH groups.

b. Yes

c. The carbon is not chiral because it is attached to two H's.

d. Yes

17.2 Each of the atoms marked with an asterisk has four different attached groups and is chiral.

a.
$$
\begin{array}{c}
\text{CH}_2\text{OH} \\
| \\
\text{C}=\text{O} \\
| \\
\text{HO}-\overset{*}{\text{C}}-\text{H} \\
| \\
\text{H}-\overset{*}{\text{C}}-\text{OH} \\
| \\
\text{CH}_2\text{OH}
\end{array}
$$

b.
$$
\begin{array}{c}
\text{CHO} \\
| \\
{}^*\text{CH}-\text{OH} \\
| \\
{}^*\text{CH}-\text{OH} \\
| \\
{}^*\text{CH}-\text{OH} \\
| \\
\text{CH}_2\text{OH}
\end{array}
$$

c.

17.3 The chiral carbon atoms (three of them) are marked with an asterisk. Thus, there are 2³, or 8, possible stereoisomers:

$$
\begin{array}{c}
\text{OH} \quad \text{OH} \quad \text{OH} \quad \text{OH} \\
| \qquad | \qquad | \qquad | \\
\text{CH}_2-\overset{*}{\text{CH}}-\overset{*}{\text{CH}}-\overset{*}{\text{CH}}-\text{CHO}
\end{array}
$$

17.4 The chiral carbon atom is the one with the attached nitrogen. The chiral carbon atom is placed at the intersection of the two lines, and the COOH is placed at the top:

17.5 In each case, look at the direction of the OH group on the bottom chiral carbon.

a. L **b.** D **c.** L

17.6 a. aldopentose

b. ketotetrose

c. aldotetrose

17.7

$$
\begin{array}{c}
\text{HOCH}_2 \quad \text{O} \\
\end{array}
$$

α

OH

OH OH

17.8 a.

$$
\begin{array}{c}
\text{CH}_2\text{OH} \\
\text{HO} \quad\quad \text{O} \\
\text{OH} \\
\text{OCH}_2\text{CH}_3 \\
\text{OH}
\end{array}
$$

acetal group

glycosidic linkage

b.

$$
\begin{array}{c}
\text{HOCH}_2 \quad \text{O} \quad \text{OCH}_3 \\
\text{CH}_2\text{OH} \\
\text{OH} \quad \text{OH}
\end{array}
$$

ketal group

glycosidic linkage

Chapter 18

18.1 One of several possibilities is the following structure:

$$
\begin{array}{c}
\text{O} \\
\parallel \\
\text{CH}_2-\text{O}-\text{C}-(\text{CH}_2)_{16}\text{CH}_3 \\
| \\
\text{O} \\
\parallel \\
\text{CH}-\text{O}-\text{C}-(\text{CH}_2)_7\text{CH}=\text{CH}(\text{CH}_2)_7\text{CH}_3 \\
| \\
\text{O} \\
\parallel \\
\text{CH}_2-\text{O}-\text{C}-(\text{CH}_2)_{14}\text{CH}_3
\end{array}
$$

(stearic acid component)

(oleic acid component)

(palmitic acid component)

18.2

$$
\begin{array}{l}
\text{CH}_2-\text{OH} \quad \text{HOOC}-(\text{CH}_2)_7\text{CH}=\text{CH}(\text{CH}_2)_7\text{CH}_3 \\
| \\
\text{CH}-\text{OH} + \text{HOOC}-(\text{CH}_2)_{14}\text{CH}_3 \\
| \\
\text{CH}_2-\text{OH} \quad \text{HOOC}-(\text{CH}_2)_{16}\text{CH}_3
\end{array}
$$

18.3

$$CH_2-OH \quad Na^{+-}OOC-(CH_2)_{16}CH_3$$
$$CH-OH + 2Na^{+-}OOC-(CH_2)_7CH=CH(CH_2)_7CH_3$$
$$CH_2-OH$$

18.4 During complete hydrogenation, all CH=CH groups become CH$_2$—CH$_2$ groups:

CH$_2$—O—C(=O)—(CH$_2$)$_7$CH$_2$CH$_2$(CH$_2$)$_7$CH$_3$

CH—O—C(=O)—(CH$_2$)$_7$CH$_2$CH$_2$CH$_2$CH$_2$CH$_2$CH$_2$CH$_2$CH$_2$CH$_2$CH$_3$

CH$_2$—O—C(=O)—(CH$_2$)$_7$CH$_2$CH$_2$CH$_2$CH$_2$CH$_2$(CH$_2$)$_4$CH$_3$

Combining all the CH$_2$ groups gives:

CH$_2$—O—C(=O)—(CH$_2$)$_{16}$CH$_3$

CH—O—C(=O)—(CH$_2$)$_{16}$CH$_3$

CH$_2$—O—C(=O)—(CH$_2$)$_{16}$CH$_3$

18.5

CH$_2$—O—C(=O)—(CH$_2$)$_{14}$CH$_3$

CH—O—C(=O)—(CH$_2$)$_7$CH=CH(CH$_2$)$_7$CH$_3$

CH$_2$—O—P(=O)(O$^-$)—OCH$_2$CH$_2$—NH$_3^+$

Chapter 19

19.1

a. H$_3$N$^+$—CH(CH$_2$—OH)—C(=O)—OH

b. H$_2$N—CH(CH$_2$—OH)—C(=O)—O$^-$

19.2

H$_3$N$^+$—CH(CH$_2$—C$_6$H$_5$)—C(=O)—NH—CH(CH$_2$—SH)—C(=O)—NH—

N-terminal residue

—CH(CH$_2$—OH)—C(=O)—NH—CH(CHCH$_3$—CH$_2$CH$_3$)—C(=O)—O$^-$

C-terminal residue

19.3 **a.** Hydrophobic attractions between nonpolar groups

b. Hydrogen bonding

c. Salt bridges between the ionic groups

Chapter 20

20.1 **a.** maltose

b. peptides

c. glucose-6-phosphate

20.2 Turnover number is the number of substrate molecules acted on per minute by one molecule of enzyme. One enzyme international unit is the amount of enzyme that catalyzes 1 μmol of substrate to react per minute.

20.3 **a.** An increase in enzyme concentration increases the rate of an enzyme-catalyzed reaction.

b. An increase in substrate concentration increases the rate until the enzyme is saturated. Above the saturation point, the rate of reaction is constant.

c. An increase in temperature increases the rate of the enzyme-catalyzed reaction until the optimum temperature is reached. Above the optimum temperature, the rate decreases.

d. An increase in pH increases the reaction rate until the optimum pH is reached. Above the optimum pH, the rate decreases.

20.4 **a.** The structure of a competitive inhibitor resembles that of the substrate. The structure of a noncompetitive inhibitor bears no resemblance to that of the substrate.

b. A competitive inhibitor binds at the active site. A noncompetitive inhibitor binds at some other region of the enzyme.

c. Increasing substrate concentration reverses the effect of a competitive inhibitor, but has no effect on a noncompetitive inhibitor.

20.5 **a.** Several enzymes are synthesized in inactive forms called zymogens. Under the proper reaction conditions, zymogens are activated to carry out their catalytic role.

b. The catalytic ability of allosteric enzymes may be either increased or decreased by the binding of modulators.

c. Certain enzymes are synthesized in greater amounts when needed and in smaller amounts when not needed.

Chapter 21

21.1

phosphodiester linkages

21.2 5' T-T-A-C-G 3' original strand

3' A-A-T-G-C 5' complementary strand

The complementary strand written in the 5' to 3' direction is C-G-T-A-A.

21.3 5' A-T-T-A-G-C-C-G 3' DNA template

3' U-A-A-U-C-G-G-C 5' mRNA

The mRNA written in the 5' to 3' direction is C-G-G-C-U-A-A-U.

21.4 fMet-His-His-Val-Leu-Cys

Chapter 22

22.1

$$\underset{\substack{| \\ CH_2-COO^-}}{\overset{\substack{OH \\ |}}{CH}}-COO^- + NAD^+ \longrightarrow \underset{\substack{| \\ CH_2-COO^-}}{\overset{\substack{O \\ ||}}{C}}-COO^- + NADH + H^+$$

22.2 **a.** $^-OOC-CH=CH-COO^- + FADH_2$

b. $CH_3-\overset{\substack{O \\ ||}}{C}-COO^- + NADH + H^+$

Chapter 23

23.1 **a.** Hexokinase regulates the first step of the glycolysis pathway. The enzyme is controlled by concentrations of glucose-6-phosphate through feedback inhibition.

b. Phosphofructokinase is an allosteric enzyme which regulates the production of fructose-1,6-bisphosphate. The enzyme is inhibited by high concentrations of ATP and citrate, and it is activated by high concentrations of ADP and AMP.

c. Pyruvate kinase controls the last step of glycolysis. It is inhibited by higher concentrations of ATP.

23.2 NAD^+ produced from the various fates of pyruvate can be used in the glycolysis pathway, enabling further glycolysis to take place.

23.3 **a.** Three: Each $FADH_2$ produces 1.5 molecules of ATP.

b. Five: Each NADH produces 2.5 molecules of ATP.

c. 20: Each acetyl CoA produces 10 molecules of ATP.

23.4 Glucose-6-phosphatase is present in the liver but not in the muscles.

Chapter 24

24.1 **a.** Each acetyl CoA contains two carbon atoms from capric acid. Thus, five acetyl CoA molecules are formed.

b. Ten carbon atoms require four trips through the fatty acid spiral. Thus, four molecules of $FADH_2$ are formed.

c. Four trips through the spiral also produce four NADH molecules and four H^+ ions.

24.2 A 16-carbon fatty acid would yield eight molecules of acetyl CoA. Seven trips through the spiral are necessary and would produce 7 $FADH_2$ and 7 NADH +7 H^+.

8 acetyl CoA × 10 ATP/acetyl CoA	=	80 ATP
7 $FADH_2$ × 1.5 ATP/$FADH_2$	=	10.5 ATP
7 NADH × 2.5 ATP/NADH	=	17.5 ATP
		108 ATP

Two high-energy bonds in ATP are consumed in the activation step.

$$\frac{-2}{106 \text{ ATP}}$$

24.3

24.4 **a.** The urea cycle converts toxic ammonia into less toxic urea for excretion from the body.

b. carbamoyl phosphate

Chapter 25

25.1 **a.** lungs

b. tissue cells

c. lungs

d. tissue cells

absolute specificity The characteristic of an enzyme that it acts on one and only one substance.

absolute zero The temperature at which all motion stops; a value of 0 on the Kelvin scale.

acetal A compound that contains the $-\overset{\displaystyle OR}{\underset{\displaystyle OR}{\overset{|}{\underset{|}{C}}}}-H$ arrangement of atoms.

acetone breath A condition in which acetone can be detected in the breath.

acid dissociation constant The equilibrium constant for the dissociation of an acid.

acidic solution A solution in which the concentration of H_3O^+ is greater than the concentration of OH^-. Also, a solution in which pH is less than 7.

acidosis A low blood pH.

activation energy Energy needed to start some spontaneous processes. Once started, the processes continue without further stimulus or energy from an outside source.

activator A substance that binds to an allosteric enzyme and increases its activity.

active site The location on an enzyme where a substrate is bound and catalysis occurs.

activity series A tabular representation of the tendencies of metals to react with H^+.

acute radiation syndrome The condition associated with and following short-term exposure to intense radiation.

addition polymer A polymer formed by the linking together of many alkene molecules through addition reactions.

addition reaction A reaction in which a compound adds to a multiple bond.

adipose tissue A kind of connective tissue where triglycerides are stored.

aerobic In the presence of oxygen.

alcohol A compound in which an $-OH$ group is connected to an aliphatic carbon atom.

alcoholic fermentation The conversion of glucose to ethanol.

aldehyde A compound that contains the $-\overset{\displaystyle O}{\overset{\|}{C}}-H$ group; the general formula is $R-\overset{\displaystyle O}{\overset{\|}{C}}-H$.

aliphatic compound Any organic compound that is not aromatic.

alkaline solution See **basic solution**.

alkaloids A class of nitrogen-containing organic compounds obtained from plants.

alkalosis An abnormally high blood pH.

alkane A hydrocarbon that contains only single bonds.

alkene A hydrocarbon containing one or more double bonds.

alkoxy group The $-O-R$ functional group.

alkyl group A group differing by one hydrogen from an alkane.

alkyl halide See **haloalkane**.

alkyne A hydrocarbon containing one or more triple bonds.

allosteric enzyme An enzyme with a quaternary structure whose activity is changed by the binding of modulators.

alpha-amino acid An organic compound containing both an amino group and a carboxylate group, with the amino group attached to the carbon next to the carboxylate group.

α-helix The helical structure in proteins that is maintained by hydrogen bonds.

alpha particle The particle that makes up alpha rays. It is identical to the helium nucleus and composed of two protons and two neutrons.

amide An organic compound having the functional group $-\overset{\displaystyle O}{\overset{\|}{C}}-\overset{\displaystyle}{\underset{|}{N}}-$.

amide linkage The carbonyl carbon-nitrogen single bond of the amide group.

amine An organic compound derived by replacing one or more of the hydrogen atoms of ammonia with alkyl or aromatic groups, as in RNH_2, R_2NH, and R_3N.

amino acid pool The total supply of amino acids in the body.

amino acid residue An amino acid that is a part of a peptide, polypeptide, or protein chain.

amphetamines A class of drugs structurally similar to epinephrine, used to stimulate the central nervous system.

anabolism All reactions involved in the synthesis of biomolecules.

anaerobic In the absence of oxygen.

anion A negatively charged ion.

anomeric carbon An acetal, ketal, hemiacetal, or hemiketal carbon atom giving rise to two stereoisomers.

anomers Stereoisomers that differ in the three-dimensional arrangement of groups at the carbon of an acetal, ketal, hemiacetal, or hemiketal group.

antibiotic A substance produced by one microorganism that kills or inhibits the growth of other microorganisms.

antibody A substance that helps protect the body from invasion by viruses, bacteria, and other foreign substances.

anticodon A three-base sequence in tRNA that is complementary to one of the codons in mRNA.

antioxidant A substance that prevents another substance from being oxidized.

apoenzyme A catalytically inactive protein formed by removal of the cofactor from an active enzyme.

aromatic hydrocarbon Any organic compound that contains the characteristic benzene ring or similar feature.

Arrhenius acid Any substance that provides H^+ ions when dissolved in water.

Arrhenius base Any substance that provides OH^- ions when dissolved in water.

atom The limit of chemical subdivision for matter.

atomic mass unit (u) A unit used to express the relative masses of atoms. One u is equal to $\frac{1}{12}$ the mass of an atom of carbon-12.

atomic number of an atom A number equal to the number of protons in the nucleus of an atom. Symbolically it is represented by Z.

atomic orbital A volume of space around atomic nuclei in which electrons of the same energy move. Groups of orbitals with the same n value form subshells.

atomic weight The mass of an average atom of an element expressed in atomic mass units.

Avogadro's law Equal volumes of gases measured at the same temperature and pressure contain equal numbers of molecules.

balanced equation An equation in which the number of atoms of each element in the reactants is the same as the number of atoms of that same element in the products.

basic solution A solution in which the concentration of OH^- is greater than the concentration of H_3O^+. Also, a solution in which pH is greater than 7; also called alkaline solution.

basic unit of measurement A specific unit from which other units for the same quantity are obtained by multiplication or division.

becquerel A physical unit of radiation measurement corresponding to one nuclear disintegration per second.

Benedict's reagent A mild oxidizing solution containing Cu^{2+} ions used to test for the presence of aldehydes.

β-oxidation process A pathway in which fatty acids are broken down into molecules of acetyl CoA.

beta particle The particle that makes up beta rays. It is identical to an electron but is produced in the nucleus when a neutron is changed into a proton and an electron.

β-pleated sheet A secondary protein structure in which protein chains are aligned side by side in a sheetlike array held together by hydrogen bonds.

binary compound A compound made up of two different elements.

biochemistry A study of the compounds and processes associated with living organisms.

biological unit of radiation A radiation measurement unit indicating the damage caused by radiation in living tissue.

biomolecule A general term referring to organic compounds essential to life.

blood sugar level The amount of glucose present in blood, normally expressed as milligrams per 100 mL of blood.

boiling point The temperature at which the vapor pressure of a liquid is equal to the prevailing atmospheric pressure.

bond polarization A result of shared electrons being attracted to the more electronegative atom of a bonded pair of atoms.

Boyle's law A gas law that describes the pressure and volume behavior of a gas sample kept at constant temperature. Mathematically, it is $PV = k$.

branched alkane An alkane in which at least one carbon atom is not part of a continuous chain.

branching chain reaction See **expanding chain reaction**.

breeder reaction A nuclear reaction in which isotopes that will not undergo spontaneous fission are changed into isotopes that will.

Brønsted acid Any hydrogen-containing substance that is capable of donating a proton (H^+) to another substance.

Brønsted base Any substance that is capable of accepting a proton (H^+) from another substance.

buffer A solution with the ability to resist changing pH when acids (H^+) or bases (OH^-) are added.

buffer capacity The amount of acid (H^+) or base (OH^-) that can be absorbed by a buffer without causing a significant change in pH.

carbaminohemoglobin Hemoglobin combined with carbon dioxide.

carbocation An ion of the form $-\overset{+}{\underset{|}{C}}-$.

carbohydrate A polyhydroxy aldehyde or ketone, or substance that yields such compounds on hydrolysis.

carbonyl group The $-\overset{\overset{O}{\|}}{C}-$ group.

carboxylate ion The $R-\overset{\overset{O}{\|}}{C}-O^-$ ion that results from the dissociation of a carboxylic acid.

carboxyl group The $-\overset{\overset{O}{\|}}{C}-OH$ group.

carboxylic acid An organic compound that contains the $-\overset{\overset{O}{\|}}{C}-OH$ functional group.

carboxylic acid anhydride An organic compound that contains the $-\overset{\overset{O}{\|}}{C}-O-\overset{\overset{O}{\|}}{C}-$ functional group.

carboxylic acid chloride An organic compound that contains the $-\overset{\overset{O}{\|}}{C}-Cl$ functional group.

carboxylic ester A compound with the $-\overset{\overset{O}{\|}}{C}-OR$ functional group.

catabolism All reactions involved in the breakdown of biomolecules.

catalyst A substance that changes (usually increases) reaction rates without being used up in the reaction.

cation A positively charged ion.

cellular respiration The entire process involved in the use of oxygen by cells.

central dogma of molecular biology The well-established process by which genetic information stored in DNA molecules is expressed in the structure of synthesized proteins.

cephalin A phosphoglyceride containing ethanolamine or serine.

chain reaction A nuclear reaction in which the products of one reaction cause a repeat of the reaction to take place. In the case of uranium fission, neutrons from fission reactions cause other fission reactions to occur.

Charles's law A gas law that describes the temperature and volume behavior of a gas sample kept at constant pressure. Mathematically, it is $V/T = k'$.

chemical changes Changes matter undergoes that involve changes in composition.

chemical properties Properties that matter demonstrates when attempts are made to change it into new substances.

chemiosmotic hypothesis The postulate that a proton flow across the inner mitochondrial membrane during operation of the electron transport chain provides energy for ATP synthesis.

chiral A descriptive term for compounds or objects that cannot be superimposed on their mirror image.

chiral carbon A carbon atom with four different groups attached.

chloride shift The movement of chloride ions from one fluid to another to maintain charge neutrality.

chromosome A tightly packed bundle of DNA and protein that is involved in cell division.

chylomicron A lipoprotein aggregate found in the lymph and the bloodstream.

cis- On the same side (as applied to geometric isomers).

citric acid cycle A series of reactions in which acetyl CoA is oxidized to carbon dioxide and reduced forms of coenzymes FAD and NAD^+ are produced.

codon A sequence of three nucleotide bases that represents a code word on mRNA molecules.

coenzyme An organic molecule required by an enzyme for catalytic activity.

cofactor A nonprotein molecule or ion required by an enzyme for catalytic activity.

cohesive forces The attractive force between particles; they are associated with potential energy.

cold spot Tissue from which a radioactive tracer is excluded or rejected.

colligative property A solution property that depends only on the concentration of solute particles in solution.

colloid A homogeneous mixture of two or more substances in which the dispersed substances are present as larger particles than are found in solutions.

combination reaction A chemical reaction in which two or more substances react to form a single substance.

combined gas law A gas law that describes the pressure, volume, and temperature behavior of a gas sample. Mathematically, it is $PV/T = k''$.

common catabolic pathway The reactions of the citric acid cycle plus those of the electron transport chain and oxidative phosphorylation.

competitive inhibitor An inhibitor that competes with substrate for binding at the active site of an enzyme.

complementary DNA strands Two strands of DNA in a double-helical form such that adenine and guanine of one strand are matched and hydrogen bonded to thymine and cytosine, respectively, of the second strand.

complete protein Protein in food that contains all essential amino acids in the proportions needed by the body.

complex carbohydrates The polysaccharides amylose and amylopectin, collectively called starch.

complex lipid An ester-containing lipid with more than two types of components: an alcohol, fatty acids—plus others.

compound A pure substance consisting of two or more kinds of atoms in the form of heteroatomic molecules or individual atoms.

compound formula A representation of the molecule of a compound, consisting of the symbols of the atoms found in the molecule. Atoms present in numbers greater than one have the number indicated by a subscript.

compressibility The change in volume of a sample resulting from a pressure change acting on the sample.

concentration The relationship between the amount of solute and the specific amount of solution in which it is contained.

condensation An exothermic process in which a gas or vapor is changed to a liquid or solid.

condensation polymerization The process by which monomers combine with the simultaneous elimination of a small molecule.

condensed structural formula A structural molecular formula showing the general arrangement of atoms but without showing all the covalent bonds.

conformations The different arrangements of atoms in space achieved by rotation about single bonds.

conjugate acid–base pair A Brønsted acid and its conjugate base.

conjugate base The species remaining when a Brønsted acid donates a proton.

conjugated protein A protein made up of amino acid residues and other organic or inorganic components.

copolymer An addition polymer formed by the reaction of two different monomers.

Cori cycle The process in which glucose is converted to lactate in muscle tissue, lactate is reconverted to glucose in the liver, and glucose is returned to the muscle.

covalent bond The attractive force that results between two atoms that are both attracted to a shared pair of electrons.

critical mass The minimum amount of fissionable material needed to sustain a critical chain reaction at a constant rate.

critical reaction A constant-rate chain reaction.

crystal lattice A rigid three-dimensional arrangement of particles.

C-terminal residue An amino acid on the end of a chain that has an unreacted or free carboxylate group.

curie A physical unit of radiation measurement corresponding to 3.7×10^{10} nuclear disintegrations per second.

cycloalkane An alkane in which carbon atoms form a ring.

cyclotron A cyclic particle accelerator that works by changing electrical polarities as charged particles cross a gap. The particles are kept in a spiral path by a strong magnetic field.

cytochrome An iron-containing enzyme located in the electron transport chain.

Daily Reference Values (DRVs) A set of standards for nutrients and food components (such as fat and fiber) that have important relationships with health; used on food labels as part of the Daily Values.

Daily Values (DVs) Reference values developed by the FDA specifically for use on food labels. The Daily Values represent two sets of standards: Reference Daily Intakes (RDIs) and Daily Reference Values (DRVs).

Dalton's law of partial pressures The total pressure exerted by a mixture of gases is equal to the sum of the partial pressures of the gases in the mixture.

daughter nuclei The new nuclei produced when unstable nuclei undergo radioactive decay.

decomposition A change in chemical composition that can result from heating.

decomposition reaction A chemical reaction in which a single substance reacts to form two or more simpler substances.

dehydration reaction A reaction in which water is chemically removed from a compound.

denaturation The process by which a protein loses its characteristic native structure and function.

density The number given when the mass of a sample of a substance is divided by the volume of the same sample.

deoxyhemoglobin (hemoglobin) Nonoxygenated hemoglobin.

deoxyribonucleic acid (DNA) A nucleic acid found primarily in the nuclei of cells.

derived unit of measurement A unit obtained by multiplication or division of one or more basic units.

dextrorotatory Rotates plane-polarized light to the right.

dialysis A process in which solvent molecules, other small molecules, and hydrated ions pass from a solution through a membrane.

dialyzing membrane A semipermeable membrane with pores large enough to allow solvent molecules, other small molecules, and hydrated ions to pass through.

diatomic molecules Molecules that contain two atoms.

diffusion A process that causes substances to spontaneously intermingle when they are brought together.

dimer Two identical molecules bonded together.

dipeptide A compound formed when two amino acids are bonded by an amide linkage.

dipolar force The attractive force that exists between the positive end of one polar molecule and the negative end of another.

diprotic acid An acid that gives up two protons (H^+) per molecule when dissolved.

disaccharide A carbohydrate formed by the combination of two monosaccharide units.

dispersed phase The substance present in a colloidal dispersion in amounts less than the amount of the dispersing medium.

dispersing medium The substance present in a collodial dispersion in the largest amount.

dispersion forces Very weak attractive forces acting between the particles of all matter. They result from momentary nonsymmetric electron distributions in molecules or atoms.

disruptive forces The force resulting from particle motion; they are associated with kinetic energy.

dissolving A term used to describe the process of solution formation when one or more solutes are dispersed in a solvent to form a homogeneous mixture.

distinguishing electron The last and highest-energy electron found in an element.

disulfide A compound containing an —S—S— group.

disulfide bridge A bond produced by the oxidation of —SH groups on two cysteine residues. The bond loops or holds two peptide chains together.

double bond The bond resulting from the sharing of two pairs of electrons.

double-replacement reaction A chemical reaction in which two compounds react and exchange partners to form two new compounds.

effective collision A collision that causes a reaction to occur between colliding molecules.

effusion A process in which a gas escapes from a container through a small hole.

electrolyte A solute that when dissolved in water forms a solution that conducts electricity.

electron capture A mode of decay for some unstable nuclei in which an electron from outside the nucleus is drawn into the nucleus, where it combines with a proton to form a neutron.

electronegativity The tendency of an atom to attract shared electrons of a covalent bond.

electronic configurations The detailed arrangement of electrons indicated by a specific notation, $1s^22s^22p^4$, etc.

electron transport chain A series of reactions in which protons and electrons from the oxidation of foods are used to reduce molecular oxygen to water.

element A pure substance consisting of only one kind of atom in the form of homoatomic molecules or individual atoms.

elemental symbol A symbol assigned to an element based on the name of the element, consisting of one capital letter or a capital letter followed by a lowercase letter.

elimination reaction A reaction in which two or more covalent bonds are broken and a new multiple bond is formed.

emulsifying agent (stabilizing agent) A substance that when added to colloids prevents them from coalescing and settling.

enantiomers Stereoisomers that are mirror images.

endergonic process A process that gains or accepts energy as it takes place.

endothermic reaction A reaction that absorbs heat.

endpoint of a titration The point at which the titration is stopped on the basis of an indicator color change or pH meter reading.

entropy A measurement or indication of the disorder or randomness of a system. The more disorderly a system is, the higher its entropy.

enzyme A biomolecule that catalyzes chemical reactions.

enzyme activity The rate at which an enzyme catalyzes a reaction.

enzyme induction The synthesis of an enzyme in response to a cellular need.

enzyme inhibitor A substance that decreases the activity of an enzyme.

enzyme international unit (IU) A quantity of enzyme that catalyzes the conversion of 1 μmol of substrate per minute under specified conditions.

equilibrium concentrations The unchanging concentrations of reactants and products in a reaction system that is in a state of equilibrium.

equilibrium constant A numerical relationship between reactant and product concentrations in a reaction at equilibrium.

equilibrium expression An equation relating the equilibrium constant and reactant and product equilibrium concentrations.

equivalence point of a titration The point at which the unknown solution has exactly reacted with the known solution. Neither is in excess.

equivalent of salt The amount that will produce 1 mol of positive (or negative) electrical charges when dissolved and dissociated.

essential amino acid An amino acid that cannot be synthesized within the body at a rate adequate to meet metabolic needs.

essential fatty acid A fatty acid needed by the body but not synthesized within the body.

ester A compound in which the —OH of an acid is replaced by an —OR.

esterification The process of forming an ester.

ester linkage The carbonyl carbon–oxygen single bond of the ester group.

ether A compound that contains a $-\overset{|}{\underset{|}{C}}-O-\overset{|}{\underset{|}{C}}-$ functional group.

eukaryotic cell A cell containing membrane-enclosed organelles, particularly a nucleus.

evaporation An endothermic process in which a liquid is changed to a gas; also called vaporization.

exact numbers Numbers that have no uncertainty; numbers from defined relationships, counting numbers, and numbers that are part of simple fractions.

exergonic process A process that gives up energy as it takes place.

exon A segment of a eukaryotic DNA molecule that is coded for amino acids.

exothermic reaction A reaction that liberates heat.

expanded structural formula A structural molecular formula showing all the covalent bonds.

expanding chain reaction A reaction in which the products of one reaction cause more than one more reaction to occur; also called branching chain reaction.

extracellular fluid Body fluid located outside cells.

extremozyme A nickname for certain enzymes isolated from microorganisms that thrive in extreme environments.

factors used in the factor-unit method Fractions obtained from numerical relationships between quantities.

family of the periodic table See group of the periodic table.

fat A triglyceride that is a solid at room temperature.

fat mobilization The hydrolysis of stored triglycerides, followed by the entry of fatty acids and glycerol into the bloodstream.

fatty acid A long-chain carboxylic acid found in fats.

feedback inhibition A process in which the end product of a sequence of enzyme-catalyzed reactions inhibits an earlier step in the process.

fermentation A reaction of sugars, starch, or cellulose to produce ethanol and carbon dioxide.

fibrous protein A protein made up of long rod-shaped or stringlike molecules that intertwine to form fibers.

first ionization energy The energy required to remove the first electron from a neutral atom.

Fischer projection A method of depicting three-dimensional shapes for chiral molecules.

fluid-mosaic model A model of membrane structure in which proteins are embedded in a flexible lipid bilayer.

formula weight The sum of the atomic weights of the atoms shown in the formula of an ionic compound.

functional group A unique reactive combination of atoms that differentiates molecules of organic compounds of one class from those of another.

furanose ring A five-membered sugar ring system containing an oxygen atom.

gamma ray A high-energy ray that is like an X-ray, but with a higher energy.

gas laws Mathematical relationships that describe the behavior of gases as they are mixed, subjected to pressure or temperature changes, or allowed to diffuse.

Geiger-Müller tube A radiation-detection device operating on the principle that ions form when radiation passes through a tube filled with low-pressure gas.

gene An individual section of a chromosomal DNA molecule that is the fundamental unit of heredity.

genome A summation of all the genetic material (chromosomes) of a cell; a person's genetic blueprint.

geometric isomers Molecules with restricted rotation around C—C bonds that differ in the three-dimensional arrangements of their atoms in space and not in the order of linkage of atoms.

globular protein A spherical protein that usually forms stable suspensions in water or dissolves in water.

glucogenic amino acid An amino acid whose carbon skeleton can be converted metabolically to an intermediate used in the synthesis of glucose.

gluconeogenesis The synthesis of glucose from noncarbohydrate molecules.

glucosuria A condition in which elevated blood sugar levels result in the excretion of glucose in the urine.

glycogenesis The synthesis of glycogen from glucose.

glycogenolysis The breakdown of glycogen to glucose.

glycolipid A complex lipid containing a sphingosine, a fatty acid, and a carbohydrate.

glycolysis A series of reactions by which glucose is oxidized to pyruvate.

glycoside Another name for a carbohydrate containing an acetal or ketal group.

glycosidic linkage The carbon–oxygen–carbon linkage that joins the components of a glycoside to the ring.

Graham's law A mathematical expression that relates rates of effusion or diffusion of two gases to the masses of the molecules of the two gases.

gray A biological unit of radiation measurement corresponding to the transfer of 1 J of energy to 1 kg of tissue.

group of the periodic table A vertical column of elements that have similar chemical properties; also called family of the periodic table.

half-life The time required for one-half the unstable nuclei in a sample to undergo radioactive decay.

haloalkane A derivative of an alkane in which one or more hydrogens are replaced by halogens; also called alkyl halide.

Haworth structure A method of depicting three-dimensional carbohydrate structures.

heat of fusion The amount of heat energy required to melt exactly 1 g of a solid substance at constant temperature.

heat of vaporization The amount of heat energy required to vaporize exactly 1 g of a liquid substance at constant temperature.

hemiacetal A compound that contains the functional group

$$-\overset{\displaystyle OH}{\underset{\displaystyle OR}{\overset{|}{\underset{|}{C}}}}-H.$$

hemiketal A compound that contains the $-\overset{|}{\underset{|}{C}}-\overset{\displaystyle OH}{\overset{|}{\underset{|}{C}}}-\overset{|}{\underset{\displaystyle OR}{\underset{|}{C}}}-$ group.

Henderson-Hasselbalch equation A relationship between the pH of a buffer, pK_a, and the concentrations of acid and salt in the buffer.

heteroatomic molecules Molecules that contain two or more kinds of atoms.

heterocyclic ring A ring in which one or more atoms are an atom other than carbon.

heterogeneous catalyst A catalytic substance normally used in the form of a solid with a large surface area on which reactions take place; also called surface catalyst.

heterogeneous matter Matter with properties that are not the same throughout the sample.

heterogeneous nuclear RNA (hnRNA) RNA produced when both introns and exons of eukaryotic cellular DNA are transcribed.

high-energy compound A substance that on hydrolysis liberates a great amount of free energy.

homeostasis A condition of chemical balance in the body.

homoatomic molecules Molecules that contain only one kind of atom.

homogeneous catalyst A catalytic substance that is distributed uniformly throughout the reaction mixture.

homogeneous matter Matter that has the same properties throughout the sample.

homologous series Compounds of the same functional class that differ by a —CH_2— group.

hormone A chemical messenger secreted by specific glands and carried by the blood to a target tissue, where it triggers a particular response.

hot spot Tissue in which a radioactive tracer concentrates.

Hund's rule A statement of the behavior of electrons when they occupy orbitals: Electrons will not join other electrons in an orbital if an empty orbital of the same energy is available for occupancy.

hybrid orbital An orbital produced from the combination of two or more nonequivalent orbitals of an atom.

hydrate A salt that contains specific numbers of water molecules as part of the solid crystalline structure.

hydrated ions Ions in solution that are surrounded by water molecules.

hydration The addition of water to a multiple bond.

hydrocarbon An organic compound that contains only carbon and hydrogen.

hydrogenation A reaction in which the addition of hydrogen takes place.

hydrogen bonding The result of attractive dipolar forces between molecules in which hydrogen atoms are covalently bonded to very electronegative elements (O, N, or F).

hydrolysis Bond breakage by reaction with water.

hydrolysis reaction Any reaction with water. For salts it is a reaction of the acidic cation and/or basic anion of the salt with water.

hydrophobic Molecules or parts of molecules that repel (are insoluble in) water.

hydroxy group The —OH functional group.

hyperglycemia A higher-than-normal blood sugar level.

hyperventilation Rapid, deep breathing.

hypoglycemia A lower-than-normal blood sugar level.

hypoventilation Slow, shallow breathing.

ideal gas law A gas law that relates the pressure, volume, temperature, and number of moles in a gas sample. Mathematically, it is $PV = nRT$.

immiscible A term used to describe liquids that are insoluble in each other.

induced-fit theory A theory of enzyme action proposing that the conformation of an enzyme changes to accommodate an incoming substrate.

inhibitor A substance that decreases reaction rates.

inner-transition element An element in which the distinguishing electron is found in an f subshell.

inorganic chemistry The study of the elements and all noncarbon compounds.

insoluble substance A substance that does not dissolve to a significant extent in a solvent.

internal energy The energy associated with vibrations within molecules.

International Unit (IU) A measure of vitamin activity, determined by biological methods.

interstitial fluid The fluid surrounding individual tissue cells.

intracellular fluid Body fluid located inside cells.

intron A segment of a eukaryotic DNA molecule that carries no codes for amino acids.

inverse square law of radiation A mathematical way of saying that the intensity of radiation is inversely proportional to the square of the distance from the source of the radiation.

invert sugar A mixture of equal amounts of glucose and fructose.

ionic bond The attractive force that holds together ions of opposite charge.

ion product of water The equilibrium constant for the dissociation of pure water into H_3O^+ and OH^-.

isoelectric point The characteristic solution pH at which an amino acid has a net charge of zero.

isoelectronic A term that literally means "same electronic," used to describe atoms or ions that have identical electronic configurations.

isoenzyme A slightly different form of the same enzyme produced by different tissues.

isomerism A property in which two or more compounds have the same molecular formula but different arrangements of atoms.

isotopes Atoms that have the same atomic number but different mass numbers. That is, they are atoms of the same element that contain different numbers of neutrons in their nuclei.

ketal A compound that contains the $-\overset{\displaystyle \text{OR}}{\underset{\displaystyle \text{OR}}{\overset{|}{\underset{|}{\text{C}}}}}-\overset{|}{\underset{|}{\text{C}}}-\overset{|}{\underset{|}{\text{C}}}-$ group.

ketoacidosis A low blood pH due to elevated levels of ketone bodies.

ketogenic amino acid An amino acid whose carbon skeleton can be converted metabolically to acetyl CoA or acetoacetyl CoA.

ketone A compound that contains the $-\overset{|}{\underset{|}{\text{C}}}-\overset{\displaystyle \text{O}}{\overset{\|}{\text{C}}}-\overset{|}{\underset{|}{\text{C}}}-$ group; the general formula is $\text{R}-\overset{\displaystyle \text{O}}{\overset{\|}{\text{C}}}-\text{R}'$.

ketone bodies Three compounds—acetoacetate, β-hydroxybutyrate, and acetone—formed from acetyl CoA.

ketonemia An elevated level of ketone bodies in the blood.

ketonuria The presence of ketone bodies in the urine.

ketosis A condition in which ketonemia, ketonuria, and acetone breath exist together.

kinetic energy The energy a particle has as a result of its motion. Mathematically, it is $KE = \frac{1}{2}mv^2$.

lactate fermentation The production of lactate from glucose.

lattice site The individual location occupied by a particle in a crystal lattice.

law of conservation of matter Atoms are neither created nor destroyed in chemical reactions.

Le Châtelier's principle The position of an equilibrium shifts in response to changes made in factors of the equilibrium.

lecithin A phosphoglyceride containing choline.

levorotatory Rotates plane-polarized light to the left.

Lewis structure A representation of an atom or ion in which the elemental symbol represents the atomic nucleus and all but the valence-shell electrons. The valence-shell electrons are represented by dots arranged around the elemental symbol.

limiting reactant The reactant present in a reaction in the least amount, based on its reaction coefficients and molecular weight. It is the reactant that determines the maximum amount of product that can be formed.

limiting-reactant principle The maximum amount of product possible from a reaction is determined by the amount of reactant present in the least amount, based on its reaction coefficient and molecular weight.

linear accelerator A particle accelerator that works by changing electrical polarities as charged particles cross gaps between segments of a long tube.

lipid A biological compound that is soluble only in nonpolar solvents.

lipid bilayer A structure found in membranes, consisting of two sheets of lipid molecules arranged so that the hydrophobic portions are facing each other.

lock-and-key theory A theory of enzyme specificity proposing that a substrate has a shape fitting that of the enzyme's active site, as a key fits a lock.

macronutrient A substance needed by the body in relatively large amounts.

major mineral A mineral found in the body in quantities greater than 5 g.

Markovnikov's rule In the addition of H—X to an alkene, the hydrogen becomes attached to the carbon atom that is already bonded to more hydrogens.

mass A measurement of the amount of matter in an object.

mass number of an atom A number equal to the sum of the number of protons and neutrons in the nucleus of an atom. Symbolically it is represented by A.

matter Anything that has mass and occupies space.

melting point The temperature at which a solid changes to a liquid; the solid and liquid have the same vapor pressure.

messenger RNA (mRNA) RNA that carries genetic information from the DNA in the cell nucleus to the site of protein synthesis in the cytoplasm.

metabolic acidosis A condition of acidosis resulting from causes other than hypoventilation.

metabolic alkalosis A condition of alkalosis resulting from causes other than hyperventilation.

metabolic pathway A sequence of reactions used to produce one product or accomplish one process.

metabolism The sum of all reactions occurring in an organism.

metals Elements found in the left two-thirds of the periodic table. Most have the following properties: high thermal and electrical conductivities, high malleability and ductility, and a metallic luster.

metallic bond An attractive force responsible for holding solid metals together. It originates from the attraction between positively charged atomic kernels that occupy the lattice sites and mobile electrons that move freely through the lattice.

metalloids Elements that form a narrow diagonal band in the periodic table between metals and nonmetals. They have properties somewhat between those of metals and nonmetals.

micelle A spherical cluster of molecules in which the polar portions of the molecules are on the surface and the nonpolar portions are located in the interior.

micronutrient A substance needed by the body only in small amounts.

mineral A metal or nonmetal used in the body in the form of ions or compounds.

mitochondrion A cellular organelle where reactions of the common catabolic pathway occur.

mixture A physical blend of matter that can theoretically be physically separated into two or more components.

moderators Materials capable of slowing down neutrons that pass through them.

modulator A substance that binds to an enzyme at a location other than the active site and alters the catalytic activity.

molarity (M) A solution concentration expressed in terms of the number of moles of solute contained in a liter of solution.

mole The number of particles (atoms or molecules) contained in a sample of an element or compound with a mass in grams equal to the atomic or molecular weight, respectively. Numerically, 1 mol is equal to 6.022×10^{23} particles.

molecular equation An equation written with each compound represented by its formula.

molecular weight The relative mass of a molecule expressed in atomic mass units and calculated by adding together the atomic weights of the atoms in the molecule.

molecule The smallest particle of a pure substance that has the properties of that substance and is capable of a stable independent existence. Alternatively, a molecule is the limit of physical subdivision for a pure substance.

monomer The starting material that becomes the repeating units of polymers.

monoprotic acid An acid that gives up only one proton (H^+) per molecule when dissolved.

monosaccharide A simple carbohydrate most commonly consisting of three to six carbon atoms.

mutagen A chemical that induces mutations by reacting with DNA.

mutation Any change resulting in an incorrect base sequence on DNA.

native state The natural three-dimensional conformation of a functional protein.

net ionic equation An equation that contains only un-ionized or insoluble materials and ions that undergo changes as the reaction proceeds. All spectator ions are eliminated.

network solids Solids in which the lattice sites are occupied by atoms that are covalently bonded to each other.

neurotransmitter A substance that acts as a chemical bridge in nerve impulse transmission between nerve cells.

neutral A term used to describe any water solution in which the concentrations of H_3O^+ and OH^- are equal. Also, a water solution with pH = 7.

neutralization reaction A reaction in which an acid and base react completely, leaving a solution that contains only a salt and water.

noble gas configuration An electronic configuration in which the last eight electrons occupy and fill the s and p subshells of the highest-occupied shell.

noncompetitive inhibitor An inhibitor that binds to the enzyme at a location other than the active site.

nonelectrolyte A solute that when dissolved in water forms a solution that does not conduct electricity.

nonessential amino acid An amino acid that can be synthesized within the body in adequate amounts.

nonmetals Elements found in the right one-third of the periodic table. They often occur as brittle, powdery solids or gases and have properties generally opposite those of metals.

nonpolar covalent bond A covalent bond in which the bonding pair of electrons is shared equally by the bonded atoms.

nonpolar molecule A molecule that contains no polarized bonds, or a molecule containing polarized bonds in which the resulting charges are distributed symmetrically throughout the molecule.

normal alkane Any alkane in which all the carbon atoms are aligned in a continuous chain.

normal boiling point The temperature at which the vapor pressure of a liquid is equal to 1 standard atmosphere (760 torr); also called standard boiling point.

N-terminal residue An amino acid on the end of a chain that has an unreacted or free amino group.

nuclear fission A process in which large nuclei split into smaller, approximately equal-sized nuclei when bombarded by neutrons.

nuclear fusion A process in which small nuclei combine or fuse to form larger nuclei.

nucleic acid A biomolecule involved in the transfer of genetic information from existing cells to new cells.

nucleic acid backbone The sugar–phosphate chain that is common to all nucleic acids.

nucleotide The repeating structural unit or monomer of polymeric nucleic acids.

nucleus The central core of atoms that contains protons, neutrons, and most of the mass of atoms.

nutrition An applied science that studies food, water, and other nutrients and the ways living organisms use them.

octet rule A rule for predicting electron behavior in reacting atoms. It says that atoms will gain or lose sufficient electrons to achieve an outer electron arrangement identical to that of a noble gas. This arrangement usually consists of eight electrons in the valence shell.

oil A triglyceride that is a liquid at room temperature.

Okazaki fragment A DNA fragment produced during replication as a result of strand growth in a direction away from the replication fork.

optically active molecule A molecule that rotates the plane of polarized light.

optimum pH The pH at which enzyme activity is highest.

optimum temperature The temperature at which enzyme activity is highest.

organelle A specialized structure within a cell that performs a specific function.

organic chemistry The study of carbon-containing compounds.

organic compound A compound that contains the element carbon.

osmolarity The product of n and M in the equation $\pi = nMRT$.

osmosis The process in which solvent flows through a semipermeable membrane into a solution.

osmotic pressure The hydrostatic pressure required to prevent the net flow of solvent through a semipermeable membrane into a solution.

oxidation Originally, a process involving a reaction with oxygen. Today it means a number of things, including a process in which electrons are given up, hydrogen is lost, or an oxidation number increases.

oxidation numbers Positive or negative numbers assigned to the elements in chemical formulas according to a specific set of rules; also called oxidation states.

oxidation states See **oxidation numbers**.

oxidative deamination An oxidation process resulting in the removal of an amino group.

oxidative phosphorylation A process coupled with the electron transport chain whereby ADP is converted to ATP.

oxidizing agent The substance that contains an element that is reduced during a chemical reaction.

oxyhemoglobin Hemoglobin combined with oxygen.

partial pressure The pressure an individual gas of a mixture would exert if it were in the container alone at the same temperature as the mixture.

Pauli exclusion principle A statement of the behavior of electrons when they occupy orbitals: Only electrons spinning in opposite directions can simultaneously occupy the same orbital.

peptide An amino acid polymer of short chain length.

peptide linkage (peptide bond) The amide linkage between amino acids that results when the amino group of one acid reacts with the carboxylate group of another.

percent A solution concentration that expresses the amount of solute in 100 parts of solution.

percentage yield The percentage of the theoretical amount of a product actually produced by a reaction.

periodic law A statement about the behavior of the elements when they are arranged in a specific order. In its present form, it is stated as follows: Elements with similar chemical properties occur at regular (periodic) intervals when the elements are arranged in order of increasing atomic numbers.

period of the periodic table A horizontal row of elements.

pH The negative logarithm of the molar concentration of H^+ (H_3O^+) in a solution.

phenol A compound in which an —OH group is connected to a benzene ring. The parent compound is also called phenol.

phenyl group A benzene ring with one hydrogen absent, C_6H_5—.

phosphoglyceride A complex lipid containing glycerol, fatty acids, phosphoric acid, and an alcohol component.

phospholipid A phosphorus-containing lipid.

phosphoric anhydride A compound that contains the

$$-O-\overset{\overset{\displaystyle O}{\|}}{\underset{\underset{\displaystyle O^-}{|}}{P}}-O-\overset{\overset{\displaystyle O}{\|}}{\underset{\underset{\displaystyle O^-}{|}}{P}}-O- \text{ group.}$$

physical changes Changes matter undergoes without changing composition.

physical properties Properties of matter that can be observed or measured without trying to change the composition of the matter being studied.

physical unit of radiation A radiation measurement unit indicating the activity of the source of the radiation; for example, the number of nuclear decays per minute.

pK_a The negative logarithm of K_a.

plasma The liquid portion of whole blood.

plasmid Circular, double-stranded DNA found in the cytoplasm of bacterial cells.

polar covalent bond A covalent bond that shows bond polarization; that is, the bonding electrons are shared unequally.

polar molecule A molecule that contains polarized bonds and in which the resulting charges are distributed nonsymmetrically throughout the molecule.

polyatomic ions Covalently bonded groups of atoms that carry a net electrical charge.

polyatomic molecules Molecules that contain three or more atoms.

polycyclic aromatic compound A derivative of benzene in which carbon atoms are shared between two or more benzene rings.

polyfunctional compound A compound with two or more functional groups.

polymer A very large molecule made up of repeating units.

polymerization A reaction that produces a polymer.

polypeptide An amino acid polymer of intermediate chain length containing up to 50 amino acid residues.

polyribosome A complex of mRNA and several ribosomes; also called polysome.

polysaccharide A carbohydrate formed by the combination of many monosaccharide units.

polysome See **polyribosome**.

polyunsaturated A term usually applied to molecules with several double bonds.

position of equilibrium An indication of the relative amounts of reactants and products present at equilibrium.

positron A positively charged electron.

potential energy The energy a particle has as a result of attractive or repulsive forces acting on it.

pressure A force per unit area of surface on which the force acts. In measurements and calculations involving gases, it is often expressed in units related to measurements of atmospheric pressure.

primary alcohol An alcohol in which the OH group is attached to CH_3 or to a carbon attached to one other carbon atom.

primary amine An amine having one alkyl or aromatic group bonded to nitrogen, as in $R—NH_2$.

primary protein structure The linear sequence of amino acid residues in a protein chain.

products of a reaction The substances produced as a result of the reaction taking place. They are written on the right side of the equation representing the reaction.

prokaryotic cell A simple unicellular organism that contains no nucleus and no membrane-enclosed organelles.

prostaglandin A substance derived from unsaturated fatty acids, with hormonelike effects on a number of body tissues.

prosthetic group The non–amino acid parts of conjugated proteins.

protein An amino acid polymer made up of more than 50 amino acids.

protein turnover The continuing process in which body proteins are hydrolyzed and resynthesized.

pure substance Matter that has a constant composition and fixed properties.

pyranose ring A six-membered sugar ring system containing an oxygen atom.

quaternary ammonium salt An ionic compound containing a positively charged ion in which four alkyl or aromatic groups are bonded to nitrogen, as in R_4N^+.

quaternary protein structure The arrangement of subunits that form a larger protein.

rad A biological unit of radiation measurement corresponding to the transfer of 1×10^{-2} J or 2.4×10^{-3} cal of energy to 1 kg of tissue.

radical An electron-deficient particle that is very reactive; also called free radical.

radioactive dating A process for determining the age of artifacts and rocks, based on the amount and half-life of radioisotopes contained in the object.

radioactive decay A process in which an unstable nucleus changes energy states and in the process emits radiation.

radioactive nuclei Nuclei that undergo spontaneous changes and emit energy in the form of radiation.

radioisotope An isotope of an element that emits nuclear radiation.

reactants of a reaction The substances that undergo chemical change during the reaction. They are written on the left side of the equation representing the reaction.

reaction mechanism A detailed explanation of how a reaction actually takes place.

reaction rate The speed of a reaction.

recombinant DNA DNA of an organism that contains genetic material from another organism.

reducing agent The substance that contains an element that is oxidized during a chemical reaction.

reducing sugar A sugar that can be oxidized by Cu^{2+} solutions.

reduction Originally, a process in which oxygen was lost. Today it means a number of things, including a process in which electrons are gained, hydrogen is accepted, or an oxidation number decreases.

Reference Daily Intakes (RDIs) A set of standards for protein, vitamins, and minerals used on food labels as part of the Daily Values.

relative specificity The characteristic of an enzyme that it acts on several structurally related substances.

rem A biological unit of radiation measurement corresponding to the health effect produced by 1 roentgen of gamma or X-rays, regardless of the type of radiation involved.

renal threshold The blood glucose level at which glucose begins to be excreted in the urine.

replication The process by which an exact copy of a DNA molecule is produced.

replication fork A point where the double helix of a DNA molecule unwinds during replication.

representative element An element in which the distinguishing electron is found in an s or a p subshell.

respiratory acidosis A condition of acidosis caused by hypoventilation.

respiratory alkalosis A condition of alkalosis caused by hyperventilation.

restriction enzyme A protective enzyme found in some bacteria that catalyzes the cleaving of all but a few specific types of DNA.

ribonucleic acid (RNA) A nucleic acid found mainly in the cytoplasm of cells.

ribosomal RNA (rRNA) RNA that constitutes about 65% of the material in ribosomes, the sites of protein synthesis.

ribosome A subcellular particle that serves as the site of protein synthesis in all organisms.

roentgen A biological unit of radiation measurement used with X-rays and gamma rays; the quantity of radiation that generates 2.1×10^9 ion pairs per 1 cm^3 of dry air or 1.8×10^{12} ion pairs per 1 g of tissue.

salt A solid crystalline ionic compound at room temperature that contains the cation of a base and the anion of an acid.

saponification The basic cleavage of an ester linkage.

satiety A state of satisfaction or fullness.

saturated hydrocarbon Another name for an alkane.

saturated solution A solution that contains the maximum amount possible of dissolved solute in a stable situation under the prevailing conditions of temperature and pressure.

scientific models Explanations for observed behavior in nature.

scientific notation A way of representing numbers consisting of a product between a nonexponential number and 10 raised to a whole-number exponent that may be positive or negative.

scintillation counter A radiation-detection device operating on the principle that phosphors give off light when struck by radiation.

secondary alcohol An alcohol in which the carbon bearing the OH group is attached to two other carbon atoms.

secondary amine An amine having two alkyl or aromatic groups bonded to nitrogen, as in R_2NH.

secondary protein structure The arrangement of protein chains into patterns as a result of hydrogen bonds between amide groups of amino acid residues in the chain. The common secondary structures are the α-helix and the β-pleated sheet.

semiconservative replication A replication process that produces DNA molecules containing one strand from the parent and a new strand that is complementary to the strand from the parent.

shape Shape depends on the physical state of matter.

shell A location and energy of electrons around a nucleus that is designated by a value for n, where $n = 1, 2, 3$, etc.

side reactions Reactions that do not give the desired product of a reaction.

significant figures The numbers in a measurement that represent the certainty of the measurement, plus one number representing an estimate.

simple carbohydrates Monosaccharides and disaccharides, commonly called sugars.

simple ion An atom that has acquired a net positive or negative charge by losing or gaining electrons.

simple lipid An ester-containing lipid with just two types of components: an alcohol and one or more fatty acids.

simple protein A protein made up entirely of amino acid residues.

single-replacement reaction A chemical reaction in which an element reacts with a compound and displaces another element from the compound.

soap A salt of a fatty acid often used as a cleaning agent.

solubility The maximum amount of solute that can be dissolved in a specific amount of solvent under specific conditions of temperature and pressure.

soluble substance A substance that dissolves to a significant extent in a solvent.

solute One or more substances present in a solution in amounts less than that of the solvent.

solutions Homogenous mixtures of two or more pure substances.

solvent The substance present in a solution in the largest amount.

specific heat The amount of heat energy required to raise the temperature of exactly 1 g of a substance by exactly 1°C.

spectator ions The ions in a total ionic reaction that are not changed as the reaction proceeds. They appear in identical forms on the left and right sides of the equation.

sphingolipid A complex lipid containing the aminoalcohol sphingosine.

spontaneous process A process that takes place naturally with no apparent cause or stimulus.

stable substance A substance that does not undergo spontaneous changes under the surrounding conditions.

standard atmosphere The pressure needed to support a 760-mm column of mercury in a barometer tube.

standard boiling point See **normal boiling point**.

standard conditions (STP) A set of specific temperature and pressure values used for gas measurements.

standard position for a decimal In scientific notation, the position to the right of the first nonzero digit in the nonexponential number.

state of equilibrium A condition in a reaction system when the rates of the forward and reverse reactions are equal.

stem cell An unspecialized cell that has the ability to replicate and differentiate, giving rise to a specialized cell.

stereochemical specificity The characteristic of an enzyme that it is able to distinguish between stereoisomers.

stereoisomers Compounds with the same structural formula but different spatial arrangements of atoms.

steroid A compound containing four rings fused in a particular pattern.

stoichiometry The study of mass relationships in chemical reactions.

strong acids and strong bases Acids and bases that dissociate (ionize) completely when dissolved to form a solution.

structural isomers Compounds that have the same molecular formula but in which the atoms bond in different patterns.

sublimation The endothermic process in which a solid is changed directly to a gas without first becoming a liquid.

subshell A component of a shell that is designated by a letter from the group s, p, d, and f.

substrate The substance that undergoes a chemical change catalyzed by an enzyme.

subunit A polypeptide chain having primary, secondary, and tertiary structural features that is a part of a larger protein.

sulfhydryl group The —SH functional group.

supercritical mass The minimum amount of fissionable material that must be present to cause a branching chain reaction to occur.

supercritical reaction A branching chain reaction.

supersaturated solution An unstable solution that contains an amount of solute greater than the solute solubility under the prevailing conditions of temperature and pressure.

surface catalyst See **heterogeneous catalyst**.

tertiary alcohol An alcohol in which the carbon bearing the OH group is attached to three other carbon atoms.

tertiary amine An amine having three alkyl or aromatic groups bonded to nitrogen, as in R_3N.

tertiary protein structure A specific three-dimensional shape of a protein resulting from interactions between R groups of the amino acid residues in the protein.

thermal expansion The change in volume of a sample resulting from a change in temperature of the sample.

thermonuclear reaction A nuclear fusion reaction that requires a very high temperature to start.

thioester bond The carbon–sulfur bond of the

functional group.

thiol A compound containing an —SH group.

titration An analytical procedure in which one solution (often a base) of known concentration is slowly added to a measured volume of an unknown solution (often an acid). The volume of the added solution is measured with a buret.

Tollens' reagent A mild oxidizing solution containing silver ions used to test for aldehydes.

torr The pressure needed to support a 1-mm column of mercury in a barometer tube.

total ionic equation An equation written with all soluble ionic substances represented by the ions they form in solution.

trace mineral A mineral found in the body in quantities smaller than 5 g.

tracer A radioisotope used medically because its progress through the body or localization in specific organs can be followed.

trans- On opposite sides (as applied to geometric isomers).

transaminase An enzyme that catalyzes the transfer of an amino group.

transamination The enzyme-catalyzed transfer of an amino group to a keto acid.

transcription The transfer of genetic information from a DNA molecule to a molecule of messenger RNA.

transfer RNA (tRNA) RNA that delivers individual amino acid molecules to the site of protein synthesis.

transition element An element in which the distinguishing electron is found in a *d* subshell.

translation The conversion of the code carried by messenger RNA into an amino acid sequence of a protein.

transuranium elements Synthetic elements with atomic numbers greater than that of uranium.

triatomic molecules Molecules that contain three atoms.

triglyceride (triacylglycerol) A triester of glycerol in which all three alcohol groups are esterified.

triple bond The bond resulting from the sharing of three pairs of electrons.

triprotic acid An acid that gives up three protons (H^+) per molecule when dissolved.

turnover number The number of molecules of substrate acted on by one molecule of enzyme per minute.

Tyndall effect A property of colloids in which the path of a beam of light through the colloid is visible because the light is scattered.

universal gas constant The constant that relates pressure, volume, temperature, and number of moles of gas in the ideal gas law.

unsaturated hydrocarbon A hydrocarbon containing one or more multiple bonds.

urea cycle A metabolic pathway in which ammonium ions are converted to urea.

valence shell The outermost (highest-energy) shell of an element that contains electrons.

vaporization See **evaporation**.

vapor pressure The pressure exerted by vapor that is in equilibrium with its liquid.

vector A carrier of foreign DNA into a cell.

vitamin An organic nutrient that the body cannot produce in the small amounts needed for good health.

volume/volume percent A concentration that expresses the volume of liquid solute contained in 100 volumes of solution.

VSEPR theory A theory based on the mutual repulsion of electron pairs. It is used to predict molecular shapes.

water of hydration Water retained as part of the solid crystalline structure of some salts.

wax An ester of a long-chain fatty acid and a long-chain alcohol.

weak (or moderately weak) acids and bases Acids and bases that dissociate (ionize) less than completely when dissolved to form a solution.

weight A measurement of the gravitational force acting on an object.

weight/volume percent A concentration that expresses the grams of solute contained in 100 mL of solution.

weight/weight percent A concentration that expresses the mass of solute contained in 100 mass units of solution.

zwitterion A dipolar ion that carries both a positive and a negative charge as a result of an internal acid–base reaction in an amino acid molecule.

zymogen or proenzyme The inactive precursor of an enzyme.

INDEX

Leucine, 613
Levorotatory, 558
Lewis, G.N., 101, 111
Lewis structures
 definition of, 102
 drawings of, 114–116
 names of, 126
 oxidation number, 149
 for polyatomic ions, 116–118
 representation, 119–121
 shapes of, 118–122
 usage, 111–112
 writing, 102
Lime, 152
Limestone, 152, 294
Limiting-reactant principle, 161–162
Linear accelerator, 339
Linear alkanes, 430–431
Linoleic acid
 characteristics of, 586–587
 description of, 706–707
 melting point of, 586–587
 structural formula of, 589
Linolenic acid, 587, 706–707
Lipases, 589
Lipids. *See also* Fats; Fatty acids; Triglycerides
 bilayer, 596–597
 classification of, 584
 complex, 584
 definition of, 584
 glycolipids, 595
 macronutrients, 705–707
 metabolism, 761
 nonsaponifiable, 584
 olive oil as, 603
 overview of, 583
 phospholipids, 592
 saponifiable, 584
 simple, 584
 sphingolipids, 594–595
 types, 584
Lipoic acid, 647
Lipoproteins
 classification of, 761–763
 description of, 625
 high-density, 605, 762
 low-density, 600, 762–763
Liquid alkanes, 379
Liquids
 concentration difference, 230
 kinetic molecular view of, 180
 physical properties of, 176–177
 shape of, 176–177
 vapor pressure, 194
Liquid state, 129, 180
L isomers, 558
Lister, Joseph, 440
Lithium batteries, 162
Lithium cobalt oxide, 162
Litmus, 277
Liver, 734, 772
L-lactic acid, 558
Lock-and-key theory, 649–650
Logarithms, 285–286
Long-chain alcohols, 429–430, 592
Long-chain fatty acid, 592
Long-chain unsaturated amino alcohol, 594

Low-density lipoprotein, 600, 762–763
Low-density lipoprotein receptors, 662
Low-energy subshells, 81
Low-level radiation, 330
Low molecular weight acids, 493
Low molecular weight amines, 521
Lowry, Thomas, 278
Lung diseases, 802
Lupus, 668
Lycopodium powder, 259
Lymph system, 761–762
Lysine, 613
Lysol, 441

M

Macronutrients, 703, 705–708
Magnesium, 713
Magnesium atoms, 58, 79, 106
Magnesium ion, 106
Magnesium metal, 153
Magnesium oxide, 153
Magnesium sulfate heptahydrate, 291
Major minerals, 712–713
Male sex hormones, 602–604
Malic acid, 159, 489
Malleability, 88
Malonate, 656
Maltose
 description of, 567–568
 digestion, 733
 in milk, 656
 solution containing, 570
 sweetness of, 561
Mammary glands, 567
Manganese, 713
Marijuana, 441
Markovnikov, Vladimir, 400
Markovnikov's rule, 400, 402
Mass
 as basic unit (metric system), 16–17
 calculations of mass percentage, 64–65
 definition of, 4
 density and, relationship between, 31
Mass number, 52
Mass spectrometers, 53
Matrix, 722
Matter
 classification of, 10–13, 35
 composition, 178
 concept summary, 35
 definition of, 4
 density, 31
 destruction, 163
 kinetic molecular theory, 178–179
 law of conservation of, 146
 model of, 8–10, 35
 properties, observation, 176–178
 shape of, 176
 states of, 175–176, 197–201
Measurement
 basis of, 14
 of blood pressure, 184
 uncertainty associated with, 22
 units of, 13–14, 35
Medical ID, 261
Medical imaging, 332
Medicines, 344

Medicines in My Home, 539
Mediterranean diet, 107, 764–765
Meitner, Lise, 340
Melanin, 412
Melatonin, 527
Melting
 description of, 196–197
 heat of fusion, 199
 heat requirements for, 133
Melting point, 130, 197
Membranes, 596–598
Memory, 344
Mendeleev, Dmitri, 73
Mental illness, 531
Menthol, 425
Menthone, 478
Mercury, 653
Mercury barometer, 181–182
Messenger RNA
 alignment of, 689
 codons, 687
 definition of, 680
 molecules, 680
 ribosomes, 690
 synthesis of, 684
 translation of, 683–684
Metabolic acidosis, 803
Metabolic alkalosis, 803
Metabolic pathway, 715
Metabolism
 ATP hydrolysis, 720
 definition of, 147
 diseases caused by abnormalities in, 595
 energy production from, 715
 protein effects on, 617
Metal atoms, 105
Metal ions
 characteristics of, 105
 inorganic ions, 647
 sulfhydryl groups, 447
Metallic bond, 130
Metallic luster, 88
Metalloids, 88
Metalloproteins, 625
Metal-nonmetal reaction, 125
Metal(s)
 activity series of, 289
 burning of, 153
 hydrogen ion reduction ability, 289
 loss of electrons, 104
 in periodic table, 88
 properties of, 88
Metathesis reactions. *See* Double-replacement reaction
Methanal, 461
Methandrostenolone, 603
Methane. *See also* Natural gas
 carbon-hydrogen bond, 357
 description of, 49, 363, 379
 structural representation of, 363
Methanol, 437
Methathesis. *See* Double-replacement reaction
Methedrine, 532
Methicillin, 655
Methionine, 613, 688
Methoxymethane, 444
Methyl alcohol, 426, 436, 497

Triple bonds, 113, 118, 406
Triprotic acid, 296–297
tRNA. *See* Transfer RNA
Trypsin, 645
Trypsinogen, 626, 659
Tryptophan, 531, 613, 688, 773
Turnover number, 650
Turnover rate, 773
Tyndall effect, 235
Tyrosine
 production of, 779
 products derived from, 773
 structural formula of, 530–531, 613

U

Ultraviolet radiation, 412
Uncatalyzed reaction, 643–644
Uncharged particles, 338
Uncombined element, 148
Unhybridized p orbital
 perpendicularity, 395, 407
 shape, comparison, 357
Units
 calculation of, 27–28
 summary of, 35
Universal gas constant, 189
Unsaturated fats, 591
Unsaturated fatty acids, 585–586,
 588, 598
Unsaturated hydrocarbons, 391
Unsaturated oils, 588
Unsaturation, 591
Unsubstituted amides, 536–537
Uracil, 670–671, 680, 687
Uranium-235, 340
Uranium-238, 341–342, 343
Urea, 644, 645, 777–778
Urea cycle, 777–778
Urease, 644, 652
Uric acid, 478
Uridine triphosphate (UTP), 747
Urinalysis test strips, 796
Urinary tract infections, 657
Urine, 796–797
UTP. *See* Uridine triphosphate (UTP)
UV radiation, 131

V

Valence electrons, 93, 118
Valence shell, 79, 85
Valence-shell electron-pair repulsion, 126
Valence-shell electron-pair repulsion (VSEPR)
 theory, 118–122
Valence-shell electrons
 pairs of, 118
 representation of, 102
 sharing of, 111
 total number of, 113
Valeric acid, 492
Valine, 613, 775
Vanadium, 713
Vanilla beans, 408

Vanillin, 478
Vanilloids, 474
Vaporization, 193–194, 199
Vapor pressure, 194
Vasodilators, 716
Vasopressin, 619, 626, 798
Vector, 695
Vegetable oils, 587, 591
Vinegar, 229, 489–490
Vinyl plastic, 436
Viral gastroenteritis, 575
Viscous fiber, 600
Vital force theory, 353–354
Vitamin B12, 647
Vitamin(s)
 coenzyme forms, 647
 deficiency of, 709
 description and functions of,
 708–709
 dietary essential, 416
 sources of, 709
Voice-Activated 911, 261
Volume
 calculation using densities, 177–178
 changes in, 185
 density and, relationship between, 31
 pressure/temperature and, relationship
 among, 184–188
 units used to express, 15–16
Volumetric flask, 224
Volume/volume percent, 222
VSPER. *See* Valence-shell electron-pair repulsion
 (VSEPR) theory

W

Water. *See also* Hydration
 alcohol miscibility with, 429–431
 alkenes reaction with, 402
 amines in, hydrogen bonding of, 520
 boiling point of, 132, 196, 200
 as Brønsted base, 281, 521–522
 concentration of, in pure water, 281–282
 consumption of, 113
 conversion of, into carbohydrates, 335
 density of, temperature function, 132
 description of, 49
 ester linkage breakage effects on, 502
 expansion of, 132
 formula for, 63–64
 freezing of, 132
 healthy beverage, 237
 human intake of, 797
 of hydration, 291
 hydrogen bonding, 132, 536–537
 hydrolysis of ATP in, 718
 ionic compounds, 217–218
 ionic substance, 216–217
 Lewis structure, 120–121
 linear alkanes solubility in, 429–430
 low molecular weight acids solubility in, 493
 -OH group interaction with, 432
 pH, salts effect on, 305

 polar solute, 217–218
 polar solvent, 464
 properties of, 132
 proton transfer, 281
 role in sinkholes, 294
 self-ionization of, 281–283
 solute solubilities, 213
 treatment of, 113
 vapor pressure, 194, 230
Water molecules
 bond formation, 432
 dipolar forces effect on placement
 of, 131
 hydrogen bonds, 132
 retention of, 291
Watson, James D., 672–675
Watson-Crick three-dimensional
 structure, 677
Waxes, 592
Weak acids
 definition of, 294–295
 Ka values for, 307
 list of, 295
 pKa values for, 307
Weak bases, 294–295
Weak Brønsted bases, 297
Weak electrolytes, 229–230
Websites, health information on, 32
Weight, 4
Weight loss, 34, 519, 617, 750
Weight/volume percent, 221–222
Weight/weight percent, 221–222
Whey protein, 716
White phosphorus, 257–258
Whole grains, 620
Wilkins, M.H.F., 673
Wintergreen, 408
Wöhler, Friedrich, 353–354
Wool, 527

X

X-rays. *See also* Radiation
 CT scan, 322
 ionization effects, 331
 radiation exposure from, 332
 usage requirements, 330
Xylose, 561

Y

Yeast fermentation, 437
Yield
 acetal, 565
 adenosine triphosphate, 744
 percentage, 163–164
 from reactions, 163–164

Z

Zeros, 23
Zinc, 188, 713
Zinc 10-undecylenate, 496
Zwitterions, 614–615, 622
Zymogens, 658–659

Features within the Text